Renewable Energy in the Service of Mankind
Vol II

Ali Sayigh
Editor

Renewable Energy in the Service of Mankind Vol II

Selected Topics from the World Renewable Energy Congress WREC 2014

Springer

World Renewable
Energy Network

Editor
Ali Sayigh
World Renewable Energy Congress
Brighton
United Kingdom

ISBN 978-3-319-18214-8 ISBN 978-3-319-18215-5 (eBook)
DOI 10.1007/978-3-319-18215-5

Library of Congress Control Number: 2015947131

Springer Cham Heidelberg New York Dordrecht London

Printed on acid-free paper

Springer International Publishing AG Switzerland is part of Springer Science+Business Media (www.springer.com)

Contents

Contributors

Kamaruddin Abdullah The Graduate School, Darma Persada University, East Jakarta, Indonesia

Shahrir Abdullah Solar Energy Research Institute, Faculty of Engineering, Universiti Kebangsaan Malaysia, Bangi, Selangor, Malaysia

Qutaiba M. Abdulmajeed Universiti Malaysia Perlis, Arau Perlis, Malaysia

Kamal A. Abed Mechanical Engineering Department, National Research Centre, Giza, Egypt

Ahmed Aboulmagd Faculty of Engineering, Mechanical Power Department, Cairo University, Cairo, Egypt

Mohd Nazari Abu Bakar Faculty of Applied Sciences, Universiti Teknologi MARA Perlis, Arau, Perlis, Malaysia

Hanane Aburideh Unité de Développement des Equipements Solaires, UDES/ Centre de Développement des Energies Renouvelables, CDER, Bou Ismail, W. Tipaza, Algérie

Giovanna Adinolfi ENEA, Agenzia nazionale per le nuove tecnologie, l'energia e lo sviluppo economico sostenibile, Research Center of Portici, Portici, NA, Italy

Kennedy Aduda Faculty of the Built Environment, University of Technology Eindhoven, Eindhoven, Netherlands

J. O. Aguilar División de Ciencias e Ingeniería, Universidad de Quintana Roo, Chetumal, Q Roo, Mexico

Jong-Bo Ahn Smart Distribution Research Center, KERI, Changwon city, Republic of Korea

Kazuyuki Aihara Institute of Industrial Science, The University of Tokyo, Tokyo, Japan

Department of Mathematical Informatics, The University of Tokyo, Tokyo, Japan

CREST, Saitama, Japan

Mounir Aksas Department of Physics, Faculty of Sciences, University of Batna, Batna, Algeria

Abduljalil A. Al-abidi Solar Energy Research Institute, Faculty of Engineering, Universiti Kebangsaan Malaysia, Bangi, Selangor, Malaysia

Fatah O. Alghoul Solar Energy Research Institute, Faculty of Engineering, Universiti Kebangsaan Malaysia, Bangi, Selangor, Malaysia

Mohammed Al-Ghoul Solar Energy Research Institute, Faculty of Engineering, Universiti Kebangsaan Malaysia, Bangi, Selangor, Malaysia

S. A. Aljunid Universiti Malaysia Perlis, Arau Perlis, Malaysia

M. Alonso-Abella CIEMAT/DER, Madrid, Spain

Zakaria Mohd. Amin Mechanical Engineering Discipline, Science and Engineering Faculty, Queensland University of Technology, Brisbane, Australia

Martin Anda Centre for Responsible Citizenship and Sustainability, Murdoch University, Murdoch, Western Australia

Felice Apicella Photovoltaic Technologies Unit, ENEA, Portici Research Centre, Portici (Naples), Italy

E. Aroudam Modeling and Simulation of Mechanical Systems Team Physics Department, Faculty of Sciences, Abdelmalek Essaadi University, Tetouan, Morocco

F. Arya School of the Built Environment, University of Ulster, Coleraine, UK

Ali H. Assi Department of Electrical and Electronic Engineering, Lebanese International University, Beirut, Lebanon

Sophie Avril CEA DEN I-tésé and ECP LGI, Gif-sur-Yvette Cedex, France

Devrim Aydın Institute of Sustainable Energy Technology, University of Nottingham, Nottingham, UK

Mervat A. Badr Mechanical Engineering Department, National Research Centre, Giza, Egypt

Ahmed G. Bahgat Faculty of Engineering, Cairo University, Giza, Egypt

A. Belghachi Laboratory of semiconductors devices physics, University of Bechar, Bechar, Algeria

Abderrahmane Belghachi Laboratory of semiconductor devices physics, (LPDS) University of Bechar, Bechar, Algeria

Dalila Belhout Unité de Développement des Equipements Solaires, UDES/Centre de Développement des Energies Renouvelables, CDER, Bou Ismail, W. Tipaza, Algérie

Randha Bellatreche Unité de Développement des Equipements Solaires, UDES/ Centre de Développement des Energies Renouvelables, CDER, Bou Ismail, W. Tipaza, Algérie

Zakarya Berkai Laboratory of semiconductor devices physics, (LPDS) University of Bechar, Bechar, Algeria

Sara Berretta Dipartimento di IngegneriaCivile e Ambientale, UniversitàdegliStudi di Perugia, Perugia, Italy

Pavel Bezrukikh Krzhizhanovsky Power Engineering Institute, Moscow, Russia

Nicola Bianco Industrial Engineering Department, University of Naples Federico II, Naples, Italy

Jean-Claude Bocquet Laboratoire de Génie Industriel, ECP LGI, Ecole Centrale de Paris, Châtenay Malabry Cedex, France

Kristian Boedecker GUNT2E Hamburg, Barsbüttel, Germany

Matteo Bortolato Dipartimento di Ingegneria Industriale, Università degli Studi di Padova, Padova, Italy

Olivier Videme Bossou University of Maroua, Maroua, Cameroon

Gert Boxem Faculty of the Built Environment, University of Technology Eindhoven, Eindhoven, Netherlands

Cristiana Brasil Maia Department of Mechanical Engineering, Pontifícia Universidade Católica de Minas Gerais, Belo Horizonte, Brazil

Martin Brueckner School of Management and Governance, Murdoch University, Perth, Western Australia

Pietro Elia Campana School of Business, Society & Engineering, Mälardalen University, Västerås, Sweden

Jean-Louis Canaletti Research Centre of Vignola, University of Corsica Pasquale Paoli, Ajaccio, Route des Sanguinaires, France

Camille Cany CEA DEN I-tésé and ECP LGI, CEA Saclay, Gif-sur-Yvette Cedex, France

Antônia Sônia Alves Cardoso Diniz Department of Mechanical Engineering, Pontifícia Universidade Católica de Minas Gerais, Belo Horizonte, Brazil

R. Castrejón-García Dirección de Desarrollo Tecnológico, Instituto Nacional de Astrofísica, Óptica y Electrónica, Tonantzintla, PUE, Mexico

Janaína Oliveira Castro Silva Department of Mechanical Engineering, Pontifícia Universidade Católica de Minas Gerais, Belo Horizonte, Brazil

Giorgio Cau DIMCM, Department of Mechanical, Chemical and Materials Engineering, University of Cagliari, Sardinia, CA, Italy

Lucia Ceccherini Nelli DIDA Department, ABITA, University of Florence, Florence, Italy

Miqdam T. Chaichan University of Technology, Baghdad, Iraq

Yefri Chan Mechanical Engineering Department, Darma Persada University, East Jakarta, Indonesia

Wei-Chen Chang Institute of Organic and Polymeric Materials, National Taipei University of Technology, Taipei, Taiwan

Institute of Nuclear Energy Research, Taoyaun, Taiwan

Pradeep Chaturvedi Indian Association for the Advancement of Science, New Delhi, India

Priyanka Chawla Department of Chemistry, C.M.P. Degree College, University of Allahabad, Allahabad, India

Hung-Shuo Chen Institute of Organic and Polymeric Materials, National Taipei University of Technology, Taipei, Taiwan

F. Chenlo CIEMAT/DER, Madrid, Spain

Daniele Cocco DIMCM, Department of Mechanical, Chemical and Materials Engineering, University of Cagliari, Sardinia, CA, Italy

H. Coch Architecture & Energy Group. Department of Architectural Technology I, School of Architecture, UPC, Barcelona, Spain

A. Constantinou Applied Energy Centre, Nicosia, Cyprus

Fabio Cosentino Dipartimento di Ingegneria Elettrica, Elettronica ed Informatica, University of Catania, Catania, Italy

Pascal da Costa Laboratoire de Génie Industriel, ECP LGI, Ecole Centrale de Paris, Châtenay Malabry Cedex, France

I. Crespo Architecture & Energy Group. Department of Architectural Representation and Visual Analysis I, School of Architecture, UPC, Barcelona, Spain

Nicolas Crettenand Ecole Polytechnique Fédérale de Lausanne (EPFL), Lausanne, Switzerland

Christian Cristofari Research Centre of Vignola, University of Corsica Pasquale Paoli, Ajaccio, Route des Sanguinaires, France

R. E. Critoph School of Engineering, University of Warwick, Coventry, UK

Pascal Da Costa ECP LGI, Ecole Centrale de Paris, Laboratoire de Génie Industriel, Châtenay Malabry Cedex, France

Mebarka Daoudi Laboratory of semiconductor devices physics, (LPDS) University of Bechar, Bechar, Algeria

Chr. Darras Research Centre of Vignola, University of Corsica Pasquale Paoli, Ajaccio, Route des Sanguinaires, France

Abdul Salam Darwish Phoenix Renewable Energy Centre, Manchester, UK

N. Davies School of Engineering, University Of Liverpool, Brownlow Hill, UK

Davide Del Col Dipartimento di Ingegneria Industriale, Università degli Studi di Padova, Padova, Italy

Pankaj Deo Technische Universität Berlin, Freiburg, Germany

Valentina Dessi' Dept DAStU, Politecnico of Milano, Milano, Italy

P. Dewallef Department of Aerospace and Mechanics, Thermodynamics Laboratory, University of Liège, Liège, Belgium

M. M. S. Dezfouli Institute of Product Design & Manufacturing (IPROM), Universiti Kuala Lumpur, Kuala, Lumpur, Malaysia

Mahadzir Hj Din Faculty of Applied Sciences, Universiti Teknologi MARA Perlis, Arau, Perlis, Malaysia

Samuel Djom Rural Electrification Agency, Yaounde, Cameroon

P. C. Eames Centre for Renewable Energy Systems Technology, Loughborough University, Loughborough, UK

M. Hamdy El-Awady Water Pollution Research Department, National Research Centre, Cairo, Egypt

Adel A. Elbaset Faculty of Engineering, Minia University, Minia, Egypt

Mahmoud A. El-Bayoumi Mechanical Engineering Department, National Research Centre, Giza, Egypt

Hamdy El-Ghetany Solar Energy Department, National Research Centre, Cairo, Egypt

Anwar El-Hadi College of Engineering and Architecture, University of Bahri, Khartoum, Sudan

Business International and Resource Development, Khartoum, Sudan

David Elliott The Open University, Milton Keynes, UK

Ahmed F. Elsafty College of Engineering and Technology, American University of The Middle East (AUM), Egaila, Kuwait

O. N. A. Esmail Faculty of Engineering, Al Azhar University, Qena, Egypt

Hilal A. Fadhil Universiti Malaysia Perlis, Arau Perlis, Malaysia

Sergio Ferlito ENEA, Agenzia nazionale per le nuove tecnologie, l'energia e lo sviluppo economico sostenibile, Research Center of Portici, Portici, NA, Italy

Tauane Shaisly Fernandes Department of Mechanical Engineering, Pontifícia Universidade Católica de Minas Gerais, Belo Horizonte, Brazil

André Guimarães Ferreira Graduate Program of Energy Engineering, Centro Federal de Educação Tecnológica de Minas Gerais, Belo Horizonte, Brazil

Igor Finelli Laboratory of Control & Conversion of Energy, Federal University of Minas Gerais, Belo Horizonte, Brazil

David Finney The Open University, Milton Keynes, UK

Giuseppe Fiorenza Photovoltaic Technologies Unit, ENEA, Portici Research Centre, Portici (Naples), Italy

G. A. Florides Cyprus University of Technology, Limassol, Cyprus

M. Pilar Gárate Departamento de Industrias, Universidad Técnica Federico Santa María, Valparaíso, Chile

S Ghazi Department of Environmental Engineering, Islamic Azad University-Parand Branch, Parand, Iran

Giorgio Graditi Photovoltaic Technologies Unit, ENEA, Portici Research Centre, Portici (Naples), Italy

ENEA, Agenzia nazionale per le nuove tecnologie, l'energia e lo sviluppo economico sostenibile, Research Center of Portici, Portici, NA, Italy

Yvonne Haigh School of Management and Governance, Murdoch University, Perth, Western Australia

Hisham F. A. Hamed Faculty of Engineering, Minia University, Minia, Egypt

Ali Hamzeh Department of Electrical Engineering, Al-Ahliyya Amman University, Amman, Jordan

Sérgio de Morais Hanriot Department of Mechanical Engineering, Pontifícia Universidade Católica de Minas Gerais, Belo Horizonte, Brazil

Indra Haraksingh Department of Physics, The University of the West Indies, St. Augustine, Trinidad

Linda Hassaine Photovoltaic, Centre de Développement des Energies Renouvelables, CDER, Algiers, Algeria

A. Hassan Faculty of Engineering and Computing, Coventry University, Coventry, UK

S. A. Hassan Solar Energy Department, National Research Center, Giza, Egypt

Hassan A. N. Hejase Department of Electrical Engineering, United Arab Emirates University, Al-Ain, United Arab Emirates

Behnoosh Bakhtiari Heleyleh Department of Environmental and Energy Engineering Science and Research Branch, Islamic Azad University Tehran, Tehran, Iran

P. Henshall Centre for Renewable Energy Systems Technology, Loughborough University, Loughborough, UK

Rainer Hinrichs-Rahlwes European Renewable Energies Federation (EREF), Brussels, Belgium

Yoshito Hirata Institute of Industrial Science, The University of Tokyo, Tokyo, Japan

Department of Mathematical Informatics, The University of Tokyo, Tokyo, Japan

CREST, Saitama, Japan

Sarah Hout Unité de Développement des Equipements Solaires, UDES/Centre de Développement des Energies Renouvelables, CDER, Bou Ismail, W. Tipaza, Algérie

Faridah Hussain National Metrology Laboratory, SIRIM Berhad, Sepang, Selangor, Malaysia

Chul-Sang Hwang Smart Distribution Research Center, KERI, Changwon city, Republic of Korea

T. Hyde School of the Built Environment, University of Ulster, Coleraine, UK

A. Isalgué Architecture & Energy Group. Department of Applied Physics, School of Architecture, UPC, Barcelona, Spain

O. A. Jaramillo Instituto de Energías Renovables, Universidad Nacional Autónoma de México, Temixco, MOR, Mexico

Hasila Jarimi Faculty of Applied Sciences, Universiti Teknologi MARA Shah Alam, Shah Alam, Selangor, Malaysia

Faculty of Applied Sciences, Universiti Teknologi MARA Perlis, Arau, Perlis, Malaysia

Jin-Hong Jeon Smart Distribution Research Center, KERI, Changwon city, Republic of Korea

A. Kaabi Nejadian Renewable Energy Organization of Iran, Ahvaz, Iran

J. K. Kaldellis Mechanical Engineering Department, Soft Energy Applications and Environmental Protection Laboratory, Technological Educational Institute of Piraeus, Athens, Greece

John K. Kaldellis Soft Energy Applications and Environmental Protection Laboratory, Mechanical Engineering Department, Technological Educational Institute of Piraeus, Athens, Greece

S. A. Kalogirou Cyprus University of Technology, Limassol, Cyprus

Soteris A Kalogirou Department of Mechanical Engineering and Materials Science and Engineering, Cyprus University of Technology, Limassol, Cyprus

Eleni Kaplani RES Lab, Mechanical Engineering Department, Technological Educational Institute of Western Greece, Patra, Greece

Sergey Karabanov Ryazan State Radio Engineering University, Ryazan, Russia

M. A. Karim Mechanical Engineering Discipline, Science and Engineering Faculty, Queensland University of Technology, Brisbane, Australia

N. Katsellis Applied Energy Centre, Nicosia, Cyprus

Hussein A Kazem Renewable Energy & Sustainable Technology Research Group/ Sohar University, Sohar, Oman

Hussein A. Kazem Sohar University, Sohar, Oman

Joseph Kenfack National Advanced School of Engineering, University of Yaounde I, Yaounde, Cameroon

Fouad Khaldi Department of Physics, Faculty of Sciences, University of Batna, Batna, Algeria

Eung-Sang Kim Smart Distribution Research Center, KERI, Changwon city, Republic of Korea

Gyeong-Hun Kim Smart Distribution Research Center, KERI, Changwon city, Republic of Korea

Daniel Klauser Meteotest, Berne, Switzerland

E. Kondili Optimisation of Production Systems Laboratory, Technological Educational Institute of Piraeus, Athens, Greece

Emilia M. Kondili Optimisation of Productions Systems Laboratory, Mechanical Engineering Department, Technological Educational Institute of Piraeus, Athens, Greece

Stefan Krauter Department of Electrical Energy Technology – Sustainable Energy Concepts, University of Paderborn, Paderborn, Germany

Photovoltaik Institut Berlin AG, Berlin, Germany

Uday Kumar N. T. CSEM-UAE Innovation Center LLC, Ras Al Khaimah, United Arab Emirates

Department of Energy Technology, KTH Royal Institute of Technology, Stockholm, Sweden

Manoj Kumar P. CSEM-UAE Innovation Center LLC, Ras Al Khaimah, United Arab Emirates

P. Kyriakou Cyprus Institute of Energy, Nicosia, Cyprus

Hailong Li School of Business, Society & Engineering, Mälardalen University, Västerås, Sweden

José M. Lobo Departamento de Industrias, Universidad Técnica Federico Santa María, Valparaíso, Chile

John Lowry Hillsborough, Bishopstone, Swindon, UK

Adnan M Faculty of Electrical and Electronics Engineering, University Malaysia Pahang, Pekan, Malaysia

Stefan C. Müller Meteotest, Berne, Switzerland

Wilson N. Macêdo Group of Studies and Development in Alternatives Energy, Federal University of Pará, Belém, Brazil

Tapas K Mallick Environment and Sustainability Institute, University of Exeter, Penryn, UK

Norain A. Manaf Faculty of Applied Sciences, Universiti Teknologi MARA Shah Alam, Shah Alam, Selangor, Malaysia

Faculty of Applied Sciences, Universiti Teknologi MARA Perlis, Arau, Perlis, Malaysia

Christine Mansilla CEA DEN I-tésé, CEA Saclay, Gif-sur-Yvette Cedex, France

Bruno Marciano Management of Alternatives Energy, Energy Company of Minas Gerais, Belo Horizonte,, Brazil

Andrew Martin Department of Energy Technology, KTH Royal Institute of Technology, Stockholm, Sweden

Sohif Bin Mat Solar Energy Research Institute, Faculty of Engineering, Universiti Kebangsaan Malaysia, Bangi, Selangor, Malaysia

Gilles Mathonnière CEA DEN I-tésé and ECP LGI, CEA Saclay, Gif-sur-Yvette Cedex, France

Hudhaifa Mazin Universiti Malaysia Perlis, Arau Perlis, Malaysia

Federico Mazzelli Department of Industrial Engineering (DIEF), University of Florence, Florence, Italy

Nesrine Mendil Laboratory of semiconductor devices physics, (LPDS) University of Bechar, Bechar, Algeria

A. Michaelides RTD Talos, Engomi, Nicosia, Cyprus

Adriano Milazzo Department of Industrial Engineering (DIEF), University of Florence, Florence, Italy

Abdulrahman Th. Mohammad Solar Energy Research Institute, Faculty of Engineering, Universiti Kebangsaan Malaysia, Bangi, Selangor, Malaysia

Ali Mohammadi Department of Environmental and Energy Engineering Science and Research Branch, Islamic Azad University Tehran, Tehran, Iran

Gowtham Mohan CSEM-UAE Innovation Center LLC, Ras Al Khaimah, United Arab Emirates

Department of Energy Technology, KTH Royal Institute of Technology, Stockholm, Sweden

Luís Monteiro Laboratory of Control & Conversion of Energy, Federal University of Minas Gerais, Belo Horizonte, Brazil

Energy Engineering Department, Catholic University of Minas Gerais, Belo Horizonte,, Brazil

R. W. Moss School of Engineering, University of Warwick, Coventry, UK

Fabrice Motte Research Centre of Vignola, University of Corsica Pasquale Paoli, Ajaccio, Route des Sanguinaires, France

Amanullah MT Oo School of Engineering, Faculty of Science, Engineering & Built Environment, Deakin University, Melbourne, Australia

Shyam S. Nandwani Heredia, Costa Rica

A Kaabi Nejadian Renewable Energy Organization of Iran, Tehran, Iran

Benno Nibeler NIEBERLE-SOLAR Company, Biessenhofen, Germany

J. E. Nielsen PlanEnergi, København SV, Denmark

T. M. Nining Dyah The Graduate School, Darma Persada University, East Jakarta, Indonesia

M. L. Nivet Research Centre of Vignola, University of Corsica Pasquale Paoli, Ajaccio, Route des Sanguinaires, France

Bassam A. Noaman College of Engineering and Technology, American University of The Middle East (AUM), Egaila, Kuwait

G. Nofuentes IDEA Research Group, University of Jaén, Jaén, Spain

Eduardo Nohme Laboratory of Control & Conversion of Energy, Federal University of Minas Gerais, Belo Horizonte, Brazil

Gilles Notton Research Centre of Vignola, University of Corsica Pasquale Paoli, Ajaccio, Route des Sanguinaires, France

Alexander Olsson School of Chemical Science & Engineering, KTH Royal Institute of Technology, Stockholm, Sweden

Hugo Osorio Departamento de Industrias, Universidad Técnica Federico Santa María, Valparaíso, Chile

Mahmod Othman Faculty of Computer Sciences and Mathematics, Universiti Teknologi MARA Perlis, Arau, Perlis, Malaysia

Mohd. Yusof Hj. Othman Solar Energy Research Institute (SERI), Universiti Kebangsaan Malaysia, Bangi, Selangor, Malaysia

Maamar Ouali Unité de Développement des Equipements Solaires, UDES/Centre de Développement des Energies Renouvelables, CDER, Bou Ismail, W. Tipaza, Algérie

Andrea Padovan Dipartimento di Ingegneria Industriale, Università degli Studi di Padova, Padova, Italy

A. Pagès-Ramon Architecture & Energy Group. Department of Architectural Technology I, School of Architecture, UPC, Barcelona, Spain

G. P. Panayiotou Cyprus University of Technology, Limassol, Cyprus

Agis M. Papadopoulos Process Equipment Design Laboratory Department of Mechanical Engineering, Aristotle University Thessaloniki, Thessaloniki, Greece

Christiana Papapostolou Optimisation of Productions Systems Laboratory, Mechanical Engineering Department, Technological Educational Institute of Piraeus, Athens, Greece

Giovanni Luigi Paparo Industrial Engineering Department, University of Naples Federico II, Naples, Italy

C. Pardal Architecture & Energy Group. Department of Architectural Technology I, School of Architecture, UPC, Barcelona, Spain

T. Parisis RTD Talos, Engomi, Nicosia, Cyprus

Minwon Park Department of Electrical Engineering, Changwon National University, Changwon city, Republic of Korea

João T. Pinho Group of Studies and Development in Alternatives Energy, Federal University of Pará, Belém, Brazil

André Quinan Laboratory of Control & Conversion of Energy, Federal University of Minas Gerais, Belo Horizonte, Brazil

Ayat. A. Ragheb Mechanical Engineering Department, National Research Centre, Giza, Egypt

K S Reddy Heat Transfer and Thermal Power Laboratory, Department of Mechanical Engineering, Indian Institute of Technology Madras, Chennai, India

Jan Remund Meteotest, Berne, Switzerland

G. Roditis Applied Energy Centre, Nicosia, Cyprus

M. A. M. Rosli Solar Energy Research Institute, Faculty of Engineering, Universiti Kebangsaan Malaysia, Bangi, Selangor, Malaysia

Faculty of Mechanical Engineering, Universiti Teknikal Malaysia Melaka, Durian Tunggal, Melaka, Malaysia

Hafidz Ruslan Solar Energy Research Institute, Faculty of Engineering, Universiti Kebangsaan Malaysia, Bangi, Selangor, Malaysia

M. Saadeldin Physics Department, Faculty of Science, Cairo University, Giza, Egypt

K. S. M. Sahari Department of Mechanical Engineering, Universiti Tenaga Nasional, Jalan IKRAM-UNITEN, Selangor, Kajang, Malaysia

Imhamed M. Saleh Ali Department of Mechanical Engineering, Sirte University, Sirte, Libya

E. Salleh Solar Energy Research Institute, Faculty of Engineering, Universiti Kebangsaan Malaysia, Bangi, Selangor, Malaysia

Mohamad reza Khaje Samakoosh Mazandran Distribution Power Company, Sari, Iran

Abubakar S. Sambo Federal Secretariat, Abuja, Nigeria

Abbas Sandouk Department of Electrical power Engineering, Damascus University, Damascus, Syria

K. Sartor Department of Aerospace and Mechanics, Thermodynamics Laboratory, University of Liège, Liège, Belgium

Hasan Saygin Engineering Faculty, Mechanical Engineering Department, Istanbul Aydin University, Istanbul, Florya, Turkey

A. M. Sayigh World Renewable Energy Network (WREN), Brighton, United Kingdom

Faramarz Separi Mazandran Distribution Power Company, Sari, Iran

Sabry Shaaban ESC La Rochelle School of Business, La Rochelle, France

GM Shafiullah School of Engineering, Faculty of Science, Engineering & Built Environment, Deakin University, Melbourne, Australia

Mohammed Sheldin Solar Energy Research Institute, Faculty of Engineering, Universiti Kebangsaan Malaysia, Bangi, Selangor, Malaysia

Jingchun Shen Department of Architecture and Built Environment, University of Nottingham Ningbo China, Ningbo, China

E. T. El Shenawy Solar Energy Department, National Research Centre, Giza, Egypt

G. S. F. Shire School of Engineering, University of Warwick, Coventry, UK

T. D. Short School of Engineering, University Of Liverpool, Brownlow Hill, UK

Selênio R. Silva Laboratory of Control & Conversion of Energy, Federal University of Minas Gerais, Belo Horizonte, Brazil

Ranj Sirwan Refrigeration and Air Conditioning Engineering Department, Erbil Technical Engineering College, Erbil Polytechnic University, Erbil, Iraq

Evgeny Slivkin Ryazan State Radio Engineering University, Ryazan, Russia

Kamaruzzman Sopian Solar Energy Research Institute, Faculty of Engineering, Universiti Kebangsaan Malaysia, Bangi, Selangor, Malaysia

Alex Stojcevski School of Engineering, Faculty of Science, Engineering & Built Environment, Deakin University, Melbourne, Australia

Herliyani Suharta Energy Technology Center (B2TE)-BPPT, Tangerang, Banten, Indonesia

M. Yusof Sulaiman Solar Energy Research Institute, Faculty of Engineering, Universiti Kebangsaan Malaysia, Bangi, Selangor, Malaysia

Dmitry Suvorov Ryazan State Radio Engineering University, Ryazan, Russia

Hideyuki Suzuki Institute of Industrial Science, The University of Tokyo, Tokyo, Japan

Department of Mathematical Informatics, The University of Tokyo, Tokyo, Japan

CREST, Saitama, Japan

Donald Swift-Hook Surrey, UK

W. Tadros Solar Energy Department, National Research Center, Giza, Egypt

Jean-Baptiste Thomas CEA DEN DIR EC, CEA Saclay, Gif-sur-Yvette Cedex, France

Zahia Tigrine Unité de Développement des Equipements Solaires, UDES/Centre de Développement des Energies Renouvelables, CDER, Bou Ismail, W. Tipaza, Algérie

Giuseppe Marco Tina Dipartimento di Ingegneria Elettrica, Elettronica ed Informatica, University of Catania, Catania, Italy

Vittorio Tola DIMCM, Department of Mechanical, Chemical and Materials Engineering, University of Cagliari, Sardinia, CA, Italy

Pedro Torres Group of Studies and Development in Alternatives Energy, Federal University of Pará, Belém, Brazil

M. Torres-Ramírez IDEA Research Group, University of Jaén, Jaén, Spain

Mridula Tripathi Department of Chemistry, C.M.P. Degree College, University of Allahabad, Allahabad, India

Zafer Utlu Engineering Faculty, Mechanical Engineering Department, Istanbul Aydin University, Istanbul, Florya, Turkey

Mehdi Barimani Varandi Mazandaran Regional Electric Company, Sari, Iran

Y. Vasiades Cyprus Organisation for Standardisation, Nicosia, Cyprus

N. Velázquez Instituto de Ingeniería, Universidad Autónoma de Baja California, Mexicali, BC, Mexico

Cristina Ventura Dipartimento di Ingegneria Elettrica, Elettronica ed Informatica, University of Catania, Catania, Italy

Joseph Voufo National Advanced School of Engineering, University of Yaounde I, Yaounde, Cameroon

Arthur A. Williams Department of Electrical & Electronic Engineering, University of Nottingham, Nottingham, UK

Ying Xu Shanghai Pacific Energy Center, Shanghai, China

Jinyue Yan School of Business, Society & Engineering, Mälardalen University, Västerås, Sweden

School of Chemical Science & Engineering, KTH Royal Institute of Technology, Stockholm, Sweden

Baharuddin Yatim Solar Energy Research Institute (SERI), Universiti Kebangsaan Malaysia, Bangi, Selangor, Malaysia

H. Yatimi Modeling and Simulation of Mechanical Systems Team Physics Department, Faculty of Sciences, Abdelmalek Essaadi University, Tetouan, Morocco

In-keun Yu Department of Electrical Engineering, Changwon National University, Changwon city, Republic of Korea

Wan-Chin Yu Institute of Organic and Polymeric Materials, National Taipei University of Technology, Taipei, Taiwan

D. Zafirakis Mechanical Engineering Department, Soft Energy Applications and Environmental Protection Laboratory, Technological Educational Institute of Piraeus, Athens, Greece

Wim Zeiler Faculty of the Built Environment, University of Technology Eindhoven, Eindhoven, Netherlands

Chi Zhang School of Chemical Science & Engineering, KTH Royal Institute of Technology, Stockholm, Sweden

Xingxing Zhang Department of Architecture and Built Environment, University of Nottingham Ningbo China, Ningbo, China

Xudong Zhao School of Engineering, University of Hull, Hull, UK

Djamila Zioui Unité de Développement des Equipements Solaires, UDES/Centre de Développement des Energies Renouvelables, CDER, Bou Ismail, W. Tipaza, Algérie

Part I
Metereology 2014

Chapter 1
Estimation of Global and Diffuse Horizontal Irradiance in Abu Dhabi, United Arab Emirates

Hassan A. N. Hejase and Ali H. Assi

Abstract This chapter presents empirical regression models for estimating the monthly average daily global horizontal irradiance in the capital city of Abu Dhabi, United Arab Emirates (UAE), based on the measurement data of 9 years. It also addresses selected models available in the literature for the estimation of diffuse horizontal irradiance and their correlation to measured data. The effect of atmospheric turbidity, due to humidity, sandstorms, and aerosols, on the solar irradiance is also addressed.

Keywords Empirical model · Clearness index · Global horizontal irradiance · Diffuse horizontal irradiance · Diffuse fraction · Turbidity factor

1.1 Introduction

Solar radiation received at the surface is of primary importance for the purpose of building solar energy systems, crop productivity and soil erosion studies, and hydraulics. Empirical and satellite-derived weather models are a viable alternative in the absence of measuring stations in the regions where solar data are required.

El Chaar and Lamont [1] performed measurements for the global horizontal irradiance (GHI) in five locations of the Abu Dhabi Emirate for 2008. They also measured the clearness index between 0.14 (minimum value January) and 0.803 (May) with the monthly average clearness index in the range 0.6–0.7 throughout the year. They indicated that the winter season (Dec–Jan–Feb) observes the lowest radiation due to dust storms and scattered muddy showers. Spring season exhibits the highest radiation due to the pleasant ambient temperature and acceptable humidity levels. The high temperatures and humidity levels near the coastal areas, where

H. A. N. Hejase (✉)
Department of Electrical Engineering, United Arab Emirates University, P.O. Box 15551, Al-Ain, United Arab Emirates
e-mail: hhejase@uaeu.ac.ae

A. H. Assi
Department of Electrical and Electronic Engineering, Lebanese International University, Beirut, Lebanon

A. Sayigh (ed.), *Renewable Energy in the Service of Mankind Vol II*,
DOI 10.1007/978-3-319-18215-5_1

80 % of the population resides, may cause a decrease in GHI due to higher turbidity factor resulting from the high temperature, humidity levels, and effect of aerosols and dust particles.

Abdalla and Feregh [2] performed GHI measurements in Abu Dhabi for 10 years (1971–1980) to establish estimation models. The maximum monthly mean daily GHI measured during the 10-year period was 6.55 kWh/m^2 during May which corresponds to a clearness index of 0.6 with 11.3 mean maximum sunshine hours (SSH) per day in May and June and the lowest values of GHI and SSH being in January with 3.38 kWh/m^2 and 7.6 h, respectively. They employed the empirical models of Page [3] and Liu and Jordan [4] to estimate the diffuse horizontal radiation in the absence of measurement data.

Elnashar [5] conducted GHI and diffuse horizontal irradiance (DHI) measurements in Abu Dhabi during 1987. The author reported that the estimated instantaneous values of clearness index, diffuse radiation, atmospheric transmittance, and extinction coefficient were strongly dependent on the air mass and month of the year. It was noted that both the clearness index and diffuse fraction had diurnal and seasonal variations with higher values in summer than in the winter season. The clearness index reached its lowest value in July and the highest value in January. This was attributed to the fact that the level of fine dust particles in the air and relative humidity were higher during the summer than the winter months. Elnashar suggested an improvement of the models of Page and Liu and Jordan by incorporating the effect of air mass in the empirical formulas.

More recently, Masdar Institute researchers [6] have performed corrections to the Heliosat-2 model in order to adapt it to the climatic environment of the Arabian Peninsula which is characterized by high concentrations of airborne dust particles and high humidity levels. The authors benchmarked remote estimation of GHI by the Heliosat-2 model against ground-based measurements in different locations of the emirate of Abu Dhabi. They noted higher turbidity levels in the summer months (May–July) for inland ground-measurement stations compared to the near-coastal stations. The high values of turbidity are attributed to high dust generation in the desert areas which experience a hot and dry summer.

The annual average daily SSH and clearness index for Abu Dhabi are 9.87 h and 0.66, respectively. These data were derived from the measured National Center for Meteorology and Seismology (NCMS) data for the years 2002–2008. The data indicate that Abu Dhabi enjoys clear skies for most of the year with lower values of clearness index during rainfall or sandstorms.

The monthly mean of the daily SSH is similar all over the emirate of Abu Dhabi, United Arab Emirates (UAE), and ranges between 8.4 h in winter to 11.6 h in summer (daily average of 9.7 h), which corresponds to an average annual solar radiation of approximately 2300 kWh/m^2, that is, 6.3 kWh/m^2 per day. Referring to the climatic atlases published by the World Meteorological Organization (WMO), the Arabian Peninsula and the African desert are the two areas that receive the highest amount of the incoming solar radiation [7].

Section 1.2 presents a comparison between measured data available from ground stations, literature, and satellite website databases. Section 1.3 addresses the estimation models for GHI and DHI and discusses the results followed by conclusions in Section 1.4.

1.2 Comparison of the Available Meteorological Data

The GHI is the total amount of shortwave radiation received by a surface horizontal to the ground. This value is of particular interest to photovoltaic installations and includes both direct normal irradiance (DNI) and DHI. DNI is the solar radiation that comes in a straight line from the direction of the sun at its current position in the sky. DHI is the solar radiation that does not arrive on a direct path from the sun, but has been scattered by molecules and particles in the atmosphere and comes equally from all directions. On a clear day, most of the solar radiation received by a horizontal surface will be DNI, while on a cloudy day, it will be mostly DHI.

The variation in radiation levels is attributed to the presence of clouds which depends on seasonal and diurnal variations. The clearness index k_T measures the solar radiation under clear sky conditions. It is defined as the ratio between the global irradiance at the site and its extraterrestrial value ($k_T = \mathrm{GHI}/G_0$). The clearness index can be used as an indication of cloudiness: high values for cloudless skies and low values for overcast skies. However, because the diffusion and absorption of solar radiation in the atmosphere depend on the sun's zenith angle, and because extraterrestrial irradiance cannot offset this dependency, the clearness index is affected by solar elevation. If the diffuse solar radiation on a horizontal surface is available, then the diffuse fraction $k_\mathrm{D} = \mathrm{DHI}/\mathrm{GHI}$ gives better information on the cloudiness than the clearness index [8].

1.2.1 Data Preparation

The weather database for the city of Abu Dhabi, UAE, is provided by the NCMS in Abu Dhabi. The results presented in this chapter make use of the mean daily SSH and mean daily GHI (in $\mathrm{kWh/m^2}$). Available weather data for Abu Dhabi span across the years 2002–2010. The 7-year data for 2002–2008 are used to construct the empirical regression-based models which are tested with the remaining 2-year data (2009–2010). Leap year days are removed to ensure uniformity of month comparison over the 7-year model period.

1.2.2 Statistical Error Parameters

The estimated and measured GHI data are correlated using the following statistical parameters: root mean square error (RMSE), mean absolute bias error (MABE), mean bias error (MBE), mean absolute percentage error (MAPE), and coefficient of determination (R^2). The statistical error parameters attest to the accuracy of the models used for estimating the monthly mean daily global horizontal irradiance (GHI). The best model should yield the lowest value of RMSE, MABE, MBE, MAPE, and the highest R^2. The formulas for the computation of the statistical parameters are found in [9]. A low value of MBE indicates good long-term prediction performance.

1.2.3 Available Measured Global Horizontal Irradiance Data

Figure 1.1 shows a comparison of monthly average daily GHI measured values for the city of Abu Dhabi. The data from the National Aeronautics and Space Administration-Surface Meteorology and Solar Energy (NASA-SSE) and the Solar Radiation Data (SoDa) satellite website databases are included for comparison [10, 11]. The data from the Masdar Institute were extracted from the UAE Solar Atlas [12] for the years 2010–2012. The measured data from different sources appear to be in good agreement for all months except for April–July where NCMS values are higher. Note that the NCMS-measured data (2002–2008) exceed all other values but are very close to data presented by Alnaser et al. [13] and are slightly higher than the NASA-SSE satellite-derived data and data provided by Islam et al. [14]. On the other hand, the average data (1971–1981) given by Abdalla and Feregh [3] are the lowest for most of the year. The annual average of monthly mean daily GHI obtained from Islam et al. [14], Abdalla and Feregh [2], Alnaser et al., NCMS

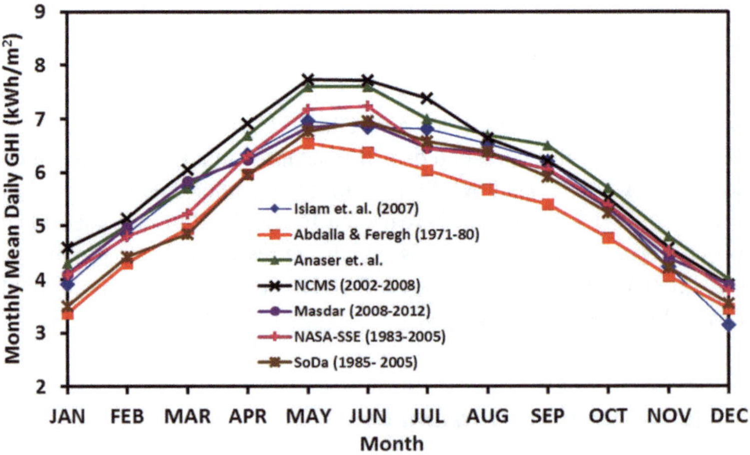

Fig. 1.1 Monthly average daily global horizontal irradiance (GHI) for Abu Dhabi from selected sources. *NCMS* National Center for Meteorology and Seismology, *NASA-SSE* National Aeronautics and Space Administration-Surface Meteorology and Solar Energy, *SoDa* Solar Radiation Data

(2002–2008), UAE Solar Atlas (2008–2012), NASA-SSE (1983–2005) and SoDa (1985–2005) are 5.58, 5.08, 5.97, 6.03, 5.62, 5.61, and 5.36 kWh/m^2, respectively. Note that, with the exception of Alnaser, the measured NCMS values are higher than all other data. GHI values by Alnaser exceed those by NCMS for the months of August–December.

The differences between the databases of the UAE Solar Atlas and NCMS could be attributed to the following factors:

- Using different types of instrumentation from different suppliers, resulting in different calibrations
- Applying different quality control criteria
- Using different data-processing approaches and formats
- Following different levels of standardization adopted by the WMO
- Type of the regular and professional maintenance performed for the weather stations recording meteorological data

Nevertheless, these weather stations provide more or less useful information that will definitely support the ongoing climatic studies of the UAE.

Furthermore, and as was concluded in recent work by Eissa et al. [6, 15], the overestimation of GHI measured by NCMS and reported by Alnaser during the months of April–July could be attributed to the higher concentrations of airborne dust particles during these months and higher humidity levels.

The computed error parameters resulting from the mutual correlation of the measured GHI data from the UAE Solar Atlas (Masdar), NCMS, NASA-SSE, and SoDa indicate that the best agreement is observed between measurements from Masdar and NASA-SSE with a coefficient of determination of 98 %, RMSE = 0.236 kWh/m^2, MBE = 0.004 kWh/m^2, and MAPE = 2.83 %.

1.3 Results and Discussions

1.3.1 Global Horizontal Irradiance Models

The daily average extraterrestrial data for SSH (S_0) and extraterrestrial solar radiation on a horizontal surface in kWh/m^2 (G_0) are computed using the geographical location of Abu Dhabi (24.43°N and 54.45°E) [9]. The monthly average daily G_0 and S_0 values are then produced.

Using the MATLAB and IBM-SPSS software packages with 7-year measured data (2002–2008), six empirical regression models for Abu Dhabi were developed for daily mean GHI and SSH. The models were tested using a 2-year data set (2009–2010) and validated against satellite-based weather data from NASA-SSE, SoDa, and Solar and Wind Energy Resource Assessment (SWERA). The estimated monthly GHI values ($G_0 * k_T$) are computed using the six developed empirical formulas in addition to the models developed by Glover and McCulloch [16] and Tiwari and

Table 1.1 Empirical regression models for Abu Dhabi. (k_T = GHI/G_0; s_T = SSH/S_0)

Model Type	Equation
Linear equation	$k_T = 0.318 + 0.417 * s_T$
Second order polynomial (quadratic)	$k_T = 0.366 + 0.295 * s_T + 0.077 * s_T{}^2$
Third order polynomial (cubic)	$k_T = 3.232 + 14.588 * s_T 18.721 * s_T{}^2 + 8.188 * s_T{}^3$
Logarithmic equation	$k_T = 0.725 + 0.327 * \log(s_T)$
Log-linear equation	$k_T = 0.307 + 0.428 * s_T 0.009 * \log(s_T)$
Exponential equation	$k_T = 0.389 * \exp(0.643 * s_T)$
Glover and McCulloch [16]	$k_T = 0.26 \cos(\phi) + 0.52 s_T$ for $\phi < 60^0$
Tiwari and Sangeeta [17]	$k_T = [-0.110 + 0.235 \cos(\phi)] + [1.772 - 0.553 \cos(\phi)] s_T - 0.694 s_T{}^2$

Sangeeta [17] which incorporate the effect of the region's latitude (ϕ in degrees). Table 1.1 shows the models used for Abu Dhabi, UAE.

All the aforementioned empirical correlations are validated against the measured data by NCMS and Masdar (UAE Solar Atlas) for the year 2010 as shown in Fig. 1.2. Data from NASA-SSE were included for comparison. Note that all the empirical models agree well with NCMS data but disagree with Masdar and NASA data during the warm season (April–September). The statistical error parameters computed from the correlation between the empirical models and Masdar 2010 data

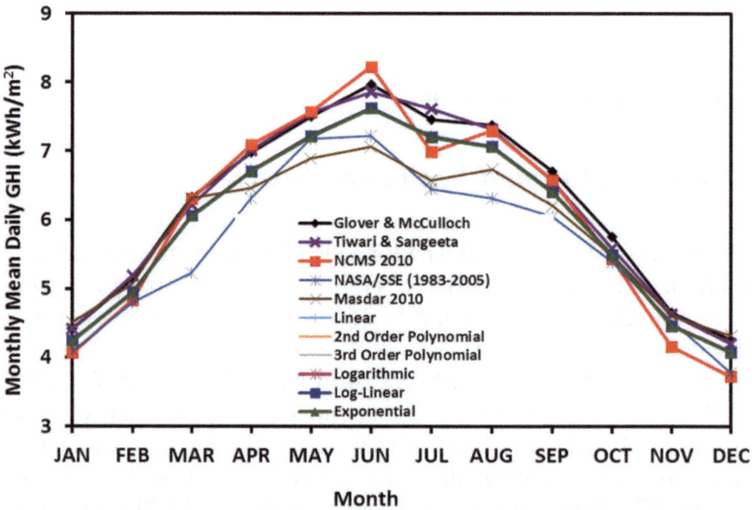

Fig. 1.2 Empirical regression models and measured global horizontal irradiance (GHI) data for Abu Dhabi (2010). *NCMS* National Center for Meteorology and Seismology, *NASA-SSE* National Aeronautics and Space Administration-Surface Meteorology and Solar Energy

yield coefficients of determination exceeding 97% for all models. The exponential empirical model outperforms the empirical models by Glover and McCulloch and Tiwari and Sangeeta as attested by the error parameters and yields RMSE = 0.444 kWh/m^2, MBE = 0.226 kWh/m^2, MAPE = 5.87%, and R^2 = 97.4%.

The drop in GHI values in a particular month may be attributed to sandstorms which have a higher frequency of occurring, while the drop in summer time (e.g., July) is due to the highly turbid atmosphere throughout the month. GHI is characteristically lower for the winter months than the summers with a distinct peak appearing in June. According to Eissa et al. [6, 15], the accuracy of satellite-based models for the estimation of solar radiation is location dependent. Over the UAE, it was observed that the Heliosat-2 method needed a recalibration before its application. Moreover, they added that annual maps generally overestimate the direct normal irradiation when compared with the available ground measurements over the UAE. That is mainly due to the unique environment of the Arabian Peninsula, with modest cloud coverage, high concentrations of airborne dust and high humidity in the coastal areas. Therefore, regionally based models are required for such environments.

The reason for the difference of NASA-SSE data from measured Masdar data could be attributed to heavy dusty days where the accumulation of dust particles on the pyranometer causes an underestimation in the measured irradiance. The uncertainty in GHI measurements may lead to more estimation errors in the developed empirical models. This may be enhanced by considering empirical models which correlate to the atmospheric turbidity in the region.

1.3.2 Diffuse Solar Radiation Models

Estimates of the DHI are needed for hardware system design parameters such as solar panel tilt, solar concentrator size, day lighting, as well as agricultural and hydrology applications. There are no known methods of estimating GHI over the globe with proven accuracy [18].

The relationship between monthly average daily DHI and GHI, named diffusion fraction (k_D = DHI/GHI), is the most significant parameter. This relationship can be found from direct meteorological observations or through an empirical relationship as studied by various researchers. Values of k_D near 1.0 indicate overcast skies, whereas values around 0.2 imply cloudless skies. At low sun elevations, the diffusion fraction is nearly 1.0 for both cloudless and overcast skies.

In the UAE, diffuse radiation measurement data from NCMS were unavailable until recently when Masdar established its UAE Solar Atlas [12]. Appropriate empirical formulas are employed, in addition to available satellite-derived data by NASA-SSE [11], for the estimation of the monthly average of daily DHI.

The DHI is computed for Abu Dhabi, UAE, using the models displayed in Table 1.2 [3–4, 19, 20–23]. Equations (1.6), (1.7), and (1.8) correspond to the non-seasonal monthly, seasonal monthly, and seasonal daily models, respectively [23].

Table 1.2 Empirical regression models for diffuse fraction in Abu Dhabi (k_D = DHI/GHI; k_T = GHI/G_0)

Model equation	Eq.
$k_D = 1 - 1.13\, k_T$	(1.1)
$k_D = 1.390 - 4.027\, k_T + 5.531\, k_T^2 - 3.108\, k_T^3$	(1.2)
$k_D = 1.4112 - 1.6956\, k_T$, for $0.34 < k_T < 0.73$	(1.3)
$k_D = 1.191 - 1.783\, k_T + 0.862\, k_T^2 - 0.324\, k_T^3$, for $0.15 < k_T < 0.8$	(1.4)
$k_D = -1.897 + 14.536\, k_T - 26.828\, k_T^2 + 14.976\, k_T^3$	(1.5)
$k_D = 1.317 - 3.023\, k_T + 3.372\, k_T^2 - 1.769\, k_T^3$, for $0.3 \le k_T \le 0.8$	(1.6)
$k_D = 1.391 - 3.569\, k_T + 4.189\, k_T^2 - 2.137\, k_T^3$ for $\omega_s \le 81.4^0$ and $0.3 \le k_T \le 0.8$	(1.7a)
$k_D = 1.311 - 3.022\, k_T + 3.427\, k_T^2 - 1.821\, k_T^3$ for $\omega_s > 81.4^0$ and $0.3 \le k_T \le 0.8$	(1.7b)
$k_D = \begin{cases} 1.0 - 0.2727\, k_T + 2.4495\, k_T^2 - 11.9514\, k_T^3 + 9.3879\, k_T^4 \\ 0.143 \end{cases}$ for $\omega_s < 81.4^0$ and $k_T < 0.715$; for $\omega_s < 81.4^0$ and $k_T \ge 0.715$	(1.8a)
$k_D = \begin{cases} 1.0 + 0.2832\, k_T - 2.5557\, k_T^2 + 0.8448\, k_T^3 & \text{for } \omega_s \ge 81.4^0 \text{ and } k_T < 0.722 \\ 0.175 & \text{for } \omega_s \ge 81.4^0 \text{ and } k_T \ge 0.722 \end{cases}$	(1.8b)

In Eqs. (1.7a, b) and (1.8a, b), ω_s is the sunset hour angle for the "monthly average day" and is given by:

$$\omega_s = \cos^{-1}[-\tan(\text{solar declination}) * \tan(\text{latitude})] \tag{1.9}$$

where

$$\text{solar declination} = 23.45 * \sin\left[6.303 * \{(284 + n)/365\}\right] \tag{1.10}$$

and n is the day number of year ($n = 1$ implies January 1). A positive value for ω_s indicates west relative to solar noon. The "monthly average day" is the day (in the month) whose declination is closest to the average declination for that month [23]. Erbs et al. [23] grouped the monthly average daily diffuse solar radiation data into three seasonal bins, following the procedure of Collares-Pereira and Rabl [24], according to the monthly average value of the sunset hour angle ω_s, where

$$\begin{array}{lcc} \text{Winter} & \text{NOV} - \text{JAN} & \omega_s < 81.4^0 \\ \text{Spring and Fall} & \text{FEB} - \text{APR; AUG} - \text{OCT} & 81.4^0 \le \omega_s \le 98.6^0 \\ \text{Summer} & \text{MAY} - \text{JUL} & \omega_s \le 98.6^0 \end{array} \tag{1.11}$$

Eissa et al. [15] suggested the following corrected formula to compute the DHI under clear sky conditions (cloud-free):

$$\mathrm{DHI}_{\mathrm{Clear}} = I_0 \varepsilon \cos\theta_z \{0.444 + [-1.124 + 0.154 * T_L(2)] * \cos\theta_z + [0.685 - 0.117 * T_L(2)] \cos^2\theta_z\} \tag{1.12}$$

In Eq. (1.12), I_0 is the solar constant (= 1367 W/m^2) and ε is the sun–earth distance correction factor which is computed from Spencer [25] as:

$$\varepsilon = 1.00011 + 0.034221 \cos\varphi + 0.00128 \sin\varphi + 0.000719 \cos 2\varphi + 0.000077 \sin 2\varphi \tag{1.13}$$

where φ is the day angle in radians (i.e., $\varphi = 2\pi \frac{n-1}{365}$) with "$n$" denoting the day of the year index ($n = 1, 2, \ldots, 365$). θ_z is the solar zenith angle, which is estimated using results from spherical trigonometry by [26–27] as:

$$\cos\theta_z = \sin\phi \sin\delta + \cos\phi \cos\delta \cos h \tag{1.14}$$

where h is the hour angle in the local solar time, δ is the current declination of the sun, and ϕ is the local latitude. A simpler formula for θ_z on a daily basis is suggested by Cardell [28]:

$$\theta_z(\text{degrees}) = 41 + 23.5 * \cos(\pi\, n / 180) \tag{1.15}$$

The turbidity factor $T_L(2)$ in Eq. (1.12) is computed by Eissa et al. [15] as a function of the measured DNI, air mass, sun–earth distance correction factor, and optical thickness of the Rayleigh atmosphere. The DHI is computed for Abu Dhabi (latitude = 24.43° and longitude = 54.45°) from Eq. (1.12) and compared with data reported by the UAE Solar Atlas [12]. Figure 1.3 shows the comparison between measured and estimated models including the suggested enhanced models for DHI in the coastal region of the UAE (Abu Dhabi). For the warm season (April–September), the SoDa data are in better agreement with the computed and measured data from Masdar and computed inland and near coast. The reported data by NASA-SSE and empirical models in Eqs. (1.1–1.8) are closer to the measured Masdar data and corrected DHI models for the mild season (October–December, January–March). A combination of both models may then be taken for best estimation of DHI in Abu Dhabi. Note that DHI almost linearly increases to peak in July and then linearly decreases to the minimum value in December.

The monthly average daily diffuse fraction (diffuse to global solar irradiance ratio—DHI/GHI) is computed for the Abu Dhabi models in Eqs. (1.4–1.11) from Table 1.2 using the GHI values of the Masdar Solar Atlas for the year 2010, as shown in Fig. 1.4. Values for the UAE Solar Atlas, NASA-SSE, SoDa, and those computed near the coastal areas of Abu Dhabi are included for reference. Note that

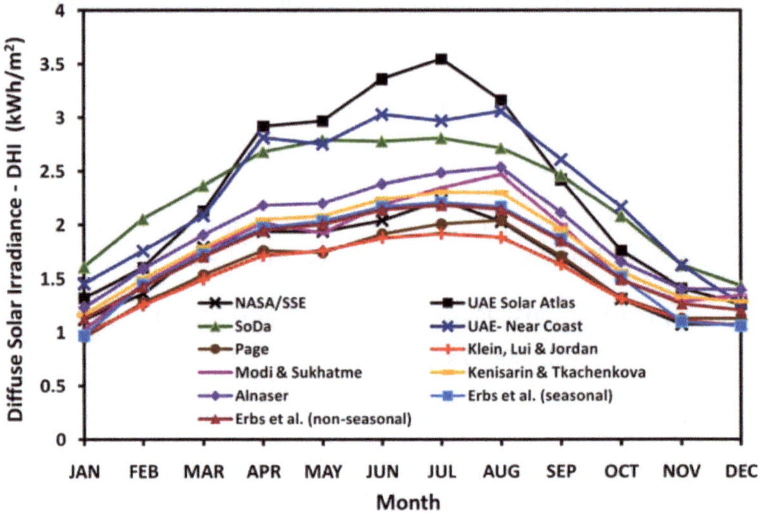

Fig. 1.3 Diffuse solar irradiance in Abu Dhabi. *DHI* diffuse horizontal irradiance, *NASA-SSE* National Aeronautics and Space Administration-Surface Meteorology and Solar Energy, *SoDa* Solar Radiation Data, *UAE* United Arab Emirates

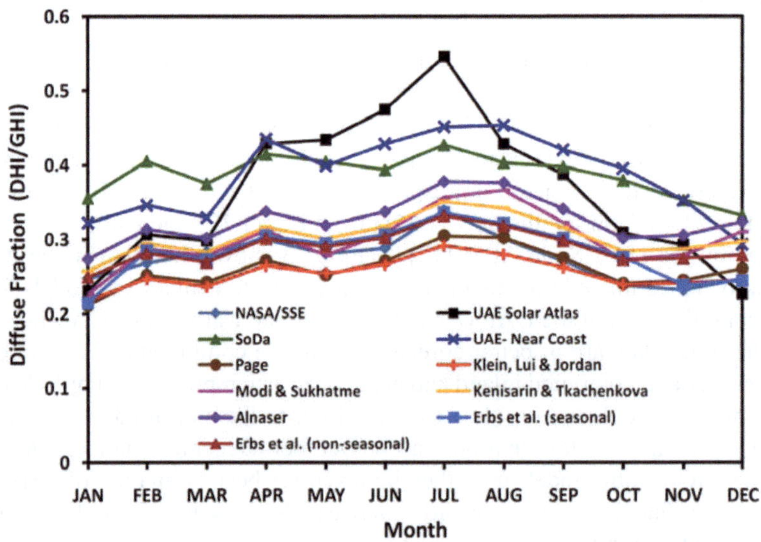

Fig. 1.4 Diffuse solar fraction in Abu Dhabi (2010). *DHI* diffuse horizontal irradiance, *GHI* global horizontal irradiance, *NASA-SSE* National Aeronautics and Space Administration-Surface Meteorology and Solar Energy, *SoDa* Solar Radiation Data, *UAE* United Arab Emirates

the diffuse fraction does not exceed 45.4 % during August (Abu Dhabi coast) and is lowest with 21 % (Page) during January. The data provided by the UAE Solar Atlas show values up to 54 % in July. The annual average diffuse fraction does not exceed

40 % thus implying that Abu Dhabi enjoys clear sky weather conditions for at least 60 % of the year. The models by Page, Klein, Liu and Jordan, and NASA-SSE underestimate the diffuse fraction for almost all the year.

1.4 Conclusion

This chapter addresses selected empirical models used to estimate the monthly mean daily GHI and DHI in the city of Abu Dhabi, UAE, based on the available measured data. The estimation models for GHI make use of the measured SSH in addition to the geographical location of Abu Dhabi, UAE (latitude and longitude). The measured data for the period 2002–2008, provided by NCMS, is used to build the models for the daily mean GHI while the available data for 2009–2010 is used to test and validate the estimation models. These models are also validated with measured GSR data available from Masdar and satellite-derived databases available from NASA-SSE and SoDa with excellent agreement with coefficients of determination exceeding 92 %.

The comparison between several well-known empirical models for DHI with recently available measured data from Masdar's UAE Solar Atlas suggest the need to correct the empirical models available in the literature for the gulf region and UAE in particular. The correction takes into consideration the turbidity factor of the atmosphere which models the effect of sandstorms, high humidity in coastal regions, and presence of aerosol particles which contribute to the reduction of the GHI and increase in DHI during the warm season in particular (April–September).

Results show that Abu Dhabi has an annual average of the monthly mean daily GHI that exceeds 6 kWh/m^2 and thus possesses a strong potential for utilization of solar energy. The monthly average clearness index throughout the year varies between 0.61 and 0.70, which implies that Abu Dhabi enjoys clear sky conditions for most of the year.

Acknowledgments The authors would like to thank the National Center for Meteorology and Seismology (NCMS), Abu Dhabi, UAE, for providing the weather data.

References

1. El Chaar L, Lamont LA (2010) Global solar radiation: multiple on-site assessments in Abu Dhabi, UAE. Renew Energy 35:1596–1601
2. Abdalla YAG, Feregh GM (1988) Contribution to the study of solar radiation in Abu Dhabi. Energy Convers Manage 28(1):63–67
3. Page JK (1961) The estimation of monthly mean values of daily total short wave radiation on vertical and inclined surfaces from sunshine records for latitudes 40°N–40°S, Proc. UN Conference on New Sources of Energy vol 4, no. 598, pp 378–390
4. Liu BYH, Jordan RC (1960) The interrelationship and characteristic distribution of direct, diffuse and total solar radiation. Sol Energy 4:1–19

5. El-Nashar A (1991) Solar radiation characteristics in Abu Dhabi. Sol Energy 47(1):49–55
6. Eissa Y, Chiesa M, Ghedira H (2012) Assessment and recalibration of the Heliosat-2 method in global horizontal irradiance modeling over the desert environment of the UAE. Sol Energy 86:1816–1825
7. Physical Geography of Abu Dhabi Emirate (2008) United Arab Emirates, Chap 5, Climate, environment agency-Abu Dhabi
8. Gordon J (2011) Solar energy: the state of the art: ISES position papers. James & James (Science Publishers) Ltd, UK (ISBN 1-902916-23–9)
9. Hejase HAN, Assi AH (2011) MATLAB-assisted regression modeling of mean daily global solar radiation in Al-Ain, UAE. In: Assi AH (ed) Engineering education and research using MATLAB (Chap 8). Intech publisher, Rijeka, pp 195–218
10. SSE (2013) NASA surface meteorology and solar energy. http://eosweb.larc.nasa.gov/sse/. Accessed 15 Oct 2013
11. SoDa (2013) Solar radiation data. http://www.soda-is.com/eng/index.html. Accessed 15 Oct 2013
12. The UAE Solar Atlas (2013) Report by the Research Center for Renewable Energy Mapping and Assessment (ReCREMA), Masdar Institute, Abu Dhabi
13. Alnaser WE, Eliagoubi B, Al-Kalak A, Trabelsi H, Al-Maalej M, El-Sayed HM, Alloush M (2004) First solar radiation atlas for the Arab world. Renew Energy 29:1085–1107
14. Islam MD, Alili AA, Kubo I, Ohadi M (2010) Measurement of solar-energy (direct beam radiation) in Abu Dhabi, UAE. Renew Energy 35(2):515–519
15. Eissa Y, Marpu PR, Gherboudj I, Ghedira H, Ouarda TBMJ, Chiesa M (2013) Artificial neural network based model for retrieval of the direct normal, diffuse horizontal and global horizontal irradiances using SEVIRI images. Sol Energy 89:1–16
16. Glover J, McGulloch JDG (1958) The empirical relation between solar radiation and hours of sunshine. Q J Royal Meteorol Soc 84(360):172–175. doi:10.1002/qj.49708436011
17. Tiwari GN, Suleja S (1997) Solar thermal engineering system. Narosa Publishing house, New Dehli
18. NASA Surface meteorology and Solar Energy: Methodology (2004) https://eosweb.larc.nasa.gov/sse/documents/SSE_Methodology.pdf. Accessed 16 Dec 2004 (revised 18 Oct 2006)
19. Klein SA (1977) Calculation of monthly average insolation on tilted surface. Sol Energy 19:325–329
20. Modi V, Sukhatme SP (1979) Estimation of daily total and diffuse insolation in India from weather data. Sol Energy 22:407–411
21. Kenisarian MM, Tkachenkova NP (1992) Estimation of solar radiation from ambient air temperature. Appl Sol Energy 28:66–70
22. Alnaser WE (1989) Empirical correlation for total and diffuse radiation in Bahrain. Energy 14(7):409–414
23. Erbs DG, Klein SA, Duffie JA (1982) Estimation of the diffuse radiation fraction for hourly, daily and monthly average global radiation. Sol Energy 28(4):293–302
24. Collares-Pereira M, Rabl A (1979) The average distribution of solar radiation-correlations between diffuse and hemispherical and between daily and hourly insolation values. Sol Energy 22(2):155–164
25. Spencer JW (1971) Fourier series representation of the position of the sun. Search 2:172
26. Jacobson MZ (2005). Fundamentals of atmospheric modeling, 2nd edn. Cambridge University Press, Cambridge. p 317. (ISBN 0521548659)
27. Hartmann DL (1994). Global physical climatology. Academic Press, San Diego p. 30. ISBN 0080571638
28. Cardell J course notes, Engineering for everyone, EGR 100, Smith College, Northampton, Massachusetts, USA http://www.science.smith.edu/~jcardell/Courses/EGR100/. Accessed 15 Oct 2013

Chapter 2
Time Series Prediction of Renewable Energy: What We Can and What We Should Do Next

Yoshito Hirata, Kazuyuki Aihara and Hideyuki Suzuki

Abstract We summarize our recent developments of time series prediction for renewable energy. Given the past parts of high-dimensional time series for renewable energy outputs, we can predict their multistep future in real time with confidence intervals. We also proposed a way to evaluate the closeness in the high-dimensional space for improving the prediction, and an index showing when the prediction is more likely to fail. In addition, it is straightforward to apply the proposed framework to predict the electricity demands. Therefore, we can generate information necessary to consider efficient unit commitments for a case where more renewable energy resources are installed.

Keywords Online multistep prediction · High-dimensional time series · Confidence intervals · Prediction credibility · Renewable energy · Electricity demands

2.1 Introduction

Because we need to reduce CO_2 mission to mitigate the impacts of the climate change, we need to introduce more renewable energy. However, because renewable energy outputs fluctuate due to the changes of weather conditions, we should predict the temporal changes of renewable energy outputs and demands, and prepare backup plants such as thermal power plants and hydroelectricity power plants so that the supplies of electricity match its demands.

Y. Hirata (✉) · K. Aihara · H. Suzuki
Institute of Industrial Science, The University of Tokyo, 4-6-1 Komaba, Meguro-ku, Tokyo 153–8505, Japan
e-mail: yoshito@sat.t.u-tokyo.ac.jp

Department of Mathematical Informatics, The University of Tokyo, 7-3-1 Hongo, Bunkyo-ku, Tokyo 113–8656, Japan

CREST, JST, 4-1-8 Honcho, Kawaguchi, Saitama 332−0012, Japan

A. Sayigh (ed.), *Renewable Energy in the Service of Mankind Vol II*,
DOI 10.1007/978-3-319-18215-5_2

When we predict renewable energy outputs and electricity demands, our prediction should be done in real time, provide estimates for multiple steps ahead simultaneously, and show uncertainties representing how reliable these predictions are. We have been developing methods of time series prediction that meet these requirements [1–4]. Because we only use the past observations of interested quantities that should be forecasted, we can provide the predictions of renewable energy outputs and electricity demands simultaneously by the method described in [1–4] when we have their past time series data.

In this chapter, we first summarize our recent developments of such time series prediction methods. Then, in Sect. 2.3, we show some examples of time series prediction for electricity demands. In Sect. 2.4, we discuss necessary techniques for controlling and/or scheduling power plants, and conclude this chapter.

2.2 Online Time Series Prediction

We discuss the method we recently proposed [1–4] for predicting renewable energy resources, which is based on the method of Kwasniok and Smith [5]. Suppose that we constructed a database of past temporal changes of renewable energy outputs. Then, at each time, the method requires us to conduct two steps: In the first step, we find close matches to the current state within the database, and follow the time evolution of these close matches to obtain confidence intervals for the future values of multisteps ahead. In the second step, we observe the next value and update the database so that we can keep the size of the database but possibly reduce the prediction errors. In Ref. [1], we proposed the basic framework for online multistep prediction. In Ref. [2], we obtained confidence intervals. In Ref. [3], we extended the method to multivariate data in such a way that the prediction errors become smaller by evaluating the p-norm between two states where $p<1$ [6] than by using the normal Euclidean distance for evaluating the closeness between these two states. In Ref. [4], we introduced an index for evaluating when the proposed method is more likely to fail by evaluating the metric of the current state to the neighboring states in the database. This new index is information that is complementary to maximal local Lyapunov exponents [7, 8], which evaluate how fast the prediction errors grow as the prediction step becomes longer.

Therefore, we can evaluate, using the methods described in Ref. [1–4], what predicted values are, how large the prediction errors will be, and how faithful the prediction is.

2.3 Example for Predicting Electricity Demands

To demonstrate the framework of our recent developments, we apply the time series prediction discussed above to the prediction of electricity demands. The dataset was downloaded from the homepage of the Tokyo Electric Power Company

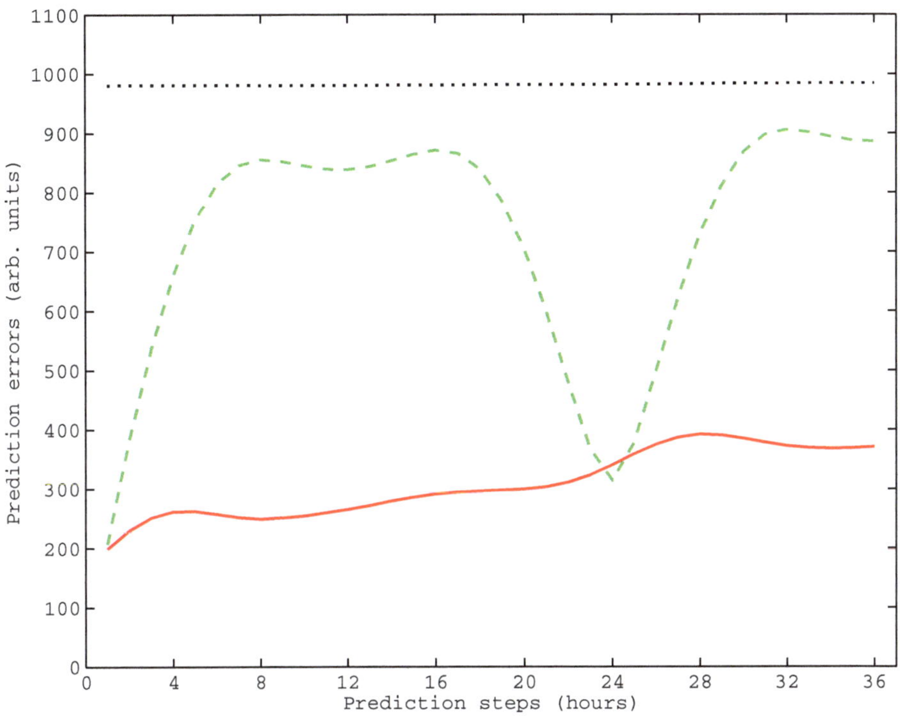

Fig. 2.1 Prediction errors for the mean for the proposed method *(red solid line)*, the persistence prediction *(green dashed line)*, and the mean prediction *(black dotted line)*

(http://www.tepco.co.jp/en/forecast/html/download-e.html). The dataset was hourly measurements between 1 January 2010 and 28 February 2011 (because the Tohoku-Oki earthquake of 11 March 2011 caused big changes in the electricity consumptions, we used the part of the dataset before the earthquake occurred).

The prediction errors are shown in Fig. 2.1. The mean for the proposed method enjoyed its advantage when the prediction step is less than 24 h compared with the persistence prediction, within which we let the current value for the prediction of q hours ahead.

We show in Fig. 2.2 an example of our prediction. Because this is the period in winter, there seemed to exist two peaks in the electricity demand within a day: morning and evening.

In Fig. 2.3, we compared the case where the metric to the datasets in the database was shorter than its median with the case where the metric was larger than its median. We clearly saw that 96 % confidence intervals had a higher confidence level when the metric was short, while the difference was statistically significant for only a limited number of prediction steps, especially those between 2 and 6 h when we evaluated the difference using the chi-square test.

Summing up, the above prediction for the electricity demands presented similar to the prediction for solar irradiation discussed in [3, 4].

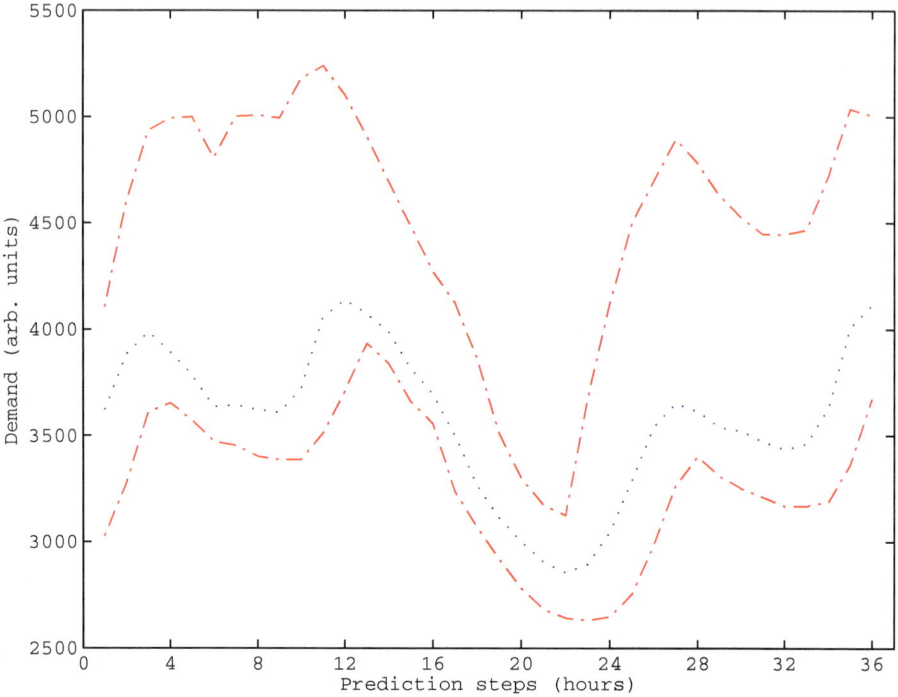

Fig. 2.2 Example of prediction for the electricity demands. The *red dot-dashed lines* show 96% confidence intervals of prediction. The *black dotted line* shows the actual value

2.4 Discussion

We described our recent developments for predicting time series. The method for the time series prediction can run online, span multiple steps ahead, and produce confidence intervals. In addition, we also proposed to use the p-norm with $p<1$ for evaluating the closeness in the high-dimensional space, which improves the accuracy of the prediction. Moreover, we derived an index for showing when the prediction is more likely to fail. Because the developed method uses the deterministic property for the underlying dynamics, we can apply the same method to predict the electricity demands. Thus, we now potentially have sufficient information to solve the problem of unit commitments, by which we can make power grid systems with much renewable energy sources more reliable. Therefore, we should try to optimize unit commitments robustly so that we can ensure the efficiency and the flexibility of our plan of starting and/or stopping backup power plants to meet supplies and demands of electricity.

Acknowledgments We thank Tokyo Electric Power Company for providing the dataset of the electricity demands used in this chapter. This research is supported by Core Research for Evolutional Science and Technology (CREST), Japan Science and Technology Agency (JST).

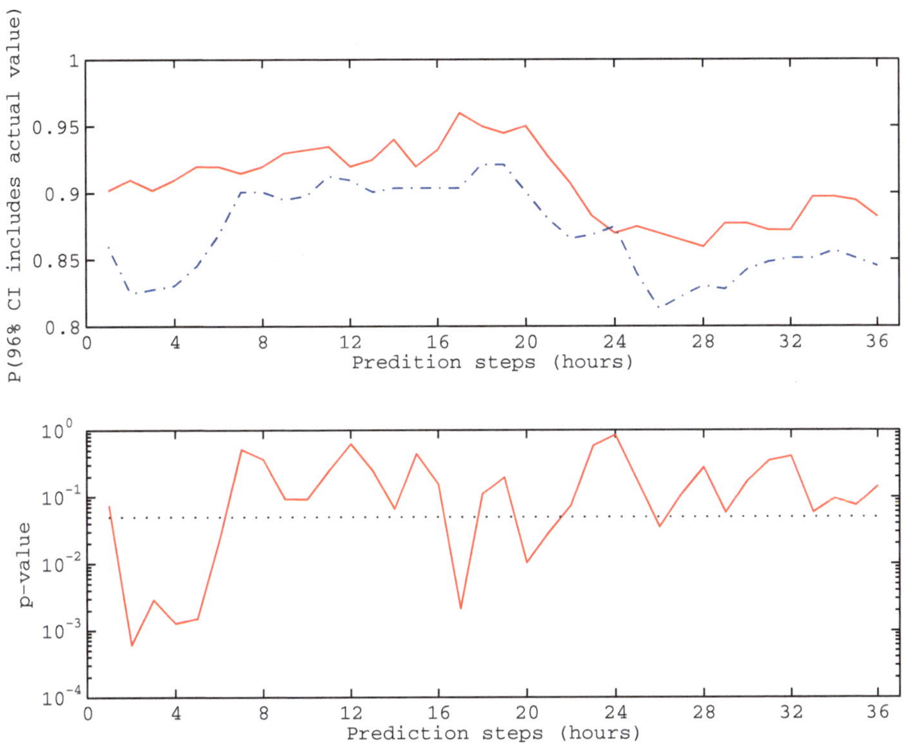

Fig. 2.3 Credibility of predicted confidence intervals. The *top* panel shows the probability that the actual values are contained within 96 % confidence intervals given a prediction step. The *red line* shows the probability that the metric was small, and the *blue dash-dotted line* shows the probability that the metric was large. The *bottom* shows the *p*-values for the difference between the cases where the metrics were small and large, respectively

References

1. Hirata Y, Yamada T, Takahashi J, Suzuki H (2012) Real-time multi-step predictors from data streams. Phys Lett A 376:3092–3097
2. Hirata Y, Yamada T, Takahashi J, Aihara K, Suzuki H (2014) Online multi-step prediction for wind speeds and solar irradiation: evaluation of prediction errors. Renew Energy 67:35–39
3. Hirata Y, Aihara K, Suzuki H (in press) Predicting multivariate time series in real time with confidence intervals: applications to renewable energy. Eur Phys J Spec Top 223:2451–2460
4. Hirata Y (2014) A fast time-series prediction using high-dimensional data: evaluating confidence interval credibility. Phys Rev E 89:052916
5. Kwasniok F, Smith LA (2004) Real-time construction of optimized predictors from data streams. Phys Rev Lett 92:164101
6. François D (2008) High-dimensional data analysis. VDM Verlag, Saarbrucken
7. Abarbanel HDI, Brown R, Kennel MB (1991) Variation of Lyapunov exponents on a strange attractor. J Nonlinear Sci 1:175–199
8. Abarbanel HDI, Brown R, Kennel MB (1992) Local Lyapunov exponents computed from observed data. J Nonlinear Sci 2:343–365

Chapter 3
Linear and Nonlinear Modeling for Solar Energy Prediction for Zone, Region and Global Areas

Hudhaifa Mazin, Hussein A. Kazem, Hilal A. Fadhil, S. A. Aljunid, Qutaiba M. Abdulmajeed and Miqdam T. Chaichan

Abstract Solar energy data provides information on the sun's potential at a location during a specific time period. These data are very important for designing and sizing solar energy systems. Due to the high cost and installation difficulties in solar measurement, solar energy data are not always available. Therefore, there is a demand to develop alternative ways of predicting solar energy data. Here, we present linear and nonlinear models for global and diffuse solar radiation. The various models have been tested in different areas. These areas are classified as zone, region and global. The zone and region models were found to be accurate and could be used to predict solar radiation, which is a really interesting achievement. However, the global models had a high percent error. Three statistical values were used to evaluate the developed models, that is, root mean square error, the mean absolute percentage error (MAPE), and mean bias error.

The results found the nonlinear models to be more accurate than the linear models—when calculating the solar energy in Malaysia using the nonlinear model, the MAPE was 6.4 %; however, when using the linear models, the MAPE was 7.3 %.

Keywords Diffuse solar energy · Clearness index · Modeling of solar energy

H. Mazin (✉) · H. A. Fadhil · S. A. Aljunid · Q. M. Abdulmajeed
Universiti Malaysia Perlis, Kampus Kubang Gajah, 02600 Arau Perlis, Malaysia
e-mail: hothyfa_mazen@yahoo.com

H. A. Kazem
Sohar University, PO Box 44, PC 311 Sohar, Oman

M. T. Chaichan
University of Technology, Baghdad, Iraq

A. Sayigh (ed.), *Renewable Energy in the Service of Mankind Vol II*,
DOI 10.1007/978-3-319-18215-5_3

3.1 Introduction

Solar energy is the light and heat we receive from the Sun. On Earth, sunlight is filtered through the Earth's atmosphere. When the direct solar radiation is not blocked by clouds, it becomes sunshine, which is a combination of bright light and radiant heat. If it is blocked by the clouds or reflects off other objects, it becomes diffused light [1]. Solar energy is secure, clean, and available on the Earth throughout the year. Solar energy technologies include solar heating, solar photovoltaics (PVs), solar thermal electricity, and solar architecture, which can make considerable contributions to solving some of the energy problems the world now faces. Solar PVs is one of best solutions for this problem, especially with regard to cost and it may also provide an excellent design with a high ratio of energy. PV technology is well-proven for producing electricity. Solar radiation data provide information on the sun's potential at a location on the Earth during a specific time period. These data are very important for designing and sizing solar PV systems. Due to the high cost and installation difficulties of measurement, these data are not always available [2, 3].

The measured data for solar energy values can be used for developing solar PV models which describe the mathematical relationship between solar energy and meteorological variables such as ambient temperature, humidity and sunshine ratio. These models can later be used to predict direct and diffuse solar energy using historical meteorological data at sites where there is no solar energy measuring device installed [3, 4].

Different models have been developed to simulate solar energy, including linear, nonlinear, artificial neural networks (ANN), fuzzy logic (FL) modeling, multiple linear regression (MLR) modeling and particle swarm optimization (PSO) models [4]. An accurate model for solar energy needs to be developed to provide a comprehensive database for the solar energy potential. Most of the previous literature focuses on linear, nonlinear and ANN models for solar energy. Therefore, we propose a regional model for solar energy based on linear and nonlinear techniques for a group of countries in a specific region—South East Asia (SEA)—using Malaysia, Thailand, Indonesia and South China as an example.

3.2 Linear Modeling of Solar Radiation

In this section, solar radiation is presented globally and regionally for different cases. By using linear models we can predict the solar energy in different places and regions. The linear model is the relationship between diffused solar energy E_d and global solar energy E_t, (E_d/E_t) with respect to the clearness index K_t [5, 6]. The values of the diffused solar energy to global solar energy E_d/E_t, have been taken from the general linear model formula $E_d/E_t = a + bK_t$, where E_d, E_T, and K_T are the diffuse solar energy, global solar energy and clearness index, respectively. Also, a and b are coefficients of the model [7, 8]. This linear expression, which represents diffuse solar energy in terms of global solar energy and clearness index is used for modeling diffuse solar energy in zones (countries) [9–18]. The average value has been taken from these published models to have a common domain of the K_t value. Since some of these models have common constraints, so the general constraint for the proposed model is K_t, $0.34 < K_t < 0.6$, including the author's proposed model.

We propose here to take the average coefficient values that belong to the different locations for the published models [11–18]. A different linear model for diffused solar energy is proposed as $E_d/E_t = 1.0707 + 1.07926K_t$, where the value of 1.0707 is the global average for the coefficient *a*, while 1.0792 is the global average for coefficient *b* for the same linear model. Both of these coefficients were collected from different global countries. This model shows little variation in accuracy from other published models. It is possible to use this model globally, generally on any country, but with some error, so that it is useful for fast prediction. It was also found that the proposed global model is similar to the model for Turkey and the surrounding region. This led us to propose another model focusing on zone and region, where zone could be a country and region could be a group of nearby countries. Since this model was tested for Malaysia and the nearby countries (Thailand, Indonesia, and South China) it was called the author's South East Asia 'SEA' model. This model shows a convergence of the predicted solar radiation with the models from the SEA region, especially Malaysia, Thailand and Indonesia. It has the shape of a curve sloping downwards, similar to all models that belong to the same region (Figs. 3.1 and 3.2).

The proposed SEA model is $E_d/E_t = 1.190935 - 1.06974K_t$, where 1.190935 is the value of coefficient *a* and −1.06974 is coefficient *b*. It is clearly seen that the

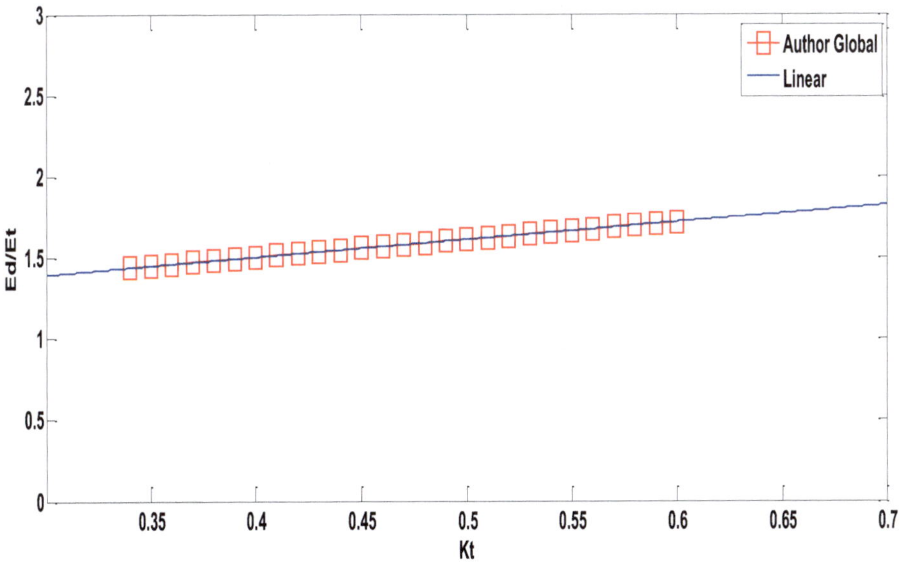

Fig. 3.1 Correlation between E_d/E_T and K_t for the proposed global linear model, with linear fitting

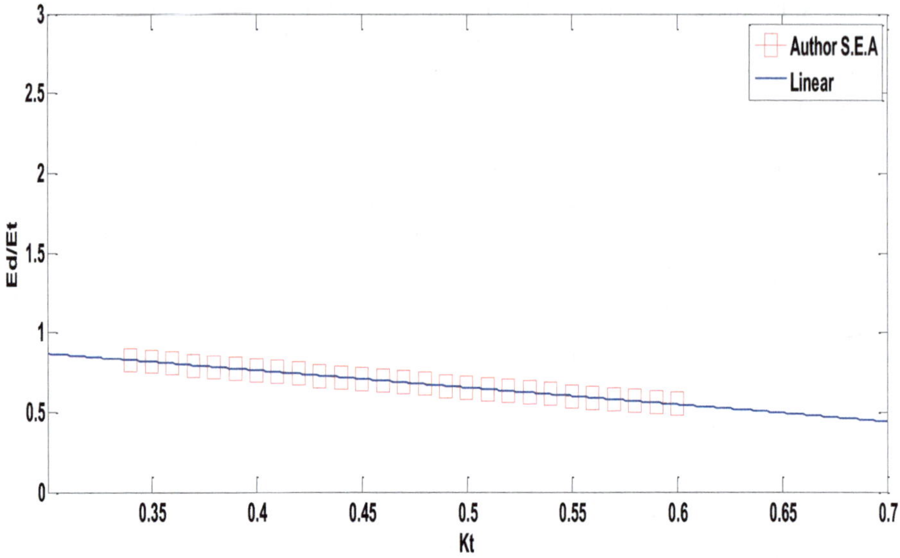

Fig. 3.2 Correlation between E_d/E_T and K_t for the author's SEA linear model, with linear fitting

variation of diffused solar energy E_d with respect to the clearness index K_t, is different from one zone (country) to another when each one is considered individually. Different published models have been tested as shown in Fig. 3.3. Figure 3.3a, b, c and d show the correlation between E_d/E_T and K_t for a linear model in Canada by Tuller [17], in Turkey by Page JK [18], in India by Modi et al. [19], and in Malaysia by Tamer et al. [4], respectively, where it can be seen in these figures that the curves have direct linear relationship.

The same procedure that had been followed to specify the Malaysia model was applied to Thailand to find coefficients *a* and *b*. These data were taken from previous models that belong to different places in Thailand as reported in reference [20]. These data were obtained from Thailand meteorological stations, and by adding the Indonesia model to the graph according to reference [21], we can see how these models are similar and how neighboring countries could have the same linear prediction attributes as shown in Fig. 3.4. The Thailand model obviously shows a convergence with Indonesia and Malaysia. A convergence can be seen between the three models which belong to South East Asia.

All three models seem to be similar since they are in the same region. Hence, the average data of Malaysia, Thailand and Indonesia can be proposed as the author's SEA model. The author's proposed SEA model has a place among models of the same zone as it is compatible with each of them, as shown in Fig. 3.5a. By comparing the SEA model with the author's proposed global model as shown in Fig. 3.5b, we can see the incompatibility between them; hence, proving that the SEA model gives an accurate prediction of solar radiation if applied to Malaysia and nearby countries in the same region.

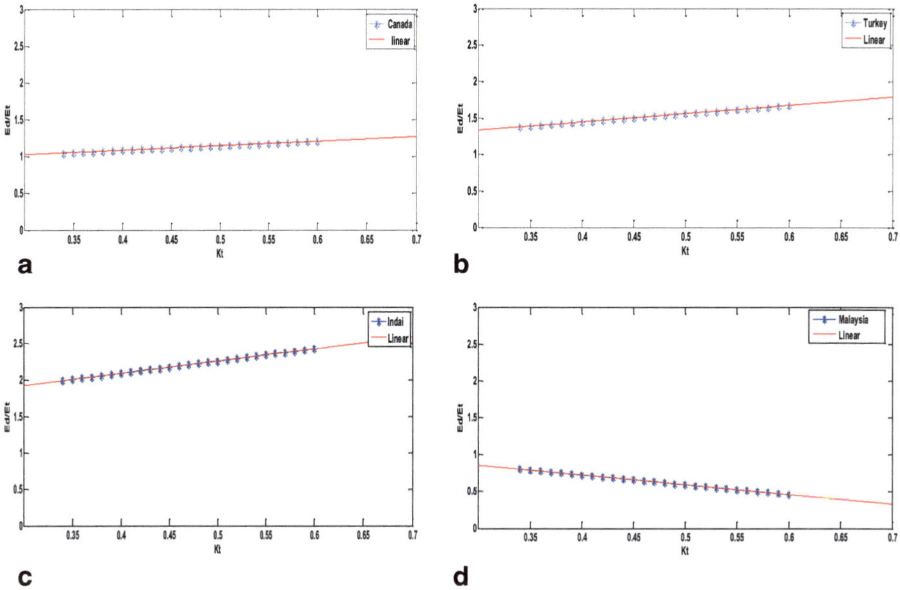

Fig. 3.3 Correlation between E_d/E_T and K_t **a** for Canada linear model, with linear fitting, **b** for Turkey linear model, with linear fitting, **c** for India linear model, with linear fitting, **d** for Malaysia average model, with linear fitting

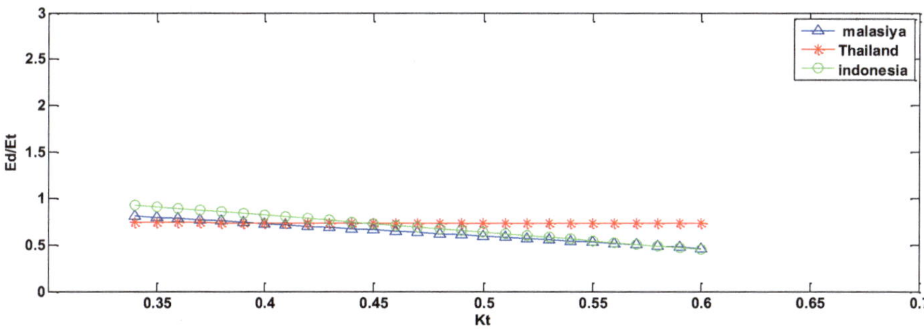

Fig. 3.4 Correlation between E_d/E_T and K_t for Thailand, Malaysia and Indonesia

The comparison between different models shows a divergence and convergence between the linear models based on the variation of the countries or regions as shown in Fig. 3.6a. This means that due to different climates and cloud cover for these regions, there are different amounts of energy. The data belong to meteorological stations in different country and states, e.g., Canada (62.2270° N, 105.3809° W), India (21.0000° N, 78.0000° E), etc. The idea is to compare linear models in different regions. The author's model showed good results in the middle of the different models. In other words, it can be clearly seen that there is a similarity in models

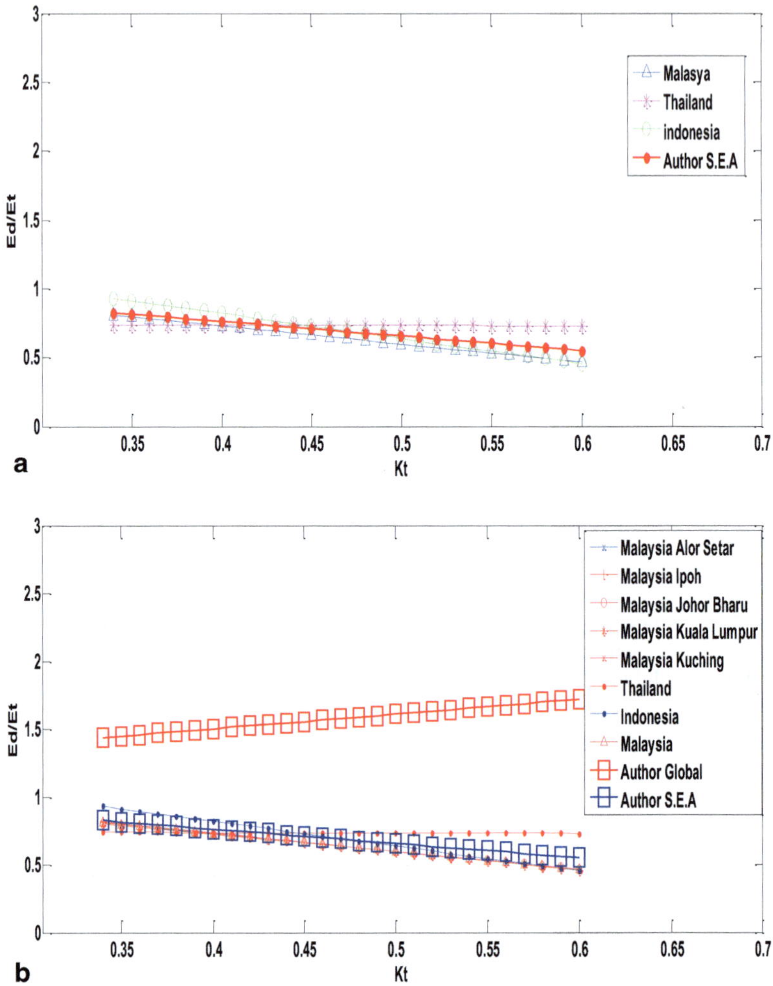

Fig. 3.5 Correlation between E_d/E_T and K_t **a** for Thailand, Malaysia and Indonesia, with the author's SEA model, **b** for five Malaysian states along with the author's SEA and global models

between the Turkey [18] model and the author's global model, which could be used for any region around Turkey but with some error.

Data from the author's global and SEA models were calculated based on the average value of *a* and *b* for all the previously mentioned models. By observing the graphs in Fig. 3.6b, a difference was seen in diffusion energy with respect to the various locations due to different cloud cover. The Malaysia model belongs to the SEA region and the proposed model seems similar.

In conclusion, we can adopt one linear model for a zone (SEA), which provides interesting and promising results and another one globally, but with some error.

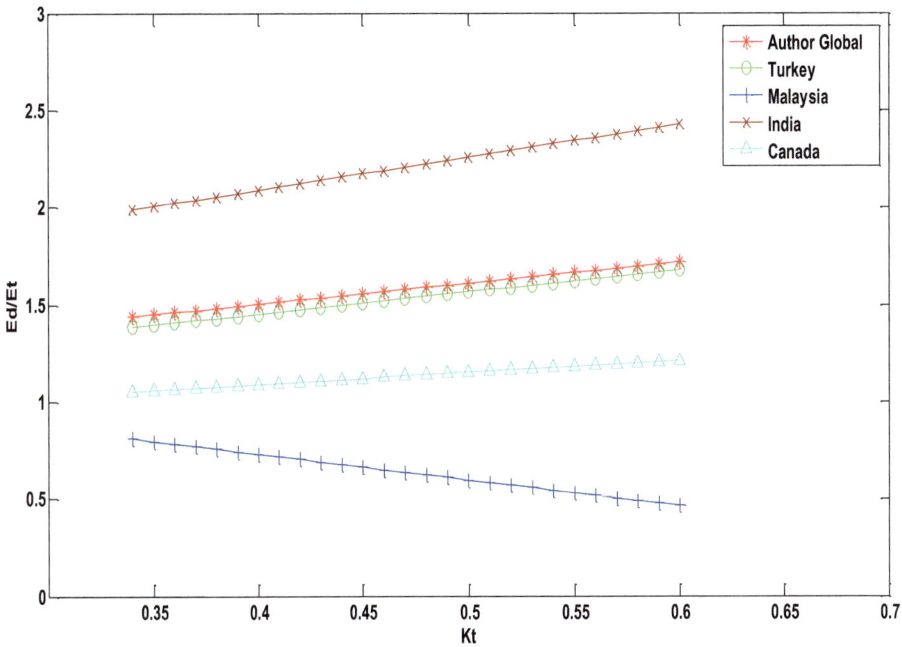

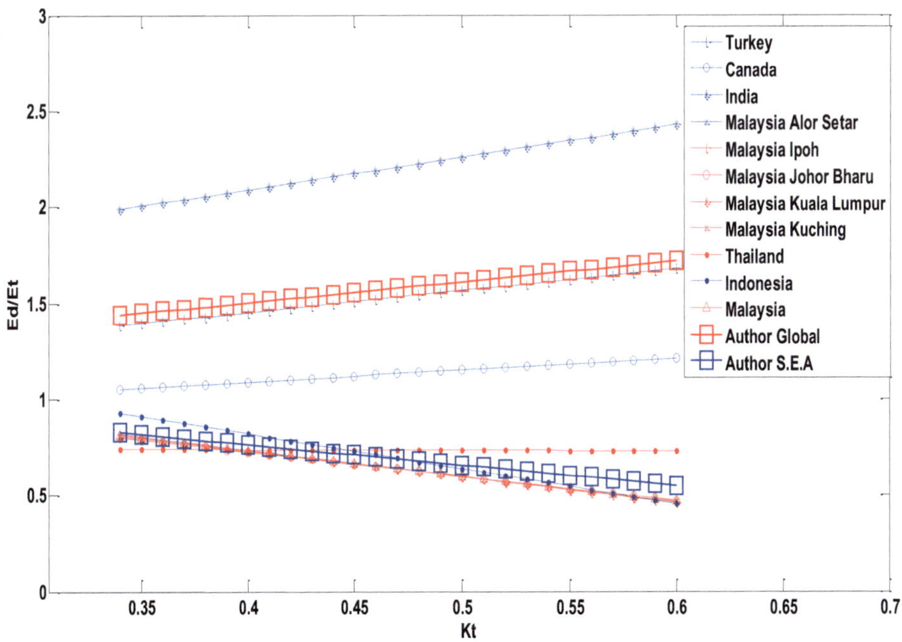

Fig. 3.6 Correlation between E_d/E_T and K_t **a** for the author's global model with different models for different places and regions, **b** for the author's global and SEA models with different models for different places and regions

3.3 Nonlinear Modeling of Solar Radiation

The same procedure was adopted to propose the linear model repeat for the nonlinear model, and the average was taken for coefficients *a, b, c* and *d* for different countries to propose the author's global and SEA models. In this section, solar radiation is presented for different regional and global case zones. Using the previous approach, the nonlinear model is the relationship between diffused solar energy E_d and global solar energy E_t, E_d/E_t with respect to the clearness index K_t. The values of diffused solar energy to global solar energy E_d/E_t, were taken from the general nonlinear model formula $E_d/E_t = a + bK_t + cK_t^2 + dK_t^3$, where *a, b, c* and *d* are the coefficients of the model [11–18], where K_t values have been taken from the common domain of the K_t values that belong to previous models $0.3 < K_t < 0.8$. This constraint is adopted for all published nonlinear models. The author's proposed global model was also compared with other nonlinear published models. It is recommended to be used for fast prediction using the nonlinear model, with some error.

Average coefficient values that belong to different locations were considered as shown in Fig. 3.7a. Different nonlinear models for diffused solar energy were used to propose the author's model as $E_d/E_t = 0.40303 + 1.89154K_t - 1.11836K_t^2 + 2.2828K_t^3$. Moreover, the proposed SEA model showed good accuracy result and is recommended for regional used. Data for the proposed SEA model were taken from the average of models held in the Malaysia zone and is derived as $E_d/E_t = 0.7014 + 0.7052K_t - 3.2929K_t^2 + 2.2828K_t^3$. The authors of reference [3] suggested one nonlinear model for Malaysia, by taking the average of the coefficients *a, b, c* and *d* for five states in Malaysia, namely Alor Setar, Kuching, Kuala Laumpur, johor Bharu and Ipoh. As shown in Fig. 3.7b, the correlation between E_d/E_t and K_t is inverse relationship. We will use this model to compare the Malaysia nonlinear model with other published nonlinear models. The figure of the Indonesia nonlinear model [21] was plotted for comparison with other models in the same zone of Malaysia and with the author's SEA model. South China was also one of the zones in the SEA region of comparison. Its model was taken from the average of 86 stations throughout China [22].

The same procedure that had been used in linear models was applied to the nonlinear models. The previous nonlinear models have been included in one graph as shown in Fig. 3.9a in order to compare all the models that belong to the same zone. The models showed good interesting results. It is clearly seen that these models are in close with each other, especially when they are plotted using a global scale on the Y-axis, that suits all nonlinear models. This scale has been tested with different nonlinear models that belong to different counters; scale Y-axis constraint is between $-2 < Y < 10$ as shown in the Fig. 3.9. By adding the author's SEA model to the previous group, it can be clearly seen that the proposed SEA nonlinear model is in harmony with them, suggesting that the proposed SEA model is compatible with the South East Asia zone.

The proposed global model is also included in Fig. 3.9b with other zone models. A deviation between the proposed global model and the other published models

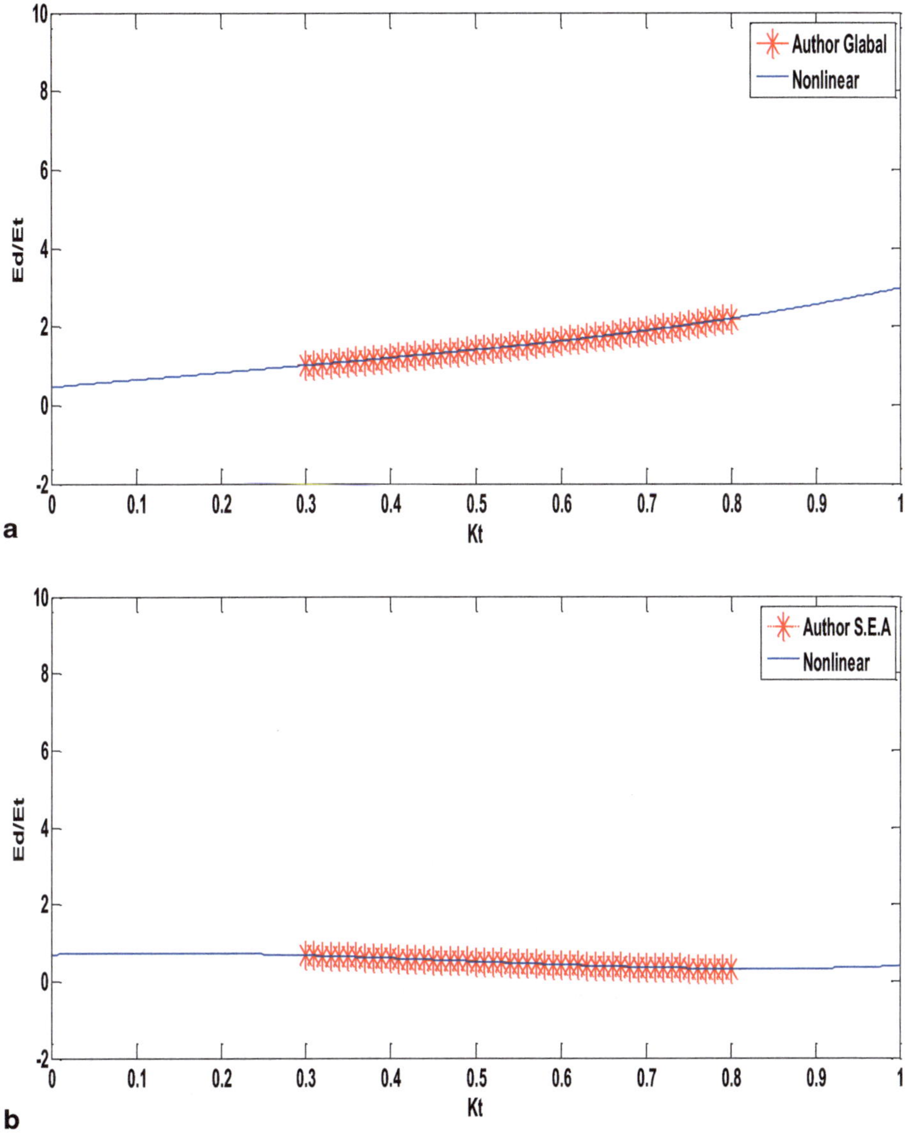

Fig. 3.7 Correlation between E_d/E_T and K_t **a** for the author's global nonlinear model, **b** for the author's SEA nonlinear model

can be clearly seen. Figure 3.10 shows a comparison between different nonlinear models; the data belong to meteorological stations in different countries and states, e.g., of the USA (62.2270° N, 105.3809° W), USSR (21.0000° N, 78.0000° E), etc. The idea is to compare nonlinear models in different regions [22–29]. Diffused solar energy prediction using nonlinear models takes a different direction for different

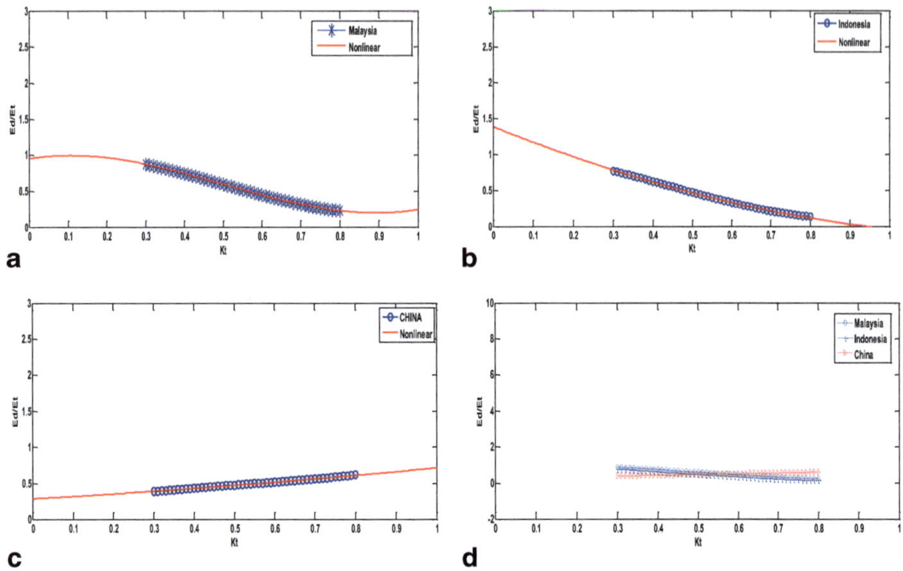

Fig. 3.8 Correlation between E_d/E_T and K_t **a** for Malaysia, nonlinear model in Malaysia, **b** for Indonesia, nonlinear model in Malaysia, **c** for China, nonlinear model, **d** for Malaysia, Indonesia and China, nonlinear model

locations, except for locations that belong to the same region which usually take the same direction as shown in Fig. 3.10.

Obviously, the proposed global model showed good results in the middle of different models, which could be used for any one of these region, but with some error. Furthermore, the proposed SEA is more suitable for the Malaysia zone with less error for nearby countries in the same region. To evaluate the proposed models, three statistical errors have been used, that is, mean bias error (MBE), the mean absolute percentage error (MAPE), and root mean square error (RMSE). The MBE is an indicator of the average deviation of the predicted values from the measured data. A positive MBE value indicates the amount of overestimation in the predicted global solar energy and vice versa. On the other hand, the RMSE provides information on the short-term performance of the models and is a measure of the variation of the predicted values around the measured data. It indicates the scattering of data around linear lines. A large positive *RMSE* implies a large deviation in the predicted value from the measured value. *MBE* and *RMSE* are given as:

$$MBE = \frac{1}{n}\sum_{i=1}^{n} I_{pi} - I_i \tag{3.1}$$

$$RSME = \sqrt{\frac{1}{n}\sum_{i=1}^{n}(I_{pi} - I_i)} \tag{3.2}$$

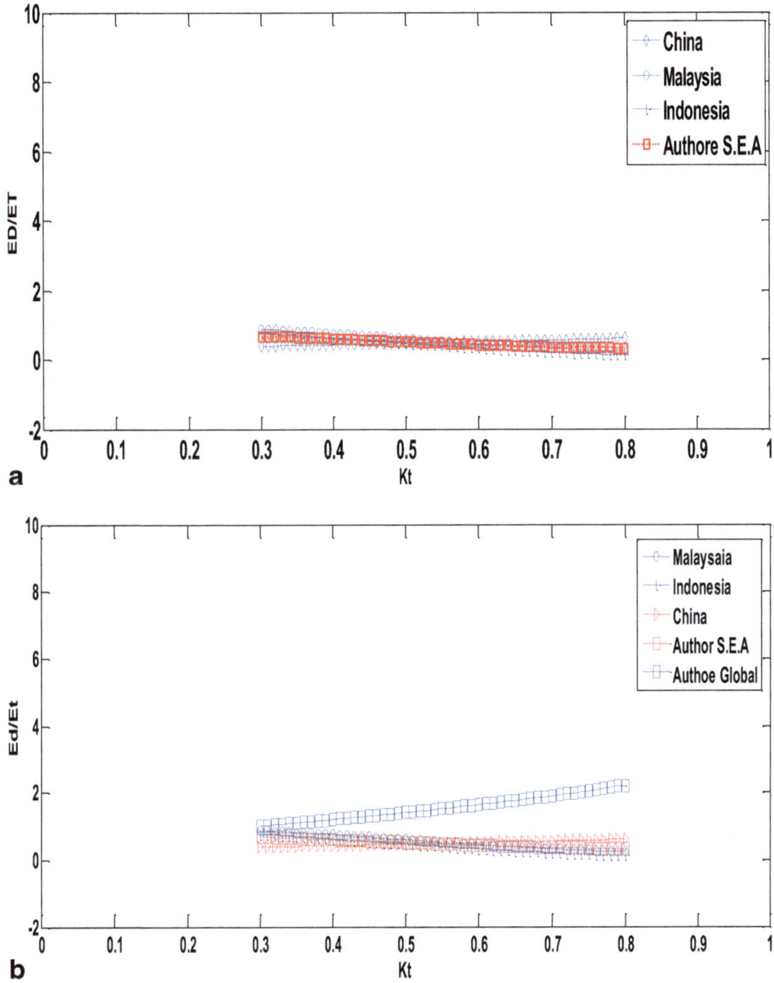

Fig. 3.9 Correlation between E_d/E_T and K_t **a** for Malaysia, Indonesia and China as well as the author's SEA nonlinear model, **b** for Malaysia, Indonesia and China as well as the author's SEA and global nonlinear models

where I_{pi} is the predicted value, I_i is the measured value, and n is the number of observations. The *MAPE* is an indicator of accuracy in which it usually expresses accuracy as a percentage and is given by

$$MAPE = \frac{1}{n}\sum_{t=1}^{n}\frac{I - I_p}{I} \tag{3.3}$$

where I is the real value and I_p is the predicted value. The calculated *MAPE* is summed for every fitted or forecasted point in time and divided again by the number

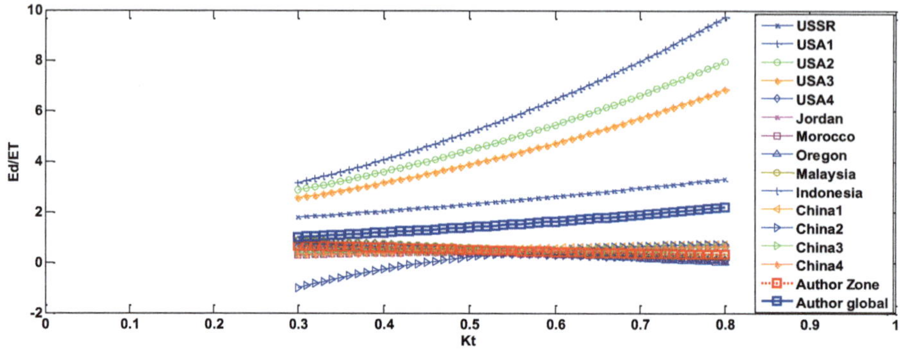

Fig. 3.10 Correlation between E_d/E_T and K_t for different nonlinear models in different places as well as the proposed SEA and global nonlinear models

Table 3.1 Evaluation of the proposed global solar energy models

	MAPE	RMSE	RMSE (%)	MBE	MBE (%)
Linear	7.311	0.4129	8.5523	−0.0545	−1.1132
Nonlinear	6.4328	0.3866	8.1002	−0.0562	−1.1276

Table 3.2 Evaluation of the proposed diffuse solar energy models

	MAPE	RMSE	RMSE (%)	MBE	MBE (%)
Linear	4.2	0.136	4.77	0.010	0.317
Nonlinear	3.77	0.133	4.69	0.008	0.166

of fitted points, *n*. Tables 3.1 and 3.2 shows the MAPE, RMSE, and MBE values for the proposed models. On the basis of the MAPE and the RMSE results shown in Tables 3.1 and 3.2, the nonlinear model is the most accurate model in calculating global solar energy among the linear models.

3.4 Conclusions

Linear and nonlinear models for calculating global and diffuse solar energy have been proposed for the regional and global areas. The proposed models have been tested for the South East Asia region. Malaysia, Thailand, Indonesia and South China have been taken as an example. The models have been evaluated on the basis of three statistical values, that is, RMSE, MAPE and MBE. The results showed that nonlinear models are more accurate than linear models in calculating global and diffuse solar energy. The MAPE values for nonlinear-based global and diffuse solar energy models are 6.4 and 3.7 %, respectively. However, the MAPE values for linear models for global and diffuse solar energy are 7.3 and 4.2 %, respectively. Finally, the proposed regional model was found to give accurate results for a group of countries as well as diffused countries.

References

1. Khatib T, Sopian K, Kazem HA (2013) Actual performance and characteristic of a grid connected photovoltaic power system in the tropics: a short term evaluation. Energy Convers Manage 71:115–119
2. Cole Boulevard Golden, Colorado Solar radiation modeling and measurements for renewable energy applications: data and model quality. NREL, (March 2003), 1617: 80401–3393
3. Bakirci K (2009) Models of solar radiation with hours of bright sunshine: a review. Renew Sustain Energy Rev 13:2580–2587
4. Khatib T, Mohamed A, Mahmoud M, Sopian K (2011) An assessment of diffuse solar energy models in terms of estimation accuracy. 2nd International Conference on Advances in Energy Engineering (ICEAA 2011)
5. Kreider JF, Kreith F (1981) Solar energy handbook. McGraw-Hill, New York
6. Kazem HA, Khatib T, Sopian K (2013) Sizing of a standalone photovoltaic/battery system at minimum cost for remote housing electrification in Sohar, Oman. Energy Build 6C:108–115
7. Jebaraj S, Iniyan S (2006) A review of energy models. Renew Sustain Energy Rev 10:281–311
8. Khatib T, Mohamed A, Sopian K (2012) A review of solar energy modeling techniques. Renew Sustain Energy Rev 16:2864–2869
9. Iqbal M (1983) An introduction to solar radiation. Academic, Orlando
10. Collares-Pereira M, Rabl A (1979) The average distribution of solar radiation: correlations between diffuse and hemispherical and between daily and hourly insolation values. Sol Energy 22:155–164
11. Khatib T, Mohamed A, Mahmoud M, Sopian K (2011) Modeling of daily solar energy on a horizontal surface for five main sites in Malaysia. Int J Green Energy 8:795–819
12. Abdalla Y, Baghdady M. (1985) Global and diffuse solar radiation in Doha (Qatar). Solar Wind Echnol 2:209–216
13. Benson R, Paris M, Sherry J, Justus C (1984) Estimation of daily and monthly direct, diffuse and global solar radiation from sunshine duration measurements. Sol Energy 32:523–558
14. Trabea A (2000) Analysis of solar radiation measurements at Al-Arish area, North Sinai, Egypt. Renew Energy 20:109–134
15. Top S, Dilma U, Aslan Z (1995) Study of hourly solar radiation data in Istanbul. Renew Energy 6:171–175
16. El-Sebaii A, Al-Hazmi F, Al-Ghamdi A, Yaghmour S (2010) Global, direct and diffuse solar radiation on horizontal and tilted surfaces in Jeddah, Saudi Arabia. Appl Energy 87:568–576
17. Tuller S (1976) The relationship between diffuse, total and extraterrestrial solar radiation. Sol Energy 18:259–263
18. Page JK (1961) The estimation of monthly ea values of daily total short wave radiation on vertical and inclined surfaces from sunshine records for latitudes 40°N–40°S. Proceedings of the UN conference on new sources of energy, paper no. 598, 4, pp 378–390
19. Modi V, Sukhatme S (1979) Estimation of daily total and diffuse insolation in India from weather data. Sol Energy 22:407–411
20. Nimnuan P, Janjai S (2012) An approach for estimating average daily global solar radiation from cloud cover in Thailand. Procedia Eng 32:399–406
21. Parangtopo H, Poseposeotjipoto (1985) Review and analysis of the global and diffuse solar radiations in Jakarta, Indonesia. Sol Wind Technol 1(3):135–152
22. Rensheng C, Shihua L, Ersi K, Jianping Y, Xibin J (2006) Estimating daily global radiation using two types of revised models in China. Energy Convers Manage 47:865–78
23. Kenisarin MM, Tkachenkova NP (1992) Estimation of solar radiation from ambient air temperature. Appl Sol Energy 28(4):66
24. Klein SA (1977) Calculation of monthly average insolation on tilted surface. Sol Energy 19:325–329
25. Erbs DG, Klein SA, Duffie JA (1982) Estimation of the diffuse radiation fraction for hourly daily and monthly-average global radiation. Sol Energy 28(4):293–304

26. Alnaser WE, Eliagoubi B, Al-Kalak A, Trabelsi H, Al-Maalej M, El-Sayed HM, Alloush M (2004) First solar radiation atlas for the Arab world. Renew Energy 29:1085–1107
27. Nagaraja Rao CR et al (1982) Solar radiation and related meteorological data for Corvallis, Oregon. Report published by Department of Atmospheric Sciences, Oregon State University, Corvallis, Oregon, April 1982
28. Huashan L, Weibin M, Lian Y, Wang X, Zhao L (2011) Global solar radiation estimation with sunshine duration in Tibet, China. Renew Energy 36:3141–3145
29. Waewsak J, Chancham C (2010) The clearness index model for estimation of global solar radiation in Thailand. Thammasat Int J Sci Technol 15(2):54–61

Chapter 4
Shortest Term Solar Energy Forecast

Jan Remund, Daniel Klauser and Stefan C. Müller

Abstract Due to the strong growth in photovoltaics (PVs), forecasting is gaining importance. At present, most forecasts are based on numerical weather prediction (NWP) models without cloud assimilation. These models lack the ability to forecast the exact position of clouds, which is needed for single-point forecasts. Nowcasting methods, based on current information about cloud positions, provide an opportunity to enhance solar forecasts. Here, we review two of four existing methods—sky camera and satellite image-based methods. Sky cameras are still at an early stage of development and much work is currently being performed. Uncertainties are in the range of 25 % for 30 s data and up to 10 min ahead. Satellite image-based methods have been in use for several years. It is possible to show that a simplified approach using NWP-based cloud vectors leads to similar accuracy as using cloud vectors based on multiple images. Satellite-based methods enhance forecasts by 30–40 % compared to NWP forecasts for forecast horizons of 15 min to 4 h. We also briefly discuss simple post-processing and aggregation methods.

Keywords Nowcasting · Solar radiation forecasts · Satellite data · PV production forecasts

4.1 Introduction

Due to the large amount of decentralized power production based on photovoltaic (PV) systems, rapidly changing cloud conditions have a major impact on total electricity production at a regional level, as well as on the stability of the power grid. Short-range forecasts for time periods of 6–72 h are usually based on statistically corrected numerical weather prediction (NWP) models. The disadvantage of NWP is that based on the chaotic behavior of the climate system, it will never be possible to forecast the exact time and position of a cloud in the future (>6 h ahead).

More recent and ongoing forecasts with high temporal as well as local resolutions are required for intraday trading or grid management. Nowcasting techniques,

J. Remund (✉) · D. Klauser · S. C. Müller
Meteotest, Fabrikstrasse 14, 3012 Berne, Switzerland
e-mail: jan.remund@meteotest.ch

A. Sayigh (ed.), *Renewable Energy in the Service of Mankind Vol II*,
DOI 10.1007/978-3-319-18215-5_4

based on sky cameras and satellite data, are able to detect the position of clouds, making it possible to forecast the position of clouds more precisely.

Until 10 years ago, global radiation was not recognized as an important parameter for NWP; however, this has now changed to a great extent. National weather services are upgrading their models and aerosol forecasts. Some countries have started programs for enhancing the quality of solar forecasts, e.g., Eweline[1] or sunshot initiative[2], and many international operations are underway. In the framework of IEA PVPS 14, a report about the state-of-the-art of solar forecasts was published in 2014[3]; however, due to fast moving technology, the report is already outdated to some extent.

Here, we review two nowcasting techniques based on sky cameras (imagers) and satellite images. Two other methods based on statistical approaches and rapid updates of NWP are not included.

4.2 Shortest Term Forecasting Based on Sky Cameras

Sky cameras deliver forecasts for the next 30 min in time steps of 1–30 s. Many university groups are working on the use and enhancement of those methods. A workshop held in the framework of COST ES1002 in June 2014 was attended by many interested parties. Forecasting with the help of a sky camera includes five steps:

1. Transformation of image
2. Detection of clouds
3. Detection of cloud vectors
4. Forecast of cloud position
5. Forecast of radiation parameters

The following points represent the current state-of-the-art (which is also quickly evolving):

- The use of 'cheap' cameras is possible (e.g., standard hemispheric webcams like Mobotix Q25)
- High Dynamic Range (HDR) or Enfuse methods (based on several pictures taken with different apertures) are needed to obtain a good cloud image and to select the optimal image per pixel
- Correction of sky brightness is needed
- Cloud movement is most often based on Kanade–Lucas–Tomasi feature tracker
- Additional measurements of the altitude of the clouds are presumably necessary (multiple cameras, lidar, ceilometers, etc.

There is little knowledge about the quality of the radiation data based on sky cameras. Yang et al. [1] give a relative root-mean-square error (RMSE) of ~25 % (for 30 s data up to 10 min ahead). Generally, the forecast horizon is in the range of 1–15 min.

[1] http://www.projekt-eweline.de

[2] http://energy.gov/eere/sunshot/sunshot-initiative

[3] http://www.iea-pvps.org/index.php?id=278

4.3 Shortest Term Forecasting Based on Satellite Images

The objective of satellite-based methods is to enhance the forecast quality between 0.5 and 6 h on a regional basis. The method used to analyze two consecutive images (using a normally visible channel at 0.6 μm) to obtain cloud motion vectors (CMVs) has been in use for several years [2].

A research project sponsored by the Swiss utility company, BKW Meteotest, tested a new simplified method between 2012 and spring 2014. The idea behind the new approach is to use two widely existing data sources—satellite radiation data and wind vectors of NWP. With the aid of satellite images, the current cloud positions are detected. In a second step, the clouds are shifted with trajectories (based on wind vectors from NWP) to future positions. Instead of using satellite maps to detect vectors, wind vectors are used. The computational time is much faster. Whereas satellite image-based CMVs are always linear, NWP vectors also have the advantage of being variable over the next 6 h (change directions and speed).

Further steps are the same as the standard CMV method—radiation is calculated based on the future position of the clouds and the sun, and with help of a clear sky model (Solis [3]) the global radiation is calculated. The forecast is updated every 15 min due to the standard update cycle of weather satellite images. The large-scale coverage of satellite images also makes it possible to create a regional forecast.

The infrared channel (3.9 μm) of the satellite provides information about the cloud situation during the night to enable a prediction for the early morning hours and can be used to detect clouds over snow, ice or salt deserts. The altitude of clouds, also detected by the satellite, is used to determine fog.

There are two critical issues to be mentioned here:

1. The transport of clouds with trajectories can lead to fraying of clouds. This effect can be suppressed by using nearest neighbor values at the destination coordinates instead of interpolation.
2. Aerosol data are extremely critical—mainly for clear sky values of direct normal irradiance (DNI). In our current system we use Monitoring Atmospheric Composition and Climate (MACC)[4] aerosol forecast of the European Center for Medium-Range Weather Forecasts (ECMWF) (for the test project climatological values have been used).

Simple post-processing is used to analyze the factors between the measurements and the forecast of the previous hours; however, this is only applied if the clearness factor is stable and either high (clear sky situation) or low (overcast situation). In mixed situations (with changing sign of the factor) no post-processing is applied.

Furthermore, for regional aggregation, a simple method was introduced and tested—as the inclination and azimuth of the installations as well as the technology are not known, a standard installation with 20° inclination (towards South) has proven to be optimal for Central Europe. The installations are summed up according to their

[4] http://www.gmes-atmosphere.eu/

weight (the installed nameplate capacities of PV are known). Each day the aggregated data of the forecasts and the measurements of the last 30 days are compared and the forecasts are corrected with the help of linear regression.

At the end of the test project, the area had been enlarged to cover the whole of Southern Europe (approximately <50°).

4.4 Validations

The test region covered an area of 120 × 120 km in Switzerland (Canton of Bern) and included 13 ground weather stations of MeteoSwiss and three additional measurement sites at PV installations. The MeteoSwiss dataset was used to validate the forecast, whereas the additional stations were used for post-processing the output and to reduce the local uncertainty. The predictions of 90 individual PV installations are aggregated to provide an overall regional production and compared with the measured total production of all PV installations in the test region. The forecasts are also compared to persistence assuming a constant clearness factor.

The improvement of this approach over the NWP Weather Research and Forecasting (WRF) direct model outputs for 3–6 h is approximately 40 %. Figure 4.1 shows the uncertainties at Neuchatel with and without post-processing.

The uncertainties are in the range of the satellite-based methods (15 min ahead) and the NWP-based methods (4–6 h ahead). The results vary a great deal for Alpine and Plateau stations and for the seasons. Summer and Plateau stations have much lower uncertainties. For Alpine and winter conditions, uncertainties are much

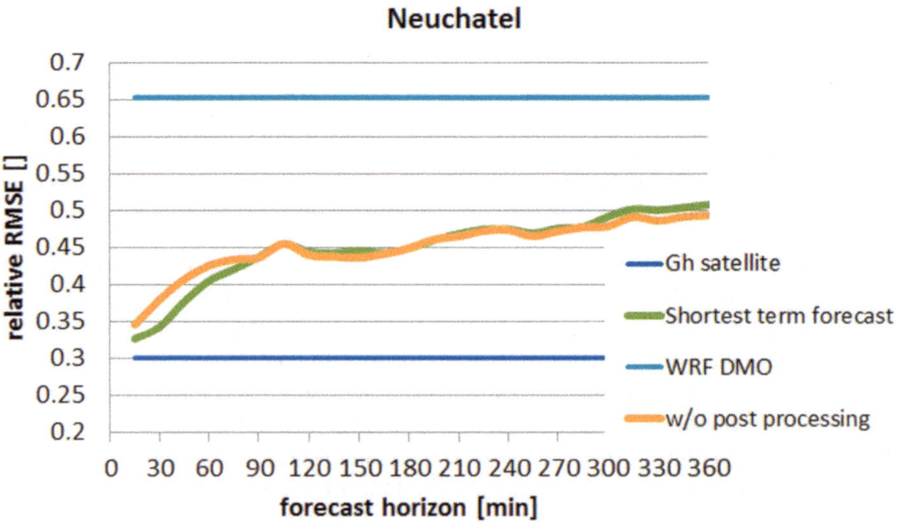

Fig. 4.1 Uncertainty of the shortest term forecast for Neuchatel. Post-processing enhances the quality for the next 90 min (for mountain stations even longer)

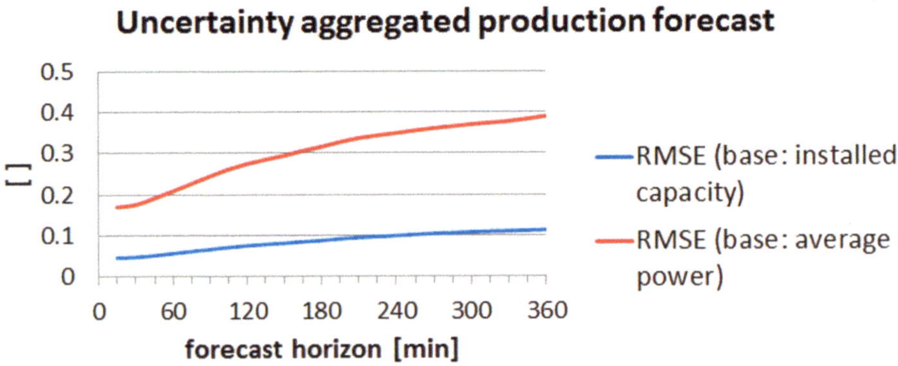

Fig. 4.2 Uncertainty of the shortest term forecast for aggregated production of 90 PV installations

higher—mostly introduced by the uncertainty of the estimation of the current cloud situation based on the satellite method.

Figure 4.2 shows the uncertainties of the aggregated forecasts.

The uncertainties for relative RMSE based on average power are in the range of 17–40 %, which is approximately 30–50 % lower than for single-point forecasts. The relative RMSE based on installed capacity (which is a common measure for PV production forecasts) is in the range of 5–10 %.

Additionally, the University of Oldenburg compared the new approach (NWP-based motion vectors) to determine the CMV using the standard approach. In a preliminary report, they concluded that the new approach showed a similar accuracy as the standard approach and that the optimal model level for cloud speed is 650 hPa (approximately 3500 m)—with a broad optimum between 600 and 800 hPa.

4.5 Conclusions and Outlook

As PV penetration is growing strongly, solar forecasting is gaining importance. However, uncertainties are high, which leads to high integration costs.

Shortest term forecasting based on sky cameras has been investigated by numerous research groups and first results are now available. Nevertheless, some technical issues have yet to be resolved and the forecast horizon is still short (10 min).

Satellite image-based shortest term forecasting has proved to be a way to lower solar forecast uncertainties. There are less uncertainties compared to NWP-based methods for forecasting horizons between 15 min and 4–6 h. For 1–2 h ahead, uncertainty is generally 30–40 % less than NWP. The short-cut of using NWP-based cloud vectors was successful and will be used for Meteotest's operational model.

A broad benchmark will be made by the IEA SHC Task 46 in 2015. In the EU FP7 project, DNICast sky cameras are being tested and NWP and satellite image-based methods will also be benchmarked. The results will be available in 2016.

References

1. Yang H, Kurtz B, Nguyen D et al (2014) Solar irradiance forecasting using a ground-based sky imager developed at UC San Diego. Sol Energy 103:502–524
2. Lorenz E (2004) Methoden zur Beschreibung der Wolkenentwicklung in Satellitenbildern und ihre Anwendung zur Solarstrahlungsvorhersage, PhD Thesis, University of Oldenburg, Germany
3. Ineichen P (2008) A broadband simplified version of the Solis clear sky model. Sol Energy 82(8):758–762

Part II
R. Integration 2014

Chapter 5
Photovoltaic-Integrated Buildings in Florence and Lucca, Italy: Case Studies

Lucia Ceccherini Nelli

Abstract When using the integrated approach, solar systems become part of the general building design. In fact, they often become regular building elements. This is due to the fact that integrating solar systems into the building envelope is often a necessity if the systems are to be economically feasible. The solar elements cannot be separate elements that are added after the building, or at least the architectural design of it, is complete. Rather, they must replace other building elements, thereby serving dual functions and reducing total costs.

The following case studies depict a coming-of-age of building-integrated photovoltaics (PVs).

These PV elements are specially designed for glass shading devices. The PVs will serve as shading elements for areas protected by the new system.

The overhanging shading roof provides adequate shade in the summer and allows for useful solar heat gain in the winter. These factors combined should help to keep the building's running costs to a minimum. In conclusion, the simulations and testing at the design stage show that the overall environmental strategy will reduce the building's running costs while optimizing visual and thermal comfort.

Integrating PVs into the architectural design offers more than cost benefits; it allows the creation of an environmentally friendly energy-efficient building.

The systems consist of crystalline PV modules integrated with a semi-transparent module. We also present an example of PV modules in thin films.

Keywords Retrofitting buildings · Building-integrated photovoltaics · Photovoltaic products · Replicability

The PV-integrated case studies described below are the atrium of the Pediatric Meyer Hospital in Florence (1), the University Library and classrooms (2) and the Physics Laboratory at Sesto Fiorentino (3), the Virtual Competence Centre ITC in Lucca (4), and the student residential buildings at the University of Florence (5).

L. Ceccherini Nelli (✉)
DIDA Department, ABITA, University of Florence,
Via S. Niccolò 93, 50125 Florence, Italy
e-mail: lucia.ceccherininelli@unifi.it

A. Sayigh (ed.), *Renewable Energy in the Service of Mankind Vol II*,
DOI 10.1007/978-3-319-18215-5_5

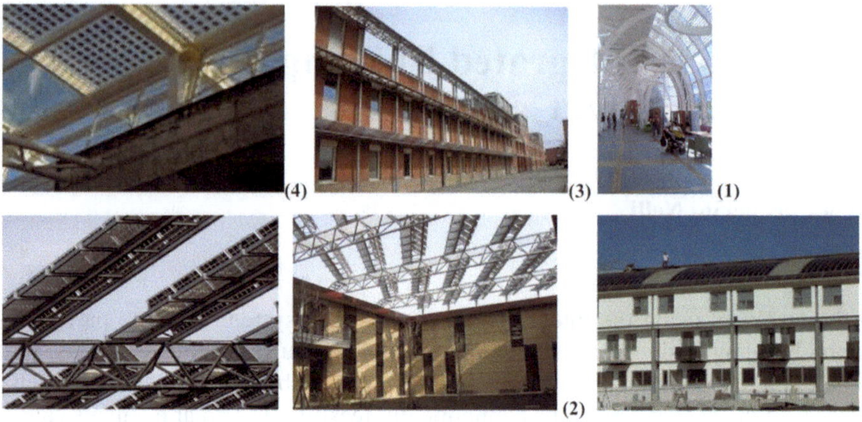

5.1 Objectives and Scientific Innovation and Relevance

The integration of photovoltaics (PVs) in buildings has huge development potential regarding both the market for building-integrated PV systems systems and the contribution to renewable electricity production. PV installation integrated into roofs and facades allows the possibility of combining energy production with other functions of the building envelope, such as shading, weather shielding and heat production.

Cost savings through these combined functions can be substantial, for example, in expensive facade systems where cladding costs may equal the costs of the PV modules. Additionally, no high-value land is required and no separate support structure is necessary. Electricity is generated at the point of use; this avoids transmission and distribution losses and reduces the utility company's capital and maintenance costs. 'Multiple integration'is perhaps the most appropriate expression. Building integration does not just mean mounting PV modules onto roofs or facades. Real integration can involve much more; it includes all the steps incorporated in the process of new construction or in retrofitting buildings, that is, from planning the production of the construction materials through to operation and recycling. Multiple integration does not produce multiple costs; however, if it is carried out in the right way, it results in multiple savings. Savings of landscape, cladding materials, engineering effort, etc. have often been mentioned in the literature. Further steps in integration have not yet been studied in detail, but it is obvious that integrating cladding materials into the existing manufacturing process should lead to further cost reductions. Integration starts at the beginning of the planning process of a building construction or renovation and continues until the building is finished.

However, the integration of PV into the architectural design offers more than cost benefits. It also allows a designer to create an environmentally friendly and energy-efficient building without sacrificing comfort, esthetics or economy.

5.2 Case Studies

Below are some examples of PV-integrated systems realized in public buildings in Tuscany over the past few years. They were designed and monitored by the researchers of the ABITA Interuniversity Centre at the University of Florence.

5.2.1 Meyer Pediatric Hospital in Florence

The system consists of 181 PV modules made with glazing of different sizes; the total output is 32,000 Wp [2, 5]. The modules have been integrated into the PV facade of the greenhouse at the main entrance of the building. Most of the PV panels measure 2.20 × 0.938 m with a power of 201 Wp. A group of conversion can transform direct current into alternating current and is laid on the roof of the greenhouse; the control panel and interface with the network is located in the center of the greenhouse inside a wooden structure.

The system is composed of three PV fields—east, west and central. Each field feeds each of the three phases of the electrical network of the hospital. In order to optimize the conversion efficiency of the modules connected to each inverter they have the same tolerance (Fig. 5.1).

1. East lot
   Fifty-five modules of 201 Wp (B1)

- Twelve modules of 88 Wp (B1/2)
- Five inverters SMA SB2500

2. Central pool

- Thirty-five modules of 188 Wp (B2)
- 12 modules of 88 Wp (B2/2)
- 1 + 1 inverter SMA SB3000 inverters SMA SB3300

3. West lot

- Fifty-five modules of 201 Wp (B1)
- Twelve modules 88 Wp (B1/second)
- Five inverters SMA SB2500

Fig. 5.1 Two main atriums with glass-glass PV panels

There are four types of PV modules (B1, B2, B1/2, B2/2) which can be grouped essentially into two types—2-m long modules with a power of 200 W and 1-m long modules with a power of 88 W. The project has achieved electrically compatible modules in order to avoid mismatching current in the strings. The modules are certified and both sides are tempered HST glass (guaranteed 20 years).

The resistance of the cables on the DC side is minimized by the size of the cables:

1. The string section in each of the conductors is 6 mm^2
2. Inverters in the framework of the interface section of the conductor are 16 mm^2
3. Framework of the interface to the general framework, the conductor cross-section is 25 mm^2.

The connections were made with compression terminals and sometimes by soft soldering.

The monitoring system has a DG 700 display which displays the values that characterize the PV system and its operation. Degree of protection IP41. $700 \times 500 \times 45$ mm/6 kg

The PV system production is controlled by checking power energy production.

5.2.2 *Virtual Competence Centre ICT Lucca*

This building has installed 3 PV-integrated systems. The PV facade system consists of 84 PV modules. 3 PV modules are arranged horizontally in each of the front panels (2.92 × 1.48 m).

The PV facade is classified as an 'integrated and intelligent system'. It has a power of 15.96 kW and an estimated production of 12.840 kWh of energy per year, from 84 modules occupying an area of 120.71m^2.

The PV array is divided into six sub-strings of two, each connected to an inverter to be installed outside in a shelter away from direct sunlight and accessible for visual inspection and maintenance activities. Each string will be equipped with an insulated and blocking diode.

Total area 120.71 m^2 modules
Inclination of the modules (tilt) 90°
Orientation of the modules (azimuth) 7th
Annual solar radiation on the surface of the modules 1.074 kWh/m^2
Technical data
Total power 15.96 kWp
Total number of modules 84
Number inverter 6
Total annul energy production 12.840 kWh
Model SANYO HIP,—Model ITALY SMA SB 3300 TL HC
Strings × 7 × 2 Modules

The PV glass roof of the greenhouse has a power of 5.76 kW and an estimated production of 6.615 kWh of energy annually. It consists of 24 transparent glass-glass PV panels (system glass room) measuring 3020 × 1620 m, arranged in parallel rows of 12 panels each, occupying an area of 117.5 m^2. In the upper part of the structure, which forms proper coverage, 24 panels are arranged in two rows. Each panel has a peak power of 240 W (produced from 96 cells). The PV array is divided into six sub-strings of one, each connected to an inverter. The inverters are installed outside and are sheltered from direct sunlight and accessible for visual inspection and maintenance activities.

The protection framework is provided with the interface device and provides surveillance of the phase voltages and network protection for minimum or maximum voltage and frequency protection and minimum and maximum frequency.

For sizing the system, two different inclinations of panels are considered—16° and 13°.

Technical details of the section with panels at an angle of 16°.
Number of surfaces available 1
Total surface area 58.70 m^2 modules
Inclination of the modules (tilt) 16°
Orientation of the modules (azimuth) 7th
Annual solar radiation on the surface of the modules 1.541 kWh/m^2
Technical data
Total power 2.88 kW
Total number of modules 12
Total number of inverters 3
Energy performance 3.327 kWh total annual energy
Form Module Module with 96 solar cells with a power of 240 W
Inverter brand—Model SMA ITALY—SB 1100 Modules strings $\times 1 \times 4$

Another part of the roof is covered with laminated PV. It covers the greenhouse and is built with PV glass-glass (6+6) modules with elements of tempered glass that meet the technical and performance audits conducted in accordance with the main terms and conditions of the site, that is, atmospheric characteristics, thermal and mechanical properties.

5.2.3 University Classrooms and Library in Sesto Fiorentino, Florence

The PV-integrated system, with a total power output of 20 kWp, occupies a surface of 300 m^2 [1–4]. The system is composed by 160 PV glass/tedlar transparent modules with a power of 125 Wp/each, divided into five subsystem of 4000 Wp each.

Each subsystem has a dedicated inverter with an interconnection box

The PV system is composed of the following:

- **Principle structure**—4 reticular beams (22 m each) positioned on the shortest side of the internal court.
- **Secondary structure**—25 beams with two steel IPE beams supported by the principal beams
- **Module structure**—modules are positioned with the longest side along the beam direction. The PV modules, transparent glass/tedlar are supported by an aluminium tripod. The modules are then sustained and screwed down to the aluminium easels with three steel L-profiles.
- Footbridges—to guarantee maintenance operations and security, grill footbridges have been installed (made of Alugril or Orsogril). These are positioned behind the PV modules on the principal and secondary beams. In the executive project we suggested handrails; however, during a function check we decided to remove the handrails to improve PV efficiency in wintertime.

PV modules are tilted 35° with southern orientation. The electricity produced in DC current by PV modules will be fed, after conversion into alternate 400 V and 50 Hz, into the building grid connected to the Medium Voltage Distribution National Grid. The energy produced will be measured through a meter installed by the grid manager and issued under the directive n.°224/00 by the National Energy Authority.

On the basis of sun values on the floor and on the module plane tilted at 35°, and assuming an average energy efficiency of 75 % of the system in the various situations, the electricity production of the system, agreed as electrical energy given to the ENEL grid is approximately 32,996 kWh/year.

Description of the components and general scheme of the system

The PV system produces energy which is connected to the electrical grid of the building in low voltage and alternate current; the system is connected electrically to the network of the client. At the connection point, the voltage is 400 Vac three-phase, 50 Hz frequency.

The energy produced will be fed into the grid, according to the technical and economic conditions of the exchange service defined by Electrical and Gas Authority directive n. 224/2000.

The principal elements that constitute the system are as follows:

- One hundred and sixty PV modules in polycrystalline silicon, 125 W peak each
- Five inverter single-phase power 3.3 kW DC, 4.2 kWp PV side
- Interface panel for grid connection (low tension) according to CEI 11.20 regulations

Conversion and electric energy delivery

Five inverters are dedicated to the five subsystems for energy conversion from DC to AC current. The inverters convert the energy of the PV generator in alternate 3-phase currents, for energy distribution into the grid.

Five inverters installed: manufactured by Sun Power Solartechnik Model SP 3100-600

5.2.4 Physic University Laboratories—Sesto Fiorentino 50 kWp

This PV system is a retrofit installation in the building of the Department of Physics of the University of Florence.

The PV modules are placed in the field with the following values of the tilt and azimuth: 1/2/3 Subfields: TILT 30.0–32° azimuth.

The project involves the construction of three PV generators connected in parallel on the general framework (QAC). The building is divided into several sections with two, three or four floors, arranged to form two quadrilaterals which enclose two interior courtyards.

The PV generator is divided in three parts, two on the south facade, as shading devices and one is installed, as a shelter, on the flat roof. A metal structure provide to support the PV glass-glass modules.

The three subfields are composed respectively of 60, 54 and 74 PV modules of variable strings divided by 9–12 modules.

The total number of panels, each divided by the PV field is equal to:

Subfield 1: 54 modules of 290 Wp—15.66 kWp
Subfield 2: 60 modules of 290 Wp—17.40 kWp
Subfield 3: 74 modules of 225 Wp—16.65 kWp

The total installed peak power is equal to: 49.71 kWp.

Total annual production of the solar field: 51.766 kWh

The PV modules are the type used at high-peak power; they are composed of 60 multicrystalline solar cells measuring 156 × 156 mm.

The body is made from an anodized aluminum frame with high resistance to corrosion, tempered solar glass with low-iron content, with maximum load equal to 550 kg/mq. Three by-pass diodes are used to minimize power loss for eventual shading or damage.

DATA FIELD FRONT PANEL
Manufacturer: SCHOTT SOLAR
Model: POLY TM 290
Peak power: 290 Wp
Number of cells: 80 connected in series
Module efficiency: 13.7 %

5.2.5 University Residential Student Building, Florence (20 kWp)

The system is integrated into the type of coverage for a building which is used as a place of residence at a University because of Mezzetta in Florence.

The system is grid-connected and the connection mode is 'three-phase low voltage'. The power is 19.92 kWp, with an estimated production of 21.320 kWh of energy per year (minimum guarantee). 94 amorphos PV modules occupy an area of 220 m^2 and 34 monocrystalline PV modules occupy an area of approximately 44 m^2 power, respectively—12.784 kWp–7,14 kWp.

Modules in amorphous Si thin film

The PV panels are made of 10/10 aluminum sheet with high corrosion resistance measuring 5700 × 467 mm, which support the laminates and are to be mounted parallel to the cover plates of the building.

The metal sheet cover must be able to adapt to the underlying curve through a system of extruded aluminum profiles and should not interfere with the disposal of water.

The profile of the sheet is integrated with a system of spacer elements and stiffening in extruded aluminum that are anchored through the cover to the support structure below.

The spacer elements are formed by two types of drawn extruded aluminum, one with an omega-shaped profile, the other with an inverted T profile. The omega spacers are anchored at the rafters of the existing structure.

The T-sections are positioned orthogonally to the omega ones and represent both the longitudinal strength of the panels and anchoring element.

The PV structure (omega and T profiles) is curved to follow the metal roof.

At the bottom of the cover a row of PV modules in hybrid technology will be arranged parallel to the roof surface.

The monocrystalline PV modules will be placed side by side and fixed to the cover by instead of special aluminum profiles and attached to the existing roof covering, exploiting the existing inclination (approximately 35° above the horizon).

This plant (consisting of 34 modules) with a rated power of 7.14 kWp, complete the PV generator with a total power production of 19.92 kWp.

5.3 Conclusions

Public actions make it necessary to stimulate integration of PV systems, to encourage stakeholder and technical managers to find new processes and technical solutions, that can stimulate industrial production, reduce PV costs and increase devices efficiency in the energy production. A great advantage of new product development—more suitable for architectural applications—is to remark as effect of the support's mechanisms until now adopted by several countries.

The most recent type of PV technology relates to integrated building systems. PV systems can contribute—according to the dimension of the plant—to the total global electric budget of the building. These applications introduce different advantages:

- Energy produced near the user has a greater value than energy furnished by the traditional electric power station;
- Electric energy production during insulation times allows demand to the net to be reduced during the day, only when there is the greatest request. Hypothesising a high development of building-integrated PV systems, it is possible to foresee a levelling of the daily peak request, usually corresponding to the more expensive kWh electrical cost. It is a more and more interesting alternative, particularly for the increasing use of air conditioning systems in residential, commercial and public buildings;
- PV installation costs could also be an avoided cost, decreasing the global building cost, because the PV modules can be constructive elements replacing tiles or facade glass;
- The adoption of these systems allows diffusion, directly among the consumers, of a great 'energetic conscience', with a positive increase of the use of electric energy produced and exchanged with the net.

It is necessary to highlight the PV systems esthetical value; the silicon cell has a pleasant aspect and a particular effect, making it an interesting material for contemporary architecture. It is possible to use different colored cells, adapting them to the several contexts.

References

1. Ceccherini Nelli L (2007) Schermature fotovoltaiche, Alinea, Firenze
2. Ceccherini Nelli L (2007) Fotovoltaico in architettura, Alinea, Firenze
3. Sala M, (a cura di), Ceccherini Nelli L (2003) Integrazione Architettonica del FOTOVOLTAICO, casi studio in Toscana, Alinea, Firenze
4. Ceccherini Nelli L (2004) Impianto fotovoltaico integrato da 20 kWp per l'edificio aule e biblioteca a Sesto Fiorentino, Alinea Firenze
5. Sala M, Ceccherini Nelli L (2006) AA.VV, CD-ROM EULEB European high quality Low Energy Buildings, Dortmund

Chapter 6
50—Energy Efficiency in Retrofitting a European Project for Training on Renewable Energy Solutions (REE_TROFIT)

Lucia Ceccherini Nelli

Abstract REE_TROFIT (www.reetrofit.eu) project (founded by the EU Commission in the Intelligent Energy Europe (IEE) program) aims to contribute to solve the shortage of local qualified and accredited retrofitting experts, as foreseen in the Energy Performance of Buildings Directive (EPBD) and its recast—and as indicated by various European countries in an assessment by the European Commission (EC)—for increasing the energy performance of the existing building stock. REE_TROFIT will use the in-house know-how and experiences of participants in carrying out vocational courses on innovative eco-building technologies. REE_TROFIT project defines best practices for institutionalization and implementation of vocational courses on renewable energy (RE) solutions and energy efficiency (EE) in retrofitting, setting up, and implementing a large-scale educational scheme and by fostering exchange of knowledge and best practices among stakeholders.

One of the major milestones of REE_TROFIT project is to raise awareness in the regional, national, and European policy-makers for the full implementation of the EPBD and its recasts. Additionally, during its lifespan, it intends to define an exploitation strategy for assuring the sustainability of training beyond the project duration and increases the local retrofitting markets.

The REE_TROFIT (newsletters: http://www.reetrofit.eu/content.php?p=nlt) training scheme is founded on an innovative educational model specifically targeted for the building professionals; the adopting REE_TROFIT training model offers the following attractive features:

Flexibility: is applicable in contexts with different regulatory frameworks: climate, landscape restrictions, qualification levels of learners, etc.

Transferability: is capable of responding to local training needs through methodologies and tools transferable at European level.

Innovation: is accessible, affordable, and capable of overcoming the problems encountered during the previous training program experimented in the partnering countries.

L. Ceccherini Nelli (✉)
DIDA Department, ABITA, University of Florence, Via S. Niccolò 93, 50125 Firenze, Italy
e-mail: lucia.ceccherininelli@unifi.it

A. Sayigh (ed.), *Renewable Energy in the Service of Mankind Vol II*,
DOI 10.1007/978-3-319-18215-5_6

Modularity: offers different training programs which are composed of independent, closed, domain-specific modules that may be activated according to the different training needs.

Brevity: offers training courses with a short duration, which are decomposed in shorter training tracks in order to ease the attendance of the targeted professionals.

Plurality: different training methods, tools, and media might be used in the training process in order to consider the needs of the trainees and to guarantee effectiveness.

Keywords Retrofitting buildings · Training · Courses · Renewable energy

6.1 REE_TROFIT Project Objectives

REE_TROFIT [1, 2] aims to contribute to solve the shortage of local qualified and accredited retrofitting experts, as foreseen in the Energy Performance of Buildings Directive (EPBD) and its recast—and as indicated by various European countries in an assessment by the European Commission (EC)—for increasing the energy performance of the existing building stock. REE_TROFIT will use in-house know-how and experiences of participants in carrying out vocational courses on innovative eco-building technologies to define best practices for institutionalization and implementation of vocational courses on renewable energy (RE) solutions and energy efficiency (EE) in retrofitting, setting up, and implementing a large-scale educational scheme in 6 Master Course (MS) for training more than 450 building professionals and by fostering exchange of knowledge and best practices among stakeholders, provide suggestions to regional, national, and European policy-makers on how to incentivize, de-bottleneck the local retrofitting markets for full implementation of the EPBD, and define an exploitation strategy for assuring the sustainability of training.

6.2 The Features of High-Quality Assessment Processes

The REE_TROFIT project developed and tested a quality assurance mechanism that best guarantees on site energy and carbon reductions during the energy-saving renovation process. This was only possible through a strong collaboration with stakeholders and policy-makers.

The activities foreseen by WP6 of the REE_TROFIT project aimed to improve knowledge and skills of the buildings workforce and the guide materials developed have proved valuable input for the EU BUILD UP Skills initiative.

Furthermore, REE_TROFIT vocational courses, delivered in each of the six partner countries, tested and improved the theoretical frameworks developed. On the whole, courses have reached over 470 supply side actors (primarily building operators) and energy advisors. Furthermore, REE_TROFIT partners have promoted cooperation with over 20 federations and governmental bodies involved in the energy and building sector in own EU partner countries.

A feedback loop between four main activities was established and maintained for the duration of the REE_TROFIT project as mentioned below:

1. Identification of existing and innovative best practices to encourage the uptake of energy-retrofitting approaches and to improve the quality of on-site energy-saving renovation activities, analysis, and development of an extensive database kept up to date during the lifetime of the project.
2. An efficiency assistant was developed, translating the best practice tools and identified techniques in each of the six EU partner countries context for best driving the uptake recommendations into a practical guide. The guide supports policy-makers and practitioners aiming to establish and/or to improve programs stimulating recommendations and encouraging action on EE in retrofitting.
3. Quality assurance tools for housing energy-saving renovations were developed, constantly updated and improved on the basis of success factors from existing best practices and according to market actors, and eventually tested within real renovation projects.
4. Many pilot projects were designed, delivered, and evaluated. With such projects, different aspects of the theoretical frameworks developed within the REE_TROFIT training were tested and improved, tailoring them to the specific policy context and market conditions in each country.

6.2.1 Main Actions of REE_TROFIT Consortium for the Quality Assessment

An overview of the six pilot training courses that have been running from 2010 to 2013 in the REE_TROFIT partner countries is reported, focusing on increasing the quality of on-site renovation activities:

Denmark—Evaluation of the impact of the activities of the Danish Knowledge Centre for Energy Savings in Buildings on the uptake of energy-efficient renovation, and detailed analysis of the Project Zero initiative for the municipalities.

The Aarhus School of Engineering (IHA), REE_TROFIT partner, is cooperating with Energitjenesten, the independent utility energy service in Denmark for the training activities using the "The Handcraft Companies Energy Forum" as a platform, and with the Architect School of Aarhus (DK) for offering courses to building professionals on energy-efficient building. The course includes passive house design, natural and mechanical ventilation, and heat recovery. IHA, with the support of these training partners and the Danish Federation of Small and Medium Sized Enterprises, will bring experience and know-how on advanced building techniques for retrofitting.

Greece—Energy-saving renovation of large-scale apartment buildings in Crete.

Technological Educational Institute of Crete (TEIC) is also involved in the following activities:

- RES School: A 2-week-long educational and training program on small-scale RE sources and energy saving.
- EPEAEK: Developing a distance learning course on "Renewable Energy Sources and Environmental Management."
- Research study for the Chamber of Engineers, Western Crete Branch, on the definition of the best practices and procedures for the energy audit and the role of professionals.

Italy—Establishment of a national stakeholder consultation forum with the Chamber of Commerce of Lucca (CCIL) to create consensus and hasten implementation of legislation in Italy, and at the practical level, to improve the energy-saving renovation process of a social housing company, and renovation for school buildings and other public buildings in Lucca. In particular, LUCENSE was able to be proactive in the promotion of the institutionalization of REE_TROFIT as training model among the Italian Chamber network to ensure the highest and broader level of institutionalization, contacting the Italian Union of Chamber of Commerce. "Unionfiliere" is the Italian Association of Chamber of Commerce aiming to develop and improve "Made in Italy" industry. The activities of *LUCENSE and CCIL* were aimed to establish a permanent committee for sustainable building in the context of Unionfiliere. LUCENSE attended the first meetings, on behalf of the CCIL, promoting the REE_TROFIT model as reference for building professionals training on the theme of sustainable building and retrofitting.

Hungary—Setting up collaborative initiatives to engage tenants and property owners in energy-saving retrofit of multi-dwelling residential buildings and public ownership with the Chamber of Commerce of Kecskemet.

Bulgaria—Improvement of REE_TROFIT recommendations and energy-saving renovation practice with a focus on residential buildings from the 1970s and 1980s in the Sofia area.

The European Labour Institute (ELI) together with *The Bulgarian Chamber of Commerce and Industry (BCCI)* established a National Center for Vocational Training (NCVT) for training professionals in various sectors. The NCVT shows in the membership over 52,000 companies and 28 regional Chambers of Commerce and Industry and 67 sector organizations.

Several dissemination channels will be used in Bulgaria toward the huge pool of professionals and organizations of NCVT for attracting participants to the training courses.

France—Development and testing of EE guidance and training material for residential energy-saving refurbishment approach in collaboration with the Chamber of Commerce of Drome (CCID).

The CCID incorporates a training organization called Neopolis, dedicated to eco-construction. An important mission of CCID-Neopolis is to fulfil the training needs of professionals for nurturing a competitive market. Neopolis is the only vocational training institution in France providing training sessions on eco-construction.

6.2.2 *Suggestions for Policy-Makers*

In order to sustain and incentivize local markets for qualified/certified professionals in building sector, suggestions for policy-makers were addressed with the following activities:

1. Establishing a *clear vision and goals* that all stakeholders could understand, buy into, and implement working together. Different pilot projects have shown the power of collaborative partnerships in driving change in the presence of a shared vision and efforts (e.g., for the Danish Project Zero: zero carbon by 2029).
2. *Valuing guidelines* and ensuring that the provided guidance is tailored, specific, and accurate.
3. Acknowledging the key role of *information and communications activities* of the Chambers of Commerce in driving action on energy retrofitting in order to ensure that builders, electricians, and installer could understand where and how they can seek support about certificate recommendations.
4. Ensuring *access to finance*, with local, regional, and national contributions, low interest loans or other means that can significantly increase the uptake of energy-saving measures. Second, the financing support should ideally be coupled with EE targets to be achieved.
5. *Training and support for supply chain actors* is therefore essential for ensuring homeowners to have access to skilled tradespeople able to deliver quality renovation works.
6. Helping builders to develop and maintain *homeowners' trust* is one of the most important links of the chain. Important factors for achieving this objective are the provision of impartial information/advice and the guarantee of quality throughout the supply chain in a way that renovations actually deliver foreseen results.
7. A further key aspect is *process and project management support,* for the homeowner, builder, and installer and/or the whole supply chain. This aspect is crucial for ensuring maintenance of the supply chain with all actors involved working together to deliver a successful low energy renovation intervention.

Clearly, the suggestion to policy-makers is not a simple "cut and paste" exercise of the REE_TROFIT model to different countries across Europe (and to the EPBD), rather all the elements need to be tailored and adapted to local context. In the below sections, we zoom in on some of the overarching and common outcomes of the REE_TROFIT courses in different countries. In fact, in order to address the key elements listed above, insight into specific parts of the whole process could thus be provided in order to create a fully functioning energy-saving renovation market.

6.3 Methodology—Strengthening the Role and Impact of REE_TROFIT Trainees

Improving skills and knowledge of technicians and professionals who attended REE_TROFIT courses includes detailed recommendations for cost-optimal energy saving, financial analysis (i.e., payback times), and technical specifications. All these points resulted to be crucial for strengthening the value of the vocational course (Fig. 6.1).

The training of energy experts and building professionals having specific competences allowed to link to the EPBD methodology, and to address grant or financial support schemes (regional and national, for instance integration of renewable energies) by imposing minimum levels for the overall energy performance of both

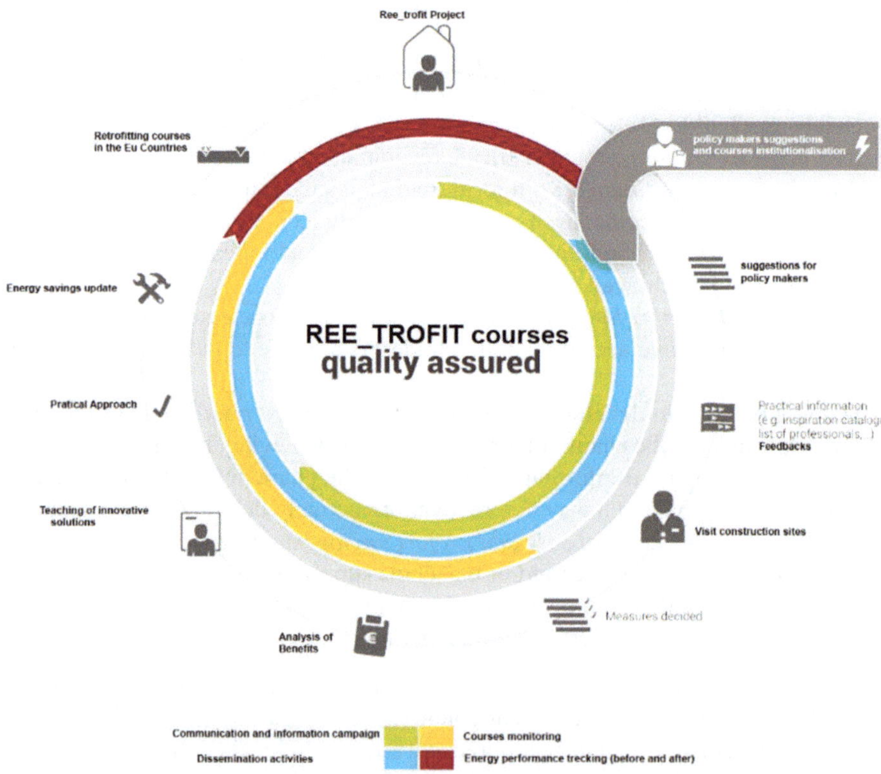

Fig. 6.1 Fully functioning energy-saving renovation REE_TROFIT training courses

public and residential buildings after renovations. This methodology thus resulted to be important to firmly embed REE_TROFIT installer and builder in the start-up phase of any energy-saving retrofit activity.

On the other hand, the existing cost differences between energy-efficient and "standard" refurbishments resulted to have a great influence on decision-making process of the property owners. Public institutions and policy-makers have thus a crucial role in encouraging investment in ambitious energy-saving retrofit interventions: Policy-makers' recommendations should be accurate, based on robust data and analysis, and the communication to target audience should be effective. Furthermore, national or regional regulation issuing body should directly interact with the supply chain, helping to create new networks and supporting the selection of services of certified experts, such as those trained by REE_TROFIT vocational courses and/or by other vocational courses aimed to train professionals in low energy retrofitting. However, any successful program driving the uptake of energy-efficient methodologies and sustainable refurbishments is highly dependent on the quality of consultancy. This need for quality consultancy goes beyond the building owners. The process management aims to ensure an integrated supply chain and a seamless customer journey, but, although it is believed to be necessary to maximize the EE potential of retrofitted buildings, this approach is not yet formalized in most of the countries. However, a number of different higher professionals (site managers, assessors, project managers) could be able to address this point.

Will follows an analysis of the principal impact strategies followed by the partner countries for the planning, organization, and institution of the vocational Retrofit courses:

The REE_TROFIT training model aims to contribute to improve the cultural level and the skills of building professionals (construction small and medium-sized enterprises (SMEs), electrical installers, thermo-hydraulic installers) in the field of EE and RE in building retrofitting, which is an area where major potential exists to reduce energy consumption and improve sustainability in buildings, with economic and environmental win-wins.

The main ten features of REE_TROFIT model are the following:

- *Flexible*: applicable in contexts with different regulatory frameworks: climate, landscape restrictions, qualification levels of learners, etc.
- *Transferable*: capable of responding to local training needs through methodologies and tools transferable at European level.
- *Innovative*: accessible, affordable, and capable of overcoming the problems encountered during the previous training program experimented in the partnering countries.
- *Modular*: the different training programs are composed of independent, closed, domain-specific modules that may be activated according to the different training needs.
- *Short*: short duration of the training courses, which are decomposed in shorter training tracks in order to ease the attendance of the targeted professionals,

capable of breaking down barriers such as lack of time, reluctance to invest in training, poor habit to listen, etc.
- *Plural*: different training methods, tools, and media might be used in the training process in order to consider the needs of the trainees and to guarantee effectiveness.
- *Interactive*: in order to ensure the active involvement of trainees.
- *Open*: possible different training environment also supported or codesigned with the leading companies in the market.
- *Pragmatic*: committed to providing high quality and accessible training opportunities to all trainees and effective skills and competence readily applicable in their workplace.
- *Effective*: allowing professionals to grow in their jobs and improve their performance.

The impact of vocational education and training (VET) is producing an innovative and sufficient model for training courses of professionals in the retrofitting sectors with concrete *platforms* for training of three types of professionals able to test it through one pilot and two further batches of trainings.

- The model proposes a new holistic approach to training in order to prepare professionals for integrated solutions.
- It uses advanced methodologies in vocational training courses—a balance between class and pragmatic activities, usage of mock-ups and practical works, site visits followed by class analyses, interactive workshops, role-playing, integral solutions' fashioning, etc.
- It crystallizes "best practices" for vocational training courses.
- It utilizes assisting tools for the expansion of the vocational training courses in the partner countries.
- Guidelines, assessment tools, questionnaires, could be developed in the web portal to assist trainees to verify their own knowledge level.
- It finishes with concrete qualitative and quantitative results and conclusions of the level of the achieved results by country and all put together.

6.4 Results—Impact on the Energy Actors at the Retrofitting Market/Involvement of New Actors in the Professional Vocational Training Processes

Italy Competent energy and financial analysis is necessary to achieve the widespread adoption of whole systems retrofits.
While the industry has grown rapidly in the past 5 years, we have been hampered by a number of issues that make it challenging to provide cost-effective and high-quality analysis.

REE_TROFIT's building energy solutions approach in the institutionalization of VET courses:

Tools and resources	Developed a set of tools and templates that will save time and increase the quality of teaching in energy savings in buildings
Training and education	Developed training and education materials that are the basis of the principal arguments involved in the VET
Building energy innovation	Direct involvement of industrial and municipal stakeholders involved in renovation of buildings

Impact on energy retrofitting for the VET

Cost-effective EE as a high-priority resource
Process in place, such as a city and/or regional collaborative, to pursue EE as a high-priority resource
Policy established to recognize EE as high-priority resource
Potential identified for cost-effective, achievable EE over the long term
EE savings goals or expected energy savings targets established consistent with cost-effective potential
EE savings goals and targets integrated into a regional energy resource plan
Developing processes to align utility and other program administrator incentives
Utility and other program administrator incentives for EE savings reviewed and established as necessary
Establishing evaluation, measurement, and verification mechanisms
Robust, transparent procedures established. Strong public education programs on EE in place
Developing region policies and incentives to ensure robust EE practices
National policies require routine review and updating of building codes
Building codes effectively enforced
National- and local government-led example programs in place

Impact on Teaching Methods Another approach of teaching will be given by LUCENSE and Abita to provide a new platform to diffuse e-learning methods of teaching, using modern technology. E-learning system offers a great potential to bring learning to the beneficiary, to their own organizations and communities.

VET in the partner countries: SMEs—the number of trainees foreseen to be trained was sometimes doubled or tripled, and their feedback showed complete satisfaction and need for continued further trainings. The impacts are as follows:

- Involvement of public authorities and energy agencies.
- Involvement in the training process, interest in further projects and training activities.
- Chambers of Commerce actively involved in the trainings, spreading information and training tools among their members.

- Training institutions, secondary schools, higher technical institutes provoked for integrating new disciplines in their curricula and offered platforms and advanced model for their integration.
- Great interest for the new products and innovation in retrofitting.

Impact on the MS VET Activities The impacts produced by the VET product model are:

- Good results on test of the model and continuous improvement for the institutionalization of the courses.
- Experienced a new training approach also with the use of technological supports and e-learning methods.
- A holistic approach.
- Worked out guidelines, plans, and other tools for easing the VET process in the MS and the further implementation of the REE_TROFIT model.
- Gaps in the VETs trainings identified and "Recommendations" for successful carrying out proposed.

Suggestions to Policy-Makers

- Create incentives and stimulate for broader, continuing training of professionals in this very important field.
- Provide funding or attractive financial models for training of the SMEs in the retrofitting sector.
- Create mechanisms for certifying short-term dynamic training courses at national level and registration of the trained professionals.
- Assure funding for training of public officers.
- Include disciplines corresponding to the needs of the retrofitting market in the compulsory educational programs in the secondary and higher schools; stimulate the creation of new professions, green collars, and jobs.
- Assure affordable funds and models for retrofitting of buildings.
- Stimulate national SMEs in producing and implementing new advanced products and technologies necessary for the EE renovation of buildings.
- Improve legislation, fashion policies for easing the implementation of RES.
- Produce levers for the intensive EE renovation of the building stock and massive application of RES.

6.5 Conclusions

The REE_TROFIT project is supporting the idea that a tight connection between the supply chain and the demand for sustainable refurbishment focusing on energy saving could successfully drive interventions toward low-energy buildings particularly by promoting vocational courses in order to improve competences of building professionals. However, public institutions and policy-makers have a crucial role in encouraging and promoting energy-saving retrofit interventions.

Currently, levels of general retrofitting activity are poorly monitored across Europe and there is virtually no monitoring of retrofit activity undertaken in response to energy-saving measures. There is, in other words, a huge potential for much better tracking and analysis to identify the remaining potential for action on EE and reduction of CO_2 emissions from European buildings. However, the most recent EU directives strongly support and promote energy performance of buildings. This is yet another important element that could support policy-makers, market actors, local authorities, and householders themselves in planning low-carbon improvement strategies.

References

1. www.reetrofit.eu
2. newsletters: http://www.reetrofit.eu/content.php?p=nlt

Chapter 7
Nuclear Power: A Promising Backup Option to Promote Renewable Penetration in the French Power System?

Camille Cany, Christine Mansilla, Pascal Da Costa, Gilles Mathonnière and Jean-Baptiste Thomas

Abstract The general 3X20 European directive proposes renewable penetration goals. In France, 27% of the electricity is to be produced by renewable resources by 2020, and this share will be continuously growing until 2050. Among these resources, some—such as wind and solar—are not dispatchable, which trigger challenges to maintain the reliability target level of the power system, both in the short and long term. Wind and solar are expected to contribute to about 12% of the French electricity production by 2020 and according to voluntaristic scenarios they could contribute to over 50% of the total electricity production by 2050.

The increase of the non-dispatchable renewable share in the power system modifies the residual demand (which is equal to the demand minus non-dispatchable renewable production) pattern to a more power ramping and higher variation in amplitudes. To keep the system balanced, backup options are numerous, though scarcely considered exhaustively in power systems modeling. Besides peaking unit production—such as gas turbines—which is usually the preferred option, storage, demand (or supply) curtailment, interconnections, and baseload (nuclear for

C. Cany (✉)
CEA DEN I-tésé and ECP LGI, CEA Saclay, 91191 Gif-sur-Yvette Cedex, France
e-mail: camille.cany@cea.fr

C. Mansilla
CEA DEN I-tésé, CEA Saclay, 91191 Gif-sur-Yvette Cedex, France
e-mail: christine.mansilla@cea.fr

P. Da Costa
ECP LGI, Ecole Centrale de Paris, Laboratoire de Génie Industriel, Grande Voie des Vignes, 92 295 Châtenay Malabry Cedex, France
e-mail: pascal.da-costa@ecp.fr

G. Mathonnière
e-mail: gilles.mathonniere@cea.fr

J. P. Thomas
5CEA DEN DIR EC, CEA Saclay, 91191 Gif-sur-Yvette Cedex, France
e-mail: jean-baptiste.thomas@cea.fr

A. Sayigh (ed.), *Renewable Energy in the Service of Mankind Vol II*,
DOI 10.1007/978-3-319-18215-5_7

instance) power modulation should also be considered, particularly in such a context of high non-dispatchable renewable penetration.

In the French context of significant production of nuclear power, nuclear modulation is investigated as a feasible opportunity to facilitate renewable energy penetration. Available capacity is assessed based on realistic scenarios and through the use of residual load duration curves. The load modulation impact on the nuclear levelized cost of electricity is estimated, and nuclear backup option is compared to gas on economic and environmental terms. Gas backup is more competitive than nuclear in each of the studied scenarios, but including a carbon tax could change the trends (as low as € 20/ton$_{CO2}$ in some cases). This advocates for such incentives to avoid the effective greenhouse gas release. Nuclear backup would be all the more competitive than power plants used are amortized.

Keywords Renewable energy penetration · Non-dispatchable renewable sources · Backup · Low-carbon sources synergy

7.1 Introduction

The 3X20 European directive proposes renewable penetration goals [1]. In France, 27 % of the electricity is to be produced by renewable resources (RES) by 2020 [2] and this share will be continuously growing until 2050. Among these resources, some—such as wind and solar—are not dispatchable, which trigger challenges to maintain the reliability target level of the power system, both in the short and long term [3, 4].

Non-dispatchable renewable resources currently benefit from a dispatch priority. The increase of their share in the power system modifies the residual demand (which is equal to the demand minus non-dispatchable renewable production) pattern to more power ramping and higher variation amplitudes. To keep the system balanced, backup options are numerous, though scarcely considered exhaustively in power systems modeling (e.g., [5, 6]). Besides peaking unit production—such as gas turbines—which is usually the preferred option, storage, demand (or supply) curtailment, interconnections, and baseload (nuclear for instance) power modulation should also be considered, particularly in such a context of high non-dispatchable renewable penetration [7].

Considering the French power system, in order to maintain its reliability target level, nuclear power could participate to system balancing thanks to characteristics compatible with power modulation [8–11]. Increasing non-dispatchable renewable penetration levels would entail higher balancing needs [12], hence lesser outputs and operational modifications for nuclear power (e.g., higher fuel use, more maintenance needs, increased material, and components ageing) [9].

This chapter investigates the overall economic impact of such an operating mode and compares nuclear and gas power load following options until 2050, based on production costs and CO_2 emission criteria. Several non-dispatchable renewable penetration level scenarios have been developed for France that take into account both wind and solar whereas most studies only focus on wind and hardly ever consider nuclear (e.g., [3, 5]). These scenarios are first detailed (Sect. 7.2). Then the methodology is presented (Sect. 7.3), followed by the results and discussion (Sect. 7.4).

7.2 Power System Scenarios

In 2013, the French government launched a broad national consultation on the energy transition [13]. In this framework, several institutions designed energy prospective scenarios for France.

The Institute for techno-economics of energy systems of the *Commissariat à l'Energie Atomique et aux Energies Alternatives* (CEA/I-tésé) was involved in the contribution of the ANCRE (*Alliance Nationale de Coordination de la Recherche pour l'Energie,* the French National Alliance for Energy Research Coordination), that proposed three contrasted scenarios that all satisfy the *factor 4,* that is, a reduction by four of the greenhouse gas emissions by 2050, compared to the 1990 level [14]. The specificity of these scenarios is a voluntaristic approach in terms of scientific and technological innovation. All these scenarios propose a similar non-dispatchable renewable penetration for a given time horizon, leading to a maximum energy penetration of 30 % by 2050 (wind and solar). As regards nuclear power, two cases are considered: a 50 % value (of the annual electricity production), in line with the French government policy, and a "high nuclear" case with a 65 % value.

In order to consider a more voluntaristic scenario in terms of non-dispatchable renewable penetration, we also considered the scenario proposed by the ADEME (*Agence de l'Environnement et de la Maîtrise de l'Energie*, the French Environment and Energy Management Agency), which examines the case of 50 % non-dispatchable renewable penetration by 2050 [15]. The ADEME also considered lower nuclear penetration (down to 20 %).

Furthermore, as regards nuclear power contribution to the French electricity mix, we also considered a "business-as-usual" value of 75 %, which corresponds to the nuclear current contribution to the French annual electricity production [16].

Finally, we established generic scenarios by crossing the renewable and nuclear penetration assumptions (cf. Tables 7.1 and 7.2), and studied each of them. For readability issues, only three contrasted cases will be presented in the results and discussion section. They provide conceivable target capacities in line with the current energy policy as regards RES development and the specific French context characterized by the significant place of nuclear power in the power mix:

1. "median scenario" (med): 16.5 % wind and solar in 2030, 30 % in 2050, and 50 % nuclear in both cases;
2. "high nuclear scenario" (high nuc): 65 % nuclear, wind and solar shares as in the median case;

Table 7.1 Renewable energy penetration scenarios

Time horizon	Wind penetration (%)	Solar penetration (%)	Total (%)	Reference scenario
2015	5	1	6	ANCRE [14]
2020	8	2	10	ANCRE [14]
2025	10	3	13	ANCRE [14]
2030	12	4.5	16.5	ANCRE [14]
2040	16	7	23	ANCRE [14]
2050	20	10	30	ANCRE [14]
2050	30	15	45	ADEME [15]
2050	35	15	50	ADEME [15]

Table 7.2 Nuclear energy scenarios

Nuclear penetration (%)	Reference scenario
75	Business-as-usual
65	High nuclear in ANCRE [14]
50	Low nuclear in ANCRE [14] or High nuclear in ADEME [15]
20	Low nuclear in ADEME [15]

3. "high non-dispatchable renewable scenario" (high ren): 30% wind and solar in 2030, 50% in 2050, 50% nuclear in both cases.

In order to assess the economic impact of nuclear power modulation operation, two scenarios have been defined as regards the French operational nuclear fleet (Pressurized water reactors (PWR) and Evolutionary power reactors (EPR™), capacity evolution is shown in Fig. 7.1.

- Current reactors are operated for 40 years and then shut down ("40 years" scenario);
- The current fleet is progressively replaced to avoid massive investment in a short period of time: 2 GWe are shut down every year ("prog."; it is the "best-estimate" scenario).

In each case, the current fleet is completed by EPR™ in order to achieve the target nuclear penetration level of the selected scenario (cf. Fig. 7.1).

These scenarios have been established based on realistic expectations (the operating life extension of nuclear power plants is currently being discussed in France) [17] and will result in different production costs according to the composition of the nuclear mix (the capex of the nuclear mix is computed from the share between amortized power plants (PWR) and the recently installed ones (EPR ™). Opex and life extension costs have also been included. They are issued from the French Court of Auditors in-depth study of the nuclear costs in the French context [24]).

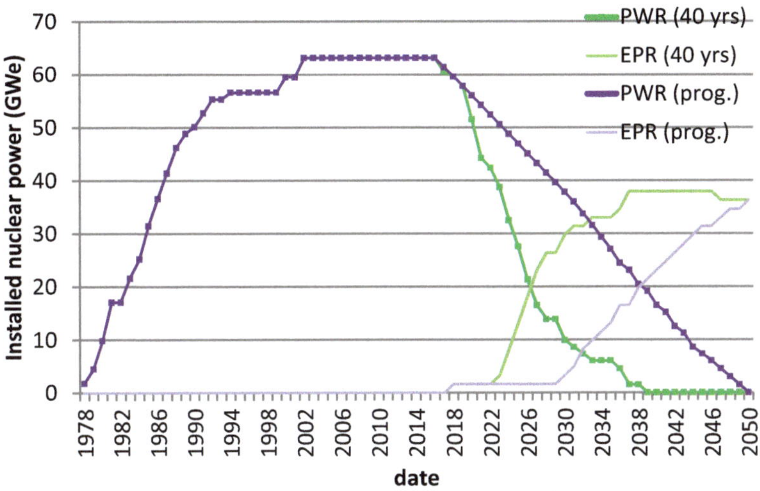

Fig. 7.1 Pressurized water reactors *(PWR)* andEvolutionary power reactors *(EPR™)* capacity evolution scenarios (for 50 % nuclear penetration), own calculus

7.3 Methodology: An Approach Based on the Construction of Annual Residual Load Duration Curves

To assess the impacts of non-dispatchable renewable resources on other power plants capacity requirements and annual productions in a power system, two major types of approaches can be distinguished. First, the construction of load net of wind and solar (called "residual" in what follows) duration curves for different wind and solar penetration levels allows determining the optimal mix of other generation resources in the long term [5, 6, 18]. Second, a minimum cost economic dispatch model can serve refining the general trends obtained from the first approach [5, 6, 19, 20]. The use of such a time-step model allows short-term supply and demand balancing in response to technical, economic and political constraints. We focus here on the first approach to put a light on general trends in the long term.

The developed methodology proceeds in several steps. On the one hand, the construction of annual residual load duration curves allows assessing the annual energy productions and optimal installed capacities, among which the needs for baseload installed power will be fulfilled by nuclear in the French case. On the other hand, nuclear cycling capacity is evaluated based on the realistic French nuclear fleet scenarios that were presented in the previous section. Thanks to the annual residual load duration curves, we evaluate and compare nuclear fleet effective full power hours of both options: nuclear power fleet operated as a baseload capacity or as a combined option— baseload and cycling capacity. We then carry out the economic comparison. Finally, through the same approach, the nuclear load following option is compared to the gas one, which is considered as the reference backup option.

7.3.1 Construction of Annual Residual Load Duration Curves

A load duration curve represents the sorted hourly load of one year, starting with the highest load hour. The residual load curve is obtained by withdrawing the wind and solar production to the total annual production (considered constant on an annual cumulated basis in this study, hourly variations have been computed into the residual load duration curve based on the historical data from the French Transmission System Operator (RTE)). Wind and solar production was estimated by calculating mean hourly power values of historical data (between 2010 and 2013) also from RTE [21], and by calibrating their hourly power to fit their assumed energy penetration level to satisfy a given demand (cf. Table 7.1). The resulting curves are displayed in Fig. 7.2.

7.3.2 Assessment of the "Optimal" Nuclear Power Generation and the Nuclear Cycling Capacity

The construction of screening curves (cf. Fig. 7.3) allows estimating the lowest total cost of technology for a given operating time. Thus, they provide the optimal annual operating time for each resource type (e.g., 7900 h for nuclear in the example shown in Fig. 7.3), which can then be copied on the residual load duration curve, to assess the corresponding optimal installed power.

Baseload installed power will be fulfilled by nuclear in the French case. Other unavoidable energy production capacities have been taken into account nonetheless (e.g., run-of-river hydroelectric plant).

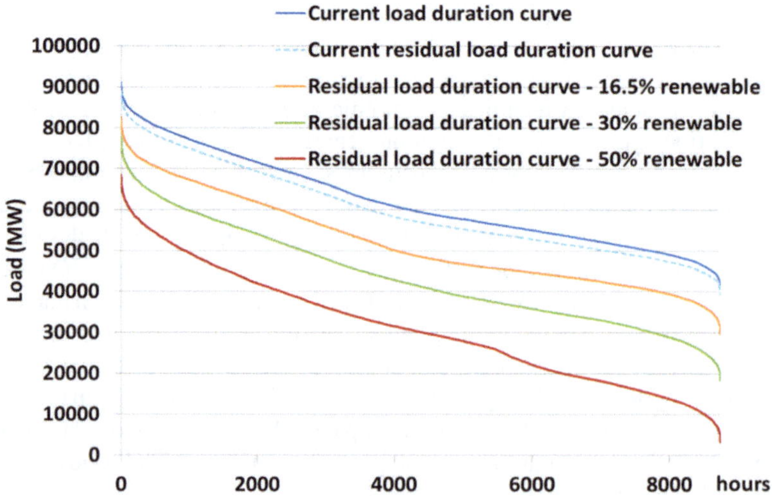

Fig. 7.2 Residual load duration curves, own calculus

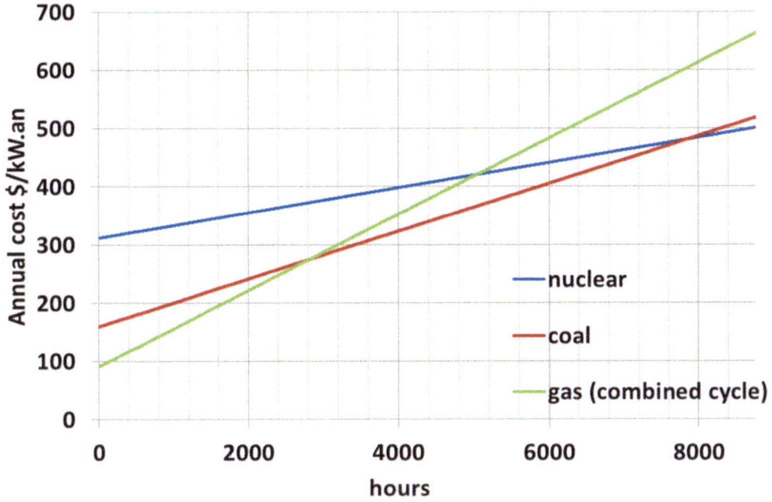

Fig. 7.3 Screening curve example (cost data from [22], 8 % discount rate, and no carbon price), own calculus

While screening curves are commonly used to establish the optimal electricity mix, our analysis showed that the results are extremely sensitive to the assumptions in terms of investment costs (are the interests during construction taken into account? What is the discount rate?), and CO_2 price, only to name a few. When considering nuclear investment (very capex intensive and low carbon), these assumptions are critical. According to our calculations, when varying the discount rate by 3 points, or when taking into account a carbon price of $ 30/tonCO_2, the optimal operating time of nuclear power could vary between 3000 and 8000 h, leading to a resulting installed capacity between 55 and 40 GW, for 16.5 % of renewable penetration for instance.

Given this uncertainty in range, we retained an empirical approach based on the historical availability factor of nuclear power in France (return on operating experience) [23]. The selected value is 7000 h. Such an approach rather minimizes the resulting installed nuclear capacity. The optimal nuclear power will be derived by confronting this economical operating duration to the residual load curve that was constructed as explained in the previous section. By comparing the optimal nuclear power that corresponds to a given total annual production and non-dispatchable renewable energy penetration, to the installed power provided by the selected scenario, the available cycling power is assessed (cf. Fig. 7.4).

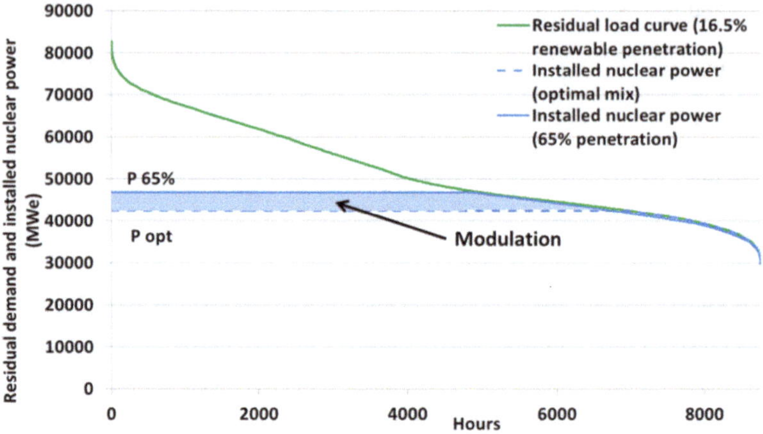

Fig. 7.4 Example of assessment of the nuclear power modulation capacity ("high nuc" 2030), own calculus

7.3.3 *Assessment of the Economic and Environmental Impacts*

The economic impact can be assessed from two viewpoints:

- From the power plant operator viewpoint: The power plant does not supply all the available energy (cf. Fig. 7.4) and cycling operation induces overcosts (to be included in future studies). This will be assessed by comparing the levelized cost of electricity (LCOE, [22]) of the installed nuclear power either operated as a baseload or as a cycling capacity as imposed by the residual load (as shown in Fig. 7.4).
- From the social welfare viewpoint: If not provided by nuclear, load following would have been carried out by other means. In this study we chose to compare to the gas power plant reference. The comparison is done by confronting the total cost of supplying the "cycling power" part (in blue in Fig. 7.4) either by nuclear or by gas. In the latter case, CO_2 emissions will also be assessed and internalized in the electricity cost.

Cost assumptions as regards nuclear plants are issued from the French Court of Auditors [24] while the Organization for Economic Co-operation and Development (OECD) assumptions have been adapted in the French case for the gas-fired power plants [22]. The open-cycle gas turbine (OCGT) was selected as the reference due to its cycling abilities.

7.4 Results and Discussion: The Impact of Renewable Penetration on Nuclear Power Generation—Cost and Environmental Issues

As explained in the previous section, the installed nuclear power was confronted to the "optimal" value for each scenario, in order to assess the cycling possibility.

First, scenarios are compared through the corresponding optimal installed nuclear power. Table 7.3 shows the share of the total electricity production that would be provided by nuclear (for an economically optimal mix), according to the renewable penetration. From this table, we can point out that, for 16.5% renewable penetration, 57% of the electricity should be supplied by nuclear in the optimal case. Hence, 50% nuclear penetration (corresponding to the "med" scenario in 2030), is sub-optimal: no nuclear cycling capacity is available.

Concerning the nuclear production cost, the two scenarios of nuclear fleet composition are compared in Fig. 7.5 ("prog" and "40 years"). Implementing a progressive shutdown first enables making profit from amortized PWRs (cycling is even less expensive than baseload operation from 2030 to 2050). It remains less expensive than the "40 years" scenario, even once amortized PWRs have

Table 7.3 Nuclear contribution to satisfy the total electricity production if the optimal mix was implemented, according to the renewable penetration

Renewable penetration (%)	Nuclear penetration (%)
16.5	57
30	41
50	17

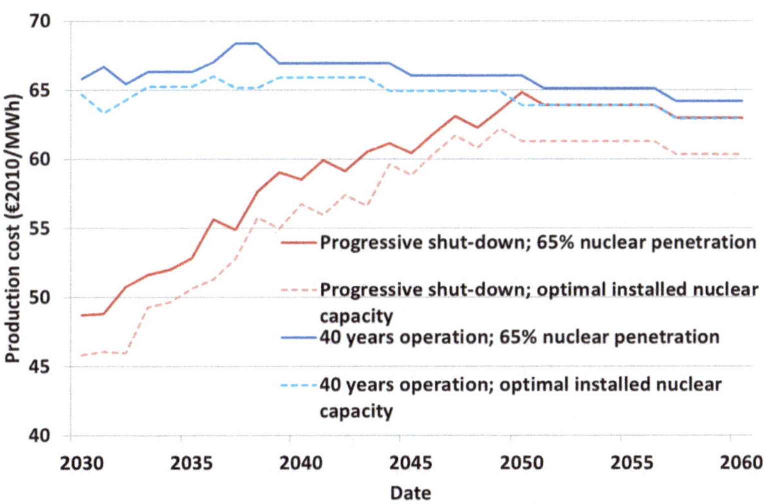

Fig. 7.5 Nuclear production cost according to the nuclear fleet composition scenario ("40 years" or "prog"), own calculus

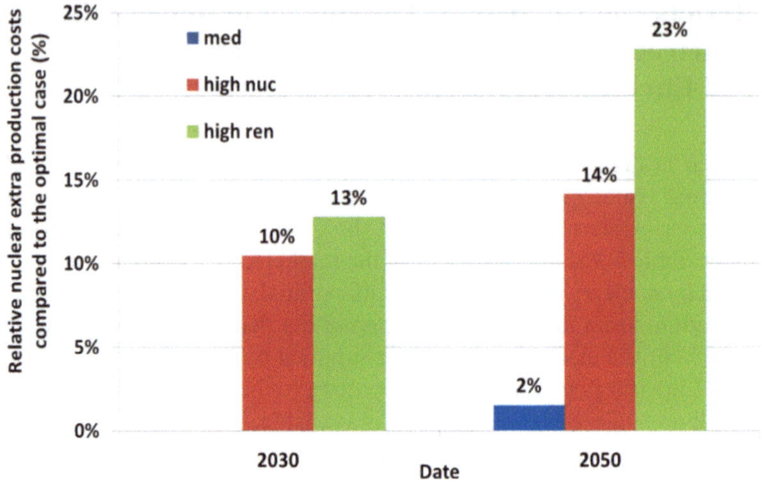

Fig. 7.6 Load modulation impact on the nuclear LCOE, own calculus

been shut down. Indeed EPRs™ that are installed later are characterized by higher availability, which results in smaller installed capacity needs, hence a reduced investment.

Since the "prog" scenario corresponds to the more realistic one in the French context, we will focus on it. For a given nuclear penetration level, the share of nuclear energy which is devoted to modulation increases more than linearly with the renewable penetration level. It is worth noticing that for the "high ren" scenario in 2050, 50 % of nuclear energy would be dedicated to modulation. Figure 7.6 shows the impact of modulation (loss of load factor) on the nuclear LCOE. The impact of the nuclear fleet composition on nuclear LCOE is pointed out by comparing the "med" 2050 case to the "high ren" 2030 case: similar nuclear and renewable penetration levels end up in different relative extra production costs. Indeed in 2050, nuclear fleets are only composed of EPR™ whatever the renewable penetration scenario, while in 2030 relative costs are strongly dependent on the PWR replacement rate according to nuclear energy penetration targets.

Finally, results show that the gas cycling option is cheaper than nuclear backup in each scenario. For a given CO_2 price value, the gas price which leads to equalize the cost of both options is determined. Conversely, the same approach is followed for a given gas price value. For all the scenarios the results are in the same order of magnitude: (i) for a CO_2 price of € 5/ton CO_2 (which corresponds to the current market value [25]), nuclear backup is competitive when the gas price exceeds € 35/MWh; (ii) for a gas price of € 31/MWh (which corresponds to the ANCRE assumption for 2030 [14]), the gas option dominates as long as the CO_2 price is lower than € 20/ton_{CO2}; (iii) if the combined cycle power plant (CCGT) was considered as the gas option (less flexible, yet probably enough), the previous results would be modified to € 50/MWh and € 110/ton_{CO2} respectively.

Overall, the gas option, even when cheaper would generate CO_2 emissions in the range of 10–50 Mt per year, depending on the scenario and the time horizon.

7.5 Conclusion and Perspectives

Nuclear modulation seems to be achievable, especially in the case of France since a large production variation could be reached only by small increments in each power plant. This would remain true as long as installed nuclear power is large enough, in terms of contribution to the global electricity mix.

In economic terms, gas backup is more competitive than nuclear in each of the studied scenarios. However a carbon tax could smooth the gap. When compared to OCGT, a value as low as € 20/ton$_{CO2}$ could be enough to change the trends, which advocates for such incentives to avoid the effective greenhouse gas release. Nuclear backup is all the more competitive than amortized power plants are utilized.

Further studies could focus on assessing the impact of an increasing electricity demand, or technological progress by also including the overcosts due to cycling operation (both for the nuclear and gas options). When increasing the non-dispatchable energy penetration, other backup options, such as storage, curtailment, or demand-side management tools, should be considered as contributors to the "backup" mix whose optimum remains to be established. This could be done through the use of a minimum cost economic dispatch model.

Finally, nuclear modulation was apprehended via the concept of load following. Another way to reach the same target could be modulating through the output. In this case, the nuclear power plant would be operated as a baseload and electricity would be supplied to the electric system when requested. For the remaining time, instead of reducing the plant load, the output could be used to produce other valuable products such as heat or hydrogen [26]. The competitiveness of these productions is still to be proven and will be the topic of future works.

Acknowledgment This work was carried out in the framework of a Ph.D. thesis. The authors want to thank J. C. Bocquet, Head of the Laboratoire de Génie Industriel, and J. G. Devezeaux, Head of the Institute for techno-economics of energy systems (I-tésé) for the fruitful discussions.

References

1. European Directive 2009/28/CE (23 Apr. 2009)
2. Ministère de l'Ecologie, de l'Energie, du Développement Durable et de la Mer (2009) Plan d'action national en faveur des énergies renouvelables. Période 2009–2020. En application de l'article 4 de la Directive 2009/28/CE de l'Union Européenne
3. Gross R, Heptonstall P, Anderson D, Green T, Leach M et al (Mar. 2006) The costs and impacts of intermittency: an assessment of the evidence on the costs and impacts of intermittent generation on the British electricity network. UK Energy Research Centre, London
4. Hart E, Stoutenburg ED, Jacobson MZ (Feb. 2012) The potential of intermittent renewables to meet electric power demand: current methods and emerging analytical techniques. Proc IEEE 100(2):322–334
5. Davis CD, Gotham DJ, Preckel PV, Liu AL (Sep. 2013) Determining the impact of wind on system costs via the temporal patterns of load and wind generation. Energy Policy 60:122–131
6. Keppler JH, Cometto M (2012) Nuclear energy and renewables: system effects in low-carbon electricity systems. Nuclear Energy Agency, OECD, Paris

7. Hand MM, Baldwin S, DeMeo E, Reilly JM, Mai T et al (2012) Renewable electricity futures study, vol.1: exploration of high-penetration renewable electricity futures. National Renewable Energy Laboratory, Golden, CO, NREL/TP-6A20-52409-1
8. Bruynooghe C, Eriksson A, Fulli G (2010) Load following operating mode at nuclear power plants and incidence on operation and maintenance costs. Compatibility with wind power variability. EC, JRC Scientific and Technical Reports, Brussels
9. Lokhov A (Jun. 2011) Technical and economic aspects of load following with nuclear power plants. NEA, OECD, Paris
10. Choho AM Major innovation in PWR load follow operations by AREVA. In Proc. 2013 Atoms for the Future Conf
11. Grard H, Thomas JB The nuclear fleet: a decisive enabler to integrate a share of renewables (in the framework of a comprehensive systems solution). In Proc. 2013 Atoms for the Future Conf
12. Wagner F (Sep. 2012) Feature of an electricity supply system based on variable input. Max Planck Institute of Plasma Physics, IPP 18-1
13. Ministère de l'Ecologie, du Développement Durable et de l'Energie (Jul. 2013) Synthèse des travaux du débat national sur la transition énergétique de la France. Conseil National du Débat
14. ANCRE (2013) Scénarios de l'ANCRE pour la transition énergétique
15. ADEME (Jun. 2013) L'exercice de prospective de l'ADEME: Vision 2030–2050
16. RTE (2013) Bilan électrique 2013
17. Nucléaire: EDF détaille les coûts de son "grand carénage", Les Echos, Fev. 21 (2014) www.lesechos.fr/entreprises-secteurs/energie-environnement/actu/0203331225826-nucleaire-edf-detaille-les-couts-de-son-grand-carenage-651911.php
18. Ueckerdt F, Hirth L, Luderer G, Edenhofer O (Dec. 2013) System LCOE: what are the costs of variable renewables? Energy 63:61–75
19. Connolly D, Lund H, Mathiesen BV, Leahy M (Mar. 2010) Modelling the existing Irish energy-system to identify future energy costs and the maximum wind penetration feasible. Energy 35:2164–2173
20. Barnacle M, Robertson E, Galloway S, Barton J, Ault G (May 2012) Modelling generation and infrastructure requirements for transition pathways. Energy Policy 52:60–75
21. RTE database http://www.rte-france.com/fr/developpement-durable/eco2mix/telechargement-de-donnees
22. Salvadores MS, Keppler JH (2010) Projected costs of generating electricity. Nuclear Energy Agency, OECD, Paris
23. Power Reactor Information System (PRIS) database IAEA http://www.iaea.org/pris/
24. Les coûts de la filière électronucléaire (Jan. 2012) Cour des comptes, Rapport public thématique, Paris, France.
25. CO_2 market. http://www.pointcarbon.com. Accessed 27 Feb 14
26. Ruth MF, Zinaman OR, Antkowiak M, Boardman RD, Cherry RS, Bazilian MD (Feb. 2014) Nuclear-renewable hybrid energy systems: opportunities, interconnections and needs. Energy Convers Manage 78:684–694

Chapter 8
Impact of Integrated Hybrid PV/Wind Generation on Harmonic Power Flow in Medium-Voltage Grid

Ali Hamzeh and Abbas Sandouk

Abstract The renewable energy sources such as wind energy or solar energy play an increasingly significant role as electric supply resources and as an integral part of the electrical grid. The negative influences on the power quality of renewable energies derive mainly from two typical characteristics of renewable sources: their randomly varying availability which can lead to frequency variation, to grid instability, and to a total or partial loss of load power supply and the presence of a static converter as output interface of the generating plants that introduces voltage and current harmonics into the electrical system that negatively affect the system power quality. As the availability of wind varies in a different manner than the sun radiation does, the hybrid wind/photovoltaic (PV) generation may increase the resultant availability and consequently reduce the negative impact on power quality of the voltage and current of the host network. One of the many aspects of power quality is the harmonic content of voltages and currents. Harmonics can be analyzed in either the frequency domain or in the time domain with post-processing using Fourier analysis. This chapter uses the Power Factory software to carry out the analysis of harmonics in the integrated power system in the frequency domain. The analysis comprises harmonic power flow including harmonic power flow according to International Electrotechnical Commission (IEC) 61000-3-6 and flicker analysis according to IEC 61400-21. Harmonic load flow calculates actual harmonic indices related to voltage or current distortion and harmonic losses caused by harmonic sources (usually nonlinear loads such as current converters). In the harmonic power flow calculation, the program carries out a steady-state network analysis at each frequency at which harmonic sources are defined. The flicker analysis performs the calculation of long- and short-term flicker disturbance factors introduced by wind turbine generators.

The effect of renewable units on the host grid depends, among other things, on the grid topology. Therefore, this impact on the power quality is studied using a typical medium-voltage (20 kV) distribution network. The test grid is a 20-bus distribution

A. Hamzeh (✉)
Department of Electrical Engineering, Al-Ahliyya Amman University, Amman, Jordan
e-mail: hamzeh.ali@gmail.com

A. Sandouk
Department of Electrical power Engineering, Damascus University, Damascus, Syria

A. Sayigh (ed.), *Renewable Energy in the Service of Mankind Vol II*,
DOI 10.1007/978-3-319-18215-5_8

system including 12 transformers, 7 cables, 7×20 kV buses, 6×0.69 kV buses, 7×0.4 kV buses, and 5 general loads. A wide range of simulations have been carried out, and the simulation results have been thoroughly discussed, and very important findings have been concluded. The conclusions would be a useful aid for planning and operation of renewable integration in medium-voltage distribution systems.

Keywords Hybrid PV/wind · Medium voltage · Harmonic power flow · Flicker analysis

8.1 Introduction

Wind and solar power generation are two of the most promising renewable power generation technologies. As these resources are highly dependent on climate and they can complement each other, a hybrid wind/photovoltaic (PV) system captures the best features of each energy resource and has great potential to provide higher quality and more reliable power to the host grid than a system based on a single resource. Renewable distributed generation units, if properly controlled and designed, can improve the power flow management on the grid and reduce the probability of grid faults, thus increasing the power quality of the energy supply [1, 2].

The negative influences on the power quality of renewable energies derive mainly from two typical characteristics of renewable sources: their randomly varying availability and the presence of a static converter as output interface of the generating plants. This kind of availability can lead to frequency variation, grid instability, and a total or partial loss of load power supply. Power electronic converters used to interface generation equipment can cause harmonic currents to flow, but conventional rotating plants (e.g., synchronous or induction generators) will alter the harmonic impedance of the network and hence its response to other harmonic sources. Further, the introduction of shunt capacitor banks used to compensate induction generators may lead to resonances [2].

As the availability of wind varies in a different manner than the sun radiation does, the hybrid wind/PV generation may increase the resultant availability and consequently reduce the negative impact on power quality of the voltage and current of the host network. One of the many aspects of power quality is the harmonic content of voltages and currents. Harmonics can be analyzed in either the frequency domain or in the time domain with post-processing using Fourier analysis. This chapter uses the Power Factory software to carry out the analysis of harmonics in the integrated medium-voltage (MV) network in the frequency domain. The analysis comprises harmonic load flow including harmonic power flow according to International Electrotechnical Commission (IEC) 61000-3-6 and flicker analysis according to IEC 61400-21. Harmonic load flow calculates actual harmonic indices related to voltage or current distortion and harmonic losses caused by harmonic sources (usually nonlinear loads such as current converters). In the harmonic load flow calculation, the program carries out a steady-state network analysis at each frequency at which harmonic sources are defined. The flicker analysis performs the calculation of long- and short-term flicker disturbance factors introduced by wind turbine generators (WTGs) [3].

8.2 Harmonic Load Flow [3]

High-frequency harmonic components in the electric system can affect the grid current and the grid voltage too. Due to the impedances of the system, current harmonic components can produce voltage harmonics and vice versa. Anyway, the primary causes of voltage and current harmonics are quite different. The harmonic components in grid current are produced by the loads equipped with electronic devices that absorb high-frequency current components. These harmonic components can be reduced only by acting directly on the loads. The voltage harmonic components are introduced in the system by the interface converter and are produced mainly by the switching of electronic components. Voltage harmonics can also be presented due to regulator malfunctioning or due to harmonic components at frequency lower than the cutoff frequency of the power bus regulator.

High levels of harmonic distortion (HD) can lead to problems for the utility's distribution system, plant distribution system, and any other equipment serviced by that distribution system. Effects can range from spurious operation of equipment to a shutdown of important plant equipment, such as machines or assembly lines.

To eliminate the grid voltage harmonics produced by the voltage harmonics on the power bus, it is necessary to guarantee the stability of the voltage level of the power bus. This can be achieved with a proper power flow control algorithm and by adding filters at the output of the other static converters connected to the power bus. Moreover, the dynamic behavior of the storage system and its converter can affect the harmonic content of the power bus voltage and, consequently, also of the grid voltage. The storage system converter regulator should maintain the energy balance between loads and sources on the common power bus.

The harmonic load flow calculation in the Power Factory program provides a vast number of results for network elements. Some of the more prominent result variables are described in this section. The HD of a current or of a voltage can be quantified in terms of the HD index. To describe the overall distortion, the total harmonic distortion (THD) index has been introduced. All distortion indices are described by their equations for the current but may be similarly described for voltage distortion [3].

$$HD_I(f_i) = \frac{|I_{fi}|}{|I_{f1}|}, \tag{8.1}$$

$$THD_I = \frac{\sqrt{I_{rms}^2 - I_{f1}^2}}{I_{\text{ref}}}, \tag{8.2}$$

where $I(f_i)$ is the component of the current at frequency f_i and I_{ref} is the reference value for the current. The total rms value is:

$$I_{rms} = \sqrt{\sum_{i=1}^{n} I_{fi}^2}. \tag{8.3}$$

The reference value I_{ref} depends on the standard used. The two possible options are the calculation according to Deutsches Institut für Normung (DIN, German Institute for Standardization) and according to the Institute of Electrical and Electronics Engineers (IEEE):

$$I_{\text{ref}}, DIN = I_{rms}, \tag{8.4}$$

$$I_{\text{ref}}, IEEE = \left| I_{f1} \right|. \tag{8.5}$$

Another value which may be of importance is the total power, which describes the power absorbed over all frequency components:

$$P_{\text{tot}} = \sum_{i=1}^{n} P_{fi}. \tag{8.6}$$

8.3 Flicker Analysis [3]

Voltage flicker describes dynamic variations in the network voltage which may be caused either by distributed generation or by loads. Traditionally, voltage flicker was of concern when the connection of large fluctuating loads (e.g., arc furnaces, rock crushing machinery, sawmills, etc.) was under consideration. However, it is of considerable significance for distributed generation, which (i) often uses relatively large individual items of plant compared to load equipment, (ii) may start and stop frequently, and (iii) may be subject to continuous variations in input power from a fluctuating energy source. Items of embedded generation plant which require assessment for potential disturbance caused by voltage flicker include:

- Connection and disconnection of induction generators
- Operation of wind turbines (WTs)
- Operation of photovoltaic generators

Flicker is usually evaluated over a 10-min period to give a "short-term" severity value Pst. The Pst value is obtained from a 10-min time series of measured network voltage using an algorithm based on the disturbance perceived by the human eye in fluctuating light. P_{st} is linear with respect to the magnitudes of the voltage change but, of course, includes the frequency dependency. Twelve P_{st} values may then be combined using a root of the sum of the cubes calculation to give a "long-term" severity value P_{lt} over a 2-h period.

Following the calculation of flicker according to IEC 61400-21, the following result variables for every node in the network are available in the single-line graphic. It should be noted that Power Factory calculates these flicker disturbance factors and relative voltage change for impedance angles with lines at 20 °C and at maximum operation temperature. The following result variables are the worst-case values in the impedance angle range, which is based on the temperature range:

- P_{st_cont}, P_{lt_cont}: short-term and long-term flicker disturbance factors, respectively, for continuous operation of the WTG(s)
- P_{st_sw}, P_{lt_sw}: short-term and long-term flicker disturbance factors, respectively, for switching operations of the WTG(s)
- d_sw: relative voltage change (as a percentage)

8.3.1 Continuous Operation

Continuous operation is defined in IEC standard 61400-21 as the normal operation of the WTG excluding start-up and shutdown operations. The short-term and long-term flicker disturbance factors during continuous operation are:

$$P_{lt} = P_{st} = c \times \frac{S_n}{S_k}, \tag{8.7}$$

where

P_{st}	Short-term flicker disturbance factor
P_{lt}	Long-term flicker disturbance factor
c[a]	Flicker coefficient for continuous operation
S_n	Rated apparent power of the WT (VA)
S_k	Short-circuit apparent power of the grid (VA)

[a] At the present time, the flicker coefficient is only known for wind energy plants. It is dependent on the network impedance angle ψ_k and on the average annual wind speed v_a

When more than one WTG exists at the point of common coupling (PCC), the resulting short-term and long-term flicker disturbance factors for continuous operation are:

$$P_{lt\,res} = P_{st\,res} = \sqrt{\sum_i^N P_{lti}^2}, \tag{8.8}$$

where N is the number of WTGs at the PCC.

For a generating plant consisting of N equal generating units, the resulting value of the flicker disturbance factor is as follows:

$$P_{\text{lt res}} = P_{\text{st res}} = \sqrt{N} \times c \times \frac{S_n}{S_k}. \tag{8.9}$$

A range of permissible limits for flicker on distribution networks is given in national and international standards. As of Bundesverband der Energie- und Wasserwirtschaft (BDEW, German association of energy and water industries), the long-term flicker disturbance factor for the MV host network should not exceed 0.46 ($P_{\text{lt}} \leq 0.46$).

8.3.2 *Switching Operations*

Switching operations are defined in IEC standard 61400-21 as start-up or switching between WTGs. In this mode of operation, the short-term and long-term flicker disturbance factors during switching operations are:

$$P_{\text{st}} = 18 * N_{10}^{0.31} * k_f\left(\psi_k\right) * \frac{S_n}{S_k}, \tag{8.10}$$

$$P_{\text{lt}} = 8 * N_{120}^{0.31} * k_f\left(\psi_k\right) * \frac{S_n}{S_k}, \tag{8.11}$$

where N_{10} and N_{120} are the number of switching operations in a 10- and a 120-min period, respectively, k_{f} is the flicker step factor, and ψ_{k} is the network impedance angle (degrees).

When more than one WTG exists at the PCC, the short-term flicker disturbance factor and long-term flicker disturbance factor under switching operations are:

$$P_{\text{st res}} = \frac{18}{S_k}\left[\sum_{i=1}^{N} N_{10,i} * (k_{f,i}\left(\psi_k\right) * S_{n,i})^{3.2}\right]^{0.31}, \tag{8.12}$$

$$P_{\text{lt res}} = \frac{8}{S_k}\left[\sum_{i=1}^{N} N_{120,i} * (k_{f,i}(\psi_k) * S_{n,i})^{3.2}\right]^{0.31}. \tag{8.13}$$

The relative voltage change (%) due to the switching operation of a single WTG is:

$$d = 100 * k_u\left(\psi_k\right) * \frac{S_n}{S_k}. \tag{8.14}$$

For the balanced harmonic load flow, the calculation of the short-circuit power, S_K, at each bus is as follows:

$$S_k = \frac{1}{|Z_{\text{bus}}|}, \tag{8.15}$$

where Z_{bus} is the impedance calculated at the bus.

8.4 Computer Simulations

8.4.1 *Test MV Grid*

The impacts of renewable units on the host grid are not generic to any network, and consequently they could be of positive, negative, or neutral influences, depending on the distribution network topology, operation mode, and load profile. Therefore, the impact on the harmonic content of grid currents and voltages is studied using a typical MV (20 kV) distribution network (Fig. 8.1), which is a modified version of an example network in the Power Factory. The test grid is a 20-bus distribution

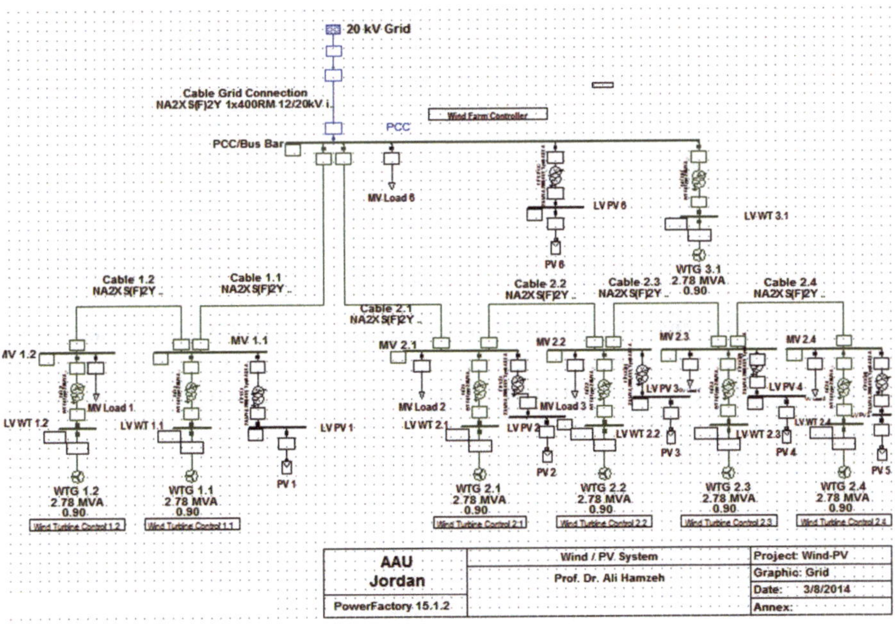

Fig. 8.1 Test medium-voltage *(MV)* distribution network (20 kV) with hybrid wind/PV generations. *PCC* point of common coupling, *WTG* wind turbine generator, *PV* photovoltaic, *LVPV* low-voltage photovoltaic, *LVWT* low-voltage wind turbine. (Modified grid based on [3])

system including a number of wind and PV generators, 12 transformers, 7 cables, 7×20 kV buses, 6×0.69 kV buses, 7×0.4 kV buses, and 6 general loads.

Each WTG is connected to a 20 kV distribution network and has a nominal active power of 2.5 MW. The machines are equipped with fully-rated converters (WT type 4). The WTG model includes the main elements of the WT (static generator and transformer) and additional dynamic models which represent the controllers. The static generator is a model of a generator which is connected to the grid via a fully-rated voltage-sourced converter (VSC). A model of the central wind farm controller is included for steady-state analysis. In this example it performs a cos ϕ *(P)* characteristic at the PCC.

8.4.2 Harmonic Current Injections and Flicker Coefficients

For the used 2.5-MW WT, the program defines implicitly the harmonic source and calculates the harmonic currents injection among others according to IEC 61000-3-6/IEC 61400-21, which includes integer and non-integer harmonics. For the PV units, the harmonics source library comprises only thyristor bridge inverters. The 12-pulse bridge inverter was selected as a PV harmonic source, whose harmonics are calculated for the balanced, phase correct harmonic source type.

The flicker analysis of the grid requires that the flicker coefficients for the generators must first be defined and assigned. The flicker coefficients of the used 2.5-MW WT are available in the program, but they are not available in the Power Factory library, and therefore they are left unassigned; the PV generator is then considered to be an ideal source for the flicker calculation.

8.4.3 Case Studies

A wide spectrum of simulation studies have been carried out to analyze the system harmonic performance under different scenarios using Power Factory 15.1.2.

Case Study 1: Wind Generators Connected to the MV Grid This case study aims to investigate the harmonic and flicker performance at PCC of the MV grid as a function of penetration level of wind generation where *Penetration Level (PL%)* is defined as the ratio of the total WT generation to the peak load of the hosting distribution network [4–6]:

$$PL\% = \frac{P_{\mathrm{WT}}}{P_{\mathrm{Load}}} \times 100. \tag{8.16}$$

The computation results show that with increasing *PL* from 0 to 125%, the THD rises from 0.24% for *PL* of 20% to 0.79% for *PL* of 125% according to an approximate linear equation ($y=0.04\,x+0.14$) as shown in Fig. 8.2. Also, all flicker

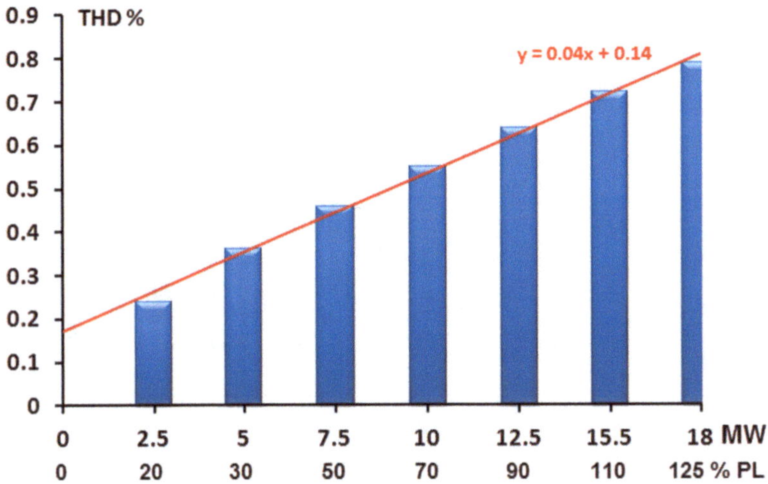

Fig. 8.2 Total harmonic distortion at point of common coupling (PCC) for wind generation with increasing penetration level. *THD* total harmonic distortion, *PL* penetration level

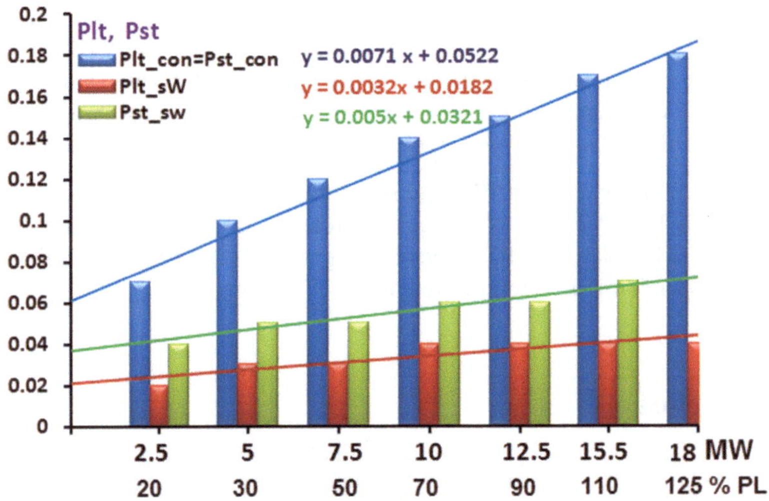

Fig. 8.3 Flicker disturbance factors at point of common coupling (PCC) for wind generation with increasing penetration levels. *PL* penetration level

factors rise linearly but with different slopes (Fig. 8.3). The long-term and short-term flicker disturbance factors during continuous operation have the larger line slope (0.0071), whereas the short-term and long-term flicker disturbance factors during switching operations have the line slopes of 0.005 and 0.0032, respectively.

Case Study 2: PV Generators Connected to the MV Grid In this case study, we use PV generators only to investigate the harmonic performance in general and to

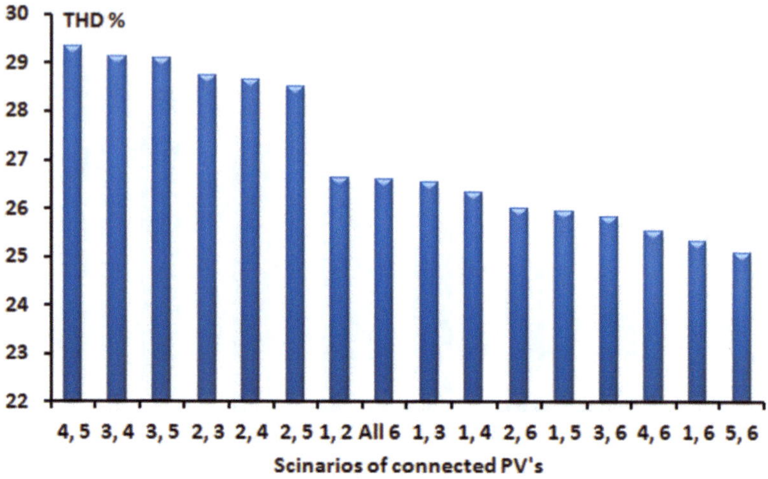

Fig. 8.4 Total harmonic distortion *(THD)* at point of common coupling (PCC) for different scenarios of photovoltaics *(PVs)* with dispersion index *(DI)* = 100 and 33 %

find out the locations of PV placement realizing the best performance. As Fig. 8.4 shows, the lowest THD% is reached if two PV plants 5 and 6 of 6 MW each are connected to the grid with a THD% of 25.07 %, whereas the maximum THD of 29.34 % belongs to the PV plants 4 and 5. It is worth mentioning that the high values of THD for the used PV systems are caused due to the selected 12-pulse bridge inverter, which is the only available inverter type in the Power Factory software.

Case Study 3: Hybrid Wind/PV The simulation results (Fig. 8.5) show that the operation of hybrid WT and PV generators on a MV grid will reduce the resultant THD at PCC. The relative THD reduction ranges between 3–5 % depending on the share and dispersion index (DI) of the PV and WT plants.

8.5 Conclusions

As the availability of wind varies in a different manner than the sun radiation does, the hybrid wind/PV generation may increase the resultant availability and consequently reduce the negative impact on power quality of the voltage and current of the host network. This chapter uses the Power Factory software to carry out the analysis of harmonics in the integrated MV network in the frequency domain. The analysis comprises harmonic load flow according to IEC 61000-3-6 and flicker analysis according to IEC 61400-21. As the impacts of renewable units on the host grid are not generic to any network, the impact on the harmonic content of grid currents and voltages is studied using a typical MV (20 kV) distribution network. The simulation results show the following conclusions:

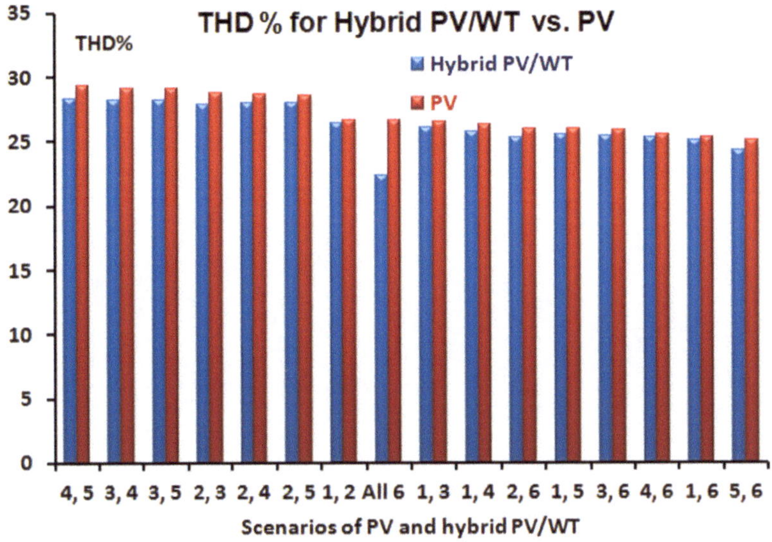

Fig. 8.5 Total harmonic distortion *(THD)* at point of common coupling (PCC) for hybrid photovoltaic *(PV)*/wind turbine *(WT)* versus PV only

- For wind generations only, and with increasing its penetration level, the THD at PCC rises linearly, and all flicker factors also rise linearly but with different slopes. The long-term and short-term flicker disturbance factors during continuous operation have the larger line slope, whereas the short-term and long-term flicker disturbance factors during switching operations have different line slopes.
- In case of PV generation only, the lowest THD% is reached if the two PV plants 5 and 6 of 6 MW each are connected with a THD% of 25.07 %, whereas the maximum THD of 29.34 % occurs for the PV plants 4 and 5. As there is no possibility in the used program to conduct flicker analysis for PV systems, the PV system is considered as an ideal flicker source.
- The hybrid operation of wind and PV will reduce the resultant THD at PCC by about 3–5 % depending on the share and DI of the two generation types.

References

1. Alliance for rural electrification (2012) Hybrid power systems based on renewable energies, Brochure
2. Gregorio R (2011) Electrical generation and distribution systems and power quality disturbances. InTech. ISBN 978-953-307-329-3 pp318. http://www.intechopen.com/books/editor/electrical-generation-and-distribution-systems-and-power-quality-disturbances
3. DIgSILENT Power Factory 15 (2012) User manual of Power Factory software V15.1.3, Germany 2012. Wind_5Harmonics-Wind_1_Base_Case
4. Hamzeh A (2010–2011) Integration of distributed generation into electric distribution systems. Damascus University Press, Damascus

5. Hamzeh A (2009) Evaluation of influence of renewable distributed generation on voltage profile and power losses in electric distribution systems. Damascus Univ J 25(1):343–391
6. Hamzeh A (2008–2009) Photovoltaic engineering, analysis and design. Damascus University Press, Damascus

Chapter 9
Implementation and Validation of Energy Conversion Efficiency Inverter Models for Small PV Systems in the North of Brazil

Luís Monteiro, Igor Finelli, André Quinan, Wilson N. Macêdo, Pedro Torres, João T. Pinho, Eduardo Nohme, Bruno Marciano and Selênio R. Silva

Abstract The increasing amount of distributed photovoltaic (PV) system components into distribution networks involves the development of accurate simulation models that take into account an increasing number of factors that influence the output power from the distributed generation systems. The modeling of PV system components in power systems and the relative control architecture is an important part of the introduction of a relevant quantity of renewable energy to the future development of the smart grid. Therefore, it is essential to have proper validated models to help operators perform improved studies and be more confident with the results. We present two energy conversion efficiency inverter models developed for two small PV systems using the Jantsch inverter models (Jantsch, Schmidt, Schmid, Results of the concerted actions on power conditioning and control. XI European Photovoltaic Solar Energy Conference, Swiss, 1992) and Sandia inverter model (King, Gonzalez, Galbraith, Boyson,Performance model for grid-connected photovoltaic inverters, Sandia National Laboratory Report, New México, 2007).

This work has been supported by the Brazilian agency CAPES.

L. Monteiro (✉) · I. Finelli · A. Quinan · E. Nohme · S. R. Silva
Laboratory of Control & Conversion of Energy,
Federal University of Minas Gerais, Belo Horizonte, Brazil
e-mail: luis.monteiro@gmail.com

W. N. Macêdo · P. Torres · J. T. Pinho
Group of Studies and Development in Alternatives Energy,
Federal University of Pará, Belém, Brazil
e-mail: wnmcedo@ufpa.br

B. Marciano
Management of Alternatives Energy, Energy Company of Minas Gerais,
Belo Horizonte, Brazil
e-mail: bruno.marciano@cemig.com.br

L. Monteiro
Energy Engineering Department, Catholic University of Minas Gerais,
Belo Horizonte, Brazil

A. Sayigh (ed.), *Renewable Energy in the Service of Mankind Vol II*,
DOI 10.1007/978-3-319-18215-5_9

The component models were implemented in MATLAB software and the simulation results were compared, firstly, with the datasheet values of the inverter Xantrex GT2.8-NA-240/208 UL-05 model and then with the microinverter Enphase® Energy (M215) model. To confirm the strong dependency on ambient conditions and to validate the simulation models, operation data from two small PV systems using the Xantrex and Enphase Energy inverters located in the north of Brazil were used and statistical analysis and comparison of the results was performed.

Keywords Conversion efficiency inverter models · System performance validation · Computational simulation · Small grid-connected PV systems

9.1 Introduction

The evaluation of energy conversion efficiency and hence the operation of inverters for photovoltaic (PV) systems connected to the grid are based on electrical parameters (voltage input and output, DC and AC power circuits, etc.) and thermal equipment. Several mathematical models of energy conversion efficiency inverters for such systems have been developed over time [1–4]. For sizing PV systems, designers need a reliable mathematical model and simple application that performs computer simulations and consequently generates estimates of energy production of PV systems with a lower margin of uncertainty in order to provide better financial return and investment in the project.

Here, we studied two energy conversion efficiency inverter mathematical models for grid-connected PV systems proposed by [1, 2]. They were chosen because they are simple to operate and because each one is based on data from the manufacturer's inverter datasheets and can also use operational data (from the field) of inverters that are coupled to small PV systems in real operation conditions. Each one was implement in a mathematical computational tool and validated, under different climatic conditions, by operating data from the Xantrex GT2.8-NA-240/280 UL-05 inverter and the Enphase® M215-60-2LL-S22-IG microinverter and data from the manufacturer's datasheets [5, 6]. In the following sections we present the cited models, their development, validation and results analysis using statistical indicators.

9.2 Methodology

The models proposed by [1, 2] were chosen and implemented using MATLAB® software R2013a. Each one was validated using datasheets from the inverter manufacturers [5, 6], and field data from two small PV systems operating under different environmental conditions. The Xantrex model (rated at 2.8 kW) is coupled to a grid-connected PV system with 3.36 kWp installed power, with m-Si modules Kyocera

KC 120. The Enphase model with 215 W nominal power is coupled to a 245 Wp PV module (m-Si Aleo S19G245) in a bus interconnected to a 220 V AC grid. Both systems were located in the test area of the Group of Studies and Development of Alternatives Energy (GEDAE) at the Federal University of Pará (UFPA) in Belém city in the north of Brazil. A detailed description of these small PV systems, their instrumentation, and monitoring and data acquisition are described in [7, 8]. The data generated by the inverters were first treated and then applied to the models. A comparison of datasheets and field data was carried out using two figures of merit commonly used in statistical literature [9, 10]—MBE[1] and RMSE[2] errors and a parallel of the models was performed. Equations (9.1) and (9.2) show the calculations of average errors where 'K' is a general variable and 'n' is the number of datasets used.

$$RMSE = \left(\frac{1}{n} * \sum_{i=1}^{n} \left(k_{\text{calculated}} - k_{\text{measured}} \right)^2 \right)^{\frac{1}{2}} \quad (9.1)$$

$$MBE = \frac{1}{n} * \sum_{i=1}^{n} \left(k_{\text{calculated}} - k_{\text{measured}} \right) \quad (9.2)$$

9.3 Energy Conversion Efficiency Inverter Mathematical Models for Grid-Connected PV Systems

9.3.1 Jantsch Model

The Jantsch model, developed by [1] proposes the calculation of the losses of the inverter depending on their power output, not taking into account other possible variations that may occur. The model considers the losses as a second-degree function and seeks to connect the terms of each order to the actual losses in the inverters. Eq. (9.3) shows total losses.

$$P_{\text{losses}} = k_0 + k_1 * P_{\text{output}} + k_2 * P_{\text{output}}^2 \quad (9.3)$$

The input power and efficiency are defined as in Eqs. (9.4) and (9.5).

$$P_{\text{input}} = P_{\text{output}} + P_{\text{losses}} = P_{\text{output}} + k_0 + k_1 * P_{\text{output}} + k_2 * P_{\text{output}}^2 \quad (9.4)$$

[1] Mean bias error (MBE) is a systematic error and is the trend of whether the data is overestimated or underestimated.

[2] Root mean square error (RMSE) indicates an average absolute error; the lower the RMSE values, the more accurate the estimate of the model.

$$\eta = \frac{P_{output}}{P_{input}} = \frac{P_{output}}{P_{output} + k_0 + k_1 * P_{output} + k_2 * P^2_{output}} \tag{9.5}$$

Having knowledge of the parameters k_0, k_1 and k_2 can set losses throughout the range of operation of the inverter, and the energy conversion efficiency. These characteristic parameters are defined by Eqs. (9.6), (9.7) and (9.8).

$$k_0 = \frac{1}{9} * \frac{1}{\eta_{inv} 1} - \frac{1}{4} * \frac{1}{\eta_{inv} 0.5} + \frac{5}{36} * \frac{1}{\eta_{inv} 0.1} \tag{9.6}$$

$$k_1 = \frac{-4}{93} * \frac{1}{\eta_{inv} 1} + \frac{33}{12} * \frac{1}{\eta_{inv} 0.5} - \frac{5}{12} * \frac{1}{\eta_{inv} 0.1} - 1 \tag{9.7}$$

$$k_2 = \frac{20}{9} * \frac{1}{\eta_{inv} 1} - \frac{5}{2} * \frac{1}{\eta_{inv} 0.5} + \frac{5}{18} * \frac{1}{\eta_{inv} 0.1} \tag{9.8}$$

where $\eta_{inv}1$, $\eta_{inv}0.5$ and $\eta_{inv}0.1$ refer to the instant efficiencies at 100, 50 and 10% of rated output power, respectively. K_0 is the loss of self-consumption, without depending on power output. K_1 refers to the linear losses in power output (e.g. voltage drops in semiconductor devices, diodes and switches, IGBT). Finally, K_2 is the quadratic loss with output power, for example, the ohmic losses.

9.3.2 Sandia Model

The mathematical model of the energy conversion efficiency for PV inverters proposed by [2], also called the Sandia/King model, is an empirical model which has good accuracy in results and good versatility, having been successfully applied to large and small PV systems. The model consists of an expression that relates AC power inverter output as a function of input power and DC bus voltage; these being the two variables needed to estimate the behavior of the conversion efficiency for an inverter. The Sandia model is described by the power AC expression, Eq. (9.9), which is complemented by Eqs. (9.10), (9.11), (9.12).

$$P_{ac} = \left\{ \left(\frac{P_{aco}}{A - B} \right) - C * (A - B) \right\} * (P_{dc} - B) + C * (P_{dc} - B)^2 \tag{9.9}$$

$$A = P_{dco} * \{1 + C_1 * (V_{dc} - V_{dco})\} \tag{9.10}$$

$$B = P_{so} * \{1 + C_2 * (V_{dc} - V_{dco})\} \tag{9.11}$$

$$C = C_0 * \{1 + C_3 * (V_{dc} - V_{dco})\} \tag{9.12}$$

where P_{ac} is the power output of the inverter, P_{dc} is the input power of the inverter, P_{aco} is rated output; P_{dco} is the input power to the inverter-rated power supply output, P_{so} is own consumption or minimum power for start-up of the inverter, V_{dco} is nominal voltage, and V_{dc} is input voltage.

9.4 Validation of Jantsch and Sandia Models

9.4.1 Analysis of Results for the Xantrex GT2.8-NA-240/208 UL-05 Inverter

Table 9.1 shows the performance of the Jantsch and Sandia models by comparing the results of MBE and RMSE using data from the manufacturer's datasheet and field data measurements.

According to Table 9.1, for output power and also efficiency (through RMSE% error), the Jantsch model showed better results compared to the Sandia model, when used datasheet data. However, the ($\Delta_{datasheet}$) difference between the numerical results for the RMSE% error is sparingly small between models, for example, 0.05 % (for output power) and 0.07 % (for efficiency). Compared to field data, the Sandia model showed better results (for the output power) and the Δ_{field} difference is equal to 0.02 %; however, for efficiency, the Jantsch model was slightly better but the Δ_{field} difference is equal to 0.03 %, which is still very low. Therefore, both models showed good quality outcomes, mean errors, and are very similar. Therefore, both models represented the inverter manufacturer Xantrex model GT 2.8 kW well.

Regarding the indicator MBE% error, for output power in Table 9.1, the Sandia model tended to underestimate (−0.17 %) but the Jantsch model tended to overestimate (0.11 %) the datasheet data ($\Delta_{datasheet}$) showing a value equal to 0.28 %, i.e., a small variation in the end result. For the field data, both models underestimated, but datasheet data ($\Delta_{datasheet}$) showed a very low result (0.03 %) between the models. Regarding MBE% for the efficiency using datasheet data, the Sandia model

Table 9.1 Results obtained for the Jantsch and Sandia models from field data and datasheet data

Sandia and Jantsch models (Xantrex GT2.8)		Data			
		Datasheet (%)	$\Delta_{datasheet}$ (%)	Field (%)	Δfield (%)
Output power	RMSE%$_{Sandia}$	0.22	0.05	0.10	0.02
	RMSE%$_{jantsch}$	0.17		0.12	
	MBE%$_{Sandia}$	−0.17	0.28	−0.06	0.03
	MBE%$_{Jantsch}$	0.11		−0.09	
Efficiency	RMSE%$_{Sandia}$	1.15	0.07	1.07	0.03
	RMSE%j$_{antsch}$	1.12		1.04	
	MBE% $_{Sandia}$	−0.70	0.85	−0.40	
	MBE%$_{Jantsch}$	0.15		−0.45	0.05

had a tendency to underestimate (−0.70 %) as opposed to the Jantsch model which overestimated their perspective (0.15 %); the ($\Delta_{datasheet}$) between models was 0.85 %. Regarding efficiency using field data, again both models tended to underestimate but ($\Delta_{datasheet}$) between models was very low; therefore, the results achieved were almost similar between the two models in this regard.

Finally, Table 9.1 shows good results (RMSE% and MBE% errors) for efficiency and output power, good equivalence and small variations in values (as mentioned above), showing the high quality of the Sandia and Jantsch models for representing a model of energy conversion efficiency to the Xantrex GT 2.8 kW inverter. Figure 9.1 shows the efficiency curves for the Jantsch model for the manufacturer's datasheet (black curve), for field measurements (red curve) and the curve of the inverter under real operation conditions (green) and Fig. 9.2 shows the efficiency curves for the Sandia models for the Xantrex GT 2.8 kW inverter. The red curve is from the datasheet, the black curve is from field measurements and the green curve is the actual curve of the inverter in real operation conditions.

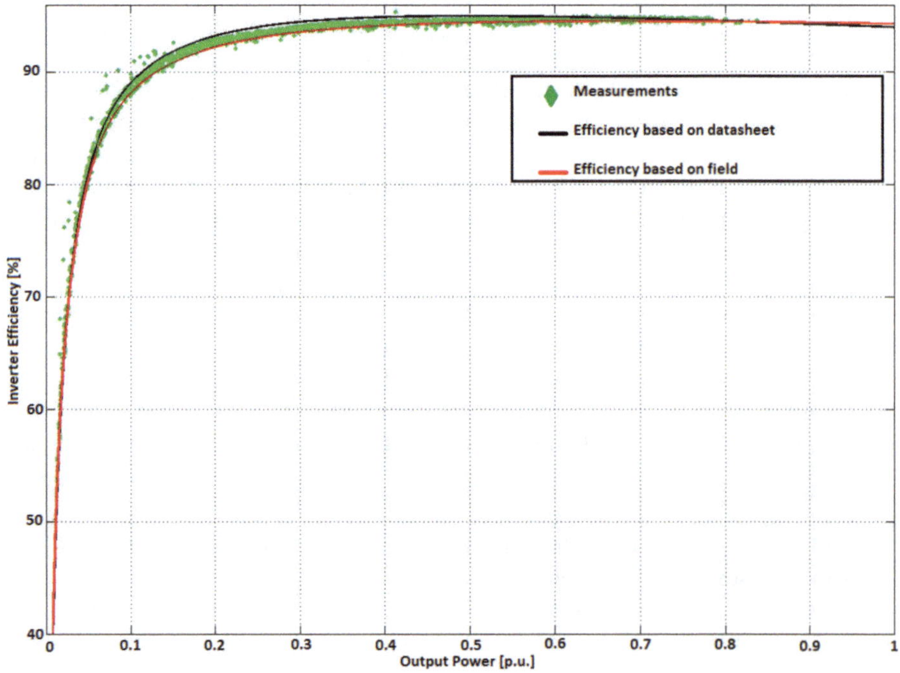

Fig. 9.1 Efficiency curves of the Jantsch model for the Xantrex GT 2.8 kW inverter using datasheet data and field data

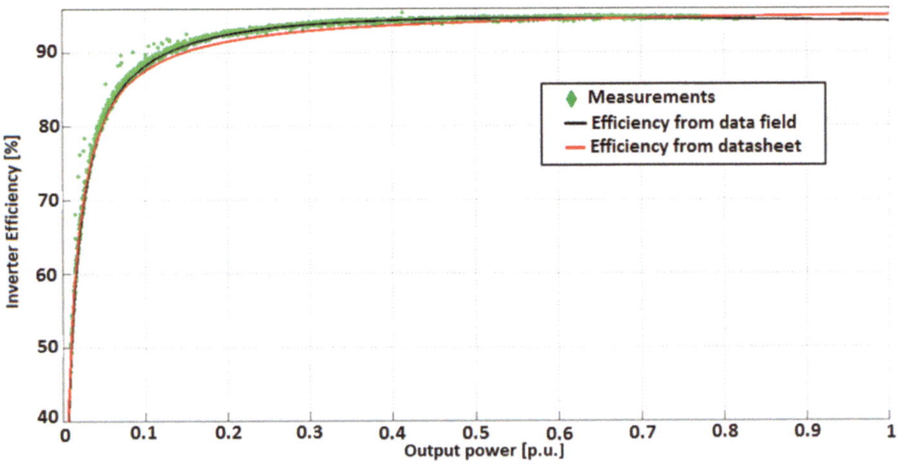

Fig. 9.2 Efficiency curves of the Sandia model for the Xantrex GT 2.8 kW inverter using datasheet data and field data

9.4.2 Analysis of Results for the Enphase Energy 215-60-2LL-IG Microinverter

Table 9.2 shows the performance of the Jantsch and Sandia models compared with the results of MBE% and RMSE% errors using data from the manufacturer's datasheet and from field measurements.

According to Table 9.2, for the output power and also efficiency (through RMSE% error), the Jantsch model showed worse results in comparison with the Sandia model, when using datasheet data. However, the ($\Delta_{datasheet}$) difference between the numerical results for the RMSE% error is very small between models, for example, 3 % (for output power). For efficiency, when using datasheet data, the

Table 9.2 Results for the Jantsch and Sandia models from field measurements and datasheet data

Sandia and Jantsch models (Enphase Energy 215 W)		Data			
		Datasheet (%)	$\Delta_{datasheet}$ (%)	Field (%)	Δ_{field} (%)
Output power	$RMSE\%_{Sandia}$	1.26	3	1.16	0.25
	$RMSE\%_{Jantsch}$	4.26		0.91	
	$MBE\%_{Sandia}$	−0.17	3.81	−0.06	0.33
	$MBE\%_{Jantsch}$	3.98		0.27	
Efficiency	$RMSE\%_{Sandia}$	13.56	8.44	4.34	0.04
	$RMSE\%_{Jantsch}$	22		4.30	
	$MBE\%_{Sandia}$	0.35	14.85	0.57	
	$MBE\%_{Jantsch}$	15.20		1.37	0.8

RMSE% error was quite high for both models with values > 10 %. Compared with field data, the Sandia model showed worse results (for the output power) and the difference (Δ_{field})with a value equal to 0.25 %; however, for efficiency, the Jantsch model was slightly better with a (Δ_{field}) difference equal to 0.04 %, which was very low. Therefore, for both models, the RMSE% error output power showed good quality for datasheet data and field data with small variations; however, for efficiency, the RMSE% error for the Sandia and Jantsch models using datasheet data was high, showing poor quality for both models. On the other hand, for efficiency, the RMSE% error was very similar (4.30 %) and therefore both models represented the microinverter Enphase Energy 215 W for field data well.

Regarding the indicator MBE% error to the power output in Table 9.2, the Sandia model tended to underestimate (−0.17 %) but the Jantsch model tended to overestimate (3.98 %) using datasheet data with $\Delta_{datasheet}$ equal to 3.81 %. The same happened for field data for both models. The Sandia model tended to underestimate (−0.06 %) while the Jantsch model tended to overestimate (0.27 %) with Δ_{field} equal to 0.33 %. Regarding MBE% error for efficiency with datasheet data, both models had a tendency to overestimate (0.35 % for Sandia and 15.20 % for Jantsch) with $\Delta_{datasheet}$ equal to 14.85 %. Regarding efficiency with field data, again both models tended to overestimate but with small values (0.57 % for Sandia and 1.37 % for Jantsch) with Δ_{field} equal to 0.8 %, showing that the difference between the models is very low; therefore, the result achieved were almost similar between the two models in this regard.

Finally, Table 9.2 shows good results; the RMSE% and MBE% errors for efficiency and output power showed good equivalence and small variations in their values (as mentioned above), showing the high quality of the Sandia and Jantsch models for representing a model of energy conversion efficiency to the Enphase Energy 215 W inverter for field measurements. Figure 9.3 shows the efficiency curves for the Jantsch model for the manufacturer's datasheet (green curve), for field measurements (red curve) and the curve of the inverter for real operation conditions (dots blue). Fig. 9.4 shows the efficiency curves for the Sandia model for the Enphase Energy 215 W inverter. The red curve is from datasheet data, the black curve is from field measurements and the green curve (dots) is the actual curve of the inverter in real operation conditions.

9.5 Conclusion

Regarding the validation of the Jantsch and Sandia models using MBE% and RMSE% errors, both showed good results on average errors, good equivalence between models and small variations between their values (as shown in Sect. 4.1). This confirmed the high quality of the Sandia and Jantsch models for representing the energy conversion efficiency for the Xantrex model GT 2.8 inverter.

For the Enphase Energy model M215-60-2LL-S22-IG microinverter, the validation of the Jantsch and Sandia models was good for power output and efficiency

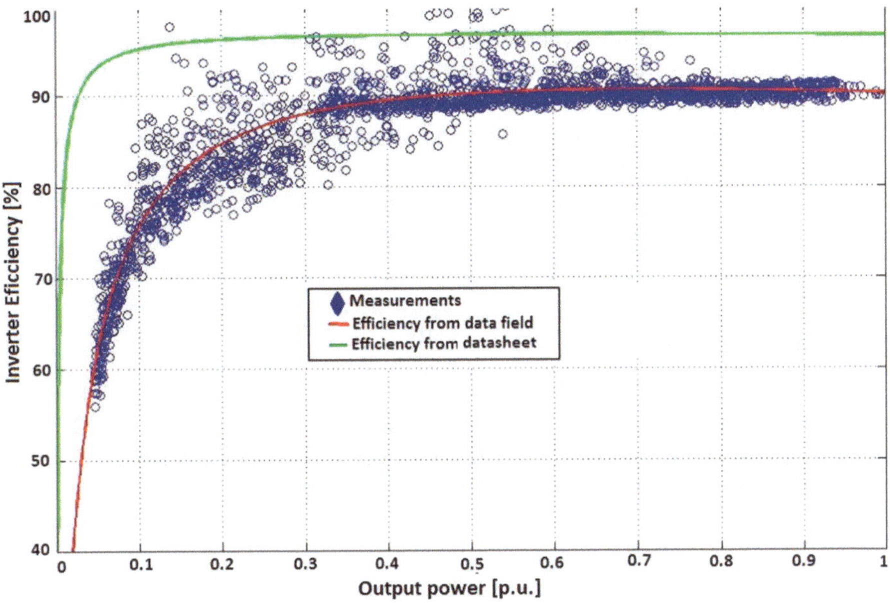

Fig. 9.3 Efficiency curves of the Jantsch model for the Enphase Energy model M215 microconverter using datasheet data and field data

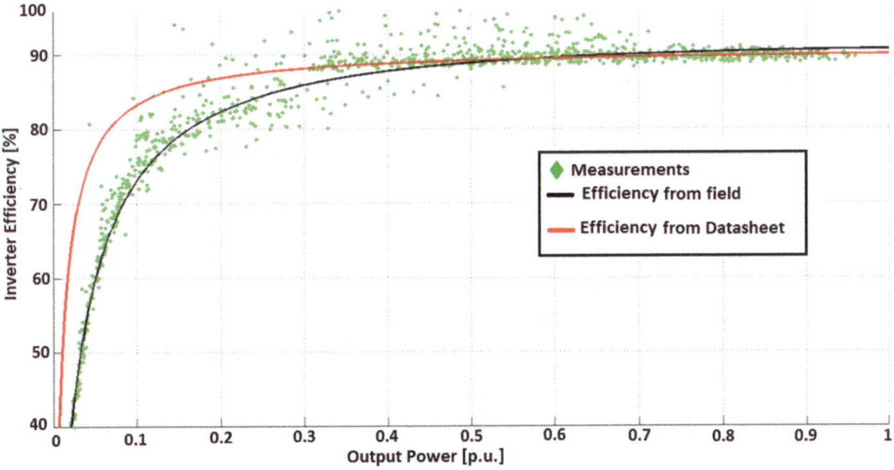

Fig. 9.4 Efficiency curves of the Sandia model for the Enphase Energy model M215 microinverter using datasheet data and field data

for data from field measurements with slight variations as shown in Table 9.2. This demonstrates the quality of both models for representing the energy conversion efficiency for the inverter. On the other hand, for datasheet data, the Sandia and Jantsch models presented very high values for RMSE% errors for efficiency. The explanation for this refers to datasheet data from the manufacturer that provides values for the efficiency curve beginning at 10 % of rated output power disregarding the values below this loading; therefore, increasing the average error and the quality of the models. Note, for MBE% errors, the Sandia model results were better than the Jantsch model. Finally, the Sandia and Jantsch mathematical models are a good and reliable for representing the energy conversion efficiency for an inverter for computer simulations and consequently to generate estimates of energy production of PV systems under different climatic conditions.

9.6 Special Thanks

This work has been supported by the Brazilian agency CAPES, CNPq, FIP PUC Minas, GEDAE/UFPA, PPGEE/UFMG and developed under the Strategic Program N°13 Call R & D Technological Electricity Sector regulated by ANEEL.

References

1. Jantsch M, Schmidt H, Schmid J (1992) Results of the concerted actions on power conditioning and control. XI European Photovoltaic Solar Energy Conference, Swiss
2. King DL, Gonzalez S, Galbraith GM, Boyson WE (2007) Performance model for grid-connected photovoltaic inverters. Sandia National Laboratory Report, New México
3. Rampinelli GA, Krenzinger A, Chenlo RF (2014) Mathematical models for efficiency of inverters used in grid connected photovoltaic systems. Renew Sustain Energy Rev 34:578–587
4. Driesse A, Jain P (2008) Beyond the curves: modeling the electrical efficiency of photovoltaic inverters. PVSC: 2008 33rd IEEE Photovoltaic Specialists Conference 1–4:1935–1940
5. XANTREX (2007) Grid Tie Solar Inverter GT2.8-NA-240/208UL-05, Datasheet, June
6. Enphase Energy (2014) Micro-inverter M215-60-2LL-S22-IG, Datasheet,
7. Macedo WN, Torres P, Braga WJ, Cavalcante RL, Galhardo M, Pinho JT (2014) Análise de eficiência de um microinversor conectado Ã rede elétrica. In: Congresso Brasileiro de Energia Solar, 2014, Recife.
8. Macêdo WN, Cavalcante RL, Júnior WB, Galhardo M, Pinho JT (2012) Contribuição Energética e Análise de Perdas de Conversão C.C.-C.A. em uma Aplicação Fotovoltaica Conectada Ã Rede Elétrica na UFPA. IV Congresso Brasileiro de Energia Solar e V Conferência Latino-Americana da ISES—São Paulo.
9. Ineichen P (2011) Global irradiance on tilted and oriented planes: model validations. Technical report. University of Geneva, February
10. Gueymard C (2009) Direct and indirect uncertainties in the prediction of tilted irradiance for solar engineering applications. Sol Energy 83(5):432–444. Erratum (1988) Sol Energy 40(2):175

Chapter 10
Mitigation Measures to Minimize Adverse Impacts of Renewable Energy Integration

GM Shafiullah, Amanullah MT Oo and Alex Stojcevski

Abstract The intermittent nature of renewable energy (RE) sources, in particular solar and wind energy, has an impact on system operations including voltage and frequency, harmonics and power quality (PQ) in general, and influences the overall performance of the power network as well as the distribution network (DN). Hence, integration of large-scale RE into the DN is one of the biggest challenges today.

Therefore, this study initially investigates the potential impacts, in particular, voltage regulation, reactive power compensation, current and voltage harmonics causes due to large-scale integration of RE into the Berserker Street Feeder, Frenchville substation under Rockhampton distribution network (DN), Central Queensland, Australia. Finally, to reduce the level of impacts observed, energy storage system and STATCOM devices (both optimised) were developed and integrated into the network that ensures a smooth power supply to the customers. Results show that integration of both optimised STATCOM and energy storage enhances the overall power quality of the Rockhampton power network as it enhances voltage regulation and improves power distribution and transformer utilisation and reduces total harmonic distortion of the power network.

Keywords Renewable energy · Distribution network · Power quality · Energy storage · STATCOM

10.1 Introduction

In general, power flows from the upstream network (the transmission network) to the downstream network (the distribution or low-voltage network). Integration of RE systems causes reverse power flows, i.e., feeding back into the grid, if the power generation from these systems is greater than the load in the local network. Therefore, RE integration introduces bidirectional power flows across distribution

G. Shafiullah (✉) · A. MT Oo · A. Stojcevski
School of Engineering, Faculty of Science, Engineering & Built Environment,
Deakin University, Geelong Waurn Ponds Campus, Melbourne, Australia
e-mail: g.shafiullah@deakin.edu.au

A. Sayigh (ed.), *Renewable Energy in the Service of Mankind Vol II*,
DOI 10.1007/978-3-319-18215-5_10

transformer (DT), and hence, DN experiences with several potential problems that include voltage fluctuations, overloading of DT, poor power factor and harmonics injection in the DN, which detracts the overall PQ of the power system network. Moreover, large-scale RE integration in the DN introduces uncertainties due to its intermittent weather-dependent nature, which causes voltage rise within the network and this voltage rise is significant in the case of single-phase PV system connections. The receiving end (customer premises) voltage may drop if RE is unable to support customer load demand, especially during peak demand periods [1–3]. Moreover, the intermittent nature of RE causes uneven generation and hence might exceed the capacity of the connected transformer. The operation of the distribution network involves reactive power due to customer loads, line impedances and RE sources, in particular induction generators used in wind turbines, which are unsafe for the smooth operation of the network [4, 5]. Inverters connected with RE sources, nonlinear customer loads, and power electronic devices introduce harmonics in the distribution network that causes overheating of transformers, tripping of circuit breakers, and reduces the life of connected equipment [6].

Significant research and development works are undertaken by various agencies throughout the world to investigate and mitigate the observed potential technical challenges to ensure reliable and uninterrupted power supply to the consumers. Albarracin and Amaris [5] investigated a voltage fluctuation model used for the evaluation of flicker assessment under sunny and cloudy conditions with photovoltaic energy sources. Results showed that irregular solar irradiation caused by cloud movement produced voltage and power fluctuations. The Gardner MA PV project [7] explores four areas: the effect on the system in steady state and during slow and cloud transients; responses of concentrated PV under fast transients; harmonic effects on the PV system; and the overall performance of distribution system, in which the total impact of the high penetration of PV was evaluated. Results showed that 37 % penetration of PV at Gardner was achieved without any significant problems. Asano et al. [8] analysed the impact of high penetration of PV on grid frequency regulation, which responds to short-term irradiance transients due to clouds. It was shown that the breakeven cost of PV is unacceptably high unless PV penetration reaches 10 % or higher. Therefore, PV integration needs to be increased and impacts to be identified and mitigated. A comprehensive study was carried out by Fekete et al. [9] that analysed the harmonic impacts in both winter and summer seasons with 10 kW PV penetration on the distribution network. Recent studies by Ergon Energy and Chant et al. [10] have explored the issues involved with small-scale PV penetration in urban networks. It was found that increased penetration exhibited increased voltage rise on LV networks, increased harmonic distortion; as a result, load rejection occurs.

From simulation results [11] it was observed that both the transient and voltage stability of the system are in a stable condition with 24.55 % wind integration level and with SVC. However, the allowable wind energy integration level for the system studied is less than 77 % [11]. From the literature [12] it was shown that DFIGs are the most efficient designs for the regulation of reactive power along with the adjustment of angular velocity to maximise the output power efficiency. Characteristics of harmonics injected into a wind energy integrated power system were investigated with a variety of configuration and operating conditions [13, 14].

Appropriate design of electrical circuits with control systems can mitigate voltage fluctuations and harmonic distortion, provide reactive power compensation and power factor improvements, and thus ensure PQ improvements in the power system. Customised power devices such as SVCs, STATCOMs, DVRs, thyristor-controlled series compensators (TCSCs), static synchronous series compensators (SSSCs), and a combination of series and shunt active power filters are the latest developments of interfacing devices between grids and consumer appliances that overcome voltage and current disturbances and improve the PQ by compensating the reactive and harmonic power [15]. A STATCOM is the best performing device for reducing voltage fluctuations and harmonics as well as improving the PQ of the power network compared to other flexible AC transmission system (FACTS) devices. STATCOMs are faster, smaller, and have better performance at low-voltage conditions, though the cost of STATCOMs is comparatively higher [16, 17].

On the other hand, energy storage plays an important role in facilitating large-scale RE integration by supporting peak load demand and peak shaving, improving voltage stability and power quality. It also maintains constant grid power and reduces GHG emissions by maximising RE utilisation. Batteries are one of the most cost-effective energy storage technologies for power applications such as regulation, protection, spinning reserve, and power factor correction [18]. Lead-acid and lithium-ion batteries are a renowned and effective storage technology today.

In order to enhance the terminal voltage quality, SVCs were used for reactive power compensation of wind power induction generators [19]. A STATCOM-based control mechanism is used to reduce the power quality problems as well as harmonics on integrating wind energy into the grid [17, 20]. The system is capable of meeting the reactive power demand from the wind generator and the load at the PCC to the grid, and maintaining the source voltage and current in-phase. Hybrid-battery super capacitor energy storage systems are expected to play a major role in power smoothing, power quality improvement, and low-voltage ride in a wind energy conversion system [12].

From the literature, it is observed that most of the available research was carried out primarily in the USA and Europe [4, 5, 8, 9]. However, the distribution network characteristic of Australia is different compared to other developed countries in many forms, and research conducted by other countries could not simply be adopted without further research in the Australian context. Therefore, this study developed a model that investigates the observed potential challenges into the grid using power system simulator PSS SINCAL [21]. The Berserker Street Feeder, Frenchville Substation, Rockhampton, was selected as a suitable site for investigation of large-scale deployment of RE into the grid. By increasing PV/wind integration into the network, the points at which distributed generation starts to have an impact on utility grid operations have been investigated with regard to network performance standards. Worst-case scenarios were identified and measures were taken to reduce the level of adverse potential impacts.

10.2 Potential Impact Analysis Model

10.2.1 Model Evaluation

To investigate the impacts of RE integration to support the loading of a feeder, this study developed a model for Berserker St. Feeder, Frenchville. Berserker Street Feeder comprises 1442 industrial and residential customers with a connected load of 4775 kVA. A 66/11 kV transformer with a capacity of 6250 kVA delivers power to the customer. Ergon Energy [22] as a local distribution network service provider (DNSP) is fully responsible for the smooth delivery of electricity to the customers in the Berserker Street Feeder.

In line with the original network facility, a 66/11 kV step-down transformer with a capacity of 6250 KVA was considered in the Frenchville Substation to deliver power supply in the feeder. The zero-sequence resistance and reactance of the step-down transformer is 1.29 and 25.84 Ω with 8 % short-circuit voltage. This case study, subdivided the feeder into two zone feeders and several zone observation points to observe adverse potential impacts on the different points of the distribution network. Berserker Street Feeder contains 37 load centres.

Three-phase HV load was transformed to three single-phase loads with an equal amount through an 11 kV/415 V DT as shown in Fig. 10.1; hence, zone-wise load

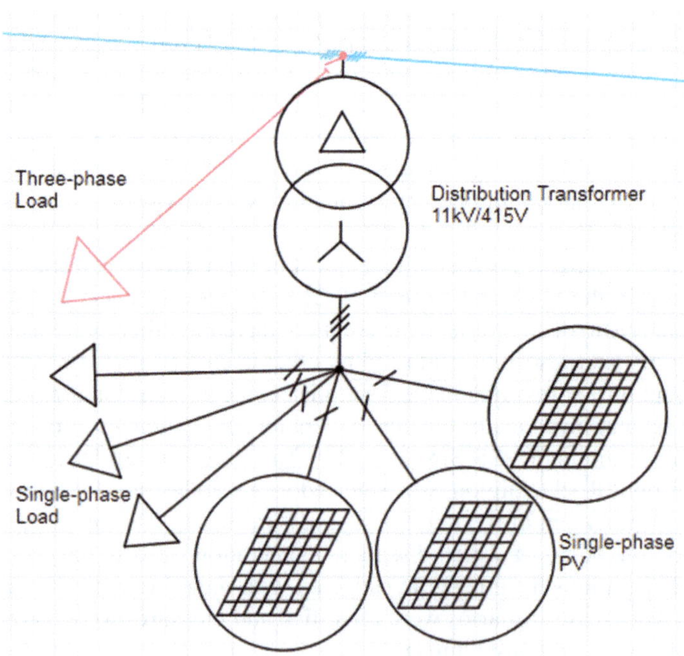

Fig. 10.1 Single-phase load and PV connection

allocation remains the same. Single-phase RE was connected in the LV side of the DT as shown in Fig. 10.1 and the amount of PV and wind depends on the percentage of load demand based on considered-case scenarios. Losses until inverter and inverter efficiency were considered 5 and 97 %, respectively, both for PV and wind.

Daily load profiles of Rockhampton for summer was used in this study in which it was shown that maximum load was measured at 2:00 p.m. A PV load profile was generated for this study in p.u. values from the hourly average summer solar radiation of Rockhampton, Australia, collected from the BOM [23] in which it was indicated that reasonable solar power is only available from 7:30 a.m. to 5:30 p.m., while peak solar radiation is at 12:30 p.m. Hourly average summer wind speed of Rockhampton, Australia, is used for load flow analysis.

10.2.2 Modelling Case Scenarios

Considering feeder load demand and future RE integration, several case scenarios were considered and accordingly developed for the model. Actual line length and impedances were considered for all of the studied case scenarios. The case scenarios considered in this case study are:

- *Case 1: grid with only load*
- *Case 2: grid with 20 % PV integration*
- *Case 3: grid with 50 % PV integration*
- *Case 4: grid with 50 % wind integration*
- *Case 5: grid with 100 % PV integration*
- *Case 6: grid with 100 % wind integration*
- *Case 7: grid with 100 % PV and 150 % load demand*
- *Case 8: grid with 100 % PV integration in centralised approach:* Instead of distributed PV connection with individual loads this case considered centralised PV connection with each of the zone observation point. The considered centralised installation was 100 m apart from the zone observation point. This centralised generation expected to deliver electricity to the local customer located around the zone observation point
- *Case 9: grid with 100 % wind integration in centralised approach:* In this case centralised wind generation was considered into the feeder and was connected to only one location in each zone
- *Case 10: grid with 200 % RE integration in both decentralised and centralised approach:* In this case, 100 % PV of total loading was connected with decentralised connection and 100 % wind of total loading was connected with centralised connection approach as shown in Fig. 10.2.

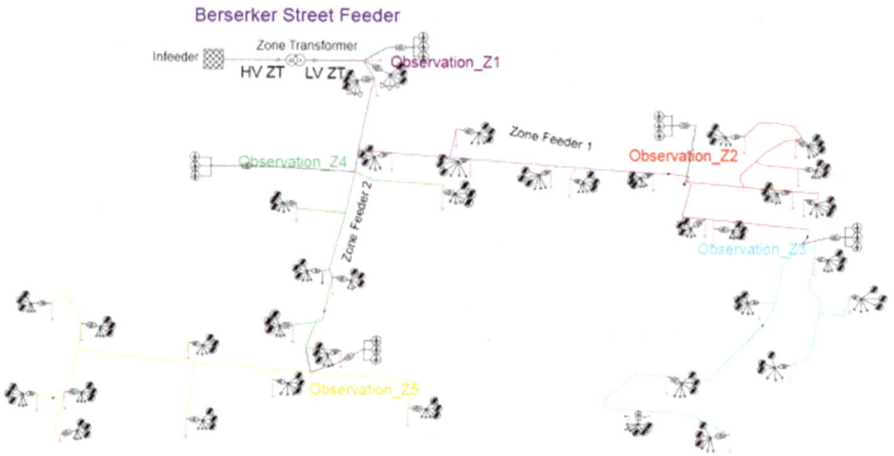

Fig. 10.2 RE integration with the feeder

10.2.3 Result Analysis

10.2.3.1 Voltage Regulation

From the model analyses, it is evident that voltage of the Berserker Street Feeder fluctuates with the increased integration of RE and causes uncertainties in the feeder as well as in the DN. From load flow analysis, it was observed that voltage gradually decreases from the source station, that is, zone substation, to the load connection point for both balanced and unbalanced systems.

Voltage regulation of cases 1 to 10 is shown in Fig. 10.3, in which it has been shown that voltage level reduces from substation to the customer end for all of the

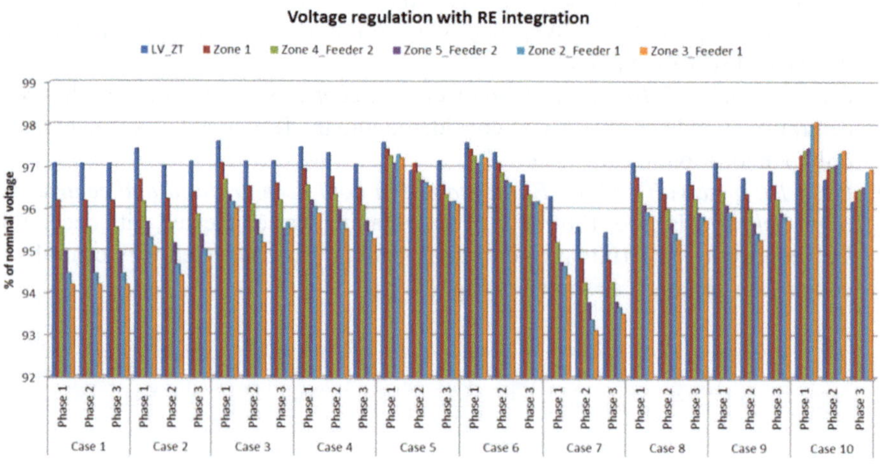

Fig. 10.3 Voltage fluctuations with RE integration

case scenarios. However, phase voltages increased with the increase in RE integration in all of the observation points as shown in cases 2 to 6. This happened due to the bidirectional power flow at the point of connection as RE sources generally connected near the customer load and in a few cases RE generation is greater than the load demand. Load demand and RE generation in a particular time also responsible for uncertainties in the network such as PV-generated maximum energy in the middle of the day, while load demand of the residential areas is minimum. Voltage of the network reduces significantly with the increased load demand as shown in case 7. In this case, voltages across zones 2 and 3 in feeder 1 and zone 5 in feeder 2 are less than 94 % of the nominal voltage; hence, the voltage limit exceeded the regulatory limit of ± 6 % of the nominal voltage. Voltage regulation of zone 3 feeder 1 and zone 5 feeder 2 with the integration of PV and wind is shown in Fig. 10.4a and b, respectively, in which it was clearly evident that voltage rises significantly with the increase in PV and wind integration into the network, though the increased voltage levels are within the allowable safety limit.

This study also considered both the centralised and decentralised RE connection approach to explore the suitability of RE integration into the Berserker Street Feeder. Figure 10.5a shows the voltage regulation of case 5, that is, 100 % PV of total loading with decentralised connection and case 8 is the same PV with centralised connection. From the figure it was shown that voltages of different observation points were decreasing in the centralised connection approach compared to the

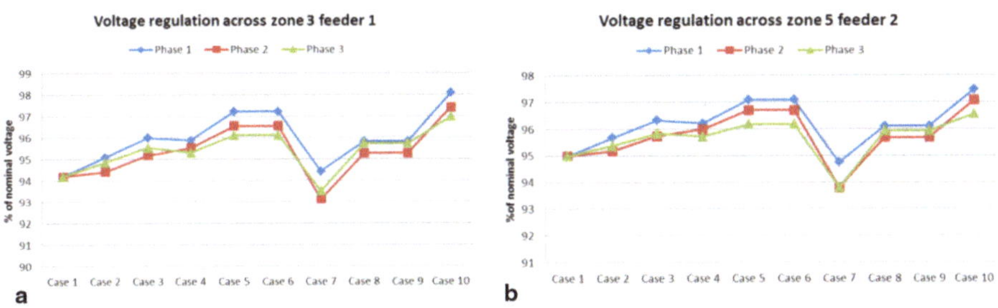

Fig. 10.4 Voltage regulation with RE integration **a** zone 3 feeder 1 and **b** zone 5 feeder 2

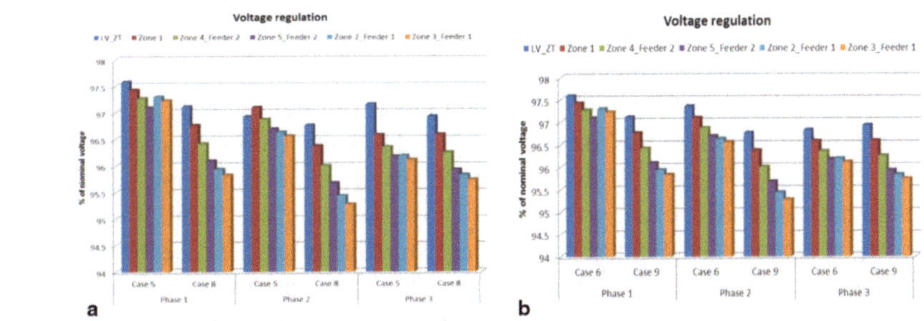

Fig. 10.5 Voltage variations due to centralised and decentralised connection of **a** PV and **b** wind

decentralised approach. One of the main reasons for this is that intermittent generation is supplied from specific nodes instead of individual load-connected nodes; hence, bidirectional power flows occur only from the centralised connection point of each zone. Figure 10.5b shows the voltage regulation of decentralised wind turbine connection with 100 % (case 6) and centralised connection (case 9) and it was shown that phase voltages decrease at different observation points in the centralised connection approach.

10.2.3.2 Power Distribution

Power generation over time was observed in Fig. 10.6a and b for cases 7 and 10. From Fig. 10.6a it was clearly indicated that from 10:10 a.m. to 2:05 p.m. customers can feed back surplus electricity to the grid as PV generation was high during this period, while at night the grid needs to supply energy to meet customer load demand as there was no PV generation. In case 10, RE could not fulfil customer demand only from 12:00 to 7:30 a.m. and 6:00 to 12:00 p.m. as PV generation was not possible and sufficient generation was not possible from wind due to low wind speed as shown in Fig. 10.6b. However, there is a huge surplus of electricity that can be fed back to the grid from 7:30 a.m. to 6:00 p.m. From model results it is evident that most of the reactive power is required for customer-connected load and wind turbine, while PV inverter has negligible impacts on reactive power. This also influences poor power factor regulation in the DN as well as affecting the PQ of the network.

10.2.3.3 Harmonics

Voltage harmonic distortion of different studied case scenarios are shown in Fig. 10.7 in which it is evident that voltage harmonics in different observation points increase with the increase in PV integration and load demand. Cases 2, 3, and 5 have the same system configurations except for the level of PV integration, and total voltage harmonics distortion in the LV network for cases 2, 3, and 4 are 2.56, 2.83, and 3.16 %, respectively. Therefore, it can be stated that the use of power electronic devices in PV inverter along with intermittent generation causes harmonics in the system. Voltage harmonic distortion increases with the increase in load demand as shown for case 7 compared to case 5 (both cases 5 and 7 have the same amount of PV connections). Voltage harmonic distortion increases significantly with the integration of wind turbine into the feeders compared to solar PV integration as shown in Fig. 10.7 for cases 4 and 6. However, a centralised connection approach reduces the voltage harmonic distortion significantly as shown in cases 8 and 9 for PV and wind energy, respectively. Voltage harmonic distortion is less in case 10 compared to cases 4 and 6, though 200 % RE of the total loading was integrated in this case (100 % PV in a decentralised approach and 100 % wind in a centralised approach).

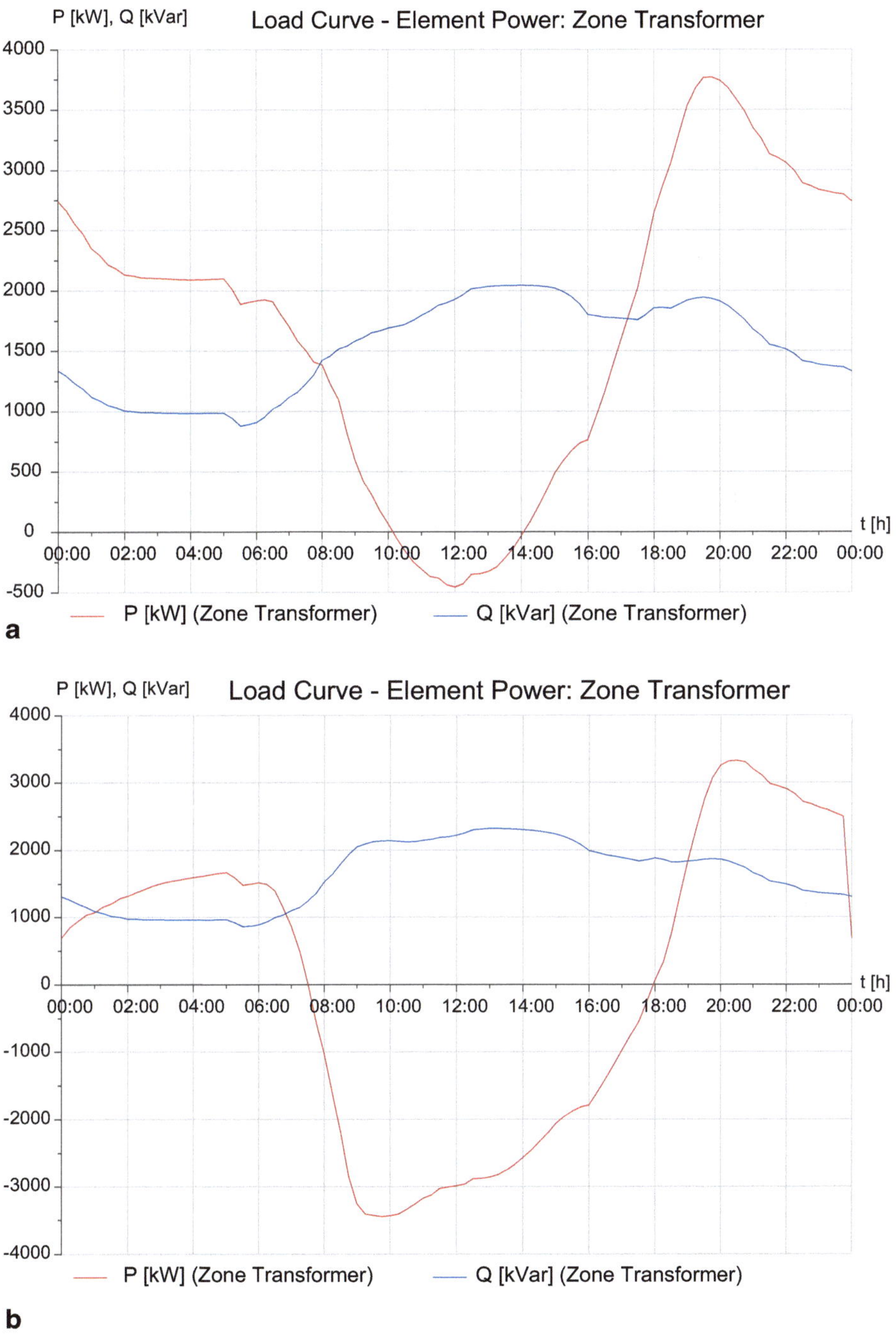

Fig. 10.6 Power distribution of the network **a** case 7 and **b** case 10

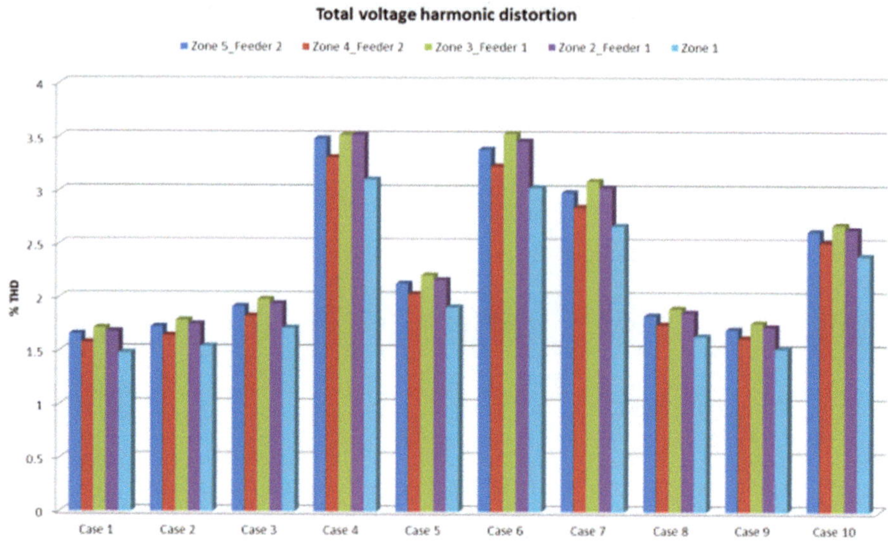

Fig. 10.7 THD for the studied cases

Voltage harmonic distortion across the HV DN and LV DN is shown in Fig. 10.8 in which it can be seen that harmonic distortion approaching the regulatory limits occurs across the LV network for cases 4, 6, and 7, while being within the safe limits in HV DN.

Harmonic current is of the greatest concern as most of the adverse effects in the distribution network are caused due to these currents. From the simulation results, it is clearly indicated that current harmonics increase with the increase in RE integration in the Berserker Street Feeder as well as into the DN. The total current

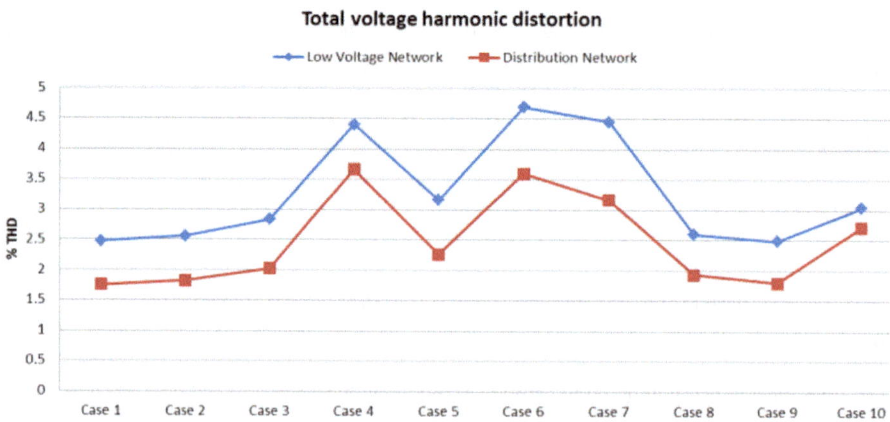

Fig. 10.8 THD across HV DN and LV DN

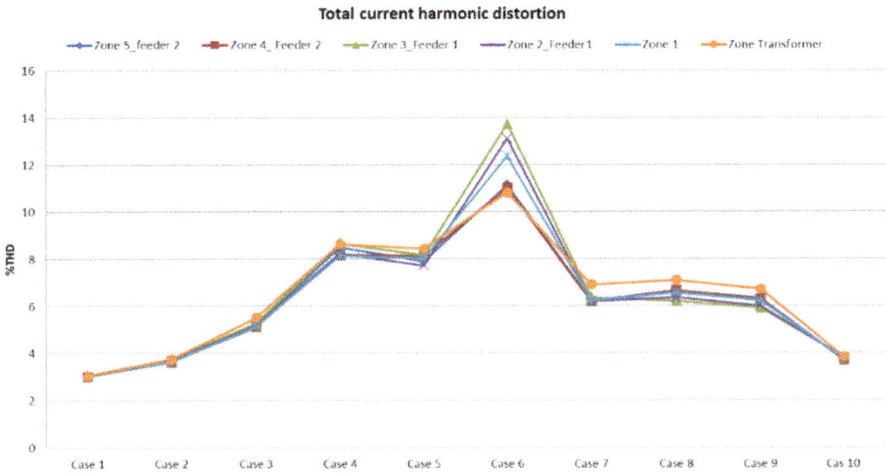

Fig. 10.9 Current harmonic distortion of feeders

harmonic distortion at different observation zone points is shown in Fig. 10.9 in which it is clearly indicated that harmonic distortion increases with the increase of PV array utilisation. Cases 2, 3, and 5 have the same conditions except for variations in the level of PV integration into the network. THDs for cases 1, 2, 3, and 5 are 3.02, 3.71, 5.49, and 8.42 %, respectively, across the zone transformer in which case 5 exceeded the regulatory standard limit as shown in Fig. 10.9. From the figure, it can be clearly seen that the current harmonic distortion is higher for wind (case 6) compared to the same amount of solar PV (case 5) integration. Current harmonic distortion for case 7 is reduced to 6.9 % due to increased load demand. Current harmonic distortion was reduced in the case of centralised PV connections (case 8) compared to decentralised PV connections (case 5). Two hundred per cent integration of total loading (case 10) reduces current harmonic distortion as this system generates more power than the household load demand and causes reverse power flow in the system. Therefore, current harmonic distortion at most of the observation points is within the allowable limits of safety as shown in Fig. 10.9, except in cases 4, 5, and 6.

From analysis of the results it was shown that current harmonic distortion exceeded the regulatory limits for 100 % PV integration and both 50 and 100 % wind integration; hence, it can be stated that wind energy injected more harmonics into the feeder than the solar PV. Moreover, harmonic distortion decreases in the centralised connection approach compared to the decentralised connection. Therefore, this study explores the possible mitigation measures that can potentially reduce the adverse impacts caused by large-scale RE integration.

To ensure optimal performances of the Rockhampton power network, an energy storage system and a STATCOM device (both optimised) were developed and integrated into the system, effectively reducing the potential impacts of integrating large-scale RE into the grid and thus ensuring reliable power supply to the customer.

10.3 Mitigation of Impacts

10.3.1 Introduction

The potential impact analysis model was further developed by integrating the optimised STATCOM and energy storage in the Berserker Street Feeder. Optimised STATCOM and energy storage devices were connected in a centralised connection approach at the observation node points and busbars. Specifications of the STATCOM and energy storage were chosen for different observation nodes based on load demand and RE integration. A trial-and-error approach was considered to optimise the specification of both the STATCOM and energy storage. To explore the mitigation outcomes, the case scenarios that had the highest generation from RE were considered as the base case.

Four case scenarios were considered to investigate the impacts on integrating the STATCOM and energy storage separately and in combination as shown below:

- Base case—Grid with 100 % PV in decentralised and 100 % wind in centralised connection approaches (case scenario 10)
- Case A—STATCOM was integrated with the base case
- Case B—energy storage was integrated with the base case
- Case C—both STATCOM and energy storage were integrated with the base case

10.3.2 Model Evaluation

A three-phase STATCOM was designed with 250 kVAR capacitive and 50 kVAR inductive reactive power in which the zero sequence resistance and reactance were assigned 0.567 and 0.453 Ω, respectively, as shown in Fig. 10.10. The lower and upper voltage limits were considered as 95 and 105 % of the rated voltage, respectively. One STATCOM was connected in each observation node, and hence a total of five STATCOMs were integrated with the feeder as shown in Fig. 10.11. In general, energy storage can be applied to generation, transmission, or various other parts of the distribution system, at the customer site or with any particular appliance. The storage system was designed using the DC-Infeeder module in PSS Sincal to allocate battery size with realistic losses between the array and the inverter (15 %) and the inverter efficiency (97 %). Initial energy storage level, maximal and minimal storage were considered as 60, 50, and 1200, respectively, as shown in Fig. 10.12. Efficiency In and Efficiency Out were considered as 0.96 and 0.97 p.u. to define two different degrees of efficiency for loading energy into the storage and removing energy from the storage. The SMA Sunny Backup SBU5000 inverter was considered with their harmonic current emission levels, which follow the EN 61000-3-12 standard [183]. For this case study, energy storage was connected in a centralised

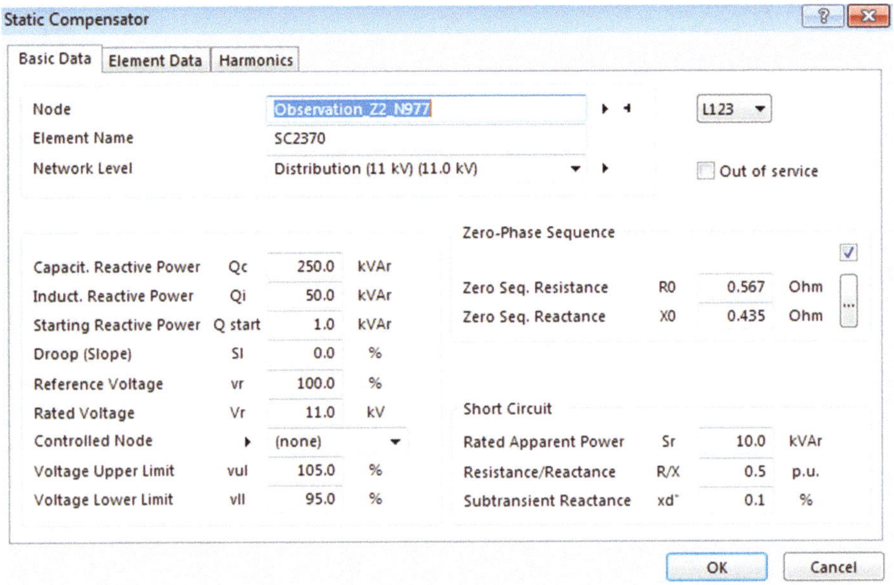

Fig. 10.10 Specifications of STATCOM for Berserker Street feeder

connection approach at observation nodes. Three single-phase 100 kW lithium-ion batteries, each with an inverter, were connected at each observation node as the energy storage device, and a total of 1500 kW of batteries were connected in the feeder as shown in Fig. 10.11.

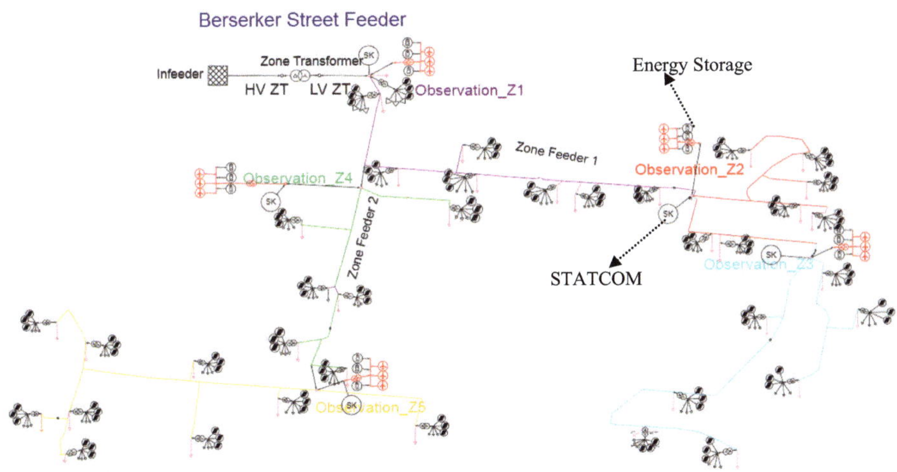

Fig. 10.11 Berserker Street Feeder with STATCOM and energy storage

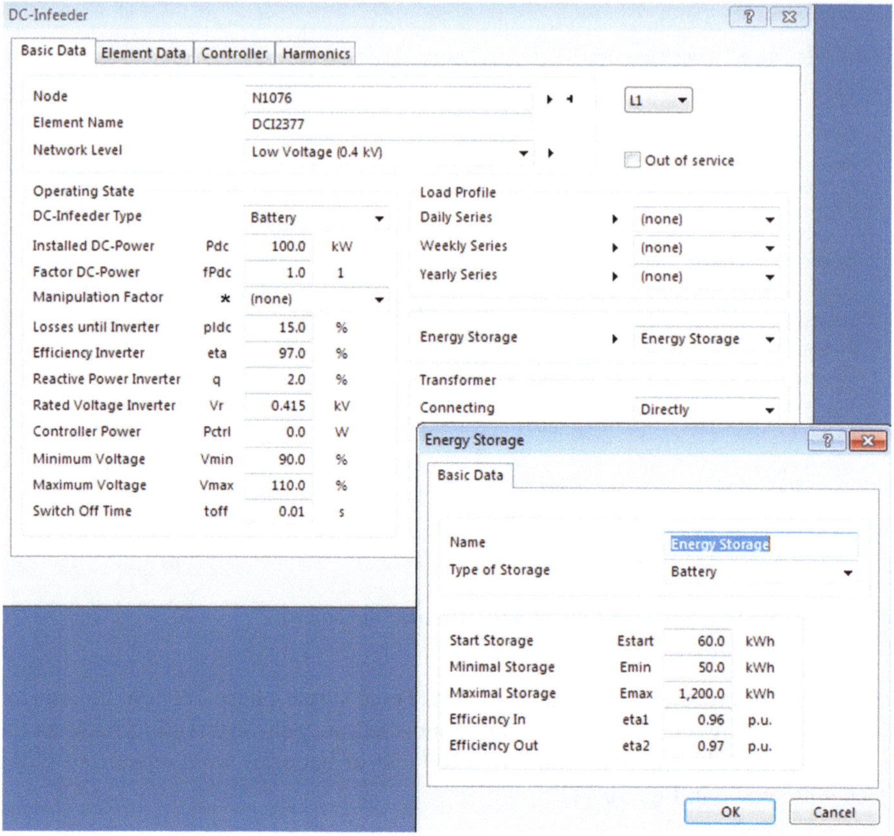

Fig. 10.12 Specification of energy storage

10.3.3 Result Analysis

10.3.3.1 Voltage Regulation

This case study investigated voltage regulation of Berserket Street Feeder after integrating STATCOM and energy storage with the base case and compared the performance of voltage regulation with the base case. Maximum feeder voltages of the studied cases are shown in Fig. 10.13 where it can be seen that STATCOM plays an active role to improve the feeder voltages. With the integration of STATCOM in case A phase to ground feeder voltage in zone 5, feeder 2 increases from 6.19 to 6.31 V. In case B, energy storage was integrated with the base case and from Fig. 10.13 it was shown that maximum feeder voltage reduces slightly from 6.19 to 6.18 V as the storage also required energy from the network for its charging. Feeder voltage increases with the integration of both energy storage and STATCOM as STATCOM improves the system voltage. The percentages of rated voltage across

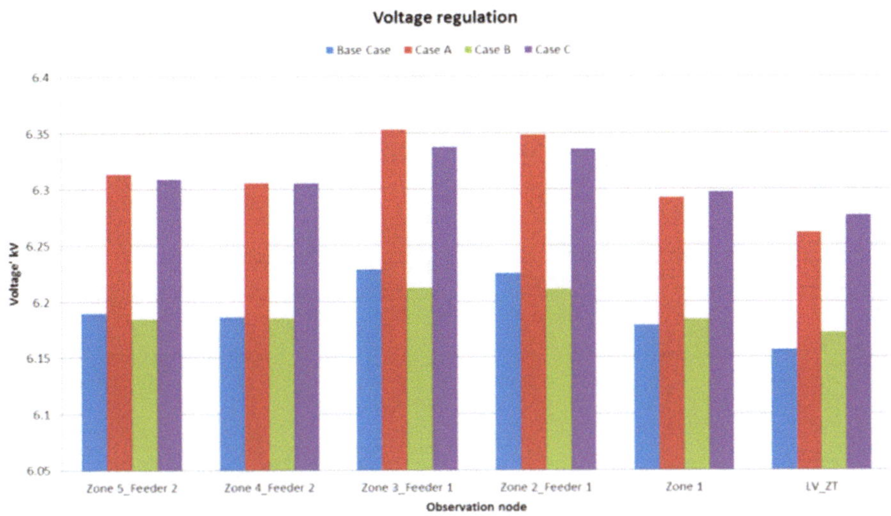

Fig. 10.13 Feeder voltages for studied cases

the zone substation are 96.94 and 98.83 % for the base case and case C, respectively. However, the load flow study doesn't explore the voltage regulation over the whole day as only the maximum voltage of the feeder was shown in the load flow analysis. Load curve analysis shows the exact scenarios of voltage regulation as both RE generation and load demand vary with time.

Figure 10.14a–d shows voltage regulation over time for the base case and cases A, B, and C, respectively. Percentages of rated voltage at 8:00 p.m. in zone 3, feeder 1 are 95.38 and 97.6 % and, at 6:00 a.m., percentages of rated voltage in zone 1 are 98.82 and 100.22 % for the base case and case A, respectively. Therefore, it can be stated that STATCOM enhances the voltage regulation of the feeder. Feeder voltage decreases at 7:45 a.m. in case B as storage starts charging with the solar PV generation and again at 5:30 p.m. as storage starts discharging with the absence of solar PV generation. Feeder voltage increases slightly for most, but not all, of the remaining period. Percentages of rated voltage in zone 3, feeder 1 for the base case and case B are approximately 97.7 and 96.8 % at 7:45 a.m. and 96.8 and 95.6 % at 5:30 p.m., respectively. Percentages of rated voltage at 8:00 p.m. are 95.38 and 95.85 % for the base case and case B in zone 3, feeder 1 as storage can also supply energy to the grid. Voltage regulation improves with the integration of STATCOM and storage as shown in Fig. 10.14d for case C. Percentages of rated voltage at 7:45 a.m. (start of storage charging) are 97.7, 96.8, and 99.2 % and at 5:45 p.m. (storage discharging) are 96.5, 95.58, and 97.84 % for the base case and cases B, and C, respectively. From the load flow and load-curve analysis, it was evident that both STATCOM and energy storage play an active role in improving voltage regulation of the Berserker Street feeder as well as in the distribution network.

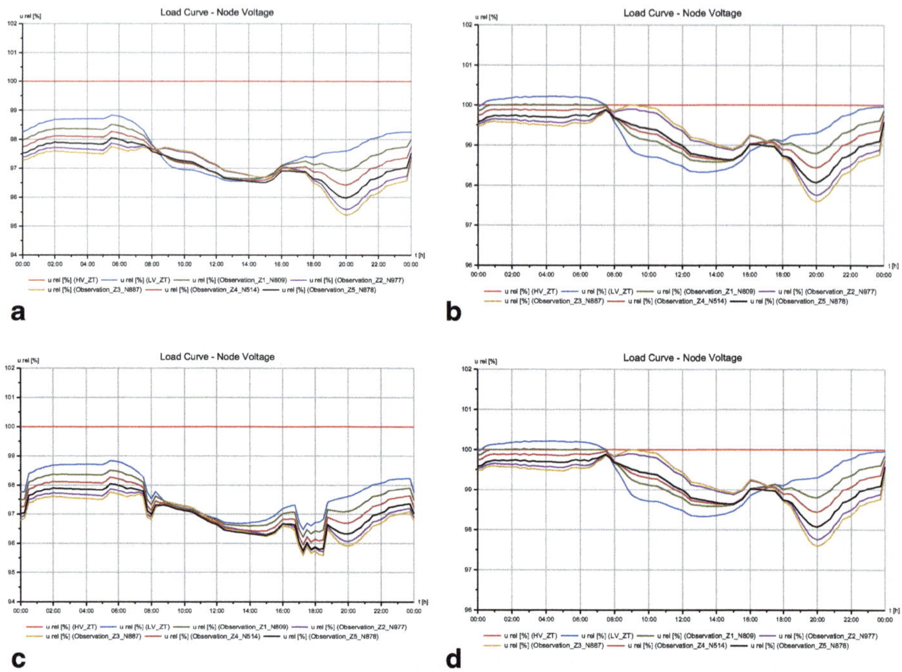

Fig. 10.14 Voltage regulation over time for studied cases **a** base case, **b** cases A, **c** case B, and **d** case C

10.3.3.2 Power Distribution

STATCOM reduces reactive power requirements from the grid significantly as it can generate reactive power that can be used to assist the reactive power demand of the network. However, it has minimum capability to control active power of the network. In the base case, reactive power demand was maximum at 1:00 p.m., which is 2317.56 kVAR as shown in Fig. 10.6b, while with the integration of STATCOM in case A, the reactive power demand is 1000 kW at the same time as shown in Fig. 10.15a. The reactive power demand at 12:00 a.m. is approximately 1300 and 0 kW for the base case and case A, respectively. Integration of STATCOM not only reduces the reactive power demand of the network but also improves the overall power factor of the network.

Feeder maximum power demand for the base case is 3326 kW at 8:15 p.m. as shown in Fig. 10.6b; however, with the integration of storage, the maximum power demand reduced to 2065 kW as storage delivered power to customers at night as shown in Fig. 10.15b. On the other hand, RE can supply a maximum of 3445 kW after meeting load demand at 9:00 a.m. with RE generation at its maximum both from solar PV and wind. However, with the integration of storage, the surplus electricity was reduced to 2220 kW as the storage required power for its charging as shown in Fig. 10.15b.

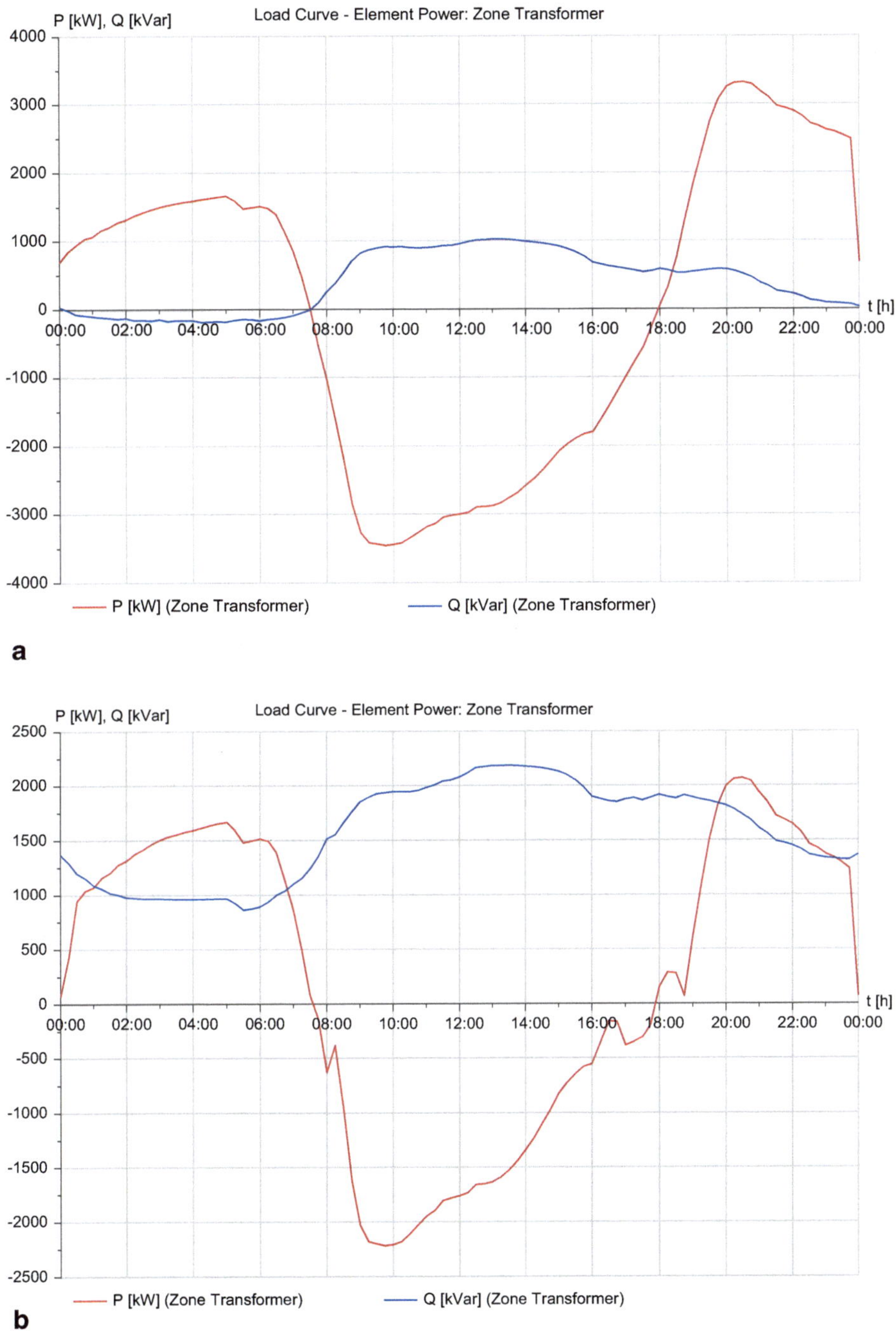

Fig. 10.15 Power distribution across the zone transformer for the studied cases **a** base case, **b** case A, **c** case B, and **d** case C

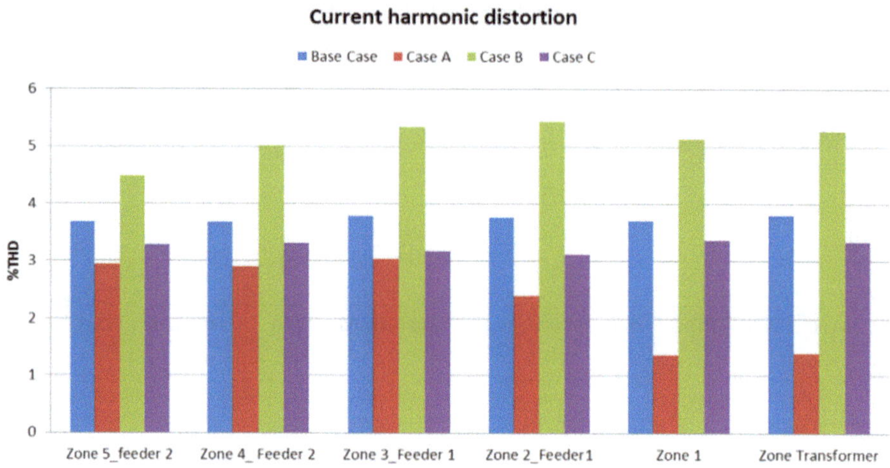

Fig. 10.16 Total current harmonic distortion for all cases

10.3.3.3 Harmonic Distortion

The total current harmonic distortion decreases with the integration of STATCOM and increases with the integration of energy storage in the Berserker Street Feeder. Integration of STATCOM decreases current harmonic distortion from 3.79 to 3.04%, while integration of energy storage increases current harmonic distortion from 3.79 to 5.35% in zone 3, feeder 1 as shown in Fig. 10.16. The total current harmonic distortion is lower in case C with the integration of both STATCOM and energy storage compared to the base case. The total current harmonic distortions across the zone transformer are 3.8, 1.39, 5.23, and 3.34% for the base case and cases A, B, and C, respectively.

Integration of STATCOM reduces total voltage harmonic distortion of the Berserker Street Feeder significantly as shown in Fig. 10.17 for case A. Voltage

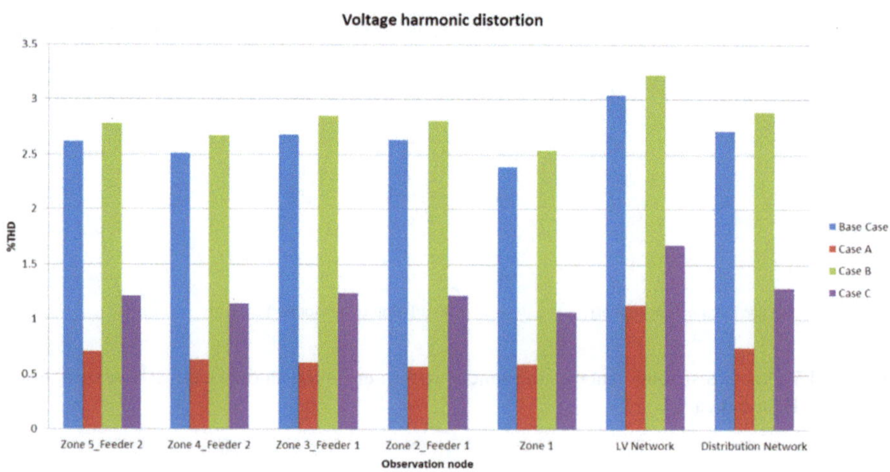

Fig. 10.17 Voltage harmonic distortion of the feeder

harmonic distortions in zone 5, feeder 2 are 2.61 and 0.71 % for the base case and case A, respectively. However, integration of energy storage into the base case increases voltage harmonic distortion, and voltage harmonic distortions in zone 5, feeder 2 are 2.61 and 2.78 % for the base case and case B, respectively. Integration of both STATCOM and energy storage reduces the voltage harmonic distortion compared to the base case. Voltage harmonic distortions in the LV network of the feeder are 3.04, 1.13, 3.22, and 1.29 % for the base case and cases A, B, and C, respectively.

Therefore, it can be concluded that STATCOM plays an active role in reducing both current and voltage harmonics in the Berserker Street Feeder as well as in the Rockhampton power network.

10.4 Conclusion

From the modelling analyses, it has been evident that the integration of RE causes uncertainties in the Rockhampton power network and causes adverse influences on voltage regulation, DT loading, phase unbalance, and harmonics injection into the network. From model analyses it is evident that unbalanced RE generation causes significant uncertainties in the feeder voltages, and for some cases exceeded the lower allowable limit of −6 %. Voltage and current harmonic distortion was less in the centralised connection approach than the decentralised connection approach. From the results, it was evident that an approach using a combination of centralised and decentralised connections is a better solution both for voltage regulation and harmonic distortion.

From the results, it was evident that the integration of an optimised STATCOM improves the voltage regulation of the distribution network and reduces the reactive power demand of the network; this in turn improves the poor power factor regulation and reduces the total voltage and current harmonic distortion significantly. On the other hand, the integration of energy storage improves the voltage regulation, reduces active power demand from the grid and improves transformer utilisation of the distribution network. Integration of both STATCOM and energy storage enhances the overall power quality of the Rockhampton power network as it enhances voltage regulation and improves power distribution and transformer utilisation and reduces total harmonic distortion of the power network. The outcome of the study is expected to be used as a guideline to the utilities for integrating large-scale renewable energy sources into the grid.

References

1. Guo Y, Hill DJ, Wang Y (2001) Global transient stability and voltage regulation for power systems. IEEE Trans Power Syst 16:678–688
2. Glover JD, Sarma MS, Overbye TJ (2012) Power system analysis and design, 5th edn. Cengage Learning, Stamford
3. Shafiullah GM, Oo AM, Stojcevski A (2013) Voltage regulation in renewable energy integration into thy distribution network. In: Proceedings of the Australasian Universities Power Engineering Conference (AUPEC), Hobart, Tasmania, Australia, 29 Sept–3 Oct 2013
4. Eltawil MA, Zhao Z (2010) Grid-connected photovoltaic power systems: technical and potential problems—a review. J Renew Sustain Energy Rev 14:112–129. (Elsevier)
5. Albarracin R, Amaris H (2009) Power quality in distribution power networks with photovoltaic energy sources. In: Proceedings of the 8th International Conference on Environment and Electrical Engineering. IEEE, Karpacz, Poland, 10–13 May 2009
6. Latheef A, Robinson DA, Gosbell VI, Smith V (2006) Harmonic impact of photovoltaic inverters on low voltage distribution systems. In: Proceedings of the 12th International Conference on Harmonics and Quality of Power, Lisbon, Portugal, 1–5 Oct 2006
7. EPRI report (1990) EL-6754 photovoltaic generation effects on distribution feeders, vol 1: description of the Gardner, Massachusetts, Twenty-First Century PV Community and Research Program, March
8. Asano H, Yajima K, Kaya Y (1996) Influence of photovoltaic power generation on required capacity for load frequency control. IEEE Trans Energy Convers 11:188–193
9. Fekete K, Klaic Z, Majdandzic L (2012) Expansion of the residential photovoltaic systems and its harmonic impact on the distribution grid. J Renew Energy 43:140–148 (Elsevier)
10. Chant TI, Shafiullah GM, Oo AMT, Harvey B (2011) Impacts of increased photovoltaic panel utilization on utility grid operations—a case study for Central Queensland. In: Proceedings of the Innovative Smart Grid Technologies Conference (ISGT 2011), Perth, Australia, 13–16 Nov 2011
11. EL-Shimy M, Badr MAL, Rassem OM (2008) Impact of large scale wind power on power system stability. In: Proceedings of 12th International Middle-East Power System Conference (MEPCON 2008), Cairo, Egypt, 12–15 March 2008
12. Muljadi E, Butterfield CP, Yinger R, Romanowitz H (2004) Energy storage and reactive power compensator in a large wind farm. In: Proceedings of 42nd AIAA Aerospace Science Meeting and Exhibit, Reno, Nevada, 5–8 Jan 2004
13. Papathanassiou SA, Papadopoulos MP (2006) Harmonic analysis in a power system with wind generation. IEEE Trans Power Deliv 21:2006–2016
14. Muljadi E, Butterfield CP (2005) Self excitation and harmonics in wind power generation. In: Proceedings of 43rd AIAA Aerospace Science Meeting and Exhibit, Reno, Nevada, 10–13 Jan 2005
15. Musunuri S, Dehnavi G (2010) Comparison of STATCOM, SVC, TCSC, and SSSC performance in steady state voltage stability improvement. In: Proceedings of North American Power Symposium (NAPS), MS, USA, 26–28 Sept 2010
16. Sannino J, Svensson TL (2003) Power-electronic solutions to power quality problems. J Electr Power Syst Res 66:71–82 (Elsevier)
17. Yuvaraj V, Deepa SN, Rozario APR, Kumar M (2011) Improving grid power quality with FACTS device on integration of wind energy system. In: Proceedings of 5th Asia Modelling Symposium (AMS), Kuala Lumpur, Malaysia, 24–26 May 2011
18. Miller NW, Zrebiec RS, Delmerico RW, Hunt G (1996) Design and commissioning of a 5 MVA, 2.5 MV battery energy storage. In: Proceedings of IEEE Power Engineering Society Transmission and Distribution conference, Los Angeles, CA, USA, 15–20 Sept 1996
19. Ra Jambal K, Umamaheswari B, Chellamuthu C (2005) Steady state analysis of grid-connected fixed-speed wind turbines. Int J Power Energy Syst 25:230–236 (ACTA Press)

20. Kook KS, Liu Y, Atcitty S (2006) Mitigation of the wind generation integration related power quality issues by energy storage. J Electr Power Qual Util 12:77–82
21. PSS SINCAL, PSS Product Suite, Siemens. http://www.energy.siemens.com/us/en/services/power-transmission-distribution/power-technologies-international/software-solutions/pss-SINCAL.htm
22. Ergon Energy A local Queensland based distribution company. http://www.ergon.com.au
23. Bureau of Meteorology, Weather Data (2011) Australian Government, Melbourne. http://reg.bom.gov.au/. Accessed 3 June 2011

Part III
Policy 2014

Chapter 11
Smart Grid and Intelligent Office Buildings: Virtual Power Plants—The Basis for the Optimal Use of Renewable Energy Sources

Kennedy Aduda, Wim Zeiler and Gert Boxem

Abstract Electricity energy generation and its supply through electricity networks are mainly organized in a top-down, centralized manner. Energy consumption can be predicted quite accurately at a high level, and this forms the basis for prescheduling the production by large power plants. Only few actors are involved in the generation, trade, and transportation of electricity, but this is changing rapidly. The increasing share of decentralized renewable energy conversion in combination with the new types of consumers will drastically alter the operation of electricity systems. Office buildings will become a potential source of energy flexibility which can be offered to the grid as a virtual power plant (VPP). In order to minimize uncertainty in the balance between energy supply and demand, it is necessary to develop realistic user behavior, installations behavior, and smart grid interaction. Monitoring the needs and preferences of users is necessary to predict future states of the demand for the smart energy systems (SES; e.g., based on weather forecasts and user behavior). The consumer of energy can become a producer due to the decentralized renewable energy conversion. To control this dynamic and interactive process automated prosumer support is needed to optimize interaction between offices and the smart grid. The chapter describes the first steps towards such systems.

Keywords Building design · Conceptual design

11.1 Introduction

As concerns grow about the environmental cost and limited supply of fossil energy resources, so does the importance to society of carefully managing the energy resources available and of developing and implementing alternative energy sources. The built environment is currently a major consumer of fossil energy, but it also has huge potential to contribute to the supply and management of renewable energy.

W. Zeiler (✉) · K. Aduda · G. Boxem
Faculty of the Built Environment, University of Technology Eindhoven, Eindhoven, Netherlands
e-mail: w.zeiler@bwk.tue.nl

A. Sayigh (ed.), *Renewable Energy in the Service of Mankind Vol II*,
DOI 10.1007/978-3-319-18215-5_11

The energy consumption of the households has decreased by 10 %; however, in contrast, the energy consumption of the office buildings has slightly increased by nearly 20 % due to higher comfort demand especially for cooling and ventilation. In the Netherlands, there are around 78,000 office buildings representing a total floor area of 46 million m^2 [1]. Offices use in total around 1420 MJprimair/m^2 compared to around 880 MJprimair/m^2 for households in the Netherlands [1].

Overall, both the electricity use and gas consumption of office buildings are increasing slightly, despite the 2020 targets set by the EU, which call for a 20 % reduction in energy use by the year 2020 [1]. Analysis of the nearly 400 projects uploaded to the CarbonBuzz database demonstrates that, on average, most buildings consume 1.5 and 2.5 times that of the declared design stage calculations, and in the case of offices this anomaly largely arises from increased electrical energy use [2]. Buildings and building services systems must be operated in a robust way, meaning that uses of the building other than those intended by occupants must not result in great variations of the energy consumption or indoor environmental conditions.

Smart adaptive control of energy consumption and generation inside (nanogrid) and around buildings (microgrid) can provide major contributions to address the imminent energy problems within the total energy infrastructure (electricity as well as the gas distribution). Large-scale integration of renewable energy sources (RES), being widely dispersed and of a highly stochastic nature, challenges the current power system in balancing power supply and demand at all times. The stochastic nature of renewable production and its negative impact on system balancing is depicted in the following figure ([3]; Fig. 11.1).

Breakthroughs need to be achieved in the field of process control for storage, demand, and distribution of heat and electricity. New approaches are needed to reduce the energy demand of existing buildings in the built environment, especially offices.

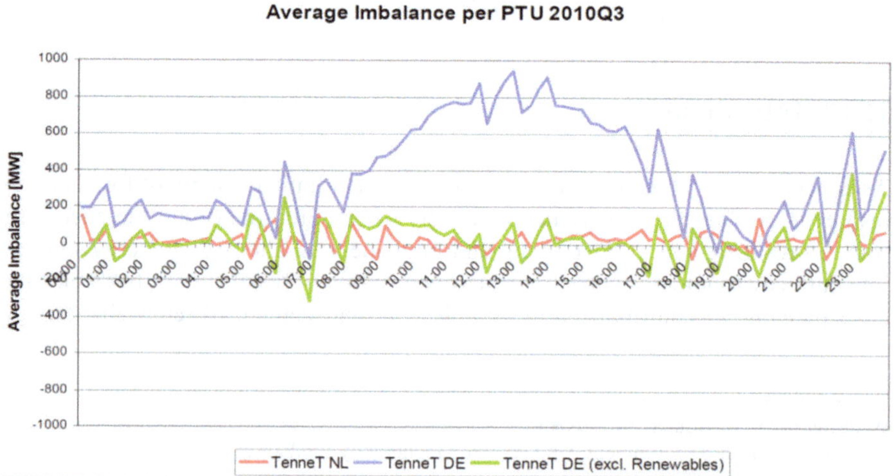

Fig. 11.1 Impact of renewable production on the average imbalance. *PTU* programme time unit. [3]

Improved process control and the use of distributed energy resources (DER), including local RES, greatly contribute to a reduction in energy demand [4]. Fluctuations of demand as well as in the supply of RES might result in instability of the supply system in specific situations of electricity production as the electrical grid has essentially no storage [5].

11.2 Methodology: Virtual Power Plant Concept

The increasing share of DER gave rise to the virtual power plant (VPP) concept [6], which aggregates many small capacity generation, storage, and demand units into one larger system. The VPP can operate as a single entity [7] collectively run by a central control system [8], while their total capacity can be comparable to a conventional power plant [9]. A VPP is primarily based on advanced information communication technology (ICT) infrastructure, providing two-way communication with the DER units for effective operation and optimal use [10, 11]. Energy storage, in the form of a local virtual power plant (LVPP), could reduce uncertainty within energy management and process control in a distributed but coordinated manner and on a critical scale, thereby addressing supply and demand balancing concerns. Instability of the electrical grid due to the supply issues as well as the unreliability of specific types of RES might lead to blackouts with enormous social and economic effects. To stabilize the smart grid, office buildings can act as LVPP [12]; see Fig. 11.2.

LVPPs can participate in the energy balancing market by employing the available RES units, storage devices, and controllable loads (heating, cooling, ventilation,

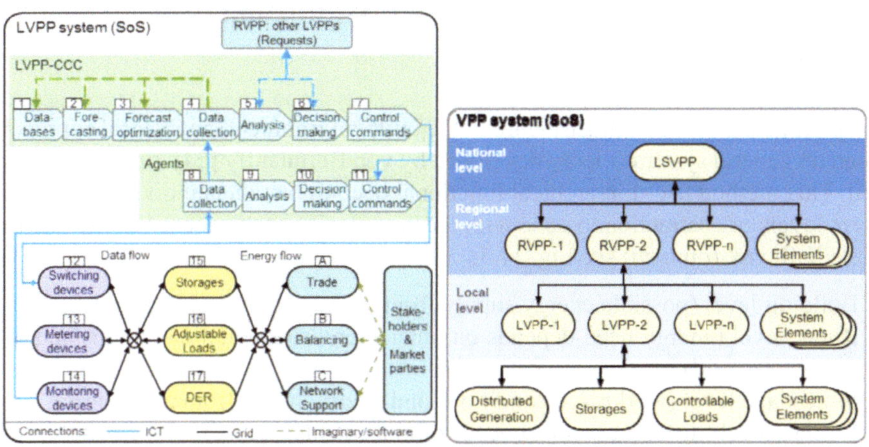

Fig. 11.2 Modular local virtual power plant (LVPP) and scalable design of LVPP system. Central Control Computer (*CCC*) closed-cycle cooling, *DER* distributed energy resources, *ICT* information communication technology, *LSVPP* large-scale virtual power plants, *RVPP* regional virtual power plant, *SoS* system of systems, *VPP* virtual power plant. [12, 13]

lighting, and power demand) on different time scales [14]. These modular and scalable LVPPs can be clustered into regional or even large-scale virtual power plants (RVPP or LSVPP); see Fig. 11.2 [13]. The LSVPP for commercial purposes, commercial virtual power plant (CVPP), is an entity that takes the aggregated DER to the market in order to buy energy but also to provide ancillary services, such as tertiary reserve or reactive power capacity.

In Europe there are a large number of on-going research projects on smart electricity distribution network solutions and ICT systems for energy efficiency, such as: Integral, flexibEL, CRISP, GREENCOM, I3RES, INERTIA, INTREPID, SMARTC2NET, SMARTHG, E-HUB, GreenerBuildings, Adapt4ee, KnohollEM, EnPROVE, ADDRESS, and STARGRID. An overview of the lessons learned and current developments can be found in the GeSI reports SMART [15] and SMARTer [16], and Giordano et al. [8, 17]. The main focus in the majority of these projects is on the combination of physical and virtual energy storage capacity in and around buildings in an attempt to stabilize the smart grid. However, a key component, a clear integral demand driven approach, is missing. An important factor of energy consumption in buildings is rather unknown: the occupancy behavior [18]. Building energy management systems (BEMS) operate on the level of occupants. However, since traditional BEMS lack intelligent reasoning to optimize process control [19], multi-agent system (MAS) technology will be used additionally in order to cope with all of the dynamic influences, internal as well as external, on buildings in LVPPs. Starting from concepts such as PowerMatcher [20, 21] or HeatMatcher [22, 23] more refined models are used to include all functional levels of buildings and building services to make optimal use of all energy storage possibilities for energy balancing.

11.3 Integral Approach: Top-down, Middle-out, and Bottom-up

To optimize the energy infrastructure in the built environment, an integral approach based on general systems theory developed by von Bertalanffy [24] is proposed [25] [26]. This method uses hierarchical functional decomposition and division into different levels of abstraction to cope with the complexity of the energy infrastructure of the built environment; see Fig. 11.3:

- Building level (possible energy supply from microgrid, nanogrid, and RES)
- Room level (energy need depends on outside environmental conditions and internal heat load)
- Workplace level (workplace conditions and energy needs from appliances)
- Human level (different comfort needs of individuals)

The process control infrastructure within office buildings is handled by BEMS. The functionality of BEMS will be extended through a middle-out approach to define necessary MAS modules. Currently, they are limited in that they do not consider the

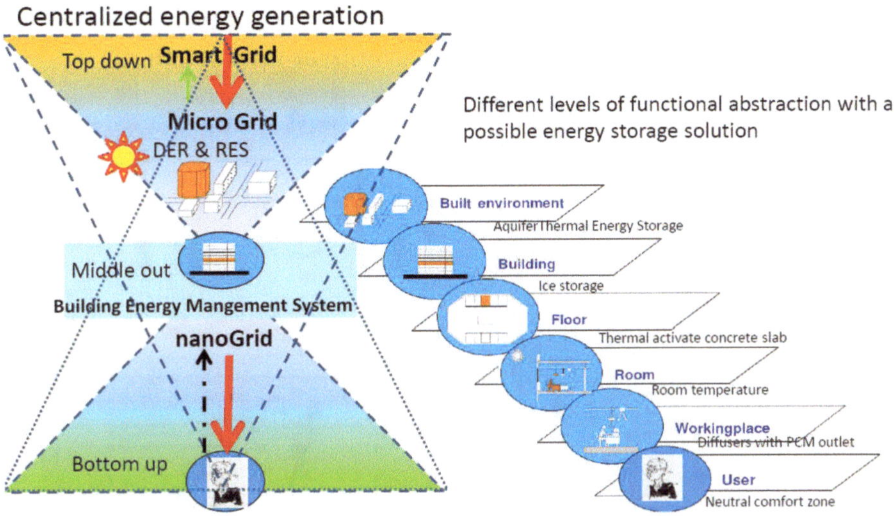

Fig. 11.3 Representation of approach for optimizing building interaction with the smart grid. *DER* distributed energy resources, *RES* renewable energy sources, *PCM* phase change material

individual end-user's consumption behavior, and it is problematic to integrate them into advanced autonomous and decentralized agent-based control technologies [27] [28]. As mentioned by Bloem and Strachan [18] a top-down approach could give the boundaries for energy consumption related to occupancy behavior. Identifying the specific building energy consumption from available measured data could support the optimization of energy balancing. In the longer term, bottom-up research should give more insight in important aspects related to occupancy behavior in a wider urban-related energy consumption context (including transport, living–work relationships) [18].

There is a different focus on the processes that occur in the building, which also depends on the strategy that is leading: bottom-up (user orientated), middle-out (building services systems orientated), and top-down (smart grid). Based on each of these approaches, the results and insights are used to specify specific functionalities for the agents of the multi-agent platform, see Fig. 11.4, a semantic representation which is based on Kolokotsa et al. [29] and Kofler et al. [30].

11.4 Discussion and Conclusions

The insights gained from the modeling of the energy demands on the different levels of a building and its surroundings lead to a concept for the monitoring and management of the energy flows in the nanogrid and microgrid in a more detailed and accurate way. The viability of these solutions will be investigated and demonstrated in real office situations. First, simple representations of the process will be made

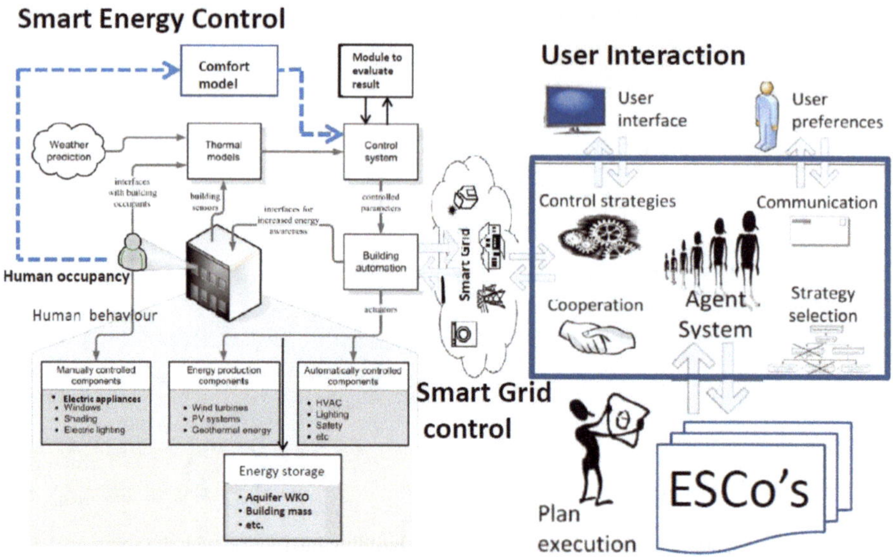

Fig. 11.4 Functional orientation of the smart energy control, smart grid control, and user interaction. *PV* photovoltaic; *HVAC* heating, ventilating, and air-conditioning; *WKO warmte/koude opslag* (heat/cold storage); *ESCo* energy savings company

to look into the interrelations between the different levels within a building system, with a specific focus on the smart grid, microgrid, and nanogrid:

- Building level (possible energy supply from the smart grid, microgrid, and renewable energy generation within the building)
- Room level (energy exchange depending on the outside environmental conditions and internal heat load)
- Workplace level (workplace conditions and energy need from appliances)
- User level (defining the different comfort needs of individuals and the resulting energy demands)

A so-called Smart Grid—Top-down Operation to Optimize distributed Energy Resources (SG-TOOSTER) agent-based platform focuses primarily on balancing local supply–demand but also on being sufficiently flexible to integrate with other smart grid functions and services. This platform can be extended with: (i) forecasting capability within the built environment, (ii) identifying flexibility of the building to compensate for the uncertainty, and (iii) enabling the LVPP concept from bottom-up to balance local power/energy supply and demand. Principled statistical methodology will be used to analyze and model the data streams emanating from sensors and actuators. The predictive models, grounded in rigorous statistical methodology so that it becomes possible to quantifiably determine the uncertainty of the predictions, will be implemented within the framework of a learning and adaptive MAS [31, 32]. The choice of a MAS is motivated by the need to ensure flexibility and robustness. Firstly, individual agents can be assigned to be in charge of distinct subsystems (e.g., based on location or objective). This simplifies the expansion of the monitoring and control system over time as the introduction of new functionalities leaves the configuration of the existing system intact and simply requires the addition of a new agent. Secondly, in view of the complexity and heterogeneity of the input data, it is advantageous to apply a wide range of modeling and learning strategies (e.g., regression and time series models as well as rule- or case-based models). Such widely different approaches are most naturally encoded by different agents, each of which communicates its results to a set of supervising agents which then, based on the estimated reliability of the individual decisions, fuse the information to reach a final decision. It is to be expected that over time it will become clear that particular strategies are more successful than others. Again, this can most easily be accommodated in a multi-agent platform setting where agents implementing unsuccessful approaches will be removed or downgraded from the pool, whereas agents with excellent performance will be boosted and diversified. Research items are explained in detail as follows:

- Forecasting: Due to the highly nonlinear and varied nature in local environments, statistical methods based on historical data and regression analysis are used to predict energy consumption. Polynomial approximations with their nonlinear mapping capability can partially address requirements of the bottom-up context. Recently, computational intelligence (CI) has been extensively investigated in many domains and is becoming a proven means for forecasting renewable energy production and energy consumption [33]. Artificial neural network (ANN) is a popular CI technique, which is able to map any nonlinear function. A properly trained ANN-based short-term forecasting tool is able to achieve very accurate approximations of the mapping of stochastic data. Learning techniques of CI can enrich local information that is highly uncertain due to end-user's behaviors and stochastic due to increasing on-site RES and DER units [34, 35].
- Flexibility will be created by the multi-agent platform by integrating a number of flexible components including thermal storage. Such devices are crucial to contribute in mitigating uncertainty within the built environment and smart energy system (SES) at large. Depending on their location, functions, and physical constraints, such devices can offer a certain degree of flexibility. Their interaction with different targets calls for multi-objective optimization. Multi-objective particle swarm optimization (MOPSO) is an advanced CI technique dedicated to such problems [36] which enables to exploit optimal potential flexibility from the built environment to reduce SES uncertainty.

The multi-agent platform acts as an aggregator and is the core element of the LVPP concept both in the built environment (smart grid and microgrid) and in buildings (nanogrid). Due to the nature of the built environment, that is, including different levels, zones, and layers, with different physical components, the overall control system of BEMS is quite complex with quite different time reposes for different functions and services. Therefore, a double control layer for the agent–aggregator platform was introduced that makes the LVPP responses with the direct demand control signal in a short time interval and with the energy storage-based control signal in a longer time interval. The former depends on occupancy detection to meet a certain comfort level of end users, whereas the latter aims to optimize the energy usage while taking into account flexibility of the built environment.

References

1. Cloquet R (2012) Kwantitatief data onderzoek Energie–innovaties, Agentschap NL projectnummer BA7156-100-100, publicatienummer BL-EH20112840, Januari 2012, versie 4
2. Kimpian Ma J (2013) Fantasy vs reality. CIBSE J June 2013:22–24
3. TenneT, 2011, Imbalance management TenneT analysis report, Version 1.0 28-04-2011
4. El Bakari KE, Myrzik JMA, Kling WL (2009) Prospects of a virtual power plant to control a cluster of distributed generation and renewable energy sources. Proceedings of the 44th International Universities Power Engineering Conference
5. MacCracken M (2010) Energy storage, providing for a low-carbon future. ASHRAE J 2010:28–34
6. Ioakimidis CS, Oliveira L, Geniskonisakis KN, Rycerski P (2013) A virtual power plant with the use of the energy box in a smart grid concept. Renew Energy Power Qual J (1)
7. Peik-Herfeh M, Seifi H, Sheikh-El-Eslami MK (2013) Decision making of a virtual power plant under uncertainties for bidding in a day-ahead market using point estimate method. Int J Electr Power Energy Syst 44:88–98
8. Giordano V, Gangale F, Fulli G, Jiménez MS (2011) Smart grid projects in Europe: lessons learned and current developments, JRC reference report, Petten, The Netherlands
9. Molderink A, Bakker V, Bosman MGC, Hurink JL, Smit GJM (2010) Management and control of domestic smart grid technology. IEEE Trans Smart Grid 1:109–119
10. Lombardi P, Powalko M, Rudion K (2009) Optimal operation of a virtual power plant. IEEE Power and Energy Society General Meeting, PES 2009, 26–30 July, Calgary
11. Asmus P (2010) Microgrids, virtual power plants and our distributed energy future. Electr J 23:72–82
12. El Bakari KE, Kling WL (2010) Virtual power plants: an answer to increasing distributed generation, 2010 IEEE PES Innovative Smart Grid Technologies Conference Europe (ISGT Europe)
13. El Bakari KE, Kling WL (2012) Fitting distributed generation in future power markets through virtual power plants, 9th International Conference on the European Energy Market (EEM)
14. Pennings LWHA (2009) Multi agent system to optimize comfort and energy flows in the built environment. MSc Thesis, Technische Universiteit Eindhoven
15. GeSI (2008) SMART 2020. http://www.gesi.org/LinkClick.aspx?fileticket=tbp5WRTHUoY%3D&tabid. Accessed 22 July 2015
16. GeSI (2012) GeSI SMARTer 2020: the role of ICT in driving a sustainable future, global e-sustainability initiative. Aisbl and the Boston consulting group, Inc. http://gesi.org/assets/js/lib/tinymce/jscripts/tiny_mce/plugins/ajaxfilemanager/uploaded/SMARTer2020.pdf. Accessed 22 July 2015
17. Giordano V, Meletiou A, Covrig CF, Mengolini A, Ardelean M, Fulli G, Jiménez MS, Filiou C (2013) Smart grid projects in Europe: lessons learned and current developments, JRC Scientific and policy reports, Luxembourg
18. Klein L, Kwak J, Kavulya F, Becerik-Gerber B, Varakantham P, Tambe M (2012) Coordinating occupant behaviour for building energy and comfort management using multi-agent systems. Autom Constr 22:525–536
19. Kok K, Warmer C, Kamphuis R (2005) PowerMatcher: multiagent control in the electricity infrastructure. Proceedings AAMAS 2005, International conference on autonomous agents and multiagent systems, volume industry track, New York

20. Kok K (2013) The PowerMatcher: smart coordination for the smart electricity grid. PhD thesis, Vrije Universiteit Amsterdam
21. van Pruissen OP, Kamphuis IG, Boxem G, Zeiler W, Wortel W, Velden JAJ van der (2009) Design of an agent architecture based on the power matcher approach. CISBAT Lausanne
22. van Pruissen OP, Kamphuis R (2010) Multi agent building study on the control of the energy balance of an aquifer. Proceedings improving energy efficiency in commercial buildings, Frankfurt a. M.
23. Blanchard BS, Fabrycky WJ (2005) Systems engineering and analysis, 4th ed. Pearson Prentice Hall, Upper Saddle River
24. Savanović P (2009) Integral design method in the context of sustainable building design. PhD thesis, Technische Universiteit Eindhoven
25. Zeiler W, Savanović P (2009) General systems theory based integral design method, proceedings ICED'09, Stanford, USA
26. Wang S (2013) Intelligent building electricity demand management and interactions with smart grid. Proceedings Clima 2013, Prague
27. Hurtado LA, Nguyen PH, Kling WL, Zeiler W, (2013) Building energy management systems-optimization of comfort and energy use. In 48th International Universities' Power Engineering Conference (UPEC) IEEE, pp 1–6
28. Bloem JJ, Strachan P (2012) Evaluating and modelling nearly-zero energy buildings; are we ready for 2018? Proceedings IEECB, Frankfurt a. M.
29. Kolokotsa D, Rovas D, Kosmatopoulos E, Kalaitzakis K (2011) A roadmap towards intelligent net zero- and positive-energy buildings. Sol Energy 85:3067–3084
30. Kofler MJ, Reinisch C, Kastner W (2012) A semantic representation of energy-related information in future smart homes. Energy Build 47:169–179
31. Ramchurn SD, Vytelingum P, Rogers A, Jennings NR (2011) Agent-based control for decentralized demand side management in the smart grid. Proceedings 10th international conference on autonomous agents and multi-agent systems (AAMAS 2011)
32. Mamidi S, Chang YH, Maheswaran R (2012) Improving building energy efficiency with a network of sensing, learning and prediction agents. Proceedings 11th international conference on autonomous agents and multi-agent systems, Valencia, Spain
33. Venayagamoorthy G, Rohrig K, Erlich I (2012) One step ahead: short-term wind power forecasting and intelligent predictive control based on data analytics. IEEE Power Energy Mag 10(5):70–78
34. Vonk BMJ, Nguyen PH, Grand MOW, Slootweg JG, Kling WL (2012) Improving short-term load forecasting for a local energy storage system. Proceedings 47th Universities Power Engineering Conference (UPEC)
35. Ahmed B, Ampatzis M, Nguyen PH, Lopes Ferreira HM, Kling WL (2013) Mitigating imbalances from wind power by using an agent-based matching mechanism. Proceedings of the 48th Universities Power Engineering Conference
36. Reyes-Sierra M, Coello C (2006) Multi-objective particle swarm optimizers: a survey of the state-of-the-art. Int J Comput Intell Res 2(3):287–308
37. Abarrategui O, Marti J, Gonzalez A (2009) Construction the active European power grid. Online J Power Energy Eng 1(4):130–133
38. Kelly S, Michael Pollitt M (2011) The local dimension of energy, January 2011, CWPE 1114 & EPRG 1103
39. Macek K, Marik K, Stluka P (2011) Ontology-driven design of an energy management system. 21th European symposium on computer aided process engineering, ESCAPE, Chalkidiki, Greece
40. Nutaro J, Protopopescu V (2009) The impact of market clearing time and price signal delay on the stability of electrical power markets. IEEE Trans Power Syst 24(3):1337–1345
41. PowerMatcher website. www.powermatcher.net
42. Taskforce Intelligente Netten (2011) Op weg naar intelligente netten in Nederland, Einddocument van de Taskforce Intelligente Netten

Chapter 12
Survey of Renewable Electricity Tariffs in Iran

A. Kaabi Nejadian, Faramarz Separi, Mehdi Barimani Varandi and Mohamad reza Khaje Samakoosh

Abstract The protective tariff policy, investment and guaranteed purchase of electricity through renewable energy sources are some of the most important tools in Iran for developing electricity production. Here, we measure and compare the marginal cost of electricity production through renewable electricity in Iran, and survey the renewable electricity tariff policy in Iran. To have balanced development in renewable energy resources, researchers have calculated and recommended different tariffs for different technologies, i.e., solar thermal, wind and small hydropower. All our research measurements were performed using COMFAR software.

Keywords Tariffs policy · Renewable electricity

12.1 Introduction

Reforms in the electricity industry and privatization are essential to achieving sustainable energy in Iran. To attract investment and encourage private investment are the main factors in enhancing renewable electricity. Guaranteed purchasing of renewable electricity is the main condition for continued investment.

A. Kaabi Nejadian (✉)
Renewable Energy Organization of Iran, Ahvaz, Iran
e-mail: Kaabi@Iranenergy.org.ir

F. Separi · M. r. K. Samakoosh
Mazandran Distribution Power Company, Sari, Iran
e-mail: fa_separi@yahoo.com

M. r. K. Samakoosh
e-mail: Samakoosh2007@yahoo.com

M. B. Varandi
Mazandaran Regional Electric Company, Sari, Iran
e-mail: m.barimani@mazrec.co.ir

A. Sayigh (ed.), *Renewable Energy in the Service of Mankind Vol II*,
DOI 10.1007/978-3-319-18215-5_12

Clause 'A' of Article 44 of the Iranian Constitutional Law emphasizes the development of the government sector and lack of development of government establishments [1]. According to this clause, the government does not have the right to perform new economic activity outside the items at the top of Article 44, and is required to grant any activity (including continuation and utilization of previous activities) which is not included in the titles at the top of Article 44 to the private sector. It generally recommends private investment.

Hence, the government should provide the necessary conditions for the development of renewable energy in Iran's energy basket by the private sector by removing obstacles and providing suitable conditions for investment, subjecting policies, and setting proper requirements (market-oriented), realization of electricity tariffs and contracting long-term agreements for the guaranteed purchasing of electricity.

It should be noted that the protective tariff policy, investment and guaranteed purchase of electricity through renewable energy sources are some of the most important tools in Iran for developing electricity production. This policy was used successfully in Turkey, Canada, Germany, Denmark, America, and Spain, and also in other countries with the aim of supporting investments in renewable energy sources. Renewable energy has attempted to create a special fund [2].

For supporting sustainable energy and also for increasing the renewable energy quota of Iran's energy basket, researchers recommend the tariff for each of these technologies and sensitivity analysis of the principal cost and marginal cost of electricity production through every renewable source.

12.2 Materials and Methods

Benefit–cost analysis was used for analyzing and simulating the data in this study. This is the usual method for evaluating the economy, for measuring the marginal cost of producing units or the cost of electricity production units, which is the specified factor used for recommending different tariffs for buying guaranteed electricity and finally the equality of B/C for comparing.

The equivalent uniform annual cost (EUAC) and, in our case, the levelized cost of energy (LCOE) is used.

$$B/C = EUAB/EUAC \tag{12.1}$$

If $B/C \geq 1$, it is suitable for private investment and

If $B/C < 1$ it is not suitable for private investment.

In this method all the annual marginal costs are measured using a discounted rate ($i = 10$) to the reference year which is then distributed during the lifetime of the project.

The loan interest rate is the most important factor in making financial decisions; all measurements and comparisons and simulations use a 7 % interest rate loan from the National Development Box of Iran [3]. Furthermore, all our research indexes

Table 12.1 The method of calculation of some of the technical–economic parameters

Tax	(Income − cost of exploitation − capital amortization − loan interest) × tax coefficient
Annual net production energy	(Access coefficient × internal consumption − 1) × nominal capacity × production coefficient × 8760
Income	Tariff * annual net production energy
Annual amortization	Investment cost × plant lifetime/(depreciation coefficient − 1)
Cash flow	Income − maintenance cost − rebating loan origin − loan interest
Total investment	Total primary cost + interest of manufacturing period
Levelized cost of energy (LCOE)	Investment annual cost + annual maintenance and operation cost + (total annual gross energy produced by plant/annual cost of consumption fuel)

are measured with the aim of studying suitable plants for investment in the private sector (with IRRE = 20 %). All the measurements in this research are performed using *COMFAR software.*

In this study, benefit−cost analysis is used to measure and compare the marginal cost of electricity production through every renewable—solar thermal, wind and small hydropower—as well as the recommended tariffs for guaranteed electricity with regard to the B/C ratio. For measuring and comparing the annual benefit−cost related to every technology, we used NPV, IRRE, DPB, NPB, IRR criteria (refer to Appendix).

In this method, the LCOE is calculated as follows (Table 12.1):

$$\text{LCOE} = \text{AC} + \text{O \& M} + P_{\text{vf}} / E_{\text{out}} \tag{12.2}$$

In the following relationship, the method of calculation for each of the variables would be as follows:

AC Annual cost of investment ($)
O&M Annual cost of maintenance and operation ($)
P_{vf} Annual cost of consumption fuel ($)
E_{out} Total annual gross electrical energy produced by plant (Kwh)

Annual Cost of investment (AC) The annual cost of investment is a uniform cost which has a constant value throughout the life of the plant. To obtain the annual cost of investment, the investment return coefficient (CRF) should be multiplied by the total cost of the primary investment.

$$\text{AC} = \text{CRF} \times \text{C} \tag{12.3}$$

In the above relationship, C is the total primary investment as engineering, procurement and construction (EPC), and

$$\text{CRF} = i/1 - (1+i)^n \tag{12.4}$$

n is the lifetime of the plant and
i is the discount rate

Maintenance and Operation (O&M) The cost of maintenance and operation is considered as a percentage of annual cost.

Annual cost of consumption fuel (P_{vf}) The amount of annual cost of fuel is calculated using the following relationship:

$$P_{\mathrm{vf}} = \left(W \times h / R_{\mathrm{a}} \times \mathrm{NHv}\right) * P_{\mathrm{g}} \tag{12.5}$$

W	The plant power (MW)
R_a	Plant efficiency
P_g	Base price of consumed fuel ($)
NHv	Heat value of net fuel (MW)
h	Amount of work hours of the plant unit in a year $\left(h = \mathrm{CR} \times 8760\right)$
CR	Coefficient of access ability

Annual gross value of plant production (E_{out}) The total annual gross energy production by plant (Kwh):

$$E_{\mathrm{out}} = W \times \mathrm{CR} \times 8760 \tag{12.6}$$

The properties and technological information in selecting each type of technology are considered by experts and professionals in the electricity industry and the Renewable Energy Organization of Iran (SUNA)—solar thermal, wind and small hydropower.

12.3 Calculations and Recommended Tariffs for Guaranteed Renewable Electricity Shopping in Iran

The first and the most successful mechanism for the development of renewable energy throughout the world is the guaranteed purchase of electricity produced from renewable energy sources. This is because private investors perform feasibility calculations for the amount of electricity based on the financial economy, that is, to save or not to invest in this sector.

Iran's first electricity tariff guaranteed the purchase of renewable energy development with a renewable target of 2008 with a 5-year contract (for peak load hours and normal 13.13 $¢/kWh, and for fewer hours 9.10 $¢/kWh) [4]. In 2010, it appeared that the tariff base guaranteed purchase price was not attractive enough for private investors to enter the renewable energy sector.

In 2013, 5 years later according to the instructions in paragraph (b) of Article 133 of Iran's Fifth Five-Year Plan and the procedures for determining the purchase power rate of new and cleaner energy sources, the basic rate for the purchase of electricity for 5-year guaranteed contracts was as follows—guaranteed purchase rate of electricity from renewable energy sources is equal to:

The average price of energy converted into the electricity market per kWh

+

Saving fuel values for liquid–combined fuel per kWh

+

Saving values due to the lack of emissions − social cost per kWh.

Base price was calculated guaranteed purchase rate was determined and communicated of electricity from sources of renewable energy 17.77 $¢/kWh [5].

In order to assess the economic justifiability of power generation through renewable sources (solar thermal, wind and small hydropower) in 2014 we asked the following three questions relating to the benefit−cost ratio of the technologies under discussion:

1. Is the production of electricity through wind power plants justified in Iran?
2. Is the production of electricity through small hydropower plants justified in Iran?
3. Is the production of electricity through solar thermal power plants justified in Iran?

According to the tariff and the cost of electricity generated from renewable sources, LCOE in 2014 is shown in Table 12.2.

It should be noted that calculating electricity production from renewable sources using a 7 % interest loan to finance the foreign exchange reserves is considered to be a private investment. Therefore, to justify the economic viability tariff cost (LCOE) by these sources, the loan interest is emphasized by 7 %. With the tariff set by the government, it can be seen that generating electricity through solar thermal power plants is not economically feasible, although it might be an appropriate tariff for the other two options.

It is very important that the cost of producing electricity through investment in various technologies must be considered when determining tariffs for the guaranteed purchase of electricity to enable different tariffs for different technologies to be determined. The same tariffs for different technologies impede balanced development of this technology; the technology is more profitable for investors when more technologies are developed.

Thus, researchers from the Economic and Strategic Studies Expert Advisory of Renewable Energy Organization of Iran found that according to the unit cost of electricity generated through renewable sources (LCOE in 2014), different tariffs should be proposed for each of the technologies with the aim of creating economic justifiability to encourage private investment in this section. The benefit−cost ratio

Table 12.2 The benefit−cost ratio of renewable resources tariff in 2014 (calculated by researchers)

	Wind	Small hydropower	Solar thermal
LCOE ($¢/kWh)	9.42	9.40	20.96
Tariff ($¢/kWh)	17.77	17.77	17.77
Benefit−cost	1.80	2.03	0.91

Table 12.3 Calculations of cost price of renewable electricity from an investor's viewpoint in 2014 (calculated by researchers)

Calculation consumptions	Wind	Solar thermal	Small hydropower
Investment cost as *EPC* ($¢/kWh)	1600	4700	2300
Maintenance and operate cost ($¢/kWh)	0.8	1.3	2%
Coefficient of net production (%)	30%	30%	50%
Plant lifetime (years)	20	30	10
Cost price components from viewpoint of private sector	*Wind*	*Solar thermal*	*Small hydropower*
Fuel cost ($¢/kWh)	0.00	0.00	0.00
Maintenance and operating cost ($¢/kWh)	0.81	1.30	1.05
Constant brought cost ($¢/kWh)	1.07	2.85	1.28
The cost of installing facilities ($¢/kWh)	7.54	16.81	7.07
Social cost	0.00	0.00	0.00
Cost price—before tax (LCOE)	*9.42*	*20.96*	*9.40*
Minimum *tariffs*	*17.90*	*43.40*	*12.70*
Cost price—after tax	10.35	24.10	9.40

and the calculations for the cost price of renewable electricity from an investor's viewpoint in 2014 was calculated as shown in Table 12.3:

It can be seen that the proposed electricity tariff for the three technologies could be economically feasible. It is hoped that due to the proposed tariff for solar thermal power plants, which is relatively high, more power can be considered to increase the installed capacity of the plant, and its growth well as reduced in the future as fees an investment's candidate be raised to produce cheap electricity.

Wind power forecasting depends on the economic and social development of wind power. The research-based purchase price is expected to reflect the value of the high potential of wind power in the technology chain and gain further support from the government for wind power plants.

Investment policy of tariff protection and guaranteed purchase of electricity from renewable energy sources are the most important tools for developing electricity production; however, this policy has also been used successfully in Turkey Canada, Germany, Denmark, America, and Spain [6].

The policy of guaranteed electricity shopping and tariffs is affected by certain factors such as upstream documents, strategic programming, and the social and political conditions of the country. In Iran, the goals determined in upstream documents of a 20-year period outlook documentation and the law of the Fifth Development Program determining 10% contribution of electricity production through renewable energies until the end of the year 2016 and determining 50,000 MW electricity production via renewable energies until the end of the year 2016, would become a lever and the reason by which the banks and National Development Box represent appropriate facilities for investment in this sector and the government would pay this tariffs, and as a result the background of development of this technology for increasing contribution of this energy in Iran's energy basket.

12.4 Summary, Conclusion and Recommendations

The price and cost price of electricity production has particular importance in different economic sectors. Lack of cost coverage through tariffs as well as the lack of competitive markets and the lack of considerable activity in the private sector in investment in the electricity sector have endangered the continuous service of the electricity industry.

The problems associated with the cost of electricity production through renewable energies are investigated from two viewpoints:

According to investigative studies from supportive policies, the first and most successful mechanism of development of the renewable energies sector throughout the world is the guaranteed purchasing of the electricity produced through renewable energy resources, because the nongovernmental sector investor could be confident about receiving payment from the electricity sales, and can perform financial and economic feasibility calculations about whether it is economical to invest in this technology. Furthermore, it is evident that alternative technologies for producing electricity as well as the cost and tariff of purchasing electricity through production resources from renewable resources require different investments as they cannot be regarded as identical. Therefore, any viewpoint or law which is based on considering the uniformity of different technologies, based on wisdom and logic, is rejected and therefore ineffective (refer to Table 12.2). Assuming we have accepted guaranteed renewable electricity shopping, lack of confident resources results in doubt for the investor as well as the banks granting these facilities and reluctance to enter into these fields. If they are sure that the required budget for purchasing electricity is guaranteed by the government (supplying confident resources), they would be eager to invest in this sector.

Researchers report the following results and suggestions for developing and improving the field of electricity production through renewable energy resources in Iran. The goal is to increase the contribution of renewable electricity to Iran's energy basket and to achieve the goals determined in the 20-year vision document and Iran's 5-year Development Plan [6].

1. Consider justifiable production of renewable electricity with a loan interest of approximately 7%. it is recommended that to encourage private investment in this field, either the National Development Box grants loans to this sector or, similar to progressive countries in this industry, establishes a 'special box' for renewable energies in Iran.
2. It is suggested that to have a balanced development in renewable energy resources, the researchers calculate and recommend different tariffs according to Table 12.4 for different technologies—17.90 $¢/kWh for wind power plants, 12.70 $¢/kWh for small hydropower plants and 43.40 $¢/kWh for solar thermal plants—to be determined and notified.

Table 12.4 Researcher-proposed tariff and the benefit–cost ratio-proposed tariff in 2014 (calculated by researchers)

	Wind	Small hydropower	Solar thermal
LCOE ($¢/kWh)	9.42	9.40	20.96
Tariff ($¢/kWh)	17.70	12.70	43.40
Benefit–cost	1.84	1.91	2.27

12.5 Appendix

12.5.1 Measurements Related to Renewable Technologies (2014)

	Interest (%)	Loan payment period (years)	LCOE ($¢/kWh)	Tariff ($¢/kWh)	IRR (%)	IRRE (%)	NPB—total (years)	DPB—total (years)	NPB—equity (years)	DPB—equity (years)	Benefit–cost ratio
Solar thermal	7	8.5	*20.96*	*43.40*	11.98	22.30	9.16	18.37	11.08	13.20	*2.27*
Small hydropower	7	8.5	*9.40*	*12.70*	13.53	20.00	7.93	14.10	10.73	13.12	*1.91*
Wind	7	8.5	*9.42*	*17.90*	12.91	20.02	7.84	13.57	10.61	12.65	*1.84*

References

1. Parliament of Islamic Republic of Iran/Research Center/Law/Constitutional Law Article-44, http://rc.majlis.ir/fa/content/iran_constitution
2. Dabiri F et al (2013) Comparative study of the legal aspects of the development and use of renewable energy inefficiently in Iran and selected countries, 28th International Power System Conference, 4–6 Nov 2013, Tehran, Iran
3. Central Bank of the Islamic Republic of Iran (2014) Key economic indicators. Public Relations Department, Tehran. http://www.cbi.ir
4. Parliament of Islamic Republic of Iran/Research Center/Law. http://rc.majlis.ir/fa/law
5. Renewable Energy Organization of Iran, Private Sectors Cooperation Office, Establishment of Non-governmental Renewable Energy power plants (2014) http://privatesectors.suna.org.ir/fa/home
6. Report DOE (2009) A comparative study of OECD country for renewable energy technology. http://www.energy.gov/conferences/winter2002/gallery/jacobsson
7. Parliament of Islamic Republic of Iran/Research Center/Law/Iran's 20-year Vision Plan (2014) http://rc.majlis.ir/fa/law/show/132295

Chapter 13
Transitions to a Post-carbon Society: Scenarios for Western Australia

Martin Anda, Martin Brueckner and Yvonne Haigh

Abstract Pathways towards a post-carbon society are being explored across all levels of government, within the scientific community and society in general. This chapter presents scenarios for cities and regions in Australia after the Age of Oil, particularly the energy-intensive state of Western Australia (WA). It argues that a post-carbon WA would ideally use technological and wider social choices to reduce carbon emissions close to zero. It focuses on policy requirements, institutional and governance arrangements and socio-technical systems to provide an industry-focussed renewable energy development plan that will help to balance ongoing and past emissions and lead to a low-carbon society.

Keywords Post-carbon society · Renewable energy · Scenarios · Western Australia

13.1 Introduction

The realisation that the Age of Oil is declining is forcing governments, science and citizens to consider options for a future without the dependency on fossil fuels that produced carbon emissions and caused ocean acidification and climate change during the Industrial Era. Having reached 'peak oil' and a critical point of 400 ppm of greenhouse gas emissions in 2013, the political thrust around global environment issues now aspires to transform the traditional fossil fuel-based economy into a low-carbon and sustainable world. A post-carbon society would achieve a balance in the global carbon cycle with additional sinks and offsets for past emissions. Ideally, a post-carbon society would have used technological and wider social choices to reduce carbon emissions to zero in different locations across the world.

In Australia, various scenarios have been proposed: solar-hydrogen energy and transport systems, nuclear fusion electric power and transportation, biomass-to-

M. Anda (✉)
Centre for Responsible Citizenship and Sustainability, Murdoch University,
90 South Street, 6150 Murdoch, Western Australia
e-mail: m.anda@murdoch.edu.au

M. Brueckner · Y. Haigh
School of Management and Governance, Murdoch University, Perth, Western Australia

A. Sayigh (ed.), *Renewable Energy in the Service of Mankind Vol II*,
DOI 10.1007/978-3-319-18215-5_13

energy where carbon is cycled through plantations, algae farms and combustion for energy production. In practice, at least for the foreseeable future, remaining and past emissions may be balanced with carbon capture and storage (CCS), biosequestration and other methods. All of these scenarios require a range of government support, variable policy setting, broad-ranging governance arrangements, institutional capability and the support of the broader public.

The Western Australian population of 2.5 million people supports diverse communities and settlement types across an area ten times greater than the UK with 1/30 of the population. The mining, housing, agriculture and land management sectors are powerful drivers of an economy that has enabled Western Australians to have one of the highest carbon footprints per capita on the planet at approximately 33 t per annum [1].

The purpose of this chapter is to provide an overview of international, national and local initiatives and approaches that are leading towards a post-carbon society. It will draw on these examples to highlight current advances and identify relevant approaches for the Western Australian context. The chapter is presented in four sections in order to highlight the imperative of the intersection between science, government policy and civil society: policy implications, governance arrangements, community-based initiatives and socio-technical systems.

13.2 Policies for Transition to Post-carbon Society

The transition to a post-carbon society requires governments, citizens and businesses work collaboratively to ensure our dependency on fossil fuels is indeed challenged. While the scientific environment highlights the raft of challenges, the policy context is integral to our move to a post-carbon society. More specifically, a post-carbon society requires policymakers to draw on lessons from a micro-policy environment, which often focuses on a localised or small-scale policy success, such as pollution licensing, tourism development [2] and marine protection areas ([3]; see also the list developed by [4]). Developing policy for a post-carbon society requires policymakers to acknowledge the fact that policy functions on three levels—macro, meso and micro. While the micro approach addresses the day-to-day policy management of programmes and policy plans, without due attention to the meso and macro levels of policy development, policy that aims to provide the basis for a post-carbon society would more than likely fail [5].

The policy advocacy coalition framework (ACF) [6] provides a sound basis to begin exploring a range of policy options that would assist with the transition to a post-carbon society. The ACF emphasises the importance of stakeholder actions—that is participants, local, state, federal and across borders in order to build collaborative frameworks that enhance policy decision-making. In particular, this approach provides a basis to develop policies that target the advancement of society due to the emphasis on the processes of human decision-making. For example, this framework provides the basis from which to embed the role of technological innovation in the policy context, such as the contribution of photovoltaic panels in the generation of electricity [4]. The ACF also provides a platform for identifying the political and social barriers that exacerbate institutional and political stagnation [7, 8]. As all policies are developed by people who operate within political, institutional and social contexts, identifying and examining barriers, bias and interdependencies is integral to developing effective policy outcomes. Moreover, as a framework for developing policy options, this framework encourages communication between governments, communities and businesses in order to engender creative solutions that will challenge our dependence on fossil fuels.

13.2.1 Policy Implications for WA

Accordingly, the policy framework for post-carbon Western Australia (WA) would function on the three levels as follows:

1. National macro policy would connect WA through the Council of Australian Governments (COAG) to the global emissions trading and reporting protocols through United Nations Framework Convention on Climate Change (UNFCCC).
2. The meso-level policy development will maintain and improve national legislative instruments such as National Greenhouse and Energy Reporting Scheme (NGERS), Mandatory Renewable Energy Target (MRET) and financing mechanisms as well as provide substantial national regulatory policies to enable inter- and intrastate transmission networks to which the new utility-scale renewable energy systems would be connected.
3. The micro-level addresses the day-to-day policy management of programmes and policy plans such as enabling public–private partnerships between local governments and the private sectors to develop localised carbon reduction infrastructures such as resource recovery for methane production and between regional governments, network operators and utilities to develop local, utility-scale renewable power stations.

13.3 Governance Structures for a Post-carbon Society

The complexity of the problems facing society's use of fossil fuels requires a collective approach that does not assume an optimal solution ([9], p. 5). In the absence of *one-best-way* for addressing climate change a selection of promising climate initiatives is presented below. These initiatives—involving a broad spectrum of different actors from across different sectors—can be seen as a part of a multitier climate governance system which ranges from multilateralism, transnational alliances and national policy solutions to regional and local self-governance experiments. It is this amalgam of interrelated approaches that can be seen as a key ingredient for future post-carbon governance; a governance system that involves a network of actors across all scales and sectors geared towards the management of the large socio-technical changes a post-carbon transition would entail.

13.3.1 Multilateral Agreements

The multilateral process offers the opportunity for progress to be monitored and for the development of accounting systems that all countries follow for measuring and reporting [10]. Negotiations under the UNFCCC, therefore, continue to be the principal vehicle for reaching a global consensus for action on climate change and for fostering collaboration between nation states [11]. Initiatives at the United Nations (UN) level can be complemented by commitments made through bilateral, regional and other multilateral forums such as the European Union (EU). To illustrate, under the UNFCCC's 1997 Kyoto Protocol, the 15 EU member countries committed to reducing their collective emissions to 8 % below 1990 levels by the years 2008–2012. The EU has recently committed to cutting its emissions by 2020 to 20 % below 1990 levels and to increase its emissions reduction to 30 % by 2020 if other high emission countries commit to reduce their emissions[12].[1] Parallel to the UN process, EU member states have instituted the European Emission Trading System (EU ETS). The EU ETS is the largest carbon trading scheme operating in the world, controlling about half of the EU's CO_2 emissions and accounting for almost 80 % of carbon credit markets in terms of the globally traded value of credits [13].

13.3.2 Transnational Approaches

Transnational forms of collaboration are organised primarily 'through cross-border networks of different configurations of actors' ([14], p. 56), transcending the boundaries of formal intergovernmental relations. The Cities for Climate Protection (CCP) is an example of, and one of the world's largest transnational governance networks; it is a global network structured through regional and national campaign offices. The CCP programme is an initiative of the International Council for Local Environmental Initiatives (ICLEI) and provides a vehicle for local authorities to develop strategies for controlling greenhouse gas emissions [15]. The programme involves more than 1000 local governments in 86 countries, including 12 mega-cities, 100 super-cities and urban regions, 450 large cities as well as 450 medium-sized cities and towns. Members of the network commit to completing five milestones [16]:

- Conducting an energy and emissions inventory and forecast
- Establishing an emissions reduction target

[1] These targets still fall well short of bringing the EU on a trajectory towards meeting their 2050 objective of reducing emissions by 80–95 % compared to 1990 levels.

- Developing a local action plan to achieve the goal
- Implementing policies and measures and
- Undertaking processes of monitoring and verifying results

To date, CCP members have been able to deliver annual emission reductions of more than 60 million tons CO_2eq, bucking the trend of rising national and global emissions [16].

13.3.3 National Approaches

Strong national leadership is a key building block of effective climate governance. Such leadership can take the form of legislating national climate protection targets, encouraging and driving investment in key infrastructure and green economy projects [4]. Government leadership provides policy certainty and longevity, which in turn creates a platform for needed long-term private sector investment. Governments can help facilitate market and social transitions through education and incentive schemes, which can encourage changes in both production and consumption patterns.

Beyond target setting, governments can use a variety of policies and instruments to create incentives for mitigation action that proves both environmentally effective and economically feasible. Policies, for example, that create a carbon price provide incentives for producers and consumers to significantly invest in low-carbon products, technologies and processes. Such policies could include economic instruments, government funding and regulation. Governments can also provide support by way of offering financial contributions (e.g. feed-in tariffs for renewable energy technologies), tax credits (e.g. for carbon management in agriculture), standard setting (e.g. new housing codes) and market creation. Measures such as these are vital for the development, innovation and deployment of new technologies [17].

The introduction of a carbon price in Australia, for example, was designed to create incentives for investments in the alternative energy sector. A ‘carbon tax’ was introduced in 2012 by the previous government, (this was not a tax, but a flexible market mechanism, an emissions cap-and-trade scheme) which was applied to the country’s largest 500 emitters; these are companies that emit more than 25,000 t of CO_2 or supply or use natural gas. The ‘tax’ was used to create financial disincentives for pollution-intensive industries and products and to offer price advantages to low-carbon alternatives [18]. This initiative helped Australia join the group of 32 countries with a carbon price currently in place covering around 850 million people, around 30 % of the global economy and 20 % of global emissions [19]. Under the new federal government, however, in 2014 the Clean Energy Act 2011 and related legislation that established the carbon pricing mechanism was repealed, jeopardising the policy stability that long-term investment decisions critically rely upon.[2]

---

[2] In addition, a carbon price alone was not going to deliver structural changes to Australia’s economy. Despite the introduction of the carbon tax in 2012 emission-intensive industries such as

The alternative programme by the new Government is the 'Direct Action Plan', which takes a 'baseline and credit' approach that creates incentives for emission reductions on the land. Soil carbon through biosequestration is considered 'the single largest opportunity for CO_2 emissions reduction in Australia' ([20], p. 16). However, there are uncertainties in the science of soil carbon sequestration. The second component of the Direct Action Plan is direct industry action through an Emissions Reduction Fund. The fund would provide financial incentives rather than liability. Using the existing National Greenhouse and Energy Reporting (NGER) scheme the new government would see the establishment of individual baselines for businesses against which their emissions could be measured, with reductions earning credits. The baseline would also enable liability for businesses to incur a financial penalty if their emissions level exceeds the given 'business as usual' levels. The dominant cycles of this plan are focused on individual business innovation towards reducing the overall level of carbon emissions. As the plan uses the Emissions Reduction Fund as its key tool, there is not a high level of penalties driving revenue. Funds will be allocated through the sale of abatement strategies to the government, which will result in firms' emissions being reduced below their individual baselines [20]. The key to effective innovation is research and development, which will provide cost-effective abatement strategies. The driver behind this scheme is the provision of positive incentives, as opposed to the mandated cap established by the emissions trading scheme (ETS) approach. A weakness in this scheme, it has been argued, will be the delay in the impact of biosequestration, which may also require additional tree plots to be planted to meet the large sequestration levels necessary. One of the implications of carbon sequestration will be the reduction in incentives for businesses to employ other strategies.

13.3.4 Local Approaches and Regional Approaches

Regional and local responses to climate change vary widely in scope and focus but commonly combine a variety of different mitigation and adaptation measures and seek to be community-centred and equity-focused. Approaches such as these include transition town concepts and urban renewal approaches as well as low-impact living and energy autonomy initiatives. They all serve the purpose of future-proofing local communities from the socioeconomic fallout from future peak oil and climate change scenarios, enhance livability and empower communities to take local action on the global climate change phenomenon. These local initiatives can also be coordinated and integrated regionally (e.g. Portland city planning; see [23]) and even become part of transnational programmes. In many cases, local climate change efforts have thus far proven more progressive than national, let alone international, initiatives.

---

mining, metals and energy continued to receive disproportionate protection, either through large government subsidies, emission exemptions or considerable funding for research into 'clean technologies' (e.g. carbon storage) [21, 22].

The Low Impact Living Affordable Community (Lilac) project in Leeds (UK) is an attempt to build an affordable, ecological cohousing project. The project pioneers low-carbon, permanently affordable, sustainable mixed urban housing solution. Similar to other British cohousing models Lilac aims at low-impact living through the use of natural building materials and the use of renewable technologies and housing designs that maximise conviviality and community interaction. At the same time, affordability is a key aspect of the Lilac initiative achieved through low-cost design features (e.g. shared areas, low-cost building materials (straw), adoption of more sustainable lifestyles (e.g. resource sharing) and unique financing mechanisms. Residents pay 35 % of their monthly income towards their equity share. When leaving, residents receive an out-payment that is linked to national wage changes as opposed to housing market fluctuations [24].

The small town of Feldheim (Germany) is an example of successful energy self-governance. In the mid-1990s, Feldheim's small agricultural community was confronted with dwindling prices for their agricultural produce and rising energy costs. In conjunction with local energy company *Energiequelle,* local farmers began to lease their land to be used for wind farms. Within 13 years, *the town reached energy autonomy using wind and solar energy as well as a biogas factory run as a joint venture between the town of Feldheim and Energiequelle*. In 2009, the local community invested, with support from the EU and Energiequelle, in the establishment of a smart grid, which has since enabled local residents to receive locally produced heat and electricity at prices determined by the community [25]. Feldheim today is Germany's only energy-independent town, and its 125 residents now pay just over half of what other Germans pay on average and are able to capitalise on the surplus electricity (99 %) they are able to sell back to the market.

13.3.5 Market and Private Sector Approaches

Despite high climate stakes, action on climate change to this day has remained slow in the political and commercial realms in the uncertain post-Kyoto policy context. Whilst rated highly as a policy priority globally, most countries to date have no direct regulation for the reduction of greenhouse gas emissions, and existing regulation varies greatly in terms of stringency, scope and mix of instruments [26]. As such no level playing field exists for the private sector where there is a relative lack of comprehensive company climate change strategies. Many strategies are still work in progress [27] or fail to deliver the kind of business models needed for a carbon-constrained future [28]. Company strategies for climate change have long been limited to the business case for sustainability, which equates good environmental outcomes to good business outcomes [29, 30].

For many years, cost and/or competitiveness arguments have been driving investments in energy efficiency and conservation, improved use of alternative energies, reforestation and improved land management as well as the development of alternative and clean technologies (Hall et al. 1990). In addition, changes in society and investor expectations [31, 32], legal compliance pressure [33] and rising environmental cost and associated risks (The Climate Institute Australia 2006) have driven attempts by companies to reduce their carbon footprint. More recently, the introduction of market mechanisms such as emission trading in Europe have given companies the option to compensate for their emissions instead of reducing them by changing their products or processes.

13.3.6 Governance Implications for WA

The key implication for WA is that it would be operating within a framework of multilateral, regional and local governance arrangements. This approach to governance would ensure that a subnational government such as WA could effectively transition to a post-carbon society. In essence, the governance arrangements would be as follows:

- A multilateral system of accounting under which all countries can follow the same rules for measuring and reporting. The UNFCCC continues to be the principal means for achieving this
- A global Emission Trading System commencing with the EU ETS as the largest carbon trading scheme in the world with WA needing to firstly develop a programme like the former NSW Greenhouse Gas Abatement Program and then join the EU scheme
- The transnational CCP programme provides the vehicle for local authorities to implement and report on local and regional carbon-reduction initiatives
- Nationally, policies are required that create a carbon price, feed-in tariffs for renewable energy technologies, tax credits for carbon management in agriculture, standard setting for energy-efficient buildings and industrial processes and market creation. The Australian government will need to rebuild them so that WA can have a stable national policy environment as well as the international platform
- At a local level the transition town and urban renewal approaches combined with industrial and agricultural redevelopment as well as supply chain reorganisation will enable the implementation of the necessary local programmes and infrastructures.

13.4 Post-carbon WA

13.4.1 Background on WA

Energy consumption for WA in 2009–2010 totalled 1025 petajoules. Most energy consumed in WA in 2009–2010 was used for mining (30 %), manufacturing (24 %) and transport (20 %) with lesser proportions for electricity generation (17 %) and residential (4 %). WA's energy consumption accounted for 17 % of total Australian net energy consumption in 2009–2010 while only having 11 % of the nation's population [34] no doubt due to higher mining and travel distance requirements.

Carbon emissions for the same period totalled 77 million t, an increase of 23 % from 1999. Energy production accounted for 75 % of total emissions followed by agriculture (16 %), industrial processes (7 %) and waste (2 %). WA produced 14 % of total Australian emissions in 2009, a lesser proportion than was the case for energy consumption due to a higher proportion of gas use than coal [35].

With a population of approximately 2.3 million in 2009 this would indicate a carbon footprint of 33 t per person, making WA one of the highest carbon emitters on the planet. The only places in the world with a larger carbon footprint would seem to be Kuwait, Netherlands Antilles, Qatar and Trinidad and Tobago while Australia as a whole averages 18 t per person compared with the UK at 8 and the USA at 17 [1].

WA is a geographically large state compared with the rest of Australia currently with a population of approximately 2.5 and 1.7 million of those in Perth, the capital city. There are several smaller regional cities with 10,000–40,000 people, numerous small country towns of around 1000 people and around 100 remote indigenous settlements with several hundred people in each. Transport between these locations is mostly by vehicles running on petrol, diesel or fossil gas. Many large industries are burning fossil gas for heat and power.

These various cities and towns are provided with electricity infrastructure through the South West Interconnected System (SWIS), the North West Interconnected System (NWIS) and 29 regional non-interconnected power systems. The SWIS electricity network grid is operated by state-owned Western Power, with generators operated by state-owned Verve Energy and other private providers and the main electricity retailer is state-owned Synergy. Horizon Power operates the regional state-owned systems and there are a number of large privately owned mining power systems.

In the year ending June 2010, the SWIS had an installed generation capacity of over 5900 MW and generated over 17,400 GWh of electricity powered by a mix of largely open cycle coal thermal and gas turbine stations. The SWIS represents 90 % of the Western Australian power market. In 2011, the SWIS was connected to 913,000 residential customers, 86,000 SME customers and 19,000 large-market customers. The large mines and port towns in the north are typically operating on a mix of open cycle gas and diesel power stations and the small remote towns operate almost exclusively on expensive diesel power generation.

Current renewable electricity sources in WA are principally wind farms and solar photovoltaic installations as well as the Ord River hydro scheme in the north of the state. These systems supplied only 6 % of total consumption in 2011/2012, with 67 % from wind farms, 14 % from solar photovoltaics, 12 % from the Ord hydro and 7 % from landfill methane gas engines. In the same year, the SWIS recorded 9.2 % of electricity sourced from renewables, mostly from wind farms [36, 37]. The larger wind farms connected to the SWIS are Collgar near Merridin at 206 MW with 111 turbines, Walkaway near Geraldton at 89 MW with 54 turbines, Emu Downs near Cervantes at 80 MW with 48 turbines, Mumbida also near Geraldton at 55 MW with

22 turbines and Albany at 35 MW with 18 turbines. There are seven other smaller operating wind farms connected to the SWIS and several other larger farms planned for construction in the near future. The Greenough River solar farm near Geraldton is 10 MW in size and could in future be expanded up to 40 MW. Connected to the SWIS, it was built by Synergy and joint venture partners GE Financial Services and is big enough to service about 3000 average homes. The US company, First Solar, provided the photovoltaic (PV) technology, using more than 150,000 thin-film PV modules to construct the solar farm.

Carnegie Wave Energy [38] signed a deal to supply power from its CETO wave energy technology to a Department of Defence submarine facility on Garden Island in WA. This 5 MW commercial Perth Wave Energy project will be the largest marine energy power plant in the country and will be completed during 2014.

Some local governments in WA have sought to achieve 'carbon neutral' status through various ICLEI, CCP, Greenpower, energy efficiency and sustainability programmes. City of Fremantle [39] was the first local government in WA and second in Australia, after City of Sydney, to achieve carbon neutral status, initially achieving this through the purchase of Greenpower from the network electricity retailer and continuing with other initiatives.

13.4.2 Pathways to Post-carbon Society in WA

There are a number of pathways that WA could take towards a post-carbon society. Consider firstly, stationary energy use.

Beyond Zero Emissions (BZE) [40] developed a Zero Carbon Australia—Stationary Energy Plan to demonstrate that 100% renewable energy is achievable and affordable. BZE designed a fully costed and detailed system of renewable energy power generation plant with commercially available and proven technologies. With a selection of concentrating solar thermal (CST) plants and large-scale wind farms in their modelling, BZE showed that their proposed system can power Australia to be 100% renewable energy within 10 years. BZE developed a two-stage installation plan over 2010–2015 and 2015–2020.

In 2011, the Australian Government announced its Clean Energy Future Plan and through the then Department of Climate Change commissioned the *Australian Energy Market Operator* [41] to undertake a study to explore two future scenarios featuring a National Electricity Market (NEM) fuelled entirely by renewable resources. The large renewable technology resources modelled were CST and PV. At utility scale, PV with rooftop and PV and EVs for storage were also considered. In decreasing capacity size as follows, geothermal, onshore and offshore wind, wave, biomass and pumped hydro and other storage options for balancing supply and demand were also modelled into the scenarios. The AEMO high-level study found that the operational issues would be manageable and would not prevent secure and reliable operability of a 100% renewable future power system. While this study did not include WA, its findings bode well for the development of a similar approach in WA and its overall costs, components and operability are in line with the BZE findings further reinforcing the viability of this pathway.

The community interest group *Sustainable Energy Now* (SEN 2013) modelled two scenarios of 100% renewable energy on the SWIS in WA by 2029 and compared this with business as usual (BAU) to 2031:

- *Scenario 1:* Solar thermal dominant with backup biomass at solar CST plants, capital cost = $ 48 billion and levelized cost of electricity (LCOE) = $ 215/MWh
- *Scenario 2:* Lower cost diverse mix with wind and solar PV dominant, capital cost = $ 43 billion and LCOE = $ 208/MWh
- *Scenario 3:* Business as usual (BAU) without carbon capture and storage or energy efficiency gains, following the fossil growth philosophy of the WA State Government's Energy 2031 Strategic Energy Initiative, LCOE = $ 203/MWh

SEN found that the two renewable scenarios will require more capital expenditure than adapting the existing fossil fuelled grid, but it will ultimately provide energy at less cost because technology costs of renewables are decreasing and there are no fuel costs, except for biomass.

13.4.3 An Industry-Focussed Scenario for a Post-carbon WA

An industry-based transition could occur alongside a variation of the preceding scenarios. Consider for example the main industry energy users in WA:

- Mining 30% (consider iron ore mining in the north of the state as an example)
- Agriculture (consider cooperative owned grain handling from wheat and grain growers by CBH across the wheatbelt in the southwest as an example)
- Manufacturing 24% (consider water and wastewater services provided statewide by the government-owned enterprise the Water Corporation of WA)
- Transport 20% (consider the fuel source for the motor vehicles across the state) and
- The building industry which is responsible for approximately 40% of emissions across the lifecycle of sourcing materials, construction and operation of homes, offices and commercial buildings

The economic sectors above provide the financial basis to implement the following strategies in a 10-year plan. The socio-technical requirements for a post-carbon scenario for WA would, therefore, be as follows:

1. Major programmes in energy efficiency upgrades and energy conservation across industrial, commercial, agricultural and domestic sectors.
2. All homes and offices achieve maximum energy efficiency through point-of-sale rating disclosure regulation and retrofit programmes designed to accelerate take-up of green skills in the workforce and widespread deployment of new 'green-collar' trades.
3. In the Kimberley region in the north of WA, tidal resources are harnessed and the Ord River hydro scheme is expanded along with distributed PV and utility-scale CST systems to provide power requirements. Expanded agricultural production on the Ord River irrigation scheme leads to large growth in food exports

to southeast Asia and this drives investment into a subsea transmission line to export solar and marine energy to Indonesia, allowing it to decommission its vast Borneo coal mines and power stations, in exchange for geothermal energy coming back to Australia.

4. Utility-scale CST power station development occurs rapidly in the Pilbara and Midwest, with the NWIS grid expanded and interconnected to all major mine-sites, with mining company support similar to the earlier Worley Parsons plan of 2008 to develop CST power stations across the north of Australia [42, 43].
5. With ongoing growth in demand for energy from expansion in population and industry in Perth and the southwest, the second gas pipeline from the north-west is built. This second pipeline, to transport the fossil gas transition fuel, is designed to withstand embrittlement from liquid hydrogen produced in the Pilbara from seawater desalination with solar energy from the new large-scale CST and PV power stations. Many of the domestic, transport and industrial gas needs are now met with hydrogen and fuel cells.
6. In coastal towns like Midwest Geraldton and Northwest Karratha large algae farms are developed for biomass and liquid biofuels production using seawater, municipal wastewater and carbon dioxide waste by-product from industrial processes. Geraldton is a suitable seasonal alternative site because in Karratha, productivity rates can be lower during summer months due to the algae's susceptibility to 'sunburn'. Such facilities can provide the alternative fuel to replace the massive fossil diesel use in mine to port locomotives. Existing mining and industries in the Pilbara alone consume diesel at approximately 2 million t/annum and a 5000 ha algae development will produce 400,000 t per annum of biofuels [44]. As with so many other technology innovations and improvements over time, algal biotechnology will become more productive through innovations such as that proposed by Moheimani and Parlevliet [45] where a semitransparent thin-film solar PV cell used as a power-generating roof over the algae ponds could be used to harvest the higher frequency end (blue and green) of the spectrum and allow the lower frequency (red) region to pass. This transmission of red light would be used by the chlorophyll and associated pigments. These facilities can then produce electrical power and the liquid biofuels needed by rail and motor vehicle transport fleets.
7. In all towns across the state the local government, Water Corporation, local community groups and private investors work together to develop firstly, systems that produce methane fuel for heat and power by anaerobic digestion of municipal solid waste, sewage biosolids and agricultural residues, for example, using a system like the locally developed Dicom/Anaeco process (www.anaeco.com).
8. In all towns the local government, Water Corporation, local community group and private investors work together to develop local food production systems using the Sundrop system of South Australia (www.sundropfarms.com) or similar where CST is providing heat and power for the production of fresh water and electricity for operation of greenhouse fans and water pumping.

9. In coastal towns in the more humid north of the state, seawater greenhouses can also contribute to fresh water and food production and in southern coastal towns with a more suitable resource Carnegie wave energy plants can be deployed for both power and fresh water production. In each of these local town initiatives above, support to local governments would be provided through the global CCP programme.
10. In the southwest, the SWIS electricity grid is converted to 100 % renewable energy through deployment of large scale CST, wind and biomass power systems, pumped hydro and EVs for storage, capacity and stabilisation as well a major rollout of distributed PV systems. This is led by utilities Western Power and Water Corporation adapting their infrastructure accordingly and with other major industry participants including WA's grain handler CBH adapting its 200 rural grain receival facilities in numerous fringe-of-grid towns to solar-storage power stations.
11. The SWIS is further stabilised by a combination of distributed generation across this network. This would include vast amounts of distributed solar PV generation installed on nearly all household and warehouse roofs; there would be large wind turbines installed at the suburban fringes of the metropolitan area, in the open spaces of coastal regions and in the rural spaces of the nearby hills known as the Darling Scarp; the SWIS itself would be upgraded to accommodate numerous localised smart grids; and there would be a proliferation of home and EV energy-storage batteries.
12. A major biosequestration programme is developed across the state as the sink for past and future ongoing emissions from the fossil gas transition fuel, ongoing biomass firing and combustion of hydrogen and liquid biofuels for transport. The oil mallee biomass power programme and other tree crops for desalinisation are expanded throughout the wheatbelt of the southwest. Major carbon farming initiatives are led by indigenous communities in remote areas throughout the state with Kimberley Land Council commencing a large-scale programme, similar to the West Arnhem Land Fire Abatement project [46, 47], in partnership with the major transition gas corporations operating in the northwest. Biodiversity revegetation programmes such as Gondwana Link project also become major carbon sinks.

13.5 Conclusions

This chapter has explained three main sectors that result in an intersection between science, government policy and civil society to yield a post-carbon scenario for WA: policy implications, governance arrangements, community-based initiatives and socio-technical systems. There were 12 socio-technical approaches identified in an industry-focussed scenario that will enable the development of a post-carbon WA.

Acknowledgements Thanks to Sharon Delmerge and Lauren O'Mahoney of the School of Arts, Navid Moheimani of the School of Veterinary and Life Sciences, Carol Warren and Alfred Tseng of the School of Management and Governance and Jonathon Whale and Tania Urmee of the School of Engineering and Information Technology that contributed to an earlier, longer version of this chapter. Their additional data will form the basis of future chapters. The authors overall are together developing a new interdisciplinary learning and teaching unit for Murdoch University undergraduate students to be called transitions to a post-carbon society.

References

1. United Nations Statistics Division (2008) Millennium development goals indicators. Department of economic and social affairs. United Nations. http://mdgs.un.org/unsd/mdg/Data.aspx. Accessed 25 April 2014
2. Hall CM (2011) Policy learning and policy failure in sustainable tourism governance: from first- and second-order to third-order change? J Sustain Tour 19(4–5):649–671
3. Weible CM (2006) An advocacy coalition framework approach to stakeholder analysis. J Pub Adm Res Theor 17:95–117
4. Wiseman J, Edwards T, Luckins K (2013) Post carbon pathways: a meta-analysis of 18 large-scale post carbon economy transition strategies. Environ Innov Soc Transit 8:76–93
5. McConnell A (2010) Policy success, policy failure and the grey areas in-between. J Public Policy 30(3):345–362
6. Sabatier PA, Jenkins-Smith HC (eds) (1993) Policy change and learning: an advocacy coalition approach. Westview Press, Boulder
7. Geva-May I (2002) Cultural theory: the neglected variable in the craft of policy analysis. J Comp Policy Anal 4(3):243–265
8. Howlett M (2009) Policy analytical capacity and evidence-based policy making: lessons from Canada. Can Pub Adm 52(2):53–175
9. Innes JE, Booher DE (2004) Reframing public participation: strategies for the 21st century. Plan Theory Pract 5(4):419–436
10. Figueres C (2012) Climate change: why we need a multilateral solution Singapore
11. Putra NA, Han E (eds) (2014) Governments' responses to climate change: selected examples from Asia Pacific. Springer, Singapore
12. Bäckstrand K, Elgström O (2013) The EU's role in climate change negotiations: from leader to 'leadiator'. J Eur Pub Policy 20(10):1369–1386
13. Klepper G (2011) The future of the European emission trading system and the clean development mechanism in a post-Kyoto world. Energ Econ 33(4):687–698
14. Andonova LB, Betsill MM, Bulkeley H (2009) Transnational climate governance. Glob Environ Politics 9(2):52–73
15. Betsill MM, Bulkeley H (2006) Cities and the multilevel governance of global climate change. Glob Gov 12(2):141
16. International Council for Local Environmental Initiatives (2014) Who is ILCEI? http://www.iclei.org/iclei-global/who-is-iclei.html. Accessed 19 March
17. Metz B, Davidson OR, Bosch PR, Dave R, Meyer LA (eds) (2007) Contribution of working group III to the fourth assessment report of the intergovernmental panel on climate change, 2007. Cambridge University Press, Cambridge
18. Clean Energy Regulator (2013) Carbon pricing mechanism. http://www.cleanenergyregulator.gov.au. Accessed 21 June 2014
19. SBS World News (2012) "Factbox: carbon taxes around the world." http://www.sbs.com.au. Accessed 21 Aug 2012

20. Coalition (2010) The reference is: the Coalition's Direct Action Plan. www.greghunt.com.au/Portals/0/PDF/TheCoalitionsDirectActionPlan2010PDF. Accessed 15 March 2014
21. Australian Conservation Foundation (2010) Australia: a clean energy superpower. http://www.acfonline.org.au/jobs-map. Accessed 14 May 2014
22. Jotzo F, Pickering J, Wood PJ (2011) Fulfilling Australia's international climate finance commitments: which sources of financing are promising and how much could they raise? Canberra
23. Peterson MN, Peterson TR, Liu J (2013) The housing bomb. Why our addiction to houses is destroying the environment and threatening our society. Johns Hopkins University Press, Baltimore
24. Chatterton P (2013) Towards an agenda for post-carbon cities: lessons from Lilac, the UK's first ecological, affordable cohousing community. Int J Urban Reg Res 37(5):1654–1674
25. Guevara-Stone L (2014) Renewables power a rural German village. http://blog.rmi.org/blog_2014_02_19_renewables_power_a_rural_german_village. Accessed 18 March
26. Organisation for Economic Cooperation and Development (2010) Transition to a low-carbon economy: public goals and corporate practices. Paris
27. Kolk A, Pinkse J (2005) Business responses to climate change: identifying emergent strategies. Calif Manage Rev 47(3):6–20
28. Ihlen Ø (2009) Business and climate change: the climate response of the world's 30 largest corporations. Environ Commun 3(2):244–262. doi:10.1080/17524030902916632
29. World Business Council for Sustainable Development (2002) Striking the balance. Earthscan Publications, London
30. Dyllick T, Hockerts K (2002). Beyond the business case for corporate sustainability. Bus Strategy Environ 11:130–141
31. Carroll AB, Shabana KM (2010) The business case for corporate social responsibility: a review of concepts, research and practice. Int J Manage Rev 12(1):85–105. doi:10.1111/j.1468-2370.2009.00275.x
32. MacLeod M, Park J (2011) Financial activism and global climate change: the rise of investor-driven governance networks. Glob Environ Politics 11(2):54–74. doi:10.1162/GLEP_a_00055
33. Wilhelm K (2013) Return on sustainability: how business can increase profitability and address climate change in an uncertain economy. Pearson, Upper Saddle River
34. Australian Bureau of Statistics (2012a) 1367.0—State and territory statistical indicators. http://www.abs.gov.au/ausstats/abs@.nsf/Lookup/by+Subject/1367.0~2012~Main+Features~Energy+Consumption~6.38. Accessed 25 April 2014. Canberra
35. Australian Bureau of Statistics (2012b) 1367.0—State and territory statistical indicators. http://www.abs.gov.au/ausstats/abs@.nsf/Lookup/by%20Subject/1367.0~2012~Main%20Features~Carbon%20Emissions~6.39. Accessed 25 April 2014. Canberra
36. Government of Western Australia (2014) Renewable energy. http://www.finance.wa.gov.au/cms/content.aspx?id=15108. Accessed 25 April 2014
37. McHenry MP (2010) Diverse, remote & innovative: prospects for a globally unique electricity network and market in Western Australia. A discussion paper prepared for Institute for resources technology, Murdoch University, South Street, Murdoch WA 6150
38. Carnegie Wave Energy (2014) Perth wave energy project, Western Australia. http://www.carnegiewave.com/index.php?url=/projects/perthproject. Accessed 26 April 2014
39. City of Fremantle (2011) Low carbon city plan 2011–2015. Fremantle
40. Beyond Zero Emissions (2010) Australian sustainable energy: zero carbon Australia; stationary energy plan. University of Melbourne
41. Australian Energy Market Operator Limited (AEMO) (2013) 100 Per cent renewables study—modelling outcomes. Department of Climate Change and Energy Efficiency (DCCEE), Canberra

42. Dawson & Schlyter (2012) Less is more: strategic scale site suitability for concentrated solar thermal power in Western Australia. Energ Policy 47:91–101 (Elsevier)
43. Robertson S (2008) Advancing solar thermal, ecogeneration—November/December 2008. http://ecogeneration.com.au/news/advancing_solar_thermal/002019/. Accessed 27 April 2014
44. Worley Parsons (2012) Pilbara algae industry study. WA Department of Regional Development and Lands
45. Moheimani NR, Parlevliet D (2013) Sustainable solar energy conversion to chemical and electrical energy. Renew Sust Energ Rev 27:494–504 (Elsevier)
46. Heckbert S, Russell-Smith J, Reeson A, James G (2011) Indigenous Australians fight climate change with fire. Solut Sustain Desirable Futur 2(6):50–56 http://www.thesolutionsjournal.com/node/1006
47. NAILSMA (2012) West Arnhem land fire abatement project. http://www.nailsma.org.au/walfa-west-arnhem-land-fire-abatement-project. Accessed 27 April 2014

Chapter 14
Future Indian Programme in Renewable Energy

Pradeep Chaturvedi

Abstract India is the world's fourth largest economy as well as the fourth largest energy consumer, importing 80 % of its oil, 18 % of its gas and now even 23 % of its coal. As the Indian economy continues to grow, so will its energy consumption. India's total energy demand, which was nearly 700 Mtoe in 2010, is expected to cross 1500 Mtoe by 2030. This is likely to increase India's dependence on energy imports from 30 % in 2010 to an unsustainable figure of over 50 % by 2030.

Keywords Insert keywords here.

India is the world's fourth largest economy as well as the fourth largest energy consumer, importing 80 % of its oil, 18 % of its gas and now even 23 % of its coal. As the Indian economy continues to grow, so will its energy consumption. India's total energy demand, which was nearly 700 Mtoe in 2010, is expected to cross 1500 Mtoe by 2030. This is likely to increase India's dependence on energy imports from 30 % in 2010 to an unsustainable figure of over 50 % by 2030.

14.1 2030 Energy Vision and Road Map for India

2030 Energy Vision is required that takes into account the future energy requirements and a sound understanding of the current and potential energy resources that are available to meet these energy requirements. This vision needs to consider and address the following questions:

1. How to fully explore and exploit the domestic resources, for example, coal, conventional and unconventional oil and gas?
2. What market pricing and taxation mechanisms need to be put in place to attract investments/private participation in the energy sector?

P. Chaturvedi (✉)
Indian Association for the Advancement of Science, New Delhi, India
e-mail: pradeepc08@gmail.com

A. Sayigh (ed.), *Renewable Energy in the Service of Mankind Vol II*,
DOI 10.1007/978-3-319-18215-5_14

3. How to accelerate the development of renewable power capacity (both wind and solar), while setting global cost benchmarks to make them economically viable?
4. What initiatives are needed to drive energy efficiency, given a vast proportion of India's infrastructure is yet to be built, for example, energy-efficient buildings, long-distance transportation and an optimal road–rail modal mix?
5. What strategic tie-ups/partnerships are possible to secure long-term supplies, especially given the changing global energy landscape?

14.2 India's Renewable Energy Programme

India launched an organised renewable energy programme implementation in 1981. With consistent efforts, though more intense recently, about 30,000 MW of grid-connected installed power generation capacity has been created. There has been a visible impact of renewable energy in the energy scenario over the past three decades. In September 2013, renewable energy-based electricity generation installed capacity contributed to 13.6% of the national capacity and about 4.7% in terms of electrical energy. Details of renewable energy capacities deployed in the country are given in Table 14.1.

The need to increase the total domestic energy production in order to reduce the import dependency combined with the need to move away from fossil fuels in the longer run in view of climate change considerations points to the need for stronger efforts to increase the supply of energy from renewables.

An important limitation on the extent to which we can shift to renewables is the high unit cost at present compared with other conventional sources. The unit cost of renewable energy, especially solar energy, is coming down, and the marginal cost of conventional energy based on fossil fuels is likely to remain high and rise in future. These trends suggest that over the next 10–12 years the unit cost of energy from renewable sources such as wind and solar may come close to the unsubsidised cost of conventional energy. Since India has a large potential of both, wind and solar energy, the exploitation of this potential should form an important part of our long-term energy strategy.

In the early 1980s, India was estimated to have a renewable energy potential of 85 GW from commercially exploitable sources, viz. (i) wind 50 GW at 50 m mast height, (ii) small hydro 15 GW, (iii) bioenergy 20 GW and (iv) solar radiation sufficient to generate 50 MW/m^2 using solar photovoltaic and solar energy. These estimates have since been revised to reflect technological advancements. Initial estimates from Centre for Wind Energy Technology (C-WET) suggest that the wind energy potential at 80 m height (with 2% availability) would be over 100 GW. Some studies have estimated even higher potential ranges up to 300 GW.

Table 14.1 Renewable energy use in India at a glance (as on 31 December 2013)

Renewable energy programme/systems	Target for 2013–2014	Development during December, 2013	Total deployment in 2013–2014	Cumulative achievement up to 31 December 2013
Power from renewables				
I. Grid-interactive power (capacities in MW)				
Wind power	2500	99.30	1096.50	20,149.50
Small hydropower	300	31.55	130.90	3763.15
Biomass power	105	6.00	21.00	1284.60
Bagasse cogeneration	300	40.40	175.45	1512.88
Waste to power—urban-industrial	20	2.40	3.00	99.08
Solar power	1100	129.09	495.13	2180.00
Total	4325	308.74	1921.98	29,989.21
II. Off-grid/captive power (capacities in MWEQ)				
Wind power—urban-industrial	10.00	2.50	4.06	119.63
Biomass (non-bagasse) cogeneration	80.00	2.10	38.54	509.69
Biomass gasifiers—rural	1.00	0.064	0.35	17.05
Industrial	9.00	0.250	2.88	141.67
Aerogenerators/hybrid systems	1.00	–	0.04	2.15
SPV systems (>1 kW)	40.00	2.52	19.70	144.38
Water mills/micro hydel	500 nos	–	–	10.18 (2547 nos)
Biogas-based energy system	2.00	2.00	–	–
Total	145.00	7.434	67.23	944.75
III. Other renewable energy systems				
Family biogas plants (number in lakhs)	1.10	0.40	0.50	47.10
Solar water heating collector areas (million sq. metre)	0.60	0.10	0.47	7.47

MWEQ megawatt equivalent *SPV* solar photovoltaic, *nos* numbers

14.3 Recent Renewable Energy Sector Application Modals

A number of initiatives have been launched in India to fulfil energy needs of the common man and rural industries. Some of these are mentioned below for their innovative approach and likely long-term impact.

1. Development of solar cities

Over 30 % of India's population lives in cities, which will increase to 50 % in the next 30 years. Urban areas have emerged as one of the largest sources of greenhouse gas (GHG) emissions, with buildings alone contributing to around 40 % of the total GHG emissions. Several Indian cities and towns are experiencing 15 % growth in peak electricity demand. The local governments and electricity utilities are finding it difficult to cope with this rapid rise in demand, and as a result most of the cities/towns are facing severe electricity shortages. There is a need to develop a framework that will encourage and assist cities in assessing their present energy consumption status, setting clear targets and preparing action plans for generating energy through renewable energy sources and conserving energy utilised in conducting urban services.

Solar City is a concept to integrate all the renewable energy projects and schemes in a city to saturate the renewable energy applications in various sectors. To achieve this, the Ministry of New and Renewable Energy has been implementing a 'Development of Solar Cities Programme' since 2009 in which 60 cities/towns are being supported for development as solar/green cities.

Solar City aims at a minimum 10 % reduction in projected demand for conventional energy at the end of 5 years, which can be achieved through a combination of energy efficiency measures and an enhanced supply of renewable energy in the city.

The specific objectives of the Solar City are:

1. To enable empower urban local governments to address energy challenges at city level
2. To provide a framework and support to prepare a master plan including assessment of the current energy situation, future demand and action plans
3. To build capacity in the urban local bodies and create awareness among all sections of civil society
4. To involve various stakeholders in the planning process
5. To oversee the implementation of sustainable energy options through public sector partnerships

Out of the selected 45 cities, master plans have been prepared for 37 cities. The Solar City programme has been successful in motivating the urban local bodies, which are now adopting renewable energy and energy efficiency in their main activities. They are also making/modifying the suitable by-laws for promoting renewable energy in their respective cities.

2. Solar heating system for poor rural households in the Himalayan region

The concept was mooted to popularise model technologies among rural masses which they may not see and use keeping in view their individualistic approach and financial constraints despite heavy government subsidy. The Himalayan Research Group (HRG) took up the pilot project of developing solar passive retrofitting for space heating and solar water heaters for houses in villages of the Mashobra Block in the district Shimla.

The community water heater has been installed at a location in the village, which is approachable by a number of households. A solar street light has been installed to facilitate women and children to take warm water when it is dark.

Solar passive retrofitting to warm housing has been designed to utilise locally available materials. Local carpenters were trained and locally available wood, glass and steel sheets were used. This is a modification blended with modern material to provide maximum solar energy for the warmth of the houses backed with scientific principles. Solar passive retrofitting, viz. thermosiphon air heating panel, trombe wall and sun spaces, was designed for existing houses keeping in view their orientation to the sun. Solar retrofitting was installed on the south-facing wall of the houses.

3. Mini grids for solar power transmission

No licence is required for generation and distribution of power in notified rural areas. Private developers have therefore taken up the challenging tasks of setting small photovoltaic-based power units and connecting houses through mini grids. At present the government's role is mainly limited to provision of capital subsidies and possibly selection of villages for such mini grids. Private developers are demanding a strong and regulated policy framework, which would enable private developers in scaling up the business and creating an efficient outreach into all rural and remote areas facing a lack of electricity.

4. Cold storage solution for villages

Postharvest wastage of food produce has been a major impediment in ensuring food security in India. Lack of electrified villages without adequate cold storage facility is only aggravating the problem. A ground-making method of running cold storage in non-electrified villages has been designed using solar energy from the sun and heat energy from a biomass-based gasifier, developed by the National Institute of Solar Energy, The Energy and Resources Institute (TERI) and Thermax Limited. Cold storages can now be installed at the farm-gate level in villages. The prototype system under development can provide cold storage facility for storing 25 t of food and vegetables, besides supplying clean power for village electrification. The cold storage facility is capable of achieving temperature as low as 0 °C which offers an opportunity to keep a wide variety of foods, vegetable and horticulture produce in it. The system also provides 50 KW of electrical power at grid quality which can be supplied for running domestic, community, commercial as well as other productive loads in a typical village. This is another initiative undertaken to develop the real-time biomass-based cold storage and village electrification.

14.4 Key Concerns in Growth of Renewable-Energy-Based Power Generation

Some of the key issues facing renewable power generation are:

Regional Concentration of Renewable Energy Potential Renewable energy is location specific and not evenly distributed. There are problems of scaling up grid-connected renewable power. For instance, wind potential is mainly confined to the wind-resource-rich states of Tamil Nadu, Maharashtra, Gujarat, Karnataka, Rajasthan, Andhra Pradesh and Madhya Pradesh. The states of Gujarat and Rajasthan have excellent solar radiation, and the other suitable states for solar power are Andhra Pradesh, Tamil Nadu, Karnataka, Madhya Pradesh, Maharashtra, Orissa and so on. Similarly, small hydropower potential is mainly available in the Himalayan states and northeastern states. The intermittent nature of solar and wind power in the absence of an adequate balancing mechanism limits the flexibility of the state grid to absorb this power.

Insufficiency and High Cost of Evacuation Infrastructure Utilisation of variable renewable energy requires a robust transmission infrastructure from remotely located generating plants to the load centres. Further, combining geographically dispersed renewable energy sources to reduce variability requires much larger, smarter and upgraded transmission networks. A recent study conducted by the Power Grid Corporation Limited has identified the requirement for strengthening both intrastate and interstate transmission systems for facilitating the transfer of renewable energy from renewable-rich potential states to other states as well as for adoption within the host states. The study has estimated that for the 12th Five-Year Plan period (2012–2017) an investment of around ₹ 30,000 crores would be required for creating a renewable power transmission infrastructure.

Regulatory Issues Renewable power, especially solar, is significantly costlier than conventional power, thus making its adoption by the cash-starved utilities difficult unless it is incentivised through the Renewable Purchase Obligation (RPO) and the introduction of Renewable Energy Certificates (REC). This would enable states to procure a fixed percentage of their power portfolio from renewable power.

Financial Barriers Renewable energy technologies require large initial capital investments, making the levelised cost of generation higher than it is for many conventional sources. These technologies need to be supported until the technologies' breakthrough and market volumes generated are able to bring the tariff down at the grid parity level. Moreover, high technology and project risk perceived by financers for renewable projects make access to low-cost and long-term funding difficult. Thus, there is a need to generate instruments for low-cost and long-term financing of such projects from both domestic and overseas resources and also banks to adopt a separate exposure limit for the renewable energy sector.

Low Penetration of Renewables for Urban and Industrial Applications Solar applications for heating water in urban, industrial and commercial applications are one of the most matured and viable renewable energy technologies available worldwide.

Better market penetration of such technologies can lead to better demand-side management for commercial as well as household usage. With already matured technology and rapidly growing industry, solar water heater installations have witnessed a massive growth throughout the world, but the installations in India remained low on account of poor adoption of high upfront costs and poor quality standards of collectors. Moreover, the binding regulation in building codes that encourage adoption of such technologies is seldom implemented, and only few states have such regulations.

Consultancy Issues The renewable energy sector has been facing serious shortage of good quality, large consultancy firms. Even the detailed project reports have been observed to be a cut and paste from other projects totally in disregard of the local conditions and situations. Very often even the resource availability data and the cost for manufacturing and installation of devices are reflected in a site-neutral manner, whereas this is really not the desirable route.

Enforcement Issues Wind-based power generation has suffered the most partly because of the lack of evacuation infrastructure in the resource-rich states and partly because of lack of enforcement mechanisms and incentives for operational performance of wind turbines. Incentives such as accelerated depreciation have not yielded the desired result. Possibly the desired route is to enforce generation-based incentives. Achievement in capacity addition has been significant for most sectors, except for waste to power.

Though most of the states have come up with the RPO, proper enforcement and monitoring is an issue.

Storage Technologies Globally, development of storage technologies has not been in line with the technology development in wind and solar energy, due to which capacity utilisation of grid-connected solar and wind energy has been relatively poor. The challenges in India are serious.

'Off-Grid Renewable Powers'—Lack of Scalable Business Models and Nonavailability of Institutional Finances The off-grid renewable sector has the advantage that it is potentially much more competitive with conventional power because it avoids investments in transmission to remote locations. Lack of scalable business models and nonavailability of institutional finance have stalled the pace of its progress. Policy interventions are required to incentivise the creation of financeable business models, such as the rice husk gasifier-based electricity generation. The issue of unwillingness of public sector banks to finance small-scale off-grid renewable-based business models need to be addressed.

14.5 Future Course

Significant progress has been made recently which is an outcome of a combination of government policies and facilitation on one hand and the private sector initiative and private sector investments on the other. Based on this combination, it has been realised that the renewable energy can play an expanding role in achieving energy security and access in the years ahead. During the 12th Five-Year Plan period (2012–2017), a capacity addition of 30,000 MW of grid-connected renewable power is proposed, of which 15,000 MW is envisaged to come from wind power, 10,000 MW from solar capacity and 5000 MW from other types of renewable sources. The areas on which attention needs to be focused during the 12th Five-Year Plan period (2012–2017), and then continued during the 13th Five-Year Plan period (2017–2022), includes the following:

- Grid-interactive and off-grid/distributed renewable power
- Renewable energy for rural application
- Renewable energy for urban, industrial and commercial applications
- Research, design and development for new and renewable energy products
- Strengthening of institutional mechanism for enhanced deployment and creation of public awareness

1. Grid-connected renewable power

A capacity addition of 30,000 MW of grid-connected renewable power is proposed during the 12th Five-Year Plan period of which 15,000 MW is envisaged to come from grid power, 10,000 MW from solar energy and 5000 MW from other types of renewable sources. Institutional mechanisms to accelerate adoption of renewable power by states in the form of RPOs are sought to be enforced by bringing in an amendment into the Electricity Act 2003. It is expected that the solar power under Jawaharlal Nehru Solar National Mission will continue falling due to enhanced indigenisation and local manufacturing. The results are really unpredictable but promising and can even lead to a boost of this capacity, should a technical breakthrough take place.

2. Renewable energy for rural application through higher-level technology products

Biogas and solar cooker programmes are finding wider applications mainly for large-scale use. Appropriate business models need to be developed, wherein mass cooking applications can be encouraged.

Some of the exciting models for providing off-grid electrification have shown notable response. Consequently, models such as a solar home lighting system through a banking mechanism, entrepreneur-based biomass gasifier models for providing electricity for lighting and mini–micro hydro systems would continue to expand if supported through policies. Effective utilisation of off-grid power will require focused efforts for development of efficient storage technologies and promoting hybrid models by balancing with other power capacities. In the long term, hybrid technology options such as gas with solar/wind need to be developed. As the cost of power through conventional generation rise in the long term and technological developments in future increase the commercial viability of hybrid options, the cumulative financial benefits realised from using these options to meet peak demand requirements may outweigh the financial push provided to them in the present scenario.

14.6 Major New Initiatives

Following are some of the new initiatives in the areas of renewable energy:

1. *National Institute of Solar Energy (NISE):* The existing Solar Energy Centre would be converted into an autonomous institution for undertaking applied research, demonstration and development in solar energy including solar hybrid areas.
2. *National Bioenergy Corporation of India (NBECI):* NBECI will be set up to implement a bioenergy mission including a cookstove programme.
3. *Renewable Energy Development Fund:* In order to address the financial constraints for the grid-connected as well as the off-grid applications of renewable sources, it is proposed to create a renewable energy development fund. The fund will plug the gap between the sector's financing needs and the amount that falls short of the banks' obligations to their lending to this priority sector.
4. *National Bioenergy Mission:* It is proposed to launch the Bioenergy Mission with an objective to create a policy framework for attracting investment and to facilitate rapid development of the commercial biomass energy market based on utilisation of surplus agro-residues and development of energy plantations.
5. *Renewable power evacuation infrastructure:* Special emphasis will be placed on creating the evacuation infrastructure and transmission facilities for renewable power in a time-bound manner to support the large expansion in consumption and production of renewable power. Judicious planning of the transmission system, that is, creating pooling substations for the cluster of renewable power generators and connecting them with the receiving station of STU/CTU at appropriate voltage level will lead to optimal utilisation of the transmission system.
6. *National Cookstove Programme:* The proposed initiative plans to universalise access of improved biomass cookstoves by providing assistance in exploring a range of technologies deployment, biomass processing and delivery models leveraging public–private partnerships.

14.7 Policy Approaches

Following policy approaches are recommended:

Subsidy Fade Out Basically the policy approach has to be such that once a critical mass in terms of manufacturing capacity is created, subsidies for new initiatives should fade away, and the equipment should survive without receiving any subsidy or fiscal incentive from the government. In keeping with this approach, the objective should be to fix time targets and move towards market-based mechanism to the extent possible. The competitive bidding process under Jawaharlal Nehru Solar Mission is an effort in this direction. The success of the same is making solar-based power supply competitive with the fossil-source-based supply.

Special Sectoral Exposure by Banks There is also a need for a policy directive so that a special sectoral exposure limit is created by the banks for the renewable energy sector. Priority sector status may also be granted to the renewable energy sector in view of the social and environmental benefits of projects. This will act as a major policy push for the off-grid applications, which face maximum barriers in receiving low-cost financing.

Augmenting Decentralised Renewable Energy Capacity in the Rural Areas Off-grid renewable energy applications have significant potential of reducing furnace oil/diesel/kerosene consumption in the country and can significantly contribute to oil import substitution. A cluster-based approach for village electrification needs to be adopted. Tariff-based bidding mechanism for such clusters inviting participation from business models would bring down the tariff by a significant amount.

14.8 Conclusion

Rapid development in the renewable energy sector has taken place because of the government creating a favourable policy framework at a fast pace and the private entrepreneurs taking the initiative to respond to the opportunity. The market mechanism had not been very favourable to the late entrants, and the international competition has created operational difficulties for established renewable energy industries. Having passed through this turmoil, the renewable energy entrepreneurs have learnt the tricks of the trade and developed innovative strategies and tactics to counter market forces.

India is hoping to cross the 100,000 MW power generation mark for grid-connected electricity from renewable sources by the year 2021–2022. The recent development efforts and the level of industrialisation in the sector also indicate that India can cross the 300,000 MW mark for grid-connected power generation through renewable sources by the year 2030 to meet its energy requirements. Appropriate policy framework, regulatory mechanism, financing and technology development are being looked into in that direction to achieve the target.

Chapter 15
Solar and Wind Energy–Present and Future Energy Prospects in the Middle East and North Africa

Abdul Salam Darwish and Sabry Shaaban

Abstract The countries of the Middle East possess an enormous potential for the development of renewable energy resources. This region of the globe captures a plentiful amount of direct sunshine, which in turn creates both wind and solar energy. Tapping into this potential will dramatically reduce fossil fuel dependency, and thus create a cleaner environment and new platforms of socio-economic growth. Although previous attempts to evaluate this renewable energy potential have been small in scale and scattered, they have demonstrated an encouraging outlook for the entire region and lead to the contention that there is a sufficient amount of renewable energy to meet all of these countries' energy needs. This chapter examines future renewable energy exploitation goals for Middle Eastern and North African (MENA) countries and explains how solar and wind resource assessment and site selection models, along with strong governmental involvement, can create success or failure in emerging renewable markets.

Keywords Middle East · Renewable Energy · Solar Energy · Wind Energy · MENA

15.1 Introduction

When people hear the term "natural resources", they often think of desirable materials that are stripped from the surface of the Earth or are mined from underground. These materials include lumber, minerals, gems, natural gas and crude oil. However, solar and wind are also part of a country's natural resource potential and just like those mentioned, they are more abundant in some areas than in others. In the Middle Eastern and North African (MENA) countries, solar and wind resources are

A. S. Darwish (✉)
Phoenix Renewable Energy Centre, Manchester, UK
e-mail: darwish@renewableenergycentre.org

S. Shaaban
ESC La Rochelle School of Business, La Rochelle, France
e-mail: shaabans@esc-larochelle.fr

A. Sayigh (ed.), *Renewable Energy in the Service of Mankind Vol II*,
DOI 10.1007/978-3-319-18215-5_15

abundant and can readily be harvested. Both wind and solar energies are regularly replenished by the sun, which shines brightly and steadily in these countries. Additionally, areas with a good number of sunshine hours appear to also have a good probability of strong and steady winds.

Tapping into renewable energy sources is not something new to the MENA region. The world's first parabolic trough solar plant was commissioned in the Egyptian suburb of Maadi, Cairo, in 1913. It was so successful that there were plans to build a massive 20,000-mile plant in the sub-Saharan desert that could have produced the same amount of power as the rest of the entire planet (at that time) [1]. If it were not for the onset of World War I, which waylaid those plans [2], solar power may have spent the last century playing as important of a role as crude oil products, and may have even displaced the petroleum era altogether.

This chapter will detail the present and future renewable energy goals for MENA region countries and will present some of the factors that either promote or obstruct the realisation of those goals.

15.2 MENA Potential, Technical Adaptations, Recent Achievements and Projected Targets

The MENA region includes Algeria, Bahrain, Djibouti, Egypt, Iran, Iraq, Israel, Jordan, Kuwait, Lebanon, Libya, Malta, Morocco, Oman, Qatar, Saudi Arabia, Syria, Tunisia, United Arab Emirates, the West Bank and Gaza and Yemen [3]. MENA has the world's highest potential for renewable energy exploitation. In the past, however, renewables did not account for more than 1 % of the region's primary energy mix [4]. Rapid and sustained increases in global energy consumption have encouraged a recent corresponding increase in adaptations to renewable energy technologies specifically suited to the MENA climate and geography, which will allow better collection of wind and solar resources. For instance, "Gulf-Spec" solar panels have been developed that are specially modified for increased efficiency in the Gulf countries' dust and humidity.

The MENA countries are positioning themselves to benefit from the development of these technologies, not only in the arena of domestic use but also as an export commodity. As a result of these types of adaptations, MENA is expected in 2015 to produce about 3.5 GW of energy from renewables, or nearly 8 % of the world's demand. The renewable energy industry in certain parts of the MENA region has grown rapidly in recent years, spurred by an increase in investment, well-organised plans and projects, government sponsorship and supportive policies. Countries that have seen the greatest successes usually have a high level of government intervention in terms of market alterations and diversions of community and industry away from systems that rely on fossil fuels.

These policies build a solid foundation for the renewable industry, which in turn attracts serious and deep-pocketed investors. Policy frameworks, coupled with renewable energy markets, have developed swiftly and today's outlook indicates that over the next 10 years, the area's diversification efforts will shift significantly, particularly in the Gulf Cooperation Council (GCC) countries. The area witnessed a marked increase in investor interest during the 2009–2012 period, with new renewable energy investments totalling more than US $ 2.9 billion in 2012. This was a 40 % increase over the year 2011 and is a 6.5 % increase since 2004. Most impressively, some of the world's largest energy players, especially national and international oil and gas companies, are now involved in the MENA solar market.

National resources are being devoted not only to the research and development of such technologies but also to their implementation and real-world application. Examples of these successful government interventions include:

- The UAE Masdar city. Aimed to develop the world's most sustainable eco-city.
- Enertech is the branch of Kuwait's National Technology Enterprises Company (NTEC), responsible for strategic investment in renewables and clean technology [5].
- Saudi Arabia's KA-CARE King Abdullah City for Atomic and Renewable Energy which focuses on policy and research.
- Saudi Arabia's King Abdullah University of Science and Technology, whose R&D focus is on water/solar and solar-powered desalinisation plants.

Moroccan projects are largely responsible for the recent 40 % jump in the region's investments between 2011 and 2012. It is in that country's city of Quarzazate that the Saudi Arabian company ACWA Power International, together with World Bank financing through the Climate Technology Fund, developed and built a US$ 1.16 billion concentrated solar plant (CSP), which boasts a 160 MW capacity. In the city of Tarfaya, a US$ 563 million, 300 MW wind farm was developed by the Moroccan company Nareva Holdings [6].

Meanwhile, Saudi Arabia houses the world's largest CSP plant, the United Arab Emirate's 100 MW Shams-1 project, valued at US$ 765 million and 20 % owned by France's Total SA. Also in Saudi Arabia, the 0.5 MW Farasan Island photovoltaic (PV) plant was developed through financing from Royal Dutch Shell's Japanese subsidiary; Showa Shell Sekiyu KK [7].

MENA countries also motivated by the European Union (EU) stated desire to transmit large amounts of solar and wind-derived power from their territories into Europe via a cross-continental power grid, with the goal of meeting as much as 20 % of the continent's energy demand. Such a transmission would save the EU about € 33 billion annually. It is further expected that global energy demand will increase 40 % by 2035 [4].

Renewable energy technology targets in the MENA region are adapting to the change in investments and the continuously increasing demand for energy. Table 15.1 below shows recently announced MENA countries targets:

Table 15.1 The most recently announced MENA countries targets [8–14]

Country	Targets	Type of Technology
Algeria	15 % 2020 40 % 2040	37 % Solar (PV and CSP) 3 % Wind
Libya	7 % 2020 10 % 2025	Solar PV
Morocco	42 % 2020	Solar PV and CSP
Tunisia	11 % 2016 25 % 2030 40 % installed 2030	Wind 1.7 GW 2030 Solar PV 1.5 GW 2030 Solar CSP 500 MW
Egypt	20 % 2020	8 % Solar PV & CSP 12 % Wind
Palestine	10 % 2020	Solar PV & CSP (460 MW Solar farms, 110 MW roof panels) Wind (800 MW)
Jordan	7 % 2015 10 % 2020	Solar PV & CSP (460 MW Solar farms, 110 MW roof panels) Wind (800 MW)
Lebanon	12 % 2020	Solar PV & CSP 100-150 MW Wind 400 MW
Syria	5 % 2025	1000–1500 MW of Wind Power 250 MW of PV Solar plant
Bahrain	5 % 2020	PV Solar Wind
Iran	10 % 2025	Solar PV & CSP Wind
Iraq	2 % 2016	240 MW Solar PV 400 MW Solar CSP 80 MW Wind
Kuwait	5 % 2015 10 % 2020 15 % 2030	Solar PV 15 % Solar CSP 40 % Wind 10 %
Oman	10 % 2020	Solar PV Solar CSP Wind
Qatar	2 % 2020	Solar PV
Saudi Arabia	50 % 2032	Total 54 GW Solar PV & CSP 42 GW Wind 9 GW
UAE	Dubai 5 % 2030 Abu Dhabi 7 % 2020 20 % 2030 (2.5 GW)	Solar PV Solar CSP Wind
Yemen	15 % 2025	400 MW Wind 100 MW CSP 4 MW Solar PV

PV photovoltaic, *CSP* concentrated solar power, *UAE* United Arab Emirates

15.3 Wind Contribution

As can be seen in Table 15.1, wind energy is projected to play nearly as large role in future industry developments as solar energy. Wind energy is an avenue that is attracting a great deal of research and investment. Countries such as Morocco and Oman boast of several locations with brilliant natural wind profiles and high wind speeds, while countries such as Iran and Turkey are planning to incorporate large wind farms in order to supplement their overall energy profile and decrease their dependence on foreign imports [1, 3]. Wind energy is one of the sources for electricity in Egypt, Morocco and Tunisia [15], and by the end of 2012 eight countries had a wind power capacity totalling 1.1 GW [10].

Additionally, Saudi Arabia is analysing wind potential for remote villages in order to decrease high energy costs that result from having long transmission lines [16]. Jordan is even considering wind turbines as the means of operating brackish water reverse osmosis desalination plants [17]. Due to MENA countries varied topography, the nature of the wind is not the same everywhere. The geographical character of a location affects wind speed, which in turn affects wind potential. The wind potential for the MENA countries is graphically presented in Fig. 15.1 [18] at a 100 m hub height:

15.4 Can MENA Meet Its Solar and Wind Targets?

There are many factors that will affect whether or not any specific MENA country is successful at meeting its stated renewable energy goals. As mentioned earlier, government's involvement and active participation in developing policies and re-directing market dependencies plays a crucial role in the initial success or failure of an emerging renewables infrastructure. When MENA governments do not or cannot aid the budding technology by protecting it against well-established fossil

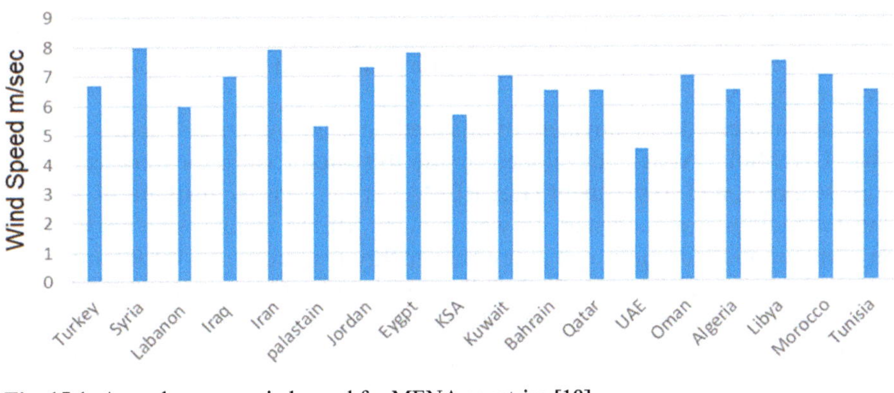

Fig. 15.1 Annual average wind speed for MENA countries [18]

fuel-based energies, it will be very difficult for renewable technologies to break through the market.

In addition to establishing an industrial infrastructure, cultural acceptance must be created through public awareness and education campaigns, and regulations must be set to standardise the new technologies. Countries suffering from a high level of internal unrest, such as Egypt, Syria, Iraq, Yemen and Libya will find it difficult to either implement or enforce such policy and market changes, and their lack of stability will additionally turn off investors. These nations are, therefore, highly unlikely to meet even modest goals, despite past achievements.

Where internal stability is not an issue, success can rest upon the depth of planning. This has been achieved to a limited degree in Morocco and Algeria; countries that are leading the region in tangible gains, by implementing highly structured regulatory codes and well organised plans for expanding renewable energy capacity, despite the lack of financial resources. Still, without securing more foreign investments, even these countries will struggle to meet their goals by the set deadlines.

On the other hand, very secure and economically stable countries such as United Arab Emirates and Saudi Arabia will rank very high and are expected to meet all of their goals, due to government support, with Saudi Arabia being expected to reach the greatest achievement with up to 54 GW of renewable energy capacity by 2032. Iran, Israel and other GCC countries are also expected to achieve their targets by 2020.

15.5 Solar and Wind Resource Assessment

One of the best ways to guarantee success in the development of renewable technologies is via proper planning and preparation, and the best way to achieve that is through the development and use of the "Solar and Wind Resource Assessment" (SWERA) program, which like any other technical project, requires planning and coordination and is constrained by budget and schedule limitations. It also demands a clear set of objectives so that the best approach is selected. A SWERA's ultimate success rests on the quality of the program's assets—sound siting and measurement techniques, trained staff, quality equipment and thorough data analysis methods [19].

Unfortunately, the deployment of this method is not yet utilised across MENA. The "UAE Solar Atlas ", developed by the Research Centre for Renewable Energy Mapping and Assessment (ReCREMA) at the Masdar Institute of Science and Technology is a good start. However, many MENA countries do not yet have any type of atlas for their own renewable energy resources. A SWERA maps out where the best locations are for building either solar plants or wind farms. Not using the SWERA program is equivalent to not having a map that tells you where all of your country's woods, mineral deposits, rivers, underground water, oil and natural gas reserves are located. However, the intergovernmental organisation International Renewable Energy Agency (IRENA), has published the Global Atlas for Renewable Energy, which is closing the gap between nations by providing access to the needed data sets, expertise and financial support they need to evaluate their national renewable energy potentials [20]. Starting with solar and wind, IRENA is expected to expand the initiative to encompass all renewable energy resources.

15.6 Solar and Wind Potential Assessment

The renewable energy resources in the MENA region can be categorised with respect to power generation as follows [21]:

- Direct solar irradiance on surfaces tracking the sun CSP.
- Direct and diffused solar irradiance on a fixed surface tilted towards the south according to the latitude angle (photovoltaic power).
- Wind speed (onshore and offshore wind power plants).

For the MENA region, the following three indicators must be examined to properly assess renewable energy potential:

- Technical indicators—Those which could potentially be exploited using current technology.
- Performance indicators—Such as the average renewable energy yield which can be potentially generated.
- Economic indicators—That will allow new plants in the medium and long term to become competitive with other renewable and conventional power sources, considering their potential technical development and economies of scales [21].

15.7 Solar and Wind Technology Options for MENA and Site Selection Factors

Concentrated solar thermal power generation is by far the greatest renewable energy resource in the MENA region. Photovoltaic technology can be widely used for small size farms as well as large complex systems. The two main characteristics that make concentrated solar power a key technology for MENA's future renewable energy supply mix are that:

- It can deliver secured power as requested and on-demand.
- It is very abundant and practically unlimited.

Other solar power technologies can also be exploited, such as those used in heating and cooling systems.

Wind energy can also be used to balance the increasing energy demand. The wind turbines that are most suitable for the MENA region fall into the 1–250 kW range for onshore sites. If a future consideration includes offshore wind power, then those turbines will fall into the 250 kW–2.5 MW range. Through the ideas and adaptations presented in this chapter, we can ease this evolution with the best possible outcomes. SMART technologies and intelligent systems are one of the key factors for the transitioning of a net neutral community in these countries [15].

In addition to solar radiation or wind speed availability, there are many parameters that must be considered in order to correctly select the right solar or wind site. For instance, choosing a site as the result of high average wind speeds alone will be insufficient. The probability of their occurrence and the height at which they are measured must also be considered. Most available figures are at a height of 10 m, which must be carefully converted to other heights in order to determine their utilisation possibility. Additionally, these figures must be supported by data derived from a remote sensing research facility. When considering solar energy sites, an important parameter for the site selection is the annual solar radiation levels at different orientations versus tilt angles. Figure 15.2 below displays the methodology to consider in order to identify appropriate solar farm sites.

When placing solar panels, it is beneficial to not only consider ecological parameters such as environmentally sensitive areas and land accessibility but also economical parameters such as energy production potential, existing transmission system and the solar power market [20]. The solar and wind energies site selection parameters are shown, respectively, in Tables 15.2 and 15.3 below:

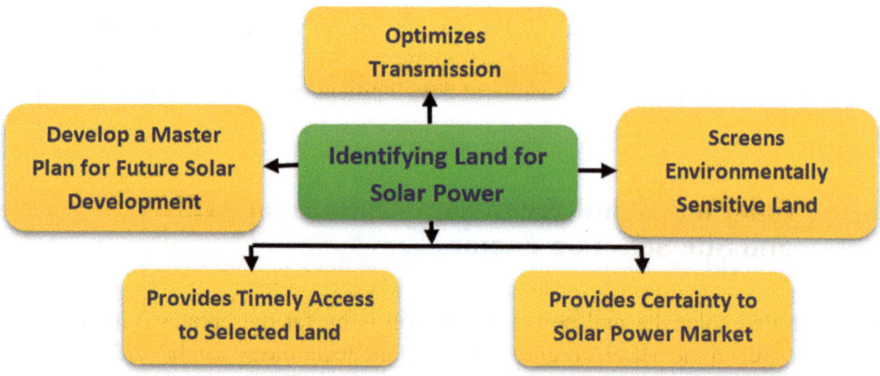

Fig. 15.2 Solar power site selection factors [20]

Table 15.2 Solar energy site selection parameters [22]

Criterion	Description
Demand and supply	Potential energy demand and generation capacity of local areas/regions, which can be assessed to determine the optimal size of the farm
Slope	The land's slope affects the direction of receiving the radiation
Direct normal irradiance/solar potential	Solar potential information to area of interest
Proximity to roads	Proximity minimises construction and maintenance costs
Distance to transmission/ power lines or pipelines	Long distances lead to loss of power and connection to network
Sand/dust/fog risk	A combination of dust, mist and fog
Access to water source	Necessary for cooling the panels
Environment sensitive areas	Effects should be specified according to the local areas
Land cover/use profile (accessibility)	Accessibility of land to another land
Visual impact	The aesthetic regarding the surrounding environment
Weather history	The amount of expected during the different seasons of the year
Solar panels specifications (to match the selected site)	Different system configurations for different sites
Natural disaster risk	Risks of flood, earthquake, dam burst, etc
Cultural heritage	Protecting the cultural heritage

15.8 Conclusion

Solar and wind energies are part of a country's natural resources and should be valued as highly as all other indigenous resources. The MENA countries have an abundance of these resources to tap at will into. The proper exploitation of these resources relies upon strong governmental involvement, which includes setting policies and regulations, markets alterations, developing detailed and thorough project plans and implementing a "(SWERA)" program. Renewable energy targets are unlikely to be met for some unstable and limited financial resource countries. International (particularly EU) support is needed to achieve the best utilization. The EU countries can get a very good source of power supply if they encourage investors to undertake projects in the MENA region. Higher public awareness and education are major factors that MENA governments need to enhance in order to achieve a net-neutral community. Community zero energy behaviour is essential. This can be materialised by setting community goals for energy use.

Table 15.3 Wind energy site selection parameters [23, 24]

Criterion	Sub-criterion
Wind energy potential	Wind frequency distribution
	Wind power density
	Annual mean wind speed
	Gusts and extreme turbulence
Land issues	Land size
	Topography and soil conditions
	Availability
	Roughness and obstacles (flat, smooth, open, coastline, mountains, tunnelling gaps, no trees, sloping)
	Conservation
	Transport (roads and railways)
	Cultural heritage and archaeology
	Distance from residential areas
	Internationally-designated areas, such as
	Special protection areas (SPA)
	Special areas of conservation (SAC)
	Ramsar site (sites are designated under European legislation and are afforded more protection under law in this respect)
	World heritage site
	Existing land use (agricultural, cattle, sheep, grazing and cutting for silage)
	Masts for communication signals
	Special scientific interest
	Distance from green belt
	Broad-leaves woodland
	Veteran trees
	Water courses
	Wetlands
	Flower-rich meadow
	Mature hedgerows
	Geological and hydrological conditions (drainage, water sedimentation, water supply issues, streams, rivers, lakes or other watercourses/aquatic habitat on or within 200–500 m)
Benefits	Proper selected height of installation
	Affordable technology
	Policy supports
Economical	Capital cost
	Operation and maintenance cost
	Land cost
	Payback period
	Incentives
	Electricity market
Social and political	Public interest and acceptance
	Government policies
	Job creation

Table 15.3 (continued)

Criterion	Sub-criterion
Wind turbine related issues	Noise
	Visual impact
	Size, scale and layout, which could affect the visibility of these turbines to a certain distance
	Shadow flicker
Risks and technical issues	Aviation
	Air defence
	Electrical connectivity
	Electro-magnetic interference
	Distance from the power network

Reference

1. Media DW (2012) Ambition vs. reality. Construct Arabia. http://www.constructarabia.com. Accessed 10 July 2014
2. Mostafaeipour A (2010) Feasibility study of harnessing wind energy for turbine installation in province of Yazd in Iran. Renew Sustain Energy Rev 14(11):93–111
3. MENA (2013) Renewable status. http://www.ren21.net/Portals/0/documents/activities/Regional%20Reports/MENA_2013_lowres.pdf. Accessed 21 July 2014
4. Bowden J (2014) Middle eastern money oils wheels of solar expansion. http://www.renewableenergyworld.com/. Accessed 20 Aug 2014
5. Frankfurt School-UNEP Centre/BNEF (2014) Global trends in renewable energy investment 2013. http://www.unep.org/. Accessed 21 Aug 2014
6. Rehman S, El-Amin IM, Shaahid S, Ahmad A, Ahmed F, Thabit T (2007) Wind measurements and energy potential for a remote village in Saudi Arabia. In: Proceedings of the IEEE PES Power Africa Conference and Exposition, Johannesburg, South Africa, 16–20 July 2007
7. Block E (2014) Middle east energy. Power Engineering International. http://www.powerengineeringint.com. Accessed 12 April 2014
8. Carbon Market Ltd (2011) Israel Govt. Sets 10 % Renewable targets. Econews. http://econews.com.au. Accessed 13 April 2014
9. Greenpeace International (2013) Jordan's future energy. Mediterranean. http://www.greenpeace.org/arabic/PageFiles/481146/Jordan_Report2013.pdf. Accessed 28 Aug 2014
10. TV, W E (2013) 12 % Renewable energy by 2020: mission possible. World Environment Magazine & TV: http://www.worldenvironment.tv. Accessed 4 June 2014
11. Mohsen R et al (2013) The role of renewable energies in sustainable development: case study Iran. Iran J Energy Environ 4:320–329
12. Nalan CB, Murat O, Nuri O (2009) Renewable energy market conditions and barriers in Turkey. Renew Sustain Energy Rev 13(6–7):1428–1436
13. Shawon MJ, El Chaar L, Lamont LA (2013) Overview of wind energy and its cost in the Middle East. Sustain Energy Technol Assess 2:1–11
14. Darwish A (2014) Eco-friendly buildings: the central factor in transitioning to a green economy. Int J Environ Sustain 3(1):54–62
15. REN21, Century RE (2014) Renewables interactive map. map.ren21: http://www.map.ren21.net. Accessed 7 July 2014
16. Habali SM, Saleh IA (1994) Design of stand-alone brackish water desalination wind energy system for Jordan. Sol Energy 52(6):525–532
17. REN21 (2013) MENA renewables status report. http://www.ren21.net/Portals/0/documents

18. Shihab-Eldin A (2014) Renewable energy in GCC. Annual Arab Energy Club Meeting, Amman, Jordan. June 2014
19. Gastli A, Charabi Y (2010) Solar electricity prospects in Oman using gis-based solar radiation maps. Renew Sustain Energy Rev 14:790–797
20. DLR (2014) Renewable energy resources in EU-MENA. DLR Aerospace Centre of Germany. http://www.dlr.de/tt/Portaldata/41/Resources. Accessed 20 May 2014
21. Dongrong L (2013) Using GIS and remote sensing techniques for solar panel installation site selection. Thesis, University of Waterloo, Canada
22. Lee AHI, Chen HH, Kang H-Y (2009) Multi-criteria decision making on strategic selection of wind farms. Renew Energy 34(1):120–126
23. Maynard JE (2010) Factors influencing the development of wind power in rural Alaska communities. MSc Thesis, University of Alaska Fairbanks. Fairbanks, Alaska

Chapter 16
The Pugwash UK 2050 High Renewables Pathway

David Elliott and David Finney

Abstract Making use of the UK Department of Energy and Climate Change (DECC) 2050 Pathways Calculator, in 2013 the British Pugwash group produced a report with three basic UK scenarios: High Nuclear, High Renewables and an Intermediate pathway. This chapter summarises the conclusions from the high renewables scenario, in which around 80 % of UK energy will be supplied from renewables by 2050, with full grid balancing, at a slightly lower cost and with lower emissions than for the Pugwash High Nuclear and Intermediate scenarios. Export of surplus electricity could earn around £ 15 billion p.a., or if a 'power to gas' conversion route was adopted an extension to near 100 % renewables was considered possible.

Keywords High Renewables · scenario DECC 2050 Pathways · UK energy options

16.1 Introduction

The UK Department of Energy and Climate Change (DECC) has made a 2050 Pathways Calculator available via the web for interested parties to use to produce their own test scenarios [1]. British Pugwash decided to take up this challenge, and in 2013 produced a report with three basic scenarios: High Nuclear, High Renewables and an Intermediate pathway [2].

The present authors produced the High Renewables scenario, which is reported below in summary form.

D. Elliott (✉) · D. Finney
The Open University, Milton Keynes, UK
e-mail: d.a.elliott@open.ac.uk

A. Sayigh (ed.), *Renewable Energy in the Service of Mankind Vol II*,
DOI 10.1007/978-3-319-18215-5_16

16.2 High Renewables Pathway

The High Renewables Pathway aimed to achieve an 80% reduction in greenhouse gas (GHG) emissions by 2050, reduce dependence on fossil fuels and foreign imports, by providing enough energy from indigenous low-carbon sources to meet foreseen demand and, in the process, ensure security of supply. It also aimed to decrease the use of nuclear power as quickly as possible, while minimising the dependence on carbon capture and storage (CCS) to reduce GHG emissions, thereby reducing the risk associated with this as yet unproven technology. In addition, it also sought to limit the fossil-fuelled backup capacity required by the intermittency of some renewables and minimise energy wastage, through the adoption of efficient uses of energy in all sectors, and, finally, it aimed to avoid biomass imports given their biodiversity and land-use issues.

Accordingly, appropriate trajectories and levels for supply and demand were selected in DECC's Pathways 2050 spreadsheet. The 'MAX' scenario set out in a 2011 Pöyry report, which supplied 94% of UK electricity from renewables by 2050, was used as a starting point for electricity supply, as shown below. GWn is the nameplate capacity, GWav is the capacity actually available over the year based on the relevant load factors used by Pöyry, that is offshore wind 37%, onshore wind 21%, photovoltaic (PV) 10%, marine/hydro 25% and biomass 50% (Table 16.1).

We did not take the view that Pöyry's scenario selections should be regarded as a 'maximum': their figures for the biomass, geothermal, wave and tidal stream contributions all looked low, and the PV figure might also be revisited, given that DECC now says PV could reach 20 GW by 2020. Furthermore, their load factor figures are lower than those now regarded by DECC as feasible. The load factors used in the DECC Pathways 2050 spreadsheet are 45% for offshore wind, 30% for onshore wind, 10% for PV, 32% (average) for marine and 80% for biomass with combined heat and power (CHP).

There could also be additional sources beyond those proposed by Pöyry, for example, tidal barrages and/or (less invasively) lagoons might contribute up to 20 GW, and could become economically viable if, by 2050, significant electricity (or hydrogen) storage capacity becomes available, or export of excess/out of phase output via a super-grid becomes feasible.

Pöyry also included interconnects in their calculations, but saw exports dominating (Pöyry calculated the net outflow as 35 TWh p.a.); so, although economically attractive, they cannot be considered as adding to UK's capacity or net UK's energy,

Table 16.1 Pöyry 2050 MAX scenario [3]

Generation	Capacity		Output
Option	GWn	GWav	TWh/y
Offshore wind	156	57	501
Onshore wind	33	7	61
Photovoltaics	38	3	27
Tidal stream/wave/hydro	31	7	63
Other/biomass and geothermal	6	1.5	13
Total renewables	264	75.5	665

GWn nameplate capacity, *GWav* capacity actually available, TWh= Terawatt-hours though they provide a useful balancing option for times when renewables are contributing low or zero power, or are generating surplus energy.

Taking account of these points, in our chosen pathway, to be cautious, we reduced the Pöyry offshore wind contribution significantly and adopted most of their other figures (again therefore being conservative given their low levels), but added in some tidal range sources (small barrages and medium-sized lagoons), as well as a range of nonelectrical sources, such as biomass for heating and transport. We also specified 15 GW of interconnections.

We then selected the nearest acceptable trajectories/levels from the DECC Pathways 2050 menu and fed the results into the DECC Calculator at http://2050-calculator-tool.decc.gov.uk/pathways/1011ot2wr1frz41303441210044144403423041023042304101 33.

Although relatively cautious, nevertheless in many cases the primary energy supply choices we made in the DECC model were above 'level 2' in DECC's classification—high, but not extreme or maximal. For example, offshore wind has been set at level 2.4 and onshore wind at level 2.9, ambitious, but not, we felt, excessive, given that wind power now has an established track record and market acceptance.

The DECC Calculator (Excel version) then gave the figures in Table 16.2 for primary energy inputs. It should be noted that the annual input figures in TWh/y are net inputs, after making appropriate adjustments for intermittency and conversion losses. Renewables make up ~70–80% depending on the content of the imports.

As can be seen, some fossil fuel use was retained mostly for transport. This was forced on us by the DECC software, which would not let us replace all the fossil fuel. It simply exported any excess electricity generation from wind, etc. We did however limit the use of CCS to just the 1.7 GWn of projects already planned.

On the demand side, we selected high levels of energy saving in most sectors, in most cases at DECC's level 4, that is very ambitious, resulting in an overall primary energy saving of around 40% by 2050. We recognised that this was challenging, but

Table 16.2 DECC Calculator results

Inputs to electricity supply in renewables	GWn	TWh/y	GWav
Offshore wind	76	*299.8*	*34.2*
Onshore wind	30.8	*81.1*	*9.3*
Photovoltaics (PV)	35.2	*30*	*3.4*
Tidal stream/wave/hydro	35.1	*89.1*	*10.2*
Tidal stream	11.9	37.6	4.3
Tidal range	9.6	19.2	2.2
Wave	9.6	19.0	2.2
Hydro	4	13.3	1.5
Geothermal	4	*28.1*	*3.2*
Non-renewable inputs			
Gas-fuelled thermal generation	1.7	*11*	*0.6*
CHP domestic electricity supply		*44*	*5.0*
CHP commercial electric supply		*16.4*	*1.9*
Electricity imports		*70*	*8.0*
Total	182.8	*669.5*	*77*

CHP combined heat and power, *PV* photovoltaic, *GWn* nameplate capacity, *GWav* capacity actually available felt it was realistic given that Germany is aiming for a 50 % primary energy reduction by 2050. In addition, DECC has suggested that 40 % savings are possible in key UK sectors by 2030, including industry [4]. We also included some storage capacity (pumped hydro) and demand side management (vehicle to grid) to help with grid balancing, along with the interconnectors for imports and exports (although the DECC spreadsheet's insistence on exporting all the excess means that more would be required than we specified).

Figures 16.1, 16.2 and 16.3 below present the results for primary energy supply and demand and electricity generation, as produced by the DECC Calculator. By 2050 emissions will be 18 % below 1990 levels, that is exceeding the 80 % reduction target, with the total cost per capita per year put at £ 4378, based on DECC's costing calculator, with the full range (low to high estimate) being £ 3257–6588.

Primary energy at 2050		
GWn	TWh/y	GWav
76	299.8	34.2
30.8	81.1	9.3
35.2	30	3.4
35.1	89.1	10.2
11.9	37.6	4.3
9.6	19.2	2.2
9.6	19.0	2.2
4	13.3	1.5
4	28.1	3.2
1.7	11	0.6
	44	5.0
	16.4	1.9
	70	8.0
182.8	669.5	77.9

GWn nameplate capacity, *GWav* capacity actually available

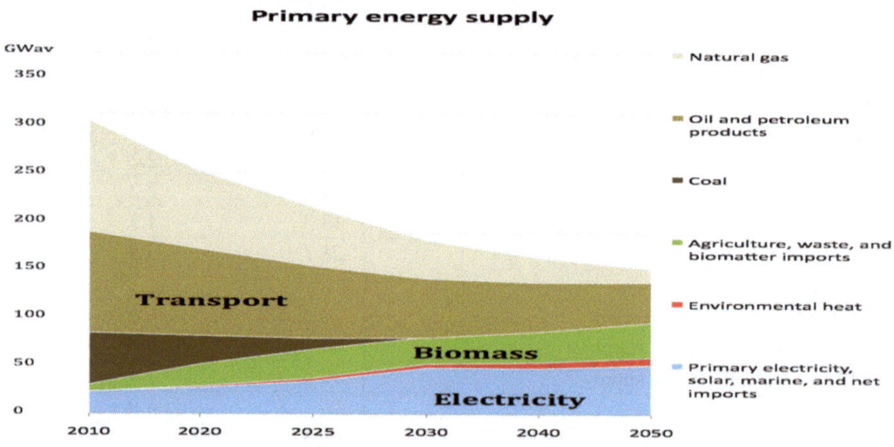

Fig. 16.1 Primary energy supply

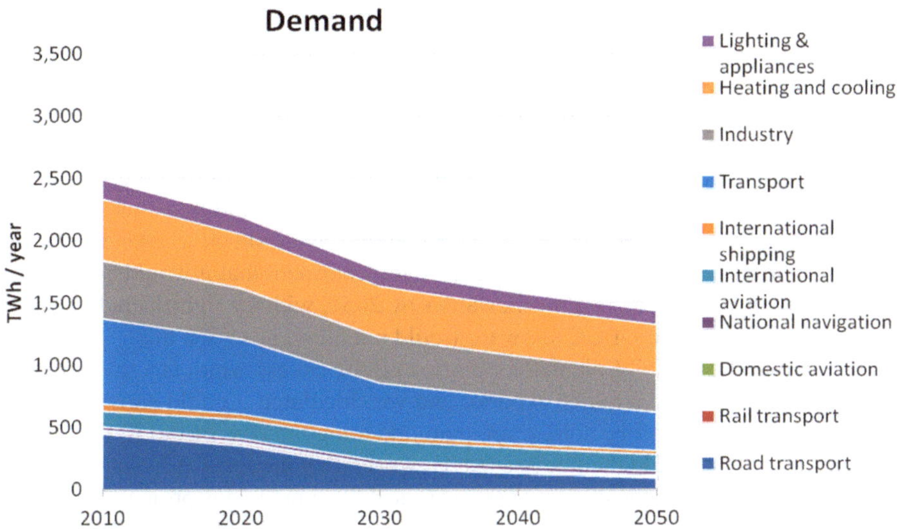

Fig. 16.2 Primary energy demand

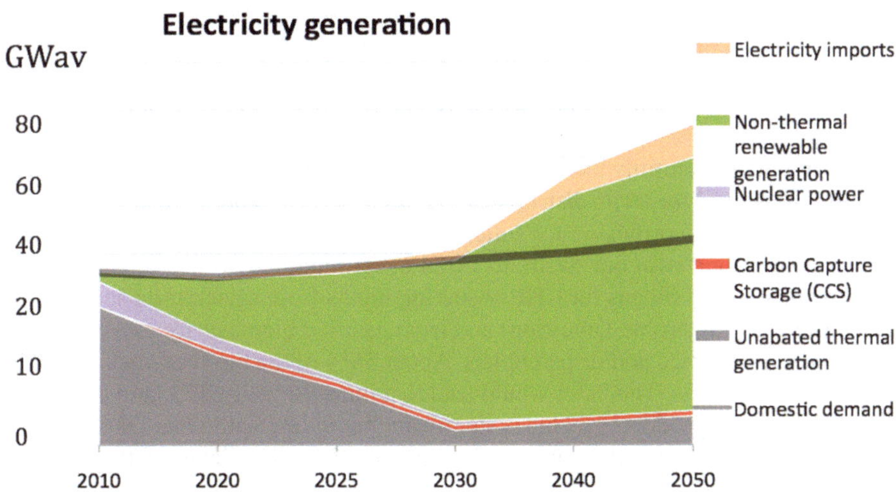

Fig. 16.3 Electricity generation

16.3 Discussion of Results

As can be seen from Fig. 16.3, electricity use expands significantly by 2050, most of this expansion being for electrified transport, although the selected supply and demand mix produces a significant electricity oversupply by 2050. We would have preferred to use some of this excess to make hydrogen via electrolysis ('wind to gas') to substitute for the retained fossil fuel and also perhaps to be used to fuel back up plants when wind inputs were low. As noted above, the model would not let us do this.

However, perhaps surprisingly, it did show that, in fact, given the very large wind and quite large biomass-fired CHP inputs and the inclusion of some storage and demand side management provision, along with imports via interconnectors, there would be no need for any extra back up capacity by 2050 as indicated by the DECC 'stress test', which looks at system robustness when wind, etc. inputs, and also temperatures, are low for 5 days. Our mix successfully passed the test, using 15 GW of imports, 7 GW of pumped hydro and some demand management, with, after 2035, no standby generators being needed to provide shock resistance. On this basis, assuming wind could be expanded, we felt confident that it would be possible in theory to move to near 100 % renewables by 2050, without significant balancing problems, although the DECC software would not allow us to test this.

Given that the DECC Calculator exported all the excess wind, etc., in that case, the UK would become a major exporter and we calculated it could earn £15.6 billion p.a. if this could all be sold, more than offsetting the cost of the 27 GW of interconnectors that would then be needed. However, that income (and cost) would of course be reduced if some of the excess was used for wind to gas projects to replace the retained fossil fuel use, most of which, under the DECC model's constraints is used for transport.

Another restriction of the Pathways 2050 spreadsheet is that, as the capacity of the biomass power stations is increased, the model automatically increases the coal burn evidently assuming cofiring. If no biomass imports are allowed, as we assumed in our selection of trajectories, the biomass power stations trajectory thus cannot exceed level 2.8 without seriously compromising the emissions reduction due to this unsolicited but enforced coal firing.

Given that, unhelpfully the DECC spreadsheet converts most of the bio resource to solid fuels used for CHP, district heating, domestic and commercial heating and export, we make no provision for solid biomass power stations—a choice that was in any case in-line with our desire to avoid using imported wood chips. However we did make use of biogas for CHP, assuming inputs from agricultural and municipal waste AD (Anaerobic Digestion) and from gasified biomass, for example using short-rotation coppice and/or perennials. About 56 % of the bio feedstock would be from grown biomass. The latter would lead to around 10 % of UK's land area being used for biomass production. This is significant but in line with what DECC says is feasible, although it could mean that farming practices and dietary patterns would have to change [5].

Of the remaining feedstock about 37 % is derived from agricultural waste, such as crop waste and manure, and the balance from other nonagricultural waste. Neither of these latter sources requires more land above 10 % indicated and no biofeedstock needs to be imported.

Our pathway is clearly based on high contributions from the main renewables, and there could be problems in achieving the targets we have set. For example, the floating variant of offshore wind technology may not turn out to be as successful as hoped: our DECC Pathways 2050 choices have the effect that from 2035, half the offshore turbines constructed each year will be floating ones. If that proves to be a problem, then the wind contribution in our scenario could be reduced. The main consequence would be to reduce the potential for net electricity export.

We did not include the Severn Barrage, in part on environmental grounds. This major project would also produce large cyclic bursts of energy, which would often be out of phase with demand. A distributed network of smaller barrages, lagoons and tidal stream turbines would be less invasive and more flexible.

Our selection of renewable contributions did not exhaust their full potential. If some contributions proved to be smaller than expected then other sources might take over. For example, the potential for wave and tidal stream generation may be significantly larger than we have assumed in our pathway, as could be the potential for PV solar.

Deep geothermal energy may also be able to contribute much more than we have indicated (including for CHP heat) which would increase balancing options and enable biomass to be diverted away from electricity generation towards vehicle fuel production to displace oil and gas. We also felt that solar-fed district heating networks linked to large inter-seasonal heat stores could have significant potential, reducing the need for biomass for heating and aiding balancing.

Overall, there would seem to be sufficient additional options, if some of the technology selections in our pathway prove to be inadequate

16.4 Conclusion

The Pugwash study looked at three 2050 pathways using the DECC Calculator. The High Renewable Pathway, in which ~70% of energy is supplied from UK renewables by 2050:

- Balances total energy supply and demand using renewables for nearly all needs with no biomass imports
- Ensures that supply meets demand with no need for fossil back up even when variable renewables are low and demand is high
- Is slightly lower cost than the Pugwash High Nuclear and Intermediate scenarios and cuts emissions by 1–2% more
- Could result in a net income from electricity exports set against imports £ 15.6 billion p.a.
- With the wind-to-gas backup option, could allow for 100% of UK's energy to be supplied from renewable sources, including imports

The Sankey diagram in Fig. 16.4 below also illustrates that system losses are low and significantly lower than those for the nuclear and intermediate scenarios, which included, by 2050, 80 and 39 GW of nuclear, respectively. The High Renewables scenario has no nuclear after 2030.

The High Renewables Pathway thus outlines a very different path from that now evidently seen as almost inevitable for the UK, with for example, Prof. John Beddington, then the government's chief scientific adviser, saying in March 2013, at

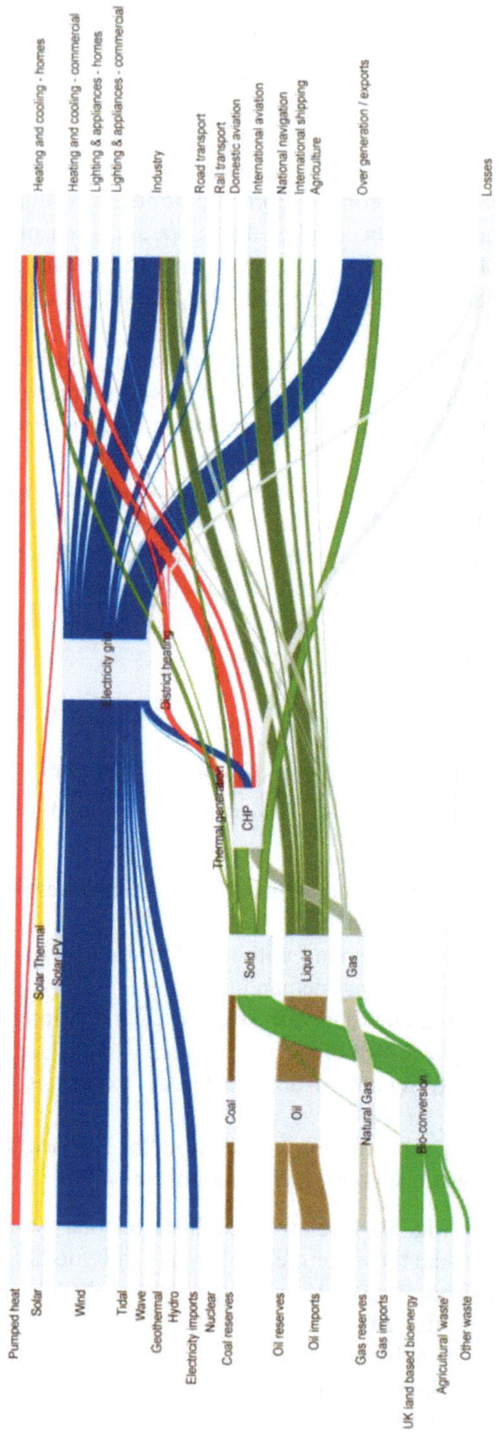

Fig. 16.4 Sankey diagram for High Renewables Pathway

the launch of the government's new nuclear strategy reports 'We really can't see a future for the UK energy sector, if we are to meet our climate change obligations and have resilience in the power sector, without a significant component of nuclear'. The new Nuclear Industrial Strategy suggests that 75 GW of nuclear might be installed by 2050, supplying 86 % of UK electricity, up from 17 % now [6]. The Pugwash study suggests this is not necessary and offers a clear viable alternative.

References

1. DECC (2012) 2050 Pathway Analysis. Department of Energy and Climate Change, London. http://webarchive.nationalarchives.gov.uk/20121217150421/http://www.decc.gov.uk/en/content/cms/tackling/2050/2050.aspx. Accessed 27 July 2015
2. Pugwash (2013) Pathways to 2050: three possible UK energy strategies. British Pugwash, London. http://britishpugwash.org/pathways-to-2050-three-possible-uk-energy-strategies/
3. Pöyry (2011) Analysing technical constraints on renewable generation to 2050. Pöyry report to the Committee on Climate Change, London, March
4. DECC (2012) Electricity Demand Reduction. Department of Energy and Climate Change, London. https://www.gov.uk/government/uploads/system/uploads/attachment_data/file/66212/Electricity_demand_reduction.pdf/
5. DECC (2012) Bioenergy Strategy. Department of Energy and Climate Change, BIS, DETR, London https://www.gov.uk/government/uploads/system/uploads/attachment_data/file/48337/5142-bioenergy-strategy-.pdf
6. BIS (2013) Nuclear Industry Strategy. Department for Business Innovation and Skills, London. www.gov.uk/government/organisations/department-for-business-innovation-skills/series/nuclear-industrial-strategy

Chapter 17
The Role of Renewable Energy Sources in Solving Energy and Water Problems of Mediterranean Sea Islands

D. Zafirakis, Gilles Notton, Chr. Darras, M. L. Nivet, E. Kondili and J. K. Kaldellis

Abstract According to the Amsterdam Treaty, declaration No. 30, "…insular regions suffer from structural handicaps linked to their island status, the permanence of which impairs their economic and social development". Considering the above, the present work aims to present different aspects of the current energy and water situation in Mediterranean Sea islands, using as case studies two representative French and Greek island regions. To this end, common problems as well as differences that call upon the elaboration of more case-specific solutions are identified. Accordingly, emphasis is given on future prospects for renewable energy sources and the role of integrated, hybrid solutions including energy storage and desalination aspects.

Keywords Island grids · Energy storage · Desalination · Hybrid energy solutions

17.1 Introduction

According to the Amsterdam Treaty, declaration No. 30, "…insular regions suffer from structural handicaps linked to their island status, the permanence of which impairs their economic and social development". In this regard, satisfaction of vital

D. Zafirakis (✉) · J. K. Kaldellis
Mechanical Engineering Department, Soft Energy Applications and Environmental Protection Laboratory, Technological Educational Institute of Piraeus, P.O. Box 41046, 12201 Athens, Greece
e-mail: jkald@teipir.gr

G. Notton · C. Darras · M. L. Nivet
Research Centre of Vignola, University of Corsica Pasquale Paoli, UMR CNRS 6134, Centre of Vignola, Route des Sanguinaires, F20000 Ajaccio, France
e-mail: gilles.notton@univ-corse.fr

E. Kondili
Optimisation of Production Systems Laboratory, Technological Educational Institute of Piraeus, P.O. Box 41046, 12201 Athens, Greece
e-mail: ekondili@teipir.gr

A. Sayigh (ed.), *Renewable Energy in the Service of Mankind Vol II*,
DOI 10.1007/978-3-319-18215-5_17

needs for most of these regions relies on strong dependence bonds with supply networks (normally being mainland based) of questionable efficiency, both in terms of supply security and costs. Of major significance in this debate, energy and water resources undergoing severe stresses on and off in many island regions [1–3] produce volatile conditions in all aspects of everyday life, damaging the local economy and development prospects of these communities. At the same time, vulnerability of the local landscape and the cultural heritage in many of these areas dictates investigation of minimum-impact and low-intensity solutions that may adapt to the local environment distinct characteristics. Application of renewable energy sources (RES) to a certain, sustainable extent may align with both these ends [4–6]; that is satisfaction of local habitants' energy and water needs under more secure terms on the one hand and preservation of the local environment on the other. In this context, although there are certain island areas that have already adopted some aspects of the proposed solution, progress met in the field is yet discouraging. In fact, current patterns of energy and water supply in most island regions largely restrict or almost eliminate the contribution of RES, despite the medium-high quality of RES potential met across several areas (e.g. in the Mediterranean). The result of this strategy is the implementation of short-sighted solutions, based almost exclusively on fuel imports [1] for covering energy needs (mainly electricity) and the operation of unorthodox water supply networks so as to cover existing water deficits that may even reach 100 % of the local water demand for certain dry island regions.

Considering the above, the present work aims to present different aspects of the current energy and water situation in Mediterranean Sea islands, using as case studies two representative French and Greek island regions. To this end, common problems as well as differences that call upon the elaboration of more case-specific solutions are identified. Accordingly, emphasis is given on future prospects for RES and the role of integrated, hybrid solutions including energy storage and desalination aspects.

17.2 Brief Presentation of the Current Energy and Water Situation

Capturing a large number of island regions of different scales, the Aegean region can be thought of as representative of the entire Mediterranean area. In fact, as also seen in Fig. 17.1, island regions of the Aegean Sea vary considerably in terms of scale, with the local power stations' size ranging from below 1 MW to even more than 100 MW for the bigger-scale islands. As already mentioned, it is these power stations that are mainly responsible for the electrification of non-interconnected Aegean islands since the contribution of RES is quite restricted, normally not exceeding 10–15 % of the total annual power generation. At the same time, owed to the variation of load demand during the different seasons of the year, it is normal for such power stations to exhibit low levels of capacity factor, which together with the need for imported fuel lead to increased costs of electricity production (especially for the smallest-scale islands) that may even reach 1000 €/MWh (see also Fig. 17.2).

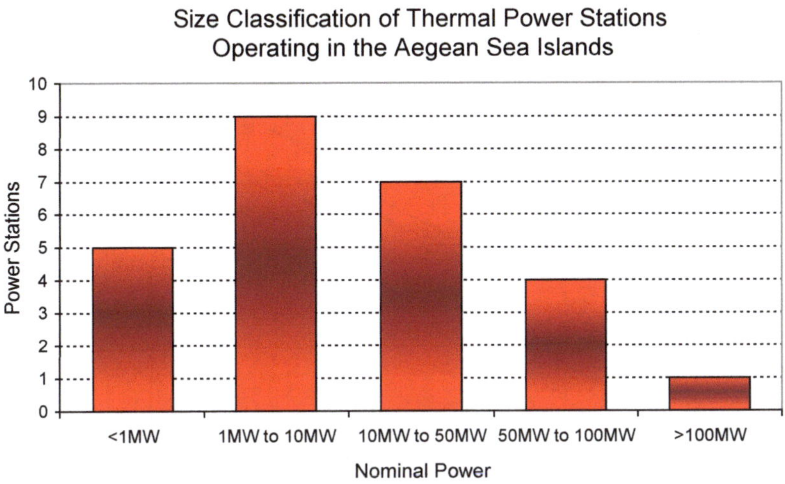

Fig. 17.1 Size classification of Aegean Sea islands thermal power stations

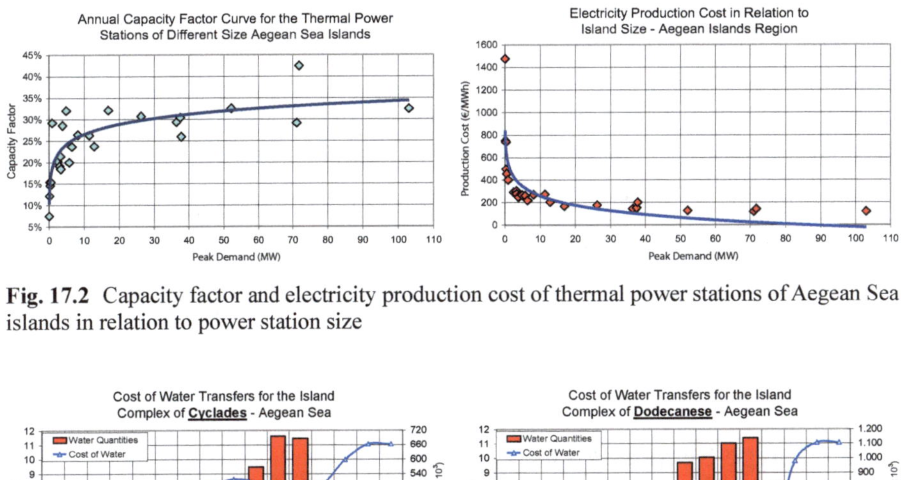

Fig. 17.2 Capacity factor and electricity production cost of thermal power stations of Aegean Sea islands in relation to power station size

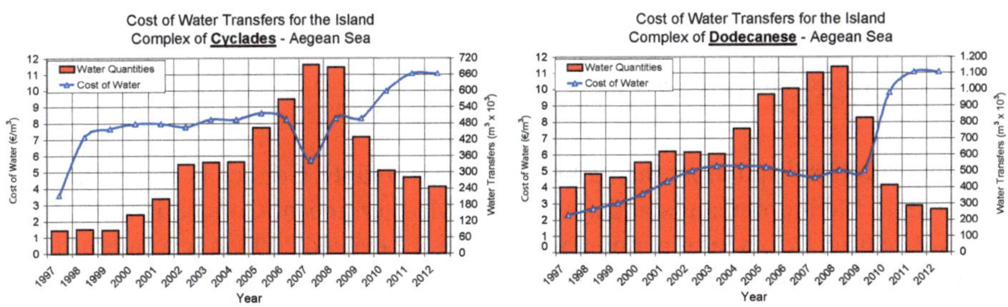

Fig. 17.3 Fresh water shipments to the island complexes of Cyclades and Dodecanese and corresponding specific cost

On top of the dominant diesel power generation pattern, in many cases island, regions also suffer from fresh water problems. To this end, requirement for water shipments (see also Fig. 17.3 for the island complexes of Cyclades and Dodecanese in the Aegean Sea) in order to cover water deficits throughout the year poses

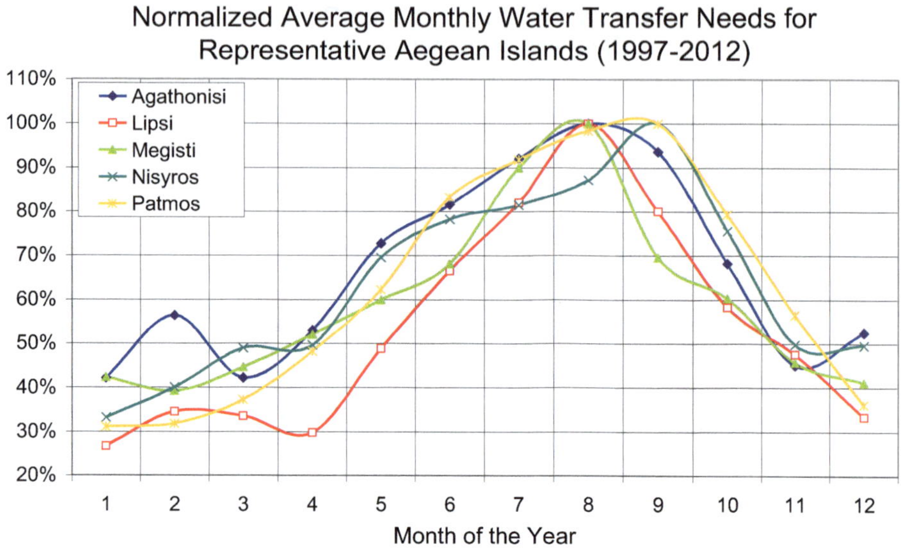

Fig. 17.4 Average needs for fresh water shipments—representative Aegean island cases

additional costs that may even exceed 10 €/m^3. These water deficits are, as expected, more pronounced during the summer season (see also Fig. 17.4) when both electricity and water demand present a remarkable increase, owed also to the arrival of numerous tourists.

Nevertheless, problems previously identified are not encountered at the same level by all island regions of the Mediterranean. On the contrary, each of the islands sets different challenges that have to be met. To underline this, in the following we give an overview of two quite dissimilar island regions, that is the large-scale island of Corsica and the small–medium-scale island of Amorgos.

17.3 Large-Scale Island—Corsica

With its 8680 km^2 and more than 300,000 local habitants, Corsica is one of the very large-scale Mediterranean islands. Although Mediterranean, Corsica is also alpine by its structure. In fact, as ten mountains of Corsica exceed 2000 m (average altitude of 568 m), the island is described as "a mountain in the sea". Furthermore, despite the fact that Corsica fits naturally into the Mediterranean climate zone, there are large differences of temperature and precipitation across its territory, largely owed to the alpine character defining the island to a large extent. Electricity production in Corsica mainly depends on the operation of thermal and hydropower installations located on the island, supported also by the contribution of interconnections with Italy and Sardinia, as well as some photovoltaics (PV) and wind power installations (see also Fig. 17.5). In this context, the local electricity production cost reaches

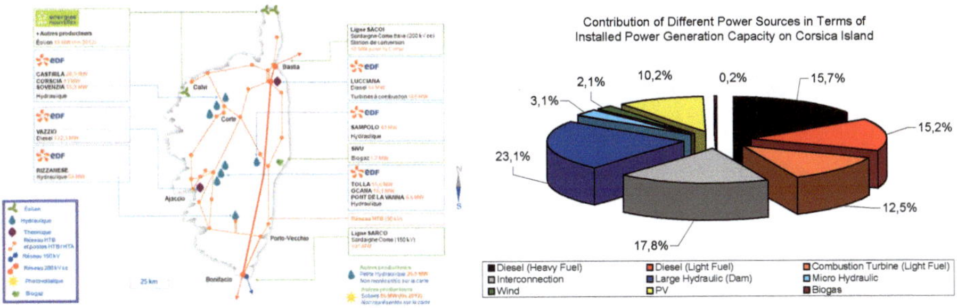

Fig. 17.5 Overview of Corsica island power stations and interconnections (total installed capacity of 840 MW). *PV* photovoltaics

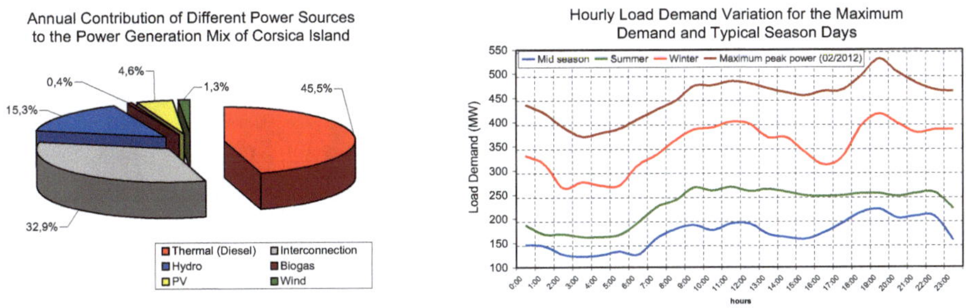

Fig. 17.6 Annual contribution of different power sources and load demand variation for Corsica. *PV* photovoltaics

110 €/MWh, which is almost double that of the French mainland. The total installed power generation capacity on the island reaches approximately 840 MW, while the respective annual power generation rises to almost 2.2 TWh.

Moreover, although the island is largely based on the delivery of electricity from thermal power generation, local energy security levels can be considered as satisfactory, especially if taking into account the diversity of power generation sources participating (Figs. 17.5 and 17.6). One other aspect of Corsica that deserves to be designated is also the fact that the maximum load demand appears during the winter season rather than the summer (Figs. 17.6 and 17.7). This is explained on the basis of year-round tourism and the fact that increased heating needs during the winter months are mainly covered by electricity sources. During summer, on the other hand, although PV generation maximizes, the contribution of hydropower minimizes, owed to the increased need of drinking water (the local population may increase to even 2 million people) and other fresh water uses covered by the same dams used for power generation.

Despite its satisfactory levels of power generation diversity, the local RES potential of Corsica (firstly hydro potential, then solar potential and excellent wind potential in the extreme south and north) is only partly exploited (see also Fig. 17.8), largely owing to the presence of interconnections and the operation of diesel plants. On the other hand, alternative schemes could still be applied on the island in order

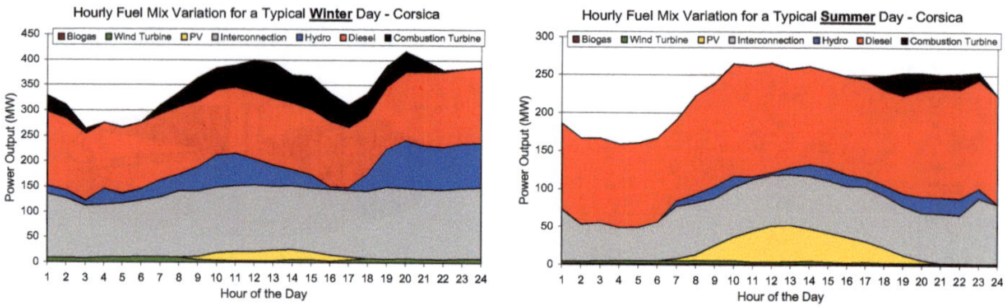

Fig. 17.7 Typical winter and summer day load demand variation for the island of Corsica

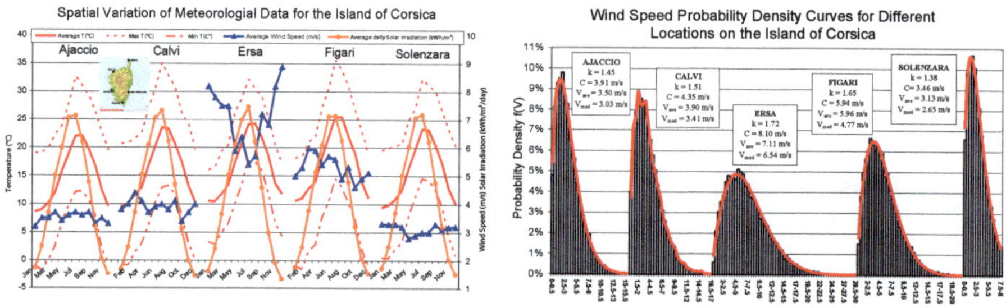

Fig. 17.8 Meteorological data for different locations on the island of Corsica

to maximize RES participation. Such schemes can take into account transformation of existing hydropower to pumped hydro storage [7] and optimum use of interconnectors facilitating maximum RES penetration and export of energy surplus. At the same time, use of biomass and cogeneration for covering both electricity and heating purposes [8] can also comprise an effective solution for the island of Corsica.

17.4 Small–Medium Scale Island—Amorgos

Unlike the island of Corsica, Amorgos is a small–medium-scale remote island found at the south side of the Aegean Sea and belonging to the island complex of Cyclades. The local population of the island reaches almost 2000 habitants, although during the summer period it may increase to reach even triple figures. At the moment, the electrical needs of the islanders, that is almost 10 GWh per annum with a peak load demand of approximately 3 MW (see Fig. 17.9), are covered by the operation of the local autonomous thermal power station. To this end, owed to the increased cost of oil imports required to operate the local thermal power station, the respective electricity production cost well exceeds 300 €/MWh (i.e. almost three times higher than that of Corsica island), having also reached in the past approximately 380 €/MWh.

At the same time, the island also encounters water supply stresses, especially during the summer period, with the total water consumption reaching approximately 140,000 m^3 per year (Fig. 17.9) and with approximately 1/3 of it being covered

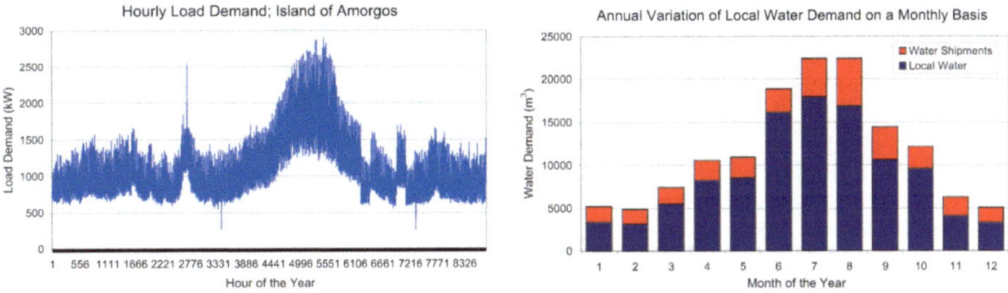

Fig. 17.9 Hourly load demand and monthly water demand for the island of Amorgos

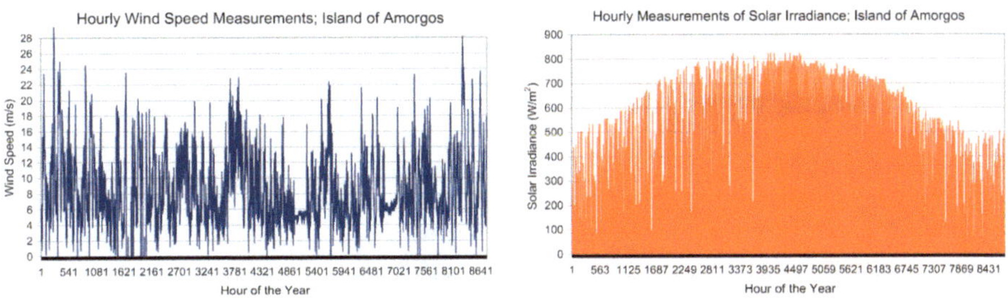

Fig. 17.10 Wind speed and solar irradiance measurements for the island of Amorgos

by water shipments. On the other hand, the current use of renewable energy on the island is rather limited, despite the fact that the entire region appreciates a considerable wind and solar potential. In fact, the local wind potential quality is determined by a mean annual wind speed in the order of 8–9 m/s (see also Fig. 17.10) which encourages investigation of wind-based energy solutions, while the local solar potential exceeds 1600 kWh/m^2/year.

Moreover, Amorgos is determined by the typical, mild Mediterranean climate (Fig. 17.11), which, opposite to the case of Corsica, implies minimum heating needs during the winter period. As a result, the electricity demand profile tends to follow the respective population distributions, peaking during the summer period. Acknowledging the above, Amorgos could benefit considerably from the establishment of alternative solutions combining RES with small–medium-scale energy storage [9, 10] and desalination aspects [11]. The most appropriate configuration to this end should also take into account the notable, seasonal complementarity between the local wind and solar potential as well as the need of maximum electricity and water demand during the summer period.

17.5 Discussion and Conclusions

Based on the brief analysis carried out in the previous paragraphs, it becomes evident that island regions of the Mediterranean present significant diversity in various aspects. To this end, we currently examined the extreme case of the very large-scale

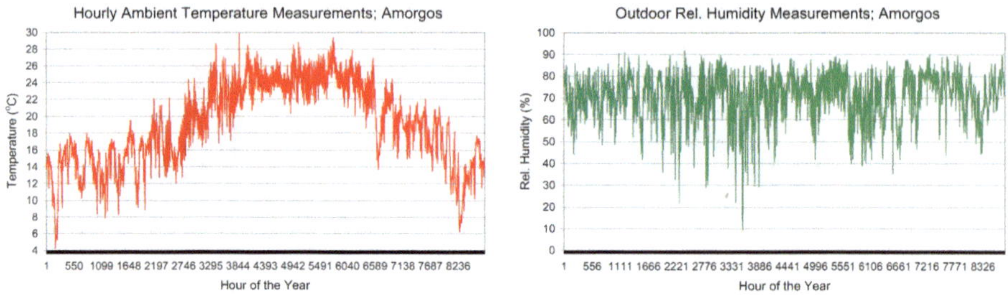

Fig. 17.11 Temperature and relative humidity measurements for the island of Amorgos

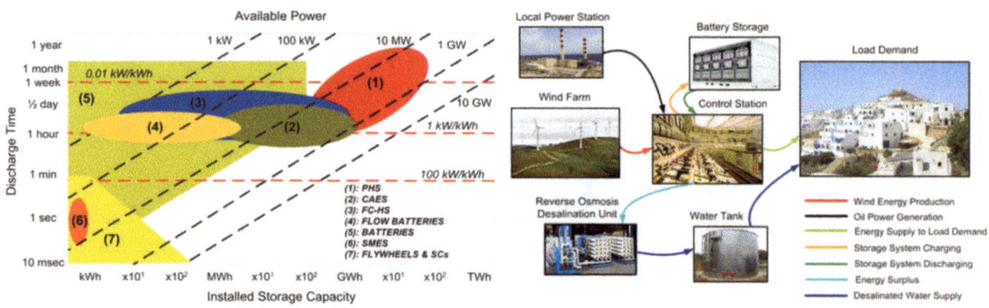

Fig. 17.12 Application map of contemporary energy storage technologies and indicative configuration for the satisfaction of local energy and water needs

island of Corsica and the small–medium-size island of Amorgos in the Aegean. A comparison between the two island regions revealed the different patterns of electricity demand and local fuel mix that call for solutions of significantly diverse characteristics. More precisely, to increase RES penetration on the island of Corsica, investigation of large-scale energy storage solutions such as pumped hydro could be put forward (see also Fig. 17.12). At the same time, owing to its increased heating needs, the island could also benefit from the promotion of combined heat and power, based, for example, on biogas. Furthermore, owing to the presence of interconnectors, increased penetration of RES could also be used for exporting purposes. To this end, combination of RES with large-scale energy storage could also support exports of guaranteed energy amounts that can produce added value for the local RES power generation.

On the other hand, the non-interconnected island of Amorgos, being almost exclusively satisfied by the operation of thermal-based power generation, presents both energy and water security problems that call for the implementation of alternative energy schemes. To this end, exploitation of the local RES potential in combination with small–medium-scale energy storage (e.g. NaS batteries) and an appropriate desalination unit can provide an effective solution that deals with both energy and water issues locally.

In conclusion, and if considering that currently energy storage offers a bundle of technological solutions that are readily available (Fig. 17.12), the role of RES in island regions of the Mediterranean becomes increasingly significant. Optimum sizing and energy management directions in this context suggest that apart from exploiting any complementarity between RES, the same should be considered for the demand side. As a result, any energy surplus should be optimally exploited, which encourages the use of deferrable loads such as desalination and leads to the development of RES-based systems (e.g. Fig. 17.12) that can achieve high levels of autonomy for both energy and water.

References

1. Kaldellis JK, Zafirakis D (2007) Present situation and future prospects of electricity generation in Aegean Archipelago islands. Energy Policy 35:4623–4639
2. Kaldellis JK, Kondili EM (2007) The water shortage problem in the Aegean archipelago islands: cost-effective desalination prospects. Desalination 216:123–138
3. Kondili E, Kaldellis JK, Papapostolou C (2010) A novel systemic approach to water resources optimisation in areas with limited water resources. Desalination 250:297–301
4. Notton G, Diaf S, Stoyanov L (2011) Hybrid photovoltaic/wind energy systems for remote locations. Energy Procedia 6:666–667
5. Notton G, Stoyanov L, Ezzat M, Lararov V, Diaf S, Cristofari C (2011) Integration limit of renewable energy systems in small electrical grid. Energy Procedia 6:651–665
6. Kaldellis JK, Kavadias KA, Kondili E (2004) Renewable energy desalination plants for the Greek islands-technical and economic considerations. Desalination 170:187–203
7. Anagnostopoulos JS, Papantonis DE (2012) Study of pumped storage schemes to support high RES penetration in the electric power system of Greece. Energy 45:416–423
8. Clement J, Martin N, Magnus B (2011) District and community heating aspects of combined heat and power (CHP) systems. In: Beith R (ed) Small and micro combined heat and power (CHP) Systems, 1st edn. Woodhead Publishing, Cambridge. pp 347–364
9. Kaldellis JK, Zafirakis D, Kavadias K (2009) Techno-economic comparison of energy storage systems for island autonomous electrical networks. Renew Sustain Energy Rev 13:378–392
10. Kaldellis JK, Zafirakis D (2007) Optimum energy storage techniques for the improvement of renewable energy sources-based electricity generation economic efficiency. Energy 32:2295–2305
11. Kaldellis JK, Kapsali M, Kondili E, Zafirakis D (2013) Design of an integrated PV-based pumped hydro and battery storage system including desalination aspects for the island of Tilos. In: 4th international conference on clean electrical power (ICCEP), Alghero, Italy, 11–13 June 2013

Chapter 18
Bias in Energy Statistics—A Review of Misinformation About Sustainable Energy

Arthur A. Williams

Abstract Some previous reviews of predictions of energy supply have shown how official statistics have tended to underplay the contribution of renewable energy to future needs. This chapter is based on a range of examples of bias within information regarding energy use, fossil fuel reserves and comparisons of energy sources. Some of these relate to the presentation of statistics in a way that misinforms, often through visual information in graphs and charts. Other biases relate to future predictions and lack of transparency regarding assumptions made. However, as shown in this chapter, even short-term predictions often turn out to be incorrect, and there is evidence that this is often due to biased use or interpretation of statistics. Some of the biases relate to prices, others to growth rates and implied limits to diffusion of sustainable energy technologies. In some cases it is possible to show in retrospect how statements promoting conventional energy over renewable sources have not in fact proved to be correct. Of course, some examples of bias in promotion of renewable energy sources also exist.

This chapter reviews examples of how energy statistics have been used to give bias (whether accidentally or deliberate) to conventional energy technologies over sustainable ones and identifies methods to assist in a critical evaluation of information sources.

Keywords Energy statistics · Energy predictions

18.1 Introduction

Energy statistics are an important tool used by energy professionals and policy makers to guide them in understanding trends and in making investment decisions. They are also used by journalists to support articles or broadcasts about energy issues. It is therefore important that such statistics are presented in a way that gives accurate

A. A. Williams (✉)
Department of Electrical & Electronic Engineering, University of Nottingham, Nottingham, UK
e-mail: arthur.williams@nottingham.ac.uk

A. Sayigh (ed.), *Renewable Energy in the Service of Mankind Vol II*,
DOI 10.1007/978-3-319-18215 5_18

information, but as this chapter shows, there are a number of ways in which statistics and prediction trends can be presented in a biased manner.

Previous studies looking at these issues include those that have looked at predictions of oil supply and predictions of "peak oil". Two recent studies on either side of the debate are from [1, 2]. Prediction of energy trends has been a theme of many publications over a long period of time. However, there has been limited review of these predictions, the assumptions on which they have been based and their accuracy, even though this is relatively easy to study in hindsight. A good review of these issues is included in a report by Rechsteiner [3] for the Energy Watch Group, which concentrates particularly on the wind energy industry. The current chapter picks up the themes raised by the Energy Watch report and looks at the record of energy predictions in the following 5-year period.

Before investigating the record of energy predictions, the first part of this chapter looks at some examples of misleading presentation of statistics. This highlights some of the problems in interpreting energy statistics. In some cases, calculations are based on assumptions that lead to unbalanced comparisons between conventional and renewable energy sources. In another example, poor presentation of graphical data has led to a bias in the presentation of statistics.

18.2 Problems with Energy Statistics

One of the basic building blocks of energy statistics is the data of "Primary Energy Supply". This presents the original source of energy before conversion and utilisation. It comes at the beginning of the annual data presented by International Energy Agency (IEA—not to be confused with the US government Energy Information Administration—EIA!). Each year IEA produces a useful book *Key World Energy Statistics,* the latest edition being 2013. Because of delays in collating data, the latest statistics for primary energy supply are for 2011. Included in these are the data for nuclear energy, hydropower and other renewable energies. According to these data, in 2011, nuclear energy supplied 5.1 % of the global energy, whereas hydropower supplied 2.3 % and other renewables (excluding biomass and waste) supplied 1.0 %. This gives the impression that nuclear energy was a more significant contributor to global energy than that all of hydropower and wind-, solar- and geothermal energies counted together.

Later, the contribution of various types of "fuel" to electrical generation is compared (p. 24). This shows that hydropower contributes 15.8 %, whereas nuclear energy only contributes 11.7 %. The figure for other renewables is 4.3 %, which in this graph includes biomass and waste. Whereas nuclear energy is claiming to supply 5.1 % of the global energy, the electrical output represents a lower contribution than hydropower, which is only 2.3 % of global supply. So is there another use of nuclear power apart from generation of electricity? No, the energy supply figure includes an allowance for the energy that is wasted as heat through the thermal conversion of nuclear fuel to electricity. Checking the small print in the glossary shows that a conversion efficiency of 33 % is used for nuclear energy, whereas no conversion losses are used for the hydropower contribution—nor for solar photovoltaic (PV) or wind energies. Theoretically the 67 % of the nuclear energy producing waste heat could be used to run a district heating scheme, but so far no such project exists! Taking out this dubious conversion efficiency would show the contribution of nuclear energy to primary energy supply as only 1.7 %—a much smaller slice of the pie chart. In this case an apparent bias against renewable energy relates to current usage, but a similar trend exists in relation to energy projections, as discussed later.

An example of statistics presented in a misleading way is the oil reserves statistics in the *BP Statistical Review of World Energy*. Using a chart in the shape of a doughnut, the regional breakdown of reserves has been indicated for 1993, 2003 and 2013 (Fig. 18.1). The diameter of the doughnut has been made approximately proportional to the quantity of reserves (values are given in 1000 million barrels). This means that the area of each section increases with the square of the reserves, giving a misleading indication of the trends.

Actually, the diameter increases slightly more than the ratio of reserves. This distortion means that the largest section of each doughnut, which represents the Middle East reserves at each stage, more than doubles in area from 1993 to 2013. In reality, the value has only increased by 22 %, from 662 to 808, as shown in Table 18.1. The table also shows that the greatest increase of reserves is in North and South America, which comes from including nonconventional resources such as Canadian and Venezuelan Tar Sands. Most of the other increases are due to deep sea discoveries or natural gas liquids—very little is "conventional" oil reserves. It is also worth noting that from 2011 to 2013 the reserves only increased 1 % per year. Although BP appears to be presenting a positive view of the future of oil supply, the reality is that most new reserves can only be extracted using more expensive technologies. In the case of tar sands, there are also significant environmental costs.

Some misleading statistics have later been corrected or removed. One example from a commercial trade association is related to bioethanol production. A report published by the Grain Farmers of Ontario [5] stated that "Global ethanol production is at 100 billion litres from 125 million tonnes of corn, or 15pc of the world's corn. This is 5 per cent of global petrol supply with 60pc used in the US." Although bioethanol represented nearly 5 % of petrol in terms of litres in 2010, the energy value of ethanol is only two thirds that of petrol, so it really contributed only 3 %. Furthermore, a significant proportion of bioethanol is produced in Brazil from sugarcane, not from corn, so the contribution of corn to energy supply was much less than implied by the report.

Table 18.1 Oil reserves (10^9 barrels) for 1993–2013

Year	1993	2003	2013
Middle East	662	746	809
S & C America	80	100	329
N America	121	225	230
Europe & Eurasia	78	116	149
Africa	61	107	130
Asia Pacific	39	40	42
World total	1041	1334	1689

Fig. 18.1 Oil reserves chart [4]

18.3 The Problems of Prediction

The difficulty of predicting trends in energy production and use is well recognized by most researchers of energy policy. Smil [6] gives a good review of this topic that points to key pitfalls in the process of predicting future energy scenarios. He cites a long record of failed predictions of energy supply (some of them his own) and highlights the fact that since these trends are the result of social, economic and technological developments, the models required to make predictions are too complex to give anything but a wide range of results. In his book published in 2003 [7], Smil makes some further comments on future energy use, including one on the "dubious" benefit of corn-derived ethanol, a policy which the USA has strongly promoted in the intervening years. Changes in energy trends due to one-off events, such as the Fukushima plant closure, are rarely taken into account even when using scenario-based predictions.

Following from the oil reserve figures above, it is interesting to look at the failure of more recent price predictions, since these have a key bearing on energy use trends. In 2004, the UK Department of Trade and Industry (DTI) gave various predictions of energy supply and price [8]. Of particular note is the price projection for oil, for which the report gave three values—low, medium and high, but even the high value is more than 50 % too low for most of the years since 2008, as shown in Fig. 18.2. Clearly, trying to predict the cost of a product with such high price volatility is not easy. In the case of the DTI, they were at least able to correctly predict the price for 2005.

In relation to the prediction of the contribution to renewable energy, there is an excellent review of wind power statistics by Rechsteiner [3]. He demonstrated how the IEA consistently under-predicted the contribution of wind power by assuming that current growth rates would not be maintained, even in the short term. Looking at the statistics since then shows that the pattern has not changed much. The "450 Scenario" outlined by the IEA in *World Energy Outlook 2009* [9]

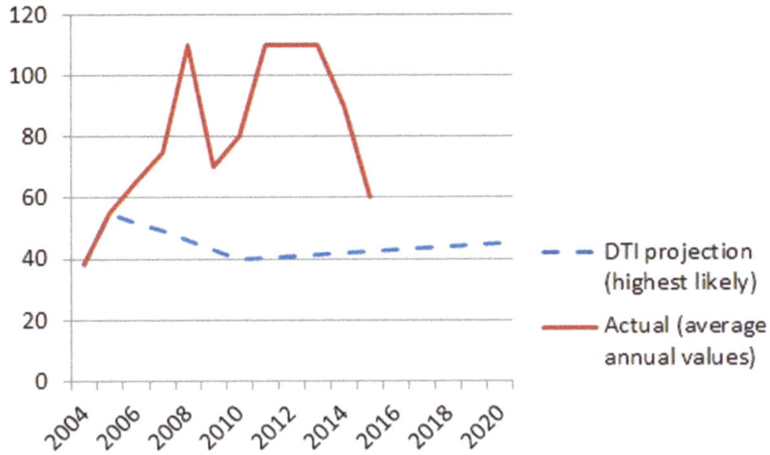

Fig. 18.2 Oil price projections by the UK government

places a much greater role for renewable energy than their "Reference Scenario". According to their predictions global capacity additions for onshore wind from 2008–2020 would be 399 GW. In reality 217 GW (over half that prediction) has already been achieved between 2008 and end of 2013. Therefore, the 399 GW target will be exceeded even with a collapse in the onshore wind energy market and zero growth from the relatively poor year of 2013, in which only 33 GW of onshore capacity was added worldwide. Even the optimistic addition of 64 GW of offshore wind globally between 2008 and 2020 is likely to be met at current growth rates.

On the other hand, predictions of the steady growth in nuclear energy have not been fulfilled. Whereas IEA [9] predicts 810 Mtoe of energy supply from nuclear energy by 2015 (based on their conversion factor), the actual output in 2011 was down compared with 2007 to 673 Mtoe.

When looking at predictions in the growth of solar PV, the IEA record is even poorer. The same 2009 publication "450 scenario" predicts 108 GW of new PV capacity from 2008 to 2020, a figure which has already been surpassed, since PV capacity has risen by a factor of 10 since the end of 2008 to reach 139 GW worldwide till the end of 2013. In a more recent publication [10], the predicted global PV market in 2020 was expected to grow to 34 GW/year, a figure that was already exceeded in 2013, within 2 years of publication. The report predicted an annual capacity addition for 2012 that would be 20 % lower than that for 2011, despite the continuing trends in cost reduction. In fact the market grew by around 2 %, according to EPIA [11]. From 2012 to 2013, the market grew again by around 25 %, and growth appears to be continuing through 2014. If 20 % growth rates for PV continue (half the historical growth rate) the market would reach around 75 GW/year by 2017, as shown in Fig. 18.3. This capacity is nearly one quarter of the current level of capacity additions to electrical power generation worldwide.

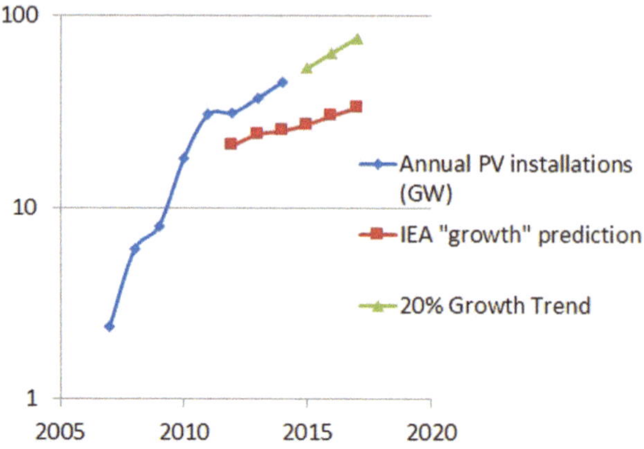

Fig. 18.3 Solar PV growth (Annual capacity additions in GW)

18.4 Conclusions

There are a number of ways in which energy statistics are presented with bias, either through the manner of presentation, or due to the assumptions behind the figures presented. In some cases, e.g. the data from the *Grain Farmers of Ontario,* it is perhaps not surprising that figures are presented to support a particular form of energy supply. However, it is more concerning when the data from trusted sources such as the IEA are found to give consistent bias. In general their data give credibility to large-scale conventional energy supply from fossil or nuclear power, while downplaying the role of renewable technologies such as solar PV and wind energies. Although the latest *World Energy Outlook 2013* [12] gives more weight to renewables, the IEA assume a decrease in new wind and solar capacity additions after 2020.

References

1. Aleklett K et al (2010) The peak of the oil age—analyzing the world oil production reference scenario in World Energy Outlook 2008. Energy Policy 38(3):1398–1414
2. McGlade CE (2014) Comment on the 'Uppsala critique'. Energy Policy 67:154–158
3. Rechsteiner R (2008) Wind power in context—a clean revolution in the energy sector. Energy Watch, Ludwig-Boelkow-Foundation
4. British Petroleum (2014) BP statistical review of world energy. BP plc, London
5. Grain Farmers of Ontario (2011) Although this report appears no longer to be available, these statistics are still quoted at another website. http://www.stockjournal.com.au/news/agriculture/general/news/new-bioproducts-market-to-outgrow-biofuels/2321929.aspx. Accessed Aug 2014
6. Smil V (2000) Perils of long-range energy forecasting: reflections on looking far ahead. Technol Forecast Soc Change 65(3):251–264
7. Smil V (2003) Energy at the Crossroads. MIT Press, Cambridge. (ISBN: 9780262693240)
8. Department of Trade and Industry (2004) DTI energy projections. DTI, London
9. International Energy Agency (2009) World energy outlook 2009. IEA, Paris

10. International Energy Agency (2012) Renewable energy: medium-term market report: market trends and projections to 2017. IEA, Paris
11. European Photovoltaic Industry Association (2013) Global market outlook for photovoltaics, 2013–2017. EPIA, Brussels
12. International Energy Agency (2009) World energy outlook 2013. IEA, Paris

Chapter 19
Equipment for Technical Education in the Field of Energy and Environment

Kristian Boedecker

Abstract GUNT Hamburg is a leading manufacturer of technical equipment for engineering education. In order to promote possible technical solutions for global challenges from climate change and progressive resource depletion, the 2E division of GUNT concentrates on didactic equipment for the field of energy and environment. The close interconnection between the fields of renewable energy, energy efficiency, and environmental technologies is a fundamental aspect of the 2E concept as indicated in the scheme below. Among the environmental technologies, the water treatment field is of special importance since it is crucial for the conservation of the human habitat. The objective of 2E is to integrate the principles of sustainability into the field of technical training. A carefully thought-out spectrum of teaching and research equipment covering high-didactic demands for the energy and environment sector has been developed.

This contribution gives an overview of the structure of the 2E curriculum, summarizes the basic didactical and technical concepts, and gives short examples of selected training units from different subject areas. The didactical concept of the development of a 2E technical training unit emphasizes the key training aspects from practice and theory with a special focus on sustainable engineering. The scope of typical experiments ranges from the commissioning and operation of important system components to the calculation and optimization of the essential key performance indicators. The energy efficiency of a process or a system is among the most important and meaningful key figures. Thus enabling the students to perform appropriate measurements for a step-by-step analysis of parameters affecting the energy efficiency is a predominant objective when preparing the instructional material.

As a typical example of this concept, our experiments in the field of photovoltaic (PV) energy conversion are presented in detail. The ET 250 trainer enables measurements on typical PV modules by means of a basic electrical circuit. Current, voltage, temperature, and irradiation intensity can be measured using a preconfigured measurement unit. Knowledge gained from these experiments serves as a fundament to understand details such as the function of maximum power point tracking systems which are typical in modern PV installations.

K. Boedecker (✉)
GUNT2E Hamburg, PO Box 1125, 22881 Barsbüttel, Germany
e-mail: Boedecker@gunt.de

A. Sayigh (ed.), *Renewable Energy in the Service of Mankind Vol II*,
DOI 10.1007/978-3-319-18215-5_19

In the next step of our PV course, methods and practical components to compensate temporal fluctuations in the energy demand and offer are investigated. To cover the associated learning objectives, we developed the ET 255 trainer. The ET 255 provides a built-in PV simulator besides PC-based measurement and analysis software. Thus the effect of varying irradiation on the behavior of systems components such as grid-connected inverters or stand-alone charge regulators can be investigated during a laboratory course without being hampered by actual weather conditions.

At the end of this contribution, some remarks on new 2E developments are given. In the biomass energy field, the laboratory biogas plant CE 642 was successfully commissioned for two customers. In the field of energy from sea waves, our ET 270 was completed which provides a laboratory system based on the principle of an oscillating water column (OWC).

Keywords Engineering education · Sustainability · Renewable energy · Environment · Curriculum · Photovoltaic · Technical education · Energy efficiency

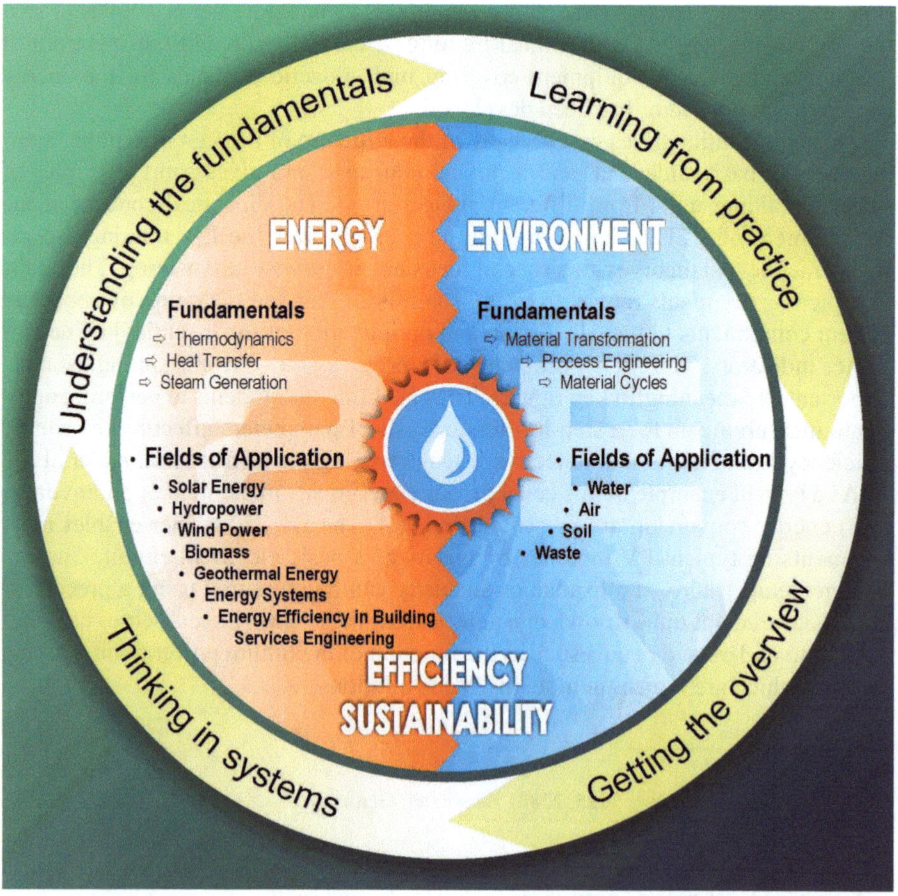

19.1 The 2E Curriculum

The essential idea behind the 2E curriculum is the close interconnection between the fields of energy and environment. This idea results from the insight that all facets of the human habitat are part of one system. To keep this system available in the future, a sustainable and efficient use of resources is necessary. For technical education, it follows that thinking in systems and use of sustainable technology must have the highest priority in didactical concepts. To implement this idea in the field of equipment for technical engineering, the 2E division of GUNT was established.

As an example of the 2E activities, some details of the two training systems from the energy branch of 2E are discussed below. The systems ET 250 and ET 255 belong to the solar energy field, which comprises PVs and solar thermal energy as subtopics. Our table of learning fields in solar energy with the assigned 2E systems is given below (Fig. 19.1):

19.2 ET 250 Solar Module Measurements

The intention for ET 250 was the preparation of a ground-level training system that enables basic measurements on PV solar modules and comprises real-world components. For the development of ET 250, the following learning objectives were considered:

- Practical aspects when using PV solar modules
- Familiarization with measurands such as short-circuit current, open-circuit voltage, maximum power
- Calculation of effective power, fill factor, and collector efficiency
- Influence of the module tilt on the electrical power
- Effect of shaded cells

The ET 250 trainer contains two standard silicon solar modules whose tilt can be adjusted. The modules can be connected in series or in parallel. A slide resistor enables varying electrical load, thus making it possible to record current–voltage curves. Sensors attached to the solar modules detect illuminance and temperature (Fig. 19.2).

The compact measuring unit provides displays for measurements of voltage, current, illuminance, and temperature. ET 250 accessories include an inclinometer and a complete set of cables. The well-structured instructional material sets out the basic principles and guides through the experiments step-by-step. Typical experimental results are shown below (Fig. 19.3).

Fig. 19.1 2E learning fields in solar energy

19.2.1 The Instruction Manuals

A detailed instruction manual is prepared for every 2E training system. Besides the construction and the production of a system, this work represents a substantial part of the development process. The preparation of the manuals includes appropriate treatment of fundamentals as well as the conduction of reference experiments.

The instruction manual starts with an introduction to the system followed by important safety instructions and detailed description of system components. After

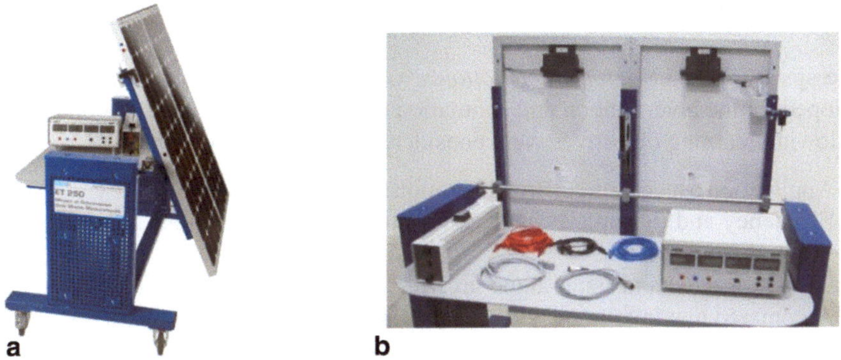

Fig. 19.2 **a** ET 250 trainer, **b** ET 250 components

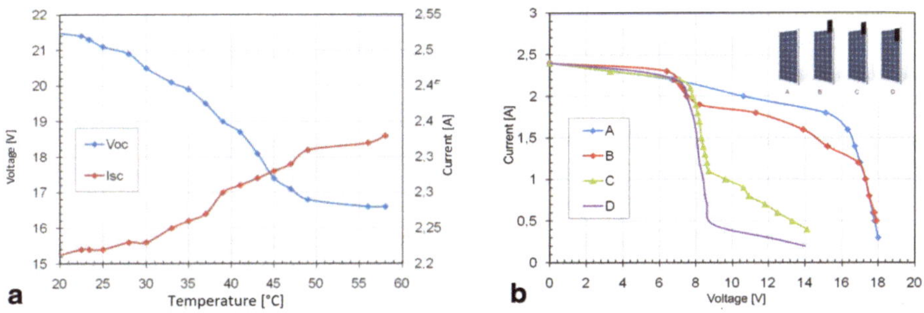

Fig. 19.3 **a** V_{oc} & I_{sc} vs temperature. The students can investigate the importance of the module temperature. The results (Fig. 19.3a) show a decrease of the open circuit voltage V_{oc} and a slight increase of the short circuit current I_{sc} with module temperature. **b** *IV* curves from shaded modules. The learning objectives for this experiment (Fig. 19.3b) also include the function of bypass diodes. Curves *A, B, C,* and *D* show how *IV* behavior changes when a part of the module is increasingly shaded

that a chapter with fundamentals and the technical background follows. The typical content of a complete instruction manual is as follows:

- Introduction and didactical information
- Safety information
- Detailed description of the device and essential components
- Basic principles
- Exercises with solutions
- Experiments with step-by-step instructions
- References, results, and technical data

It is strongly recommended that the students prepare themselves using the listed exercises before starting the practical experiments. The instruction manuals contain carefully compiled information and reference data for each recommended experiment. Step-by-step instructions ensure the success of each experiment.

19.2.2 *Exploring PV System Components with ET 255*

The system ET 255 *Using Photovoltaics: Grid connected or Stand-alone* was developed for the investigation and simulation of PV systems and its components. The following learning objectives were considered during the development:

- Application aspects of electrical components for PV systems
- Efficiency and dynamic behavior of system components in grid-connected and stand-alone operations
- Grid-connected and stand-alone inverters
- Automated maximum power point tracking
- Function of inverters
- Function of solar batteries and charge controllers
- Behavior of components under varying temperature and illuminance

ET 255 comprises typical components for the usage of PV current. The PV current can be used for feeding into a public power grid (grid-connected operation) or for local consumption (stand-alone operation). ET 255 can be run by both, with actual solar modules (such as ET 250) and with the built-in PV simulator. Operation and parameterization of the PV simulator is carried out via a dropdown menu in the software program. With relatively little effort, the PV simulator makes it possible to investigate the effects of changing illuminance and temperatures (Fig. 19.4).

A variety of software functions are available for capturing and displaying the measurement data. The efficiency and dynamic behavior of the electrical system components can be studied by analyzing these results. Actual readings for current,

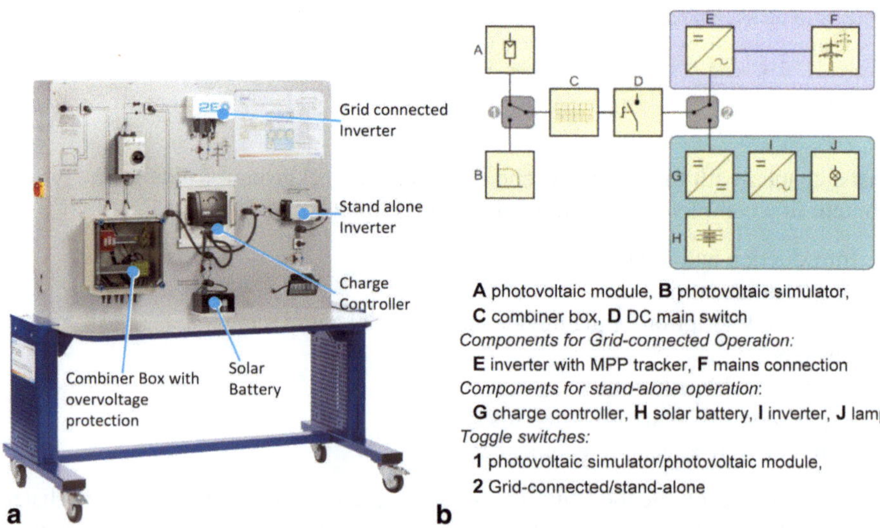

Fig. 19.4 **a** ET 255 trainer, **b** ET 255 schematic. *A* photovoltaic module, *B* photovoltaic simulator, *C* combiner box, *D* DC main switch. *Components for grid-connected operation*: *E* inverter with MPP tracker, *F* mains connection. *Components for stand-alone operation: G* charge controller, *H* solar battery, *I* inverter, *J* lamp; *Toggle switches: 1* photovoltaic simulator/photovoltaic module, *2* Grid connected/stand-alone

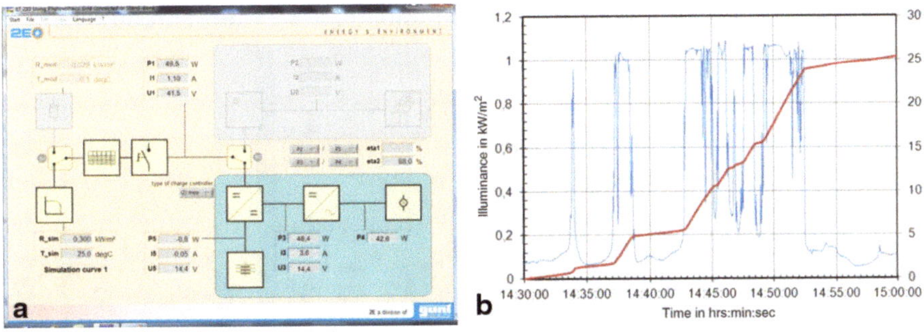

Fig. 19.5 **a** Screenshot of the ET 255 software, **b** Results from ET 255 with ET 250

voltage, and electrical power are displayed in the system schematic view of the software (Fig. 19.5a).

The ET 255 software can also be used to record and transfer measured data from longer measurement periods. Results from the combination of ET 255 with ET 250 during a cloudy day are displayed in Fig. 19.5b as an example. The results show a time resolved measurement of the illuminance in kW/m^2 (blue curve) and the supplied electrical energy in Wh (red curve). The data were recorded under natural sunlight with some clouds passing during the measurement. From the diagram it can be clearly seen how the clouds affect the PV generation of electrical energy. The analysis of the system performance under varying illumination and load conditions is discussed in detail within the results section of the instructions manual.

19.3 Other Training Systems from the Energy Section of 2E

19.3.1 ET 270 Wave Energy Converter (Fig. 19.6)

19.3.2 CE 642 Biogas Plant (Fig. 19.7)

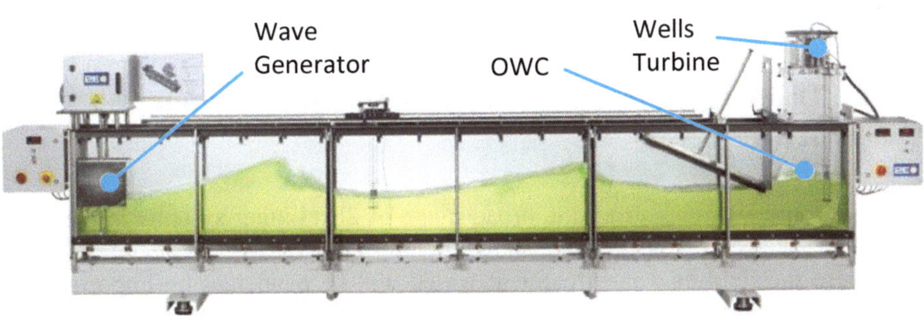

Fig. 19.6 Wave energy converter. The ET 270 gives an example for further training systems of the 2E curriculum. The ET 270 wave energy converter belongs to the 2E hydropower section. The system demonstrates the conversion of energy from sea waves into electrical energy by using the so-called oscillating water column (OWC)

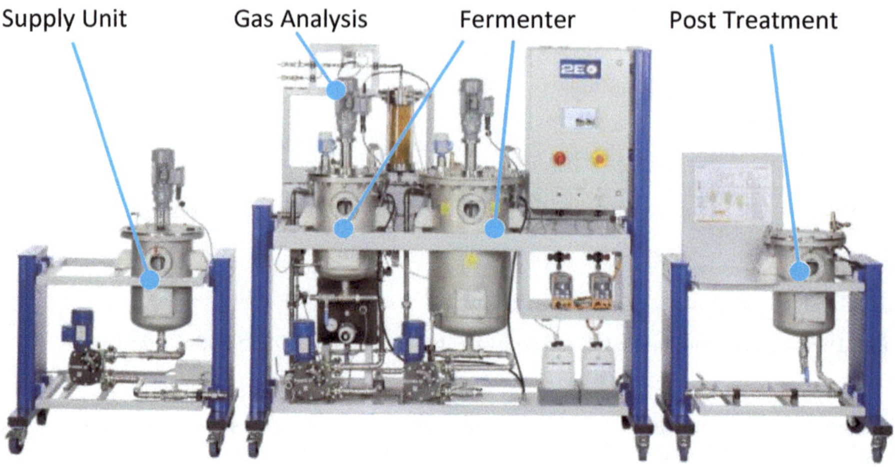

Fig. 19.7 Biogas plant. The system CE 642 "biogas plant" enables the fermentation of biomass to produce biogas. CE 642 expands the 2E programme in the biomass section. CE 642 is equipped with industrial standard components and programmable logic controller (PLC) to provide optimized process conditions

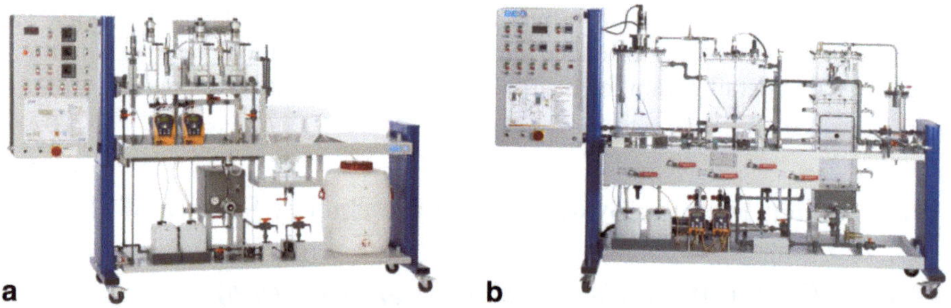

Fig. 19.8 **a** Activated sludge process. The system CE 705 "activated sludge process" demonstrates the most important biological process in water treatment. CE 705 represents a wastewater treatment plant in laboratory scale. **b** Anaerobic water treatment. CE 702 "anaerobic water treatment" enables the degradation of organic substances. The system comprises a so-called "*U*pflow *A*naerobic *S*ludge *B*lanket" reactor

19.4 Training System from the Environment Section of 2E

To give some examples from the broad range of 2E training systems in the environmental sector, two systems are shown below: The systems CE 702 and CE 705 belong to the field of water treatment which will be of increasing importance in sustainable technical education (Fig. 19.8).

19.5 Summary

Experiences from vocational schools, technical colleges, and universities confirm repeatedly that practical work with experimental setups is crucial for a successful technical education. In the field of energy and environment, the usage of sustainable technologies represents a fundamental requirement for modern engineering education. It becomes obvious that technical systems cannot be optimized in a sustainable way unless the complete material and energy balance of the system during its lifecycle is considered. In this context the 2E curriculum combines the areas of energy and environment and provides appropriate guidelines for engineering education. GUNT Hamburg established the 2E division to implement this concept permanently within development and production of technical equipment for engineering education.

As shown here for the PV sector, the identification of learning fields and related sets of learning objectives are prerequisites for a comprehensive portfolio of successful technical training systems. The ease of use, a clearly predefined range of experiments, and the use of modern industrial components are essential aspects for the development process. A further conclusion that can be drawn from experiences of more than 30 years is that carefully prepared instructional material is an indispensable part of a successful technical training system. More information on 2E and GUNT can be found under www.gunt2e.de or www.gunt.de.

Chapter 20
The Development of Renewable Energy in Russia

Sergey Karabanov, Pavel Bezrukikh, Evgeny Slivkin and Dmitry Suvorov

Abstract The beginning of the twenty-first century has been marked by quick growth of renewable energy sources market in the world, especially, in Europe. In Russia this market is extremely small; big energy projects with the application of renewable energy sources are absent. The present chapter provides the data on the potential of renewables in Russia; the analysis of the factors affecting the application of renewable energy sources.

In 2012 the Russian government took a number of steps for the development of renewable energy, first of all, photovoltaics (PV), wind energy, and small hydropower. The chapter analyzes these measures and data on the quantity of installations of PV, wind energy, and small hydroelectric power stations. The information on a number of projects of renewable energy sources use in houses, urban facilities is provided. The additional initiatives promoting the use of renewable energy in Russia are offered.

Keywords Renewable energy sources · Solar power · Wind energy · Small hydropower · Initiatives

20.1 Introduction

Russia has all kinds of renewable energy sources, and the majority of the regions has two and more renewable sources. For example, the Krasnodar Krai has the economic potential for all renewable energy sources: geothermal and solar power, wind energy, hydropower of small rivers and watercourses, low-potential power of the sea, ambient air and technical water supply drains. The technical potential of the renewable energy sources in the region is two times higher than the present energy

P. Bezrukikh (✉)
Krzhizhanovsky Power Engineering Institute, Moscow, Russia
e-mail: bezruky@yandex.ru

S. Karabanov · E. Slivkin · D. Suvorov
Ryazan State Radio Engineering University, Ryazan, Russia
e-mail: pvs.solar@gmail.com

A. Sayigh (ed.), *Renewable Energy in the Service of Mankind Vol II*,
DOI 10.1007/978-3-319-18215-5_20

Table 20.1 The estimation of renewable energy sources potential in Russia

Resource	Gross potential, 10^6 t (c.e.)/year	Technical potential, 10^6 t (c.e.)/year	Economical potential, 10^6 t (c.e.)/year
Wind energy	44,326	2216	11
Small hydropower	402	126	70
Solar power	2,205,400	9695	3
Biomass energy	467	129	69
Geothermal energy (hydro-thermal resources)	[a]	11,869	114
Low potential heat	563	194	53
Total	*2,251,158*	*24,229*	*320*

[a] Gross potential of hydrothermal energy amounts to 29.2×10^{12} t (c.e.)

consumption in the territory. That is why solar heating systems are widely spread just in this region. The economic potential of renewable energy sources in Russia is about 320 million tons of reference fuel (t (c.e.)), that is about 30 % of the domestic consumption of the energy sources in 2010 (970×10^6 t (c.e.)) (Table 20.1) [1].

In spite of such possibilities the application of renewables in Russia is small. According to the data of the annual report of the Federal State Statistics Service about technical and economic indices and consumption of reference fuel at power stations, the share of renewables in power engineering of the Russian Federation in 2002–2012 was 0.66 % (Y2012) (Table 20.2).

Until 2012 there were the following renewable energy stations in Russia:

- *Heat power stations using biomass*:

  Quantity—36. Total power—1355 MW.
  The share of biomass in the fuel is 54.5 %. Nominal capacity—738 MW.

- *Small hydroelectric power stations*

  a. Quantity—58 (belong to Energo JSC). Installed capacity—618 MW.
  b. The quantity of stations not belonging to Energo JSC—17. Installed capacity—118,045 MW.

Total: 75 small hydropower stations. Installed capacity—735,964 MW.

- *Geothermal power stations*

  Quantity—5. Installed capacity—86.5 MВт.

- *Wind power stations*

  Quantity—7. Working—5. Installed capacity—9.27 MВт.

- *Solar power stations*

  Quantity—4. Installed capacity—150 кВт.

The chapter considers the situation of renewable energy sources application in Russia and legislative initiatives as far as encouragement of renewables use concerns.

Table 20.2 Power generation on the basis of renewable energy sources in 2002–2012 in Russia, including small hydroelectric power stations, 10^9 kWh

Periods, years	2002	2003	2004	2005	2006	2007	2008	2009	2010	2011	2012
Wind power station	0.007	0.009	0.014	0.010	0.008	0.007	0.005	0.004	0.004	0.005	0.004
Geothermal power station	0.15	0.31	395.1	0.39	0.46	0.48	0.46	0.46	0.47	0.47	0.48
Small hydroelectric power station	2.42	2.42	2.75	2.78	2.55	2.72	2.87	3.32	2.85	2.73	2.87
Biomass heat power station	2.44	2.62	2.82	2.71	2.91	2.82	3.12	2.96	2.99	3.30	3.72
Total	*5.02*	*5.36*	*5.98*	*5.89*	*5.93*	*6.03*	*6.46*	*6.75*	*6.32*	*6.50*	*7.07*
All power stations in Russia	891.3	916.3	931.9	953.1	931.4	1008.3	1040.4	990.0	1025.4	1052	1066
Renewables share, %	*0.56*	*0.59*	*0.64*	*0.62*	*0.64*	*0.60*	*0.62*	*0.68*	*0.62*	*0.62*	*0.66*

Table 20.3 Marginal amounts of capital costs for renewable energy sources, RUB/1 кW

Renewable energy stations	2014	2015	2016	2017
Wind power stations	65,762	65,696	65,630	65,565
Solar power stations	116,451	114,122	111,839	109,602
Small hydroelectric power stations	14,600	14,600	14,600	14,600

Table 20.4 Target values for localization of different renewable energy sources

Renewable energy stations	Start-up	Localization index, %
Wind power stations	2014	35
	2015	55
	From 2016 to 2017	65
Solar power stations	From 2014 to 2015	50
	From 2016 to 2017	70
Small hydroelectric power stations	From 2014 to 2015	20
	From 2016 to 2017	45

20.2 Analysis of the Legal Basis for Renewable Energy Sources Application

The Russian Government issued the Decree No. 449 of May 28, 2012 "On the mechanism of stimulation of renewable energy sources application at the wholesale market of electric power." The law touches upon three types of renewables:

- Solar power stations (not less than 5 MW)
- Wind power stations (not less than 5 MW)
- Small hydroelectric power stations (up to 25 MW)

The mechanism of power sale from generating facilities was chosen as stimulation for the renewable energy sources application. The rules of power pricing for solar, wind, and small hydroelectric power stations were set.

a. The operating costs values have been established, for example, the values of the unit operating costs in 2012 were the following:
   - For solar power stations—RUB 170,000.00 (~ US$ 5100.00)/MW per month.
   - For wind power stations—RUB 118,000.00 (~ US$ 3500.00)/MW per month.
   - For small hydroelectric power stations—RUB 100,000.00 (~ US$ 3000.00)/MW per month.
b. The marginal amounts of capital costs for 1 kW installed capacity have been established (Table 20.3).
c. The target values of the production localization have been established (Table 20.4).
d. The planned quantities of power takeoff for renewable energy sources have been established (Table 20.5).

Table 20.5 Planned quantities for power takeoff, MW

Renewable energy stations	2014	2015	2016	2017	Σ
Wind power stations	100	250	250	250	850
Solar power stations	120	140	200	250	710
Small hydroelectric power stations	18	26	124	124	292
Total					*1752*

Table 20.6 Energy volumes of the selected projects, MW

Renewable energy stations	2013	2014
Wind power stations	105	51
Solar power stations	394.198	505
Small hydroelectric power stations	–	20.64

The payback of investments is carried out at the expense of the price for the delivered power.

The methods for calculation of the price components for the power of renewable energy stations which assures the payback of capital and operating costs are developed.

20.3 Results of the Decree Realization

According to the established rules, the renewable energy projects are selected every year. Table 20.6 shows the energy volumes of the projects selected in 2013 and 2014. It can be seen that the decree plays a big role in the development of renewable energy.

The criterion of the localization degree has a significant effect on the development of the renewable energy industry. We can see it through the example of solar power stations. Photovoltaic (PV) modules and grid-connected inverters are necessary for the station construction. Full-scale production of the following items is not available in Russia:

- Polycrystalline silicon
- Solar cells
- Inverters

The production of PV modules is limited. The similar situation is with wind power stations. Currently the measures on creation and extension of production facilities are taken.

20.4 Stand-Alone Renewable Energy Stations

One of the lines for renewable energy source researches, getting traction in the Russian Federation, is power supply of residential houses and lighting.

Fig. 20.1 Multifamily residential house with hybrid power supply

20.5 Residential Houses

At construction of residential houses different design and technological concepts which allow to reduce building operation costs by application of power-efficient technologies and renewable energy sources are used.

One of such projects realized in the central region is building of a two-storeyed residential house with the total area of 713.5 m^2. A PV power plant, thermal pumps, solar collectors were used for the house power supply [2] (Fig. 20.1).

Specifications of Renewable Energy Sources Used in the Project

1. PV power plant
   Peak power—7.5 kW
   PV modules—RZMP-235-T, 32 pcs.
   Inverter—TripleLinx 8k, 1 pc.
   Storage battery—200A2/12Bx32
2. Thermal pump
   Corsa-25 (Russia) with the heat power of 24.4 kW.
3. Solar collectors

EE-SHS solar water heaters (360 pcs.) and Cordivari boilers were used as solar collectors.

In case of optimal installation of water heaters the system gives 2100 L of hot (55 °C) water during 6 months per year: from May to September.

The use of a PV power plant provides uninterrupted year-round power supply for common areas and lighting of courtyard areas. The application of solar collectors ensures seasonal hot water supply for the house. A thermal pump provides year-round heat supply.

It has been established that the operating costs of living quarters is 1.9–2.1 times lower than that of standard house maintenance.

The house operation with the use of renewable energy sources (thermal pumps, solar heat, PV power plant) and modern energy saving materials (low-emissivity glass, construction materials, etc.) has shown that the effective application of such technologies is possible in Central Russia.

20.6 Lighting

Wide application of light emitting diode (LED) lamps in recent years and considerable price decrease of lamps and PV modules allow to use PV systems for public lighting. Some objects, such as playgrounds, parks allow to use lighting during shorter period of time; lighting is also possible with the use of motion sensors. It can reduce significantly the requirements to the energy necessary for lighting [3].

20.7 Structure of the System for Sports Grounds Lighting

There are two options of the lighting system creation—stand-alone (when each lamp has a PV module, a storage battery, and balance of system (BOS) components) and locally centralized (a PV generator is placed in one point of the illuminated territory, the most convenient one from the point of view of solar resources availability. The container with control and converting electronics, accumulators are placed there too). The power wiring to the lamps located in this territory is carried out by an alternating current to decrease the losses. Such an option can be used at illumination of courtyard territories or internal building premises (PV modules are grouped on a building roof where the probability of their shading is considerably less and the area, not used for other purposes, is available). Besides, the locally centralized option allows to place larger quantity of PV modules on the object, without being limited to strength and stability requirements of an individual lamp support that is especially important for regions with the limited potential of solar irradiance.

For the sports ground lighting, the systems on the basis of the following equipment were used:

- 215Wp RZMP-215-T PV modules—12 pcs.
- ILLUMINEX LED lamps, (power—38 W, supply voltage—175–240 W, luminous flux—3200 lm, fluorescent temperature—5000 K, provide even horizontal illuminance at the level of coating of 30 Lx at the spot with the diameter of about 15 m at lamp location height of 5 m)—6 pcs.
- Storage battery 200 Ah—6 pcs.
- Inverter 24 V, 600 W—1 pc.
- Charge controller 60 A—2 pcs

The specified set provides stand-alone operation of the illumination system within 5 h per day at the most unfavorable climatic conditions during the winter period

Fig. 20.2 View of the lighting system installed on a sports ground in Moscow (2011)

with the minimum duration of daylight hours (November–January). The storage batteries installed have a capacity that provides the load consumption of 0.23 kWh for 6 days at total lack of their charge from PV modules.

The system (Fig. 20.2) generates about 6000 kWh/year, the real power consumption for sports ground lighting during the period of September–April makes up 280 kWh. The rest of electric power, generated and not consumed for the ground lighting, is used for standby illumination of a nearby apartment house entrances.

20.8 Basic Trends of Scientific Researches

1. Researches in the field of high-efficiency units are based on PV modules with solar concentrators and precision tracking for the Sun. Ground systems with the efficiency of more than 35 % and the lifetime of not less than 25 years were created [4].
2. Researches in the field of space solar power plants based on A_3B_5 compounds [4].
3. Researches on the development of high-voltage planar cascade solar cells on the basis of silicon multilayer structures [4].
4. Researches in the field of $A_3B_5$and A_4B_6 nanoheterostructures production [4].
5. Researches in the field of silicon production based on chlorine-free technologies:
   - Plasma chemical purification of metallurgical-grade silicon [5].
   - Purification by the method of impurities extraction from a solid phase [6].

20.9 Factors Encouraging Further Development of Renewable Energy in Russia

The adopted decree does not solve all the problems in the application of renewable energy sources. First of all, it is the limited use of renewables: wind, solar, and small hydroelectric power stations. So, it is necessary to take the following measures:

1. Extend the decree for the other sources: geothermal, bioenergetics, etc.
2. Stimulate the use of renewable energy sources at the regional levels. In Russia there are 85 regions, and each region has the potential to use minimum two types of renewable energy sources.
3. Take power generated by renewable energy sources to the grid without limits.
4. Provide financial support for the development of the equipment production for renewable energy by creating federal and regional programs and taxation reduction.
5. Stimulate the development of scientific researches in the field of renewable energy sources and to train scientific and engineering personnel.

20.10 Conclusion

The creation of new conditions—adoption of the law for stimulating the application of renewable energy sources—plays a big role in the development of renewable energy in the Russian Federation.

The chapter gives the analysis of measures stimulating the renewable energy development. The factors encouraging further growth of the share of renewable energy in the country energy balance are considered. The basic trends of science development in PV are given.

References

1. Karabanov SM, Kazakova NV, Kuropov MV (2011) The prospects for photovoltaics development in Russia. Proceedings 26th European photovoltaic solar energy conference
2. Evseenkov OV, Karabanov SM, Kirakovskiy VV (2013) Hybrid power supply of a residential house in Central Russia using a PV Station. Proceedings 28th European photovoltaic solar energy conference
3. Karabanov SM, Kukhmistrov YV, Arushanov KA, Osipov DV (2012) Assessment of PV street lighting application in Russia. Proceedings 27th European photovoltaic solar energy conference
4. Proceedings 1st International Renewable Energy Forum "Towards raising energy and economic efficiencies" (REENFOR-2013), Moscow, October 22–23, 2013
5. Karabanov S, Dshkhunyan V, Yasevich V (2012) Patent of the Russian Federation № 2465202
6. Karabanov SM, Suvorov DV, Slivkin EV, Sazhin BN (2013) The analysis of silicon purification efficiency by the method of extraction from a solid phase. Proceedings 28th European photovoltaic solar energy conference

Chapter 21
Future Energy Without Oil and Fossil Fuel

John Lowry

Abstract Problems caused by global warming due to carbon dioxide released from combusting fossil fuels are now generally accepted. Fossil fuels are a finite resource that will start to run out during this century, we have, for example, around 40 years of oil supply at current usage rates. It is shown in the chapter that world can run successfully without fossil fuel sources using the current technology.

21.1 Introduction

Problems caused by global warming due to carbon dioxide released from combusting fossil fuels are now generally accepted by the scientific community. These problems include rise in sea levels, melting of ice in Greenland and Polar regions, and an increase in tornadoes, droughts and floods. The volume of Antarctic ice detected to be melting into the ocean has doubled in the past decade. Since the start of 2013 alone we have experienced major typhoons in the Philippines and India, hurricanes in Mexico, tornadoes in the USA and flooding in the UK—all of which are considered to have been caused or made worse by global warming. Coupled to this is the fact that fossil fuels are a finite resource that will start to run out during this century.

It is definitely time for dynamic action. It is quite possible with current technology to run society without using fossil fuels and this is the main context of this chapter.

To do this, we need to understand three things. First, how much energy we derive from fossil fuels, second where we use this energy and third how we can realistically adapt the power sources using today's technology to use the energy from non-fossil fuel sources.

J. Lowry (✉)
Hillsborough, Bishopstone, Swindon, SN6 8PW, UK
e-mail: John1000000@live.com

A. Sayigh (ed.), *Renewable Energy in the Service of Mankind Vol II*,
DOI 10.1007/978-3-319-18215-5_21

21.2 How Much Energy Do We Use and Where Do We Use This

The world currently uses some 10,000 million tons of oil equivalent per annum or 116,000 TWh of energy.

The annual percentage breakdown of different energy sources for 2008 is illustrated in Fig. 21.1.

An example of where this energy is used in the USA is shown in Fig. 21.2.

An approximate breakdown of energy used by different forms of transport is illustrated in Fig. 21.3. This is based on worldwide emission figures.

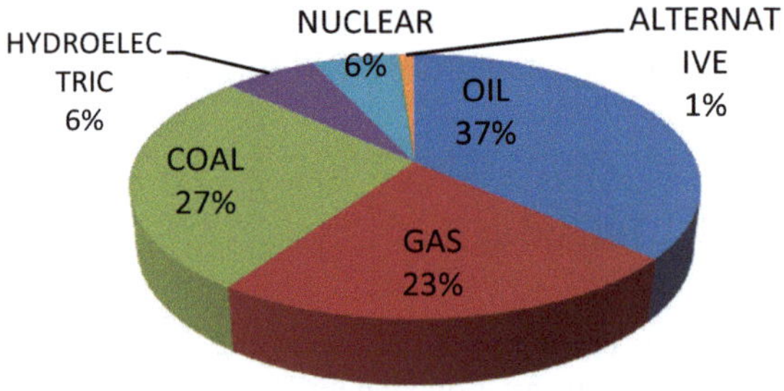

Fig. 21.1 Percentage breakdown of worldwide energy for 2008

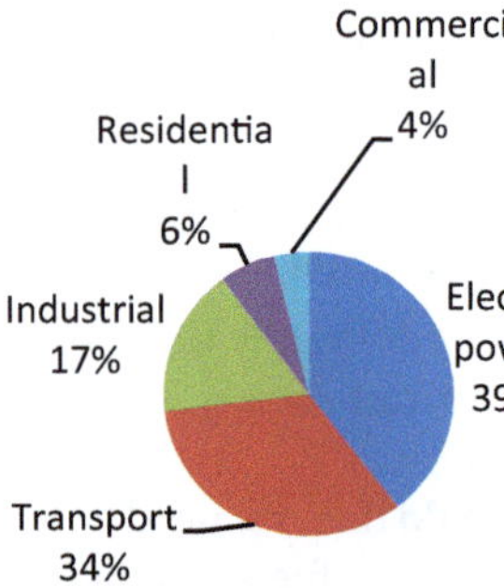

Fig. 21.2 Electricity generated from fossil fuels in the USA in 2005. (Source http://www.greenenvironmentnews.com/energy_co2_emissions_by_sector.php)

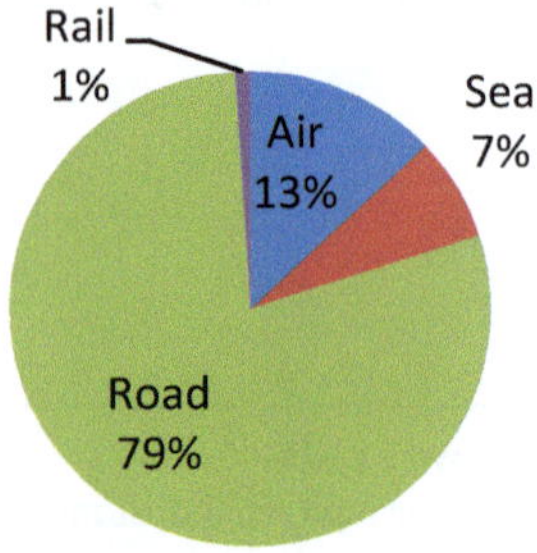

Fig. 21.3 Breakdown of how fossil fuels are used in worldwide transport. (Source http://www.unep.org/climateneutral/Topics/Transport/tabid/154/Default.aspx)

21.3 Non-Fossil Fuel Alternatives

There are two options if we wish to produce energy from non-fossil fuel sources. Either we need to obtain energy from non-fossil fuel sources or we need to derive fuels from non-fossil fuel sources.

The existing established technologies for producing non-fossil fuel energy that does not release carbon dioxide include nuclear (fission) energy, solar energy, wind energy, hydro energy, tidal energy and geothermal energy. Alternative fuels that do not rely on fossil fuels include biofuels, hydrogen electrolysed from water and fuels produced from algae.

Nuclear fission is well established as a power source already providing most of the electricity in France and around 20 % of the electricity in the USA at a cost that is considered to be economic. The safety of nuclear power stations is a highly controversial issue and is certainly being addressed. It must be borne in mind that while there are safety issues relating to nuclear energy, the consequences of continuing to use fossil fuels may itself result in a series of disasters that will be brought about by climate change.

Sources of sustainable or alternative energy are well known and these include hydro, solar, wind and tidal power. Hydro power is already well established, currently providing about 6 % of electricity worldwide. Among others solar and wind are perhaps the most established.

Solar has the biggest potential. It is abundant and available worldwide. Every year the sun irradiates the land with 0.22 billion TWh. Only a fraction of this would satisfy the world's present energy needs currently supplied by fossil fuels. Put differently, the amount of solar energy falling on the earth in less than an hour could power the planet for 1 year. In 2008, the earth used around 116,000 TWh of fossil fuel energy, a fraction of a per cent of the solar energy available.

Clearly, there are adequate amounts of solar energy available now and for the future. The problems of using solar energy have been the cost and availability of technology to capture it. The amount of solar radiation varies both with the time of day and with the weather and rarely matches the requirements of users. However, there has been a quiet revolution in the cost of solar power.

The cost of photovoltaic panels has fallen consistently since the 1980s and will continue to fall. The cost is predicted to fall below US$1000 peak kilowatt in the near future. The cost of energy generated by solar photovoltaic power has already reached cost parity with electric grid costs when using solar photovoltaics in sunny regions of southern Europe and is predicted to reach parity with less sunny places such as the UK and Germany within this decade.

It is estimated that the average cost of solar power in the USA will fall below the current average retail electricity price of 12 cents per kilowatt-hour in around 2020. In fact, given that retail electricity prices are rising by a few per cent each year, prices will probably cross earlier, around 2018 for the country as a whole, and as early as 2015 for the sunniest parts of America.

Solar radiation in itself has a low energy density in terms of energy per square metre and in countries such as the UK where the land per head of population is in relatively short supply, sufficient land will not be available to generate enough energy to replace the energy generated from fossil fuels. However, there are vast areas of land such as the Sahara Desert in North Africa, which have plenty of land and plenty of solar radiation. Power generated from solar plants in the desert could be transmitted back to countries in Europe via the supergrid. For the night-time use in Europe the energy would need to be transmitted from further afield.

In the Sahara, the average sunshine radiation on a horizontal surface is 273 W/m^2. Solar photovoltaic panels have an efficiency of about 15 % (electrical energy generated/solar energy). Therefore, in the Sahara 1 m^2 of panel will produce an average of 41 W and 359 kWh per annum. The UK, for example, uses 2461 TWh of fossil fuels. Assuming an average conversion efficiency of 33.3 %, an average of 820 TWh of electricity would be needed to replace this. In order to generate this much power, an area of photovoltaic panel 2284 km^2 would be required. The area of the Sahara is 9.3 million km^2 so the photovoltaic panels would occupy 0.025 % of the total area. This is little more in proportion to a small rug on a football pitch. If the solar panels are tracked to face the sun at all times, nearly 600 kWh per square metre of electricity would be obtained every year.

As a simple example consider solar photovoltaics costing US$1000 installed per peak kilowatt—i.e. the photovoltaics will produce 1 kWh with a solar radiation level of 1000 W/m^2. In practice a much lower radiation figure will be achieved and in the solar radiation averaged across 8769 h in a year in locations such as the Sahara will be 273 W/m^2. The annual electricity produced by photovoltaics rated for 1 kW peak will therefore be 2390 kWh per annum. Spread over a life of 20 years and ignoring interest, this photovoltaic panel will cost US$50 per year and produce electricity costing 2.1 cents per kWh. This figure needs to be increased to account for 10 % loss in performance over the life of the photovoltaics and a further 20 % to account for inefficiencies in inverters and transmission. This still brings the cost of electricity to less than 3 US cents per kilowatt—an attractive figure.

Electricity from solar power stations in desert areas would need to be transmitted back to areas where the power was needed using the supergrids mentioned earlier. The cost of high-voltage transmission lines is relatively less compared with the cost of the solar panels.

By using supergrids with solar power stations distributed around the globe and connected by high-voltage, high-efficiency electric transmission lines, solar power would provide solar electricity throughout the day and night.

Solar power stations are not of course limited to the Sahara. The best location is in the "sun belt," located roughly between the 40th parallels north and south of the equator, between southern Spain and South Africa, for example. This vast area of the earth's surface would include parts of Spain, France, Italy, the USA, India and Australia.

There are no insurmountable technical problems with solar photovoltaic power stations or indeed with solar thermal power stations. Bearing in mind the almost limitless availability of solar radiation and that the cost of solar will fall continuously—

whereas the capital cost of virtually every other power source is likely to increase—solar power is an extremely attractive option for the future.

An alternative to photovoltaics are solar thermal systems that are also considered to be economic, such as wind, tidal, geothermal and marine currents. Wind energy currently produces less than 1 % of the energy provided by fossil fuels. Wind energy, if fully developed, could produce up to 40 %.

Tidal energy production such as that produced by the Rance Tidal Power Station in France has proved to be successful. There is plenty of scope for increasing the use of tidal power including a Bristol channel tidal scheme. Extreme tidal schemes such as damming the Irish Sea alone could produce over 1 % of the energy produced by fossil fuels. Geothermal energy prevalent in Iceland and marine current energy can also provide substantial energy.

An alternative to using fossil fuels is to obtain energy from non-fossil fuel resources. Such fuels include biofuels, hydrogen, algae-based fuels or fuels synthesised from carbon dioxide and water. It is important that synthesised fuels use resources and energy from non-fossil sources, it is also important that biofuels do not detract from land currently used for growing food.

It is quite possible to produce fuels from carbon dioxide, in other words reversing the process of combustion. A firm known as Air Fuel Synthesis Ltd is working on developing commercial units to produce a fuel similar to petrol and is developing a plant that will produce a kerosene substitute for use in conventional jet engines. Studies by Air Fuel Synthesis indicate that 3 kWh electrical energy is needed to produce 1 kWh jet fuel.

Hydrogen can be produced by electrolysing water with electricity generated from non-fossil fuel sources such as nuclear of solar. The efficiency of electrolysing hydrogen can be as high as 85 %.

Producing fuels from algae is another possibility.

21.4 Use of Electrical Supergrids

Nuclear and alternative energy are not as effective in producing power as they might be. Nuclear has a high base load and electricity is often wasted, as it is hard to regulate nuclear power stations to match demands. Renewable energy is often readily available in one place but needed in another. As a simple example, wind energy may be abundantly available from offshore wind farms in the North Sea when it is needed in Spain at times of low wind there. High solar energy may be produced in the Sahara when more northerly climates may not be able to produce much solar energy in mid-winter. It is hard to store energy in meaningful quantities using modern technology.

A global energy network, or a supergrid, would allow electricity to be passed around the world and this would make non-fossil fuel energy generation more viable, allowing us to make much of our electricity from renewable and nuclear

Fig. 21.4 The worldwide supergrid is shown on a Dymaxion, or Fuller map projection, looking down

sources. The world needs a series of ring mains or supergrids, the advantages of which would be huge.

The idea was first mooted in the 1970s by Richard Buckminster Fuller. With the development of high-voltage direct current (DC) transmission, which allows very efficient transmission of electricity, the electrical supergrid is now possible.

The use of a high-voltage DC transmission system has already been started in Europe. High-voltage DC transmission is probably the only way of connecting the diverse electrical systems used in different countries.

This can be expanded into a worldwide supergrid as originally proposed by Buckminster Fuller (Fig. 21.4).

The use of supergrid networks has been started but they need substantial further development. Their cost is not a prohibitive factor as they would bring with them substantial benefits.

21.5 Transport

Figure 21.5 compares the energy use for different forms of transport in terms of energy use per passenger kilometre.

Comparing various modes of transport is difficult as a lot depends on the percentage occupancy of the mode concerned. Assuming 100 % occupancy, a Boeing jet uses the most energy with the two types of trains using a much lower amount (Fig. 21.5).

It should be noted that the conventional electric train at 100 mph uses 4 % of the energy of the Boeing 747 and that the high-speed train uses 7.4 % of the energy of

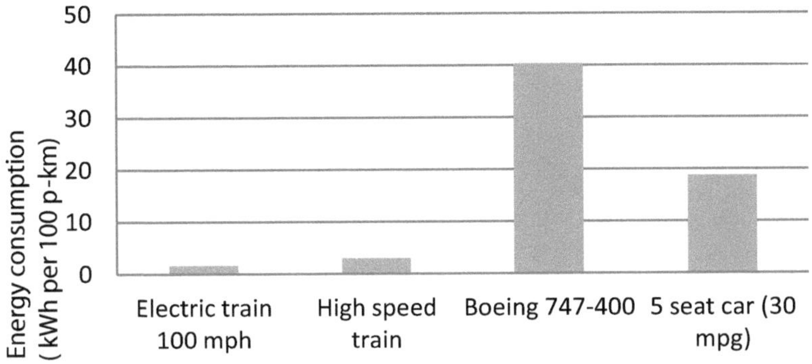

Fig. 21.5 Comparison of different forms of transport with 100% occupancy. (Compiled from figures in *Sustainable energy – Without the hot air*)

the Boeing 747. The electric train at 100 mph uses 7.5% of the energy of a fully loaded car and the high-speed train uses 16% of the energy of a fully loaded car. Some caution is needed when using these figures as the efficiency of converting fuel into energy is used for the calculation of the Boeing 747 and for the car, whereas the high-speed train figure is calculated for the output from the power station, probably nuclear energy. Nevertheless, the conventional train at 100 mph and the high-speed train are considerably more energy efficient in terms of kWh per person per kilometre than both the aeroplane and the car.

High-speed electric trains are an important development as not only they are energy efficient, consuming only 7% of the energy of airliners, but also they are starting to reach speeds where they can become competitive with air transport for inter-city transport overland, particularly where new tracks are used. Initially, this is true for shorter routes but speeds are increasing all the time. Maglev trains can already reach speeds of up to 360 mph (580 km/hr) a speed not far below that of an airliner. Trains have added advantages in that they can travel to and from city centres, whereas aeroplanes have to land at airports often placed at considerable distances from the city centre. In addition, trains are quicker to load and unload than aeroplanes. In countries such as France where electricity is largely generated by nuclear power, electric trains do not rely on fossil fuels and produce no carbon emissions.

Developing road transport that can be free from fossil fuels is considerably more complex than that for electric trains that can be connected to the electric grid by supply lines using electricity generated from fossil fuel-free sources—as is the case in France. Until recently, there were no effective road vehicles that could be considered suitable for transporting passengers and goods at reasonable speeds over reasonable distances.

Perhaps the biggest change in electric vehicles in the last few years has been the development of the lithium battery, which has a reasonable specific energy and a more reasonable charge time than previous batteries. This has ultimately led to the introduction of a series of commercial vehicles such as the Tesla S, a battery electric

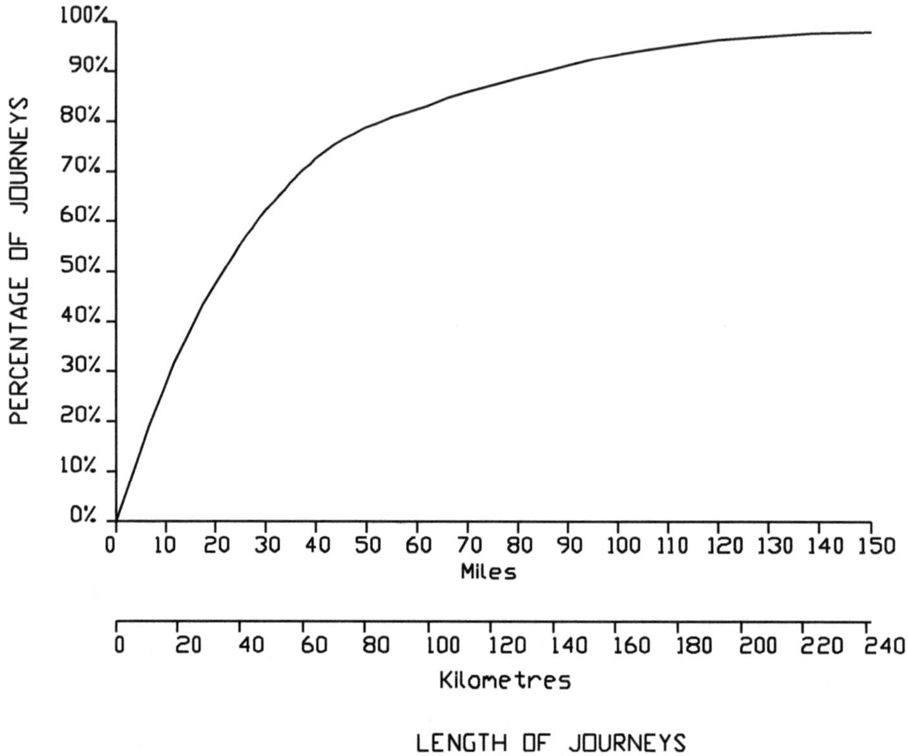

Fig. 21.6 Typical journey lengths in the USA. (Adapted from data http://www.hybridconsortium.org/dot_data.html)

vehicle (BEV) produced by Tesla Motors in the USA, that uses lithium batteries and has a range of up to 300 km. Typical journey lengths in the USA are shown in Fig. 21.6 where it can be seen that the majority of journeys are within the range of small commercial electric cars.

Lithium batteries continue to develop rapidly, and future batteries are likely to have considerable higher specific energy than those at present. The range can be extended by the use of electrical pickups while driving.

An alternative to using BEV is to use fuel cell vehicles propelled by hydrogen. Prototype cars have extensive ranges and can be refuelled quickly. At present the cost of the fuel cells is prohibitive but this is coming down.

If aircraft are to be run using fuels that are non-fossil based, there are only two realistic options. Aircraft should be fuelled by either hydrogen or synthesised fuels. Running conventional aircraft on hydrogen would be problematic. To obtain a reasonable range, we need to store the hydrogen cryogenically and due to its low density very large fuel tanks would be needed. This would need a substantial redesign of aircraft, which would be bulky due to the volume required for fuel tanks. A more attractive way of running aircraft without fossil fuels would be to use fuels that have been synthesised from water and carbon dioxide using energy from non-fossil fuel sources. Of course substituting air travel with high-speed trains on medium overland routes would achieve non-fossil fuel high-speed travel at considerably lower energy use and avoid many of the environmental problems associated with air travel.

Nuclear energy is one alternative for shipping and has been well tested. Nuclear submarines, for example, are used by many of the world's leading navies and have covered vast distances successfully for over half a century.

Nuclear cargo and passenger ships such as nuclear-powered cargo ship the *Otto Hahn* have been trialled successfully, as has the *NS Savannah*, which was the first nuclear-powered cargo–passenger ship.

As with aircraft, the application of synthesised fuels would allow us to use conventional ships that were effectively fossil fuel free. Again modern agriculture could be made fossil fuel free with the use of synthesised fuels.

21.6 Resources

There are adequate supplies of uranium for nuclear power and silicon for solar cells. The availability of lithium for batteries is a greater problem. According to a 2011 study conducted at Lawrence Berkeley National Laboratory and the University of California Berkeley, the currently estimated reserve base of lithium should not be a limiting factor for large-scale battery production for electric vehicles, as the study estimated that about 1 billion 40 kWh lithium-based batteries could be built with current land-based reserves. Another 2011 study by researchers from the University of Michigan and Ford Motor Company found that there are sufficient lithium resources to support global demand until 2100, including the lithium required for the potential widespread use of electric vehicles. In addition, lithium is recyclable. The above analysis is based on current lithium reserves based on land reserves of 10–20 million tons.

There is 230 billion ton of lithium in the sea—over 10,000 times that of the land reserves. South Korea is building a plant to extract lithium from seawater and by 2014 this plant is predicted to extract 33 tons of lithium per year. In total it is expected to produce 20,000–100,000 tons of lithium from the plant. Such plants can be replicated.

21.7 The Future

The world is reaching the point where oil and other fossil fuels have a limited time before they are totally used up. We can argue about precisely when this will happen but it would seem that there is little over 40 years of oil remaining at current usage rates. We have slightly longer to use up all our natural gas and over 100 years of coal supplies left. We currently rely on fossil fuels to produce our food, and to run our homes, industry and transport systems.

Concerns about global warming due to carbon release continue and our ice caps continue to melt. Not everyone accepts global warming is due to carbon release or the potentially dire consequences of this. If we do not accept the argument that global warming is caused by carbon release and are prepared to gamble on the considerable consequences, we still have to find a way of running the world without fossil fuels if for no other reason than there will be none left.

Various future technologies hold considerable promise, but promises are not always honoured. Nuclear fusion holds the promise of clean inexhaustible energy as do superconductors, which work at ambient temperatures. This is not to say that we should stop working on these technologies; indeed we should do quite the opposite. However we cannot yet plan them into a successful system until they are fully proven. A host of inventions will arrive in the future, some of which will be even better than what we are dreaming of now.

All we can do is continue to work from sound technical principles to develop technology that can produce a worldwide power system that does not rely on fossil fuels. There are perfectly good technical solutions to living without fossil fuels—we do not have to return to a seventeenth-century lifestyle of relying on animal power and sail boats. We should be able to achieve as good or ideally better standards of living without having to burn fossil fuels.

The basic key to surviving without fossil fuels is threefold. First, we have to generate power in the quantities we need without using fossil fuels; second we have to transmit this power to where it is needed in the world; and third we have to develop transport and agricultural machines that can use energy from non-fossil fuel sources.

We have adequate existing methods of generating power without fossil fuels using both nuclear fission energy and alternative energy. Nuclear fission power stations have been in use for over half a century. It is true that new designs need to be able to avoid meltdown to avoid disasters such as the one caused by the recent tsunami in Japan—but nuclear power simply cannot be written off. We should bear in mind the disaster of running out of fossil fuels and not having contingency plans.

There are a range of alternative energy sources including hydro power, which already supplies 3.4 % of the world's total energy sources, and solar, wind, tidal, undersea currents, wave energy and geothermal energy. It is emphasised that development of alternative power sources should follow sound economic principles.

If there has been a revolution in alternative energy, it has been in solar energy where the price has fallen to levels where solar electricity can be generated economically in the so-called solar belt.

The key to providing widespread electricity from non-fossil fuel systems is undoubtedly a worldwide electric supergrid to allow electricity to be passed around the planet. A supergrid could be used to connect diverse power sources such as nuclear power stations, solar plants, wind farms and tidal systems. This would allow electricity to be passed from places generating large amounts of power to places where power was needed. The electric supergrid is likely to consist of a very high-voltage DC cable that will transmit electricity with very little loss across the globe. The supergrid enables the national grids of different countries to be connected.

A worldwide supergrid would also allow nuclear power stations to transmit the electricity they generate at off-peak periods to other countries with peak demand at different times. This would also be of considerable benefit to the economics of nuclear electricity generation. Transmission of electricity for use in households and in industry is not a particular problem as electricity will be connected to national grid systems.

Transport and agriculture are particular problems as much of their needs rely on oil—the fossil fuel likely to run out first. Train transport should be encouraged as it is energy efficient and can already be run from non-fossil fuel-generated electricity, as is the case in France. High-speed trains use only 7 % of the energy per passenger-mile in comparison to an aeroplane. High-speed train travel to replace air transport use would save a considerable amount of energy, and the energy it does use can be supplied from non-fossil fuel-generated electricity. Considerable over-land air travel is used in the USA and elsewhere, and there is scope for considerable savings in the use of fossil fuels and the resulting carbon release by encouraging the use of high-speed rail travel. It would be impractical to replace all air travel with high-speed trains. The remaining air transport will need to use fuel synthesised from carbon dioxide and water, algae-based fuels or hydrogen if fossil fuels are to be avoided. The former is a simpler solution as it would avoid the need for a major redesign of aeroplanes. Using biofuels grown on agricultural land is considered to be an unacceptable solution at the time of potential food shortages.

Road transport is likely to continue to embrace the new electric technologies of BEV and battery/internal combustion hybrids. Until the cost of fuel cells is reduced, the number of vehicles using this technology is less likely to grow, although it is still a strong possibility for the future. The adoption of pick-up systems to recharge vehicles while driving is another option that has advantages including reducing battery size and giving an extended range.

The question is why, if all these technologies work, should we take action now to plan for the future when the fossil fuel runs out? The answer is that there is considerable work to be done to integrate all of these technologies. If we examine the history of technical developments, the time taken to implement known technologies can be lengthy. Setting up electrical supergrids or massive solar power stations is going to take years of planning and international agreements but while these are arguably in everybody's interest, international politics is rarely so simple.

It is up to scientists, engineers, politicians and virtually all thinking people to take up responsibility for running the world without fossil fuels.

References

1. Lowry J Life without oil and other fossil fuels
2. Larminie J, Lowry J (2012) Electric vehicle technology explained, 2nd edn. Wiley
3. Mackay DJC (2009) Sustainable energy without the hot air. UIT Cambridge
4. The Economist (2008) Case history: in search of the perfect battery
5. http://environment.nationalgeographic.com/environment/global-warming/gw-effects/
6. http://www1.eere.energy.gov/hydrogenandfuelcells/pdfs/32405b27.pdf
7. http://www.itm-power.com/tech/electrolyser-stacks/
8. http://www.airfuelsynthesis.com/
9. http://en.wikipedia.org/wiki/Lithium
10. http://www.scientificamerican.com/article.cfm?id=how-long-will-global-uranium-deposits-last
11. http://www.inference.phy.cam.ac.uk/sustainable/refs/nuclear/TroubleLithium.pdf

Chapter 22
Problematic Integration of Fatal Renewable Energy Systems in Island Grids

Gilles Notton

Abstract The islands are often not or partially connected to continental networks and manage their own energy supplies in reaching alone the supply/demand balance and in assuring the quality of the delivered electricity. The low consumption and the lack of interconnections induce energy costs higher than in other regions. The islands are good locations for using new technologies of energy production and storage. For increasing their security, they often prefer to use indigenous sources as renewable ones but the intermittence and stochastic character of these "fatal" energy sources make them more difficult to control and manage and more again in small island networks. Most islands have good renewable energy (RE) resources which are underused in comparison with their real potential. A particular attention is paid to the problem induced by the renewable energy sources (RES) utilization in an island electrical grid on the electrical management.

Keywords Islands · Renewable energy · Integration rate · Stochastic production

22.1 Introduction

In Europe, there are about 300 islands (6 % of the territory) for 14 M-inhabitants, that is, higher than the population of some European countries, and more than 100,000 islands of all sizes are scattered in the world at all latitudes with almost 500 M-inhabitants [1]. The total islands area is one sixth of the earth area.

Already it is not an easy task for an electric network manager to maintain at each moment a production/consumption balance with controllable energy production means in a large interconnected network; but when some fatal energy systems are integrated in the network and if, moreover, the network is not interconnected, the task is even more complicated.

G. Notton (✉)
Research Centre of Vignola, University of Corsica Pasquale Paoli, UMR CNRS 6134, Centre of Vignola, Route des Sanguinaires, F20000 Ajaccio, France
e-mail: gilles.notton@univ-corse.fr

A. Sayigh (ed.), *Renewable Energy in the Service of Mankind Vol II*,
DOI 10.1007/978-3-319-18215-5_22

22.2 Electricity Production/Consumption Balance

The electricity is used immediately after being produced! The electricity manager must dispatch the electricity and ensure a perfect electricity supply/demand balance. In a power system, power balance must be maintained, that is, load demand must be balanced by generation supply at all times. The electrical network must work at a stabilized frequency (50 Hz), thus the production systems must at each moment adapt their production to the power consumption either in the electrical station or in the dispatching. If the production decreases suddenly due to the loss of a production mean or a cloud passage above a photovoltaic (PV) plant, this balance is weakened and the frequency falls below 50 Hz. In this case, a rapid increase of the production by connected means must occur by an increase of the motor power working at partial load or starting a new production mean is needed. But the rise speed of the power and the starting time is not instantaneous as shown in Table 22.1

If an increase of the consumption and consequently the starting of new energy mean were not anticipated, the imbalance can continue and it will be necessary to shed a fraction of the load. If the load shedding is too slow or insufficient, there is a risk of blackout. Similarly, if the load is lower than the production, the frequency increases and this presents danger for the electrical machines; the security system stops the electrical machine and leads to power failures in the network.

Table 22.1 Characteristics of electricity production plants [2]

Type	Size (MW)	Minimum power capacity (%)	Rise speed in power per min (%)	Start-up time (hours)
Nuclear power plant	400–1300 per reactor	20	1	40 h (cold) 18 h (hot)
Steam thermal plant	200–800 per turbine	50	0.5–5	11–20 h (cold) 5 h (hot)
Fossil-fired power plants	1–200	50–80	10	10 min–1 h
Combined-cycle plant	100–400	50	7	1–4 h
Hydropower plant	50–1300			5 min
Combustion turbine (light fuel)	25			15–20 min
Internal combustion engine	20	65		45–60 min

22.3 Particular Case of Island Territories

The insular electrical systems are governed by the same electro-technical rules as those that are applied in mainland network but the absence or the limitation of an interconnection to a large network provide it some specificity recognized by the competent European and French authorities [3]. As islands are not interconnected (Fig. 22.1), the energy manager has to reach the supply/demand balance without the assistance of external production means. Islands have a structural fragility: A short circuit on an electrical system will generate a voltage drop in all the islands [4]; the low inertia implies a high-frequency variability with voltage consequences [3]; the previous problems are compounded by the high unit size of an electrical generator in comparison with the peak power. Voltage and frequency drops are more numerous and deeper in islands than in mainland: The number of shedding in an islander grid is important (on average, 2.1 shedding by month on the 5 last years [5] between 0.3 in Corsica and 3.5 in Guadeloupe). In Corsica, before the partial alternate current (AC) interconnection with Sardinia, more than 200 failures per year occurred on the transmission network with voltage and frequency dips [6].

Islands have often a small population, a low and variable energy consumption that prohibits the use of high-rated power production means and requires the utilization of small electrical plants for a better adaptation to the load and for limiting the disturbances due to the loss of a electrical plant [4]. An electrical unit power must not exceed 25 % of the average power [6] (in the European grid with a 300 GW peak power, the loss of a 1.3 GW nuclear unit corresponds to a ratio of 0.4 %; in an insular grid of 200 MW, the loss of the most powerful unit of 40 MW represents a

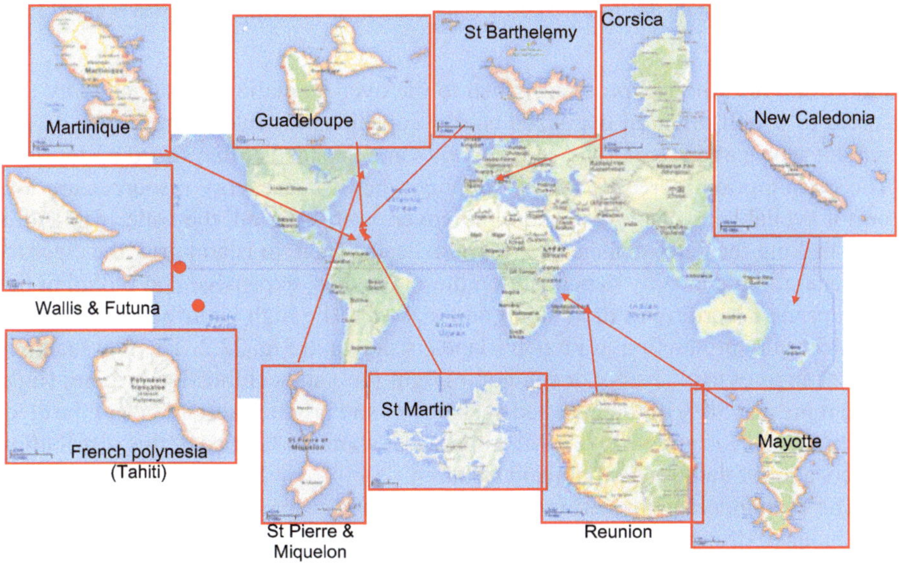

Fig. 22.1 Some French islands all over the world

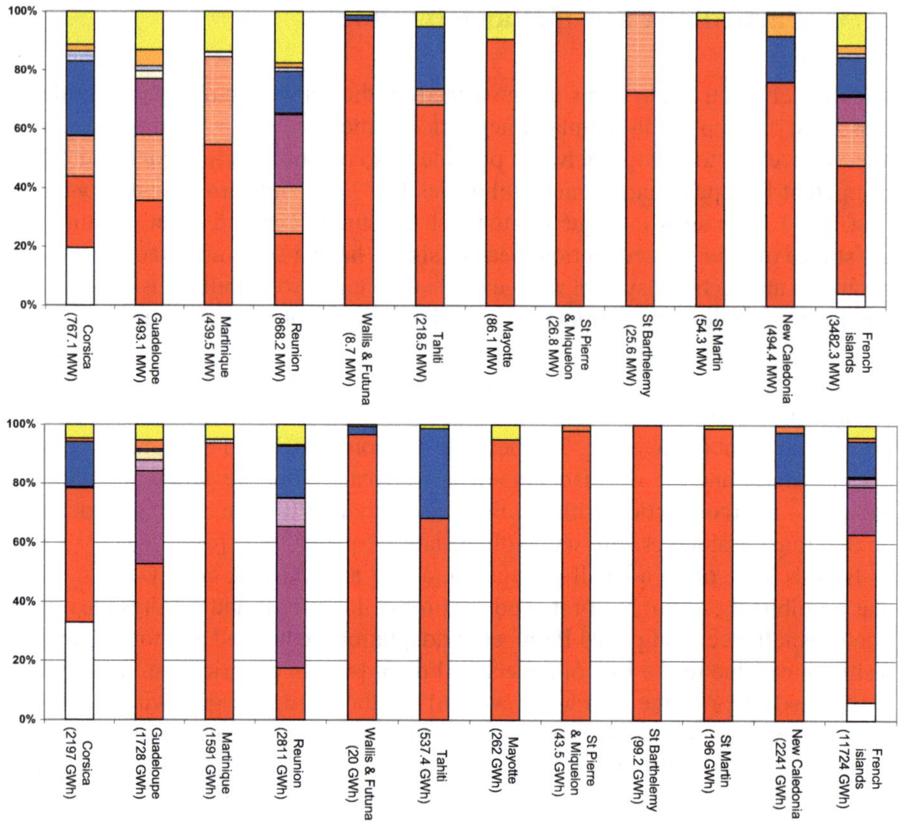

Fig. 22.2 Repartition of the installed power and energy production in French islands in 2012

ratio of 20 %). Another problem is lied to the over-specialization of economy that forces to install often an oversized energy capacities to cover factors such as important seasonal demand. The economy of some islands is often based on tourism. The consumption increases during summer when some energy resources are low or not available as water resources; moreover, in this period, the water demand is high making the dam management more delicate. Then, in island grid, the electricity generation is heavily dependent on diesel engines, expensive and polluting but appropriate according to their small unit size [1]. The high contribution of fuel in the electricity production in French islands is shown in Fig. 22.2. The fuel facilities (diesel engines using heavy or sometimes light fuel) and combustion turbines (light fuel) represent 58 % of the installed peak power and produce about 57 % of the electricity (in some islands 100 %) with a further of about 16 % of coal; this distribution differs a lot of the French mainland situation.

The preponderant fuel utilization in the islands was confirmed by Liotard [7] and induces strong dependence and increasing sensibility to the oil barrel price. In

Corsica and overseas departments, the cost of electricity is in the best state, at least twice the purchase price secured by the electricity tariff balancing out in all the French territory. The European Union defined a special status for the "small isolated networks" for which the members of states can adopt specific measures different from the European continent [8]. The French law [9] identifies some "area not interconnected to the continental network" called ZNI (Fig. 22.1): French Polynesia with Electricity of Tahiti (Suez) with 11 islands, New Caledonia with ENERCAL and EEC, Wallis and Futuna (EEWF Suez), Corsica, French overseas collectivities (FOC; Saint Pierre and Miquelon, Saint Barthelemy and Saint Martin) and French overseas departments (FOD; Guadeloupe, French Guyana, Martinique, Mayotte, and Reunion) where EDF (Electricity of France) ensures the public service (production, single buyer, transport, distribution, and commercialization).

The French laws impose [8–11] the electricity tariff equalization over all the French territory and take place in the Contribution of the Public Service of Electricity (CPSE) which is paid by all the electricity consumers and serves particularly to compensate the high gap in insular area between the production cost and the regulated sale price of electricity. This CPSE is controlled and attributed by the Energy Regulatory Commission (CRE). The FOC and FOD must produce in whole their electricity and largely in Corsica (partially connected). For the CRE, the MWh cost was, in 2010, around 122 and 315 € for a regulated sale price at 51.7 €/MWh [12]. For EDF [13], in France, the production cost is 45 €/MWh against 110 €/MWh in Corsica and Reunion, 140 € in Martinique and Guadeloupe, and 160 € in Guyana. Thus, the regulated sale price is less than 50 % of the cost price of the electricity produced locally.

These ZNI are characterized by an energy production growth higher than in mainland (+3.8 % for islands and 1 % for French mainland). Other technical factors increase the production cost as material corrosion, more modulated production peak, and more expensive maintenance. Another specificity is the unusually high gap between the produced and sold energy due to technical and nontechnical losses: In 2012, the technical and nontechnical losses were: 316 GWh (14.4 %) in Corsica [14], 196 GWh (11.3 %) in Guadeloupe [15], 169 GWh (10.6 %) in Martinique [16], 284 GWh (10.1 %) in Reunion [17], 3.48 GWh (17.6 %) in Wallis and Futuna [18], and 51.2 GWh (9.5 %) in Tahiti [19].

The islands must reach more constraining objectives about energy management and the three criteria [20]: accessibility to electricity at an acceptable price, availability in term of continuity and quality of service, and social and environmental acceptability.

22.4 Integration of Intermittent and Stochastic RES

The power system is a dynamic system, subjected to continuously changing conditions, some of which can be anticipated and some of which cannot. The primary function of the power system is to serve a continuously varying customer load.

From a control perspective, the load is the primary independent variable—the driver to which all the controllable elements in the power system must be positioned and respond. There are annual, seasonal, daily, minute-to-minute, and second-to-second changes in the amplitude (and character) of load served by the system. The reliability of the system then becomes dependent on the ability of the system to accommodate expected and unexpected changes and disturbances while maintaining quality and continuity of service to the customers.

Wind and sun resources are variable and largely undispatchable, and they impact the planning and operation processes. The variability of these sources affects the electrical network management due to which the penetration level of the electrical systems using these sources is high. When the renewable energy is lower or higher than the demand, other generation units have to compensate the difference. This implies that a system with a significant amount of installed renewable energy systems must be operated with a sufficient reserve margin, as most traditional generation units require a considerable start-up time [21]. Using solar or wind energy systems induces two main constraints [4, 22, 23]:

1. Constraint related to the source: no guaranteed power due to the intermittence of the source; thus, we must provide other means of production in reserve and compensate the variations of power by other means of production having some reserves and able to react immediately (in increasing or inversely in decreasing their power rapidly); these "fatal" sources are not always available when required and sometimes are present when the network does not require them, their production being then in excess.
2. Technical constraint: these systems have no inertia and are sensible to voltage and frequency drops which cut the production and aggravate the incidents: Moreover, they do not participate in the electrical network stability and quality. These systems do not participate in the "system service" (voltage and frequency regulation, switch-on in autonomous mode or black-start) and behave as a passive generator from an electrical point of view.

The first constraint implies that the conventional energy system must work at partial load to be able to react rapidly to a quick variation of fatal production. Then, the efficiency of the conventional motor is reduced by implying an increase in the consumption of fuel and production cost. Thus, it is impossible to optimize the operating mode of the conventional energy means [23].

The second constraint implies to have a voltage and frequency regulation made by conventional means and consequently, a given amount of conventional means with inertia must work permanently.

The two constraints show the difficulties or even impossibility to use "fatal" renewable systems alone without a part of the production coming from conventional energy means or guaranteed renewable energies. Integrating fatal renewable systems in the energy mix allows to reduce the proportional cost in the production cost from thermal plant (fuel, maintenance, etc.) but not the fixed costs [5].

The penetration rate of fatal energy systems, that is, the power generated by these systems compared with the total production must be limited in view to guaranty the electrical grid stability and the production/consumption balance. Some feedbacks in Denmark show that for a penetration rate up to 20 or 30 % some stability problems can occur [22]. A French order of the April 23, 2008 states that any facility (>3 kW) whose power is at least 1 % of the minimum power circulating in the network and implementing randomness fatal energy can be disconnected from the public distribution network by the network operator when the active power from these plants reached 30 % of the total active power in the network [24]. The same maximum penetration rate of renewable electrical systems is also used in other countries as Canary Archipelago (Spain). In other words, in small systems particularly sensitive to disturbance and with a low electric inertia—as in islands, it is necessary to limit the introduction of wind and solar power to 30 % of the power generated at any time.

When and how the disconnection happens? The PV wind systems are disconnected according to the first-come, first-served principle: The last arrived is disconnected first and the last authorized to be reconnected. The order is established according to the date of dispatch of the connection request form to the electricity supplier.

Even, if the part of the solar and wind energy production in the ZNI in France is still low, this limit of 30 % is rapidly reached. In Guadeloupe and Reunion, this limit of 30 % has been reached [5]: Corsica and Guadeloupe reach 34 %, La Réunion 35 %, and Martinique 22 %.

This limit of 30 % is more easily reached in a small territory as an island because the "profusion" of the renewable energy means at an island scale is very small. The "profusion" means that the random fluctuations of the wind or solar production systems were statistically reduced when the systems are spread over a large territory and their production are injected in the same grid. In particular, in cases of significant expansion, production will neither be zero nor maximum.

Small island perimeters require a balanced geographical distribution of projects on the island, indeed, if projects are geographically concentrated and the wind fall or the sky is cloudy, a significant portion of power is suddenly lost inducing important problem for the electrical management; but also, as we wrote previously, if suddenly the sky is cloudless, the production of the PV systems will increase suddenly and the total production can exceed the load consumption.

If the production of a wind farm is intermittent, the electrical wind production at a national French level (with a wide distribution of wind turbines across a territory) is not or at least less intermittent because the geographical profusion, and the presence of three uncorrelated wind regimes (oceanic, continental, and Mediterranean) allows an attenuation of the production variation.

Such an effect is illustrated in Fig. 22.3 where we see that the production of an alone PV system can suffer sudden variations, at a region scale and a fortiori at France scale; the PV production is smoothed by "profusion effect" [25]. This smoothing is strongly reduced in a small territory and more often, the production of PV system or wind turbines spread over the island area will vary in the same direction, complicating the work of the electrical manager.

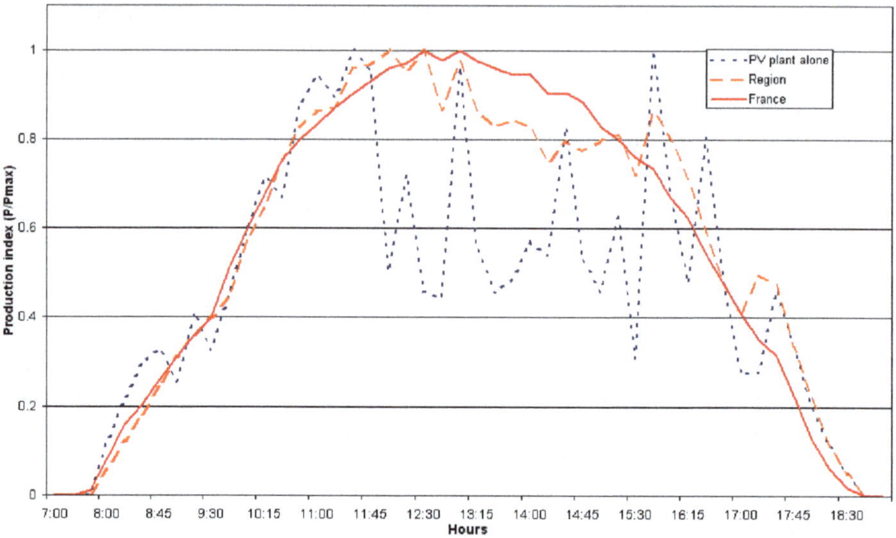

Fig. 22.3 Illustration of the profusion effect for PV production in France (15/03/2013) [25]

This limitation of the "profusion" in an island has also an influence on the load consumption: In a mainland, the profusion of loads due to the different electrical demands in the various regions allows to predict the load more easily with a best accuracy and the programming of the production mean is more easy than in an island where this profusion is very limited [26].

The small penetration rate limits the development of such energy systems and the contribution of renewable energy systems to the global energy balance of the islands. Consequently, and as said by Duic and Carvalho [27], the higher penetration of renewable energy sources (RES) in islands is limited by its intermittent nature, which can be increased if some kinds of energy accumulation are used.

22.5 Renewable Energy Development in the French Islands

Managing an electrical grid in islands is more complex than in inland for all the reasons previously exposed and this problem becomes even more complicated when renewable energy systems with stochastic and intermittent production are integrated into this grid; however, in these areas, the rate of fatal renewable energy (leaving aside hydraulic energy and biomass) in the production reached 5.5 % in 2012 (14 % in installed power), that is, more than in the French mainland territory (3.5 % in energy and 8.5 % in installed power). The part of renewable energy in the electrical production is shown in Fig. 22.4 [14–19]. Despite the difficulties for integrating

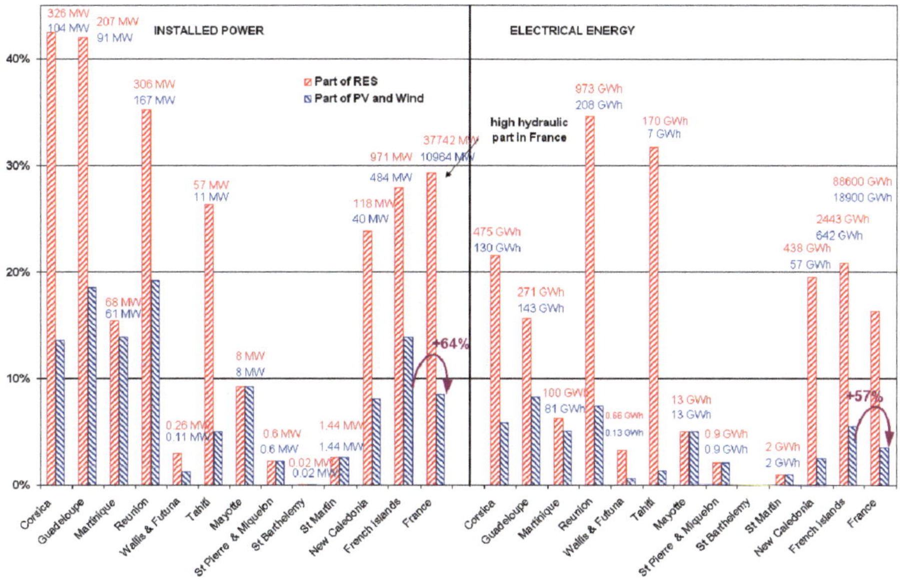

Fig. 22.4 Part of the renewable energy in the installed electrical power and electrical energy

fatal resources in island grids, the islands reached a higher level of integration of renewable energy than in mainland.

The part of the renewable power in France is higher than in the islands because the rate of hydraulic is high in France (second position after nuclear). The gap between the islands is important: in term of installed renewable power, from 42 % for Corsica and Guadeloupe to around 3 % for Saint Martin, Saint Pierre and Miquelon, and Wallis and Futuna and practically nil for Saint Barthelemy and in term of electrical energy with a maximum of 35 % for Reunion. The fatal energy part reaches 19 % of the total installed power in Guadeloupe and Reunion and up to 8.3 % in energy for Reunion.

This strong interest for fatal renewable energy systems in the islands is mainly due to the high energy production cost underlined in the third paragraph. The utilization of costly fossil fuels due to the fluctuations in crude oil prices, the small size of the production means, and a more expensive maintenance due to the remoteness cause such high costs and encourage the emergence of innovating solutions as the development of renewable energy and the setting up of a policy of electricity demand reduction and energy management.

Lacoste [5] showed that the "fatal" renewable energy (unguaranteed energy) is cheaper than heavy fuel and of course than domestic fuel (used in combustion turbine): wind energy, small hydraulic energy, and sometimes PV plants; moreover, the guaranteed renewable energy is less expensive than the fuel in basis utilization: as geothermal, biogas, and close to fuel cost as biomass.

22.6 Conclusion

The islands have everything to gain by developing their use of renewable energy from an economic point of view but also for ecological reasons. In view to delete or to decrease the constraint of the integration limits, three solutions must be developed together:

- To predict the production of the fatal sources for achieving by anticipation of an optimal switchover towards conventional plants, the prediction of these productions is generally less accurate in the insular networks
- To develop smart grids for transport and management of production means. The network will be managed with more flexibility for reaching the supply/demand balance in controlling not only the production but also the consumption
- To develop energy storage means in order to absorb the surplus energy and to release it when the consumption requires

References

1. Duic N, Alves LM, Chen F, Da Graça Carvalho M (2003) Potential of Kyoto protocol clean development mechanism in transfer of clean energy technologies to small island developing states: case study of Cape Verde. Renew Sustain Energy Rev 7:83–98
2. Saguan M (2007) L'analyse économique des architectures de marché électrique. Application au market design du "temps réel". PhD dissertation, University Paris Sud XI
3. EDF (2008) Technical reference: key parameters for the insular electrical systems safety. SEI REF 05, 01/12/2008. http://corse.edf.com/fichiers/fckeditor/Commun/SEI/corp/SEI_REF_05_V1.pdf. Accessed 12 April 2014 (in French)
4. EDF (2009) Intermittent random renewable energy in the insular energy system: why a maximum integration rate of 30%, EDF report, 19/05/2009, CCP SEI. http://sei.edf.com/fichiers/fckeditor/Commun/SEI/corp/Pourquoi-30-dans-SEI.pdf. Accessed 11 April 2014
5. Lacoste J (2013) Les systèmes électriques insulaires. IEEE P & ES, 25/04/2013, RTE, La Défense, Paris. http://ewh.ieee.org. Accessed 12 Feb 2014 (in French)
6. Barlier Y (2000) The situation of electricity generation in Corsica. Proceeding of dissemination of the advanced control technologies and SCADA systems for the isolated power networks with increased use of renewable energies, Ajaccio, France, 2000 (pp 27–35)
7. Liotard C (2008) L'électricité dans les îles: vers une production durable? In: Guerassimoff G, Maïzi N (eds) îles et énergie: un paysage de contrastes. Presses de l'école des Mines; Collection Libres Opinions
8. Directive 2003/54/EC of the European Parliament and of the Council of 26 June 2003 concerning common rules for the internal market in electricity and repealing Directive 96/92/EC. Official J. European Union, L176–37-L176–55, 15/01/2003
9. French Act 2000–108 (10/02/2000) on the modernisation and development of the public electricity service; reinforced the 01 January 2012
10. French Act SPEGEEG 2004–803 (09/08/2004) relating to the public service of the supply of electricity and gas and to other electric and gas companies; reinforced the 01 January 2013
11. Energy Program Act POPE 2005–781 (13/07/2005) establishing France's energy policy priorities; reinforced the 01/06/2011
12. Energy Regulatory Commission (CRE) (2013) www.smartgrids-cre.fr. Accessed 11 April 2014

13. EDF-SEI Corsica (2013) http://www.corse-energia.fr. Accessed 11 April 2014
14. EDF SEI Corse (2013) Generation adequacy report on the electricity supply—demand balance. http://sei.edf.com. Accessed 09 April 2014 (in French)
15. EDF SEI Guadeloupe (2013) Generation adequacy report on the electricity supply—demand balance. http://sei.edf.com. Accessed 09 April 2014 (in French)
16. EDF SEI Martinique (2013) Generation adequacy report on the electricity supply—demand balance. http://sei.edf.com. Accessed 09 April 2014 (in French)
17. EDF SEI La Réunion (2013) Generation adequacy report on the electricity supply—demand balance. http://sei.edf.com. Accessed 09 April 2014 (in French)
18. IEOM (Institut d'Emission d'Outre Mer) (2013) Wallis et Futuna. Report ISSN 1635–2262. http://www.ieom.fr. Accessed 01 Feb 2014 (in French)
19. EDT (Electricité de Tahiti) (2013) EDT in brief. www.edt.pf. Accessed 09 April 2014 (in French)
20. Guerassimoff G, Maïzi N (2008) Iles et énergie: un paysage de contrastes. Presses de l'école des Mines, Collection Libres Opinions
21. Van Hertem D (2009) The use of power flow controlling devices in the liberalized market. Catholic University of Leuven report, January 2009. ISBN 978–94-6018–024-8
22. Robyns B, Davigny A, Saudemont C, Ansel A, Courtecuisse V, Francois B, Plumel S, Deuse J (2006) Impact de l'éolien sur le réseau de transport et la qualité de l'énergie. J3eA, 5, 2006. http://www.j3ea.org. Accessed 11 April 2014
23. Pestourie J (2013) Les EnR intermittentes: cas des réseaux insulaires. IEEE P & ES, 25/04/2013, RTE, La Défense, Paris. http://ewh.ieee.org. Accessed 12 Feb 2014 (in French)
24. Legislative French Assembly (2008) French Act n°2008–386 of April 23, 2008 relating to the general technical prescriptions for the conception and the operation for the connection of production installations to the public electrical networks, 2008. (in French)
25. RTE (Réseau de Transport d'Electricité) (2012) Overview of renewable energies 2013. www.rte-france.com. Accessed 11 April 2014
26. Wang Y (2012) Evaluation de la Performance des Réglages de Fréquence des Eoliennes Ã l'Echelle du Système Electrique: application Ã un Cas Insulaire. PhD dissertation, Ecole Centrale de Lille, 20/11/2012
27. Duic N, Carvalho MG (2004) Increasing renewable energy sources in island energy supply: case study Porto Santo. Renew Sustain Energy Rev 8:383–399

Chapter 23
Renewable Energy Development in Africa: Issues, Challenges and Prospects

Abubakar S. Sambo

Abstract The solar radiation climatology of Africa is such that more than half of the continent has solar radiation intensities of more than 7000 kWh/m^2/day. The hydropower resource of the continent is huge especially in the central and eastern parts of the continent with the Grand Inga of Democratic Republic of Congo alone of having a potential of about 50,000 MW. There is a significant geothermal potential along the East African Rift Valley spanning Djibouti, Ethiopia, Kenya and Tanzania. High-wind potentials exist along the coastal and mountainous parts of the continent. Biomass in the form of fuelwood is the dominant fuel in the household sector in the whole continent and most especially in the tropical zones. Biodiesel from Jatropha is gaining more recognition in many African nations.

The optimal utilization of the renewable energy resources of Africa will call for a continent-wide increase in energy access in order to uplift the continent's socio-economy. This will in turn call for the practical elimination of the challenges bedevilling the sector. This can be done through the adoption of strong political will by African national governments to drive renewable energy development. There is also the need for sound policies and plans as well as institutional and legal frameworks in addition to the enthronement of pragmatic capacity building in renewable energy technologies.

Keywords Renewable energy · Solar energy · Wind energy · Biomass/biofuels

23.1 Introduction

Africa is a continent of 55 countries, covering a land area of about 30.3 million sq km constituting about 20.3 % of the total land area on Earth, and with a population of about 0.9334 billion people. The continent is endowed with energy resources of oil,

A. S. Sambo (✉)
Federal Secretariat, Abuja, Nigeria
e-mail: assambo@gmail.com; assambo@yahoo.com

A. Sayigh (ed.), *Renewable Energy in the Service of Mankind Vol II*,
DOI 10.1007/978-3-319-18215-5_23

natural gas, coal, hydropower, solar, wind, geothermal, biomass, tidal and wave energy in its territorial waters, which are unevenly distributed. Notwithstanding the availability of energy resources in Africa, the continent today has the highest poverty of modern energy services culminating in massive power outages and long queues in petrol service stations. Given the fact that conventional energy resources are available in a limited number of African nations and the fact that their utilization leads to global warming and climate change problems, the sustainable energy development pathway for Africa is to ensure large-scale deployment of renewable energy sources which exist in sizeable magnitudes in at least one form all over the continent.

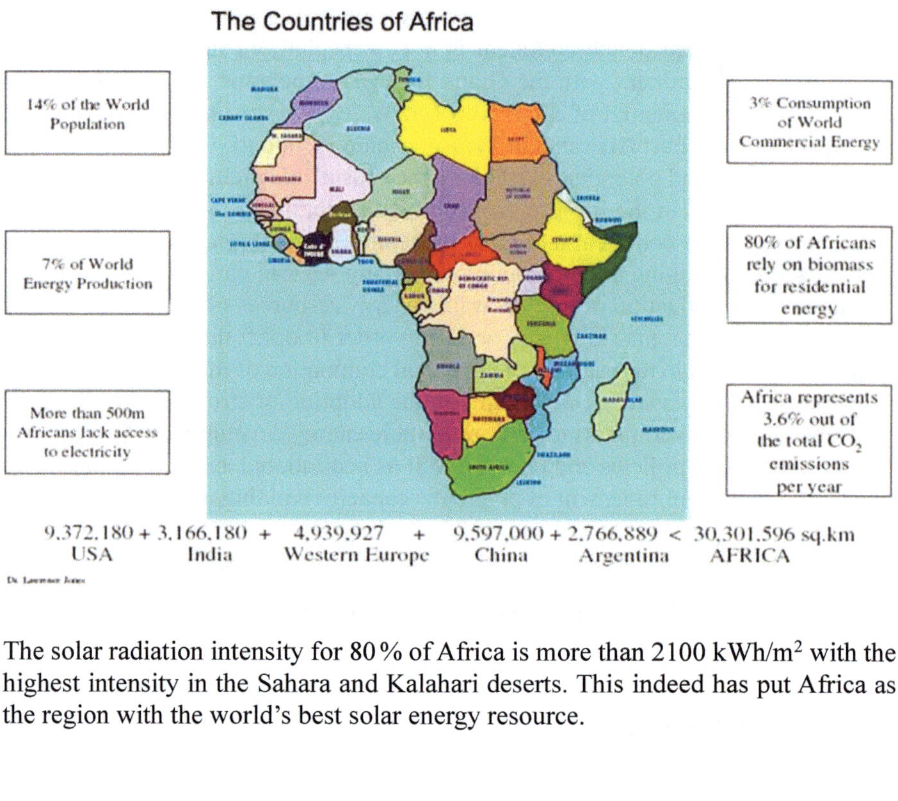

The solar radiation intensity for 80 % of Africa is more than 2100 kWh/m^2 with the highest intensity in the Sahara and Kalahari deserts. This indeed has put Africa as the region with the world's best solar energy resource.

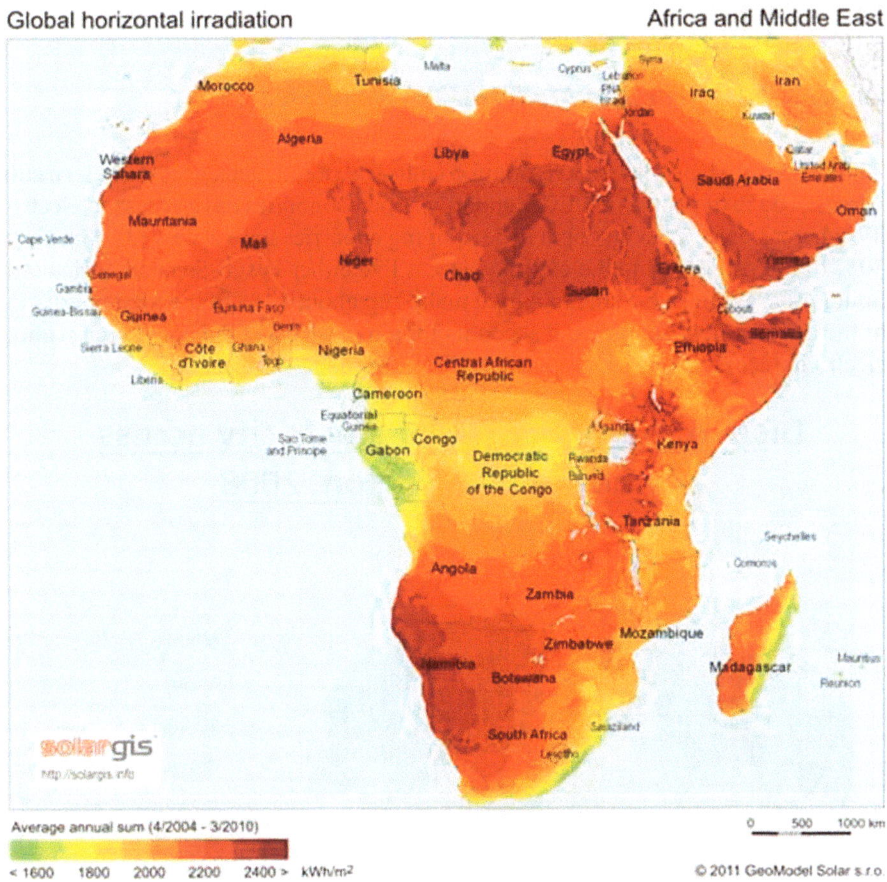

The highest wind regimes are along the northern and southern coasts of the continent. High wind regimes are noticeable around the mountainous parts of Africa. The wind potential for Africa was estimated as 10,600 TWh per year assuming that 10% of the land area has average wind speeds exceeding 5.1 m/s at a height of 10 m. Africa's technically exploitable hydropower capability is estimated to be in excess of 1,917 TWh/year which is about 13 % of the World's hydropower potential [1]. It is said that the economically exploitable hydropower is at least 1,100 TWh/year [2]. The geothermal energy potential of Africa is estimated to be in the range of 2.5–6.5 GW. African desert vegetations make it rich in traditional biomass in the form of fuelwood and agricultural residue. Ethanol, from sugar cane, is also produced as an addition to gasoline in some African nations.

23.2 Major Energy Issues

23.2.1 Low Access to Modern Energy Services

About 1.5 billion people in developing countries currently lack access to electricity. Almost 100 % of people in OECD and transition economies have access to electricity, whereas only 72 % of people in developing countries have access [3]. More than 80 % of the people without electricity access live either in sub-Saharan Africa or in South Asia. While sub-Saharan Africa makes up about 14 % of the total population of the development countries, it accounts for almost 40 % of the population without electricity access.

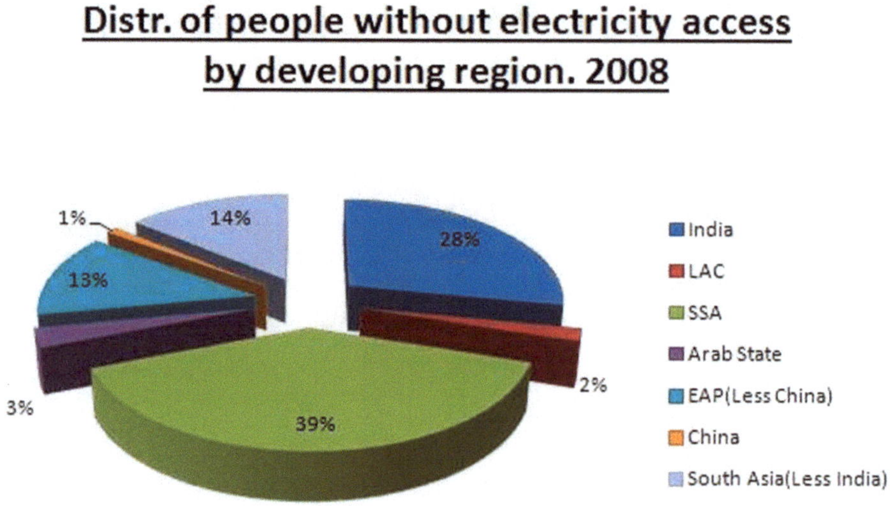

23.2.2 Over Dependence on Fuelwood

Today, with the exception of South Africa, where only 16 % of the population depends on traditional biomass energy, almost 80 % of the population in sub-Saharan Africa depends on biomass resource for cooking [4]. Firewood and charcoal burn inefficiently, resulting in major energy losses (up to 85 %). Furthermore, when burned indoors, they emit pollutants such as carbon monoxide, small particulates, nitrogen oxide, benzene, butadiene, formaldehyde, poly-aromatic hydrocarbons and many more other health damaging substances. It is estimated that 396,000 deaths in sub-Saharan Africa in 2002 were due to indoor air pollution from the burning of biomass fuels [5]. It is particularly dangerous to women and children as they do most of the cooking.

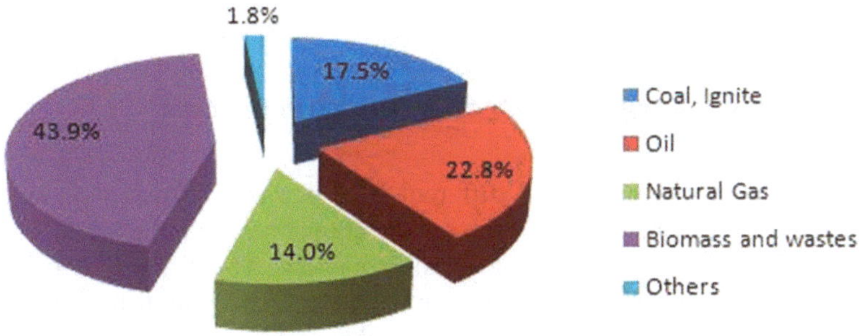

23.3 Challenges

23.3.1 Poor Infrastructure

Although energy resources are abundant in some regions of sub-Saharan Africa, the energy infrastructure to transport, distribute, transform and efficiently utilize them is lacking. Poor delivery mechanisms associated with poor services have hindered the sustainable development of energy services in the region. The centralized system like pipeline and national grids have not reached majority of the people and cost more to the poor when it finally reaches them. In most countries, the energy distribution network is overstretched resulting in power outages and shortages in fuel supply.

23.3.2 Insufficient Skilled Manpower

Developing a competitive energy sector requires large numbers of highly skilled people in many specialized areas, which Africa lacks. Consequently, many of the specialized energy infrastructural development in Africa are carried out by foreign personnel, resulting in significant financial drain.

Areas in which this is a problem includes the conventional areas of oil and gas exploration and development; power generation, transmission and distribution; and financial investment decision-making related to energy. These areas encompass energy-planning, analysis and modelling based on national, subregional and regional consideration. Of course skilled manpower in all aspects of renewable energy, for all African countries, is grossly inadequate.

23.3.3 Weak Energy Cooperation and Integration

Africa is made up of countries of different sizes and is bestowed with an uneven distribution of energy resources. Weak energy cooperation framework in Africa has seriously affected energy production in the continent. The difficulties for an individual country to mobilize the huge capital investment required for energy infrastructure projects justifies the need for regional energy cooperation and integration. A notable step has been made in the right direction with the creation of power pools in the four major regions of the continent. But the power pools need to be made operational in order for them to achieve the objectives for which they were created.

23.4 Prospects

The current large-scale energy poverty bedeviling Africa can be effectively addressed with the large-scale deployment of renewable energy technologies, especially those based on solar energy and hydropower. There is prospect of significant improvement of the quality of life of Africans, and there is also prospect of serious business transactions for the African entrepreneurs along with their foreign business partners.

In most African countries today, one will find at least one type of renewable energy system. Such systems were mostly installed as pilot or demonstration projects often from donor countries or international development agencies. One will find solar photovoltaic systems in the forefront followed by wind electricity converters and biomass/biogas/biofuel plants.

There is a huge potential for renewable energy market in Africa based on the fact that one form of renewable energy resource or two or even three are available in all African countries. This implies that the energy demand for national economies and for all the major energy utilizing sectors of industry, transport, household and services can indeed be met by renewable energy-based technologies.

23.5 The Way Forward

To ensure large-scale utilization of renewable energy there is need for African countries to conduct a comprehensive compilation of total energy supply and consumption on the one hand and the country's energy resources endowments on the other.

Again, there is the need to produce a comprehensive scenario-based energy demand projections using modern energy modelling tools on short-, medium- and long-term time horizons and covering the major economic sectors.

African countries should form strong international cooperation with technologically advanced countries to ensure technology transfer to Africa and capacity building of Africans.

All African countries should produce comprehensive energy supply strategies using the energy demand projections and to begin with, for the period 2015–2030 and to key into the United Nation's Sustainable Energy for All Initiative that calls for:

a. Universal access to modern energy services by 2030
b. Doubling the share of renewable energy in national energy mix by 2030
c. Doubling the energy efficiency thereby reducing energy intensity by 2030

For the period 2030–2040 it is recommended that African nations should implement the recent UNFCCC/IPCC resolution of tripling the uptake of renewable energy.

For the period beyond 2040, African countries should aim at making renewable energy to contribute at least 50 % of overall national energy supplied.

Other national aspirations, such as Nigeria's commitment to be amongst the top 20 strongest economies of the world by the year 2020, should be considered while developing the energy supply strategy.

The creation of Ministries of Renewable and Alternative Energy by African countries will strongly promote the use of RE on large scale as it is the current situation for countries like India.

There is also the need to review energy law(s) to ensure that remote rural communities are catered for but essentially to reform the energy sectors to make them market oriented with ample encouragement of the organized private sector to invest in the development of the national energy infrastructure. There is also the need for African nations to:

a. Adopt appropriate frameworks to promote the practical adoption of new energy technologies. This will include incentives to users and producers of new energy technologies as well as feed-in-tariffs, where necessary.
b. Strengthen the national energy regulatory frameworks to ensure orderly development of the sector and also to ensure that international best practices are adopted

on the issue of licences for new plants as well as evolvement of both cost-reflective tariffs and practical provisions for indigent groups.

c. Ensure that the reviewed energy policies are passed into law by parliaments. This is necessary for the majority of African countries in view of the frequent unnecessary policy changes from one elected administration to the next one.

23.6 Conclusion

The provision of modern energy services for the Africa is far below the level for meaningful socio-economic growth, largely due to inadequate energy planning as well as policy and legislative barriers that have been curtailing the exploitation of renewable energy.

To significantly expand energy access in Africa using large-scale renewable energy technologies there is need for a fresh initiative consisting of:

a. Production of scenario-based energy demand projections covering all sectors of national economies and on short-, medium- and long-term time horizons.
b. Production of comprehensive energy supply strategies using the demand projections with phased increase of renewable energy from doubling to tripling the rate of uptake of RE and to the last phase of making total energy supply to be 50% solar and other renewable energy based.
c. Making national energy policies robust, market oriented but with provisions for the under-privileged and with practical incentives and feed-in-tariffs where necessary.
d. The reviewed energy policies should be passed into law to minimize disruptions from one elected government to the next.
e. Establishment of Ministries of Renewable and Alternative Energy.
f. Establishment of international cooperation with advanced nations for technology transfer and capacity building.

References

1. WEC (2005) Untapped hydro potential in the world energy book, Issue 1, London 24–25
2. ICSU (2007) Sustainable energy in sub-Saharan Africa. Science plan of the international council for science (ICSU) Regional office for Africa
3. IEA (2009) World energy outlook. International energy agency. Paris
4. Kerekezi S (2006) African energy sector: Priority issues. A background paper for the forum of energy ministers in Africa for CSD-14
5. WHO (2006) Fuel for life: Household energy and health. WHO report WA 754

Chapter 24
Perspectives for Renewable Energy in Europe

Challenges on the Way Towards a Stable and Reliable Policy Framework Until and Beyond 2020

Rainer Hinrichs-Rahlwes

Abstract The Renewable Energies Directive (RED) of the European Union (EU), enacted in 2009, which established a binding target of a 20 % share of renewables for the EU and differentiated binding targets for each Member State by 2020, is facing significant implementation challenges. It appears that, due in part to a range of policy adaptations in the interim, the EU and many of its Member States will fail to achieve the targets. A European Commission-supported consortium of scientists and European and national renewable energy associations tasked with tracking progress towards the established targets has developed policy recommendations that would enable the EU and all Member States to achieve or even overshoot their 2020 targets. In order to get back on track, and succeed in 2020 and beyond, the consortium has developed an overarching six-point recommendation, highlighting the need for policy stability, and a focus on the key sectors of transportation and heating and cooling, among other pressing concerns.

Keywords Climate and Energy Package · Policy framework · Targets · European debate

24.1 Progress Towards the European Union's 2020-Renewables Targets

The European Union's (EU's) Climate and Energy Package 2020[1] was politically agreed by the Heads of State and Government in March 2007 and enacted by a set of directives and resolutions, which entered into force in 2010. The package comprises

[1] There is plenty of literature about the Climate and Energy Package. I mention here only my own book (with contributions from some colleagues), published in mid-2013, which describes and analyses the decision process until the agreement and the included directives and other decisions.

R. Hinrichs-Rahlwes (✉)
European Renewable Energies Federation (EREF), Avenue Marnix 28, 1000 Brussels, Belgium
e-mail: rainer.hinrichs@bee-ev.de

A. Sayigh (ed.), *Renewable Energy in the Service of Mankind Vol II*,
DOI 10.1007/978-3-319-18215-5_24

a target of 20 % energy efficiency (EE) increase until 2020, a share of at least 20 % renewable energy (RE) in gross final energy consumption (i.e. in all the three sectors: electricity, heating and cooling, and transport), and a target of reducing greenhouse gas (GHG) emissions by 20 % compared with 1990 levels (or 30 % if other relevant countries agree to similar efforts). The most effective part of the package is the Renewable Energies Directive (RED), the main instrument to achieve the 20 % renewables target. It sets binding targets (and indicative trajectories up to 2020) for every Member State (MS) and establishes transparency and clear administrative regulations, including National Renewable Energy Action Plans (NREAPs). MSs are responsible for reaching their national targets by domestic action (i.e. national support mechanisms designed by the MSs) and/or by voluntary cooperation with neighbours or 'third' (i.e. non-EU) countries. NREAPs have to outline the policies which will be applied to reach the mandatory targets.

The national targets for each MS are differentiated according to resources, economic potential and some other factors. Achievement is closely monitored by regular reporting, official statistical data and—among others—a project consortium of scientists and European and national renewable energy associations, supported by the European Commission, branded as 'keep-on-track'.[2] The consortium has published a second 'Tracking Roadmap' in summer 2014 and will publish an updated version in 2015.

Evaluation of 2012 data provides a sobering picture. Several MSs have backed away from their original NREAPs and have lowered their ambitions. A wide range of policy adaptations is disturbing legal certainty and risking economic viability of projects, which were relying on the stability of existing policies and frameworks. Some changes (e.g. in Czech Republic, Spain, Bulgaria) were even introduced retrospectively. Data evaluation and scenario calculations suggest that additional policies are necessary. Without them, the EU as a whole will probably fail to achieve the 20 % RE target in 2020, and so will many MSs.

According to the consortium's calculations, (see Fig.24.1) the EU will only reach 17.9 % RE in 2020 (30.9 % in the power sector, 18.8 % in heating and cooling and 5.5 % in transport). Only 9 out of 27[3] MSs seem to be well on track towards their national targets. For four MSs (Finland, Germany, Ireland and Slovakia) there are serious doubts about whether their targets will be reached without additional measures. The remaining 14 are clearly classified as 'not well on track'. Most MSs will have to undertake additional efforts and adjust their policies to achieve the 2020 targets and to keep the EU on track.

It also deals with the implementation status and the upcoming discussions and challenges. A long list of references leads to all major primary and secondary sources, which might be relevant in this context: *Rainer Hinrichs-Rahlwes* (with contributions from Christine Lins, Jan Geiss, Markus Kahles and Thorsten Müller), *Sustainable Energy Policies for Europe: Towards 100 % Renewable Energy*, CRC Press/Balkema, Leiden NL http://www.crcpress.com/product/isbn/9780415620994.

[2] www.keepontrack.eu All publications mentioned in the article can be downloaded from the website.

[3] The most recent 28th Member State, Croatia, will only be included in the 2015 publications.

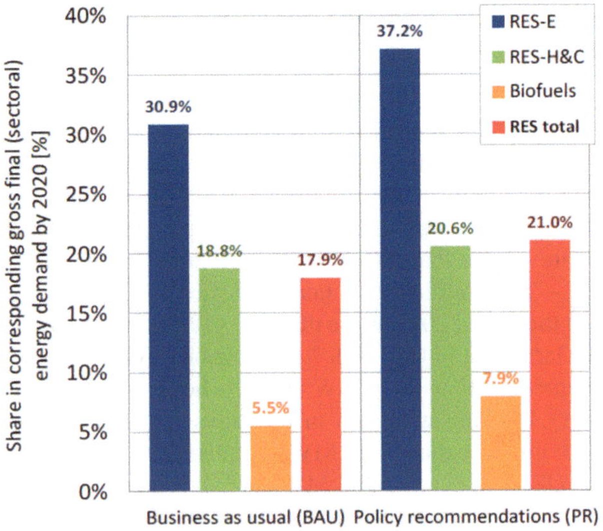

Fig 24.1 Sector Specific RES-Shares 2020 according to assessed cases. (*Source*: Keep-on-track! EU Tracking Roadmap 2014

The consortium has developed policy recommendations, which would enable the EU and all MSs to achieve or even overshoot their 2020 targets. France, Luxembourg, Malta and the UK would need to use the cooperation mechanisms of the RED. All others could reach their targets by domestic action. The EU's 2020 share of renewables in the gross final energy consumption would amount to 21 %, with the highest share (37.2 %) in the electricity sector, 20.6 % in the heating and cooling sector and 7.9 % in the transport sector.

The policy recommendations developed by the consortium provide detailed proposals for each MS. Also, they are summarised in six overarching recommendations for all MSs and the EU. The most important ones deal with regulatory and policy stability, and with reliability of framework conditions. The first recommendation refers to the discussions about a post-2020 framework and related targets. The consortium underlines the need to 'adopt an ambitious binding renewable energy target for 2030, including binding national targets, alongside EE and GHG emissions targets'. The second recommendation is as clear as the first one: 'Ensure a predictable and stable legislative framework for RES at the national level and in particular to avoid any retroactive changes to existing support schemes'. Furthermore, a stronger focus on the heating and cooling and transport sectors is recommended—both of which are still 'strongly dependent on the existence of a supportive and comprehensive framework'. In particular, for the transport sector the 'current policy vacuum' needs to be removed. Finally, the 'focus on the removal of administrative barriers' needs to be maintained.

24.2 Renewable Sources Dominating New Installations in the Power Sector

RE has been very well developed in Europe—particularly in the power sector. In 2013, 72% of new power capacity installed in the EU was on RE technology, wind and solar energy providing by far the largest contribution. Other renewables contributed at a lower percentage. Also, this was not a single year only, but it was the sixth consecutive year that the share of renewables exceeded all other sources (Source: EWEA—Wind in Power: Annual Statistics 2013).

Altogether, at the end of 2013, more than 117 GW of windpower capacity existed in the EU, with Germany (34 GW—29%) and Spain (23 GW—20%) still far ahead of all other countries (UK with 11 GW and 9% holds the third position). But the balance is shifting. Looking at the new installations in 2013 only, Germany still was the number one with more than 3200 MW installed (29% of the market), but number two was the UK with nearly 1900 MW (17%), followed by Poland with nearly 900 MW (8%), and Sweden, Romania, Denmark and France with around 6% market share each. The European windpower market is expected to continue expanding, but—depending on policy decisions and international competition—probably at a slower pace.

Since 2010, with a peak of more than 22 GW of new capacity in 2011 and down to 11 GW in 2013, the solar photovoltaic (PV) market had experienced tremendous growth—increasing five times from an installed capacity of close to 17 GW in 2010 to 81.5 GW in 2013. Around 75% of Europe's cumulative PV capacity today is in Germany with more than 36 GW and in Italy—the two countries that dominated the EU PV market in 2010, 2011 and 2012, dropping to less than 50% in 2013 (Source: EPIA, Global Market Outlook for Photovoltaics 2014–2018). Depending on policy decisions and related market development, the European Photovoltaic Industry Association (EPIA) predicts another reduction of the market size in 2014 to below 7 GW followed by a gradual increase to 8 GW in 2018 or an uptake (13 GW) already in 2014, increasing to 17 GW in 2018.

Globally, similar tendencies can be observed: Renewables are providing major shares of new power capacities—contributing 56% worldwide in 2013 (Source: REN21—Global Status Report 2014). Non-European regions are accelerating deployment. China and the USA have overtaken Europe in terms of installed capacity from renewables, and others—e.g. the BRICS-countries—are following suit. This is another reason, why Europe needs to develop an ambitious and efficient policy framework for the 2030 time horizon—as soon as possible in order not to lose the competitive edge on growing global markets.

24.3 The 2030 Challenge: What Is at Stake?

Considerable growth of renewables and an increasing number of enabling policies can be observed[4] outside Europe. The picture is dramatically changing. A few years ago, Europe was the most important global hub for RE development, primarily for the power sector—particularly with regard to successful policy development. Meanwhile, the epicentre of global growth of REs is shifting to other regions—e.g. to China, India and other Asian countries, and also to the USA and to several South American countries; more recently to Africa, where renewables provide major opportunities for clean growth and access to energy. Europe's market shares are shrinking and major investment flows are shifting towards other parts of the world.

Many reasons for this shift are due to global developments, but some are home-made failures and policy deficiencies. Renewables are domestic energies, which considerably reduce import dependencies and spending for fossil fuel imports. They increase security of supply significantly. And given that renewables have steep and rapid learning curves resulting in cost decreases (many of which were induced by supportive frameworks in Germany, Denmark, Spain and a few other European countries), it is not surprising that more and more countries and regions around the world have decided (and are developing enabling policies) to tap the vast potentials of clean and sustainable renewables.

The global uptake could and should spur competition in order to further facilitate development and deployment of renewables. European policymakers are just about failing to take necessary decisions until and beyond 2020. It is obvious that—without a reliable policy framework with ambitious targets for 2030—Europe's renewable energy development will slow down or even stagnate, to the detriment of Europe's economy and global competitiveness.

For more than a year, the 28 MSs of the EU and the European Parliament have tried to develop a consensus about the post-2020 Climate and Energy Framework. There is a verbal consensus about the development of a framework that is in line with global GHG reduction targets and the objective of limiting global warming to a maximum of 2 °C by the end of the century. But there is no consensus about policies and instruments to reach these objectives—due to the fact that some countries (particularly, the UK and France) are trying to uphold nuclear energy as an option, while others (Poland and Czech Republic) are reluctant to reduce coal and fossil fuel use and are therefore advocating low targets and carbon capture and storage (CCS) and shale gas. The debate about the 2030 framework was not focussed on spurring EE and growth of renewables, but was creating ambiguity instead—by overstating costs rather than benefits and at the same time downplaying costs and risks of conventional and nuclear energy.

[4] For details see the 'Renewables Global Status Report', which is published annually by REN21—the Renewable Energy Policy Network for the twenty-first century. The latest available version evaluated for the article is of 2014 http://www.ren21.net/REN21Activities/GlobalStatusReport.aspx and REthinking Energy published by IRENA—the International Renewable Energy Agency: Towards a new power system http://www.irena.org/rethinking/.

24.4 Member States' Positions and European Commission's Proposals

The different priorities of the 28 MSs are reflected in their opinions regarding the elements the 2030 Climate and Energy Framework should include. Whereas all seem to agree on the need of stability and reliability including the objective of completing the internal energy market (IEM), there are different opinions about the ambition level of 2030 targets for GHG reduction and even more so about targets for EE and renewables. Poland (in line with other Visegrad-countries) is aggressively opposing all targets (including a meaningful GHG-reduction target), if the agreement does not include a range of exceptions and benefits in favour of their coal-fired power plants and shale gas projects. The UK, France and some others are advocating a single (GHG) target only—arguing that only such an approach would lead to consensus and comply with the MS' prerogative of their energy mix. Following the good example of the 2020 framework, Germany and some others are asking—though not very outspoken—for three targets (GHG, EE and RE). The RE sector as well as major environmental and climate change NGOs is advocating three mutually reinforcing binding targets, including a RES target of at least 45 % in 2030, underpinned by binding national targets in order to maintain a stable and ambitious framework for sustainable growth. In contrast, the conventional energy sector, e.g. represented by EURELECTRIC, is striving for a GHG target only, explicitly leaving the energy mix open—and thus supporting those MSs that are planning to slow down RE development and subsidise coal and/or nuclear energy instead.

These different positions of the MSs are the background for the European Commission's proposal for a 'Climate and Energy Framework 2030' (CEF 2030), presented in January 2014.[5] It was widely welcomed by the conventional energy sector and strongly criticised by the renewables sector and environmental NGOs. The European Parliament, too, asked for more ambitious targets, including for RE and EE. The main elements of the proposal are a GHG-reduction target of 40 % compared to 1990 levels, which should be achieved 'domestically', and a RES target of 27 % in 2030, which should be 'binding on the EU level'. The proposal did not include an EE target, but envisaged to evaluate the existing directive and other legislation first. In addition, a 'new governance structure' was proposed, which is intended to replace the existing binding targets and mandatory national action plans for EE and RE, allegedly leaving more flexibility for MS to determine their energy mix and pick the GHG options they prefer. This 'new governance' was widely criticised as not being sufficiently precise to have any real impact.

[5] 'A policy framework for climate and energy in the period from 2020 to 2030' http://eur-lex.europa.eu/legal-content/EN/TXT/PDF/?uri=CELEX:52014DC0015&from=EN.

24.5 Letting 'the Market' Decide

Discussions about the 2030 framework are not the only threats to reliability and stability of the policy framework. Feed-in tariffs have dominated the policy landscape in the RES electricity sector in Europe for many years. They have been very effective to support renewables development. On the other side they have been criticised for disturbing the market, particularly by those who are interested in limiting the growth of RE, and by those who are interested in limiting competition from new and independent market entrants, which were and are gaining market shares, and thus are threatening incumbent business models and those utilities that are not willing or able to adapt to increasing RES shares and the resulting needs of system change. Opponents to feed-in tariffs have long since advocated quota systems with tradable certificates and/or tendering procedures in order to control better the growth of renewables. Many of them have pushed for harmonising support systems for renewables on the basis of quota or tendering systems. A major attempt in this direction was rejected by the European Parliament and the European Council in 2009. As a result, the RED[6] as a major element of the 2020 Climate and Energy Package unequivocally states that MSs are responsible for achieving their national RE targets and have the right to design and implement support schemes of their choice, including the liberty to restrict support to domestically produced renewable electricity. Cooperation between MSs is an option, but limited to voluntary agreements between them.

Parts of the European Commission, and particularly EURELECTRIC and other incumbent stakeholders have never stopped campaigning for 'more Europeanisation' of renewables support, accelerating convergence of national support schemes and also for 'harmonisation' as opposed to the present legislation, which was recently (July 1, 2014) confirmed by the European Court of Justice in the Åland case.[7] Despite this ruling and the strong political consensus about MS' right to define their energy mix and their support schemes, the European Commission agreed on new Environmental and Energy Aid Guidelines (EEAG),[8] which could severely undermine the MS' prerogative. Among various other regulations, the guidelines include a limited variety of conditions, under which MS may grant operating aid (state aid) for producers of energy from renewable sources.

Except for very small installations and for exceptions requiring detailed reasons, the EEAG exclude feed-in tariffs from the list of lawful state aid (i.e. permissible state aid, considered not to be distorting markets). Legally, guidelines can only

[6] Directive 2009/28/EC on the promotion of the use of energy from renewable sources http://eur-lex.europa.eu/legal-content/EN/TXT/PDF/?uri=CELEX:32009L0028&from=EN.

[7] In a case about the Swedish support system, the European Court of Justice decided that MSs have the right to limit access to their support systems to domestic producers—For details see the ECJ's press release with further links: http://curia.europa.eu/jcms/upload/docs/application/pdf/2014-07/cp140090en.pdf.

[8] Guidelines on State aid for environmental protection and energy 2014–2020 (EEAG) http://eur-lex.europa.eu/legal-content/EN/TXT/PDF/?uri=CELEX:52014XC0628(01)&from=EN.

facilitate the application of existing legislation and cannot create new legislation, which these guidelines obviously do by limiting the MS' choice of support schemes. The European Commission argues that—if they provide good reasons—MSs are free to pick other mechanisms. This is, however, only a theoretical possibility, if legal certainty for beneficiaries of the support is taken into account.[9] They would risk repayment of the support, if a MS implemented legislation, before receiving state aid clearance from the European Commission. As a result, MSs have little choice but to apply the EEAG, in order to receive swift clearance from the European Commission.

24.6 Another Missed Opportunity: European Energy Security Strategy

The crisis in Ukraine and the tensions with Russia should have been a major incentive for reducing Europe's dependence on fossil and nuclear imports and thus increased energy supply security. For obvious reasons, renewables should be a major part of the solution, which should have been included in the European Commission's 'European Energy Security Strategy' (EESS),[10] which was presented in May 2014.

It includes suggestions for short-term diversification of energy supply, particularly alternatives to Russian gas. It suggests the need for improving energy infrastructure and EE. Unfortunately, it almost completely fails to include RE. The potentials and opportunities are hardly mentioned. The focus remains on fossil fuels (including unconventional sources) and nuclear energy (misinterpreting uranium as a domestic source). According to the EESS, renewables need to achieve the 2020 targets 'in the context of a market-based approach' and with 'improved coordination of national support schemes'. There is no reference to the potentials of various RES, including for heating and cooling and transport sectors. Moreover, there is no reference to any policy framework or targets beyond 2020—as if the 2030 debate were not taking place in parallel.

Although the EESS acknowledges the EU's increasing energy import dependency, it is widely oblivious of consequences to be derived from these facts. For more than a decade, the EU's domestic energy production from fossil fuels has constantly been decreasing, whereas the production from renewables has been

---

[9] This is why in September 2014 EREF filed an Annulment plea against the EEAG to the European Court of Justice, accusing the European Commission of exceeding its powers by issuing guidelines which are apparently in conflict with existing legislation including the TFEU.

[10] http://ec.europa.eu/energy/doc/20140528_energy_security_communication.pdf. The document was discussed and basically endorsed by the Heads of State and Government of the EU in their June Council Meeting.

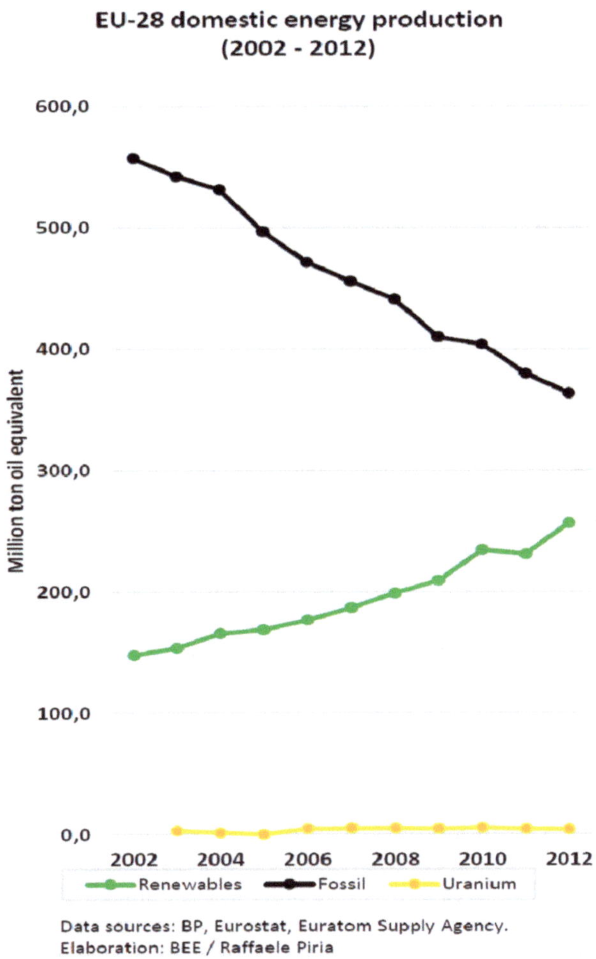

Fig. 24.2 EU-28 domestic energy production (2002–2012). (*Source*: EU Energy Security Strategy—Tackling the causes, not just the symptoms, BEE—German Renewable Energy Federation, June 2014)

smoothly increasing (Fig 24.2).[11] This tendency is even more obvious for the individual sources (Fig 24.3). Whereas renewables are 100% domestic (with exceptions for biomass, where imports play a certain role), coal, gas, oil and uranium are increasingly imported from outside Europe. From 2001 to 2012, the domestic part of coal use fell from 65 to 55%. For natural gas, the figures are even more worrying: from 52% domestic in 2001 to only 34% in 2012. Instead of trying to find more

[11] This has been analysed and assessed in a study published by the German Renewable Energy Federation und June 2013 http://www.bee-ev.de/Publikationen/BEE2014_EU_Energy_Security_Strategy-Tackling_thecauses_not_just_the_symptoms.pdf.

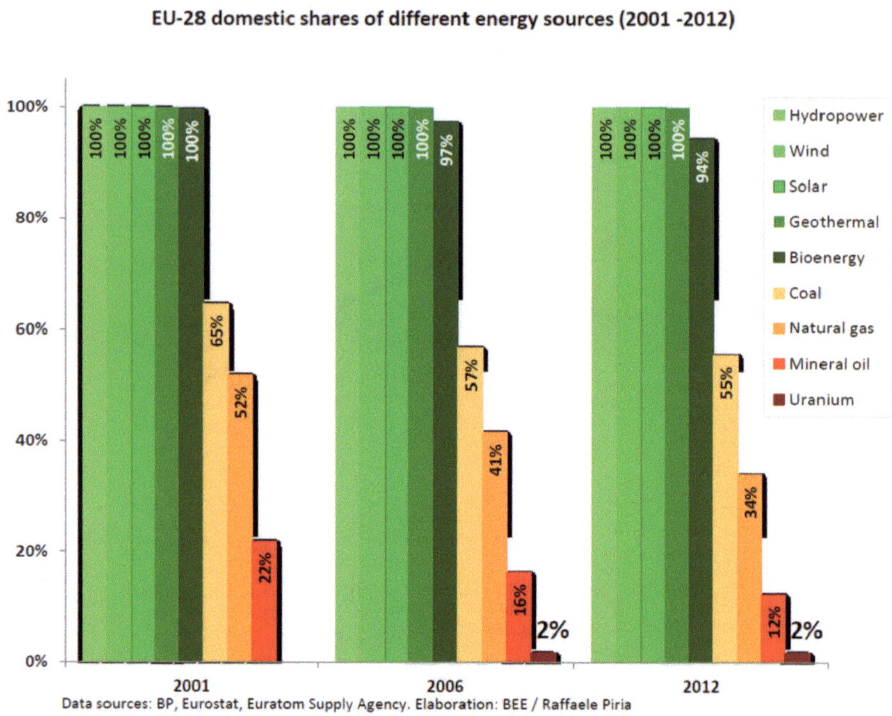

Fig 24.3 EU-28 domestic shares of different energy sources (2001–2012). (*Source*: EU Energy Security Strategy—Tackling the causes, not just the symptoms, BEE—German Renewable Energy Federation, June 2014)

sellers of fossil fuels and/or build more pipelines and liquefied natural gas (LNG) terminals, the focus on increasing renewable sources would be the smarter option. Replacing fossil fuel imports by renewables would also save billions of Euros. In 2013 alone, the EU spent more than 540 billion €.

24.7 Council Conclusions and the Next Steps

The major elements of the CEF 2030 proposal were endorsed by the European Council Meetings in March and June 2014, and—after adding a non-binding EE target in the context of the Ukraine crisis—became the official input to the European Council on October 23/24, 2014, where the Heads of State and Government had planned to agree on the main principles and ambition levels of the 2030 Climate and Energy Framework. When the conclusions were eventually presented to the public [1], the ambition levels had further decreased and advantages for coal users had been added. The Council once more did not bother mentioning renewables at least as part of the solution for energy supply security.

The Council endorsed 'a binding EU target of an at least 40% domestic reduction in greenhouse gas emissions by 2030 compared to 1990'. The target will be 'delivered collectively by the EU in the most cost-effective manner possible, with the reduction in the Emissions Trading System (ETS) and non-ETS sectors amounting to 43% and 30% by 2030 compared to 2005, respectively'. The reform proposals for reviving the ETS, however, are so weak and unspecific that ETS will probably not have a relevant impact on carbon prices and thus on decarbonisation. The conclusions include loopholes for countries with high emissions, for example, continued free allocation of certificates (allegedly to prevent carbon leakage).

The renewables and efficiency part of the conclusions is even more disappointing, because there had been some hope that the good arguments about reducing energy dependence, creating wealth and reducing GHG emissions would have some impact. At the end of the day, the Council agreed on an 'EU target of at least 27% ... for the share of renewable energy consumed in the EU in 2030. This target will be binding at EU level. It will be fulfilled through Member States contributions guided by the need to deliver collectively the EU target without preventing Member States from setting their own more ambitious national targets and supporting them, in line with the state aid guidelines, as well as taking into account their degree of integration in the internal energy market'. Underling that support for renewables will have to be 'in line with the state aid guidelines' is another problematic decision. The Council thereby de facto ratifies guidelines, which the European Commission had imposed in partial breach of the RED and also the European Treaties and thus depriving the Council and Parliament of their constitutional rights of legislation.

The Council also agreed on an 'indicative target at the EU level of at least 27% ... for improving energy efficiency in 2030 compared to projections of future energy consumption based on the current criteria'. Apart from the fact that this is a non-binding target and a very weak definition of efficiency, the target level was lowered from 30 to 27% at the very last moment.

The conclusions underline that the targets 'will not be translated into nationally binding targets', thus renouncing from an effective tool to foster and eventually enforce compliance, if MSs are not developing and implementing appropriate policies.

The Council reiterated its commitment to improve EU-wide transmission grids and gradually increase interconnection capacity between MSs from the existing low levels via 10% of installed capacity to higher levels.

The Council sets up some governance elements, the most concrete ones being the requirement of 'a reliable and transparent governance system'. The system will 'build on the existing building blocks, such as national climate programmes, national plans for renewable energy and energy efficiency. Separate planning and reporting strands will be streamlined and brought together'. By highlighting the intention to 'keep all elements of the framework under review' and to 'continue to give strategic orientations as appropriate, notably with respect to consensus ...' the Council not only underlines the importance of these conclusions and the related targets and frameworks, but also—implicitly—undermines the European Parliament's role in Europe's legislation and rules out decisions taken by Qualified Majority Vote as defined in the Lisbon Treaty, the latest version of the EU Treaties.

In the coming months, the *Council Conclusions* will be further detailed by legislative and procedural proposals from the European Commission and eventually be transposed into legislation at European and national levels. According to the EU Treaties, the legislative process will have to result in a consensus (or a qualified majority on both sides) between MSs (acting through the European Council) and the European Parliament. This leaves some leeway for improving or at least further specifying the insufficient Council agreement.

Acknowledgements The original paper analysing the status of and challenges for renewables in Europe was presented at the WREC XIII in Kingston/London on August 5, 2014. Regarding the challenges and perspectives, it could therefore only reflect the policy debate until the summer break 2014. The written version of the paper was finalised at the end of October 2014, just a few days after the European Council (i.e. the European Heads of State and Government) decided on the basic principles and targets of Europe's Climate and Energy Framework for 2030, thus providing an answer—though an insufficient one—to the challenges and questions raised in my August presentation. Until the paper was published in 2015, further development of European energy policies took place. Meanwhile the issues raised in the paper have become part of the "Energy Union" process" with legislatition on governance, compliance, target development, energy market design and other elements of the CEF 2030 being prepared in detail.

References

1. European Council (23 and 24 October 2014) Conclusions on 2030 climate and energy policy framework. http://www.consilium.europa.eu/uedocs/cms_Data/docs/pressdata/en/ec/145356.pdf. Accessed 21 Aug 2015

Chapter 25
Examination of Energy Usage of Electrical House Applications in Terms of Energy Efficiency

Zafer Utlu and Hasan Saygin

Abstract This study evaluates the energy utilization efficiency of the Turkish residential-commercial sector (TRCS) and electrical house applications in 2011 by using energy and exergy analyses. Of the total energy input, 28.15 % was produced in 2011, while the rest was obtained by imports. In 2011, 34.47 % of Turkey's total end-use energy was consumed by the residential-commercial sector. Annual fuel consumptions in space heating, water heating, and cooking activities as well as electrical energy used by appliances are determined for the period 2000–2011. It is clear from this figure that the energy efficiencies in the years studied range from 57.05 to 65.53 %, while the exergy efficiencies vary from 8.02 to 10.07 %. In addition, researches on EHA efficiency indicated an efficiency of 81 % for the first law of thermodynamics and 22 % for the second law of thermodynamics.

Keywords Electrical household appliances · Energy · Exergy · Efficiency

25.1 Introduction

Considering the increase in energy production and consumption of Turkey in the last decade, energy consumption, which was 3527.2 pJ in 2000, has increased to 4859.16 pJ in 2011. The rate of meeting consumption through production is 28.1 %; thus, rest of the 71.9 % consumption is met through imports in 2011. This situation indicates that substantial amount of national revenues of the country are spent on energy imports. When this situation is investigated in terms of electricity consumption in parallel with technological advancement and economic growth, 128.295 GWh consumption in 2000 increased to 229.395 GWh in 2011 [1].

Z. Utlu (✉) · H. Saygin
Engineering Faculty, Mechanical Engineering Department,
Istanbul Aydin University, Florya, Istanbul, Turkey
e-mail: zaferutlu@aydin.edu.tr

H. Saygin
e-mail: hasansaygin@aydin.edu.tr

A. Sayigh (ed.), *Renewable Energy in the Service of Mankind Vol II*,
DOI 10.1007/978-3-319-18215-5_25

Investigations on electricity consumption on sectorial basis show that 47.3 % of the consumption in 2011 belongs to industrial sector and the remaining sectorial consumption is distributed among dwellings as 23.8 %, business organizations as 16.4 %, government offices as 3.9 %, lighting as 2.2 %, agricultural irrigation as 2.0 %, and other as 4.4 %. It is known that electricity energy consumption of dwellings is caused by electrical household appliances (EHA). Considering that EHA are used in other electricity-consuming sectors, it is understood that electrical consumption of EHA corresponds to 28–32 %. This situation becomes clear with the annual sales figures of new electrical household appliances, which increased two fold in the last decade. Considering that energy consumption of dwellings is substantially based on the use of EHA, increasing energy efficiency in such products and market dominance of products with high energy efficiency would lead to substantial amount of energy saving. In addition, the total greenhouse gas emission of Turkey, which was 298 million t in 2000, increased by 5.9 % and reached to 422.41 million t in 2011. In 2011, 71 % of the total emission is caused by energy sector, 9 % is from wastes, 13 % is from industry, and 7 % is from agriculture sectors.

Earlier studies conducted on the energy and exergy analyses of Turkey are based on the values of the past years. The present study analyzes the energy and exergy use in the Turkish residential-commercial sector (TRCS) and EHA in 2011. This analysis is also done based on the actual data for 2011.

25.2 Total Input Configuration of Turkey

Turkey, with a population of 74,724,269 and with dwelling units of 19,481,678 on 800,000 km^2 of land is located between 35°50′ and 42°06′ N latitudes and 25°40′ and 44°48′ E longitudes. Most of Turkey is in Asia. The far north-western part of the country is in Europe and is separated from the rest of the country by the Dardanelles and Bospherous Straits and the Sea of Marmara [2, 3]. The values of energy and exergy inputs for TRCS in 2011 according to energy carriers are illustrated in Table 25.1. As can be seen in the table, the total energy and exergy inputs to the Turkish sectors were 4859.28 and 4724.79 pJ, respectively, while for residential-commercial sector, they were determined as 1243.79 and 1190.87 pJ in 2011, respectively. Of the total energy input, 35.05 % was produced in 2000. It was also determined that 28.15 % of the total energy input was produced in 2011, while the rest was obtained by imports.

In 2011, 34.47 % of Turkey's total end-use energy was consumed by the residential-commercial sector. The other sectors were determined as listed: the industrial sector with 35.45 %; the transportation sector with 18.34 %; the agricultural sector with 6.6 %; and out of energy with 5.1 %. In this year, in terms of consumption of the 13 largest energy sources, natural gas had the largest share, with 34.99 %, followed by petroleum, with 26.22 %. Figure 25.1 illustrates energy and exergy flows in a macrosystem for Turkey's whole and TRCS.

Table 25.1 Energy and exergy inputs to TRCS during 2011

			Residential-commercial			Turkey	
Energy carrier toe/q[a]			Input (pJ)	Sector (%)	Turkey (%)	(pJ)	(%)
Hard coal	0.61	Energy	172.70	13.88	3.55	668.76	13.76
	1.03	Exergy	177.88	14.94	3.77	688.82	14.64
Lignite	0.21	Energy	61.23	4.92	1.26	648.98	13.36
	1.04	Exergy	63.68	5.35	1.35	674.94	14.35
Asphaltite	1.03	Energy	6.56	0.53	0.14	15.54	0.32
	0.97	Exergy	6.76	0.57	0.14	15.07	0.32
Petroleum	1.05	Energy	52.01	4.18	1.07	1274.17	26.22
	0.99	Exergy	51.49	4.32	1.09	1261.43	26.81
Natural gas	0.91	Energy	426.18	34.26	8.77	1700.68	35.00
	0.92	Exergy	392.08	32.92	8.30	1564.63	33.26
Wood	0.30	Energy	102.06	8.21	2.10	102.25	2.10
	1.05	Exergy	107.17	9.00	2.27	107.36	2.28
Biomass	0.23	Energy	42.89	3.45	0.88	45.61	0.94
	1.05	Exergy	45.03	3.78	0.95	47.89	1.02
Hydropower (electricity)	0.09	Energy	316.54	25.45	6.51	188.14	3.87
	1.00	Exergy	316.54	26.58	6.70	188.14	4.00
Geothermal	0.86	Energy	45.19	3.63	0.93	86.09	1.77
	0.29	Exergy	13.10	1.10	0.28	24.97	0.53
Solar	0.86	Energy	18.43	1.48	0.38	26.33	0.54
	0.93	Exergy	17.14	1.44	0.36	24.49	0.52
Wind	0.09	Energy	0.00	0.00	0.00	16.98	0.35
	1.00	Exergy	0.00	0.00	0.00	16.98	0.36
Coke	0.70	Energy	0.00	0.00	0.00	68.99	1.42
	1.05	Exergy	0.00	0.00	0.00	72.44	1.54
Petrocoke	0.77	Energy	0.00	0.00	0.00	16.64	0.34
	1.04	Exergy	0.00	0.00	0.00	17.31	0.37
Total		Energy	1243.79	100.00	25.60	4859.16	100
		Exergy	1190.87	100.00	25.21	4724.79	100

[a] The upper values are conversion factor to tons oil of equivalent (toe), while the lower values are quality factor

25.3 Energy Utilization in the TRCS

The TRCS includes space heating, water heating, cooking, and electrical appliances for energy consumption. In the following subsections, the utilization of energy and exergy in the TRCS during the year 2011 is analyzed. The specific applications for energy and exergy consumptions were determined for 2011 and shown in Fig. 25.1. In 2011, of Turkey's end-use energy, 34.47% was used by the residential sector. Table 25.1 illustrates the use of energy and exergy as well as the shares of the resources in this sector for the year 2011. Share of the energy utilization in the

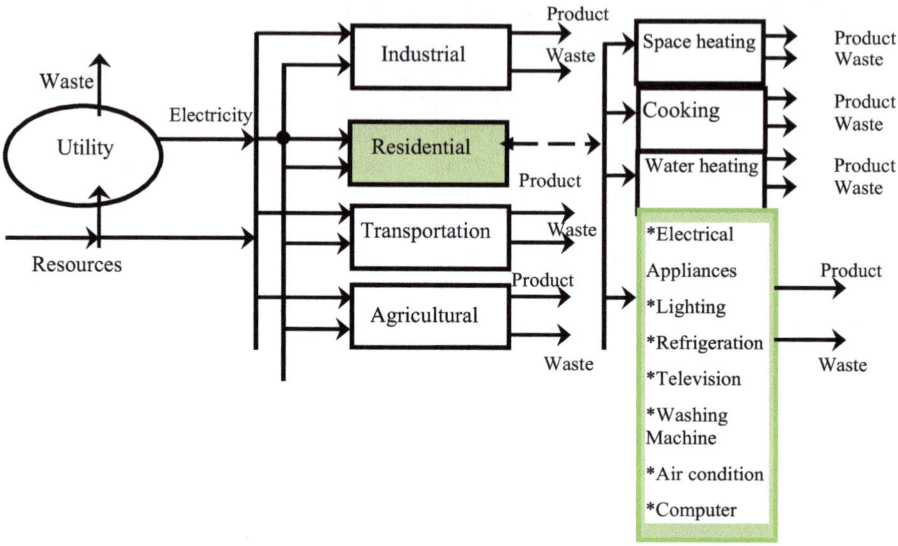

Fig. 25.1 An illustrative presentation of the energy flows in a macrosystem and residential for Turkey

residential-commercial modes is as follows: space heating with 42 %, water heating with 26 %, cooking with 12 %, and electrical appliances with 20 % in the studied year. These values are determined for 2011. Table 25.1 shows energy and exergy utilization values for the year studied in the TRCS. The highest contributions came from renewable resources (including wood) with 38.70 %, fuel with 35.85 %, and electricity with 25.45 % in 2011. In 2011, the highest contributions came from wood with 102.40 pJ. However, natural gas usage has continuously increased in the TRCS for space heating, water heating, and cooking purposes in several cities.

Natural gas constituted 103.73 pJ of used energy in this sector in 2011. In addition, utilization of renewable energy is spread in the TRCS, especially from sunlight for water heating, from geothermal for water heating and space heating, and from biowaste for general usage.

25.4 Efficiency Analysis for Electric Utilization

As living standards rise, the use of electrical appliances is increasing fast and boosting electricity demand. Increasing use of air conditioning, especially in the Mediterranean region, has shifted the peak hours of electricity demand to noon in the summer. According to a survey conducted by Turkey Statistical Institute (TSI) in 2011, Turkey has a total of 19,481,678 residences. In Turkey, the average home appliances used in a residential are: fridge, washing machine, dishwasher, television, ironing board, vacuum cleaner, computer, oven, and microwave oven.

Table 25.2 Energy utilization values of residential-commercial service in Turkey in 2011

	Electrical (%)	Fuels (%)	Renewables[a] (%)
Component			
Lighting	36		
Incandescent	40		
Fluorescent	60		
Refrigeration	35		
Water heating	2	30	54
Cooking	3	18	
Space heating	2	50	46
Washing machine	2		
Vacuum cleaner	2		
Air conditioning	4		
Television	7		
Iron	1		
Computer	4		
Miscellaneous	2	2	
	100	100	100

[a]Renewables include biomass, wood, geothermal, and solar

Table 25.3 Saturation values of electrical appliances in1998, 2000, and 2011

Type of appliances	1998[a] (%)	2000 (%)	2011 (%)
Lighting	100	100	100
Incandescent	80	70	30
Fluorescent	20	30	70
Refrigeration	97.38	99	99
Television	96.59	99	99
Washing machine	78.97	86.20	95
Dishes machine	14.49	30.30	65.45
Vacuum cleaner	77.21	89.19	9240
Air-conditioning	1.26	1.45	9.25
Clothes drying	92.54	96.60	98.75
Electrical furnace	74.57	90.00	95.00
Hair drying machine	61.62	82.40	89.00
Computer	3.6	3.9	24

[a]These datas were taken from references [2,7,9]

Electrical energy is used for various purposes such as lighting, refrigeration, television, washing machine, etc. in this sector. Energy utilization values for the TRCS are indicated in Table 25.2, while the saturation values of electrical appliances are given in Table 25.3.

Refrigeration requires the largest fraction of electricity with 33–40 % in the year studied, followed by lighting with 35 %. The overall efficiency and effectiveness values for electric utilization were estimated as follows [4–6]:

Table 25.4 Annual energy use of lighting and electrical values in housing sector

Type of appliances	Consumption (kW/year)	Consumption share (%)
Fridge	917.28	27.9
Washing machine	187.2	5.7
Dishwasher	171.6	5.2
Television	296.4	9.0
Television 2	192.4	5.8
Iron	130	4.0
Broom	78	2.4
Computer	109.2	3.3
Bakery	260	7.9
Microwave oven	12.48	0.4
Lighting	936	28.4
Total	3290.56	100

Table 25.5 Values of energy efficiency index according to energy efficiency class

Energy efficiency index (I)	Energy efficiency class
$I < 55$	A
$55 \leq I < 75$	B
$75 \leq I < 90$	C
$90 \leq I < 100$	D
$100 \leq I < 110$	E
$110 \leq I < 125$	F
$125 \leq \mathrm{I}$	G

25.4.1 Energy Usage Rate of Lighting and Electrical Appliances

Table 25.4 shows the annual energy use of lighting and electrical values in housing sector; besides, the energy consumption share is also given in Table 25.4.

25.4.2 Energy Efficiency Index (I)

Depending on the values of energy efficiency index, which allows for determination of the energy efficiency, class values are given in Table 25.5.

Energy efficiency index (I) is calculated as follows and expressed in percent (%).

$$I = E/\mathrm{EST} \tag{25.1}$$

where I is the energy efficiency index; E is the annual energy consumption of the device in accordance with the standards specified, expressed in kWh/year terms (24 h consumption $\times$ 365), will be provided and it will be measured according to TS-EN 153; Est is the standard annual energy consumption of device, expressed in kWh/year.

$$\mathrm{Est} = M \times \mathrm{DH} + N \tag{25.2}$$

Table 25.6 Values of energy efficiency index according to energy efficiency class

Class of device	M	N
Domestic refrigerator	0.233	245
Fridge-chiller	0.233	245
Starless fridge	0.233	245
One-star refrigerator[a]	0.643	191
Double-star refrigerator[b]	0.450	245
Three-star refrigerator[c]	0.657	235
Fridge-freezer[a]([c])	0.777	303
Vertical freezer	0.472	286
Horizontal freezer	0.446	181
Other cooling devices	(1)	(1)

where DH is the adjusted net volume (liter); *M* and *N* values are based on Table 25.6, which is as below.

M and *N* values for these devices are determined by considering low-temperature chamber and the number of stars according to the temperature of the coldest part. Refrigerators consist of various compartments maintained at different temperatures. These compartments will affect energy consumption in different ways. Refrigerator's electricity consumption is a function of the corrected volume. Adjusted volume is the weighted sum of the volumes of the different compartments.

Thus, the adjusted volume (DH) may be expressed as:

$$\mathrm{DH} = V_c \times W_c \times F_c \times C_c \ldots \quad (\text{for each chamber}) \tag{25.3}$$

where V_c is the net volume of each compartment (l); $W_c = (25 - T_c)/20$; T_c is the design temperature of each compartment (°C); F_c is the factor that cools by internal forced air circulation and taken as 1, 2 for chamber icing and 1 for other chambers. C_c is the coefficient for tropical climates class of refrigerator that depends on compartments features of the refrigerator given below. Partition coefficient of variation is given by C_c in Table 25.7.

A++—Energy Class Products; the difference between the classes A, A+, and A++ is due to energy consumption, and the class more efficient than class A are

Table 25.7 Values of energy efficiency index according to energy efficiency class

Compartment	C_c
Pantry compartment	1.35
Fresh food compartment	1.30
0° compartment	1.25
Starless compartment	1.25
One-star[a] compartment	1.20
Two stars[b] compartment	1.15
Three[c] and four stars[d] compartment	1.10

[a]high energy consumption value
[b]medium energy consumption value
[c]low energy consumtion value
[d]lowest energy consumption value

Table 25.8 Daily electricity consumption according to energy class

Energy class	Daily electricity consumption (kWh/24 Saat)
B	1.70
A	1.23
A+	1.07
A++	0.50

named either A+ or A++. In Table 25.8 the energy consumption of different classes is compared.

A++ energy level consumed 0.5 kWh energy in a day. This 0.5 kWh consumption is equivalent to 40 W lighting lamp used for 12.5 h in a day.

25.5 Results and Discussion

In this study, Turkey's energy and exergy utilization efficiencies in 2011 were analyzed for the TRCS as well as EHA. The energy and exergy inputs were also compared, while losses and efficiencies were identified. Based on the estimations for energy and exergy consumptions, it is expected that annual energy consumption in the TRCS will increase by 6.5% from 2000 to 2011.

General relations used in the modeling and some of the key aspects of thermodynamics in terms of energy and exergy are taken from references [2–7].

25.5.1 Lighting

Approximately 35–38% of all electricity use was for lighting [6]. Electrical energy consumption value for lighting was 138 kWh per dwelling unit annually in 1990 [7]. Annual electricity consumption of dwelling unit for lighting is assumed to change linearly from 138–180 kWh in 2001 to 235 kWh in 2011 [8, 9]. Lighting is assumed to be 60% incandescent and 40% fluorescent, with first- and second-law efficiencies of about 5 and 4.5% This year is 1998. and 20 and 18.5%, in 2011, respectively. Utilization ratio of fluorescent in lighting is expected to increase from 20% in 2000 to 50% in 2011, as given in Table 25.9. Combining the relevant first- and second-law efficiencies for lighting, we calculated $\eta = 9.5 - 15.5\%$ and $\varepsilon = 8.70 - 14.3\%$ for the years considered, as indicated in Table 25.9.

25.5.2 Refrigeration

Refrigerators consume huge share of electricity in RCS as 35–40 % [4, 6]. By adopting new technology for refrigerators, electricity consumption can be decreased. Electricity consumption is projected to decrease by using new technologies

Table 25.9 Energy and exergy efficiencies values of electrical appliances

Component	2000		2011	
	η	ε	η	ε
Lighting	9.5	8.7	15.5	14.3
(Incandescent)	5	4.5	5	45
(Fluorescent)	20	18.5	20	18.5
Refrigeration	100	10.6	150	15.7
Water heating	90	10.8	90	10.8
Cooking	80	17.2	80	17.2
Space heating	98	7.3	98	7.3
Washing machine	80	80	90	90
Vacuum cleaner	70	70	80	80
Air conditioning	200	14	200	14
Television	80	80	80	80
Iron	98	30	98	30
Miscellaneous	70	65	90	70
Overall efficiencies	80.98	22.17	86.03	24.35

for refrigerators. Average annual consumption was calculated to be 346 kWh in 1990, 328 kWh in 1995, 300 kWh in 2001 and 2005, and 285 kWh in 2011 [7–9].

The second-law efficiency of refrigeration was calculated from

$$\varepsilon_2 = \varepsilon_1 \left[\left(\frac{T_0}{T_3} \right) - 1 \right] \tag{25.4}$$

It is assumed that the temperatures inside freezers and refrigerators are approximately −8 °C, the coefficient of performance (COP) is 1.0, and the room temperature near the refrigerator coil is 20 °C. Using Eq. (25.4), these assumptions yield $\varepsilon = 15.70\,\%$ in 2011, while second-law efficiencies are given in Table 25.9 for 2011.

25.5.3 Water Heating

In 2011, 4 % of all electrical use was for water heating. First-law efficiency of electrical use for water heating is assumed to be 90 % [4]. It is assumed that hot water and ambient temperatures are 60 and 20 °C, respectively, while quality factor (q_{fuel}) is 1.0 for electrical use. The second-law efficiency of water heating was calculated from Eq. (25.4) and was found to be 10.8 %, as given in Table 25.9.

25.5.4 Cooking

A total of 3–4 % of all direct electrical use was for cooking. It is assumed that first-law efficiency of electrical use for cooking is 80 %, and the cooking and ambient temperatures are 120 and 20 °C, respectively [5]. Using Eq. (25.4), these assumptions yield $\varepsilon = 17.2\,\%$ for cooking.

25.5.5 Space Heating

In all, 2–3 % of all direct electrical use was for space heating. It is assumed that first-law efficiency for space heating is 98 %, the supply temperature for the space heating equipment is 50 °C, and the ambient temperature is 20 °C [4, 6]. Using Eq. (25.4), the numerical values and the assumed first-law efficiency, we found $\varepsilon = 7.3\,\%$ for space heating.

25.5.6 Air Conditioning

Assuming that the COP value of the electric air conditioning unit is 2, this unit extracts heat from air at 14 °C and the outside temperature is 35 °C, and using Eq. (25.4) in a similar manner, we found $\varepsilon = 14\,\%$ in the years studied.

25.5.7 Television

Six to seven percent of all electrical use was for television and computers. Annual electricity consumption for television has increased compared with previous years due to an increase in the number of TV channels, a longer daily broadcast period, and utilization of color TVs. First- and second-law efficiencies are assumed to be 80 %.

25.5.8 Others

Electricity consumption values of other electrical appliances, for instance, washing machine, dish washer, iron, computer, and vacuum cleaner were selected and estimated as in Tables 25.2 and 25.3, while the first and second-law efficiencies of these appliances are listed in Table 25.9.

Substituting the relevant numerical values into Eqs. (25.5) and (25.6), we found $\eta = 66.16$, 86.03 % and $\varepsilon = 18.66$, 23.35 % for electrical use in 2000 and 2011, respectively.

Overall, the first- and second-law efficiencies ($\eta_{,orc}$ and $\varepsilon_{,orc}$) for the entire residential-commercial sector were calculated by aggregating both purchased electrical energy and direct fuel use as follows:

$$\varepsilon_{1,orc} = \frac{\left(\varepsilon_{1e} \times e_{rc} + \varepsilon_{1of} \times f_{erc}\varepsilon_{1r} \times r_{rc}\right)}{\left(e_{rC} + f_{erc} + r_{rc}\right)} \tag{25.5}$$

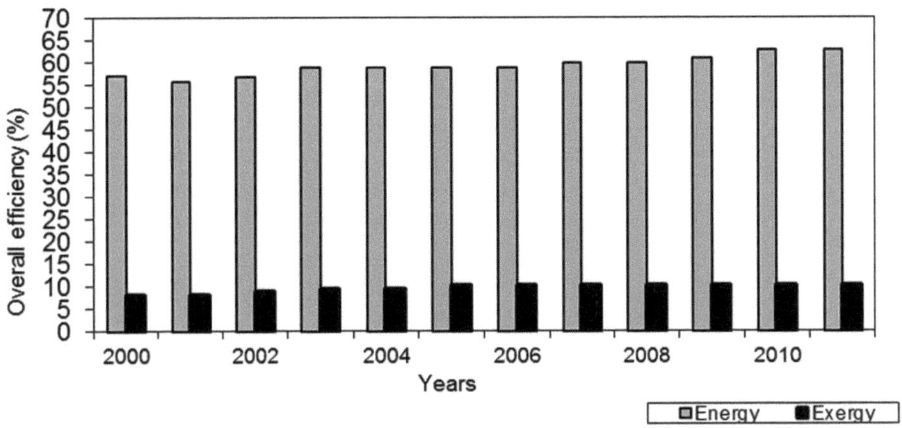

Fig. 25.2 TRCS energy and exergy efficiencies improvement over the period from 2000 to 2011 in Turkey

$$\varepsilon_{2,\mathrm{orc}} = \frac{\left(\varepsilon_{2\mathrm{e}} \times e_{\mathrm{rc}} + \varepsilon_{2\mathrm{of}} \times f_{\mathrm{erc}} \varepsilon_{2\mathrm{r}} \times r_{\mathrm{rc}}\right)}{\left(e_{\mathrm{rc}} + f_{\mathrm{exrc}} + r_{\mathrm{exrc}}\right)} \tag{25.6}$$

Using the numerical values given in Tables 25.2 and 25.3, the weighted mean overall energy and exergy efficiencies for the entire residential-commercial sector were found to be $\eta_{\mathrm{orc}}=63.53\,\%$, and $\varepsilon_{\mathrm{orc}}=9.02\,\%$ in 2011. It should also be noted that energy and exergy efficiencies for the TRCS are almost similar, as also denoted by Rosen and Dincer [4]. Furthermore, a variation of the overall mean energy and exergy efficiencies for the TRCS for 2011 is shown in Fig. 25.2.

25.5.9 Development of Energy and Exergy Efficiencies in the TRCS

As can be seen in Fig. 25.2, a comparison of energy and exergy efficiency values for the TRCS is illustrated. It is clear from this figure that the energy efficiencies in the years studied range from 57.05 to 65.53 %, while the exergy efficiencies vary from 8.02 to 10.07 %. This sector, the TRCS shows considerably important and comparable losses of energy and exergy. In terms of exergy losses, this sector ranks rather differently, accounting for about 89 to 91 % of all exergy losses.

The present study indicates that exergy utilization in Turkey was even worse than energy utilization. In other words, Turkey represents a big potential for increasing the exergy efficiency. It is clear that a conscious and planned effort is needed to improve exergy utilization in Turkey. Considering the existence of energy-efficient technologies in the world, the major problem is delivering these technologies to consumers or, in other words, using effective energy-efficiency delivery mechanisms, as reported in detail elsewhere [3–7]. These results indicate the need of sav-

Table 25.10 Energy and exergy efficiency values of TRCS in 2011 (%)

	2011	
	η (%)	ε (%)
Space heating ($\eta_{sh.of}$; $\varepsilon_{2sh.of}$)[a]	61.08	4.95
Water heating ($\eta_{wh.of}$; $\varepsilon_{2wh.of}$)[a]	63.58	7.56
Cooking ($\eta_{c.of}$; $\varepsilon_{2c.of}$)[a]	56.60	11.75
Overall sector for fuel use (η_{ofrcs}; $\varepsilon_{2.ofrcs}$)[a]	61.02	6.96
Overall sector for electrical energy use (η_{oercs}; $\varepsilon_{2.oercs}$)	86.03	23.35
Entire sector(η_{orcs}; $\varepsilon_{2.orcs}$)	65.53	10.07

[a] Fuel and renewable includes

ing the energy use and to improve habits of energy use in this sector and its subsectors. The space heating constitutes the biggest energy loss, followed by the water heating and cooking activities in the TRCS. From the evaluation of the results given in Table 25.10, it may be concluded that the TRCS has about equal and fairly high energy efficiencies, while it indicates a very poor performance in terms of its exergy efficiency values.

25.6 Conclusions

Exergy clarifies the situation by exposing the losses of a process and the emissions to the environment. The exergy concept assigns a number to these flows, which are to be minimized in order to meet sustainable conditions. Thus, exergy is a suitable and necessary concept in the development of a sustainable society, and future research of exergy and its applications must be further directed toward the development of a sustainable society [7].

- The present study has clearly shown the necessity of the planned studies toward increasing exergy efficiencies in the sector studied.
- It may also be concluded that the analyses reported here will provide the investigators with knowledge about how much effective and efficient a country uses its natural resources.
- This knowledge is also needed for identifying energy efficiency and/or energy conservation opportunities as well as for dictating the energy strategies of a country or a society.
- This result indicates that exergy and energy inefficiencies in the TRCS are caused by currently available techniques.
- Using promising building technologies, such as ground-source heat pumps, heat pump water heaters, high-efficiency chillers and air-conditioners, combined cooling, heating, and power systems, advanced reflective window films, integrated cooling and hot water systems, thermal energy storage (TES) systems, may be basic solutions for this.

- Especially, the use of TES systems has been attracting increasing interest in several thermal applications, e.g., active and passive solar heating, water heating, cooling, and air-conditioning, while TES is presently identified as the most economical storage technology for building heating, cooling, and air-conditioning applications [8].
- As efficiency labeling for electrical appliances air-conditioners (as well as for many other appliances) is nonexistent or is only now under preparation, there is room for significant improvement.

Acknowledgments This project is being supported by the Grant Programme under Market Transformation of Energy Efficient Appliances in Turkey Project (EVÜDP) implemented by the General Directorate for Renewable Energy (GDRE) of the Ministry of Energy and Natural Resources and United Nations Development Programme (UNDP) with the financial assistance of Global Environment Facility (GEF).

References

1. SIS, State Institute Of Statistics (2013) Prime Ministry Republic of Turkey, http://www.die.gov.tr. access date 20 March 2014
2. Hepbasli A, Utlu Z (2004) Evaluating the energy utilization efficiency of Turkey's renewable energy sources during 2001. Renew Sustain Energy Rev 8(3):237–255
3. Utlu Z, Hepbaşlı A (2007) Parametrical investigation of the effect of dead (reference) state on energy and exergy utilization efficiencies of residential–commercial sectors: a review and an application. Renew Sustain Energy Rev 11(4):603–634
4. Rosen MA, Dinçer I (1997) Sectoral energy and exergy modeling of Turkey. Trans ASME 119:200–204
5. Utlu Z, Hepbaşlı A (2007) A review on analyzing and evaluating the energy utilization efficiency of countries. Renew Sustain Energy Rev 11(1):1–29
6. Utlu Z, Hepbaşlı A (2006) Estimating the energy and exergy utilization efficiencies for the residential–commercial sector: an application. Energy Policy 34(10): 1097–1105
7. State Planning Organization (2012) Electrical Energy Special Commission Report, Eighth Development Plan, Ankara, Turkey, (in Turkish)
8. Hepbaşlı A, Utlu Z (2004) Comparison of Turkey's sectoral energy utilization efficiencies between 1990 and 2000. Part 2. Residential–commercial and transportation sectors. Energy Sources 26:1345–1355
9. SIS, State Institute of Statistics (2013) Energy Consumption in Residences in 1998, Prime Ministry of Turkey, Ankara

Chapter 26
The Role of Decentralized Energy for Widening Rural Energy Access in Developing Countries

S Ghazi

Abstract Decentralized energy (DE) is an inevitable part of universal energy access, particularly in developing countries. It is not just an absolute solution for widening energy in off-grid rural areas but it is important for remote and deprived areas connected to the centralized electricity network those suffer from inefficient energy due to the high grid network losses. Moreover, it covers waste and resource management, carbon and energy efficiency, and socioeconomic sustainability which can integrate sustainable consumption and production into the business models. Modern bioenergy in the form of biomass, ethanol, biodiesel, or biogas has the highest potential mainly because the technology is mature and is a relatively easy competitive substitute for fossil fuels. Despite many advantages and benefits of DE in rural areas, there are important barriers which prevent the increase of access to modern energy services; all of them are associated with social, technical, managerial, and institutional issues. This chapter states a record of DE statement in the world by emphasis on the present barriers. Therefore, to tackle these barriers, a holistic approach which covers the whole dimensions of a DE model is needed. In this regard, this interdisciplinary model should be flexible and should consider the benefits of all stakeholders, including national and international authorities, private sectors, developers, investors, small and medium enterprises, and local communities.

Keywords Decentralized energy · Bioenergy · Rural energy

26.1 Energy Access in Developing Countries

Despite many efforts made by international bodies to eliminate energy poverty around the world, there is still a clear distinction between the millennium development and the achieved goals of widening energy access, particularly in developing countries. The latest record of international energy agency shows that approximately one third

S. Ghazi (✉)
Department of Environmental Engineering,
Islamic Azad University-Parand Branch, Parand, Iran
e-mail: s.ghazi@brighton.ac.uk; sanaz_ghazo2001@yahoo.com

A. Sayigh (ed.), *Renewable Energy in the Service of Mankind Vol II*,
DOI 10.1007/978-3-319-18215-5_26

of the world's population does not have access to electricity [1]. Moreover, 2.7 billion people around the world still rely on traditional biomass for cooking and heating and 1.3 billion people do not have access to electricity [1]. Most of these people live in developing and underdeveloped areas mainly with low population density unsuitable for justifying grid access or connection [1].

On the other hand, economic development and diversification are hindered in many of the developing countries owing to the lack of electrical power.

As shown in Fig. 26.1, it is obvious that many people in India and South Africa do not have access to electricity and they heavily rely on biomass for production of their required energy.

Presently, there are 460 million people in China and India alone without access to modern energy systems. This lack of energy access contributes to shortened life expectancy, reduced health, lower educational levels, and degradation of the environment. In India, only about one third of rural households are electrified, and in Kenya access to electricity stands at 15 %. Governments of these countries have set targets to increase access to electricity and implement measures to achieve these, for instance, through rural electrification programs. The most successful rural electrification program was implemented in Thailand, where the scheme increased rural energy access in the country from 20 % in 1974 to 98 % today [1].

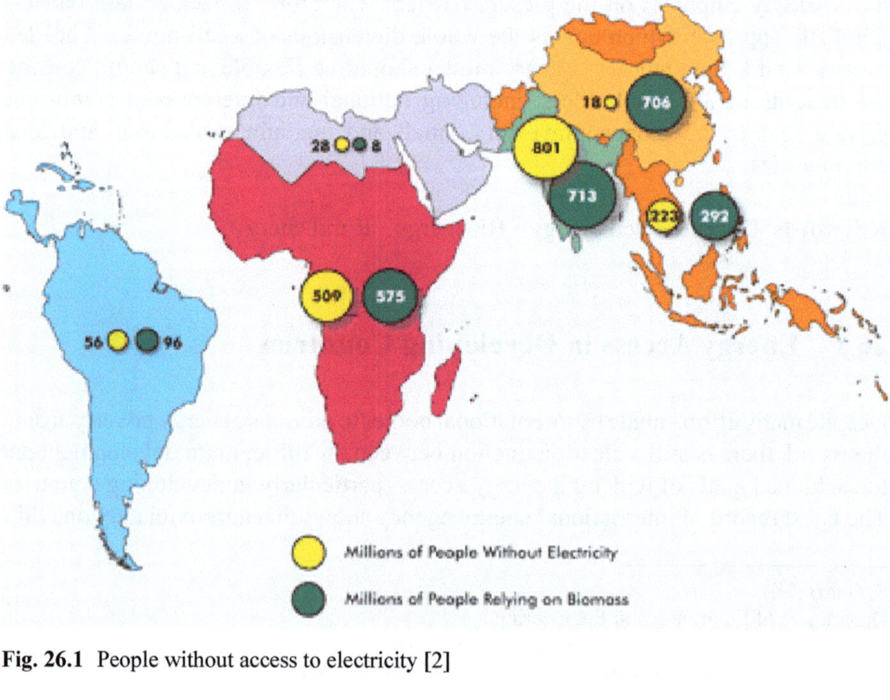

Fig. 26.1 People without access to electricity [2]

26.2 Decentralized Energy (DE): An Absolute Solution for Widening Energy Access in Rural Areas

The lack of energy access for billions of poor people in the world is a big concern among the world's major development agencies (World Bank, UN, and others) [3]. It is felt that without access to modern energy people cannot escape from poverty and the Millennium Development Goals will not be achieved. The financial gap to provide energy access for the poor is large [3]. The majority of people in energy poverty live in rural areas in South Asia and sub-Saharan Africa those are more vulnerable to the impacts of climate change. Moreover, the use of alternative energies has not been progressed in these areas mainly because of high installation cost. In order to widening access to energy and poverty alleviation, small decentralized solutions have huge potential for providing clean reliable, sustainable, and affordable energy services, particularly electricity in rural areas of developing countries [4]. In this context, DE can play an important role in changing perspectives on resource management, carbon, and energy efficiency. In fact, this sophisticated model provides a strategic approach to energy use, resource efficiency, and waste management which can integrate sustainable consumption and production into the business models (Fig. 26.2) [2, 5].

There is no doubt DE brings numerous benefits in rural areas, including cost-effective electricity, heat generation, sustainable socioeconomic development, increased use of renewable and lower carbon emissions, poverty alleviation, job opportunities, gender equality, higher efficient system, lower installation cost than traditional power plant, local independence, productive uses, sizable according to the local needs, using local resources, reducing the dependence on fossil fuels, greater energy security, and greater awareness of energy issues through community-based energy systems [7].

DE options in rural areas mainly cover solar photovoltaics (PVs), solar water heater, micro-wind turbine, micro-hydropower, biogas, and biomass systems. Among

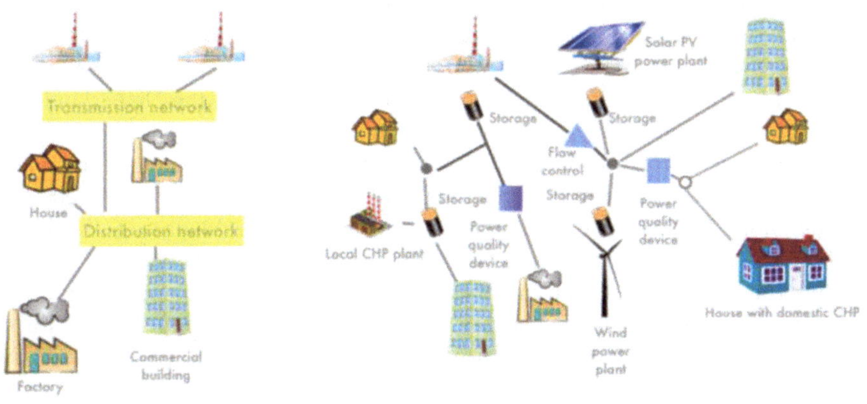

Fig. 26.2 Schematic picture of centralized and decentralized network [6]

various options, low-carbon micro-combined heat and power (CHP) systems can be considered for a wide range of applications. Modern bioenergy in the form of biomass, ethanol, biodiesel, or biogas has the highest potential for expansion among renewable energy technologies (RETs), mainly because the technology is mature and is a relatively easy substitute for fossil fuels. Bioenergy amounts to about 50 % of the primary energy supply in Africa. The proportion of bioenergy in some developing countries exceeds 90 % of their primary energy supply [8]. Sub-Saharan Africa has the greatest bioenergy potential of all world regions while the use of modern bioenergy in Africa has generally been limited to a few industries such as timber and sugar factories [9].

Modern biomass technologies, such as biogas and improved cooking stoves, could be used as substitutes for traditional cooking stoves in the household sector. Bioenergy, in the form of bio-ethanol and biodiesel, could serve as a substitute for petroleum products in the transport sector. Improving household energy services across Africa can provide various socioeconomic and environmental co-benefits. These include reduced deforestation and greenhouse gas (GHG) emissions, reduced dependence on fossil fuels, low-cost energy, improved sanitation, reduced indoor pollution, saving of time by women and children, rural employment, bio-slurry fertilizers, reduced pressure on rangelands, and, in general, rural development. Some biomass sources include residues from agro-industries, wood wastes from forestry and industry, animal manure, postharvest crop waste, grasses, and municipal solid waste [10].

Part of the attraction of these energy sources is the fact that liquid biofuels require little or no changes to today's vehicles and infrastructure and have strong potential as near-term alternative fuels [9]. Brazil has the most advanced biofuels program. The current Brazilian policy requires a 20–25 % blend of ethanol and 3 % blend of biodiesel in all motor fuels [11]. A number of African countries, such as Kenya, South Africa, Ethiopia, and Zambia, have started to blend biofuels into their national fuel mix. They have also begun to include diesel fuel in their national energy and transport policies. These policies are beginning to bring together domestic actors from agriculture and industry. Biomass combustion for heat and electricity is a well-established commercial technology. Despite the obvious advantages of cogeneration, it remains untapped in Africa, perhaps because of low heat demand in households due to the hot weather and/or to the low level of industrial development, while about 30 % of total electricity production in Denmark, the Netherlands, and Finland is cogenerated [3].

However, a few African countries have started to use cogeneration in their industrial sector. For example, Mauritius meets 40 % of its electricity needs from cogeneration. Following the success of cogeneration in Mauritius, the United Nations Environment Programme (UNEP) and African Development Bank have jointly promoted cogeneration in Africa with a 2007 initiative planned to run for 6 years [9]. The initiative, using bagasse, is being implemented in the agro-industrial sector of seven eastern and southern African countries (Kenya, Ethiopia, Malawi, Sudan, Uganda, Tanzania, and Swaziland).

Biogas technologies are increasingly being used for power generation as well. Some of the promising technologies include anaerobic digestion and landfill digestion. Recently, landfill electricity generation projects have been funded under the Clean Development Mechanism (CDM) of the Kyoto Protocol. However, out of the 130 landfill CDM projects worldwide, only 10 registered projects exist in Africa [9]; another 8 are at the validation stage. However, the CDM methodologies are very conservative and tend to under-calculate Certified Emission Reductions (CERs). China, India, Nepal, Thailand, Germany, USA, and Denmark have a long experience in the development of biogas programs and projects. However, experiences in developing countries have been largely limited to small-scale applications of anaerobic digestion in rural areas for cooking and for running gas generators [10].

Grid-connected, large-scale biogas programs are uncommon in developing countries. A few countries, including Tanzania, have implemented an integrated approach by using large-scale biogas to produce grid-connected electricity and organic fertilizers [10].

The most remarkable examples related to the implementation of decentralized renewable energies consist of 150,000 in Kenya, 100,000 in China, 85,000 in Zimbabwe, 60,000 in Indonesia, and 40,000 in Mexico; around 150,000 PV and wind systems for health clinics, schools, and other communal buildings worldwide. Over 45,000 small-scale hydro schemes are operational in China, providing electricity to more than 50 million people. Over 100,000 families in Vietnam use very small water turbines to generate electricity from hydro schemes, while more than 50,000 small-scale wind turbines provide electricity in remote rural areas in the world [10].

Moreover, recent samples of decentralization energy in Africa have been established in Ghana and Mozambique [10].

The Government of Ghana (GoG) intends achieving universal access to electricity by 2020 [10]. With barely 10 years to go, about 40 % of the Ghanaian populace are still without access to electricity. While the electrification rate reaches 85 % in urban areas, in rural areas it is about 23 % [10]. The majority of those without access are located in remote areas and island communities where extending the national grid is difficult and costly. The two strategic targets for this area are to increase the share of renewables in the energy mix to 10 % by 2020 (excluding large hydro) and to increase the use of decentralized renewable energy systems for electricity use to 30 % in rural areas [9].

Although Mozambique is a large hydroelectricity producer, its electricity sector is export oriented, and its national population electrification rate, about 12 %, is among the lowest in the continent [10]. While the electrification rate reaches 21 % in urban areas, in rural areas it is about 6.3 % mainly served by diesel generators in isolated systems [10]. According to the Ministry of Energy's Management Strategy for the Energy Sector (2008–2012), the objective of the government is to reach an electrification rate of 15 % by 2019 (and 20 % by 2020) [10].

26.3 Barriers to Develop DE in Developing Countries

Despite many advantages and benefits of DE in rural areas, there are important barriers which prevent the increase of access to modern energy services; all of them are associated with social, technical, managerial, and institutional issues. The greatest challenge is to bridge the existing gap between the technology providers and the end-users in developing countries [6]. The main barriers include low acceptance of the local communities, lack of market demand and fragile local economies, lack of bankability, no guarantee returns on investment, unwillingness among private sectors to invest, lack of proper regulation, allocation of much less funding to smaller scale projects, too expensive installation cost of RETs for poor rural people, customization of technology, inadequate equipment standards or training, lack of proper market characteristics, lack of skills and a supporting market, lack of knowledge/skills relevant to new business opportunities, the knowledge gap in the field of energy, particularly among small and medium enterprises business opportunities and high investment costs, low public capacity to create an enabling environment, insufficient transparency and accountability, and the lack of an adequate government policy promoting RETs [3–5, 7, 10]. Table 26.1 indicates information on detailed present barriers to promote DE mostly in developing countries.

Table 26.1 List of barriers to develop DE in rural areas [3–5, 7, 10]

Barrier	Description
Local barriers	Low acceptance of the local communities: local communities may not use new energy applications even when they are available, as they prefer to stick to traditional methods. They may not also feel responsible for maintaining and repairing new devices, especially if they are not participating in the decision-making process from the outset. In some cases, a technology may lack cultural acceptance
Financial barriers	Such projects are less "bankable" than major, grid-connected projects, which often serve large industries. Services to more vulnerable, low-income population groups appear far riskier in terms of potential returns on investment
	Lack of capital to invest in new machinery
	Lack of market demand and fragile local economies: from end-user's perspective, DE service delivery should be affordable, whereas for the service provider the delivery model must be economically viable. Accommodating both the end-user's and service provider's intentions introduces complexity into DE services delivery in many developing countries
	The amount of funding allocated to smaller scale projects has been much less than that allocated to the other priority areas
	The low level of investment from the private sector as cofounders of these projects reflects that the private sector is not yet confident to invest its own money in energy access in poor countries. In developing countries, it is not attractive to the private sector, and more work is needed on enabling the environment through better laws, regulations, subsidies, and good governance in order to attract private funds for poor groups of the population in developing countries
	In most cases, the installation cost of RETs is too expensive for poor rural inhabitants. Only in some places do microcredit schemes exist that allow people to take out a loan to pay for the installation of an RET such as a solar PV system

Table 26.1 (continued)

Barrier	Description
Technological barriers	DE production systems supply energy just like any other centralized energy system, but constituent technologies need to match the local resources (such as energy sources and income) and end-user needs. This customization of technology is a key challenge in DE services delivery from the technical perspective
	Inadequate equipment standards or training are likely to increase the risks associated with faulty or poorly installed or maintained systems
Market barriers	The exiting market conditions of energy in many developing countries can be characterized by subsidies, distorted prices, and market regulation
	Lack of skills and a supporting market: communities in many of the developing countries do not have the appropriate skills and training to successfully adapt, install, and maintain technologies. Several decentralized RE projects have failed to deliver energy after a few years when equipment fell into disrepair and spare parts could not be obtained. For example, the success of the Grameen Shakti solar home systems initiative in Bangladesh is largely due to the hands-on, long-term support it provides to customers and the training of local engineers and technicians
Capacity building and awareness barriers	Lack of knowledge/skills relevant to new business opportunities created
	At the macro-level, the knowledge gap in the field of energy is a major concern in developing countries. The gap is affecting small and medium enterprises that are delivering or want to deliver energy services
Legal and governmental barriers	Low public capacity to create an enabling environment: the public sector's low capacity to develop and enforce laws, policies, and regulations to promote energy access, the ability to monitor progress and enforce laws, as well as the competence to measure progress has proved an obstacle to both the introduction and scale-up of RE technologies
	Insufficient transparency and accountability: technology suppliers and/or installers are often imposed with a grant or loan which can result in a lack of accountability if there are problems with the equipment or service and may not result in the best value for money for the customers
	The lack of an adequate government policy promoting RETs can be a significant barrier to their uptake
	In most cases, rather than seeing RETs as complementary to a national grid system, governments often continue to prioritize grid expansion
Institutional barriers	Small- and medium-sized enterprises (SMEs) are key players for delivering DE services due to the dispersed nature of the market. The main challenges in the case of SMEs in developing countries are lack of awareness about business opportunities and high investment costs. In addition, the current market models of SMEs are still evolving and are yet to attract external investments. These local business opportunities create local employment in developing countries

DC decentralized energy, *RET* renewable energy technology, *PV* photovoltaic, *RE* renewable energy

Conclusion and Recommendations To eliminate the present barriers and scale up DE in rural areas of developing countries, a comprehensive support by international and donor countries under recent international commitments is significantly crucial. Other related activities in this field include appropriate management models (bottom-up, demand side), political will, national policies including integration of decentralized renewable energy into their energy policies, developing policy guidelines for the sustainable development of the biofuels industry and direct targeted support including subsidies and incentives, financial schemes such as appropriate tariff, microcredit, rural energy supply companies (RESCOs), community-/nongovernmental organization (NGO)-based approaches, flexibility, monitoring, and evaluation in project design, demand analysis, private participation in small-scale infrastructure provision, productive uses of electricity, capacity building, raising the possibility use of climate-related funds in form of CDM, technology development, technology transfer and pilot projects to reduce costs of small DE, donor flexibility and consistency with national policies, consideration of local needs and involvement of local communities in the planning and implementation, particularly women, adequate training in areas such as installation, operation and maintenance to build local capacity among both technology suppliers and users, and participation of private sector in developing standards and accreditation schemes [5, 7, 10, 11].

Some other remarkable suggestions raised from executed DE projects are as follows:

- The NGO sector has shown a strong commitment to energy access for the poor; therefore, both governments and international bodies should continue supporting and extending the participation of NGOs in future projects. This may require extra efforts in terms of capacity building, but may be worth the effort to maximize the reach of projects to the poor, and rural and isolated people [3].
- In many developing countries, water is an increasingly scarce resource and competition over it is set to intensify with the onset of climate change. For poor communities, access to clean water is especially critical—and a priority over energy—so it is crucial that energy projects are thoroughly evaluated for their impacts on clean water availability.
- Improve policy and planning by prioritizing energy access as part of national development priorities, set national targets for universal energy access and establish plans, institutions and an enabling environment to deliver them [8, 9].
- Integrate decentralized renewable energy into their energy policies and planning alongside conventional grid extension, with clear interim goals for progress that are monitored.
- Integrate energy planning with planning in other sectors and to target energy access support to all geographical areas with economic and/or employment potential that can enable multiplier benefits [8, 9]. In China, increased policy focus on domestic use of PV as part of the Renewable Energy Development Program (REDP) program significantly motivated domestic private PV manufacturers to consider domestic markets [11].
- Develop policy guidelines for the sustainable development of the biofuels industry and direct targeted support, including subsidies and incentives in developing energy.
- Investments need to be scaled up considerably to achieve the UN target of universal access to energy by 2030 [9].
- Renewable energy investments will require a range of capacity-building activities and innovative financing mechanisms, which should be supported by the EU. Several ongoing initiatives by the EU or EU Member States promote local small- and medium-sized enterprise (SME) operation to scale up delivery of energy services to the poor and increase developmental benefits [3].

- Existing market conditions need to be improved for promoting sustainable DE service delivery through the market. Improvement is essential in order to overcome competitive disadvantages introduced by the distorted energy prices of centralized energy service delivery [11].
- At the macro-level, the knowledge gap in the field of energy is a major concern in developing countries. The gap is affecting small and medium enterprises that are delivering or want to deliver energy services [10]. Adequate training in areas such as installation, operation, and maintenance—as well as learning and awareness-raising activities—are key to developing the local knowledge required for effective and sustainable RET use.
- In addition, development agencies need to promote DE services delivery as an opportunity rather than a development perspective.

In this context, a holistic approach which covers the technical, economic, social, environmental, and institutional dimensions of the project is necessary. Moreover, decentralized and bottom-up (demand) approaches may be the most effective for delivering energy to the poorest, and should receive greater attention in energy plans and investments. To reach the poorest, existing market failures must be overcome through public policy and finance, social enterprises and businesses focused on serving low-income energy markets and multi-sector partnerships.

References

1. Arranz-Piera P, Velo E (2011) Energy for all 2030. Institute for sustainability research group on cooperation and human development. Universitat Politècnica de Catalunya, Spain
2. Energy access in Developing Countries (2014) World Alliance for Decentralized Energy (WADE). http://localpower.org/ben_access.html
3. United Nations (2010) Renewable energy technologies for rural development. United Nations conference on trade and development
4. Deve (2013) Access to energy on developing countries, directorate general for external policies. http://www.europarl.europa.eu/activities/committees/studies.do?language=EN
5. Cecelski E (2002) Enabling equitable access to rural electrification: current thinking on energy, poverty, and gender. Energia News 5(3)
6. http://www.ilsr.org/graphics-from-the-report-democratizing-the-electricity-system/
7. The UK Government's Business Taskforce on Sustainable Consumption and Production (2013) Decentralised energy business opportunity in resource efficiency and carbon management
8. United Nations Economic Commission for Africa African Climate Policy Centre (2011) Critical issues in bioenergy for Africa: exploring the opportunities, constraints, and Trade-OFFS. http://www.uneca.org/acpc/publications
9. Workshop Report (2006) Energy policy options for the poor in Africa and Latin America, Hotel Silken Berlaymont, Boulevard Charlemagne 11–19, B1000 Brussels, Belgium, 27th & 28th March 2006
10. Yadoo A (2012) Delivery models for decentralised rural electrification case studies in Nepal, Peru and Kenya. International Institute for Environment and Development. www.iied.org/pubs
11. Sanchez T (2013) Energy for all 2030, rural electrification. Universitat Politècnica de Catalunya, Spain

Chapter 27
Renewables Are to Save Fuel

Donald Swift-Hook

Abstract All countries, especially the biggest fuel consumers, want to save fuel. Fuel efficiency improvements have had significant success but renewables are also an option to save fuel and they are making great strides: operating wind power capacity overtook nuclear power in the middle of 2014.

China, the USA and India are leading the way, not only in the amount of coal they are burning (nearly three quarters of the total) but also in installing wind power (over half the total) and other renewables to save coal.

If additional capacity were needed to stop the UK's lights going out, cheap plant would be needed to cover just the peaks, not a capital intensive plant like wind or solar or nuclear. In fact, no new capacity can be justified because there is plenty of spare including plant that has been mothballed. Subsidies for new plant such as wind and solar and possibly nuclear are justified by their fuel saving.

A 20 % minority of countries, including all the European Union (EU) and some other Organisation for Economic Co-operation and Development (OECD) countries, are prepared to limit their emissions of greenhouse gases to tackle the dangers of global warming as set out by the Intergovernmental Panel on Climate Change (IPCC) but most countries are not. They regard the dangers of global warming as of secondary importance to their own prosperity.

One of the most feared dangers, as promulgated by the media, is that the ice caps will melt and we shall all be inundated. They are excited to see the sea ice melting in the Summer but they forget that it refreezes in the Winter and in any case in accordance with Archimedes' principle, floating ice does not change the water level when it melts. The IPCC predict sea level rises of around 0.4 m over the next 100 years, which seem very moderate and unfrightening for governments when compared with the tidal ranges of many metres that they experience around the world every 12 h.

So carbon abatement to tackle global warming does not provide a strong case for installing renewables but the urgent need to save costly fuel imports strongly supports the huge expansion of renewables that we see today.

D. Swift-Hook (✉)
Woking, Surrey GU21 4XX, UK
e-mail: donald@swift-hook.com

A. Sayigh (ed.), *Renewable Energy in the Service of Mankind Vol II*,
DOI 10.1007/978-3-319-18215-5_27

Keywords Global warming · Renewables · Wind power · Fuel saving · Rising sea-levels

27.1 Introduction

All countries around the world have to rely on coal, gas and oil to fuel their economies and feed the power stations which generate their electricity, whatever other means they may also have. Most countries have to import the fossil fuels they need and that gives them a great incentive to save fuel. Saving fuel improves their balance of trade, balance of payments and their security of supply.

Even countries which export fuel are interested in saving it to conserve their reserves and to increase the amount available for export.

27.2 Energy Efficiency

An obvious way to save fuel is to improve energy efficiency. Although this is not as exciting as other methods, it has, slowly and almost unnoticeably, been making considerable inroads.

Microwave ovens (in which only the food is heated) have replaced the ones in which the whole oven is heated and the electrical consumptions of electronic systems have reduced as fast as the systems themselves have multiplied. When I recently needed to buy a refrigerator to keep a bottle of milk a day from going sour in a rather sunny studio, I discovered that the smallest and cheapest refrigerator that can be bought on the EU High Street today has an efficiency rating of class A: the highest rating is the lowest available! Excellent insulation is responsible for that and this is not so easy to achieve on small items which necessarily have large surface-to-volume ratios. On a larger scale, building regulations ensure that modern houses are all well insulated so they require far less heating.

It used to be said that electricity consumption was an indication of the standard of living and there was a clear correlation between the two. That is no longer the case. In the UK, although the standard of living has relentlessly increased, the demand for electricity is 4 % less today than it was 10 years ago, according to the energy statistics for the UK published annually by the UK Department of Energy and Climate Change (DECC). Primary energy consumption in the UK in 2013 was 10 % less than it was at the turn of the century, down to 203 TWh from 239 TWh in 2000, and today it is actually less than it was in 1970!

These achievements are a clear indication of the dramatic success of energy-saving measures, which are sometimes looked down upon as rather boring, since they do not involve any particularly high levels of technology.

27.3 Renewables Save Fuel

Renewables also save fuel and, by contrast, they are seen as far more exciting than pedestrian energy efficiency projects and their successes (shown in Fig. 27.1) are widely greeted with loud appreciation. Sometimes competing sources of energy, such as nuclear, voice their disapproval but rarely do these rapid developments of renewable energy go unnoticed.

In 2014, 51 GW of wind farms were built bringing the total generating capacity to 370 GW of wind power (according to the Global Wind Energy Council, GWEC). Only 333 GW of nuclear capacity was actually operational in the middle of 2014. By the end of 2014, wind farm capacity operating worldwide was more than nuclear capacity, so some envious reaction from the nuclear lobby is understandable. (Wind already significantly exceed nuclear capacity in the UK in 2012 and in the EU in 2013.)

Solar generation has been expanding even more rapidly than wind with more than 50 % of new plant being installed each year for the past 9 years, that is from 5.1 GW at the end of 2005 to 177 GW at the end of 2014 (according to the European Photovoltaic Industries Association).

The countries installing all these renewables are those that are producing and/or burning huge quantities of coal. Figure 27.2 shows that China, the USA and India are leading the way, not only in burning coal but also in installing wind power to save coal. (Wind data from GWEC, coal data from BP Statistical Review of World Energy 2014).

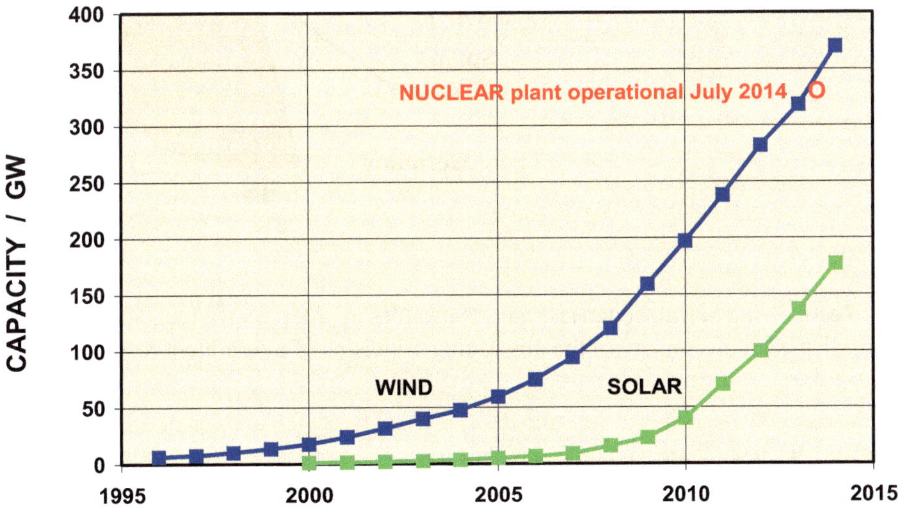

Fig. 27.1 Successful expansion of wind and solar power. Operating wind power capacity will overtake nuclear in the middle of 2014

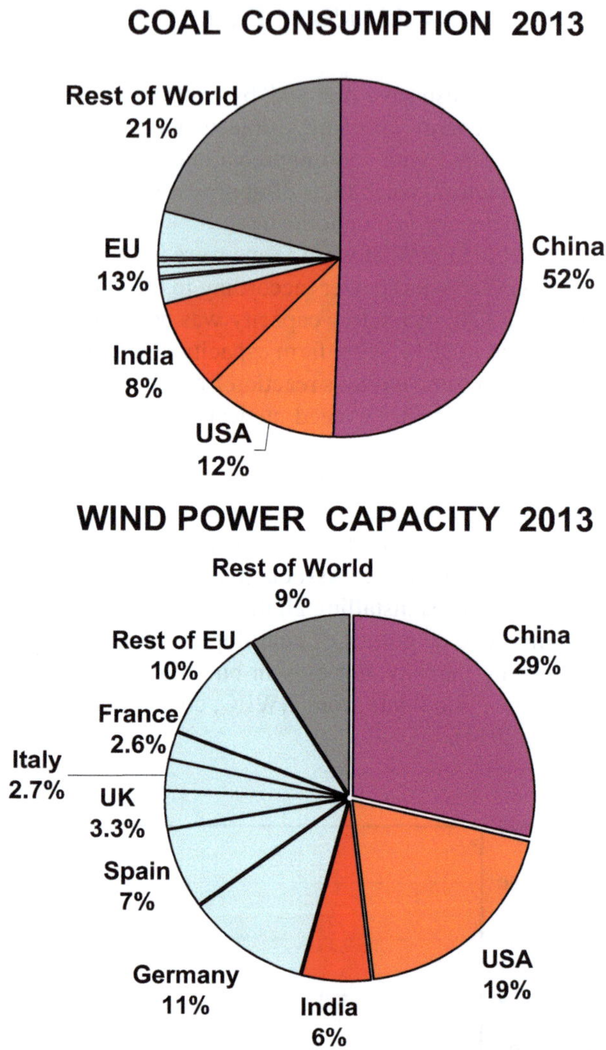

Fig. 27.2 China, the USA and India lead the way both in consuming coal and in installing wind power to save coal

Recognising that all countries want renewables in order to save fuel, not to cut their emissions, counters many of the misleading criticisms of intermittent renewables.

27.4 To Stop the Lights Going Out

It is often asserted that new capacity is needed in the UK "to stop the lights going out", an emotive phrase much loved by the media. Wind is criticised because it is intermittent and it is often erroneously suggested that 100 % standby plant will need to be installed alongside every wind farm for when the wind does not blow.

Once it is clearly accepted that wind is installed to save fuel, not to provide generating capacity, the whole perspective changes and it can be seen how ludicrous the abovementioned suggestion is. What you do when the wind stops blowing is you stop saving fuel, that is to say you carry on burning fuel as you did before you installed plant to save it. Nobody suggested that you would save fuel all the time and you just go back to burning fuel. When you are burning that fuel, to suggest that you need to install 100 % standby plant is ludicrous.

If there were a danger of the lights going out, the danger would only arise for short periods of peak demand. Peaking plant needs to be low on capital cost and need not be very efficient because it will only need to operate and burn fuel for shortpeak periods. Suggestions that high capital cost plant, such as wind and nuclear, should provide such peak capacity totally and foolishly misunderstand the problem.

Similarly, advocating more nuclear power because it will provide additional base-load capacity makes little sense. Times of base-load are when the power system has the least need for more generating capacity and when the most of the spare capacity is standing idle. The value of more capacity at that time is negligible. Wind power and nuclear power are to save fuel, not to provide generating capacity.

By the same token, it is foolish to advocate installing storage in conjunction with wind to give wind base-load capability and to cover periods which need least covering.

27.5 Spare Capacity on the UK System

However, the idea of the lights going out in the UK and the need for more generating capacity is itself mistaken and misunderstood. Far more generating capacity is available today than is needed to ensure system security.

When old coal and oil plant was replaced by more efficient gas turbine power plant, much of it was mothballed rather than permanently closed. This means that it is still available for generation with due notice (and at a cost). According to EUROSTAT, the UK has 95 GW of installed capacity which is not permanently closed.

Peak electricity demand in the UK has stayed virtually level around 60 GW for the past 10 years, according to DECC statistics, despite huge increases in the amount of electrical equipment in use, especially in the electronics arena.

This means that there is more than 50 % spare capacity available on the UK power system, whereas only 23 % is regarded as necessary to provide adequate power system reliability (a few failures in 100 years to meet maximum demand due to lack of installed capacity).

At all events, with so much spare available, it is clear that the UK does not need more generating capacity of any sort right now "to stop the lights going out", even when facing the shut-down of a fair amount of old coal and nuclear plant. Indeed, several operational gas turbine stations have recently been shut down. However, the

UK does have an urgent need to save fuel imports right now. That is what justifies subsidising the installation of large amounts of wind and solar power and possibly nuclear.

27.6 Global Warming: 80 % Versus 20 %

Apart from saving fuel, renewables do not have any emissions so they save the carbon dioxide and other greenhouse gases that would be emitted by the fossil fuels they save and replace.

All countries recognise that there is global warming and that it is caused by man-made emissions of greenhouse gases but most countries regard climate change as a matter of secondary importance. This includes the USA, China, India and all the states around the Gulf, nearly 80 % of all countries in total. They do not have direct carbon abatement policies to limit their carbon emissions because that would seriously damage their economies. They regard world poverty—and their own prosperity—as of primary importance. This was clearly set out in the 15th Conference of the Parties (COP 15) Declaration agreed—or at least received—after the United Nations Climate Change Conference held in Copenhagen in December 2009.

The USA, China and India find it unthinkable to shut down huge sections of their economies. They prefer to face the consequences of global warming in 100 years time; when their economies will have developed to a fitter state and they will have more funds to deal with whatever problems arise then than they would have if they impoverished themselves now. They place their reliance on accommodation rather than mitigation.

A 20 % minority of countries, notably the EU and a few Organisation for Economic Co-operation and Development (OECD) countries take a different view. They fear that the consequences set out by the United Nations Intergovernmental Panel on Climate Change (IPCC) will be sufficiently damaging over the next 100 years to justify spending large sums of money today to limit the effects and hopefully to limit global warming. They are leading the way with carbon abatement policies in the hope that everyone else will follow their example.

27.7 The IPCC

The IPCC is the body that is recognised as collecting together and providing the most authoritative views on global warming. In their deliberations they deal with both accommodation (how to cope with the effects of global warming when the time comes, perhaps 100 years from now) and mitigation (controlling emissions) but, perhaps understandably, the immediate requirements of mitigation and emission controls receive far more attention in the media and by today's politicians than what will be required in 100 years time.

In balancing the pros and cons of carbon abatement and emission limitation, countries have to consider their own circumstances. The direct effect of global warming in simply warming the atmosphere does not seem particularly difficult to cope with. Indeed, extra warmth may be quite advantageous. More people die from cold than from heat stroke, so an increase in the world's average temperature of a degree or two might actually be beneficial. It could also have significant advantages for agriculture, as could increased levels of carbon dioxide, as long as the fears of desertification can be countered by attention to irrigation. It is suggested that, in 50 years time, the climate of London could be the same as that of the French Riviera is today. Most Londoners I speak to would welcome such a prospect!

The most serious concerns expressed are that the polar ice caps will melt and inundate us all. There are other possible effects of climate change that are threatened, such as increased incidence of extreme events, more rainfall, less rainfall, etc., but none of these has the same broad and emotional impact as rising sea levels.

27.8 Melting Ice Caps

This fear of melting ice caps is not to be lightly disregarded. By the time the last Ice Age ended, only 7000 years ago—which is the blink of an eye on a geological timescale—sea levels had risen 100 m, see Fig. 27.3.

The English Channel, the North Sea, the Baltic Sea had all previously been dry. It was possible to walk from Russia to Alaska (and there is evidence that both people

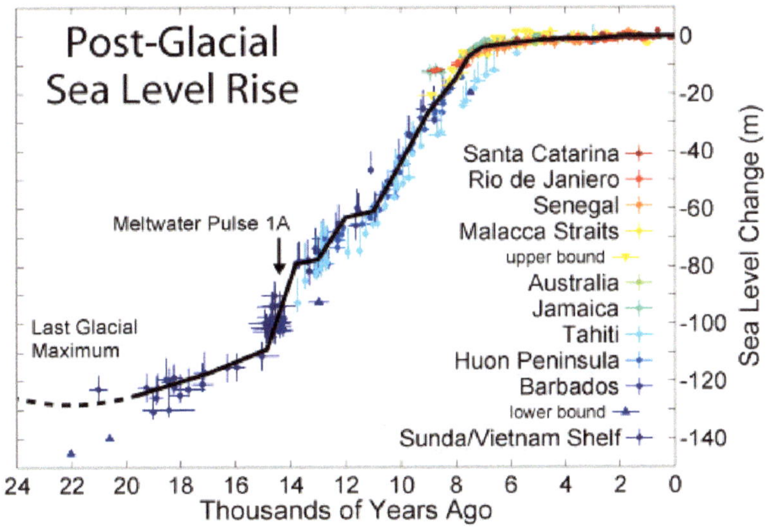

Fig. 27.3 Rises in sea levels when the ice melted at the end of the last Ice Age

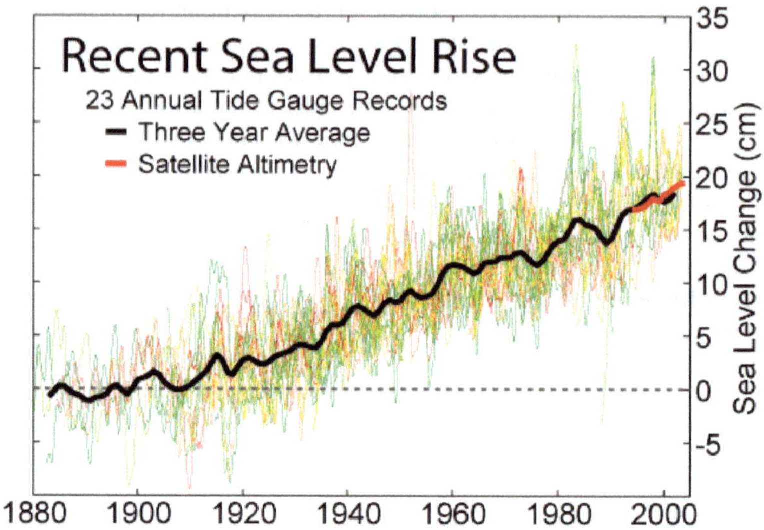

Fig. 27.4 Rises in sea levels over the last hundred years have been around 0.2 m. Over the next century, the IPCC predict an increase to 0.4 m +/−0.2 m

and animals did so). Australia was joined to New Guinea and the Gulf was just a river valley. Another similar sea rise would have disastrous consequences.

Fortunately, the rates of rise have slowed considerably and sea levels have only risen by 0.2 m in the last century. The IPCC says that this rate is likely to increase substantially over the next 100 years, perhaps as much as doubling to 0.4 m, although this estimate is hedged with an uncertainty of +/−0.2 m (Fig. 27.4).

The intrepid Londoner who is prepared to face the rigours of a Riviera climate may be forgiven for thinking that no action is necessary, when threatened by the IPCC with a gradual sea-level rise between 0.2 m and 0.6 m over the next 100 years, as the River Thames rises and falls 5 m twice every day at Tower Bridge! Table 27.1 shos the tidal ranges in various locations around the globe.

Table 27.1 Tidal ranges around the world

London	5 m
Cardiff	14 m
Edinburgh	5 m
Belfast	3.5 m
Mumbai	4 m
Boston	5 m
San Francisco	14 m
Shanghai	5 m
Hong Kong	3.5 m
Abu Dhabi	2 m
Hamburg	3.6 m

Such are the realistic threats to be faced by different countries in their own parts of the world and, perhaps not surprisingly, the great majority are prepared to meet such problems in 100 years time rather than institute costly carbon controls today.

27.9 Media Excitement

The media pay little attention to responsible and well-attested estimates which are not at all exciting (like those from the IPCC). They publish wildly exaggerated and totally invalid estimates of sea-level rises as high as 6 m or even 10 m and they are often guilty of the most elementary errors.

For instance, a standard video clip shows huge cliffs of sea ice collapsing into the sea to form icebergs. The accompanying commentary implies that this goes on all the time, which is half right—this actually goes on half the time. About half of the north polar ice cap melts every year during the Arctic summer when the sun never sets—and it then refreezes every year during the Arctic winter when the sun never rises.

Of course, a video of the sea refreezing in the dark would lack the excitement of iceberg formation, with its thunderous splashes—it would have the restful quality of watching paint drying (if some illumination could be provided)—but it is ironical that the year after the iconic and much repeated video clip was recorded in the South Atlantic, Antarctic sea ice reached its greatest extent of coverage on record, another less advertised fact. (This "on record" length of time is not perhaps quite as impressive as it sounds, because satellite records only began in 1979, so it is only the greatest ice cover in 30 years, but it certainly belies the message behind the video clip.)

Such media horror stories are clearly intended to be frightening but the threat they represent is purely fictional, which would be amusing if the implications were not so serious. Sea ice is by definition floating and any school going child is familiar with the cry of "Eureka" that Archimedes proverbially gave as he leapt from his bath to proclaim his principle: The ice weighs the same as the water displaced. This means that when the ice remelts, there is no change in the water level. If every bit of sea ice melts, the sea level would stay just the same!

It would, of course, be more worrying if the ice cover over Greenland and Antarctica were to melt. Irresponsible journalists—and some scientists who should know better—delight in estimating the significant sea rise that would be involved. Such calculations are purely hypothetical and fortunately, the IPCC do not find this a realistic proposition.

The IPCC set out their estimates in a major report which they published in 2007. A major update was eagerly anticipated for 2013 and many global warming enthusiasts hoped for major new predictions. That was rather unwise because any major changes which cast doubts on previous estimates would inevitably raise question marks about the whole approach and so would reflect on the new estimates too.

Fortunately there are no great changes and the latest 2013 report confirmed all the important estimates published previously.

So, carbon abatement to tackle global warming does not provide a particularly strong case for installing renewables as far as most countries are concerned.

Fortunately, saving costly fuel imports is an urgent and immediate need which strongly supports the huge expansion of renewables that we see today.

27.10 Conclusions

Renewables are to save fuel, not to provide reliable capacity and most countries pursue them with that in mind. Conveniently, renewables save emissions and the 20 % minority of countries who are committed to limiting their emissions say they are subsidising renewables for that reason, although they do not have carbon taxes or much support for carbon capture and sequestration.

The majority of countries, led by China, USA and India on both fronts, burn increasing amounts of coal and install wind power and other renewables to save the coal they need to burn and stretch it further. This is why wind and solar and other renewables are surging ahead.

Chapter 28
Renewable Energy Against an Oil and Gas Background: Challenges, Drivers, and Case Studies

Indra Haraksingh

Abstract Renewable energy (RE) development has always been a difficult proposition in oil- and gas-rich countries. Since it has not been the main economic driver for these countries, inadequate emphasis has been placed in this area. But most countries have realized that fossil fuel reserves are exhaustible and that in order to ensure future sustainable development, more importance must be attached to developing indigenous renewable sources of energy.

In an oil and gas backdrop, huge challenges are being faced on an ongoing basis. Therefore, much more effort has to be placed on education and sensitization programs. The average person is more concerned with his present economic status and therefore pays less attention to the future benefits that can be derived from using RE technologies.

Ensuring sustainable socioeconomic development is challenging in societies where substantial subsidies continue to be placed on fossil fuel electricity generation as opposed to green technologies. But governments are fast realizing that their economies will not be sustained when oil and gas reserves are reducing. Emphasis is being placed on international cooperation to intensify RE project development.

Many oil- and gas-rich countries are now moving toward a "Green Energy Revolution." Some classic examples include Abu Dhabi's carbon-neutral Masdar City, Qatar's Solar Schools' project, and Saudi Arabia's solar heating plant in Riyadh. The Caribbean Community is also moving toward a 50 % RE in its energy mix. Even Trinidad and Tobago, a net exporter of natural gas, is forging ahead with its plans to develop RE technologies, in particular solar and wind energy.

The goal is therefore to improve overall economic advantage and accelerate the transition from fossil fuels to sustainable technologies for future environmental preservation.

Keywords Renewable energy · Oil and gas · Challenges · Drivers · Case studies

I. Haraksingh (✉)
Department of Physics, The University of the West Indies, St. Augustine, Trinidad
e-mail: Indra.Haraksingh@sta.uwi.edu; iharaksingh@yahoo.com

A. Sayigh (ed.), *Renewable Energy in the Service of Mankind Vol II*,
DOI 10.1007/978-3-319-18215-5_28

28.1 Introduction

The oil and gas sector and the renewable energy (RE) sector have traditionally been viewed as competitors, in particular in the power sector. While both natural gas and RE appear to be in direct competition, given the current state of global greenhouse gas (GHG) emissions, it is in every country's interest to transition to a significant percentage usage of RE in its energy mix as a matter of urgency. This has been a difficult situation for many oil- and gas-producing countries since their economies have been heavily based on these industries. Nonetheless, these countries are quickly realizing that not only because of serious environmental consequences and climate change, but also due to the fact that the petroleum-based resources are finite and they must promote the efficient use of the remaining resources for their own energy security. In addition to this, these countries must conform to international conventions and treaties on climate and energy. Consequently, there is a concerted drive to promote the use of various and appropriate forms of RE.

28.2 Challenges

It is generally accepted that the energy sector is the largest contributor to GHG emissions. The petroleum industry is energy intensive and as a result, is usually one of the main contributors to climate change. According to the Intergovernmental Panel on Climate Change (IPCC), in 2010 the energy supply sector was responsible for 35 % of anthropogenic GHG emissions. Figure 28.1 shows the prediction for change in sea ice thickness from the 1950s, showing the sea ice thickness to be 100 % of the 1955 volume, as opposed to the 2050s, showing only 54 % of the 1955 volume. This is a drastic change with serious implications for sea-level rise, among many other problems, especially for Small Island Developing States.

Figure 28.2 shows the global climate change by GHGs and aerosols since 1750 with factors above zero having a warming effect while those below zero having a cooling effect. It is clear that the net effect is warming. A vertical line between two data points indicates scientific uncertainty regarding the estimated contribution of a particular factor.

Figure 28.3 shows the contribution of natural factors and human activities in radiatively forcing climate change. Radiative forcing values are for the year 2005, relative to the preindustrial era (1750). The contribution of solar irradiance to radiative forcing is 5 %, the value of the combined radiative forcing due to increases in the atmospheric concentrations of carbon dioxide (Fig. 28.4).

Despite international treaties, such as the United Nations Framework Convention (UNFCCC) and the Kyoto Protocol, GHG emissions rose more rapidly between 2001 and 2010 than in the preceding decade (IPCC). In fact, about half of cumulative anthropogenic carbon dioxide (CO_2) emissions between 1750 and 2010 occurred in the past 40 years. Globally, both economic and population growth

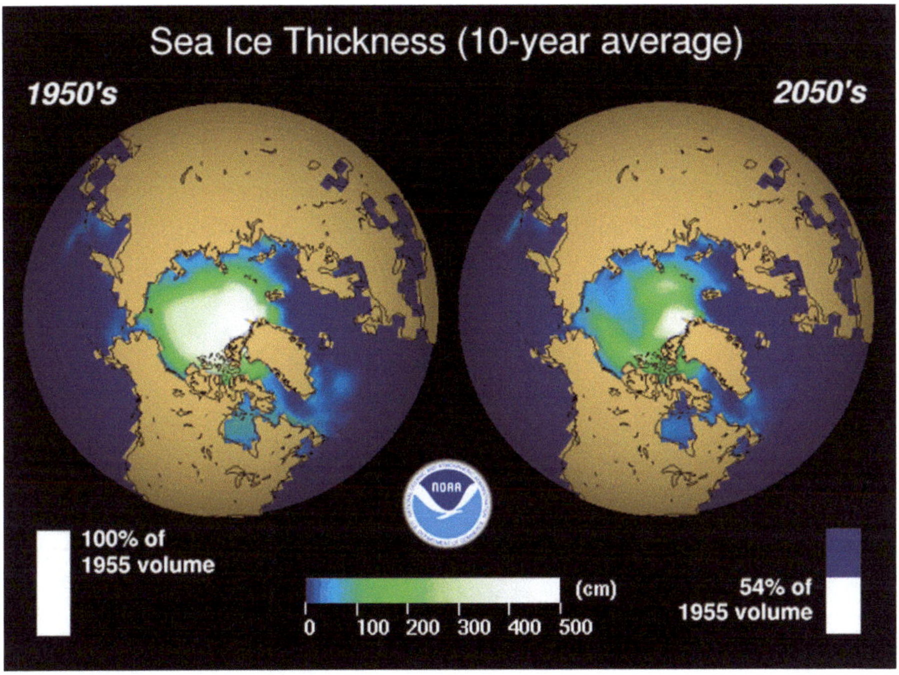

Fig. 28.1 Change in sea ice thickness over a 100-year period

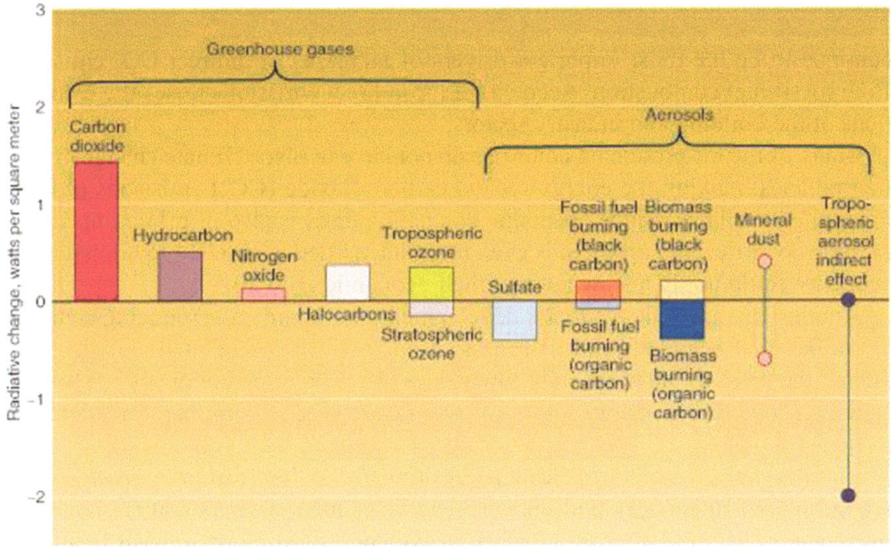

Fig. 28.2 Global climate change due to greenhouse gases and aerosols since 1750

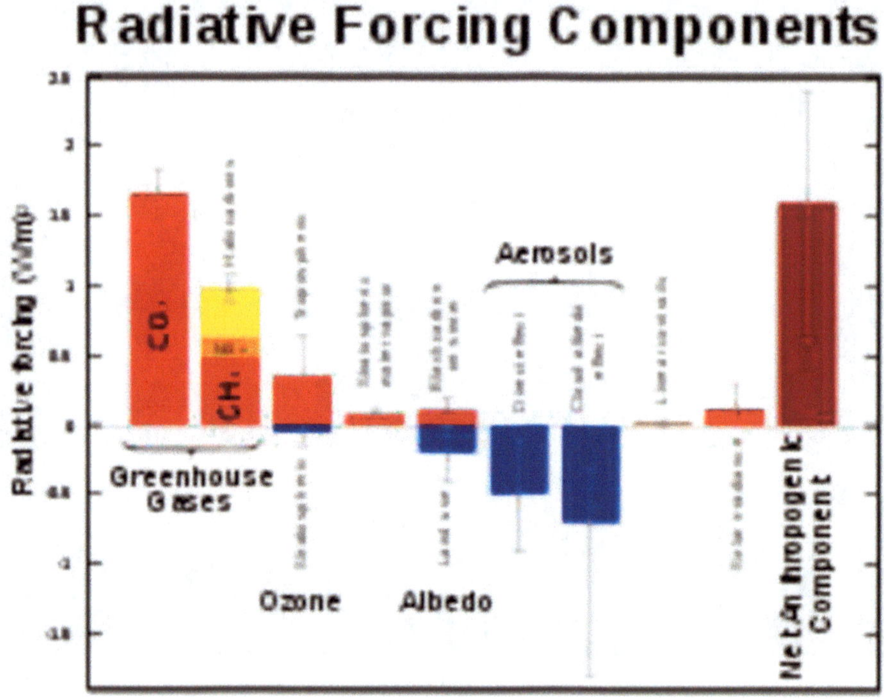

Fig. 28.3 Contribution of natural factors and human activities to radiative forcing of climate change

continue to be the most important drivers of increases in indirect CO_2 emissions from fossil fuel combustion. Accounting for indirect emissions raises the contributions of the building and industry sectors.

Many of the oil-producing countries do not have in place climate change mitigation policies, making the energy-related carbon dioxide (CO_2) emissions problem more uncontrollable. Since economic competitiveness is also of prime importance to these countries, the problem is even more intensified as moving to a low-carbon economy could imply a slowdown of their economic growth.

Another major challenge to RE development in oil- and gas-producing countries is the fact that heavy subsidies are placed on fossil fuel-generated electricity, reducing the price to unrealistically low values. This situation continues to stifle the growth of renewable energy sources in such countries, making the transition very difficult.

There are other challenges faced in transitioning to a low-carbon economy. There are technological barriers, and financial and investment barriers that are not easily overcome given the long history of extensive concentration on a fossil fuel-based economy. There are also cultural, institutional, and legal barriers to be broken in the transitioning. This will have even more impact on poorer countries which are more

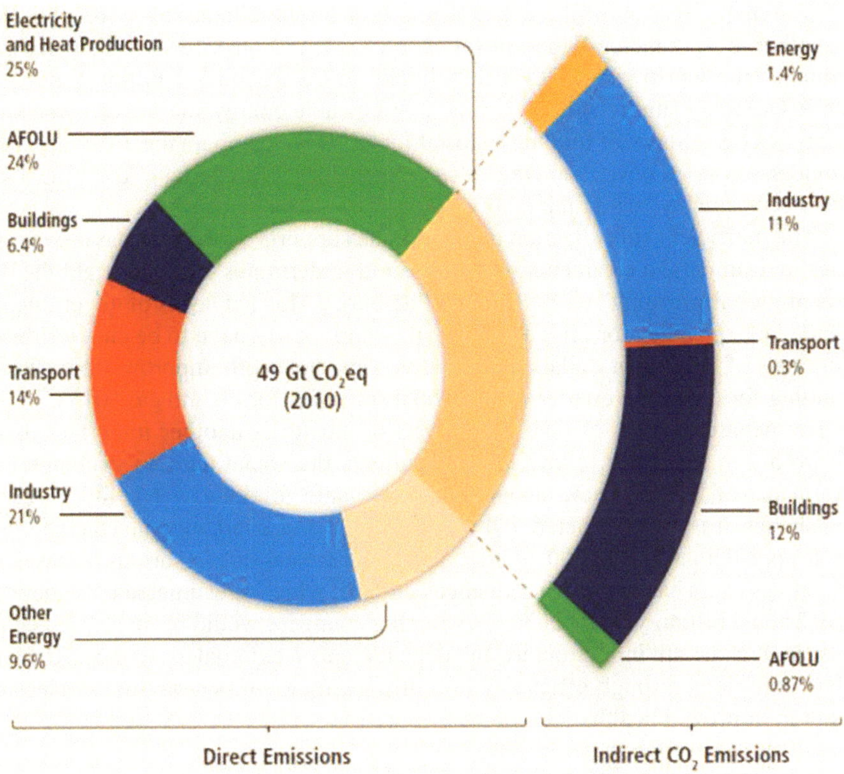

Fig. 28.4 Greenhousegas emissions by economic sectors (Source: IPCC)

vulnerable to the implications of transitioning since this cannot be achieved without the necessary financial backing. Such a transition is also dependent upon the extent of capacity building and the rate at which this is achieved.

28.3 Drivers

The major driver working in favor of RE development is the realization that fossil fuel reserves of countries are finite, and there is compelling need to preserve such reserves in the economic interest of the countries. Indeed, preservation of the resources can lead to great economic advantage in years to come. These countries have a dire need to continue to remain world leaders. However, these same countries are very much aware of their status as being the world's largest carbon emitters per capita. Being signatories to international treaties, such as Kyoto Protocol, these countries must preserve their image and therefore conform to the terms of the treaties. Hence, moving to a low-carbon strategy must be a priority for them.

Awareness creation of the extent of the cost on the climate by GHG emissions is another driving factor. Economist Swantja Küchler from Green Budget Germany estimates the worldwide cost on the climate to be around 2.7 billion € per year. While Germany supports the transition to renewable energy sources, it gives about five times the amount of funding to fossil fuels. "If we compare the different energy providers on price now, then we can see renewables will overtake fossil fuels suppliers very quickly" said Steiner Teymurov.

"The inefficient subsidization of fossil fuels supports wasteful behavior, complicates investments in clean energy sources, and undermines efforts to fight the dangers of global warming" (G20 Summit, Pittsburg). This led heads of government to decide that subsidies for oil, gas, and coal would slowly have to be cut worldwide. Hundreds of billions of dollars are spent on subsidies, with importing countries accounting for close to a quarter of the global subsidies for oil and gas (IEA).

The negative effect of GHG emissions on health is another important driver. According to the IPCC, climate change affects the social and environmental determinants of health. These include clean air, safe drinking water, food security, and shelter. It is also predicted that between 2030 and 2050, climate change would cause approximately 250,000 additional deaths per year due to various diseases and heat stress. This has direct consequences on cost, which is estimated to be between US$ 2 and 4 billion/year by 2030. Developing countries would be most affected due to their weak health infrastructure and dependence on external assistance. Reducing GHG emissions through efficiency in transport, food, and energy-use choices can result in improved health.

28.4 Case Studies

United Arab Emirates A brief look at some oil- and gas-producing countries shows that while many of these do not have high penetration of renewable energy sources into their energy mix, they have already started the transition and some actually have significantly large RE projects. The United Arab Emirates (UAE) lies in a belt of high solar irradiance and therefore has a wealth in solar energy availability. The UAE and other Gulf countries are increasingly investing heavily in green energy. The Saudi royal family has intentions of moving to 100% renewable and low-carbon energy generation in the following decades.

The UAE has recently signed the United Nations' Sustainable Energy for All initiative led by UN Secretary-General Ban Ki-moon. This initiative aims to mobilize all sectors of society to support the goals by 2030: universal access to modern energy services, doubling the rate of energy efficiency gains, and doubling the share of RE in the global energy mix. The International Renewable Energy Agency (IRENA) has also found its base in the UAE, reinforcing the commitment of the UAE to the de-carbonization transition.

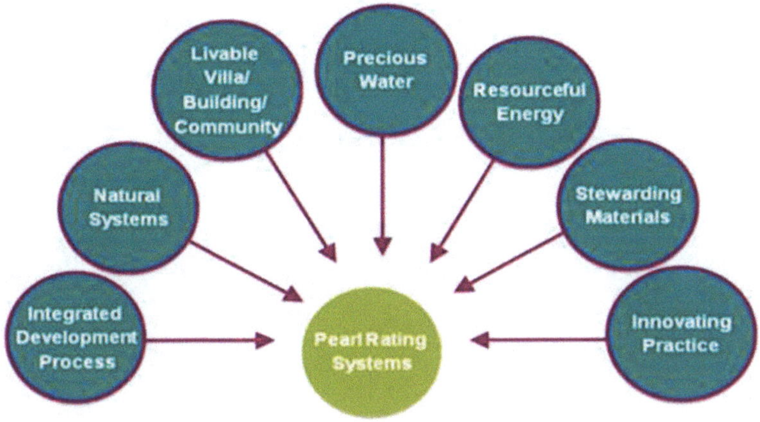

Fig. 28.5 Pearl Rating System

The UAE has committed to building a sustainable, stable economy through its vision 2030. It hopes to achieve this through diversifying and expanding its enterprise base across many sectors of the economy. Owing to the large share of energy use in buildings, development of infrastructure will be given priority in order to sustain the projected levels of economic growth in Abu Dhabi, where 50–60 % of electricity generated is consumed by buildings, as opposed to 30–40 % in the global average. This move is strategic as both environmental and sustainability benefits are to be derived. Estidama has been established for governance and community development in the Middle East. The purpose of Estidama is to create a new sustainability framework that will direct the current course while allowing adaptation. Estidama seeks to ensure the main tenets of sustainability: environmental, economic, social, and cultural. Creation of new tools, resources, and procedures necessary for 2030 vision has taken on a greater level of importance. To achieve some of the goals, workshops and training sessions are organized to create awareness and understanding of the Pearl Rating System, which is a framework for sustainable design, construction and operation of communities, buildings, and villas to advance Estidama.

Figure 28.5 shows the Pearl Rating System that is organized into seven categories, some of which provide both mandatory and optional credits. All the mandatory credit requirements must be met in order to achieve a 1 Pearl rating. To achieve a higher Pearl rating (2–5 Pearls), all the mandatory credit requirements must be met in addition to a minimum number of credit points. All new projects must achieve a minimum 1 Pearl rating to receive approval from the planning and permitting authorities. Government-funded buildings must achieve a minimum 2 Pearl rating.

Abu Dhabi Abu Dhabi has traditionally been a leader as a hydrocarbon producer and in global energy markets, having 8 % of the global crude oil reserves. It is estimated that the Emirate has enough hydrocarbon reserves for another 100 years. But Abu Dhabi is trying to create a balance between hydrocarbons and RE by increasingly moving toward RE. Abu Dhabi Economic Vision 2030 is guiding its

Fig.28.6 Masdar City

Fig. 28.7 Masdar's residential block

aspiration to be an international hub for RE and low-carbon energy technologies. This would certainly enhance its image in the international arena. Abu Dhabi and Dubai have set targets of 7 and 5 %, respectively, of their total energy needs from RE by 2020. One of the big achievements for Abu Dhabi is Masdar, a carbon-neutral town (Fig. 28.6). Figures 28.7 and 28.8 reveal details of Masdar's residential block and city skylight. This is a classic example of what a big oil producer can do to create a balance between hydrocarbons and RE, while at the same time focusing on climate change and energy security.

Fig. 28.8 Masdar City skylight

Fig. 28.9 Saudi Arabia's solar thermal plant

Saudi Arabia The world's largest solar thermal plant was opened in Riyadh, Saudi Arabia, doubling the size of what was previously the largest solar thermal facility located in Denmark of collector area 36,000 m^2. It is designed to generate enough power to heat water for the Princess Noura Bint Abdulrahman University for Women in Riyadh comprising 40,000 students and 20,000 female staff. GREENTecONE,

Fig. 28.10 Qatar's School's project

an Austrian solar design company, supplied the solar panels for the project target of 10 % of energy from solar source by 2020 (Fig. 28.9).

Qatar Qatar's school project is involved in powering Qatari schools using solar energy. The project was launched on Earth Day. Qatar has the highest emission per capita in the world and this was a major driver to efforts in transitioning to low carbon generation. Their action involved building a strong body of RE resources and a knowledge center for solar and REs. Targeting young people in schools was one of their strategies. Figure 28.10 shows some of the activities of the schools' project as well as the students involved.

The Caribbean Moving closer to home, the Caribbean Community (CARICOM) is aggressively pursuing development of RE as a priority. The energy ministers of the government have signed an agreement for a target of 47 % RE by 2027 at the Council for Trade and Economic Development (COTED) meeting in Trinidad in March 2013. Each country in CARICOM has a country target as well.

The CARICOM comprises 15 countries with a total population of approximately 16 million. Of these Haiti has 9.3 million (58 %), Jamaica 2.7 million (17 %), Trinidad and Tobago 1.3 million (8.7 %), and the Organization of Eastern Caribbean States (OECS) 2.3 million (16.3 %). Its geographical area is 465,000 km^2. Annual fuel import bill was US$ 10 billion in 2008 and annual food bill was US$ 3 billion. Oil imports amounted to 93 million Bbls/year in 2008. Electricity capacity in 2009 was 4800 MW (approx.). Electricity consumption was 18,000 GWh in 2009. Oil production (Trinidad and Tobago, Barbados, Belize, and Suriname) totaled 120,000 Bbls/day (approx.). Oil-refining capacity was 200,000 Bbls/day (approx.), Trinidad and Tobago—160, Jamaica—35, and Suriname—7. Electricity tariff ranged from US$ 0.04/kWh to US$ 0.50/kWh (2008). Impact of oil price volatility was a rise of US$ 10 in the price of oil and increase in tariffs of US3.5 cents/kWh (worst case in 2008).

However, RE potential is large. Hydro, solar, wind, geothermal, biomass, and marine energy resources are available to be harnessed. Currently, the percentage RE in the electricity sector is approximately 8 %. The percentage RE in the total energy

Table 28.1 Oil and gas production and utilization in Trinidad and Tobago

	Total
Crude oil production	*81,157 BOPD*
Natural gas production	*4144* MMSCF/D
Natural gas utilization	*3854* MMSCF/D
Production of LNG	*32,700,292* m^3

supply is 3%. Persons without access to modern forms of energy are approximately 7 million. Countries with draft or approved energy policy are in total 13, with a Regional Energy Policy established in March 2013 at the meeting of energy ministers in Trinidad.

28.4.1 Regional Initiatives

There are many initiatives taken by the CARICOM in recent years. These include:

- Caribbean Renewable Energy Development Programme
- Implementation of a Regional Energy Policy
- Caribbean Sustainable Energy Road Map and Strategy (C-SERMS) and Platform
- Caribbean Renewable Energy Capacity Support (CRECS) Project
- Sustainable Energy Technical assistance (SETA-OECS/CDB) Project
- Caribbean Sustainable Energy Program (CSEP)
- CARICOM Energy Awareness Initiatives

All these initiatives are established to assist the region in a smooth transition to low-carbon economy and to reach the 2027 target of 47% RE.

Trinidad and Tobago Placing Trinidad and Tobago in the Caribbean context is important as this establishes the uniqueness of this country with respect to energy. Trinidad and Tobago is the only net producer and exporter of oil and gas in the English-speaking Caribbean. It has a huge advantage over the other Caribbean islands. Table 28.1 shows some figures for oil and gas production and utilization. Electricity prices range from US$ 0.04 to US$ 0.06, whereas electricity tariffs in most other Caribbean islands range from US$ 0.20 to US$ 0.45.

Notwithstanding the country's wealth in hydrocarbon resources, the government is moving ahead with plans for development of its RE resources to complement its hydrocarbon resources. The Government of Trinidad and Tobago is currently developing a national energy policy green paper that recognizes RE combined with energy efficiency and utilization of compressed natural gas in the transportation sector, as important strategies for the promotion of sustainable growth. The Ministry of Energy and Energy Affairs is currently spearheading a number of initiatives in this regard. Figure 28.11 shows the incentives offered by the government to stimulate the growth of renewable energy in the country.

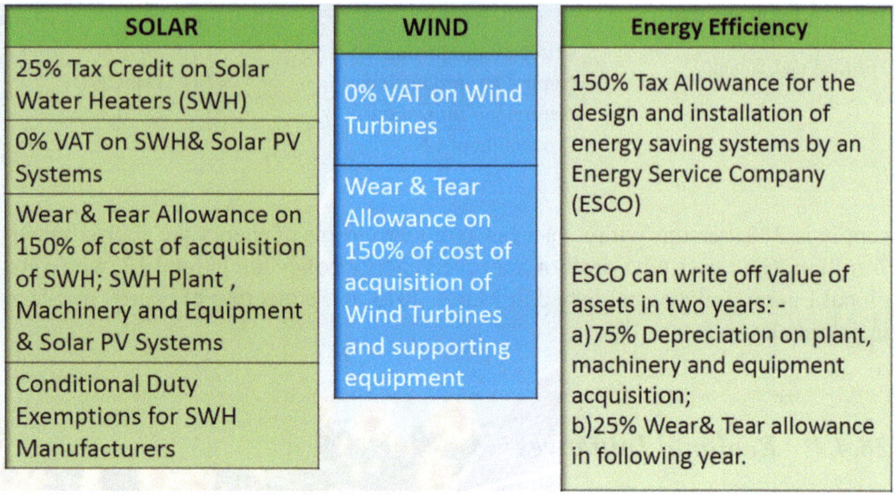

Fig. 28.11 Finance Act No. 13 of 2010

The legislative agenda includes review of the Trinidad and Tobago Electricity Commission (T&TEC) and Regulated Industries Commission (RIC) Acts. Key issues are open access, grid interconnection, feed-in tariff, net metering, and net billing. The country is also collaborating with UNEP to develop a framework for policy and legislation to govern feed-in tariffs. The Government is currently investigating possibilities for setting up a photovoltaic manufacturing industry and is preparing for a wind resource assessment to establish sites for wind farm installation in conjunction with the University of the West Indies. Trinidad and Tobago is also earmarked to be the home of the Centre for Excellence in Renewable Energy, with support from the US Department of Energy.

28.5 Conclusion

Despite the numerous challenges faced by oil and gas producers, the many benefits to be derived from transitioning to alternatives, given the serious environmental consequences, is unquestionable. There are numerous reasons why a hydrocarbon-producing country should invest in RE. This will prolong the life of the hydrocarbon reserves, reduce risk, enhance profits through diversification, and ensure long-term growth and energy security, while reinforcing its position as a global/regional energy leader. This will encourage a country to gain international recognition and secure a sustainable future, in line with the Sustainable Energy for All principles (SE4ALL). Moreover, it will work toward preservation of the environment and the fragile ecosystem.

The economic analysis of solar energy shows that the cost of power generation from RE is less expensive than that from fossil fuels if the hidden costs of environmental and public health costs are considered. A reduction in GHG emissions reduces environmental pollution and saves expenditure on health care. While petrol and electricity should always be cheaper in oil- and gas-producing countries, unrealistically reduced prices simply encourages waste and inefficiency in transportation. More efficient public transport will help to reduce carbon emissions.

It is important to increase cooperation with international organizations, foreign companies, and donor countries to get a head start on the movement toward clean energy. It is now crucial to advance economic policy activities with respect to RE, and to formulate legislative frameworks and establish government structures. Reducing subsidy on fossil fuels and increasing subsidy on RE will go a long way in transitioning to a low-carbon economy.

References

1. http://www.aee-intec.at
2. http://www.caricom.org
3. http://www.climate.noaa.gov
4. http://www.greenonetec.com
5. http://millenniumenergy.co.uk
6. https://str.llnl.gov/str/April03/Chuang.html
7. http://www.unep.org
8. http://www.upc.gov.ae/estidama-and-pearl-rating-system/pearl-rating-system.aspx

Chapter 29
Development of an Optimisation Model for the Evaluation of Alternative Energy and Fuel Supply Chains

Christiana Papapostolou, Emilia M. Kondili and John K. Kaldellis

Abstract Energy and fuel supply chains (SCs) and their optimisation not only in techno-economic terms but also accounting for environmental and social implications imbedded in each supply chain option render a very interesting research topic, especially in areas with limited energy resources and areas where strict air quality limitations apply. Considering that the aim of this chapter is to present a developed methodology and the assorted modelling tool for the optimisation of alternative SCs considering sustainability dimensions in the design and operation on top of technical limitations. To evidence that a small-scale power planning problem will be assessed, that is, an isolated consumer, Milos island in Greece, for a specific timeframe, proofing the implementation possibilities and flexibility of the model to change over different optimisation targets and limitations applicable in each specific energy decision strategy.

Keywords Energy planning · Evaluation framework · MILP · Power generation

C. Papapostolou (✉) · E. M. Kondili
Optimisation of Productions Systems Laboratory, Mechanical Engineering Department, Technological Educational Institute of Piraeus, P. Ralli St. and Thivon 250, 122 44 Athens, Greece
e-mail: cpap@teipir.gr

E. M. Kondili
e-mail: ekondili@teipir.gr

J. K. Kaldellis
Soft Energy Applications and Environmental Protection Laboratory, Mechanical Engineering Department, Technological Educational Institute of Piraeus, P.O. Box 41046, 12201 Athens, Greece
e-mail: jkald@teipir.gr

A. Sayigh (ed.), *Renewable Energy in the Service of Mankind Vol II*,
DOI 10.1007/978-3-319-18215-5_29

29.1 Introduction

The transition towards a sustainable energy supply is rather complicated and shaped from a wide spectrum of parameters and actors affecting the decision-making process. Energy planning, on top of techno-economic criteria, requires the contemplation of social implications, such as awareness, perception, opposition, local development and environmental considerations like the minimisation of the carbon footprint of the energy systems, in the evaluation of alternative energy and fuel supply chains (SCs). Furthermore, the security of energy supply that must be equally met, has made it necessary for us to address and examine the problem under a holistic perspective, accounting simultaneously for economic, social and environmental criteria in each policy decision, assessing possible uncertainties, under the goals and the priorities set.

The aim of this chapter is to present an approach according to which a mixed integer linear programming (MILP) model has been developed for the optimal design and operation of alternative energy and fuel SCs with special focus on electricity generation.. The energy system of an isolated area consists of a set of energy inputs—supply sources, storage facilities as well as a certain profile of energy demand. The optimisation model may accommodate various criteria such as cost minimisation, environmental benefits maximisation, micro and macro social benefits maximisation and a set of constraints expressing the structure and the limitations of the system.

The developed methodology, along with the corresponding integrated tool for the economic, social and environmental evaluation of various alternative energy and fuel SCs, comprises the formulation of a mathematical model, suitable performance indexes as well as a representative case study of an isolated area in Greece. Under a predetermined geographical region (Milos island, Greece) along its characteristics, the proposed model will optimise the alternative electricity SCs taking into account simultaneously, social environmental and techno-economic criteria, on top of satisfying the different end users' needs.The novelty of the present work relies on the flexibility of the optimisation criterion, which can differ with respect to the social, environmental, technical or economic objectives of the operation of the system. Due to its generic nature, the proposed methodology maybe equally applied to other energy systems modelling and optimisation like water and energy.

29.2 Current Modelling Approaches and Position of the Energy Planning Problem

Energy planning, design and operation for meeting specific energy needs and demand priorities constitute a composite problem, because various resources and technological options with local characteristics, impacts and availabilities exist.. The majority of the available models have been developed and formulated with special application and reference to very specific energy systems' design and simulation and/or optimisation, in response to the problem of meeting future demands under

under various uncertainties and stochastic parameters, identifying the most cost-effective solution. However, energy planning in general is rendered particularly difficult as different timescales of various and conflicting decisions, that is, investment versus operation, environmental performance versus economic profitability, there exist on top of assorted stakeholders' and communities' interests, preferences and needs over the time horizon examined [1–4]. In addition, the multi-attribute parameters that are embedded in each alternative energy SC option, assign a degree of uncertainty in the related decisions.

Addressing to that problem, special emphasis has been placed from early years, by the researchers, on power and heat SCs optimisation in terms of strategic and/or operational planning or even at the level of detailed or discrete scheduling with respect to their technical feasibility and economic viability (profit maximisation and/or cost minimisation [5–7]), and/or with regard to sustainability issues, in terms of life cycle analysis (LCA), considering also the possibility of these energy SCs to be competitive to primary needs of food production [8–10]. However, recently, emphasis has been put on the assessment of more holistic SC considerations to incorporate more sustainability dimensions (economic and/or environmental and/or social implications) [11, 12]. Nevertheless, the integration of all these matrices and decisions in respective models of energy and fuel SCs across a comparative basis, reflecting the multidimensional space of the problem environment, has not yet been achieved (Fig. 29.1).

In response to that, this work cites an integrated approach for the evaluation of alternative energy and fuel SCs. The developed evaluation framework [13, 14] consists of:

- A suitably selected representation tool based on the resource-task network approach.
- A generic mathematical model (reflecting the constraints and the system variables)—MILP (presented in the following section).
- A set of appropriately selected and quantified technical, economic, environmental and social indexes.

The types of decisions and alternatives that are supported by the current optimisation problem include:

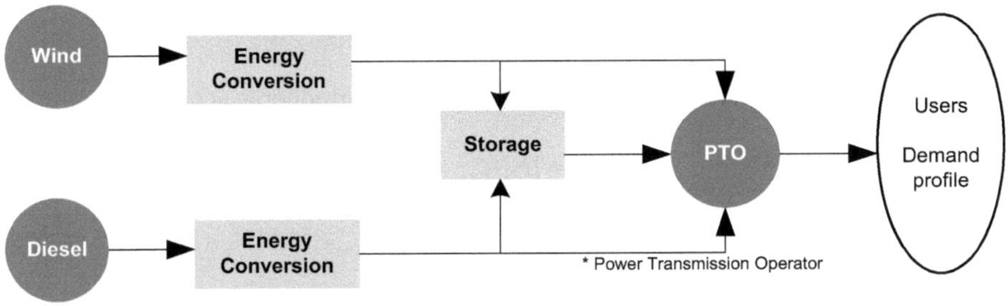

Fig. 29.1 The Resource-Task Network (RTN)-generic representation of the energy planning problem under consideration

- Selection of specific resources/energy inputs/fuel mix.
- Selection of production capacity (domestic or not) and/or expansion of existing capacity and/or consideration of imports (binary types of decisions, 0–1)
- Selection of specific SC configuration according to local (regional-specific) priorities and needs.
- Possibility of switching over different optimisation targets depending on the regional character of the proposed configuration and the stakeholders' interests.

29.3 The Developed Mathematical Model

The developed mathematical model (which is formulated as a MILP problem, seeking to combine both discrete and continuous variables and decisions in the complex planning and operating energy SCs) seeks, for a predetermined area under consideration with specific load demand characteristics and resources availability, to identify the optimal set of energy supply options accounting simultaneously for technical, economic, social and environmental criteria, of different energy resources, for time-varying needs and priorities.

Thus, for a preset time horizon H, a set of resources R, storage stations S, interconnection options I and a set of users U to be satisfied, the optimisation criterion is the maximisation of the total value of the Energy Supply Chain (ESC) taking into account economic and social as well as environmental benefits:

$$\mathrm{Max} = \sum \left\{ a_1 \times \mathrm{ECONV}_{r,s,i} + a_2 \times \mathrm{ENVV}_{r,s,i} + a_3 \times \mathrm{SOCV}_{r,s,i} \right\} \tag{29.1}$$

where a_1, a_2, a_3 are appropriate weighted normalisation factors of the different dimensions considered.

The economic value considers the total costs (TC) of the electricity generation (Investment INV, maintenance MAIN and operational costs OPER) (€):

$$\mathrm{ECONV}_{r,s,i} = \mathrm{TC}_{r,s,i} \forall r,s,i \in R,S,I \tag{29.2}$$

where

$$\mathrm{TC}_{r,s,i} = \mathrm{INV}_{r,s,i} + \mathrm{MAIN}_{r,s,i} + \mathrm{OPER}_{r,s,i}. \tag{29.3}$$

For each time step t and for $r,s,i \in R,S,I$, the cost components accordingly are:

$$\mathrm{INV}_{r,s,i} = \mathrm{INV}_{r,s,i} \times \exp_{r,s,i}, \tag{29.4}$$

$$\mathrm{MAIN}_{r,s,i} = \mathrm{MAIN}_r \times \mathrm{EG}_r + \mathrm{MAIN}_s \times \mathrm{ES}_{r,s} + \mathrm{MAIN}_i \times \mathrm{EIN}_i, \tag{29.5}$$

$$\mathrm{OPER}_{r,s,i} = \mathrm{OPER}_r \times \mathrm{EG}_r + \mathrm{OPER}_s \times \mathrm{ES}_{r,s} + \mathrm{OPER}_i \times \mathrm{EIN}_i. \tag{29.6}$$

Where exp is an expansion factor considering the creation or not of new electricity generation and storage infrastructure if needed $\exp = 0,1$, EG_r is the electricity generated for the selected time step t from resource r, $ES_{r,s}$ is the electricity stored at storage stage s for the selected time step t generated from resource r and EIN_i is the electricity imported from the interconnection option i for the selected time step t.

The environmental value considers the environmental positive impact of the ESC in LCA compared to an environmental friendly Renewable Energy Resources (RES)-based ESC with a "typical-average" impact on the environment (e.g. the wind SC) (€). Thus, for each time step t and for $\forall r,s,i \in R,S,I$, we have:

$$ENVV = (ENVF_r \times EG_r + ENVF_{r,s} \times ES_{r,s} + ENVF_i \times EIN_i) \times PCO_2 \quad (29.7)$$

$$ENVF_{r,s,i} = LCREF - LCENVF_{r,s,i} \quad \forall r,s,i \in R,S,I \quad (29.8)$$

where PCO_2 is the current commercialisation price of CO_2 (€/kWh), LCREF is the reference ESC environmental value and $LCENVF_{r,s,i}$ is the environmental footprint (in LCA terms) of each ESC.

Finally, the social value of the ESC considers micro (MiSOC) and macro (MaSOC) economic benefits from the implementation and operation of the different ESC configurations (€):

$$SOCV = MiSOCV + MaSOCV \quad (29.9)$$

and,

$$MiSOC = EMPL_r \times EG_r + EMPL_s \times ES_{r,s} \quad (29.10)$$

$$MaSOC = (SEC_r \times EG_r + SEC_{r,s} \times ES_{r,s} + SEC_i \times EIN_i) \times PEX \quad (29.11)$$

where EMPL is the employment yield—benefit in Euros from each kilowatt hour of electricity produced, SEC is the energy security index for each ESC (1 for RES-based plants, 0.5 for domestic produced coal, -1 for imported diesel and natural gas), $PEX(€/toe)$ is the equivalent exchange losses in Euros spent in each case for electricity generation from any type of imported energy (and resources).

Model constraints account for:

- Energy production and capacity limitations of each energy supply source:

$$EG_r = CAP_r \times AF_r \times n_r \text{ (for wind, PV, oil and LNG resource)} \forall r \in R \quad (29.12)$$

$$EI_t = INT_i \times AF_i \text{ (for interconnection)} \forall i \in I \quad (29.13)$$

$$CAP_r \le CAP_r^{max} \forall r \in R \quad (29.14)$$

$$ESS_s^{min} \le ESS_s \le ESS_s^{max} \forall s \in S \quad (29.15)$$

$$\mathrm{INT}_i \leq \mathrm{INT}_i^{\max} \forall i \in I, \tag{29.16}$$

where CAP is the capacity, AF is the availability factor and n is the efficiency of each electricity generation infrastructure/plant.

Storage balance rate of charge of the storage stations in each time step t:

$$\mathrm{ES}_{r,s,t} = \mathrm{ES}_{r,s,t-1} + \mathrm{EG}_{r,t} - \mathrm{D}_{u,t} + \mathrm{EIN}_i. \tag{29.17}$$

- Land availability limitations for the set of selected energy supply options:

$$\mathrm{LF}_{r,s} \times \mathrm{CAP}_{r,s}^{\max} \leq \mathrm{Amax} \tag{29.18}$$

whilst the interconnection solution is considered to have zero land footprint.

- Direct emissions (in CO_{2eq}) from each energy supply resource for the selected horizon H (t=1, 2…12) in years, months, days, hours:

$$\sum\nolimits_1^t \mathrm{EG}_{r,s} \times \mathrm{EMF}_{r,s} \leq \sum\nolimits_1^t \mathrm{EG}_{r,s} \times \mathrm{EMF}_{\mathrm{ceiling}} \tag{29.19}$$

where $\mathrm{EMF}_{r,s}$ is the emission factor of each energy SC/resource and $\mathrm{EMF}_{\mathrm{ceiling}}$ is the emission factor ceiling that is considered "allowable" when compared to a conventional but clean electricity generation plant and infrastructure (e.g. natural gas).

29.4 The Case of Milos Island

Model evaluation is carried out through an exemplar case study in solver excel, for Milos island in Greece. Milos is positioned in the Cyclades island complex, and is characterised by an intense terrain along with a history on volcanism and a very good wind potential (Fig. 29.2). In the existing power system of Milos, the electrical load is provided by thermal generators (diesel, heavy oil) and by wind power

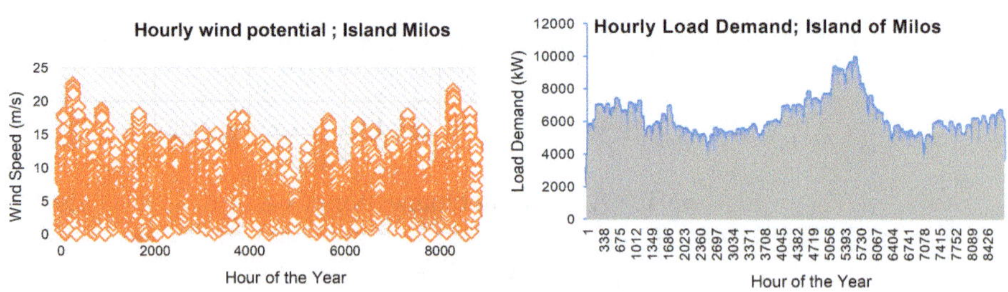

Fig. 29.2 Typical hourly wind potential and load demand variation for the island of Milos (for 2013)

generation, with peak load demand reaching approximately 10.00 MW whilst minimum demand is around 1.8 MW (Fig. 29.2).

Model implementation is performed for a small (due to solver's limitations) but detailed timeframe (one typical day in January—24 h) for a wind–diesel energy SC configuration considering also the energy storage option (typical set of lead acid batteries), acknowledging model implementation possibilities and operational issues. Diesel unit is treated as a base load power station constantly operating at its minimum capacity. Load demand variation is met by wind energy generation and by storage operation in cases of energy deficit. The data used for model validation were the following (Table 29.1):

The considered scenario accounts technical, economic, environmental and social implications simultaneously (a_1: 0.4, a_2: 0.4, a_3: 0.2). Results evidence that although wind is a more expensive energy source, due mainly to its environmental perfor-

Table 29.1 System characteristics and selected values for model optimisation

CAP_r^{max}	Maximum capacity of each electricity production plant r (MW)	Wind: 10, diesel: 2
CAP_r^{min}	Minimum capacity of each electricity production plant r (MW)	Diesel: 1
ESS_s^{min}, ESS_s^{max}	Minimum and maximum capacity of the storage station s (MWh)	Min: 1–max: 10
$INV_{r,s,i}$	Investment cost of the selected resource, storage station or interconnection option (€/kW)	Wind: 1100, diesel: 500, energy storage: 200
$MAIN_{r,s,i}, OPER_{r,s,i}$	Maintenance and operational cost of the selected resource, storage station or interconnection option (€/kWh)	Wind: 0.02, diesel: 0.03, energy storage: 0.01
LCREF	Reference life cycle environmental footprint of selected ESC (in CO_2 eq/kWh)	Wind: 65
$LCENVF_{r,s,i}$	Life cycle environmental footprint of each energy supply option (r, s, i) (in CO_2 eq/kWh)	Wind: 65, diesel: 770, energy storage: 275
$EMPL_{r,s}$	Employment yield (€/kWh)	Wind: 0.097, diesel: 0.113, energy storage: 0.030
$LF_{r,s}$	Land footprint of each resource r, s capacity installation (in km^2/kW)	Wind: 0.8, diesel: 0.5, energy storage: 0.6
Amax	Maximum land being available for the installation of each resource r, s (in km^2)	10 % of the island area (160 km^2)
EMF_r	Direct emission factor from each energy resource r, s (in kg CO_2eq/kWh)	Wind: 0.0, diesel: 0.8, energy storage: 0.0
$EMF_{ceiling}$	Emissions ceiling factor for the electricity generated (in kg CO_2eq/kWh)	Natural gas: 0.4
PEX	Exchange losses from imported energy (and resources) (€/kWh)	0.078
PCO_2	Current commercialisation price of CO_2 (€/kWh)	15

mance and to emissions' limitations ceiling, it is almost fully exploited to meet the demand and charge the energy storage system (Fig. 29.3).

29.5 Discussion and Conclusions

In order to prove the sensibility and the adaptability of the model to different optimisation strategies and goals, two alternative scenarios are considered: One with special emphasis on environmental dimension of electricity generation, ignoring cost issues (a_1: 0.0, a_2: 1.0 a_3: 0.0) Fig. 29.4, and a second one, a more moderate scenario for which, under the current techno-economic considerations, no emissions' ceiling is applicable for power production and the goal is to eliminated to demand fulfilment ($EMF_{ceiling} = 1(a_1 : 1.0, a_2 : 0.0, a_3 : 0.0)$. Diesel power operation cost = 0.02 €/kWh; Fig. 29.4).

Results in the first case show that under mandates for air quality limitations with special focus on the environmental performance of energy systems, particularly in small, tourist attractive islands, RES-based plants are always the optimal solution,

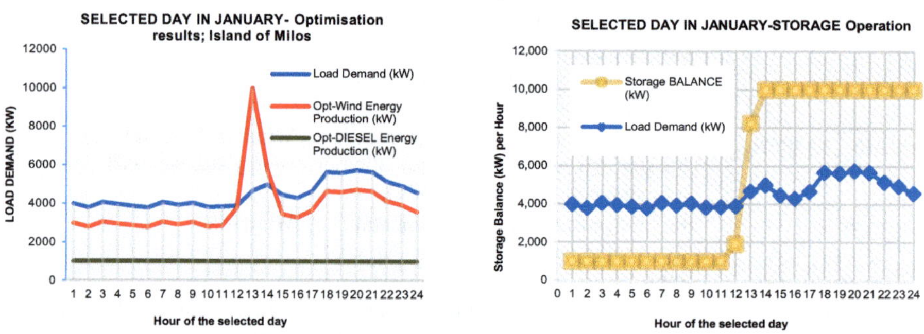

Fig. 29.3 Load demand profile, wind and diesel power generation and storage balance for the selected solution

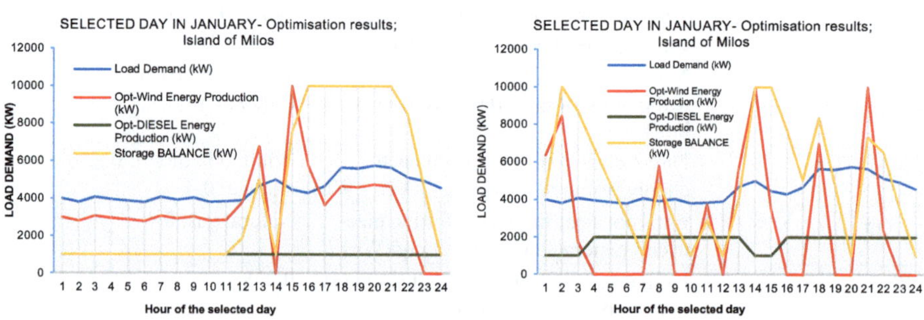

Fig. 29.4 Load demand profile, wind, diesel power generation and storage balance for the selected solutions

when constantly treated with an appropriately sized energy storage system. However, under the current techno-economic considerations with no provision of strict environmental standards and RES penetration, diesel power stations will remain the preferable solution (Fig. 29.4), exhausting their available per case capacity.

Recapitulating, in the present work, the developed optimisation framework for holistic evaluation of alternative energy and fuel SCs was introduced on top of the concept of energy and fuel SCs and their generic (uniform) representation. By accounting the sustainability dimensions that need to be incorporated in the energy planning problem, the optimisation criterion is set under the goal of maximising the total value of the SC subjected to both social and environmental constraints. The innovative crisps of the present work lie primarily on the introduction of the social and environmental function as problem/system constraints, and secondly on the flexibility of the optimisation criterion to be treated in a hybrid way depending on the energy planning strategy that the decision-maker seeks to implement.

In the next stage of the present and ongoing research, different criteria weights along with a more extensive time horizon planning will be assed.

Acknowledgements This work is part of the PhD research thesis entitled "Social, environmental and economic impacts of alternative energy and fuel supply chains" which is financially co-supported by the "Bursary Program of individualized assessment for the academic year 2012–2013" from resources of the operational programme "Education and Lifelong Learning" of the European Social Fund and from the National Strategic Reference Framework (2007–2013).

References

1 Lee JH (2014) Energy supply planning and supply chain optimization under uncertainty. J Process Control 24:323–331
2 Løken E (2007) Use of multicriteria decision analysis methods for energy planning problems. Renew Sustain Energy Rev 11:1584–1595
3 Pereira MVF, Pinto LMVG (1991) Multi-stage stochastic optimization applied to energy planning. Math Program 52:359–375
4 Pohekar SD, Ramachandran M (2004) Application of multi-criteria decision making to sustainable energy planning—a review. Renew Sustain Energy Rev 8:365–381
5 Almansoori A, Shah N (2009) Design and operation of a future hydrogen supply chain: multiperiod model. Int J Hydrog Energy 34:7883–7897
6 Hamedi M, Farahani RZ, Husseini MM, Esmaeilian GR (2009) A distribution planning model for natural gas supply chain: a case study. Energy Policy 37:799–812
7 Papapostolou C, Kondili E, Kaldellis JK (2011) Development and implementation of an optimisation model for biofuels supply chain. Energy 36:6019–6026
8 Pimentel D, Patzek T, Siegert F, Giampietro M, Haberl H (2008) Biofuel in question. The New Scientist 197:18
9 Su CL, Lee YM (2009) Development status and life cycle inventory analysis of biofuels in Taiwan. Energy Policy 37:754–758
10 Papapostolou C, Kondili E, Kaldellis JK (2008) Modelling, optimisation and life cycle analysis of biofuels supply chain. In: Proceedings of the World Renewable Energy Congress (WREC X), Glasgow, United Kingdom, 2008
11 Seuring S, Müller M (2008) From a literature review to a conceptual framework for sustainable supply chain management. J Clean Prod 16:1699–1710

12 Seuring S, Müller M (2008) Core issues in sustainable supply chain management—a Delphi study. Bus Strategy Environ 17:455–466
13 Papapostolou C, Kondili E, Kaldellis JK (2013) Energy supply chain optimisation: special considerations for the solution of the energy planning problem. Comput Aided Chem Eng 33(2014):1525–1530
14 Papapostolou C, Früh W-G, Kondili E, Kaldellis JK (2014) Evaluation framework of energy and fuel supply chains: a methodological approach, accepted for publication in Fresenius Environmental Bulletin, F-2014-451_MESAEP

Part IV
PV 2014

Chapter 30
An Innovative Dynamic Model for the Performance Analysis of a Concentrating Photovoltaic/Thermal (CPV/T) Solar Collector

Giuseppe Fiorenza, Giovanni Luigi Paparo, Felice Apicella, Nicola Bianco and Giorgio Graditi

Abstract A zero-dimensional dynamic model has been developed in order to simulate both electrical and thermal performance of a real concentrating photovoltaic/thermal (CPV/T) solar collector, namely the Absolicon X10 PVT prototype. This model calculates the instantaneous average temperatures of the main components of the system by handling a system of nonlinear differential equations, obtained via the application of the first law of thermodynamics, which is introduced in a MATLAB routine. Input parameters of the model include all the values characterising both the weather conditions and the main physical, geometrical and dimensional properties of the collector. Model validation was performed, firstly, via a steady-state simulation according to the EN 12975 standard in order to compare the thermal efficiency provided by the model with the solar collector efficiency curve. The difference among these values was constantly limited around 1 %. As a second step, the model was verified in a daily transient simulation by using the experimental data collected during the tests carried out at the ENEA Portici Research Centre, where an Absolicon X10 PVT collector is currently in operation. The measured and simulated thermal and electric yields were compared, thus achieving the confirmation that model can provide estimations consistent with the expected results.

Keywords Concentrating PV/T collector · Zero-dimensional dynamic model · Thermal efficiency · Performance analysis

30.1 Introduction

A photovoltaic/thermal hybrid solar system (or PV/T system for simplicity) is a combination of photovoltaic (PV) and solar thermal (T) components able to produce both electricity and heat simultaneously. In comparison with the use of

G. Fiorenza (✉) · F. Apicella · G. Graditi
Photovoltaic Technologies Unit, ENEA, Portici Research Centre, Portici (Naples), Italy
e-mail: giuseppe.fiorenza@enea.it

G. L. Paparo · N. Bianco
Industrial Engineering Department, University of Naples Federico II, Naples, Italy

A. Sayigh (ed.), *Renewable Energy in the Service of Mankind Vol II*,
DOI 10.1007/978-3-319-18215-5_30

two separate PV and solar thermal systems, the advantage of the combined approach is fundamentally the saving in space and, prospectively, in cost on equal energy production. Different kinds of hybrid PV/T collectors have been proposed in the past, mostly assembled via a flat-plate water-heating collector inserted with a PV module [1]. On the other hand, the operating temperature of such systems is rather limited (typically below 60 °C). The implementation of the concentration concept allows one to improve the hybrid collector performance. By combining advanced solar thermal and photovoltaic technologies (multi-junction solar cells), concentrating PV/T collectors (simply known as CPV/T collectors) can reach significantly higher overall efficiencies in comparison with a common flat PV/T collector. Furthermore, the operating temperature can be raised up to 90 °C or even more, therefore CPV/T collectors can be combined to desiccant wheels or absorption chillers in order to achieve a solar cooling system [2].

The technically easiest method to validate the efficiency of a CPV/T collector is to run it through one full season. Energy production is continuously measured so that the global performance can be evaluated by the sum of the observed values. This approach is unfortunately very time-consuming and the outcome cannot be considered as a standard result, being applicable only to the location where the collector is tested. In addition, laboratories that can handle advanced long-term testing are in a limited number, and it is expensive to obtain such an efficiency assessment of the collector. Alternatively, the performances of a CPV/T collector on an annual basis can be calculated from its efficiency curve, which can be obtained through a reduced number of tests according to the European Standard [3]. Again, international laboratories that can perform this type of tests are limited and the characterisation procedure is both time- and money consuming.

A smarter evaluation of the thermal and electrical performances consists of defining a mathematical model of the physical phenomenon, expressing the energy output as a function of the thermal state of the collector and its surrounding conditions. By following this approach, the evaluation of the collector efficiency is not related to a particular season and location. Although different multidimensional models for predicting the performance of a CPV/T collector are currently available in the literature, a simplified shape for the collector is mostly assumed, since remarkable computational efforts are implied in case of complex geometry, in particular when calculating annual electrical and thermal production. On the other hand, some examples of zero-dimensional models, which are also available, are conceived for stationary applications, while a CPV/T collector fundamentally works under dynamic conditions. A steady-state model is not able to predict operational temperatures of the PV cells and the heat-removal fluid during periods of fluctuating solar irradiance or intermittent fluid flow [4]. For a better simulation of the real performances of a CPV/T solar collector a dynamic model is needed [5].

In the current chapter, a zero-dimensional dynamic model is presented that is able to simulate both the electrical and thermal performance of a CPV/T collector, namely the Absolicon X10 PVT prototype, considering its real physical, geometrical and dimensional characteristics.

30.2 Experimental Work

Figure 30.1 shows the Absolicon X10 PVT prototype, which has been installed at the ENEA Research Centre of Portici, near Naples, in order to investigate the performance of a commercial CPV/T collector under real operative conditions. This type of collector consists of a parabolic concentrator with a single-axis tracker that focuses the incident beam of solar radiation onto a central receiver [6]. The collector has a gross area of 10.91 m^2 and an aperture area of 10.37 m^2. The PV module is composed of 166 mono-crystalline solar cells in series. The size of each cell is 32×110 mm, therefore, the total module area is 0.584 m^2 and the resulting concentration ratio is 17.8. The certified values of the main collector performance parameters, according to the technical bulletin provided by the manufacturer, are listed in Table 30.1. The tracking of the sun is carried out by rotating the structure around an axis oriented in the east–west direction. As it is known, concentrating solar systems can only exploit a fraction of the diffused light. In contrast, non-concentrating systems such as standard PV modules and flat-plate collectors use the global radiation [7].

The correct modelling of the receiver's cross section is fundamental in order to develop a consistent simulation tool. The absorber of the collector under investigation is roughly triangular with a double aluminium section bar structure (substrate). Each bar is provided with a fluid channel aiming at achieving a countercurrent flow configuration. The two receiver sides opposite to the parabolic concentrator are

Fig. 30.1 The Absolicon X10 PVT collector operating at the ENEA Portici Research Centre

Table 30.1 Certified values of the main performance parameters of the Absolicon X10 PVT collector

Zero-loss coefficient, η_0	0.548
Heat-loss coefficient, a_1	0.849
Second degree heat-loss coefficient, a_2	0.003
Zero-loss coefficient calculated with	*Absorber efficiency factor, $F'(\tau\alpha)_{en}$: 0.617*
	Incidence angle modifier, diffuse radiation, K(θ,d): 0.294
STC nominal power	1000 W (±5%)
Power drop	−0.4%/°C

STC measured at cell temperature of 25 °C and irradiance of 1000 W/m^2 with air mass 1.5 (AM1.5) spectrum

equipped with laser groove buried contact (LGBC) solar cells, whereas the top side of the receiver is covered with a thermal absorber. In Fig. 30.2, the cross section of the receiver without top absorber is shown together with a sketch where its different components are identified.

The Absolicon X10 PVT prototype installed at the ENEA Portici Research Centre is in operation since the beginning of 2013. During the first period of performance monitoring, maximum temperature of collector was limited to 50 °C. In Table 30.2, the values of the main operating parameters (inlet fluid temperature and fluid flow rate), as well as weather data (air temperature, both diffuse and global radiation), are reported, with reference to a suitable day of spring 2013. Experimental results for outlet fluid temperature and electrical power are also shown, limited to the part of the day characterised by an adequate global solar radiation (>100 W/m^2).

Based on the data in Table 30.2, heat and power production during the day were calculated [8]. The variation in these quantities over the day is shown in Fig. 30.3. It can be deduced that peak values of thermal and electrical efficiency are both around 85% of the expected values according to the collector-certified characteristics, as listed in Table 30.1. In addition to the possible imprecision correlated to

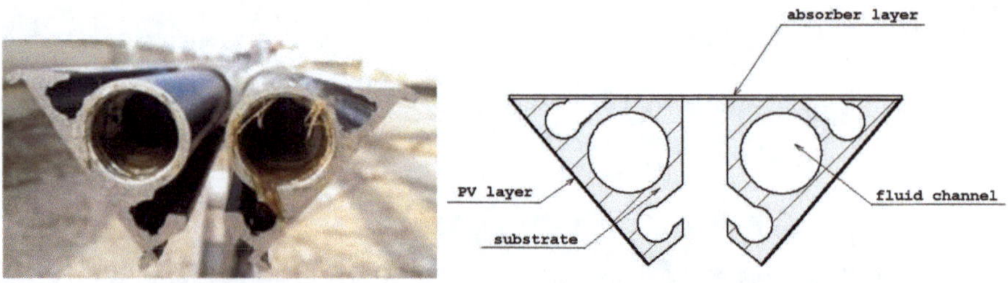

Fig. 30.2 Picture and representation of the receiver of the Absolicon ×10 PVT collector

Table 30.2 Operating parameters, weather and experimental data during a suitable spring day (21 March 2013)

Time	Inlet fluid temperature (°C)	Outlet fluid temperature (°C)	Fluid flow rate (l/min)	Air temperature (°C)	Diffuse solar radiation (W/m^2)	Global solar radiation (W/m^2)	Electrical power (W)
07:30	13.55	13.80	11.12	7.60	51.78	101.43	11.00
08:30	20.68	22.17	11.37	9.25	103.16	494.94	71.17
09:30	25.42	27.67	11.47	9.91	134.84	566.28	189.33
10:30	34.09	37.96	11.67	11.40	127.02	831.50	444.00
11:30	41.62	47.20	11.79	10.78	107.77	1055.92	685.00
12:30	43.43	49.33	11.83	11.00	105.20	1072.36	720.17
13:30	42.38	47.94	11.91	11.24	103.80	994.28	652.33
14:30	38.59	43.24	11.77	11.21	99.29	839.02	508.33
15:30	32.23	35.44	11.55	11.13	85.62	621.45	289.33
16:30	23.92	25.49	11.50	10.51	64.65	350.16	17.33

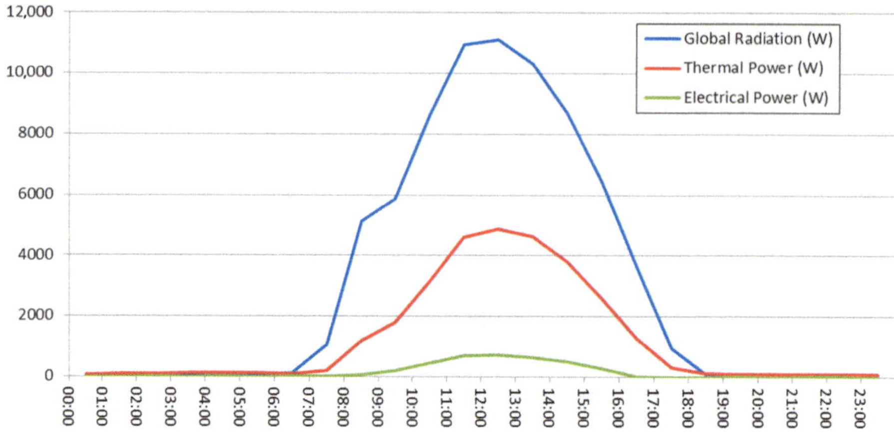

Fig. 30.3 Heat and power production during a suitable spring day (21 March 2013)

the industrial class accuracy of the collector field instrumentation, this difference is presumably due for the most part to the imperfect tracking of the sun by the collector, which was observed during the period under investigation.

30.3 Theoretical Model

The basic assumptions for the development of the theoretical model are thermodynamic equilibrium, negligibility of the kinetic and gravitational terms and solar radiation uniformly concentrated along the PV area. Furthermore, the temperature gradients in both PV film and substrate are supposed to be negligible because of the low thickness of the former and the high conductivity of the latter. It is to be stressed that PV and substrate temperatures are definitely uniform; as a consequence, a zero-dimensional approach guarantees adequate accuracy along with the reduction in the computational effort [9]. Finally, pure water is assumed as the heat transfer fluid.

The concentration ratio, C, is defined as:

$$C = \frac{A_{\mathrm{ap}}}{A_{\mathrm{PV}}},$$

where A_{ap} and A_{PV} are the aperture area of the collector and the area of the PV cells, respectively.

As it is commonly assumed in concentrating systems, only beam incident radiation, I_{b}, is considered in the energy balance on the PV cells. Anyway, the amount of solar radiation reaching the absorber surface is affected by a number of optical effects due to the collector geometry and material properties. In instantaneous power calculations, these effects are taken into account by considering the optical efficiency (zero-loss coefficient), η_0, calculated with direct radiation normal to the collector surface. An incidence angle modifier, *K(θ),* correlating normal to any given incidence angle optical efficiencies, is introduced in the energy balances as well. Therefore, the radiation incident on the PV surface, I_{PV}, is expressed by the quantity [10]:

$$I_{\mathrm{PV}} = CA_{\mathrm{PV}} I_{\mathrm{b}} \cos\theta K(\theta).$$

A system of equations in six unknown average temperatures, $T_{\mathrm{fluid,outlet}}$ (or $T_{\mathrm{fluid,inlet}}$), T_{cell}, $T_{\mathrm{substrate}}$, $T_{\mathrm{top\ absorber}}$, $T_{\mathrm{concentrator}}$, $T_{\mathrm{covering\ glass}}$, is presented. This system of equations is highly nonlinear because of the radiative terms and the correlations for evaluating the heat transfer coefficients introduced in the energy balances; therefore, it may be solved through the conventional numerical iterative techniques. It is important to point out that the aforementioned average temperatures, except for $T_{\mathrm{concentrator}}$ and $T_{\mathrm{covering\ glass}}$, are split into two values (denoted with the subscript *in* and *out,* respectively), related to each of the two aluminium profiles, which form the absorber, since the heat transfer fluid runs through them in sequence. This leads to a system of ten ordinary differential equations to be solved.

Considering the control volumes that delimit the two fluid channels, the following energy balances are obtained:

$$A_{\mathrm{ex}} h_{\mathrm{f}} (T_{\mathrm{sub_{in}}} - T_{\mathrm{f_{in}}}) - \dot{m} c_{\mathrm{f}} (T_{\mathrm{int}} - T_{\mathrm{in}}) = V_{\mathrm{ch}} \frac{d}{dt} (\rho_{\mathrm{f}} c_{\mathrm{f}} T_{\mathrm{f_{in}}}) \tag{30.1a}$$

$$A_{\mathrm{ex}} h_{\mathrm{f}} (T_{\mathrm{sub_{out}}} - T_{\mathrm{f_{out}}}) - \dot{m} c_{\mathrm{f}} (T_{\mathrm{out}} - T_{\mathrm{int}}) = V_{\mathrm{ch}} \frac{d}{dt} (\rho_{\mathrm{f}} c_{\mathrm{f}} T_{\mathrm{f_{out}}}) \tag{30.1b}$$

$$T_{\mathrm{int}} = \frac{T_{\mathrm{in}} + T_{\mathrm{out}}}{2} \quad T_{\mathrm{f_{in}}} = \frac{T_{\mathrm{in}} + T_{\mathrm{int}}}{2} \quad T_{\mathrm{f_{out}}} = \frac{T_{\mathrm{int}} + T_{\mathrm{out}}}{2}$$

The symbol A_{ex} denotes the fluid channel lateral area. The fluid heat transfer coefficient, h_{f}, is calculated using the following correlation provided by Gnielinski (1976) [11]:

$$Nu_{\mathrm{f}} = \frac{\mathrm{Pr_f}\, fr/8(\mathrm{Re_f} - 1000)}{\left(1 + 12.7(fr/8)^{0.5} (Pr_{\mathrm{f}}^{2/3} - 1)\right)} \quad \text{where} \quad fr = (0.790 \ln \mathrm{Re_f} - 1.64)^{-2}.$$

The thermo-physical properties necessary to determine Re and Pr numbers are evaluated at the average inlet/outlet fluid temperature. Nu_{f} is calculated at each integer value of the temperature between fluid triple and boiling point; then the coefficients of the polynomial correlation between h_{f} and T_{f} are determined through curve fitting. Thermophysical properties ρ_{f} and c_{f} are assumed constant as the fluid temperature is always lower than its boiling value.

With reference to the control volumes delimiting the boundaries of the two pieces of substrate, two more energy balances are deduced:

$$A_{\text{cell}}\frac{T_{\text{cell}_{\text{in}}}-T_{\text{sub}_{\text{in}}}}{R_{\text{cell-sub}}}+A_{\text{top}}\frac{T_{\text{top}_{\text{in}}}-T_{\text{sub}_{\text{in}}}}{R_{\text{top-sub}}} \tag{30.2a}$$

$$-A_{\text{ex}}h_{\text{f}}(T_{\text{sub}_{\text{in}}}-T_{f_{\text{in}}})-A_{\text{sub}_{\text{conv}}}h_{\text{sub}}\left(T_{\text{sub}_{\text{in}}}-\frac{T_{\text{sub}_{\text{in}}}+T_{\text{sub}_{\text{out}}}}{2}\right),$$

$$+CA_{\text{sub}_{\text{rad}}}K(\theta)\tau_{\text{glass}}I_{\text{b}}\alpha_{\text{sub}}=V_{\text{sub}}\frac{d}{dt}(\rho_{\text{sub}}c_{\text{sub}}T_{\text{sub}_{\text{in}}})$$

$$A_{\text{cell}}\frac{T_{\text{cell}_{\text{out}}}-T_{\text{sub}_{\text{out}}}}{R_{\text{cell-sub}}}+A_{\text{top}}\frac{T_{\text{top}_{\text{out}}}-T_{\text{sub}_{\text{out}}}}{R_{\text{top-sub}}}-A_{\text{ex}}h_{\text{f}}(T_{\text{sub}_{\text{out}}}-T_{f_{\text{out}}})$$

$$-A_{\text{sub}_{\text{conv}}}h_{\text{sub}}\left(T_{\text{sub}_{\text{out}}}-\frac{T_{\text{sub}_{\text{in}}}+T_{\text{sub}_{\text{out}}}}{2}\right). \tag{30.2b}$$

$$+CA_{\text{sub}_{\text{rad}}}K(\theta)\cos\theta\,\tau_{\text{glass}}I_{\text{b}}\alpha_{\text{sub}}=V_{\text{sub}}\frac{d}{dt}(\rho_{\text{sub}}c_{\text{sub}}T_{\text{sub}_{\text{out}}})$$

The last term of equation first member represents the solar radiation that passes through the surface of the substrate that is not covered by any PV cell or absorber layer. $R_{\text{cell-sub}}$ and $R_{\text{top-sub}}$ are the conductive resistances calculated considering the metallic substrate coupled with the cells surface and the top absorbing surface, respectively [9]. In both cases the conductivity as well as the density and the specific heat of the substrate is assumed to be constant in the investigated range of temperatures.

The third couple of energy balances refers to PV cells control volumes:

$$CA_{\text{cell}}K(\theta)cos\theta\,\tau_{\text{glass}}I_{\text{b}}\alpha_{\text{cell}}=A_{\text{cell}}\frac{T_{\text{cell}_{\text{in}}}-T_{\text{sub}_{\text{in}}}}{R_{\text{cell-sub}}}+A_{\text{cell}}h_{\text{cell}}\left[T_{\text{cell}_{\text{in}}}-\left(\frac{T_{\text{glass}}+T_{\text{conc}}}{2}\right)\right],$$

$$+CA_{\text{cell}}K(\theta)cos\theta\,\tau_{\text{glass}}I_{\text{b}}\alpha_{\text{cell}}\eta_{\text{cell}}+A_{\text{cell}}\sigma\varepsilon_{\text{cell}}(T_{\text{cell}_{\text{in}}}{}^{4}-T_{\text{conc}}{}^{4}) \tag{30.3a}$$

$$CA_{\text{cell}}K(\theta)cos\theta\,\tau_{\text{glass}}I_{\text{b}}\alpha_{\text{cell}}=A_{\text{cell}}\frac{T_{\text{cell}_{\text{out}}}-T_{\text{sub}_{\text{in}}}}{R_{\text{cell-sub}}}+A_{\text{cell}}h_{\text{cell}}\left[T_{\text{cell}_{\text{out}}}-\left(\frac{T_{\text{glass}}+T_{\text{conc}}}{2}\right)\right].$$

$$+CA_{\text{cell}}K(\theta)cos\theta\,\tau_{\text{glass}}I_{\text{b}}\alpha_{\text{cell}}\eta_{\text{cell}}+A_{\text{cell}}\sigma\varepsilon_{\text{cell}}(T_{\text{cell}_{\text{out}}}{}^{4}-T_{\text{conc}}{}^{4}) \tag{30.3b}$$

Equation first member is the fraction of the incident radiation, which is absorbed by the cells, while the second term in equation second member is the convective thermal flux related to the interaction between the cells and the still air inside the collector. In particular, h_{cell} is evaluated through the expression [12]:

$$Nu = 0.59Ra^{\frac{1}{4}},$$

which is suitable for natural convection over vertical plate surfaces (assumed the perfect symmetry of the system) in the field from 10^4 to 10^9 Ra. An interpolating curve has been determined in order to explicit the correlation between h_{cell} and temperature. The third term in equation second member is the gross electrical power produced by the PV module. In particular η_{cell} is calculated as temperature linear-dependent considering the STC efficiency and a power drop, as reported in Table 30.1. Finally, the radiative term is obtained by assuming all surfaces as "grey bodies" and PV surface area as negligible if compared to the collector aperture area (the model for radiation from a small body in a large cavity is adopted).

Two additional energy balances are defined referring to the top absorber control volumes:

$$A_{top}\tau_{glass}I_{tot}\alpha_{top} = A_{top}\frac{T_{top_{in}} - T_{sub_{in}}}{R_{top\text{-}sub}} + A_{top}\sigma\varepsilon_{top}(T_{top_{in}}{}^4 - T_{glass}{}^4) + A_{top}h_{top}[T_{top_{in}} - T_{glass}], \tag{30.4a}$$

$$A_{top}\tau_{glass}I_{tot}\alpha_{top} = A_{top}\frac{T_{top_{out}} - T_{sub_{out}}}{R_{top\text{-}sub}} + A_{top}\sigma\varepsilon_{top}(T_{top_{out}}{}^4 - T_{glass}{}^4) + A_{top}h_{top}[T_{top_{out}} - T_{glass}]. \tag{30.4b}$$

Equation first member represents the absorbed fraction of the total incident radiation. The radiative term is calculated by assuming involved surfaces as "grey bodies" and the top absorber surface area as negligible in comparison with covering glass surface area. Finally, the second term in equation second member is the convective heat flux associated with the top absorber and still air inside the collector, which has been determined via the following Jakob's correlation (1949) [13]:

$$Nu = 0.068Ra^{\frac{1}{3}},$$

which is suitable for natural convection in horizontal rectangular enclosures in the range of 4×10^5–10^7 Ra. A quadratic fitting curve has been individuated to determine the h_{top} *as a* function of the temperature.

The ninth energy balance is about the concentrator control volume:

$$\begin{aligned} &A_{conc}\alpha_{conc}\tau_{glass}I_{tot} + A_{cell}\sigma\varepsilon_{cell}\left(\left(\frac{T_{cell_{in}} + T_{cell_{out}}}{2}\right)^4 - T_{conc}^4\right) \\ &\qquad + \frac{A_{glass}\sigma(T_{glass}^4 - T_{conc}^4)}{\dfrac{1}{\varepsilon_{glass}} + \dfrac{A_{glass}}{A_{conc}}\left(\dfrac{1}{\varepsilon_{conc}} - 1\right)} \\ &= A_{conc_{ext}}\sigma\varepsilon_{cell}(T_{conc}^4 - T_{sky}^4) + A_{conc,back}h_{conc,back}(T_{conc} - T_{amb}) \\ &\qquad + A_{conc,front}h_{conc,front}\left[T_{conc} - \left(\frac{T_{conc} + T_{glass}}{2}\right)\right]. \end{aligned} \tag{30.5}$$

The first term in equation first member is the absorbed fraction of the total incident radiation, whereas the second term is the radiative term, which already has been presented in (30.3). The third term represents the incoming radiative heat flux estimated with the model of the surface in an enclosure (both assumed as “grey bodies”), which can be considered valid, since absorber and cells areas are negligible if compared to glass and concentrator areas. The first term in equation second member is the external radiative thermal flux on the back side, while the second and third ones are convective terms taking into account the interactions between concentrator and the environment and concentrator and the air filling in the cavity, respectively. The coefficient $h_{conc,back}$ is evaluated through a multiple regression as a function of T_{film} and v_{air}.

The last energy balance refers to covering glass control volume:

$$\begin{aligned} A_{glass}\alpha_{glass}I_{tot} &= A_{glass}\sigma\varepsilon_{glass}(T_{glass}^4 - T_{sky}^4) + \frac{A_{glass}\sigma(T_{glass}^4 - T_{conc}^4)}{\dfrac{1}{\varepsilon_{glass}} + \dfrac{A_{glass}}{A_{conc}}\left(\dfrac{1}{\varepsilon_{conc}} - 1\right)} \\ &\quad + A_{glass}h_{glass,ext}(T_{glass} - T_{amb}) + A_{glass}h_{glass,int}\left[T_{glass} - \left(\frac{T_{glass} + T_{top}}{2}\right)\right] \end{aligned} \tag{30.6}$$

The first term in first member in the equation represents the absorbed radiation fraction, while the first term in second member provides the radiative heat exchange between the covering glass and the environment under “grey bodies” hypotheses and negligibility of the glass surface area if compared to sky extension. Third and fourth terms determine the convective heat exchange driving from the interactions

Table 30.3 Thermal conductive resistance values at various channel wall temperatures and concentrated incident radiation

Thermal conductive resistance (m^2K/W)	Temperature (°C)	Concentrated incident radiation	
		6000 W/m^2	0 W/m^2
$R_{cell-sub,out}$	50	2.71×10^{-5}	2.76×10^{-4}
	80	3.74×10^{-5}	3.07×10^{-4}
$R_{cell-sub,in}$	10	1.29×10^{-5}	3.81×10^{-4}
	40	2.35×10^{-5}	8.55×10^{-4}
$R_{top-sub,out}$	50	1.51×10^{-5}	7.96×10^{-6}
	80	1.33×10^{-5}	7.93×10^{-6}
$R_{top-sub,in}$	10	1.10×10^{-5}	8.18×10^{-6}
	40	1.86×10^{-5}	8.28×10^{-6}

between glass and the environment and glass and the air filling in the cavity, respectively. In particular $h_{glass,ext}$ is evaluated through the expression [12]:

$$Nu = 0.037\,\mathrm{Re}^{\frac{4}{5}}\,\mathrm{Pr}^{\frac{1}{3}},$$

which is suitable for the parallel flow over flat plates in the range 5×10^5–10^7 Ra. In the present study, a linear regression plus an interaction term of $h_{glass,ext}$ as a function of T_{film} and v_{air} has been determined.

A further step concerns the evaluation of both thermal conductive resistances $R_{top-sub}$ and $R_{cell-sub}$ by carrying out a two-dimensional simulation of heat transfer in solid bodies.

Thanks to the zero-dimensional hypothesis, the temperature of each element is assumed uniformly distributed in its own domain. As a consequence, an equivalent electric network can be introduced. It consists of four thermally interacting nodes at the potentials of T_{sub}, T_{top}, T_{cell} and T_{air}, respectively. It must be stressed that T_{air} is evaluated as a linear average on the uncovered side of the substrate boundary. The net heat fluxes in each temperature node are calculated as total line-averaged heat flux magnitudes on the interfaces among elements whose interaction is being analysed. In the attempt to determine the aforementioned thermal conductive resistances, the temperature differences among thermal potentials are compared to the relative heat fluxes.

The problem is solved after imposing the following conditions:

- Uniformity of temperature on fluid channels boundaries
- Negligibility of radiative and convective heat fluxes on all other boundaries

A sensitivity analysis is performed in order to estimate $R_{top-sub}$ and $R_{cell-sub}$ maximum values, considering that the thermal resistance of the convective heat exchange between fluid and substrate is roughly 10^{-3} W/m^2K. Different conditions have been examined in terms of wall temperatures of the channels and concentrated incident radiation on the absorber, including the condition of high wall temperature with no incident radiation, which has no physical meaning and is shown only in order to provide a complete description of the trend in thermal conductive resistances. The results of these calculations are reported in Table 30.3.

It can be noticed that the thermal resistance values related to both cell/substrate and absorber/substrate heat exchange are at least two orders of magnitude lower than the fluid/substrate one, with the exception of $R_{cell-sub}$ under the scarcely significant condition of no incoming radiation. Therefore, it is possible to neglect these resistances assuming that PV cells, top absorber layer, and aluminium substrate are all at the same thermal potential in order to simplify the calculations, by reducing the number of the equations in the system from ten to six. Moreover, it must be considered that the practical heat transfer fluid is a mixture of ethylene glycol and water. The addition of glycol to the water causes a decrease in the convective heat transfer coefficient and, as a result, an increase in the thermal convective resistance, thus making this simplification to be even more suitable for the physics of the phenomenon.

30.4 Results and Discussion

The differential equations system described in the previous section has been solved via its implementation in a MATLAB routine. In order to validate the model, the first step is the comparison between the simulated stationary thermal efficiency and the collector efficiency curve obtained from the data reported in Table 30.1. As input data, a water flow rate of 0.010 kg/s/m^2 of collector area, a global solar radiation of 1000 W/m^2 and an air temperature of 30 °C are assumed, according to the test conditions defined by the European standard [3]. In particular, thermal efficiency is calculated for different values of the normalised operational temperature, which is obtained by fixing the values of both global radiation and air temperature according to the testing method and by varying fluid inlet temperature, under the condition that fluid outlet temperature must not exceed 80 °C. The achieved results are presented in Fig. 30.4, whereas the variance between expected efficiency according to collector curve and corresponding values provided by the model is stressed in

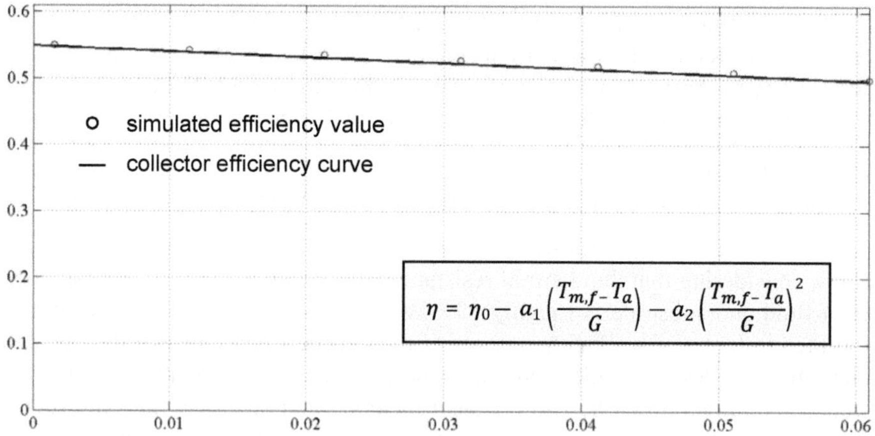

Fig. 30.4 Comparison between model calculations and collector efficiency curve

Table 30.4 Difference between expected efficiency according to collector curve and calculated efficiency

Normalised temperature [m^2K/W]	0.0015	0.0114	0.0213	0.0312	0.0411	0.051	0.0609
Expected efficiency according to collector curve	0.547	0.538	0.530	0.521	0.513	0.505	0.496
Efficiency values provided by the model	0.550	0.542	0.535	0.527	0.518	0.509	0.499
Variance [%]	+0.55	+0.74	+0.93	+1.14	+0.97	+0.79	+0.60

Table 30.4. The accuracy of model predictions is evident, with variances restricted around 1 % in the whole practical temperature range.

As a second and final step, model accuracy has been verified under transient conditions. For this purpose, a daily simulation has been carried out in order to evaluate thermal and electrical production. Average 10-min values for 1-min interval data are assumed for all the input parameters. Both thermal and electrical energy production are calculated every 10 min as well; subsequently, the daily total amount is inferred. With reference to the data acquired on 21 March 2013 and already reported in Table 30.2, the model supplies heat and power production values of 36.1 and 5.39 kWh, respectively, while the corresponding experimental values are 29.6 and 3.59 kWh. Therefore, model results overrate the test data of 22 and 50 %, respectively. The lower accuracy in the prediction of electrical energy production is explained by the absence in the current model version of a corrective factor, taking account of the performance decline of a single-axis tracker CPV collector when the sun moves away from solar noon. According to the experimental surveys, a decrease in the effective PV module area of around 10% per hour is observed. Therefore a corrective factor for daily power production of about 0.8 must be adopted. This figure is coherent with literature data relevant to the difference in gain between one-axis and two-axis trackers [14]. By implementing this correction, the model value of power production drops to 4.31 kWh and the overestimation reduces to 20%. The same result is obtained, if only the period of the day with a high direct solar radiation, that is quite close to the solar noon, is considered. The residual overestimation of around 20% affecting model predictions of both thermal and electrical productivity was expected, given the proved imperfection in the collector tracking system, which was already discussed in Sect. 30.2.

30.5 Conclusions

- CPV/T collectors allow combined production of heat and power with higher efficiency with respect to flat PV/T collector with an operating temperature suitable for thermally driven cooling technologies.
- The performance of a commercial CPV/T collector is being monitored at the ENEA Research Centre of Portici: experimental thermal end electrical efficiency

peak values are 15 % lower than the expected results according to the manufacturer data, thus, confirming the necessity of better tuning collector tracking system.
- In order to achieve an accurate CPV/T collector performance prediction as an alternative to cost- and money-consuming test campaigns, a dynamic model based upon effective receiver geometry and size has been developed; this model can be used as a valuable tool for system optimisation as well.
- An analysis of thermal conductive resistances, which regulate the heat exchange among receiver components, demonstrates that the model system of equations can be significantly simplified.
- Model validation under stationary conditions has been accomplished with encouraging results since the variance between simulated and expected efficiency according to collector curve is limited around 1 %.
- Model results under dynamic conditions are close to the expected values concerning thermal energy production, while the implementation of a correction factor associated to the efficiency drop when the sun moves away from solar noon is necessary for an improved estimation of electrical energy production.

Acknowledgements The authors would like to express their gratitude to the Italian Ministry of the Environment, Land and Sea for funding a research project concerning the development and testing of CPV/T systems, thus, allowing the accomplishment of this work.

References

1. Chow TT (2010) A review on photovoltaic/thermal hybrid solar technology. Appl Energy 87:365–379
2. Sabatelli V, Marano D, Fiorenza G (2007) Techno-economic evaluation of solar-assisted air-conditioning systems in Europe. In: Proceedings of ESTEC 2007, 3rd European solar thermal energy conference, Freiburg, Germany, 19–20 June 2007, pp 431–437
3. CEN (2006) EN 12975-2:2006. Thermal solar systems and components—solar collectors—part 2: test methods. CEN Ed, Brussels
4. Othman MYH, Yatim B, Sopian K, Bakar MNA (2005) Performance analysis of a double-pass photovoltaic/thermal (PV/T) solar collector with CPC and fins. Renew Energy 30:2005–2017
5. Chow TT (2003) Performance analysis of photovoltaic-thermal collector by explicit dynamic model. Sol Energy 75:143–152
6. Bernardo LR, Perers B, Håkansson H, Karlsson B (2011) Performance evaluation of low concentrating photovoltaic/thermal systems: a case study from Sweden. Sol Energy 85:1499–1510
7. Winston R, Miñano JC, Benitez P (2005) Nonimaging optics. Academic, New York. ISBN: 0-12-759751–759754
8. Duffie JA, Backman WA (1991) Solar engineering of thermal processes, 2nd edn. Wiley, New York. ISBN: 0-471-51056–51054
9. Calise F, Vanoli L (2012) Parabolic trough photovoltaic/thermal collectors: design and simulation model. Energies 5:4186–4208
10. Carvalho MJ, Horta P, Farinha Mendes J, Collares Pereira M, Carbajal WM (2007) Incidence angle modifiers: a general approach for energy calculations. In: Proceedings of ISES world congress vol. I–V, Beijing, China, 18–21 Sept 2007, pp 608–612

11. Gnielinski V (1976) New equation for heat and mass transfer in turbulent pipe and channel flow. Int Che Eng 16:359–368
12. Cengel YA, Ghajar AJ (2010) Heat and mass transfer: fundamentals and applications, 4th edn. McGraw-Hill, Blacklick. ISBN: 0077366646
13. Jakob M (1949) Heat transfer, vol 1. Wiley, New York. ISBN: 0-471-43824-3
14. Mousazadeh H, Keyhani A, Javadi A, Mobli H, Abrinia K, Sharifi A (2009) A review of principle and sun-tracking methods for maximizing solar systems output. Renew Sustain Energy Rev 13:1800–1818

Chapter 31
A Key Player Towards a Sustainable Energy Mix

Sophie Avril, Christine Mansilla, Pascal da Costa and Jean-Claude Bocquet

Abstract The energetic issues are on the top of the political agenda in many countries for environmental reasons, for its driving role in all the economic sectors, and for the energetic independence concerns. These problems are stressed by the increasing weight of renewable intermittent power sources in the global mix. In particular, due to its high potential and the strong national policy support it beneficiated, solar photovoltaic energy is now a key player in the world energy mutation and the way it is integrated into the global mix should be carefully performed.

To deeply understand the way the solar energy can penetrate and transform the forthcoming energy framework, we adopted a three-level strategy to provide some answers to the following questions: (i) How can we situate the photovoltaic power role in future energy mixes? (ii) Due to its main drawback, intermittency, could we provide an optimal design of a system combining storage devices? (iii) What is the efficiency of the incentive policies that are or have been implemented to accelerate its deployment?

That is why, in a first part, the peculiar position of the solar photovoltaic energy in the energy mix is analyzed. After recalling the general issues of future global energetic mixes, we propose a brief description of the different photovoltaic technologies and their promising evolutions in terms of technical improvements and cost reductions. Then, we describe the fast growing photovoltaic market and its consequences both on the electricity mix and the industry sector.

In the second part, we investigate the problematic of integrating such intermittent energy in the electricity mix, by developing a multi-criteria optimization

S. Avril (✉)
CEA DEN I-tésé and ECP LGI, CEA Saclay, 91191 Gif-sur-Yvette Cedex, France
e-mail: sophie.avril@cea.fr

C. Mansilla
CEA DEN I-tésé, CEA Saclay, 91191 Gif-sur-Yvette Cedex, France
e-mail: christine.mansilla@cea.fr

P. da Costa · J.-C. Bocquet
Laboratoire de Génie Industriel, ECP LGI, Ecole Centrale de Paris, Grande Voie des Vignes, 92 295 Châtenay Malabry Cedex, France
e-mail: pascal.da-costa@ecp.fr

J.-C. Bocquet
e-mail: jean-claude.bocquet@ecp.fr

A. Sayigh (ed.), *Renewable Energy in the Service of Mankind Vol II*,
DOI 10.1007/978-3-319-18215-5_31

methodology that simulates a system composed of photovoltaic panels and storage devices. Applications on a real case in the *Cirque de Mafate* (*L'île de la Réunion,* France) are provided to illustrate the interest of our method.

Finally, we question the efficiency of different public supports to the photovoltaic technologies in the most relevant countries. We focus on correlating the installed power capacity with the spent public money and the electricity prices.

In these works, we put into relief the necessity to consider energy issues through the prism of technical basis. Indeed, solutions that cannot be efficient should not be implemented in the system. However, a solely technical treatment of energy challenges is obviously insufficient, since energy is core point for the economy, for the citizen and so for the politician. Such a combined approach needs to remain anchored on concrete data close to the reality of the technical devices and keeping in mind the financial feasibility of the proposed solutions, when designing a new energy landscape and thus a new societal model.

Keywords Photovoltaic energy · Sustainable energy mix · Optimal design · Support policy

31.1 Introduction

Undoubtedly, energy has played an important role in human development and continues to do so. Thus, W. Fred Cottrell wrote "The energy available to man limits what he can do, and influences what he will do" [1, 2]. According to some historians [3, 4], the link between energy and human development is correlated to the domestication of the fire about 400,000 years ago. Access to energy has first allowed satisfying basic needs, such as heating or cooking. The energy applications have gradually diversified, thus developing agriculture and industry, allowing people and goods to move on thousands of miles away, offering access to drinking water, health and education for instance.

Today, the global energy mix is strongly based on fossil fuels, since the modern economic development is mainly based on mineral resources and not on renewables. This is due to the extraordinary properties of oil which is present everywhere in our societies [5]. Thus, in 2012, more than 80 % of the global primary energy demand was met by fossil fuels [6]. Oil provides 33 % of the world energy needs, followed by coal (22 %) and gas (21 %). The nuclear energy covers 6 % of primary energy needs. Finally, the renewable energy sources represent 13 % of the global supply with a strong weight for hydraulics (10 %).

For some years now, there has been an increasing interest in the problem of the release of greenhouse gases (GHG) and air pollutants. It began with the Summit of Rio in 1992, during which the *United Nations Framework Convention on Climate Change* was adopted. The first strong policy measures against the CO_2 emissions were signed in 1998 with the Kyoto protocol and consisted in the establishment of targets of emission reductions in relation to those of 1990 [7].

The energetic issues are on the top of the political agenda in many countries for environmental reasons, for its driving role in all the economic sectors, and for the energetic independence concerns. These problems are stressed by the increasing weight of renewable intermittent power sources in the global mix. In particular, due to its high potential and the strong national policy support it beneficiated, solar photovoltaic (PV) energy is now a key player in the world energy mutation and the way it is integrated into the global mix should be carefully performed.

To deeply understand the way the solar energy can penetrate and transform the forthcoming energy framework, we adopted a three-level strategy to provide some answers to the following questions: (i) How can we situate the photovoltaic power role in future energy mixes? (ii) Due to its main drawback, intermittency, could we provide an optimal design of a system combining storage devices? (iii) What is the efficiency of the incentive policies that are or have been implemented to accelerate its deployment? The paper structure will adopt this approach in what follows.

31.2 Solar Energy in the Energy Mix

31.2.1 Solar PV Technologies

Current technologies and possible future options have been categorized from current first-generation to future third-generation technologies [8]. Performances differ from one technology to another, depending on the different maturities (cf., Fig. 31.1).

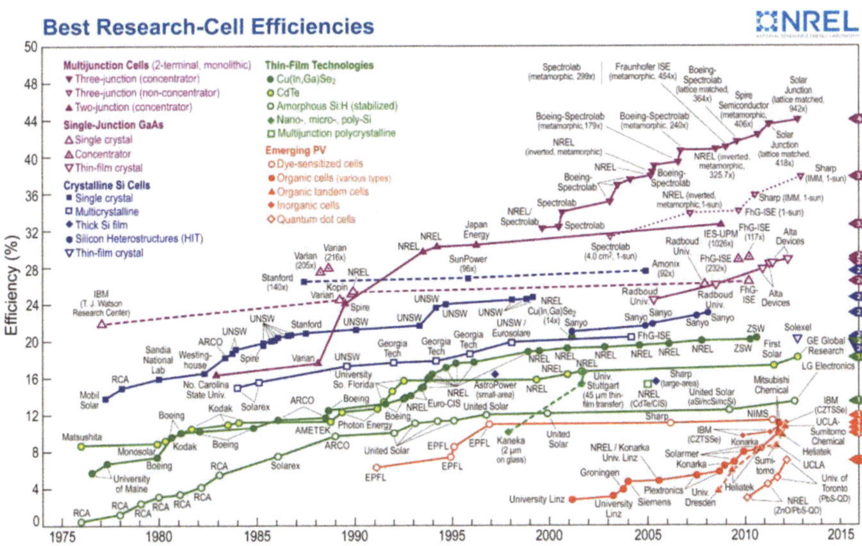

Fig. 31.1 Best research-cell efficiency from 1974 to 2012 [12]

The current PV market consists of arrangement of technologies including wafer-based silicon and a variety of thin-film technologies. More precisely, the first generation of solar cells based on single-junction mono- and multi-crystalline silicon wafers currently account for 90 % of PV production.

To reduce the cost-to-power ratio of the PV technology, the next step consists in removing the unnecessary material from the cost equation by using thin-film devices. In this approach, thin layers of semiconductor material are deposited onto a supporting substrate, or superstrate, such as a large sheet of glass. This typically requires a 100–1000 times thinner slice than the thickness of a silicon wafer of the first generation.

In the wake of the second-generation thin-film technologies, due to the reduction of active material cost, the low-cost substrate tends to become the cost limiting factor and improved efficiencies of the solar devices are required to pursue the cost-to-power ratio reduction trend. That is why the third generation of PV technologies consists in using multiple junctions [9, 10]. Here, a stack of different solar cells with various bandgaps is able to use the entire solar spectrum. Due to their quite expensive cost, these technologies are nowadays only used in "niche" applications such as satellite power generation [11].

31.2.2 Solar PV Market

The photovoltaic market significantly grew these past 10 years. The global PV cumulative installed capacity was 102,156 MW at the end of 2012 (it was only 2235 MW in 2002, which represents almost 50 % growth each year during 10 years) [13]. This market has been driven by Europe, mainly for political reasons (to limit the energy dependence and the GHG emissions) and thanks to strong public supports. Thus, Germany is the world's top PV market with 7.6 GW of newly connected systems and a cumulative installed capacity of 32.4 GW at the end of 2012 [14]. Then, there is Italy with 16.25 GW (second position), the USA with 7.2 GW installed (third position), and China and Japan with 7 GW installed each (fourth position). France is seventh with 4 GW installed at the end of 2012, 1 GW having been installed in 2012 (Fig. 31.2).

However, we can now observe a slowdown in the capacities installed in Europe. According to the European Photovoltaic Industry Association (EPIA), this is due to a stabilization phase after the previous boom. This slowdown can also be explained by the fact that this development is expensive especially for the countries those have developed PV thanks to feed-in tariffs (FIT) which generate high prices of electricity which are not appropriate in the actual context of financial and economic crisis.

The major consequence of this high deployment of PV technologies is its impact on the electricity mix. Indeed, whereas new capacities were off-grid systems at the beginning of this deployment, almost all new installed systems are connected to the grid, especially in Europe. This electric production begins to be significant in some countries, representing more than 5 % of the annual electricity demand, on average, for Germany and Italy.

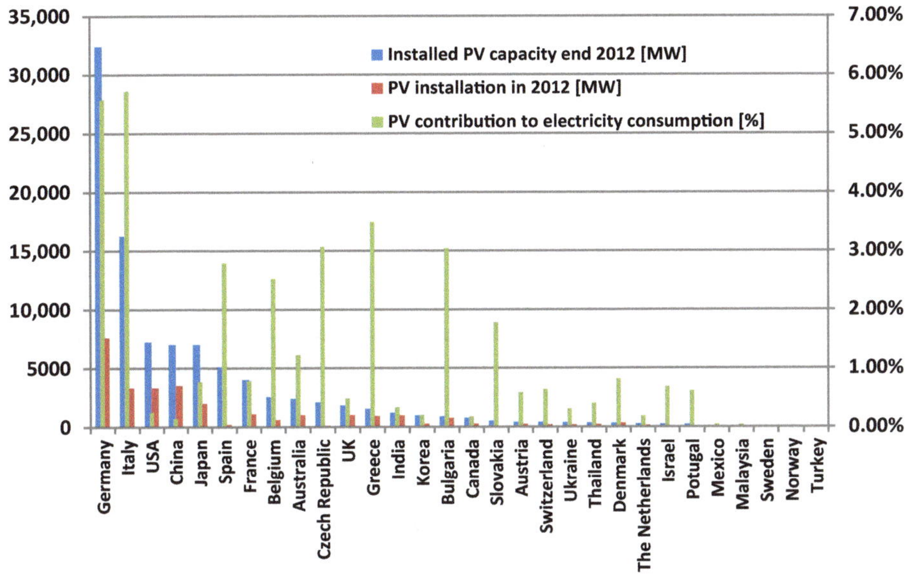

Fig. 31.2 Installed PV capacity at the end of 2012 and PV installation in 2012 in MW (*left* axis) and PV contribution to electricity consumption in 2012 [%] (*right* axis) [14]

This has an impact on the electric production (less use of fossil fuels but more intermittency) and consequently on the electricity market. It also has consequences on the industry sector since the PV market is very profitable and fast growing. Thus, problems of overcapacities and dumping were observed; they have been profitable for Chinese companies, rising up the issue of the national support to the PV market.

Considering the challenges of the energy sector, PV is meant to have a strong place in the future energy mix due to its high potential, the strong public support devoted to it (which accompanies its development) and with technologies that are still improving. However, there is a technical challenge to wider its development: its intermittency, a solution to which being coupling PV to storage devices. This is the theme of the following section.

31.3 Integrating Intermittent Energy Source in the Mix: Development of a Multicriteria Approach to Design a PV-Storage Integrated System

To evaluate the economic relevance of a PV system, the aim of this section is first to present a methodology for multi-objective optimization of a whole chain of electricity production from PV panels (based on the total levelized cost and other criteria), and then to present a possible application of the developed methodology on an interesting case in *La Réunion* island.

31.3.1 Methodology

In order to optimize a system coupling energy storage to PV, we propose to proceed in several steps:

I. We must choose the right devices both in terms of technological and economical potentials, regarding the selected place to install it. The whole system is characterized by different parameters and behavioral equations related to each technology and to environmental data (meteorological data and electrical consumption). This is the techno-economic model of the system.
II. For a given configuration, we are able to calculate some variables, such as the total levelized cost of the whole system output or the occupied area for instance. We can also take into account constraints which can be intrinsic or extrinsic.
III. The final step is to implement the optimization process through a code which operates the simulation for a given configuration of the system and also performs multicriteria optimization under constraints. This lead us to the creation of the MOMuS code (*Modélisation et Optimisation MUlticritère du Stockage d'énergie*—Multicriteria modeling and optimization of energy storage), in strong relationship with our colleagues of the LGLS (*Laboratoire de Génie Logiciel et Simulation*—Laboratory of software engineering and simulation). The optimizations can be performed on the variables (calculated in phase ii) as well as on the parameters (defined in phase i) of the system, depending on what has to be tested, respecting the constraints (defined in phase ii).

31.3.2 Application on a Real Case: The Cirque de Mafate (L'île de la Réunion, France)

Satisfying the energy demand has now become a sensitive topic in the world, especially in insular areas, in addition to the need to limit fossil fuel consumption for both sustainability and energy self-sufficiency issues. Thus, the European Commission has set a goal of 22 % of electricity production from renewable energies in 2010 for the whole electricity consumption in Europe [15]. Likewise, in the particular case of the French island *La Réunion,* an electrical self-sufficiency in 2025 is targeted [16]. Thus, we decided to test our model to optimize renewable electricity production coupled to storage technologies in the *La Nouvelle* village [17, 18].

The site of *La Nouvelle* in the cirque of Mafate on the French island *La Réunion,* in the Indian Ocean, is only accessible on foot or by helicopter. The annual consumption of the 750 inhabitants is provided by the Regional Energy Agency in *La Réunion* (ARER) [19], presenting a peak of electrical consumption of 140 kW. We assumed that it is weakly connected to the grid (at the present time, the electricity is produced by diesel generators). To reach the objective of energy independence by 2025, our goal is to satisfy the electricity demand of the consumers (the load) by producing electricity thanks to PV panels only. Indeed, due to green energy pressure and the guiding rules predicted by the PRERURE (Regional Plan of Renewable Energies and of Rational Use of Energy—Plan Régional des Energies Renouvelables et d'Utilisation Rationnelle de l'Energie), the fossil fuels are planned to be replaced. Of course, PV electricity production depends on the meteorological data of the site land. These fluctuating data are also provided by the ARER [19]. The load cannot be fully satisfied without storage technologies and we have evaluated the relevancy of a storage system based on batteries coupled with a H_2 chain as described in Sect. 31.2.

Our goal is to test different storage devices in order to minimize both the total levelized cost *TC* of the whole system and its degree of intermittency which is represented thanks to the grid connection S_{max} (allowed in the 1–10 kW range), while granting at least a 97% consumer satisfaction rate. The computational results are presented at the short and medium terms (namely, "2012" and "2020").

In the precise case of *La Nouvelle,* it appeared that:

- For the short term, the most economical solution is to use Pb-A batteries coupled with few H_2 bottles as in [20]
- For the medium term, this is still the cheapest solution but with 10% of profit compared to a solution using Ni-Cd batteries coupled with H_2 storage
- Coupled with Pb-A batteries, the H_2 is used as a seasonal storage which is not the case when coupling it with Ni-Cd batteries
- Using Pb-A batteries and H_2 storage leads to use fewer PV panels than when using Ni-Cd batteries and H_2 storage. Indeed, due to the high cost of Ni-Cd batteries, H_2 storage is more utilized than with Pb-A batteries. Thus, due to the efficiency of the global energy chain which is better for the batteries than for H_2 devices, using Pb-A batteries and H_2 leads to a smaller occupied surface than using Ni-Cd batteries and H_2 storage.

So, for the short term, we have to select Pb-A batteries. For the medium term, the choice is quite more complex, since it is then not only cost dependent. For instance, if environmental issues are considered, the use of batteries can be avoided and H_2 storage solution should be preferred. On the contrary, the H_2 storage solution should be avoided when intending to limit the dedicated surface (Fig. 31.3).

These results demonstrate the practical utility of the developed design method. It also has to be noticed that despite the fact that those simulations were based on the *La Nouvelle* concrete case, the results can be extrapolated to many remote areas presenting the same constraints. We can, thus, assert that in the short-term Pb-A batteries have an edge on other analyzed technologies.

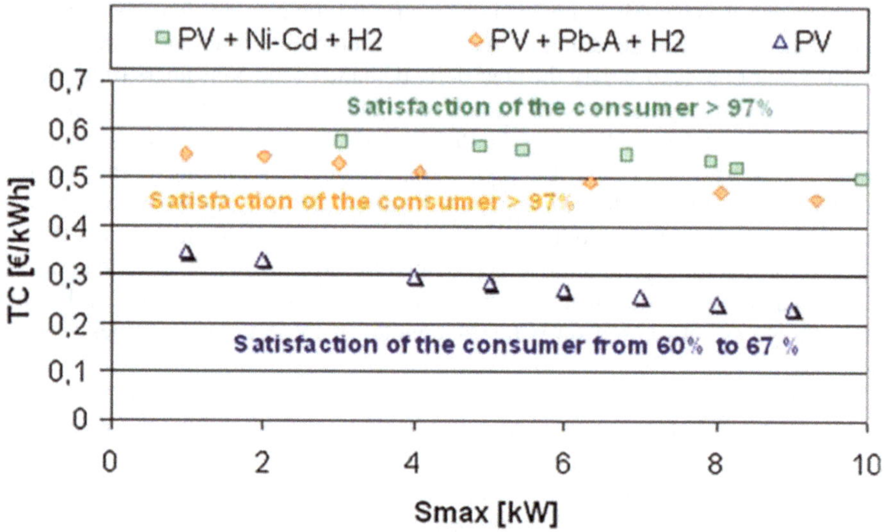

Fig. 31.3 Total levelized cost for the medium term versus the connection to the grid for three cases: (i) PV without storage, (ii) PV & Pb-A batteries & H_2 chain, and (iii) PV & Ni-Cd batteries & H_2 chain

31.4 PV Support Policy Assessment

Last decade saw PV technology emerge as one of the most promising technologies for power generation in the world, with an average annual global growth rate of around 47% over the last 10 years (between 2002 and 2011) [21], 75% of the installed capacities being located in Europe. This spectacular growth, despite the high levelized costs of PV electricity generation which were in 2010 in the range of $ 333–600/MWh in the Organization for Economic Co-operation and Development (OECD) countries for a 10% discount rate [22], was accompanied by strong national policy supports. Indeed, renewable energy sources are receiving increasing support worldwide from public authorities because of the environmental advantages they procure in comparison with conventional energy sources, especially when considering the risk of climate change [23, 24]. Thus, many countries have set targets for PV deployment [15, 25]. The possibility to achieve them at a lower cost, which was not until now a central issue given that the objectives were limited, has now become a major concern, making it necessary to examine the efficiency of the instruments used to promote PV [26, 27].

Public support can be categorized into direct and indirect subventions, but also—and we will focus on this distinction—into supply or demand policies. In general, supply policies, also called technology-push, aim at developing the products portfolio, while demand-pull policies correspond to Keynesian policies that act on demand in order to re-launch employment and production [28]. In the case of new energy technologies, demand-pull policies target the learning-by-doing effect which is particularly marked for PV [29]. Traditionally, R&D support is categorized as a technology-push measure. Public support for R&D is crucial since a major risk of underinvestment exists in comparison to what the public interest would be [30 *in* 28]. Indeed, financial risks underlie R&D investment [31 *in* 28], risks that some companies are unwilling to take. This is especially true for environmental R&D [32 *in* 28]. As regards PV, technology-push measures would facilitate the PV cost decrease through the technology improvement (reduction of raw material use for instance) or through breakthroughs (new cheapest technology).

Technology-push measures are most often accompanied by demand-pull ones in order to drive the consumers to adopt the developed innovations, as it will be detailed hereafter as regards PV support measures. Such measures gather FIT [26], tax abatement, value added tax (VAT) rate reduction and/or regional aids. The aim is to incite consumers to adopt technologies that are beneficial for the global social welfare, despite their higher costs compared to competing technologies. Technology-push policies usually tend to prevail during the first innovation phases, while demand-pull measures are predominant at the end when the technology becomes more mature [33 *in* 28]. Overall both effects may coexist, as suggested by models with the so-called two-factor learning curve [34]. The two factors are cumulative experience ("learning by doing" [35–37] and accumulated knowledge ("learning by searching"). As shown by Neuhoff [38 *in* 39], strategic deployment needs to be coupled with increased R&D amounts. In the PV field, the US PV manufacturing program is an example of this policy [40, 41 *in* 39].

As a matter of fact, previous studies demonstrated that the reduction in PV cost is the result of diverse factors such as module efficiency, plant size, silicon cost, silicon consumption, yield, wafer size [42, 43]. Based on empirical data on the considered period, the three factors that were identified as most important in explaining cost declines were the plant size, cell efficiency, and the cost of silicon [42]. While some of these factors very much depend on the installed PV capacity, others—such as the module efficiency—are more likely the results of the R&D outcomes. Hence, a policy only lying on demand-pull measures would not ensure continuous progress since a part of it can only be brought thanks to R&D.

When calculating the cumulative cost of the different policies to promote PV at the end of 2010 and their sharing in terms of market incentives (investment subsidies, loans, and FITs) and technologies incentives (R&D and demonstration programs), the differences are highlighted (see Fig. 31.4). Thus, the average cost spent per Wp until 2010 (i.e., cumulative costs divided by the cumulative installed capacity) was calculated at around €2.6 for France, €1.1 for Germany, €1.1 for Spain, €1.0 for Japan, and €2.5 for the USA. The best scheme until today seems to be the Japanese one. With their large demonstration programs (more than 25% of the total incentives at the end of 2010), they were able to choose the size, the location, and the kind of installations to subsidize as well as the installed capacity per year. In parallel, the part dedicated to R&D was significant. Germany, which also has a low average cost in 2010, led a totally different policy to support PV, with a strong effort on FIT. If the average cost is not too high today due to the important capacity installed every year, it will be different in the future and will heavily impact the

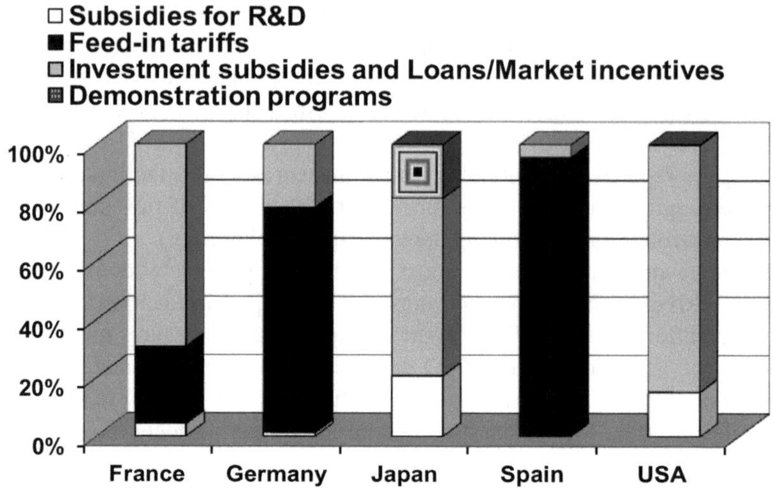

Fig. 31.4 Sharing out of the cumulative cost at the end of 2010among demonstration program costs, FIT, market incentives (other than FIT), and subsidies for R&D in France, Germany, Japan, Spain, and the USA (R&D is unknown for Spain)

electricity mix price. With an average cost spent per Wp until 2010 of €1.1, Spain is far from being a model. Indeed, the very attractive FIT of 2007–2008 which led to a spectacular boom in terms of installed capacity (3345 MWp cumulative installed PV capacity at the end of 2008) will imply high public expenses during 25 years. As for the USA, they began to support PV comparatively late. The result is a cumulative installed capacity 30 % lower than the Japan's one. At the present time, the amounts dedicated to R&D are significant, but they are decreasing in favor of market incentives which comprise FIT. So, as for Germany and Spain, the average cost spent per Wp is intended to increase. Finally, the French policy appears to be the least efficient one with the highest average cost. The main spending concerns the investment subsidies for nearly 70 %. However, as for the other European countries FIT are going to weigh more and more in the future.

31.5 Conclusion

In these works, we put into relief the necessity to consider energy issues through the prism of technical basis. Indeed, solutions that cannot be efficient should not be implemented in the system. However, a solely technical treatment of energy challenges is obviously insufficient, since energy is core point for the economy, for the citizen and so for the politician. Such a combined approach needs to remain anchored on concrete data close to the reality of the technical devices and keeping in mind the financial feasibility of the proposed solutions, when designing a new energy landscape and thus a new societal model.

Acknowledgments This work was carried out in the framework of a PhD thesis. The authors want to thank J. G. Devezeaux, Head of the Institute for techno-economics of energy systems (I-tésé) for the fruitful discussions.

References

1. Cottrell F (1955) Energy and society: the relation between energy, social change and economic development. McGraw-Hill Book Company, New York
2. Cottrell F (2009) Energy & society: the relation between energy, social change, and economic development. AuthorHouse, Bloomington
3. de Lumley H (2006) Il y a 400 000 ans: la domestication du feu, un formidable moteur d'hominisation. CR Palevol 5(1):149–154
4. Goudsblom J (1987) The domestication of fire as a civilizing process. Theory Cult Soc 4(2):457–476
5. Amyx JW, Bass DM, Whiting RL (1960) Petroleum reservoir engineering: physical properties, vol 1. McGraw-Hill College, New York
6. CAS 2012 Rapport Energies 2050, Rapport du groupe de travail présidé par Jacques Percebois
7. Protocol K (1997) United Nations framework convention on climate change. Kyoto Protocol, Kyoto
8. Green MA (2000) Photovoltaics: technology overview. Energy Policy 28(14):989–998
9. Green MA (2003). Third generation photovoltaics: advanced solar energy conversion. Springer, Berlin
10. Yoshimi M, Sasaki T, Sawada T, Suezaki T, Meguro T, Matsuda T, Yamamoto K (2003) High efficiency thin film silicon hybrid solar cell module on 1 m/sup 2/-class large area substrate. In: Photovoltaic energy conversion, 2003. Proceedings of 3rd world conference on, vol 2. IEEE, Japan, pp 1566–1569
11. Brown GF, Wu J (2009) Third generation photovoltaics. Laser Photon Rev 3(4):394–405
12. National Renewable Energy Laboratory (NREL) (2013) Figure representing the best research-Cell Efficiency from 1974 to 2012. http://a141.idata.over-blog.com/3/69/76/43/Partheme/Environnement/Chine-Centrale-solaire/Cellules-solaires-efficacite-rendement-top-recherche.jpg. Accessed Sept 2013
13. EPIA (2013) Global market Outlook for photovoltaics 2013–2017. http://www.epia.org/index.php?eID=tx_nawsecuredl&u=0&file=/uploads/tx_epiapublications/GMO_2013_-_Final_PDF_01.pdf&t=1372326209&hash=a894a7bc758d6136c99a30e8b60ec531bf691c56. Accessed 26 June 2013
14. IEA-PVPS (2012) PVPS Report. A snapshot of global PV 1992-2012. Preliminary information from the IEA PVPS Programme. Report IEA-PVPS T1-22:2013. ISBN 978-3-906042-10-7
15. Directive EU (2001) Directive 2001/77/EC of the European Parliament and the Council of 27 September 2001 on the promotion of electricity produced from renewable energy sources in the internal electricity market. Official J L 283:33–40
16. PRERURE (2003) Regional Plan of Renewable Energies and of Rational Use of Energy (Plan Régional des Energies Renouvelables et d'Utilisation Rationnelle de l'Energie). www.prerure.org (*in French*)
17. Avril S, Arnaud G, Colin H, Montignac F, Mansilla C, Vinard M (2010) Optimization of a remote photovoltaic system coupled to storage in an island context, 5th international renewable energy and storage (IRES 2010), 22–25 November 2010, Berlin
18. Avril S, Arnaud G, Colin H, Montignac F, Mansilla C, Vinard M (2011) Cost and surface optimization of a remote photovoltaic system for two kinds of panels' technologies. J Power Sources 196(19):8166–8169
19. Lacassin G, Vinard M (2008) Study of setting up a micro grid at La nouvelle (Etude test de mise en place d'un micro réseau électrique Ã La Nouvelle), ARER—Agence Régionale Energie Réunion, November 2008 (in French)

20. Dufo-Lopez R, Bernal-Agustin JL (2008) Multi-objective design of PV–wind–diesel–hydrogen–battery systems. Renew Energy 33(12):2559–2572
21. European Photovoltaic Industry Association (EPIA) (2011) Market Report 2011, May 2010. http://www.epia.org/. Accessed 4 April 2012
22. International Energy Agency (IEA) (2010) Projected costs of generating electricity, 2010, 218 p, ISBN 978-92-64-08430-8
23. Nakicenovic N, Alcamo J, Davis G, de Vries B, Fenhann J, Gaffin S, Dadi Z (2000) Special report on emissions scenarios: a special report of Working Group III of the Intergovernmental Panel on Climate Change (No. PNNL-SA-39650). Pacific Northwest National Laboratory, Richland, WA (US), Environmental Molecular Sciences Laboratory (US)
24. Stern NNH (ed) (2007) The economics of climate change: the Stern review. Cambridge University Press, Cambridge
25. California Public Utilities Commission (CPUC) (2010) Renewables Portfolio Standard, quarterly Report, Second quarter 2010. http://www.cpuc.ca.gov/NR/rdonlyres/66FBACA7-173F-47FF-A5F4-BE8F9D70DD59/0/Q22010RPSReporttotheLegislature.pdf>. Accessed 4 Aug 2010
26. Couture T, Gagnon Y (2010) An analysis of feed-in tariff remuneration models: implications for renewable energy investment. Energy Policy 38(2):955–965
27. Menanteau P, Finon D, Lamy ML (2003) Prices versus quantities: choosing policies for promoting the development of renewable energy. Energy Policy 31(8):799–812
28. Gallié EP (2011) Rapport d'activité du projet POLINOTEN
29. Kersten F, Doll R, Kux A, Huljic DM, Görig MA, Berger C, Wawer P (2011) PV learning curves: past and future drivers of cost reduction. In: Proceedings of the 26th European photovoltaic solar energy conference, pp 5–9
30. Arrow K (1962) Economic welfare and the allocation of resources for invention. In: The rate and direction of inventive activity: economic and social factors. NBER, pp 609–626
31. Martin S, Scott JT (2000) The nature of innovation market failure and the design of public support for private innovation. Res Policy 29(4):437–447
32. Jaffe AB, Newell RG, Stavins RN (2002) Environmental policy and technological change. Environ Res Econ 22(1–2):41–70
33. Dosi G (1988). Sources, procedures, and microeconomic effects of innovation. J Econ Lit 26(3):1120–1171
34. Miketa A, Schrattenholzer L (2004) Experiments with a methodology to model the role of R & D expenditures in energy technology learning processes; first results. Energy Policy 32(15):1679–1692
35. Argote L, Epple D (1990) Learning curves in manufacturing. Science 247(4945):920–924
36. Dutton JM, Thomas A (1984) Treating progress functions as a managerial opportunity. Acad Manag Rev 9(2):235–247
37. Yelle LE (1979) The learning curve: historical review and comprehensive survey. Decis Sci 10(2):302–328
38. Neuhoff K (2005) Large-scale deployment of renewables for electricity generation. Oxford Rev Econ Pol 21(1):88–110
39. Mitchell C, Sawin JL, Pokharel GR, Kammen D, Wang Z, Fifita S, Jaccard M, Langniss O, Lucas H, Nadai A, Trujillo Blanco R, Usher E, Verbruggen A, Wüstenhagen A, Yamaguchi K (2011) Policy, financing and implementation. In: IPCC Special report on renewable energy sources and climate change mitigation. Edenhofer O, Pichs-Madruga R, Sokona Y, Seyboth K, Matschoss P, Kadner S, Zwickel T, Eickemeier P, Hansen G, Schlömer S, von Stechow C (eds) Cambridge University Press, Cambridge, United Kingdom, New York, NY, USA
40. Mitchell RL, Witt CE, King R, Ruby D (2002) PVMaT advances in the photovoltaic industry and the focus of future PV manufacturing R&D. In: Conference record of the 29th IEEE photovoltaic specialists conference. New Orleans, LA, USA, 19–24 May 2002, pp 1444–1447

41. Jayanthi S, Witt EC, Singh V (2009) Evaluation of potential of innovations: a DEA-based application to US photovoltaic industry. IEEE T Eng Manage 56(3):478–493
42. Nemet GF (2006) Beyond the learning curve: factors influencing cost reductions in photovoltaics. Energy Policy 34(17):3218–3232
43. Nemet GF (2006) How well does learning-by-doing explain cost reductions in a carbon-free energy technology? Fondazione Eni Enrico Mattei (FEEM), Italy

Chapter 32
On-Grid Photovoltaic Water Pumping Systems for Agricultural Purposes: Comparison of the Potential Benefits Under Three Incentive Schemes

Pietro Elia Campana, Alexander Olsson, Chi Zhang, Sara Berretta, Hailong Li and Jinyue Yan

Abstract Electricity self-consumption represents a key issue to shorten the photovoltaic (PV) systems' payback period (PBP). To achieve high self-consumption rates, load control or storage systems are commonly used. Water pumping for irrigation is an interesting application that can largely increase the share of PV electricity self-consumption. The main objective of this chapter is to analyze the potential benefits of on-grid photovoltaic water pumping (PVWP) systems compared to those of conventional PV installations without water pumping. Three countries with different incentive schemes for PV installations have been studied: China with feed-in tariff, Italy with investment subsidies and net metering, and Sweden with tax reduction, investment subsidies, and green certificates. Wheat, maize, and potatoes have been chosen as reference crops to investigate the feasibility of PVWP systems as they are some of the most irrigated crops in China, Italy, and Sweden, respectively. The results show that the grid-connected PVWP systems lead to a lower PBP, between

P. E. Campana (✉) · H. Li · J. Yan
School of Business, Society & Engineering, Mälardalen University, SE-72123 Västerås, Sweden
e-mail: pietro.campana@mdh.se

H. Li
e-mail: hailong.li@mdh.se

A. Olsson · C. Zhang · J. Yan
School of Chemical Science & Engineering, KTH Royal Institute of Technology, SE-10044 Stockholm, Sweden
e-mail: alols@kth.se

C. Zhang
e-mail: chz@kth.se

J. Yan
e-mail: jinyue.yan@mdh.se; jinyue@kth.se

S. Berretta
Dipartimento di IngegneriaCivile e Ambientale, UniversitàdegliStudi di Perugia, Via G. Duranti 93, 06125 Perugia, Italy
e-mail: saraberrettamail@gmail.com

A. Sayigh (ed.), *Renewable Energy in the Service of Mankind Vol II*,
DOI 10.1007/978-3-319-18215-5_32

1 and 2 years, compared to PV installations without pumping. Owing to the low cost of electricity and feed-in tariffs, China presented the longest PBP. The PVWP systems in Italy have the shortest PBP mostly due to the more favorable incentive scheme compared to China and Sweden. The PBP of PVWP systems in Sweden is extremely affected by the uncertainty to benefit of the investment subsidies.

Keywords Self-consumption · Photovoltaic · Water pumping · Agriculture

32.1 Introduction

In 2013, the capacity of worldwide installed PV systems reached almost 140 GW_p. Residential, commercial, industrial, and ground-mounted segments accounted for 22, 27, 17, and 34 %, respectively. Germany, Italy, and Spain were leading the European PV market whereas China was the top market worldwide with the capacity of 11.8 GW_p PV systems installed [1]. The rapid growth of the PV market during the past decade has been boosted by the supporting renewable policies, especially subsidies, feed-in tariffs, and net metering. In general, self-consumption is always primarily encouraged to promote the onsite-produced electricity consumption [2]. For example, Germany and Italy introduced special premium tariffs for self-consumed electricity in addition to the feed-in tariffs [2, 3]. Although PV systems have reached the grid parity in most of the countries where incentive schemes are adopted, the gradual abandon of supporting policies makes the electricity self-consumption an important issue worth to be addressed for new installations [4]. Several studies have investigated how to maximize PV electricity self-consumption. *Widén* [2] studied how to increase the electricity self-consumption in 3–9 kW_p PV systems with optimal daily scheduling of household appliances. Castillo-Cagigal et al. [5] presented a control technique to maximize PV energy consumption based on deferrable and non-deferrable household loads. Thygesen et al. [6] studied the effects of storing PV electricity in batteries and hot water storage tanks to achieve high levels of self-consumption in the residential sector. However, there is still a strong need to explore new ways to further enhance the self-consumption. This study investigates the feasibility of using water pumping systems for irrigation to increase the self-consumption. The PV water pumping (PVWP) technology has been considered as one of the most promising technologies for irrigation due to the match between water supply and crop water requirements [7–9] and its technical and economic feasibility of PVWP has been proven. Commonly, PVWP technology is used in the remote areas where there is no grid access. However, if PVWP systems are connected to the grid, using the power output for the purpose of water pumping can largely increase the self-consumption and achieve more benefits. The main objective of this chapter is to compare the conventional on-grid PV systems with the grid-connected PVWP. This work will provide a better understanding about the combination of PV and water pumping from the perspective of enhancing the self-consumption of

PV power output. To estimate the economic benefit, three countries with different incentive schemes for PV systems have been studied: China, Italy, and Sweden.

32.2 Methodology

This section describes the methodological approaches used to design and simulate the PVWP systems. The economic analysis carried out in this chapter uses the payback period (PBP) as an indicator to compare grid-connected PVWP systems and conventional grid-connected PV installations. The chosen sites for this work are Taiyuan (37°52′10″N 112°33′37″E) in China, Piacenza (45°03′00″N 9°42′00″E) in Italy, and Lund (55°42′N 13°12′E) in Sweden. The choice of these three sites is determined by the high annual amount of water used for irrigation. Winter wheat, maize, and potatoes are used as reference crops for China, Italy, and Lund, respectively, due to local production circumstances [10].

32.2.1 PV Water Pumping System Design

The design of the grid-connected PVWP systems has been made following the same approach used for off-grid PVWP installations. The approach is based on the calculation of the PV array power peak for the designing month. Typically, the design month is the month marked out by the highest ratio between monthly crop water requirements and available solar irradiation [11]. The PVWP system power peak has been calculated with the following equation [12]:

$$P_p = \frac{0.0027}{f_m[1-\alpha_c(T_{cell}-T_0)]\eta_{MP}} \frac{IWR_p TDH}{E_S} \qquad (32.1)$$

where P_p is the array power peak (kW_p), 0.0027 is a conversion factor, IWR_p is the peak of the daily irrigation requirements during the irrigation season ($m^3/ha/day$), TDH is the total dynamic head assumed equal to 80 m, f_m is the matching factor equal to 0.9, α_C is the PV module temperature coefficient equal to 0.0045 %/°C, T_{cell} is the cell temperature (°C), T_0 is the reference temperature equal to 25 °C, and E_S is the daily solar irradiation hitting the array for the designing month ($kWh/m^2/day$). The peak irrigation water requirements have been estimated from the reference evapotranspiration ET_0 using the Penman-Monteith approach [13]. The pump working head and the irrigated area have been assumed to be equal to 80 m and 3 ha, respectively. The effect of future climate change on the crop water demand is also addressed in this chapter taking into consideration the Intergovernmental Panel on Climate Change (IPCC) scenarios A2 and B1 [14]. Comparison between PVWP and PV systems has been made assuming the same PV power capacity. To assess the benefits of both PVWP and PV systems in terms of self-consumed electricity, we have taken both power requirements for water pumping and farm building electricity consumption into consideration. The farm building electric load has been evaluated through simulations of IDA ICE assuming a surface of 180 m^2 and heat pump as heating and cooling device [15]. The hourly simulations of PV power output, farm load consumption, and electricity consumption for irrigation have been performed to evaluate the PV power production, the share of electricity self-consumed, and electricity fed into the grid.

32.2.2 PV System Simulations

The simulations of the PV system have been executed considering the PV array behavior depending on the solar irradiation hitting the PV modules. The power output from the photovoltaic array has been simulated using the following relation [16]:

$$P_{PV} = \eta_{PV} A_{PV} G_{g,t} \tag{32.2}$$

where η_{pv} is the efficiency of the PV module, A_{PV} is the array area, and $G_{g,t}$ is the total global irradiation hitting the tilted PV module surface. The annual PV module power output decay has been assumed to be equal to 0.3 %. For the PVWP system, the simulation of the pumping unit has been performed by taking into account its characteristic curve (power input versus water flow) [17]. The on-grid PVWP system is integrated with a novel control system proposed by Campana et al. [18], which can prioritize pumping and can dispatch the generated power for different purposes.

32.2.3 Economic Analysis

The PBP has been used as criteria for comparison. It can be calculated as [19]:

$$\text{PBP} = -\frac{\text{In}\left(1 - \dfrac{i\ \text{ICC}}{B_a - m\ \text{ICC}}\right)}{\text{In}(1+i)} \tag{32.3}$$

where i is the real discount rate, ICC is the initial capital cost, B_a is the annual revenue, and m is the percentage of the annual operation and maintenance costs compared to the ICC. The ICCs both for PVWP systems and for PV system have been taken from two manufacturing companies assuming a PV module price of 0.75 €/ W_p and project implementation costs equal to 30 % of the components cost [20, 21]. Currently, actors on the Chinese electricity market can benefit from feed-in tariffs depending on the size and location of the PV installation [22]. The incentive for distributed PV power generation systems is 0.049 €/kWh (0.42 CNY/kWh), whereas

Table 32.1 Economic analysis assumptions

	China	Italy	Sweden
Feed-in tariff	0.049	–	–
Subsidies (% ICC)	–	50 (up to 96 k€ and 20 kW_p)	35
Electricity price (buying) (€/kWh)	0.093	0.2	0.15
Electricity price (selling) (€/kWh)	0.053	0.1	0.05
Tax reduction (€/kWh)	–	–	0.07 (up to 30,000 kWh)
Green certificates (€/kWh)	–	–	0.0217
Real interest rate (%) [28]	6.41	2.45	3.69

the tariffs for power plants range from 0.10 to 0.12 €/kWh [23]. Feed-in tariff incentives finished in Italy in 2013. Currently, the Italian government is incentivizing PV systems with subsidies equal to 50% of the ICC in the form of tax reduction up to 96,000 € and power peak lower than 20 kW_p [3, 24]. In Sweden, the government proposed a tax reduction of 0.07 €/kWh up to 30,000 kWh per connection point with the constraint that the power fed into the grid should be lower than the power consumed. In this chapter, we have assumed that this law will be implemented [25]. PV installation put in operation before December 31st 2016 can also receive investment subsidies of 35% of the ICC up to 130,000 € (1.2 million SEK) or maximum 4000 € including VAT (37,000 SEK) per kW_p, but the amount is capped and this subsidy is uncertain [26]. Moreover, green certificates are given to producers of renewable electricity as one certificate for each MWh produced, and the price is market based and fluctuates from about 15 to 35 €/MWh (150–350 SEK/MWh) [27]. The assumptions made regarding the current electricity prices, feed-in tariffs, and subsidies for PV systems in the investigated countries are summarized in Table 32.1.

32.3 Results and Discussions

32.3.1 Crop Water Requirements and PVWP System Design and Simulation

For the PVWP system, the self-consumption of electricity is sensitively dependent on the water requirement of crops. Figure 32.1 shows the assessment of crop water requirements. The size of the resulting PVWP systems is 12, 20, and 24 kW_p for Taiyuan, Piacenza, and Lund, respectively. It has to be pointed out that the system design carried out in this chapter is based on a single crop. Nevertheless, an accurate design of the PVWP system should be based on the crop rotation planning during the years to estimate the absolute peak crop demand. The watering of different crops during the year, an option that can sensitively increase the PV electricity self-consumption, should also be considered in the design of the PVWP system.

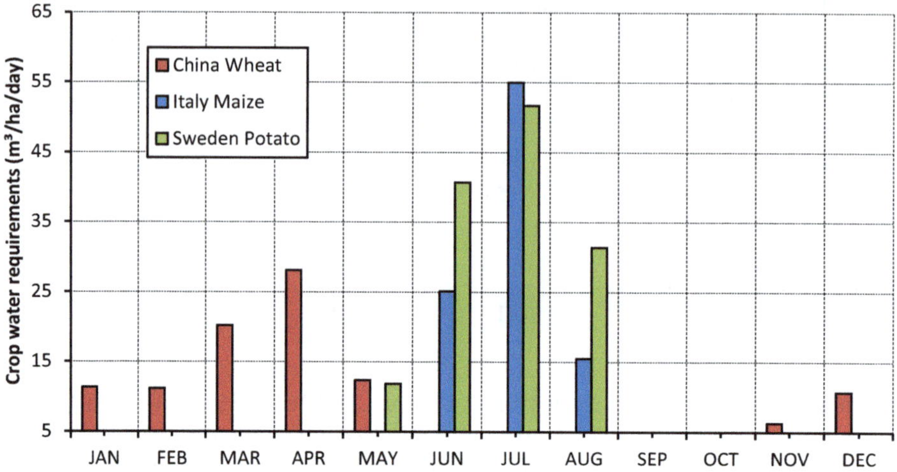

Fig. 32.1 Crops water requirements

32.3.2 Payback Period Analysis

The results of the PBP are depicted in Fig. 32.2. In all cases, PVWP systems present lower PBP compared to conventional PV systems due to the high self-consumption rates of electricity. Irrigation allows reaching high self-consumption rates since water pumping represent a controlled load. The simulation results executed for China, Italy, and Sweden show that the PV self-consumed electricity varies between 30 and 50 %. If the PVWP system is used for watering multiple crops, especially winter

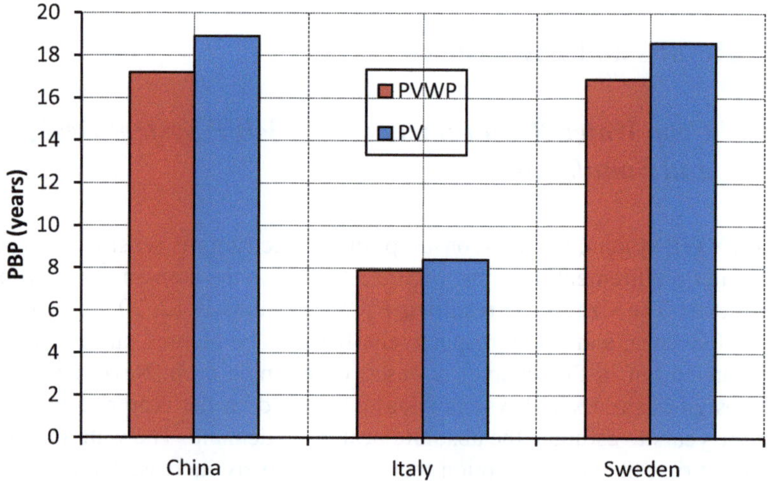

Fig. 32.2 Comparison between PVWP and conventional PV grid-connected systems

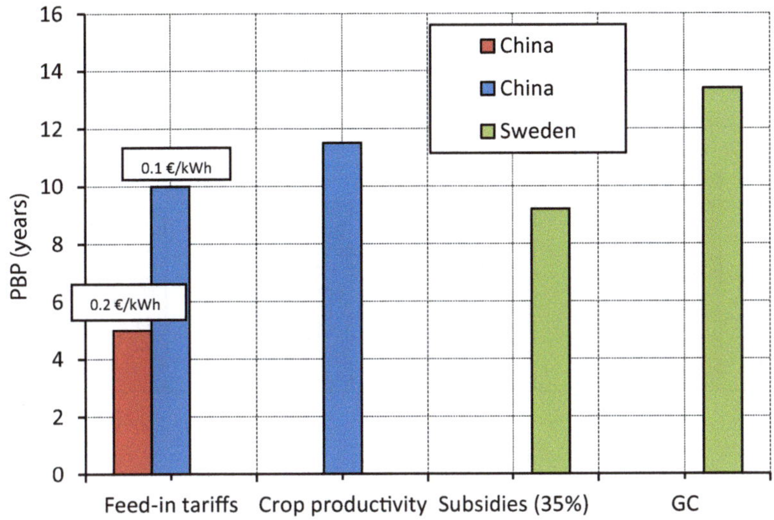

Fig. 32.3 Effect of different scenarios on the PVWP system PBP in China and Sweden

and summer crops, the PV electricity self-consumption can reach 60–70 %. China presents a longer PBP compared to Italy mainly for two reasons: first, the lack of subsidies that can cover 50 % of the initial capital cost; and second, the prices of electricity in China is lower than Italy and thus reduces the benefits of self-consumption and selling surplus electricity. We have not included investment subsidies in the Swedish case here and that explains the long PBP of PVWP and PV systems in Sweden compared to Italy.

Different scenarios regarding the incentive schemes have been analyzed for China and Sweden. The effects on the PBP of feed-in tariff variation and combining the revenues of electricity production and crop yield have been taken into consideration for China. The effect of subsidies and green certificates on the PBP has been investigated for Sweden. Figure 32.3 shows the PBP results for the scenarios investigated. To achieve a competitive PBP for PVWP systems in China between 5 and 10 years, the feed-in tariffs should range between 0.1 and 0.2 €/kWh. Considering feed-in tariffs for the distributed PV system equal to those for large-scale PV power plants (0.12 €/kWh ≈ 1 CNY/kWh) allows reducing the PBP of 50 % compared to the reference case. This assumption has been made since, unlike most of the countries that adopted feed-in tariffs for supporting the PV market (France, Germany, Italy, and Spain), China is supporting large power plants with higher subsidies compared to distributed PV installations [23, 29]. PVWP systems can cover electricity consumption for water pumping and farm buildings and inject the surplus power into the grid. The aim of the multipurpose use of PVWP, electricity production, and crop productivity, is to shorten the PBP. The average yield and price of winter wheat in 2013 were 5.52 t/ha and 290 €/t, respectively [30]. If 10 % of the gross profit is invested in the PVWP system, the PBP can be shorten to 11 years. It has been

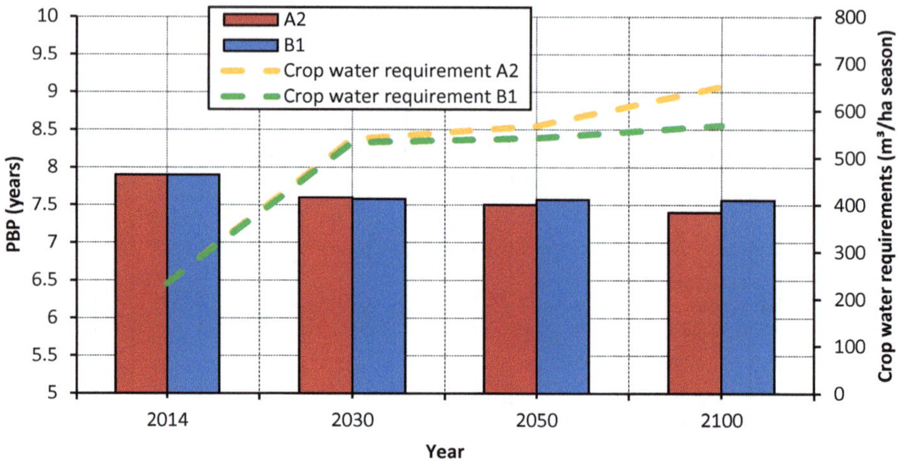

Fig. 32.4 PBP reduction due to the increase of crop water requirements

pointed out that the profitability of the system is extremely sensitive to the peak of crop water requirements, operational hydraulic head, and crop yield and price. Depending on the site-dependent parameters, PVWP systems can achieve PBP as low as 3 years [31]. The Swedish governmental investment subsidy for PV systems reduces the PBP from 16.9 to 9.2 years, making the investment more profitable. The possibility of combining tax reduction and green certificates brings the PBP down to 13.4 years, about 20 % less than the reference case.

The size of PVWP system varies from site to site due to different crop water requirements, hydraulic head, and solar irradiation. To fairly compare the three incentive schemes, we have taken the Swedish case (farm building load and water pumping electricity requirements) as the reference case. The resulting PBPs are 20, 8.5, and 17 years for China, Italy, and Sweden (reference case), respectively.

The increase of temperatures and the decrease of precipitations are some of the direct consequences of climate change [14]. Temperature and precipitation are the main parameters affecting the crop water requirements to be provided through irrigation and thus power requirements for water pumping. The effects of climate change have been addressed in this work taking into account the variation of water demand induced by the future temperature and precipitation variations. The variations of temperature and precipitation have been modeled on the basis of the IPCC scenario A2 (high-emission scenario) and B1 (low-emission scenario) in 2030, 2050, and 2100 [14]. Figure 32.4 shows the reduction of the PBP due to the augmentation of crop water requirements for Italy, assuming the same electricity prices of 2014. The increase in crop water requirements during the irrigation season directly increases the share of electricity that is self-consumed, and thus the benefits of the PVWP systems. In particular, the increase of crop water requirements in 2030 is significant and results in a reduction of the PBP of about 6 %. It is clear that higher crop water requirements have the negative effect of making the system undersized, especially in meeting the peak water demand. Although the crop water requirements increase to 130 %, the share of PV electricity self-consumption has a small increase without sensitively affecting the PBP. This aspect is also crucial in case of drought periods where the peak water demand might be substantially higher and investors might want to size the PVWP system to tackle the risk of lower yields compared to higher investment cost. Both future variations of temperature and precipitation, and droughts should be considered in a detailed design of PVWP systems.

32.4 Conclusions

Self-consumption represents a key issue to achieve high PV system profitability. Water pumping for irrigation purposes can increase the share of PV electricity self-consumption. This study investigates the benefits of grid connected PVWP systems. The PBP has been assessed under different incentive schemes for PV installations in three countries. The following conclusions can be drawn:

- Grid-connected PVWP systems allow us to increase the share of electricity that is self-consumed. Correspondingly, the increased self-consumption results in shorter PVWP system PBPs.
- Currently, the Italian subsidies (tax reduction) represent the best incentive scheme for grid-connected PVWP systems with PBP of 7.9 years, not considering the revenues due to crop sale.
- The multipurpose use of PVWP systems can increase the revenues produced by its operation, shortening the PBP.
- The increase in future crop water requirements due to climate change scenario projections can increase the benefits of PVWP systems connected to the grid, increasing the share of self-consumed electricity. This makes PVWP systems an important climate change adaptation technology for the agricultural sector.

References

1. EPIA Global Market Outlook for Photovoltaics 2014–2018. http://www.epia.org/fileadmin/user_upload/Publications/EPIA_Global_Market_Outlook_for_Photovoltaics_2014–2018_-_Medium_Res.pdf. Accessed 1 July 2014
2. Widén J (2014) Improved photovoltaic self-consumption with appliance scheduling in 200 single-family buildings. Appl Energy 126:199–212
3. Tudisca S, Di Trapani AM, Sgroi F, Testa R, Squatrito R (2013) Economic analysis of PV systems on buildings in Sicilian farms. Renew Sustain Energy Rev 28:691–701
4. Talavera DL, de la Casa J, Muñoz-Cerón E, Almonacid G (2014) Grid parity and self-consumption with photovoltaic systems under the present regulatory framework in Spain: the case of the University of Jaén Campus. Renew Sustain Energy Rev 33:752–771

5. Castillo-Cagigal M, Gutiérrez A, Monasterio-Huelin F, Caamaño-Martín E, Masa D, Jiménez-Leube J (2011) A semi-distributed electric demand-side management system with PV generation for self-consumption enhancement. Energy Convers Manage 52:2659–2666
6. Thygesen R, Karlsson B (2014) Simulation and analysis of a solar assisted heat pump system with two different storage types for high levels of PV electricity self-consumption. Sol Energy 103:19–27
7. Bayrakc AG, Kocar G (2012) Utilization of renewable energies in Turkey's agriculture. Renew Sustain Energy Rev 16:618–633
8. Mahmoud MM, Kukhun WR, Daud A-K, Efficiency improvement of a dual PV water pumping system on a desert well by solar matched load control. Int J Energy Eng 3(5):151–157
9. Campana PE, Li H, Yan J (2013) Dynamic modeling of a PV pumping system with special consideration on water demand. Appl Energy 112:635–645
10. Faostat http://faostat3.fao.org/faostat-gateway/go/to/home/E. Accessed 1 July 2014
11. Glasnovic Z, Margeta J (2007) A model for optimal sizing of photovoltaic irrigation water pumping systems. Sol Energy 81:904–916
12. Khatib T (2010) Design of photovoltaic water pumping system at minimum cost for Palestine: a review. J Appl Sci 10(22):2773–2784
13. Allen RG, Pereira LS, Raes D, Smith M (1998) Crop evapotranspiration. Guidelines for computing crop water requirements. FAO, Rome
14. IPCC IPCC special report-Emissions scenarios. https://www.ipcc.ch/pdf/special-reports/spm/sres-en.pdf. Accessed 1 July 2014
15. Berretta S (2013) Feasibility study of using wind power driven heat pump to supply heat for a single house. Master Thesis at Mälardalen University. http://www.diva-portal.org/smash/get/diva2:679376/FULLTEXT01.pdf. Accessed 1 July 2014
16. Duffie JA, Beckman WA (2006) Solar engineering of thermal processes, 3rd ed. Wiley
17. Abella MA, Lorenzo E, Chenlo F (2003) PV water pumping systems based on standard frequency converters. Prog Photovolt Res Appl 11:179–191. doi:10.1002/pip.475
18. Campana PE, Zhu Y, Brugiati E, Li H, Yan J (2014) PV water pumping for irrigation equipped with a novel control system for water savings. Proceedings of the 6th International Conference on Applied Energy—ICAE 2014
19. Mathew S (2006) Wind energy-fundamentals, resource analysis and economics. Springer
20. Solartech http://www.solartech.cn/. Accessed 1 July 2014
21. Wholesale Solar http://www.wholesalesolar.com/solar-panels.html. Accessed 1 July 2014
22. Ministry of Finance People's Republic of China Subsidy for small-scale distributed photovoltaic grid system. http://jjs.mof.gov.cn/zhengwuxinxi/tongzhigonggao/201307/t20130731_971420.html. Accessed 1 July 2014
23. Sun H, Zhi Q, Wang Y, Yao Q, Su J (2014) China's solar photovoltaic industry development: the status quo, problems and approaches. Appl Energy 118:221–230
24. Law (27 December 2013) n. 147. http://www.gazzettaufficiale.it/eli/id/2013/12/27/13G00191/sg%20. Accessed 1 July 2014. (In Italian)
25. Swedish Government Lagrådsremiss, Skattereduktion för mikroproducenter av förnybar el (Tax reduction for micro producer of renewable electricity) in Swedish. http://www.regeringen.se/content/1/c6/23/30/28/6194aef4.pdf. Accessed 1 July 2014
26. SEA (Swedish Energy Agency) https://www.energimyndigheten.se/Hushall/Aktuella-bidrag-och-stod-du-kan-soka/Stod-till-solceller/. Accessed 1 July 2014
27. Ekonomifakta http://www.ekonomifakta.se/sv/Fakta/Energi/Styrmedel/Elcertifikat/. Accessed 1 July 2014
28. The World Bank http://data.worldbank.org/indicator/FR.INR.RINR. Accessed 1 July 2014
29. Campoccia A, Dusonchet L, Telaretti E, Zizzo G (2009) Comparative analysis of different supporting measures for the production of electrical energy by solar PV and wind systems: four representative European cases. Sol Energy 83:287–297
30. Trading economics http://www.tradingeconomics.com. Accessed 1 July 2014
31. Campana PE, Olsson A, Li H, Yan J An economic analysis of photovoltaic water pumping irrigation systems. Proceedings of the 5th International Conference on Applied Energy—ICAE2013

Chapter 33
Enhancement in Conversion Efficiency by Surface Modification of Photoanode for Natural Dye-Sensitized Solar Cell

Mridula Tripathi and Priyanka Chawla

Abstract To avoid the negative effects of current energy systems on environmental pollution and global warming, research have been focused on renewable energy sources for the future. Among the renewable energy sources, solar cells have attracted a great interest as a solution to this problem. As a result, the conversion of solar energy into different forms has been the core of research for the past few years. The conventional solid state silicon-based solar cells, though highly efficient, are yet to become popular for mass applications as they are highly expensive. Therefore, developing low-cost devices for harvesting solar energy is most desirable. Dye-sensitized solar cells (DSSCs) have been considered as one of the most promising photovoltaic technologies because they are generally made from inexpensive components and have a simple designed structure. Numerous metal complexes and organic dyes have been utilized as sensitizers so far; the highest efficiency of DSSCs sensitized by ruthenium complex and osmium complex compounds absorbed on nanocrystalline TiO_2 reached 11–12 %. The major drawbacks of ruthenium are its rarity, high cost and the complicated synthesis of ruthenium complexes. The natural dyes anthocyanin/betacyanin obtained from fruits, flowers and leaves can be easily extracted by simple procedures and their cost-effectiveness, non-toxicity and complete biodegradation; therefore, the use of natural dyes in DSSCs has been a major focus of research. Among various wide-bandgap semiconducting oxides, nanocrystalline TiO_2 (ns-TiO_2) is the most suitable material for the photoanode of DSSC but TiO_2 can utilize only 6 % of the total solar irradiation in photocatalysis. Doping with metal oxides has been considered a promising way for improving the photocatalytic efficiency of ns-TiO_2. The increased photoactivity of CuO–TiO_2 may be attributed to the improvement of the light absorption properties and the slowdown of the recombination between the photoexcited electrons and holes during the photoreaction. We have studied the performance of DSSCs sensitized with anthocyanin pigments extracted from black grapes. The solar cell was assembled using CuO–TiO_2 thin film on ITO-coated glass with anthocyanin dye, liquid electrolyte system with LiI: I_2 as a redox couple. The obtained solar conversion efficiency was 4.8 % using an irradiation of 100 mW/cm^2.

M. Tripathi (✉) · P. Chawla
Department of Chemistry, C.M.P. Degree College, University of Allahabad, Allahabad, India
e-mail: ccmpau@gmail.com

A. Sayigh (ed.), *Renewable Energy in the Service of Mankind Vol II*,
DOI 10.1007/978-3-319-18215-5_33

Keywords Natural dye-sensitized solar cell · Anthocyanin · ns-CuO–TiO_2 · Conversion efficiency

33.1 Introduction

Increasing energy demands and concerns about global warming have encouraged scientists to develop cheap and easily accessible renewable energy sources. Among all renewable energy sources, solar energy has a fair amount of consistency due to its abundance and because it does not cause pollution. As a result, the conversion of solar energy into different forms has been the core of research for the past few years. The conventional solid state silicon-based solar cells, though highly efficient, are yet to become popular among mass applications as they are highly expensive. Therefore, it is necessary to develop low-cost devices for harvesting solar energy [1, 2]. Dye-sensitized solar cells (DSSCs) have been widely investigated as a next-generation solar cell because of their simple structure and low manufacturing cost [3, 4]. A DSSC is composed of a nanocrystalline porous semiconductor electrode, sensitizer, counter electrode and an electrolyte containing iodide and triiodide ions. The use of dye molecules as light harvesters in DSSCs for future clean energy is most attractive and successful when the dye is chemo-adsorbed on a porous network of interconnected nanometer-sized crystallites of a wide-bandgap semiconductor. The best examples of studies are the ruthenium- and osmium-based dyes, which show a high conversion efficiency of about 11–12% [5, 6]. Although such DSSCs have provided a relatively high efficiency, there are several disadvantages of using noble metals in them, which are limited in amount, hence their production is costly. On the other hand, organic dyes are not only cheaper but have also been reported to reach a high efficiency. However, synthetic organic dyes have often presented problems as well, such as complicated synthetic routes and low yields. The use of natural pigments may be a convenient alternative to ruthenium complexes and synthetic organic dyes, despite the lower efficiency because they are easy to obtain [7, 8]. In nature, some fruits, flowers and so on show various colours and contain several pigments that can be easily extracted and then employed in DSSCs. Therefore, unlike synthetic dyes, the natural ones are available, easy to prepare, low in cost, non-toxic, environmentally friendly and fully biodegradable. The anthocyanin and betacyanin dyes are natural pigments responsible for the red and blue colours of many fruits and leaves and may serve as photoprotective agents, antioxidants and osmotic regulatories. Moreover, their absorption spectra have favourable overlap with the solar spectrum [9–11].

Nanocrystalline TiO_2 (ns-TiO_2) is the most suitable material for the photoanode of DSSCs because the edge of the conduction band is located where it allows electron injection from the excited state of the dye. It is not only sensitive to light and corrosion–resistant but also inexpensive as an industrial material [12]. Basically, ns-TiO_2 can only utilize 6% of the total solar irradiation in photocatalysis due to the large band gap of anatase ns-TiO_2 (3.2 eV), but doping techniques shift the activity of ns-doped TiO_2 from the ultraviolet (UV) region to the visible light region [13]. It has been reported that doping with non-metal ions, alkali metals, transition metal ions and lanthanides (rare earth metal ions) has been considered a promising way to improve the photocatalytic efficiency of nanosized TiO_2 [14]. The wide range of applications of CuO as a gas sensor, photochromic and electrochromic material make it an interesting candidate for the modification of the TiO_2 photoelectrode [15, 16].

We present in this chapter our studies of an anthocyanin dye obtained from a black grape skin-sensitized CuO–TiO_2 photoanode solar cell. The properties of the natural solid state DSSC have been studied by measuring their short-circuit photocurrent density (J_{sc}), open-circuit voltage (V_{oc}), fill factor (*ff*) and conversion efficiency.

33.1.1 Experimental

The CuO and TiO_2 nanopowders were prepared by the sol–gel process. To prepare a sol–gel of TiO_2, $Ti[OCH(CH_3)_2]_4$ was added slowly to propanol, drop by drop. Deionized water was slowly added under vigorous stirring conditions for the duration of 10 min. During addition, a white precipitate was formed, and then 1 ml of 70 % HNO_3 was added to the mixture. The mixture was then stirred for 15 min at 80 °C. The propanol together with some water was allowed to evaporate during this time. In this way, a stable TiO_2 colloidal solution resulted.

The aqueous solution of $CuCl_2.6H_2O$ (0.2 M) is prepared in a cleaned round-bottom flask, and 1 ml of glacial acetic acid is added to the aqueous solution and heated to 100 °C with constant stirring. Then 8 M NaOH is added to the heated solution till pH value reaches 7. The large amount of precipitate is formed immediately, resulting in a CuO colloidal solution.

For the preparation of the CuO–TiO_2 admixed nanopowder, CuO solution was slowly added to TiO_2 colloidal solution under vigorous stirring conditions for 6 h. The resulting gel was dried and calcined at 450 °C to get CuO–TiO_2 nanopowder.

Fresh black grapes (*Vitis vinifera*) were purchased from the local market of Allahabad, India. The anthocyanin dye (Fig. 33.1) extracted with ethanol was obtained by the following steps: fresh fruits were washed with water and dried at 60 °C. After

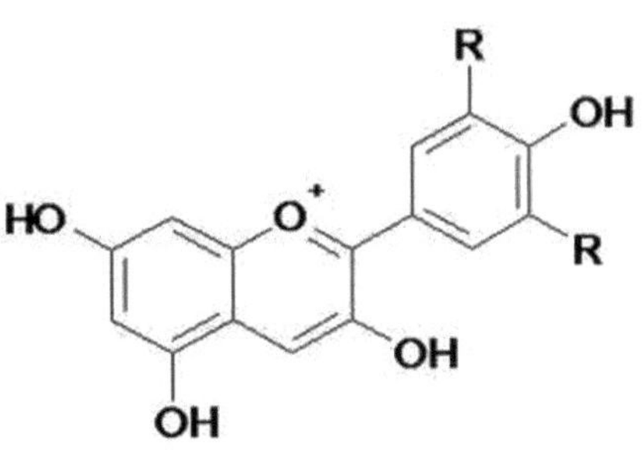

Fig. 33.1 The anthocyanin dye

crushing, these materials were immersed in absolute ethanol at room temperature in the dark for 1 week. Then the solids were filtrated out, and the filtrates were concentrated in Rotavapor at 40 °C, and these ethanolic extracts were refined by the chromatogram method. After that, the natural anthocyanin dye sensitizer alcohol solution was prepared.

Indium tin oxide (ITO) conductive glass with a sheet resistance of 15–20 Ω/cm^2 was first cleaned in a detergent solution using an ultrasonic bath for 15 min, rinsed with water and ethanol and then dried. CuO–TiO_2 nanopowder pastes were deposited on the ITO conductive glass by the doctor-blading technique in order to obtain CuO–TiO_2 film. The film on the substrate was annealed in the oven at 150 °C for 5–10 min.

The CuO–TiO_2-based electrode was immersed in an ethanol solution containing a natural anthocyanin dye for 10–12 h. Figure 33.4 shows a cell diagram that consists of both dye-sensitized and platinum (Pt) counter electrodes, which were assembled to form a solar cell by sandwiching a redox (I^-/I_3^-) electrolyte solution. The electrolyte solution was composed of 0.5 M LiI, 0.05 M iodine in acetonitrile. The cell voltage observed under illumination corresponds to the difference ΔV between the quasi fermi level of TiO_2 under illumination and the electrochemical potential of the electrolyte. The latter is equal to the Nernst potential of the redox couple (R/R^-) used to mediate charge transfer between the electrodes.

All electrochemical measurement was carried out by using Biologics SP-150 Potentiostat and Galvanostat. A xenon–mercury lamp (Oriel Corporation, USA) was used as the illumination source. The intensity of incident radiation was adjusted and fixed at 100 mW/cm^2. The fill factor *(ff)* and conversion efficiency *(η)* of the DSSC were determined according to $ff = (I \times V)_{max} / (I_{sc} \times V_{oc})$ and $\eta = I_{sc} \times V_{oc} \times ff / \text{Intensity (mW/cm}^2)$ based on *I–V* curve.

33.2 Results and Discussion

Figure 33.2 shows X-ray diffraction (XRD) patterns of the CuO–TiO_2 materials. Analysis of XRD patterns revealed that alloying ns-TiO_2 with CuO did not lead to the formation of any new composite material. However, a slight change in the lattice parameter was invariably found. The XRD also confirmed the presence of CuO–TiO_2 nanopowder.

With the help of a scanning electron microscope (SEM), the microstructural characteristics of the synthesized mixed oxide CuO–TiO_2 nanopowder was carried out (Fig. 33.3). The observed nanostructured characteristics showed a very fine-grained structure suggestive of a nanocrystalline-like matrix. By employing Scherrer's equation for the powder diffraction peaks and the SEM, the average grain size of the CuO–TiO_2 photoelectrodes was found to be 50–100 nm respectively.

The black grapes contained the anthocyanin dye. The chemical absorption of these dyes is generally accepted to occur because of the condensation of alcoholic

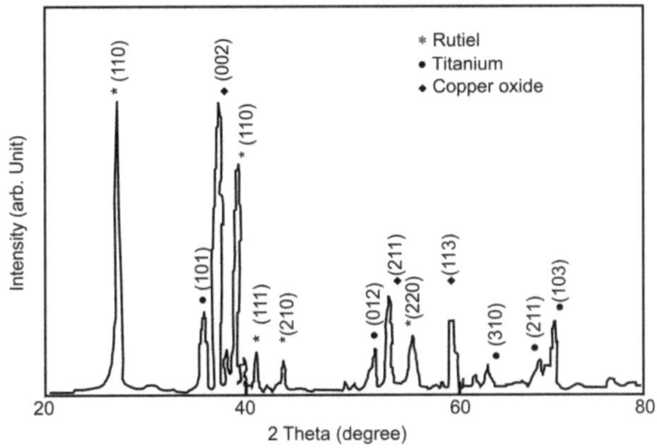

Fig. 33.2 X-ray diffraction pattern of CuO–TiO_2 nanopowder

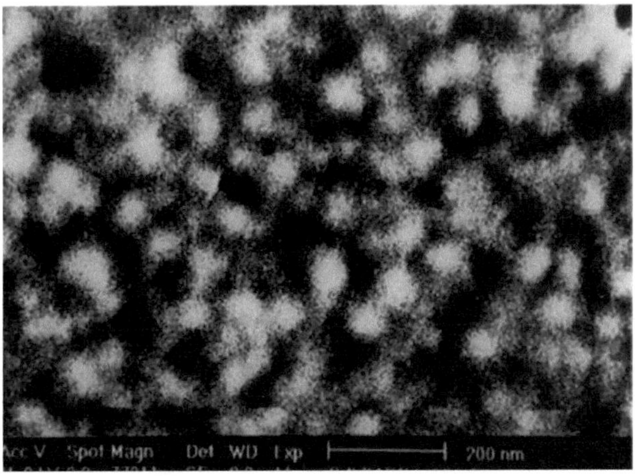

Fig. 33.3 Scanning electron micrograph of nanocrystalline (ns) CuO–TiO_2 nanopowder

bound protons with the hydroxyl group on the surface of nanostructured CuO–TiO_2 porous film and thus results in the high photoelectric conversion effect. Figure 33.4 shows the absorption spectrum of anthocyanin dye that covers the entire solar spectrum. It exhibits an intramolecular charge transfer (ICT) band at 540 nm and shows broad absorption spectra with an extended absorption edge in the solid state.

Figure 33.5 shows the resulting photocurrent, photovoltage curve for the DSSC-fabricated photoelectrode, anthocyanin dye-modified film from the suspension of the CuO–TiO_2. The photocurrent density (J_{sc}) and photovoltage (V_{oc}) was found to be 12.8 mA/cm^2 and 400 mV, respectively. The enhancement in the photocurrent with CuO modification is thought to be due to the improved spectral response of the photoanode.

By providing an inherent energy barrier that led to a decrease in recombination, the value of fill factor (*ff*) and conversion efficiency (η) was found to be 52 and 4.8 %, respectively, without any heat treatment or static press processing.

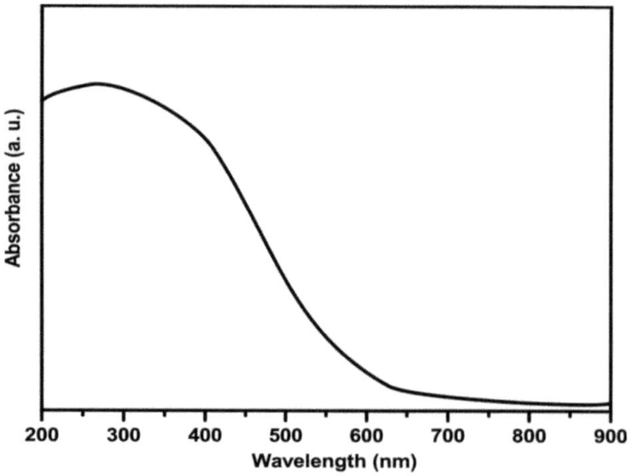

Fig. 33.4 Absorption spectra of anthocyanin dye-coated CuO–TiO_2 photoelectrode on indium tin oxide (ITO) glass substrate

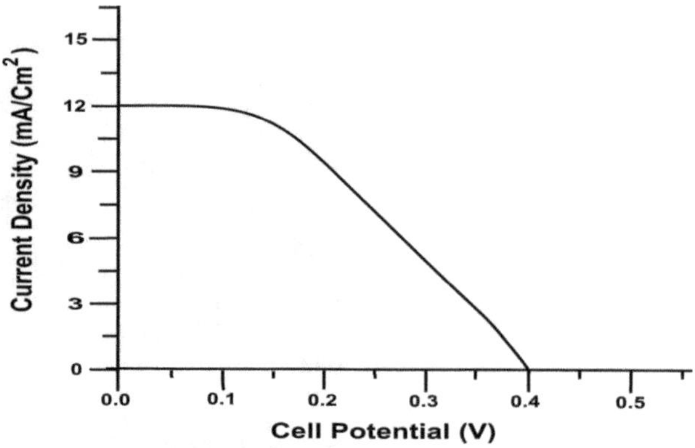

Fig. 33.5 Current–voltage characteristic curve of anthocyanin dye-coated CuO–TiO_2 photoelectrode

The light absorption behaviour of anthocyanin dye-coated CuO–TiO_2 photoelectrodes is elucidated by studying the absorption characteristics of the two photoelectrodes; a clear enhancement in a photoabsorption range (Fig. 33.4) is thought to be due to higher interaction between CuO–TiO_2 and anthocyanin dye, which leads to a better charge transfer and lower recombination rate.

Admixing of CuO species on TiO_2 can significantly increase the surface acidity. Due to this increased acidity, CuO–TiO_2 can absorb more anthocyanin dye on its surface. Thus, significant improvement of the photocatalytic activity has been achieved in comparison to the bare TiO_2 photocatalysts.

Moreover, anthocyanin in the black grapes extract has a shorter distance between the dye skeleton and the point connected to the oxide surface of photoanode. This could facilitate an electron transfer from anthocyanin in the black grapes extract to the CuO–TiO_2 surface and could be accounted for a better performance of anthocyanin dye sensitization.

33.3 Conclusions

In this chapter, we applied CuO–TiO_2 to natural DSSCs as a photoelectrode to reduce the recombination rate by providing an energy barrier, and we also explore the use of natural dyes, anthocyanin obtained from black grapes, which are characterized by a high content of red-purple pigments with a high absorption coefficient in the visible part of the solar spectrum. The dye absorption spectrum, CuO–TiO_2 electrical properties and DSSC efficiency are then investigated. It may be pointed out that the development of CuO–TiO_2 DSSCs as studied in the present investigation represents the first work of its type. From these studies, it can be concluded that the modification of TiO_2 with CuO deposition enhances cell performance due to the reduced recombination of photoinjected electrons. The anthocyanin dye has higher photosensitized performance due to better charge transfer between the anthocyanin dye molecule and the oxide surface, which is related to a dye structure. The enhancement of photoactivity can also be thought to be due to dye sensitization with high conductivity and wide absorption range (200–540 nm).

Photoelectrochemical parameters of 0.5-cm^2 active photoelectrode area devices sensitized by anthocyanin dye were constant for 8 weeks of continuous evaluation. Therefore, the anthocyanin dye-sensitized CuO–TiO_2 DSSC opens up a perspective of commercial feasibility for inexpensive and environmentally friendly dye cells.

Acknowledgements We are thankful to the Board of Research in Nuclear Sciences, Bombay, India (2012/34/33/BRNS) for the financial support.

References

1. Sastrawan R, Beier J, Belledin U, Hemming S, Hinsch A, Kern R, Vetter C, Petrat FM, Prodi-Schwab A, Lechner P, Hoffmann W (2006) A glass frit-sealed dye solar cell module with integrated series connections. Sol Energy Mater Sol Cells 90:1680–1691
2. Rowe RD, Lang CM, Chestnut LG, Latimer D, Rae D, Bernow SM (1995) ESEERCO. New York state environmental externalities cost study. Oceana Publications, New York
3. O'Regan B, Gratzel M (1991) A low cost, high efficiency solar cell based on dye sensitized colloidal TiO_2 films. Nature 353:737–740
4. Gratzel M (2001) Photolectrochemical Cells. Nature 414:338–344
5. Nazeeruddin MK, Pechy P, Liska P, Renouard T, Zakeeruddin SM, Humphry-Baker R, Comte P, Cevey L, Costa E, Shklover V, Spiccia L, Deacon GB, Bignozzi CA, Gratzel M (2001) Engineering efficient panchromatic sensitizers for nanocrystalline TiO_2-based solar cells. J Am Chem Soc 123:1613–1624
6. Altobello S, Argazzi R, Caramori S, Contado C, Da Fre S, Rubino, Chone C, Larramona G, Bignozzi CA (2005) Sensitization of nanocrystalline TiO_2 with black absorbers based on OS and Ru polypyridine complexes. J Am Chem Soc 127:15342–15343
7. Hao S, Wu J, Huang Y, Lin J (2006) Natural dyes as photosensitizers for dyesensitized solar cell. J Sol Energy 80:209–214
8. Kim S, Lee JK, Kang SO, Ko JJ, Yum JH, Fantacci S, De Angelis F, DiCenso D, Nazeeruddin MK, Grätzel M (2006) Molecular engineering of organic sensitizers for solar cell applications. J Am Chem Soc 128:16701–16707

9. Calogero G, Di Marco G, Cazzanti S, Caramori S, Argazzi R, Di Carlo AD, Bignozzi CA (2010) Efficient dye-sensitized solar cells using red turnip and purple wild Sicilian prickly pear fruits. Int J Mol Sci 11:254–267
10. Garcia CG, Polo AS, Murakami Iha NY (2003) Fruit extracts and ruthenium polypyridinic dyes for sensitization of TiO_2 in photoelectrochemical solar cells. J Photochem Photobiol A 160:87–91
11. Calogero G, Di Marco G (2008) Red Sicilian orange and purple eggplant fruits as natural sensitizers for dye-sensitized solar cells. Sol Energ Mater Sol Cells 92:1341–1346
12. Nowotny J, Bak T, Nowotny MK, Sheppard LR (2007) Titanium dioxide for solar hydrogen. Intl J Hydrog Energy 32:2609–2629
13. Zhang XT, Udagawa K, Liu Z, Nishimoto S, Xu C, Liu Y, Sakai H, Abe M, Murakami T, Fujishma A (2009) Photocatalytic and photoelectrochemical studies on N-doped TiO_2 photocatalyst. J Photochem Photobiol A- Chem 202:39–47
14. Hambrock J, Becker R, Birkner A, Weiß J, Fischer RA (2002) A non-aqueous organometallic route to highly monodispersed copper nanoparticles using $[Cu(OCH(Me)CH_2NME_2)_2]$. Chem Commun 1:68–69
15. Park BK, Jeong S, Kim D, Moon J, Lim S, Kim JS (2007) Synthesis and size control of monodisperse copper nanoparticles by polyol method. J Colloid Interface Sci 311:417
16. Balamurugan B, Mehta BR (2001) Optical and structural properties of nanocrystalline copper oxide thin films prepared by activated reactive evaporation. Thin Solid Films 396:90–96

Chapter 34
Study of Trap Density Effect on Current Voltage Characteristics of SubPc and C_{60} Organic Semiconductors for Photovoltaic Application

Mebarka Daoudi, Nesrine Mendil, Zakarya Berkai and Abderrahmane Belghachi

Abstract Our work is focussed on the effect of trap density on the current density–voltage characterization of organic active layers chloro(subphthalocyaninato)boron (SubPc) or fullerene (C_{60}), in which SubPc and C_{60} were applied as the donor and acceptor materials, respectively. The results show that at high voltage (>0.17 V) the trap density affected J–V dramatically. Our results are in good agreement with those by Holmes et al. (Adv Funct Mater 22:617–624, 2012)

Keywords SubPc and C_{60} organic semiconductors · Carrier mobility · Charge carrier current density · Trap density

34.1 Introduction

Most of the power generated nowadays is produced using fossil fuels, which emit tons of carbon dioxide and other pollutants every second [1]. In order to make the development of our civilization sustainable and cause less harm to our environment, scientists are looking for new sources of substitute clean energy. As the photovoltaic (PV) solar energy industry is one of the fastest growing forces in the market, the research for new technology which uses cheap PV materials as well as fabrication methods to collect solar energy becomes increasingly important. In 1986, the first organic thin-film solar cell with reasonable efficiency (up to 1 %) was created and

---

M. Daoudi (✉) · N. Mendil · Z. Berkai · A. Belghachi
Laboratory of semiconductor devices physics, (LPDS) University of Bechar, PB n°417, 08000 Bechar, Algeria
e-mail: mebarkadaoudi@yahoo.fr; lpds2000@gmail.com

N. Mendil
e-mail: nesrine_mendil@yahoo.fr

Z. Berkai
e-mail: Berakai_zakarya@yahoo.com

A. Belghachi
e-mail: abelghachi@yahoo.fr

A. Sayigh (ed.), *Renewable Energy in the Service of Mankind Vol II*,
DOI 10.1007/978-3-319-18215-5_34

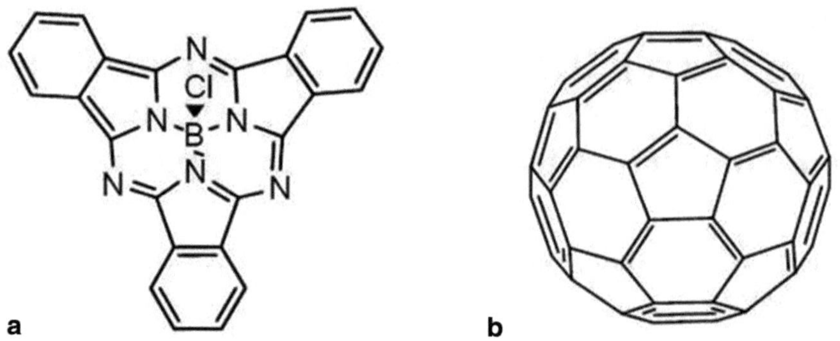

Fig. 34.1 Molecular structures of **a** SubPc and **b** C_{60} [2]

reported by Tang [1]. Since then, great interest has been devoted to the advantages of using organic materials in the solar cell field. Organic photovoltaic (OPV) cells based on donor–acceptor thin films have been demonstrated as the most promising systems for solar energy conversion. Chloro(subphthalocyaninato)boron (SubPc) and fullerene (C_{60}) are the most frequently used donor and acceptor materials, respectively [2] (Fig. 34.1).

In contrast to the rapid progress made in device performance, there is limited research on the charge transport properties with or without light interaction, which is mainly due to the complexity of the disorder nature in organic semiconductors [3]. In this study, we investigate the charge carrier transport properties in SubPc and C_{60} films by analysing the current density–voltage (J–V) characteristics in darkness for neat films at various trap densities.

34.2 Theoretical Model

We investigated the transport properties of charge carriers in organic material properties using a mathematical simulation taking into account the effect of trap density. In this context, the current density–voltage (J–V) characteristics of metal/(SubPc or C_{60})/metal diodes could be helpful in understanding and optimization of the active layer for organic solar cells. Carrier current density investigations of neat films have been studied by simulating hole-only and electron-only diodes using a model of space-charge-limited (SCL) current. The SCL current density *(J)* can be written as a function of the applied voltage *(V)* as [4–6]:

$$J = \frac{9}{8} \varepsilon \mu_{\mathrm{eff}} \frac{V^2}{d^3} \tag{34.1}$$

where ε is the dielectric constant of the organic thin film, d is the organic film thickness, and μ_{eff} is the effective charge carrier mobility, where it is calculated using the Gaussian disorder model governed by the width of Gaussian density of states σ in the following equation [7–10]:

Table 34.1 Parameters of SubPc and C_{60} single-carrier devices

Parameters	Symbol	Values (SubPc/ C_{60})	References
HOMO[a] (eV)	E_v	5.6/6.2	[11]
LUMO[b] (eV)	E_c	3.6/3.6	[11]
Trap band (eV)	E_t	σ^2/K_BT	
Effective density of state (cm^{-3})	N_c/N_v	10^{21}	[12]
Trap density of state (cm^{-3})	N_t	10^{17}	
Infinity mobility (cm^2/Vs)	μ_∞	$1.5\times10^{-7}/0.25$	
Dielectric constant	E	4/3.5	[13]
Gaussian density width (eV)	σ	0.03/0.06	[13]
Boltzmann constant (eV/K)	k_B	8.62×10^{-5}	
Inter-site spacing (cm)	a	$12\times10^{-8}/14.2\times10^{-8}$	[14,15]

[a] Highest occupied molecular orbital
[b] Low unoccupied molecular orbital

$$\mu_{eff} = \vartheta\left[\mu_\infty exp\left(-\left(\frac{3\sigma}{5k_BT}\right)^2 + 0.78\left(\left(\frac{\sigma}{k_BT}\right)^{1.5} - 2\right)\sqrt{\frac{eaV}{\sigma d}}\right)\right] \tag{34.2}$$

where

$$\vartheta = \frac{N_c}{N_t} exp\left[-\frac{(E_C - E_t)}{k_BT}\right] \tag{34.3}$$

where μ_∞ is the mobility as the temperature goes to infinity, a is the inter-site spacing k_B is the Boltzmann constant. N_c is the density of states, N_t the density of shallow trap and E_C and E_t are the localized states and specific energy of trap, respectively. All the parameters used in Eqs. (34.1)–(34.3) are shown in Table 34.1.

34.3 Results and Discussions

The simulation results of charge carrier current density of the molecular hole conductor SubPc and the electron conductor C_{60} are presented in the Figs. 34.2 and 34.3 as a function of applied voltage compared with those extracted from the literature. In this case, we note that the trap effect is neglected.

We observe that below a critical voltage (applied voltage <0.17V) the current density varies linearly, which corresponds to an ohmic regime. Above the critical voltage, the current density strongly increases which is characteristic for SCL. The transition between the two different regimes is sharp for trap levels located in a single energy level. In addition, it is possible to achieve a good fit between experimental and theoretical values.

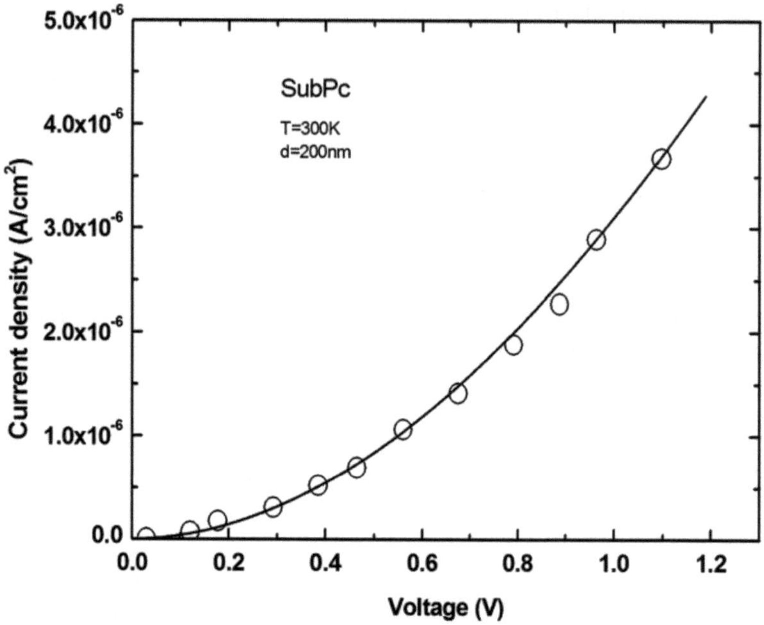

Fig. 34.2 *J–V* characteristics for SubPc single-carrier diode; the *circles* are experimental results extracted from Ref. [1]

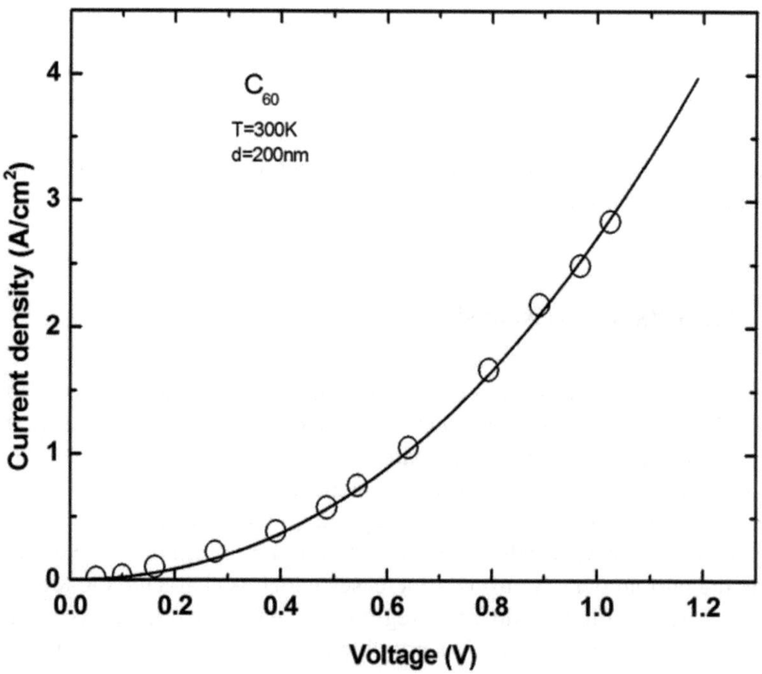

Fig. 34.3 *J–V* characteristics for C_{60} single carrier-diode; the *circles* are experimental results extracted from Ref. [1]

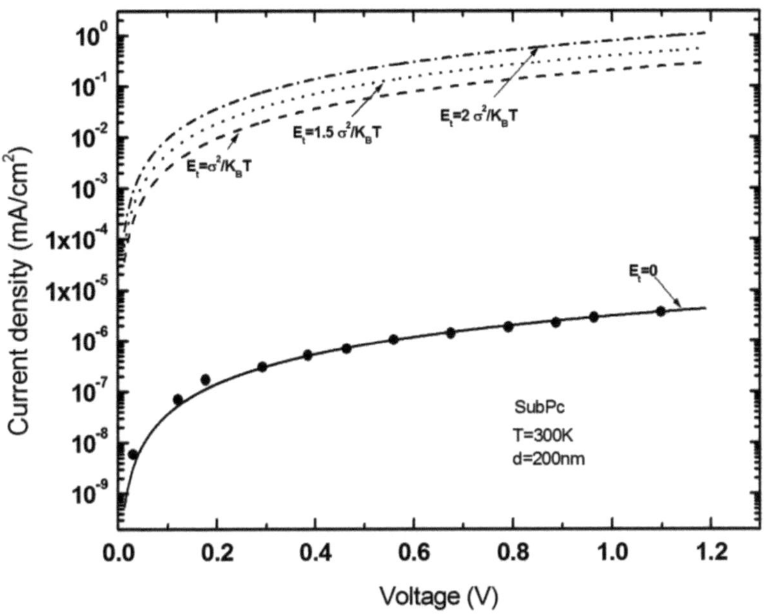

Fig. 34.4 *J–V* characteristics for SubPc single-carrier diodes for different values of trap energy; the *circles* are experimental results extracted from Ref. [1]

Impurities can have a major impact on transport properties by creating regions with different energetic levels. Such energetic levels act as traps since they represent a situation which is more thermodynamically stable. For the charge carrier, the lifetime of a hole or electron in a trap state is a function of the trap depth. We can distinguish shallow traps with a depth in the order of K_BT and deep traps with depths much larger than the energy [12]. In this study, a simple model incorporating deep traps is presented to understand this behavior and its impact on single-layer diodes. The characteristic current density–voltage as a function of the trap depth is shown in Figs. 34.4 and 34.5 which is appropriate for SubPc and C_{60} organic single crystals, respectively.

Traps play an important role in the electron and hole transport process of organic disorder materials. We find that the hole transport in SubPc with trap energy is higher than that of SubPc without of trap energy as the J–V characteristics show in the Fig. 34.4. However, in Fig. 34.5 the increasing of the trap energy results in a higher hopping barrier, which is responsible for the lower electron mobility in the C_{60} thin film. The electron transport initially increases from low trap energy and then decreases at higher trap energy ($>2\sigma^2/KT$). In high trap energy, the Fermi level moves toward the center of the DOS distribution and will lead to further decreases in carrier mobility.

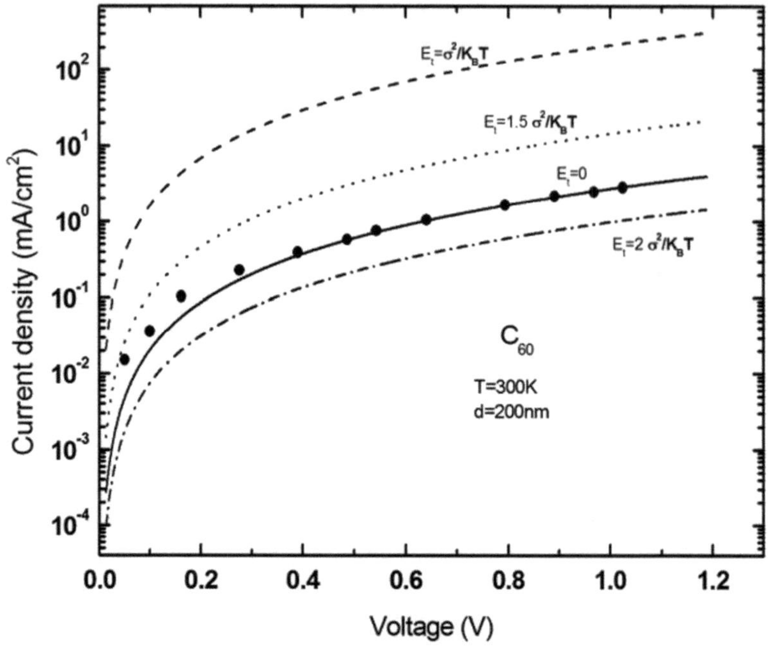

Fig. 34.5 *J–V* characteristics for C_{60} single-carrier diodes for different values of trap energy; the *circles* are experimental results extracted from Ref. [1]

34.4 Conclusion

Organic thin-film solar cells are very attractive for flat-plate terrestrial applications because they can be fabricated using potentially low-cost methods, which can process large-area substrates in an economic way. In this study, we have rigorously explored the electrical J–V characteristic of neat films of SubPc and C_{60} organic semiconductors for PV application. For the two layers, the J–V decreases with increasing trap energy. It is apparent that the charge transport in pure organic SubPc or C_{60} is limited by trap density and scattering the charge carriers. The charge carriers are multiply trapped by and thermally released from trap states associated with the defects.

Acknowledgments This work is partly supported by Algerian ministry of higher education and research (CNEPRU) project N° D03820140006.

References

1. Holmes J, Pandey R, Gunawan AA, Andre Mkhoyan K (2012) Efficient organic photovoltaic cells based on nanocrystalline mixtures of boron subphthalocyanine chloride and C_{60}. Adv Funct Mater 22:617–624
2. Jason UW (2009) Carrier mobility in organic charge transport materials: methods of measurement, analysis, and modulation. PhD thesis, University of Rochester Rochester, New York
3. Junbo Wu (March 2011) Organic photovolatic cells and grapheme transparent conductors. PhD thesis, Stanford University
4. Gommans H, Verreet B, Rand BP, Muller R, Poortmans J, Heremans P, Genoe J (2008) On the role of bathocuproine in organic photovoltaic cells. Adv Funct Mater 18:3686–3691
5. Kang M-G, Xu T, Park HJ, Luo X, Jay Guo L (2010) Efficiency enhancement of organic solar cells using transparent plasmonic Ag nanowire electrodes. Adv Mater 22:4378–4383
6. Opitz A, Bronner M, Bru¨tting W (2008) Charge carrier injection and ambipolar transport in C_{60}/CuPc organic semiconductor blends. J Phys Conf Ser 100:082043
7. Bru¨tting W, Bronner M, Gotzenbrugger M, Opitz A (2008) Ambipolar blends of Cu-phthalocyanine and fullerene: charge carrier mobility, electronic structure and their implications for solar cell applications. Macromol Symp 268:38–42
8. Schon JH, Batlogg B (January 1999) Modeling of the temperature dependence of the field-effect mobility in thin film devices of conjugated oligomers. Appl Phys Lett 74(2):11
9. Letizia JA, Rivnay J, Facchetti A, Ratner MA, Marks TJ (2010) Variable temperature mobility analysis of n-channel, p-channel, and ambipolar organic field-effect transistors. Adv Funct Mater 20:50–58
10. Coropceanu V, Cornil J, da Silva Filho DA, Olivier Y, Silbey R, Brédas J-L (2007) Charge transport in organic semiconductors. Chem Rev 107:926–952
11. Yang F (September 2008) Thin film solar cells grown by organic vapor phase deposition. PhD thesis, Princeton University
12. Wang JC et al (2011) Charge accumulation induced S-shape J–V curves in bilayer heterojunction organic solar cells. Organic electronics 12:880–885
13. Yu X (April 2012) Monte Carlo simulation of charge transport in organic solar cells. Thesis, Duke University
14. Pevzner B (December 1995) Transport and dielectric properties of thin fullerene (C_{60}) films. Thesis, Massachusetts Institute of Technology
15. Morse GE (2012) Engineering boronsubphthalocyanine for organic electronic applications, University of Toronto

Chapter 35
PV Cell and Module Degradation, Detection and Diagnostics

Eleni Kaplani

Abstract With crystalline silicon photovoltaic (PV) modules being in the market for over three decades, investigation into usual causes and extent of module degradation after prolonged exposure in field conditions is now possible. Degradation phenomena vary significantly between cells, modules and installations, giving rise to different power degradation rates reported. The main defects observed in the field-aged PV modules include EVA browning, degradation of the antireflective coating, delamination between the glass-encapsulant and the cell-encapsulant interfaces, humidity ingress, corrosion of busbars and contacts, shunt paths, cracks/micro-cracks in the cell, damage of the glass and the back sealing and bypass diode failure. This study presents severe degradation effects observed in PV modules operating outdoors for over 20 years. In many of the cases investigated, different defects were found to coexist within the same cell or module, leading to more severe effects of optical/physical, thermal and electrical degradation phenomena, significantly reducing the PV power output. Other modules that exhibited extensive optical/physical degradation showed milder degradation in performance. Detection of module degradation was carried out in this study first through visual inspection and I-V curve analysis. Further, nondestructive diagnostic techniques were used such as infrared thermography for the identification of hot spots and these were seen to be mainly linked to resistive busbars and contacts and electroluminescence imaging for the identification of shunts and other defects. The detection, diagnosis and monitoring of such defects is of great importance for a deeper understanding of the complex ageing mechanisms that take place after prolonged PV exposure in field conditions, and the identification of underlying causes, assisting the early identification of defects and the extension of the energy life of PV systems.

Keywords PV degradation · Diagnostics · Infrared thermography · Electroluminescence

---

E. Kaplani (✉)
RES Lab, Mechanical Engineering Department, Technological Educational Institute of Western Greece, Meg. Alexandrou 1, 26334 Patra, Greece
e-mail: ekaplani@teiwest.gr

A. Sayigh (ed.), *Renewable Energy in the Service of Mankind Vol II*,
DOI 10.1007/978-3-319-18215-5_35

35.1 Introduction

A deeper understanding of degradation mechanisms that take place from the very early to the late stages of PV module life during outdoor operation is essential for a more accurate estimation of module lifetime and prognosis of possible future faults. The PV module qualification tests, as for instance the IEC 61215 for crystalline silicon PV modules, are based on accelerated stress tests such as thermal cycling, damp heat test, humidity-freeze cycling, UV test, static and dynamic mechanical load test, hot spot test, etc., in an attempt to replicate failure modes that may be observed in the field. These tests identify design and material faults that would lead to premature field failure, but do not address failure mechanisms or quantify lifetime for the intended application, system configuration and climate [1]. Thus, studies on PV module operation in field conditions for many years can provide valuable information on the reliability and durability of PV modules in real life. A review of PV degradation rates reported in the literature from field testing gave a skewed distribution with a mean degradation of 0.8 % per year and a median of 0.5 % per year, with 78 % of the reported data having a degradation rate below 1 % per year [2]. Initial rapid light-induced degradation accounts for 1–5 % loss in short-circuit current [3].

In this study, degradation phenomena observed in PV modules operating in the field for up to 25 years are presented. Nondestructive diagnostic techniques including I-V curve analysis, infrared (IR) thermography and electroluminescence (EL) are employed here to assess PV performance and identify the defects. Modules that have undergone different stages of degradation, from initial rapid light-induced degradation to mild and severe degradation phenomena, are analysed, and degradation estimates are given for several of these modules tested, revealing the need for a deeper understanding of PV degradation phenomena that occur under real conditions of operation.

35.2 Diagnostic Techniques

Visual inspection is a simple and significant procedure for the identification of defects and early signs of module failure mechanisms. A close examination of PV modules can reveal early signs of browning of the ethylene-vinyl-acetate (EVA) encapsulant, degradation of the antireflective (AR) coating, delamination, cracks in the cells, burn marks, etc. Illumination of the cells with UV light at 375 nm can assist the identification of EVA browning [4]. Figure 35.1 shows the digital image of a cell with mild browning of the EVA and the UV-illuminated image where the EVA browning is clearly revealed, making possible the assessment and monitoring of the extent of the browning. This may also be achieved through digital image processing [5].

IR thermography is a powerful tool used to locate defected cells or cell areas that exhibit higher or lower localised temperature than the remaining part of the module or cell, respectively. Figure 35.2a shows the IR image of several modules of a PV generator operating in field conditions for 20 years. The IR image is captured using

Fig. 35.1 **a** Digital image of PV cell with mild EVA browning, **b** the same cell illuminated with UV light. The module operates outdoors in the RES Lab for the past 20 years

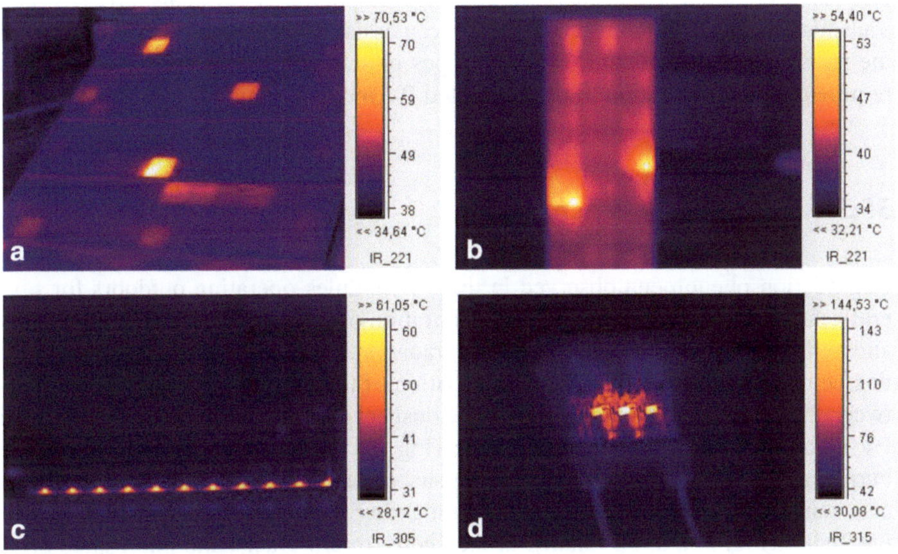

Fig. 35.2 IR thermography of **a** an sc-Si PV generator operating at load in the RES Lab for 20 years, **b** an sc-Si PV module forward biased in the dark within the lab, **c** a flexible a-Si PV module reverse biased in the lab, **d** the junction box of a pc-Si PV module reverse biased in the lab

an IR camera with spectral range of 7.5–14 μm. Hot cells operating at temperatures of 20–30 °C higher than the average temperature of the remaining part of the module are visible. Such large temperature differences signify defected cells. IR thermography may also be recorded in the dark with the module in forward bias mode using a DC power supply. Figure 35.2b shows an sc-Si PV module with 20 year outdoor operation tested in the lab. The IR image reveals two cells having a hot spot at the busbar. By reverse biasing the module in the dark, it is possible to check diode functionality. Figure 35.2c shows the IR image of a flexible a-Si PV module with bypass diodes connected across each of the 11 cells, and Fig. 35.2d shows the junction box of a pc-Si reverse-biased PV module showing the operation of the three diodes.

EL imaging is a sensitive technique for deeper investigation into failure causes, assisting in the identification of cracks, broken metallization and regions of higher resistance and shunts [6]. It is carried out in the dark with the PV module in forward-bias mode, and is based on the light emission of the PV module at around 1150 nm as a result of the radiative recombination of carriers. A near infrared (NIR) sensitive Si-CCD or a shortwave infrared (SWIR) camera is required. An SWIR camera with InGaAs sensor and a spectral range from 0.9 to 1.7 μm was used in this study for capturing the EL image. Photoluminescence (PL) is a contactless diagnostic technique with a module at open-circuit mode which involves SWIR image capture. It is based on the PL emission due to radiative recombination after optical excitation with light at 800 nm. Additional information may be obtained through separation of band-to-band emission images at around 1150 nm from the defect-band emission images at around 1550 nm using cut-off filters [7].

Microscopic techniques such as scanning electron microscopy (SEM), including electron beam induced current (EBIC), energy-dispersive X-ray analysis (EDX) and transmission electron microscopy (TEM) can assist in the in-depth investigation of the physical defects, such as different types of shunts [8]. Destructive techniques require a proper dissection of the laminated PV module, which is a difficult task.

35.3 Degradation Phenomena

Degradation phenomena observed in the PV modules operating outdoors for several years in the RES Lab include discolouration of the EVA with different patterns and yellowness index (Fig. 35.3a–d), degradation of the AR coating with different degree of severity (Fig. 35.3b–d), degradation of the cell-encapsulant interface between the gridlines in the proximity of the busbars (Fig. 35.3b) and in a flexible a-Si PV module in the form of delamination (Fig. 35.3f). Furthermore, corrosion and burn marks were observed mainly on the busbars and cell interconnections, usually also visible at the backside of the module together with bubbles, cracks and tears of the Tedlar (Fig. 35.3d, e). Finally, broken cells, micro-cracks and interrupted gridlines (Fig. 35.3g h) were also observed.

35.4 PV Degradation and Diagnosis

Several modules that have been operating outdoors in the RES Lab from only several hours up to 25 years have been examined. Defect detection and diagnosis has been carried out via means of visual inspection, digital image with UV light illumination, I-V curve analysis, IR thermography and EL imaging.

Fig. 35.3 Degradation phenomena observed in PV modules operating outdoors in the RES Lab, **a** EVA browning, **b** deterioration of the cell-encapsulant interface and other effects, **c** degradation of the AR coating and other effects, **d** corrosion, burn mark, EVA browning and other effects, **e** corrosion, burn marks, tear and bubble at the back of the module, **f** delamination in a flexible a-Si module, **g** broken cell, **h** broken gridlines

The visual inspection of a new sc-Si PV module that has operated outdoors only for several hundred hours and has experienced initial rapid light-induced degradation has revealed slight deterioration of the cell-encapsulant interface at the proximity of the busbar (Fig. 35.4a). EL with the module forward biased at near V_{oc} revealed a crack in one cell with a large region appearing dark in the EL image (Fig. 35.4b). The EL image of part of the module reveals several grid finger interruptions and a rather inhomogeneous EL emission in some cells (Fig. 35.5a). The IR thermography revealed some cells exhibiting about 10 °C higher temperature than the average temperature of the module, while some correlations between the IR image and EL image were observed (Fig. 35.5b). I-V curve analysis and translation to standard test conditions (STC) gave a 3.7 % reduction in power output.

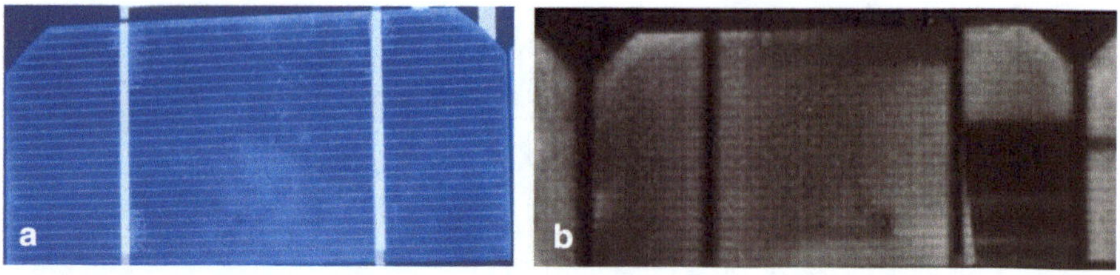

Fig. 35.4 A new sc-Si PV module with only several hundred hours of outdoor operation in the RES Lab. **a** Digital image of a cell exhibiting slight deterioration in the cell-encapsulant interface, **b** EL image of the same cell revealing a crack

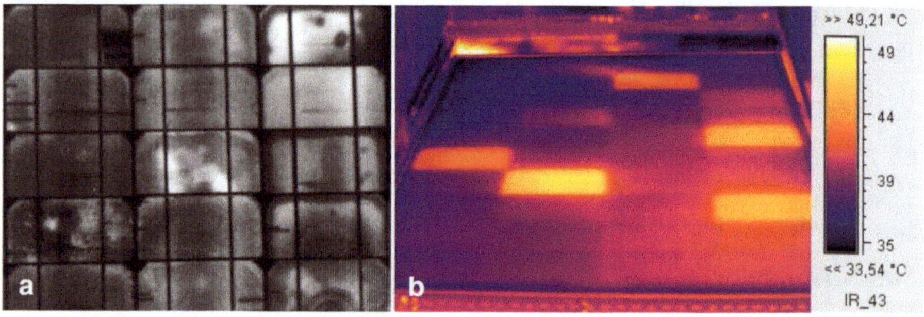

Fig. 35.5 Image from part of a new sc-Si PV module, with only several hundred hours of outdoor operation in the RES Lab. **a** EL image **b** IR image

Fig. 35.6 sc-Si PV module operating outdoors in the RES Lab for 15 years exhibiting mild signs of degradation of the **a** AR coating, **b** the cell-encapsulant interface

An sc-Si PV module operating outdoors for 15 years exhibits mild signs of AR deterioration and degradation in the cell-encapsulant interface at the proximity of the busbars, see Fig. 35.6. IR thermography shows a rather uniform temperature distribution of the module with only one cell exhibiting higher temperature by about

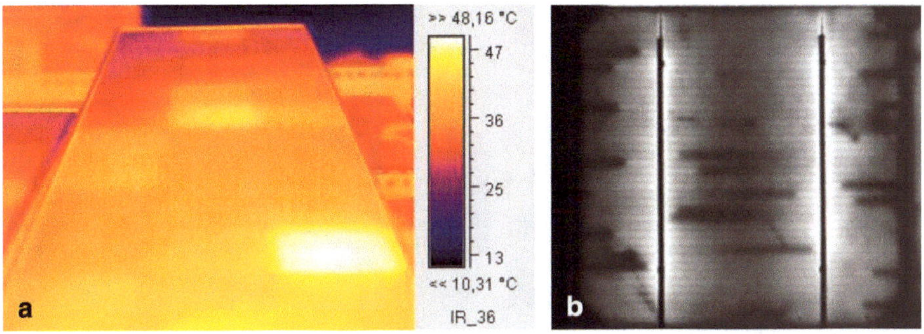

Fig. 35.7 sc-Si PV module operating outdoors in the RES Lab for 15 years exhibiting mild signs of degradation. **a** IR image **b** EL image of one of the cells

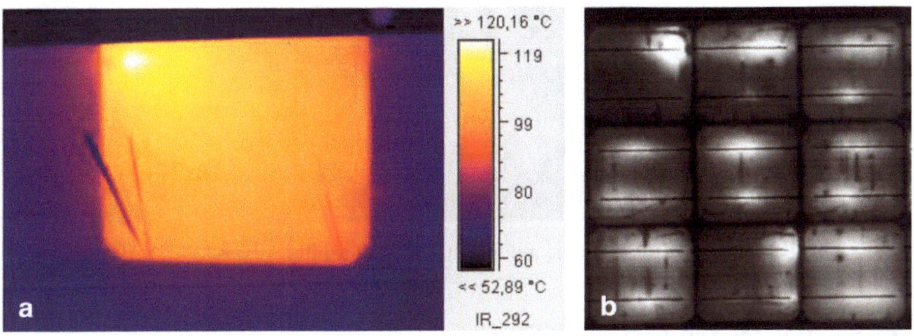

Fig. 35.8 sc-Si PV module operating outdoors for 20 years, **a** IR image at the backside of the module at the hot cell, **b** EL image of part of the module

10 °C from the average temperature of the module (Fig. 35.7a). EL imaging further reveals cells with several interrupted gridlines (Fig. 35.7b).

Several sc-Si PV modules operating in the RES Lab for 20 years have experienced degradation through induced shading effects for a prolonged period of several months [5]. These modules now exhibit severe optical and electrical degradation phenomena mainly associated to the affected cells [9]. As an example, the IR image of one of these modules reveals a hot cell with a hot spot on the busbar exhibiting a temperature difference greater than 40 °C from the average temperature of the module (Fig. 35.8a). The cell and spot temperature exceeds 100 °C, which is detrimental for the operational life of the cell and module. The EL image of part of this module (Fig. 35.8b) reveals inhomogeneous EL emission, several interrupted gridlines, shunts and micro-cracks.

Other sc-Si PV modules operating outdoors in the RES Lab for 24 and 25 years, and have aged naturally, exhibit a rather uniform temperature distribution (Fig. 35.9a). Further investigation through EL reveals the existence of cracks in several cells, some of which isolate a part of the cell, inactive areas and inhomogeneous EL emission (Figs. 35.9b and 35.10).

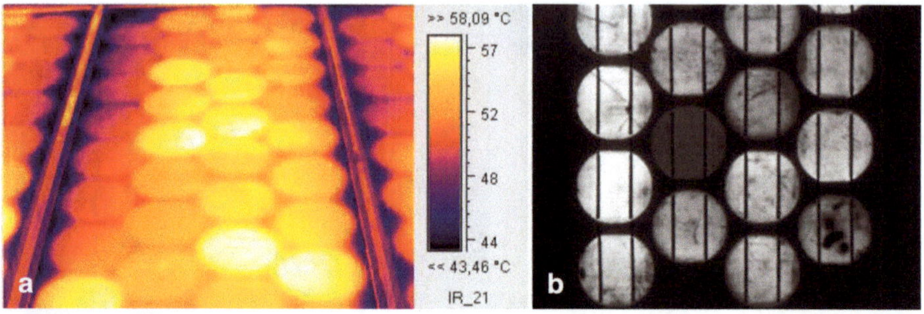

Fig. 35.9 sc-Si PV module operating outdoors in the RES Lab for 25 years—naturally aged, **a** IR image, **b** EL image

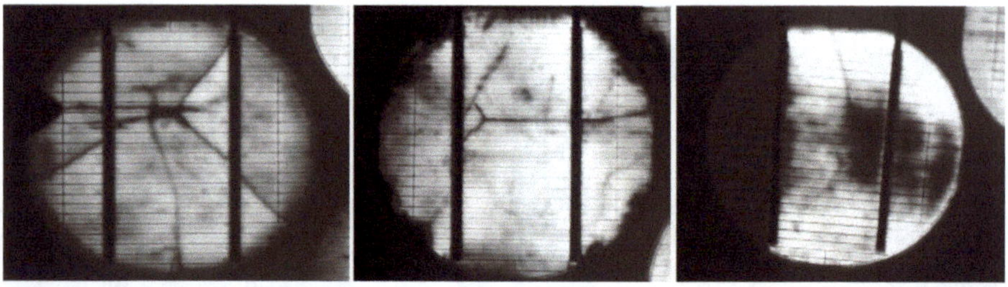

Fig. 35.10 EL image of cells from the module in Fig. 35.9 exhibiting cracks

35.5 PV Degradation Estimates

An estimate of the degradation of the 20 and 25-year-old PV modules experiencing induced and natural ageing respectively is carried out through I-V curve analysis. The STC normalised I-V curves of the modules that experienced induced shading effects in the past are displayed in Fig. 35.11a. Large deviations in the I-V curve characteristics are observed in these modules, some of which reveal the existence of some optically degraded cells, while others reveal an increased series resistance due to corrosion, as regarded in the slope of the curve at open-circuit voltage. The relative reduction in power output ranges between 25 and 41 %, in the short-circuit current 6–15 %, in the open-circuit voltage 3–5 % and in the fill factor 12–34 %. On the other hand, the naturally aged 24–25 year-old modules have non-deformed I-V curve characteristics, with relative reduction in the fill factor of only 4–5 %. The relative reduction in power output ranges between 20 and 25 %, which is mostly attributed to the short-circuit current with a relative reduction of 13–15 %. This is not only partly due to the optical losses caused by the severe EVA browning that affects all cells of the module (Fig. 35.3a) but is also due to the several cracks and inactive areas in the cells revealed by the EL image. The reduction in the open-circuit voltage is in the range of 5–7 %.

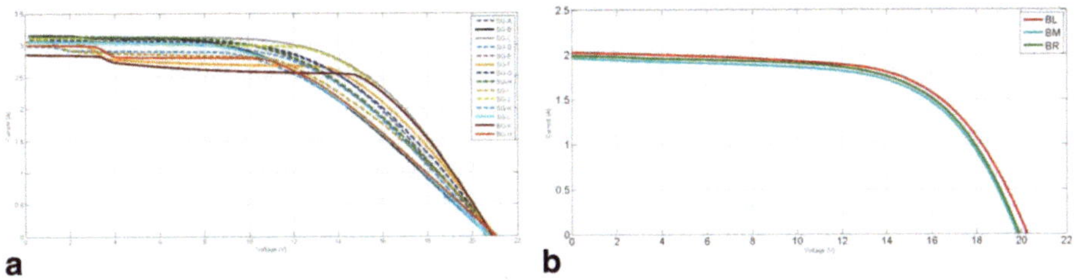

Fig. 35.11 STC normalized I-V curves. **a** 20-year-old modules with induced shading effects, **b** 24–25 year-old modules with natural ageing

35.6 Conclusions

The ageing of PV modules is a complex process that may involve a cascade of events based on different effects caused by a chain of chemical reactions. It was shown to significantly differ between modules and cells of the same module. Several modules operating outdoors from several hours to as many as 25 years have been examined and defect detection and diagnosis was carried out through visual inspection, also with the assistance of UV light illumination, I-V curve analysis, IR thermography and EL. A combination of all these techniques was found important for an in-depth analysis of PV degradation and underlying defects. Furthermore, EL imaging gave significant insights to the actual physical condition of the modules, especially in cases where I-V curve analysis and IR thermography did not show abnormal behaviour due to the uniformity of the effects within the module. IR thermography and EL imaging can serve as important tools not only for diagnosis but also for the prognosis of possible future defects in the modules.

References

1. Wohlgemuth J (2012) IEC 61215: What it is and isn't. 2012 PV Module Reliability Workshop. NREL/PR-5200-54714
2. Jordan DC, Kurtz SR (2012) Photovoltaic degradation rates—an analytical review. NREL/JA-5200-51664
3. Sharma V, Chandel SS (2013) Performance and degradation analysis for long term reliability of solar photovoltaic systems: a review. Renew Sustain Energy Rev 27:753–767
4. Meyer EL, van Dyk EE (2004) Assessing the reliability and degradation of photovoltaic module performance parameters. IEEE Transact Reliab 53(1):83–92
5. Kaplani E (2012) Detection of degradation effects in field-aged c-Si solar cells through IR thermography and digital image processing. Intl J Photoenergy 2012:1–11. doi:10.1155/2012/396792. (Art. no. 396792)
6. Sander M, Henke B, Schweizer S, Ebert M, Bagdahn J (2010) PV module defect detection by combination of mechanical and electrical analysis methods. Proceedings of theIEEE 35th Photovoltaic Specialists Conference, 20–25 June 2010, Honolulu

7. Yan F, Johnston S, Al-Jassim M, Zaunbrecher K, Sidelkheir O, Blosse A (2011) Defect-band emission photoluminescence imaging on multi-crystalline Si solar cells. Proceedings of the 37th IEEE Photovoltaic Specialists Conference, June 19–24 2011, Seattle, Washington. NREL/CP-5200-50725
8. Breitenstein O, Rakotoniaina JP, Al Rifai MH, Werner M (2004) Shunt types in crystalline silicon solar cells. Prog Photovolt: Res Appl 12:529–538
9. Kaplani E (2012) Degradation effects in sc-Si PV modules subjected to natural and induced ageing after several years of field operation. J Eng Sci Technol Rev 5(4):18–23

Chapter 36
Practical Identification of the Photovoltaic Module Parameters

E. T. El Shenawy, O. N. A. Esmail, Adel A. Elbaset and Hisham F. A. Hamed

Abstract As the amount of the energy radiated to the earth by the sun exceeds the annual energy requirement of the world population, making use of this inexhaustible energy source for our everyday electricity requirement is the great challenge of the present and the future. The generation of electric energy from the sun via photovoltaic (PV) generation is one of the most appropriate candidates in this manner. For PV applications, the accurate simulation of the PV module performance, which is an important point in the accurate design of these systems, depends mainly on the electrical parameters of the PV module (series resistance, reverse saturation current, short-circuit current temperature coefficient, and ideality factor). Also, the knowing of the effect of the ageing on the PV module performance after a certain period requires an accurate identification of these parameters. This chapter presents accurate methods to get the PV module electrical parameters depending on practical measurements for its current–voltage (I–V) characteristics under different operating conditions (different solar radiation intensity and module temperature) using an accurate data acquisition system. Then, the chapter compares the theoretical performance of the PV module using the identified parameters with the measured curves to arrive at the accurate determination of the PV module parameters.

Keywords PV module parameters · PV measurements · Photovoltaic simulation

E. T. E. Shenawy (✉)
Solar Energy Department, National Research Centre, El Behooth St., Dokki, Giza, Egypt
e-mail: essamahame@hotmail.com

O. N. A. Esmail
Faculty of Engineering, Al Azhar University, Qena, Egypt

A. A. Elbaset · H. F. A. Hamed
Faculty of Engineering, Minia University, Minia, Egypt

A. Sayigh (ed.), *Renewable Energy in the Service of Mankind Vol II*,
DOI 10.1007/978-3-319-18215-5_36

36.1 Introduction

The strong demand for alternatives to fossil fuel-based energy sources and growing environmental concerns have increased interest in solar (photovoltaic, PV) cells as a long-term, exhaustless, environmentally friendly, and reliable energy technology. Continuous efforts to develop various types of solar cells are being made in order to produce solar cells with improved efficiencies at a lower cost, thereby, taking advantage of the vast amounts of free energy available from the sun. Due to rising costs of conventional energy and their limited resources, PV energy becomes a promising energy for the future of the world [1]. The PV system is a generating system using the photoelectric effect that changes light energy into electric energy. Also, the performance of PV system is dependent on radiation and temperature [2].

Most solar cell parameters can be obtained from simple current–voltage (I–V) measurements. Figure 36.1 shows the I–V characteristics of a typical solar cell under forward bias and illumination. The short-circuit current (I_{sc}) is the current through the solar cell when the voltage across the solar cell is zero. The open-circuit voltage (V_{oc}) is the voltage across the solar cell when the current through the solar cell is zero and it is the maximum voltage available from the solar cell. The maximum power point (MPP) is the condition under which the solar cell generates its maximum power (P_{max}); the current and voltage in this condition are defined as I_{max} and V_{max}, respectively. The fill factor (FF) and the conversion efficiency *(η)* are metrics used to characterize the performance of the solar cell. The FF is defined as the ratio of P_{max} divided by the product of V_{oc} and I_{sc}. The conversion efficiency is

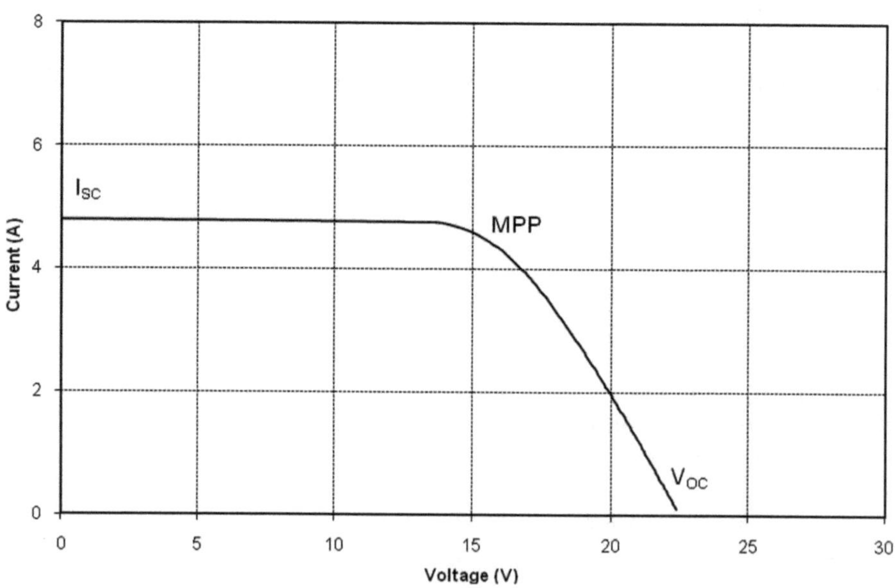

Fig. 36.1 I–V characteristics of a typical solar cell

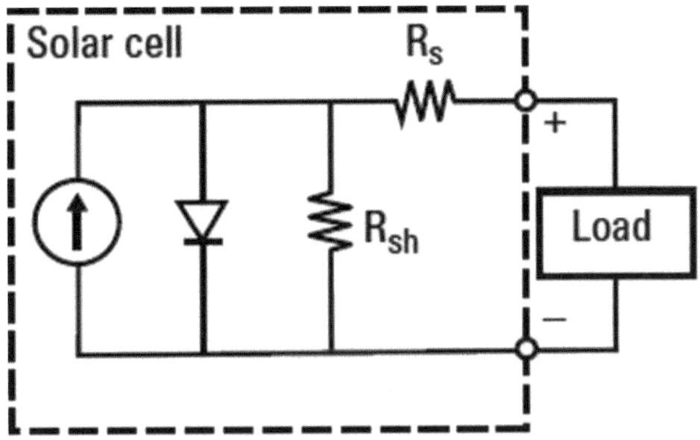

Fig. 36.2 Equivalent circuit of a typical solar cell

defined as the ratio of P_{max} to the product of the input light irradiance and the solar cell surface area. Therefore, it is of prime importance to measure the I–V characteristics with high accuracy under natural environmental conditions [3].

The equivalent circuit used to model the DC behaviour of a typical solar cell is shown in Fig. 36.2. The equivalent circuit, which describes the static behaviour of the solar cell, is commonly composed of a current source, a pn junction diode, and a shunt resistor (R_{sh}) in parallel along with a series resistor (R_s). The current source models electron injection from light. R_s is the total Ohmic resistance of the solar cell, which is essentially the bulk resistance. Smaller R_s values equate to increased solar cell efficiencies. R_{sh} accounts for stray currents, such as recombination currents and leakage currents around the edge of devices. In this case a larger R_{sh} value equates to increased solar cell efficiency, since it means that the stray currents are reduced.

From the equivalent circuit of the typical solar cell, we can write the cell current as follows:

$$I_c = I_{ph} - I_d - I_{sh} \tag{36.1}$$

$$I_{sh} = \frac{V_c + I_c R_s}{R_{sh}} \tag{36.2}$$

Since the shunt resistance of the PV cell is much greater than its series resistance, the current I_{sh} becomes much smaller compared with the other currents in the cell. Therefore, this current will be neglected as it will not cause a large error in the PV cell simulation model [4]. Then, Eq. (36.1) can be written as follows:

$$I_c = I_{ph} - I_d \tag{36.3}$$

Also, I–V characteristics of a PV cell can be expressed in terms of the cell current and voltage as follows [4]:

$$I_c = I_{ph} - I_o \left\{ e^{\left[\frac{q}{AKT}(V_c + I_c R_s)\right]} - 1 \right\} \tag{36.4}$$

$$V_c = \frac{AKT}{q} \ln\left(\frac{I_{ph} + I_o - I_c}{I_o} \right) - I_c R_s \tag{36.5}$$

The PV cell photoelectric current and reverse saturation current will be calculated, respectively, using the following equations [5]:

$$I_{ph} = [I_{scr} + K_i (T - 298)] \frac{G}{1000} \tag{36.6}$$

$$I_o = I_{or} \left(\frac{T}{T_r} \right)^3 e^{\left[\frac{qE_{Go}}{AK} \left(\frac{1}{T_r} - \frac{1}{T} \right) \right]} \tag{36.7}$$

Since the photovoltaic module consists of a series of connected solar cells, the I–V characteristics of the whole module can be derived by scaling the characteristics of one cell with a factor equal to the number of cells in series. Then Eqs. (36.4) and (36.5) will be modified to the following pattern:

$$I = I_{ph} - I_o \left\{ e^{\left[\frac{q}{n_s AKT}(V + n_s I R_s)\right]} - 1 \right\} \tag{36.8}$$

$$V = \frac{n_s AKT}{q} \ln\left(\frac{I_{ph} + I_o - I}{I_o} \right) - n_s I R_s \tag{36.9}$$

Equations (36.8) and (36.9) simulate the I–V characteristics of the PV module, provided we know the PV module parameters such as series resistance, reverse saturation current, short-circuit current temperature coefficient, and module ideality factor.

Several methods for the determination of photovoltaic module parameters are proposed by several authors [6–9]. Accurate knowledge of the photovoltaic module parameters is essential for the design, quality control of solar modules and for simulation of their performance. These parameters are often determined using experimental data under a given illumination and temperature [10].

The present chapter presents an accurate method to estimate the PV module parameters such as module series resistance, reverse saturation current, short-circuit current temperature coefficient, and ideality factor based on practical measurements of I–V curves for the PV module under different operating conditions.

36.2 Practical Measurements

Practical measurements of the I–V curves of the PV module depend mainly on the data acquisition system (DAS). DAS is used to record the signals from the different sensors that are used for measuring the different physical parameters of the PV system. These measurements include the incident solar radiation on the surface of the PV module, PV module surface temperature, module voltage, and module current. These parameters can be measured and recorded via a PC driven by Analog to Digital (AD) card [11]. Figure 36.3, shows the electronic circuit diagram for the proposed DAS.

As shown in Fig. 36.3, using the DAS, PV module voltage can be measured accurately by using LV 25-P voltage transducer, with galvanic isolation between the primary circuit (high voltage) and the secondary circuit (low voltage), while a current transducer LA 25-NP is used to measure the PV module current. A thermopile pyranometer, mounted at the PV module structure and parallel to the module, of type Kipp and Zonen (model CM5-774035) is used to measure the solar radiation intensity. A type K thermocouple is used to measure the PV module surface temperature.

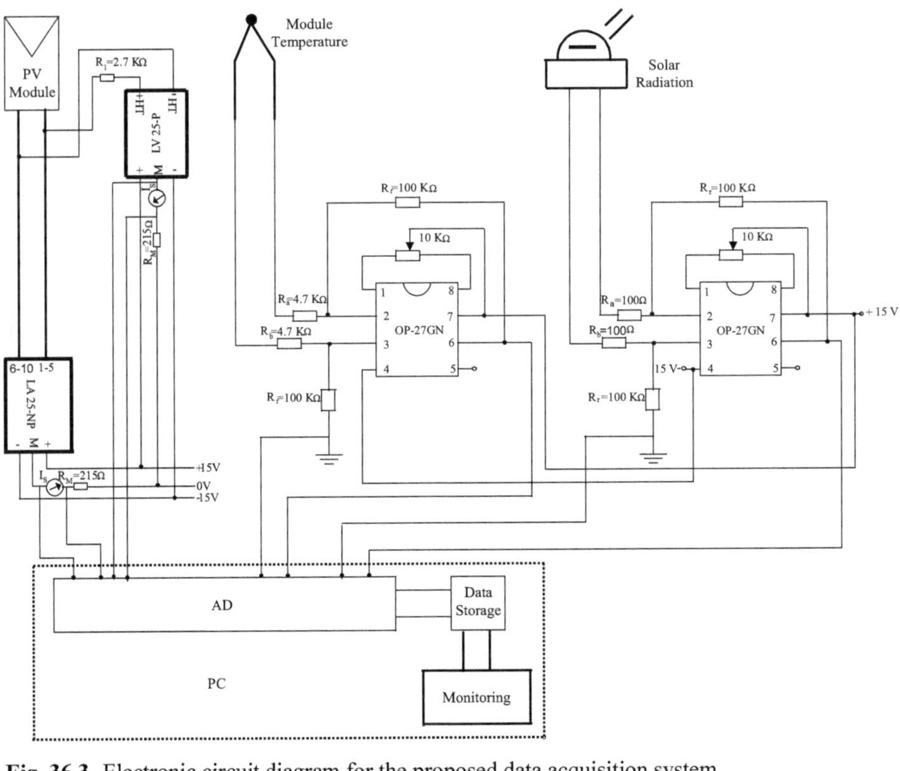

Fig. 36.3 Electronic circuit diagram for the proposed data acquisition system

Table 36.1 PV module characteristics at STC

Maximum power	64 W
Operating temperature	47 °C
Open-circuit voltage	27.1 V
Short-circuit current	4.8 A
Dimensions in cm^3	137 cm × 74.1 cm × 3.2 cm
Weight	9.17 kg

36.3 PV Module

The PV module is a thin film triple junction silicon solar cell technology with maximum output of 64 W (at Standard Test Conditions (STC) 25 °C ambient temperature and 1000 W/m^2 global radiation). The complete specifications of the PV module are listed in Table 36.1.

36.4 Determination of the PV Module Parameters

The PV module parameters that needed to be determined are the module series resistance (R_s), reverse saturation current (I_o), short-circuit current temperature coefficient *(K_i)*, and ideality factor *(A)*. The next sections describe the methods of detecting these parameters.

36.4.1 PV Module Series Resistance

The series resistance modifies slightly the shape of the I–V characteristics of the PV module. It causes a reduction in the slope of the I–V characteristics in the region of the peak power point. For more efficient PV modules, its series resistance must be very low [12].

To obtain more accurate estimates of R_s, a series of forward-biased I–V measurements are made using different values of input light irradiance (see Fig. 36.4). First, a forward-biased I–V measurement under an arbitrary light irradiance is made and a value V_1 that is slightly higher than V_{max} (as shown in Fig. 36.4) is selected. Next a value of ΔI is calculated, where $\Delta I = I_{sc1}(0) - I_{sc1}(V_1)$. This process is then repeated more times using lower values of light irradiance as shown in Fig. 36.4. Finally, R_s is estimated by averaging R_1, R_2, and R_3, as shown below [12]. Figure 36.5 shows the group of practical measured I–V curves used in these calculations.

$$R_s = \frac{R_1 + R_2 + R_3}{3}, \tag{36.10}$$

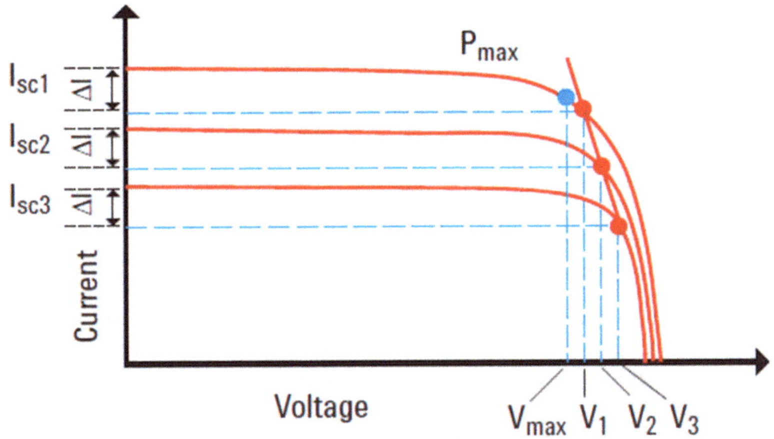

Fig. 36.4 Determination of the PV module series resistance

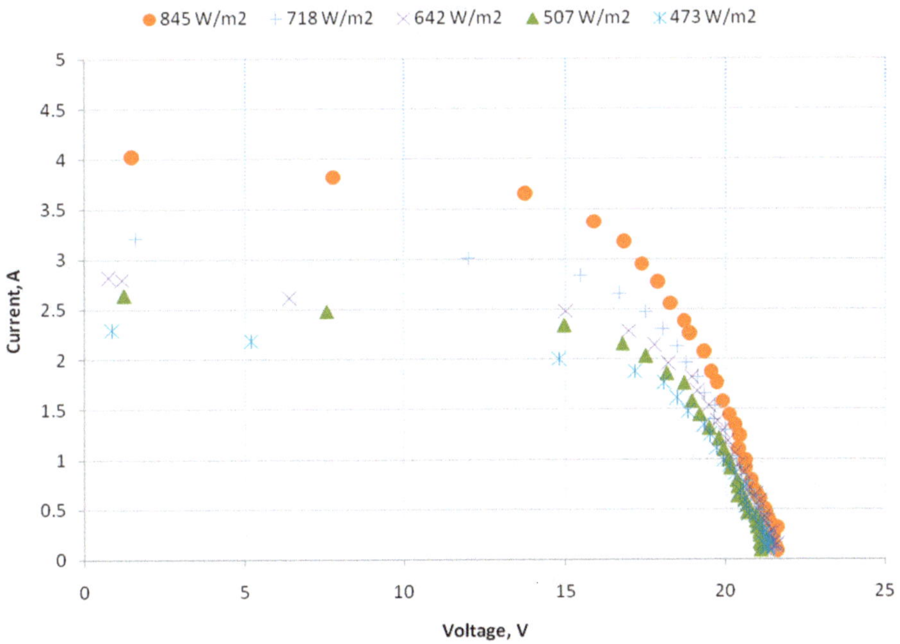

Fig. 36.5 Measured I–V curves of the PV module at different radiation levels (temp. 45 °C)

where

$$R_1 = \frac{V_2 - V_1}{I_{sc1}(V_1) - I_{sc2}(V_2)}, R_2 = \frac{V_3 - V_2}{I_{sc2}(V_2) - I_{sc3}(V_3)}, R_3 = \frac{V_3 - V_1}{I_{sc3}(V_3) - I_{sc1}(V_1)}. \quad (36.11)$$

36.4.2 Other PV Module Parameters

The PV module light-generated current, I_{ph} as calculated in Eq. (36.6) slightly differs from the practical PV module short-circuit current. This difference resulted from the temperature coefficient of the short-circuit current, K_i, as presented in Eqs. (36.8) and (36.9). Also, the I–V characteristics of the PV module presented in Eqs. (36.8) and (36.9) is valid only for a certain solar radiation and module temperature. If the solar radiation or the module temperature is changed, then another I–V characteristic for this module can be drawn. Equations (36.8) and (36.9) contain two important parameters that are PV module saturation current, I_o, and module ideality factor, A. These parameters can be calculated by curve fitting for the measured I–V characteristics of the PV module, as shown in Fig. 36.5, with the theoretical model in Eqs. (36.8) and (36.9), based on the least-square method.

36.5 Test the Estimated PV Module Parameters

The obtained parameters of the PV module from the previous sections are listed in Table 36.2. These parameters will be used in the simulation of the PV module characteristics at different temperature and radiation levels.

Figures 36.6 and 36.7 show the measured and the calculated I–V curves of the PV module at different operating conditions. From the figures, it is clear that there is good agreement between the calculated and the measured values of the PV module characteristics, which indicates the accuracy of a method used to get the PV module parameters based on practical measurements.

Table 36.2 The calculated PV module parameters

Series resistance, R_s	1.055 Ω
Reverse saturation current, I_o	1.9955×10^{-9} A
Short-circuit current temperature coefficient, K_i	0.0014 A/°C
Ideality factor, A	1.48

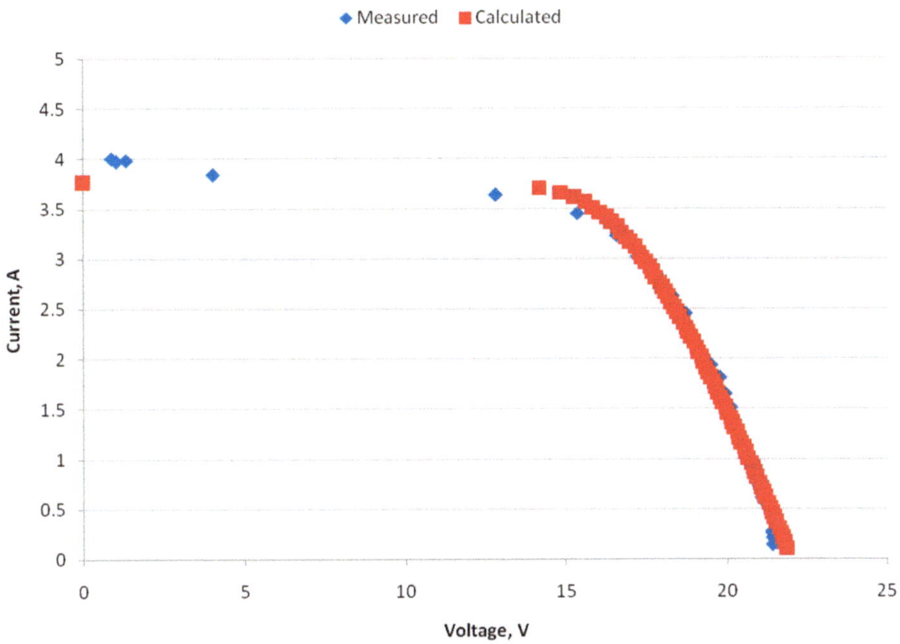

Fig. 36.6 Measured and calculated I–V curves of the PV module at 778 W/m^2, 45 °C

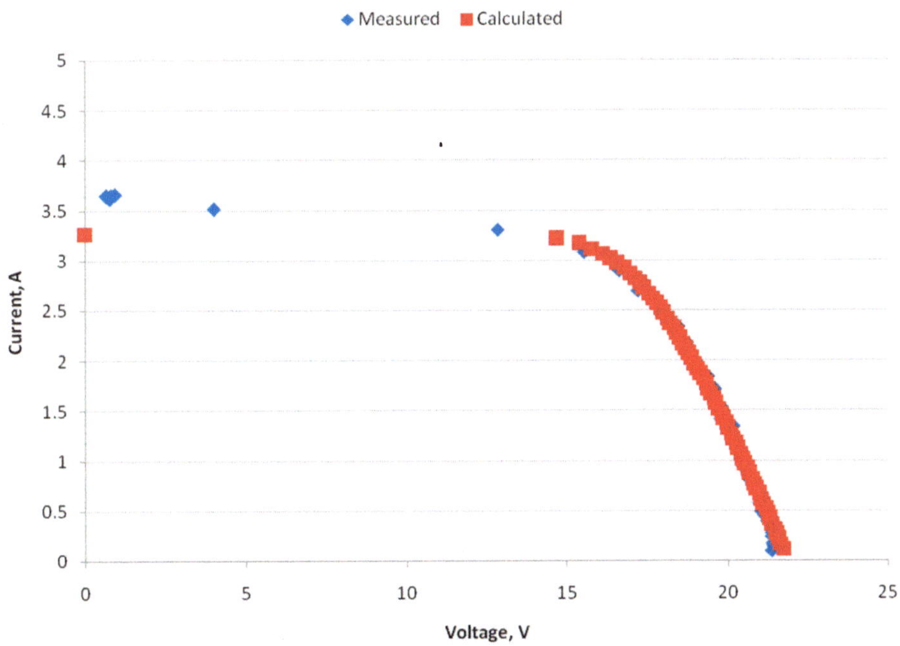

Fig. 36.7 Measured and calculated I–V curves of the PV module at 676 W/m^2, 45 °C

36.6 Conclusion

The present chapter introduces an effective method used for determining PV module series resistance, reverse saturation current, short-circuit current temperature coefficient, and ideality factor based on practical measurement of I–V curves for the PV module under different operating conditions. The determination process depends mainly on the accurate measurement of the module characteristics and calculation of the module parameters via slope of the I–V curve and curve fitting with the least-square method. The agreement between the measured and the calculated I–V curves of the PV module explains the quality of the method used to get the module parameters.

36.7 Nomenclature

A	Ideality factor
E_{Go}	Band gap for silicon (1.1 eV)
G	Solar radiation on the PV cell surface (W/m^2)
I_d	Diode current *(A)*
I_o	PV cell reverse saturation current *(A)*
I_{or}	PV cell reverse saturation current at reference temperature, T_r *(A)*
I_{ph}	PV cell light-generated current *(A)*
I_{scr}	PV cell short-circuit current at STC *(A)*
I_{sh}	Shunt current *(A)*
K	Boltzmann's constant (1.38×10^{-23} N m/ K)
K_i	PV short-circuit current temperature coefficient (A/°C)
n_s	The number of cells in series in the PV module (22)
Q	Electron charge (1.6×10^{-19} C)
R_s	PV module series resistance (Ω)
R_{sh}	PV module shunt resistance (Ω)
T	PV module surface temperature (K)
T_r	Reference temperature (298 K)

References

1. Khemissi L et al (2012) Low cost and high efficiency of single phase photovoltaic system based on microcontroller. Sol Energy 86:1129–1141
2. Ko JS, Chung DH (2012) Control of a novel PV tracking system considering the shadow influence. J Electr Eng Technol 7(4):524–529
3. Mahmoud MM (2006) Transient analysis of a PV power generator charging a capacitor for measurement of the I-V characteristics. Renew Energy 31:2198–2206

4. Altas IH, Sharaf AM (1992) A fuzzy logic power tracking controller for a photovoltaic energy conversion scheme. Electr Power Res 25:227–238
5. Ro K (1997) Two loop controller for maximizing performance of grid connected photovoltaic fuel cell hybrid power plant, Ph. D. Thesis, Virginia University, April 1997
6. Tivanov M, Patryn A, Drozdov N, Fedotov A, Mazanik A (2005) Determination of solar cell parameters from its current-voltage and spectral characteristics. Sol Energy Mater Sol Cells 87:457–465
7. Ortiz-Conde A, Javier F, Sanchez G, Muci J (2006) New method to extract the model parameters of solar cells from the explicit analytic solutions of their illuminated I-V characteristics. Sol Energy Mater Sol Cells 90:352–361
8. Thongpron J, Kirtikaraa K, Jivacate C (2006) A method for the determination of dynamic resistance of photovoltaic modules under illumination. Sol Energy Mater Sol Cells 90:3078–3084
9. Macabebe EQB, Van Dyk EE (2008) Parameter extraction from dark current-voltage characteristics of solar cells. S Afr J Sci 104:401–404
10. Bencherif M, Chermitti A (2012) New method to assess the losses parameters of the photovoltaic modules. Int J Eng Adv Technol (IJEAT) 2:2249–8958
11. The manual of the data acquisition system (PCL-711S) (1999)
12. El-Azaly GEA (1996) Design and modernizing testing equipment for photovoltaic pannels, Ph. D. Thesis, Ain Shams University, Cairo, Egypt, 1996

Chapter 37
Experimental Results of Computer Monitoring of PV-based Energy System

Kamal Abed, Ahmed Bahgat, Mervat Badr, Mahmoud El-Bayoumi and Ayat. Ragheb

Abstract Renewable energy emerges as a cleaner and sustainable alternative for generating electricity. This chapter proposes development of an experimental monitor and control scheme for stand-alone photovoltaic (PV) systems. The proposed system comprises a set of PV panels, a number of storage batteries, a charge controller, DC/AC inverter to feed electrical load, and a backup generator. A monitoring and control unit was designed and implemented to operate the system using a specially developed computer program based on LabVIEW software. The main task of the proposed scheme is to manage and protect the batteries from over discharging that may destroy batteries. To do so, the developed control scheme disconnects the load from PV and battery system, using switching control, when state of charge (SOC) of batteries falls to 35 % or less, and switching on a petrol engine/AC generator unit. The control panel that employs a USB data acquisition card (DAQ) connected to the PC was built, installed, operated, and calibrated with the system. The cost of generated electricity was not taken into consideration as the system is built on experimental scale. Results of the study indicate a significant decrease in the battery useful capacity with higher discharged current (load requirements). It is also possible to use the I-V curves to identify the batteries SOC, particularly at low SOCs to protect the batteries when SOCs are lower than 35 %. A significant output is the possibility for continuous, through monitoring of the system performance setting an accurate archive of generated power, battery performance, effect of electrical load, and control of the backup generator to ensure the continuity of electrical load supply.

Keywords Integrated renewable energy systems · PV modules · Control strategy · Batteries

K. A. Abed (✉) · M. A. Badr · M. El-Bayoumi · A. A. Ragheb
Mechanical Engineering Department, National Research Centre, Giza, Egypt
e-mail: kaabed3@hotmail.com

A. Bahgat
Faculty of Engineering, Cairo University, Giza, Egypt

A. Sayigh (ed.), *Renewable Energy in the Service of Mankind Vol II*,
DOI 10.1007/978-3-319-18215-5_37

37.1 Introduction

Significant developments in the design, analysis, and installation of integrated energy systems (IES) have been achieved in the past decade. Stand-alone integrated renewable energy systems usually incur lower costs and demonstrate higher reliability than PV or wind systems supplying electricity for off-grid applications, especially in remote areas with difficult access to electricity grid. Control and optimization of the hybrid systems are usually complex tasks. Most frequent hybrid systems consist of PV generator and/or wind turbines and/or diesel generator, with energy storage unit. Batteries used in PV applications have different performance characteristics compared with batteries used in more traditional applications. The shortened battery life contributes significantly to the cost of a PV system. In some PV systems, the battery accounts for more than 40 % of the life cycle cost [1]. The life of a lead-acid battery can be extended by avoiding critical operating conditions such as overcharge and deep discharge.

Most of IES studies are theoretical based on simulation software packages. From solar radiation data and load profiles, the performance of the PV centralized system was simulated using a time step scheme [2]. The advantages of this system are large charging rate of power, high efficiency, and low cost compared with conventional individual PV systems. The economic study shows that the life cycle cost and the price of kilo watt hour generated in the centralized system is lower than that for the individual system.

Size optimization of solar array and battery in a stand-alone photovoltaic (SPV) system was investigated and applied on a case study in Malaysian weather conditions. This SPV system typically consists of a solar array, a controller with maximum power point tracker (MPPT), a battery, an inverter, and loads. The controller with an MPPT helps in extracting maximum power from the solar array regardless of the variation of solar radiation and temperature as well as protects the battery from overcharging and under-discharging [3]. Constant-voltage and current-limited control were realized by parallel controller to meet the energy storage of the system requirement and to protect the battery. Super capacitor was used to filter the PV cell output and to reduce the small charging and discharging cycle of the battery [4].

Dispatch strategies for the operation of a PV/diesel/battery hybrid power system using "set points" were introduced including the determination of the optimum values of set points for the starting and stopping of the diesel generator to minimize the overall system cost [5]. A computer program for a typical dispatch strategy was developed to predict the long-term energy performance and the life cycle cost of the system. The power output of PV systems can be made stable utilizing online on/off control function of power conversion devices for PV systems [6]. In that study, a remote monitoring and control system has been developed obtaining measurements and allowing for remote control operations. One of the experimental results showed that sizing of a PV/diesel hybrid system by taking into account the solar radiation and the load/demand profile of a typical area may lead the diesel generator to operate near its optimal point (70–80 % of its nominal power) [7].

37.2 PV Energy System

PV energy systems comprise a number of PV modules, battery bank, charge controller, inverter, and a controller. The battery meets the daily load fluctuation and the diesel generator takes care of the long-term fluctuation. The current experimental setup consists of 900 W PV modules with six lead-acid, gel-batteries 100 AH each, charge controller, and a 1200 W inverter to power AC loads.

37.2.1 Characteristics of System Components

37.2.1.1 PV Characteristics

There are three classic parameters that are very important on the PV characteristics, namely, short-circuit current (I_{sc}), open-circuit voltage (V_{oc}), and the maximum power point (I_{mp};V_{mp}). The power delivered by a PV cell attains a maximum value at the points (I_{mp};V_{mp}). The classical points are shown in Fig. 37.1 and are usually given as part of a manufacturer's data sheet for a PV module.

Another important parameter of the PV characteristics is called the fill factor (FF), which describes how the curve fills the rectangle that is defined by (V_{oc}) and (I_{sc}). It gives an indication of the quality of a cell's semiconductor junction and measures of how well a solar cell is able to collect the carriers generated by light. It is defined as:

$$FF = V_{mpp} I_{mpp} / V_{oc} I_{sc} \tag{37.1}$$

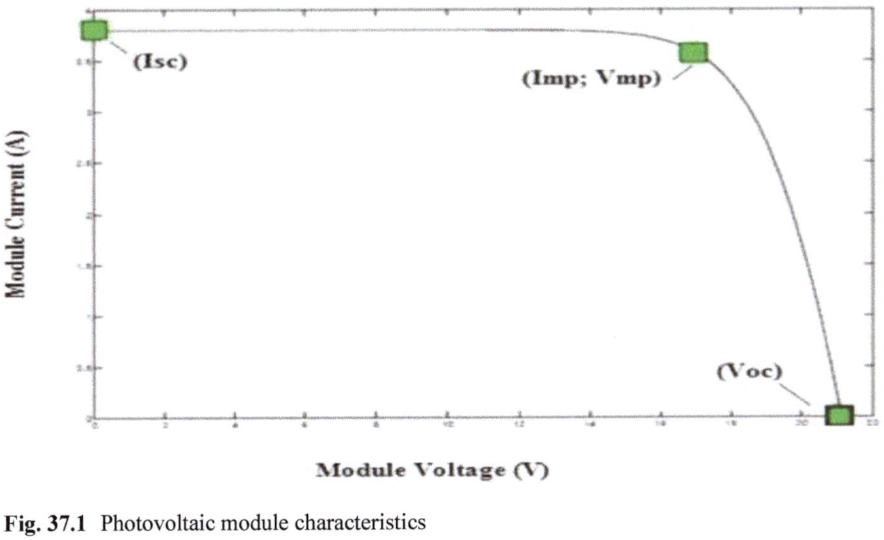

Fig. 37.1 Photovoltaic module characteristics

or:

$$V_{oc}I_{sc}\cdot \mathrm{FF}=V_{mpp}I_{mpp}=P_{max}. \tag{37.2}$$

The efficiency η is defined as the ratio of the maximum output power p_{max} to the solar power received by the cell surface: P_L

$$\eta = V_{mpp}I_{mpp}/P_L. \tag{37.3}$$

37.2.1.2 Battery Characteristics

The battery capacity is an elementary performance characteristic that allows specifying the amount of stored energy. Battery capacity is the total charge that can be discharged from the battery, and it is measured in ampere hour (Ah). As shown in Fig. 37.2, the percent of the battery capacity to be consumed according to hourly discharging current (I).

Battery state of charge (SOC) is the cumulative sum of the daily charge/discharge energy transfers. When solar radiation decreases due to the weather conditions, less energy is supplied by the PV array. This causes the battery SOC to reduce accordingly, [8]. When the battery SOC is low, the system will still attempt to supply the electrical load demand causing daily cycles around the deep discharge threshold. More intelligent management system would monitor the SOC and gradually reduce the energy taken from the battery to help prevent continuous operation at a low SOC. The minimum SOC or deep discharge protection (DDP) is often implemented by measuring the battery voltage, which is not a true measure of the SOC. In winter, the battery may experience a low SOC for extended periods due to the seasonal variation in solar irradiation. A sever low SOC reduces the life of the battery, [9].

Increasing the battery capacity cannot provide sufficient storage capacity for the worst season of lowest solar irradiation because of the battery self-discharge. One

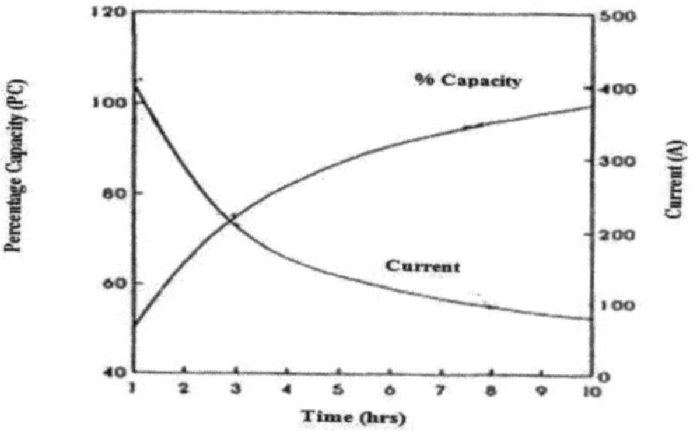

Fig. 37.2 Relation between battery discharging current and percentage capacity [10]

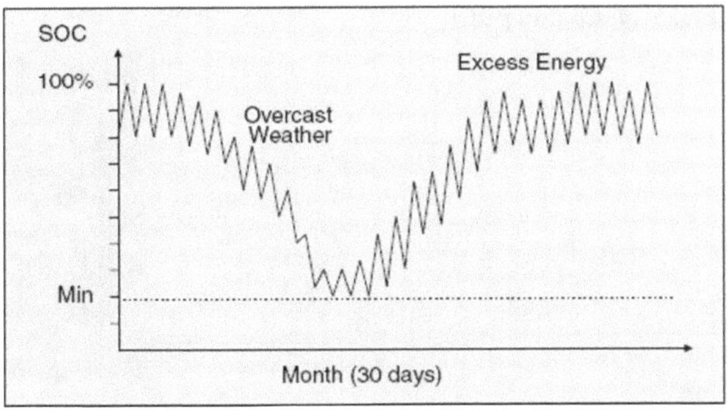

Fig. 37.3 Typical battery cycling pattern in a PV system

of the objectives of the proposed battery management system is to measure the SOC and use it to minimize extended periods of low SOC. SOC (defined as the ratio between the difference of the rated capacity and the charge balance divided by the rated capacity) can be seen in Eq. (37.4), [10].

$$SOC = \left[\frac{C_{\text{nom}} - Q_{\text{Bat}}}{C_{\text{nom}}} \right] \tag{37.4}$$

$$\text{where; } Q_{\text{Bat}} = \int_c Idt \tag{37.5}$$

C_{nom} The nominal capacity of the battery (Ah).
Q_{Bat} The Ah-balance (i.e. net Ah discharged or charged since the last full state of charge), and I is the main reaction current (A).

In renewable energy systems, the charge of solar batteries is cycled while providing energy on a daily basis. A photovoltaic (PV) system's output power fluctuates according to the weather conditions, season, and geographic location. Usually, charge controller is used to regulate the PV output power fluctuations. Figure 37.3 shows the typical battery cycling pattern in a PV system.

37.3 Development Computer Unit

37.3.1 Control Panel

The current system relies on a specially built, dedicated control panel. This control panel collects data regarding the installed system performance through a data-acquisition system. The data collected is used for evaluation purposes of the system performance and decision making of the need for the generator. The control panel consists of the following units as shown in Fig. 37.4.

Fig. 37.4 Control panel

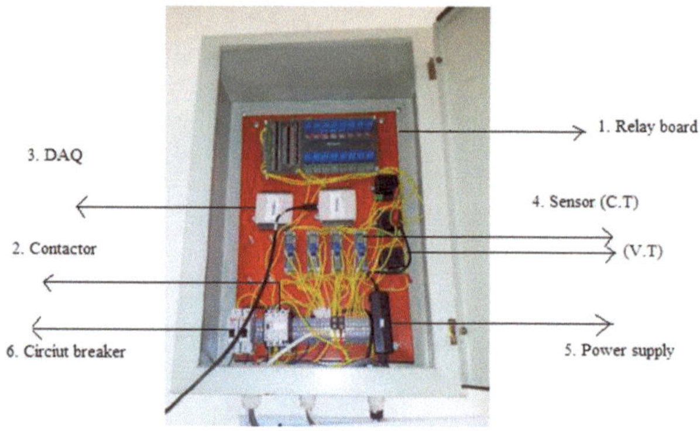

1. Relay board: 12 V DC, contact current: 5 Amp,
   - Driving relay for contactors on/off
   - Switching on/off between generator and inverter
2. Contactors: coil volt: 220 V AC and protection for load on/off
3. DAQ: The fundamental task of a DAQ system is the measurement of real-world physical signals. Before a computer-based system can measure a physical signal, a sensor or transducer must convert the physical signal into an electrical signal
4. Sensors/transducers and voltage sensors:
   - Current transducer converts current to signal volt (0–5) V DC
   - Voltage transformer convert 220–12 VAC divided by resistance to be measured by DAQ
5. Power supply: 12 V DC exciting relay board and sensors
6. Circuit breaker: on/off main power supply of the control panel

37.3.2 Monitoring and Control Strategy

The system operation control is the strategy by which the supervisory controller efficiently coordinates and dispatches the available energy of each generator and storage unit to meet the demand and supply system requirements. The major concern of the system operation control concepts is to utilize the available energy of each component in an efficient way to meet the load requirements. Some parameters that constrain the functionality of the control strategies are mainly economics of operation, environmental impacts, reliability of supply, and quality of the delivered energy [10]. Regardless of the system configuration, all of the implemented strategies have agreed on utilizing first the output of the renewable energy converters (e.g. PV, etc.) to cover the load. In case of an insufficiency of the renewable energy, different strategies are proposed to fill the void. One suggests that the rest of the demand (defined as net-load), which is not covered by the renewable should be met first by the storage batteries. The remaining load will be covered by the diesel engine, if any, or it is simply considered as shortage in supplying the load (loss of load probability).The strategy affects the operation lifetime of the system components, which consequently influence the energy generation costs.

The control strategy is based on:

- Using a PC for data logging, processing, display, and control
- Collection of available signals and status using transducers and switches
- Using DAQ to interface between PC and signals
- Real-time data is graphically displayed
- Real-time data is analyzed and control signals are generated in form of switching actions
- Logged data is used for IES performance analysis

The analog data signals are conditioned and sent from the system to the PC through the DAQ unit of the control panel. The DAQ unit is controlled by a special developed interface, using the LabVIEW software. The DAQ unit converts analog signals representing system parameters to digital data and sends it to the PC, where the LabVIEW dedicated program displays and stores it on the PC disk and then processes it further.

37.3.3 PC/Software

The employed PC is an IBM compatible computer with a display and a number of available USB connectors. The USB connectors are used to connect the PC to the DAQ of the control panel. The PC is equipped with sufficient storage space for data collection. The PC is capable of receiving and processing the operation's data in timely manner. Software package LabVIEW is used to generate the code for implementing the current thesis. The software communicates with the DAQ unit to receive the system parameters. These parameters are displayed on the monitor in an easy way in the form of charts and numerical values. Charts are divided into two graphic display charts one for current and one for voltage.

37.3.4 Configuration of SW Package

- LVSW Package is used to perform the two following tasks:

1. *Data acquisition*
   This part concerns the acquisition of data collected by the national instrument DAQ card. Data are stored in the form of time tagged files. Data analysis is performed based on given functions. Data are prepared for report generation. Control archives are sent via DAQ to a PC.

2. *Data visualization and control actions*
   A group of graphic displays based on available library have been prepared. The graphic display for currents showed real-time value of the current of PV, batteries, and inverter (load), taken every 2 s. Also, the graphic display for voltages shows the voltage of batteries, inverter DC input, and inverter RMS of output. On the same graph, the real time values of these voltages are displayed, taken every 2 s. In both graphic display of currents and voltages, there are two active buttons, the first one is to start/stop the program and the second active button connect/disconnect load (inverter output). Real-time values are extracted from acquired data or processed data in part I and then displayed. Another function is to receive manual actions and setting given by the operator, as there are two text boxes the first gives the date/time of the run and the second box gives the initial state of charge. These actions and setting are transferred to the real-time data base (part I).

For the purpose of calibration, a set of measuring devices were employed and the test was performed at fixed and known conditions as possible. The test results were compared to measured levels and necessary adjustments were made to the system to reflect the true system signals.

37.4 Results

A set of experiments were performed using the developed experimental setup. These experiments were conducted under different environmental conditions. The results are presented in the following sections.

37.4.1 Effect of Battery State of Charge (SOC) on the Capacity of the System

For the current work, the effect of batteries SOC (%) on the capacity of the system to supply load demand was evaluated. A special test was carried out in which the PV modules were covered by an opaque sheet to ensure zero PV current and the load was supported only by the batteries. The power delivered to the load was evaluated at different loads and different battery SOC. The results of this test, Fig. 37.5, show that the SOC (%) does not significantly affect the battery output at different loads (the five curves are almost coinciding), as long as the current demand is not higher than battery maximum current.

- **Control system's evaluation of batteries SOC**

The SOC of batteries was calculated by the means of integration or summation of the battery output current (Ah). For this purpose, batteries were fully charged (i.e. 100 % SOC), then it was discharged; using a fixed load 400 W, down to 35 % SOC

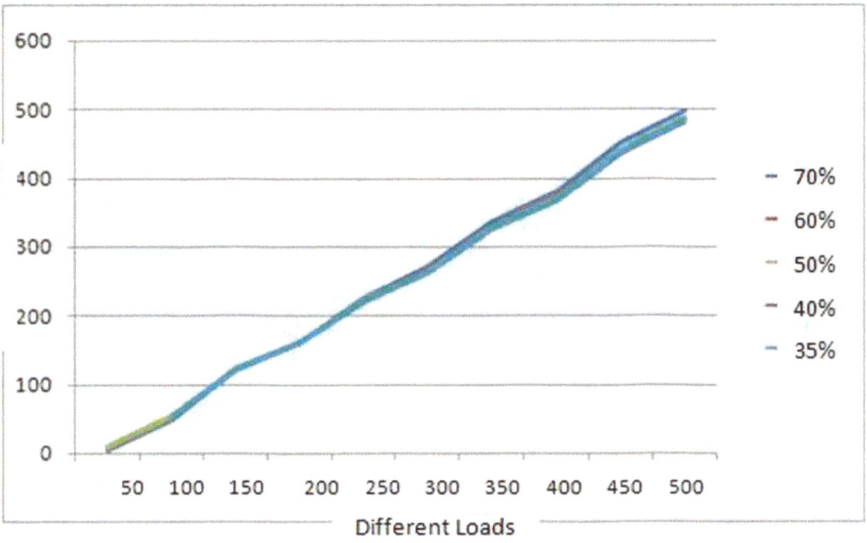

Fig. 37.5 Output power at different loads and different battery SOC

and the Ah was integrated using the following equation. The summation was used to calculate the battery capacity at 100, 90, 80, 70, 60, 50, and 40 % SOCs, when each case was discharged down to 35 % SOC. The results of this test are shown in Fig. 37.6.

This Test Provided Data for the System to Carry Out Two Tasks

- **The first task** of the control system is to evaluate the SOC of batteries so if the control recognizes that the batteries SOC has reached the alarm level of 35 %, then the system instructs the operator to switch on the backup generator as shown in Fig. 37.6.

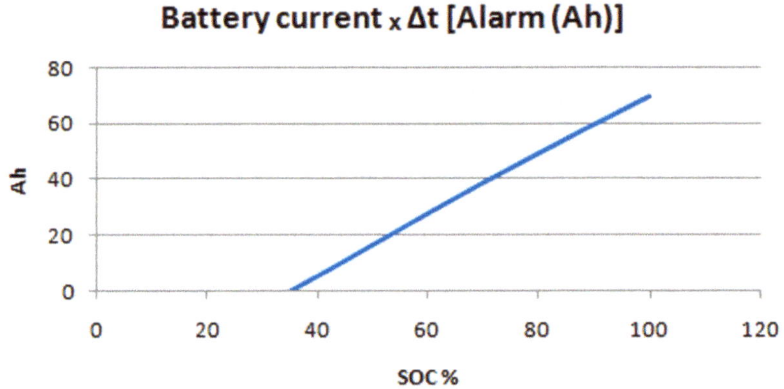

Fig. 37.6 The results of battery capacity (Alarm Ah)

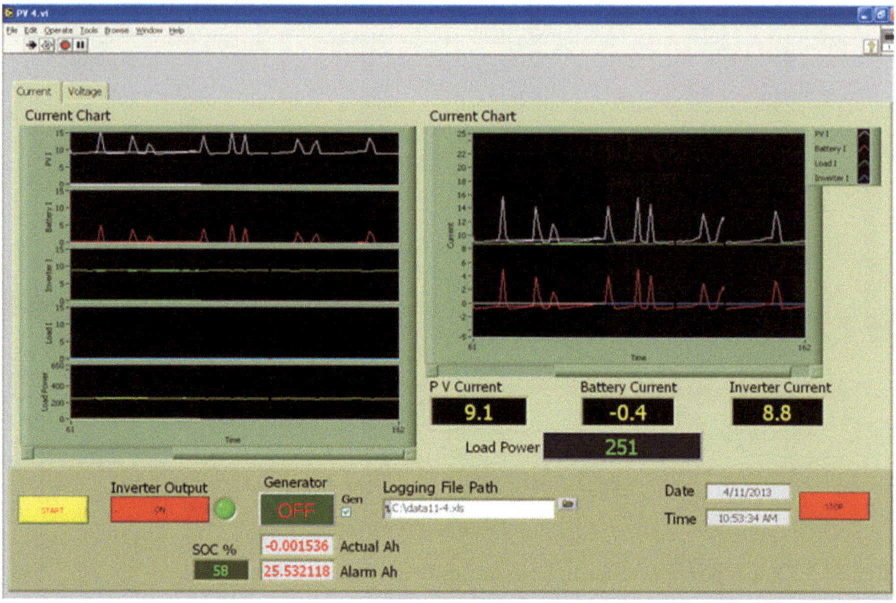

Fig. 37.7 Program screen showing load supported by PV and batteries

- **The second task** is to enable the system to start at any SOC defined from the batteries' charge controller. The obtained value is required as an input to the software; in case that the system was shut down for service or any other reason. The program would use this SOC to evaluate the battery capacity available and deduce the integrated Ah (Actual Ah) withdrawn in order to evaluate the 35 % SOC. Figure 37.7 shows the system starting at initial SOC 58 %. The control system calculates the available Ah as 25.5 Ah, set it as the "Alarm Ah," and starts summation of discharge Ah (Actual Ah). The screen shows "Engine Off" as there is no need to start the engine.

37.4.2 Effect of Load on Battery Capacity

A set of runs were conducted to evaluate battery capacity during which the summation of the discharged current (actual Ah) were calculated from 100 to 35 % SOC, measured on the charge controller, using different fixed loads. The results in Fig. 37.8 show a significant decrease in battery useful capacity at higher discharge current. At a load of 400 W, the battery capacity was around 66 Ah compared with typical capacity of 100 Ah/20 h. This indicates poor battery efficiency if discharged at higher rates.

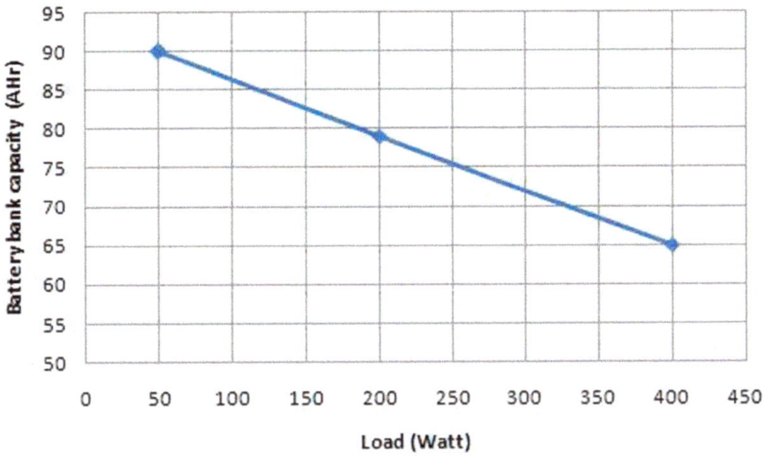

Fig. 37.8 Variation of battery bank capacity at different consumption rates

37.4.3 Collected Results

The following section is an illustration of a sample of the collected results during several months.

Data of 3-12-2012 PV current is not enough to supply load but once inverter is closed, the PV start to charge the battery as shown in Fig. 37.9.

Data of 13-1-2013 Figure 37.10 shows the battery volt variation when subjected to a constant load of 450 W and variable PV current. When the PV current is not sufficient ,the battery supplies the difference resulting in battery voltage reduction as shown in Fig. 37.10.

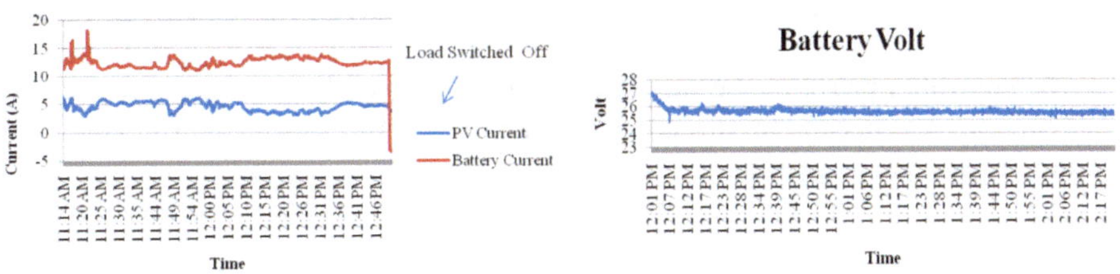

Fig. 37.9 Collected data at 3-12-2012

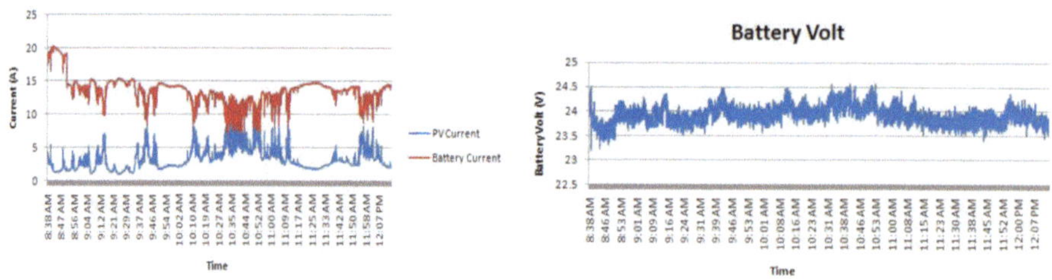

Fig. 37.10 Collected data at 13-1-2013

37.5 Conclusions

A control scheme for a PV-based energy system is designed and installed. The major system components include PV modules, storage batteries, charger controller, DC/AC power inverter, and an electrical load. The control panel employs a USB DAQ connected to a PC and a number of signal conditioning devices and transducers to control the system using specially built LAbVIEW software. After calibration, a set of results were taken for the current system at fixed and variable loads to evaluate its performance. Results of the study indicate that:

- The developed control scheme could be considered efficient to perform control of IES for all operating conditions and is capable of operating and applying different control strategies.
- SOC (%) does not significantly affect the battery output at different loads, as long as the current demand is not higher than the battery maximum current.
- Significant decrease in battery useful capacity (efficiency) with higher discharged current.
- It is also possible to use the I-V curves to identify the batteries SOC, particularly at low SOCs to protect the batteries when SOCs lower than 35 %.
- Maximum PV output is around noon except for cloudy periods.
- Noise capture signal of the battery voltage is eliminated by using a small capacitor.
- The controller proved to be successful in detecting conditions where the pack-up generator was needed for disconnecting the PV and batteries system and instructing the generator ignition.

References

1. Bilal BO, Sambou V, Kébéa CMF, Ndiaye PA, Ndongo M (2012) Methodology to size an optimal stand-alone PV/wind/diesel/battery system minimizing the levelized cost of energy and the CO_2 emissions. Energy Procedia 14:1636–1647
2. Bernal-Agustın JL, Dufo-Lopéz R (2009) Simulation and optimization of stand-alone hybrid renewable energy systems. Renew Sustain Energy Rev 13:2111–2118

3. Dakkak M, Hirata A, Muhida R, Kawasaki Z (2003) Operation strategy of residential centralized photovoltaic system in remote areas. Renew Energy 28:997–1012
4. Wang C, Chena W, Shao S, Chena Z, Bin Z, Li H (2011) Energy management of stand-alone hybrid PV system energy. Energy Procedia 12:471–479
5. Ashari M, Nayar CV (1999) An optimum dispatch strategy using set points for a photovoltaic (PV)/diesel/battery hybrid power system. Sol Energy 66(1):1–9
6. Funabashi T, Hayashi T, Okuno Y (2008) On/off control method to stabilize the power output of photo voltaic generation system. Proceedings of 4th European conference on PV-hybrid systems and mini-grids Athens, Greece, 29th and 30th May 2008
7. Yamegueu D, Azoumah Y, Py X, Zongo N (2011) Experimental study of electricity generation by solar PV/diesel hybrid systems without battery storage for off-grid areas. Renew Energy 36:1780–1787
8. Duryea S, Islam S, Lawrance W (2001) A battery management system for stand-alone photovoltaic energy systems. IEEE Industry Applications Magazine, May/June 2001
9. Ibrahim MI (2002) Decentralized hybrid renewable energy systems: control optimization and battery ageing estimation based on fuzzy logic. Ph. D. Thesis. Kassel University, Germany
10. El-Mallah AA, Shalaby MA, Badr MA (1994) A stochastic simulation model for a hybrid renewable energy system supplying remote community. Proceedings of 3rd world renewable energy congress, Reading, UK

Chapter 38
New Renewable Energy Promotion Approach for Rural Electrification in Cameroon

Joseph Kenfack, Olivier Videme Bossou, Joseph Voufo, Samuel Djom and Nicolas Crettenand

Abstract Rural electrification in Cameroon was in the past planned by the Government through the Ministry of Water and Energy. This issue was addressed through grid extension using three-phase medium voltage, a single wire earth return solution, or off-grid thermal plants for remote localities. Because the architecture of the electricity grid is radial isolated, this leads to many technical problems and a limited rate of electrification, especially in remote areas. The country learned a lot from failures and realized that the rural electrification solutions were yet to come. In order to really tackle the rural electrification problem, a dedicated Agency for Rural Electrification was created, bringing new approaches.

Given the cost of fossil fuel in the country, thermal plants are very expensive from the operating and maintenance point of view, leading to high cost of electricity, especially in places with low income and poor communication and infrastructures. Grid extension is constructed with specially treated wooden poles with limited life expectancy. It hence requires power lines with frequent maintenance under the threat of bushfire. Furthermore, there is high voltage drop due to very long distances and heavy lightning problems during the rainy season. On the other hand, Cameroon has huge hydro, biomass, and solar potential, allowing the country to envisage the development of these sources of renewable energy in most remote areas.

In order to address the issue, the Rural Electrification Agency has launched a new approach since 2009 based on the promotion of renewable energy to electrify the rural areas. This is based on the recommendation of the World Bank which

J. Kenfack (✉) · J. Voufo
National Advanced School of Engineering,
University of Yaounde I, P.O. Box 7048, Yaounde, Cameroon
e-mail: joskenfack@yahoo.fr

O. V. Bossou
University of Maroua, Maroua, Cameroon

S. Djom
Rural Electrification Agency, Yaounde, Cameroon

N. Crettenand
Ecole Polytechnique Fédérale de Lausanne (EPFL), Lausanne, Switzerland

A. Sayigh (ed.), *Renewable Energy in the Service of Mankind Vol II*,
DOI 10.1007/978-3-319-18215-5_38

has provided initial funds to subsidy the initial investment costs. Objective criteria have been set up to ensure promotion of solar hybrid energy and micro hydro in remote areas, including capacity building of all stakeholders. The study is based on the institutional reforms of the sector, the previous rural electrification programs, and the ongoing organization and projects in the framework of renewable energy promotion.

Keywords Rural electrification · Renewable energy · Energy policy

38.1 Introduction

The issue of rural electrification is particularly crucial in developing countries in Africa in general and in Cameroon in particular where there is huge hydro [1], solar [2, 3], and biomass [4] potential. Many different types of solutions are put in place in many countries to address rural electrification. Among the initiatives, the creation of a rural electrification agency and later a dedicated rural electricity fund to sustain the action is one of the solutions adopted by some countries as depicted in Table 38.1.

In the framework of a long-term development plan, Cameroon has brought out many papers on the issue since a decade. These papers include energy policy in general and dedicated action towards electrification and rural electrification, in particular. The main papers are Cameroon's Strategy for Growth and Employment (DSCE) [6], Poverty Reduction Strategy (DSRP) [7], National Energy Action Plan for Poverty Reduction (PANERP) [8], and the Cameroon vision 2035 [9].

Despite the papers mentioned above and the creation of a Rural Electrification Agency in 1999 and later the Rural Energy Fund (FER) in 2009, rural areas still have a very poor electrification rate. The current evolution cannot permit Cameroon to attain the millennium goals by 2015, not even 2020. The previous trend of the budget allocated to rural electrification for the past years depicted in the figure shows that tremendous effort is needed (Fig. 38.1).

In fact, the rural population in 2012 was 11 million dispatched in 13.634 localities. From the data given by the utility AES-SONEL, the global rural electrification rate to date is 19 %, giving electricity access to only 800,000 households [5].

Table 38.1 Countries with a dedicated structure funding rural electrification [5]

	Burkina Faso	Cameroon	Congo (Brazza)	Madagascar	Mauritania	Senegal
Structure	Electrification Development Fund	Rural Electrification Fund	Electricity Sector Development Fund	National Electricity Fund	Investment Fund for Rural Electrification	Rural Electricity Fund

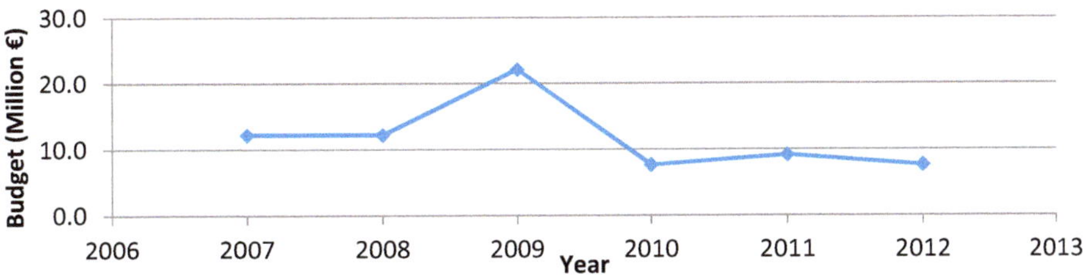

Fig. 38.1 Evolution of government budget allocation for rural electrification

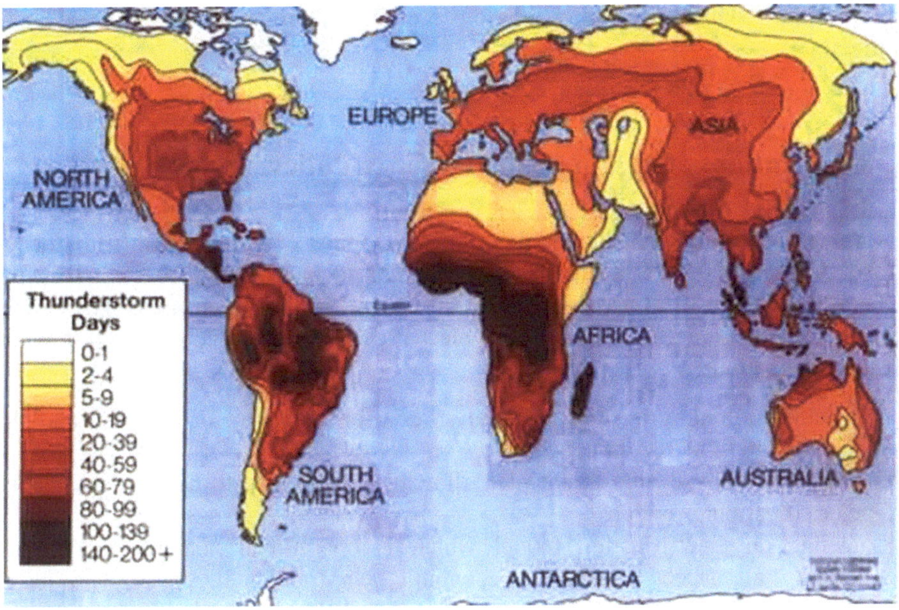

Fig. 38.2 World lightning map, average world thunderstorm in days/year [11]

38.2 Network Architecture and Strategy

Central Africa is among the places in the world under heavy lightning threats as depicted in Fig. 38.2. As the network structure is radial, long-distance power lines need to be constructed as the main solution for rural electrification through grids. This can only lead to very long distance distribution lines, high voltage drop and less energy security. Besides, the single wire earth return solution adopted since the past three decades has limited capacity (less than 25 kVA) per medium to low voltage transformer. Furthermore, standalone system initiatives still have many issues to be addressed [10].

38.3 Technologies Used in Rural Electrification

Two models are currently used to electrify remote areas: grid extension and stand-alone systems.Grid extension is the first model and is used when the locality is not very far from the interconnected grid. The breakeven cost admitted should be less than 500 € per subscriber [5].

The second model is the standalone systems that are used when areas are very far from the grid. To ensure the sustainability of the project, the recommended primary source used is renewable energy (hydro, solar, biomass, wind or hybrid) [12]. From the local economic environment, solar remains the most expensive solution. Other renewable solutions are much more competitive, hence cost-effective.

38.4 National Ongoing Rural Energy Policy

Rural energy policy is challenging for developing countries in all continents [13]. In order to allow 1.2 million households to have access to electricity by the year 2020 in rural areas and 400,000 households in urban areas, Cameroon has adopted, since 2012, a new rural electrification plan based on the following recommendations:

- Establish a long-term plan and clear criteria for the choice of localities and the technology to be recommended
- Define a realistic investment plan allowing the country to have a sustainable electrification growth [5]
- Define an electrification model which is operational and sustainable including all aspects (technical, economic, environmental, social, and institutional) with a clear identification of all stakeholders and their operational mode,
- Define a business plan with a good share of initial investments compatible with technologies to be implemented including hybrid systems for energy access.

The objective of this new strategy is to boost the new approach of rural electrification planning and fund-raising initiatives.

The national energy policy for the next 10 years is depicted in Cameroon's DSCE paper. This vision aims at reducing poverty and meeting the targets stated in the PANERP paper.

In particular, the Government of Cameroon has pointed out some key issues to be addressed in order to enhance energy access for all. Those points are the implementation of the PANERP paper, the development of rural electrification, the increase of electricity access, and the development of other forms of energy.

38.5 Investments Estimates for Each Technology Approach

Global investment estimates made for electricity generation, transport, and distribution should be around 447 million € for a total generation capacity of 220 MW. For grid extension projects, global investment estimates are140 million €, the total capacity being 100 MW. Standalone systems are estimated to be 307 million € for a capacity of 120 MW. The cost is high because the Government of Cameroon has chosen to develop solar standalone systems under an OTC agreement [5].

Promotion of renewable energy through the development of decentralized systems is an opportunity for clean electricity generation in a low-income environment. This will contribute to a sustainable fight against poverty and deserves to be encouraged by the government. Beyond wealth generation, standalone systems will certainly improve the service quality by ensuring:

- Local service enchantment
- Development of a national network layout (open medium voltage loop technology)
- Capacity building of local authorities and municipalities in the management of energy
- New job creation

38.6 Assessment of Current Situation

The suburban and rural electrification sector lacks investors. To date, more than 20 memorandums of understanding have been signed by local and foreign private potential investors, but all projects are behind schedule because of some core issues still need to be addressed. Those identified are classified in three main categories and are depicted in Table 38.2.

38.7 Proposed Ongoing Solutions for Identified Issues

From a swot analysis of the current institutional sector reform, core issues to address have been identified. The issues are pertained in Table 38.3.

From the assessment of the Cameroon energy sector [14] and given the new organization of the rural electrification sector, each key stakeholder has a specific role for others as shown in Table 38.4.

Table 38.2 Identified issues to be addressed

Financial problems	Governance problems	Technical problems
Liquidity shortage for pilot projects	Poor rural electrification planning	Need of capacity building on how to run a rural electrification project
Rural electrification is not prioritized enough when raising bilateral or multilateral funding	Insufficient coordination of rural electrification projects	Poor statistics on rural electrification
Cost-effectiveness of rural electrification projects	Mutual agreements for key role projects that are noncost-effective	Distribution network is old and sometimes faulty in rural areas
Poor access to local investors to funds for rural electrification compared to neighboring countries	Nonrealization of rural electrification project by the utility	Need for specific norms in rural areas
	Many institutions are separately involved in the promotion of rural electrification	Absence of specific objective targets to meet in terms of technology
	Human resources needs capacity enhancement and development of local expertise	

Table 38.3 Proposed solutions for identified issues

Financial issues	Governance issues	Technical issues
Public funding and foreign funding allowed by decree	Assign a clear mission to all stakeholders and get them to respect them	Ensure the implementation of energy services in order to attain the millennium goal
Support with subsidized pilot projects that are technical sustainable, socially acceptable, and environmentally sound	Thoroughly revise the rural electrification national planning at the national level	Promote good energy use for the wellbeing and economic development of villagers

Table 38.4 Role matrix of key stakeholders in Cameroon

	AER/FER	ARSEL	Project developers	Municipalities	Banks
AER/FER			Funding agreement	Assistance agreement	Partnership agreement
Regulatory board (ARSEL)			Autorisation agreement	Autorisation agreement	
Project developers	Funding agreement	Autorisation agreement		Public-private-partnership	Funding agreement
Municipalities	Assistance agreement	Autorisation agreement	Public-private-partnership		
Banks	Partnership agreement		Funding agreement		

38.8 Investment Plans and Current Funding Issues

38.8.1 Investment Plans

447 million € should be raised from the following several financial institutions:

- Multilateral institutions; World Bank, Islamic Development Bank (143 million €)
- African Development Bank (80 million €)
- Long-term loan banks; French Development Agency, Eximbank (118 million €).

This action plan is developed under the supervision of the FER. In order to achieve its goals, this institution should collaborate with many other partners and be transformed into a real financial institution with the capability to optimally raise funds and manage projects. Given the specificity of the rural energy sector (low income, nonviability of projects, remote areas, etc.), a dedicated approach for rural electrification is required. It is in this framework that the Rural Electricity Fund was created with the first US$ 40 million contribution from the World Bank.

38.8.2 Current Funding Issues

The Rural Electricity Fund (FER) is the funding instrument for rural electrification. Created in December 2009, this fund intends to fill the gap in rural areas 10 years after the liberalization of the electricity sector. The main purpose is to ensure that all projects in rural areas will be cost-effective. The FER initiative will encourage private investors, banks, microfinance institutions, small and medium size enterprises and local authorities.

The Rural Electricity Fund is dedicated to increasing transparency, good governance, and proper planning and selection of projects as well as the objective award of grants. The structure of the Rural Electricity Fund supports public–private partnership and is hence adapted to provide electricity to remote areas. In order to attain its goal, the Rural Electricity Fund has set up two approaches: priority rural electrification projects (PPER) and local community initiated projects (PILER). PPER are linked to the national electrification strategy whereas local community-initiated projects are initiated by local associations, villagers, local authorities or private investors and nongovernmental organizations.

38.9 Institutional Organization of the Rural Electricity Fund

38.9.1 Initial Organization

The Rural Electricity Fund was set up with (US$ 40 million) from the World Bank. It was initially dedicated to run the development of projects chosen according to criteria from the World Bank, basically those from renewable energy source. Because of crucial delays due to the lack of feasibility studies for standalone renewable energy projects, all potential projects were behind schedule. The available funds were hence transferred to finance the grid extension projects in order to avoid fund closing. Despite the willingness of the government and other stakeholders, it has been difficult to raise funds at the national and international levels without any guaranty from the government. This can only lead to poor conditions for local institutions willing to sustain rural electrification.

38.9.2 Current Organization

It has a planning committee chaired by the ministry in charge of electricity. The committee is composed of the Ministry of Finance, Ministry of Economy, Planning and Regional Development, sponsors, municipalities, Electricity Sector Regulatory Agency (ARSEL) in charge of delivering authorizations and Rural Electrification Agency, which is the execution unit. The board of Rural Electricity Fund running in the Rural Electrification Agency is in charge of managing all projects. A special bank account has been opened in the Bank of Central African States (BEAC Central Bank) where all funds are located. Conventions with local commercial banks and microfinance institutions are also ready to complete the fund-raising by allowing credit to project initiators.

38.10 Rural Electricity Fund Resources

All government investment resources are considered, namely the public investment budget, debt relief under the heavily indebted poor countries (HIPC) initiative, the Special Council Support Fund (FEICOM), National Community-Driven Development Program in Cameroon (PNDP), Cameroon's Hydrocarbon Prices Stabilization Fund (CSPH) and available funds from donors. In case of cofinancing with international donors such as ACP-EU Energy Facility for instance, the public investment budget automatically provides the counterpart fund. The country has also developed an investment incentive scheme through the general tax code and an exemption from value-added tax. The private sector is also expected to play an important role by contributing to fund-raising to support investors.

38.11 How Does One Benefit from the Program?

To benefit from the program you have to write officially to the executive director of the Rural Electrification Agency to request for the grant for extension of power or development of a power plant in your area. Selection criteria are based on the national ongoing rural energy policy and project developers should express a willingness to co-share with FER (30/70) and endorse consent in writing. Project developers should select a cost-effective solution (grid connected or standalone) best suited for the requirements in the tender documents.

38.12 How Do We Improve

From the institutional structural analysis of the rural electrification area in Cameroon, the improvement of the electrification rate in the country should take into account the following points:

- Long-term planning taking into consideration the local context and all technical options
- Define a sustainable institutional and operational model dedicated for rural electrification
- Design a specific economic model based on the cost of service (long-term electricity supply) regardless of the technology
- New financial solutions adapted to the context combining local and foreign funds for optimum investments that ensure cost-effective electricity access

38.13 Fundraising

The country needs technical and financial assistance to improve the rural electrification rate and attain its mid- to long-term goals. Because of the state of poverty and some poor government at the national level, the help of international institutions is really needed. The following international institutions are contributing to rural electrification in Cameroon according to several funding scenarios as shown in Table 38.5.

Table 38.5 Funding scenario in African Development Bank (ADB), World Bank (WB), Islamic Development Bank (IDB), Rural Energy Fund (FER), Commercial Banks (CB)

	Scenario 1	Scenario 2	Scenario 3	Scenario 4
Type of funding	Public subsidy (70%) commercial bank (30%)	Funding of distribution 100%, CB production	Private loan	Funding of distribution and production of public share in production
Sector	Production and distribution	Production	Production and distribution	Production
Banks	WB, IDB, CB	European Union	FDA, ADB, Eximbank	FER, WB, IDB, ADB,

38.14 Case Study

The case study is focused here on a group of five villages in the western part of Cameroon to be electrified using a micro hydro plant [15]. We have stressed here on the implementation of the new financial approach. Pictures and drawings are from Ekom Nkam village [16] (Table 38.6).

The funding of the project is depicted in Table 38.6 (Figs. 38.3, 38.4, 38.5 and Table 38.7).

38.15 Conclusion

The current global electrification rate in Cameroon is 19.8 %. In order to target 41.5 % by the year 2020, Cameroon needs to raise 447 million € from funding institutions to develop 120 MW grid connected and 100 MW standalone systems from renewable sources. The country is really aware that the projects need to be subsidized in order to be cost-effective in rural areas. Because the funds are not always available, the contribution of national and international banks is really needed. This will require an adequate organization of the existing Rural Electricity Fund.

Given the specificity of the remote areas and taking into account the important impact of energy in general and its contribution to poverty alleviation in particular, there is a need to specifically address the rural electrification issue. Regardless of the cost-effectiveness of projects, the government has taken the necessary measures to ensure the future wellbeing of the remote area dwellers and the sustainability of rural electrification with the help of the World Bank through a new organization of the FER and the choice to promote renewable energy projects.

Table 38.6 Financial structure of the ADEID project. (Source: ADEID)

		Amount in millions €			
N°	Technology	Loan above 30 years, 3 % (WB, IDB)	Loan 50 years 2 %, ADB	Loan 20 years (Eximbank, FDA)	Long-term loan, 4 %, (Eximbank)
1	Grid extension	76.2	64.0	0.0	0.0
2	Standalone priority systems	15.2	15.2	15.2	15.2
3	Hydro and minigrid	15.2	0.0	15.2	18.3
4	Biomass and minigrid	4.6	0.0	0.0	10.7
5	Solar with grid feeding	7.6	0.0	22.9	15.2
6	Standalone solar plant	15.2	0.0	48.8	30.5
7	Solar home systems	7.6	0.0	15.2	15.2
8	Autonomous low consumption lamps	1.5	0.0	0.8	0.8
	Total	143.3	79.3	118.1	104.4

Fig. 38.3 Micro hydro plant layout

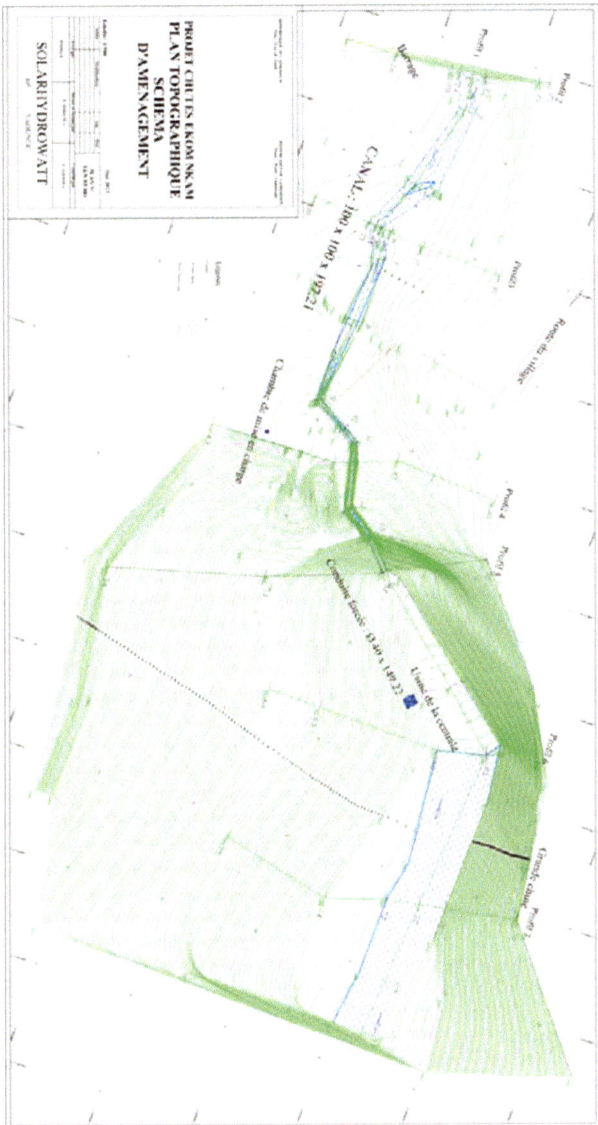

Fig. 38.4 Network design

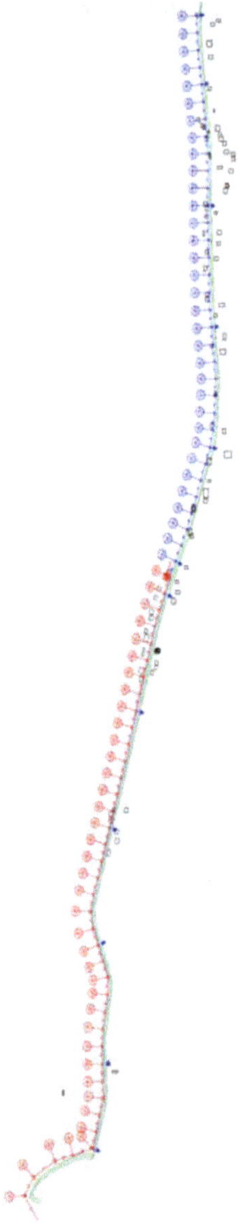

Waterfall

Fig. 38.5 Typical house in Ekom Nkam

Table 38.7 Funding of the ADEID project

	Euros	Ration (%)	Type of funding	Situation
Energy facility from European Union	439.3	23	Subsidy	Available
NGO ADEID (owner)	19.9	1	Private loan	Ready soon
Municipalities	370.3	19	Subsidy	Ongoing
Rural Electrification Fund FER (AER)	1079.8	57	Subsidy	Available
Total	1909.3	100		

References

1. Kenfack J, Lejeune A, Fogue M, Tamo TT (2004) Hydro potential and development in Cameroon, Hydropower and Dams, congress Hydro, Porto
2. http://www.soda-is.com
3. Muller J-C (2004) Ecole d'été, Electricité solaire pour les zones ruralesUNESC0– Paris
4. Millington AC, Critchley RW, et al (1994) Evaluation de la biomasse ligneuse en Afrique Subsaharienne. Publications de la Banque Mondiale
5. Directorate of Energy (2012) Clarification for rural electrification strategy and its funding, Rural Electrification Agency of Cameroon
6. Government of Cameroon (2003) Cameroon's strategy for growth and employment paper (DSCE)
7. Government of Cameroon (2009) Poverty reduction strategy paper
8. Government of Cameroon (2005) National energy action plan for poverty reduction paper
9. Government of Cameroon (2009) Vision 2035
10. Kenfack J, Bossou O, Voufo J, Djom S (2014) Addressing the current remote area electrification problems with solar and microhydro systems in Central Africa. Renew Energy 67
11. http://www.lightningsafety.com/nlsi_info/lightningmaps/worldlightning.html
12. Kenfack J, Neirac FP, Tatietse TT, Mayer D, MédardFogue, Lejeune A (2009) Microhydro-PV-hybrid system: sizing a small hydro-PV-hybrid system for rural electrification in developing countries. Renew Energy 34(10)
13. Gurung, Prakash Gurung O, Eun Oh S (2011) The potential of a renewable energy technology for rural electrification in Nepal: a case study from Tangting. Renew Energy 36(11)
14. Kenfack J, Bignom B, Crettenand N (2014) Overview of institutional structure reform of the Cameroon electricity sector and assessments. International J Hydropower Dams (1)
15. Maher P, Smith NPA, Williams AA (1998) Pico hydro power for rural electrification in developing countries. International J Ambient Energy 19
16. ADEID (2013) Developing renewable resources for poverty alleviation in the rural areas of Bare Bakem, Kekem, Koutaba, Foumbot and Massangam

Chapter 39
Photovoltaics and the Energy System: Adaptation of Layout and Load

Stefan Krauter

Abstract In the foreseeable future sustainable electrical energy systems should be capable of handling 100 % renewable energy across electrical distribution grids. Due to the recent drop in the price of photovoltaic (PV) systems, PVs will play a major role in the global renewable energy portfolio. The technical implications of this energy transition from fossil fuels to renewables are manifold and will affect all sectors of electricity supply (generation, transmission, distribution, operation management), and consequently lead to a new system design. The major challenge facing the effective integration of PVs into the grid is variable generation as a result of positioning to the sun, weather conditions and load level. Here, we address generation issues by suggesting that PV power plants be adapted to load requirements to achieve peak power output during periods of high demand using azimuth and tilt angle modifications, thermal conditioning of PV modules, smart site selection and a combination of complementary energy sources. Moreover, we also address several possibilities regarding load management such as load shifting, inducements, load adaptation and appropriate market design. A brand new, non-technical method is to shift the holiday season; this has the potential to achieve the most effective results for a small cost within the shortest time period.

Keywords Energy transition · Photovoltaics · Demand-side management · Plant modification

39.1 Introduction

The technical implications surrounding the transition to a renewable energy electrical grid impacts all aspects of electricity supply (generation, transmission, distribution, load management). The cumulative impacts will eventually lead to an entirely

---

S. Krauter (✉)
Department of Electrical Energy Technology–Sustainable Energy Concepts,
University of Paderborn, Pohlweg 55, 33098 Paderborn, Germany
e-mail: Stefan.Krauter@upb.de

Photovoltaik Institut Berlin AG, Wrangelstr. 100, 10997 Berlin, Germany

A. Sayigh (ed.), *Renewable Energy in the Service of Mankind Vol II*,
DOI 10.1007/978-3-319-18215-5_39

new system design, a process often referred to as *energy transition, energy revolution, or energy system 2.0.*

Changes to the generation sector are manifold. On one hand, large grid systems are being fed increasingly from generating sources at distant locations such as offshore wind farms or large hydropower plants from Scandinavia and the Alps. On the other hand, energy grids are also being supplied by a significant increase in smaller, distributed generation systems, particularly, photovoltaics (PVs). The integration challenges facing solar PVs are generation intermittency/variability (solar radiation exposure) and load levels. In order to reduce the requirements for distribution, storage and additional generation units, electrical grids should be adapted to manage both load and generation levels. That issue is presently treated with high priority by system operators, utilities and government entities and will be overcome in the foreseeable future. Here, we focus on the potential contributions of three measures—(1) how PV power plants can be adapted to load requirements, (2) the possibilities pertaining to load management, and (3) adequate market system design [1].

We take the situation in Germany as an example, but most of the results are applicable to other countries, given the appropriate adjustments in the parameters.

39.2 Measures for PV Generation

On sunny days, in countries with a high PV share, grid oversupply occurs as a result of the combination of the PV feed and inflexible supply from fossil fuel power plants. In terms of basic economics, combined oversupply often depresses the market price of electricity. Because fixed feed-in tariffs have to be paid, even during periods of oversupply, a considerable amount of money is spent on excess variable energy.

Figure 39.1 shows the irradiance and the spot-market prices over the course of a sunny day [2]. The system would be more balanced with a shift toward more PV power supplied for morning and/or evening power demands, with any corresponding high prices to appear on the spot-market. Such a shift could be facilitated via aligning the PV arrays in a more eastern or western orientation—with results as shown in Figs. 39.2 and 39.3.

39.2.1 Adaptation of Location of PV Power Plant

While fossil and nuclear power plants depend on a fuel supply infrastructure and cooling water intake, the application of PVs within Germany (and most other countries) is relatively independent from its location. Throughout Germany, the variation of solar irradiance and corresponding yield is ± 15 %. In order to minimize power transmission costs, it is useful to install generating capacity as close as possible to the end user. However, from the perspective of the PV investor who receives a

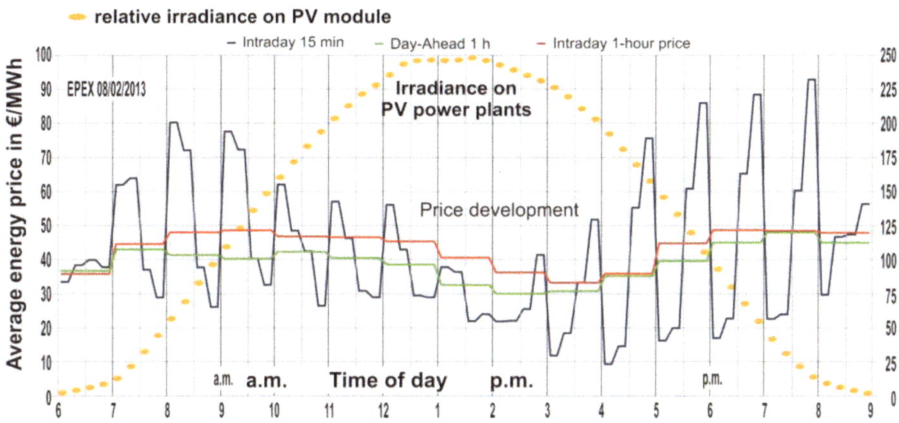

Fig. 39.1 Typical irradiance on a sunny day in Germany and EPEX price development for electricity (day-ahead and intraday trading prices [2])

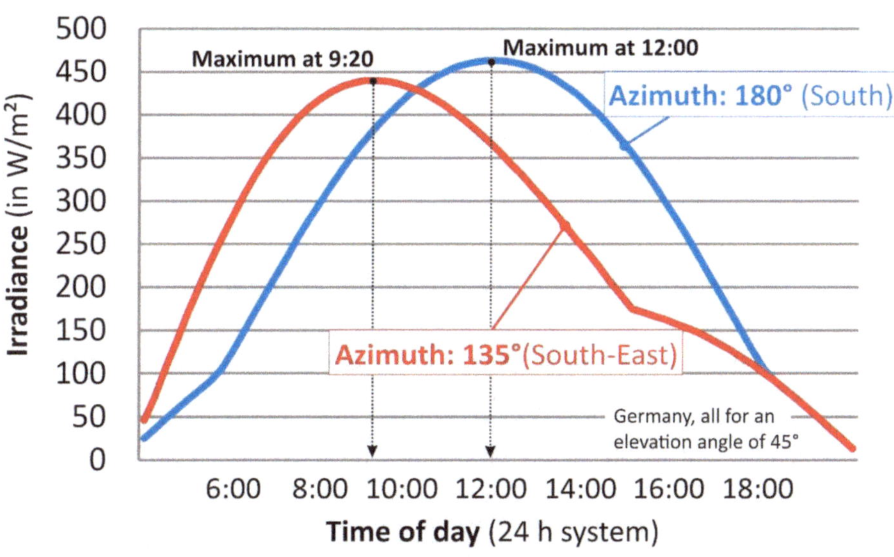

Fig. 39.2 PV generation management via azimuthal deviation from South (standard configuration) to 45° South-East: more PV power is available during the morning hours (at higher spot-market prices)

fixed feed-in tariff regardless of the location, the costs for land and local workforce often take precedence. Therefore, a lot of PV capacity has been installed far from the majority of consumers in regions with less economic activity and less consumption than more densely populated regions. As a result, system operators have been forced to invest in additional transmission capacity and grid stabilizing measures—all of which ultimately increase the total cost of electrical energy. Siting PV arrays at former nuclear power plants can often be cost beneficial for both PV investors

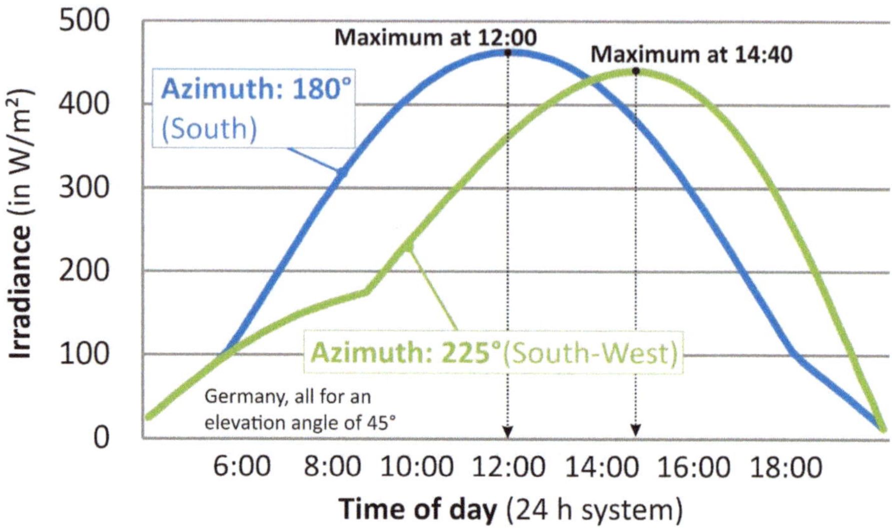

Fig. 39.3 PV generation management via azimuthal deviation from South (standard configuration) to 45° South-West: more PV power is available during the morning hours (at higher spot-market prices)

and system operators since they combine excellent transmission infrastructure with at least 1000 MW of capacity and cheap land.

39.2.2 Adaptation to Load via Azimuthal Orientation of PV

One useful way to improve the adaptation of PV generation to the load profile is to transition away from the tradition of a strictly southern orientation of arrays. While a southern orientation (for all sites north of the equator) does provide for the highest annual energy yield and maximum profitability for fixed feed-in tariffs, the actual prices on the electricity spot-market paint a different picture. At the beginning of PV development, peak prices were observed around noon which meant that expensive peak electricity was replaced by solar power. In order to persuade investors to transition to new optimization objectives, either feed-in tariffs should be modified to account for actual spot-market price development or financial compensation in return for slightly reduced energy yield might serve as sufficient incentives.

39.2.3 Adaptation to Seasonal Load Through Modification of PV Array Elevation Angles

Electricity demand in Germany (as is the case with all other countries that are located at 40–90° of latitude) is higher in the winter than in the summer; however, present PV arrays are optimized for a maximum annual energy yield , Which results

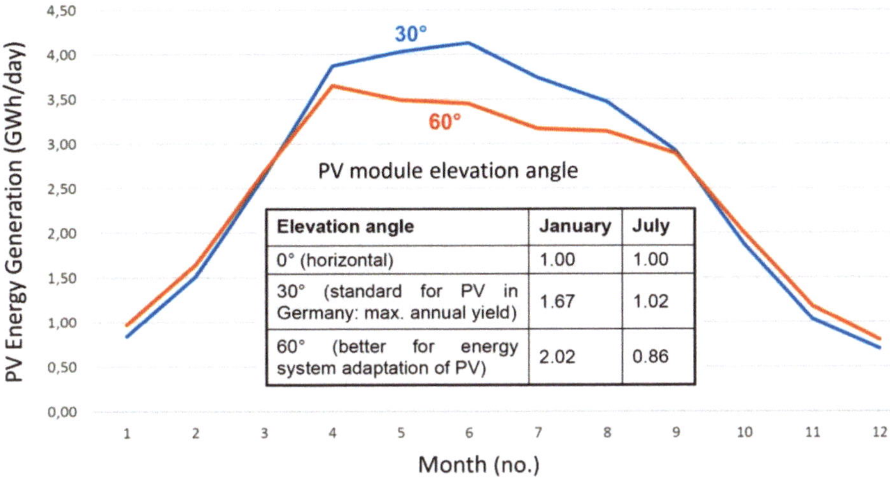

Elevation angle	January	July
0° (horizontal)	1.00	1.00
30° (standard for PV in Germany: max. annual yield)	1.67	1.02
60° (better for energy system adaptation of PV)	2.02	0.86

Fig. 39.4 Average daily electrical energy output of a 1-GW PV power plant located in Berlin (latitude: 52.5°N) for 30° and 60° module elevation angles. The included table shows the relative energy output for elevation angles of 0° (ref), 30° and 60° during the months of January and July

in a layout optimized for the summer months with a corresponding an average module elevation angle of 30° only (within Germany's latitude range of 47.5–54.8°). A steeper PV array angle would reduce large power discrepancies found between summer and winter months and also generate more electricity during winter when there is higher demand for it. An increase in the elevation angle to 60° would better capture winter sun and diminish oversupply during summer. As the graph in Fig. 39.4 shows, the relative increase in energy generation during winter is barely visible, so the relative generation data for average summer and winter months are provided in the table included. To compensate for considerable reductions in terms of annual energy harvest, adequate compensation should be provided in order to create an incentive for PV investors to implement the measures described.

39.2.4 PV Tracking Systems

PV tracking systems provide users with the greatest flexibility to reduce solar variability or to increase power generation throughout the day. Prior to 2008, tracking systems were frequently employed because they helped to generate an additional 30–40% increase in energy yield. However, since 2008, as PV modules have become more efficient and less expensive, the gain in energy harvest provided by tracking systems, which carry higher installation and operating costs, are not as economically beneficial as they once were. However, with the inclusion of actual spot-market prices and cost of grid system services, a renaissance in tracking technologies seems possible.

39.2.5 Thermal Conditioning of PV

Opportunities for increased efficiency also arise through the use of latent heat storage attached to PV modules. When the warming of PV modules is delayed, PV arrays operate with higher conversion efficiencies and increased power output during the morning hours (see also [3], other cooling strategies [4]). Similar to the tracking systems mentioned in Sect. 2.4, the cooling measures carry higher installation costs and are not as economically beneficial as they once were (from the energy gain point of view). However, with the inclusion of actual spot-market prices and grid system services, there could also be a renaissance in cooling technologies.

39.2.6 Optical Enhancement

The use of angular-selective/anti-reflection surface layers or structures can reduce optical reflection, particularly in the morning and in the evening. This is demonstrated for a 90°V-structure on the module surface which leads to increased optical transmission for incidence angles of 36° to 50°, see Fig. 39.6, and therefore to higher PV power output (see Fig. 39.5) during periods of high spot-market prices from 9 to 10 am (see Fig. 39.4).

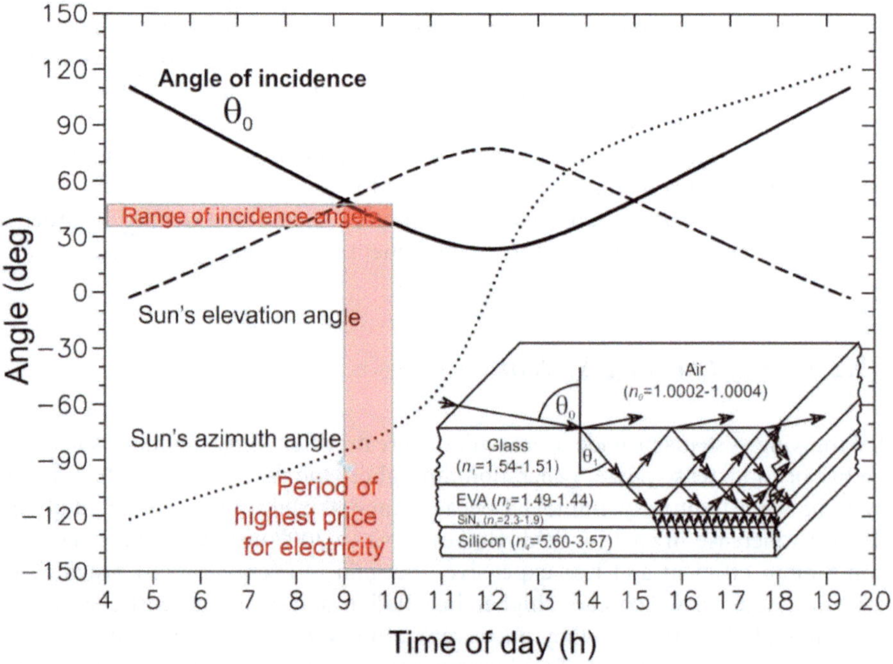

Fig. 39.5 Angle of incidence *(bold line)* during the course of the day; the *red color* area represents a period of high spot-market prices and the accordingly relevant angles of incidence. The *inset* shows a cross-section of a PV module encapsulation with the optical path of the incident irradiance

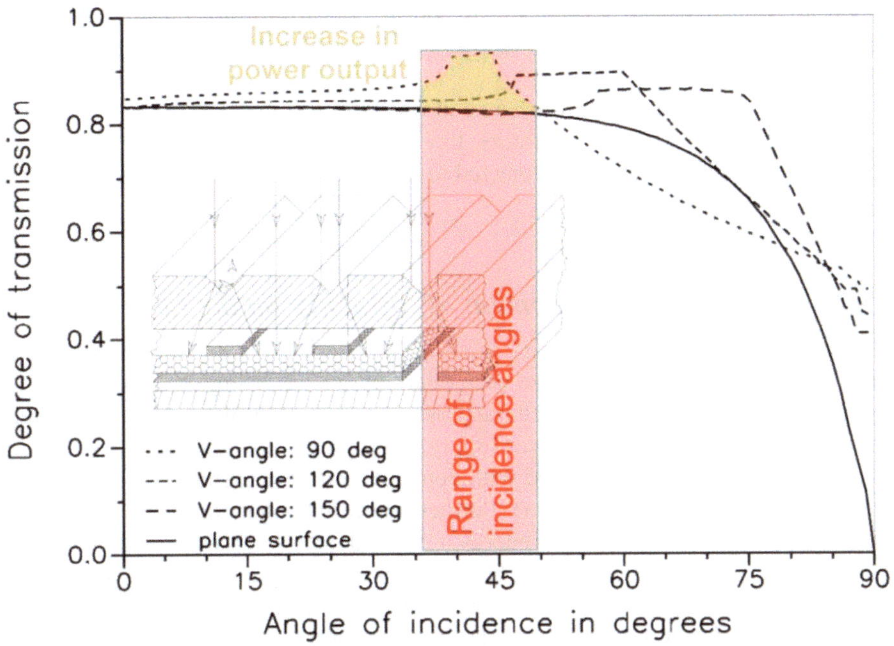

Fig. 39.6 Optical transmission through a typical three-layer module as a function of incidence angle. Plane surface (*bold line;* see also inside image of Fig. 39.5) and different V-structured surfaces (see inside image with a V-angle of 90°) together with the relevant incidence angles (36° to 50°) that correlate to high spot-market prices, as identified in Fig. 39.5. The increase in transmission and power output *(olive color)* is displayed for an advantageous V-angle of 90°

While the overall PV energy generation gain via V-structures is approximately 3–4% only [1], that gain is achieved during periods of the day with flat incidence angles and high spot-market prices, so the value of that increase may double (depending on the actual spot-market price variation and its future development). A similar effect in optical enhancement plus an additional energy gain from thermal conditioning of up to 10% can be achieved via a surface water film which has been investigated in [4].

39.3 Load Management

It is often assumed that the load profile is fixed and constant, with little prospect for customization; however, upon closer inspection, macro- and micro-scale opportunities are available for load distribution. Examples, with a focus on modifications that are relatively easy to implement, are shown below. Possibilities for load adaptation to PV power output are numerous:

- Flexible tariffs to trigger use of appliances during favorable periods.
- Flexible consumers/producers controlled by grid load, PV output and predictive use.

- Cheap storage located in proximity to end-users. Various forms of storage capacity may include:
  - Local heat storage (dishwashers and washing machines).
  - Local cold storage (refrigerators, freezers).
  - Parked electric vehicles connected to the grid.

39.3.1 Historical Development of Influence on the Load Curve

Since the 1960s, load curves have been adapted to an existing or desired electricity generation structure with its characteristic generation profiles. During the very early development of electric power supply, hydropower plants were popular which allowed generation to be matched with demand relatively easily via the control of water flow (reaction time $\leq$1 min). Furthermore, this balancing measure operates quite efficiently with little loss of energy. Coal and later, nuclear generation sources (known as *base-load power plants*), became increasingly popular around the middle of the twentieth century; however, their flexibility to throttle up/down is limited and it requires significant time frames to do so (up to 8 h).

Actual electricity requirements and corresponding load profiles in the 1950s and 1960s were characterized by high demand during the day and low demand during the night. The most expensive electricity, during peak hours, had to be supplied by additional power plants around midday and in the evening. In order to achieve uniform capacity from base-load power plants during the day and night ('load shaving': to reduce the amplitude of the load profile), attractive overnight tariffs were offered, often in combination with local heat storage options for the consumption units. Some of the most prominent examples are electric night storage heaters and electrical heat pumps. The effect of differentiated overnight tariffs can be observed in Fig. 39.7 with an increase in night consumption being the most apparent. In other words, load curve adjustments are possible and have been successfully achieved in the past. Sometimes the term 'load shaping' is used to describe measures to modify the load profile. Due to an increase in solar generating capacity, the load shaping performed over recent decades should be reversed (towards more pronounced day versus night cycles—the opposite of 'load shaving') in order to better coordinate PV generation with consumption demands and therefore reduce costs.

39.3.2 Load Control by 'Large Virtual Consumers' (LVCs)

An example of a *large virtual consumer* (LVC) might be end users that employ numerous refrigerators equipped with latent cold storage in combination with a common control mechanism. The main effect is the release of *control* (or *balancing*)

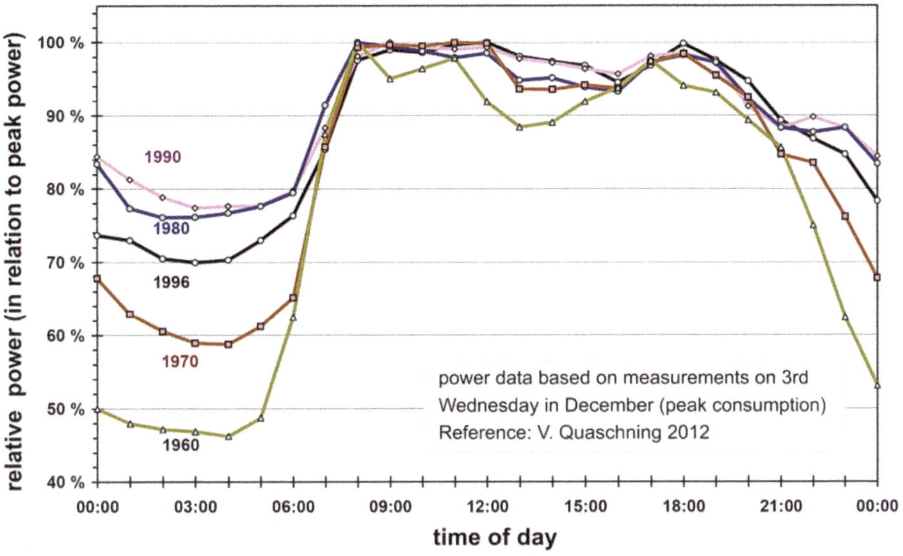

Fig. 39.7 Evolution of the relative diurnal electrical load profile in Germany from 1960 to 1996 [5]. Until recently, night load has been increased to reduce the day–night variation ('load shaving')

energy by LVCs for use by the electrical grid. Each of the refrigeration units with latent cold storage is also equipped with a communication device that is able to receive direct secure messaging (DSM) via the GSM mobile phone network from an electrical grid system operator who may ask for a load release *(minimal load)*, load increase *(maximum load)* or allow for control as required by the local device *(standard)* (Fig. 39.8).

DSM was selected to allow for regional control of the broadcasting message, to avoid individual control of each single refrigerator and to maintain privacy. Depending on the operating state of the refrigerator, the command to switch off will be

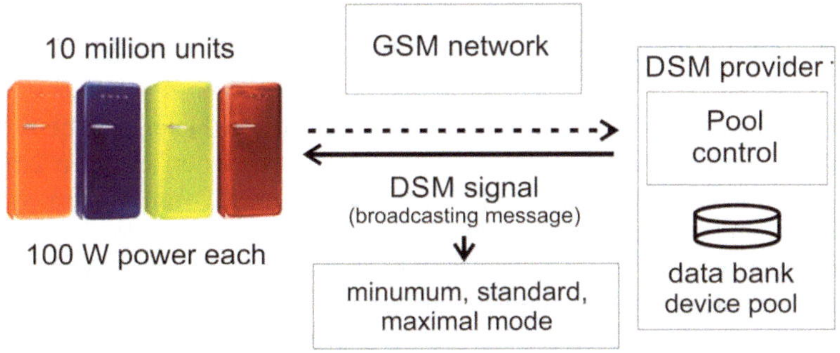

Fig. 39.8 Example of distributed load management: communication structure with DSM signals for influencing load by cooling units (equipped with latent cold storage)

ignored if the products being cooled are at risk of high temperatures; however, due to the condition of latent storage, this seldom happens and a majority of the refrigerators do participate and effectively contribute to load management.

In Germany, refrigerators and freezers account for 3.6 GW of power which could be theoretically available for load management if they were all equipped with suitable control units. If only 10 million new refrigerators were equipped (each unit consumes approximately 100 W during its active operation period), this would account for 1 GW of controllable load with high availability at little extra cost.

39.3.3 'Power-to-Gas'

Power-to-Gas is a concept in which excess electricity is used to produce hydrogen via electrolysis, and CO_2 is then used to convert the hydrogen into synthetic methane. For storage and distribution of methane (and to some extent, hydrogen) existing natural gas infrastructure such as pipelines, underground storage, and gas stations, can be used. Methane could serve as mobile fuel or as an energy source for heat generation. To produce the gas, excess amounts of electricity from PV may be used. Hydrogen and methane can be converted back into electrical energy (via a fuel cell or a gas turbine), and so serve as as (indirect) electical energy storage. Through power-to-gas, the energy sectors are becoming more integrated and can also support an increase in the proportion of renewable energies for transport and heat generation. Power-to-gas is an effective method to counteract renewable power fluctuations in the network and thus balance generation and load management as well as prevent network congestion. Siting power-to-gas plants close to generation and gas infrastructure can reduce electrical grid loads and other associated costs. Energy transport would be shifted from the electrical to the gas grid with its higher transmission capacities (e.g., 30 GW for a single pipeline). A model of a power structure operating with 100 % renewable energy using the power-to-gas concept is depicted in Fig. 39.9.

39.3.4 Electric Mobility to Influence the Load and for Storage

By 2020 it is estimated that at least 1 million electric vehicles will be operating in Germany. In the private sector, most vehicles are used for less than 2 h a day. Therefore, if a sufficient number of bi-directional power terminals are installed in parking lots and garages, vehicle batteries could be available for load management and grid control services for approximately 20 h a day. Since the charging necessary for a vehicle is completed much quicker than the parking duration (at least for normal operation), load management can be carried out via adjusting the speed of the charge (within the capability of the battery).

Assuming that 70 % of all vehicles have an average storage capacity of 20 kWh and an average charge of 50 %, an additional 7 GWh of capacity could be available

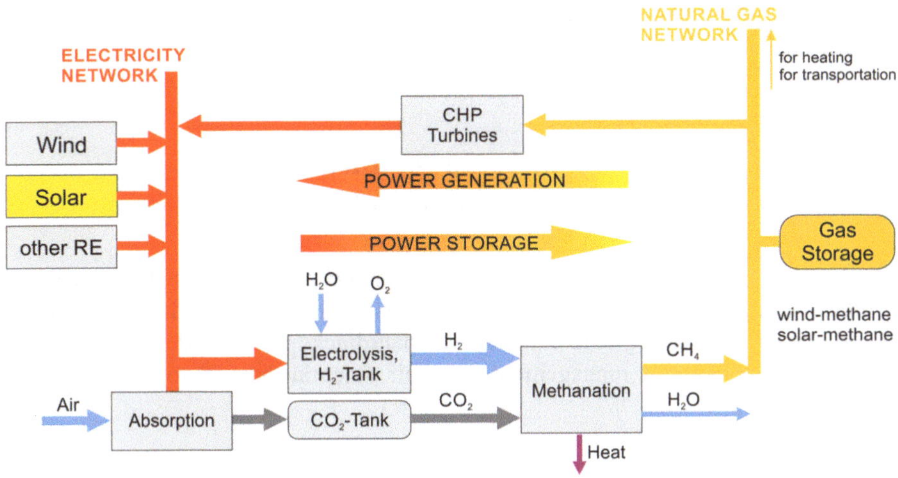

Fig. 39.9 "Power-to-Gas" concept [6] using natural gas infrastructure for storage and distribution of energy from PV and other renewable sources

to control/balance energy demands. For further expansion of electric mobility, in a scenario where 90 % of all vehicles in Germany are electric, the available balancing energy would increase to 277 GWh.

Control could be maintained either directly by the user, bundled via *aggregators* (see Fig. 39.10) or via the grid/system operator. It should be noted that utilizing vehicles as storage can be quite expensive and, thus, load management should mainly

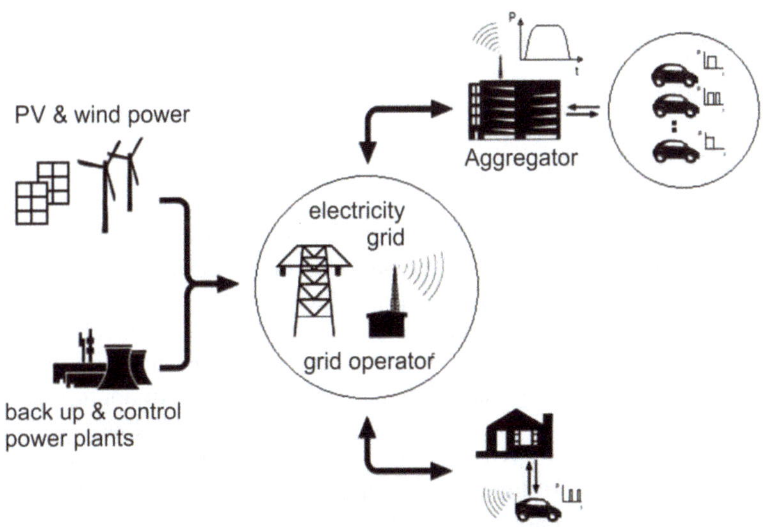

Fig. 39.10 Electric mobility with controlled charge for load management and, in exceptional cases, for intermediate storage

be employed for the charging process—only in emergency cases should vehicle batteries be used as a form of distributed storage.

39.3.5 Load Management in the Industrial Sector and Information Technology (IT)

Load management can be further employed in industrial sectors (heavy electric grinders and saws), and also for transportation, refrigeration and heating. Many IT processes such as big-data calculations, data backup and rendering could also serve as a form of load management but their potential has yet to be sufficiently determined.

39.3.6 Seasonal Load Adaptation via a Shift in the Holiday Season

Unfortunately, most electrical energy consumption does not occur in summer (when more PV power is available) but in the winter (at lest for colder climates). While this may be explained, in part, by the demand for more heat and light during the winter, holiday behaviour also plays a role. For example, a considerable portion of Germany's economic activity (and of many other countries) slows down during the summer holiday season, causing a decrease in consumption which leads to overproduction of PV supply. A shift from traditional summer holidays to wards winter holidays could have a significant impact on the adaptation of load to PV power availability. Research on this subject is being conducted within the Department for Sustainable Energy Concepts at the University of Paderborn.

39.4 Marketplace

In order to encourage investment in the concepts shown above, financial incentives are necessary. These incentives can be provided either through government subsidies or through access to the electricity price fluctuations in spot-markets. Profitable participation in a spot-market requires an open and easily reachable marketplace along with an effective communication infrastructure. Currently, pre-qualification hurdles are prohibitively high and thus limit participation. An appropriate market design example is shown in Fig. 39.11 and discussed further below.

In this model, participants offer their services (generation, storage, transport, energy management, load shifts) and articulate their needs (power/energy requirements, backup) to a common platform that is accessible to the end user. A deal is made when the participants agree on a price. Market regulation and ratings systems

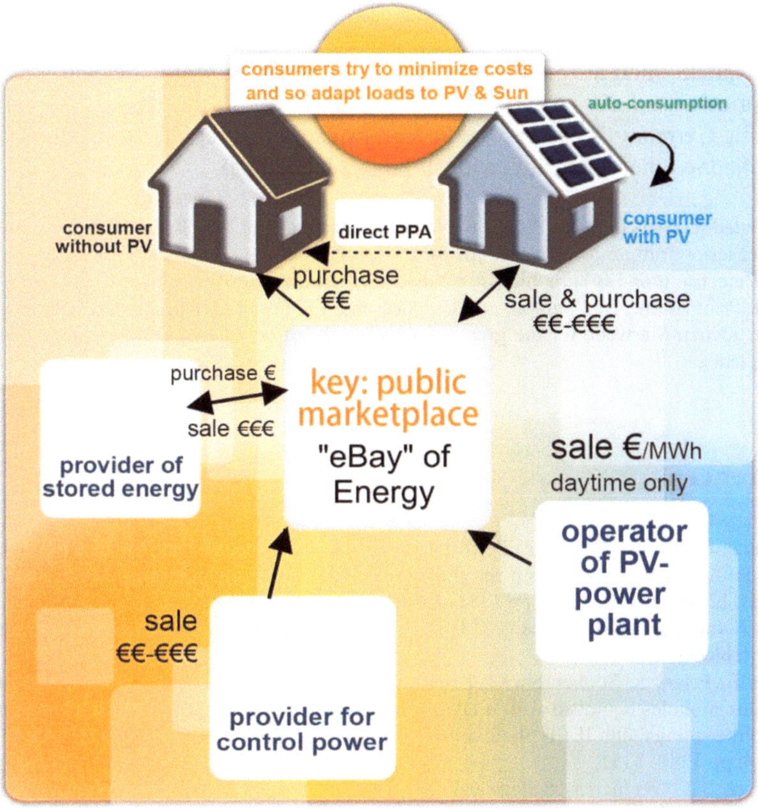

Fig. 39.11 Common market platform through which all participants are able to offer and trade their energy services and needs [7]

(e.g., for confidence of delivery), similar to the eBay platform, could be employed between entities. Direct sales from one participant to another via a power purchase agreement (PPA) should be possible. The transport of generated, sold, purchased or used electricity is becoming more and more simple, accessible and market-based.

In order to fully achieve a PV energy transition, there are administrative and legislative challenges to overcome before an open market platform is implemented. Without effective market access, many innovative solutions to integrate PV in the energy matrix will not be realized.

39.5 Conclusions

For effective integration of PV into the energy matrix, a two-way strategy should be employed. One strategy will require different adaptations to the generation process and the other will require adaptations to the demand. The latter can be achieved via changes in consumer behavior or/and the utilization of appliance equipment that is capable of carrying out load management, e.g., via local storage units. To achieve results quickly and at low costs, a change of behavior on the demand side seems appropriate.

While Germany is used as the basis for data in this report, these conditions can be applied to other countries (with suitable adaptations).

Acknowledgements Sincere thanks to Mr. Keith Parsons and Mr. Stefan Brill for proofreading the manuscript draft, to Mr. Ewald Japs for the graphics of the spot-market prices, Mr. Ali Ameli for the 'electric mobility' graphics, Mr. Yassin Bouyraaman for the 'power-to gas' graphics and Dr Dirk Prior of the 'LVC' graphics, Mrs. Alessandra Simplicio for the 'market-place' graphics, and Mr. Jochen Marwede for the graphics showing the effect of east/west orientation of the PV power plants.

References

1. Krauter S, Hanitsch R (1992) Performance of a partly structured surface at a PV-module. In: Proceedings of the11th European photovoltaic solar energy conference, Montreux (Switzerland), 12–16 October 1992, pp 1351–1354
2. EPEX Intraday Market Data (2013) http://www.epexspot.com/en/market-data/intraday/intraday-table/-/DE
3. Japs E, Peters S, Sonnenrein G, Krauter S (2014) Energy-economic comparison of photovoltaic modules equipped with a layer of conventional and improved phase-change material. In: Proceedings 40th IEEE photovoltaic specialist conference, Denver, CO (USA), 8–13 June 2014, pp 1348–1352
4. Krauter S (2004) Increased electrical yield via water flow over the front of photovoltaic panels. Sol Energy Mater Sol Cells 82:137–137
5. Quaschning V, Communication private (2012) Regenerative Energiesysteme, 8th edition. Carl Hanser 2013. (ISBN:978-3-446-43526-1, German version of graphics is included in the book)
6. Sterner M Bioenergy and renewable power methane in integrated 100% renewable energy systems. Limiting global warming by transforming energy systems. Renewable energies and energy efficiency 14. Kassel University Press. (ISBN:978-3-89958-798-2)
7. Krauter S (2012) PV 3.0. In: Proceedings of the 27th European photovoltaic solar energy conference, Frankfurt a. M., Germany, 24–28 September 2012, pp 4520–4522

Chapter 40
Evaluation of the Characteristics of the PV Module Considering Effects of Real Climatic Conditions

H. Yatimi and E. Aroudam

Abstract The photovoltaic (PV) model is used in a simulation study to validate the system design of a PV system. This work presents the modeling and the simulation of the PV module using the Matlab environment taking into consideration the measurements carried out under real working conditions in Tetouan (northern Morocco). The model is developed based on the mathematical model of the PV module. A particular PV module is selected for the analysis of the developed model. The three-dimensional profile of solar radiation used in this work is presented. The output current and power characteristic curves, which highly depend on some climatic factors such as solar radiation and temperature, are obtained by simulation for the selected module and discussed. The instantaneous power production characteristics of the 12 months of the year are presented and interpreted.

Keywords PV module · Modeling · Simulation · I–V and P–V curves

40.1 Introduction

Currently, the sources of energy and the harmful increase in greenhouse gases are problems of actuality. The additional danger is that excessive consumption of natural resources has reduced the reserves of this type of energy in a dangerous way for

H. Yatimi (✉) · E. Aroudam
Modeling and Simulation of Mechanical Systems Laboratory,
Physics Department, Faculty of Sciences, Abdelmalek Essaadi University,
Sebta Ave., Mhannech II BP 2121, Tetouan 93002, Morocco
e-mail: yatimi.hanane@gmail.com

E. Aroudam
e-mail: aroudam_hass@hotmail.com

A. Sayigh (ed.), *Renewable Energy in the Service of Mankind Vol II*,
DOI 10.1007/978-3-319-18215-5_40

future generations, mainly with the industrial development. In order to preserve our natural living spaces and protect their resilience, a renewed compatibility would require a suitable form of alternative energy sources that should be independent and easily accessible. Moreover, with industrial developments, the use of power generation technology with renewable energy sources is developing rapidly, and the application of photovoltaic (PV) solar energy has become popular. Renewable energy has advanced all over the world in environment protection since it is clean, operates silently, has long lifetime and absence of fuel cost, and is inexhaustible [1, 2]. Renewable energy, and particularly the power generation from solar energy using PV panels, has emerged in the past decades since it has the aforesaid advantages and requires less maintenance.

The photovoltaic energy is the electrical energy produced from the solar radiation by the solar panels. The increase in the number of PV systems installed all over the world brings the need of proper supervision and control algorithms as well as modeling and simulation tools for researchers and practitioners involved in its application. The modeling and electrical characterization of panels currently marketed are needed to optimize the operation of PV systems using these PV panels. This may considerably reduce the cost of the PV systems [3] and increase the efficiency of PV generators. The major problem of the production of electrical energy by this technique is the optimal functioning of the PV panels (modules). However, the development of profitable and economically viable conversion systems pass necessarily by the understanding of different components of the system in which the fundamental unit is the solar PV module. The latter is composed of several solar cells that require study and understanding. A precise knowledge of the current–voltage (I–V) and power–voltage (P–V) characteristic curves of PV modules is required to study PV systems in an efficient manner.

The main aim of this chapter is to provide the reader with the fundamental knowledge of modeling and simulation of PV generator blocks based on mathematical equations using real meteorological data collected from Tetouan [4]. The electrical modeling of real (practical) PV cells is presented, and the mathematical equations are obtained from the modeling circuit. Afterward, the electrical characteristic data of the reference model are presented. Additionally, the meteorological data sets used in this chapter, each one of which consists of daylong data from sunrise to sunset, are presented. The three-dimensional profile of solar radiation is also presented. Afterward, the effects of solar radiation and temperature, which are locally measured for a typical clear day of 1 month of every season on the output current and power characteristics obtained by simulation for the selected module, are considered and analyzed. Finally, the instantaneous power production characteristics of the 12 months of the year are presented and interpreted.

40.2 PV Module Modeling

40.2.1 Modeling Using Single Diode Model

As the PV module is composed of a group of cells, its model is based on that of a PV solar cell. The equivalent circuit of a PV solar cell is shown in Fig. 40.1. Series resistor R_s and parallel resistor R_p that limit the performance of the cell are added to the model to take into account the dissipative phenomena (internal losses) in the cell [5].

R_s: Series resistor, mainly due to losses by Joule effect through grids collection and to the specific resistor of the semiconductor as well as bad contacts (semiconductor, electrodes). R_p: Parallel resistor comes from the recombination losses mainly due to the thickness, the surface effects, and the non-ideality of the junction.

The I–V characteristic equation is given as follows [6]:

$$I = I_{\mathrm{ph}} - I_0\left[\exp\left(\frac{q}{akT}(V + IR_s)\right) - 1\right] - \frac{V + IR_s}{R_p}, \tag{40.1}$$

where I_{ph} is the light-generated current or photocurrent, I_0 is the reverse saturation current of the diode, q is the electron charge ($1.60217646 \times 10^{-19}$ C), k is the Boltzmann constant ($1.3806503 \times 10^{-23}$ J/K), and T is the temperature of the p–n junction in K. Variable a represents the diode ideality factor. It depends on recombination mechanisms in the space charge zone. In the ideal case, R_s tends towards 0 and R_p to infinity. And in the real case, these resistors provide an assessment of the imperfections of the diode, considering that the resistance R_s has a low value, the slopes of the I–V characteristics are calculated at $I=0$ open circuit and V = 0 short circuit and give the inverse of series and shunt resistance values, respectively.

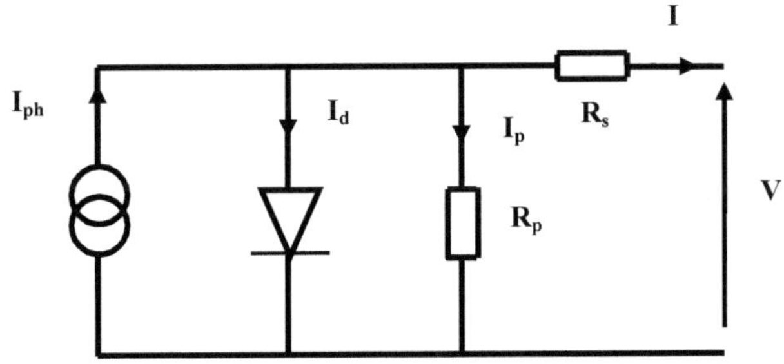

Fig. 40.1 Equivalent circuit of a real solar cell

Table 40.1 Electrical characteristic data of the PV module at STC

Parameters	Value
Maximum power (P_{max})	90 W
Voltage at maximum power (V_{mp})	47.4 V
Current at maximum power (I_{mp})	1.90 A
Open-circuit voltage (V_{oc})	60.5 V
Short-circuit current (I_{sc})	2.06 A

40.2.2 *Reference Model*

In order to apply these concepts to the developments of the solar cell model, the First Solar FS Series 3 Black PV module has been chosen for modeling [7]. The performance of the solar cell is normally evaluated under the standard test conditions (STC), where an average solar spectrum at air mass (AM) 1.5 is used, the solar radiation is normalized to 1000 W/m^2, and the cell temperature is defined as 25 °C. The key specifications are shown in Table 40.1.

40.3 Results and Discussions

In order to have a model to simulate the operation of our associated cells, we developed a model in Matlab environment by using the mathematical equation given in the previous section, and the First Solar FS Series 3 Black PV module is taken as reference module.

40.3.1 *Meteorological Data Sets*

The results of prediction of the operating current from four parameter models were analyzed for a number of meteorological data sets. Each set consisted of daylong data, from sunrise to sunset. The important statistics of the meteorological data used are presented in Table 40.2.

40.3.2 *Profile of Solar Radiation*

As the solar radiation varies over the course of the year, it is important to pay particular attention to the solar radiation profile. This daily profile (Fig. 40.2) is obtained by extrapolation of locally measured data at regular intervals (1 h; [4]) throughout the typical clear day of each month of the year in northern Morocco (Tetouan).

Table 40.2 Important statistics of the daily data set measured from experimental setup used in the current chapter

Data set	Date	Local time of day (p.m.)	Solar radiation (maximum) (W/m^2)	Temperature (°C)
1	January 5,1990	13.00	461.3340	16.7117
2	February 22,1990	13.00	637.1327	19.9941
3	March 14,1990	12.00	720.8477	19.1060
4	April 30,1990	12.20	865.0566	21.4096
5	May 11,1990	12.00	888.6379	23.8182
6	June 22, 1990	13.00	903.6524	27.0578
7	July 3,1990	12.30	873.0143	27.1189
8	August 26,1990	12.30	800.7225	29.8530
9	September 7,1990	12.00	692.8818	27.2954
10	October 26, 1990	12.00	711.9397	25.0706
11	November 23,1990	12.00	592.3771	26.9733
12	December 01, 1990	12.50	490.1700	15.6431

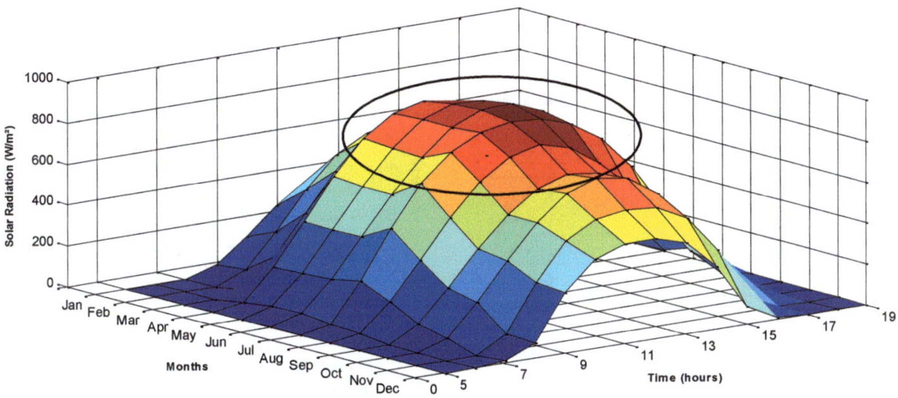

Fig. 40.2 Northern Morocco solar radiation profile

The figure above shows the three-dimensional profile of solar radiation. From the surf view, the solar radiation reaches its maximum from April to August and from 9 a.m. to 4 p.m.; during this period the power production is maximal. In the figure, we note that during June (summer) the solar radiation is maximum (903.65 W/m^2), during April (spring) it decreases (865.1 W/m^2), then in October (autumn) it is still decreasing (711.9 W/m^2), and finally in December (winter) the solar radiation is minimum (490.17 W/m^2).

40.3.3 Solar Radiation Effects on PV Module Characteristics

The response of a PV cell at different levels of solar radiation and at a constant temperature of 25 °C shows that solar radiation has a significant effect on the short-circuit current (Fig. 40.3), while the effect on the voltage in the open circuit is quite low. Regarding the power, we can clearly see the existence of the maximum on the power curves (Fig. 40.3), corresponding to the maximum power point P_{max}.

This figure represents the curves of the generated current and power by one PV module for a typical clear day of 1 month of each season of the year.

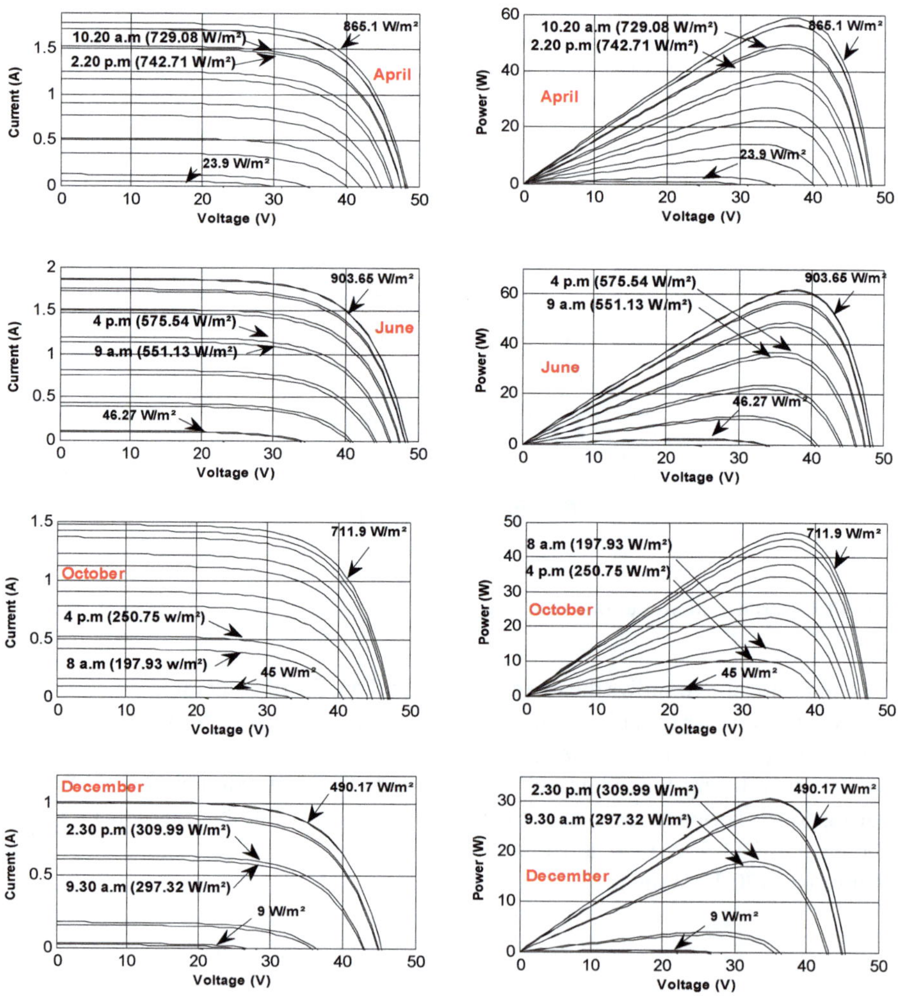

Fig. 40.3 Current–voltage (I–V) and power–voltage (P–V) characteristics at constant temperature and variable solar radiation for a typical clear day of 1 month of each season—Tetouan

Table 40.3 Maximum power point *(MPP)* under the influence of solar radiation for a typical clear day of 1 month of each season

Months	Solar radiation (W/m^2)	Constant temperature T=25°C		
		P_{mpp} (W)	V_{mpp} (V)	I_{mpp} (A)
April, spring	865.1	61.74	12.67	1.86
June, summer	903.65	58.73	12.61	1.78
October, autumn	711.9	47.03	12.36	1.47
December, winter	490.17	30.70	12.02	1.012

When the solar radiation is high, the cell generates more power. The PV power increases during the first hours of the day and gradually decreases, as we can see in the figures above. Table 40.3 gives the digital value of the maximum power point (MPP) of the PV module at constant temperature and variable solar radiation.

40.3.4 Temperature Effects on PV Module Characteristics

Temperature is an important parameter in the behavior of solar cells. It has an influence on the characteristics of a PV generator. In Fig. 40.4, the curves show the variation of the characteristics of a PV module at different levels of temperature and at a given solar radiation (fixed at the maximum of a typical clear day of 1 month of each season of the year). These curves represent the generated current and power by one PV module of a typical clear day of 1 month of each season.

From these curves, we can clearly see that the temperature has a very important effect on the open-circuit voltage and does not have a remarkable effect on both the short-circuit current and the power of the module as shown in Table 40.4. When the temperature is low, the cell generates more power.

Table 40.4 gives the MPP of the PV module for a typical clear day of each month for the maximum and the minimum values of the temperature and at a given solar radiation.

40.3.5 Power Production of the 12 Months of the Year

The daily powers for each month of the year are represented in the curves in Fig. 40.5 and Table 40.5.

After analyzing the results for the daily powers of PV, we note that the period of operation of PV is closer during the winter months and is approximately between 8 a.m. and 5 p.m., 9 h/day (November, December, and January), and vice versa for the summer period. Production by PV is zero or very low during the periods of peak power consumption (between 8 p.m. and 10 p.m.), which requires either the use of other generators or stored energy, and this is according to the peak value and the classification of the day in the year.

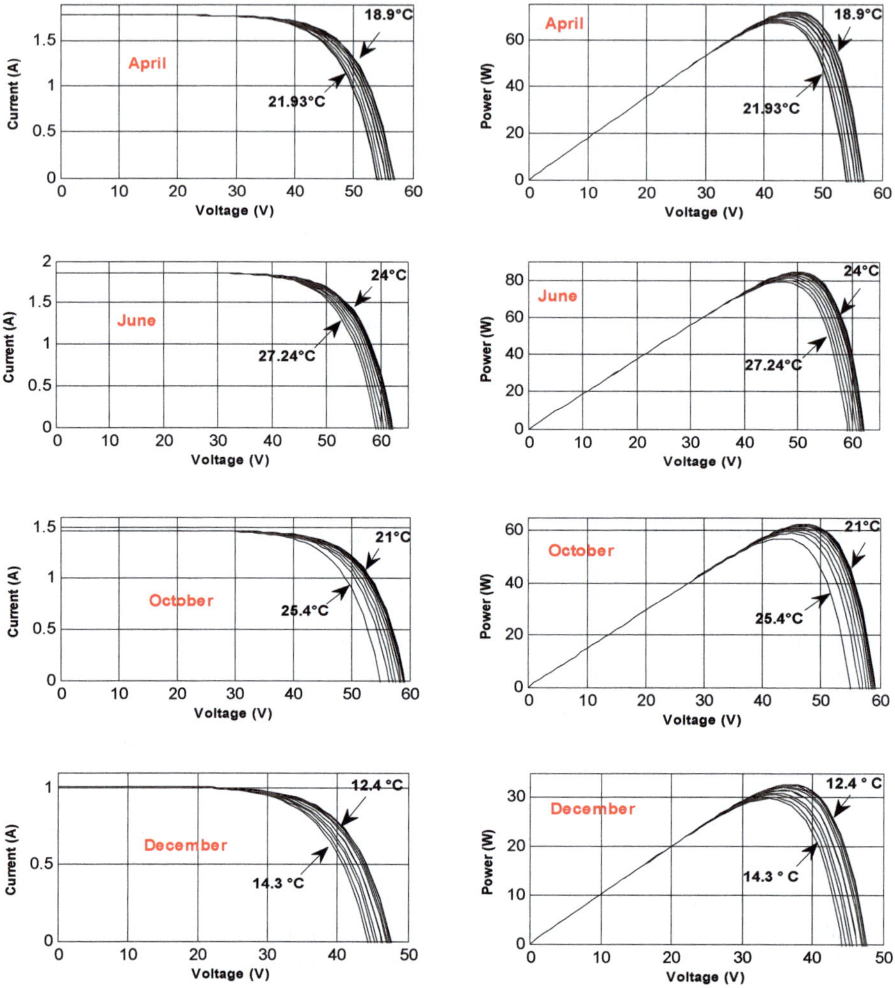

Fig. 40.4 Current–voltage (I–V) and power–voltage (P–V) characteristics at constant solar radiation and variable temperature for a typical clear day of 1 month of each season—Tetouan

Figure 40.5 shows that the instantaneous power produced in April is 58.73 W (at 00.20 p.m.) and in June 61.74 W (at 1 p.m.). These two powers, which are highly dependent on solar radiation, are very close, and this is due not only to the high value of solar radiation in June (903.65 W/m^2), which is normal since it is summer, but also to its high value in April (865.1 W/m^2), which is a month of spring where the atmosphere is clean with less dust and particles, thanks to the rain of winter.

Table 40.4 Maximum power point *(MPP)* under the influence of temperature for a typical clear day of 1 month of each season

Months	Temperature (°C)	Constant soar radiation (fixed at the maximum of a typical clear day of each month) (W/m²)		
		P_{mpp} (W)	V_{mpp} (V)	I_{mpp} (A)
April, spring	18.9	72.41	56.9	1.783
	21.93	67.52	53.91	1.783
June, summer	24	84.5	62.25	1.861
	27.24	79.27	59.14	1.861
October, autumn	21	62.58	59.25	1.467
	25.4	56.95	55	1.467
December, winter	12.4	32.69	47.63	1.011
	14.3	29.73	44.26	1.011

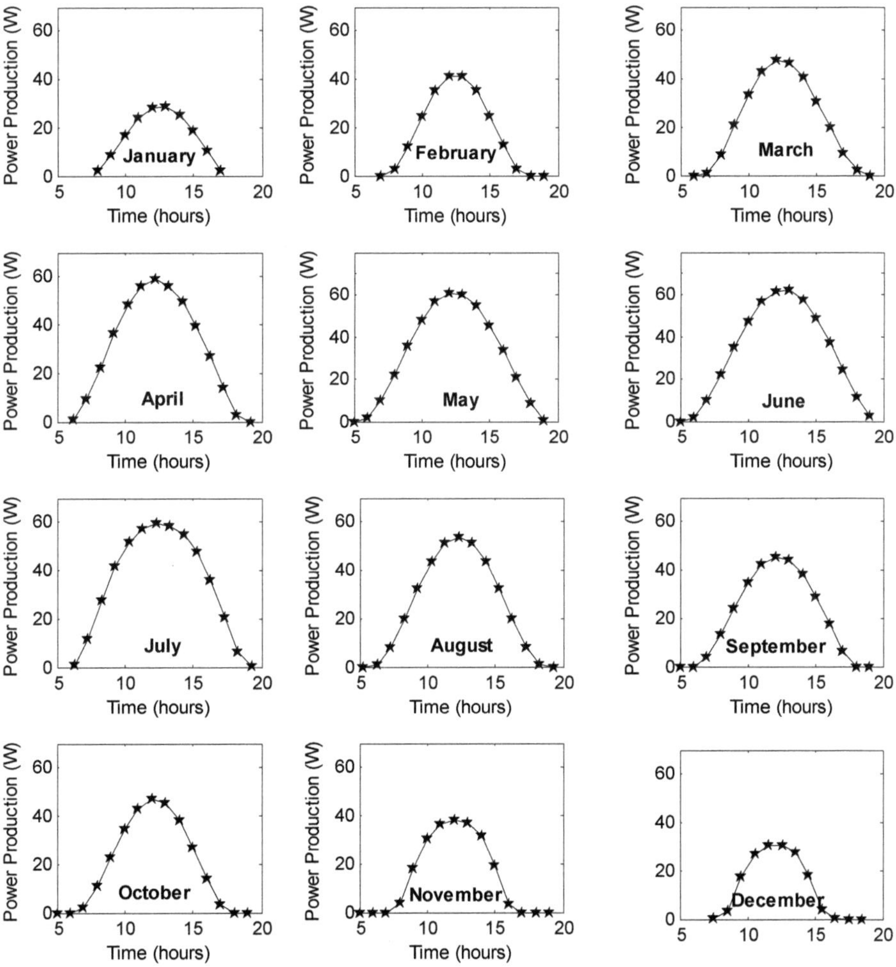

Fig. 40.5 Daily power production of the 12 months of the year

Table 40.5 Daily power (MPP) of each month of the year

Months	January	February	March	April	May	June	July	August	September	October	November	December
Power (W)	163.37	232.31	302.99	420.36	455.49	475.73	475.43	365.397	298.76	287.54	218.42	158.97

40.4 Conclusions and Perspectives

In this chapter, we presented a modeling and simulation of the PV module to understand its electrical operation depending on real climatic factors. The simulation results showed that the I–V and P–V characteristic curves are highly influenced by temperature and solar radiation that are locally measured, and also the curves of daily power production of the 12 months of the year are highly affected by solar radiation. The following of this work is based on optimizing the performance of PV modules using commands and techniques to minimize the influence of these meteorological parameters.

References

1. Mozaffari Niapour SAKH, Danyali S, Sharifian MBB, Feyzi MR (2011) Brushless DC motor drives supplied by PV power system based on Z-source inverter and FL-IC MPPT controller. Energy Convers Manag 52:3043–3059
2. Reza N, Gevorg B (2013) Gharehpetian, an investigation on combined operation of active power filter with photovoltaic arrays. Int J Electr Power Energy Syst 46:392–399
3. Mrabti T, El Ouariachi M, Tidhaf B, Kassmi K, Chadli E, Kassmi K (2009) Modeling of electrical properties and characterization of solar panels, modelisation des proprietes electriques et caracterisation des panneaux photovoltaïques. Renew Energy Rev 12(1):105–116
4. Aroudam El H (1992) Evaluation du Gisement solaire dans la région de Tétouan. Doctoral Thesis, Faculty of Sciences, Abdelmalek Essaadi University, Morocco
5. Akassewa Tchapo SINGO (2010) Système d'alimentation photovoltaïque avec stockage hybride pour l'habitat énergétiquement autonome. Doctoral Thesis, Henri Poincaré University, Nancy-I
6. Villalva MG, Gazoli JR, Filho ER (2009) Comprehensive approach to modeling and simulation of photovoltaics array. IEEE Trans Power Electron 5:1198–1208
7. The Datasheet of module First Solar FS Series 3 Black PV module (2013) [Online] http://www.firstsolar.com/Company-Documentation. November 2013

Chapter 41
Temperature Dependence of Carrier Mobility in SubPc and C_{60} Organic Semiconductors for Photovoltaic Applications

Nesrine Mendil, Mebarka Daoudi, Zakarya Berkai and Abderrahmane Belghachi

Abstract Carrier mobility is an important parameter in determining device performance in optoelectronics. Here, we study the charge carrier mobility of C_{60} and SubPc neat layers for different temperatures calculated in darkness. The results show that temperature affects the disorder energy (σ) of organic semiconductors; when the temperature is <289 K, this parameter is around 0.03 eV and it is double when the temperature is >289 K. In these conditions, the mobility of electrons and holes are around 0.14 and 10^{-7} Cm^2/Vs, respectively. Our results are in good agreement with experimental data reported by Pandey et al., Adv Funct Mater 22:617–624. 2012)

Keywords Organic semi conductors (C_{60}, SubPc) · Effect of temperature · Charge carrier mobility · Characteristic current density–voltage

41.1 Introduction

One new type of photovoltaic (PV) technology, organic solar cell technology, is based on fullerene (C_{60}), boron subphthalocyanine chloride (SubPc) and molecules. In organic solar cells, it is essential to deduce and understand the laws that determine the charge transport and the parameters which affect the charge transport as carrier mobility, charge distribution and temperature. These parameters play a critical role in determining the fluency and efficiency of the organic solar cells. In order

N. Mendil (✉) · M. Daoudi · Z. Berkai · A. Belghachi
Laboratory of semiconductor devices physics, (LPDS) University of Bechar, PB n°417, 08000 Bechar, Algeria
e-mail: mendil_nesrine@yahoo.com

M. Daoudi
e-mail: Mebarkadaoudi@yahoo.fr

Z. Berkai
e-mail: Berakai_zakarya@yahoo.com

A. Belghachi
e-mail: abelghachi@yahoo.fr

A. Sayigh (ed.), *Renewable Energy in the Service of Mankind Vol II*,
DOI 10.1007/978-3-319-18215-5_41

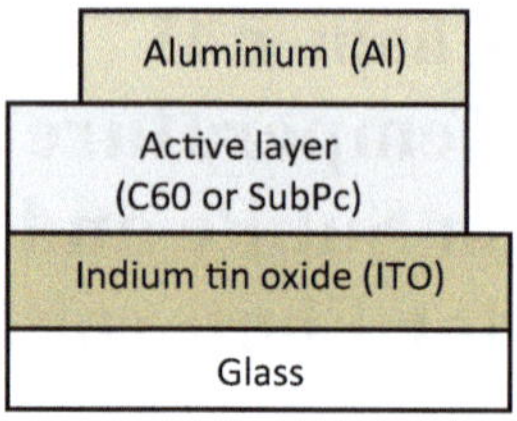

Fig. 41.1 Organic (C_{60} or SubPc) Schottky diode structure

to do this, the use of sophisticated modeling to simulate the charge transport is necessary. In contrast to the situation with crystalline and amorphous organic materials, there has been little research on charge mobility in liquid crystals. Here, we initiated a study on the mobility and current density of charge carrier results obtained for different temperatures of C_{60} or SubPc layers calculated without illumination. The results present nonlinear behavior.

The structure of a solar cell is shown in Fig. 41.1. Starting from the bottom, there is a glass substrate (the indium tin oxide [ITO] layer) which is a transparent layer which serves as the hole contact. In the middle is the active layer which is based on C_{60} or SubPc.C_{60} organic semiconductors, and is used as the organic electron-acceptor due to its high stability and high carrier mobility [2]. SubPc organic semiconductors are used as organic hole-donor semiconductors. Lastly, the aluminum layer is the electron contact.

The aim of our work is to investigate the effect of temperature on mobility and current density–voltage characteristics for both C_{60} and SubPc Schottky solar structures.

41.2 Theoretical Model

The key factor is to decide the efficiency of the carrier transport and collection process by the electrodes. Therefore, carrier mobility is a critical factor in determining the efficiency. Some theoretical models have been established for organic solar cells. In the absence of traps, the space charge limited current density *(J)* can be written as a function of the applied voltage *(V)* as [3]:

$$J = \frac{9}{8}\varepsilon\mu\frac{V^2}{d^3} \tag{41.1}$$

where ε is the relative dielectric constant of the organic thin film and is approximated as $\varepsilon=4$ [1], V is the applied voltage, μ is the charge carrier mobility of material, and d is the organic film thickness.

The mobility dependence of temperature as determined from the various electron-only or hole-only devices is given by the Gaussian disorder model. It is governed by the width of a Gaussian density-of-states as follows [4]:

$$\mu(E) = \mu_{\infty} exp\left[-\left(\frac{3\delta}{5K_BT}\right)^2 + 0.78(\left(\frac{\delta}{K_BT}\right)^{1.5} - 2)\sqrt{\frac{eaE}{\delta}}\right] \tag{41.2}$$

where μ_{∞} is the mobility as the temperature is going to infinity, δ is the Gaussian disorder, a is the inter-site spacing, E is the electric field, and K_B is the Boltzmann constant.

41.3 Results and Discussion

In this work we investigated carrier mobility and the effect of applied voltage on the current densities of the neat electron layer (C_{60}) and neat hole layer (SubPc), each one sandwiched between a cathode and an anode as illustrated in Fig. 41.1.

Figure 41.2 shows the electron-only (a) and hole-only (b) current density−voltage characteristics using Eq. 41.2, with the corresponding parameters for C_{60} and SubPc organic semiconductors in darkness shown in Table 41.1.

It is clear that the *J–V* characteristics exhibit two regions. When the applied bias is <0.2 V, the electron and hole current densities are almost zero. Above this region, the current densities continue to increase as the voltage increases. The overall behavior of these characteristics is nonlinear. Our results are in a good agreement with the experiment data.

We can see that our results in Fig. 41.3a exhibit two regions for range of voltage (0.3, 0.4, 0.6, 0.8 and 1.2 V) with various temperature values. The first one is from 120 K to T_c = 185 K (the critical temperature); the electron mobility increases with increasing temperature and voltage. In contrast, when the temperature is above T_c (the second region) we observe that the electron mobility increases with increasing temperature and decreasing voltage. In Fig. 41.3b the electron mobility increases with increasing temperature from 300 K to T_c = 371.98 K and decreasing voltage; however, after T_c, the electron mobility increases with increasing temperature and voltage.

Figure 41.4a shows two regions for range of voltage (0.3, 0.4, 0.6, 0.8 and 1.2 V) with various values of temperature. In Fig. 41.4a, the first region is from 120 K to T_c = 187 K (the critical temperature). We can see when the bias is between 0.3 and 0.4 V the hole mobility increases with increasing of temperature. In contrast, when the bias is >0.4 V mobility decreases. In the second region we observe that the hole mobility decreases with increasing temperature and voltage. As a result, in graph (b) when voltage is >0.08 V, the hole mobility increases with increasing temperature except for 1.2 V. However, in the second region, after T_c the hole mobility decreases with increasing temperature and decreasing voltage. The effect of voltage has a strong influence on the carrier mobility of both SubPc and C_{60}.

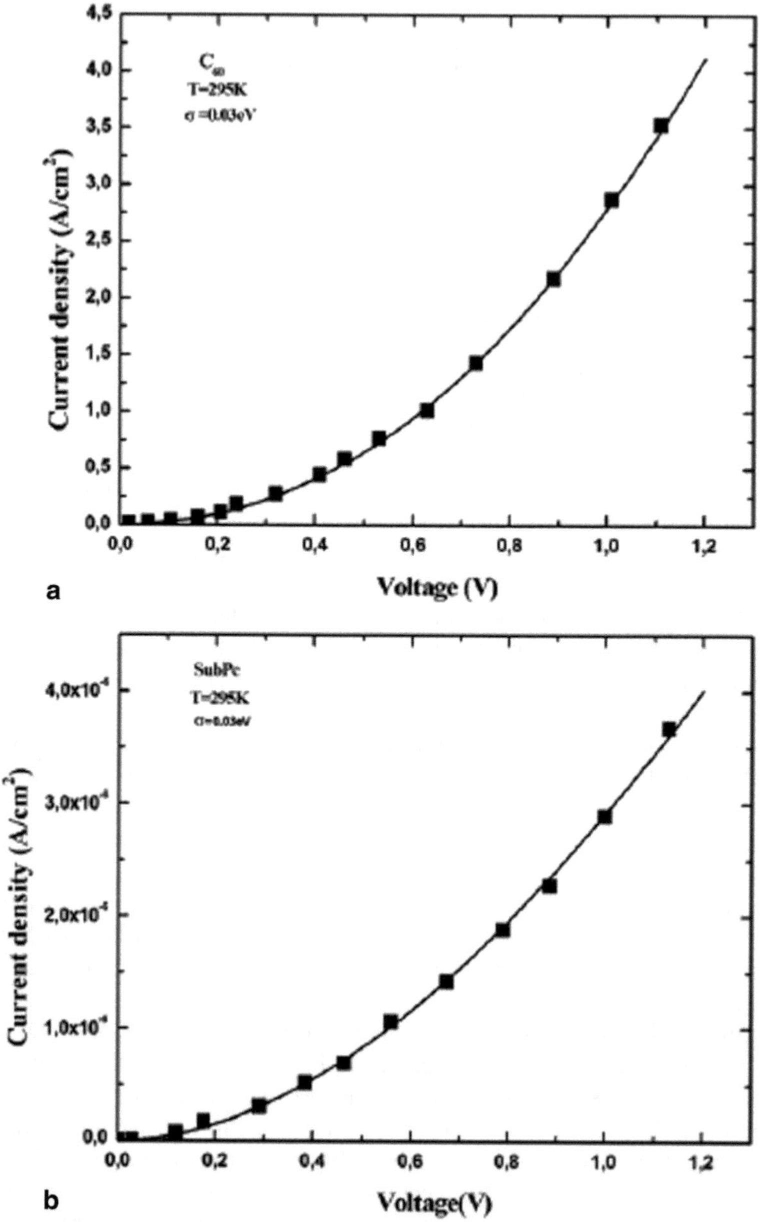

Fig. 41.2 Dependence of the current density–voltage characteristic versus bias voltage; for electron transport in neat C_{60} (**a**) and hole transport in neat SubPc (**b**) at room temperature 298 K. The solid lines represent our results and black squares represent experimental data taken from ref [1]

Table 41.1 Material properties of C_{60} and SubPc

	a (10^{-8} cm)	μ_∞ (Cm²/Vs)	Δ (eV)	
			$T \leq 289$ K	$T > 300$ K
C_{60}	14.17 [5]	0.3	0.03 [6]	0.06 [7]
SubPc	12 [8]	1.5×10^{-7}	0.03	0.06

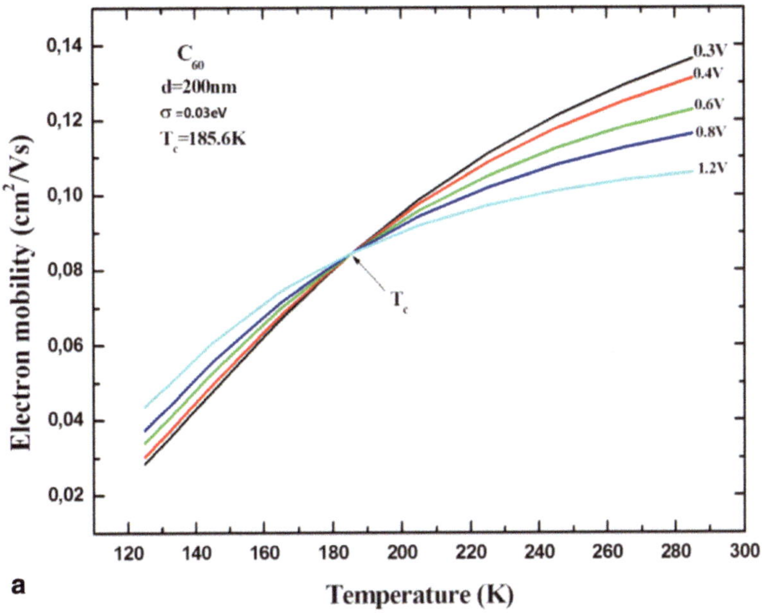

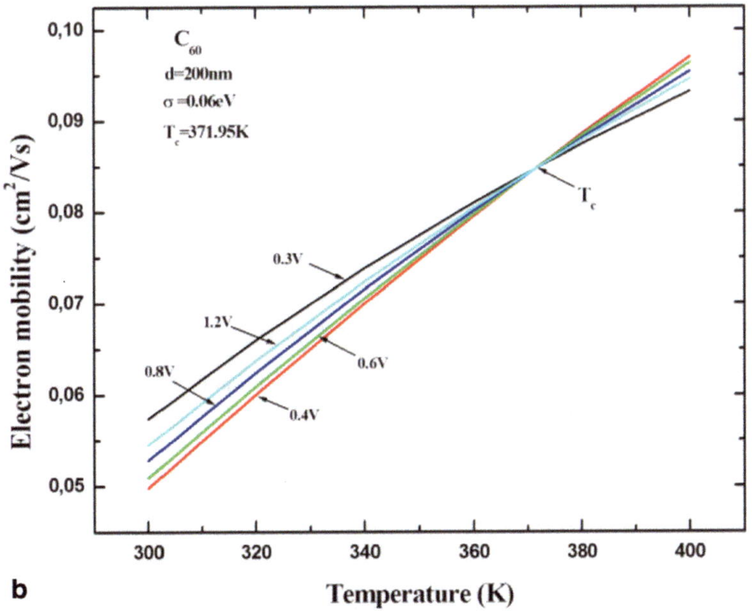

Fig. 41.3 Electron mobility dependence of temperature in neat C_{60} at various voltages. Two cases of Gaussian disorder **a** 0.03 eV and **b** 0.06 eV

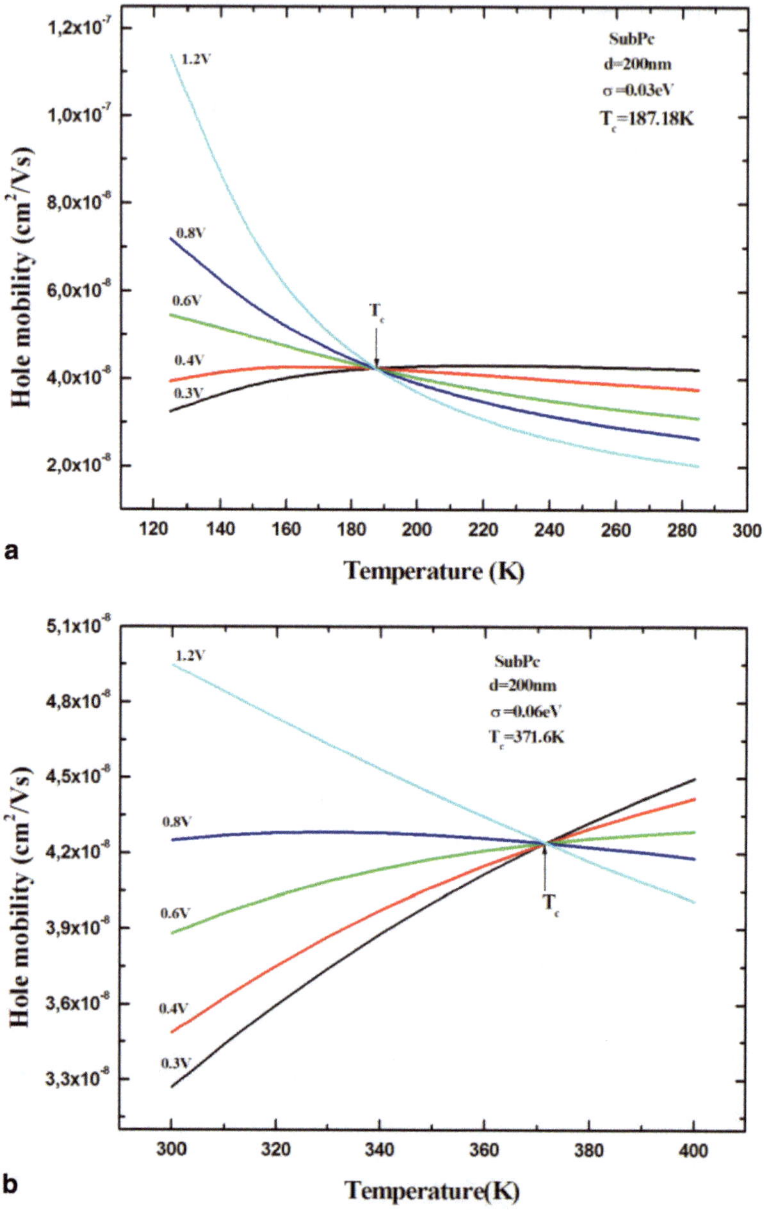

Fig. 41.4 Hole mobility as a function of temperature in neat SubPc for various voltage values. Two cases of Gaussian disorder **a** 0.03 eV and **b** 0.06 eV

41.4 Conclusion

In this study, we investigated electron mobility in neat C_{60} n-type Schottky diodes and neat hole mobility in SubPc p-type Schottky diodes using a recent analytical model assuming a Gaussian density-of-states distribution. Our results show good agreement with experimental results. Furthermore, the temperature has a dramatic influence on the ability of carrier transportation on device performance. For each characteristic there are two distinct critical temperature regions.

References

1. Pandey R, Gunawan AA, Mkhoyan KA, Holmes RJ (2012) Efficient organic photovoltaic cells based on nanocrystalline mixtures of Boron subphthalocyanine chloride and C_{60}. Adv Funct Mater 22:617–624. (WILEY-VCH Verlag GmbH & Co. KGaA, Weinheim)
2. Wallace Jason U (2009) Thesis PHD carrier mobility in organic charge transport materials: methods of measurement, analysis and modulation. Rochester, New York
3. Martinelli N (2011) Charge transport in organic conjugated materials: impact of lattice dynamics and Coulombic interaction effects. Dissertation, 26 Sept 2011
4. Lenes M, Shelton SW, Sieval AB, Kronholm DF, (Kees) Hummelen JC, Blom PWM (2009) Electron trapping in higher adduct fullerene-based solar cells. Adv Funct Mater 19:3002–3007. (WILEY-VCH Verlag GmbH & Co. KGaA, Weinheim)
5. Hu JY, Niu NN, Piao GZ, Yang Y, Zhao Q, Yao Y, Gu CZ, Jin CQ, Yu RC (2012) Carbon 50:5458–5462. (Elsevier Ltd. All rights reserved)
6. Copyright 2008 by Yang, Fan Fan Yang Santa Clara, CA May, 2008
7. Deibel C, Strobel T arXiv:0906.2486v2 [cond-mat.mtrl-sci] 16 Jul 2009
8. Trelka M, Medina A, Ecija D, Urban C, Groning O, Fasel R, Gallego JM, Claessens CG, Otero R, Torres T, Miranda R (2011) Subphthalocyanine-based nanocrystals. Chem Commun 47:9986–9988. (This journal is The Royal Society of Chemistry 2011)

Chapter 42
An Assessment on the Impact of the Solar Spectrum on Different PV Materials in Sunny Sites by Using Different Time Scales

G. Nofuentes, M. Alonso-Abella, M. Torres-Ramírez and F. Chenlo

Abstract This chapter is addressed at analyzing how the performance of some photovoltaic (PV) materials is influenced by the solar spectrum distribution according to the months of the year. Spectral responses of four different PV technologies—amorphous silicon, cadmium telluride, copper indium diselenide, and monocrystalline silicon—have been used to develop this study. Spectra and incident global irradiance were scanned at 5-min intervals in two inland sunny sites located in Spain during a 12-month experimental period. Regarding the results, it was concluded that amorphous silicon and cadmium telluride PV modules undergo the highest differences of monthly spectral gains over the year. Much slighter seasonal variations of these gains are perceptible in the other two considered PV technologies. Specifically in the two locations analyzed, spectral gains range from approximately −12 % (January) to around 2 % (June) for the amorphous silicon (a-Si) PV module while such gains range from values close to 0 % (April to September, inclusive) to values slightly lower than −2 % (December) for the copper indium diselenide (CIS) and monocrystalline silicon (m-Si) PV module. Prevailing "red-rich" and "blue-rich" spectra in winter and summer, respectively, could explain these results.

Keywords PV technologies · Energy yield · Spectral effects · Solar spectra measurement

42.1 Introduction

The influence of the solar spectrum distribution on photovoltaic (PV) devices has been widely explored in literature. PV devices based on thin-film technologies are especially sensitive to the spectral distribution of the irradiance. Despite the existing valuable contributions addressed at dealing with this topic, the impact of the

G. Nofuentes (✉) · M. Torres-Ramírez
IDEA Research Group, University of Jaén, Campus de Las Lagunillas, 23071 Jaén, Spain
e-mail: gnofuen@ujaen.es

M. Alonso-Abella · F. Chenlo
CIEMAT/DER, Avda. Complutense, 22., 28040 Madrid, Spain

A. Sayigh (ed.), *Renewable Energy in the Service of Mankind Vol II*,
DOI 10.1007/978-3-319-18215-5_42

spectral irradiance as a function of the considered time scale has been scarcely addressed.

Quantifying how much the solar spectrum distribution of a given site increases or decreases the electricity produced by a specific PV technology over periods longer than an hour or a day is crucial. Because of this fact, if spectral effects are not taken into account in order to estimate this electricity, noticeable errors can be made when only irradiance data collected provided by pyranometers (flat response sensors) are considered [1, 2].

In this sense, this chapter is aimed at looking at the gains or losses—only due to spectral effects—that some kinds of solar cells of different technologies experience on monthly periods of integration.

42.2 Materials and Methods

A 12-month experimental campaign in which spectral irradiances have been periodically recorded in two sunny inland sites located in Spain—Madrid (latitude 40.41 °N, longitude 3.73 °W) and Jaén (latitude 37.77 °N, longitude 3.80 °W)—have provided the necessary experimental data. Then, the relative spectral responses of four different PV solar cells have been selected, so as to represent the following technologies: amorphous silicon (a-Si:H), cadmium telluride (CdTe), copper indium diselenide (CIS), and monocrystalline silicon (m-Si).

An EKO™ MS-700 spectroradiometer measured the spectral irradiance in Madrid and Jaén. This instrument recorded solar spectral distributions within a range from 350 to 1050 nm. The incident irradiance data were obtained by means of a Kipp & Zonnen™ CMP 21 pyranometer in Madrid and Jaén. Both instruments were facing due south with a tilt angle of 35° for each site. The experimental campaign was run from April 2011 to March 2012, inclusive. Five-minute recording intervals between 10:00 and 14:00 were considered during this campaign. This time window of measurement implies the decrease of the angle of incidence dependence because of the sun elevation, which is then higher than that of the rest of the day. Measurements recorded at values of G below 300 W m^{-2} were also dropped, due to their little relevance in sunny locations in terms of electricity production.

According to the IEC 60904-7 standard [3], the spectral factor *(SF)* is defined as follows:

$$SF = \frac{\int_{\lambda_3}^{\lambda_4} E^*(\lambda)d\lambda \int_{\lambda_1}^{\lambda_2} E(\lambda)SR(\lambda)d\lambda}{\int_{\lambda_3}^{\lambda_4} E(\lambda)d\lambda \int_{\lambda_1}^{\lambda_2} E^*(\lambda)SR(\lambda)d\lambda}, \tag{42.1}$$

where λ_1 (nm) and λ_2 (nm) are the lower and upper limits within which the absorption takes place in the PV device, λ_3 (nm) and λ_4 (nm) are the wavelength limits within which the reference device is spectrally sensitive, $SR(\lambda)$ (A W^{-1}) is the spectral response of the PV device, $E(\lambda)$ (W m^{-2} nm^{-1}) is the spectral irradiance of the actual solar spectrum, and $E^*(\lambda)$ (W m^{-2} nm^{-1}) is the spectral irradiance of the standard AM 1.5G spectrum, as stated in the IEC 60904-3 standard [4].

Thus, values of *SF* higher than 1 imply spectral gains, which means that the PV device has a better performance under the actual solar spectral distribution than that under the standard solar spectrum AM 1.5G. On the other hand, values of *SF* lower than 1 correspond to spectral losses.

SF values were determined from each solar spectral distribution registered over 5-min time intervals. The monthly spectral factor (SF_m), as defined in [5] is a meaningful parameter that is obtained by means of integrating Eq. (42.1) over a month. The four spectral responses tested have been used to produce the evolution graphs for the calculated monthly spectral gains that are going to be shown hereafter. In this sense, the monthly spectral gains experienced by a PV technology, in terms of percentage, may be written as follows:

$$\text{Monthly spectral gains (\%)} = 100 \cdot (SF_m - 1). \tag{42.2}$$

In order to provide a better understanding of the solar spectral influence over the outdoor performance of a PV technology, monthly spectral gains are shown and analyzed in the following section, instead of *SF*.

42.3 Results and Discussion

Over 16,000 and 10,000 spectra were measured in Madrid and Jaén, respectively, together with the same number of measured samples corresponding to *G*. Figure 42.1 shows the evolution of the monthly spectral gains in Madrid. As it can be appreciated, the maximum variations of monthly spectral gains take place for the a-Si:H PV cell, ranging from − 13.4 % (January) to 1.3 % (June) in Madrid. These spectral gains lie in the interval − 11.5 % (January) to 2.0 % (June) for the same cell in Jaén, as depicted in Fig. 42.2. The behavior of both CIS and m-Si PV solar cells is extremely similar and less sensitive to the spectrum distribution for both sites. Specifically, the span between the month with the lowest and the highest spectral gains integrated over a month equals some 4.5 and 3 % for Madrid and Jaén, respectively. Finally, CdTe monthly behavior is similar to the CIS and m-Si during the April to September period. The largest spectral gains in the CdTe take place during the winter season (February), around − 8 and − 5 % for Madrid and Jaén, respectively.

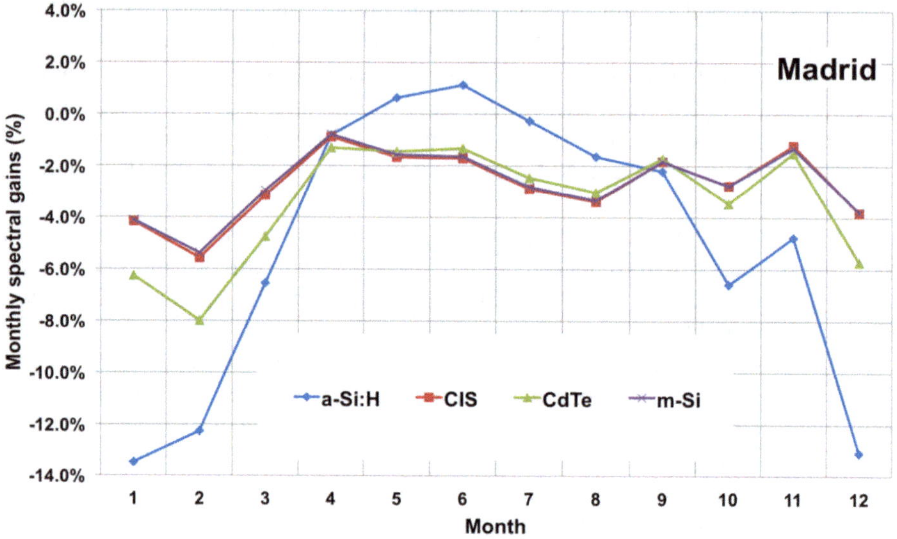

Fig. 42.1 Monthly spectral gains in the city of Madrid for the four considered photovoltaic (PV) solar cell spectral responses. *a-Si* amorphous silicon, *CIS* copper indium diselenide, *CdTe* cadmium telluride, *m-Si* monocrystalline silicon

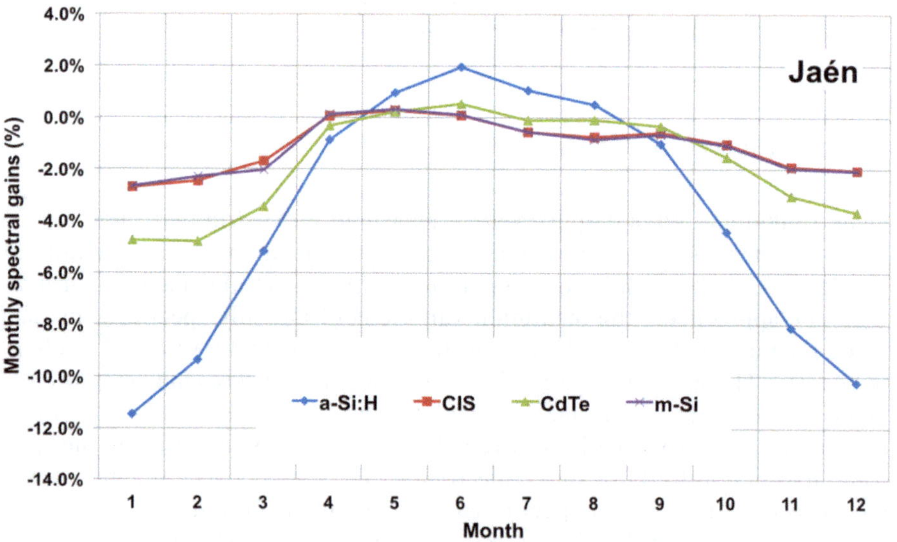

Fig. 42.2 Monthly spectral gains in the city of Jaén for the four considered photovoltaic (PV) solar cell spectral responses. *a-Si* amorphous silicon, *CIS* copper indium diselenide, *CdTe* cadmium telluride, *m-Si* monocrystalline silicon

42.4 Conclusions and Future Research Lines

This chapter presents an effort to analyze the influence of the solar spectral effect on four PV technologies: a-Si:H, CdTe, CIS, and m-Si, bearing in mind that this influence is especially noticeable in thin-film technologies due to the peculiarities of their spectral responses. Results of the influence of the solar spectrum distribution on the outdoor behavior of some thin-film technologies in the cities of Madrid and Jaén (Spain) have been presented. The prevailing environmental conditions related to these two locations correspond to those of sunny and inland climates.

Results show that in general, monthly spectral gain values present seasonal oscillations, reaching their maxima in winter months. PV technologies based on a-Si:H, CdTe, CIS, and m-Si are ordered from highest to lowest impact of the solar spectral irradiance distribution on the outdoor performance of these ones.

The prevailing "red-rich" spectra of both sites help explain the results obtained. Anyway, the outcome of this work clearly shows a conclusion raised in some previous good works: The longer the scale of integration, the smaller the impact of the spectrum, whatever the considered PV technology is. The above conclusions should be taken cautiously given the limited amount of data from which these conclusions have been drawn. In fact, the experimental campaign is still ongoing in Madrid and Jaén.

Acknowledgments This work was supported by the Spanish Science and Innovation Ministry and the European Regional Development Fund (ERDF) within the frame of the project "Estimación de la energía generada por módulos fotovoltaicos de capa delgada: influencia del espectro" under expedient code ENE2008-05098/ALT and *Fundación Iberdrola* within the frame of the project "Contribución al modelado del comportamiento eléctrico en exterior de módulos fotovoltaicos de capa delgada."

One of the authors (M. Torres-Ramírez) greatly acknowledges *Fundación Iberdrola* for granting an "Energy for Research" Scholarship.

References

1. Nofuentes G, García-Domingo B, Muñoz JV, Chenlo F (2014) Analysis of the dependence of the spectral factor of some PV technologies on the solar spectrum distribution. Appl Energy 113:302–309
2. Ishii T, Otani K, Takashima T, Itagaki A, Utsunomiya K (2013) A simplified methodology for estimating solar spectral influence on photovoltaic energy yield using average photon energy. Energy Sci Eng 1(1):18e26
3. IEC 60904-7 (Ed. 3.0) (2008) Photovoltaic devices—Part 7: computation of the spectral mismatch correction for measurements of photovoltaic devices
4. Intenational Electronical Commission (1989) IEC 60904-3, ed. 2. Photovoltaic devices—Part 3: measurement principles for terrestrial photovoltaic (PV) solar devices with reference spectral irradiance data; 2008
5. Alonso-Abella M, Chenlo F, Nofuentes G, Torres-Ramírez M (2014) Analysis of spectral effects on the energy yield of different PV (photovoltaic) technologies: the case of four specific sites. Energy 67:435–443

Chapter 43
Optimizing Photovoltaic Water Pumping Systems for Developing Countries Through the Addition of a Novel Induced-flow Subsystem

N. Davies, T. D. Short and A. Hassan

Abstract Water supply and sanitation in developing countries represents an issue responsible for millions of deaths each year. Despite the recent advances in photovoltaic (PV) water pumping systems, high set-up costs, lack of local expertise and reliability issues have prevented many people from benefiting from this potentially life-saving technology. This is particularly unfortunate given that PV pumps have been shown to be cheaper than conventional diesel pumps over the product lifetime. This research has been undertaken to reduce these high implementation costs and other stability issues by developing the theory towards a self-regulating, sustainable pumping system based on the use of an additional flow subsystem modelled on a DC boost-converter circuit from power electronics.

Initial experimental results demonstrate a good proof of concept. Using an induced-flow test rig, based on the use of a high-speed ON/OFF valve, oscillations in the flow through the system have been observed as having waveforms which are in line with those anticipated from theory. Additionally, a boost in output pressure was seen as the active time (or duty cycle) of the valve was increased to achieve flow to greater pumping heads and increased output flow with improved efficiency. In conclusion, these results reveal a potential to develop a one-size-fits-all, sustainable PV pumping system based on a centrifugal pump. Ultimately, this could offer a cheaper and more reliable supply of a resource, which is vital to any developing community: *clean water*.

Keywords Photovoltaic water pumping · Induced flow · Electric–hydraulic analogy · Renewable energy

---

N. Davies (✉) · T. D. Short
School of Engineering, University Of Liverpool, Brownlow Hill L69 3GH, UK
e-mail: n.davies@liv.ac.uk

T. D. Short
e-mail: tdshort@liv.ac.uk

A. Hassan
Faculty of Engineering and Computing, Coventry University, Coventry CV1 5FB, UK
e-mail: ab6922@coventry.ac.uk

A. Sayigh (ed.), *Renewable Energy in the Service of Mankind Vol II*,
DOI 10.1007/978 3 319 18215 5_43

43.1 PV Pumping Systems and Developing Communities

Access to clean water is among one of the most basic human needs, yet it is still not fully met in many developing countries. Of the 884 million people worldwide without access to safe drinking water [1], more than 5 million people in developing countries die each year as a direct result of diseases caused by unsafe drinking water or lack of clean water for sanitation [2]. In the year 2000, the United Nations introduced a series of goals to address access to clean water for drinking and sanitation purposes in an effort to reduce the global mortality rates from water-related diseases [3]. Studies into the impact of the installation of a basic borehole pump within a community have shown a drastic improvement in health and sanitation in the area, with a 50% reduction in illness due to a clean water source [4]. Sustainable water pumps, such as those involving PV power sources, can be one way of providing a clean, sustainable water source for drinking, sanitation, irrigation and agriculture [5]. This has been seen especially in projects such as the 'Programme Regionale Solaire' (PRS) [6], which installed 626 PV pumps in the Sahel region of Africa.

Pumps that make use of a sustainable energy source have the inherent benefit that they have considerably lower operational costs than comparative fuel-based systems—namely diesel generator pumps [7, 8]. They are doubly suited to 'off-grid' locations where high levels of solar irradiance are prevalent [9], making them ideally suited to most developing communities. Furthermore, PV water pumping holds its place on an environmental platform outside of developing communities as a means of reducing the amount of fossil fuels used in pumping applications globally [10].

Unfortunately, PV pumps are not without their flaws. Targets set nearly three decades ago to implement 10 million PVP's [11] saw only 0.6% of this figure realized by the end of the twentieth century [12], many of which were shown to be nonoperational by 2001 [13]. This failure may be attributed in large part to the installation of the pumps. Conditions under which a sustainable pump must operate are not expected to be constant and fluctuations in the operating head of a pump can result in decreased, or no performance. Ordinarily, it is not possible to compensate entirely for changes in environmental conditions, such as high or low rainfall, save implementing a pump with a good range of pumping heads based on an area survey or site-specific assessment/installation. This results in a higher level of expertise than usually available at the local level for correct set-up and often equates to an installation cost of up to 500% more than an equivalently sized diesel generator [8].

It is proposed that through the addition of an induced-flow subsystem (IFS) to a standard centrifugal pump, its performance may be decoupled from its operating conditions, allowing the pump to operate continuously at its 'Best Efficiency Point' (BEP) irrespective of fluctuations in environmental conditions or improper installation. Much in the same way a maximum power point tracker (MPPT) is currently used with PV power supplies to optimize their power output [14], an IFS pump would provide optimum hydraulic power, enabling maximum power draw avail-

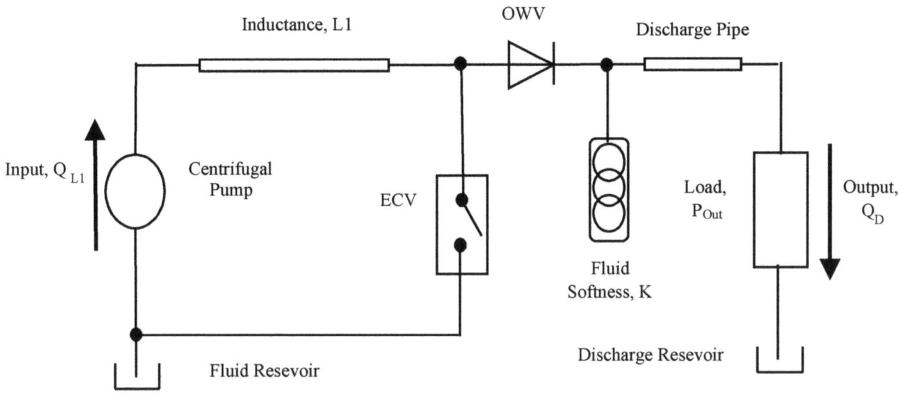

Fig. 43.1 Hydraulic circuit with IFS

able from a PV power source. This would support the self-regulation of a system—compensating for any fluctuations in operating conditions and thereby improving the reliability of the pump. While the concept of using an IFS in conjunction with positive displacement pumps has been the subject of many studies for almost four decades now, the principles have not been applied to pumping regimes focused around a centrifugal pump as a means of maintaining BEP. Given that it has been demonstrated that the addition of such a subsystem to positive displacement PV pumping regimes can be beneficial to their operation [15–18], it stands to reason that the same could hold true for the cheaper and more widely available centrifugal pump. Such a system would also lend itself to an auto-setup (one size fits all) capability, eliminating the need for site-specific assessment, which in turn would lower the cost of installation.

43.2 Principles and Operation

The proposed PV water-pumping system is shown in Fig. 43.1 and utilizes the same design as that of the DC boost converter circuit as observed in Fig. 43.2. Water is transferred from the reservoir/well to the fluid inductance, L, via the centrifugal pump with a flow rate $Q_{L1}(t)$. Flow is then diverted either back to the fluid reservoir via the operation of an electronically controlled valve (ECV) or to a fluid 'softness', K via a one-way check valve (OWV). The ECV operates with period T:

$$T = t_o + t_c, \tag{43.1}$$

where t_o and t_c represent the open and closed times of the ECV, respectively, and vary in accordance with the duty cycle, D, for the pumping system such that:

$$t_o = D \cdot T \tag{43.2}$$

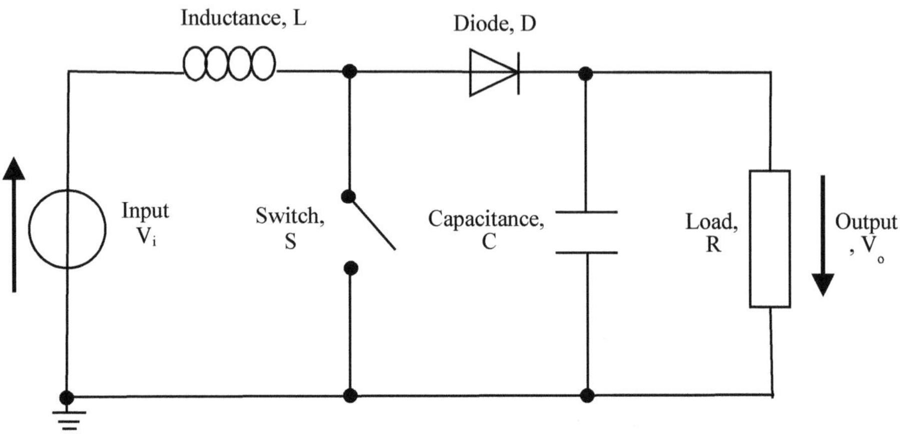

Fig. 43.2 DC boost-converter circuit

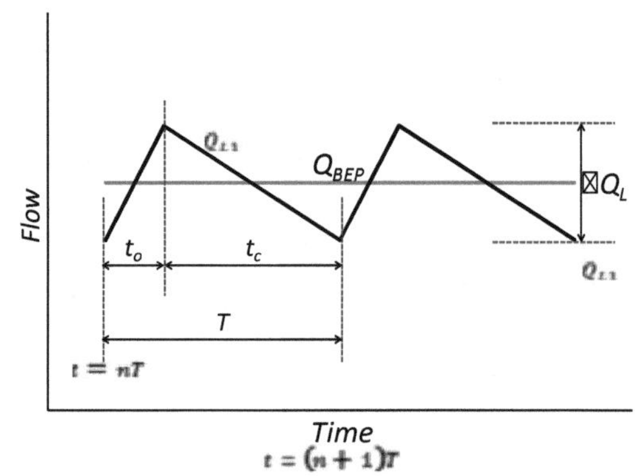

Fig. 43.3 Periodic oscillations in flow through the inductance pipe w.r.t. time

Considering the periodic operation of the ECV, it is clear that there will be two distinct time conditions, shown in Fig. 43.3, under which the system operates; $Q_{L1}(t) = Q_{L1}(t)_{ON}$ and $Q_{L1}(t) = Q_{L1}(t)_{OFF}$. Here, the total change in flow from a certain time $nT \rightarrow (nT + t_0)$ must be equal and opposite to the change in flow from $(nT + t_0) \rightarrow (n+1)T$ ('n' denoting the number of the period over which the flow is oscillating, from $0 \rightarrow \infty$). Thus $Q_{L1}(nT)_{ON} = Q_{L1}(nT)_{OFF}$ and $Q_{L1}(nT + t_0)_{ON} = Q_{L1}(nT + t_0)_{OFF}$. During time t_o the ECV provides a path of least resistance from the inductance to the reservoir causing the flow to rise linearly from minimum flow, $Q_{\min}$, to maximum flow, $Q_{\max}$. During time t_c, flow rate falls back to $Q_{\min}$ and is diverted to supply the fluid capacitance and discharge flow, Q_D, at the system outlet. These oscillations in inductance flow interact with the IFS to create oscillations in Q_D. The duty cycle varies depending on the operating head of the pumping system, H_{Out}, and is given by the DC boost converter circuit equation:

$$D = 1 - \frac{H_{BEP}}{H_{OUT}}. \tag{43.3}$$

H_{BEP} is the BEP operating head and the original voltage terms in the equation have been replaced with hydraulic pressure terms using the electric–hydraulic analogy. This equation is validated for the hydraulic circuit in Fig. 43.1 by considering the fluid flow after the inductance. From continuity:

$$Q_{L1}(t) = Q_{ECV} + Q_{OWV}. \tag{43.4}$$

Assuming all flow is diverted through the ECV during t_o:

$$Q_{ECV} = DQ_{L1}(t) \tag{43.5}$$

and;

$$Q_{OWV} = (1-D)Q_{L1}(t). \tag{43.6}$$

From the Steady Flow Energy equation:

$$\Delta E = \sum \dot{m}\left(U + gH + \frac{c^2}{2} + gz\right) = 0, \tag{43.7}$$

where mass flow $\dot{m} = \rho Q$, change in internal energy $U = 0$, fluid velocity $c = Q / A$, and negligible changes in height, z, between pump inlet and EVC. Assuming all pipe internal areas, A, are equal:

$$gH_{L1} + \frac{Q_{L1}{}^2}{2A^2} = (1-D)\left(gH_{Out} + \frac{(1-D)^2 Q_{L1}{}^2}{2A^2}\right) + D\left(gH_{Sub} + \frac{D^2 Q_{L1}{}^2}{2A^2}\right), \tag{43.8}$$

where H_{Out} and H_{Sub} are the output and reservoir submergence heads, respectively. Rearranging and simplifying by ignoring the submergence head and noting $Q_{L1} = Q_{BEP}$, likewise $H_{L1} = H_{BEP}$:

$$\Delta H = \frac{DH_{BEP}}{1-D} + \frac{3\rho DQ_{BEP}{}^2}{2gA^2} \tag{43.9}$$

with $\Delta H = (H_{Out} - H_{BEP})$. This provides a definition of D numerically equal to Eq. 43.3. From the definition of duty, it is seen there is no theoretical limit to the operating head the system can deliver to, as $D < 1$, illustrated in Fig. 43.4. Thus, the pump is decoupled from its operating conditions and maintains its BEP.

Now, considering the total average flow across a single time period:

$$\sum\nolimits_0^T Q_{AV(L1)} = Q_{BEP} \cdot T. \tag{43.10}$$

Fig. 43.4 ECV duty cycle

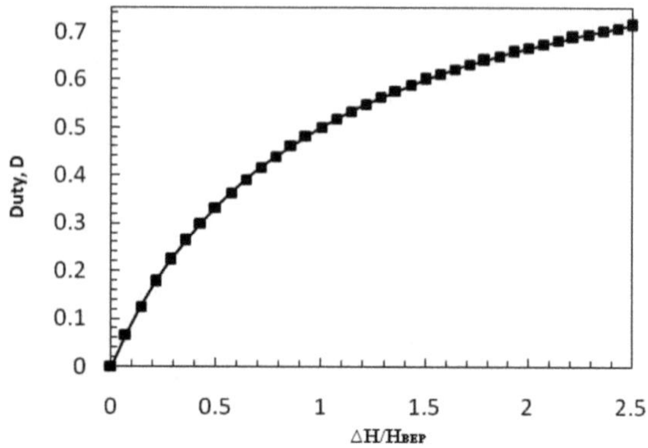

Discharge flow, $Q_D(t)_{OFF}$, across the $Q_{L1}(t)_{OFF}$ period is similarly defined:

$$\sum_{TD}^{T} Q_D(t)_{OFF} = \frac{1}{2}(Q_{MAX} + Q_{MIN}).T(1-D). \tag{43.11}$$

This value will be equal to the total output flow over a single time period as by definition of the operation of the valve, there is no flow through the ECV during time t_c. It is also assumed that any additional flow from the capacitance chamber, Q_C, during time t_o will be equal to the loss in flow during t_c as the capacitance chamber is re-pressurized. Consequently, the average discharge flow from the system for any given head is:

$$Q_{AV(D)} = \frac{1}{2}(Q_{MAX} + Q_{MIN})(1-D). \tag{43.12}$$

43.3 Method and Experimental Results

43.3.1 Test Rig

The novel IFS test rig was completed as illustrated by Fig. 43.5. The main construction was simple, primarily comprising a centrifugal pump, a rigid stainless steel inductance pipe, electronically controllable high-speed ON/OFF valve, one-way check valve, capacitance chamber and an adjustable head output pipe. Full test rig specifications can be found in Table 43.1. For the purpose of testing, a mains variable power source was used as opposed to a PV array. Given restrictions on pumping head and lab space, a small-scale pump was selected for ease of use and to allow for the greatest range of test data. There is nothing to suggest that this is not representative of a larger pumping system as all components should scale up well.

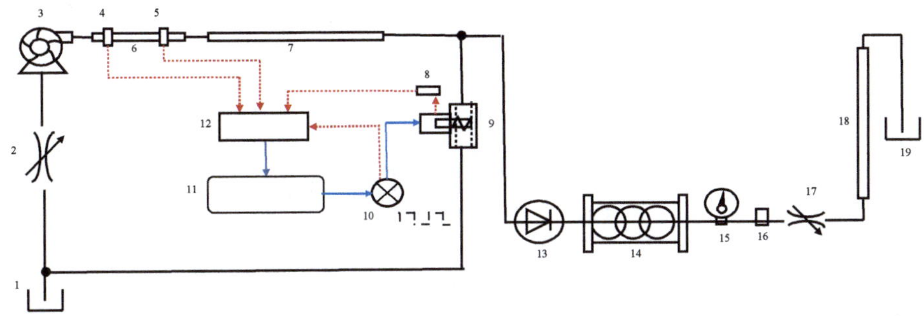

Fig. 43.5 Schematic of experimental test rig. *1* fluid reservoir, *2* throttle valve, *3* centrifugal pump, *4* pump-side pressure transducer, *5* valve-side pressure transducer, *6* 0.15 m inductance pipe length, *7* additional 3.55 m inductance pipe length, *8* piezo-electric transducer for measuring physical valve timing, *9* high-speed ON/OFF solenoid valve and coil, *10* Arduino chip valve controller, *11* computer, *12* data acquisition unit, *13* one-way check valve, *14* capacitance chamber, *15* pressure gauge, *16* 5-bar safety pressure release valve, *17* throttle valve, *18* adjustable discharge pipe, *19* output fluid reservoir

Table 43.1 Experimental test rig specifications

Test rig detail	Symbol	Value	Unit
Inductance total length	l	3.7	m
Inductance internal diameter	d_i	15×10^{-3}	m
Fluid 'Softness', K	K	2.31	GPa.m^{-3}
Fluid density	ρ	1000	kg.m^{-3}
Inductance pipe area	A	1.77×10^{-4}	m^2
ECV minimum opening time	–	20	ms
ECV minimum closing time	–	30	ms
Steel pipe wave celerity	a	1360	m.s^{-1}
Data acquisition sample rate	–	1	ms
Data acquisition sample period	–	2	S

The inductance pipe was split into two equal diameter sections, the first of which contained a pair of pressure transducers with a 0.15-m spacing to measure fluctuations in flow. The second section supplied the additional length to make up the complete inductance pipe. This inductance length of 3.7 m was selected to ensure correct valve operation over a complete range of heads, based on the definition of time period in relation to the natural frequency of the IFS:

$$T = \frac{2\pi}{r\omega_n} \tag{43.13}$$

with frequency ratio, $r < 1$ for over-resonant pumping conditions and natural frequency defined by:

$$\omega_n = \sqrt{\frac{KA}{\rho l}}, \tag{43.14}$$

Fig. 43.6 IFS test rig in the University of Liverpool hydraulics lab

where A is the inductance pipe internal cross-sectional area, ρ the fluid density and ι the inductance length. Although substantially longer than would be anticipated for practical use, the inductance pipe allowed the high-speed ON/OFF valve to operate at lower frequencies and so test across a complete range of duties. While it is envisaged that a higher performance valve would be required to improve the overall performance of the test rig and scale down its size, the valve selected for lab testing was sufficient to provide a good 'Proof Of Concept' (POC). Figure 43.6 shows the fully constructed rig in the University of Liverpool hydraulics lab.

A commercially available Danfoss™ series EV210-10B solenoid valve was used with a minimum cycling time of 50 ms, comfortably enabling a basic operating frequency at (frequency ratio) $r = 2$, of $3.3Hz$. A piezo-electric transducer was fixed directly to the solenoid housing as a means of measuring the physical valve timing, and the digital timing of the valve was controlled by a simple Arduino chip circuit using a software interface for programming; allowing for timings to be corrected to match calculated duties. Valve duties were selected and programmed based on the varying output head according to Eq. 43.3.These duties were varied according to different input voltages, and thus BEP's, to the pump to represent changing levels of solar irradiance.

The output pipe was built in segments to accommodate a maximum head of 5 m in increments of 0.5 m limited only by the height restrictions of the test space. Positive pressure through the system was provided by a 24 V magnetically coupled centrifugal pump. Experimental testing of the pump rated it to maximum head of 4.2 m with a BEP of 2.88 m. The range of test duties was maximized by running the centrifugal pump at voltages as low as 6 V at which maximum head was only 1.7 m. In this way the output flow of the IFS rig could be monitored over a variety of operating conditions.

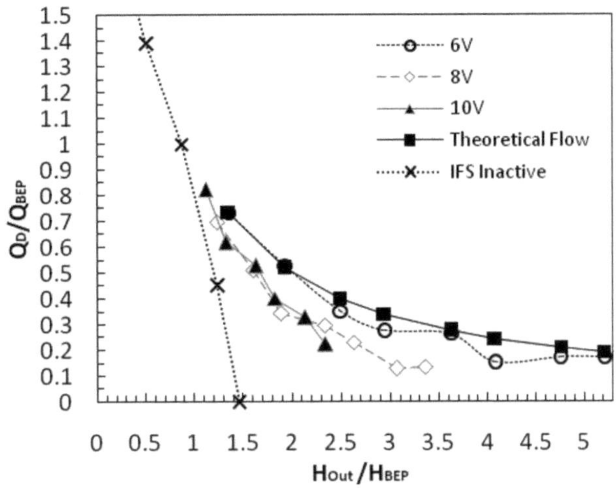

Fig. 43.7 Comparison of IFS operation at varying levels of pump operational power, r = 2

Experiments were conducted by starting the pump and allowing pressure to stabilize before engaging the high-speed ON/OFF valve. A data acquisition unit (DAQ) was used to capture the voltage and current traces from both the high-speed valve and the pump, the pressure transducer outputs and the piezo-electric output from the solenoid valve housing. Adequate sample time of 1 ms was allowed to provide an accurate trace of the system and flow behaviour, relative to the operating frequency. Average flow output from the system was calculated from the increase in collected water mass across 1 min.

43.3.2 Output Flow and Hydraulic Power

For each set of results, the flow at the outlet pipe of the fluid circuit was plotted against the control curve based on input voltage with no IFS. All results are presented in nondimensional terms, relative to associated Q_{BEP} and H_{BEP} at each voltage. Figure 43.6 presents the comparison in output flow, both with and without IFS, under increasing output head over a range of input voltages for over resonant condition, $r = 2$. On activation of the IFS, in each test case, there was an increase in average discharge flow above that of the unmodified pumping system. Additionally, the system was able to pump to a higher head with the IFS active. The relationship between increasing duty and output pressure was validated as the boost in pressure allowed the test rig to produce an output flow even after the maximum operating head of the pump had been reached and exceeded for each operating voltage. It is also seen that the flow trends are a good match to the theoretical data, with a closer match being achieved at lower pump voltages. This is discussed in Sect. 43.4. Initial tests were repeated for a variety of valve frequency ratios to determine its effect on system discharge flow.

As can be seen in Figs. 43.7 and 43.8 higher ratios achieved discharge flows closer to those predicted theoretically. Tests were also conducted using a con-

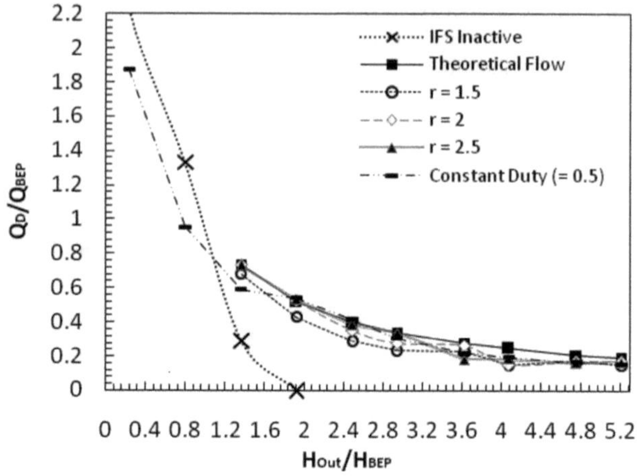

Fig. 43.8 Comparison of flow rates for ECV operational frequency ratios at 6 V pump power

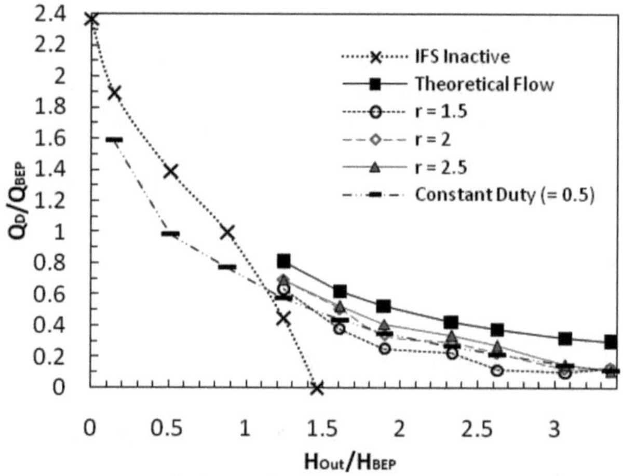

Fig. 43.9 Comparison of flow rates for ECV operational frequency ratios at 8 V pump power

stant operational duty of 0.5 at $r = 2$ to determine how detrimental to the systems performance any deviation in theoretical duty is. Average discharge flow was reduced, seen more clearly in Fig. 43.8, but the difference between flows was minimal and there was seen to be good grouping with the other r value tests. Figures 43.9 and 43.10 present hydraulic power, P_H, against increasing head. It is more clearly seen from these figures the degree to which tests at lower operational voltages more closely match the theoretical BEP line. A significant boost in hydraulic power on activation of the IFS with a more even distribution in r value variance can be seen in the higher operational voltages. This is especially relevant at the lower heads above H_{BEP} where there exists a greatly improved discharge flow and hydraulic power, demonstrating an improved use of available energy (Fig. 43.11).

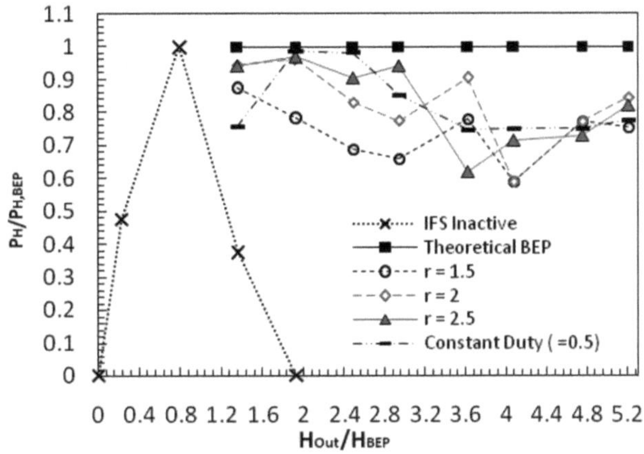

Fig. 43.10 Comparison of hydraulic power outputs for ECV operational frequency ratios at 6 V pump power

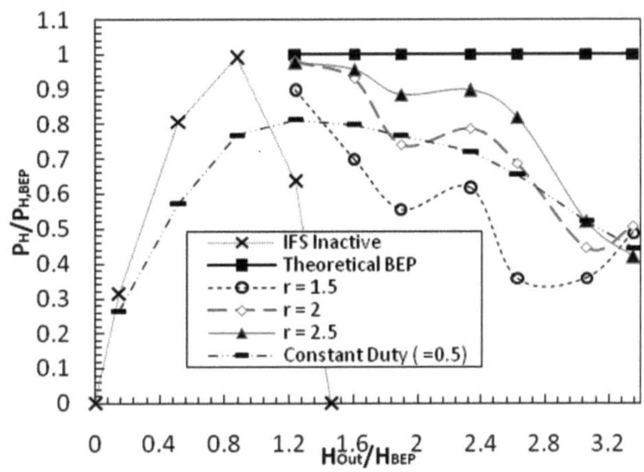

Fig. 43.11 Comparison of hydraulic power outputs for ECV operational frequency ratios at 8 V pump power

43.4 Discussion

43.4.1 Pumping Power

Experimental results agree well with the flow trend obtained from theory. Although results show improved delivery head and output flow on operation of the IFS, they also hold relevance in the context of PV pumping through the optimization of available electrical power, P_E. Plotting the relationship between pumping head and operating power (from the product of pump current draw and operating voltage), it is seen that there is a greater energy draw by the pump while the IFS is active. This is shown in Fig. 43.12. While similar to the increased pumping capabilities of reciprocating pumps using IFS, the results from centrifugal pumps differ by means of efficiency. As has been previously shown by Burton and Short [16], reciprocating pumps, while providing good matching to renewable energy sources through in-

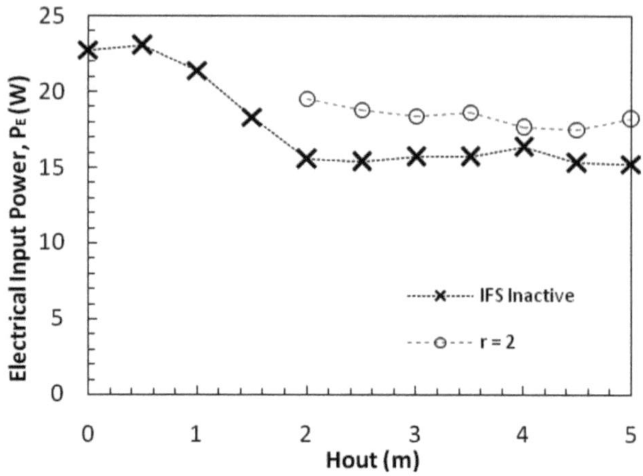

Fig. 43.12 Comparison of power usage with and without IFS

creased output flow and greater energy usage, prove less efficient than without IFS. This differs from centrifugal pumping systems as inspection of Fig. 43.12 shows an increase in 'wire-to-water' efficiency when compared to the hydraulic power trends given in Figs. 43.9 and 43.10. It is seen that for heads above BEP conditions the IFS pumping system makes greater use of the energy available to the pump while also providing increased hydraulic power. This demonstrates the use of such subsystems to be suitable for use with PV power sources.

43.4.2 Flow Trends

The duty equation based on a lumped parameter analysis of the pumping system and compared to the DC boost converter equation has shown good matching to the experimental results. However, the flow rates were consistently lower than theoretically predicted, especially at higher operational voltages. This could be at least in part to the frictional and component losses in the pumping system. Further tests are currently ongoing to fully establish the cause of these drops in flow at higher pumping powers.

43.5 Conclusions

The addition of an IFS to a standard centrifugal pump for the purpose of PV water pumping systems has been shown to maximize the use of available energy by enabling the pump to maintain a close to maximum hydraulic power point based on BEP. Furthermore, the current theory being developed suggests that the presence of this subsystem results in no upper limit on the head to which the pump can deliver water and an increased output flow at conditions beyond those expected of the pump, increasing its effective operating range and reliability. This demonstrates that such a system would lend itself to an auto-setup capability, reducing costs and widening accessibility to sustainable PV pumping.

Acknowledgements This chapter and the research presented within were made possible with the support of Dytecna Ltd., the EPSRC, the Technology Strategy Board and the University of Liverpool.

References

1. Assembly UNG (2010) The human right to water and sanitation. A/RES. 64
2. Gleick PH (2002) Dirty water: estimated deaths from water related diseases 2000–2020. Pacific Institute Research Report
3. WHO (2005) Minimum water quantity needed for domestic use in emergencies. Technical note, World Health Organisation (WHO)
4. Nogier A (1998) Setting up an organisational PV pumping scheme in Nigeria. International workshop on PV water supply issues, Marrakech
5. Short TD, Oldach R (2003) Solar powered water pumps—the past, the present and the future? J Sol Energy Eng 125(1):76–82
6. Kabore F (1998) Le programme régional solaire: L'hydraulique villageoise et pompage solaire au service du développement duarble dans les pays du Sahel. Lecons et impacts sur le terrain. International workshop on PV water supply issues, Marrakech
7. Solar Electric Light Fund SELF (2008) A cost and reliability comparison between solar and diesel powered pumps. Solar electric light fund publication. http://self.org/solarvsdiesel. Accessed Jan 2012
8. Odeh I, Yohanis YG, Norton B (2006) Economic viability of photovoltaic water pumping systems. Sol Energy 80(7):850–860
9. Jafar M (2000) A model for small-scale photovoltaic solar water pumping. Renew Energy 19(1–2):85–90
10. McNelis B, Derrick DA (1989) Solar water pumping: clean water for rural areas. In: Proceedings of the 2nd ASEAN science and technology week, p. 19
11. Barlow R, McNelis B, Derrick A (1993) Solar pumping: an introduction and update on the technology, performance, costs and economics. World Bank technical paper, Intermediate Technology Publications Ltd. p. 68
12. Derrick A (1998) 20 years of PV water pumping—successes and problems. International workshop on PV water supply issues, Marrakech
13. Kaunmuang P (2001) Assessment of photovoltaic pumping systems in Thailand—one decade experience. Sol Energy Mater Sol Cells 67(1–4):529–532
14. Saadi A, Moussi A (2007) Optimisation of back-boost converter by MPPT technique with a variable reference voltage applied to photovoltaic water pumping system under variable weather conditions. Asian J Inf Technol 6(2):222–229
15. Burton JD, Short TD (1999) Induced flow reciprocating pumps: part 1. Proc Inst Mech Eng Part A J Power Energy 213:363–373
16. Burton JD, Short TD (1999) Induced flow reciprocating pumps: part 2. Proc Inst Mech Eng Part A J Power Energy 213:375–389
17. Burton JD, Lobo-Guerrero UJ (1972) Induced flow reciprocating pump theory supported by new performance data. J Basic Eng 94(4):706–714
18. Wade NS, Short TD (2007) The performance of a new positive displacement pump targeted to improve rural water supplies. Proc Inst Mech Eng Part A J Power Energy 221(A8):1163–1171

Chapter 44
Power Converters and Control of Grid-Connected Photovoltaic Systems

L. Hassaine

Abstract Photovoltaic (PV) installations have had an exponential growth mainly due to the governments and utility companies that support programs that focus on grid-connected PV systems (Haeberlin H (2001) Evolution of inverters for grid-connected PV systems from 1989 to 2000, in Proceedings of the 17th European Photovoltaic Solar Energy Conference, pp. 426–430. Munich). In an effort to use solar energy effectively, a great deal of research has been done on the grid-connected PV generation systems. Out of the total PV power installed in the Europe, 98.7% are grid-connected PV systems and 1.3% are off-grid PV systems. Therefore, in grid-connected PV systems, the inverter which converts the output direct current (DC) of the solar modules to alternating current (AC) is receiving increased interest in order to generate power to utility. A key consideration in the design and operation of inverters is how to achieve high efficiency with power output for different power configurations. The requirements for inverter connection include maximum power point, high efficiency, control power injected into the grid, and low total harmonic distortion of the currents injected into the grid. Consequently, the performance of the inverters connected to the grid depends largely on the control strategy applied. Many topologies are used for this purpose.

An overview of power inverter topologies and control structures for grid-connected PV systems are presented. The first section describes the various configurations for grid-connected PV systems and power inverter topologies. The following sections report and present control structures for single-phase and three-phase inverters. Some solutions to control the power injected into the grid and functional structures of each configuration are also proposed.

Keywords Grid-connected photovoltaic system · Power conditioning · Inverters · Control

L. Hassaine (✉)
Photovoltaic, Centre de Développement des Energies Renouvelables, CDER, BP 62 Route de l'Observatoire Bouzaréah, 16340, Algiers, Algeria
e-mail: l.hassaine @cder.dz

A. Sayigh (ed.), *Renewable Energy in the Service of Mankind Vol II*,
DOI 10.1007/978-3-319-18215-5_44

44.1 Introduction

With the increasing concern for global environmental protection, the need to produce pollution-free natural energy such as solar energy has received great interest as an alternative source of energy for the future since it is clean, pollution-free, and inexhaustible.

Photovoltaic (PV) installations have had an exponential growth mainly due to the governments and utility companies that support programs that focus on grid-connected PV systems [1]. In an effort to use the solar energy effectively, a great deal of research has been done on the grid-connected PV generation systems, 98.7 % PV power has been connected to the grid in Europe as shown in Fig. 44.1.

In grid-connected PV systems, the inverter that converts the output direct current (DC) of the solar modules to the alternating current (AC) is receiving increased interest in order to generate power for utility.

44.2 Topologies Structures for Grid-Connected PV Systems

44.2.1 Central Inverters

Central technology illustrated in Fig. 44.2 was based on centralized inverters that interfaced a large number of PV modules to the grid [2–5]. The PV modules were divided into series connections (called a string), each generating sufficiently high voltage to avoid further amplification. These series connections were then connected in parallel through string diodes in order to reach high power levels, 10–250 kW [5]. This centralized inverter includes some severe limitations such as high-voltage DC cables between the PV modules and the inverter, power losses due to a centralized MPPT, mismatch losses between the PV modules, losses in the string diodes, and a nonflexible design where the benefits of mass production could not be reached. The grid-connected stage was usually line commutated by means of thyristors, involving many current harmonics and poor power quality.

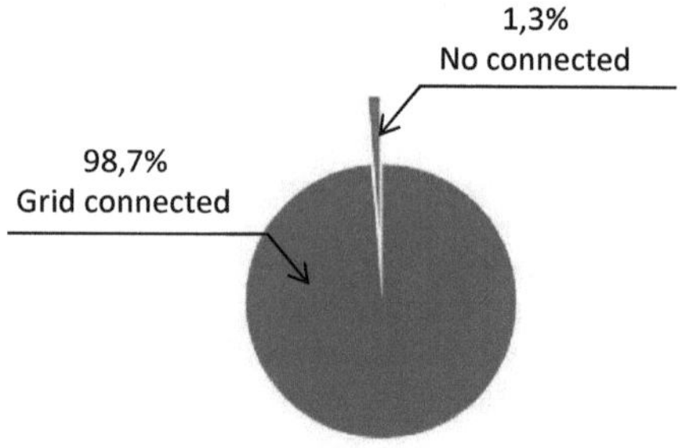

Fig. 44.1 PV power installed in Europe

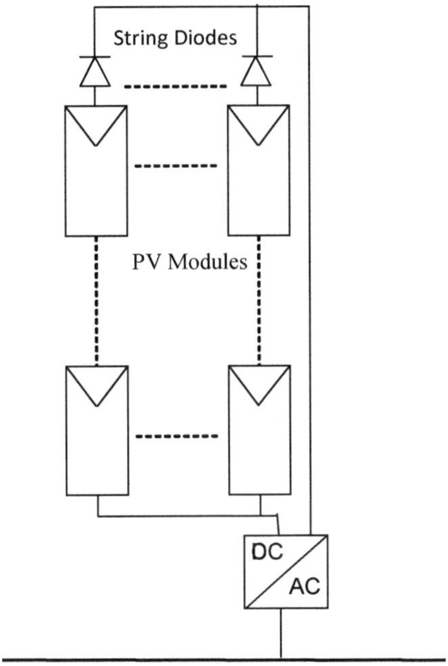

Fig. 44.2 Centralized technology [3]

44.2.2 String Inverters

The string inverters shown in Fig. 44.3 are a reduced version of the centralized inverter, where a single string of PV modules is connected to the inverter [2, 3, 6]. The input voltage may be high enough to avoid voltage amplification. There are no losses associated with string diodes and separate MPPTs can be applied to each string. This increases the overall efficiency compared to the centralized inverter and reduces the price due to mass production.

44.2.3 Multi-string Inverters

The multi-string inverter depicted is the further development of the string inverter, where several strings are interfaced with their own DC–DC converter to a common DC–AC inverter [1, 3, 6]. This is beneficial compared to the centralized system, since every string can be controlled individually. A general structure for distributed systems is illustrated in Fig. 44.4. The range of power is 3–10 kW [4].

44.2.4 AC Modules

The AC module depicted in Fig. 44.5 is the integration of the inverter and PV module into one electrical device [1]. It removes the mismatch losses between PV

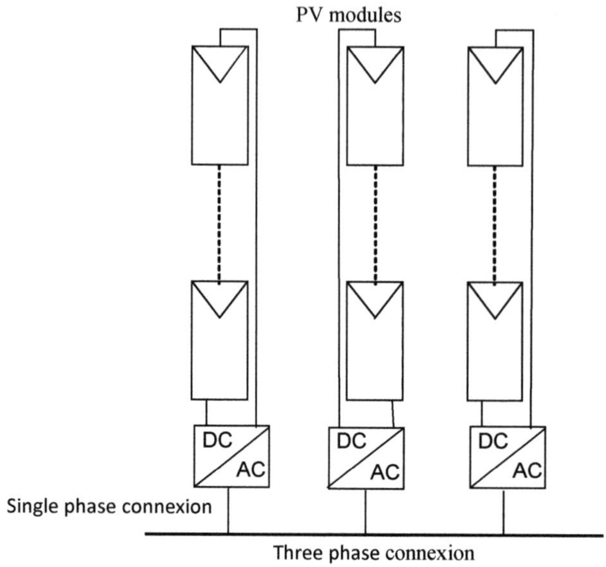

Fig. 44.3 String technology [3]

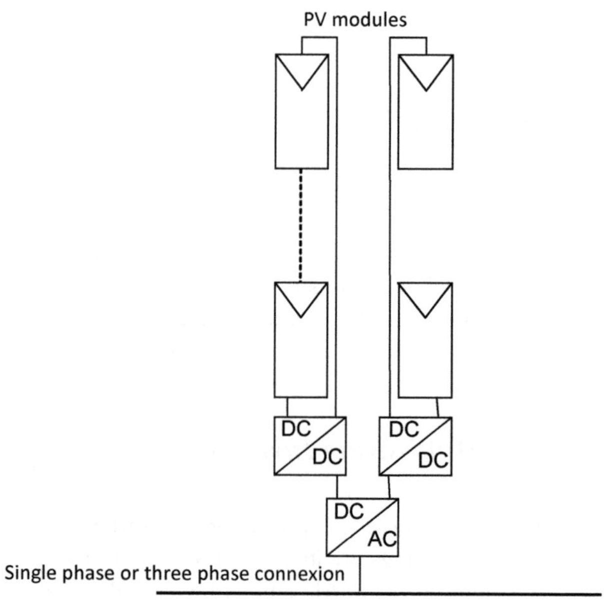

Fig. 44.4 Multi-string technology [3]

modules, since there is only one PV module, as well as supports optimal adjustment between the PV module and the inverter and, hence, the individual MPPT. It includes the possibility of an easy enlarging of the system due to the modular structure. The necessary high-voltage amplification may reduce the overall efficiency

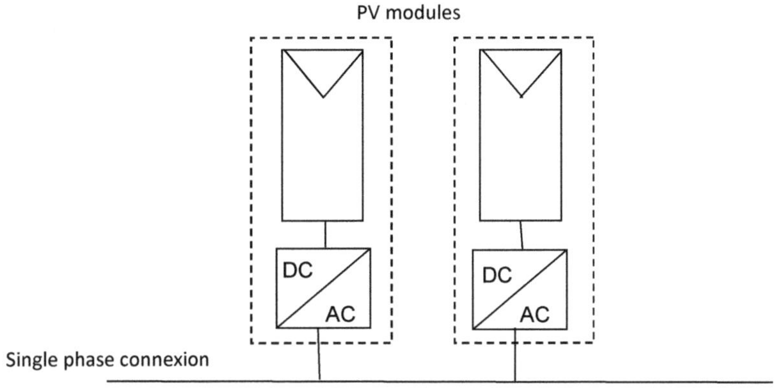

Fig. 44.5 AC module [3]

and increase the price per watt because of more complex circuit topologies. The present solutions use self-commutated DC–AC.

44.3 Control Structures for Grid-Connected PV Systems

The DC–AC converters inject sinusoidal current into the grid controlling the power factor. Therefore, the inverter converts the DC power from the PV generator into AC power for grid injection. One important part of the PV-connected grid system is its control. The control can be divided into two important parts.

1. MPP controller, with the main property to extract the maximum power from the input source (PV module).
2. Inverter controller, which ensures the control of active and reactive power generated to the grid; the control of DC-link voltage; high quality of the injected power; and grid synchronization.

The control strategy applied to the inverter mainly consists of two cascaded loops. Usually, there is a fast internal current loop, which regulates the grid current, and an external voltage loop, which controls the DC-link voltage. The current loop is responsible for power quality issues and current protection; thus, harmonic compensation and dynamics are the important properties of the current controller. The DC-link voltage controller is designed for balancing the power flow in the system. Usually, the design of this external controller aims the optimal regulation and stability of systems having slow dynamics. This voltage loop is designed for a stability time higher than the internal current loop by 5–20 times. The internal and external loops can be considered decoupling; therefore, the transfer function of the current control loop is not considered when the voltage controller is designed [5–8].

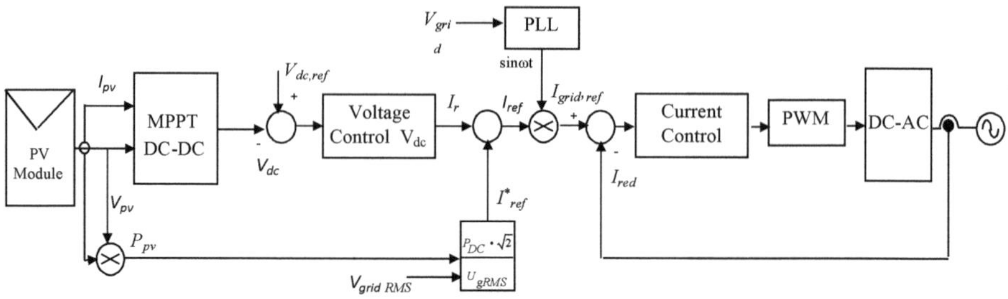

Fig. 44.6 Control structure with DC–DC converter [8]

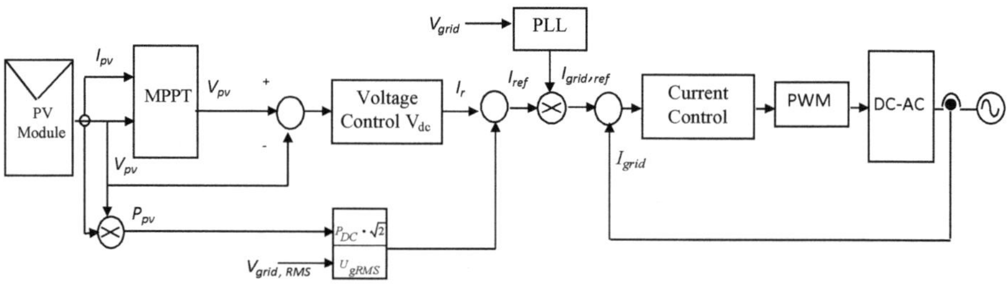

Fig. 44.7 Control structure without DC–DC converter [8]

In some works, the control of the inverter connected to the grid is based on a DC-link voltage loop cascaded with an inner power loop instead of a current one. In this way, the current injected into the grid is indirectly controlled.

44.3.1 Control Structure for Single-Phase with DC–DC Converter

The control structure for single phase with DC–DC converter proposed in [6–9] is shown in Fig. 44.6. The most common control structure for the DC–AC grid converter is a current-controlled H-bridge PWM inverter having a low-pass output filters. Typically, L filters are used but the new trend is to use LCL filters that have a higher order which leads to more compact design.

- Control of instantaneous values current
- Current is injected in phase with gird voltage (PF = 1)
- Use PLL for synchronization of current I_{grid} and V_{grid}

44.3.2 Control Structure for Single Phase Without DC–DC Converter

In Fig. 44.7, the control structure for single phase without DC–DC converter proposed in [5, 6, 9] is shown.

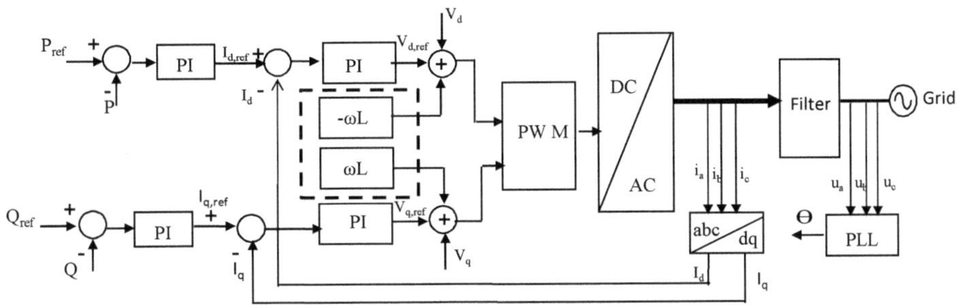

Fig. 44.8 General structure for dq control [9]

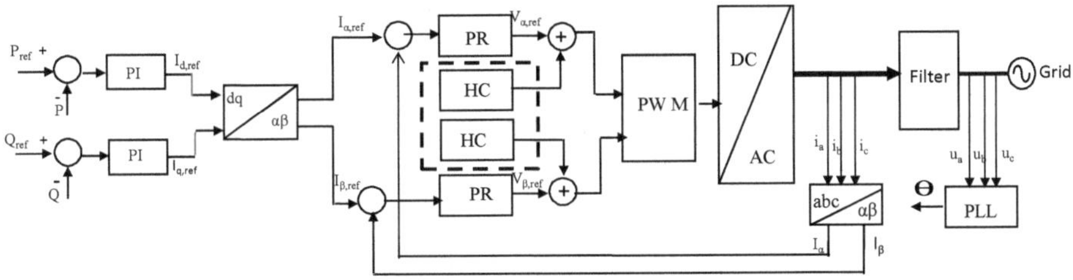

Fig. 44.9 General structure for αβ control [9]

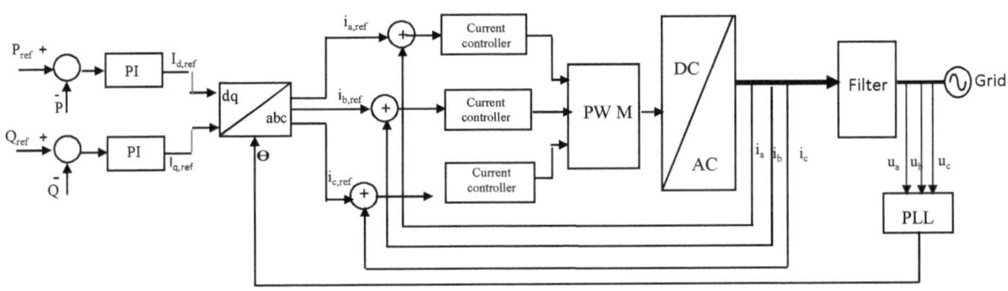

Fig. 44.10 General structure for abc control [9]

44.3.3 *Control Structure for Three-Phase Inverter Connected to the Grid*

In Fig. 44.8, the control structure for three-phase inverter proposed in [9] is shown. Synchronous reference frame control, also called *dq* control, uses a reference frame transformation module, $abc \rightarrow dq$.

Another possible way to structure the control loops proposed by [9] is to use the implementation in stationary reference frame, as shown in Fig. 44.9; in this case, the grid currents are transformed into stationary reference frame using the $abc \rightarrow \alpha\beta$ module.

The *abc* control is to have an individual controller for each grid current; however, the different ways to connect the three-phase systems are shown in Fig. 44.10.

44.4 Conclusion

This chapter presented solutions, where large areas of PV modules were connected to the grid by means of centralized inverters. This included many shortcomings for which reason the string inverters emerged. A natural development was to add more strings, each with an individual DC–DC converter and MPPT to the common DC–AC inverter; thus, the multi-string inverters were brought to light. This is believed to be one of the solutions for the future. Another trend seen in this field is the development of the AC module, where each PV module is interfaced to the grid with its own DC–AC inverter.

In the next part, some control structures for single-phase and three-phase inverters has been discussed. Control structures for single-phase inverter with DC–DC converter and without a DC–DC converter. Different implementation structures for three-phase inverter such as *dq* and stationary and natural frame control structures were presented, and their major characteristics were pointed out.

References

1. Annual Report (2012) Photovoltaic industry association, EPIA
2. Haeberlin H (2001) Evolution of inverters for grid connected PV-systems from 1989–2000, in Proceedings of the 17th European Photovoltaic Solar Energy Conference, pp. 426–430. Munich
3. Kjaer SB, Pedersen JK, Blaabjerg F (2005) A review of single-phase grid-connected inverters for photovoltaic modules. IEEE Trans Ind Appl 41(5):1292–1306
4. Blaabjerg F, Chen Z, Kjaer SB (2004) Power electronics as efficient interface in dispersed power generation systems. IEEE Trans Power Electron 19(5):1184–1194
5. Blaabjerg F, Teodorescu R, Liserre M, Timbus AV (2006) Overview of control and grid synchronization for distributed power generation systems. IEEE Trans Ind Electron 53(5):1389–1409
6. Hassaine L, olías E Implementación de un Control Digital de Potencia activa y Reactiva para inversores. Aplicación a Sistemas Fotovoltaicos conectados a Red, Doctoral Thesis, Carlos III University Madrid, Spain
7. Ciobotaru M, Teodorescu R, Blaabjerg F (2005) Control of single-stage PV inverter, Proceedings of PELINCEC, CDROM
8. Teodorescu R, Blaabjerg F (2006) Overview of renewable energy system, ECPE Seminar renewable energy, ISET, Germany
9. Hassaine L, Olías E, Quintero J, Salas V (2014) Overview of power inverter topologies and control structures for grid connected photovoltaic systems. Renew Sustain Energy Rev 30:796–807

Chapter 45
Monitoring and Diagnostics of Photovoltaic Power Plants

Giuseppe Marco Tina, Fabio Cosentino and Cristina Ventura

Abstract Photovoltaic (PV) systems should be monitored in order to control their production and detect any possible faults. Different possibilities exist for data analysis. Some perform it yearly, analyzing the performance of the PV system over a significant time period of operation and comparing it with similar systems. This shows that a system is performing poorly, but it has the disadvantage that it cannot be used to explain the causes of this underperformance. Others use high-resolution monitoring (minutely to hourly intervals) to analyze the performance of the system; with this higher resolution, a fault diagnosis procedure can be executed. These systems can detect general faults like constant energy losses, total blackout, and short-time energy losses, and the best can also detect shading; however, they cannot identify the exact cause. With the introduction of distributed maximum power point tracking (DMPPT) systems—power optimizers and micro-inverters—a new level of PV system monitoring is possible. Since these systems require the monitoring of the modules' operating voltage and current (for the maximum power point tracking (MPPT) algorithm), the use of voltage and current sensors for each module is at no extra cost. It is only necessary to add a communication module since it is not directly incorporated in the DMPPT board.

Based on the use of such appliances, a wireless sensor network (WSN) that allows monitoring, at panel level, the efficiency of PV panels has been proposed; nodes of the WSN, which are installed on each PV module, are equipped with voltage, current, irradiance, and temperature. Acquired data are then transferred to a management center which is in charge of estimating efficiency losses and correlated causes at the level of the single module.

This research has been further developed in this chapter. The authors propose the possibility of using the DC/DC converter inside the system for MPPT. It allows

G. M. Tina (✉) · F. Cosentino · C. Ventura
Dipartimento di Ingegneria Elettrica, Elettronica ed Informatica,
University of Catania, Catania, Italy
e-mail: giuseppe.tina@dieei.unict.it

F. Cosentino
e-mail: fab.ing@libero.it

C. Ventura
e-mail: cristina.ventura@dieei.unict.it

A. Sayigh (ed.), *Renewable Energy in the Service of Mankind Vol II*,
DOI 10.1007/978-3-319-18215-5_45

to span the PV module voltage within a certain range, to measure a partial current–voltage (I-V) curve under the assumption that failures can be detected mostly by the parameter variation of R_S, R_{SH}, and diode factor η in the I-V curve based on equivalent circuit equation. These parameter variations should be calculated from I-V curve variation. The failure pattern would be presumed from the parameter variations caused by I-V curve variation, if the field data were accumulated. Due to the fact that I-V is partially available, new algorithms, analytical and metaheuristic, for parameter identification have been developed.

The calculated parameters are used not only to detect long-term faults (e.g., aging, soling, delamination, and so on), but also to build an I-V curve reference when an impromptu fault happens (e.g., shading, breakdown diode, and so on).

An experimental hardware and software (in LabVIEW/MATLAB environment) setup has been realized, so many measurement campaigns have been done. The proposed algorithms for parameter identification have been checked against real outdoor conditions. The chapter contains several graphs and charts which explain the relationship between I-V curve shapes and type of faults. A full literature survey will also be included in this study.

45.1 Introduction

The energy produced by a photovoltaic (PV) system depends on various factors such as nominal characteristics of the system components, electrical and geometrical configurations, weather conditions of the installation site, shadowing, PV plant availability, and faults that may occur during normal operations [1]. A certain number of different problems may therefore cause energy losses in the PV plant. Some losses can be related to the PV plant, including errors in tracking the maximum power point, dispersion module parameters (mismatch), wiring losses, and aging [2]. Others are influenced by environmental characteristics such as operating temperature and solar radiation level. In addition, other losses are related to the power conditioning unit (DC/DC and DC/AC converters).

In this context, a very promising research field is the monitoring of PV systems, which may include display operation, data monitoring, protection against theft, detection and correction of failures, safety shutdown, and reducing mismatch losses [3]. The monitoring and control are essential to supervise the performance of PV systems in order to prevent electrical failures and malfunctions [4].

Diagnosis methods can be generally categorized in two groups: nonelectric methods such as visual (discoloration, browning, surface soiling, delamination) and thermal (thermal extraordinary heating), and electric methods (dark/illuminated current–voltage (I-V) measurement, transmittance line diagnosis, radio frequency (RF) measurement). Among these methods, the nonelectric methods need frequent visual checks of the PV array in order to observe the color changes of the modules or the thermal properties such as hot spots. These methods need thermal cameras or other equipment in front of the array. Contrarily, the electrical methods need only the measurements of the output electrical signature such as voltage and current [5–7]. Thus, fault diagnosis using electrical methods is more advantageous and promising for the monitoring and diagnostic PV systems [4].

Recently, many monitoring and diagnostic systems have been proposed [8–13]. Most of them use local sensors to measure temperature and irradiance [8, 9]; some instead use satellite data [10, 11], which, however, appear to be less accurate. Another common point of most of the monitoring systems is that they analyze data from the PV generator, limiting the range of possible faults which can be detected. Moreover, many of these analyze only the energy production [8, 10]. These systems allow the detection of general type of failures, such as constant energy losses, total blackout, the energy losses in the short term [10] and, in the best cases, are able to detect cases of shadowing [8]. However, such systems are not able to detect the exact cause of the fault.

With the introduction of the concept of distributed maximum power point tracking (DMPPT), where power optimizers and micro-inverters are used at the single panel level, it is possible to achieve a new level of control and supervision of PV systems, in which the exact point of the fault and the causes of power losses can be determined [4, 12, 13]. In this context, some researcher introduced PV systems modeling and fault diagnosis based on PV module parameter estimation [14–16].

In this chapter, an approach for the monitoring and diagnostic of PV systems based on the PV array I-V characteristic is proposed; by exploiting information from the I-V characteristic, in fact, the detection and localization of defects can be achieved. The method is mainly based on the real-time estimation of the equivalent circuit parameters of a PV module/array.

The PV module parameters can be evaluated by means of either numerical methods, which minimize the difference between a measured I-V curve and the one calculated by the model [16], or just using some points of the I-V characteristic that can be measured and, referring to standard test conditions (STC), are also provided by that manufacturer of the PV module [17]. Most of these methods need the measurement of three remarkable points of the I-V characteristic: open circuit (V_{OC}), short circuit (I_{SC}), and maximum power point (V_{MPP} and I_{MPP}), besides the PV panel temperature (T) and the irradiance (G). However, it must be observed that usually PV plants are managed by inverters which perform maximum power point tracking (MPPT). In such a way, controlling the MPPT voltage window of any commercial inverter, it is possible to easily measure V_{MPP}, I_{MPP}, and V_{OC}, while I_{SC} is usually not achievable during the normal operation of a PV panel. The novelty of the method here proposed is that it allows the estimation of the PV module parameters using data obtainable using a commercial inverter, besides the PV panel temperature and the irradiance that are usually available in a real PV plant. Wireless sensor network (WSN) nodes installed on each PV module can be used to measure these data [18]. The estimation of the PV module parameters can be used for both short- and long-term diagnosis analyses. They allow, in fact, the estimation of the I-V characteristic of a PV module. A reliable estimation of the I-V characteristic is strategic for both assessment of PV panel performance and improvement of the power generation system efficiency. Furthermore, it can be compared with the partial measured I-V curve in order to detect possible causes of fault. Moreover, keeping track of the key PV module parameters would reflect changes in the PV generator's health state [19–21]. In this work, a LabVIEW interface has been implemented in order to test the proposed approach in real time on a PV model under real operating conditions.

The chapter is organized as follows: first, the architecture for the monitoring and the long- and short-term diagnosis is analyzed, then the method used to find the key parameters starting from the partial measure of the I-V characteristic is presented, finally the experimental data are reported to demonstrate the potentiality of the adopted solutions.

45.2 Architecture of the Monitoring and Diagnostic System

The general approach to assess the electrical performance of a PV system is based upon the capability of analytically describing the I-V characteristic of the PV component for each operative temperature and solar radiation. Traditionally, the analytical models used in the study of these phenomena evaluate the behavior of the PV cell by assimilating it into an equivalent electrical circuit that includes some non-linear components [16]. These electrical equivalent circuits are based on some unknown parameters, from three to seven, depending on the complexity of the model.

The most common and used of them is the single diode-based equivalent circuit model, which is characterized by five unknown parameters. First of all, I_{PH} (light or photo-generated current) represents the charge carrier generation in the semiconductor layer of the PV cell caused by the incident radiation. Due to the presence of the diode, there is one unknown parameter, I_0 (reverse-bias saturation current for the diode) and η (the ideality factor of the diode). R_{SH} (parallel or shunt resistance) expresses the losses due to the high current path through the semiconductor throughout the mechanical defects and the leakage of current to the ground. Finally, R_S (series resistance) represents the internal losses due to current flow and the connection between cells.

The I-V curve, and therefore the parameters, of a PV panel varies with irradiance and cell temperature. Anyhow, the reference value of each parameter can be calculated starting from its value in the operating conditions and vice versa [22]. Once the parameters have been obtained, the I-V characteristic can be estimated by model (45.1):

$$I = I_{PH} - I_o \cdot \left(e^{\frac{V + I \cdot R_S}{n \cdot V_t}} - 1 \right) - \frac{V + I \cdot R_S}{R_{SH}}. \tag{45.1}$$

The procedure here proposed for monitoring and diagnostic purposes is based upon the comparison between the measured I-V curve in real working conditions and the one approximated starting from the estimated parameters. In particular, the methodology adopted can be described by the following steps:

- Calculation of the five parameters starting from data provided by manufacturers, which are referred to as STC
- Measurement of solar radiation and cell temperature (in the real working conditions)
- Calculation of the five parameters in the real working conditions through relations proposed in [22]
- Estimation of the I-V characteristic in real working conditions using the calculated parameters
- Measurement of the partial I-V curve
- Calculation of the error (*mse*) between the measured and the estimated I-V curve: if the *mse* is less than a threshold (mse_{th}), there are no anomalies in the measured I-V characteristic, no faults are detected, the five parameters can be estimated and stored in order to develop a long-term diagnostic analysis; whereas, if the *mse* is higher than the threshold (mse_{th}), a fault is detected, the estimated five parameters are not stored and a short-term diagnostic procedure can be launched to identify the location and the type of fault.

The block diagram of the algorithm here developed is shown in Fig. 45.1.

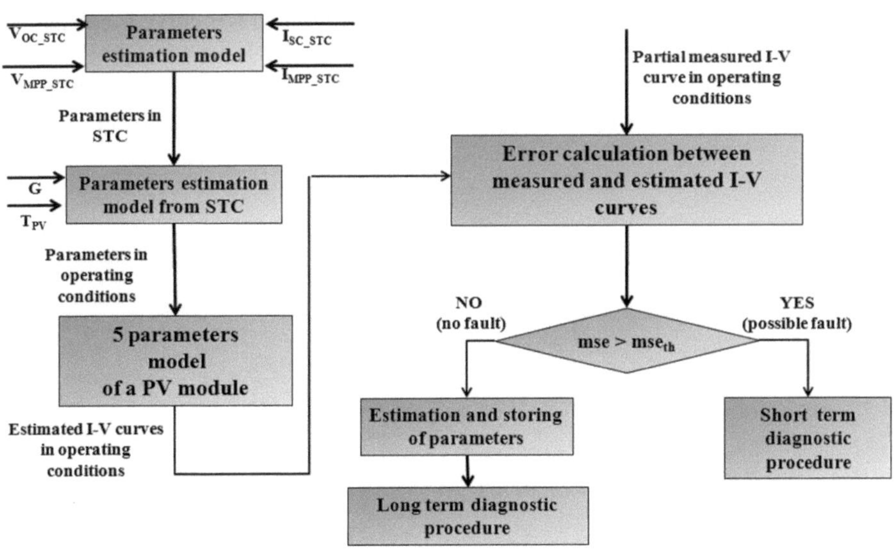

Fig. 45.1 Block diagram of the procedure here proposed for monitoring and diagnostic purpose

45.3 Parameter Evaluation Based on Partial I-V Curve

The five parameter model is relative to the equivalent circuit representative either of a PV cell or a PV module.

To find the five parameters, the analytical method proposed in [17] has been used, which allows to find the five parameters by the three remarkable points of the I-V curve: open circuit (V_{OC}), maximum power (V_{MPP}, I_{MPP}), and short circuit operating conditions (I_{SC}). Moreover, a metaheuristic algorithm [16] has been also used for parameter identification, the AB–DE. This algorithm is derived from the artificial bee colony (ABC), which is inspired by the foraging behavior of a bee swarm which searches honey sources; movement search and information exchange between the bees are inspired by differential evolution (DE).

Anyway, as aforementioned, it must be observed that usually PV plants are connected to inverters which perform MPPT. In such cases, V_{MPP}, I_{MPP}, and V_{OC} can be easily measured, while I_{SC} is usually not achievable during the normal operation of a PV panel. Therefore, an algorithm that allows us to find the five parameters with a good accuracy, without measuring I_{SC}, has been developed. The algorithm starts from the value of the current *(I_{C_tr})* correspondent to a voltage value which can be set depending on the inverter that is used in the real application *(V_{C_tr})*, for example, the value of the voltage can be 60 % of V_{OC}. The obtained measured I-V characteristic is shown in Fig. 45.2.

The procedure here proposed can be described by the following steps:

1. The five parameters are calculated using the five parameters model [17] setting the short circuit current *(I_{C_SC})* equal to I_{M_tr} and the I-V characteristic is estimated.
2. The current (I_{C_tr}) correspondent to V_{C_tr} in the estimated I-V curve is calculated.
3. The value of I_{C_tr} is compared with the value of I_{M_tr}.
4. If I_{C_tr} is greater than or equal to I_{M_tr}, it means that $I_{SC}=I_{C\_tr}$ and the algorithm is stopped because the value of I_{C_SC} is the closest one to the real I_{SC},

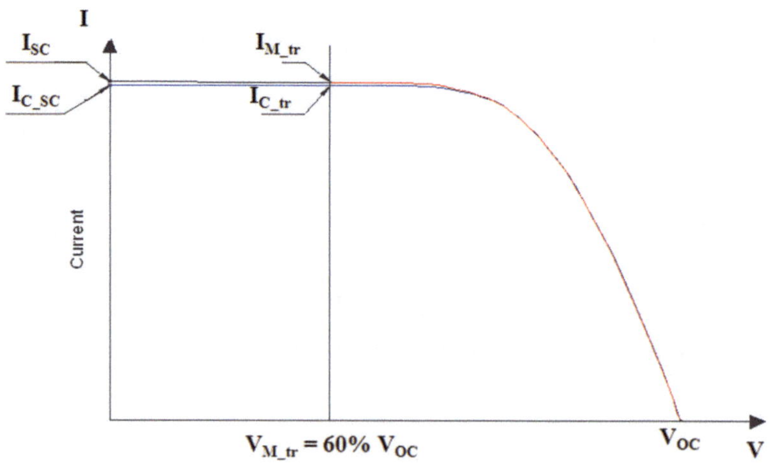

Fig. 45.2 Example of partial curve *(read line)* that can be used to estimate the I_{SC} parameter

5. Otherwise, the five parameters are estimated setting $I_{c\_sc}(t) = I_{c\_sc}(t-1) + \frac{I_{M\_tr} - I_{c\_tr}}{2}$ and the I-V characteristic is estimated; the algorithm continues from point 3.

The block diagram of the algorithm used to calculate I_{SC} is shown in Fig. 45.3.

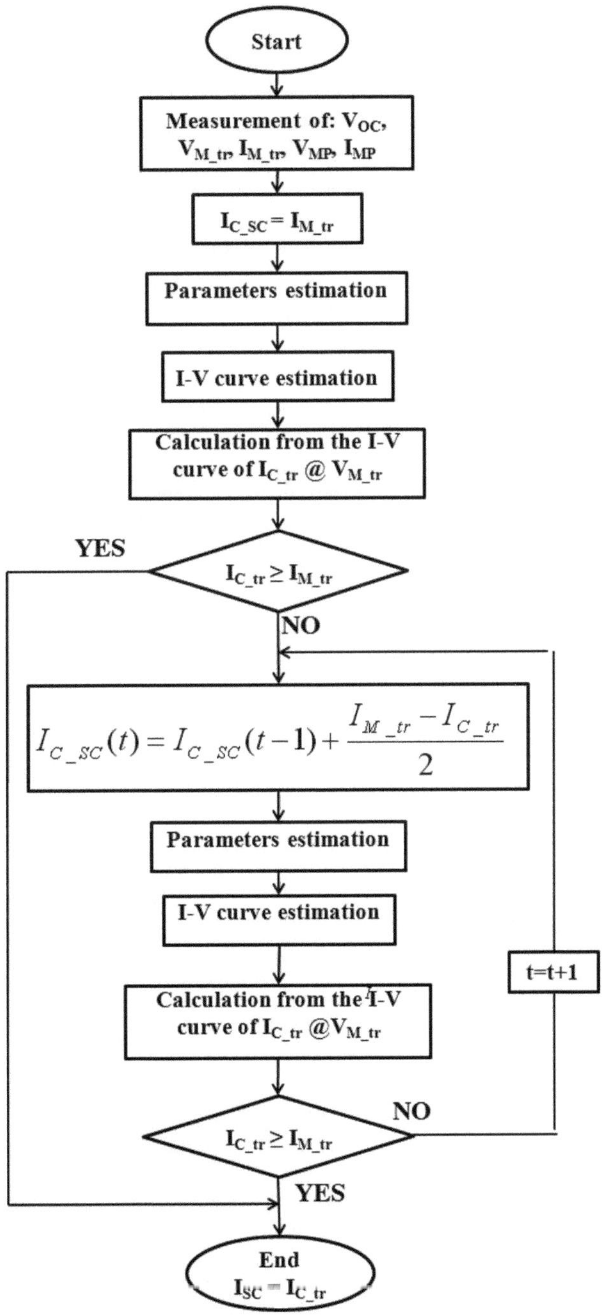

Fig. 45.3 Block diagram of the algorithm developed to find I_{SC}

45.4 Experimental Results

45.4.1 Experimental Setup

To perform experimental tests of the developed platform, a glass–glass PV module composed by 45 Bosch Solar Cell M 3BB by Wp 4.24/each (5 strings of 9 cells each). Electrical parameters in STCs ($G=1000$ W/m^2, $T=25$ °C, air mass $=1.5$) are: $P_{max}=168{,}809$ W, $V_{MPP}=21{,}850$ V, $I_{MPP}=7726$ A, $V_{OC}=2721$ V, $I_{MPP}=8324$ A.

To measure the I-V characteristic of the PV module, an electronic load (6063B 250 W DC Electronic Load by Agilent) in the constant voltage mode has been used. An irradiation sensor (SPEKTRON 210 by TRITEC), placed in the plane of the PV module, has been used to measure the solar radiation, while an NTC thermistor has been used to measure the temperature of the PV module. Signals coming from sensors have been acquired using an NI cDAQ-9188 data acquisition board.

The block diagram of the experimental setup is shown in Fig. 45.4. All data are acquired, stored, and elaborated through a LabVIEW interface.

45.4.2 Experimental Results

First, the I-V characteristics of the glass–glass PV module have been measured, then the five parameters have been calculated starting from STC values using

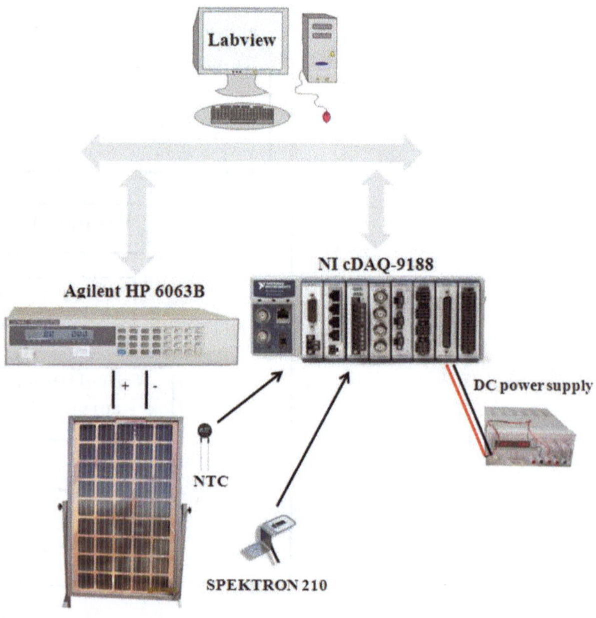

Fig. 45.4 Block diagram of the experimental setup

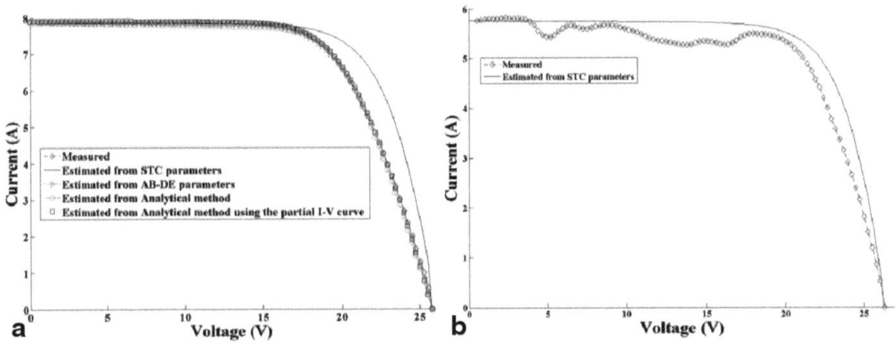

Fig. 45.5 Example of I-V curve. **a** The calculated five parameters can be stored and a long-term diagnostic analysis can be applied. **b** The five parameters cannot be calculated or stored and the measured I-V curve can be used for the short-term diagnostic analysis

relations proposed in [22] and using the three different methods here proposed; in all cases, the I-V characteristics have been estimated. Figure 45.6a and b shows the I-V curves measured, the I-V curves estimated using the parameters calculated from STC values, the I-V curves estimated using the parameters calculated through the AB-DE algorithm, the I-V curves estimated using the parameters calculated through the analytical algorithm, and the I-V curves estimated using the parameters calculated through the analytical algorithm starting from the partial I-V characteristic. In particular, Fig. 45.5a shows an example of I-V curve when no fault occurred, therefore the calculated five parameters can be stored and a long-term diagnostic analysis can be applied; Fig. 45.5b shows an example of I-V curve when a fault occurred, therefore the calculated five parameters cannot be stored and the measured I-V curve should be used for the short-term diagnostic analysis.

Moreover, in the case of no fault, the five parameters in the operating condition have been stored and compared with their values estimated starting from the correspondent values in STC [22]. Figure 45.6a shows the results obtained for I_0, Fig. 45.6b shows the results obtained for η, Fig. 45.6c shows the results obtained for R_S, Fig. 45.6d shows the results obtained for R_{SH}. The value of I_{PH} is very close to I_{SC}.

The results demonstrate how the algorithms here proposed allow the estimation of the five parameters with good accuracy in case irradiance is greater than 500 W/m^2. This suggests the addition of another block that allows the estimation and storage of the parameters after a check on the irradiance value.

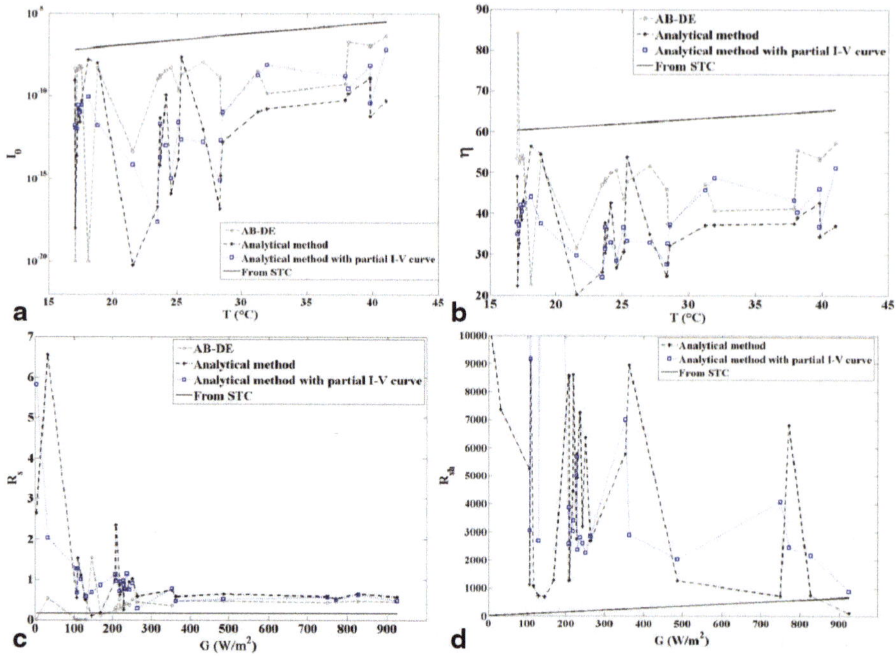

Fig. 45.6 Comparison between the parameters obtained from the ones estimated in STC [22], with the AB-DE method, with the analytical method, and with the analytical method using a partial I-V curve: **a** I_0, **b** η, **c** R_S, **d** R_{SH}

45.5 Conclusions and Future Work

PV systems are subject to different sorts of failures, and a monitoring and diagnostic system must be present to identify different causes of failure. In this chapter, a procedure that allows evaluation if a fault occurs in a PV system is proposed. The algorithm is based on the estimation of the key parameters of a PV module. In particular, a metaheuristic method and an analytical method are proposed. These methods are based on the measurement of the current and the voltage in short circuit, in maximum power point and in open circuit. However, since a DC/AC converter performs MPPT in a PV system, it is not possible to measure the current in short-circuit conditions; in this chapter, a procedure that allows a precise estimation of this parameter has been developed. The algorithm here proposed allows the detection and location of a fault that occurred in a PV module by comparing the partial measured I-V characteristic with the one estimated using the parameters estimated in the operating condition starting from their values in STC. If there are no anomalies in the measured I-V characteristic, no faults are detected, the five parameters can be estimated and stored in order to develop a long-term diagnostic analysis; if a fault is detected, the estimated five parameters are not stored and a short-term diagnostic procedure can be launched to identify the location and the type of fault. Results demonstrate that the proposed algorithms used to estimate the key parameters allow good accuracy if the irradiance is greater than 500 W/m^2.

Acknowledgments This work is developed under the ICARO project funded by the PO. FESR 2007–2013—Sicilia—Linea di Intervento 4.1.1.1.

References

1. Almonacid F, Rus C, Pérez-Higueras P, Hontoria L (2011) Calculation of the energy provided by a PV generator. Comparative study: conventional methods vs. artificial neural networks. Energy 36:375–384
2. Nordmann T, Jahn U, Nasse W (2004) Performance of PV systems under real conditions. In: Workshop on Life Cycle Analysis and Recycling of Solar Modules - the "Waste" Challenge; Brussels, Belgium
3. Burger B, Goeldi B, Rogalla S, Schmidt H (2010) Module integrated electronics. An overview, 25th European photovoltaic solar energy conference and exhibition/5th world conference on photovoltaic energy conversion, pp 3700–3707
4. Davarifar M, Rabhi A, El Hajjaji A, Dahmane M (2013) Real-time model base fault diagnosis of photovoltaic panels using statistical signal processing. International conference on renewable energy research and applications. IEEE ICRERA, Madrid
5. Takashima T, Yamaguchi J, Ishida M (2008) Disconnection detection using earth capacitance measurement in photovoltaic module string. Prog Photovolt Res Appl 16(8):669–677
6. Merhej P, Dallago E, Finarelli D (2010) Effect of capacitance on the output characteristics of solar cells. In: Ph. D. research in microelectronics and electronics (PRIME), 2010 conference on, 18–21 July 2010 pp 1–4
7. Ducange P, Fazzolari M, Lazzerini B, Marcelloni F (2011) An intelligent system for detecting faults in photovoltaic fields. In: Intelligent systems design and applications (ISDA), 2011 11th international conference on, 22–24 Nov. 2011. pp 1341–1346
8. Firth SK, Lomas KJ, Rees SJ (2010) A simple model of PV system performance and its use in fault detection. Sol Energy 84(4):624–635
9. Chouder A, Silvestre S (2010) Automatic supervision and fault detection of PV systems based on power losses analysis. Energy Convers Manage 51(10):1929–1937
10. Drews A et al (2007) Monitoring and remote failure detection of grid-connected PV systems based on satellite observations. Sol Energy 81(4):548–564
11. Stettler S, Toggweiler P, Remund J (2006) SPYCE: satellite photovoltaic yield control and evaluation. In: Proceedings of the 21st European photovoltaic solar energy conference, pp 2613–2616
12. Solórzano J, Egido MA (2013) Automatic fault diagnosis in PV systems with distributed MPPT. Energy Convers Manage 76:925–934
13. Prieto MJ, Pernía AM, Nuño F, Díaz J, Villegas PJ (2014) Development of a wireless sensor network for individual monitoring of panels in a photovoltaic plant, Sensors, 14(2), pp2379–2396
14. Chao K-H, Sheng-Han H, Wang M-H (2008) Modeling and fault diagnosis of a photovoltaic system. Electr Power Syst Res 78(1):97–105
15. Chang C-H, Zhu J-J, Tsai H-L (2010) Model-based performance diagnosis for PV systems. Proceedings of SICE annual conference, pp 2139–2145, 18–21 Aug. 2010
16. Hachana O, Hemsas KE, Tina GM, Ventura C (2013) Comparison of different metaheuristic algorithms for parameter identification of photovoltaic cell/module. J Renew Sustain Energy 5(5):053122
17. Tina GM, Ventura C (2013) Evaluation and validation of an electrical model of photovoltaic module based on manufacturer measurements. Sustainability in energy and buildings, smart innovation, systems and technologies Vol 22, pp 15–24, Springer. DOI: 10.1007/978-3-642-36645-1_2
18. Ando B, Baglio A, Pistorio A, Tina GM, Ventura C (2013) SENTINELLA: A WSN for a smart monitoring of PV systems at module level. IEEE international workshop on measurements and networking proceedings (M & N) 2013, pp 36–40, October

19. Sera D (2010) Series resistance monitoring for photovoltaic modules in the vicinity of MPP. 25th European photovoltaic solar energy conference and exhibition, pp 4506–4510
20. Barbato M, Meneghini M, Giliberto V, Giaffreda D (2012) Effect of shunt resistance on the performance of mc-silicon solar cells: a combined electro-optical and thermal investigation. Photovoltaic specialists conference
21. Kang G-H, Kim K-S, Park C-H, Waithiru L, Yu G-J, Ahn HK, Han D-Y (2006) Current-voltage characteristics with degradation of field-aged silicon photovoltaic modules. 21st European Photovoltaic Solar Energy Conference, 4–8 September 2006, Dresden, Germany
22. Boyd MT, Klein SA, Reindl DT, Dougherty BP (2011) Evaluation and validation of equivalent circuit photovoltaic solar cell performance models. J Solar Energy Eng 133(2)
23. Barbato M, Meneghini M, Giliberto V, Giaffreda D (2012) Effect of shunt resistance on the performance of mc-silicon solar cells: a combined electro-optical and thermal investigation. Photovoltaic specialists conference

Chapter 46
AC Power Short-term Forecasting of a Thin-film Photovoltaic Plant Based on Artificial Neural Network Models

Giuseppe Marco Tina, Cristina Ventura, Giovanna Adinolfi, Sergio Ferlito and Giorgio Graditi

Abstract Solar and PV forecasts for time horizons ranging from a few minutes ahead to several days ahead are generally significant for planning the operations of power plants which convert renewable energies into electricity. With the increasing penetration of photovoltaic (PV) power systems into the grid, the problems caused by the fluctuation and intermittence of PV power output are gaining interest. The power output fluctuations impact the power system's stability. For this reason, an accurate forecast of PV production is necessary to consider PVs a reliable energy source.

Different techniques can be used to generate solar and PV forecasts, depending on the forecast horizon—very short-term forecasts (0–6 h ahead) perform best when they make use of measured data, while numerical weather prediction models are essential for forecast horizons beyond approximately 6 h.

The aim of this study is to evaluate models for PV AC power short-term forecast using different artificial intelligence-based techniques and using ahead values of solar radiation and ambient temperature as data sources for forecasting. The data refer to a 1-kWp experimental micromorph silicon modules plant located at ENEA

C. Ventura (✉) · G. M. Tina
Dipartimento di Ingegneria Elettrica, Elettronica ed Informatica,
University of Catania, Catania, Italy
e-mail: cristina.ventura@dieei.unict.it

G. M. Tina
e-mail: giuseppe.tina@dieei.unict.it

G. Adinolfi · S. Ferlito · G. Graditi
ENEA, Agenzia nazionale per le nuove tecnologie,
l'energia e lo sviluppo economico sostenibile, Research Center of Portici,
80055 Portici, NA, Italy
e-mail: giovanna.adinolfi@enea.it

S. Ferlito
e-mail: sergio.ferlito@enea.it

G. Graditi
e-mail: giorgio.graditi@enea.it

A. Sayigh (ed.), *Renewable Energy in the Service of Mankind Vol II*,
DOI 10.1007/978-3-319-18215-5_46

Portici Research Centre in southern Italy. A large dataset consisting of data measured every 5 min and acquired from 2006 to 2012, is used for the training/testing of the artificial neural networks (ANNs) proposed here. The AC power production evaluation is based upon data measured by a commercial inverter used for plant connection to the grid. This kind of inverter allows acquisition of operative data during their functioning hours, which are usually the central hours of the day. Therefore, when commercial inverters are used to acquire data, the use of ANNs is the best method for forecasting.

In order to verify the effectiveness of the forecast data, measured and predicted data have been compared and the errors have been calculated by means of the relative mean bias error, the relative root mean square error and the correlation coefficient. Experimental data are reported to demonstrate the potentiality of the adopted solutions and to compare the different techniques proposed here. Furthermore, an algorithm that allows classification of a day as variable, cloudy, slightly cloudy or clear has been used for verification as forecast uncertainty depends on the meteorological conditions.

Keywords Photovoltaic system model · Artificial Neural Network · Clear Sky Model

46.1 Introduction

At the moment, photovoltaics (PVs) cover 3 % of the electricity demand and 6 % of the peak electricity demand in Europe. As the share of PVs in the electricity mix increases, grid and market integration challenges are becoming more and more important for the future development of PVs [1]. Unfortunately, the major shortcoming when using PV systems is that solar panels generate an amount of electric power that strongly depends on weather and shadow conditions. Therefore, the total power production coming from PV plants in a specified future time period cannot be determined precisely, as it is a non-deterministic and stochastic process. This power output fluctuation will impact the power system's stability. However, if the energy produced by PV systems can be forecasted with sufficient precision, it will be possible to make PVs a more reliable source of electricity.

An interesting and in-depth overview of the state-of–the-art of solar PV and solar forecasting is offered in [2] and in [3]. The authors emphasize how different forecasting methods can be used depending on the use of PV forecasts, i.e., if they are focused on the output power of systems or on the rate of change, and on the tools and information available to forecasters, such as data from weather stations and satellites. Moreover, forecasting methods can be mainly divided in two groups—physical or statistical. The physical approach uses solar and PV models to generate PV forecasts, whereas the statistical approach relies primarily on past data to 'train' models, such as autoregressive or artificial intelligence models, with little or no reliance on solar and PV systems [2]. Diverse resources can be used to generate solar and PV forecasts including measured weather and PV system data, satellite and sky imagery observations of clouds, and numerical weather prediction (NWP) models which form the basis of modern weather forecasting. The forecast horizon considered depends on the data availability. If only measured data are available, the best performance is achieved for very short-term forecasts (0–6 h ahead). Intra-day forecasts are an important component of the integration of variable renewable resources into the electric grid [4]. On the other hand, the most appropriate forecasting technique depends on the forecast horizon required—NWP models perform best for horizons of 1 or 2 days ahead [2], whereas statistical models based on local ground measurements, possibly combined with satellite or sky imagery data of cloud movements, are more appropriate for short-horizons of <6 h [5].

In the literature there are several reports on PV forecasting; some publications describe forecasting models applied to PV power [6–13], while others report on the prevalence of irradiation forecasting models [14, 15].

For short-horizon forecasting, the techniques used in the models include time series ARMA/ARIMA models [11], artificial neural network (ANN)-based models [7, 10], fuzzy-based models [17], genetic algorithms [13], AR/ARX/NARX approaches [8, 9] and hybrid models [12, 14]. However, conventional forecasting methods (empiric, analytic, numeric simulation and statistic approaches such as AR, ARMA, ARIMA, and Markov chain, etc.) as well as estimation and modelling of the meteorological data and of the power output of a PV plant, cannot be used in the following problems [16]:

1. Long-term forecasting and modelling of the the data;
2. Data missing from the database;
3. Predicting data in locations where measurement instruments are not available.

In [6], several forecasting techniques using no exogenous inputs for predicting the solar power output of a 1-MWp, single-axis tracking PV power plant operating in Merced, California, are evaluated and compared. The authors demonstrate that ANN-based forecasting models perform better than other forecasting techniques and that the accuracy of all models depends strongly on the seasonal characteristics of solar variability.

Here, we propose an ANN approach for forecasting the power production of a 1-kWp experimental micromorph silicon modules plant which uses a commercial inverter for plant connection to the grid. Such commercial DC-AC converters allow the acquisition of operative data during their operating hours. This involves missing samples in the data during the periods of time when the inverter does not work. Considering the described scenario, ANNs seems to be a suitable method to obtain forecasts, as shown by the approach proposed here [16].

In PV plants, AC output production data as well as solar radiation and ambient/PV module temperature can be available. In this context, in order to consider data availability, various ANNs have been developed using a different set of inputs. Moreover, clear sky radiation, that can be easily calculated using appropriate models, has been used as input for all the developed ANNs. A large dataset consisting of data measured every 5 min and acquired from 2006 to 2012, is used for the training/

testing of the ANNs (single hidden layer multilayer perceptron [MLP] neural networks). One-hour ahead and 2-h ahead forecasting have been implemented. Analysis of the results has been reported using 5-min or hourly data.

As stated in [17], the accuracy of solar and PV forecasting depends mostly on climate type and weather conditions. Therefore, a method to classify each minute of a day as variable, cloudy, slightly cloudy or clear has been used [17]. In this way it is possible to understand the correlation between the percentages of the minutes of a specific day that belong to each class and the error performed on that day. Here, we present the input data analysis, the structure of the model, the training procedures and simulation results.

46.2 Experimental PV Plant Description and Data Analysis

The developed ANN models makes use of data provided by a data acquisition system connected to an experimental plant consisting of micromorph silicon modules (amorphous and microcrystalline).

The rated power of the plant is 1.05 kWp. The PV system architecture is made up of 28 modules which are divided into 14 strings (each made by 2 modules). The single module efficiency is close to 9% [18]. The above-described PV plant is connected to the grid by means of a DC-AC converter with a nominal efficiency of 92%. The used modules are installed at 20° tilt angle with south orientation and are assembled on a wooden structure to simulate a typical real roof installation.

Using a data logger, it is possible to monitor and acquire the analog signals coming from the solar radiation sensors (pyranometer) and ambient temperature sensors (Pt 100 probes) that are installed on the PV generator.

The PV plant has been monitored since 2006, so a large dataset of experimental data are available to train and test the ANN models. In particular, 6 years of data sampled with a 5-min time step have been used for ANN training and valuation. To make the data coherent and physically acceptable they have been filtered and interpolated. During the interpolating process, data relative to days with > 1 h of missing data or with a number of available data < 4 h have been eliminated; this is the reason why available data after interpolation are decreased. The number of available data for each year, before and after the filtering process and after interpolation, is reported in Table 46.1. Theoretical data are calculated considering the number of samples that should be available from 1 h after the theoretical sunrise until 1 h before the theoretical sunset [17].

Cumulative curves of the available irradiance data (Gi) for each year compared with clear sky radiation are shown in Fig. 46.1; this allows possible anomalies (anomalous shape) in the measured data to be identified.

Moreover, since PV plant power production is rarely forecasted in situations with inhomogeneous clouds as it results in lower accuracy in output predictions, an

Table 46.1 Number of available data before and after filtering for each year

Year	Theoretical availability	Data before filtering	Data after filtering	Data after filtering and interpolation
2006	38,740 (from 01/02/2006 7:15)	34,322	33,765	21,788
2007	41,392	36,823	36,351	21,313
2008	41,477	22,531	19,945	11,465
2009	41392	22,501	20,422	13,430
2010	41,392	35,394	25,269	16,064
2011	41,392	37,746	37,203	22,155
2012	30,245 (until 08/09/2012 9:20)	14,334	14,120	7934

algorithm that calculates the percentage of time steps that can be classified as clear, slightly cloudy, cloudy or variable during a day [17] has been adopted. The algorithm has been applied to the data available for the different years in order to verify the particular characteristics of each year. For example, if a particular year has been characterized by clear sky weather conditions, the clear and slightly cloudy percentages should be higher than the cloudy and variable ones. In this case the accuracy of the forecasting is expected to be better than a year characterized by variable weather conditions. Results are shown in Fig. 46.2.

All these consideration have been used to choose the best year to train the ANN. In this context, 2007 was chosen since it had sufficient data, a cumulative curve that did not present anomalies and it seemed to be a year which was characterized by balanced weather conditions.

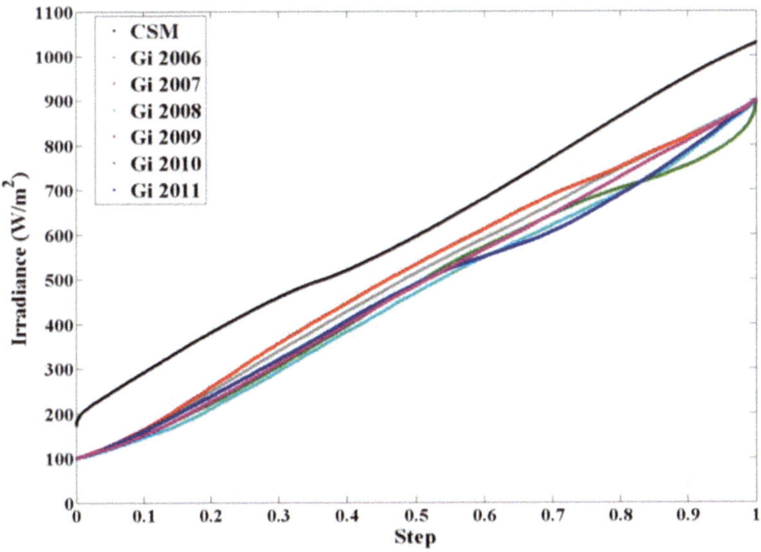

Fig. 46.1 Cumulative curves for each year of available data

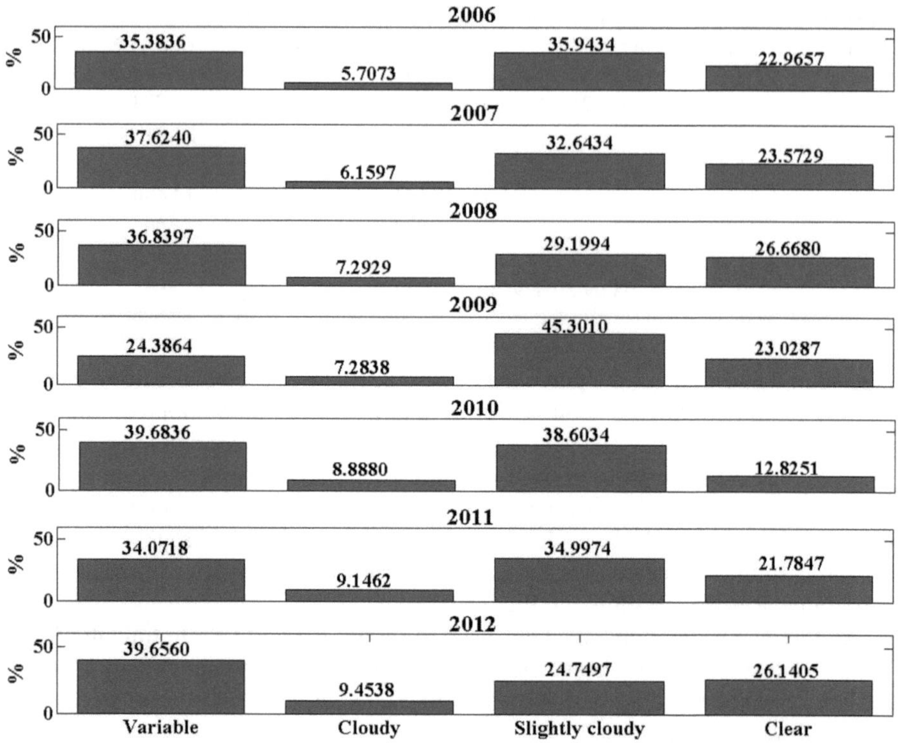

Fig. 46.2 Classification of days according to weather conditions—clear, slightly cloudy, cloudy or variable for each considered year

46.3 Neural Network to Forecast the AC Power Generation

Artificial intelligence techniques are becoming useful as an alternate approach to conventional techniques or as components of integrated systems. They have been used to solve complicated practical problems in various areas and are becoming more popular nowadays [16].

Different ANNs have been implemented for PV AC power short-term forecasting, considering different input data. Moreover, 1-h ahead and 2-h ahead forecasting has been implemented.

First, the minimum availability of data is considered. Therefore, only the AC output power (Pac) of 1- or 2-h ahead and the clear sky radiation (CSM) [17] are considered as input. Clear sky models estimate the terrestrial solar radiation under a cloudless sky as a function of the solar elevation angle, site altitude, aerosol concentration, water vapor, and various atmospheric conditions. It stands as a reference for the daily variance of irradiance and the upper limit for it.

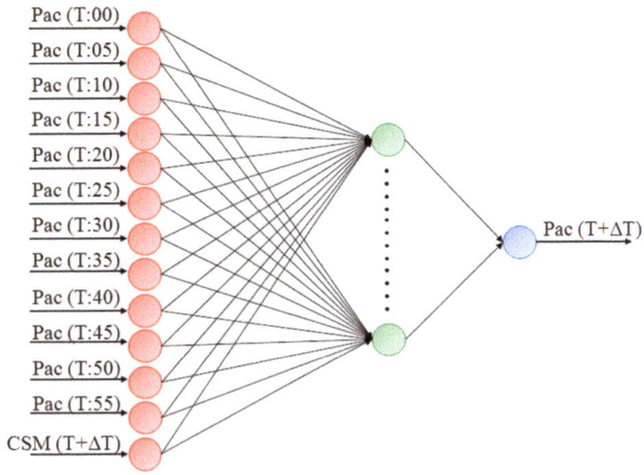

Fig. 46.3 Block diagram of the ANN used for the short-time forecasting of the AC power using as inputs the clear sky solar radiation and the Pac power data relative to 1 h, and 1-h (ΔT=1) and 2-h ahead (ΔT=2)

A block diagram of an example of the neural network implemented for short-time forecasting of the AC power using as inputs the AC power data relative to 1 h, and 1-h (ΔT=1) or 2-h ahead (ΔT=2), and the clear sky solar radiation, is shown in Fig. 46.3.

Then, assuming that environmental measurement data (the solar radiation and the ambient temperature) are also available, other inputs are considered—the measured solar radiation (Gi) and the ambient temperature (Tamb) relative to 1 h, and 1- or 2-h ahead.

In order to simplify the structure of the neural network and using the environmental data as inputs, another solution has been implemented. First, a neural network that allows Gi forecasting starting from data relative to its values 1- or 2-h ahead and the CMS was implemented. The same procedure was used to forecast Tamb and, finally, the forecasted data were used, together with the CSM, to estimate Pac [18].

The characteristics of the different neural networks implemented here, using the inputs of 1-or 2-h ahead, are summarised in Table 46.2.

In detail, the proposed ANN is made up of a single hidden layer (the number of neurons are different for each developed ANN), and training is evaluated using the mean squared error (MSE) as a performance parameter with a maximum of 500 epochs. Moreover, a hyperbolic tangent sigmoid transfer function is used as a neuron activation function of the hidden layer, while the learning algorithm used to train the ANN is the Levenberg-Marquardt backpropagation. Using this algorithm, weights and bias values are updated according to Levenberg-Marquardt optimization [19]. The values of the learning period, amount of data used in training, neurons on the hidden layer and learning rate were set on a trial and error basis.

Table 46.2 Main characteristics of the different neural networks implemented here using the inputs of 1- (ΔT=1) or 2-h (ΔT=2) ahead with respect to the output

ANN	Number of inputs	Neurons in the hidden layer	Inputs	Output
Pac ANN1	13	10	Pac (T:00): Pac (T:55), CSM (T+ΔT:00)	Pac (T+ΔT:00)
Pac ANN2	37	15	Pac (T:00): Pac (T:55), Gi (T:00):Gi (T:55), Tamb (T:00):Tamb (T:55), CSM (T+ΔT:00)	Pac (T+ΔT:00)
Gi	13	20	Gi (T:00):Gi (T:55), CSM (T+ΔT:00)	Gi (T+ΔT:00)
Tamb	13	20	Tamb (T:00):Tamb (T:55), CSM (T+ΔT:00)	Tamb (T+ΔT:00)
Pac ANN3	14	15	Forecasted Gi (T+ΔT:00), Forecasted Tamb (T+ΔT:00), CSM (T+ΔT:00)	Pac (T+ΔT:00)

46.4 Experimental Results

In order to appreciate the effectiveness of the ANN estimation, some statistical indicators as suggested in [20] were considered. The relative mean bias error (MBE/A) (Eq. 46.1), the relative root mean square error (RMSE/A) (Eq. 46.2) and the correlation coefficient (CC) (Eq. 46.3) were calculated using the following expressions:

$$MBE/A = \frac{\sum_{n=1}^{N}\left(P_{y,n} - P_{x,n}\right)}{N \cdot \overline{P_x}} \cdot 100 \tag{46.1}$$

$$RMSE/A = \frac{\sqrt{\frac{1}{N} \cdot \sum_{n=1}^{N}\left(P_{y,n} - P_{x,n}\right)^2}}{\overline{P_x}} \cdot 100 \tag{46.2}$$

$$CC = \frac{\sum_{n=1}^{N}\left(P_{y,n} - \overline{P_y}\right) \cdot \sum_{n=1}^{N}\left(P_{x,n} - \overline{P_x}\right)}{\sqrt{\sum_{n=1}^{N}\left(P_{y,n} - \overline{P_y}\right)^2 \cdot \sum_{n=1}^{N}\left(P_{y,n} - \overline{P_x}\right)^2}} \tag{46.3}$$

where $P_{x,n}$ represents the measured data, $P_{y,n}$ represents the approximated data, N is the number of available samples P_x and P_y, and are the mean values of the measured and approximated data, respectively.

The methodology adopted to evaluate the performance of the MLP ANN can be described by following steps:

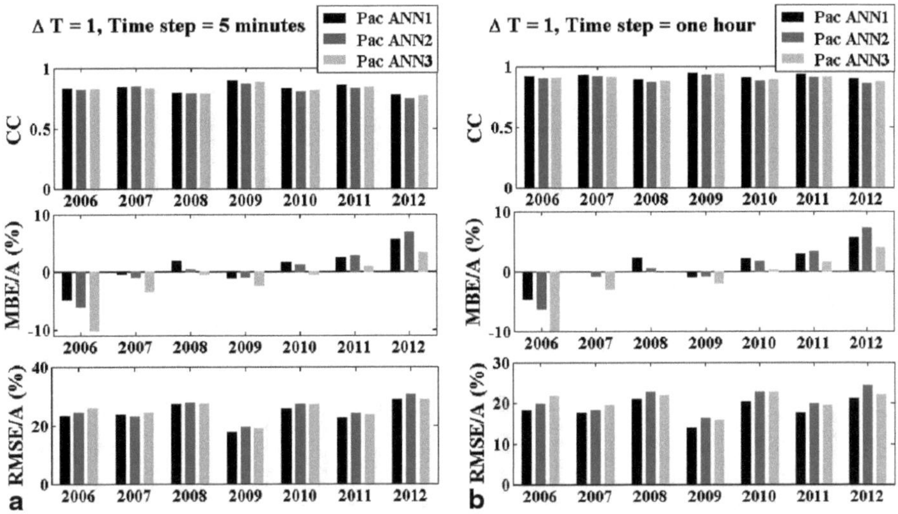

Fig. 46.4 Results obtained when the AC power is forecast using data relative to 1-h ahead (ΔT=1) using the ANN approaches proposed here (Pac ANN1, Pac ANN2, Pac ANN3). The values of statistical coefficients are calculated for data: **a** with 5-min time steps; **b** with 1-h time steps

- Divide the entire available dataset into subsets, each regarding a specific year;
- Select a specific year for ANN phases of training/evaluation and testing, according to a percentage division; typically 98 % for training, 1 % for testing and 1 % for evaluation;
- Train the network using data relative to the year chosen for the training phase;
- Simulate the network output utilizing all remaining years and compute the statistical indexes for each year, as described before.

The designed MLP ANNs were trained using 2007 and simulated with all the other available years (2006, 2008, 2009, 2010, 2011, 2012).

Figure 46.4a and b show the results obtained when the AC power is forecasted considering data relative to 1-h ahead (ΔT=1) using the ANN approaches proposed here (Pac ANN1, Pac ANN2, Pac ANN3). The values of statistical coefficientsare calculated for data with 5-min time steps (Fig. 46.4a) and of 1 h (Fig. 46.4b).

Figure 46.5a and b show the results obtained when the AC power is forecast using data relative to 2-h ahead (ΔT=2).

Moreover, Fig. 46.6 shows an example of the relationship between the weather conditions and the correlation coefficients (RMSE/A in Fig. 46.6) when Pac ANN 1 is used to forecast data 1-h ahead. It is possible to note that better results can be obtained when the variable or the cloudy percentages are higher than the slightly cloudy or clear ones.

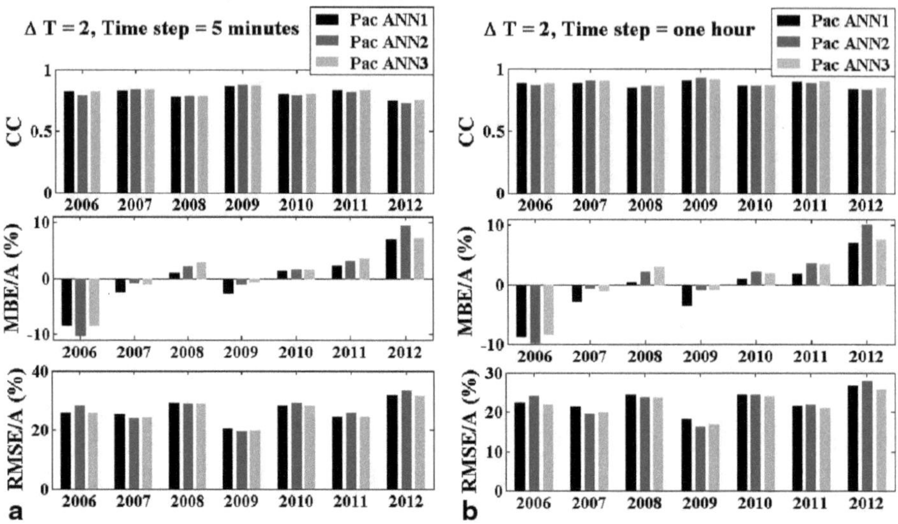

Fig. 46.5 Results obtained when the AC power is forecast using data relative to 2-h ahead ($\Delta T = 2$) using the ANN approaches proposed here (Pac ANN1, Pac ANN2, Pac ANN3). The values of statistical coefficients are calculated for data: **a** with 5-min time steps; **b** with 1-h time steps

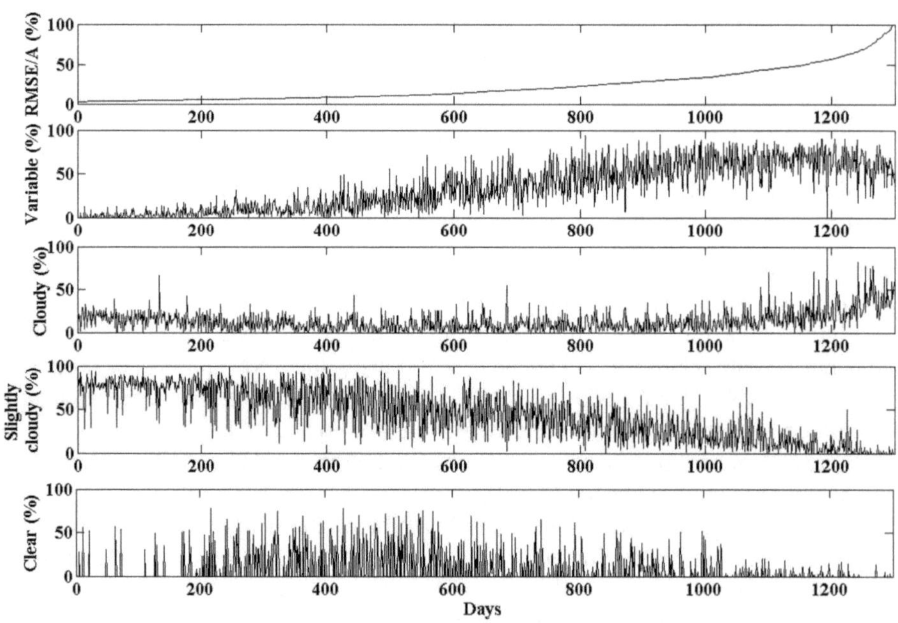

Fig. 46.6 Relationship between the weather conditions and the RMSE/A when the Pac ANN 1 is used ($\Delta T = 1$)

46.5 Conclusions

Here, we proposed an ANN approach to forecast the AC power generation of a PV thin-film photovoltaic plant 1- and 2-h ahead. In particular, three different ANNs were used for the forecasting. The first one used as input only the data measured by a commercial inverter and the clear sky radiation that can be easily estimated. Apart from the clear sky radiation, the second one used as input the measured solar radiation and ambient temperature. In the third one, the AC output power was forecast from predicted values of solar radiation and ambient temperature.

To evaluate the effectiveness of the estimations performed using the ANNs in the evaluation of AC power, three statistical coefficients—the RMSE/A, the MBE/A and the CC—were evaluated. In particular, for forecasting relative to 1-h ahead and with 5-min time steps of the data, the RMSE/A results were between 17.75 and 27.46 %, the MBE/A were between − 8.20 and 7.70 %, while the CC were between 0.77 and 0.89. Using hourly data, the RMSE/A results were between 15.32 and 22.92 %, the MBE/A were between − 7.58 and 8.71 %, and the CC were between 0.88 and 0.94. In particular, for forecasting relative to 2-h ahead and with 5-min time steps of the data, the RMSE/A results were between 19.46 and 31.73 %, the MBE/A were between − 11.12 and 9.24 %, and the CC were between 0.72 and 0.88. Using hourly data, the RMSE/A results were between 16.41 and 27.68 %, the MBE/A were between − 9.14 and 12.08 %, and the CC were between 0.81 and 0.92. These results were obtained using the ANN which uses the minimum availability of data; this is due to the fact that when using a complicated structure of the ANN, worse results can be achieved. Moreover, an algorithm was used that allowed classification of days by means of four percentages—variable, cloudy, slightly cloudy or clear. This classification allowed us to demonstrate how forecast uncertainty depends on the meteorological situation, i.e., inhomogeneous cloud situations are generally more difficult to forecast and show a lower accuracy than forecasts for clear sky days.

References

1. Gaëtan M, Sinead O, Manoël R (2014) Global market outlook for photovoltaics 2014–2018. European Photovoltaic Industry Association (EPIA)
2. Pelland S, Remund J, Kleissl J, Oozeki T, De Brabandere K (2013) Photovoltaic and solar forecasting: state of the art. IEA PVPS Task 14, Subtask 3.1 Report IEA-PVPS T14-01: 2013, October 2013
3. Diagne M, David M, Lauret P, Boland J, Schmutz N (2013) Review of solar irradiance forecasting methods and a proposition for small-scale insular grids. Renew Sustain Energy Rev 27:65–76
4. Jones LE (2011) Strategies and decision support systems for integrating variable energy resources in control centers for reliable grid operations-global best practices, examples of excellence and lessons learned. Report under contract no. DE-EE0001375 of U.S. Government, December 2011
5. Lorenz E, Hurka J, Heinemann D, Beyer HG (2009) Irradiance forecasting for the power prediction of grid connected photovoltaic systems. IEEE J Sel Top Appl Earth Obs Remote Sens 2(1):2–10. doi:10.1109/JSTARS.2009.2020300

6. Pedro HTC, Coimbra CFM (2012) Assessment of forecasting techniques for solar power production with no exogenous inputs. Sol Energy 86(7):2017–2028
7. İzgi E et al (2012) Short–mid-term solar power prediction by using artificial neural networks. Sol Energy 86(2):725–733
8. Bacher P, Madsen H, Nielsen HA (2009) Online short-term solar power forecasting. Sol Energy 83:1772–1783
9. Tao C, Shanxu D, Changsong C (2010) Forecasting power output for grid-connected photovoltaic power system without using solar radiation measurement. Power Electronics for Distributed Generation Systems (PEDG), 2010 2nd IEEE International Symposium on 2(1):773–777
10. Chupong C, Plangklang B (2011) Forecasting power output of PV grid connected system in Thailand without using solar radiation measurement. Energy Procedia 9:230–237
11. Hassanzadeh M, Etezadi-Amoli M, Fadali MS (2010) Practical approach for sub-hourly and hourly prediction of PV power output. North American Power Symposium (NAPS), pp. 1–5, 26–28 Sept. 2010
12. Wu YK, Chen CR, Rahman HA (2014) A novel hybrid model for short-term forecasting in PV power generation. Int J Photoenergy 2014:9 (Article ID 569249, http://dx.doi.org/10.1155/2014/569249)
13. Wu J, Chan CK, Zhang Y, Xiong BY, Zhang QH (2014) Prediction of solar radiation with genetic approach combing multi-model frame work. Renew Energy 66:132–139
14. Voyant C, Muselli M, Paoli C, Nivet ML (2012) Numerical weather prediction (NWP) and hybrid ARMA/ANN model to predict global radiation. Energy 39(1):15
15. Chaabene M, Ammar MB (2008) Neuro-fuzzy dynamic model with Kalman filter to forecast irradiance and temperature for solar energy systems. Renew Energy 33(7):1435–1443
16. Mellit A, Soteris AK (2008) Artificial intelligence techniques for photovoltaic applications: a review. Prog Energy Combust Sci 34(5):574–632
17. Tina GM, De Fiore S, Ventura C (2012) Analysis of forecast errors for irradiance on the horizontal plane. Energy Convers Manage 64:533–540
18. Graditi G, Ferlito S, Adinolfi G, Tina GM, Ventura C (2014) Performance estimation of a thin-film photovoltaic plant based on an artificial neural network model. IREC 14, 25–27 March 2014
19. Chen Y, Zhang S (2012) Research on EEG classification with neural networks based on the levenberg-marquardt algorithm. Commun Comput Inf Sci 308:195–202
20. Notton G, Paoli C, Ivanova L, Vasileva S, Nivet ML (2013) Neural network approach to estimate 10-min solar global irradiation values on tilted planes. Renew Energy 50:576–584

Chapter 47
Digital Signal Processor-Based Power Management System Implementation for a Stand-Alone Microgrid on a Small Island in Korea

Chul-Sang Hwang, Jong-Bo Ahn, Jin-Hong Jeon, Gyeong-Hun Kim, Eung-Sang Kim, Minwon Park and In-keun Yu

Abstract This chapter presents a power management system (PMS) for the control of a stand-alone microgrid that was installed in Mara Island's microgrid system in Korea. Most stand-alone microgrids can control or confine the start, stop, or output of each distributed generator using the energy management system, but these functions cannot guarantee stability against disturbances that occur transiently and unexpectedly due to their reliance on communications. Therefore, Mara Island's microgrid system developed and applied a PMS that does not rely on communications to secure the stability of the transient system. The PMS controls controllable active and reactive power, particularly that of energy storage systems, for the controllable active power of a diesel engine. First, we performed a controller hardware-in-the-loop simulation to verify the PMS performance. The microgrid system was modeled in a real-time digital simulator, connecting the PMS designed by a digital signal processor. Second, the PMS was installed in Mara Island's microgrid system in Korea, which includes a photovoltaic power generation system, a diesel engine,

---

I.-k. Yu (✉) · M. Park
Department of Electrical Engineering, Changwon National University, Changwon city, Republic of Korea
e-mail: yuik@changwon.ac.kr

M. Park
e-mail: paku@changwon.ac.kr

C.-S. Hwang · J.-B. Ahn · J.-H. Jeon · G.-H. Kim · E.-S. Kim
Smart Distribution Research Center, KERI, Changwon city, Republic of Korea
e-mail: hcs1006@keri.re.kr

J.-B. Ahn
e-mail: jbahn@keri.re.kr

J.-H. Jeon
e-mail: jhjeon@keri.re.kr

G.-H. Kim
e-mail: kgh1001@keri.re.kr

E.-S. Kim
e-mail: eskim@keri.re.kr

A. Sayigh (ed.), *Renewable Energy in the Service of Mankind Vol II*,
DOI 10.1007/978-3-319-18215-5_47

a battery energy storage system, and the PMS. These systems are connected to a 380-V, one-feeder distribution subsystem. The loads of the microgrid usually vary from 40 to 120 kW. The results show that the proposed PMS helps to improve the stability of the stand-alone microgrid. The stability and utilization of the system can be increased by utilizing the PMS in accordance with the purpose of the microgrid system.

Keywords Controller hardware-in-the-loop simulation · Digital signal processor · Power management system · Real-time digital simulator

47.1 Introduction

Microgrids are systems that integrate distributed generators (DGs), battery energy storage systems (BESS), and controllable loads on a low-voltage network [1, 2]. There are two different types of microgrids: a grid-connected and a stand-alone microgrid. In the case of a stand-alone microgrid, it always operates in islanding mode because it is usually installed in remote areas such as mountainous districts or islands without power grids. A microgrid, in general, consists of a diesel engine, micro-turbines, wind generators, photovoltaic (PV) generation systems, local loads, and so on. These DGs are managed by a power management system (PMS) and an energy management system (EMS), which offer many advantages such as peak load shaving, stability maintenance against the utility faults, and high power quality [4–8]. An important consideration when implementing optimal generators is system stability. When a microgrid is operated in stand-alone mode, its dynamics are strongly dependent on the connected sources and on the power regulation control of the converter interfaces. This is similar to a conventional grid, where the system stability is largely influenced by the synchronous generators. For droop-controlled microgrids, which offer advantages in terms of their autonomous operation, analysis has shown that the parameters that determine generator power sharing have a significant effect on stability in stand-alone operation [9, 10]. It is quite costly to operate a diesel engine due to the high fuel cost, and it also causes environmental issues, such as green-house gas emissions [11]. For such reasons, renewable energy sources like a PV generation system and a wind turbine have been considered as a solution to the high operation cost and environmental issues of diesel engines. However, the high penetration of intermittent renewable energy sources into a small stand-alone system may cause problems such as instability, poor power quality, and low operation efficiency owing to the intermittency of renewable energy sources [12–14].

Therefore, in this chapter, we developed and applied a PMS that has several functions, such as measurement, digital input/output (I/O), analog output, and communication. The PMS controls active and reactive power, particularly that of energy storage systems, to achieve the controllable active power of a diesel engine.

In this chapter, the stand-alone microgrid on Mara Island was first modeled in a real-time digital simulator (RSCAD/RTDS), connecting the PMS designed by a digital signal processor (DSP). The microgrid is composed of a 380-V, one-feeder distribution subsystem, which is the stand-alone system far away from the utility network. It consists of a diesel engine, a PV generation system, a BESS, a PMS, and loads; these systems are connected to a 380-V, low-voltage line. The loads of the microgrid usually vary from 40 to 120 kW. The PMS and RTDS are connected using an interface I/O card for controller hardware-in-the-loop simulation (CHILS). The CHILS was conducted in advance to verify the PMS performance. A typical synchronous generator model in the RSCAD/RTDS library was used to represent the diesel engine, with a simplified first-order model of an exciter and governor. The BESS was also modeled by the current source. The power reference of the BESS was calculated by the PMS. The PMS compensates for diesel engine power with the function of the charge and discharge of the BESS. Second, the results of the simulation study were applied to a real microgrid. The PMS was installed in Mara Island's microgrid system in Korea. The purposes of using PMS control algorithms are to control the BESS output power and to determine the reference power calculation for compensating for diesel engine power. The PMS calculates the power reference of the BESS to maintain the power of the diesel engine.

We confirmed that the reliability and running efficiency of the stand-alone microgrid with the PMS were improved, and the control strategy in the system responded appropriately.

47.2 Configuration of a Stand-Alone Microgrid on a Small Island in Korea

Figure 47.1 shows the configuration of Mara Island's microgrid in Korea. The microgrid is a stand-alone microgrid type that has a diesel engine as a main source that is not connected to the Korea Electric Power Corporation (KEPCO). It is a common alternating current (AC)-type power grid that consists of two diesel engines of 150 and 250 kVA, two PV arrays with a 75 kW capacity, and a BESS with 50 and 600 kWh capacities. Mara Island has a large proportion of a floating population due to its large number of tourists and has a unique power load characteristic, with a maximum load of 120 kW during the day and 40 kW at night, which is why it requires an optimum operation method.

47.3 PMS Control Strategy

Stand-alone microgrids can control or confine the start, stop, or output of each DG using the EMS, but these functions cannot guarantee stability against disturbances that occur transiently and unexpectedly due to their reliance on communications. Therefore, in this chapter, we developed and applied a PMS that does not rely on communications to secure the stability of the transient system. The needs for the

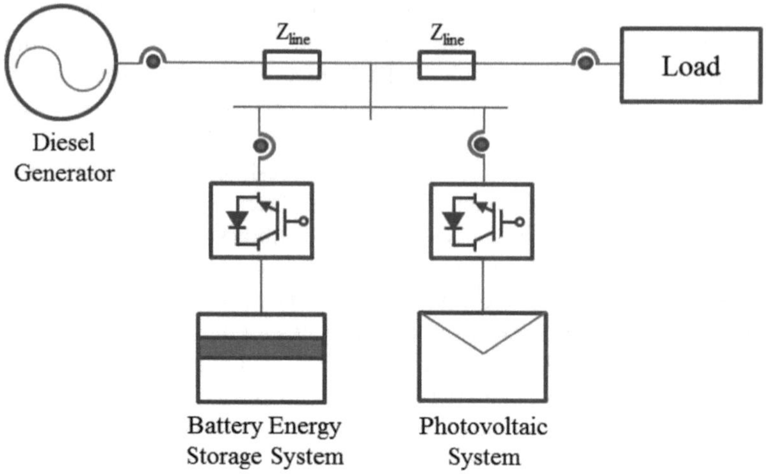

Fig. 47.1 Configuration of Mara Island's microgrid in Korea

PMS can be summarized as follows. First, the system controls active and reactive power, particularly that of energy storage systems. This function particularly ensures operations to safeguard against disturbances such as accidental trips in renewable energy sources or rapid output changes. Second, the EMS, which monitors and controls microgrids, is the computer system connected with each equipment via communications and does not have direct input and output control functions. However, existing devices that do not offer communication functions, such as circuit breakers and dummy loads, can also be used. The control of these devices in an integrated manner requires a means capable of their direct control by receiving orders via communications from the EMS, and the PMS can provide this function. Third, microgrids consist of various devices such as DGs and control systems, and as an automation system linked via communications, these devices require a function to monitor system failures and perform backup operations during system failures. The PMS can perform this independent function.

The PMS was designed as follows to perform the above functions. Since there is no need to measure the individual outputs of multiple DGs installed within microgrids, electric power was measured by adding up the diesel engine, the PV generation systems, loads, and so on. Three CTs were installed in areas where unbalanced currents flow, such as the diesel engine and loads. On the other hand, currents were detected with one or two CTs in areas where three-phase balanced currents flow, such as inverters. For voltages, one area of bus voltages was detected.

- Measurement function: Other forms of measurement such as digital power meters were not used because such measuring instruments cannot detect outputs quickly as they are internally designed to have large filters or time delays for safe detection. The calculation of power realized a maximum measuring speed of below one cycle by applying either the discrete Fourier transform or direct-quadrature (DQ) transformation method.

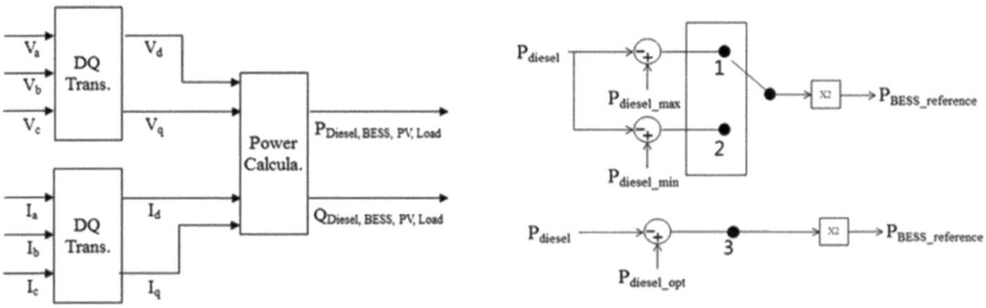

Fig. 47.2 Block diagram of the power calculation and control in the power management system (PMS). *DQ* direct-quadrature, *BESS* battery energy storage system, *PV* photovoltaic

- Digital I/O: The system provides digital output ports to control circuit breakers or dummy loads and generates outputs in the form of contacts of point to facilitate interfaces. It enables the quick identification of fault conditions without relying on communications by receiving information on the operating state of each DG via digital inputs.
- Analog output: The system was designed to transmit signals using the analog method to deliver charge–discharge orders speedily to energy storage systems capable of controlling the output of active power. It also enabled the electrical insulation of signals and the transmission of current signals to be resilient against noises.
- Communication: The system provides communication functions to operate through cooperation with the EMS and exchanges auxiliary information such as operating modes and conditions with the EMS.

Figure 47.2 shows a block diagram of the power calculation and control in the PMS.

47.4 Modeling and Simulation

47.4.1 RTDS-Based Modeling and the Simulation

The capacity of each system is shown in Table 47.1. Figure 47.3 shows the standalone microgrid system modeled in an RSCAD/RTDS. The microgrid system consists of a PV, a BESS, a diesel engine, and loads.

47.4.2 CHILS-Based Simulation

A stand-alone microgrid was modeled in the RSCAD/RTDS. The PMS was designed by a DSP. Figure 47.4 shows the configuration of the CHILS. A typical synchronous generator model in the RSCAD/RTDS library is used to represent the

Table 47.1 Capacity of each system

System	Diesel engine	PV array	BESS	Load
Capacity	150 kW	75 kW × 2	50 kW × 2	40–120 kW

PV photovoltaic, *BESS* battery energy storage system

diesel generator, which is equipped with a simplified first-order model of an exciter and governor. The BESS is also modeled by using a current source. The PMS is connected to the RTDS through interface boards. A personal computer is used to program and debug the software. Through an emulator, the computer can communicate with the PMS, in which the real-time running program resides. The voltage and current signals of the microgrid from the RTDS are sent to an analog-to-digital converter (ADC) of the PMS through a gigabit transceiver analog output (GTAO) card. The power reference of the BESS signal from the PMS is sent to the ADC of the RTDS through a gigabit transceiver analog input (GTAI) card.

47.4.3 Simulation Conditions and the Results

In the simulation model, the diesel engine is controlled to a minimum value of 10 kW and a maximum value of 20 kW. The PMS determines the reference of the BESS that controls the active power of the diesel engine. The active power output of the BESS fluctuates according to the power of the diesel engine, and the positive value of the output power means that the BESS is discharged. The load power condition occurs after 4 s, as shown in the simulation. The loads of the microgrid usually vary from 10 to 40 kW. Figure 47.5 shows the power of the PV generation system, the diesel engine, the load, and the BESS output. During the microgrid operation, the BESS output power is charged or discharged by control of the PMS. When the diesel engine power is less than 10 kW, the BESS is charged, and at more than 20 kW the BESS is discharged.

47.5 Experiment (Fig. 47.6)

The test shown in Fig. 47.7 indicates a resulting waveform of the minimum-, maximum-, and optimum-efficient operations of the diesel engine utilizing the PV generation system and the BESS in the stand-alone operation in which real loads of Mara Island are connected. This is an operational result, as a minimum and maximum region of the diesel engine was set to 25 and 45 kW, respectively, with a 1830 s interval as the operating condition. This is a waveform that is compensated for by producing more power stored in the BESS to minimize the output of the diesel engine under a given condition. After 1855 s, the output of the diesel engine

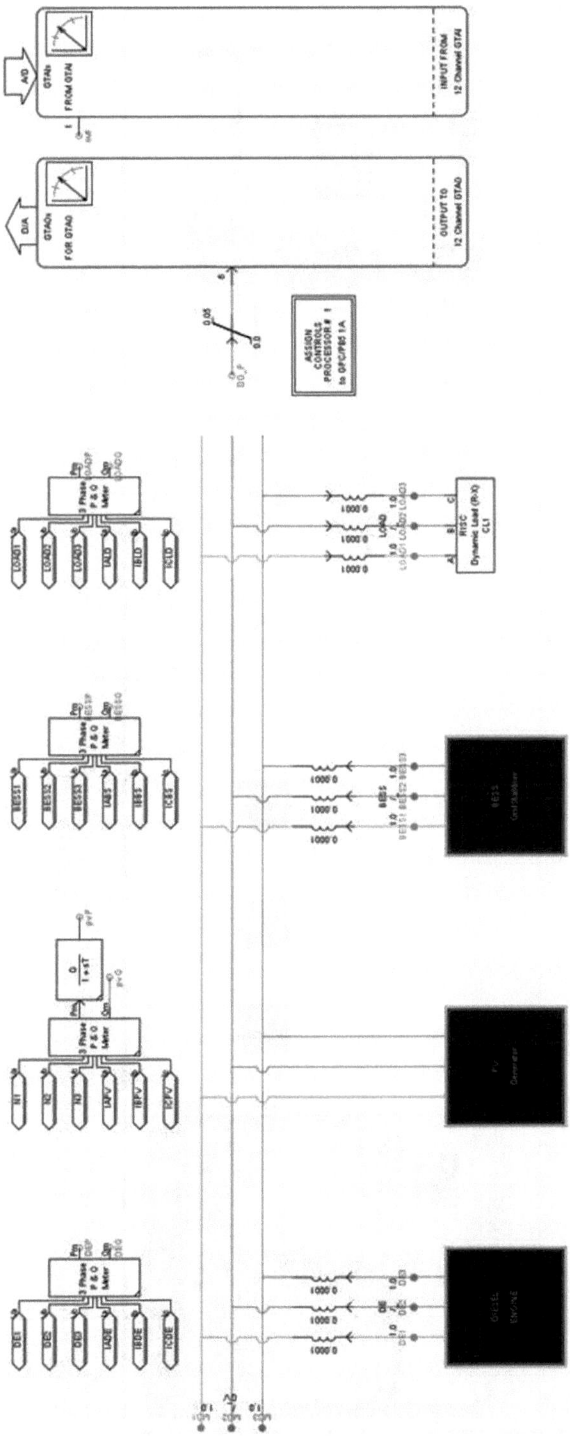

Fig. 47.3 System setup in RSCAD/RTDS

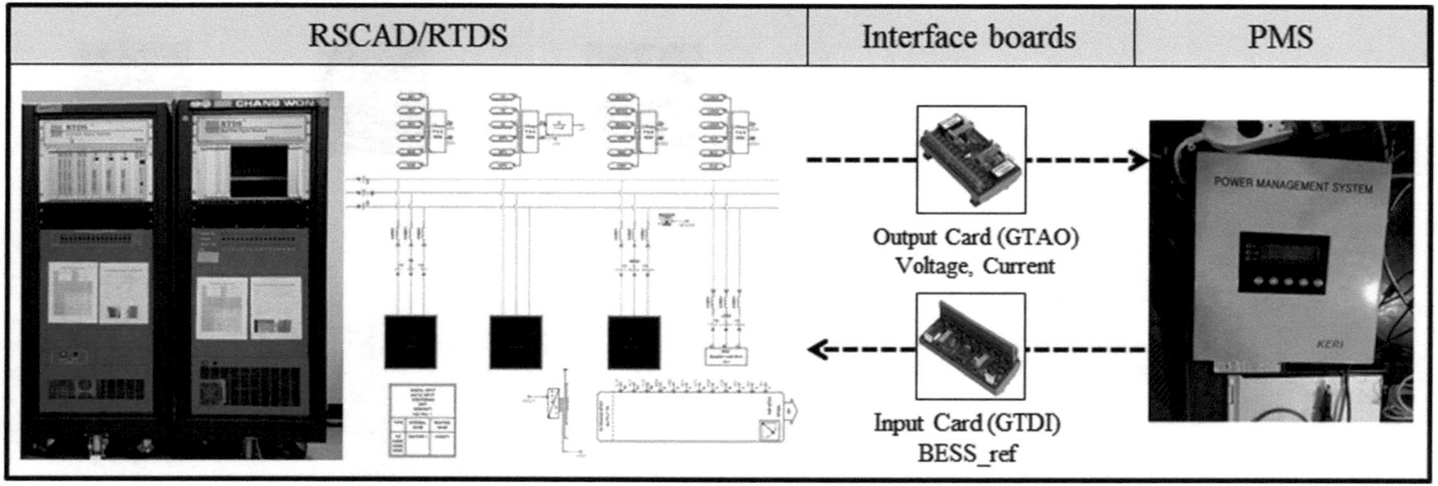

Fig. 47.4 Configuration of the controller hardware-in-the-loop simulation (CHILS). *PMS* power management system, *GTAO* gigabit transceiver analog output, *GTAI* gigabit transceiver analog output, *BESS* battery energy storage system

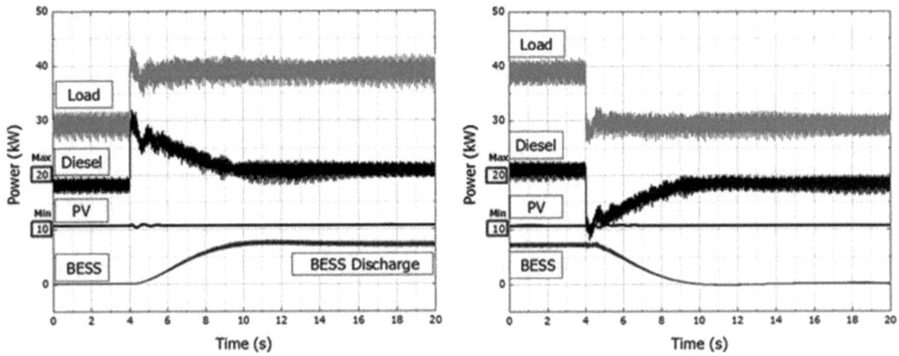

Fig. 47.5 Power of a load with the battery energy storage system *(BESS)* controlled by the power management system (PMS). *PV* photovoltaic

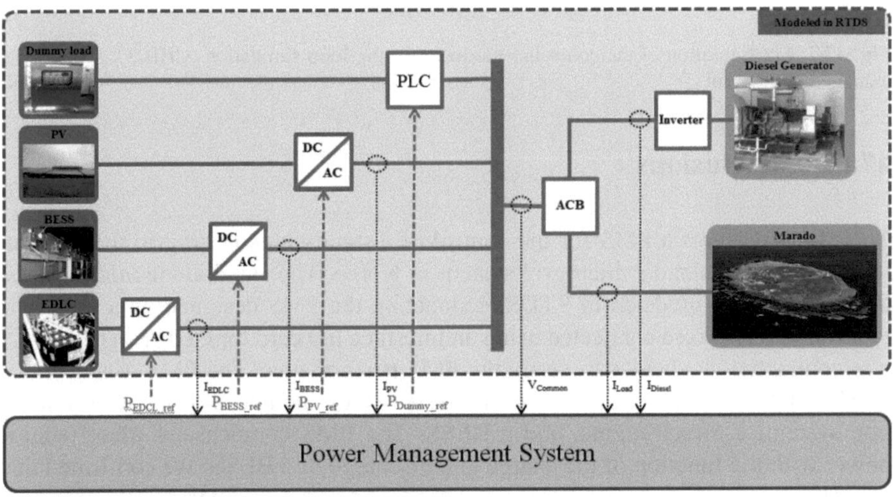

Fig. 47.6 Circuit diagram of Mara Island's system with the power management system (PMS). *PV* photovoltaic, *BESS* battery energy storage system

was controlled artificially to obtain the same effect of variable loads, even under a circumstance that cannot change the load in real towns on Mara Island so that the optimum efficient operation characteristics of the BESS were tested under the assumption of variable load. We verified that optimum efficient operations were performed correctly as the BESS was able to store redundant power when the diesel engine output increased, whereas energy storage in the BESS was decreased when the generator output decreased. Moreover, the PV generation showed the characteristic of a base load power station that supplies power continuously under the constraint of power output as 25 kW since the load rate in towns was not significantly high.

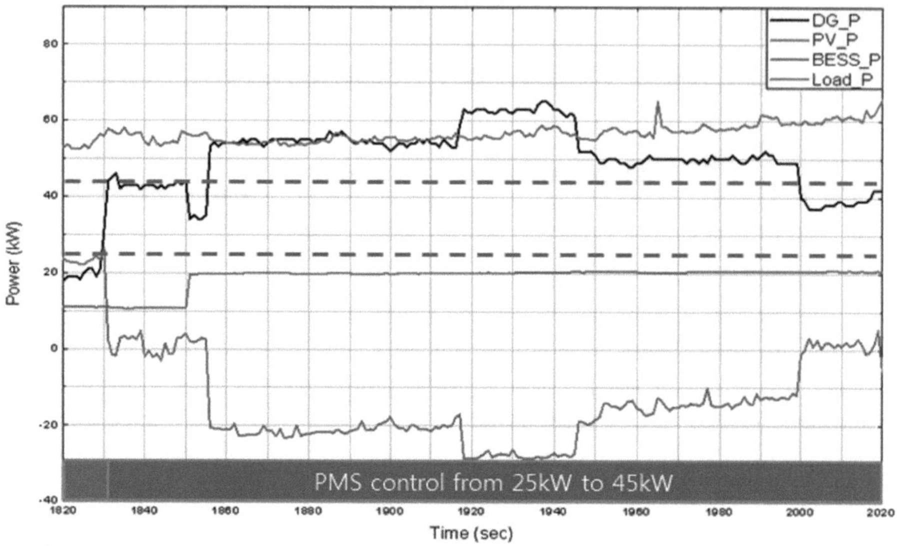

Fig. 47.7 Configuration of the controller hardware-in-the-loop simulation (CHILS). *PMS* power management system

47.6 Conclusions

This study presents a PMS for the control of a stand-alone microgrid that was installed in Mara Island's microgrid system in Korea. The stand-alone microgrid on Mara Island was modeled in RTDS, connecting the PMS designed by a DSP. The PMS and RTDS were connected using an interface I/O card for CHILS. The CHILS was conducted in advance to verify the PMS performance. The PMS was installed in Mara Island's microgrid system in Korea, which includes a PV power generation system, a diesel engine, and a BESS. The PMS compensates diesel engine power with the function of the charge and discharge of a BESS. We confirmed that the reliability and running efficiency of the stand-alone microgrid with PMS was improved, and the control strategy in the system had an appropriate response. The stability and utilization of the system can be increased by utilizing the PMS in accordance with the purpose of the microgrid system. An issue of voltage and current measurement in the PMS was discovered due to the unbalance load, which was not predicted for Mara Island. As a result, the control cannot be properly achieved because of the incorrect power calculation. The functions of the PMS have been improved to solve the unbalance load issue.

Acknowledgments This work was supported by the Korea Electrotechnology Research Institute and in part by Changwon National University in 2013–2014.

References

1. Lasseter RH, Eto JH, Schenkman B, Stevens J, Vollkommer H, Klapp D, Linton E, Hurtado H, Roy J (2011) CERTS microgrid laboratory test bed. IEEE Trans Power Syst 26(1):325–332
2. Pereira Barbeiro PN, Moreira CL, Soares FJ, Almeida PMR (2010) Evaluation of the impact of large scale integration of micro-generation units in low and medium voltage distribution networks. In: IEEE conference on CITRES 265–272. doi:10.1109/CITRES.2010.5619801
3. Logenthiran T, Srinivasan D, Khambadkone AM, Raj TS (2010) Optimal sizing of an islanded microgrid using evolutionary strategy. In: IEEE conference on PMAPS 12–17. doi:10.1109/PMAPS.2010.5528840
4. Katiraei F, Iravani R, Hatziargyriou N, Dimeas A (2008) Microgrids management. IEEE Power Energy Mag 6(3):54–65
5. Dimeas AL, Hatziargyriou ND (2005) Operation of a multiagent system for microgrid control. IEEE Trans Power Syst 20(3):1447–1455
6. Tsikalakis AG, Hatziargyriou ND (2008) Centralized control for optimizing microgrids operation. IEEE Trans Energy Convers 23(1):241–248
7. Peas Lopes JA, Moreira CL, Madureira AG (2006) Defining control strategies for microgrids islanded operation. IEEE Trans Power Syst 21(2):916–924
8. Guerrero JM, Blaabjerg F, Zhelev T, Hemmes K, Monmasson E, Jemei S, Comeh MP, Granadino R, Frau JI (2010) Distributed generation: toward a new energy paradigm. IEEE Ind Electron Mag 4(3):52–64
9. Pogaku N, Prodanovic M, Green TC (2007) Modeling, analysis and testing of autonomous operation of an inverter-based microgrid. IEEE Trans Power Electron 22(2):613–625
10. Katiraei F, Iravani MR (2006) Power management strategies for microgrid with multiple distributed generation units. IEEE Trans Power Syst 21(4):1821–1831
11. Rahman S (2003) Going green. IEEE Trans Power Energy 16–18
12. Pudjianto D, Strbac G, van Overbeeke F, Androutsos AI, Larrabe Z, Saraiva JT (2003) Investigation of regulatory, commercial, economic and environmental issues in microgrids. IEEE Trans Power Syst 6. doi:10.1109/FPS.2005.204223
13. Handschin E, Neise F, Neumann H, Schultz R (2006) Optimal operation of dispersed generation under uncertainty using mathematical programming. IEEE Trans Power Syst 28(9):618–626
14. Pogaku N (2006) Application of inverter-based distributed generators for harmonic damping throughout a distribution network. In: IEEE conference on power electronics 1922–1927. doi:10.1109/PESC.2005.1581894

Chapter 48
Techno-Economic Assessment of Photovoltaic Systems in Oman: Review Article

Hussein A Kazem

Abstract Here, we present a review of the feasibility of photovoltaic (PV) systems in Oman. Different PV system configurations, for example, stand-alone and grid connected, have been reviewed and discussed for Oman. The effects of environmental parameters were taken into consideration when evaluating the performance of the PV systems. The systems were evaluated in terms of technical and economic criteria. The outcome of this work is deemed important since the cost of energy of PV systems was found to be cheaper than the cost of energy generated by fossil fuel without government subsidies. Furthermore, the technical and economic evaluation showed that the grid-connected PV system is feasible and promising in Oman. This study contains worthwhile information for those interested in PV system investment in Oman and the neighboring hot weather countries.

Keywords Feasibility study · Stand-alone and grid-connected PV systems · Sizing of PV systems · Oman

48.1 Introduction

Population and industrial growth has led to an increase in peak electricity demand in Oman from 2773 MW in 2007 to approximately 5691 MW in 2014 as shown in Fig. 48.1. The annual growth rate is high, at approximately 9 %, and peak demand is envisaged to continue increasing [1]. Electricity generation is predicted to reach 24.0 TWh at the end of 2014, and electricity shortages are expected to occur in the near future if the current trends continue [2]. The Omani government accounts for 19 % of total gas production, while the remainder is used in oil production and for export, and up to 92 % of the natural gas is domestically used for producing electricity. If Oman continues to build power stations which utilize gas for electricity production there will be a need to import, rather than export natural gas. As all power generation facilities at present are dependent on fossil fuels (nonrenewable),

H. A. Kazem (✉)
Renewable Energy & Sustainable Technology Research Group/Sohar University,
PO Box 44, Sohar PCI 311, Oman
e-mail: h.kazem@soharuni.edu.om

A. Sayigh (ed.), *Renewable Energy in the Service of Mankind Vol II*,
DOI 10.1007/978-3-319-18215-5_48

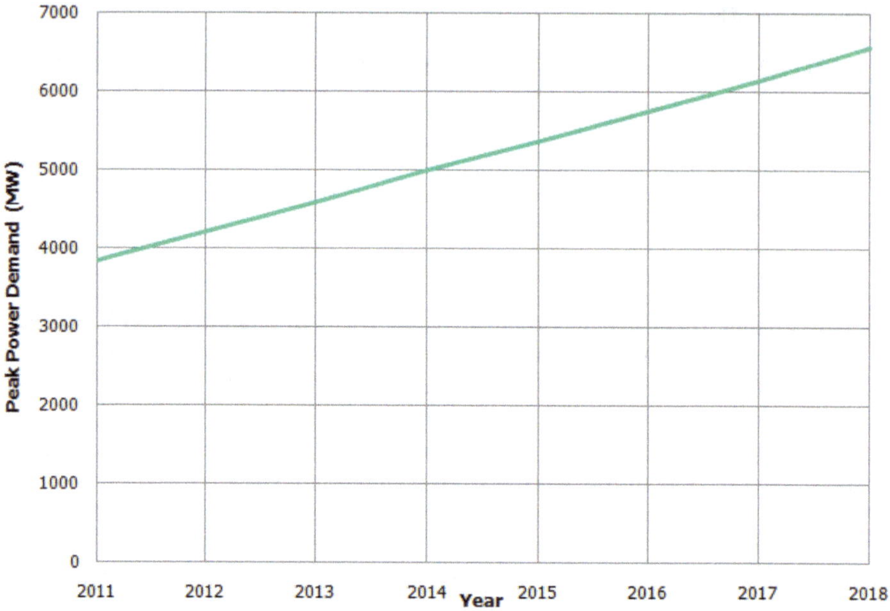

Fig. 48.1 Maximum electrical demand in Oman till 2013

it is strongly advised to seek alternative sources of energy. The three most important factors in selecting new energy sources are that they must be renewable, locally available and environmentally friendly.

Solar energy fulfills all of these requirements. It is renewable for as long the sun is shining, free to collect, and it does not harm the environment. This study aims to investigate the feasibility of using solar energy via photovoltaic (PV) technology to generate electricity in Oman. The main interconnected grid electricity system exhibits a distinct seasonal trend—demand in summer months is very much higher than in winter, perhaps as much as 6-fold [3]. Air conditioning loads in summer rise in response to higher temperatures resulting in a strong positive correlation between monthly peak demand and maximum monthly ambient temperature. The potential for producing electricity using PV systems is highest during the summer which coincides with the period of peak electricity demand in Oman. Using solar energy to produce electricity also emits zero greenhouse gases (GHGs) such as CO_2. The reduction in GHG emissions (carbon) would be approximately 175 kg/MWh where renewable energy replaces natural gas. Replacing at least part of Oman's conventional fossil fuel plants with PV technology could significantly reduce GHG emissions in Oman, as well as reducing the country's dependence on nonrenewable energy sources. The important issue is finding an appropriate starting point and development scenario for the use of renewable energy in Oman that may lead to the zero carbon scenario described above.

Solar power can be collected to produce electricity by a variety of methods. Among these methods, PV systems have shown great success for many reasons; it is the oldest and most thoroughly researched method, and has been tested and

Table 48.1 PV system specification

PV array	
PV module rated power (24 modules)	140 Wp (3.36 kWp)
Maximum voltage	17.7
Maximum current	7.91
Open circuit voltage	22.1
Short circuit current	8.68
Efficiency	13.9 %
Temperature coefficient of Voc	−0.36 %/k
Temperature coefficient of Isc	0.06 %/k

implemented for a long time throughout the world. Implementation of PV systems has shown that their reliability and efficiency depends on many factors, the dominant being location (latitude, longitude, and solar intensity), environment (temperature, wind, humidity, pollution, dust, rain, etc.) and the type of PVs used. Thus, before committing to a large PV project in Oman, a thorough investigation of the above factors is essential.

The Authority of Electricity Regulation (AER) in Oman published their study of the potential of renewable energy in Oman in 2008 and they claim that solar and wind could be the priority and best choice for generating electricity [2]. The Public Authority for Electricity and Water (PAEW) was awarded a consultancy contract in 2010 for some international companies to advise on how to implement a large-scale solar power project in Oman. More recently, in 2011, the Research Council of Oman funded a project by Sohar University (SU) to investigate the technical and economic feasibility of PV systems in Oman [4]. Here, we review the results of this study and focus on the main important conclusions and recommendations related to the technical and economic issues.

In this research, 24 PV panels were installed in SU. Each panel has a power output of 140 W (total of 3.36 kW, specification shown in Table 48.1). Different PV system configurations were designed, installed and evaluated as shown in Fig. 48.2a. A key aspect of the implemented work involved measuring the environmental parameters mentioned above using weather stations as shown in Fig. 48.2b, and their effects on system performance. Connection of the PV modules was made

Fig. 48.2 **a** Installed PV systems; **b** Installed weather station

both off-grid and grid-connected; this is to allow further research on maximizing the power delivery to the grid from PVs. The performance of the PV system was optimized for the Omani environment, especially in the heavily populated North Al-Batinah region centered around Sohar. Investigation of PV performance parameters was also carried out using solar monitoring stations located at different places in Oman. The outcomes from this project are a pillar point for the government and companies planning to use PV in future power generation in Oman.

Three PV system configurations were investigated—grid-connected, stand-alone, and tracking systems. Furthermore, different technologies were used [multi-crystalline (multi-c), mono-crystalline (mono-c), and amorphous silicon (a-Si)] to test the effect of Oman's environment on the performance of PV systems and technologies. Here, we describe the most important achievements in this project.

48.2 Techno-Economic Assessment of the Grid-Connected PV System

In this section, a techno-economic methodology is presented to evaluate the productivity of a grid-connected PV system (see Fig. 48.3) at a specific site in Oman called Sohar [5]. The methodology is based on three factors namely capacity factor (CF), yield factor (YF), and cost of energy (CoE). CF and YF were used to evaluate the PV system technically. The CF is the ratio of the actual annual energy output to the amount of energy the PV array would generate if it operated at full power for 24 h per day for a year. Generally, the typical CF for a PV array is in the range of 15–40 %. YF is defined as the annual, monthly or daily net AC energy output of the system divided by the peak power of the installed PV array under standard test conditions (STCs). On the other hand, the cost of energy produced by the system is used as economic evaluation criteria. The analysis is performed by MATLAB software using hourly meteorological data and a model of the grid-connected PV system.

Analysis of the obtained meteorological data showed that the average daily solar energy in the Sohar zone is 6.182 kWh/m^2.day. On the other hand, the result showed

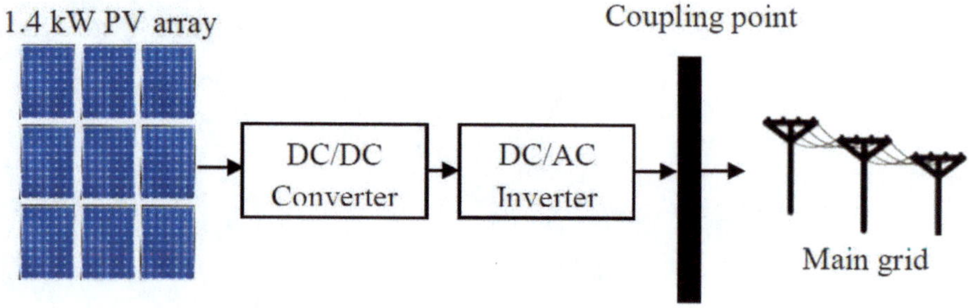

Fig. 48.3 Grid-connected PV system

that the annual YF of the system is 1696.6 kWh/kWp. Meanwhile, the CF of the proposed system is 19.46 %. The cost of the energy generated by the proposed system is 0.1975 USD/kWh.

48.3 Sizing of Stand-Alone PV/Battery System at Minimum Cost

This section presents the investigation for optimal sizing of a stand-alone PV system for remote areas in Sohar as shown in Fig. 48.4 [6]. The PV array tilt angle as well as the size of the system's energy sources are optimally designed for better performance and lower energy cost. Numerical methods for optimization of the PV module tilt angle, PV array size and storage battery capacity are implemented using MATLAB and hourly meteorological data and load demand [7].

The authors recommended optimal sizing of a stand-alone PV system for Sohar zone. The recommendations included PV module/array optimum tilt angle, optimum PV array size and optimum capacity of the storage battery. Furthermore, a comparison between some methods for designing a stand-alone PV system as well as the HOMER software and the proposed numerical method was performed in order to show the significance of the proposed method.

The results show that for the Sohar zone, the tilt angle (β) of a PV array must be adjusted twice a year. The PV array must be slanted at 49° during the period of

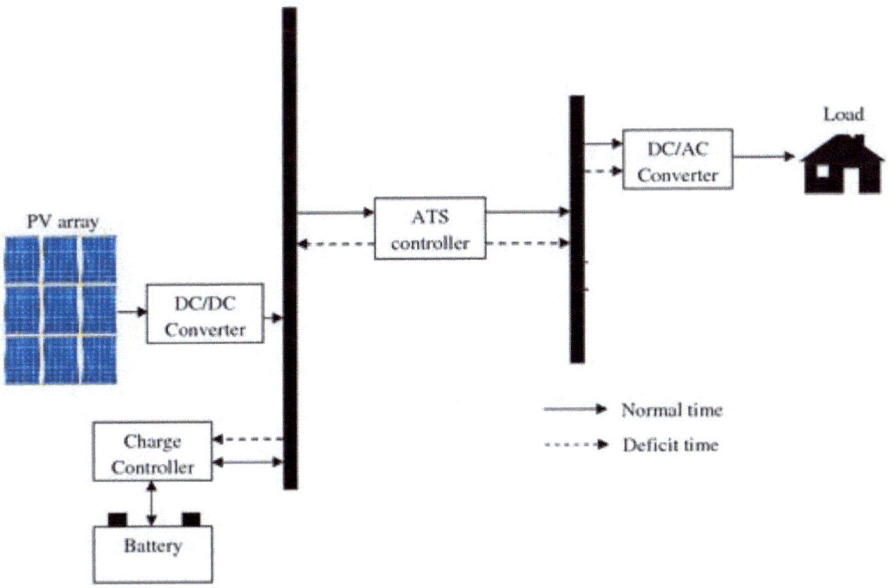

Fig. 48.4 Stand-alone PV system

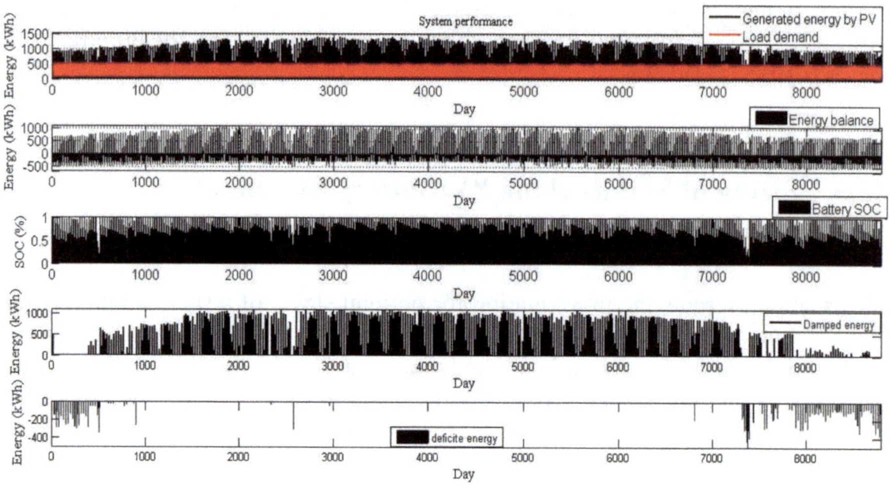

Fig. 48.5 Designed system performance

21/09–21/03 (n=255–81), and it must be horizontal (tilt angle is zero) during the period of 21/03–21/09 (n=81–255). This adjustment practice increases the energy collected by a PV array by 20.6%. This is the first time that the optimum tilt angle for PV systems in Oman has been presented monthly and seasonally (two and four seasons).

The availability of the proposed system is greater than the system designed by the intuitive method and HOMER despite the fact that the storage battery of the first system is much higher than the proposed system. However, Fig. 48.5 shows the performance of the proposed system over a 1-year period. From the figure it can be seen that the battery is used intensively used throughout the year, which enables it to supply the load on some extremely cloudy days. However, it can be seen that most of the energy deficit cases occur in December and January. Finally, the PV array for Oman was found to be 1.33 while the sizing ratio for battery was 1.6. It is worth mentioning that the same proposed system needs to be applied for different zones of Oman (Table 48.2, 48.3, 48.4).

48.4 Optimal Sizing of a Hybrid PV/Wind/Diesel Generator/Battery

In this section, the author presents a method for determining the optimal size of PV array, wind turbine, diesel generator and storage battery installed in a building-integrated system as shown in Fig. 48.6 [8]. The objective of the proposed optimization is to design a system that can supply a building load demand at minimum cost and maximum availability. Mathematical models for the system components as well

Table 48.2 Monthly tilt angle optimization results for Sohar

Period	β	E_{TILT} (kWh/m^2)	E_0 (kWh/m^2)	Energy gain
Jan	57	247.1	144.3	71.2%
Feb	48	217	152.3	50%
Mar	32	234.7	203.4	15.4%
Apr	10	199	196.3	1.3%
May	0	229.4	229.4	0
Jun	0	220.8	220.8	0
Jul	0	213.3	213.3	0
Aug	3	247	203.9	21.1
Sep	25	220.5	145.5	51.5%
Oct	44	233.3	189.2	23.3%
Nov	55	226	145.8	55%
Dec	60	257.3	181.6	41.6%
Total		2750	2230	23.3%

Table 48.3 Results for seasonal tilt angle optimization for Sohar—four seasons

Period	β	E_{TILT} (kWh/m^2)	E_0 (kWh/m^2)	Energy gain
21/03–21/06 (n=81–163)	1	581.3	581.2	≈0
21/06–21/09 (n=163–255)	0	628.4	628.4	0
21/09–21/12 (n=255–346)	47	688	492.6	39.7%
21/12–21/03 (n=346–81)	51	785	524.2	49.8
Total		2680	2230	20.2%

Table 48.4 Results for seasonal tilt angle optimization for Sohar—two seasons

Period	β	E_{TILT} (kWh/m^2)	E_0 (kWh/m^2)	Energy gain
21/03–21/09 (n=81–255)	0	1209.6	1209.6	0
21/09–21/03 (n=255–81)	49	1478.8	1016.8	45.4%
Total		2688.4	2230	20.6%

as meteorological variables such as solar energy, temperature, and wind speed are employed for this purpose. The results of the case study showed that the optimum sizing ratios for the PV array, wind turbine, diesel generator and battery for a system located in Sohar are 0.737, 0.46, 0.22 and 0.17, respectively. The results have been validated by comparing the proposed optimization method with the method using HOMER software. The superiority of the proposed optimization method compared to the method using HOMER software is also highlighted [8].

Based on the proposed algorithm and proposed system design in Sects. 48.2, 48.3, and 48.4, a MATLAB-based user-friendly software tool called REPS.OM (see Fig. 48.7) has been proposed for optimal sizing of PV systems in Oman [9].

The developed REPS.OM software tool aims to design a PV system at minimum cost for Oman. The step-by-step procedures and calculations involved in the REPS. OM are as follows:

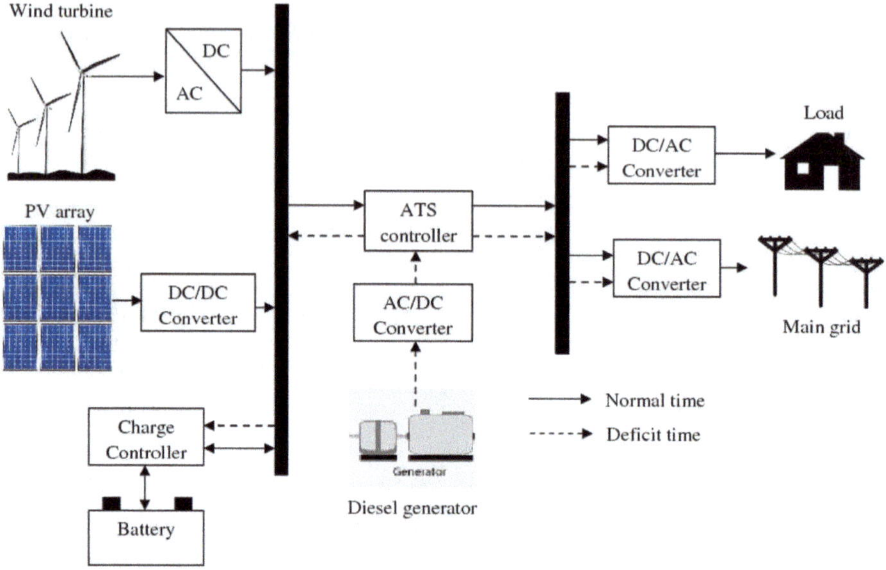

Fig. 48.6 Building-integrated hybrid PV/wind/diesel generating system

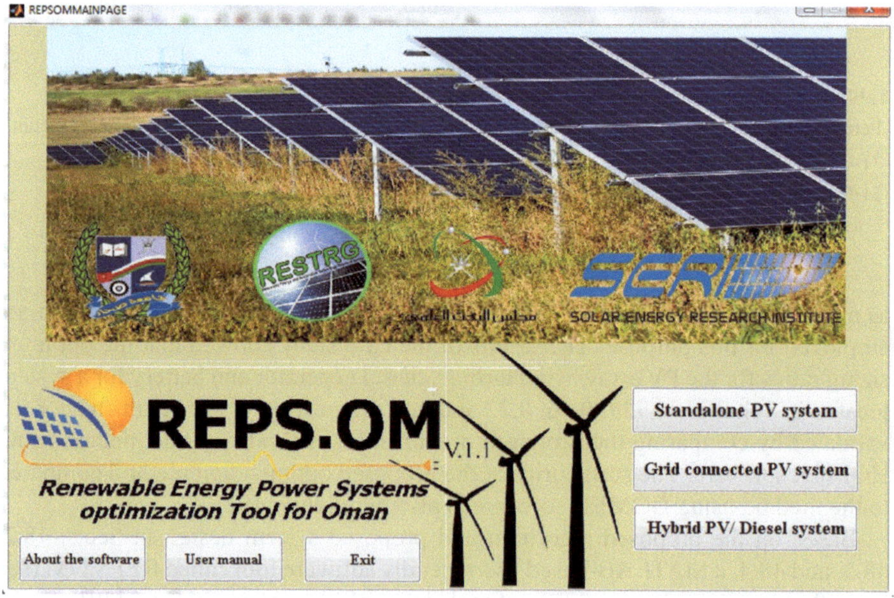

Fig. 48.7 The main page of the REPS.OM software tool

- Define a specific type of PV system under study and its location coordinates.
- Perform optimization of the PV system by determining the optimum PV array/module tilt angle, PV array battery size, wind turbine, diesel generator, and inverter capacities.
- Calculate the optimum tilt angle by first modeling the solar energy on a tilt surface.
- Determine the optimum capacity of a PV battery and diesel generator based on the type of PV system (stand-alone, grid-connected, or hybrid).
- Calculate the optimum inverter size for the PV system.
- Simulate the optimized PV system in order to investigate its performance for a 1-year period.

48.5 Effect of Dust and Temperature on PV

This section investigates the effect of dust on the PV module (multi-c) performance. The degradation of PV performance due to the deposition and accumulation of different pollutant types has been investigated. Experiments concerning the effects of air pollutants including red soil, ash, sand, calcium carbonate, and silica on the power generated have been conducted and analyzed. The results show that the reduction in PV voltage and power strongly depends on pollutant type and deposition level. The results show that ash pollutant is the most effective dust particle on the PV module voltage compared to other dust pollutants. The highest reduction in PV voltage (25 %) was recorded with ash pollutant. The author presented technical information about the performace of PV systems under Oman's desert climate [10]. It was also found that a-Si performed better than mono-c and multi-c in a dusty environment. In Sohar-Oman the collected dust deposited on the PV system was analyzed and found to contain 0.2, 36.1, 44, and 19.7 % of C. sand, F. sand, silt, and clay, respectively. It did not contain any ash particles according to the accumulated dust and cleaning is recommended every 4–5 months.

48.6 Conclusion

Here, we have reviewed the feasibility of PV systems in Oman. Different types of PV systems (stand-alone, grid-connected and tracking systems) were investigated, designed and evaluated in term of economic and technical criteria. Different PV technologies were also tested to check the correlation between PV performance parameters and weather conditions in Oman. A different design technique (intuitive, numerical, and analytical) was used to design PV systems.

It was found that PV systems are suitable for electricity generation in grid-connected and stand-alone power system in Oman. We conclude that using PV systems in Oman is justified on economic and technical grounds. This research contains worthwhile technical information for those who are interested in PV technology investment in Oman.

Acknowledgments The research leading to these results received Research Project Grant Funding from the Research Council of the Sultanate of Oman (Research Grant Agreement no. ORG SU EI 11 010). The authors would like to acknowledge support from the Research Council of Oman. Gratitude is also expressed to the Faculty of Engineering and Research and Industry Collaboration office at Sohar University.

References

1. Kazem HA, Reyah A, Feras H, Ali HA (2011) Prospects of potential renewable and clean energy in Oman. Int J Electron, Comput Commun Technol (IJECCT), Malaysia 1(2):25–29
2. Authority for Electricity Regulation in Oman (2008) Study on renewable resources. Final Report, Oman, pp. 14
3. Kazem HA (2011) Renewable energy in Oman: status and future prospects. Elsevier-Renew Sustain Energy Rev 15:3465–3469
4. The Research Council of Oman (2010). www.trc.gov.om. Accessed 20 Jan 2014
5. Kazem HA, Khatib T (2013) Techno-economical assessment of grid connected photovoltaic power systems productivity in Oman. Elsevier-Sustain Energy Technol Assess 3:61–64
6. Kazem HA, Khatib T, Sopian K (2013) Sizing of a standalone photovoltaic/battery system at minimum cost for remote housing electrification in Sohar, Oman. Elsevier-Energy Build 6C:108–115
7. Kazem HA, Khatib T, Alwaeli AAK (2013) Optimization of photovoltaic modules title angle for Oman. 7th IEEE international power engineering and optimization conference PEOCO2013, Malaysia, pp. 700–704
8. Kazem HA, Khatib T (2013) A novel numerical algorithm for optimal sizing of a photovoltaic/wind/diesel generator/battery microgrid using loss of load probability index. Int J Photoenergy 2013:8 pages
9. Kazem HA, Khatib T (2013) Photovoltaic power system prospective in Oman, technical and economical study, 1st edn, ISBN: 978–3659372957, LAP LAMBERT Academic Publishing, Germany
10. Kazem HA, Khatib T, Sopian K, Buttinger F, Elmenreich W, Albusaidi AS (2013) The effect of dust deposition on the performance of multi-crystalline photovoltaic modules based on experimental measurements. Int J Renew Energy Res (IJRER) 3(4):850–853

Chapter 49
The Photovoltaic Project of Sudan to Reach More than One Million Homes: An Exercise in Policy Making, Finance, Strategies, Education and Sustainability

Anwar El-Hadi

Abstract Sudan is enjoying the highest incidence of solar power in the world, and has a surface area of 1,890,000 km^2. Electricity from fossil fuel and hydro sources reaches 34 % of the total population—covering 57 % of the urban and 16 % of the rural population. There are plans to reach 95 % of the urban population by 2016. Despite this plan, however, no more than 28 % of the rural population can be covered. This is a clear challenge for the implementation of renewable energy sources.

A project for the promotion of solar energy was launched in 2001 comprising workshops, exhibitions and lectures. Implementation started in the following years resulting in a photovoltaic (PV) electricity supply reaching 7000 families in the rural areas. 150 healthcare units, schools, water supply, groceries and village cultural clubs were served. By 2005/2006 these efforts had succeeded in supplying electricity to 1000 villages, including in-house equipment for the service units.

A target was set to supply PV electricity to 1,100,000 homes in the rural areas over the following 20 years. The scheme was launched in 2012 in the most northerly States.

The work is perceived as a case study in the field of implementation, covering policy making, finance, strategies, education and sustainability of a PV electricity supply for rural areas.

A novel feature is the launch of a project called 'Barrier Removal to Secure PV Market Penetration in Semi-Urban Sudan'. The approaches used to carry out the various activities were discussed and evaluated and found to be highly effective for achieving the objectives.

The socio-economic impacts were clearly noticed from the start.

The success encouraged the project workers to fulfill the challenge of securing PV electricity for more than one million homes.

Keywords Sustainability · Policy · Renewable energy

A. El-Hadi (✉)
College of Engineering and Architecture, University of Bahri, Khartoum, Sudan
e-mail: anwareladi87@yahoo.com

Business International and Resource Development, Khartoum, Sudan

A. Sayigh (ed.), *Renewable Energy in the Service of Mankind Vol II*,
DOI 10.1007/978-3-319-18215-5_49

49.1 Introduction

Electric energy has been commercially available in Sudan since 1908 using fossil fuel. In 1926, the Sennar Dam, also known as 'Sinnar', was built for crop irrigation. With a height of 40 m, the first hydro-electric power station was in operation.

Following Sinnar, many dams were built—Jabal Awlia in 1937, Khashm Al-Girba in 1964, Roseires in 1966, and more recently, Merowe in 2009. The dams were multi-purposely imitated and are now producing 56.8 % of generated electricity with more projects under construction to enhance the output.

Generally, the output of generated electricity has been increasing at the rate of 12 % per annum since 2002 and reached 9,339,8 GWh in 2012 with hydro constituting 71 % [1].

A key factor for the realization of the Merowe Dam project, with 1250 MWh generating capacity, was the ability of Sudan to export oil in commercial quantities in 1999/2000.

With an electrification ratio of 34 % and 283 kWh/year electricity use per capita, Sudan has a long way to go to fulfill its aspirations in this field, especially at a time when the proceeds from petroleum are dwindling due to the secession of South Sudan.

Other renewable energy sources were surveyed with potential capacities and locations, which resulted in the formation of the following:

- Wind Atlas
- Solar Atlas
- Waste to energy feasibility studies
- Geothermal energy studies

A few projects emanating from the above were identified after appropriate investigation to feed into the national grid.

However, with its vast area of 1,890,000 km^2, extending from north to south at a length of 1950 km and from east to west at a length 2200 km, this constitutes a rural population to reach approximately 60 % of the total population.

As most of the electric power generated is along the sides of the River Nile where most of the urban population is situated, the national grid embodies only a total length of 8560 km of high voltage transmission lines.

Thus, it is clear that the majority of the population is rural, utilizing most of the land for animal raising and rain-fed agricultural development schemes. As such, the major contributor to the country's economy is situated far from the current energy sources and is isolated from the grid.

Hence, a Rural Electrification Project was launched which aims to supply the rural residents in off-grid areas with electricity through Solar Home Systems (SHS) [1], with the following features:

> To target 1.1 million SHS, benefiting 5.5 million people (i.e. 16 % of the population) between 2013 and 2032.

Table 49.1 Some of the projects that preceded the PVP

Time	Funded by	Executed by	Scope
1976–1978	UNDP	Energy Research Institute	Solar pumps global project
1982–1984	Germany	Energy Research Institute	Sudan project for renewable energy
1992–1998	UNDP/Sudan	Energy Research Institute	Rural development through solar energy
1996–1997	Sudan	Solarman Co. Ltd	Sudanese customs authority

NB Other projects followed which coincided with the PVP, resulting in close coordination

The estimated budget is 700 million USD and the project will be executed by the government-owned Sudanese Electricity Distribution Company (SEDC). It is worth noting that this project came about as a result of preliminary work which started in 1999; without it, the project may not have come into existence. The initiative is called 'Barrier Removal to Secure PV Market Penetration in Semi-Urban Sudan' or 'PVP'. It targeted 1000 villages with a total budget of 2.5 million USD. Its execution removed barriers, increased confidence and led to the rural electrification PV project with a budget of 700 million USD.

We outline the PVP project, which had a huge impact on the prevailing strategies that followed, and discuss the possibility of it being reproduced elsewhere under similar conditions. It enhanced South/South cooperation, paved the way for a smooth transition, and minimized mistakes and time lapse. It made use of international organisations and professional agencies.

49.2 The PVP: Barrier Removal to Secure PV Market Penetration in Semi-Urban Sudan

Solar PV applications in Sudan started as early as 1970. These projects were executed through foreign financial support. Table 49.1 presents some of these projects.

Mosques, khalwas, community centres and water resources were targeted. The PVP was initiated to remove the barriers that hindered market penetration of PV applications in general and households in particular.

The PVP is a project funded jointly by the United Nations Development Programme (UNDP), the Global Environment Facility (GEF), and the Ministry of Energy and Mining [2]

> It is supposed to meet the growing suppressed demand for electric energy in semi-urban Sudan through reliable win-win domestic PV systems as a substitute for fossil-based generating units. This is to be achieved by removing a number of critical barriers that inhibited the use of PV systems, thus enabling market penetration of this technology in the targeted areas.
>
> It was assumed that the barriers removed would contribute to the establishment of a sustainable foundation for the use of this proven technology as a viable option for rural and semi-urban electrification, resulting in the long-term reduction of carbon-dioxide emissions.

The main objective of the PVP was to remove the five identified critical barriers that inhibited market penetration of PV technology. These are:

1. Market infrastructure and policy barriers
2. Information and awareness barriers
3. Technical and know-how barriers
4. Financial barriers
5. Institutional barriers

An independent entity was formed to implement the project, assisted by a technical committee, which met regularly; thus, keeping stakeholders informed of the progress. It operated through 16 focal offices covering 13 states.

49.3 Policy and Institution Barriers

Energy use in Sudan relies heavily on traditional fuels, mainly charcoal and firewood. Oil is also used as a commercial source of energy, especially diesel generators in rural areas.

The goal of the PVP was to improve the quality of life both environmentally and economically and to remove any barriers that prevented this goal being achieved.

Fossil fuel-generating electricity systems in rural areas receive a 30% reduction in import taxes compared to zero tax for PV systems. In addition to the taxes, solar panels are usually produced in developed countries and then transported through difficult terrains to semi-urban areas in developing countries, which adds to the cost and logistic problems.

As PV systems incur high start-up costs coupled with lack of financing, banks were not traditionally involved in financing PV projects.

A Solar Act which encouraged easing regulations regarding the use of renewable energy technologies in general, and PV in particular, was prepared in cooperation with the Energy Committee of the National Assembly to favour PVP in general and encourage its promotion in particular.

The Ministries of Physical Planning and Public Utilities in the different States established renewable energy divisions. The focal points include these divisions.

The Act, together with the formation of these divisions, helped in the removal of most of the barriers.

The private sector was encouraged to import PVP equipment through the formation of companies and involvement in focal points.

49.4 Information and Awareness

Television, radio stations, seminars and newspapers were all utilized to spread and deepen awareness of renewable energy technology (RET) and PVP among the rural population.

In fact, the first meetings took place at the National Assembly (Parliament), and then involved members from the rural areas in the activities. Exhibitions and video releases were used.

Approximately 4500 different activities took place. Some previous projects financed by foreign agencies, which had been neglected, were encouraged by the new policy and were endorsed in the plans. Stakeholders as well as policy makers were also involved.

The activities demonstrated the role of solar energy in refrigeration in health centres and in powering water pumps for humans and animals, and underlined the fact that PV applications were economically viable in many applications if properly sized.

49.5 Technical and Know-How

Engineers, technicians and those in vocational training centres were trained at the focal points in the technical, economic and environmental aspects of RET in general and PV in particular. Approximately 853 people were trained, both from the private sector and the state. Most importantly a curriculum was established for vocational training centres (VTCs), and some centres close to rural areas were equipped with the necessary training laboratory and workshop equipment.

It should be noted that graduate studies in the field were already established in some universities; however, the real need for rural implementation lay on the shoulders of the vocational trainees.

49.6 Financial Barriers

The PVP established a specific financing mechanism. A simple credit procedure taking into account the low financial capabilities of the targeted communities was formed. A Memorandum of Understanding for the guarantee fund was signed between the Saving and Social Development Bank, the Agricultural Bank, and the Ministry. A special credit mechanism allowed individuals and group users to purchase PV systems. A microfinance system was established at federal and state levels for a credit mechanism.

49.7 Results as Evaluations and Assessments

The PVP, which was subject to scrutiny and evaluation from independent professional expertise in all elements, paved the way for rural PV electrification and other developments.

When the socio-economic and environmental impact of the PV applications was assessed [3], the reaction of the rural population to the use of solar systems was found to be positive. Villagers were able to organise meetings in the evenings, children's performance at schools improved, and health care was extended to deal with night emergencies. Awareness of what is going on the outside world increased because of television sets. Agricultural and commercial businesses were enhanced through telecommunications.

Reference [3] developed indicators to measure and assess the long-term socio-economic impact of PV system applications. These applications included health, education, telecommunication, water and community centres. Their general format is shown in Table 49.2.

Removal of the financial barriers was assessed through the Guarantee Fund Financing Mechanism [4]. It was found in some offices that the disbursement exceeded 100 % of the allotted fund by recycling the repayments and utilizing bank resources.

The Agricultural Bank increased its contribution to the Guarantee Fund from 50 to 150 %. The Saving Development Banks increased its contribution from 50 to 75 %. More bank branches showed interest in future financing.

PV training in VTCs was evaluated [5]. It was found that the new curriculum on PV technology was in agreement with the objectives and contents of the existing engineering curriculum of the VTC. Training materials were designed. It goes without saying that the establishment of relevant PV training at VTCs is a key element to sustainability.

Many other assessment reports were prepared.

A final evaluation for the project [6] noted many details, some of which are mentioned below.

At the moment the contribution of the PVP system to the carbon-free environment is negligible but will be more important with future development and further applications.

PV system applications proved to be:

- Beneficial to social and economic development.
- An ideal solution to energy problems in the vast rural areas of Sudan.
- An important factor in poverty reduction.

Sustainability of these applications is easily achieved through:

- The importance of the role of project partners and stakeholders.
- The capacity of the local institutions to manage the process of market development and to ensure a full ownership of such processes.
- The sustainability of the microfinance for facilitating the access of rural communities and individuals to PV systems and clean energy.

Most important, the possibility of replication or scale-up of the PVP experience enhanced and will continue to enhance sustainability.

Table 49.2 Indicators to measure and assess the long-term socio-economic impact of PV system applications

Key issues	Indicators	Data needed	Method of data collection/frequency/responsibility
a. Assessment of benefits on health services	% Increase in EPI coverage; % increase in the antenatal and postnatal attendance; % decrease in maternal mortality; % decrease in infant mortality; % average length of stay of health personnel in health facilities	Number of immunizations; number of antenatal and postnatal visits; number of maternal deaths; number of infant deaths; number of years health personnel in health facilities	Health centre records were collected on a daily basis and collated on a monthly basis by the health staff; information on the health personnel in the health facilities can be obtained from the community leaders in the villages and localities
b. Assessment of social benefits (including dispute/conflict management)	% Increase in the number of pupils in the school; % increase in the number of pupils passing examinations; % average number of years teachers are staying in school; % increase in the number of people attending adult education in centres with PV systems; % increase in the number of villages with TV using PV systems; % increase in the amount of water used in the houses; % increase in disputes/conflicts resolved by village elders	Number of pupils in the schools; number of pupils passing their examinations; number of years teachers are staying in the schools; number of villages with TV using PV systems; number of houses using PV for lighting; amount of water used by households from wells using PV systems; number of disputes/conflict resolved at the village level	School records kept by the school administration; information on the number of years teachers serve in the school can be obtained from the community leaders and the localities; records kept by village community leaders; survey to determine domestic water consumption patterns
c. Assessment of economic benefits	% Increase in the number of farmers using PV systems in irrigations; % increase in the range of vegetables and fruits available in the village market; % increase in the number of communication equipment using PV systems; % increase in the types of income generating activities in the villages using PV systems	Number of farmers using PV systems for irrigation; types of vegetables and fruits currently available in the village markets locally grown; number and types of communication equipment in the villages; number and types of income generating activities in the villages with PV systems	Local market surveys by local consultants and personnel in the localities

Table 49.2 (continued)

Key issues	Indicators	Data needed	Method of data collection/frequency/responsibility
d. Assessment of sustainability	Amount of money collected for the purchase of PV systems and their maintenance at the village level; number of trained personnel in the villages able to maintain the systems; number of equipment installed and still operating	Amount of money collected for the purchase of PV systems and their maintenance at the village levels; number of trained personnel in the villages able to maintain the systems; number of equipment installed and still operating	Records from the community leaders
e. Assessment of energy policy	Policy statements by the federal and state authorities	Records of the policy statements by federal and state authorities	Collection of documents by the project office
f. Assessment of environment impact	% Increase in the number of centres using PV systems for lighting instead of wood; % increase in households with stoves using solar energy	Records and household surveys	

Government policy makers encouraged the State Government to adopt solar PV in their annual development plans, and support non-grid connected areas with PV services for agriculture, social services, schools offices and portable water.

[6] reported the findings of the Auditor General Report regarding the different PVP accounts, that is, United Nations Development Programme, governments and banks involved. The audit report included Adequacy of Financial Operations and Controls; Adequacy of Management Structure and Control; Non-Expendable Equipment; the Bank Guarantee Fund; Monitoring and Evaluation Report; and the Performance Audit and Executive Summary of audit findings and recommendations. It concluded that the findings were satisfactory.

49.8 Rural Electrification; the Photovoltaic Project

The PVP symbolizes the foundation of rural electrification. All previously mentioned efforts and strategies were scaled up to cater for this 700 million USD project. In addition to the previously mentioned community and economic services, solar home systems reach 1.1 million families providing:

- 100 W: for lightning three lamps, TV, radio and mobile charge
- 50 W: for lighting three lamps

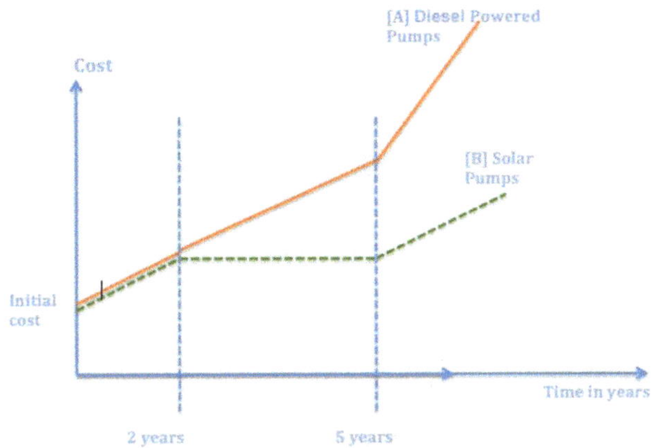

Fig. 49.1 The initial costs and operating costs are shown for pump (A), diesel power and pump (B) solar power. A water well with a 67.5-m head irrigates 10 acres of land. The costs are the same for discharge.
Pump (A): total cost paid in cash. Maintenance cost approximated as linear.
Pump (B): 50 % paid cash; 50 % paid in equal installments over 2 years. Maintenance free for 5 years. After 5 years, the operating cost of (A) increased tremendously while the cost of (B) increased but at a lesser rate.

The financial sponsors are the government, national banks and foreign banks. The loan period is 4 years with a downpayment of 10 %. Job creation is estimated at 10,000—both direct and indirect. After-sales service has created a new business. Reduction in carbon dioxide emission is estimated to be a thousand tons.

The rural electrification project extends from 2012 to 2031 in 5-year phases subject to evaluation and improvement with developments in the field of PVs.

One of the earliest achievements of this project is a simple experiment conducted on a farm with real data, as shown in Fig. 49.1.

49.9 Conclusive Remarks

The authorities have successfully ensured the sustainability of rural electrification through PV technology. First, the barriers were identified, analysed and a plan was formulated to remove them. Implementation was started, monitored and provided feedback. Assessments and evaluations were carried out and were followed by replication and scaling up.

This took place when the cost of PV systems was dropping exponentially and their use and output power was rising exponentially.

This can be considered as a contribution to South/South cooperation by replicating and scaling up these efforts elsewhere.

It is suggested that a regional conference for WREC/WREN be held in Khartoum to share the experiences of the region, and to narrow the gap between the developing and developed countries in the field of renewable energy technology, especially solar energy.

Acknowledgments The author would like to acknowledge the help and assistance received from Sudan's Minister of Water Resources and Electricity and the Directorate of Renewable Energy. The help received from the Director of Energy Affairs, and the Ministry of Petroleum, Policies and Evaluation was invaluable. The great help of Ingi, Randi and Rinad in editing and formatting the outcome are well appreciated.

References

1. Saied YA (2014) Renewable energy project, renewable energy directorate, ministry of water resource and electricity, Sudan
2. UNDP/GEF (1998) Ministry of energy and mining. Project document. SUD/99/G35 and SUD/00/013–Barrier Removal to Secure PV Market Penetration In Semi Urban Sudan–(PVP)
3. Gore PW (2002) University of Khartoum. Assessment of socio-economic and environmental impacts of the PV applications. Barrier Removal to Secure PV Market Penetration in Semi-Urban Sudan, December
4. Salih Gibreil, Awadia Fadel Moula AB, Khalaf Alla EL Bahi EL Sheikh, Asim EL-Haj A/Gadir (2005) Guarantee fund financing mechanism. Barrier Removal to Secure PV Market Penetration in Semi-Urban Sudan. Assessment report
5. Sid Ahmed MO Evaluation of PV training in VTC's, SUD/99/G35
6. Hood AH (2006) Report of the final evaluation. Barrier Removal to Secure PV Market Penetration in Semi-Urban Sudan, SUD/99/G35 and SUD/00/013, UNDP, GEF, GOS, May

Chapter 50
Enhanced Performance of Dye-sensitized Solar Cells Aided by Olive-shaped ZnO Nanocrystallite Aggregates as the Light-scattering Layer

Wei-Chen Chang, Hung-Shuo Chen and Wan-Chin Yu

Abstract Olive-shaped ZnO nanocrystallite aggregates were synthesized for dye-sensitized solar cells (DSSCs). The submicron-sized hierarchical nanostructure is composed of highly crystalline ZnO nanoparticles about 20 nm in diameter and has an overall dimension of approximately 150×300 nm. An economical and environment-friendly aqueous solution method was developed to synthesize the olive-like aggregate. This template-free self-assembly method involved the mixing of zinc nitrate and sodium hydroxide aqueous solutions at a low temperature (80 °C) and aging the mixture for a particular length of time. We employed a low-temperature (150 °C for 1 h) thermal treatment process for the fabrication of bilayer photoelectrode, with commercial ZnO nanoparticles (~20 nm) as the underlayer and submicron-sized structures as the light-scattering overlayer. The N719-sensitized DSSCs containing the aggregate overlayer reached a power conversion efficiency of 4.4 %, 33 % higher than that attained by DSSCs incorporating large solid particles (200–500 nm) as the scattering layer. The enhanced overall conversion efficiency of aggregate-based cells was correlated with a prominent increase in the short-circuit current density. Optical and dye-loading investigations show that this improvement can be attributed to the dual functionality of the olive-shaped nanocrystallite aggregates, which have excellent light-scattering ability to enhance photon capture while providing a large surface area for sufficient dye adsorption.

Keywords Zinc oxide · Aggregates · Dye-sensitized solar cell · Light scattering

W.-C. Yu (✉) · W.-C. Chang · H.-S. Chen
Institute of Organic and Polymeric Materials,
National Taipei University of Technology, Taipei, Taiwan
e-mail: wanchin@ntut.edu.tw

W.-C. Chang
Institute of Nuclear Energy Research, Taoyaun, Taiwan
e-mail: changpeter@iner.gov.tw

A. Sayigh (ed.), *Renewable Energy in the Service of Mankind Vol II*,
DOI 10.1007/978-3-319-18215-5_50

50.1 Introduction

Dye-sensitized solar cells (DSSCs) have stimulated great interests in the past two decades due to their appealing properties, such as transparency and multicolor options, potentially low production costs, compatibility with flexible substrates, and ability to function under indoor lighting [1, 2]. A critically important component of DSSC is the photoanode, which consists of a mesoporous oxide semiconductor film covered with dye molecules that capture photons from light during device operation. Photoexcitation of the dye molecules then leads to the injection of electrons into the oxide film. Therefore, the photoanode structure has great impact on the photovoltaic performance of DSSCs. Conventional photoanode films are constructed from nanoparticles approximately 20 nm in diameter to ensure a large interfacial surface area for dye adsorption. However, the high transparency of the nanocrystalline film leads to prominent optical loss, thus limiting the generated photocurrent. An important strategy that has been adopted to boost optical absorption in DSSCs is light scattering [3]. The basic principle of the light-scattering method is to confine light propagation and extend the traveling distance of light within the oxide film. As s result, the possibility of photon absorption by the dye molecules is increased, and so is the power conversion efficiency. In conventional DSSCs, submicrometer-sized solid particles are introduced into the photoanode film to serve as light scatterers; these solid particles are either mixed into [4] or form a scattering layer over the nanocrystalline film [5–7]. However, mixing large particles into the nanocrystalline film unavoidably causes a decrease in the specific interfacial surface area of the film, whereas placing large particles on top of the nanocrystalline film brings about a limited increase in the interfacial surface area of the film. To overcome these drawbacks, a class of multifunctional hierarchical nanostructure called nanocrystallite aggregates has been developed, which are submicrometer-sized structure assembled from nanometer-sized crystalline building blocks [8–9]. Because of their porous nature and submicron size, not only can the aggregates provide a large interfacial surface area for dye adsorption, they can also improve optical absorption by generating effective light scattering. Additionally, nanocrystallite aggregates may offer better electron transport capacity than dispersed nanocrystals, because of the well-interconnected nanocrystallites within the aggregates [8, 10].

In the present study, an easy self-assembly procedure was developed to synthesize olive-shaped nanocrystallite aggregates. The environment-friendly synthesis procedure was performed at a relatively low temperature (80 °C) without the use of any shape-directing agents or organic solvents. To investigate the applicability of the olive-shaped aggregates as a dual-function light-scattering material, bilayer photoanode consisting of a nanocrystalline underlayer and a light-scattering overlayer was prepared using a low-temperature (150 °C for 1 h) heat-treatment process. The olive-shaped aggregates were demonstrated to be a promosing photoanode material, leading to a power conversion efficiency of 4.4 % for aggregate-containing DSSCs, a 33 % improvement compared to that attained by solid scatterer-based cells. The improvement in cell efficiency can be attributed to the synergetic effects of enhanced light scattering and improved dye loading brought about by the olive-shaped aggregates.

50.2 Experimental

To synthesize olive-shaped ZnO nanocrystallite aggregate, equal volumes of 0.05 M $Zn(NO_3)_2 \cdot 6\,H_2O$ and 0.1 M NaOH aqueous solutions were brought to 80 °C, thoroughly mixed, and allowed to age at the same temperature for 4 h. The resulting white precipitate was centrifugally collected, washed with pure water, and dried in an oven for later characterization and photoanode fabrication. To fabricate photonaode films, ZnO pastes, consisting of ZnO nanomaterials dispersed in an aqueous solution of tert-butanol (volume ratio of tert-butanol to water was 2:1), were applied onto the fluorine-doped tin oxide (FTO) substrates (Nippon Sheet Glass, 8–10 Ω/γ, 2.2 mm-thick) by doctor-blade technique using adhesive tape as the frame and spacer (active area = 0.25 cm^2). The electrodes were then heat-treated at 150 °C for 1 h to remove organic materials from the pastes. The thickness of the nanoparticle underlayer was kept at 20 μm, while the thickness of the light-scattering overlayer was varied, in order to determine the optimal thickness for each type of light-scattering material.

After cooling to room temperature, the ZnO photoelectrodes were immersed into a solution made by dissolving 0.5 mM cis-bis (isothiocyanato) bis-2,2′-bipyridyl-4,4′-dicarboxylato)-ruthenium (II) bis-tetrabutylammonium (N719, Solaronix) in equal portions of acetonitrile and tert-butanol. The dye-adsorption time was varied from 2 to 4 h depending on film thickness. The electrodes loaded with N719 were then washed with acetonitrile and dried in the air before cell assembly. The counter electrode of DSSCs was made of FTO glass onto which the nanocrystalline Pt catalysts were deposited by decomposing H_2PtCl_6 **at** 400 °C for 20 min. The ZnO photoelectrode and the Pt counter electrode were sandwiched together with a 60 μm-thick hot-melting spacer (Surlyn, DuPont, Wilmington, DE, USA) in between, and the space between the electrodes was filled with an acetonitrile-based electrolyte containing 0.1 M lithium iodide (LiI, Sigma-Aldrich), 0.6 M 1, 2-dimethyl-3-propylimidazolium iodide (PMII, Merk), 0.05 M I_2 (Sigma-Aldrich), and 0.5 M tert-butylpyridine (TBP, Sigma-Aldrich) in acetonitrile.

The morphologies and structures of ZnO nanomaterials were characterized by both field emission scanning electron microscopy (FE-SEM, FEI Nova230) and transmission electron microscopy (TEM, H-7100). X-ray diffraction (XRD) patterns were obtained by using a diffractometer (PANalytical X'Pert PRO) with Cu Kα radiation. The film thicknesses of photoelectrodes were measured by a micro-figure-measuring instrument (Kosaka Laboratory, Surfcorder ET3000). Dye loading of the photoelectrode was estimated by desorbing the dye in a 10 mM NaOH aqueous solution and then measuring the absorbance of the solution by UV-vis spectroscopy (V-570, Jasco). The diffuse reflectance spectra of undyed films were determined by a UV-Vis spectrophotometer (V-570, Jasco) equipped with an integrating sphere. The incident photon-to-current conversion efficiency (IPCE) was measured as a function of wavelength by using a monochromator (Forter Tech). The energy conversion efficiency was measured under a white light source (Yamashita Denso, YSS-100 A) that gave an irradiance of 100 mW cm^{-2} (the equivalent of AM 1.5 one sun) on the surface of solar cell. The irradiance of simulated light was calibrated using a silicon photodiode (BS-520, Bunko Keiki).

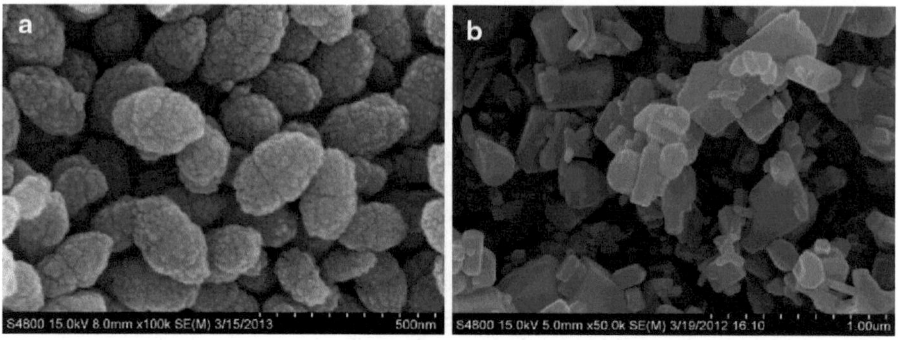

Fig. 50.1 FE-SEM images of **a** olive-shaped aggregates and **b** solid light-scattering particles

Fig. 50.2 TEM image of olive-like aggregates

50.3 Results and Discussion

Figure 50.1 displays representative FE-SEM images of the as-synthesized aggregates and the commercially available solid light-scattering particles. As shown, the nanocrystallite aggregates have an olive-like morphology with an overall dimension of approximately 150×300 nm. The size of the aggregate is comparable to the wavelengths of visible light, a characteristic likely to give the olive-like aggregate light-scattering capability. The TEM image of the aggregate (Fig. 50.2) exhibits an oval-shaped morphology, consistent with the olive-like structure observed in Fig. 50.1a. Figure 50.2 also demonstrates that the aggregates are composed of nanocrystals approximately 20 nm in size.

Figure 50.3 shows power XRD patterns of ZnO nanoparticles, as-synthesized aggregates, and heat-treated aggregates. All three samples exhibit similar diffraction

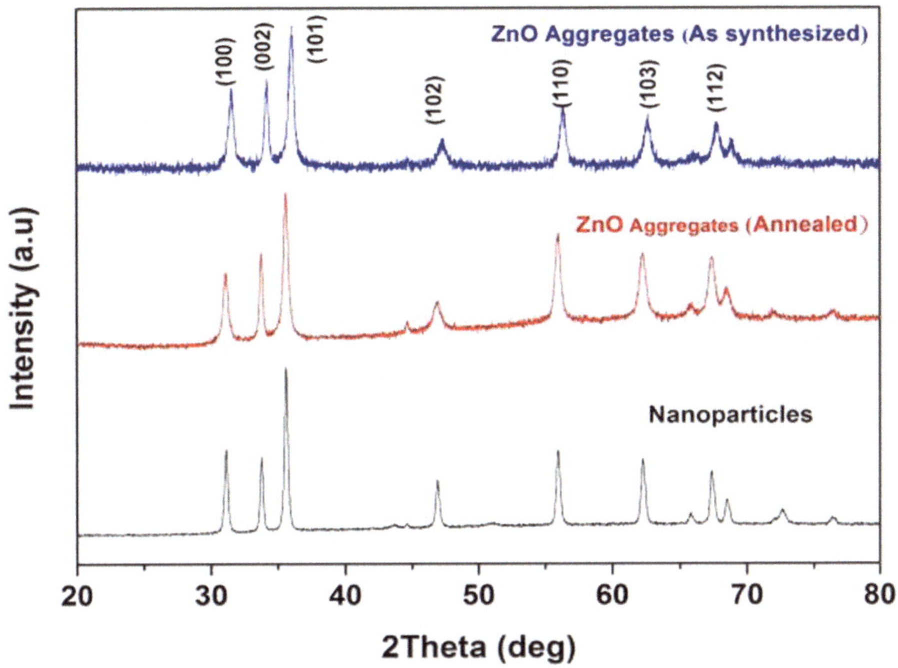

Fig. 50.3 Power XRD patterns of ZnO nanomaterials

patterns readily indexed to the wurtzite hexagonal phase of pure ZnO. The peaks at 2θ=31.7°, 34.5°, 36.3°, 47.4°, 56.6°, and 62.7° correspond to the (100), (002), (101), (102), (110), and (103) planes of hexagonal ZnO, respectively (JCPDS Card No. 36-1451). No other diffraction peaks were detected in these patterns, demonstrating that the olive-like nanostructures are composed of pure ZnO nanocrystals.

DSSCs based on bilayer-structured photoanode with commercial ZnO nanoparticles (~20 nm) as the underlayer and submicron-sized structures (aggregates or commercial solid light scatterers) as the light-scattering overlayer were fabricated. Reference DSSCs based on single-layered nanoparticle photoanode were also constructed for comparison. The thickness of the nanoparticle underlayer was kept at 20 μm, the optimal thickness for the single-layered nanoparticle photoanode. The thickness of the light-scattering layer was varied in order to determine its optimal value. As shown in Table 50.1, the optimal thickness of the solid light scatterer was 5 μm, whereas 20 μm was optimal for the olive-shaped aggregate. The J–V curves of the best-performing cells based on different photoanode configurations are summarized in Fig. 50.4a.

The photocurrent action spectra of DSSCs with different photoanodes are given in Fig. 50.4b, which displays the wavelength distribution of incident monochromatic photon-to-current conversion efficiency (IPCE). The IPCE values reach a maximum at around 530 nm for all samples, but the aggregate-containing cell exhibits the highest IPCE of 76%, consistent with the trend of J_{SC} value observed in

Table 50.1 Effect of photoanode structure on DSSC performance

Photoanode structure	J_{sc} (mA/cm²)	V_{oc} (V)	FF	η(%)
20 μm nanoparticles	6.21	0.67	0.75	3.09
+5 μm solid scatterer	8.74	0.61	0.67	3.57
+10 μm solid scatterer	7.83	0.61	0.69	3.30
+5 μm aggregates	8.78	0.60	0.67	3.65
+10 μm aggregates	8.97	0.64	0.7	4.02
+20 μm aggregates	10.54	0.61	0.69	4.43

Fig. 50.4a. The cell lacking a scattering layer had the lowest IPCE that likely resulted from the high transparency of the nanoparticle film. Introducing a light-scattering layer, whether made of submicrometer-sized solid particles or aggregates, the corresponding IPCE was greatly improved, likely due to enhanced light harvesting brought about by the light-scattering effect.

To gain more insight into the origin of the enhanced J_{SC} of the aggregate-based cell, dye loadings and diffuse reflectance were determined, because J_{SC} is closely related to these two factors [9]. As shown in Fig. 50.5a, the aggregate-based film

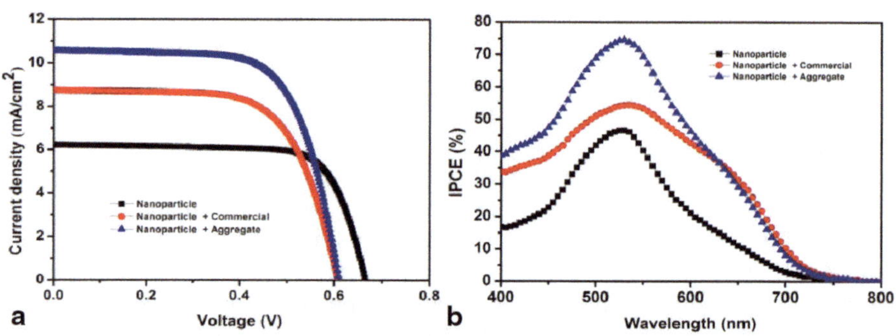

Fig. 50.4 Current density–voltage curves and photocurrent action spectra of DSSCs

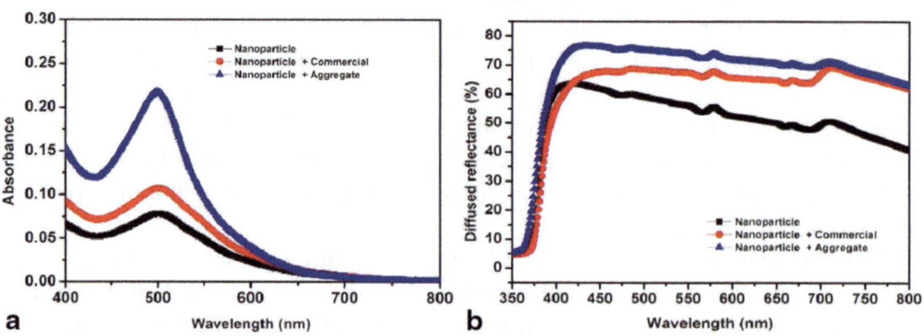

Fig. 50.5 **a** Absorption spectra of N719 solutions resulted from desorbing the dye from sensitized films and **b** diffused reflectance of undyed photoanode films

had significantly higher dye loading than the other two films. The aggregate-based film also exhibited the highest diffuse reflectance in the range of 400–800 nm, the main wavelength range that DSSC dyes capture photons. The result confirms that the olive-shaped aggregates have good light-scattering ability. The enhanced J_{SC} of the aggregate-based cell can thus be attributed to the synergetic effects of superior light scattering and improved dye loading brought about by the olive-like aggregates. The strong light-scattering effect of aggregate-based film extended the travelling distance of incident light within the photoanode film and increased the probability of photon harvesting by dye molecules. Increased dye loading may also improve light harvesting. Consequently, the IPCE and J_{SC} were greatly improved in the aggregate-based devices.

50.4 Conclusions

Olive-shaped ZnO nanocrystallite aggregates were synthesized through an environmental-friendly aqueous-solution method and adopted for the fabrication of bilayer-structured photoanode that contained a nanoparticle underlayer and a light-scattering overlayer. The olive-shaped aggregates outperformed the conventional light-scattering solid particles due to the dual functionality of the aggregates, which generated light scattering while providing a large surface area for dye adsorption. The aggregate-based DSSCs reached a power conversion efficiency of 4.4 %, 33 % higher than that attained by solid scatterer-based DSSCs.

References

1. Oregan B, Grätzel M (1991) Letters to nature. Nature 353:737
2. Yella A, Lee HW, Tsao HN, Yi CY, Chandiran AK, Nazeeruddin MK, Diau EWG, Yeh CY, Zakeeruddin SM, Grätzel M (2011) Porphyrin-sensitized solar cells with cobalt (II/III)–based redox electrolyte exceed 12 percent efficiency. Science 334:629
3. Zhang Q, Myers D, Lan J, Jenekhe SA (2012) Applications of light scattering in dye-sensitized solar cells. Phys Chem Chem Phys 14:14982
4. Kang SH, Kim JY, Kim HS, Koh HD, Lee JS, Sung YE (2008) Influence of light scattering particles in the TiO_2 photoelectrode for solid-state dye-sensitized solar cell. J Photochem Photobiol A 200:294
5. Hore S, Vetter C, Kern R, Smit H, Hinsch A (2006) Influence of scattering layers on efficiency of dye-sensitized solar cell. Sol Energy Mater Sol Cells 90:1176
6. Ito S, Murakami T, Comte P, Liska P, Grätzel C, Nazeeruddin M, Grätzel M (2008) Fabrication of thin film dye sensitized solar cells with solar to electric power conversion efficiency over 10 %. Thin Solid Films 516:4613
7. Qiu Y, Chen W, Yang S (2010) Double-layered photoanodes from variable-size anatase TiO_2 nanospindles: a candidate for high-efficiency dye-sensitized solar cells. Angew Chem Int Ed 49:3675
8. Yan K, Qiu Y, Chen W, Zhang M, Yang S (2011) A double layered photoanode made of highly crystalline TiO_2 nanooctahedra and agglutinated mesoporous TiO_2 microspheres for high efficiency dye-sensitized solar cells. Energy Environ Sci 4:2168

9. Zhang Q, Park K, Xi J, Myers D, Cao G (2011) Recent progress in dye-sensitized solar cells using nanocrystallite aggregates. Adv Energy Mater 1:988
10. Sauvage F, Chen DH, Comte P, Huang FZ, Heiniger LP, Cheng YB, Caruso RA, Grätzel M (2010) Dye-sensitized solar cells employing a single film of mesoporous TiO_2 beads achieve power conversion efficiencies over 10%. ACS Nano 4:4420

Chapter 51
Better Than Optimum: Integrated. The Integration of Renewable Energy in Architecture as an Optimization Factor

H. Coch, C. Pardal, A. Pagès-Ramon, A. Isalgué and I. Crespo

Abstract This chapter presents the importance that the embodied energy of the support material of a photovoltaic installation can have in relation to the energy produced and, therefore, in relation to other efficiency factors in the design of photovoltaic installations. The importance of these costs cannot be neglected in the overall system efficiency.

The monitoring of photovoltaic panels after a few years of operation under real conditions demonstrates that a good integration of photovoltaic systems in constructive elements, even if their inclination or orientation is not the recommended one, may be better than just overlapping them in an optimal position. The embodied energy of the added constructive materials used when the system is not integrated may be more than the loss of production as a result of a nonoptimal photovoltaic modules disposal.

This chapter illustrates the potential that architectural design applied to photovoltaic modules integration could represent, based on the experience of the photovoltaic pergola built in 2004 at the Forum esplanade in Barcelona. The ultimate goal is to consider the architectural design as another, important parameter when sizing photovoltaic systems, together with the orientation, tilt, temperature and other parameters regularly used.

Keywords Architectural design · Architectonic integration · Photovoltaic systems

H. Coch (✉) · C. Pardal · A. Pagès-Ramon
Architecture & Energy Group. Department of Architectural Technology I,
School of Architecture, UPC, Avda. Diagonal, 649, 08028 Barcelona, Spain
e-mail: helena.coch@upc.edu

A. Isalgué
Architecture & Energy Group. Department of Applied Physics, School of Architecture,
UPC, Avda. Diagonal, 649, 08028 Barcelona, Spain

I. Crespo
Architecture & Energy Group. Department of Architectural Representation and Visual Analysis I,
School of Architecture, UPC, Avda. Diagonal, 649, 08028 Barcelona, Spain

A. Sayigh (ed.), *Renewable Energy in the Service of Mankind Vol II*,
DOI 10.1007/978-3-319-18215-5_51

51.1 Introduction

It is well-known that the production of energy from renewable sources is essential on the path to sustainability. In the case of solar photovoltaic installation, the energy that could be provided depends on several factors such as the orientation and tilt of the modules, operating temperature, etc. All these factors are taken into account to maximize its performance. However, the embodied energy of the materials used on the construction of the structure or any supports needed to hold them in optimal positions is rarely evaluated. The assessment is usually based only on economic cost and payback time.

This situation leads to a main subject: the importance of integrating photovoltaic systems in the building instead of overlapping them over an already finished roof or façade. Integration means that the photovoltaic cells are incorporated in any of the building basic elements: tiles, window glasses, façade coatings, louvers or any other that may be sun-exposed in their natural position.

Although an extensive scientific bibliography on solar photovoltaic systems has been developed, for this chapter, publications with this precise approach on architectural considerations will be considered even though others have been consulted.

The purpose is to contemplate a wide range of variables related with the architectural design of solar photovoltaic systems instead of just focusing on the well-known technical performance parameters.

As Azadian 2013 [1] noted, we can classify the problems faced by architects and engineers when they incorporate solar photovoltaics in buildings in four general areas:

a) Institutional barriers
b) Public acceptance
c) Economic barriers
d) Technical barriers

Within those *technical barriers*, we find *architectural considerations*.

In this chapter, we would like to discuss *architectural considerations*, regarding them not only as *technical barriers* but also from the point of view of their *public acceptance* and *economic barriers*. This new approach will lead to a different way of evaluating the efficiency of solar photovoltaic systems.

The discussion, therefore, is focused on the concept of architectural integration. Accordingly, the importance of integration is presented by evaluating the photovoltaic pergola installed in the area of the Forum in Barcelona that was widely diffused between media. This evaluation considers the costs of the photovoltaic system—cells and support—in energy terms and CO_2 emissions.

51.2 Active Energy Production Systems: Incorporated or Integrated in Architecture

There are many systems that take advantage of renewable energy sources. These devices, normally developed by skilled engineers seeking their optimal performance in order to maximize energy production, are designed as a whole system by itself and too many times they do not consider architectural design and its final location in the building.

For a long time, architecture has had little or nothing to do in the development of these devices, not due to a lack of consideration on the use of renewable energies but because this discipline has its own architectonic mechanisms to take advantage of energy to improve the building's comfort conditions: greenhouses, galleries, patios, inertia walls, etc.

Vernacular architecture has always incorporated energy management in its definition and comfort acquisition is a value intrinsic to architecture. It was not until the Industrial Revolution that a new socioeconomic model emerged, extremely dependent on fossil energies.

Nowadays, guaranteeing user comfort without consuming fossil fuels and reducing CO_2 emissions is a must. The current requirements to reduce the consumption of fossil fuels are so strict that an efficient management of the energy through passive systems is not enough to guarantee acceptable results.

Therefore, active production systems are required to provide the comfort needed by the user without an excessive consumption of fossil fuels and without emitting greenhouse gases. Consequently, buildings are turning themselves into energy producer devices.

Solar photovoltaics are incorporated in the building only considering the optimal energetic performance. At the same time, those systems are becoming part of our architecture without having given time to architects to assimilate them as a new element of the design repertoire. This new situation leads us to raise the issue of the integration of active systems in general, and specifically the photovoltaic ones. They should not be camouflaged on the façades and roofs of buildings but be given the possibility to become part of the formal and material language of architecture [2, 3].

The incorporation of photovoltaic cells in architecture may have different approaches and degrees of integration.

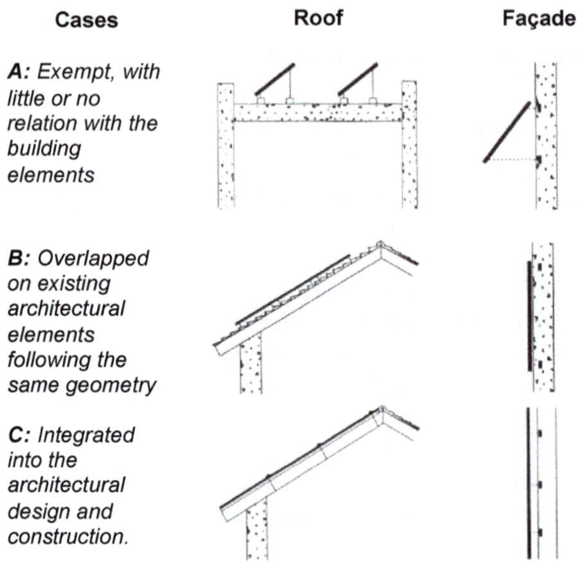

Case A includes the cases where solar photovoltaic modules are just fixed to the building with the same criteria as incorporated air conditioning machines, antennas, etc. The modules are considered gadgets of the building where they are placed on.

Case B is formed by all those installations where solar photovoltaics are placed over any of the architectural elements that compose the building, either in the roof or the façade. The module follows the architecture geometry and is overlapped in the best possible way.

The last case, *Case C*, is the only one that can be considered integration. Solar cells are designed as an architectural constructive element itself. It is not possible to talk about panels, as the elements are glasses, tiles, louvers or watertightness sheets, all of them with a common feature: their rendering is formed by solar photovoltaic cells.

A: Exempt B: Overlapped C: Integrated

This last decade, industry has developed a vast amount of photovoltaic construction elements paving the way for an integration of energy production devices in buildings. However, some designers still do not contemplate the possibility as it is considered less efficient—the roof slope or the façade plane may not be in a suitable position to obtain an optimal energy production.

51.3 Active Photovoltaic Integrated Systems Net Energy

In fact, the energy provided by the system depends on factors such as the orientation and inclination of the panels and the operating temperature. These factors are taken into account to maximize their performance. When photovoltaic systems are implemented in a building, the parameters which are considered, like the disposal of the array of panels are, in most of the cases, the same as in any other installation, without taking into account the properties of the building. Only in some emblematic buildings are architectural variables considered.

Despite the main goal of those installations being energy production, the embodied energy of the materials being used is seldom considered, not only in the case of those materials needed to make the modules but also those employed in all the elements designed to fix the system to the building. Therefore, the energetic balance does not always contemplate all the variables, embodied energy versus generated energy. However, recent considerations point out the relevance of the part of the photovoltaic installation besides the photovoltaic module, in the "Balance of System", or BOS, of the total cost [4, 5].

The aim of this chapter is to show how important it is to consider the embodied energy of those peripheral elements. This cost cannot be neglected because it could be a negative value in the general energy balance of a system considered optimal in terms of orientation of the panels and operating temperature.

Fig. 51.1 Analysed volume of the photovoltaic pergola

51.4 Case Study: Photovoltaic Pergola in the Forum of Barcelona

The pergola of the Forum in Barcelona has been taken as an example of case A, in which the embodied energy in the construction of the support of the photovoltaic system is much higher than the energy produced by the photovoltaic system.

This element situated near the sea in the Forum of Barcelona, designed by the architects Martínez Lapeña and Elías Torres in 2004, connected to the grid about 2700 photovoltaic modules, each with an array of 3×36 photovoltaic cells.

Before the opening exhibition of Forum 2004, the media published that the energy generated would be enough to supply 1000 dwellings and that 440 t of annual CO_2 emissions would be saved. However, the calculations developed in the following work prove that this balance wouldn't be achieved until two decades after, when the same amount of energy consumed for the fabrication of the building materials would have been produced. The calculations compare the embodied energy—and the CO_2 emissions—in the construction of the pergola with the amount of energy supplied.

51.4.1 Considerations and Calculations of Embodied Energy and CO_2 Emissions Generated in the Fabrication of the Photovoltaic Pergola

For the following calculations, only the energetic repercussions of the structure supporting the pergola situated in the Forum square have been considered; the foundations or the existing building below this level weren't contemplated (Fig. 51.1).

The description and amount of materials that constitute the pergola (excluding the photovoltaic modules) have been obtained from certain measurements of the project. Therefore, it is possible to know the real quantity of materials and mobilized resources, even if it doesn't correspond to the amount of material in the

Table 51.1 Energetic costs and emissions of CO_2 related with the fabrication of the photovoltaic pergola

Materials forming the photovoltaic pergola	Quantity (kg)	Embodied energy (MJ/kg) Source: [6]	Embodied energy (MJ)	CO_2 emissions ($kgCO_2$/kg) Source: [6]	CO_2 emissions ($kgCO_2$)
Concrete	3,040,000	0.6	1,781,844	0.1	330,478
Steel (concrete's reinforcement)	304,000	35.0	10,640,000	2.8	857,280
Steel	790,000	35.0	27,650,000	2.8	2,227,800
Neoprene	77	120.0	9240	17.7	1364
Mortar	842	2.4	1979	0.2	185
2686 photovoltaic panels	43,191	60.5	2,611,625	7.1	304,709

completed element. This last value is usually lower, as it doesn't take into account the losses during the building phase. The energetic equivalences and the amount of CO_2 per kilogram of material either are provided by Institut de Tecnología de la Construccio de Catalunya (ITEC) [6] or they are approximate values obtained from the comparative with similar materials in the same database.

The module used for the pergola is the I-165 of Isofoton. The description and mass of the main materials of the module are obtained from the datasheet of this model [7] and information acquired from the company Isofoton.

The wires system and the interconnections are not contemplated in the calculations. Considering that these elements are part of any regular generating and transporting system of electric energy, these components cannot be understood as a specific feature of photovoltaics production. In Mariska's works [8], it is demonstrated that the energetic expenses and the carbon footprint of the wires system are approximately around 1 % of the total of the photovoltaic modules.

About the energy consumed by the fabrication of the pergola (Table 51.1), it is important to note that, from the 42,685 GJ, only 8.2 % are related to the fabrication of materials that forms the photovoltaic modules, the elements that justify the whole construction.

51.4.2 Considerations and Calculations of the Energy Produced and the Savings of CO_2 Emissions During the Running Period of the Photovoltaic Pergola

The photovoltaic modules, with measures of 1310 × 969 × 39.5 mm, 16.5 kg and an array of 3 × 36 photovoltaic cells, are south-oriented with a tilt angle of 35° and without obstructions, the position that receives more radiation at the end of the year. Taking into account that the power of the panel is 165 W-pic and the peak sun hours (PSH) in Barcelona are 4.62 kWh/m^2 day; the theoretical energy generated in 1 year by one module in the indicated position would be 278 kWh/year (or 1002 MJ/year). Therefore, the whole pergola, which is composed of 2686 photovoltaic modules,

Table 51.2 Losses of performance, theoretical and real produced energy and CO_2 emissions saved. (Source: [12])

Year	Performance losses (%)	Energy generated by 1 panel (MJ)	Energy accumulated by 1 panel (MJ)	Energy accumulated by the pergola (GJ)	Emissions avoided by 1 panel ($kgCO_2$)[a]	Emissions avoided accumulate by 1 panel ($kgCO_2$)[a]	Emissions avoided accumulated (tCO_2)[a]
Theoretical	0	1001.7	0	0	109	0	0
1	4.5	956.6	957	2569	104	104	279
5	7.8	923.9	4701	12,627	100	510	1369
10	11.8	883.0	9198	24,706	96	997	2678
15	15.9	842.2	13,491	36,236	91	1462	3928
20	20.0	801.3	17,579	47,217	87	1906	5118

[a] It has been considered the Spanish electric mix during the period 2004–2009

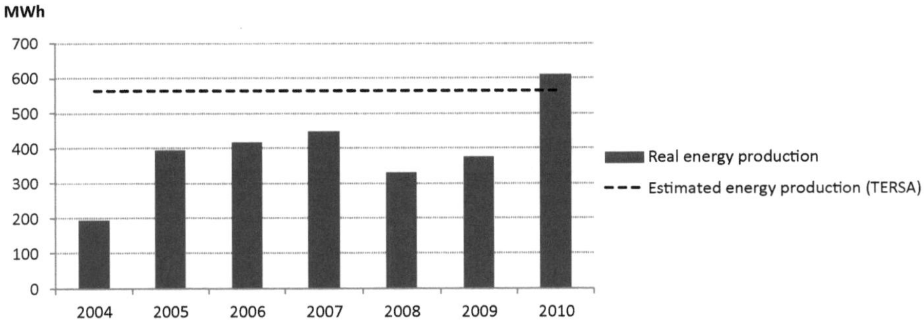

Fig. 51.2 Production of the photovoltaic pergola. (Source: own elaboration with TERSA data)

would have an annual energy production of 2690 GJ. This value is 30 % higher than the production value estimated by Treatment and Waste Selection (TERSA), the current manager of the pergola's energy.

Isofoton states that the performance of the modules decreases approximately about 4 % during the first running hours and about 0.5 % annually during the rest of its life [7]. This company guarantees an 80 % performance during 20 years. After this period of guarantee, the performance can have a wide range of variations depending on the conditions of the installation. One of these conditions is the dust accumulation on the modules, which effect has been studied with models and experiments and arrived at that reduces significantly the performance [9–11].

51.5 Discussion

Taking a look at the energetic contribution and the saved emissions, Table 51.2 exposes that one photovoltaic module can reduce between 800 and 1000 MJ in 1 year and it can avoid between 87 and 109 kg of CO_2. Therefore, if we compare these values with the results of Table 51.1, it is possible to state that in about a year and a half, this module balances the embodied energy—and CO_2 emissions—of its fabrication.

On the other hand, the case of the whole pergola is clearly different. To compensate the embodied energy in its main building materials, it is necessary to have the system running for, at least, 17.9 years; to compensate the CO_2 emissions of these materials fabrication, 14.2 years are necessary. Therefore, only 10 % can be considered net energy and only 27 % can be considered as saved emissions based on the energy that will be produced by the pergola during the first 20 years (the time of guarantee).

In addition, this period can be longer if the real annual contribution of the pergola is taken into account (Fig. 51.2). In the period 2004–2010, the amount of energy expected was produced only during the last year. The differences between the forecasted and real values are due to the combination of two factors: technical problems, and the variation of running conditions from the conditions used by the manufacturer to calculate the performance presented in the data sheet. In total, the energy produced during this period is 39 % lower than that estimated by our own calculations and 24 % lower than that estimated in TERSA's data.

Taking into account the real production between 2004 and 2010 and considering that the average annual future production would be the same, the energy payback time of the pergola would be reached after 29 years and the payback time of the emissions after 23 years.

51.6 Conclusions

This case study is a clear example of the least favourable situation of case A. The embodied energy to produce the support for the photovoltaic panels is much higher than the energy generated by these panels in the first 18 years of their lifespan. Evidently, when the support is already built, with an optimal position for the panel's performance, the system in itself is effective. However, the energetic costs of its construction in the beginning cannot be neglected.

Accordingly, it is necessary to incorporate all the costs in the general view. By doing so, it is possible to appreciate the benefits of integrating the photovoltaic elements in the architecture, even if it means having to reduce some of the optimum parameters defined in an abstract case. The integration should be incorporated in the design criteria as another extra parameter, with the value of reducing the embodied energy cost of the overlapping elements, specially designed to host or support the photovoltaic modules. If the building skin was the direct support of the cells or was an element of the solar system, we could talk about integration.

This could be a good challenge for architects to face a photovoltaic system design: to consider that architectonic integration could be a better option than the optimum nonintegrated solution.

Acknowledgements This paper is supported by the Spanish Ministry under project BIA2013-45597-R.

References

1. Azadian F, Radzi MAM (2013) A general approach toward building integrated photovoltaic systems and its implementation barriers: a review. Renew Sustain Energy Rev 22:527–538
2. Jellea BP, Breivikb C (2012) State-of-the-art building integrated photovoltaics. Energy Procedia 20:68–77
3. Jelle BP, Breivik C (2012) Hilde Drolsum Røkenes. Building integrated photovoltaic products: a state-of-the-art review and future research opportunities. Sol Energy Mater Sol Cell 100:69–96
4. Stylos N, Koroneos C (2014) Carbon footprint of polycrystalline photovoltaic systems. J Clean Prod 64:639e645
5. Beylot A, Payet J, Puech C, Adra N, Jacquin P, Blanc I, Beloin-Saint-Pierre D (2014) Environmental impacts of large-scale grid-connected ground-mounted PV installations. Renew Energy 61:2–6

6. ITec—Institut de Tecnologia de la Construcció de Catalunya (2010) BEDEC PR/PCT. Data Base of Construction Materials. Barcelona, ITeC
7. Isofotón (2002) I-165 Módulo Fotovoltaico. Madrid, Isofotón
8. De Wild-Scholten, Mariska MJ (2013) Energy payback time and carbon footprint of comercial photovoltaic systems. Sol Energy Mater Sol Cell 119(SI):296–305
9. Hegazy AA (2001) Effect of dust accumulation on solar transmittance through glass covers of plate-type collectors. Renew Energy 22:525–540
10. Beattie NS, Moir RS, Chacko C, Buffoni G, Roberts SH, Pearsall NM (2012) Understanding the effects of sand and dust accumulation on photovoltaic modules. Renew Energy 48:448–452
11. Adinoyi MJ, Syed AM (2013) Said. Effect of dust accumulation on the power outputs of solar photovoltaic modules. Renew Energy 60:633–636
12. Pages-Ramon A (2012) Caracterització del sector de la edificació des del punt de vista de les emissions de gasos d'efecte hivernacle. PhD thesis, UPC. [www.tdx.cat]

Part V
ST 2014

Chapter 52
Heat Removal Factor of an Unglazed Photovoltaic Thermal Collector with a Serpentine Tube

M. A. M. Rosli, Kamaruzzman Sopian, Sohif Bin Mat, M. Yusof Sulaiman and E. Salleh

Abstract Heat removal factor (F_R) is a vital parameter in determining the thermal efficiency of a photovoltaic thermal (PVT) system. As a main factor of thermal performance, F_R represents the ratio of the actual heat transfer to the maximum yield of heat transfer. In this study, F_R of an unglazed PVT with serpentine tube collector was determined, with focus on the flat-plate thickness of the flat-plate serpentine tubes. Thermal modeling was used to estimate the overall heat losses of the unglazed PVT. The highest F_R value of 0.88 was obtained in the 0.015-m-thick flat plate, followed by 0.84 in the 0.010-m-thick flat plate. The difference in F_R between the two designs was only 4.54 %, which can be considered within the acceptable range for the flat-plate thickness. This consideration was based on an economic point of view and the handling issues of PVT systems.

Keywords Serpentine · Photovoltaic thermal (PVT) · Flat-plate collector · Unglazed

---

M. A. M. Rosli (✉) · K. Sopian · S. B. Mat · M. Y. Sulaiman · E. Salleh
Solar Energy Research Institute, Faculty of Engineering, Universiti Kebangsaan Malaysia, 43600 Bangi, Selangor, Malaysia
e-mail: afzanizam@utem.edu.my

M. A. M. Rosli
Faculty of Mechanical Engineering, Universiti Teknikal Malaysia Melaka, Hang Tuah Jaya, 76100 Durian Tunggal, Melaka, Malaysia

K. Sopian
e-mail: ksopian@eng.ukm.my

S. B. Mat
e-mail: drsohif@gmail.com

M. Y. Sulaiman
e-mail: myusof.sulaiman@gmail.com

E. Salleh
e-mail: eelsall06@gmail.com

A. Sayigh (ed.), *Renewable Energy in the Service of Mankind Vol II*,
DOI 10.1007/978-3-319-18215-5_52

Nomenclature

Symbols

C_p Fluid specific heat
D_i Tube inside diameter
D_o Tube outside diameter
$h_{f,i}$ Fluid-to-tube heat transfer coefficient
h_i Heat transfer coefficient of insulator
h_o Heat transfer coefficient of glass to ambient
h_T Conductive heat transfer through the tedlar
k Plate thermal conductivity
K_G Conduction of glass
K_i Conduction of insulator
K_T Conduction of tedlar
L_G Thickness of glass
L Length of one serpentine segment
L_i Thickness of insulator
L_T Thickness of tedlar
$\dot{m}$ Fluid mass flow rate
N Number of segments
U_b Overall heat transfer coefficient from water to ambient
U_L Overall heat loss coefficient
U_T Conductive heat transfer coefficient from solar cell to water through tedlar
U_t Conductive heat transfer coefficient from solar cell to ambient through glass cover
U_{tT} An overall heat transfer coefficient from glass to tedlar through solar cell
U_{tW} An overall heat transfer coefficient from glass to water through solar cell and tedlar
V Wind speed
W Distance between tubes

Greek letter

δ Plate thickness

52.1 Introduction

The thermal performance of a photovoltaic thermal (PVT) collector can be determined using three important parameters, namely, the heat removal factor (F_R), the overall losses (U_L), and the effective transmittance absorptance $(\tau\alpha)_{eff}$ that cover the PVT system as follows [1]:

$$\eta_{th} = F_R(\tau\alpha)_{eff} - F_R U_L \left(\frac{T_i - T_a}{G} \right). \tag{52.1}$$

F_R is one of the crucial parameters in determining thermal efficiency [2]. F_R is defined as the ratio of useful energy to the maximum heat transfer. F_R has to be carefully determined because it is a gain factor for both $\tau\alpha$ and U_L. Based on Eq. (52.1), $\tau\alpha$ should be maximum, while U_L should be minimum for a PVT collector system to achieve high thermal efficiency.

F_R is affected by several factors such as material selection, dimensions, geometry, and operational variables. These parameters must be appropriately determined for the engineer to design a highly efficient and inexpensive collector [3, 4].

F_R in various collector geometries has been investigated extensively [5]. One collector geometry type is the serpentine arrangement, the attributes of which are different from that of the conventional tube collector. Consisting of one source of water that flows from the inlet to the outlet, the serpentine arrangement is considered as a complex geometry that needs to be expressed in simple equations. It is typically made of metal materials, such as copper, stainless steel, and aluminum.

The serpentine tube collector has been investigated since the late 1970s [3, 6, 7]. The analytical study agrees well with single-bend cases, and a difference of only 5 % was found for cases with any number of bends [3]. They concluded that the analytical model is applicable to large numbers of tubes, and the results are acceptable with a small error.

The current study aims to analyze the thermal performance of a sheet-and-tube unglazed PVT collector with serpentine tube configuration. The key objective is to determine F_R considering the variation in flat-plate thicknesses of the plate absorber. Moreover, the optimum thickness for flat plates with the given operating parameters is proposed in this study.

52.2 Thermal Modeling

An analytical study was conducted to evaluate the F_R of the PVT system. The energy balance of the system was used to estimate U_L of the unglazed PVT collector. Figure 52.1a, b, and c shows the cross section of the unglazed PVT collector, the balance of heat flow over an element length Δx underneath the tedlar, and the thermal resistance circuit of the PVT system, respectively. The following assumptions were considered to simplify the analysis [8, 9]:

I. The PVT collector is in a quasi-steady state.
II. The transmittance of ethylene vinyl acetate (EVA) is 100 %.
III. An average temperature is assumed through each layer.
IV. The flow of water between the tedlar and insulation material (force mode) is uniform.
V. Heat conduction is one-dimensional.

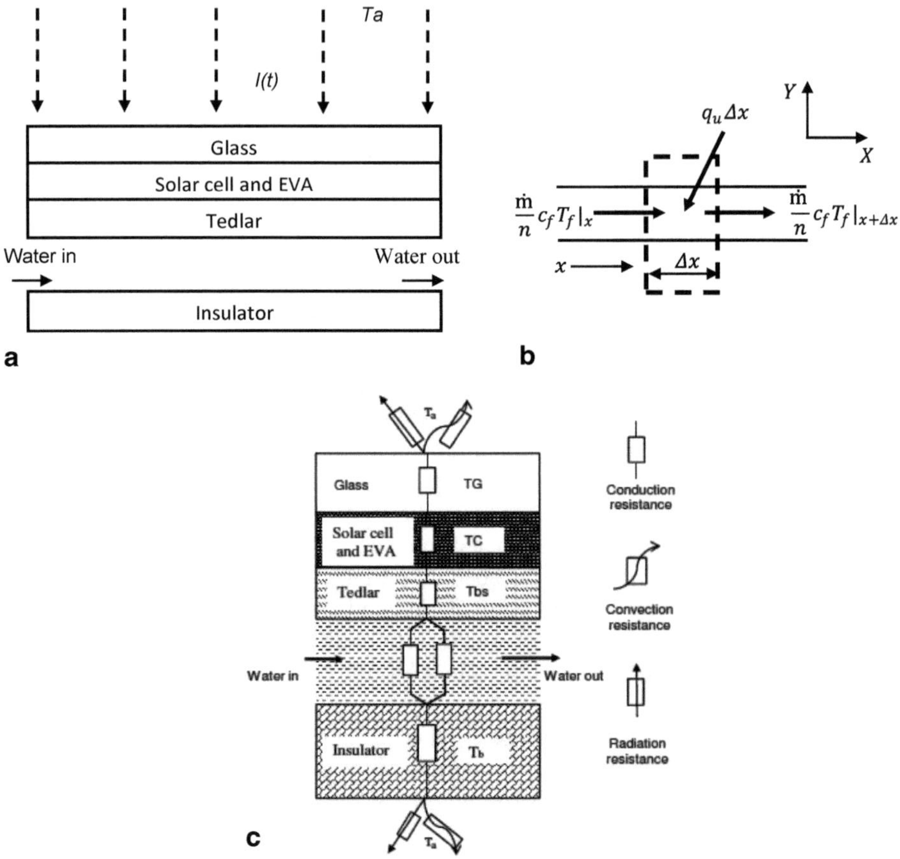

Fig. 52.1 **a** Cross section of the unglazed PVT collector. **b** Heat flow balance over an element Δx length below the tedlar. **c** Thermal resistance circuit of the PVT system

U_L can be expressed as follows [10]:

$$U_L = U_{tw} + U_b. \tag{52.2}$$

Table 52.1 shows the PVT parameters used to estimate U_L based on the thermal energy balance in equation [10].

F_R of the serpentine tube collector was analyzed after obtaining U_L for the PVT system. In this study, bond conductance (C_b) was neglected to simplify the analysis. Figure 52.2a, b and Table 52.2, respectively, show the serpentine tube arrangement (without plate), the schematic diagram of the serpentine tube collector, and the parameters used for the calculations.

The equations below were used for the analysis [3]. Figure 52.3 shows the solution of F_R in terms of three dimensionless parameters known as F_1, F_2, and the dimensionless capacitance rate.

Table 52.1 Parameters for the PVT collector

Parameter	Value	Unit
hi	5.8	W/m² K
ho	9.5	W/m² K
h_T	500	W/m² K
K_G	1	W/m K
Ki	0.035	W/m K
K_T	0.033	W/m K
LG	0.003	m
Li	0.05	m
L_T	0.0005	m
U_b	0.65	W/m² K
U_L	8.63	W/m² K
U_T	66	W/m² K
Ut	9.24	W/m² K
U_{tT}	8.1028	W/m² K
V	1	m/s

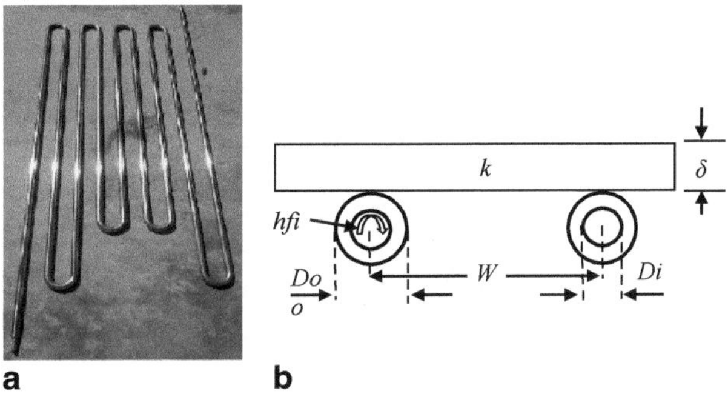

Fig. 52.2 **a** Serpentine tube arrangement (without plate). **b** Schematic diagram of the serpentine tube collector

The parameters F_1 and F_2 can be obtained as follows [3]:

$$F_1 = \frac{K}{U_L W} \frac{KR(1+\gamma)^2 - 1 - \gamma - KR}{[KR(1+\gamma) - 1]^2 - (KR)^2}, \tag{52.3}$$

$$F_2 = \frac{1}{KR(1+\gamma)^2 - 1 - \gamma - KR}, \tag{52.4}$$

where

$$k = \frac{(k\delta U_L)^{1/2}}{\sin h\left[(W-D)\left(U_L/k\delta\right)^{1/2}\right]}, \tag{52.5}$$

Table 52.2 Parameters of the serpentine collector

Parameter	Value	Unit
L	1.2	m
W	0.06	m
N	9	–
δ	0.005, 0.010, 0.015	m
Do	0.015	m
Di	0.009	m
k	211	W/m K
U_L	8.63	W/m² K
$\dot{m}$	0.01	kg/s
Cp	4910	J/kg K
hfi	1000	W/m² K

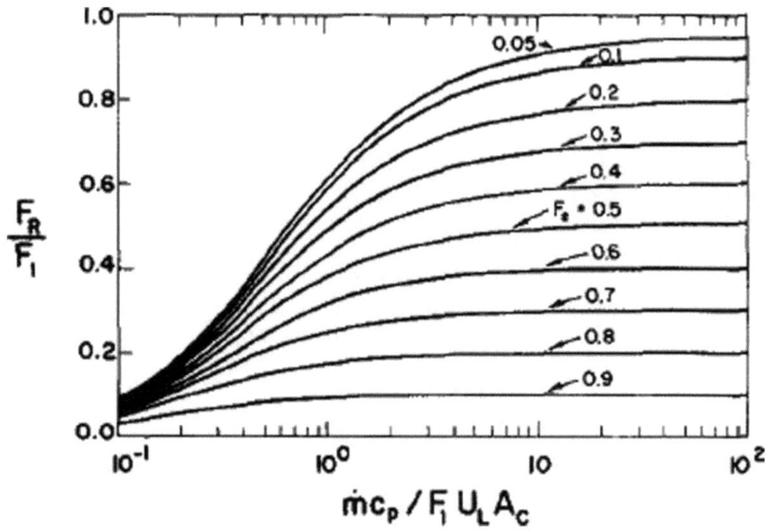

Fig. 52.3 Heat removal factor (F_R) for the serpentine collector plate [3]

$$\gamma = -2\cos h\left[(W - D)\left(\frac{U_L}{k\delta}\right)^{1/2}\right] - \frac{DU_L}{k}, \tag{52.6}$$

$$R = \frac{1}{C_b} + \frac{1}{(\pi D_i h_{f,i})} \tag{52.7}$$

The dimensionless capacitance rate is given by

$$\frac{\dot{m}C_p}{F_1 U_L A_C}, \tag{52.8}$$

Table 52.3 F_R for various flat-plate thicknesses, δ(m)

δ (m)	$\frac{\dot{m}Cp}{F_1 U_L A_C}$	F_2	F_R/F_1	F_R
0.005	0.500	0.944	0.041	0.72
0.010	0.409	0.954	0.039	0.84
0.015	0.378	0.957	0.038	0.88

where the collector area (A_C) is given by

$$A_c = N \times W \times L. \quad (52.9)$$

52.3 Results

The F_R results for the various plate thicknesses are presented in Table 52.3. The highest F_R of 0.88 was obtained using the 0.015-m-thick flat-plate collector, followed by 0.84 and 0.72 for the 0.010- and 0.005-m-thick flat plate collectors, respectively. The flat plates with the lowest and highest thickness had an approximately 18 % difference in F_R. However, F_R of the 0.010-m-thick flat-plate collector and the highest F_R had a difference of only less than 5 %.

52.4 Conclusions

F_R is a crucial factor in determining the thermal efficiency performance of a serpentine collector. The thickness of a flat-plate serpentine collector linearly increases with F_R. Considering the economic aspect and handling issues, thickness values between 0.010 and 0.015 m are considered within the acceptable range, with only a 4.54 % difference in F_R. A thicker flat plate indicates greater cost and weight, which, in turn, leads to handling issues. Thermal modeling and parametric study are important in determining an economically competitive PVT design.

Acknowledgments The authors would like to thank the Solar Energy Research Institute of the Universiti Kebangsaan Malaysia and the Universiti Teknikal Malaysia Melaka for sponsoring this project.

Appendix

The unknown parameters in Eq. (52.2) are derived as follows:

$$U_{tw} = \frac{h_T U_{tT}}{U_{tT} + h_T},\ U_{tT} = \frac{U_T U_t}{U_T + U_t},\ U_t = \left[\frac{L_G}{K_G} + \frac{1}{h_o}\right]^{-1},\ h_T = \left[\frac{L_T}{K_T}\right]^{-1},$$

$$U_b = \left[\frac{L_i}{K_i} + \frac{1}{h_o}\right]^{-1},\ h_o = 5.7 + 3.8V \text{ and } U_T = \frac{K_T}{L_T}.$$

References

1. Florschuetz LW (1979) Extension of the Hottel-Whillier model to the analysis of combined photovoltaic/thermal flat plate collectors. Sol Energy 22:361–366
2. Zondag H (2008) Flat-plate PV-thermal collectors and systems: a review. Renew Sustain Energy Rev 12(4):891–959
3. Abdel-Khalik SI (1976) Heat removal factor for a flat-plate solar collector with serpentine tube. Sol Energy 18:59–64
4. Rosli MAM, Mat S, Ruslan H, Sopian K (2013) Parametric study on water based photovoltaic thermal collector. In: 7th international conference on renewable energy sources (RES'13), 8th edn., pp 135–140
5. William JAD, Beckman A (1980) Solar engineering of thermal processes. Sol Eng Therm Process 44:197. (Wiley, New York)
6. He-Fei Zhang ZL (1985) Thermal performance of a serpentine absorber plate. Sol Energy 34(2):175–177
7. Akgiln MA (1988) Heat removal factor for a serpentine absorber plate. Sol Energy 41(1):109–111
8. Dubey S, Tiwari GN (2008) Thermal modeling of a combined system of photovoltaic thermal (PV/T) solar water heater. Sol Energy 82(7):602–612
9. Dubey S, Tiwari GN (2010) Fundamentals of photovoltaic modules and their applications. RSC Publishing, Cambridge
10. Tiwari A, Sodha MS (2006) Performance evaluation of hybrid PV/thermal water/air heating system: a parametric study. Renew Energy 31(15):2460–2474

Chapter 53
An Experimental Investigation of Solar-Assisted Heat Pump Combined with A Latent Heat Storage

Devrim Aydın and Zafer Utlu

Abstract The building sector is one of the leading sectors in energy consumption. For this reason, utilizing renewable energy sources in building air conditioning systems has significant importance. Taking advantage of using heat storage systems integrated with renewable energy-sourced air conditioning systems makes it possible to benefit from cyclic renewable sources, such as solar energy, for longer periods of time. In this way, increasing solar fraction and decreasing dependency on fossil fuels in heating systems will be possible.

In this study we present an experimental analysis of a hybrid heating system in Istanbul. Analyses are performed according to the first and second law of thermodynamics using real data obtained from a prototype structure built as part of a project. The main components of the system are solar collectors, latent heat storage (LHS) as a low-temperature heat source, and a heat pump. Furthermore, a buffer tank placed in the system provides a continuous supply of hot water to a wall heating cycle. By storing input and output heat from the heat pump, a significant reduction in consumption of compressor energy was achieved.

During the testing period an average of 0.895 kW energy was gained from solar collectors, of which 0.77 kW was stored in LHS. As a result, the charging time (average 402 min) of LHS was far higher than the discharging time (97.8 min). A large amount of heat was transferred to the heat pump and average coefficient of performance (COP) of the heat pump was calculated as 4.38 for the whole testing period.

On average, 2.42 kW of exergy destruction took place during the experiment. Solar collectors and heat pumps are promising components in terms of exergy destruction; showing 1.15 and 1.09 kW, respectively, in our study. A high percent of heat loss took place from the heat pump (1.83 kW), followed by the accumulator-

D. Aydın (✉)
Institute of Sustainable Energy Technology,
University of Nottingham, Nottingham NG7 2RD, UK
e-mail: ezxda4@nottingham.ac.uk

Z. Utlu
Engineering Faculty, Mechanical Engineering Department,
Istanbul Aydin University, Florya, Istanbul, Turkey
e-mail: zaferutlu@aydin.edu.tr

A. Sayigh (ed.), *Renewable Energy in the Service of Mankind Vol II*,
DOI 10.1007/978-3-319-18215-5_53

wall heating cycle (0.42 kW). On the contrary, the highest breakdown of exergy loss occurred in the accumulator-wall heating cycle (0.28 kW).

The discharging energy efficiency of LHS is high throughout all system components. Furthermore, the heat pump showed a significant efficiency of 78 %. The exergy efficiency of system components investigated during the discharging period was similar (approximately 32 %). However, the efficiency of solar collectors and charging of LHS was 2.3 and 7 %, which is relatively low.

Our study results indicate that the performance of solar collectors and LHS units in terms of energy is adequate; however, the system seems open to improvement in terms of exergy. Increasing the total surface area of solar collectors can contribute to an improvement of system performance. In this prototype, the total area of the solar collectors was 3.24 m^2. As a result of simulations, if the collector area is increased to 10 m^2, a significant amount of excess storable energy will be gained which will increase solar fraction in winter months.

Keywords Latent heat storage · Heat pump · Solar collectors · Renewable energy

Nomenclature

A	Area
c	Specific heat (kJ/kgK)
$\dot{E}$	Energy (W, kW)
$\dot{E}x$	Exergy (W, kW)
h	Enthalpy (kJ/kg)
I	Solar radiation (W/m^2)
$\dot{m}$	Flow rate (kg/s)
T	Temperature (°C, K)
W	Work (W, kW)
LHS	Latent heat storage
SHS	Sensible heat storage
HP	Heat pump
AT	Accumulator tank
WH	Wall heating system
HRES	Hybrid renewable energy system
PCM	Phase-change material
GSHP	Ground source heat pmp
YREB	Yildiz renewable energy building

Greek Symbols

Q	Thermal energy (W, kW)
Ψ	Exergy to energy ratio for radiation
η	Efficiency
Δ	Difference

Indices

I	Input
O	Output
s	Stored
a	Accumulated
d	Destruction
l	Loss
s,r	Solar irradation
k	Heat source
c	Collector
t	Total
n	Net
0	Ambient conditions
p	Pump
sp	Supply
g	Generation
comp	Compressor
tr	Transfer

53.1 Introduction

High fossil fuel prices and greenhouse emissions are the driving force behind renewable energy technology research [1]. For most people throughout the world, solar energy is considered as a primary energy source for the future. Whilst research about new and renewable energy resources is vital, improving energy storage systems is as important as benefitting from new energy resources. Research is continuing to find new and renewable energy resources. Furthermore, energy storage systems are as important as using new energy resources. Storage of energy in an appropriate way and developing systems to become available in the case of necessity is the key misson of today's researchers. Storing energy in a suitable way not only provides equilibrium between supply and demand but also increases the performance and utility of energy systems. Energy storage systems enable economical usage of fuels. As a result, they reduce the amount of waste energy. In this way, these systems improve the performance of energy systems in an economical way [2].

53.2 Literature Review

Over the past decade significant improvements in renewable energy technologies have increased utility of renewable sources. However, unstable heat loads from buildings and variable heat generation from renewable sources seem to be an obstacle to benefitting from these resources. Latent heat storage (LHS) is a promising technology for dealing with the fluctuations of heat load and storing the excess energy. Research is underway regarding usage of heat storage systems combined with solar energy and heat pumps.

There are several numerical and experimental studies about the storage of solar energy in heat form for heating applications. Ucar and Inalli [3], Koca et al. [4], Öztürk [5], Alkilani et al. [6], Fabrizio [7], Kousksou et al. [8], Aghbalou et al. [9], Kenisarin et al. [10], Hussain et al. [11] and Argiriou et al. [12] developed various designs for storing solar energy for long or short time periods.

Deng et al. [13] carried out experimental research on seasonal heat storage coupled with solar-assisted heat pumps. In another study, Yumurtas and Unsal [14] modelled and analytically investigated underground heat storage integrated with solar-assisted heat pumps. Benli and Durmus [15] used a ground source heat pump (GSHP)–LHS combination for experimental greenhouse heating applications. Han et al. [16] developed a numerical model for a hybrid system consisting of LHS, solar collectors and a GSHP in China. Wang et al. [17] experimentally investigated a solar GSHP system where energy was stored underground. Fernandez-Seara et al. [18] experimentally tested the heat storage performance of a 300 L solar home system (SHS) combined with a heat pump and solar collectors. Esen [19] carried out similar research for LHS and analyzed heat pump performance and temperature distribution inside a heat storage tank. Wu et al. [20] investigated heat storage materials and reported that a concentration of solutions and composites had a significant effect on heat storage potential. Long et al. [21] carried out an experimental and numerical study about usage of LHS in solar-assisted heat pump systems. Zhou et al. [22] used a thermoeconomic approach for solar energy-sourced LHS systems. On the other hand, Henchoz et al. [23], Sharma et al. [24] and Zalba et al. [25] presented comprehensive reviews related to the classification of phase-change materials (PCMs) and usage of them in air conditioning applications. Dincer et al. [26] carried out a similar study for SHS materials and analysed SHS system methods. Jagedheeswaran et al. [27] demonstrated exergy analysis methods for investigating the second law of thermodynamics of LHS systems. An experimental and numerical comparison between sensible, thermochemical and LHS systems in terms of energy–exergy performance, environmental impact and sustainability was presented by Çalışkan et al. [28] and Dinçer [29]. Furthermore, Ezan et al. [30] experimentally analysed the first and second law of thermodynamic efficiencies of LHS. Erek et al. [31] investigated entropy enhancement in cylindrical LHS. Pinel et al. [32] Conducted research relevant to heat storage methods and heat storage materials. Bedecarrats et al. [33] analysed the charging and discharging performances of a solar-sourced LHS consisting of capsulized PCMs. A similar purpose study was performed by Benli [34] for greenhouse heating using GSHP as a heat source.

53.3 System Description

53.3.1 Yildiz Renewable Energy Building

In this study, experimental heating system of Yildiz Renewable Energy Building (YREB) situated at Davutpasa Campus was investigated, Within this study, experimental energy and exergy analysis of system components has been presented. A diagram of the entire renewable energy-sourced heating system used for meeting the heating demand of YREB is shown in Fig. 53.1.

In the experimental setup, data was recorded instantaneously with sensors and data loggers installed on the system. In the system, two flat plate solar collectors each with a surface area of 1.62 m^2 was used to supply energy to the LHS unit during the charging period. The solar collectors were connected to the LHS serially and energy gained through the collectors was transferred to the LHS within a closed cycle. The thermal energy transferred from the collectors to the LHS had a theoretical efficiency of 85 %. The flow of heat transfer fluid (HTF) is provided with a 100-W pump which is controlled with an on–off control system. The mode of the pump, thus, the working conditions of the cycle, depends on the conditions given in inequalities Eqs. (53.1) and (53.2).

$$T_i > T_{LHS} + 3^{\circ}C \quad \text{On} \tag{53.1}$$

$$T_i \leq T_{LHS} + 3^{\circ}C \quad \text{Off.} \tag{53.2}$$

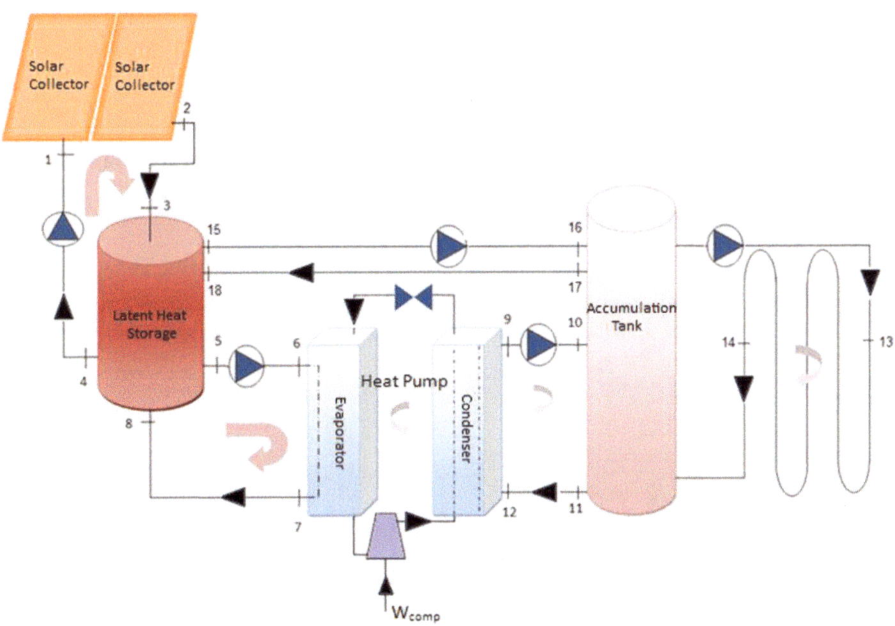

Fig. 53.1 Heating system diagram

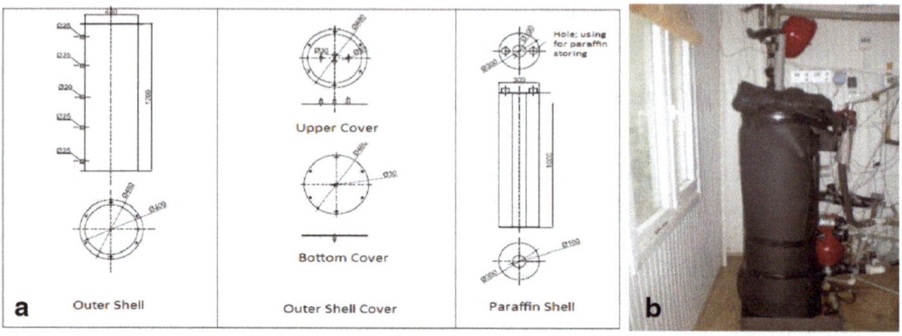

Fig. 53.2 **a** Technical drawing of LHS *(left)*. **b** Insulation *(right)*

53.3.2 Latent Heat Storage

The LHS used in the system is made of steel and is painted black to increase heat gain. The LHS is the system component where the collected energy is stored. During the experiment, paraffin which melts between 42 and 44 °C and easy to find in Turkey was used. The paraffin was stored in a shell with a volume of 0.063 m^3 and the paraffin shell was located in an outer shell with a volume of 0.22 m^3. HTF circulates inside the gap between the inner and outer shells. There is also a 0.1-m diameter gap in the middle of the inner shell to increase the heat transfer surface area. A technical drawing of the LHS and the insulation are shown in Fig. 53.2a and b, respectively.

53.3.3 Methodology

Measured temperatures of various parts belonging to system components as well as the environmental conditions can be viewed via the computer software DALI (Data Acquisition and Logging Interface) and Telecontrol. These softwares have a large databases allowing long-term records. The interfaces used for DALI and Telecontrol programs are shown in Fig. 53.3a and b.

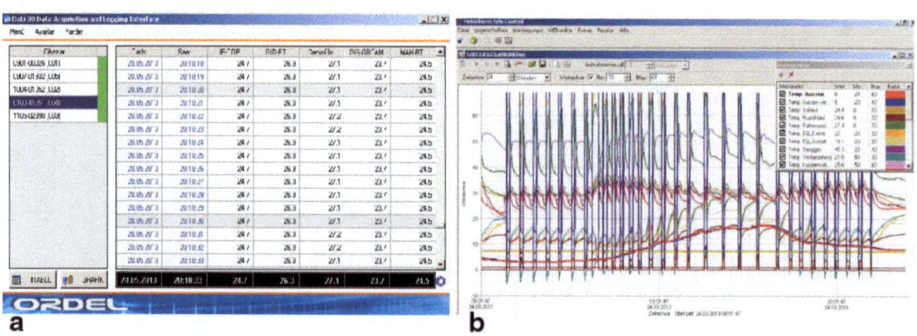

Fig. 53.3 **a** DALI interface *(left)*. **b** Telecontrol interface *(right)*

53.4 Thermodynamic Analysis

53.4.1 Charging

During the experimental period average daily solar radiation varied between 84 and 832 W/m^2. Therefore, the energy gain of solar collectors ranged from 0.183 kW to 1.839 kW. The amount of exergy 'fall down' to the surface of the collectors and the amount of exergy gain from the collectors was in the range of 256–2520 W and 3.4–72.1 W, respectively.

The efficiency of solar collectors with and without considering pump work is represented as total and net efficiency. The second law of thermodynamic efficiencies varied between 1 and 4 % due to exergy destruction while the first law of thermodynamic efficiencies were in the range of 50–80 %.

The average energy gained from the collectors (0.895 kW) was transferred to LHS (0.770 kW) with 86 % efficiency. Therefore, heat loss through the piping to the environment was calculated as 0.125 kW. Heat loss in LHS was low and the high amount of input energy stored was in the range of 0.18−1.5 kW. Energy–exergy balances and efficiencies of LHS for the charging period are shown in Eqs. (53.3), (53.4) and (53.5), (53.6)

$$\dot{E}_3 = \dot{E}_4 + \dot{Q}_l + \dot{E}_S \tag{53.3}$$

$$\dot{m}_{HTF} \cdot c_{HTF} \cdot \left[(T_3 - T_4) - T_0 \cdot \ln\left(\frac{T_3}{T_4}\right) \right] = \left[\dot{m}_{HTF} \cdot c_{HTF} \cdot (T_3 - T_4)\right] \cdot \left[1 - \left(\frac{T_0}{T_{PCM}}\right)\right] + \dot{Q}_{loss}\left[1 - \left(\frac{T_0}{T_{PCM}}\right)\right] + \dot{E}X_d \tag{53.4}$$

$$\eta_{I_{net}} = \frac{E_s}{E_3 - E_4 + W_p} \tag{53.5}$$

$$\eta_{II_n} = \frac{\dot{Q}_{PCM}\left(1 - \frac{T_0}{T_{PCM}}\right) - \dot{Q}_{loss}\left[1 - \left(\frac{T_0}{T_{PCM}}\right)\right]}{\dot{m}_{HTF} \cdot c_{HTF} \cdot \left[(T_3 - T_4) - T_0 . \ln\left(\frac{T_3}{T_4}\right)\right] + \dot{W}_P} \tag{53.6}$$

On the other hand, 27 W exergy was transferred to LHS with 78.2 % efficiency, where a total of 34.5 W was gained from the collectors.

53.4.2 Discharging

The average stored energy in the heat storage tank varied between 0.8 and 1.5 kW; however, absorbed energy from LHS was in the range of 6.2–8.8 kW. On the contrary, the charging time (average 402 min) was far longer than the discharging time (average 97.8 min.). This condition reflects the function of LHS on storing low-density heat and supplying high-density heat when the heating demand of the building increases. Hence, LHS has an important role in system performance. The average ambient temperature was measured as 284.3 K during the experimental period. Energy–exergy balances and efficiencies of LHS for the discharging period are given in Eqs. (53.7), (53.8) and (53.9), (53.10)

$$\dot{m}_{HTF}\, c_{HTF}\left(T_5 - T_8\right) = \dot{E}_{sp} - \dot{Q}_l \tag{53.7}$$

$$\begin{aligned} &\dot{m}_{HTF} \cdot c_{HTF} \cdot \left(T_5 - T_8\right) \cdot \left[1 - \left(\frac{T_0}{T_{PCM}}\right)\right] - \dot{Q}_l \cdot \left[1 - \left(\frac{T_0}{T_{PCM}}\right)\right] \\ &= \dot{m}_{HTF} \cdot c_{HTF} \cdot \left[\left(T_5 - T_8\right) - T_0 . \mathrm{In}\left(\frac{T_5}{T_8}\right) + T_0 . \dot{S}_g\right] \end{aligned} \tag{53.8}$$

$$\eta_{I_n} = \frac{\dot{E}_5 - \dot{E}_8}{\dot{E}_{sp} + \dot{W}_p} \tag{53.9}$$

$$\eta_{II_n} = \frac{\dot{E}x_5 - \dot{E}x_8}{\dot{E}x_{sp} + \dot{W}_p}. \tag{53.10}$$

The average input and output exergy to and from the heat pump is 0.07 and 0.56 kW. On the other hand, exergy destruction varied between 1 and 1.27 kW with the highest amount taking place in the compressor. Moreover, the COP value of the heat pump was 4.38 on average. Analysis results showed that the heat pump has a high performance in terms of efficiency and COP. Input and output exergy to the accumulator-wall heating system were in the range of 0.35–0.77 and 0.12–0.37, respectively. During the experimental period, the average daily stored exergy in the accumulator showed a minimum of 0.17 kW and maximum of 0.31 kW.

53.4.3 Overall Efficiency of LHS

According to Jagedheeswaran et al., for LHS units charged and discharged in different time frames, the overall efficiency of LHS is obtained by multiplying the charging and discharging efficiencies [27]. Regarding this study, the average overall total energy and exergy efficiencies of LHS were calculated as 72 and 8.4%,

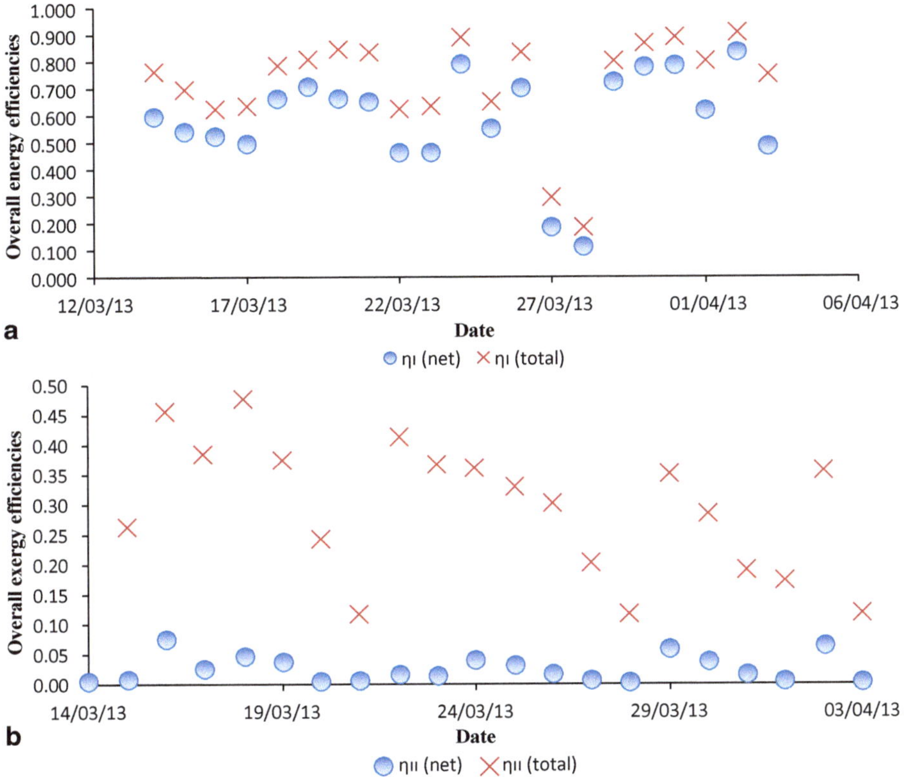

Fig. 53.4 **a** Daily average overall energy efficiencies of LHS. **b** Daily average overall exergy efficiencies of LHS

respectively. When considering pump energy consumption, the efficiencies fell to 58 and 2.3 %, respectively (Fig. 53.4a and b).

53.4.4 Comparison of Control Volumes

A comparison of heat loss, exergy loss, exergy destruction and efficiencies of heating system components is shown in Table 53.1.

According to a comparison shown in Table 53.1;

- A high percent of heat loss took place in the heat pump (1.83 kW), followed by the accumulator-wall heating cycle (0.42 kW).
- On the contrary, the highest breakdown of exergy loss occurred in the accumulator-wall heating cycle (0.28 kW).
- An average of 2.42 kW exergy destruction took place during the experiment. The solar collectors and heat pump are promising components in terms of exergy destruction at 1.15 and 1.09 kW, respectively.

Table 53.1 Performance criteria of control volumes

Control volume	Q_{loss}(kW)	Ex_{loss}(kW)	$Ex_{destruction}$(kW)	ηI (%)	ηII (%)
Solar collectors	0.212	0.014	1.150	0.630	0.023
LHS (charging)	0.080	0.002	0.017	0.620	0.070
LHS (discharging)	0.121	0.003	0.014	0.930	0.320
HP	1.837	0.095	1.096	0.786	0.319
Accumulator-water heating cycle	0.425	0.285	0.143	0.429	0.315

With regard to control volume efficiencies shown in Table 53.1;

- The discharging energy efficiency of LHS is high throughout all system components. Furthermore, the heat pump had a promising efficiency of 78 %, as shown in the figure.
- The exergy efficiency of the system components investigated during the discharging period were close (approximately 32 %). However, the efficiency of solar collectors and charging of LHS were 2.3 and 7 %, respectively, which are relatively low.

53.5 Results and Discussion

According to the experimental analysis results:
In the charging process

- 0.690 kW of 0.895 kW energy gained by the collectors was stored in LHS. Meanwhile, 0.10 kW pump work was given to the system.
- Similarly, only 0.01 kW of 0.034 collected exergy was stored in LHS.

In the discharging process

- 3.17 kW of 7.69 kW energy which was gained from LHS was transferred to the building and during operating times averaged 3.86 kW energy, collected in the accumulation tank. Stored energy in the accumulator enables low energy consumption by the heat pump when solar energy is unavailable. The accumulator tank is an important stability element of the system. However, decreasing its size will have a positive effect on the system performance.
- From an average of 0.075 kW exergy gained from LHS, 0.207 kW was transferred to the building. Meanwhile, 0.36 kW pump work and 1.679 kW compressor work was given to the system.

For the charging process, the energy and exergy efficiencies were determined as 0.68 and 0.53, respectively. For the discharging process, the values were calculated as 0.78 and 0.29, respectively. As can be seen, energy efficiencies are much higher than exergy efficiencies. This is because the exergy gained from solar collectors is low and exergy destruction in the heat pump and accumulator-wall heating cycle are high.

The functions of system components can be summarized as follows:

- Solar collectors provide heat gain to the system when solar energy is available.
- LHS provides energy to the system when solar energy is unavailable.
- The heat pump increases the exergy of thermal energy.
- The accumulator provides heat supply balance by storing excess energy.

Using these components together increases utility of the heating system as well as enabling the heating demands of the building to be met without the need for a fossil fuel heating system.

In conventional air-sourced heat pump systems, because the heat source (air) is unstable and has a relatively low temperature, the energy consumption of the compressor is high. Furthermore, variable air temperatures make design conditions difficult. On the other hand, benefitting from ground heat brings some advantages such as more stable system design, lower energy consumption and higher COP values. However, high space requirement and investment costs are the main drawbacks of GSHPs.

Compared with the other heat sources mentioned above, our study results show that utilizing LHS as a heat source in heat pump applications brings significant advantages. Due to the existence of LHS unit and accumulator tank, both the working time of the heat pump, as well as the power requirement of the compressor falls. Hence, electricity consumption of the compressor is very low. LHS works as an energy supplier for the evaporator. As a result of a significant amount of heat gain by the heat pump from LHS, compressor energy consumption decreases substantially. Furthermore, well-insulated accumulator tank store the hot water and the heat pump only works when the water temperature in the accumulator tank is below the setting value. This control mechanism decreases the working time of the heat pump, as well as providing a more stable working condition. Consequently, when compared with different energy-sourced heat pump systems, utilization of LHS and an accumulator tank, improves efficiency and decreases fossil fuel consumption by enhancing solar fraction in renewable-sourced heating systems.

Solar energy is a compatible energy source when combined with LHS systems. Usage of PCMs melting in the range of 35–40 °C will allow solar energy to be stored more efficiently, as well as making phase-change processes possible on cloudy days while solar radiation intensity is low.

Even though some research has been carried out, analysing the performance of LHS systems in different working conditions, together with more detailed analysis of these systems, will be the driving force to benefitting from cyclic renewable energy sources for longer periods of time and more efficiently in space heating applications. Considering that the heating sector has a significant breakdown in energy consumption, this will be the key for decreasing fossil fuel consumption in the near future.

References

1. Chiasson A-D, Spitler J-D, Rees S-J, Smith M-D-A (2000) Model simulating the performance of a pavement heating system supplemental heat rejecter with closed-loop ground-source pump systems. J Sol Energy Eng 122:183–191
2. Hepbasli A, Akdemir O (2004) Energy and exergy analysis of a ground source (geothermal) heat pump system. Energy Convers Manage 45:737–753
3. Ucar A, İnalli M (2008) Thermal and economic comparisons of solar heating systems with seasonal storage used in building heating. Renew Energy 33:2532–2539
4. Koca A, Oztop F-H, Koyun T, Varol Y (2008) Energy and exergy analysis of a latent heat storage system with phase change material for a solar collector. Renew Energy 33:567–574
5. Ozturk H-H (2005) Experimental evaluation of energy and exergy efficiency of a seasonal latent heat storage system for greenhouse heating. Energy Convers Manage 46:1523–1542
6. Alkilani M-M, Sopian K, Alghoul A-M, Sohif M (2011) Review of solar air collectors with thermal storage units. Renew Sustain Energy Rev 15:1476–1490
7. Fabrizio E (2012) Energy reduction measures in agricultural greenhouses heating: envelope, systems and solar energy collection. Energy Build 53:57–63
8. KousksouT, Strub F, Lasvignottes C-J, Jamil A, Bedecarrats P-J (2007) Second law analysis of latent thermal storage for solar system. Sol Energy Mater Sol Cells 91:1275–1281
9. Aghbalou F, Badia F, Illa J (2006) Exergetic optimization of solar collector and thermal energy storage system. Heat Mass Transf 49:1255–1263
10. Kenisarin M, Mahkamov K (2007) Solar energy storage using phase change materials. Renew Sustain Energy Rev 11:1913–1965
11. Hussain M-M Dincer İ, Zubair M-S (2004) A feasibility study of using thermal energy storage in a conventional air conditioning system. Int J Energy Res 28:1–13
12. Argiriou A, Klkitsikas N, Balaras A-C, Asimakloupos N-D (1997) Active solar space heating of residential buildings in Northern Hellas–a case study. Energy Build 26:215–221
13. Qi Q, Deng S, Jiang Y (2008) A simulation study on a solar heat pump heating system with seasonal latent heat storage. Sol Energy 82:669–675
14. Yumurtas R, Unsal M (2012) Energy analysis and modeling of a solar assisted house heating system with a heat pump and an underground energy storage tank. Sol Energy 86:983–993
15. Benli H, Durmus A (2009) Evaluaiton of ground source heat pump combined latent heat storage system performance in greenhouse heating. Energy Build 41:220–228
16. Han Z, Zheng M, Kong F, Wang F, Li Z, Bai T (2008) Numerical simulation of solar assisted ground-source heat pump heating system with latent heat energy storage in severely cold area. Appl Therm Eng 28:1427–1436
17. Wang H, Qi C, Wang E, Zhao J (2009) A case study of underground thermal storage in a solar ground coupled heat pump system for residental buildings. Renew Energy 34:307–314
18. Fernandez -S.J, Pineiro C, Dopazo A-J, Fernandes F, Sousa B-X-P (2012) Experimental analysis of a direct expansion solar assisted heat pump with integral storage tank for domestic water heating under zero solar radiation conditions. Energy Convers Manage 59:1–8
19. Esen M (2000) Thermal performance of a solar–aided latent heat storage for space heating by heat pump. Sol Energy 69:15–25
20. Wu H, Wang S, Zhu D (2007) Effects of impregnating variables on dynamic sorption characteristics and storage properties of composite sorbent for solar heat storage. Sol Energy 81:864–871
21. Long Y-J, Zhu S-D (2009) Numerical and experimental study on heat pump water heater with PCM for thermal storage. Energy Build 40:666–672
22. Zhou D, Zhao Y-C, Tian Y (2012) Review on thermal energy storage with phase change materials in building applications. Appl Energy 92:593–605
23. Henchoz S, Buchter F, Fevrat D, Morandin M, Mercangöz M (2012) Thermoeconomic analysis of a solar enhanced energy storage concept based on thermodynamic cycles. Energy 45:358–365

24. Sharma A, Tyagi V-V, Chen R-C, Buddhi D (2009) Review on thermal energy storage with phase change materials and applications. Renew Sustain Energy Rev 13(318):345
25. Zalba B, Marin M-J, Cabeza F-L, Mehling H (2003) Review on thermal energy storage with phase change: materials, heat transfer analysis and applications. Appl Therm Eng 23:251–283
26. Dincer I, Dost S, Li X (1997) Performance analysis of sensible heat storage systems for thermal applications. Int J Energy Res 21:1157–1171
27. Jagedheeswaran S, Pohekar D-S, Kousksou T (2010) Exergy based performance evaluation of latent heat thermal storage system: a review. Renew Sustain Energy Rev 14:2580–2595
28. Çalışkan H, Dincer I, Hepbasli A (2012) Thermodynamic analysis and assessments of various thermal energy storage systems for buildings. Energy Convers Manage 62:109–122
29. Dincer I (2002) Thermal energy storage systems as a key technology in energy conversion. Int J Energy Res 26:567–588
30. Ezan A-M, Ozdoğan M, Gunerhan H, Erek A, Hepbasli A (2010) Energetic and exergetic analysis and assessment of a thermal energy storage (TES) unit for building applications. Energy Build 42:1896–2001
31. Erek A, Dincer I (2008) An approach to entrophy analysis of a latent heat storage modul. Int J Therm Sci 47:1077–1085
32. Pinel P, Cruickshank A-C, Morrison B-I, Wills A (2011) A review of available methods for seasonal storage of solar thermal energy in residential applications. Renew Sustain Energy Rev 15:3341–3359
33. Bedecarrats P-J, Lesvignottes C-J, Strub F, Dumas P-J (2009) Study of a phase change energy storage using spherical capsules. Part 1: experimental results. Energy Convers Manage 50:2527–2536
34. Benli H (2011) Energetic performance analysis of a ground source heat pump system with latent heat storage for a greenhouse heating. Energy Conver Manage 52:581–589

Chapter 54
Experimental Study of Modified Absorption Cooling Systems by Adding Ejector–Flash Tank Unit

Ranj Sirwan, Kamaruzzman Sopian and Mohammed Al-Ghoul

Abstract In this study, an absorption cooling system with and without ejector–flash tank units is proposed by testing an experimental rig at Universiti Kebangsaan Malaysia in Bangi, Malaysia. A combined ejector–flash tank unit with absorption cooling system has been designed and installed. The major components of the system consist of the generator, ejector, condenser, flash tank, evaporator, absorber, and solution heat exchanger. Experimental results show that the COP of the absorption cooling systems with and without an ejector–flash tank unit was, respectively, 0.87 and 0.64 under the same operating conditions. In addition, the cooling effect in the former also increased, which may be caused by the improvement in the quality of ammonia entering the evaporator. This study revealed the possibility of optimizing the performance of absorption cooling systems through the addition of ejector–flash tank units.

Keywords Absorption cooling system · With and without adding ejector–flash tank unit · Experimental performance

Nomenclature

Symbols

COP	Coefficient of performance
C_p	Specific heat at constant pressure ($kJ.kg^{-1}.K^{-1}$)
m	Mass flow rate (kg/s)
Q	Heat rate (kW)
T	Temperature (°C)
W	Work rate (kW)

R. Sirwan (✉)
Refrigeration and Air Conditioning Engineering Department, Erbil Technical Engineering College, Erbil Polytechnic University, Erbil, Iraq
e-mail: ranjsirwan@gmail.com

K. Sopian · M. Al-Ghoul
Solar Energy Research Institute, Faculty of Engineering, Universiti Kebangsaan Malaysia, 43600 Bangi, Selangor, Malaysia

A. Sayigh (ed.), *Renewable Energy in the Service of Mankind Vol II*,
DOI 10.1007/978-3-319-18215-5_54

Subscripts

abs	Absorber
acwi	Absorber cooling water in
acwo	Absorber cooling water out
cond	Condenser
ccwi	Condenser cooling water in
ccwo	Condenser cooling water out
chwi	Chilled water in
chwo	Chilled water out
cw	Cooling water
chw	Chilled water
evp	Evaporator
gen	Generator
ghotwi	Generator hot water in
ghotwo	Generator hot water out
hotw	Hot water
p	Pump

54.1 Introduction

Because of energy crisis and environmental pollution, absorption refrigeration system is recommended strongly these days. The optimization of absorption cooling systems remains a challenging task because of the poor coefficient of performance (COP) of this cooling system [1]. Previous theoretical and experimental investigations focused on improving the performance of absorption cooling systems by using different working fluids, cycles, and technologies. Those studies found that the overall performance of the basic cycle ranges from 0.4 to 0.7 [2–6]. Other studies have improved the basic single-stage absorption cycle by adding a generator–absorber heat exchanger. The COPs of these cycles could increase up to 40 % [7–15]. Other studies have also introduced multi-pressure absorption cooling cycles, such as the multi-effect cycle. The multi-effect cycle can generate cooling effect and can obtain high COP that ranges from 0.8 to 1.4 [16–22]. The addition of an ejector to the absorption system results in potential COP improvements [23]. According to –previous studies, the COP of ejector–absorption systems increased by 20–49 % compared with that of basic absorption systems [1]. The increase in COP in the ejector–absorption system is caused by the ability of the system to work under triple pressure levels [24]. Many approaches have been studied regarding the configuration of combined ejector–absorption refrigeration cycles [23, 25–29]. For instance, Sirwan et al. [30] conducted a mathematical performance study by using the first law of thermodynamics to evaluate the effect of adding a flash tank between the condenser and evaporator of an absorption–ejector system. Furthermore, Sirwan et al. [31] applied a second low analysis to evaluate the effect of adding flash tank to the system performance. The authors conducted an exergy analysis of a modified absorption–ejector cooling system with a flash tank. Results show that the COP and exergetic efficiency (COP_{exe}) values of the modified cycle have been improved. Adding a flash tank to the combined system can be considered a novel system enhancement. These results encouraged the authors to perform an experimental study to test the performance of the absorption cooling system before and after modification by using ammonia–water solution as a working fluid.

54.2 Experimental Setup

54.2.1 Description of the System

An existing absorption cooling system with 5 kW cooling capacity [32] that is based on available commercial components was modified by adding a flash tank between the condenser and the evaporator. Figure 54.1 describes the schematic diagram of the absorption cooling system after adding the ejector–flash tank unit. In this system, ammonia was used as refrigerant, whereas water was used as an absorbent (ammonium solution). The absorption cooling system in this study comprised four cycles: hot water cycle (solar thermal collector), cooling water cycle, chilled water

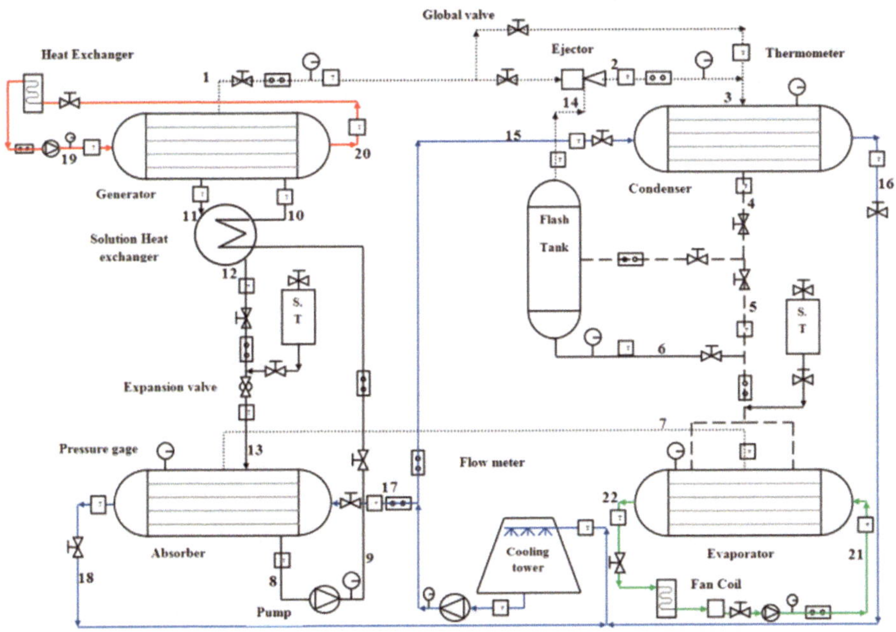

Fig. 54.1 Schematic diagram of the existing combined ejector–flash tank–absorption system after modification

Fig. 54.2 Modified combined ejector–absorption cooling system

cycle, and solution cycle. The system included a generator, an ejector, a condenser, a flash tank, an evaporator, an absorber, a solution heat exchanger, pumps, and expansion valves. The flash tank was fabricated according to the specific design capacity of existing system components. The flash tank vessel was designed to work under flexible operating conditions [33]. Figure 54.2 shows the modified experimental test rig on site.

54.2.2 Instrumentation Description

As shown in the schematic diagram in Fig. 54.1, complete instrumentation equipment was installed and calibrated to determine the heat balance and performance of the modified combined ejector–absorption system. The following instruments were used in this study:

- Temperature sensors were connected to the data acquisition system
- Pressure gauges were used to detect pressure variations at different system levels
- Flow meters were placed to measure the liquid and vapor flow rates in specific positions.

All instruments and devices were calibrated to reduce experimental uncertainties.

54.3 Experimental Methodology and Data Analysis

The experimental test rig, which is located in Green Energy Technology Innovation Park, Universiti Kebangsaan Malaysia, Bangi, Malaysia, can work at different modes by using different control valves. The performance of the prototype was measured, evaluated, and compared under the following modes:

1. Basic absorption cycle without ejector–flash tank;
2. Modified absorption cycle with ejector–flash tank.

The uncertainty measurements are calculated. The overall COP errors ranged from ±5.72 % to ±10.69 %, which are relatively reasonable. The duration of each test is related to the steady state of the system under specific operating conditions. For both cycles, each test was taken after the system operated for at least 20–30 min, which is the period required by cycles to reach the steady state. Temperatures were recorded automatically and then saved every minute. The operating conditions depended on the environment and available solar thermal heat because the generator is operated by the hot water cycle provided by the solar thermal collectors. To calculate the thermal performance of the cycles, the temperature and flow rates of the hot water, cooling water, and chilled water cycles that enter and leave the generator, condenser, absorber, and evaporator were measured. The thermal energy of the main components in both cycles was evaluated by using the following equations:

$$Q_{gen} = m_{hotw} C_p (T_{ghotwi} - T_{ghotwo}) \tag{54.1}$$

$$Q_{cond} = m_{cw} C_p (T_{ccwi} - T_{ccwo}) \tag{54.2}$$

$$Q_{abs} = m_{cw} C_p (T_{acwi} - T_{acwo}) \tag{54.3}$$

$$Q_{evp} = m_{chw} C_p (T_{chwi} - T_{chwo}) \tag{54.4}$$

Equation (56.5) is an energy balance equation for the heat exchanger components during the absorption cycle. Results show that the proposed system experiences heat losses to the environment because of the imperfections of the insulated pipes used.

$$Q_{gen} + Q_{evp} = Q_{cond} + Q_{evp} + Q_{loss} \tag{54.5}$$

COP was determined from the ratio of the cooling effect (evaporator load) and heat supplied (generator load):

$$COP = \frac{Q_{evp}}{Q_{gen} + W_{p.}} \tag{54.6}$$

54.4 Results and Discussion

54.4.1 Operating Conditions

Table 54.1 shows the operating range of the fixed and controlled parameters in the experiment. The performance limitations of the system were estimated according to existing operating conditions. The system cooling capacity was designed at 5.5 kW on the basis of the mass flow rate and available operating conditions. The realistic range of operation conditions was dictated by the experimental and practical limitations. System capacity and flow can be adjusted manually by using different valves.

Table 54.1 Operating ranges of fixed and controlled variables

Parameter	Working range
Heat source temperature	65–110 °C
Generator pressure	8–16 bar
Condenser temperature	25–40 °C
Evaporator pressure	3–8 bar
Absorber temperature	25–40 °C
Solution NH_3 mass fraction	30–60 %
Cooling water mass flow rate [condenser]	6–15 L/min
Cooling water mass flow rate [absorber]	6–15 L/min
Chilled water mass flow rate [evaporator]	5–10 L/min
Hot water flow rate [generator]	7–15 L/min
Solution mass flow rate	0.196–0.491 kg/s

54.4.2 Experimental Comparison Between Basic and Modified Absorption Systems

A test rig was experimentally tested under different operating conditions for both modified combined ejector–absorption system and basic absorption system (without ejector and flash tank). The tests demonstrated the effects of adding a flash tank with ejector on the overall performance of the system and on the cooling capacity of the evaporator. The flash tank increased the quality of liquid ammonia, which is used to generate the cooling process inside the evaporator. Figure 54.3 compares the enthalpy and temperature values entering the evaporator for the basic and modified cycles. Figure 54.4 compares the thermal cooling effect between the two systems. The ejector–flash tank unit improved the evaporator cooling effect compared to the basic system under the same operating temperature.

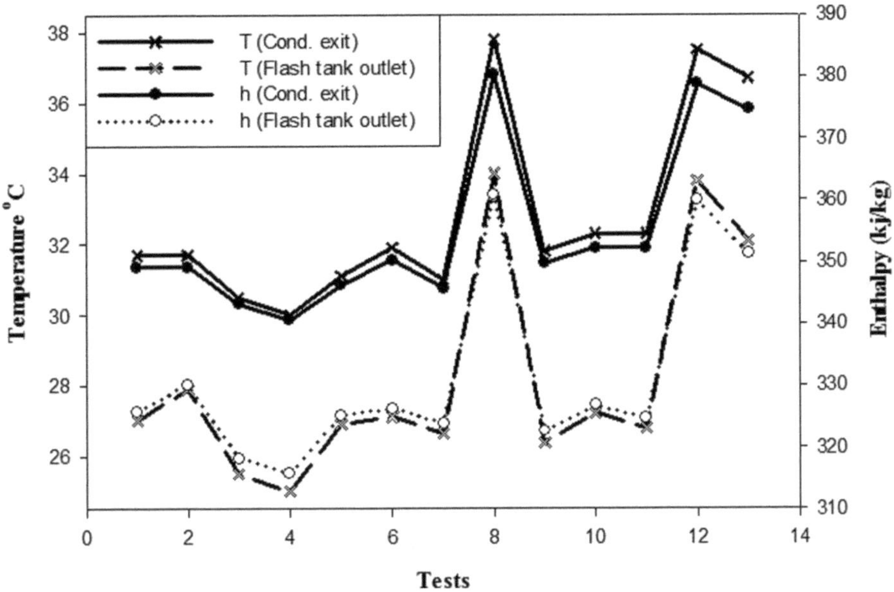

Fig. 54.3 Enthalpy and temperature variation of experimental tests in both cycles

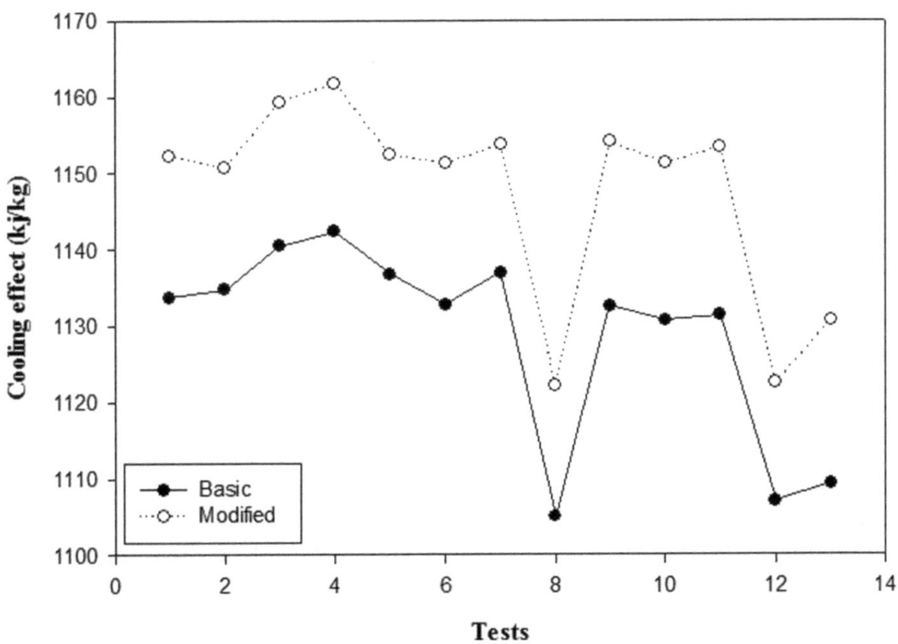

Fig. 54.4 Compression of the cooling effect between two cycles (modified and basic)

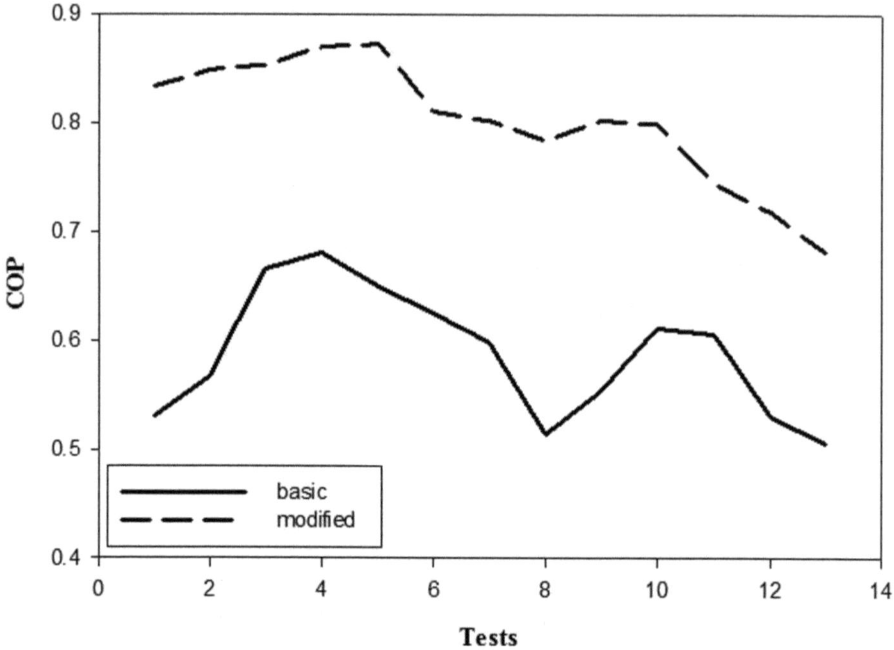

Fig. 54.5 COP compression between the two cycles (modified and basic)

Figure 54.5 compares the COP between two absorption–cooling systems under the same operating and design conditions. The modified absorption cooling system has higher COP and has better performance stability compared with the basic absorption cooling system.

54.5 Conclusion

A modified combined ejector–absorption cooling system was tested experimentally in this study, and an experimental installation was used to evaluate the performance of absorption cooling systems with and without ejector–flash tank units.

Measurement and data acquisition instruments were first installed and calibrated. Several tests were also conducted to study system behavior under different operating conditions. The experimental results showed that the COP of the modified absorption cooling system with and without an ejector–flash tank unit was 0.87 and 0.64, respectively, under the same operating conditions. The cooling effect in the system with an ejector–flash unit increased because the addition of the ejector–flash unit results in quality improvement of the liquid ammonia entering the evaporator. This study shows that adding an ejector–flash tank unit optimizes the performance of the absorption cooling system. Future developments on this experimental research would involve measuring the concentrations of the ammonium solution and determining the effects of varying ammonia concentration on system performance.

References

1. Srikhirin P, Aphornratana S, Chungpaibulpatana S (2001) A review of absorption refrigeration technologies. Renew Sustain Energy Rev 5:343–372
2. Sun D-W (1997) Thermodynamic design data and optimum design maps for absorption refrigeration systems. Appl Therm Eng 17:211–221
3. Sun D-W (1998) Comparison of the performances of NH3-H2O, NH3-LiNO3 and NH3-NaSCN absorption refrigeration systems. Energy Convers Manage 39:357–368
4. Joudi KA, Lafta AH (2001) Simulation of a simple absorption refrigeration system. Energy Convers Manage 42:1575–1605
5. Assilzadeh F, Kalogirou SA, Ali Y, Sopian K (2005) Simulation and optimization of a LiBr solar absorption cooling system with evacuated tube collectors. Renew Energy 30:1143–1159
6. Stoecker WF, Reed LD (1971) Effect of operating temperatures on the coefficient of performance of aqua-ammonia refrigerating systems. ASHREA Trans 77:163–170
7. Kandlikar SG (1982) A new absorber heat recovery cycle to improve COP of aqua ammonia absorption refrigeration system. ASHRAE Trans 88:141–158
8. Kang YT, Akisawa A, Kashiwagi T (1999) An advanced GAX cycle for waste heat recovery: WGAX cycle. Appl Therm Eng 19:933–947
9. Engler M, Grossman G, Hellmann HM (1997) Comparative simulation and investigation of ammonia-water: absorption cycles for heat pump applications. Int J of Refrig 20:504–516
10. Velázquez N, Best R (2002) Methodology for the energy analysis of an air cooled GAX absorption heat pump operated by natural gas and solar energy. Appl Therm Eng 22:1089–1103
11. Park CW, Koo J, Kang YT (2008) Performance analysis of ammonia absorption GAX cycle for combined cooling and hot water supply modes. Int J of Refrig 31:727–733
12. Gómez VH, Vidal A, Best R, García-Valladares O, Velázquez N (2008) Theoretical and experimental evaluation of an indirect-fired GAX cycle cooling system. Appl Therm Eng 28:975–987
13. Ramesh kumar A, Udayakumar M (2008) Studies of compressor pressure ratio effect on GAXAC (generator–absorber–exchange absorption compression) cooler. Appl Energy 85:1163–1172
14. Kang YT, Hong H, Park KS (2004) Performance analysis of advanced hybrid GAX cycles: HGAX. Int J Refrig 27:442–448
15. Jawahar CP, Saravanan R (2010) Generator absorber heat exchange based absorption cycle—A review, renew, and sustain. Energy Rev 14:2372–2382
16. Ma WB, Deng SM (1996) Theoretical analysis of low-temperature hot source driven two-stage LiBr/H2O absorption refrigeration system. Int J Refrig 19:141–146
17. Boer D, Valles M, Coronas A (1998) Performance of double effect absorption compression cycles for air-conditioning using methanol–TEGDME and TFE–TEGDME systems as working pairs: Performances de cycles Ã compression absorption Ã double effet pour le conditionnement d'air utilisant les couples méthanol–TEGDME ou TFE–TEGDME. Int J Refrig 21:542–555
18. Xu GP, Dai YQ, Tou KW, Tso CP (1996) Theoretical analysis and optimization of a double-effect series-flow-type absorption chiller. Appl Therm Eng 16:975–987
19. Xu GP, Dai YQ (1997) Theoretical analysis and optimization of a double-effect parallel-flow-type absorption chiller. Appl Therm Eng 17:157–170
20. Arun MB, Maiya MP, Murthy SS (2001) Performance comparison of double-effect parallel-flow and series flow water–lithium bromide absorption systems. Appl Therm Eng 21:1273–1279
21. Kaushik SC, Arora A (2009) Energy and exergy analysis of single effect and series flow double effect water–lithium bromide absorption refrigeration systems. Int J Refrig 32:1247–1258
22. Gomri R (2010) Investigation of the potential of application of single effect and multiple effect absorption cooling systems. Energy Convers Manage 51:1629–1636

23. Göktun S (1999) Optimal performance of a combined absorption and ejector refrigerator. Energy Convers Manage 40:51–58
24. Jelinek M, Levy A, Borde I (2002) Performance of a triple-pressure-level absorption cycle with R125-N, N′-dimethylethylurea. Appl Energy 71:171–189
25. Sun D-W, Eames IW, Aphornratana S (1996) Evaluation of a novel combined ejector-absorption refrigeration cycle—I: computer simulation. Int J Refrig 19:172–180
26. Eames IW, Wu S (2000) A theoretical study of an innovative ejector powered absorption–recompression cycle refrigerator. Int J Refrig 23:475–484
27. Levy A, Jelinek M, Borde I (2002) Numerical study on the design parameters of a jet ejector for absorption systems. Appl Energy 72:467–478
28. Sözen A, Özalp M (2003) Performance improvement of absorption refrigeration system using triple-pressure-level. Appl Therm Eng 23:1577–1593
29. Wang J, Dai Y, Zhang T, Ma S (2009) Parametric analysis for a new combined power and ejector–absorption refrigeration cycle. Energy 34:1587–1593
30. Sirwan R, Algoul MA, Sopian K, Ali Y, Abdulateef JM (2013) Evaluation of adding flash tank to solar combined ejector-absorption refrigeration system. Sol Energy 91:283–296
31. Sirwan R, Algoul MA, Sopian K, Ali Y (2013) Thermodynamic analysis of an ejector-flash-tank-absorption cooling system. Appl Therm Eng 58:85–97
32. Abdulateef JM (2010) Combined solar-assisted ejector absorption refrigeration system. Doctoral thesis, Univeristi Kebangssan Malaysia, Malaysia
33. Watkins RN (1967) Sizing separator and accumulators. Hydrocarb Process 46:253–256

Chapter 55
Temperature Prediction for an Integrated Solar Collector with Spherical Phase-Change Material

Fatah O. Alghoul, Kamaruzzman Sopian, Mohammed Al-Ghoul, Shahrir Abdullah, Mohammed Sheldin and Adnan M

Abstract Phase-change material (PCM) systems play a particularly important role during off peak periods in improving thermal storage systems. In this study, we aim to predict output temperature in integrated solar heating using a radial basis function rural network (RBFNN). First, we present a brief PCM numerical model including geometrical and operational parameters and discuss the effect of this parameter on the output temperature. The working fluid temperature is uniformly distributed and equal to collector output temperature. The learning data is theoretical, generated by using a standard solar collector integrated with 95 PCM modules. The RBFNN model has three input nodes representing spherical size, duct length and number of balls and two output nodes representing the first and last layer PCM temperatures. Simulation result shows the predicted PCM temperatures at the first and last layer closely match the analytical data from computational fluid dynamics.

Keywords Phase-change material · Thermal energy storage · RBFNN

55.1 Introduction

With rising energy costs there is an increasing demand for renewable energy sources as an effective method for utilization of clean energy sources, especially solar energy. During the day the collected solar energy can be stored as sensible heat, latent heat, heat of reaction or a combination of these in materials such as water and rocks for passive and active heating applications. Therefore, efficient and economical phase-change materials (PCMs) are the main factor in the utilization of solar energy in many applications such as heating buildings, egg incubation, drying vegetables, meats, fruits, and other industrial purposes.

F. O. Alghoul (✉) · K. Sopian · M. Al-Ghoul · S. Abdullah · M. Sheldin
Solar Energy Research Institute, Faculty of Engineering,
Universiti Kebangsaan Malaysia, 43600 Bangi, Selangor, Malaysia
e-mail: fatah667@gmail.com

A. M
Faculty of Electrical and Electronics Engineering,
University Malaysia Pahang, Pekan 26600, Malaysia

A. Sayigh (ed.), *Renewable Energy in the Service of Mankind Vol II*,
DOI 10.1007/978-3-319-18215-5_55

In recent years, thermal energy storage has attracted much attention as a primary energy source for the application and modeling of domestic, commercial and industrial processes [1]. A transient behavior model was developed [2] to study the performance of phase-change energy storage units using both air and liquid as a working fluid based on three assumptions—axial conduction in the flow mode is negligible, the Biot number is very low so that temperature variations normal to the flow can be neglected, and heat loss from the unit can be ignored. The temperature levels and mechanisms of heat transfer dictate the geometry and design details of the system such as walls and roofs filled with PCM [3–5], control of electronic components, and thermal protection and control [4]. Empirical methods are used for sizing PCM units for air-based solar heating systems for many locations during heating seasons [6]. Tao [7] presented a method for the analysis of solidification in cylindrical and spherical geometries based upon a fixed-grid approach. Shih et al. [8] used iterative method investigation for successive approximation to study the solidification process inside spherical geometry for large values of the Stefan number and small Biot numbers. Additional numerical and experimental studies used for solving phase-change problems are the enthalpy methods and temperature-based equivalent heat capacity methods [9–16].

In this study, a prediction of solar collector output temperature using a radial basis function rural network (RBFNN) was proposed. A mathematical model was briefly described including geometrical and operational parameters of a spherical PCM unit packed in cylindrical tubes inside a solar collector. We then investigated the optimum physical properties of the PCM on thermal performance under a variety of loads and control strategies. A MATLAB computer program has been developed to compute the air temperature, spherical in each cylinder along the duct, freezing time for each cylinder, and the time required to discharge all the thermal energy.

55.2 Related Work

55.2.1 Physical Model

This system consists of three essential parts—a single transparent glass, an isolated duct and a storage unit which consists of a single row of cylinders containing a PCM; the cylinders are placed in the cross flow of a forced air stream. This unit has two goals—to absorb and to store solar energy. The design takes a number of parameters into consideration, such as integration with the PCM storage unit, the simplicity of construction, dismantling, and handling the PCM unit, the length of the collector which was proposed by Alkilani et al. [17], and the number of spherical PCMs in the number of rows controlled by the collector length. A solar simulator charges the collector by thermal energy until the spherical PCM enters the liquid phase in order to investigate the output air temperature by assuming there will be no conflict between charging and discharging at different times due to the discharge process (Fig. 55.1)

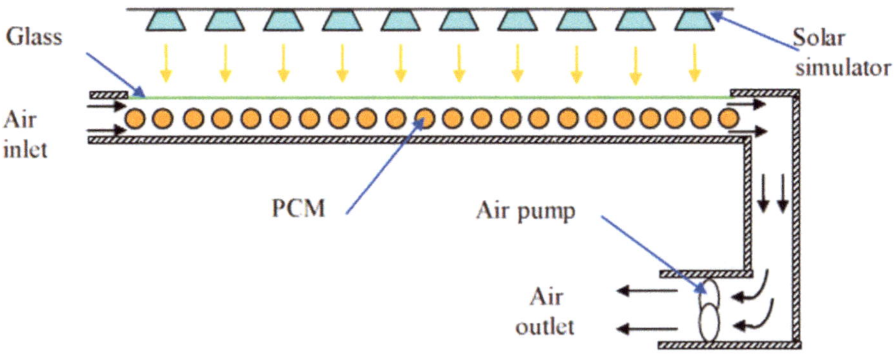

Fig. 55.1 Cross-section of the solar air collector with PCM spheres

for heat transfer factor (HTF):

$$\varepsilon \rho_f A_c L c_f \left(\frac{\partial T_f}{\partial t} + V_{max} \frac{\partial T_f}{\partial x} \right) = h_s a_p \left(T_p - T_f \right) + h_W a_W \left(T_W - T_f \right) \tag{55.1}$$

where:

A_c	The storage tank cross-sectional area
c	Specific heat
L	Height of the storage tank
h_s	Surface heat transfer coefficient between HTF and PCM
T	Temperature
t	Time
V_{max}	HTF velocity
ε	Void fraction
x	Axial direction of the storage tank

The temperature of a planar surface subject to solar radiation, assuming that the surface behaves as a black body and that it does not lose heat to its surroundings by convection or conduction, is given as:

$$q_{solar} a_w - 2\sigma T_w^4 = 0$$

where

q_{solar}	Solar radiation flux
σ	Stefan–Boltzmann constant (5.67×10^{-8})

where a typical value of solar radiation flux is 1.140 Wm^{-2}.

For PCM:

$$(1-\varepsilon) \rho_p A_c L C_p \left(\frac{\partial T_p}{\partial t} \right) = h_s a_p \left(T_f - T_p \right) \tag{55.2}$$

where

ρ: density

Initial conditions

$$T_{\mathrm{f}}(t=0,x)=T_{\mathrm{f\_ini}}$$
$$T_{\mathrm{p}}(t=0,x)=T_{\mathrm{p\_ini}} \tag{55.3}$$

Boundary conditions

$$T_{f}(t,x=0)=T_{f\_\mathrm{inlet}}$$
$$\frac{\partial T_{\mathrm{f}}(t=0,x)}{\partial x}=0 \tag{55.4}$$

Governing equations

Equations (55.1) and (55.2) are re-arranged to give the governing equations for computer programming:

$$\frac{\partial T_{\mathrm{f}}}{\partial t}=\frac{\left[h_{\mathrm{s}}a_{\mathrm{p}}\left(T_{\mathrm{p}}-T_{\mathrm{f}}\right)+h_{\mathrm{W}}a_{\mathrm{W}}\left(T_{\mathrm{W}}-T_{\mathrm{f}}\right)\right]}{\varepsilon\rho_{\mathrm{f}}A_{\mathrm{c}}Lc_{\mathrm{f}}}-\left(v_{\max}\frac{\partial T_{\mathrm{f}}}{\partial x}\right) \tag{55.5}$$

and

$$\frac{\partial T_{\mathrm{p}}}{\partial t}=\frac{h_{\mathrm{s}}a_{\mathrm{p}}\left(T_{\mathrm{f}}-T_{\mathrm{p}}\right)}{(1-\varepsilon)\rho_{\mathrm{p}}A_{\mathrm{c}}LC_{\mathrm{p}}} \tag{55.6}$$

55.2.2 Radial Basis Function Neural Network (RBFNN)

RBFNNs have increasingly attracted interest for engineering applications due to their advantages over traditional multilayer perceptions, that is, faster convergence, smaller extrapolation errors, and higher reliability. Over the past few years, more sophisticated types of neurons and activation functions have been introduced in order to solve different sorts of practical problems [18]. In particularly, RBFNNs have proved very useful for many systems and applications [19]. A RBFNN is defined as a kind of artificial neural network (ANN) that has radial activation functions on its intermediary layer. RBFNNs were robustly used in the context of neural networks as linear and nonlinear function estimators and indicated their interpolation capabilities (see Broomhead and Lowe) [20]. Park and Sandberg, [21] proved that RBFNNs are capable of approximating any function with arbitrary accuracy. The neural network is a mapping between its inputs and outputs based on a number of known sample input-output pairs. In general, the more samples available to train the

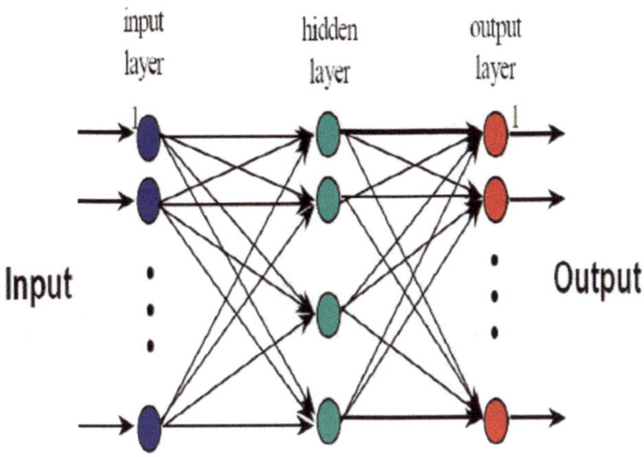

Fig. 55.2 Radial basis function neural networks

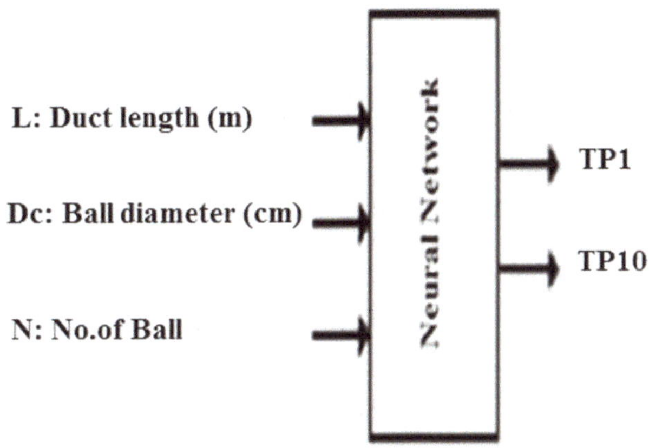

Fig. 55.3 The proposed model

network, the more accurate the representation of the real mapping will be. These samples are obtained by solving the direct problem (times). In its simplest form, a RBFNN consists of three layers of neurons as shown in Fig. 55.2. The first layer acts as the input layer of the ANN. The second layer is a hidden layer with a high-scale dimension, which promotes a linear transformation of input space dimension by computing radial functions in their neurons. The third layer, the output layer, outputs the ANN response, promoting a linear transformation of the intermediary layer high-scale dimension to low-scale dimension [22].

55.3 The Proposed Method

The ANN structure and training algorithms have been proposed as shown in Fig. 55.3. The basic form of RBFNN architecture involves three entirely different layers. The input layers is made up of three nodes (duct length, L; diameter, Dc; and number of spherical balls) of source nodes while the second layer is a hidden layer

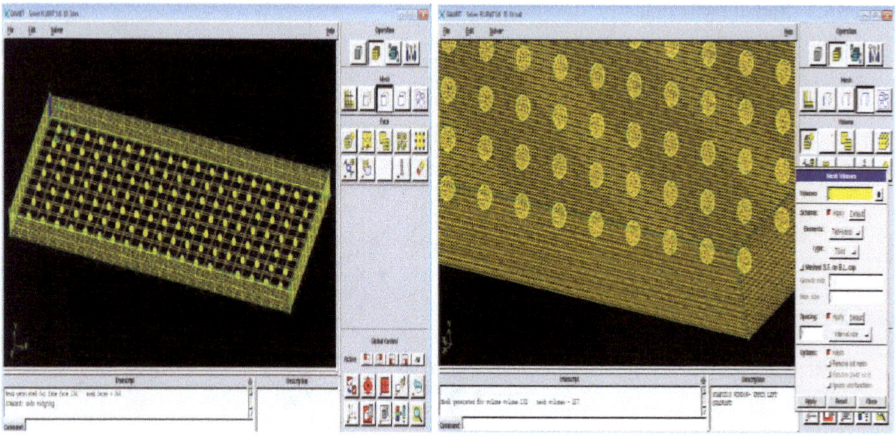

Fig. 55.4 Spherical PCM CFD simulation model

of high-dimension which senses a different purpose from that in a multilayer perception. The output layer is made up of two nodes (PCM temperature) which supply the response of the network to the activation patterns applied to the input layer. The transformation from the input layer to the hidden layer is nonlinear whereas the transformation from the hidden layer to the output layer is linear.

The major steps of the RBFNN design and training to determine maximum air velocity are summarized as follows:

1. A set of PCM patterns are generated by varying selected important parameters.
2. For each set generated in step (i) the PCM modal analysis is performed and TP1 and TP10 are calculated.
3. The RBFNNs are designed and trained by 3 different input patterns for each model and are generated as shown in Fig. 55.3.
4. In the RBFNN, the target output is TP1 and TP10 for each input pattern which is computed by running the contour program.
5. Training of the RBFNNs using the input/output patterns developed in Steps (iii) and (iv) is carried out.
6. Finally, the model outputs have been compared to check the sensitivity of maximum air velocity based on four input parameters.

55.4 Modelling Guides Using Computational Fluid Dynamics (CFD) Simulation

A model with boundary conditions was converted to algebraic equations by means of finite-volume techniques with fully implicit temporal differentiation, using a three-dimensional spherical capsule in a staggered arrangement. Delusive terms were evaluated using a central difference scheme as shown in Fig. 55.4.

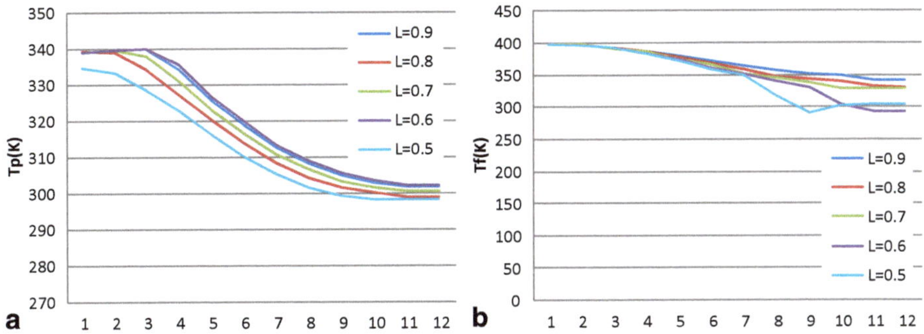

Fig. 55.5 Variable bed length. **a** PCM temperature. **b** Flow temperature

In practice, PCM characteristics and solution accuracy requirement together determine the best simulation strategy. The simulated system consists of a solar collector that can be opened or closed which is connected with an inside opaque wall, insulated on the side facing the air space and with an external spherical-granite covering panel on the facade.

55.5 Results and Discussion

A computer program was developed using MATLAB10 software to compute the PCM and mass flow temperature, layer by layer along the duct. In this study we supposed that the room temperature was 28 °C, and different parameter PCM.

Figures 55.5a and b shows the effect of increasing the duct length from 2.5 to 2.9 m on PCM temperature. In this simulation, all parameters were kept constant except the system flow rate, which was set between 0.05 and 0.19 kg/s along same sized 170 spherical PCM arranged in 10 layers. Figure 55.5a shows that the first layer has a higher temperature which decreases through the other layer. In addition, the effect of increasing the duct length on increasing the temperature of the first layer is clear. Figure 55.5b shows that flow temperature decreases by increasing the duct length.

In the second simulation, all parameters were kept constant, that is, duct length 2.7 m, and size of each ball, the number of spherical PCM was arranged in 14, 15, 16, and 17 for each 10 layers. Figures 55.6a and b shows that there is almost no significant effect on PCM and flow temperature.

Figures 55.7a and b shows the effect of increasing spherical PCM diameter from 2.5 to 4.5 cm on PCM temperature. In this simulation, all parameters were kept constant such as number of balls and duct length and 150 spherical PCM were arranged in 10 layers. Figure 55.7a shows that the first layer has a higher temperature which decreases through the other layer. In addition, the effect of increasing the diameter on increasing the temperature of the first layer is clear. Figure 55.7b shows almost no significant effect on flow temperature decrease by increasing the ball diameter.

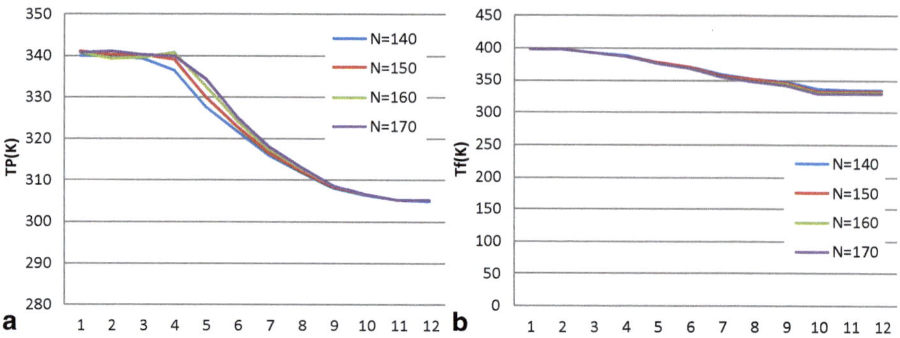

Fig. 55.6 Increased number of spherical PCMs. **a** PCM temperature. **b** Flow temperature

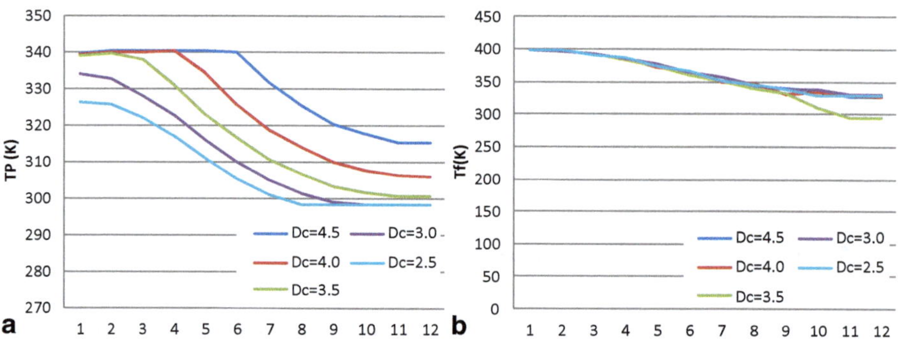

Fig. 55.7 Increased ball diameter. **a** PCM temperature. **b** Flow temperature

Finally, we studied the effect of each parameter on maximum airflow where each case kept one constant and change the others. Figure 55.8 shows the effect of ball diameter, bed length and number of balls on maximum airflow. The airflow was not affected by changing the number of PCM balls because of assuming uniform air distribution. Figure 55.8 shows airflow is significantly affected by ball diameter compared with bed length.

Finally, from the above 95 Model carried out to get the detailed of PCM temperature. Network architecture was required to train ANN models with the results of the MATLAB10 software. First, the entire training data file was randomly divided into training and testing data sets. Approximately 90 % of the data patterns were used to train the different network architectures where the remaining eight random patterns were used for testing to verify the prediction ability of each trained NN model.

Table 55.1 shows comparison sample of first and last layer PCM for a set of models calculated by RBFNN and CFD techniques. It can observed that the accuracy of the RBFNN method was slightly superior compared to the CFD method on account of mean average error (MAE).

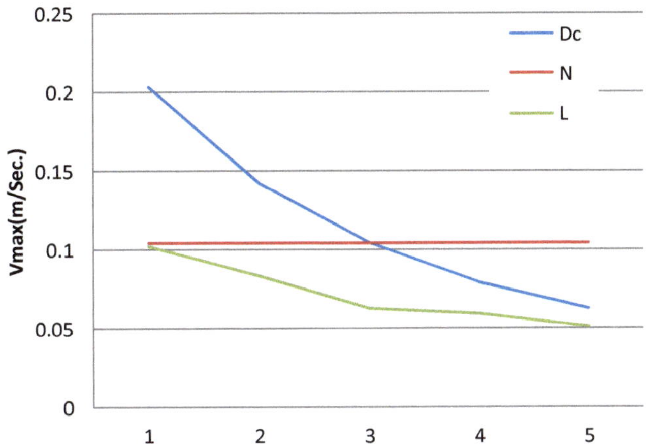

Fig. 55.8 Maximum airflow with variable PCM number and diameter, and bed length

Table 55.1 Results of sample testing of temperature

Input layer			CFD		Output		
No. of balls	DC (cm)	L (m)	TP1 (K)	TP10 (K)	TP1 (K)	TP10 (K)	Average error
140	2.5	2.55	325	302	328.9	307.134	4.517
150	2.6	2.78	339	311	343.068	314.421	3.7445
170	3.4	2.83	341	325	339.636	329.225	2.7945
160	3.8	2.67	331	299	328.4513	302.289	2.91885
140	4.1	2.75	329	307	332.0597	304.9124	2.57365
170	2.9	2.89	336	322	340.0656	319.4562	3.3047
150	3.65	2.53	338	313	342.1574	311.0594	3.049
160	4.3	2.61	327	306	331.3818	308.85192	3.61686
				Maximum absolute error			4.517
				Mean average error (MAE)			3.314883

55.6 Conclusion

In this study, the RBFNN was used to predict first and last PCM layer temperature. The simulation data from MATLAB software and CFD has been used for training and testing. A mathematical model was developed for PCM integration with the duct system was also explained briefly. The simulation results show that a better performance was obtained with RBFNN for reduction of the effort and time required determining the temperature in PCM compared with CFD, which usually deals with only a single problem for each run. In addition, the performance analysis clearly shows that PCM temperature depends on ball diameter.

References

1. Taha MA, Babus Hag RF, Callagfan PWO, Probert SD (1995) Design feasibility of an intermittent domestic energy store. Appl Energy 51(3):277–290
2. Kenisarin MM, Kenisarina KM (2012) Form-stable phase change materials for thermal energy storage. Renew Sustain Energy Rev 16(4):1999–2040
3. Mawire A, McPherson M (2009) Experimental and simulated temperature distribution of an oil-pebble bed thermal energy storage system with a variable heat source. Appl Therm Eng 29(5–6):1086–1095
4. Kurnia JC, Sasmito AP, Jangam SV, Mujumdar AS (2013) Improved design for heat transfer performance of a novel phase change material (PCM) thermal energy storage (TES). Appl Therm Eng 50(1):896–907
5. Sarı A, Karaipekli A (2009) Preparation, thermal properties and thermal reliability of palmitic acid/expanded graphite composite as form-stable PCM for thermal energy storage. Sol Energy Mater Sol Cells 93(5):571–576
6. Jurinak JJ, Abdel-Khalik SI (1979) Sizing phase-change energy storage units for airbased solar heating systems. Sol Energy 22:355–359
7. Tao LC (1967) Generalized numerical solutions of freezing a saturated liquid in cylinders and spheres. AIChE J 13:165–169
8. Shih YP, Chou TC (1971) Analytical solution for freezing a saturated liquid inside or outside spheres. Chem Eng Sci 26:1787–1793
9. Cárdenas B, León N (2013) High temperature latent heat thermal energy storage: phase change materials, design considerations and performance enhancement techniques. Renew Sustain Energy Rev 27:724–737
10. Ismail KAR, Abugderah MM (2000) Performance of a thermal storage system of the vertical tube type. Energy Conversat Manage 41:1165–1190
11. Zivkovic B, Fujii I (2001) An analysis of isothermal phase change of phase change material within rectangular and cylindrical containers. Sol Energy 70:51–61
12. Sari A, Kaygusuz K (2002) Thermal performance of a eutectic mixture of lauric and stearic acids as PCM encapsulated in the annulus of two concentric pipes. Sol Energy 72:493–504
13. Pinel P, Cruickshank CA, Beausoleil-Morrison I, Wills A (2011) A review of available methods for seasonal storage of solar thermal energy in residential applications. Renew Sustain Energy Rev 15(7):3341–3359
14. Shuangmao W, Fang G, Chen Z (2012) Discharging characteristics modeling of cool thermal energy storage system with coil pipes using n-tetradecane as phase change material. Original research article. Appl Therm Eng 37:336–343
15. Sharma A, Tyagi VV, Chen CR, Buddhi D (2009) Review on thermal energy storage with phase change materials and applications. Renew Sustain Energy Rev 13(2):318–345
16. Kuravi S, Trahan J, Goswami YD, Rahman MM, Stefanakos EK (2013) Thermal energy storage technologies and systems for concentrating solar power plants. Energy Combust Sci 39(4):285–319
17. Alkilani MM, Sopian K, Mat S, Alghoul MA (2009) Output air temperature prediction in a solar air heater integrated with phase change material. Eur J Sci Res 27(3):334–341
18. Kumar S (2005) Neural networks: a class room approach. ISBN:007-124672-X
19. Kurban T, Beşdok E (2009) A comparison of RBF neural network training algorithms for inertial sensor based terrain classification. Sensors 9:6312–6329
20. Broomhead DS, Lowe D (1988) Multivariable function interpolation and adaptive networks. Complex Syst 2:321–355
21. Pandya AS. (1995) Pattern recognition with neural networks in C++
22. Park J, Sandberg IW (1993) Universal approximation and radial basis function networks. Neural Comput J 5:305–316

Chapter 56
Design of Various Hybrid Single-Pass Photovoltaic–Thermal (PV/T) Solar Collector

Mohd. Yusof Hj. Othman, Faridah Hussain, Kamaruzzman Sopian, Baharuddin Yatim and Hafidz Ruslan

Abstract In this chapter, five different designs of hybrid photovoltaic–thermal (PV/T) solar collectors are fabricated and their performance are evaluated. All designs are based on the concept of single-pass collector. Designs of heat exchangers for the collectors include honeycomb, stainless steel wool, V-groove, *∇-groove,* and rectangular groove. The advantages of the collectors are their capability to generate both electricity and heat simultaneously. The objective of this study is to compare the performance of all PV/T collectors. All systems were tested at irradiance of ~ 800 W/m^2 with mass flow rate spanning from 0.01 to 0.13 kg/s.

Keywords Photovoltaic-thermal · Solar collector · Single pass system · Heat exchanger

56.1 Introduction

The concept of photovoltaic/thermal (PV/T) system had been studied, discussed, and published for almost 4 decades now. Review papers related to research work on PV/T collectors were written by the authors of [1]–[4]. Both water and air are suitable to be used as the cooling fluid to cool the PV module in order to avoid the drop of electrical efficiency. However, air-based PV cooling system is simpler and economical due to its minimal usage of material and low operating cost. A high thermal conductivity material together with natural or forced flow of air removes the heat from the system.

M. Y. Hj.Othman (✉) · B. Yatim
Faculty of Science and Technology, Universiti Kebangsaan Malaysia, 43600 Bangi, Selangor, Malaysia
e-mail: myho@ukm.edu.my

K. Sopian · H. Ruslan
Solar Energy Research Institute, Faculty of Engineering, Universiti Kebangsaan Malaysia, 43600 Bangi, Selangor, Malaysia

F. Hussain
National Metrology Laboratory, SIRIM Berhad, 43900 Sepang, Selangor, Malaysia
e-mail: faridahh@sirim.my

A. Sayigh (ed.), *Renewable Energy in the Service of Mankind Vol II*,
DOI 10.1007/978-3-319-18215-5_56

Extensive research work focusing on designing various hybrid PV/T was carried out from time to time before it can be mass produced for commercialization. Hegazy [5] carried out comparison study of four models of the PV/T collector. Both single-pass and double-pass system were evaluated. Heat balance equation was identified and solved for each model. Performance of a double-pass PV/T solar collector suitable for solar drying system was investigated [6]. A study on nine different design concepts of combined PV/T water and air solar collector system was carried out [7]. The design concepts were divided into four different groups. Two types of PV panels were investigated, namely, the conventional opaque PV panel and the transparent PV panel. Another study was on two low-cost improvement designs of heat remover placed on a channel of a PV/T system [8]. The design was improved by introducing a thin metal sheet and attaching a fin to the back wall of the channel.

A performance study was carried out on a double-pass air-based PV/T system with fins attached at the back of the absorber plat of the PV module [9]. The experiment concluded that using fins as the integral part of the PV module increases the overall efficiency of the system. Sopian et al. [10] evaluated the thermal efficiency of the double-pass PV/T system with porous media at the lower channel. Experimental results proved that introducing the porous media at the lower channel increased the heat transfer area which led to an increase in the thermal efficiency of the system from 60 to 70 %. A direct coupled outdoor PV/T system in Kerman, Iran, was tested [11]. The design of the system involves the use of a thin aluminum sheet placed in the middle of the air channel as the heat exchanger to cool the PV panels. Another PV/T solar collector with an active cooling system was fabricated to increase the electrical efficiency of a PV module [12].

The objective of this chapter is to evaluate the performance for various designs of hybrid PV/T. Single-pass air-based PV/T solar collector combined with various heat exchangers were studied. In order to remove the heat from the back of the PV module, five different designs of heat exchanger, honeycomb, stainless steel wool, V-groove, *∇-groove,* and rectangular groove were installed horizontally into the channel located at the back side of the PV module. Under similar setup of operational condition, each heat exchanger was tested one by one to observe the system performance.

56.2 Designs of PV/T Collector

The performance of each PV/T solar collector was tested indoor using a laboratory-fabricated solar simulator. Under similar setup of operational condition, the system was tested one by one at irradiance of 817 and 828 W/m^2 with mass flow rate span from 0.01 to 0.13 kg/s.

The first design of the PV/T collector is shown in Fig. 56.1 [13]. It is a collector with honeycomb heat exchanger. Locally purchased aluminum sheet was made into corrugated sheet. The thickness of the sheet is ~0.2 mm. Five pieces of aluminum corrugated sheets were joined together to fabricate a piece of compact honeycomb

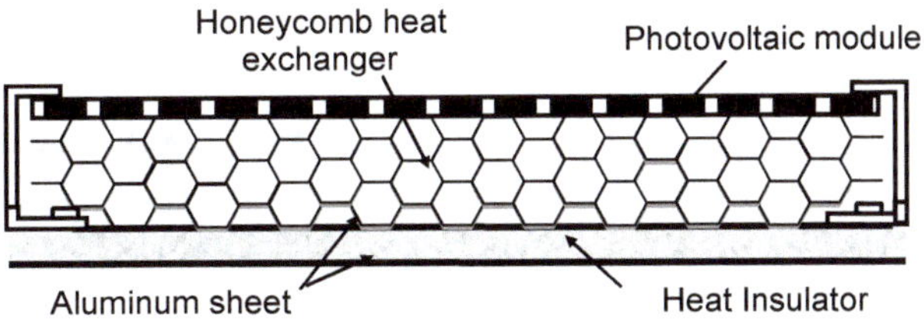

Fig. 56.1 PV/T collector with honeycomb heat exchanger

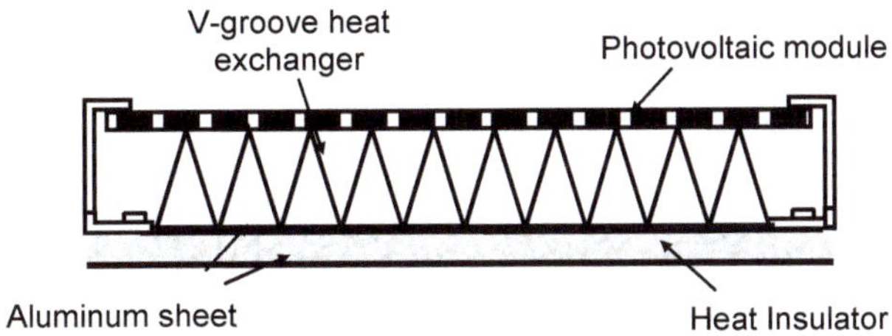

Fig. 56.2 PV/T collector with V-groove heat exchanger

with hexagonal geometry. The honeycomb was installed horizontally into the channel located at the back side of the PV module.

Figure 56.2 illustrates the second design of the PV/T collector with V-groove heat exchanger. V-groove was fabricated using locally purchased aluminum sheet. The thickness of the aluminum sheet was ~0.5 mm. The V-groove was also installed at the back of the PV module using the same method.

The third design of the PV/T collector is shown in Fig. 56.3. Locally purchased stainless steel wool was used as the heat exchanger. The stainless steel wool was tested to prove that it will not rust over time. The stainless steel wool was attached to a piece of aluminum sheet. The complete design was installed onto the back of the PV module using the same method as shown in Fig. 56.3.

The fourth design was a single-pass PV/T system with aluminum ∇-grooved absorber plate [14]. The thickness of the aluminum was 0.7 mm, attached to the

Fig. 56.3 PV/T collector with stainless steel wool heat exchanger

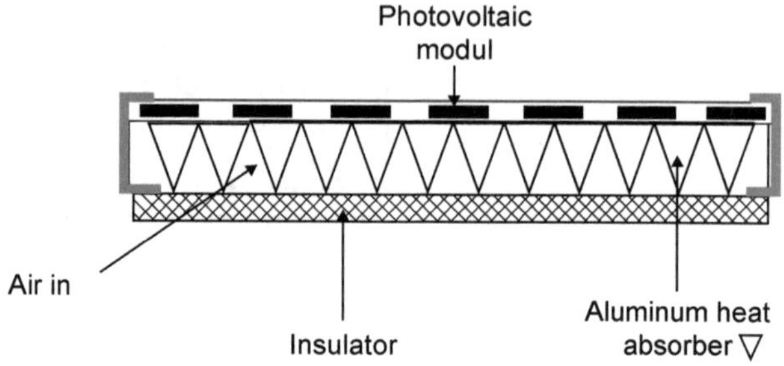

Fig. 56.4 PV/T system with aluminum ∇-grooved absorber plate

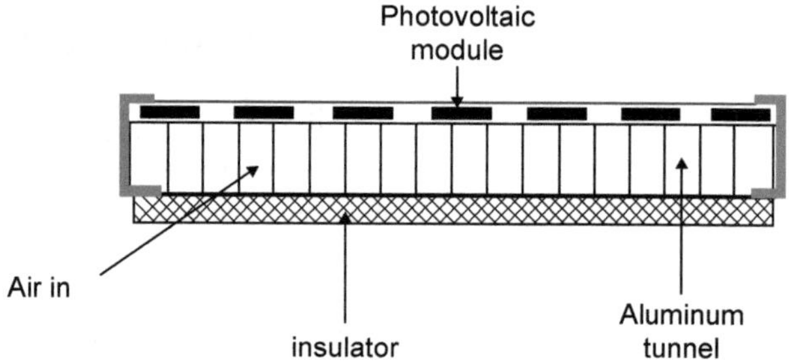

Fig. 56.5 PV/T system with rectangular tunnel heat exchanger

back of the PV module. Figure 56.4 shows the cross-sectional diagram of the PV/T system.

The fifth design of heat exchanger was a rectangular tunnel [15]. Rectangular tunnel bars of size ($1.2 \times 2.5 \times 120$) cm (width x height x length) were installed onto the back of the PV module. The material of the rectangular tunnel was aluminum. Figure 56.5 shows the design of the PV/T system. The effectiveness of PV/T collector with each design of heat exchanger was studied to identify their overall performance. The performance of the collectors was evaluated and compared in terms of its electrical and thermal efficiencies.

56.3 Experimental Procedures

Honeycomb, V-groove, and stainless steel wool were tested at the National Metrology Laboratory (NML), SIRIM Berhad, Sepang. The ∇-grooved and rectangular tunnels were tested at UKM, Bangi. The experiment was run under two similar conditions of irradiance and mass flow rate. At NML-SIRIM, the PV/T system was tested indoor at 828 W/m^2 with mass flow rates of air through the system set to 0.021, 0.042, 0.085, 0.110, and 0.128 kg/s. A monocrystalline (SolarWorld, SW 85) photovoltaic module was used to harvest the electrical energy. The thermal system consisted of a blower attached to a galvanized ducting with a length of 2.5 m. A voltage regulator was used to control the air speed of the blower between 0.4 and 1.5 m/s flow through the system. A heater controlled by a voltage regulator was installed into the ducting in order to ensure that the inlet temperature is equivalent to the ambient temperature.

At UKM, Bangi, the PV/T system was tested indoor at 817 W/m^2 with mass flow rates of the air through the system set to 0.0110, 0.0287, 0.0409, 0.0552, and 0.0754 kg/s. A solar panel with model SHARP NE-80E2EA was used to harvest the electrical energy. The thermal system consists of a blower attached to a ducting, a voltage regulator to control the air speed of the blower, and halogen lamps controlled by a voltage regulator to ensure that the inlet temperature is equivalent to the ambient temperature.

Type-T thermocouples were placed at various locations of the system. Inlet temperature, panel temperature, outlet temperature, and temperature at the back of the thermal insulator were measured and used to calculate the thermal efficiency of the system. At each mass flow rate setting, time of not less than 90 min was allowed for temperature stabilization. In the first step, the PV/T collector system efficiency was studied without any heat exchanger installed into the system. Then, the system efficiency was tested with heat exchanger one by one separately.

56.4 Results and Discussion

All parameters involved in the calculation of thermal and electrical efficiency were measured in detail. Maximum current, I_m (A), and maximum voltage, V_m(V), were determined from I/V curve measurement of the PV module using digital multimeter and rheostat. A_c is the area of the solar cell and S is the average irradiance value during the experiment. Equation (56.1) was used to calculate the electrical efficiency, π_{el}, of a PV/T system. Measurements of short-circuit current, I_{sc}(A), open-circuit voltage, V_{oc} (V), can be done by direct connection between the multimeter and the PV module.

$$\pi_{el} = \frac{I_m \, x \, V_m}{A_c \, \text{x} \, S} x \, 100\% \tag{56.1}$$

Inlet temperature, T_i, panel temperature, T_s, outlet temperature, T_o, and temperature at the back of the thermal insulator, T_b, were measured and used to calculate the thermal efficiency of the system. Air velocity, V_{av}, at input location was calculated. Mass flow rate, *m,* of the air flow through the system was calculated using Eq. (56.2).

$$m = \rho A V_{av} \tag{56.2}$$

Equation (56.3) was used to calculate the thermal efficiency, π_{th}, of the developed PV/T system. C_p is specific heat of air, and A_p is the area of the PV/T collector

$$\pi_{th} = \frac{mC_p(T_o - T_i)}{A_p S} x100\% \tag{56.3}$$

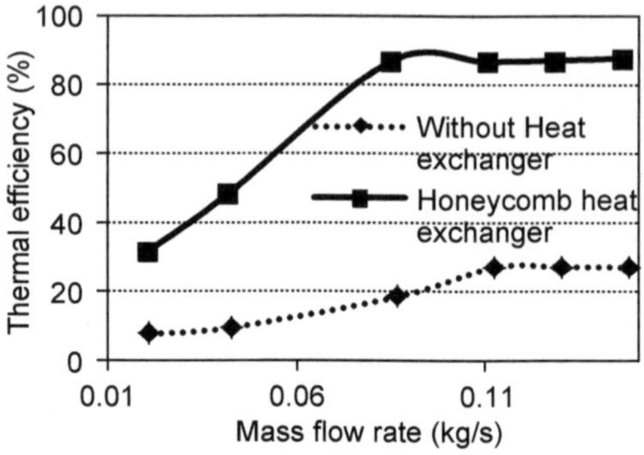

Fig. 56.6 Thermal efficiency curve for honeycomb heat exchanger

Graph in Fig. 56.6 shows the plot of the thermal efficiency for collector with honeycomb heat exchanger. The experimental result shows that the thermal efficiency, π_{th}, increases with the increase in mass flow rate up to 0.11 kg/s. Above mass flow rate of 0.11 kg/s, the thermal efficiency, π_{th}, remains stable. The maximum thermal efficiency, π_{th}, of the collector with the honeycomb is ~87.0 % and without is only 27.0 %. Graph in Fig. 56.7 shows the plot of the electrical efficiency, π_{el}, for collector with honeycomb heat exchanger. Overall, collectors show an increasing trend for electrical efficiency, π_{th}, with the increase in the fluid mass flow rate. The electrical efficiency improves by 0.2 % with the usage of honeycomb heat exchanger.

Graph in Fig. 56.8 shows the plot of the thermal efficiency for collector with V-grooved heat exchanger. Result shows that the thermal efficiency, π_{th}, increases with the increase in mass flow rate. The maximum thermal efficiency, π_{th}, is ~75.0 % and without is only 27.0 %. Graph in Fig. 56.9 shows the plot of electrical efficiency, π_{el}, for the system. The difference of electrical efficiency for collectors with and without heat exchanger is not so significant.

Graph in Fig. 56.10 shows the plot of the thermal efficiency for collector with stainless steel wool heat exchanger. The thermal efficiency, π_{th}, increases with the increase in mass flow rate. The maximum thermal efficiency, π_{th}, is ~87.0 % and without is only 27.0 %. Graph in Fig. 56.11 shows the plot of for electrical efficiency, π_{el}, for the system. Usage of stainless steel heat exchanger reduces the electrical

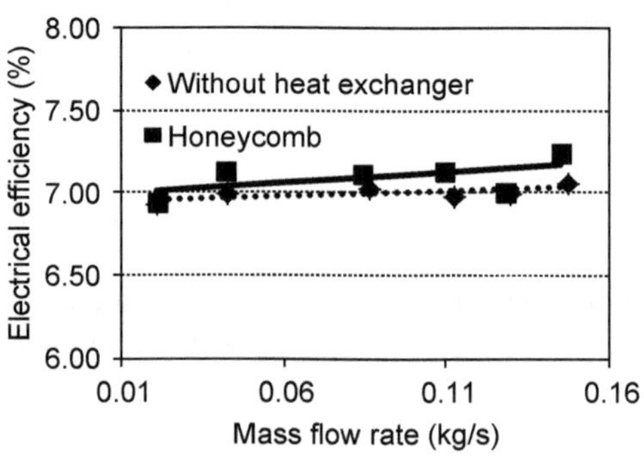

Fig. 56.7 Electrical efficiency curve for honeycomb heat exchanger

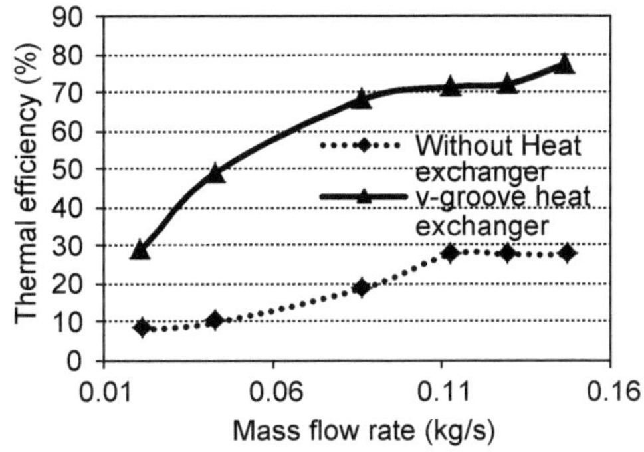

Fig. 56.8 Thermal efficiency curve for V-grooved heat exchanger

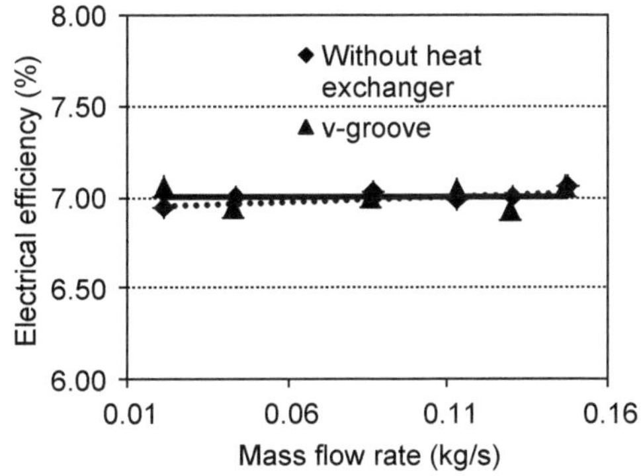

Fig. 56.9 Electrical efficiency curve for V-grooved heat exchanger

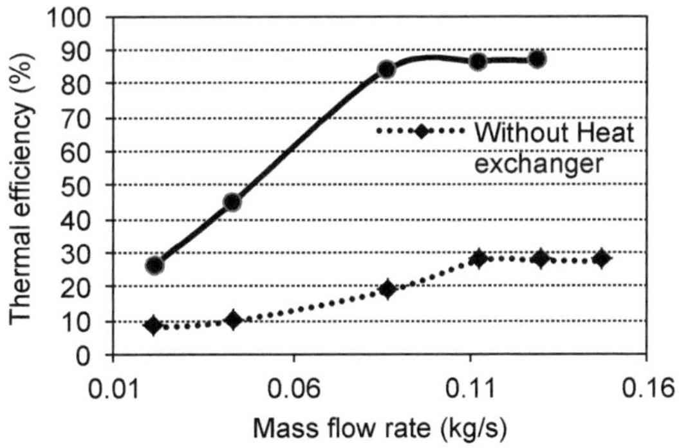

Fig. 56.10 Thermal efficiency curve for stainless steel wool heat exchanger

efficiency collectors by 0.1 % compare to without heat exchanger. The structure of the stainless steel wool is not uniform to allow air to flow freely through it from inlet to outlet. It caused a pressure drop to the flowing air and is not suitable to be used to remove heat from the back of the PV module.

Graph in Fig. 56.12 shows the plot of the thermal efficiency for collector with ∇-grooved heat exchanger. Same as other design, the thermal efficiency, π_{th}, increases

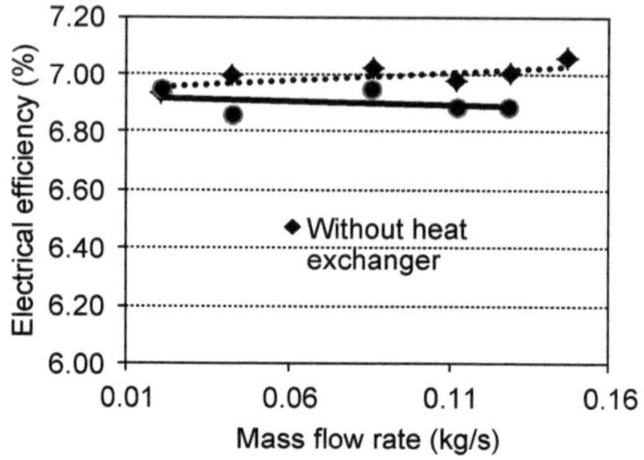

Fig. 56.11 Electrical efficiency curve for stainless steel wool heat exchanger

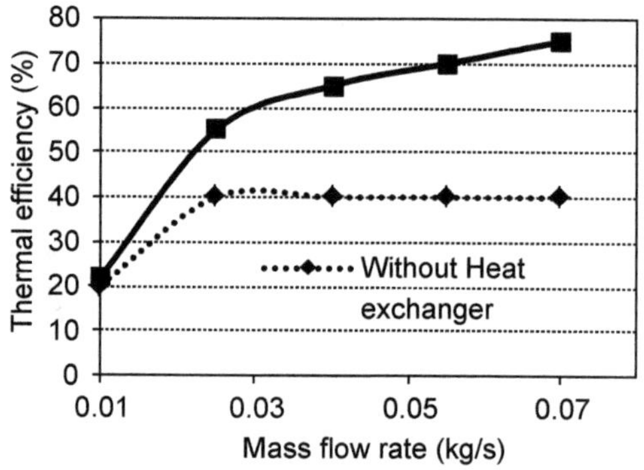

Fig. 56.12 Thermal efficiency curve for ∇-grooved heat exchanger

with the increase in mass flow rate. The maximum thermal efficiency, π_{th}, is ~75.0% and without heat exchanger is 40.0%. Graph in Fig. 56.13 shows the plot for electrical efficiency, π_{el}, for the system. Overall, collectors show an increasing trend for electrical efficiency, π_{el}, with the increase in the fluid mass flow rate. The electrical efficiency improves by 2.0% with the usage of ∇-grooved heat exchanger.

Finally graph in Fig. 56.14 shows the plot of the thermal efficiency for collector with triangular tunnel heat exchanger. The thermal efficiency, π_{th}, increases with the

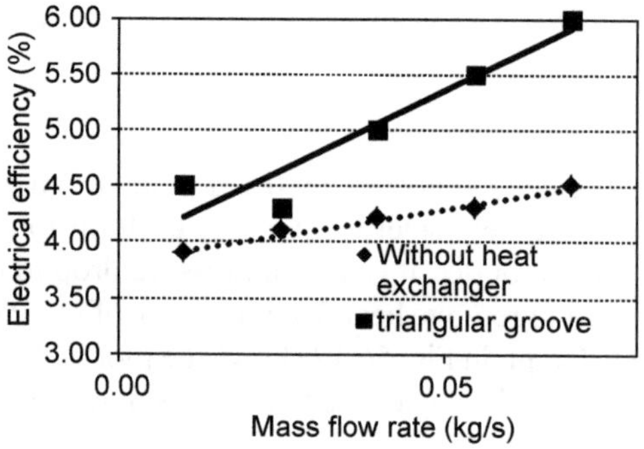

Fig. 56.13 Electrical efficiency curve for ∇-grooved heat exchanger

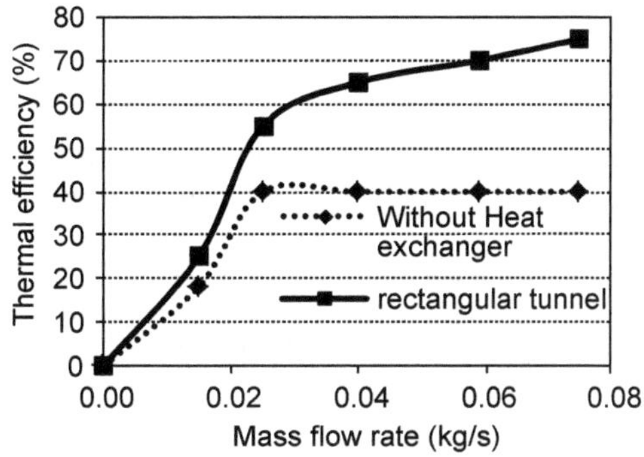

Fig. 56.14 Thermal efficiency curve for rectangular tunnel heat exchanger

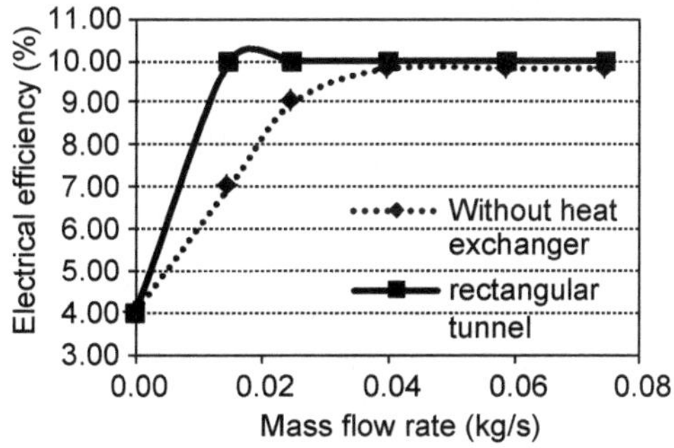

Fig. 56.15 Electrical efficiency curve for rectangular tunnel heat exchanger

increase in mass flow rate. The maximum thermal efficiency, π_{th}, is ~75.0% and without heat exchanger is 40.0%. Graph in Fig. 56.15 shows the plot of electrical efficiency, π_{el}, for the system. The collectors show an increasing trend for electrical efficiency, π_{el}, with the increase in the fluid mass flow rate. The electrical efficiency improves by 0.2% with the usage of triangular tunnel heat exchanger.

56.5 Conclusion

The performance of various designs of heat exchanger for hybrid single-pass PV/T collector is evaluated and explained in this chapter. Every design had its own capability to harvest solar energy and converted it to thermal energy and electrical energy. Overall, PV/T design with heat exchangers show improved efficiency for both electrical and thermal energy, except for PV/T with stainless steel wool which shows decreased electrical efficiency due to its non-uninform structure. PV/T with honeycomb heat exchanger is capable of increasing thermal efficiency by ~50% and electrical efficiency by 0.2%. PV/T with V-groove is capable of enhancing thermal efficiency by ~40%, and no significant improvement is observed for electrical efficiency. PV/T with stainless steel shows ~50% thermal efficiency improvement.

PV/T with ∇-grooved heat exchanger shows increasing value of ~35 % for thermal efficiency and 2.0 % for electrical efficiency. Finally, PV/T with triangular tunnel heat exchanger shows increasing value of ~35 % for thermal efficiency and 0.2 % for electrical efficiency. Findings of this experiment can be further investigated to be applied in solar drying system and space heating in future.

Acknowledgment The experiment described in this chapter was financially supported by SIRIM Berhad, Malaysia. The technical supports are received from UniversitiKebangsaan Malaysia under research grants PRGS/1/11/TK/UKM/01/12. The authors would like to express sincere appreciation to both organizations for their valuable support.

References

1. Hasan MA, Sumathy K (2010) Photovoltaic thermal module concepts and their performance analysis: a review. Renew Sustain Energy Rev 14:1845–1859
2. Kumar R, Rosen MA (2011) A critical review of photovoltaic–thermal solar collectors for air heating. Appl Energy 88:3603–3614
3. Tiwari GN, Mishra RK, Solanki SC (2011) Photovoltaic modules and their applications: a review on thermal modeling. Appl Energy 88:2287–2304
4. Hussain F, Othman MYH, Sopian K, Yatim B, Ruslan H, Othman H (2013) Design development and performance evaluation of photovoltaic/thermal (PV/T) air base solar collector. Renew Sustain Energy Rev 25:431–441
5. Hegazy AA (2000) Comparative study of the performances of four photovoltaic/thermal solar air collectors. Energy Convers Manag 41:861–881
6. Sopian K, Liu HT, Kakac S, Veziroglu TN (2000) Performance of a double pass photovoltaic thermal solar collector suitable for solar drying systems. Energy Convers Manag 41:353–365
7. Zondag HA, Vries DWd, Helden WGJv, Zolingen RJCv, Steenhoven AAV (2003) The yield of different combined PV-thermal collector designs. Sol Energy 74:253–269
8. Tripanagnostopoulos Y (2007) Aspects and improvements of hybrid photovoltaic/thermal solar energy systems. Sol Energy 81:1117–1131
9. Othman MY, Yatim B, Sopian K, Bakar MNA (2007) Performance studies on a finned double-pass photovoltaic-thermal (PV/T) solar collector. Desalination 209:43–49
10. Sopian K, Alghoul MA, Alfegi EM, Sulaiman MY, Musa EA (2009) Evaluation of thermal efficiency of double-pass solar collector with porous–nonporous media. Renew Energy 34:640–645
11. Shahsavar MA, Ameri M (2010) Experimental investigation and modeling of a direct-coupled PV/T air collector. Sol Energy 84:1938–1958
12. Teo HG, Lee PS, Hawlader MNA (2012) An active cooling system for photovoltaic modules. Appl Energy 90:309–315
13. Hussain F, Othman MY, Yatim B, Ruslan H, Sopian K, Anuar Z, Khairuddin S, Performance of a single pass air base photovoltaic/thermal solar collector with and without hexagonal honeycomb heat exchanger, World Renewable Energy Forum, WREF 2012, Including World Renewable Energy Congress XII and Colorado Renewable Energy Society (CRES) Annual Conferen 2, (2012) 859–864
14. Othman MYHJ, Ruslan H, Sopian K, Jin GL (2009) Performance study of photovoltaic-thermal (PV/T) solar collector with ∇-grooved absorber plate. Sains Malaysiana 38(4):537–541
15. Jin GL, Ibrahim A, Chean YK, Daghigh R, Ruslan H, Mat S, Othman MY, Ibrahim K, Zaharim A, Sopian K (2010) Evaluation of single-pass photovoltaic-thermal air collector with rectangle tunnel absorber. Am J Appl Sci 7:277–282

Chapter 57
Optimizing Solar Hot Water Systems (Closed Systems) for Air-Conditioning Cycles in TRNSYS

A Kaabi Nejadian, Ali Mohammadi and Behnoosh Bakhtiari Heleyleh

Abstract Parallel and series solar hot water systems (closed systems) were simulated and optimized. The system is designed to supply the required heat for solar air-conditioning cycles such as heat generators in ejector cycle, absorption cycle generator, regenerator in liquid desiccant cycle and regenerating absorbent of the desiccant wheel, and solid desiccant cycle. A model of solar hot water system was simulated in TRNSYS and then, based on solar fraction, the components of the system were optimized. For optimization of the cycle, the dynamic performance of the system in supplying the required temperature for a cycle of Three Tons Refrigeration was analyzed under different situations. The results were compared with the experimental results to obtain consistency. The effects of the parameters of the system, supplied heat capacity, the energy used by the heater, and the effect of the environment were surveyed regarding feasibility of achieving higher solar energy. Finally, the solar hot water system was optimized for Bushehr city. The results showed that the optimization of the solar system is influenced by environmental parameters such as radiation intensity, dry bulb temperature, relative moisture; the way the system is utilized (residential and institutional); the term of operation of the system; and, most importantly, the set point of the auxiliary hot water system. The optimized system was featured as follows: solar fraction: 0.5, area of collector: 86 m^2, angle of collector: 31°, set point of the auxiliary hot water system: 75°C, capacity of tank: 4 m^3, solar collector discharge: 0.25 kg/s, and mass discharge rate of the thermal converter: 0.2 kg/s.

Keywords Solar hot water system · Flat-plate collector · Storage tank · Optimization · Solar fraction

A. Kaabi Nejadian (✉)
Renewable Energy Organization of Iran, Tehran, Iran
e-mail: Kaabi@iranenergy.org.ir

A. Mohammadi · B. Bakhtiari Heleyleh
Department of Environmental and Energy Engineering Science and Research Branch, Islamic Azad University Tehran, Tehran, Iran
e-mail: ali_makran60@yahoo.com

B. Bakhtiari Heleyleh
e-mail: b.bakhtiari.h@gmail.com

A. Sayigh (ed.), *Renewable Energy in the Service of Mankind Vol II*,
DOI 10.1007/978-3-319-18215-5_57

Nomenclature

Q_{load}	Thermal capacity of the solar hot water system [kW]
T_{in}	Temperature of the fluid of the solar hot water system entering the exchanger [c]
T_{out}	Temperature of the fluid of the solar hot water system exiting of the exchanger [c]
T_{set}	Regulated temperature of the auxiliary hot water system [c]
C_{pw}	Water special temperature [kJ/kg]
M_w	Mass flow of the fluid inside the solar hot water system [kJ/kg]
COP	Coefficient of performance [-]
SF	Solar fraction [-]
A	Total collector array aperture or gross area [m^2]
A_a	Aperture area of a single collector module [m^2]
A_r	Absorber area of a single collector module [m^2]
a_1	Negative of the first-order coefficient in the collector efficiency equation [kJ/h m^2 k]
a_2	Negative of the second-order coefficient in the collector efficiency equation [kJ/h m^2 k]
b_0	Negative of the first-order coefficient in the incident angle curve fit equation [-]
b_1	Negative of the second-order coefficient in the Incidence Angle Modifier (IAM) curve fit equation [-]
F_R	Overall collector heat removal efficiency factor [-]
F_{av}	Modified value of F_R Overall collector heat removal efficiency factor when the efficiency is given in terms of T_{av}, not T_i [-]
F_o	Modified value of F_R Overall collector heat removal efficiency factor when the efficiency is given in terms of T_o, not T_i [-]
I	Global (total) horizontal radiation [kJ/h m^2]
I_d	Diffuse horizontal radiation [kJ/h m^2]
I_T	Global radiation incident on the solar collector (Tilted Surface) [kJ/h m^2]
I_{bT}	Beam radiation incident on the solar collector [kJ/h m^2]
M	Flow rate at use conditions [kg/h]
m_{test}	Flow rate in test conditions [kg/h]
N_S	Number of identical collectors in series [-]
T_a	Ambient (air) temperature [c]
T_{av}	Average collector fluid temperature [c]
T_i	Inlet temperature of the fluid to the collector [c]
T_o	Outlet temperature of the fluid from the collector [c]
U_L	Overall thermal loss coefficient of the collector per unit area [kJ/h m^2 k]
$U_{L/T}$	Thermal loss coefficient dependency on T [kJ/h m^2 k]
α	Short-wave absorptance of the absorber plate [-]
β	Collector slope above the horizontal plate [°]
θ	Incidence angle for beam radiation [°]
ρ_g	Ground reflectance [-]
T	Short-wave transmittance of the collector cover [-]
$(\tau\alpha)$	Product of the cover transmittance and the absorber absorptance [-]

$(\tau\alpha)_b$	For beam radiation (depends on the incidence angle θ) [-]
$(\tau\alpha)_n$	At normal incidence [-]
$(\tau\alpha)_s$	For sky diffuse radiation [-]
$(\tau\alpha)_g$	For ground-reflected radiation [-]
C_p	Specific heat of the fluid [kJ/kg k]
H_a	Height of the auxiliary heater above the bottom of the tank [m]
H_t	Height of the tank [m]
H_{th}	Height of the thermostat above the bottom of the tank [m]
k	Thermal conductivity [W/m^2 k]
m_h	Mass flow rate of the hot stream entering the tank [kg/h]
m_L	Mass flow rate of the load [kg/h]
Q_{aux}	Rate of auxiliary energy input to the tank [kJ/h]
Q_{env}	Rate of energy loss from the tank [kJ/h]
Q_{he}	Maximum rate of energy input to the tank by auxiliary [kJ/h]
Q_{in}	Rate of energy input to the tank from the hot fluid stream [kJ/h]
Q_{sup}	Rate of energy supplied to the load by the tank [kJ/h]
r_i	Ratio of insulation thickness of the top to the sides of the upright tanks [-]
T_D	Temperature of water delivered by the tank to the load [c]
T_{env}	Environmental temperature for losses [c]
T_h	Temperature of the hot fluid entering the tank [c]
T_I	Temperature of its segment [c]
T_L	Temperature of the load stream entering the tank [c]
T_R	Temperature of the fluid return to the heat source [c]
T_{set}	Thermostat set temperature [c]
T	Average tank temperature [c]
V_h	Volume of the fluid entering the tank from the heat source over a time interval Δt [m^3]
V_L	Volume of the fluid entering the tank from the load over a time interval Dt [m^3]
V_t	Tank volume [m^3]
ΔE	Change in the internal energy storage [kg m^2/s^2]
UA	Overall Heat Losses value of the tank [kJ/h]
Δt	Simulation time step [s]
ΔT_{db}	Thermostat temperature dead band [s]
ρ	Fluid density [kJ/kg k]
γ_{heater}	Optional control function input (0 to 1) that disables or enables the auxiliary heater [-]

57.1 Introduction

One of the greatest challenges today is world energy supply. For that reason, great efforts have been made by a number of developed countries to rationalize energy use by means of research seeking more energy-efficient systems or by using other energy sources such as the sun, wind, biomass, and wave and tide. Due to the fast development of economics and the tremendous energy demand in Iran and other developing countries, solar energy has been looked at as an attractive substitute energy source for space heating in recent years.

Riffat et al. [1] constructed a thin-membrane, flat-plate, heat-pipe solar collector and developed an analytical model that was used to simulate heat transfer processes occurring in the collector and calculated its efficiency. Bong et al. [2] presented a validated theoretical model to determine the efficiency, heat removal factor, and outlet water temperature of a single collector and an array of flat-plate heat-pipe collectors. Ezekwe [3] analyzed the thermal behavior of solar energy systems using heat-pipe absorbers and compared them with systems using conventional solar collectors. Bojic et al. [4] modeled and simulated the performance of a forced circulation solar water-heating (SWH) system using a time-marching model.

Jones and Golshekan [5] studied the desertification of a packed bed consisting of river pebbles along with several additional properties, and the experimental data of this study were found to be mutually consistent and were in good agreement with a previously published work. Sorour investigated a small-sized sensible heat energy storage unit constructed out of gypsum rocks and indicated the optimal design and operating conditions for maximum thermal efficiency of this small unit [6].

Different computational tools have been developed to numerically evaluate the long-term performance of solar systems and to study the effect of design parameters. TRNSYS is an extensive software for transient simulation of solar systems (thermal or photovoltaic (PV)), low-energy solar multi-zone buildings, renewable energy systems, fuel cells, and their related equipment. This program has been widely used to study and optimize solar systems. There have been several studies that used TRNSYS to evaluate the effect of different design parameters and operating conditions on the performance of a thermosyphon solar water-heating (TSWH) system.

TRNSYS is a complete and extensible simulation environment for the transient simulation of systems, including solar thermal systems. However, not much work has been done concerning the use of TRNSYS simulation for solar air heating. The present chapter has focused on the use of TRNSYS to analyze a solar air-based heating system for a public building in North China. An investigation was performed to observe the design parameters of the system and to determine their optimum values. The annual and monthly solar fractions (SFs) of the entire system were used as the optimization parameters.

Buckles and Klein [7] compared the performance of different configurations of forced circulation systems. Michaelides and Wilson [8] optimized the design criteria of an active SWH system for hotel application. Wongsuwan and Kumar [9] studied the performance of forced circulation systems experimentally and numerically. The numerical simulations were conducted using TRNSYS and artificial neural network. The results from both numerical models were found to be in good agreement with the experimental values. For instance, Furbo et al. [10] recommended the design of a large SWH as a low-flow system with hot water tanks, external heat exchangers, and stratification inlet pipes; Nayak and Amer [11] evaluated nine dynamic test procedures for the evaluation of flat-plate collectors; Tsilingiris [12] developed a simple simulation model for large SWH systems; Kikas [13] studied laminar flow distribution of reverse and direct return circuits in solar collectors; Prapas et al. [14] studied the thermal behavior of large central SWH; Fanney and Klein [15] investigated the influence of flow rate and also the incorporation of an auxiliary heat exchanger on the performance; Chiou [16] developed a numerical method to determine the variation in performance due to nonuniform flow distribution; and Klein and Beckman [17] presented a general design method for closed-loop SWH systems. The present work is focused on using TRNSYS to analyze a forced circulation SWH system for a single-family residential unit in Montreal, Canada. A comprehensive study has been conducted to study all design parameters of the system and to determine their optimum values. The monthly and annual SFs of the entire system are used as the optimization parameters.

The objective of this chapter is to develop a TRNSYS simulation model for forced-circulation SWH systems with flat-plate tube collectors and validate the model using measured field performance data. The validated model will be useful for long-term performance simulation under different weather and operating conditions. The model could also be used for system optimization under different load profiles.

57.2 Solar Hot Water System

As pictured in Fig. 57.1, the model comprises two loops. One includes a solar collector, a pump, and a tank; and another includes a tank, an auxiliary hot water system, a three-way valve, a pump, and a balance valve.

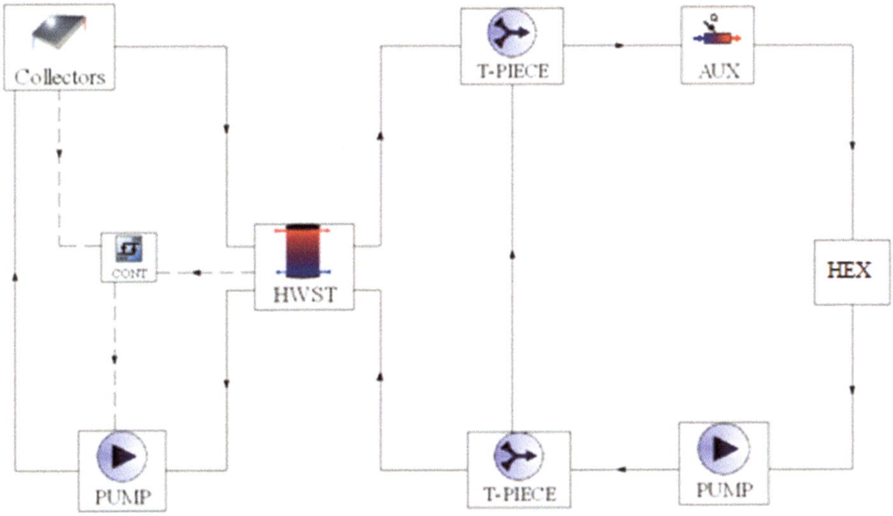

Fig. 57.1 Solar hot water system model in TRNSYS. *AUX* Auxiliary; *HEX* Heat Exchanger; *HWST* Hot Water Storage Tank

Throughout the first loop, the temperature of the fluid (water) rises by passing the solar collector before being stored in the tank. By losing its heat (supplying warm water), the water is pumped to the cycle. When the temperature of water at the outlet is higher than that of the inlet, the pump is turned off by the controller so that energy waste is prevented.

The second loop is featured with water flow from the tank to the three-way valve and inflow to the auxiliary hot water system, when the temperature of the discharge water from the tank is less than the preset temperature of the auxiliary hot water system; it heats the water up to T_{set} temperature.

When the temperature of the outflow tank is less than T_{set}, the auxiliary hot water system is turned off. Then, the water enters the shell and pipe convertor (to increase wet-bulb temperature) and is pumped into the circulation pipe after being cooled down. The discharged fluid from the pump enters the balance valve. If the temperature of the inflow fluid to the pipe is less than that of the tank's outlet, the fluid is led to path 2, otherwise to the tank.

57.3 Equations

57.3.1 Flat-Plate Collector

The energy collected by the collector is obtained from Hottel–Whillier equations. The equations of the flat-plate collector are as follows:

The amount of energy received by the collector:

$$\dot{Q}_u = \frac{A}{N_s}\sum_{j=1}^{N_s} F_{\mathrm{R},j}\left(I_{\mathrm{T}}(\tau\alpha) - U_{\mathrm{L},j}\left(T_{\mathrm{i},j} - T_{\mathrm{a}}\right)\right) \tag{57.1}$$

$$F_{\mathrm{R},j} = \frac{N_{\mathrm{s}} m_c C_{\mathrm{pc}}}{A U_{\mathrm{L},j}}\left[1 - \exp\left(-\frac{F U_{\mathrm{L},j} A}{N_{\mathrm{s}} m_c C_{\mathrm{pc}}}\right)\right]. \tag{57.2}$$

Heat transfer coefficient:

$$h_w = 5.7 + 3.8W\left(\frac{W}{m^2} - k\right) \tag{57.3}$$

$$f = \left(I - 0.04h_w + 0.0005h_w^2\right)\left(I + 0.09N_G\right) \tag{57.4}$$

$$C = 365.9\left(I - 0.00883\beta + 0.0001298\beta^2\right). \tag{57.5}$$

Transfer coefficient and the total absorption coefficient:

$$(\tau\alpha) = \frac{I_{\mathrm{bT}}(\tau\alpha)_{\mathrm{b}} + I_{\mathrm{d}}\left(\frac{1+\cos\beta}{2}\right)(\tau\alpha)_{\mathrm{s}}}{I_{\mathrm{T}}} + \frac{\rho I\left(\frac{1-\cos\beta}{2}\right)(\tau\alpha)_{\mathrm{g}}}{I_{\mathrm{T}}}. \tag{57.6}$$

Outlet temperature of each component:

$$T_{\mathrm{o},j} = \frac{AF_{\mathrm{R},j}[I_{\mathrm{T}}(\tau\alpha) - U_{\mathrm{L},j}(T_{\mathrm{i},j} - T_{\mathrm{a}})]}{N_{\mathrm{s}}\dot{m}_c C_{\mathrm{pc}}} + T_{\mathrm{i}}. \tag{57.7}$$

The fluid flow is zero:

$$T_{\mathrm{p}} = \frac{I_{\mathrm{T}}(\tau\alpha)}{U_{\mathrm{L}}} + T_{\mathrm{a}}. \tag{57.8}$$

The waste coefficient of the solar collector is obtained by highly complicated equations including those developed by Klein [18].

57.3.2 Hot Water Storage Tank

Figure 57.2 represents the idea of a plug-flow tank, which is divided into four sections with capacities of V_{i} and T_{i} [18]. The temperature order is not represented by temperature inversion. For a specific time period, heat is extracted from a specific volume of fluid (V_{h}), which is equal to $(\dot{m}_{\mathrm{h}})\Delta t / \rho$ at temperature T_{h}.

Assuming $T_{\mathrm{h}} > T_{\mathrm{L}}$, a new section is added to the top of the tank, and the profile is transferred to the new section. At the same time, the inflow volume from the section "Load" is equal to $(\dot{m}_{\mathrm{h}})\Delta t / \rho$ at temperature T_{L}. When $T_{\mathrm{L}} < T_4$, a section is added to the bottom of the tank, and the profile is retransferred. The transfer of the profile network in the tank is equal to the difference between the total volume of the temperature generator and the volume of the consumed fluid, or $(\dot{m}_{\mathrm{h}} - \dot{m}_{\mathrm{L}})\Delta t / \rho$.

The mathematical equations of the plug flow of the storage tank are as follows:

Average water temperature inside the tank to be used:

$$T_{\mathrm{D}} = \frac{V_{\mathrm{h}}T_{\mathrm{h}} + (V_{\mathrm{L}} - V_{\mathrm{h}})T_1}{V_{\mathrm{L}}}. \tag{57.9}$$

The overall mortality rate of tanks:

$$\dot{Q}_{\mathrm{env}} = \sum_{i=1}^{N}(UA)_i(T_{\mathrm{i}} - T_{\mathrm{env}}). \tag{57.10}$$

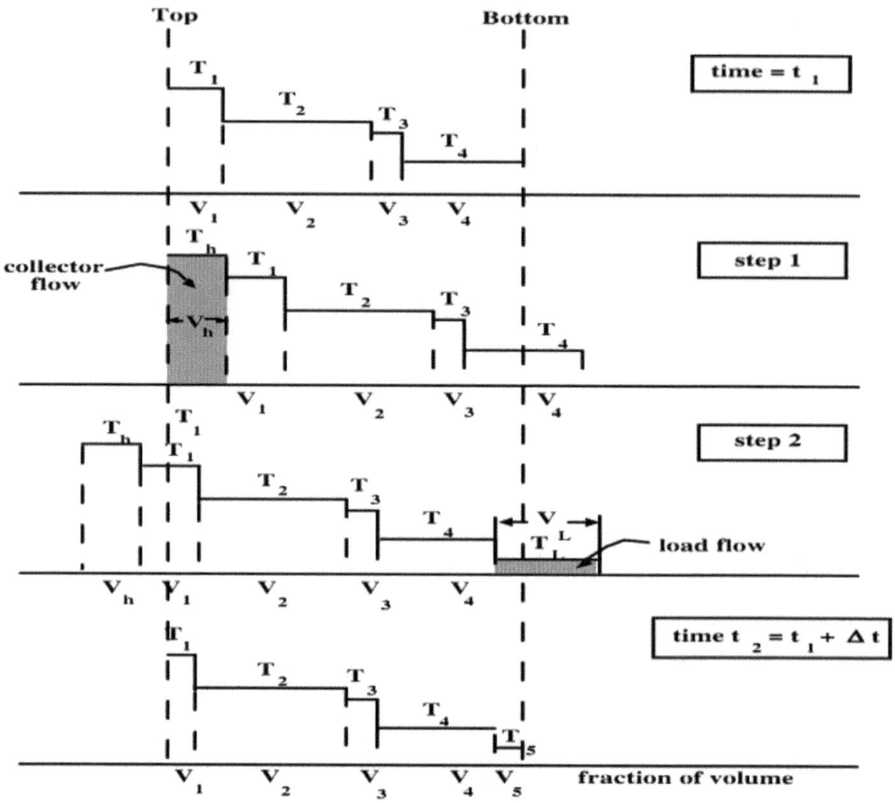

Fig. 57.2 Theoretical model of a storage tank with fixed volume

Energy input to the tank from the hot fluid flow:

$$\dot{Q}_{\text{in}} = \dot{m}_{\text{h}} C_{\text{p}} (T_{\text{h}} - T_{\text{R}}). \tag{57.11}$$

Energy available for consumption:

$$\dot{Q}_{\text{sup}} = \dot{m}_{\text{L}} C_{\text{p}} (T_{\text{D}} - T_{\text{L}}). \tag{57.12}$$

Tank internal energy changes:

$$\Delta E = \rho C_{\text{p}} \left(\sum_{i=1}^{N} T_i V_i - \sum_{i=1}^{N} T_i V_i \,\Big|_{t=\text{TIME}0} \right). \tag{57.13}$$

57.4 Definition

A definition of SF is introduced before discussing the optimization problem.

57.4.1 Solar Fraction

This value is defined as the contribution of collector and thermal storage tank in total energy needed by cooling system; notably, the electric power of the pump is omitted.

In general, different works have introduced different definitions of SF. The definition proposed in this work is as follows:

$$\text{Solar fraction} = \frac{q_{\text{load}} - q_{\text{aux}}}{q_{\text{load}}} \tag{57.14}$$

$$q_{\text{load}} = m_{\text{w}} C_{\text{pw}} \left(T_{\text{in}} - T_{\text{out}}\right). \tag{57.15}$$

57.5 Optimum Diagram of a Solar Hot Water System

57.5.1 Structural Optimum (Series/Parallel)

The optimization of a solar hot water system can be approached in two ways: (1) optimization of the structure of the system (type of tank; multi-layer, single-layer) and (2) optimization of other components and features such as fluid discharge, area of collector, and volume of storage tank (Fig. 57.3).

The structure of the system was divided into five models for optimization. The basis of optimization is the SF. The five models are as follows:

1. Multi-layer tank with parallel auxiliary hot water system
2. Single-layer tank with series auxiliary hot water system
3. Full-mix tank with parallel auxiliary hot water system
4. Full-mix tank with series auxiliary hot water system
5. Multi-layer tank with inside auxiliary hot water system

The optimum model is assumed to have minimum energy consumption by the auxiliary hot water system or maximum SF.

Other details of the system (i.e., area of the collector, discharge rate, storage tank capacity, and capacity of the auxiliary hot water system) are assumed unchanged.

Table 57.1 lists the results of the five models based on SF for a year in the city of Tehran.

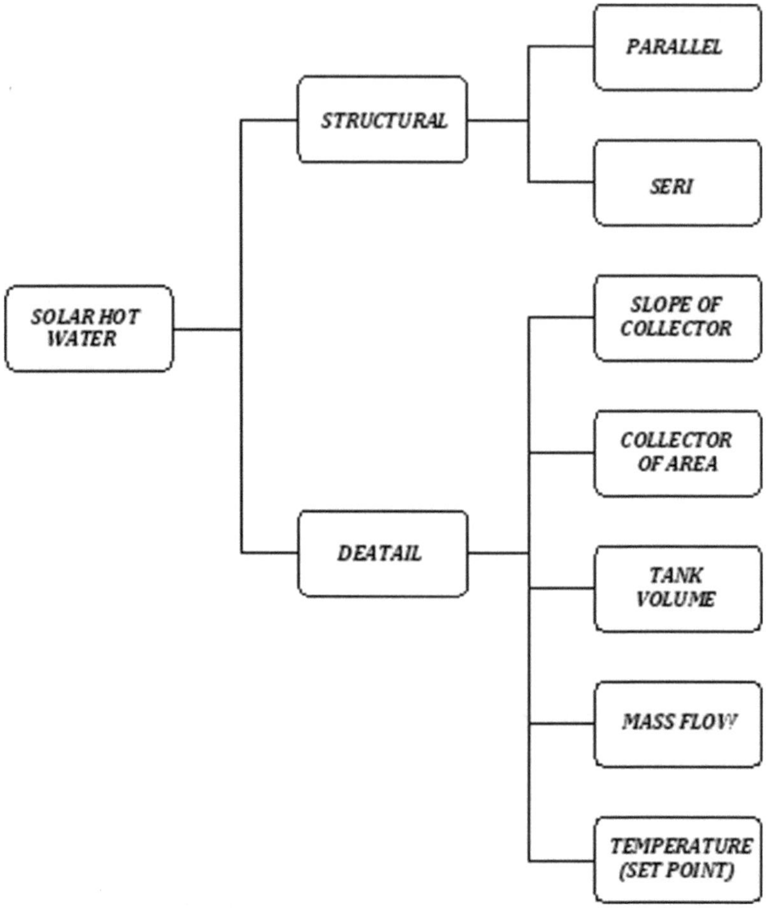

Fig. 57.3 Optimum diagram of a solar hot water system

Table 57.1 Solar fraction for the different models

Model	1	2	3	4	5
Solar fraction	0.93	0.86	0.92	0.85	0.87

As indicated in Table 57.1, maximum SF was observed in model 1 and with small difference in model 3. Model 1 was not a good option given that the cost of building or buying a multi-layer tank is much more than that of a single-layer tank or a full-mix tank. Furthermore, the difference of SF between the two models was trivial. Model 3 was adopted as the optimum structure of a solar hot water system (Fig. 57.1). Here, having the structure of the solar hot water system optimized, the area of the collector, the discharge rate, the capacity of the storage tank, and the set point temperature (TSP) of the auxiliary hot water system are optimized.

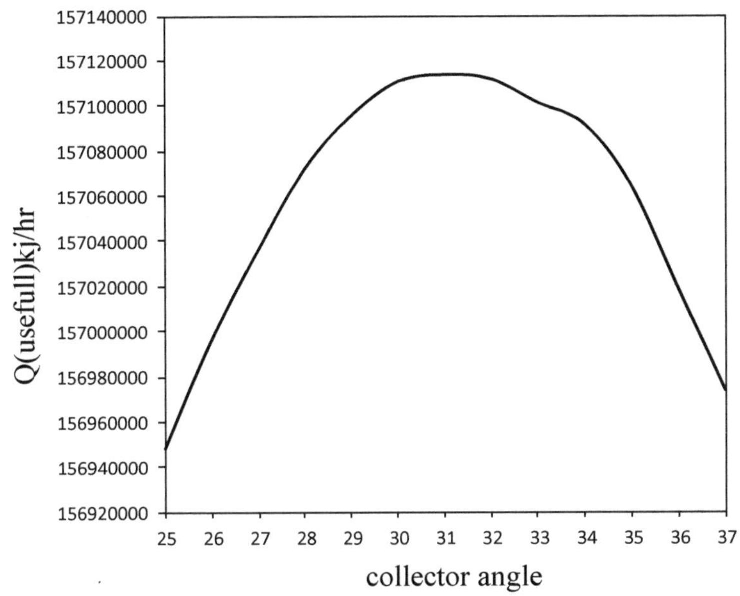

Fig. 57.4 Absorbed heat based on the collector slope angle (Bushehr city)

57.5.2 Optimization of the Components

57.5.2.1 Optimization of the Collector Angle

The slope angle of the collector was set based on the absorbed temperature by the collector throughout the cooling phase for the climate of Bushehr city in the latitude of 30° (Fig. 57.4). Based on the experiment, the slope angle of the collector ranges from 1° to 5° higher than the latitude of the site. Clearly, the optimum angle is 31°, which is consistent with the above.

57.5.2.2 Optimization of the Pump Mass Discharge Rate

Two parameters need to be adjusted for the optimization of the mass discharge rate: SF and regeneration. The discharge rate cannot be optimized by keeping one of the parameters fixed as one parameter considerably influences the other. Figure 57.5 shows the optimized mass discharge of the right loop.

The mass discharge rate of the solar collector was optimized using SF measure, after the optimization of the discharge rate of the right loop—comprising auxiliary hot water system, convertor, and storage tank. Figure 57.6 illustrates the variation in SF based on mass discharge rate. The maximum SF of the solar hot water system is obtained for the discharge rate of 0.4 kg/s. It is notable, however, that the left loop was optimized first.

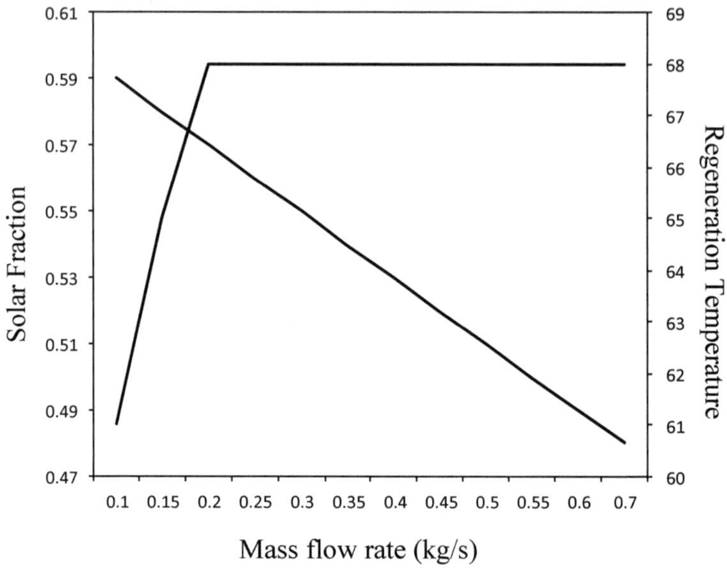

Fig. 57.5 Effect of mass discharge on regeneration temperature and solar fraction

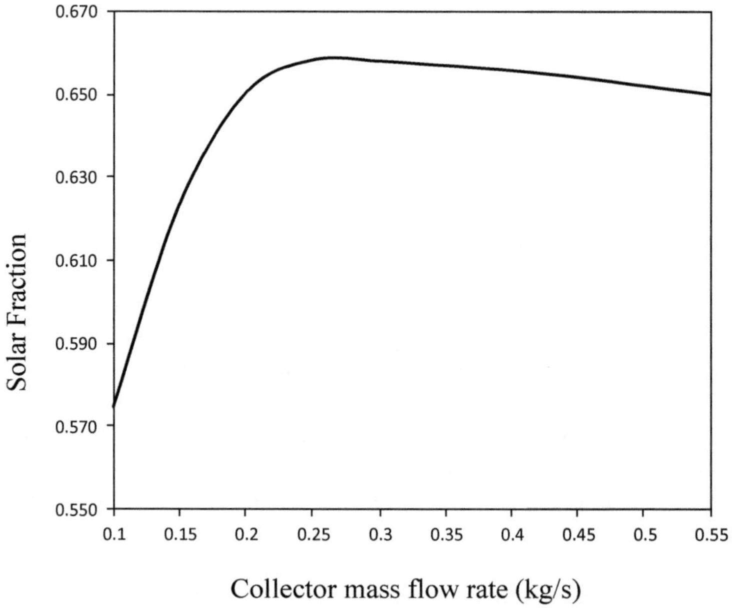

Fig. 57.6 Effect of mass discharge rate of the collector on solar fraction

As shown in the figure, the variation in mass discharge rate is trivially effective on SF, while SF is the basis of optimization in the solar desiccant cycle.

57.5.2.3 Optimization of the Area of the Collector and the Capacity of the Storage Tank

There are two ways to optimize the area of the collector and the capacity of the storage tank. First, both components are optimized at the same time, so that the area of the collector is set and then the SF is obtained based on the capacity of the tank. Afterward, the area is adjusted and the process is repeated. The area is adjusted until the minimum capacity of the SF tank is obtained.

In the second method, which is more accurate, the area is adjusted based on SF (by assuming a capacity for the storage tanker). The area for which the maximum area of SF is obtained is recorded, and then the capacity of the storage tank is adjusted based on the obtained area. Afterward, the maximum SF of the storage tank is recorded, and then the capacity of the new storage obtained in the last stage is replaced with the new capacity of the tank, and the area is readjusted to obtain the maximum area of SF. Then, using the optimum area and adjusting the capacity of the storage tanker, the maximum SF is obtained. The process is repeated until the difference between the obtained capacity of the storage tanker of two consecutive turns is close to zero (Figs. 57.7–57.12). At this point, the obtained area and capacity are optimum. It is notable that we started by choosing a preliminary discharge rate,

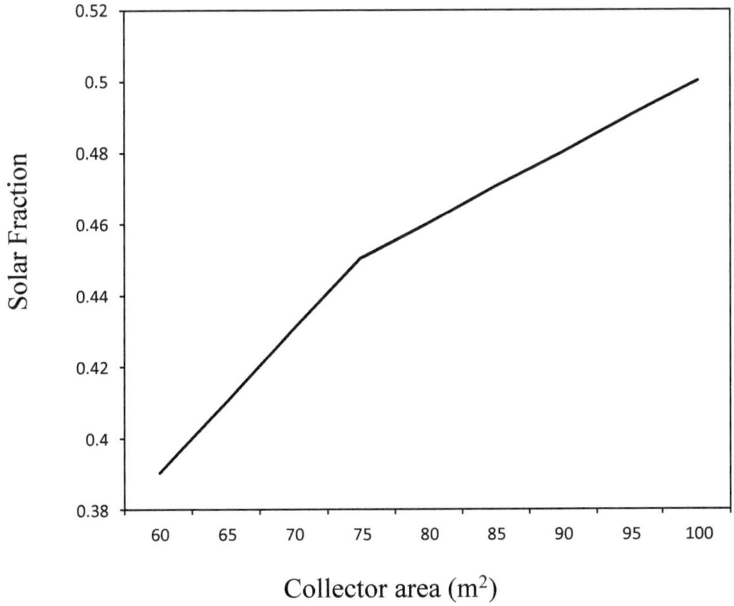

Fig. 57.7 Solar fraction based on the area of the collector

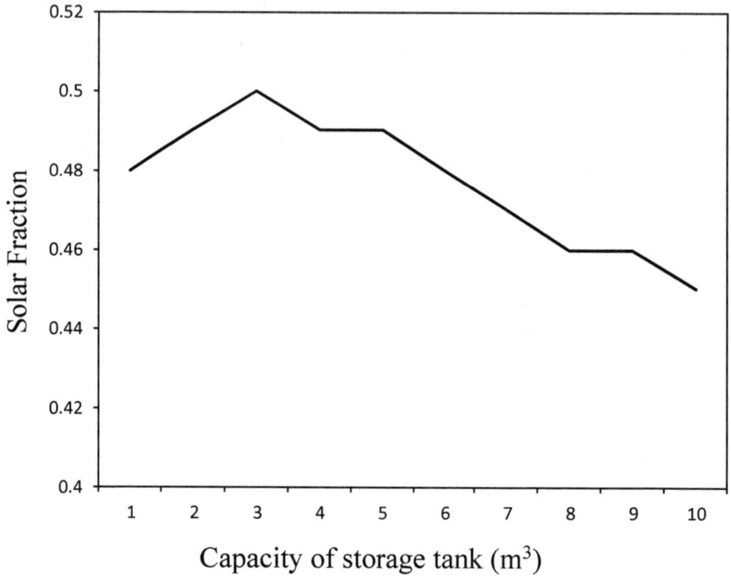

Fig. 57.8 Solar fraction based on the capacity of the storage tank

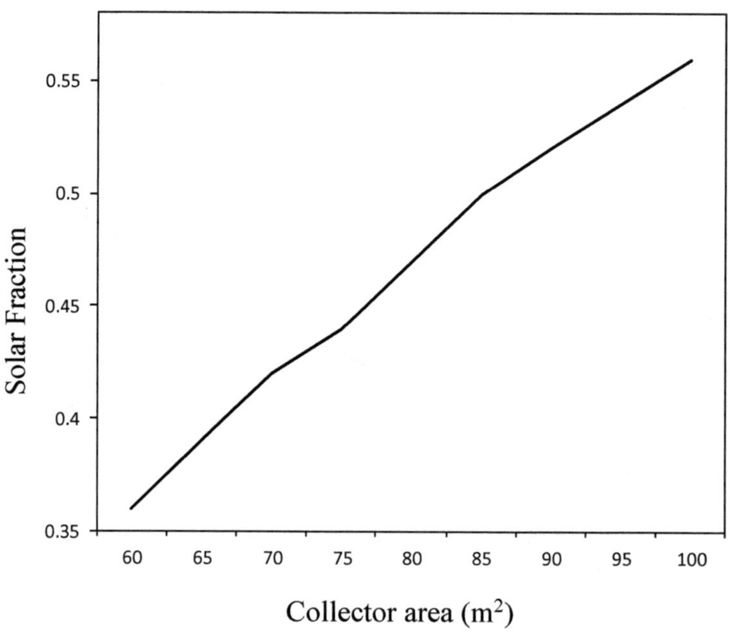

Fig. 57.9 Solar fraction based on the area of the collector

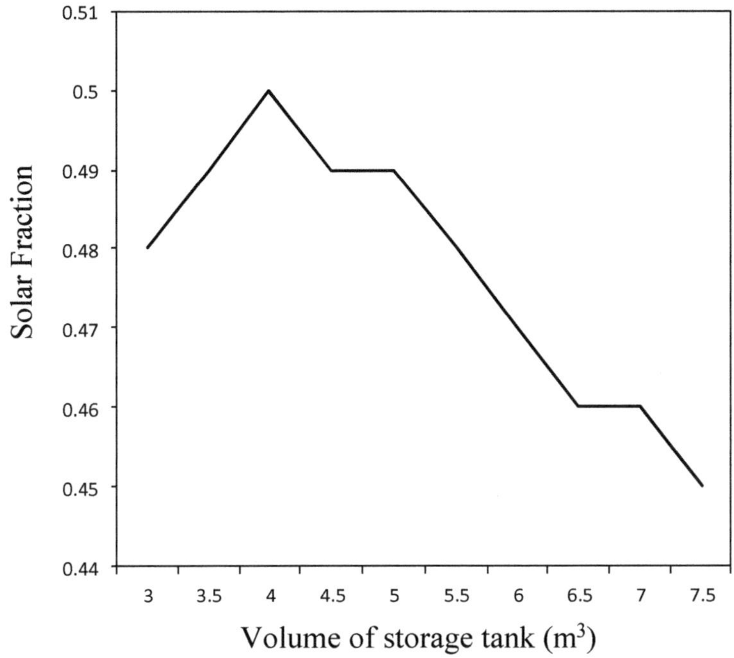

Fig. 57.10 Solar fraction based on the volume of the storage tank

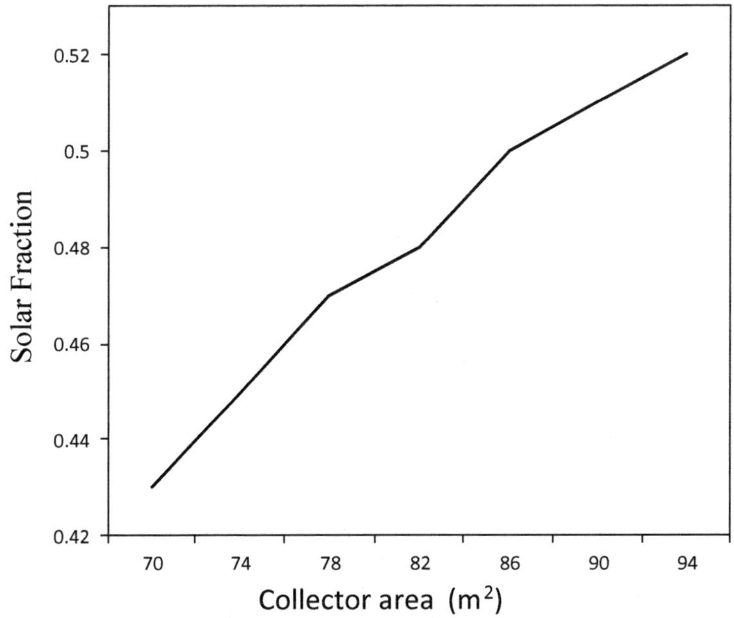

Fig. 57.11 Solar fraction for the area of the collector

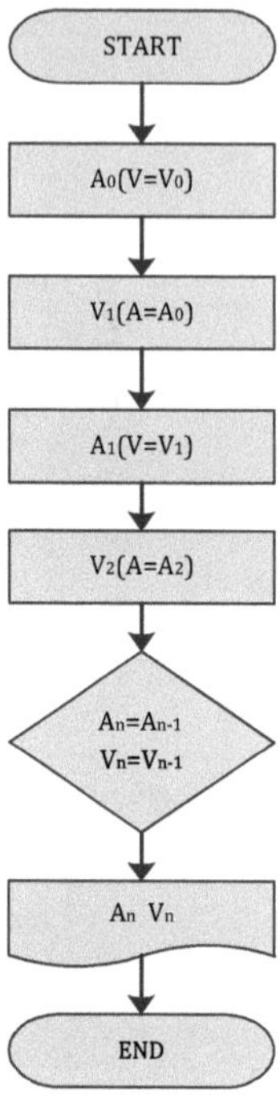

Fig. 57.12 Optimization flowchart of the area of the collector and the capacity of the tank based on solar fraction

which is almost relative to the discharge rate of the fluid on the desiccant cycle, and SF is set to 0.5. First, the optimum area is obtained for a storage capacity of 2 m^3.

With an area of 100 m^2, the optimum capacity for the tank is obtained (Fig. 57.8).

The area is recalculated for a capacity of 3 m^3 (Fig. 57.9).

In Fig. 57.10, the capacity of the tank is obtained for a collector area of 85 m^2.

For the third turn, the area is obtained for a capacity of 4 m^3 (Fig. 57.11).

Thus, the optimum area and capacity are 84 m^2 and 4 m^3, respectively, as shown in Fig. 57.12.

57.5.2.4 Optimization of TSP of the Auxiliary Hot Water System

Optimizing TSP is one of the critical and most effective steps in the optimization of solar hot water systems. Taking into consideration that economic optimization is not a concern here, SF was set to 0.5, based on which the temperature of the auxiliary hot water system (Fig. 57.13) was obtained.

However, the designer cannot make an accurate optimization regarding regeneration temperature as there are cycles with nonsolar regeneration, and there is always a fixed temperature. The generator of heat is either gas or an electric torch, which enables the designer to set the best temperature. On the other hand, the solar temperature generator varies throughout the day and the only way of modification by the designer is to set the TSP, which is the minimum required temperature. Another serious challenge ahead of the designer is the high cost of a solar hot water system.

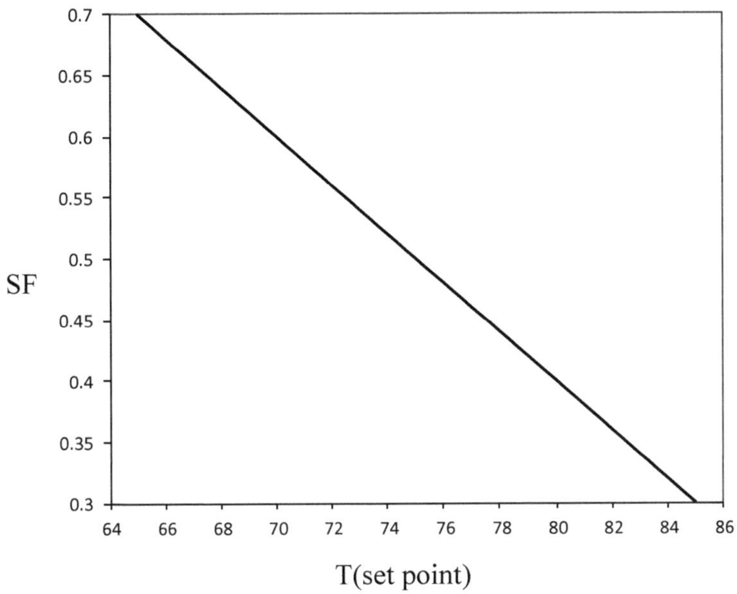

Fig. 57.13 Optimization of the set point temperature of the auxiliary hot water system based on solar fraction *(SF)*

57.6 Simulation Results

A solar hot water system was simulated and analyzed regarding dynamic and time-to-time performance in TRNSYS. As mentioned above, the optimization of the system was discussed.

Here, the performance of different components of the solar hot water system for an apartment located in Bushehr city with an area of 115 m^2 for 48 h in August is analyzed. The performance of the cycle after optimization is as follows:

Area of collector: 86 m^2
Angle of collector: 31°
Water discharge rate of the right loop: 0.2 kg/s
Water discharge rate of the left loop: 0.25 kg/s
Capacity of the storage tank: 4 m^3
Auxiliary hot water system temperature ($T_{setpoint}$): 75°C

References

1. Riffat SB, Zhao X, Doherty PS (2005) Developing a theoretical model to investigate thermal performance of a thin membrane heat-pipe solar collector. Appl Therm Eng 25(5e6):899e915
2. Bong TY, Ng KC, Bao H (1993) Thermal performance of a flat-plate heat-pipe collector array. Sol Energy 50(6):491e498

3. Ezekwe CI (1990) Thermal performance of heat pipe solar energy systems. Sol Wind Technol 7(4):349e354
4. Bojic M, Kalogirou S, Petronijevi K (2002) Simulation of a solar domestic water heating system using a time marching model. Renew Energy 27(3):441e452
5. Jones BW, Golshekan M (1989) Destratification and other properties of a packed bed heat store. Int J Heat Mass Transf 32(2):351–359
6. Sorour MM (1988) Performance of a small sensible heat energy storage unit. Energy Convers Manage 28(3):211–217
7. Buckles WE, Klein SA (1980) Analysis of solar domestic hot water heaters. Sol Energy 25(5):417–424
8. Michaelides IM, Wilson DR (1997) Simulation studies of the position of the auxiliary heater in thermosyphon solar water heating systems. Renew Energy 10(1):35–42
9. Wongsuwan W, Kumar S (2005) Forced circulation solar water heater performance prediction by TRNSYS and ANN. Int J Sustain Energy 24(2):69–86
10. Furbo S, Vejan NK, Shah LJ (2005) Thermal performance of a large low flow solar heating system with a highly thermally stratified tank. ASME J Sol Energy Eng 127:15–20
11. Nayak JK, Amer EH (2000) Experimental and theoretical evaluation of dynamic test procedures for solar flat-plate collectors. Sol Energy 69(5):377–401
12. Tsilingiris PT (1996) Solar water-heating design—a new simplified dynamic approach. Sol Energy 57(1):19–28
13. Kikas NP (1995) Laminar flow distribution in solar systems. Sol Energy 54(4):209–217
14. Prapas DE, Veliannis I, Evangelopoulos A, Sotiropoulos BA (1995) Large DHW solar system with distributed storage tanks. Sol Energy 55(3):175–184
15. Fanney AH, Klein SA (1988) Thermal performance comparisons for solar hot water systems subjected to various collector and heat exchanger flow rates. Sol Energy 40(1):1–11
16. Chiou JP (1982) The effect of non-uniform fluid flow distribution on the thermal performance of solar collectors. Sol Energy 29(6):487–502
17. Klein SA, Beckman WA (1979) A general design method for closed loop solar energy systems. Sol Energy 22:269–282
18. Klein SA et al (2004) TRNSYS Version. 16, Solar Energy Laboratory, University of Wisconsin-Madison

Chapter 58
Analytical Model Development for Efficient Solar Desalination System (SDS)

Bassam A. Noaman and Ahmed F. Elsafty

Abstract Supply of adequate quantities of fresh potable water is one of the most serious problems confronting human, especially when we know that one third of the world population are suffering from water shortage and it is expected to reach two thirds in the near future. Therefore, desalination, as a non-conventional water resource, has become one of the most interesting alternative water sources to partially face the freshwater scarcity in the near future. This chapter presents a parametric study to simulate for an optimal design of a transportable solar desalination system (SDS) in order to achieve maximum thermal performance. The study reveals that changing the solar intensity, reflector parabolic height and width, evaporation area, wind velocity, saline water depth, and absorbing tube shape changes the unit productivity.

Keywords Solar · Water · Desalination · Mathematical model

58.1 Introduction

With the expansion in the population and, in turn, in the industry development, the search for a solution to produce freshwater becomes very eminent. Therefore, many experimental and theoretical researches have been conducted to produce freshwater by means of solar energy and desalination [1].

A general mathematical model was developed for a solar still that uses parabolic reflector tube absorber desalination technology [2]. This model solves the governing heat and mass transfer action which occurs during the desalination operation. Experimental work on different desalination systems was designed and implemented to study the influence of concentrators on productivity [3] and compare the results with a theoretical model [4].

The purpose of this study is to develop a computer program and perform a parametric study on the solar desalination system (SDS) in [2] by solving the nonlinear

B. A. Noaman (✉) · A. F. Elsafty
College of Engineering and Technology,
American University of The Middle East (AUM), Egaila, Kuwait
e-mail: bassam.noaman@aum.edu.kw

A. Sayigh (ed.), *Renewable Energy in the Service of Mankind Vol II*,
DOI 10.1007/978-3-319-18215-5_58

heat balance equations and investigate the effect of the relevant parameters on the thermal performance and total production of the SDS. The modeled SDS is investigated to maximize the absorbed energy by water and reduce the loss of heat to glass cover, therefore increasing the total production. This chapter investigates the effects of a few of the following influential parameters on SDS: (1) environmental parameters, which correspond to the geographical locations and its weather conditions and are summarized as solar intensity, ambient air temperature, and wind velocity; (2) design parameters, which correspond to dimensions, material properties, and coating colors and are summarized as reflector aperture area, parabolic reflector height and width, efficiency of reflection, condenser thickness, condenser emissivity, and absorbing tube shape; and (3) operational parameters, which are summarized as evaporation area and saline water depth.

A computer program is developed to analyze and simulate for maximum thermal performance and total still production with optimal design and operation parameters.

58.2 Methodology

The main components of the system that take part in the heat transfer operation are shown in Fig. 58.1 and Table 58.1. Each component forms an energy balance equation, and the overall equations for all components of the system are solved simultaneously to compute their temperatures and finally the overall water production. A number of assumptions have been made to ease the computation and are listed as follows:

- Solar intensity and ambient temperature are approximated to a sine curve [5].
- Both absorptivity and emissivity of glass cover are constant.

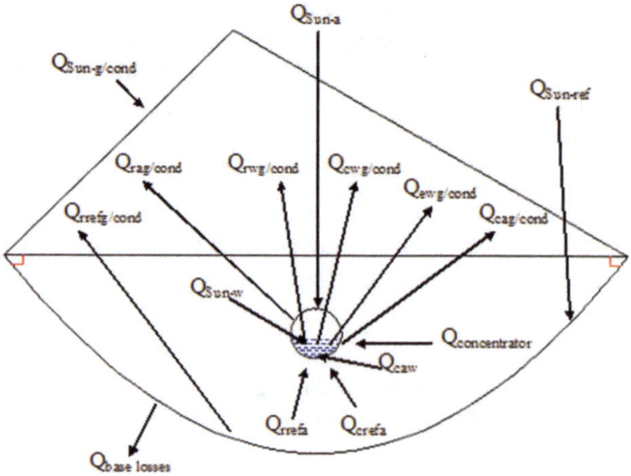

Fig. 58.1 The principles of still component's energy balance

Table 58.1 Desalination system's energy exchange mechanisms

Distillation components	Heat energy gain (W)	Heat energy loss (W)
Glass condenser	Sun ($Q_{Sun-g/cond}$) Water [*radiation* ($Q_{rwg/cond}$), *convection* ($Q_{cwg/cond}$), *evaporation* ($Q_{ewg/cond}$)] Absorber [*radiation*($Q_{rag/cond}$), *convection* ($Q_{cag/cond}$)] Reflector [*radiation* ($Q_{rrefg/cond}$)]	Sky *(radiation)* Air (*convection*)
Water	Sun (Q_{Sun-w}) Absorber [*convection* (Q_{caw})]	Glass/condenser *(radiation, convection, evaporation)*
Absorber	Sun (Q_{Sun-a}) Reflector [*radiation* (Q_{rrefa}), *convection* (Q_{crefa})] Parabolic concentrator ($Q_{concentrator}$)	Glass/condenser *(radiation, convection)* Water *(convection)*
Reflector	Sun ($Q_{Sun-ref}$)	Base ($Q_{base\ losses}$) Absorber [*radiation, convection*] Glass/condenser *(radiation)*

- All physical and thermal properties are constant with temperature.
- There is no leakage of water vapor from the still.
- Flow rate is very small; therefore, the level of water in the absorber is assumed to be constant.
- There is no temperature gradient in glass cover.

The principles of energy exchange mechanisms in the developed solar still are illustrated and summarized in Fig. 58.1 and Table 58.1, which are as follows:

- A very large part of solar energy falling on the still is absorbed by the absorber.
- The energy absorbed by the absorber is largely transferred to the water in the still, and a small part is lost to the surrounding by convection and radiation.
- Energy is transferred from the absorber and water to the glass cover by the water vapor evaporating from the water surface and then losing its heat of vaporization to the glass cover during condensation.
- Small reflection loss occurs at the glass and reflector base.
- Heat is transferred to the glass cover from water by radiation and by free convection on the trapped air in the still.
- A small part of incident solar energy is absorbed by the glass cover, and some energy is lost to the atmosphere by convection and radiation.

Different environmental, design, and operational parameter variations are investigated to study their effect on the optimal performance and water production. The environmental parameters are those related to the geographical site where the system is located and its corresponding weather. The investigated parameter is the solar intensity. The design parameters are those related to the still configurations such as dimensions, material properties, and coating colors. The investigated parameters include the parabolic reflector height and width and the absorbing tube shape. The operational parameters are those which can be easily changed by the operator during the operation in order to increase the total productivity for a certain still design and environmental conditions. The investigated parameters include evaporation area and saline water depth.

58.3 Results

Results are shown with the different parameter variations of the desalination system. The theoretical results are computed from sunrise time (6:00) to sunset time (18:00). Figure 58.2a shows the temperature variations of each of the main components of the system, with an absorber tube having a diameter of 76 mm and 10 holes, and each hole has a diameter of 15 mm. The wind speed is 2 m/s, the height of the parabolic reflector is 0.5 m, and the maximum solar intensity is 800 W/m^2, which may be received on earth surface for a given location [5]. The figure shows the temperature curves of water and receiver to be very close. This is expected since the area of evaporation is very small. The small evaporation areas, which correspond to the areas of the 10 holes, lead to trap the generated vapor inside the receiver tube, which in turn increases the water temperature. The water temperature curve intersects the receiver temperature curve after midday since the time needed by water for releasing its energy is more than the time needed by the receiver tube. The water loses its energy with respect to its heat capacity. The area of the glass is larger than the condensation area, and this is reflected in the condenser temperature where it is lower than the glass cover temperature. The ambient temperature curve is plotted versus time with maximum temperature input of 40°C. Figure 58.2b shows the effect of solar intensity on the water temperature.

The effect of water depth and volume is shown in Fig. 58.3a and b with different maximum solar intensities 400 and 800 W/m^2, respectively. It is evident from

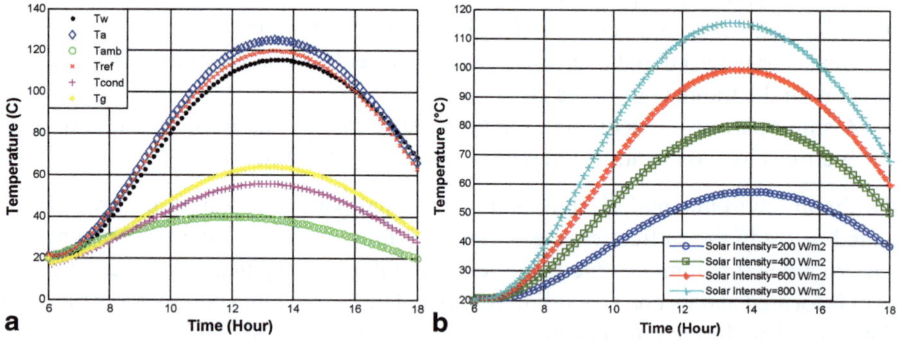

Fig. 58.2 **a** Temperature variations of water *(Tw)*, absorber *(Ta)*, ambient *(Tamb)*, reflector *(Tref)*, condenser *(Tcond)*, and glass *(Tg)* and **b** the effect of solar intensity on water temperatures

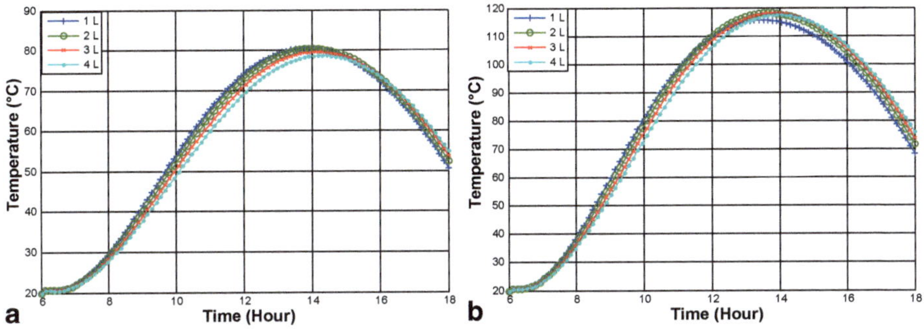

Fig. 58.3 **a** The effect of water volume on water temperatures with maximum solar intensity = 400 W/m^2 and **b** maximum solar intensity = 800 W/m^2

the figure that increasing the water depth/volume leads to a decrease in water temperature. Furthermore, the higher the water volume, the higher the time required to increase its temperature. This is expected and is clearly shown in Fig. 58.3a and b since the higher the water depth/volume, the higher the water heat capacity.

The effect of the tube diameter on water temperature is shown in Fig. 58.4a and b with different maximum solar intensities 400 and 800 W/m^2, respectively. For a given solar energy, increase in the tube diameter leads to an increase in the absorber surface area, which in turn leads to less absorbance and water temperature. This is clearly evident when increasing the tube diameter area as shown in Fig. 58.4a and b.

Figure 58.5a shows the effect of wind speed on glass temperature. It is expected that the increase in wind speed results in energy loss in glass and decrease in glass temperature. Figure 58.5b shows the effect of the number of holes on water temperature. It is shown that increasing the number of holes leads to a decrease in water temperature. This is expected since holes correspond to evaporation areas, and increasing the evaporation areas leads to more vapors generated from water, consequently losing water energy and decreasing water temperature. Figure 58.6a and b shows the effect of increasing the area of evaporation on the water productiv-

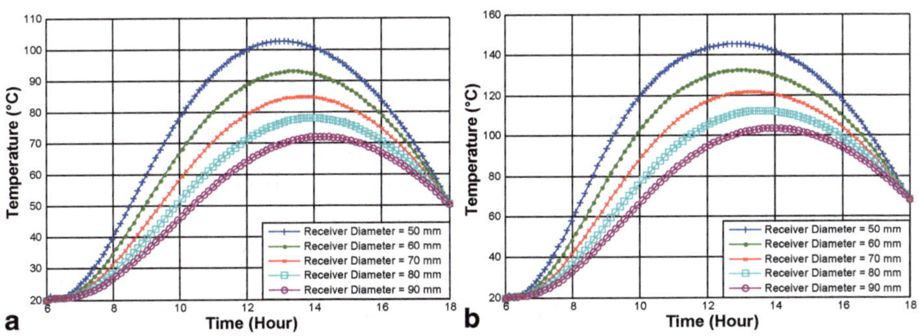

Fig. 58.4 **a** The effect of receiver's diameter on water temperatures with maximum solar intensity = 400 W/m^2 and **b** maximum solar intensity = 800 W/m^2

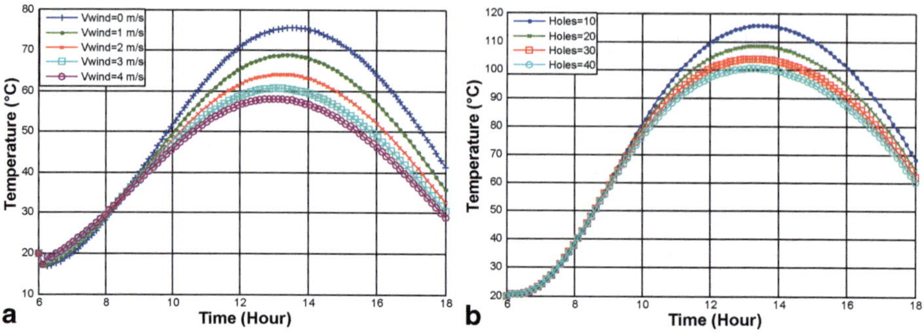

Fig. 58.5 **a** The effect of wind speed on glass temperatures for maximum solar intensity = 800 W/m^2 and **b** the effect of number of holes on water temperatures for maximum solar intensity = 800 W/m^2

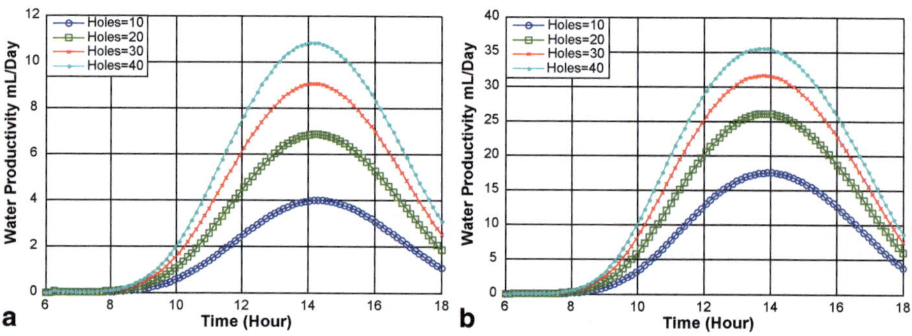

Fig. 58.6 **a** The effect of the number of holes on water productivity for maximum solar intensity = 400 W/m^2 and **b** maximum solar intensity = 800 W/m^2

ity for maximum solar intensities 400 and 800 W/m^2, respectively. It is shown that increasing the evaporation area leads to an increase in water productivity due to the more water vapor and consequently more condensation.

Figure 58.7a and b shows the effect of the parabolic reflector height and width on the reflector temperature, respectively. It is a principle of geometry that a parabolic reflector facing the sun will reflect parallel rays of light to the focal point of the parabola, which in this system is the location of the absorbing tube. It is shown that for a given aperture reflector area, increasing the parabolic reflector height leads to a decrease in the energy directed to the reflector and in turn a decrease in its temperature. Figure 58.7b shows that increasing the reflector width results in an increase in energy and temperature in the reflecting surface.

Figure 58.8a and b shows the effect of the shape of the absorbing tube on the water and absorbing tube temperatures and the water productivity, respectively. Modifying the absorbing tube from cylindrical to elliptically cylindrical shape for a given volume leads to a decreasing surface area of the absorbing tube. As a result, the

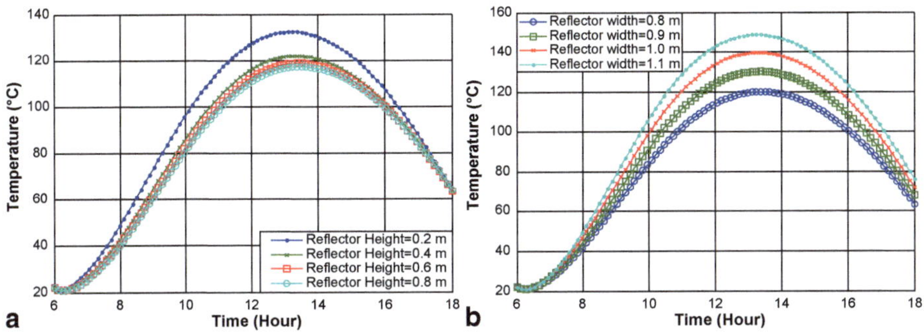

Fig. 58.7 **a** The effect of the parabolic reflector height and **b** reflector width on reflector temperature for maximum solar intensity = 800 W/m^2

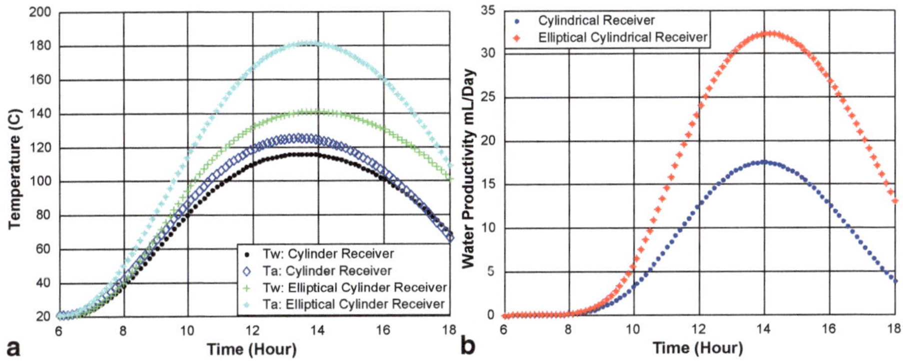

Fig. 58.8 **a** The effect of cylinder receiver and elliptical cylinder receiver on water *(Tw)* and absorber *(Ta)* temperature and **b** on water productivity for maximum solar intensity = 800 W/m^2

water and absorber's temperature increases, which increases the water productivity as shown in Fig. 58.8a and b.

58.4 Conclusion

A computer program is developed to study the effect of different environmental, design, and operational parameters on a transportable SDS performance and productivity. This system may be used in remote areas where solar energy and the source of saline water are abundant. These system parameters include solar intensity, wind speed, parabolic reflector height and width, absorbing tube shape, and water depth/volume. The investigation of the aforementioned parameters reveals that evaporation surface area, water depth, parabolic reflector height and width, absorbing tube shape, and wind speed have significant effects on water productivity in water desalination system.

References

1. Xiao G, Wang X, Mingjiang N, Wang F, Zhu W, Luo Z, Cen K (2013) A review on solar stills for brine desalination. Appl Energy 103:642–652
2. Elsafty AF, Fath HE, Amer AM (2008) Mathematical model development for a new solar desalination system (SDS). Int J Energy Convers Manage 49:3331–3337
3. Elsafty AF, Abbas MM (2013) Experimental investigation for a new solar desalination system. Energy Proced 36:676–686
4. Nematollahi F, Rahimi A, Gheinani T (2013) Experimental and theoretical energy and exergy analysis for a solar desalination system. Desalination 317:23–31
5. Radosavljević J, Đorđević A (2001) Defining of the intensity of solar radiation on horizontal and oblique surfaces on earth. Work Living Environ Prot 2(1):77–86

Chapter 59
A Model to Estimate Ambient Conditions and Behavior of the Airflow Inside a Solar Chimney

Janaína Oliveira Castro Silva, Tauane Shaisly Fernandes, Sérgio de Morais Hanriot, Antônia Sônia Alves Cardoso Diniz, André Guimarães Ferreira and Cristiana Brasil Maia

Abstract This chapter presents models to estimate the environmental conditions and the behavior of the airflow within a prototype of a small-scale solar chimney located in Belo Horizonte, Brazil. A correlation from the literature for diffuse radiation, based on clearness index and global radiation, was evaluated. A model from the literature was used to estimate the ambient temperature of the device. The results of both parameters were compared with the experimental data. An energy balance was applied to find the heat interactions between the ground, airflow, coverage, and environment, based on the estimated incident solar radiation and ambient temperature. Literature correlations were used to estimate the convective heat transfer coefficients. Consolidated correlations were then applied to estimate the mass airflow rate and outlet temperature of the airflow inside the prototype. The results were compared with the experimental data for 4 days in autumn and good agreements were found. The model was then used to estimate the airflow parameters for 1 year. The analysis performed was transient, with results provided

C. Brasil Maia (✉) · J. O. Castro Silva · T. S. Fernandes · S. d. M. Hanriot · A. S. A. Cardoso Diniz
Department of Mechanical Engineering, Pontifícia Universidade Católica de Minas Gerais, Belo Horizonte, Brazil
e-mail: cristiana@pucminas.br

J. O. Castro Silva
e-mail: janainajocs@hotmail.com1

T. S. Fernandes
e-mail: tshaisly@gmail.com1

S. d. M. Hanriot
e-mail: hanriot@pucminas.br1

A. S. A. Cardoso Diniz
e-mail: asacd@pucminas.b1

A. G. Ferreira
Graduate Program of Energy Engineering, Centro Federal de Educação Tecnológica de Minas Gerais, Belo Horizonte, Brazil
e-mail: agferreira@deii.cefetmg.br2

A. Sayigh (ed.), *Renewable Energy in the Service of Mankind Vol II*,
DOI 10.1007/978-3-319-18215-5_59

for each hour of the day, for 365 days. These results showed good accordance with the experimental data. The greater differences were found in the mass flow rate at night.

Keywords Solar chimney · Mathematical model · Ambient temperature · Solar radiation

59.1 Introduction

The past few decades have been marked by increased energy demand, associated with population growth and increasing energy consumption. The world energy matrix is mainly based on fossil fuels that, although cheaper, feature a number of drawbacks and are being depleted at an accelerated rate. In the 1980s, when environmental issues have become a global concern, the quest for sustainable energy sources began. The main goal is the search of renewable sources of energy and the pursuit of the highest possible energy efficiency. Solar energy is an abundant, inexpensive, and environmentally friendly source, and has been the subject of theoretical and experimental studies. Solar chimney is a promising technology, which absorbs direct and diffuse solar radiation to generate a hot airflow, used to drive a turbine or to dry agricultural products (Fig. 59.1). The airflow parameters, such as mass flow rate and outlet temperature, depend mainly on ambient conditions.

Bernardes et al. [1] developed a model to estimate the energy production in solar chimneys and examined the effects of different environmental conditions and structural dimensions on the power output. Larbi et al. [2] presented the performance analysis of a prototype and analyzed the influence of physical and geometrical parameters on the output power of the chimney, performing an overall analysis. Nizetic and Klarin [3] showed a mathematical model based on a

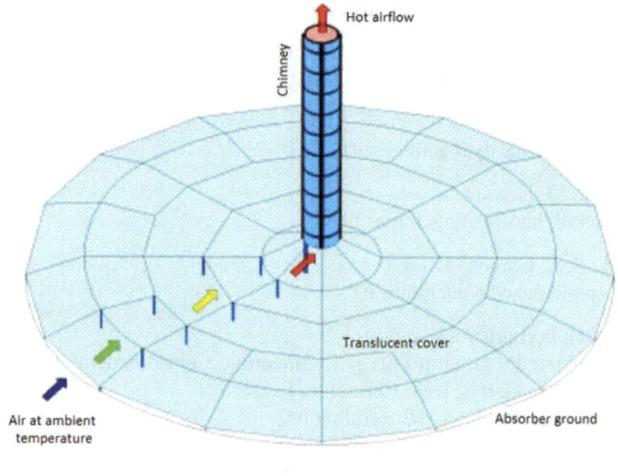

Fig. 59.1 Schematics of solar chimney

global analysis of a solar chimney to estimate the ideal factor of the turbine pressure drop. The authors stated that even for constant solar radiation the potential pressure is not fixed, but rather a function of an increase in air temperature in the collector. Hamdan [4] provided a mathematical model to evaluate the performance of a solar chimney power plant based on their dimensions and operating conditions. Pretorius and Kröger [5] evaluated the influence of a correlation coefficient for heat transfer convection with a coefficient of loss of turbine inlet, as well as various types of glass collectors on the performance of a large-scale solar chimney. Hamdan [6] developed a model of simplified thermodynamic analysis to analyze the feasibility of implementing solar chimneys. Cao et al. [7] showed a model of heat transfer that is used to compare the performance of a conventional solar chimney and two inclined solar chimneys. The results indicated that a larger angle provides better performance in the collector in winter but results in lower performance in the summer.

This chapter aims to develop a mathematical model to estimate the mass airflow rate and the outlet temperature in a prototype of a solar chimney. Literature models were used to predict the incident solar radiation and the ambient temperature. These parameters were then used as input data for the airflow model.

59.2 Mathematical Model

The total solar radiation incident I on a surface is the sum of direct radiation I_b, diffuse radiation I_d, and ground-reflected radiation. The incident solar radiation is predicted considering an isotropic sky and the average values are obtained from the literature for the clearness index (Table 59.1).

The radiation flux absorbed by the ground can be calculated by using Eq. (59.1), where $(\tau\alpha)$ is the transmittance–absorptance product for the parcels of direct and diffuse radiation, subscribed by b and d, respectively, and reflected by the ground. β represents the angle formed between the cover and a horizontal surface ($\beta = 0$) and ρ_{solo} represents the ground reflectance [9].

$$S = I_b R_b (\tau\alpha)_b + I_d (\tau\alpha)_b \left(\frac{1+\cos\beta}{2}\right) + \rho_{\text{ground}} I (\tau\alpha)_{\text{ground}} \left(\frac{1-\cos\beta}{2}\right) \quad (59.1)$$

The model for ambient temperature subdivides the day into five intervals (Eq. 59.2). The model uses the daily maximum Tai, $d_{\max}$, minimum Tai, $d_{\min}$, sunrise $\text{Tai}_{t,\text{sunrise}}$, and sunset $\text{Tai}_{\text{sunset}(j)}$ temperatures [10], given by:

Table 59.1 Ratio of monthly clearness index to Belo Horizonte [8]

Month	Jan	Feb	Mar	Apr	May	June	July	Aug	Sept	Oct	Nov	Dec
$\overline{K_T}$	0.46	0.45	0.51	0.53	0.55	0.60	0.64	0.53	0.48	0.50	0.45	0.41

$$\begin{aligned}
&\mathrm{Tai} = \mathrm{Tai}_{\mathrm{sunrise}} + \mathrm{slp}_{\mathrm{before},K_{x,\max}} K_x; && t_{\mathrm{sunrise}} < t < t_{K_x,\max} \\
&\mathrm{Tai} = \mathrm{Tai}_{t,\max} - \mathrm{slp}_{\mathrm{after},K_{x,\max}} \left(K_{x,\max} - K_x\right); && t_{K_x,\max} < t < t_{\mathrm{sunset}} \\
&\mathrm{Tai} = \mathrm{Tai}_{t,\mathrm{sunset}} - \mathrm{NCR}(j)\left(t - t_{\mathrm{sunset}}\right); && t_{\mathrm{sunset}} < t < t_{\mathrm{midnight}} \\
&\mathrm{Tai} = \mathrm{Tai}_{t,\mathrm{sunrise}} - 0.67\mathrm{NCR}(j)\left(t - t_{\mathrm{sunrise}}\right); && \mathrm{t}_{\mathrm{midnight}} < t < 2/3\left(t - t_{\mathrm{sunrise}}\right) \\
&\mathrm{Tai} = \mathrm{Tai}_{t,\mathrm{sunrise}} - 0.5\mathrm{NCR}(j)\left(t - t_{\mathrm{sunrise}}\right); && 2/3\left(t - t_{\mathrm{sunrise}}\right) < t < t_{\mathrm{sunrise}}
\end{aligned} \quad (59.2)$$

The model starts from the premise that there is a response of the air temperature at the inlet of the solar radiation, K_x. This term is the ratio of the amount of solar radiation received by the ground from sunrise and the amount of solar radiation that a surface perpendicular to the solar rays has received during the same period (Eq. 59.3). G_o and G_{sc} refer to the maximum solar radiation incident on a horizontal surface and the solar constant, respectively.

$$K_x = \frac{\int_{\mathrm{sunrise}}^{t} G_o \, \mathrm{d}t}{\int_{\mathrm{sunrise}}^{t} G_{sc} \, \mathrm{d}t} \quad (59.3)$$

More details about the model can be seen in [10].

An energy balance was applied to determine the heat $\left(S \right)$exchanges in the cover. There is a convection heat transfer between the cover and the airflow inside the chimney ($q''_{\mathrm{conv},3}$) and a radiation heat transfer between the cover and the ground ($q''_{\mathrm{rad},1}$). This heat is lost to the external environment by convection ($q''_{\mathrm{conv},2}$) and radiation ($q''_{\mathrm{rad},2}$). The heat transmitted by conduction to the deeper layers of the ground ($q''_{\mathrm{cond},0}$) is determined as the difference between the absorbed solar radiation (S), the flow of energy transferred to the ground by convection flow ($q''_{\mathrm{conv},1}$) , by radiation to the coverage ($q''_{\mathrm{rad},1}$), and to the external environment $\left(q''_{rad,3}\right)$ [Eq. (59.4)].

$$S - q''_{\mathrm{conv},1} - q''_{\mathrm{rad},1} - q''_{\mathrm{rad},3} = q''_{\mathrm{cond},0} \quad (59.4)$$

The radiation between the ground and coverage can be approximated by radiation between two infinite flat plates [9]. The net radiation flux exchanged with the coverage is given by Eq. (59.5), where $\varepsilon_{\mathrm{col}}$ and $\varepsilon_{\mathrm{ground}}$ represent the emittance of coverage and ground, respectively. T_{ground} and T_{col} represent the temperatures of the ground and coverage, respectively.

$$q''_{\mathrm{rad},1} = \frac{\sigma\left(T_{\mathrm{ground}}^{4} - T_{\mathrm{col}}^{4}\right)}{\frac{1}{\varepsilon_{\mathrm{col}}} + \frac{1}{\varepsilon_{\mathrm{ground}}} - 1} \quad (59.5)$$

The convective heat transfer between the coverage and the outer airflow is given by Eq. (59.6).

$$q''_{conv,2} = h_{col,amb}(T_{col} - T_{ai}) \tag{59.6}$$

The convective heat transfer coefficient $h_{col,amb}$ can be estimated by Eq. (59.7), where T_m is the average temperature between the coverage and the ambient temperature. ΔT is the difference between the temperature inside the coverage and the ambient temperature. ρ, μ, C_p, and k represent, respectively, density, dynamic viscosity, specific heat at constant pressure, and thermal conductivity of air evaluated at T_m . V_{amb} represents the velocity of the air outside the coverage [1, 5, 11].

$$h_{col,amb} = \frac{0.2106 + 0.0026 V_{amb} \left(\frac{\rho T_m}{\mu g \Delta T} \right)^{\frac{1}{3}}}{\left(\frac{\mu T_m}{g \Delta T C_p k^2 \rho^2} \right)^{\frac{1}{3}}} \tag{59.7}$$

The heat transfer by radiation between the coverage and the sky can be approximated by the radiation from a small convex object surrounded by a large surface, Eq. (59.8). T_{sky} is the effective sky temperature.

$$q''_{rad,2} = \sigma \varepsilon_{col}({T_{col}}^4 - {T_{sky}}^4) \tag{59.8}$$

The convective heat transfer between the ground and the airflow in the coverage is given by Eq. (59.9), where $\overline{T_{ao}}$ represents the average air temperature inside the coverage.

$$q''_{conv,1} = h_{ground,flow}(T_{ground} - \overline{T_{ao}}) \tag{59.9}$$

Pretorius [12] proposes to use Eq. (59.10) to determine the convective heat transfer coefficient between ground and flow, where Pr is the Prandtl number and V is the mean flow velocity.

$$h_{ground,flow} = 3.87 + 0.0022 \left(\frac{V \rho C_p}{\mathrm{Pr}^{\frac{2}{3}}} \right) \tag{59.10}$$

According to Maia [13] coverage transmits approximately 41 % of the radiation emitted by the ground. The radiation flux exchanged between the ground and the environment can be estimated by

$$q''_{\mathrm{rad},3} = 0.41\sigma\varepsilon_{\mathrm{col}}(T_{\mathrm{ground}}{}^{4} - T_{\mathrm{sky}}{}^{4}) \tag{59.11}$$

The temperature on the surface of the ground can be determined by Eq. (59.12) [13]:

$$T_{\mathrm{ground}}(t) = T_i + \frac{2}{\sqrt{\pi}}\frac{\alpha'}{k_{\mathrm{ground}}}\int_{t'=0}^{t}\frac{q''_{\mathrm{cond},0}(tt')}{\sqrt{4\alpha'(t-t')}}\,\mathrm{d}t' \tag{59.12}$$

The convective heat exchanged between the coverage and the internal flow is given by Eq. (59.13) where T_{ao} represents the airflow temperature.

$$q''_{\mathrm{conv},3} = h_{\mathrm{col,flow}}(T_{\mathrm{col}} - T_{\mathrm{ao}}) \tag{59.13}$$

The convective heat transfer coefficient between the cover and the airflow $h_{\mathrm{col,flow}}$ can be estimated using Eq. (59.14) where Re_x is the Reynolds number along the coverage [1, 5, 11].

$$h_{\mathrm{col,flow}} = \frac{\left(\frac{f}{8}\right)(\mathrm{Re}_x - 1000)\mathrm{Pr}}{1 + 12.7\left(\frac{f}{8}\right)^{\frac{1}{2}}\left(\mathrm{Pr}^{\frac{2}{3}} - 1\right)}\left(\frac{k}{d_h}\right) \tag{59.14}$$

The Darcy friction coefficient f is obtained from Eq. (59.15) [14]:

$$f = (1.82\,\log_{10}\mathrm{Re}_x - 1.64)^{-2};\quad \mathrm{Re}_x < 3.4\times10^{6} \tag{59.15}$$

Koonsrisuk et al. [15] proposed an expression for mass flow and temperature, disregarded the cross-sectional area of the tower in the calculation of the collector area. This simplification cannot be used in this chapter, so the expression for the mass flow rate was rewritten in the form of Eq. (59.16).

$$\dot{m} = \sqrt[3]{\frac{\frac{\rho^2\beta' g q''_{\mathrm{conv},1}\pi^3}{8Cp_{\mathrm{ao}}}H_t\left[R_{\mathrm{c}}^{2} - \left(\frac{D_t}{2}\right)^2\right]}{\frac{4F_yH_t}{D_t^{5}} + \frac{\frac{F_x}{64}}{R_{\mathrm{c}}h_{\mathrm{c}}^{3}} + \frac{1}{D_t^{4}}}} \tag{59.16}$$

H_t represents the height of the chimney, R_{c} the radius of coverage, D_t the diameter, and h_{c} height of the tower roof from the ground. β' is the coefficient of volumetric expansion for real gases, and the terms F_x and F_y represent the friction factors for collector and tower, respectively, given by Eqs. (59.17) and (59.18) [15]:

$$F_x = 0.046\mathrm{Re}_x^{-\frac{1}{5}}; \quad 5000 < \mathrm{Re}_x < 2\times10^5 \tag{59.17}$$

$$\frac{1}{\sqrt{F_y}} = 1.5635\ln\left(\frac{\mathrm{Re}_x}{7}\right); \quad 4000 < \mathrm{Re}_y < 10^7 \tag{59.18}$$

The outlet temperature of the chimney is calculated by Eq. (59.19) [15]:

$$T_{\mathrm{ao}} = \frac{q''_{\mathrm{conv},1}\pi}{\dot{m}C_p}\left[R_{\mathrm{c}}^2 - \left(\frac{D_t}{2}\right)^2\right] + T_{\mathrm{ai}} \tag{59.19}$$

The mathematical model was run for each hour of the day, for 365 days, allowing the determination of the parameters for the whole year, with a 1-hour interval. Calculations were performed using the commercial software Engineering Equation Solver.

59.3 Results and Discussion

Solar radiation and ambient temperature were modeled for the city of Belo Horizonte, Brazil (latitude 19°55'S and longitude 43°56'W). The daily maximum, minimum, sunrise, and sunset temperatures were defined using the experimental data provided by the Brazilian National System for Environmental Data (SINDA). The results were compared with the data provided by SWERA project for a typical meteorological year (TMY) and with the experimental data of Maia [13] and Ferreira [16], for 4 days in May. The airflow parameters were determined for the dimensions of a prototype of a solar chimney, built in Belo Horizonte. The diameter and height of the tower were 1 and 12.3 m, respectively; and the diameter and height of the coverage were 25 and 0.5 m, respectively [17]. Numerical results were compared with the experimental data of Maia [13] and Ferreira [16], for 4 days in May.

The diffuse and global components of solar radiation are shown in Fig. 59.2a. The values obtained using the proposed model were compared with the data from TMY and with the experimental data for the period between May 8 andMay 12. It can be seen that, although the average differences are significant, the proposed model is able to represent the general behavior of the diffuse and global components of solar radiation. It is noteworthy that the TMY tends to represent an average year, and great differences are found when the experimental data are compared with the TMY data.

Figure 59.2b shows the ambient temperature, in comparison to TMY data and experimental data for the same period presented in Fig. 59.2a. The model was able

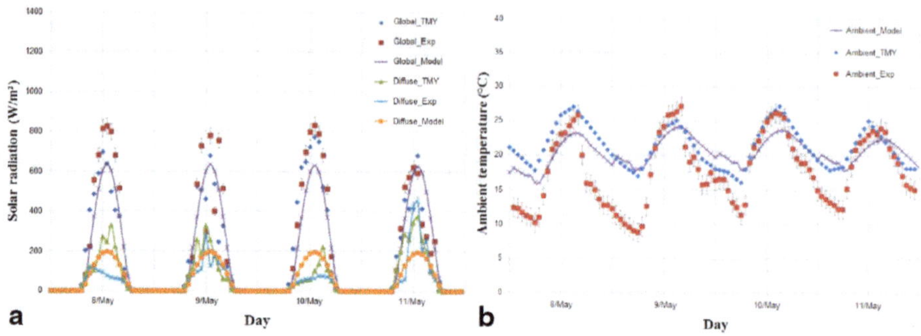

Fig. 59.2 Estimation of solar radiation (**a**) and ambient temperature (**b**)

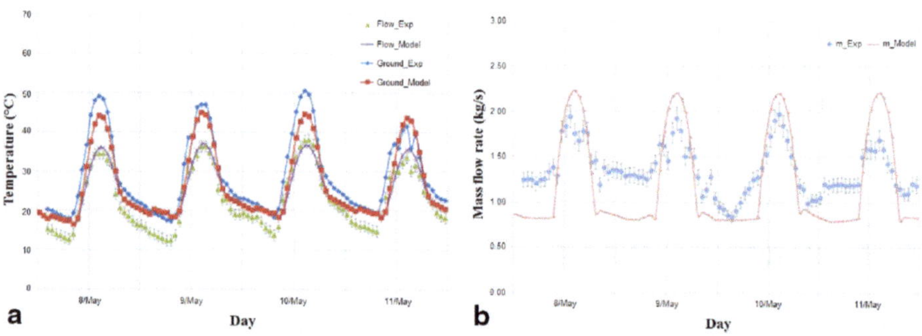

Fig. 59.3 Estimated temperatures of the flow conditions (**a**) and mass flow (**b**)

to predict the general behavior of the ambient temperature with reasonable accuracy (average difference of about 8 %), having good sensitivity for the maximum and minimum values reached during the day.

Using the results for solar radiation and ambient temperature, we predicted the ground temperature, the airflow temperature at the outlet, and the mass flow rate. Figure 59.3a presents a comparison between the temperatures obtained using the proposed model and the experimental data from the prototype. The discontinuities observed at night can be attributed to those observed in ambient temperatures. It can be seen that the model was similar to the experimental data, with the average difference of 15 % for the period evaluated. The ground temperature showed lower difference, of about 9 %.

Figure 59.3b shows a comparison between mass flow rates obtained with the model and the experimental data. Substantial differences were found (average difference of 25 %), especially at night. These differences can be attributed to the difficulties encountered during the determination of the ambient temperatures at night and of the heat transfer rates. These problems led to differences on the ground and outlet airflow temperatures. Furthermore, an equation was used to determine the mass flow rate obtained from the literature for large-scale solar chimneys, which

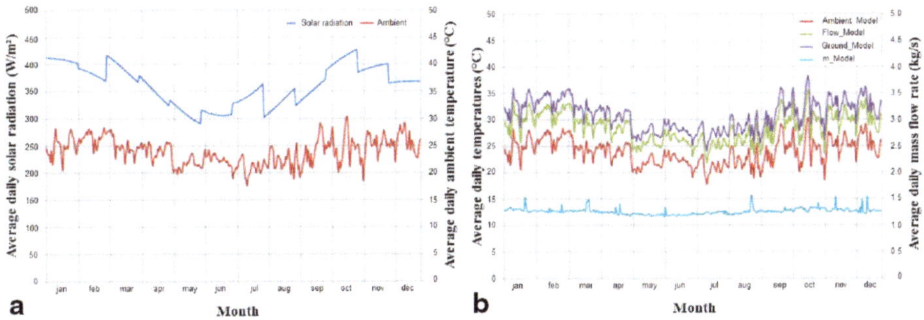

Fig. 59.4 Average daily environmental conditions (**a**) and the flow (**b**)

showed high pressure drops along the tower. Since this parameter is used for the evaluation of the mass flow rate and small-scale solar chimneys have lower pressure drops (especially at night), great differences were found.

The results were presented only for a 4-day period, in order to compare results obtained using the model with the experimental data. Nevertheless, the main objective of the model is to predict the airflow parameters for the whole year. Figure 59.4 presents the daily average ambient conditions and the daily average airflow parameters for 1 year. The discontinuities observed in solar radiation are due to the assumption used for the clearness index. A constant value was assumed for each month. The lower values of solar radiation and ambient temperature are found during the winter in southern hemisphere (from June to August). Figure 59.4b shows the airflow temperatures and mass flow rates. All the temperatures show similar behaviors, with higher values found for the ground temperature and lower values found for ambient temperature. It can also be seen that the mass flow rate does not significantly vary along the year.

59.4 Conclusions

This chapter proposed to model the ambient temperature and the incident solar radiation for the city of Belo Horizonte, Brazil. The results were used to estimate the ground temperature, outlet airflow temperature, and the mass flow rate for a prototype of solar chimney.

The results obtained under ambient conditions were compared with the data from the TMY for the city and also with the experimental data obtained for a 4-day period, available in the literature [13, 16]. The model was able to predict the parameters with good accuracy, considering that the ambient conditions are subject to random variations from year to year. The ground and outlet airflow temperatures were reasonably close to experimental data. Therefore, it can be concluded that the model is able to represent the airflow temperatures.

When the results for 1 year are evaluated, it can be seen that the general behavior of the parameters is well described by the models. The solar radiation and temperature are consistent, as well as the airflow temperatures and mass flow rates. The development of a mathematical model for the prediction of the airflow inside solar chimneys has fundamental importance on the study of their technical and economic feasibility. With the model results, flow conditions can be estimated without the need for construction of the device.

Acknowledgments The authors acknowledge the financial support provided by FAPEMIG and PUC Minas.

References

1. Bernardes MAS, Voβ A, Weinrebe G (2003) Thermal and technical analyses of solar chimneys. Sol Energy 75:511–524
2. Larbi S, Bouhdjar A, Chergui T (2010) Performance analysis of a solar chimney power plant in the southwestern region of Algeria. Renew Sustain Energy Rev 14:470–477
3. Nizetic S, Klarin B (2010) A simplified analytical approach for evaluation of the optimal ratio of pressure drop across the turbine in solar chimney power plants. Appl Energy 87:587–591
4. Hamdan MO (2013) Analysis of solar chimney power plant utilizing chimney discrete model. Renew Energy 56:50–54
5. Pretorius JP, Kröger DG (2006) Critical evaluation of solar chimney power plant performance. Sol Energy 80:535–544
6. Hamdan MO (2011) Analysis of a solar chimney power plant in the Arabian Gulf region. Renew Energy 36:2593–2598
7. Cao F, Zhao L, Li H, Guo L (2013) Performance analysis of conventional and sloped solar chimney power plants in China. Appl Therm Eng 50:582–592
8. Guimarães APC (1995) Estudo solarimétrico com base na definição de mês padrão e seqüência de radiação diária. Dissertação (Mestrado). Universidade Federal de Minas Gerais—Curso de Pós Graduação em Engenharia Mecânica
9. Duffie JA, Beckman WA (2006) Solar engineering of thermal processes, 3 edn. Wiley, New York
10. METEONORM (2013) Handbook part II: Theory
11. Li J, Guo P, Wang Y (2012) Effects of collector radius and chimney height on power output of a solar chimney power plant with turbines. Renew Energy 47:21–28
12. Pretorius JP (2007) Optimization and control of a large-scale solar chimney power plant. Tese (Doutorado)—University of Stellenbosch, ÁfricadoSul
13. Maia CB (2005) Análise Teórica e Experimental de uma Chaminé solar: Avaliação Termofluidodinâmica. Tese (Doutorado)—Universidade Federal de Minas Gerais, Departamento de Engenharia Mecânica, Brasil
14. Kröger DG, Burger M (2004) Experimental convection heat transfer coefficient on a horizontal surface exposed to the natural environment. Proceedings of the ISES EuroSun2004 International Sonnenforum 1:422–430
15. Koonsrisuk A, Lorente S, Bejan A (2010) Constructal solar chimney configuration. Int J Heat Mass Transf 53:327–333
16. Ferreira AG (2004) Estudo de Viabilidade Técnica da Utilização de uma Chaminé solar como Secador de Alimentos. 172 f. Tese (Doutorado)—Universidade Federal de Minas Gerais, Departamento de Engenharia Mecânica, Belo Horizonte

17. Maia CB, Ferreira AG, Valle RM, Cortez MFB (2009) Theoretical evaluation of the influence of geometric parameters and materials on the behavior of the airflow in a solar chimney. Comput Fluids 38:625–636
18. Collares-Pereira M, Rabl A (1979) The average distribution of solar radiation-correlations between diffuse and hemispherical and between daily and hourly insolation values. Sol Energy 22:155–164

Chapter 60
Comparative Investigation of Solar Photovoltaic (PV) and Photovoltaic/Thermal (PV/T) Systems by both Laboratory and Field Experiments

Xingxing Zhang, Jingchun Shen, Xudong Zhao, Ying Xu and Benno Nibeler

Abstract Photovoltaic (PV) semiconductor degrades in performance due to temperature rise. Therefore, a super thin-conductive thermal absorber was developed to retrofit the existing PV panel into a photovoltaic/thermal (PV/T) panel with multiple benefits including dual outputs of increased electricity and additional hot water, and potential savings in installation cost and space. The thermal absorber regulates the PV temperature by creating trade-off between PV efficiency and thermal output. This chapter presents the parallel comparative investigation of the PV and the PV/T panels through laboratory and field experiments. The laboratory evaluation consisted of one PV and one PV/T panel, while the overall field system involved 15 stand-alone PV panels and 15 retrofitted PV/T panels. The total electric installation capacity of the field system was around 6 kWp, and all the PV or PV/T panels were connected to the national grid through an electric inverter. The laboratory testing results demonstrated that this PV/T panel could achieve an electrical efficiency of PV cells at about 16.8% (about 5% increase), and produce an extra amount of heat with thermal efficiency of nearly 65% under standard testing conditions. The nominal mass flow rate of the working fluid is recommended at 50 $Lh^{-1}m^{-2}$. The thermal absorber was measured at an extremely low pressure drop of less than 20 Pa. The field-testing results indicated that the hybrid PV/T panel could essentially improve the electrical return of PV panels by nearly 3.5% in practice, and meanwhile increase the overall energy output (both electricity and heat) by nearly 324.3%. Such synergetic integration of PV and thermal absorber not only results in improved PV efficiency but also generates more energy per unit area when com-

X. Zhang (✉) · J. Shen
Department of Architecture and Built Environment, University of Nottingham Ningbo China, 315100, China
e-mail: Xingxing.zhang@nottingham.edu.cn

X. Zhao
School of Engineering, University of Hull, Hull HU6 7RX, UK

Y. Xu
Shanghai Pacific Energy Center, 200001 Shanghai, China

B. Nibeler
NIEBERLE-SOLAR Company, 87640 Biessenhofen, Germany

A. Sayigh (ed.), *Renewable Energy in the Service of Mankind Vol II*,
DOI 10.1007/978-3-319-18215-5_60

pared with stand-alone PV panel. To replace with conventional electric water heating system, the 15 PV/T panel's payback periods were estimated at less than 5 years and the corresponding CO_2 emission reduction was about 440 t throughout their life span of 25 years. Further opportunities and challenges in the built environment were discussed from aspects of different PV/T stakeholders to accelerate the development of such technology. It is expected that such a dedicated technology could become a significant solution to yield more electricity, offset heating load freely and reduce carbon footprint in contemporary energy environment.

Keywords PV · Thermal absorber · Retrofit · PV/T

60.1 Introduction

Solar thermal and photovoltaic (PV) are the fundamental pillars to assist in transition from the traditional fossil fuel energy structure to a renewable energy system in contemporary built environment. PV semiconductor degrades up to 43 % in efficiency performance with its temperature rise, leading to a subsequent reduction in annual electricity generation. Therefore, the PV/thermal (PV/T) technology was developed to regulate the temperature of PV cells and make advanced utilization of the heat trapped from PV simultaneously. Such synergetic integration of PV and thermal absorber not only results in improved PV efficiency [1] but also generates more energy per unit area when compared with stand-alone PV panel or solar thermal collector. Additional characteristics of the PV/T technology lie in potential saving in material use, reduction in installation cost and homogeneous facade appearance. It is now becoming a significant solution to yield more electricity and offset heating load freely in contemporary energy environment.

Technologies for this purpose have been developed substantially, but meanwhile exhibited some inherent problems such as complex structure, high cost, low efficiency, unsafe operation and incompatibility between tubing absorber and flat-plate PV panel. Currently, the most common way to develop a PV/T panel is to attach the PV module onto a classical flat-plate thermal collector [2]. Such method is inconvenient and expensive and involves the classical manufacturing process of thermal collectors. The conventional thermal absorbers are normally in the geometry of cylinder tubing, which is absolutely unmatched with the flat-plate PV panel and therefore largely reduces the overall working efficiency. As a result, this chapter aims to develop a new type of flat-plate thermal absorber with high heat transfer coefficient, low cost and low pressure drop that can easily retrofit the existing PV panels into the PV/T panels in a rapid way.

Moreover, an experimental evaluation of such PV/T system under either the laboratory or the real climatic conditions has not yet been fully examined. This work retains certain challenges, however, as there are several uncertain factors, including dynamic weather conditions, thermal adaptability and system robustness, etc.

60.2 Description of the New PV/T Technology

Figure 60.1 illustrates the schematic structure of the thermal absorber and its associated approach to retrofit the existing PV panels. The thermal absorber is composed of two parallel thin flat-plate metal sheets, one of which is extruded by machinery mould to formulate arrays of micro-corrugations, while another sheet remains smooth to attach beneath PV panel through a series of U-shaped resilient metal clips. A laser-welding technology is applied to join them together, forming up the built-in turbulent flow channels. Such unique compact structure engenders not only high heat transfer capacity but also convenience in rapid PV/T transformation.

Such thermal absorber offers a very easy way to retrofit an existing PV panel by the following steps: (1) take off the installed PV panel from the roof; (2) insert the thermal absorber; (3) fix the absorber with the PV panel by the U-shaped clips (each clip can hold maximum pressure at about 1.8×10^6 Pa); (4) enclose the thermal insulation layer when needed; (5) install the retrofitted PV/T panel back to the roof surface and connect the standard piping system for thermal collection. All these procedures can be accomplished in just a few minutes.

There are respectively two fluid inlets at the bottom and two fluid outlets on the top with the standard piping joints for a symmetrical fluid distribution. The overall thickness of the thermal absorber is less than 5 mm with a flexible dimension subject to the sizes of PV panels in practice. The absorber has a weight of about 10 kgm^{-2} which can hold fluid of nearly 3.5 Lm^{-2}. The nominal flow rate is designed at 50 $Lh^{-1}m^{-2}$ and the maximum operating pressure is designed at approximately 3×10^5 Pa (3 bar). The laboratory evaluation consisted of one PV and one PV/T panel respectively. During the field testing, the overall system involved 15 stand-alone PV panels and 15 retrofitted PV/T panels. The total electric installation capacity was around 6 kWp and all the PV or PV/T panels were connected to the national grid through an electric inverter. Table 60.1 presents the electrical parameters of single PV module under standard testing conditions. Table 60.2 lists some experimental instruments used in the laboratory testing. Each PV or PV/T module has an effective area of 1.28 m^2. Figure 60.2 shows the schematic system design. The demonstration project was continuously operated and recorded in real climate over 2 weeks from 1st May to 18th May 2014 in Jiangsu, China (32.9 °N, 119.8 °E).

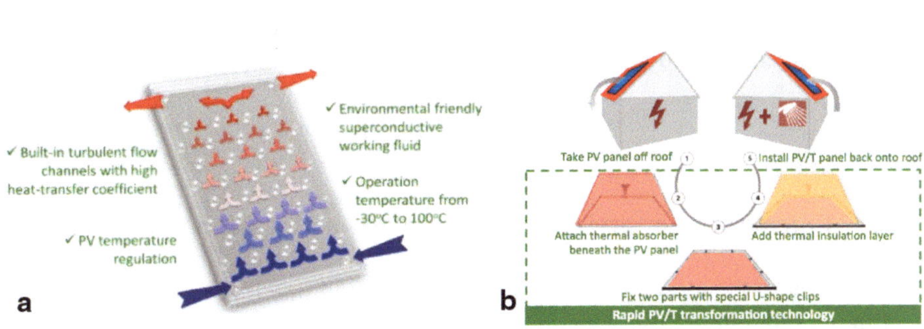

Fig. 60.1 **a** Schematic of thermal absorber. **b** Approach to retrofit the existing PV panels

Table 60.1 Photovoltaic characteristics of single PV module under standard testing conditions

At short-circuit current	I_{SC} = 8.67 A, V_{SC} = 0 V
At open-circuit voltage	I_{OC} = 0 A, V_{OC} = 30.24 V
At the maximum power point	I_{mp} = 8.1 A, V_{mp} = 24.7 V (P_{mp} = 210 W, η_o = 16.1 %)

Table 60.2 List of the experimental testing and monitoring devices

Devices	Specification	Value
OEM pressure sensor Model: Tecsis-P3297	Accuracy	≤1.0 %
Thermocouple T type	Min/max temperature sensed [°C]	−200 ~ 350
Data logger DataTaker-TD500	Record data with computing unit	–
Power sensor WB1919B35-S	DC output	≤0.2 %
Pyranometer Model: Hukseflux-LP02-TR-05	Calibration uncertainty	<1.8 %
	Sensitivity [μV/(Wm^{-2})]	14.45
	Transmitted range [W/m^2]/[mA]	0 ~ 1600/(4 ~ 20)
Pocket anemometer Model: Skywatch-Xplorer 1	Air velocity resolution [kph]	0.1
	Best air velocity accuracy	±3 %
	Maximum air velocity [kph]	150
Solar simulator system Model: SolarConstant4000	Similar global radiation to CIE Publ.85	IEC 60904-9/class B
	Radiation intensity [W/m^2]	1000/[280–3000 mm]
	Homogeneity	±10 % or better (class C)
	Control system	Control with touch panel

The testing was fully operated and the testing data were recorded automatically at 5-s interval. The on-site project is displayed in Fig. 60.3.

60.3 Results and Discussion

60.3.1 Laboratory Evaluation

To Investigate the effect of the thermal absorber to the PV panels (baseline), the varying percentage is defined as the equation below:

$$X = \frac{X_{PV/T} - X_{PV}}{X_{PV}} \times 100, \quad (60.1)$$

where X represents the varying element such as temperature, power and thermal energy.

The experimental evaluation of one PV panel and one PV/T panel was carried out in the laboratory. The solar simulator was adjusted upwards or downwards to

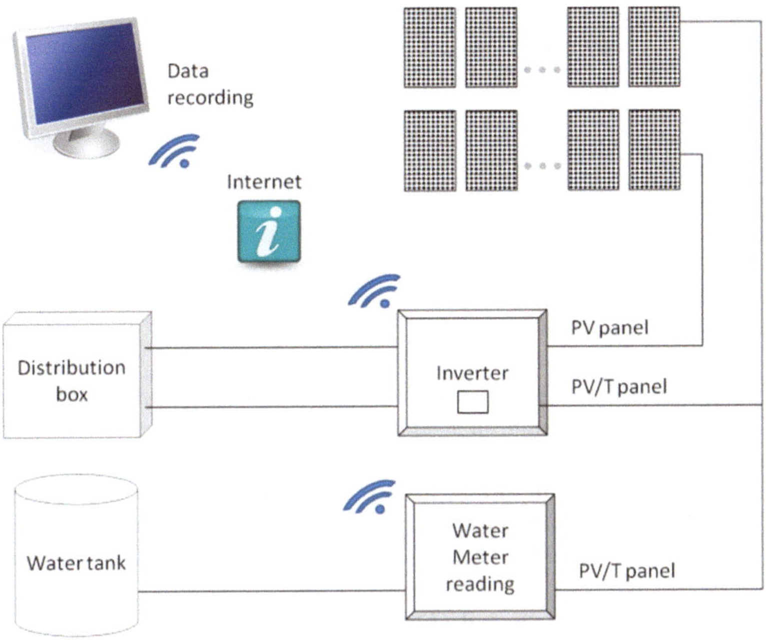

Fig. 60.2 System schematic design

Fig. 60.3 On-site diagram of laboratory evaluation and field testing

evenly simulate the solar radiation on either the PV or the PV/T panel. Table 60.3 presents some results under the standard testing conditions, i.e. 1000 W/m^2 of solar radiation, 25 °C of air temperature and 30 °C of inlet temperature. The electrical efficiency of the PV/T panel was about 16.8 %, which increased nearly 5 % when compared to the PV panel. The initial optical thermal efficiency and the linear coefficient were determined by linear regression analysis of the testing results at 73.5 % and 15.8 Wm^{-2}K^{-1} respectively, leading to the thermal performance of the thermal absorber as given in Eq. 60.2. The linear coefficient (heat loss coefficient) was a little bit high mainly because there was no thermal insulation attached at the absorber back. This needs to be considered especially in cold climate regions in practice.

Table 60.3 Testing results of the thermal absorber-based PV/T panel

Parameters	Symbol	Value	Parameters	Symbol	Value
Initial thermal efficiency	η_o	73.5%	Fluid volume	V	4.5 L
Linear coefficient	U_L	15.8 $Wm^{-2}K^{-1}$	Pressure drop	ΔP	19.7 Pa
Collecting area	A	1.28 m^2	Nominal mass flow	M	50 $Lh^{-1}m^{-2}$
Thermal output	Q_{th}	890 W	Maximum mass flow	m_{max}	230 $Lh^{-1}m^{-2}$
Electrical output	q_{th}	168 W	Joint diameter	D	0.022 m
Thermal efficiency	η_{th}	65%	Outlet temperature	T_o	42 °C
Electrical efficiency	η_e	16.8%	PV temperature	T_{PV}	45 °C

The temperature difference between the PV layer and the absorber outlet was only about 3 °C, which demonstrates that such thermal absorber had a very small contact thermal resistance when applied onto the back of PV panels. Owing to the special design of flow channel, the pressure drop in the thermal absorber was less than 20 Pa, consuming only little energy from the pump. The thermal and the overall efficiency of the combined PV/T panel achieved were 65 and 81.8%, respectively.

$$\eta_{th} = 0.735 - 15.8(T_i - T_a) / I \tag{60.2}$$

60.3.2 *Field Testing*

A parallel field testing between the 15 PV panels and the 15 PV/T panels was implemented and the comparative results were discussed from the aspects of temperature at PV cells, electrical power output and additional thermal output by the PV/T panels. Figures 60.4, 60.5 and 60.6 illustrate the parallel comparison rustles of the PV and the PV/T panels during the 2-week field-testing period. The mean daily cell temperatures of the PV panels and the PV/T panels were approximately 41.1 and 39.8 °C respectively, resulting in a decrease in the PV temperature to about 3.0% daily on average by the thermal absorber. However, the temperature difference was not too large mainly due to the moderate solar radiation in May and the thermal insulation layer at the back of the PV/T panels. Thanks to the temperature reduction of the PV layer, the mean daily powers of the PV panels and the PV/T panels were nearly 926.7 W and 959.2 W, respectively. As a result, the thermal absorber increased the power output to 3.5% on average, which was slightly lower than that in the laboratory owing to the varying weather conditions and the impact of incident solar angle. The PV/T panels could generate 11.3 kWh electricity and 35.3 kWh heat daily (46.6 kWh energy in total), while the PV panels could only produce 10.9 kWh electricity in total. Thus, an increase in the overall energy output of nearly 324.3% was achieved by the PV/T panels when compared with the PV panels.

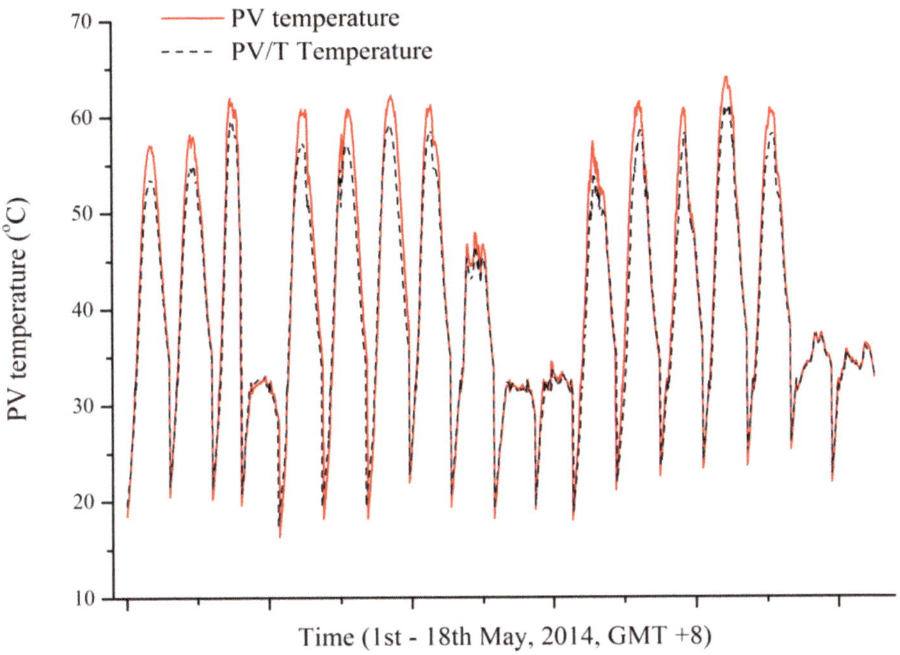

Fig. 60.4 Daily PV temperature of the PV and the PV/T panels

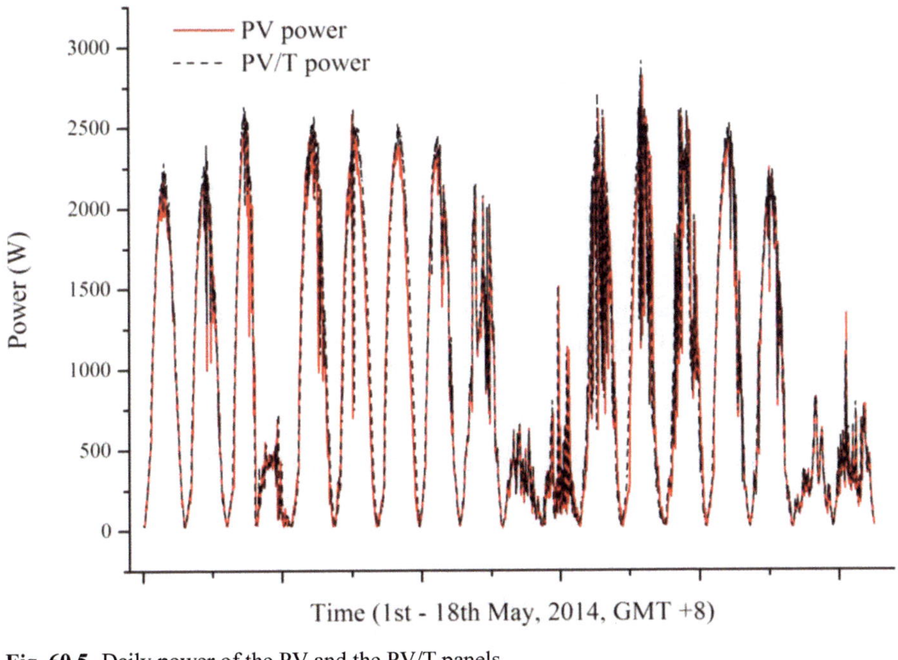

Fig. 60.5 Daily power of the PV and the PV/T panels

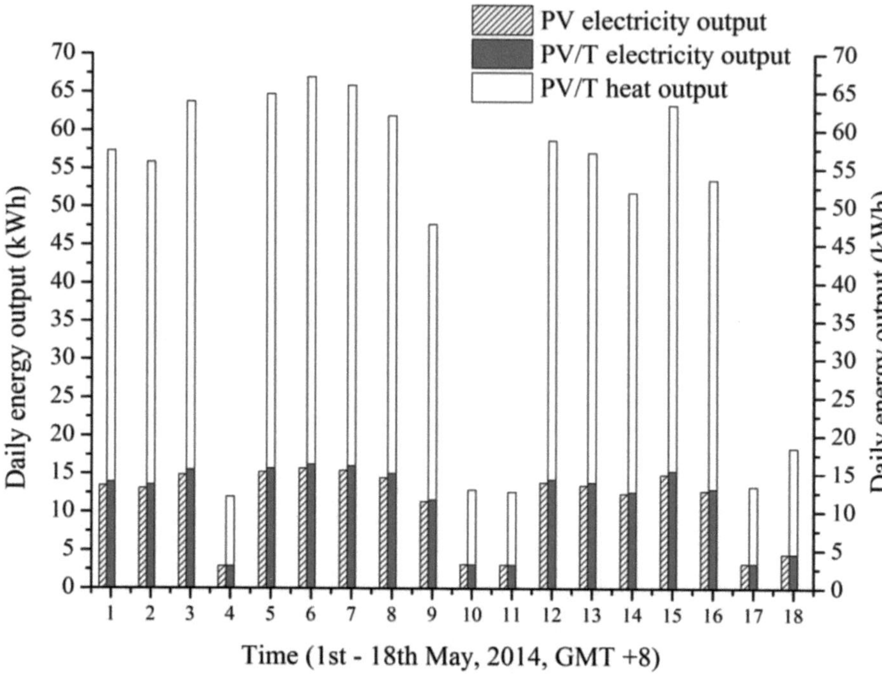

Fig. 60.6 Daily energy output of the PV and the PV/T panels

60.3.3 Economic and Environmental Benefits

A simple analysis of the economic and environmental benefits of the PV/T water heating system was carried out using the mean values from the above field-testing results. The electricity price from national grid is CNY¥0.6/kWh [3] and from the distributed PV/T panels in Jiangsu is CNY¥0.82/kWh [4]. Capital cost of the 15 PV/T panels is about CNY¥57,750. To replace with the conventional electric water heating system of 90 % heating efficiency, the system's payback periods were estimated at less than 5 years in Jiangsu, China. The CO_2 emission reduction was estimated at about 440 t by the electricity-to-CO_2 conversion factor (0.997 kg CO_2/kWh-heat in China [5]) throughout their life span of 25 years.

60.3.4 Opportunities and Challenges in Future Built Environment

To enable a widespread deployment of such hybrid solar technology, the explicit benefits and challenges for PV/T stakeholders have been identified and summarized in Table 60.4 [1], which needs to be elaborated in the coming years. The establishment of an interdisciplinary working group seems necessary, which can attain a

Table 60.4 Opportunities and challenges for PV/T stakeholders

Stakeholders	Opportunities	Challenges
R&D institutes	Quest for new technological solutions	Performance and reliability standards
		Increased system performance
Engineering consultants	Innovative and high-profile technology	Design tools development
		New system concepts development
Architects	New solutions for integration	PV/T integrated with building design
		New building concepts
Installers	Reduced installation effort	Plug-and-play integration in comfort systems
		Combination of two professional specialisms
Building industry	Increased energy performance	Integration of module into building facade
		Prefabrication possibilities
Manufacturers	Enlarged markets	Cost-effective production
		Plug-and-play systems
Policymakers	More effective path to renewable targets	Building regulations, market and R&D support

sound information exchange of results related to R&D, design specifications, design tools, test methods, installation barriers, market surveys, policy development, etc. This interdisciplinary cooperation may lead to a clear understanding of various problems over the PV/T developments.

60.4 Conclusion

This chapter discussed the development of a super thin-conductive thermal absorber to retrofit the existing PV panel into the PV/T panel with multiple benefits, including dual outputs of increased electricity and additional hot water, and potential savings in installation cost and space. It compared the PV panel and the PV/T panel in parallel through both laboratory and field testing. The laboratory testing results demonstrated that the PV/T panel could achieve an electrical efficiency of PV cells at around 16.8 % (with about 5 % increase), and produce an extra amount of heat at thermal efficiency of nearly 65 % under standard testing conditions. The nominal mass flow rate of working fluid was recommended at 50 $Lh^{-1}m^{-2}$. The thermal absorber was measured with an extremely low pressure drop at less than 20 Pa. The field-testing results indicated that the PV/T panel could essentially improve the electrical return of PV panels by nearly 3.5 % in practice, and meanwhile increase the overall energy output (both electricity and heat) by nearly 324.3 %. Such synergetic integration of PV and thermal absorber not only results in improved PV efficiency but also generates more energy per unit area when compared with stand-alone PV panel. To replace with the conventional electric water heating system, the 15 PV/T panel's payback periods were less than 5 years and the corresponding CO_2 emission reduction was about 440 t throughout their life span of 25 years. Explicit benefits and future challenges in the built environment were discussed for PV/T stakeholders to accelerate the development of such technology. An interdisciplinary working group was suggested to be established in the future to attain a sound information exchange in R&D, design specifications, design tools, test methods, practical installation barriers, market surveys, policy development, etc. The dedicated technology is expected to become the next-generation solar-driven heating system and enable a significant reduction in a building's carbon footprint.

Acknowledgements The authors would like to acknowledge their sincere appreciation to the financial supports from the Ningbo Natural Science Foundation (2015A610039) and EU Marie Curie International Research Staff Exchange Scheme (R-D-SBES-R–269205).

References

1. International Energy Agency (2010). Technology roadmap: solar photovoltaic energy. http://www.iea-pvps.org. Accessed 18 May 2014
2. Zhang X et al (2012) Review of R&D progress and practical application of the solar photovoltaic/thermal (PV/T) technologies. Renew Sustain Energy Rev 16(1):599–617
3. State grid (2014). Electricity price. www.js.sgcc.com.cn. Accessed 21 May 2014
4. Jiangsu Development and Reform Commission (2013). The standard for distributed photovoltaic award. www.jsdpc.gov.cn. Accessed 21 May 2014
5. China5e (2013). Electricity-to-CO_2 conversion ratio. http://www.china5e.com. Accessed 21 May 2014

Chapter 61
Performance of a Recirculating-Type Solar Dryer

Yefri Chan, T. M. Nining Dyah and Kamaruddin Abdullah

Abstract A novel recirculation-type integrated solar collector drying chamber solar dryer has been designed, constructed, and tested. The solar dryer comprises a feed hopper, a centrifugal blower, a pneumatic conveyor, and a transparent structure acting as drying chamber containing a hopper with a vortex at the top. The transparent structure has a dimension of 3 m in diameter and height of 3 m. The blower used has a capacity of maximum 1 kW, 220 V, and a supplying air velocity of 30.3/s. A series of tests were conducted outdoors where solar radiation was available. A test with 104 kg of rough rice indicated that the drying time required to reduce the moisture content (m.c.) of rough rice from 28.4 % wb to the final m.c. of 14.3 %wb was 5 h. During the test, the drying temperature was kept constant at 50.1 °C and the relative humidity (RH) was 21.73 %. The required power for the pneumatic conveyor was 581 W, with a total energy input of 210.7 MJ, including LPG and solar radiation. The resulting drying efficiency was 22.4 % with specific energy of 15.2 MJ/kg of water evaporated. Another test using 200 kg with initial m.c. of 27.6 %wb of rough rice has shown that the required drying time to achieve the final m.c. of 14.3 % was 8 h. During the test, the drying temperature was kept constant at 46.9 °C, and the RH was 21.7 %. The drying efficiency was 31.7 % with specific energy of 10.7 MJ/kg of water evaporated. By knowing the amount of rough rice retained in the receiving hopper and the rate of recirculation, it was possible to measure the drying time of each cycle until the drying process completed. By using the sphere model, we were able to estimate the drying curve, where it was found that a good agreement exists between the theoretical and experimental data using the value of drying constant k=2.75 (1/hr) and M_e of 7 % db. As the drying time increased toward the end of drying process, it was observed that the variation in rough rice m.c. became smaller and smaller and was better than that found in the previous study.

K. Abdullah (✉) · T. M. Nining Dyah
The Graduate School, Darma Persada University, Jl. Radin Unten II, Pondok Kelapa, East Jakarta 13450, Indonesia
e-mail: Kamaruddinabd@gmail.com

Y. Chan
Mechanical Engineering Department, Darma Persada University, Jl. Radin Unten II, Pondok Kelapa, East Jakarta 13450, Indonesia

A. Sayigh (ed.), *Renewable Energy in the Service of Mankind Vol II*,
DOI 10.1007/978-3-319-18215-5_61

Keywords Drying efficiency · Pneumatic conveyor · Solar dryer · Specific energy

Notations

Symbols

A_c	Surface area of hopper exposed to solar radiation (m^2)
C_p	Specific heat (kJ/kg °C)
CV	Calorific value (kJ/kg)
D_v	Mass diffusivity (m^2/s)
ΔH_{fg}	Latent heat of vaporization (kJ/kg)
I_{rad}	Solar radiation (W/m^2)
k	The drying constant (1/h)
m	Mass (kg)
M	Moisture content (% db)
Pw	Electric power (Watt)
S_E	Total specific energy (MJ/kg of water evaporated)
t	Time (h)
W	Mass (kg)
X	Moisture content (% wb)

Subscript

D	Drying
e	Equilibrium
g	Gas
h	Hopper
i	Initial
f	Final
o	Initial
p	Rough rice
r	Recycling

Greek Letter

θ_R	Recycling time
η_D	Drying efficiency

61.1 Drying Curve

61.1.1 Drying Process

If the grain can be assumed as a spherical body, then the drying process will follow the equation given [1]:

$$\frac{\partial M}{\partial t} = D_v \left(\frac{\partial^2 M}{\partial r^2} + \frac{2}{r} . \frac{\partial M}{\partial r} \right) \tag{61.1}$$

which can be solved under the following initial conditions (IC) and boundary conditions (BC).

IC: at $t=0$, $0<r<R$, $M=M(0)$

BC: at $t>0$, $r=\pm R$, $T=$constant, RH$=$constant, $M=M_e$

The solution of Eq. (61.1) in a simplified form is given in [2], taking the first term only

$$\frac{\bar{M}-M_e}{M_0-M_e}=A\ \exp(-Dv\pi^2 t/r^2) \quad (61.2)$$

Here A is the shape factor (−), r is the equivalent radius of the grains (m), and k is the drying constant. Under the turbulent drying process, the value of k was found to be 2.75 (1/hr) and $M_e=7$ (% db) [3].

The amount of drying during each recycling process could be estimated from the recycling time, θ_R, calculated from the known amount of grain remaining in the hopper, W_h, and the recycling rate of the grain, W_R, within the dryer, such that

$$\theta_R=\frac{W_h}{W_R} \quad (61.3)$$

where from previous study by Kamaruddin et al. [3], the value of W_R can be estimated using the following relation :

$$W_R=1.152\ Pw+7.52$$

61.1.2 Drying Efficiency

The drying efficiency, η_D can be defined as the ratio of useful energy to dry the grain to total energy input. Therefore

$$\eta_D=\frac{(m_{pf}C_{pf}T_{pf}-m_{pi}C_{pi}T_{pi})+\left\{\frac{(X_i-X_f)W}{(1-X_f)}\right\}\Delta H_{fg}}{Pwt_D+W_g CV_g+I_{rad}A_c} \quad (61.4)$$

The total specific energy, S_E is defined as the total energy input to amount of water vapor evaporated.

$$S_E=\frac{Pwt_D+W_g CV_g+I_{rad}A_c}{\left\{\frac{(X_i-X_f)W}{(1-X_f)}\right\}} \quad (61.5)$$

61.2 Experiment

Figure 61.1 shows the experimental setup. It shows the main component of the solar dryer which comprises (1) a blower, (2) a feed hopper, (3) a liquified petroleum gas (LPG) stove, (4) a transparent structure, (5) a pneumatic conveyor, and (6) a receiving hopper. The transparent structure will simultaneously act as a solar heat collector as well as the drying chamber. Before filling the hopper with grains, the temperature sensors (type k thermocouples; Chromel-Alumel) were placed at specific locations at the pneumatic conveyor inlet and outlet, in the drying chamber, at the hopper wall, and at the inlet and outlet of the LPG stove. The relative humidity (RH) meter (Lutron-type BG-UT-02P) was placed within the drying chamber, while a digital pyranometer (Tenmarr-type TM 206) was placed in horizontal position to measure global solar radiation. In addition, a voltmeter and an ampere meter were used to measure the electricity consumption during the drying process. A Lutron-type DW 6060 anemometer was used to measure the rate of air flow at the blower inlet. During the experiment, rough rice was filled into the feed hopper to a certain amount while keeping the blower operating until the grains had been collected in the receiving hopper. During this operation, the valve below the receiving hopper was closed. The hot air from the LPG stove was supplied into the blower inlet, keeping the valve below the feed hopper closed while the valve below the receiving hopper open. In this way, the grains flow from the receiving hopper into the pneumatic conveyor and into the transparent structure, which also functions as the drying chamber, and finally fall into the receiving hopper. This process continued until the final moisture content (m.c.) of the grain was achieved.

Fig. 61.1 Blower (1), feed hopper (2), LPG stove (3), transparent structure (4), pneumatic conveyor (5), and receiving hopper (6)

61.3 Results and Discussion

Figure 61.2 shows the drying curve of under 104 kg load of rough rice. It took 5 h to dry from 28.4 %wb to the final 14.3 %wb. Total gas consumption was 4.3 kg, total electricity consumption was 2.905 kWh, and the average power consumption was 580.9 W. The total solar radiation received was 8.724 kJ. Under this condition, the drying air temperature was at 50 °C, and hopper temperature was 54.5 °C, with drying chamber RH at 21.73 % (Figs. 61.3 and 61.4).

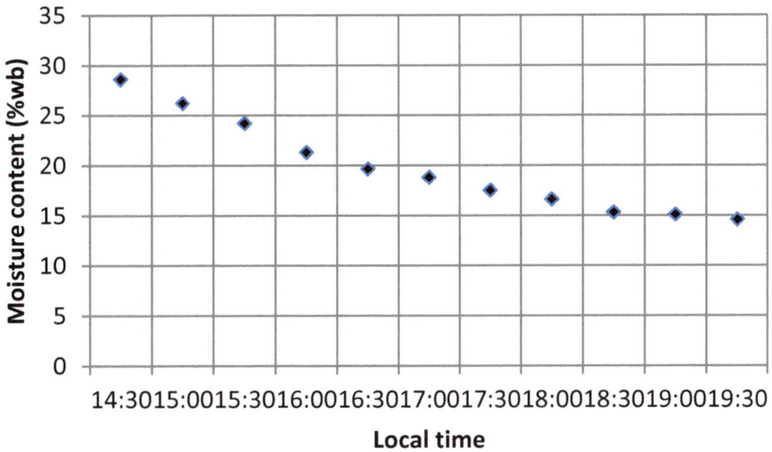

Fig. 61.2 The drying curve of 104 kg load of rough rice

Fig. 61.3 Variation in drying chamber and hopper temperature

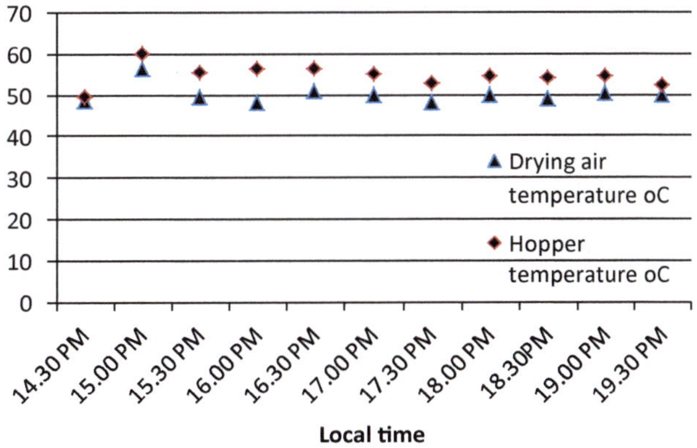

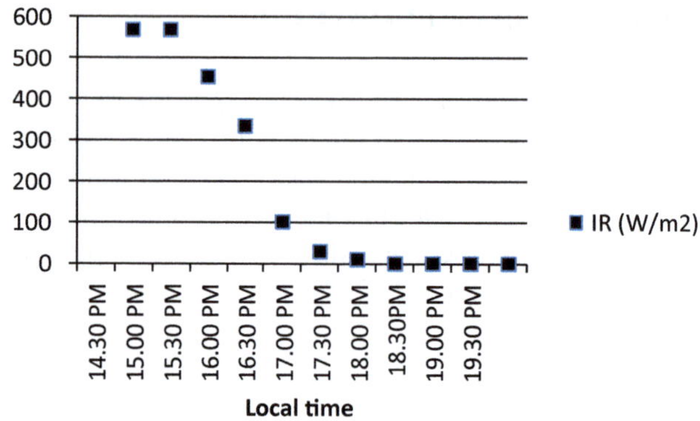

Fig. 61.4 Variation of solar radiation during the experiment

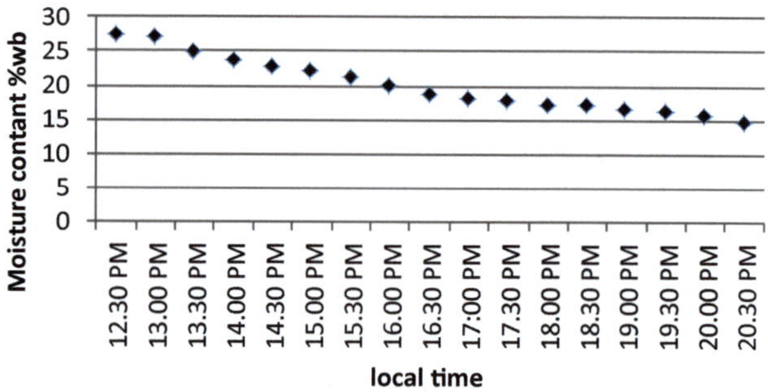

Fig. 61.5 The drying curve for the 200 kg load test

61.3.1 Drying Efficiency

The drying efficiency was calculated using Eq. (61.4) and was found to be 23.6 %, while the total specific energy calculated using Eq. (61.5) was found to be 14.6 MJ/kg of water evaporated. Figure 61.5 shows the drying curve for the 200 kg load test. It took 8 h to dry the m.c. of rough rice from 27.3 % wb to 14.6 % wb. The electricity consumption was 578.2 W on average and the LPG consumption was 6 kg with total of 10.487 kJ of solar radiation. The average drying temperature was 47 °C.

On the basis of these data, the calculated drying efficiency was found to be 35.7 %, while the total specific energy (including solar energy input) was found to be 9.475 MJ/kg of water evaporated. Figure 61.6 shows the comparison between the drying curves prepared using Eqs. (61.2) and (61.3). It shows that the model used predicted the obtained experimental data quite well. Here the value of a parameter *A*, the shape factor, *A* was given a value of unity larger than for the case of spherical body, which is $6/\pi^2$.

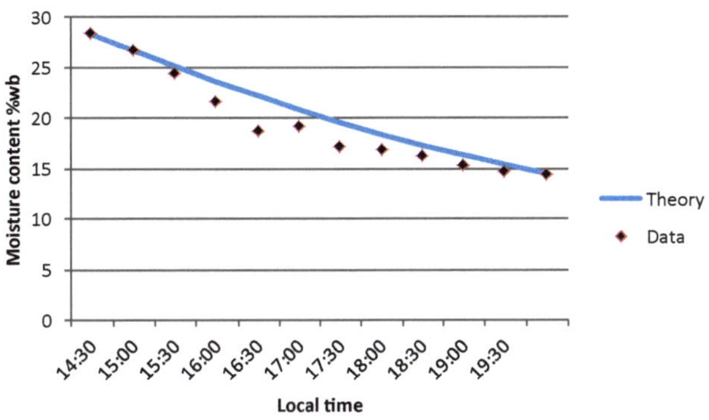

Fig. 61.6 Comparison between the theoretical and experimental data for the 104 kg drying test

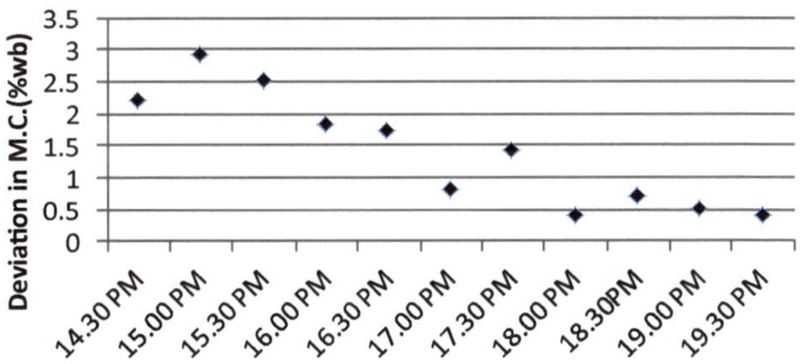

Fig. 61.7 Deviation in moisture content %wb

61.3.2 Comparison with Theory

Figure 61.6 shows the comparison between the drying curves prepared using Eq. (61.2) with coefficient $k = 2.75$ (1/hr) and $M_e = 7$ (% db), and Eq. (61.3) for the rate of recirculation. During the test, the amount of rough rice within the hopper was 55 kg, which indicated that the recirculation time was 0.8136 h/cycle.

61.3.3 Homogeneity in Moisture Content

As is described in the earlier section, by using a pneumatic conveying system, the drying process will proceed in a turbulent manner, resulting in homogenous final m.c. Five samples from each measuring time indicated that the variation in m.c. of the sample reduced as the drying process continued toward the end. It varied from 2.9 % at the earlier stage of drying to 0.4 % at the final drying stage. These results show a much lower percentage than using a flat-bed dryer for the drying process, even with reverse air flow being applied [4] (Fig. 61.7).

61.4 Conclusions

1) With 104 kg load of rough rice, the solar dryer was able to reduce the initial m.c. from 28.4 % wb to final m.c. of 14.3 % wb within 5 h. The drying air temperature was 50 °C and the RH was 21.73 %. The electric power used was 580.9 W while the LPG consumed was 4.3 kg with solar radiation of 8.724 kJ.
2) The drying efficiency of the dryer was 23.6 % whereas the total specific energy (including solar radiation) was 14.6 MJ/kg of water evaporated.
3) Under 200 kg load of rough rice, the required drying time to reduce the initial m.c. of 27.3 to 14.6 % wb was 8 h. The drying process was conducted under drying air temperature of 47 °C, where the electric power used was on the average of 578.2 W, the amount of LPG consumed was 6 kg, and solar radiation was 10.487 kJ.
4) The efficiency of the dryer was 35.7 % whereas the total specific energy (including solar radiation) was 9.475 MJ/kg of water evaporated.
5) Comparing the predicted drying curve and data shows that all data have lower values, although the deviation is small.
6) The variation in m.c. during the drying process was smaller than that using flatbed drying varying from 0.4 % to 2.9 % wb.

Acknowledgment The authors express their gratitude to the Directorate General of Higher Education, of Indonesia for providing research grant under contract no. HB 018/K3/KM/SPK/2013, May 13, 2013.

References

1. Crank J (1986) The mathematics of diffusion. Clarendon Press, Oxford
2. Pirasteh G, Saidur R, Rahman SMA, Rahim NA (2014) A review on development of solar drying applications. Renew Sustain Energy Rev 31:133–148
3. Kamaruddin A, Aep SU, Yefri C (2014) Performance of a continuous flow solar dryeing system, International Energy J 14(1)
4. Hien PH, Tam NH, Xuan NP (2003) Paper presented at the seminar on agricultural engineering and agro-product processing toward mechanization in rural areas, Nong-Lam University, Ho Chi Minh City, 11–12 December

Further Reading

1. Fudholi A, Sopian K, Yazdi MH, Ruslan MH, Gabbasa M, Kazem HA (2014) Performance analysis of solar drying system for red chili. Sol Energy 99:47–54
2. Henderson SM, Perry RL (1979) Agricultural process engineering, 3rd ed. AVI, Textbook series, Westport
3. Prakash O, Kumar A (2014) Solar green house drying: a review. Renew Sustain Energy Rev 29:905–910
4. Şevik S (2013) Design, experimental investigation and analysis of a solar drying system. Energy Convers Manage 68:227–234

Chapter 62
Thermo-Energy Transfer Optimization of a Solar Distiller with Energy Storage Under Bou-Ismail Climatic Conditions

Randha Bellatreche, Dalila Belhout, Maamar Ouali, Djamila Zioui, Zahia Tigrine, Hanane Aburideh and Sarah Hout

Abstract Seawater desalination techniques are sustainable, but their use is limited in rich countries. However, in recent years, the capacity of desalination stations has greatly increased and production costs by m3 experienced a sharp decline. Seawater and brackish water desalination using solar energy witnesses a growing interest in some developing countries. Solar distillation is a technically feasible, simple, profitable, and operational method for the production of freshwater. This thermal process is a key to future water problems in the water poorest regions, namely arid and isolated sites in southern Algeria as it permits to desalinate brackish water and treated wastewater reuse concentrated salt. The aim of this study is to evaluate the effect of temperature and the internal components of solar still on its operation. An experimental examination of a solar distiller performance is realized and designed by a team of research of UDES. The daily instantaneous and global yield will be examined to show the temperatures' influence on the different distiller elements as well as on the quantity of distillate produced. Also, a comparison of different geometries with energy storage has been undertaken in order to study the influence of geometry on the production of distilled water.

Keywords Desalination · Seawater · Solar energy · Solar distillation

R. Bellatreche (✉) · D. Belhout · M. Ouali · D. Zioui · Z. Tigrine · H. Aburideh · S. Hout
Unité de Développement des Equipements Solaires, UDES/Centre de Développement des Energies Renouvelables, CDER, Bou Ismail, 42415 W. Tipaza, Algérie
e-mail: bellatreche_randa@yahoo.fr

D. Belhout
e-mail: belhout.dalila@yahoo.fr

D. Zioui
e-mail:ziouidjamila@yahoo.fr

A. Sayigh (ed.), *Renewable Energy in the Service of Mankind Vol II*,
DOI 10.1007/978-3-319-18215-5_62

62.1 Introduction

Distillation is one of the oldest and most rustic desalination techniques. In fact, it reproduces the natural process of the industrial desalination. Solar distillation is a thermal process that represents a sustainable solution to water shortages in the world.

The desalination technique by solar energy, based on the principle of greenhouse, is not new. This process of evaporation–condensation was used for the first time on a large scale in 1872 at Las Salinas, the north of Chilean desert, to supply drinking water to construction sites, mines of nitrates. Since then, the principle has not changed, although improvements have been obtained in the form of construction materials and methods of work [1].

This operation is based on heating (solar energy), or on reducing pressure, and it turns saltwater into steam free from impurities dissolved in the liquid. By condensation, the steam results in water of high purity that may require remineralization when intended for consumption [2]. Several greenhouse solar stills exist, many of which were advanced and industrialized [3–5], including the distiller plan, waterfall, wicking, multi-effects, and spherical.

At present, a large number of experimental and theoretical works are made to improve the performance of solar still in different configurations [6–8]. According to several studies made in this field, the performance of a solar still depends on different external and internal parameters, and the quantity of distillate produced is different according to the distiller type. [9] made a systematic comparison of a simple solar still and a solar water distiller heater to improve the productivity of a solar greenhouse hot-box-type still. The results indicate that the supplied water temperature affects the daily production of the still, and therefore, it has an impact on its efficiency. [10] studied experimentally the effectiveness of a solar distiller at different angles 15, 25, 35, 45, and 55°. They found that the maximum freshwater production is achieved with the optimum angle of inclination of 35°. It was proved that water salinity affects the distillate production, even at low concentrations. [11] used a mathematical model to predict the performance of a simple solar distiller submitted to various climatic operational and design parameters, in the region of Oman. It was founded that the optimal design conditions of a solar still tend to give an average annual solar yield of 4.15 kg/m^2 per day. Further studies on solar stills doped with internal–external reflectors were made by [12, 13] studying three types of absorber (plate bare black tissue and black fleece) and show that the performance is closely linked to the type of surface used. [14] have made a theoretical and experimental study by making changes to the distiller (wing, sponge, and wick) in order to analyze the effect of these materials on the system performance. It was found that the experimental analysis is in agreement with the theoretical results with an error of about 10 %; productivity has improved by 48 %. In this context, an experimental examination of a solar distiller performance, directed and designed by a team of research of UDES, is proposed. The daily instantaneous and global yield will be examined, for two geometries (square and rectangle) to show the temperatures' influence of the different distiller elements on the quantity of distillate produced.

62.2 Solar Still Operation

The operating principle of a greenhouse solar still involves a water vapor phase change, which can separate the salts from water. The result of this phase change is distilled water and salts in the form of deposits. In fact, under the action of solar flux striking the transparent surface, the water heats and evaporates, the produced vapor condenses and trickles down on the inside face inclined glass into gutters which actually lead it to a storage tank, leaving the deposit dissolved salts and other residues in seawater. Formatting the internal circulation of water vapor is thermosiphon which arises due to the temperature difference between the glass and the water mass to be distilled.

The different heat fluxes involved in the heat exchanges that occur inside the distiller are explicit by many physical phenomena.

The various modes of heat transfer exchanged between the brine and the glass are made by convection, by radiation, and by evaporation. The distiller is a system that exchanges heat with the outside through the glass by convection and by radiation. The insulating bottom and sidewall are made by conduction and by radiation. In what follows the main heat transfer inside and outside the concrete solar distiller is determined (Fig. 62.1).

62.3 Description of the System

A new technology distillation greenhouse effect with energy storage has been developed to desalinate seawater and brackish water and demineralize wastewater treated. The energy storage occurs in the concrete mass, which comprises the base and the serpentine heat exchanger.

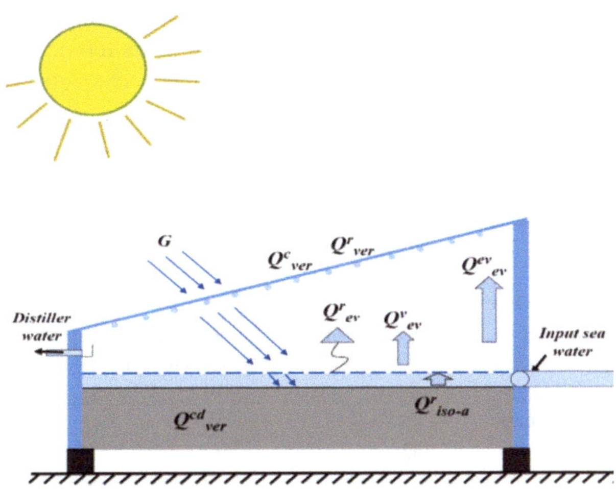

Fig. 62.1 Operating principle and heat exchange of a concrete solar still

Fig. 62.2 General overview of the experimental setup

Our equipment is characterized by using selective coating covering the surface of the exchanger, to improve the heat absorption phenomena. In order to maximize efficiency, our distiller is equipped with simple control devices, automatic control, and stabilization of the water level. The heat exchangers have been added in order to maintain the production of distilled water overnight, for parts of this thermodynamic conditions evaporation/condensation, and therefore, more advantageous is a significant increase in equipment performance. The heating of the water in the heat exchanger and concrete base is made by plan solar collectors, one for each.

The two distillers are composed mainly of a transparent glass cover, which allows the passage of solar radiation ($0.4 < \lambda < 0.8$ μm) and becomes opaque to infrared radiation (Fig. 62.2). It is placed on a basin made of concrete, wherein seawater is located, and the lower surface portion is represented by $s = 1\ m^2$. A black paint was applied on the distiller platform as an absorber for absorbing maximum solar radiation, which results in an increase in its temperature. The absorber is covered with a seawater layer of 1.5–4.5 cm depth, characterized by the parameters shown in the table below (Table 62.1). The sidewalls are made of aluminum isolated by a coating of polystyrene.

Table 62.1 Physicochemical characteristics of seawater used

Seawater	Salinity [g/L]	Electrical conductivity σ [ms/cm]	pH	Temperature [°C]
Bou-Ismail	29.3	45.5	8.20	24.9
Fouka	32	50.3	8.08	25.7

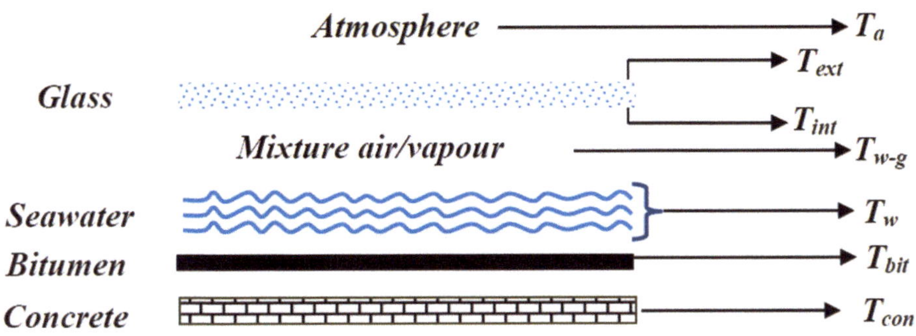

Fig. 62.3 Thermocouples' position in a solar distiller

62.3.1 Experimental Setup and Instrumentation

The absorber is covered by seawater stratum of 3 cm depth, characterized by the parameters shown in the table below (Table 62.1).

Different temperatures are measured by thermocouples of K type connected to a data logger Fluke type for data acquisition, which are recorded on a memory card, scheduled for a time interval of 1 h. Figure 62.3 shows the thermocouples' disposition inside the distiller.

In order to determine the physicochemical properties of distilled water, tests were performed, namely conductivity, salinity, TDS, and pH using an inoLab (Cond Level 1) conductivity meter. The volume of the distilled water collected at each time is measured using a graduated cylinder. The global solar radiation on a horizontal plane is measured using a pyranometer type CMP11 "Kipp & ZONEN" its sensitivity range is between 7 and 14 V/W m^2. The radiation data were recorded in the weather station of UDES.

62.4 Results and Discussion

The systematization of experimental tests allowed presenting the evolution of different temperatures versus time, and the daily quantity of distillate for two different geometries (square and rectangle) is represented by curves.

62.4.1 Solar Radiation Evolution as a Function of Time

The variation of the solar radiation received by a glass surface inclined 13° during the experimental tests is shown in Fig. 62.4. Observations show that the solar radiation becomes dominant and more intense between 11:30 am and 14:30 pm and it

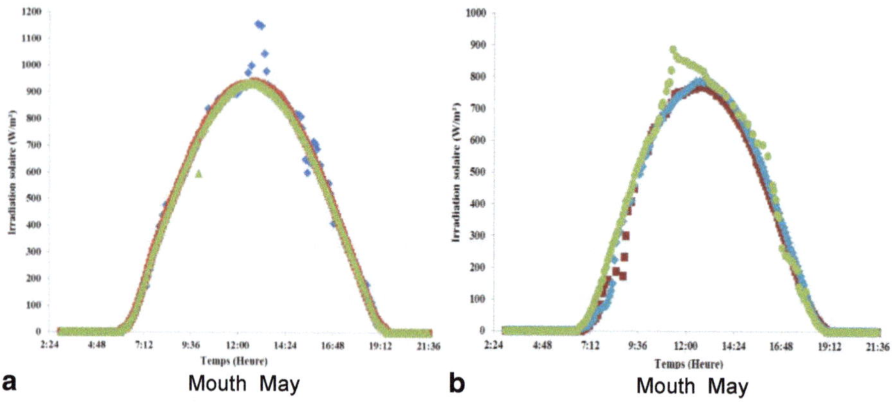

Fig. 62.4 Evolution of solar radiation versus time (**a**) April and (**b**) May

decreases from 16 h because Bou-Ismail is a coastal region which has a long insolation time during the period between April and July.

It was found that the distilled water production depends strongly on the incident solar energy.

62.4.2 *Evolution of the Temperature Versus Time*

The evolution of different temperatures as a function of time, that is the seawater temperature Tw, the interior glass pane temperature T_{int}, the exterior glass pane temperature T_{ext}, and medium temperature between the water and the glass T_{w-g} obtained in the experiments for the two geometries square and rectangle, with and without absorber, is shown in Figs. 62.5 and 62.6.

These figures show that the temperature of the different concrete distiller elements for the distiller with absorber is greater than the temperature of the distiller elements without absorber, because black color absorbs more solar radiation. The most significant temperature is that of the water, and medium temperature between the water and the glass 55–71 °C, the other temperatures are comprised within the range [40–49] °C.

62.4.3 *Evolution of the Distillate Production Versus Time*

The histogram below shows a comparison of the overall production for 24 h, for both geometries square and rectangle, during the month of March, April, and May, respectively. A better yield is obtained for the rectangle geometry, for volume of 25 l with an average value of the global production = 5250 ml and an average value of the global production = 4700 ml for the square geometry for the same volume.

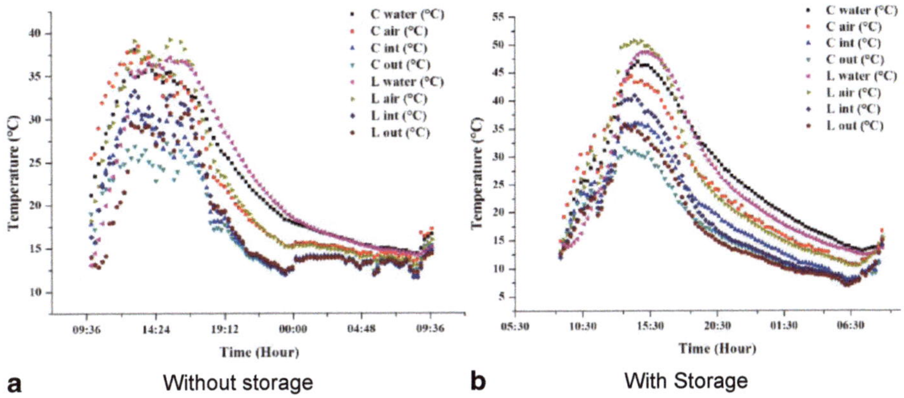

Fig. 62.5 Evolution of the daily production for a distiller without absorber

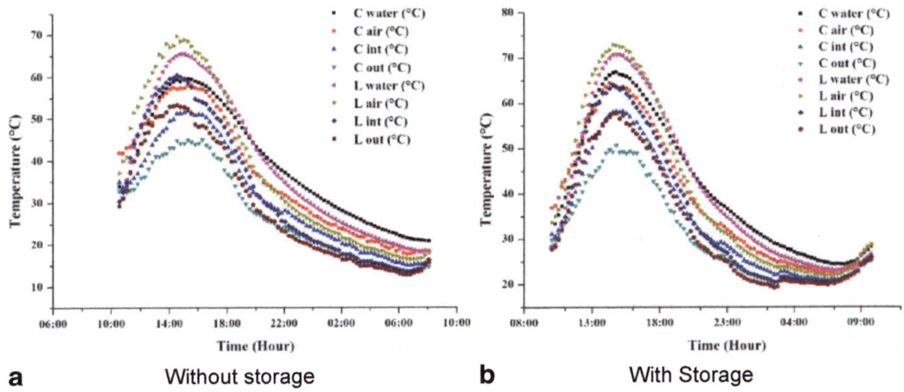

Fig. 62.6 Evolution of the daily production for a distiller with absorber

The difference between both productions is explained by two reasons. The first is due to the influence of weather conditions, in which the rectangle geometry is more adapted to receive the sun's rays (Fig. 62.8b). Thus, the rectangle geometry produced more during the night than the square one. The second is that absorber is very important for better absorption.

In fact, it was deduced that the distillate production is influenced by the presence of wind. It was observed as well that this external parameter deprived the production rate during the morning. However, it was noted that from 15 h, the quantity of distillate produced is decreased due to the decrease in temperature at the end of the day (Fig. 62.7)

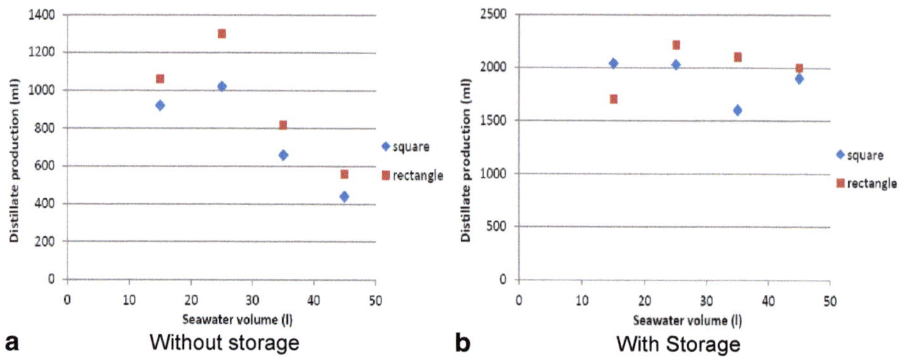

Fig. 62.7 Evolution of the daily production for a distiller without absorber

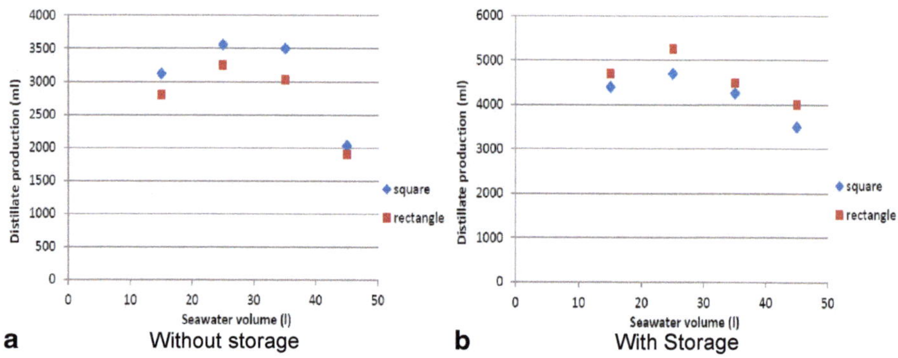

Fig. 62.8 Evolution of the daily production for a distiller with absorber

62.5 Distillate Quality

The pure water recovered was analyzed to determine the pH and electrical conductivity. Figures below show the evolution of the electrical conductivity and the pH measured of distilled water produced for the two distillers considered (Figs. 62.9, 62.10, 62.11, 62.12).

From these figures, it can be seen that the electrical conductivity is more better for the distillate produced with absorber and without storage, and it keeps practically the same value without storage (between 7 and 12). For the distiller with absorber and with storage, we note that the electrical conductivity is slightly more elevated which is around 15 µs/cm for the rectangle geometry and is around 24 µs/cm for the square one, and this can be explained by important quantity of the distillate which may influence the quality of the distilled water at its collection. On the other hand, the pH present values which are close and they do not exceed a value of 7, this is due to carbon dioxide dissolves in water.

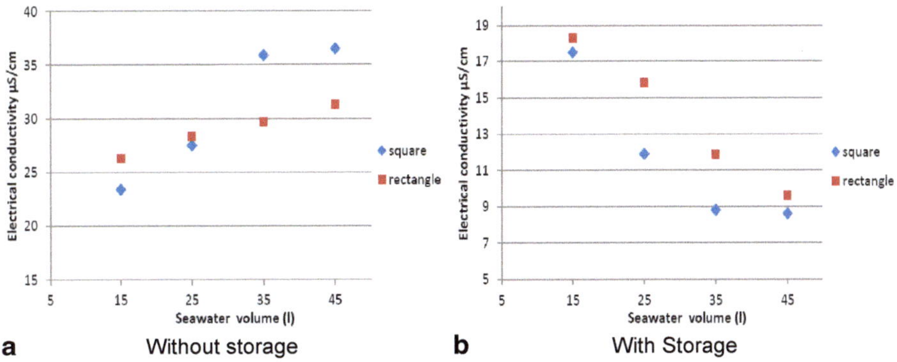

Fig. 62.9 Evolution of the electrical conductivity for two distillers without absorber

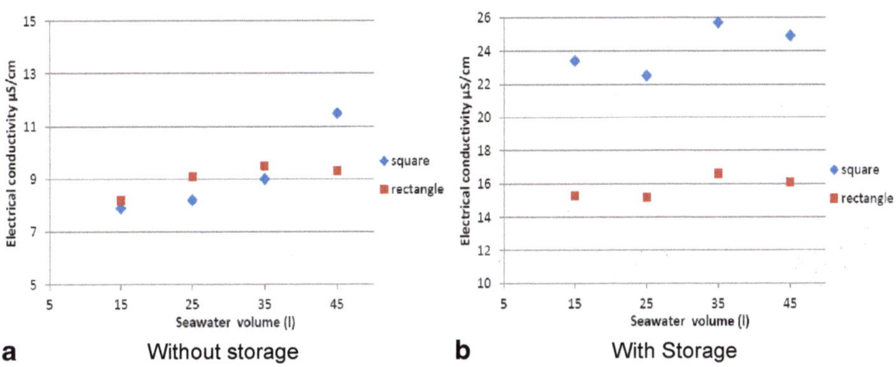

Fig. 62.10 Evolution of the electrical conductivity for two distillers with absorber

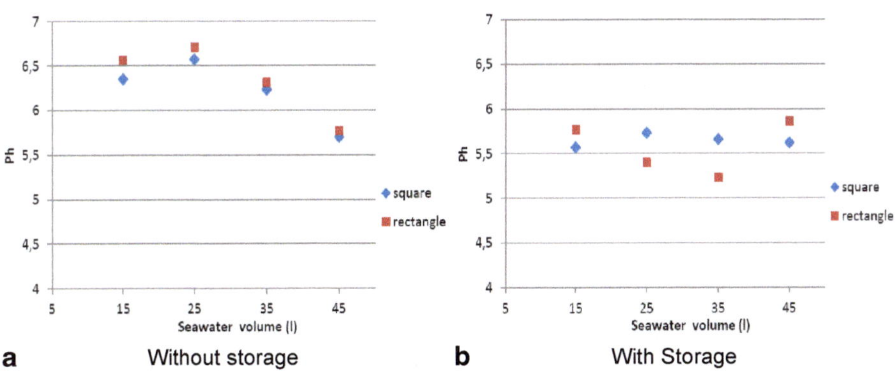

Fig. 62.11 Evolution of the pH for two distillers without absorber

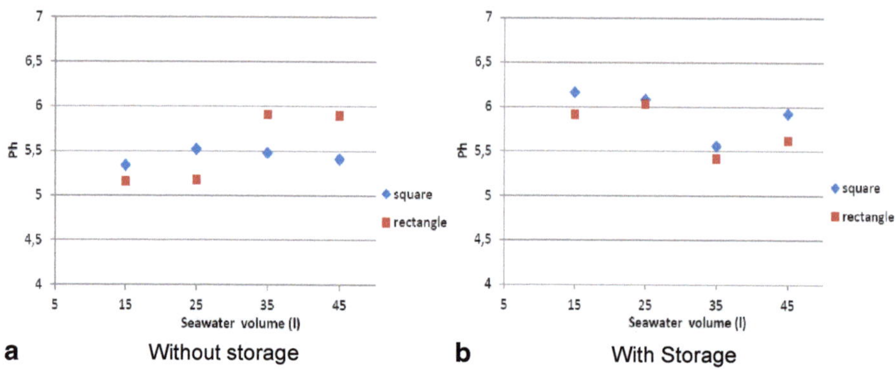

Fig. 62.12 Evolution of the pH for two distillers with absorber

62.6 Conclusion

This experimental study highlights the influence of the geometry and energy storage on the quantity and quality of the distillate water. According to this analysis, the process of seawater desalination by solar distillation appears to be sensitive to different effects, which affect the performance of the system, such as the solar energy intensity, the distilled water temperature, and the thickness of the water layer of energy storage.

The experiments carried out with seawater have allowed noticing changes related to the evolution of the distilled water quantity produced for the two geometries studied (square and rectangle). We found that the absorber is very important in the operation of distillation. In addition to the absorber, the energy storage improves the production of distilled water.

However, it was found that the effect of wind and salt deposit decreased the production of distilled water.

Against this, the incident solar energy and the temperature gradient between seawater and the ambiance tend to accelerate the physical phenomena of solar distillation, where the local time varies between 12:30 and 15:30 pm. According to this qualitative study, it is confirmed that seawater solar distillation is a good option for water treatment.

References

1. Akash BA, Mohsen MS, Nayfeh W (2000) Experimental study of the basin type solar still under local climate conditions. Energy Convers Manage 41:883–890
2. Al-Hayek I, Badran OO (2004) The effect of using different designs of solar stills on water distillation. Desalination 169:121–127
3. Al-Hinai H, Al-Nassri MS, Jubran BA (2002) Effect of climatic, design and operational parameters on the yield of a simple solar still. Energy Convers Manage 43:1639–1650

4. Aybar HS, Atikol FE (2005) An experimental study on an inclined solar water distillation system. Desalination 180:285–289
5. Aybar HS (2006) Mathematical modeling of an inclined solar water distillation system, water distillation system. Desalination 190:63–70
6. Bouidiba M et al (1987) Etude d'un distillateur solaire Ã effet de serre de type cascades. Jith 2:801–810
7. Cherrared L (1999) Amélioration du Rendement d'un Distillateur Solaire Ã Effet de Serre. Energ. Ren. Spécial issue, Volume Valorization, pp. 121–124
8. Dunkle RV (1961) Solar water distillation: the roof type still and a multiple effect diffusion still. International Developments in Heat Transfer, Int. Heat Transfer Conference, University of Colorado. pp 895–902 Part 5
9. Kaabi A, Smakdj N (2007) Impact of temperature difference (water-solar collector) on solar-still global efficiency. Desalination 209:298–305
10. Khedim A, Klemens S, Christian F, Christoph M (2004) Production décentralisée de l'eau potable Ã l'énergie solaire. Desalination 168:13–20
11. Pages J, Chaboub C, Laloë F, Sagna P, Sow I (1989) Un Point d'Eau Douce dans les Iles du Saloum? Examen des Possibilités des Distillateurs Solaires. Revue Sénégalaise des Recherches Agricoles et Halieutiques—Vol. 2– n°34
12. Rubio-Cerda E, Porta-Gandara MA, Fernandez-Zayas JL (2002) Thermal performance of the condensing covers in a triangular solar still. Instituto de Ingenierı´a, UNAM, Cd. Universitaria, Me´xico, D.F. 04510, Mexico. Renew Energy 27:301–308
13. Tanaka H (2009) Experimental study of a basin type solar still with internal and external reflectors in winter. Desalination 249:130–134
14. Velmuruga V, Gopalakrishnan M, Raghu R, Srithar K (2008) Single basin solar still with fin for enhancing productivity. Department of Mechanical Engineering, Infant Jesus College of Engineering, Thoothukudi 628–851, Tamilnadu-India. Energy Convers Manage 49:2602–2608

Chapter 63
Solar Thermal Collectors with Low and High Concentration

Matteo Bortolato, Ahmed Aboulmagd, Andrea Padovan and Davide Del Col

Abstract This chapter describes the performance analysis of different concentrating technologies through experimental and numerical modeling activities. Two solar thermal systems with different designs and, accordingly, different concentration ratios have been studied. The first solar device is a stationary compound parabolic concentrator (CPC) collector: it is provided with truncated or full CPC reflectors and evacuated tubes. Each evacuated tube is composed of an outer glass envelope and a glass absorber with selective coating in thermal contact, via absorber fin, with a U-shaped channel for the liquid flow. The second system is a parabolic trough concentrator (PTC) with two-axis solar tracking: the primary optics consists of a segment of parabolic cylinder which concentrates the direct normal irradiance (DNI) on a linear receiver. In this system, two types of flat receivers have been tested. One receiver has been designed for thermal energy extraction, and it consists of a canalized roll-bond plate provided with a semi-selective coating. The other receiver has been designed for cogeneration of electricity and heat (CPVT), and it is equipped with triple-junction photovoltaic cells, which are actively cooled by an aluminum roll-bond heat exchanger. The performance of the described collectors has been experimentally characterized at the Solar Energy Conversion Laboratory of the University of Padova (45.4°N, 11.9°E), Italy. The collectors have also been mathematically modeled, and the numerical data have been validated against the experimental measurements.

Keywords CPC · Parabolic trough · Concentrating collectors

D. Del Col (✉) · M. Bortolato · A. Padovan
Dipartimento di Ingegneria Industriale, Università degli Studi di Padova,
Via Venezia 1, Padova, Italy
e-mail: davide.delcol@unipd.it

A. Aboulmagd
Faculty of Engineering, Mechanical Power Department, Cairo University, Cairo, Egypt

A. Sayigh (ed.), *Renewable Energy in the Service of Mankind Vol II*,
DOI 10.1007/978-3-319-18215-5_63

63.1 Introduction

Solar collectors are distinguished as low-, medium-, or high-temperature heat exchangers. High temperatures can be obtained by concentrating solar irradiance via reflecting surfaces [1]. Furthermore, solar collectors are classified according to the geometrical concentration ratio (C), defined as

$$C = \frac{A_{ap}}{A_{abs}} \tag{63.1}$$

where A_{ap} is the aperture area of the reflector and A_{abs} is the absorber area. For instance, $C=1$ for flat plate collectors and $C>1$ for concentrating collectors. Hsieh [2] conducted a theoretical study for the evacuated tube collector with U-tubes and a full CPC reflector. Mathematical formulation of optical and thermal processes in CPCs was performed. In his model, any beam of radiation incident within the acceptance angle is assumed to reach the absorber. Convection heat transfer coefficient of the working fluid was constant, and the contact thermal resistance was neglected. In the work of Kumar et al. [3], theoretical performance of single-pass evacuated tube collector with a CPC reflector was analyzed under different operating conditions. In their model, the energy balance equations for each component (cover, receiver envelope, receiver, and fluid inside the cylindrical tube) were solved by using the finite difference technique. Kim and Seo [4] proposed a theoretical and experimental model of evacuated tube collectors with finned tubes and U-tube designs. In their model, the performances of the collectors were studied to find the best shape of the absorber tube. Beam, diffuse irradiation, and shade due to adjacent tubes were also taken into account. In linear concentrators, the case of PTC, the thermodynamic limit for the concentration ratio is around 213, as demonstrated by Rabl [5]. In electronics, the mini-channel technology has proved to be reliable and effective in removing high heat fluxes. Its application as active cooling system in concentrated photovoltaic (CPV) systems with densely packed photovoltaic cells appears very interesting and promising, since compactness and very low thermal resistance are necessary requirements. Furthermore, it gives the possibility to cogenerate electrical energy and heat at temperature up to 100 °C if triple-junction solar cells are employed. Mini channels can be also employed in the design of thermal receivers of solar concentrators for the production of heat at medium temperature range.

In the present work, two types of concentrating solar collectors are introduced: CPC with truncated and full geometry and PTC with two different receivers along the focal line. The working conditions for both models are at zero incidence angles of solar rays (normal to collectors' apertures, $\theta=0$) with a wide range of operating temperatures. A more comprehensive optical and thermal analysis is proposed. Theoretical analysis and numerical solution procedures are available, and simulation outputs are compared to experimental results.

63.2 Compound Parabolic Concentrator Model

63.2.1 Description of the CPC Collector

Evacuated tube solar thermal collectors are made out of glass with all the air evacuated to effectively prevent the heat loss caused by convection between the absorber and the glass. The absorber is coated with a selective material having high absorptivity and low emittance [6]. Figure 63.1 shows the test rig [7] of the CPC glass evacuated tube collector and description of the heat transfer fluid (water) flow within U-tubes. The receiver consists of a copper U-tube inside a glass-vacuumed tube. The copper tube is surrounded by a cylindrical aluminum fin pressed on it, which has the role of enhancing the heat transfer area between the inner glass absorber surface and the U-tube. Two reflector designs of the CPC glass evacuated tube collector are modeled, the truncated CPC and the full CPC (Fig. 63.2a).

The aperture width of the full CPC reflector (with $C=1.9$) is 2.5 times that for the truncated CPC (with $C=0.75$). Part of the solar radiation falls directly on the absorber surface, while the rest is reflected by the CPC on the absorber glass tube. The outer cylindrical glass transmits the rays to the inner glass tube, which conducts

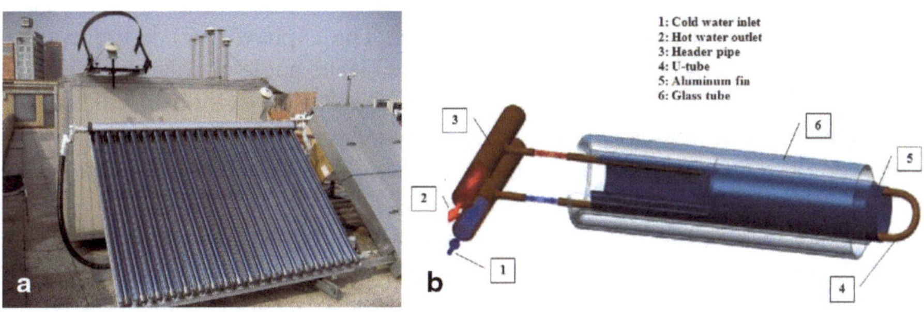

Fig. 63.1 CPC glass evacuated tube collector with U-tubes. **a** test rig [7]. **b** Fluid flow in U-tubes

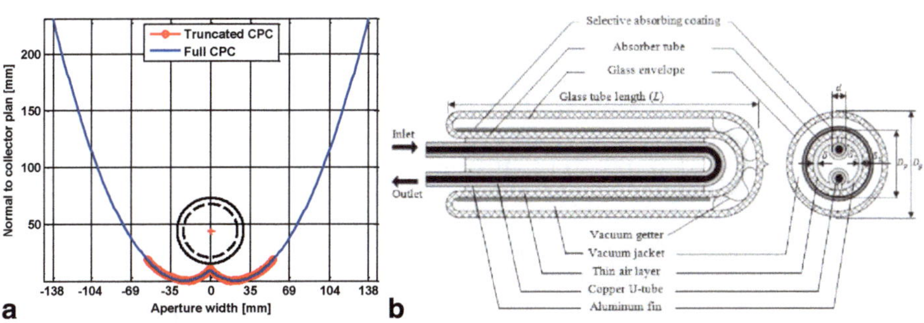

Fig. 63.2 **a** CPC reflector with tubular receiver (truncated and full geometry). **b** Illustration of the CPC glass evacuated tube solar collector with U-tube

the energy to the absorber fin [8]. The energy transformed into heat is conducted by the fin to the copper U-tube and finally absorbed by the working fluid. The detailed illustration of the evacuated tube and its cross-sectional view are given in Fig. 63.2b. Analytical model parameters considered in the present study are as follows: D_g=0.058 m, D_p=0.047 m, d=0.0063 m, L=1.56 m, δ_1=0.0016 m (glass thickness), δ=0.0008 m (fin thickness). Absorptivity and emissivity of the selective absorptive coating are 0.94 and 0.06, respectively, while the glass transmittance is 0.92. The analysis is performed for a single evacuated glass tube, whereas the collector is composed of 20 glass tubes with parallel flow channels.

63.2.2 Optical and Thermal Performance of the CPC Collector

The optical efficiency (η_0) depends on the optical properties of the materials used (reflectivity, transmissivity, and absorptivity), the geometry of the reflector, the space between collector tubes, and incidence angles of solar rays. The optical efficiency model used in this study is obtained from [9] as

$$\eta_o = \rho^{\bar{n}} \tau \alpha \gamma \tag{63.2}$$

where ρ is the reflectivity of the CPC reflector, τ is the transmissivity of the glass tube, and α is the absorptivity of the absorber. The exponent ($\bar{n}$) is the average number of reflections, which depends on the incidence angle of solar radiation (θ) and half acceptance angle of the CPC reflector (θ_a). The behavior of $\bar{n} = 0.5$ is discussed by [5], it decreases with increasing the half acceptance angle and also with decreasing the incidence angle of solar radiation. In the present study, a value of $\bar{n} = 0.3$ is obtained for the truncated CPC and $\bar{n} = 0.5$ for the full CPC. The parameter γ is the intercept factor, defined as the fraction of solar radiation accepted by the collector absorber. This factor depends on collector geometry, incidence angle of solar rays, and atmospheric conditions. The values of γ are taken as 0.92 for the truncated CPC and 0.95 for the full CPC. The thermal efficiency (η_{th}) of the collector is defined as the ratio of the useful heat gained by the heat transfer fluid (q_u) to the absorbed solar power (q_{abs}) which are given, respectively, by

$$\eta_{th} = \frac{q_u}{q_{abs}} \tag{63.3}$$

$$q_u = F' A_{abs} \left[\eta_o IC - U_L \left(T_f - T_a \right) \right] = \dot{m}_f \, c_{p,f} \left(T_{f,out} - T_{f,in} \right) \tag{63.4}$$

$$q_{abs} = \eta_o I A_{ap} \tag{63.5}$$

F' is the collector efficiency factor, I is the total solar irradiance, U_L is the overall heat loss coefficient, T_f is the mean fluid temperature inside the U-tube, and T_a is the ambient air temperature. $\dot{m}_f$ is the mass flow rate of the working fluid, $c_{p,f}$ is the specific heat, and $T_{f,in}$ and $T_{f,out}$ are the inlet and outlet fluid temperatures, respectively. The overall efficiency of the collector is given by

$$\eta_c = \frac{q_u}{IA_{ap}} = F'\eta_o - \frac{U_L F'}{C} T_m^* \quad (63.6)$$

where $T_m^* = (T_f - T_a)/I$ is the reduced temperature difference.

63.3 Parabolic Trough Concentrator Model

63.3.1 Description of the PTC Collector

The primary optics of the PTC is made up of a semi-parabolic trough with a focal length of 1810 mm and a rim angle of 78°. The reflectance of the primary mirrors measured at normal incidence is 96.1% as reported by the manufacturer. The total aperture area of the solar concentrator is equal to 6.857 m^2. In this prototype (Fig. 63.3a), the length of the focal line is 2.4 m. The system moves about two axes (azimuthal and zenithal motions) to have the beam radiation normal to the aperture area. The motion is governed by a solar algorithm when approaching the sun and by a sun sensor when achieving the best receiver alignment. Two different receivers have been installed in the parabolic trough concentrator described above. According to the shape of the primary optics, each receiver is located so that the normal to its frontal surface forms an angle of 45° with the normal to the aperture area. The first receiver has been engineered for heat production (Fig. 63.3b), while the second one has been designed for the cogeneration of electricity and heat (CPVT). The thermal receiver consists of an aluminum plate manufactured using the roll-bond process and is provided with four parallel channels with a hydraulic diameter of 2 mm. A semi-selective coating (0.87 absorptance and 0.35 emittance) is used to cover the receiving surface to enhance the absorption of solar radiation. The hybrid CPVT receiver is provided with triple-junction PV cells to achieve high electrical efficiency even at temperatures up to 100 °C. The cells have been soldered on a ceramic substrate which in turn is in thermal contact with an active cooling system including a cana-

Fig. 63.3 **a** Prototype of the PTC collector. **b** The thermal receiver

Table 63.1 Values of the optical efficiency for the CPVT receiver obtained by SolTrace

Total optical error (mrad)	2.5	3	3.5	4
Optical efficiency (%)	74.1	71.2	68.0	64.8

lized roll-bond heat exchanger. On the front side of the CPVT module, a secondary optics composed of two flat aluminum mirrors has been mounted to reduce the optical losses.

63.3.2 Optical and Thermal Performance of the PTC Collector

The optical performance of the PTC with the thermal module and the hybrid CPVT receiver has been evaluated by using SolTrace 2012.7.9 [10] as Monte Carlo ray-tracing tool. The optical efficiency in this case is given by Eq. (63.7). The optical model of the linear focus PTC with the thermal receiver is very simple and includes the reflective surfaces of the primary mirrors composing the semi-parabolic cylinder and the absorber. According to the receiver geometry, the optical efficiency is independent of the optical error in the range between 2.5 and 4 mrad and it is equal to 96 %. The geometrical concentration ratio in this case is around 50. The steady-state method described in the EN 12975-2 standard [11] is adopted here for the assessment of the thermal performance of the tested receivers. When considering the parabolic trough, the input power is the direct normal irradiance (DNI), which is measured by a pyrheliometer mounted on a high-precision sun tracker. Measurements are repeated using the experimental apparatus described in [12] at different inlet water temperatures in order to produce a set of thermal efficiency data, defined as the ratio of the useful heat flow rate to the input power from the sun, which is given by Eq. (63.8).

Table 63.1 reports the values of the optical efficiency for the CPVT receiver obtained using SolTrace as a function of the total optical error. The parabolic trough with the hybrid CPVT receiver exhibits a geometrical concentration ratio of 130. During the tests with the CPVT module, its electrical terminals are connected to a rheostat and a power analyzer that measure the output electrical power (P_e) which is averaged on a time frame of 10 min, just like the output thermal power. The rheostat was adjusted so that the hybrid module worked at the maximum power point. The electrical efficiency can be expressed by Eq. (63.9).

$$\eta_o = \frac{q_{abs}}{DNI \cdot A_{ap}} \tag{63.7}$$

$$\eta_{th} = \frac{q_u}{DNI \cdot A_{ap}} = \frac{m_f \, c_{p,f} (T_{f,out} - T_{f,in})}{DNI \cdot A_{ap}} \tag{63.8}$$

$$\eta_{el} = \frac{P_e}{DNI \cdot A_{ap}} = \frac{E_e I_e}{DNI \cdot A_{ap}} \tag{63.9}$$

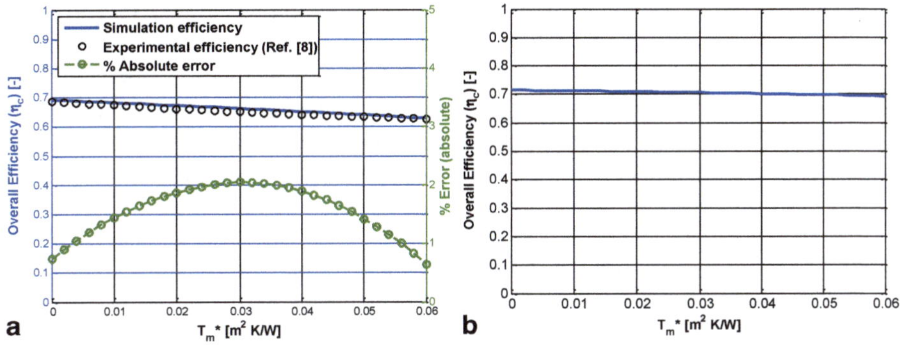

Fig. 63.4 Overall efficiency versus T_m^* (T_a=20℃ and I=1000 W m^{-2}). **a** Truncated CPC: comparison between simulation output and experimental results. **b** Full CPC: simulation output

E_e and I_e are the voltage and current of the electrical resistive load, respectively. The overall efficiency of the system can be defined as follows:

$$\eta_c = \eta_{th} + \eta_{el} = \frac{q_u + P_e}{DNI \cdot A_{ap}} \tag{63.10}$$

63.4 Simulation and Results

63.4.1 Outputs for the CPC Collector

The performance of the CPC collector is evaluated by simulations under the following working conditions: $T_a = 20$℃, $\dot{m}_{f,c} = 0.07 kg s^{-1}$ (total mass flow rate for the collector), $u_a = 2 ms^{-1}$ (wind speed), $I = 1000 W m^{-2}$, and T_m^* ranging from 0 to 0.06 Km m^{2}W^{-1}. Figure 63.4 shows the collector overall efficiency, and Table 63.2 gives a summary of the averaged simulation outputs in both cases, truncated CPC and full CPC. Comparison between simulation output and experimental results obtained from [7] for the overall efficiency with the associated percentage absolute error is provided for the case of truncated CPC collector. It can be noticed from the simulation outputs that the effect of increasing the aperture width and CPC height (full CPC geometry) is increasing the absorbed solar power and hence the collector

Table 63.2 Average values of different performance parameters for truncated CPC and full CPC (under zero incidence angle)

η_c [%]	q_{abs} [W]	q_u [W]	η_{th} [%]	η_o [%]	η_c [%]
Truncated CPC	134	114	84.9	78.0	66.2
Full CPC	344	303	88.1	80.1	70.6

heat gain. The mean absorber temperature in case of a full CPC is 51.1 °C, while it is 49.1 °C in case of a truncated CPC. In addition, an increase of the values of thermal, optical, and overall efficiencies of 3.2, 2.1, and 4.4 %, respectively, is obtained. This analysis is conducted by neglecting the component of diffuce irradiance and under zero incidence angle. In order to keep the advantages of the full CPC configuration under general operating conditions, a tracking system should be added.

63.4.2 Outputs for the PTC Collector

During the experimental tests with the thermal receiver, the mass flow rate of water in single-phase flow was set at 0.02 kg s^{-1}/m^2 of the aperture area of the PTC. The DNI ranged within 615 and 780 W m^{-2}, and the ambient air temperature was between 21.5 and 27.5 °C. The inlet water temperature has been set at three different operating values of 50, 65, and 80 °C. As illustrated in Fig. 63.5, the thermal efficiency slightly decreases with the reduced temperature difference and it is around 75 % at T_m^*=0.08 K m^2 W^{-1}. A simple model of the thermal receiver has been developed. The receiver has been discretized in many elements, and a lumped capacitance scheme has been considered for each element. The thermal balance associated with each node in steady-state conditions has been implemented to calculate the temperature distribution, the useful heat flow rate, and the thermal efficiency. The model has been validated against experimental data, showing good agreement, as depicted in Fig. 63.5. The test runs with the hybrid CPVT module have been performed with the DNI ranging between 700 and 830 W m^{-2}, and the ambient temperature between

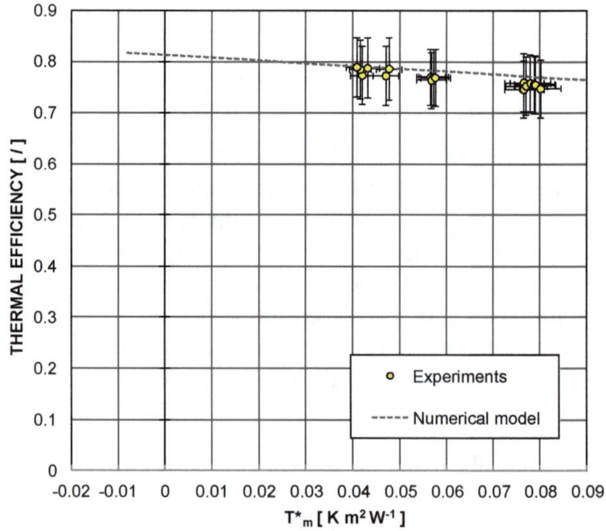

Fig. 63.5 Experimental thermal efficiency versus T_m^* obtained during the tests of the parabolic trough concentrator with the roll-bond thermal module

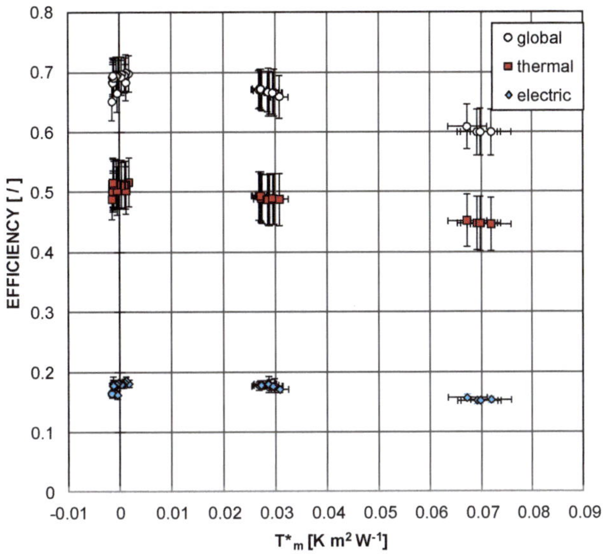

Fig. 63.6 Experimental thermal efficiency versus T_m^* obtained during the tests of the PTC with the CPVT module (electrical and thermal mode)

29 and 34 °C. The global efficiency obtained during tests with electrical load ranges between 0.7 and 0.6 as the reduced temperature difference T_m^* varies between 0 and 0.072 K m^2 W^{-1} (Fig. 63.6).

63.5 Conclusions

Numerical modeling and experimental activities have been developed for two types of solar collectors, a CPC with low concentration and a PTC with high concentration and two receiver types. Optical and thermal characterizations of the collectors are introduced. For the CPC collector, a more detailed approach is used to predict various performance parameters (temperatures, thermal powers, and efficiencies). When the analysis extended from truncated CPC to full CPC, the mean absorber temperature increased as well as the absorbed solar power. The effect on reducing the truncation also resulted in an increase in collector optical and thermal efficiencies. The comparison with experimental data measured at the same conditions, in case of truncated CPC, shows the good predicting accuracy of the model. For the PTC, with higher concentration, higher temperatures can be produced with thermal receivers, and this is interesting for many applications such as solar cooling, direct steam generation, and industrial process heat. A simple numerical model for the thermal receiver has been developed, revealing a good agreement with experiments. It has been also demonstrated that using the CPVT receiver, a cogeneration of two forms of powers suitable for electrical and thermal applications (up to 90–100 °C) can be obtained.

References

1. Foster R, Ghassemi M, Cota A, Energy S (2010) Renewable energy and the environment CRC Press, Taylor & Francis Group, 6000 Broken Sound Parkway NW, Suite 300 Boca Raton, FL, pp 33487–32742
2. Hsieh CK (1981) Thermal analysis of CPC collectors. Sol Energy 27(1):19–29
3. Kumar R, Kaushik SC, Garg HP, Vatsa N (1999) Numerical analysis of a tubular solar collector with cusp reflectors. Int J Ambient Energy 20(3):149–158
4. Kim JT, Ahn HT, Han H, Kim HT, Chun W (2007) The performance simulation of all-glass vacuum tubes with coaxial fluid conduit. Int Commun Heat Mass Transf 34(1):587–597
5. Rabl A (1975) Comparison of solar concentrators. Sol Energy 18(1):93–111
6. Kim YS, Balkoski K, Jiang L, Winston R (2013) Efficient stationary solar thermal collector systems operating at a medium-temperature range. Appl Energy 111(1):1071–1079
7. Zambolin E, Del Col D (2012) An improved procedure for the experimental characterization of optical efficiency in evacuated tube solar collectors. Renew Energy 43(1):37–46
8. Soriga I, Neaga C, 2012 Thermal analysis of a linear Fresnel lens solar collector with black body cavity receiver, U.P.B. Sci. Bull., Series D, vol. 74, no. 4, ISSN 1454-2358
9. Rabl A (1976) Optical and thermal properties of compound parabolic concentrators. Sol Energy 18(1):497–511
10. National Renewable Energy Laboratory (NREL) (2012) SolTrace version 2012.7.9. http://www.nrel.gov/csp/soltrace. Accessed Jan 2014.
11. Comité Européen de Normalisation (CEN) (2006) EN 12975-2: Thermal solar systems and components—solar collector—part 2: test methods, Brussels, Belgium
12. Zamboiln E, Del Col D (2010) Experimental analysis of thermal performance of flat plate and evacuated tube solar collectors in stationary standard and daily conditions. Sol Energy 84(1):1382–1396

Chapter 64
Building-Integrated Solar Thermal Systems

Soteris A. Kalogirou

Abstract With buildings accounting for 40% of primary energy requirements in the EU and the implementation of the Energy Performance of Buildings Directive, developing effective energy alternatives for buildings is imperative. The increasing role for renewables implies that solar thermal systems (STSs) will have a main role as they contribute directly to the heating and cooling of buildings and domestic hot water. Meeting building thermal loads will be primarily achieved through an extensive use of renewables, following standard building energy-saving measures. These systems are typically mounted on building roofs with no attempt to incorporate them into the building envelope creating aesthetic challenges, space availability issues, and envelope integrity problems. This chapter aims to give a survey of possible solutions of STS integration on the building roofs and façades. The advantages of integration are quantified, and suggestions are given to address the possible problems created.

Keywords Solar Collectors · Building integration · Façade integrated collectors · Roof integrated collectors

64.1 Introduction

The Renewable Energy Framework Directive sets a 20% target for renewables by 2020. Buildings account for 40% of the total primary energy requirements in the EU [1]. Therefore, developing effective energy alternatives for buildings, used primarily for heating, cooling, and the provision of hot water, is imperative. One way to reduce fossil fuel dependence is the use of renewable energy systems (RES) which are generally environmentally benign. In some countries, such as Cyprus, RES and in particular solar water heating are used extensively. The benefits of such systems are well known, but one area of concern has been their integration. Most

S. A. Kalogirou (✉)
Department of Mechanical Engineering and Materials Science and Engineering, Cyprus University of Technology, PO. Box 50329, 3603 Limassol, Cyprus
e-mail: Soteris.kalogirou@cut.ac.cy

A. Sayigh (ed.), *Renewable Energy in the Service of Mankind Vol II*,
DOI 10.1007/978-3-319-18215-5_64

solar components are mounted on building roofs, and they are frequently seen as a foreign element on the building structure. Due to this fact alone and irrespective of the potential benefits, some architects object to this use of solar energy systems. It is therefore necessary to find ways to better integrate solar systems within the building envelope, which should be done in a way that blends into the aesthetic appearance and form of the building architecture in the most cost-effective way.

The Energy Performance of Buildings Directive (EPBD) requires that RES are actively promoted in offsetting conventional fossil fuel use in buildings. A better appreciation of solar thermal systems (STSs) integration will directly support this objective, leading to an increased uptake in the application of renewables in buildings, which is expected to rise dramatically in the coming years. This is further augmented by the recast of EPBD, which specifies that by the year 2020, the buildings in the EU should be of nearly zero energy consumption. Meeting building thermal loads will be primarily achieved through an extensive use of renewables, following standard building energy-saving measures, such as good thermal insulation, advanced glazing systems, etc. STSs are expected to take a leading role in providing the thermal energy needs, as they can contribute directly to the building heating, cooling, and domestic hot water requirements.

64.2 Building Integration of Solar Thermal Systems

Among the renewable energy resources, solar energy is the most essential and prerequisite resource of sustainable energy because of its ubiquity, abundance, and sustainability. STSs can supply thermal energy for space heating, cooling, and the provision of hot water for the needs of a house/building. The advantages of building integration of STSs are that more space is available on the building for the installation of the required area of the STSs and that the traditional building component is replaced by the STS, which increases the economic viability of the systems.

In the case that this concept is employed, coupled with aesthetic and architectural challenges of building integration, many practical issues need to be resolved, such as rainwater sealing and protection from overheating (avoiding increased cooling loads during summer). The extra thermal energy can also be used for the heating of the building in winter. As STSs are latitude dependent, with respect to façade application and solar incidence angle effects, these need to be considered, as countries near the equator have high-incidence angles (the sun is higher on the sky), but more energy is available compared to higher-latitude countries.

The adoption of building integration of STSs can fundamentally change the accepted solar installation methodologies that affect residential and commercial buildings throughout the world. Maybe, the single most important benefit originating from this idea is the increased adoption of STSs in buildings.

A solar energy system is considered to be building integrated, if for a building component this is a prerequisite for the integrity of the building's functionality. If the building-integrated STS is dismounted, dismounting includes or affects the adjacent building component, which will have to be replaced partly or totally by a conventional/appropriate building component. This applies mostly to the case of structurally bonded modules but applies as well to other cases, such as in the case of replacing a wall-leaf in a double-wall façade with a building-integrated solar thermal system (BISTS). Therefore, building integration must provide a combination of the following:

1. Mechanical rigidity and structural integrity
2. Weather impact protection from rain, snow, wind, and hail
3. Energy economy, such as useful thermal energy, but also shading and thermal insulation
4. Fire protection
5. Noise protection

The building integration of solar thermal systems can pose a number of problems that will need to be considered such as:

1. Amount of thermal energy collected and at what temperature range
2. Resistance to wind-driven rain penetration
3. If the underlying base layer is transparent, calculation of light and solar energy characteristics
4. Calculation of thermal resistance and thermal transmittance characteristics of the construction (overall heat transfer coefficient)
5. Fire protection classification and fire protection from hot components in contact with flammable materials
6. Noise attenuation

64.3 Collector Systems that can be Building Integrated

The solar-collecting methodologies that can be applied in buildings are the simple thermosiphonic units, forced circulation systems employing flat plate collectors, integrated collector storage units, evacuated tube collector systems, and various low-concentration compound parabolic units [2]. In some countries, such as Cyprus, RES and in particular solar water heating are used extensively, with 93 % of all domestic dwellings currently equipped with such a system [3].

The benefits of solar water heating systems are well known, but one area of concern has been their integration. Most solar collecting components are mounted on building roofs with no attempt to incorporate them into the building envelope. In many instances, they are actually seen as a foreign element on the building roof. Many architects, irrespective of the potential benefits, object to this use of RES due to this fact alone. The problem will be even more serious, when solar space heating and cooling systems are used, as they required much more solar collectors. It is therefore necessary to find ways to better integrate solar collectors within the building envelope and/or structures, which should be done in a way that blends into the

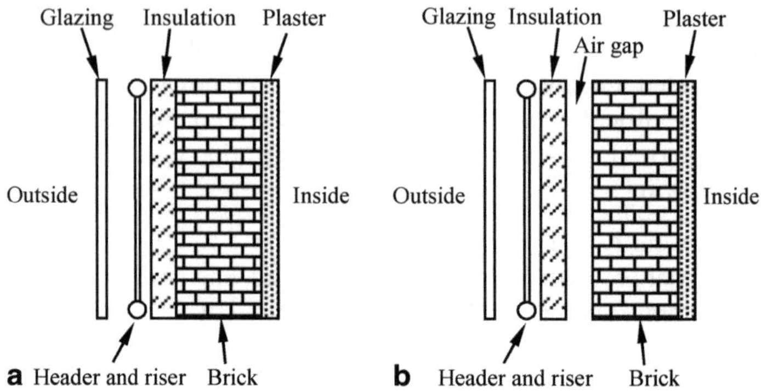

Fig. 64.1 Two solutions of façade building-integrated flat plate collectors. **a** Solar thermal system (STS) adjacent to the brick. **b** STS with air gap

aesthetic appearance and form of the building architecture in the most cost-effective way.

As was seen above, various solar heating systems can be installed in buildings and each one of them has to be considered by itself when building integration is considered. Evacuated tube collectors can lead to serious rain penetration problems when integrated in buildings unless a special construction is done behind the collector to keep rain out of the building structure. Two solutions of building-integrated flat-plate collectors are shown in Fig. 64.1 as examples of this application. The collector consists of the usual parts found in stand-alone systems without the casing, and the whole construction is set up in front of the brick of the normal brick wall. The collector can be installed by leaving an air gap between the insulation and the brick, as shown in Fig. 64.1b, according to the prevailing conditions that exist in the area of installation and the necessity to avoid migration of moisture into the building. In both cases, good insulation is used to avoid transferring unwanted heat into the building, especially during the summer months. The same construction can be used for sloping roof applications in which case the brick is replaced by a concrete slab.

In view of the EPBD, which requires also the extensive use of thermal insulation, the above solutions can be viewed, especially for retrofitting applications, as external insulation applied to the external wall surface, protected with glazing. So the only extra element required is the header and riser assembly, storage tank and piping (not shown), and glazing in order to convert the system into a thermal energy collection system. Of course, the ideas and systems to be used are not limited to the ones shown in Fig. 64.1 but are extended to various other ones as shown subsequently. Additionally, many ready-made products already appear on the market, such as roof singles, façade coverings, etc., all of which are also solar thermal collectors. Some typical examples are shown in Fig. 64.2.

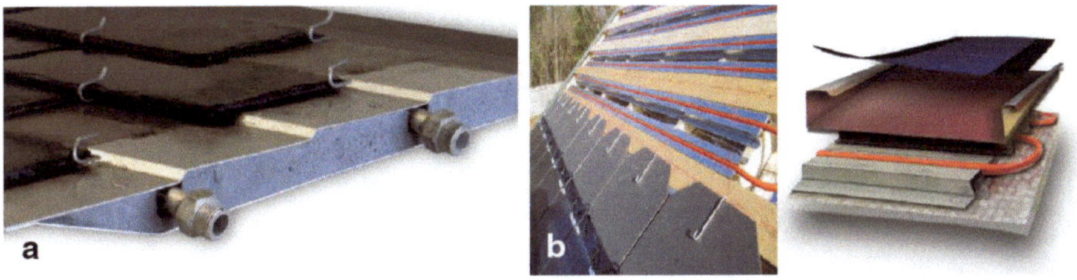

Fig. 64.2 Commercial building-integrated solar thermal system (BISTS) products. **a** Thermoslate. **b** Englert Solar Sandwich

64.4 BISTS Classifications/Systems

BISTSs have been classified across a range of operating characteristics and system features and mounting configurations. The main classification criteria of all STSs are based on the method of transferring collected solar energy to the application (active or passive), the energy carrier (air, water, water-glycol, oil, etc.), and the final application for the energy collected (hot water and/or space heating, cooling, process heat, or mixed applications).

For BISTSs, additionally the architectural integration quality based on structural, functional, and aesthetical variations have to be classified. The collector as a central element in the integration has to fulfill in some cases many more specifications than the ordinary "add-on" collectors.

The majority of BISTSs can be classified as being either passive or active, for example, in the first case using thermal buoyancy for fluid transport (natural convection or circulation) or no transport at all, and in the second case utilizing pumps or fans to circulate the thermal transfer fluid to a point of demand or storage (forced convection or circulation). A number of systems are however hybrids, operating in part through a combination of natural and forced transport methods. Many façade solar air heaters use thermal buoyancy to induce an air flow through the vertical cavities that can be further augmented with in-line fans (and heating), if necessary. The BISTS delivers thermal energy to the building, but additionally other forms of energy may contribute to the buildings energy balance. For instance, daylight comes through a transparent window or façade collector, or photovoltaic/thermal (PV/T) systems will also deliver electrical power, which may be used directly by any auxiliary electrical services. Heated air or water can be stored or delivered directly to the point of use. Although the range of applications for thermal energy is extensive, all of the evaluated studies demonstrate that the energy is used to provide one or a combination of the following:

64.4.1 Space Heating

Thermal energy produced by a BISTS may reduce the space heating load of a building by adding solar gains directly (e.g., by a passive window) or indirectly

Fig. 64.3 An indirect solar-comb construction building-integrated solar thermal system (BISTS)

(e.g., by transferring heat from the collector via a storage to a heating element) into the building as shown in Fig. 64.3.

64.4.2 Air Heating and Ventilation

Thermal heat may be used also to preheat fresh air needed in the building. Air is heated directly or indirectly (in a secondary circuit) and using forced flow or thermosiphonic action is used to provide space air heating and/or ventilation to the building as shown in Fig. 64.4. In some instances, an auxiliary heating system is used to augment the heat input because of comfort reasons.

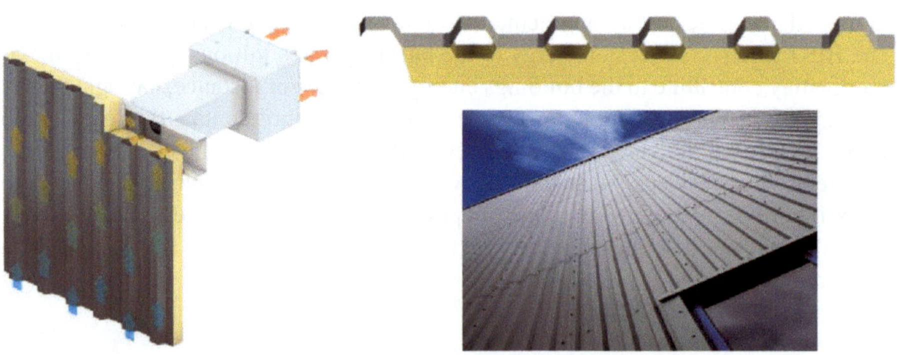

Fig. 64.4 Solar air heating façade building-integrated solar thermal system (BISTS) with auxiliary heating system

Fig. 64.5 Roof-integrated flat plate building-integrated solar thermal system (BISTS) for solar water heating

64.4.3 Water Heating

Hot water demand in the building is the most popular application. In the majority of water-heating BISTSs, a customized heat exchanger or integrated proprietary solar water heater is used to transfer collected heat to a (forced) heat transfer fluid circuit and on to an intermediate thermal store and/or directly to a Domestic Hot Water (DHW) application. In most instances, an auxiliary heating system is used to augment the heat input. An example is shown in Fig. 64.5.

64.4.4 Cooling and Ventilation

In cooling-dominated climates, buildings most of the time have an excess of thermal energy, and there BISTS can also be a technology to extract heat from a building. There are a number of methods described in providing a cooling (and/or ventilation) effect to a building: shading vital building elements, desiccant linings, supplying

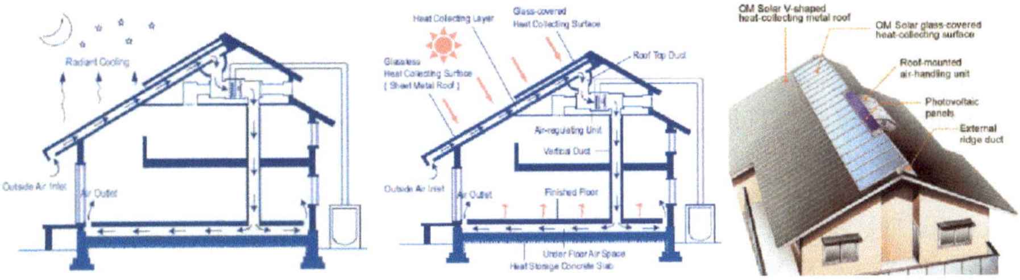

Fig. 64.6 Radiant cooling via a reversed building-integrated solar thermal system (BISTS) [4]

Fig. 64.7 Building-integrated solar thermal system (BISTS) balustrade/railing feature

heat directly to "sorption" equipment, induced ventilation through a stack effect and reverse operation of solar collecting elements for nighttime radiation cooling as illustrated in Fig. 64.6 [4].

With reference to Fig. 64.6, during the solar heating mode, fresh outdoor air enters a channel under the roof and flows upward. The air is heated on contact with the metal roof sheet, passing through an upper glazed section (to improve collection) whereupon the heated air enters the rooftop duct and is mechanically forced through the air-regulating unit. The temperature-controlled air is directed down into the space to be heated via underfloor channels between the floor and the concrete slab before finally being diffused into the room through the floor diffusers. In summer-cooling mode, outdoor air is drawn through the roof channels at nighttime, subcooled using radiant cooling, and as with the heating mode, directed into the space to be cooled via the underfloor channels [4].

64.4.5 Other

The majority of BISTSs documented are mounted on the façade or roofing structures, but a significant number can be classified as being "other." This embraces a multitude of mounting options, from shading devices to balcony balustrades such as that shown in Fig. 64.7.

An additional classification can relate to the mode of installation: new build, refurbishment, or retrofit which is often related to the form of the design or components utilized be they proprietary/prefabricated or customized. Further subsection classification can be related to features such as optical enhancements or indirect benefits associated with the BISTS, such as weatherproofing, acoustic attenuation, or thermal insulation.

A COST action is running to investigate the building integration of solar thermal systems (BISTSs). The COST identification number of the action is TU1205, and the action belongs to the Transport and Urban Development (TUD) domain. More details can be obtained from COST's Web page (www.cost.eu/domains_actions/tud/Actions/TU1205) or the project Web page (http://www.tu1205-bists.eu/).

64.5 Conclusions

The EPBD requires that RES are actively promoted in offsetting conventional fossil fuel use in buildings. A better appreciation of PV and STS integration will directly support this objective, leading to an increased uptake in the application of renewables in buildings. This uptake in RES in buildings is expected to rise dramatically in the next few years. This is further augmented by a recast of the directive, which specifies that the buildings in the EU should consume nearly zero energy (residential and commercial buildings by the year 2020 and public buildings by 2018). Meeting building thermal loads will be primarily achieved through an extensive use of renewables, following standard building energy-saving measures, such as good insulation or advanced glazing systems. STSs are expected to take a leading role in providing the electrical and thermal energy needs, as they can contribute directly to the building heating, cooling, and domestic hot water requirements.

As can be seen from the solutions presented in this chapter, a number of ideas have been tried, and others are just at the concept stage, and generally more R&D effort is needed. It is believed that in the coming years more and more of these solutions/ideas will find their way into the market in view of the implementation of the directives imposed by the EU.

Acknowledgments The author is grateful to the EU COST Action TU1205: "Building integration of solar thermal systems (BISTS)" for its sponsorship.

References

1. European Commission (2005) Doing more with less, Green Paper on energy efficiency. Accessed 22 June 2005 COM
2. Kalogirou SA (2009) Solar energy engineering: processes and systems. Academic, Elsevier, New York
3. Maxoulis CN, Kalogirou SA (2008) Cyprus energy policy: the road to the 2006 world renewable energy congress trophy. Renew Energy 33:355–365
4. OM Solar Association. http://www.omsolar.net/en/omsolar2/roof.html. Accessed 28 July 2014

Chapter 65
Experimental Study on Regenerator Performance of a Solar Hybrid Liquid Desiccant Air-Conditioning System

Sohif Bin Mat, Kamaruzzman Sopian, M. Yusof Sulaiman, Abdulrahman Th. Mohammad and Abduljalil A. Al-abidi

Abstract This chapter presents an experimental study on the performance of the liquid desiccant regenerator of a hybrid solar air-conditioning system. Lithium chloride (LiCl) solution is used as the working desiccant material. The effects of air temperature, air humidity ratio, and solution temperature on the performance of the regenerator are disused. The experimental results showed that the moisture removal rate (MRR) and effectiveness of the regenerator increase slowly as a function of the air inlet temperature. It was found that the MRR and effectiveness increased about 0.79 and 1.1 %, respectively. The moisture removal rate decreased with increasing air inlet humidity ratio and increased with desiccant inlet temperature.

Keywords Desiccant · Regenerator · MRR · Effectiveness

Nomenclature

m_a	Dry air mass flow rate at the inlet of the regenerator (kg sec^{-1})
MRR	Moisture removal rate in the regenerator (g sec^{-1})
P	Total pressure of air above the solution (Pa)
P_{wz}	Partial pressure of water vapor in the solution (Pa)
T_s	Solution temperature (°C)
W_{in}	Inlet air humidity ratio (kg/kg $_{dry\ air}$)
W_{out}	Outlet air humidity ratio (kg/kg $_{dry\ air}$)
W_{out_eq}	Humidity ratio of the air in equilibrium with the desiccant solution (kg/kg $_{dry\ air}$)
X	Solution concentration (%wt)

Greek symbols

ε	Effectiveness of the regenerator (%)

K. Sopian (✉) · S. B. Mat · M. Y. Sulaiman · A. T. Mohammad · A. A. Al-abidi
Solar Energy Research Institute, Faculty of Engineering,
Universiti Kebangsaan Malaysia, 43600 Bangi, Selangor, Malaysia
e-mail: ksopian@eng.ukm.my; sopian@yahoo.com

A. Sayigh (ed.), *Renewable Energy in the Service of Mankind Vol II*,
DOI 10.1007/978-3-319-18215-5_65

65.1 Introduction

The effectiveness and condensation rate are mostly used as an index to determine the performance of the liquid desiccant dehumidifier and regenerator. Based on the previous studies, modeling techniques for the prediction of dehumidifier/regenerator performance can be classified into theoretical analysis model, experimental model, and artificial intelligence model.

Experimental models can be developed by using conventional approaches such as statistical techniques. Many studies, such as Chung et al. [1] and Kessling et al. [2], have been carried out on the performance of a packed tower under various parameters such as the inlet mass flow rate, temperature, and air humidity ratio as well as the inlet mass flow rate, temperature, and concentration of the desiccant solution. A study was conducted by Moon et al. [3] to present new mass transfer performance data of a cross-flow liquid desiccant dehumidification system, and a new empirical correlation was developed for the dehumidification effectiveness, which fitted the experimental data within ± 10 %. Sanjeev et al. [4] developed an experimental setup to study the performance of the liquid desiccant dehumidification system under different operating conditions (hot water temperature, inlet air conditions, and solution concentration). Lithium chloride and calcium chloride were used as desiccants in the system. The results show that the effectiveness of the dehumidifier was found to range between 0.25 and 0.44, while that of the regenerator was between 0.07 and 0.80. Liu et al. [5] compared the mass transfer performance of lithium chloride and lithium bromide aqueous solutions. The experimental results show that the dehumidifcation performance of the LiCL was better than that of LiBr in the condition of having the same desiccant flow rate. Liu et al. [6] investigated a heat and mass transfer model, which was validated using experiments and numerical results to analyze the uncoupled heat and mass transfer process, relative humidity, and the enthalpy difference between the solution and air. Liu et al. [7] conducted an experimental study on the mass transfer performance of a cross-flow configuration dehumidifier. Celdek-structured packing was used in the dehumidifier, and lithium bromide was used as the liquid desiccant.

Artificial intelligence models have become the preferred trend for some researchers to predict the performance of heat and mass transfer processes in the dehumidifier and regenerator because of their ability to learn and adapt to change with little human interaction. In order to see how capable the artificial neural network is as a technique for modeling the liquid desiccant dehumidification, several studies are reviewed. Mohammad et al. [8] proposed an artificial neural network (ANN) model for predicting the performance of a liquid desiccant dehumidifier in terms of the water condensation rate and dehumidifier effectiveness. A MATLAB code was designed to study feed-forward-back propagation. The results show that the maximum percentage difference between the ANN and the experimental values for the water condensation rate and dehumidifier effectiveness were 8.13 and 9.0485 %, respectively. In the current study, an experimental analysis was performed to evaluate the performance of a lithium chloride liquid desiccant counter-flow dehumidifier at varying solution and air inlet conditions. Moisture removal rate and effectiveness were used as performance indicators.

65.2 System Description

The schematic diagram of the proposed system is shown in Fig. 65.1. The proposed system consists of three main components: dehumidifier, regenerator, and refrigerant cycle. The refrigerant cycle is used to precool the liquid desiccant on the dehumidification side and preheat it on the regeneration side. There are also some auxiliary fittings such as two centrifugal fans, two solution pumps, two valves, and two flow meters.

The dehumidifier and the regenerator are the core of the system. The structured packing has a section area of 0.24 m^2 and a height of 0.85 m. Two solution sumps, with a capacity of 40 L each, are located at the bottom of the system (one for storing

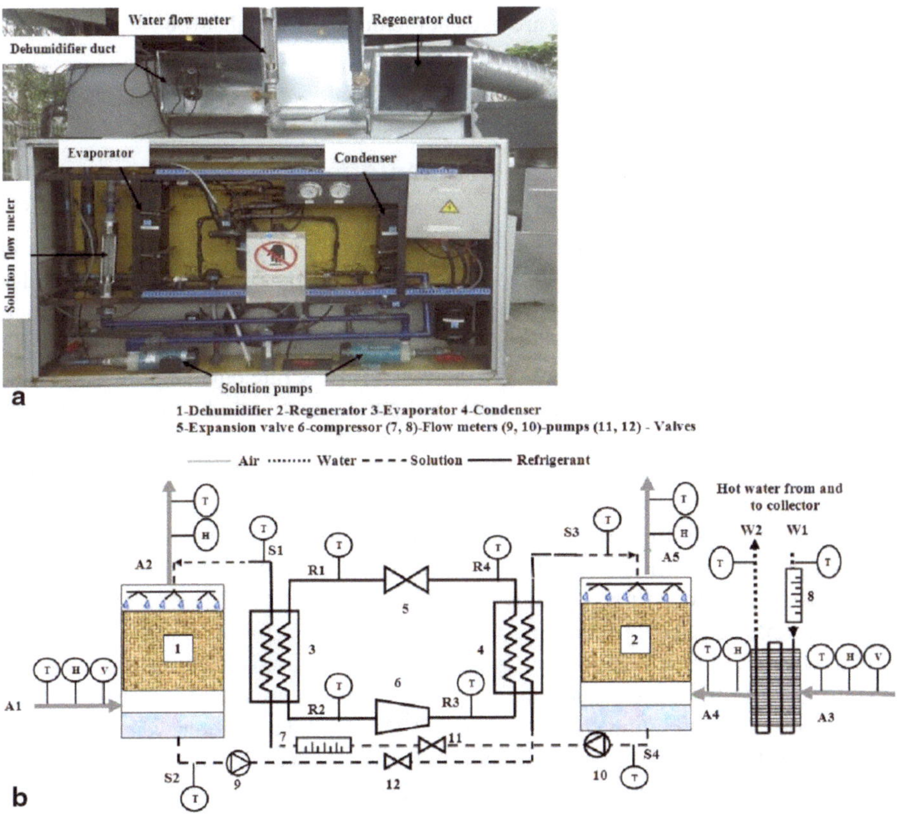

Fig. 65.1 Schematic diagram of the system **a** Photograph of the liquid desiccant unit. **b** Experimental rig

the weak solution and the other for storing the strong solution). A pump was used to circulate the hot, strong desiccant solution through the evaporator to precool it before entering the dehumidifier using the refrigerating output from the vapor compression unit. Ambient air is drawn by a centrifugal fan into the bottom of the dehumidifier and is in direct contact with the strong solution in a counter-flow configuration. Once the strong solution has attracted moisture from the air, the processed air is dehumidified, and its temperature depends on the temperature of the desiccant solution. The dehumidified warm air leaves the dehumidifier and enters the direct evaporative cooling unit. The desiccant solution in the dehumidifier sump becomes weak and pumps into the condenser, where it is preheated by the exhaust heat from the condenser of the vapor compression unit before entering the regenerator. Ambient air is also preheated in the heating coil (state A4) by solar-heated water (states W1 to W2) using the evacuated tube solar collectors, which are located about 20 m from the system. The total collector area was 34 m^2 and can produce hot water at about the maximum temperature of 95 °C. After preheating, the air stream enters the regenerator (state A5) to carry over the moisture from the hot desiccant solution. The hot and wet air leaving the regenerator subtracts into the ambient air. Air filters are used at the entrance to the dehumidifier/regenerator to prevent dust from entering the system. Two eliminators are used at the outlet of the absorber/regenerator to prevent the carryover of the desiccant solution particles into the ducts.

65.3 Measurements and Instrumentation

To measure the air-dry bulb temperature and relative humidity at different points in the system, five thermocouples with an accuracy of ±0.1 °C and TR-RH2W humidity sensors with an accuracy of ±3 % RH are installed at points A1, A2, A3, A4, and A5 as shown in Fig. 65.1. Two anemometers (AM-402) are installed at points A1 and A3 to measure the air velocity and then the air flow rate entering the dehumidifier and the regenerator. The tested desiccant parameters included temperature and flow rate. Four thermocouples with an accuracy of ±0.1 °C are installed at points S1, S2, S3, and S4 to measure the temperature before and after entering the dehumidifier and regenerator. A glass flow meter with an accuracy of ±10 l/h is installed to control the mass flow rate of the desiccant solution entering the dehumidifier. The temperatures of the precooling and preheating of the evaporator and condenser are measured by four thermocouples installed at points R1, R2, R3, and R4. Two thermocouples are used to measure water temperature before and after the heating coil at points W1 and W2, and one glass flow meter is used to control the solar hot water flow rate from the storage tank. All these sensors are coupled with a PC system to store the data in the computer. An ADAM-4018 data logger was used to collect sensor data. This data-acquisition system has a capability of two single-ended (six differential) voltage channels. Various single-channel inputs are available in the ADAM data acquisition: thermocouple, mV, V, and mA. The data-acquisition

Table 65.1 Specifications of the different measuring devices

Parameters	Devices	Accuracy	Operational range
Air velocity	Anemometer AM-402	±2% m/s	0.4–30 m/s
Air and solution temperature	Thermocouples type K	±0.1 °C	0–1370 °C
Air relative humidity	Humidity sensors TR-RH2W	±3% RH	10–95% RH
Solution and water flow rate	Glass flow meter	±10 l/h	200–2600 l/h
Solution concentration	Analysis method (ALPHA 3111-B)		

system samples data every 5 min. The main characteristics of the different measuring devices are shown in Table 65.1.

65.4 Liquid Desiccant Regenerator Performance Parameters

The performance of the regenerator is evaluated by calculating the moisture removal rate and its effectiveness. The moisture removal rate is calculated as follows [9]:

$$\mathrm{MRR} = m_{\mathrm{a}} \times (W_{\mathrm{out}} - W_{\mathrm{in}}) \tag{65.1}$$

where m_{a} is the dry air mass flow rate at the regenerator inlet, W_{in} is the inlet air humidity ratio, and W_{out} is the outlet air humidity ratio.

The effectiveness of the regenerator is evaluated as the ratio between the change in actual humidity in the air and the maximum change in humidity possible and can be represented as follows:

$$\varepsilon = {W_{\mathrm{out}} - W_{\mathrm{in}}} \big/ {W_{\mathrm{out\_eq}} - W_{\mathrm{in}}} \tag{65.2}$$

where $W_{\mathrm{out\_eq}}$ is the humidity ratio of the air in equilibrium with the desiccant solution under the following operating conditions, as shown by Kinsara et al. [10]:

$$W_{\mathrm{out\_eq}} = 0.62185 \times P_{\mathrm{wz}} \big/ (P - (P_{\mathrm{wz}}) \tag{65.3}$$

where P is the total pressure of air above the solution and P_{wz} is the partial pressure of water vapor in the solution. For lithium chloride, the pressure required to regenerate the weak desiccant for the temperature range 25–80 °C and the concentration range 5–45 % is given by [11].

$$P_{\mathrm{wz}} = -5.4442 + 0.87X + (T_{\mathrm{s}} / 15.844) \tag{65.4}$$

where X *and* T_{s} are the concentration and temperature of the solution, respectively.

65.5 Results and Discussion

The variations in moisture removal rate and effectiveness as a function of air inlet temperature are shown in Fig. 65.2. The figure indicates that the MRR is nearly constant and is unaffected by air inlet temperature, whereas regenerator effectiveness increases slightly from 44.4 to 45 %. Increasing the air inlet temperature may increase the system temperature through sensible heat transfer between the liquid desiccant and air, which may also affect the MRR and regenerator effectiveness.

Figure 65.3 shows the trend of the effect of air inlet humidity ratio on MRR and regenerator effectiveness. The MRR decreases from about 1.7 to 1.38 g/sec, whereas the regenerator effectiveness remains unchanged with increasing air humidity ratio. This result may be explained as follows: Increasing the air inlet humidity ratio causes a decrease in the mass transfer potential within the regenerator; hence, the MRR decreases although the air outlet humidity ratio increases. Increasing both inlet and outlet humidity ratio does not affect the regenerator effectiveness.

Figure 65.4 shows the impact of the solution inlet temperature on the MRR and regenerator effectiveness. The moisture removal rate increases from 0.8 to 1.5 g/sec by increasing the solution inlet temperature, whereas the regenerator effectiveness decreases slightly with the increase of the desiccant inlet temperature. An increase in solution inlet temperature causes an increase in vapor pressure on the solution surface rather than in the air stream; hence, the mass transfer potential within the

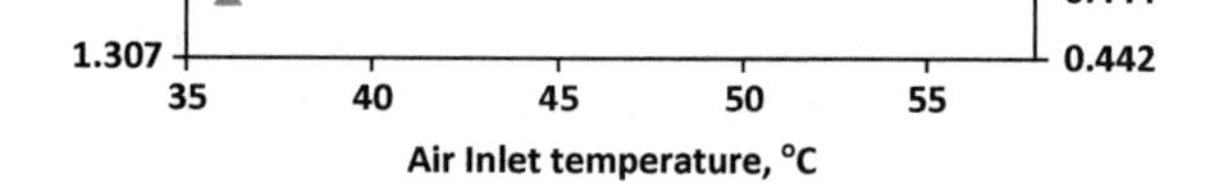

Fig. 65.2 Influence of the air inlet temperature on the MRR and the regenerator effectiveness

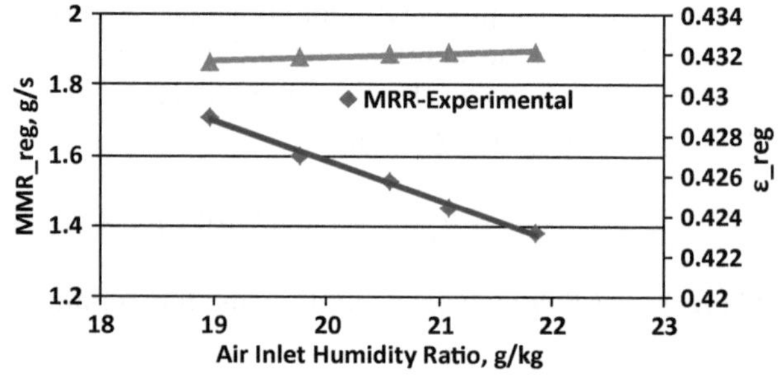

Fig. 65.3 Influence of the air inlet humidity ratio on the MRR and the regenerator effectiveness

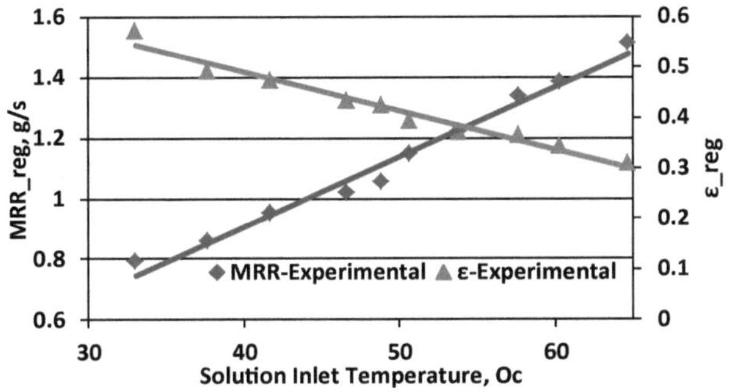

Fig. 65.4 Influence of the solution inlet temperature on the MRR and the regenerator effectiveness

regenerator increases. Both equilibrium humidity ratio on the solution surface and outlet air humidity ratio from the regenerator increase by increasing the solution input temperature. The increase in the former is less obvious than that in the latter, which results in a decrease in regenerator effectiveness.

65.6 Conclusions

An experimental analysis was performed to evaluate the performance of a lithium chloride liquid desiccant counter-flow regenerator at varying solutions and air inlet conditions. Based on the experimental results, the following conclusions can be drawn:

- The MRR and the effectiveness of the regenerator increase slowly as a function of the air inlet temperature. It was found that the MRR and the effectiveness increased about 0.79 % and 1.1 %, respectively.
- The moisture removal rate decreased with increasing air inlet humidity ratio and increased with the desiccant inlet temperature. The regenerator effectiveness increased very little with increasing air inlet humidity ratio and decreased with the desiccant inlet temperature.

References

1. Chung TW, G. TK, Hines AL (1993) Dehumidification of air by aqueous lithium chloride in a packed column. Sep Sci Technol 28:533–50
2. Kessling W, Laevemann E, Kapfhammer C (1998) Energy storage for desiccant cooling systems component development. Sol Energy 64:209–221
3. Moon CG, Bansal PK, Jain S (2009) New mass transfer performance data of a cross-flow liquid desiccant dehumidification system. Int J Refrig 32:524–533
4. Sanjeev J, Sagun T, Rajat SD (2011) Experimental performance of a liquid desiccant dehumidification system under tropical climates. Energy Convers Manage 52:2461–2466
5. Liu XH, Yi XQ, Jiang Y (2011) Mass transfer performance comparison of two commonly used liquid desiccants: LiBr and LiCl aqueous solutions. Energy Convers Manage 52:180–190

6. Liu XH, Jiang Y (2008) Coupled heat and mass transfer characteristic in packed bed dehumidifier/regenerator using liquid desiccant. Energy Convers Manage 49:1357–1366
7. Liu XH, Zhang Y, Qu KY, Jiang Y (2006) Experimental study on mass transfer performances of cross flow dehumidifier using liquid desiccant. Energy Convers Manage 47:2682–2692
8. Mohammad ATh, Mat SB, Sulaiman MY, Sopian K, Al-abidi AA (2013) Implementation and validation of an Artificial Neural Network for predicting the performance of a liquid desiccant dehumidifier. Energy Convers Manage 67:240–250
9. Liu XH, Jiang Y, Chang XM, Yi XQ (2007) Experimental investigation of the heat and mass transfer between air and liquid desiccant in a cross-flow regenerator. Renew Energy 32:1623–1636
10. Kinsara AA, Elsayed MM, Al-Rabghi OM (1996) Proposed energy-efficient air-conditioning system using liquid desiccant. Appl Therm Eng 16:791–806
11. Al-Farayedhi AA, Gandhidasan P, Younus Ahmed S (1999) Regeneration of liquid desiccants using membrane technology. Energy Convers Manage 40:1405–1411

Chapter 66
Integration of Concentrated Solar Power Plant and Coal-Fired Power Plants for Block Size of 100 MW

Pankaj Deo

Abstract Development of renewable energy sources has not reached a stage where it can completely replace the conventional power plants. However, renewable energy technologies like wind energy, solar photovoltaic or concentrated solar plant (CSP) can join hands with conventional power plants to address the energy crisis and develop themselves to make this replacement possible in near future. This chapter explores the possibility of integrating CSP with coal power plant (CPP) for reducing the installation cost of CSP, which in turn would also reduce coal consumption and carbon emission from coal plants. Moreover, it would also boost up the implementation of different government renewable energy policies.

Based on the proposed technique in the chapter, CSP can be installed in existing coal power plants (CPP). Auxiliary drives in coal power plant consume approximately 8 % of total power generation in CPP. These auxiliary drives can run on the thermal energy provided by CSP. Important concerns like identification of land in coal plant for CSP installation, total land requirement, process of installation and integration of CSP with CPP are elaborated in this chapter.

This chapter also investigates the advantages and challenges for the integration of CPP and CSP. CSP is more effective in regions where direct normal irradiation (DNI) is more than 1900 kWh/m^2/year which limits the use of suggested integration method. On the other hand, availability of power block for CSP would result in significant cost reduction in CSP plant installation. Moreover, cost required for resources like man power for operation and maintenance would be reduced considerably. These plants would not be dependent on any kind of power purchase agreement; hence, it would encourage CSP project developers for new installations all over the world.

Keywords Concentrated solar power (CSP) · Auxiliary power · Carbon emission · Coal power plant (CPP) · Power generation

P. Deo (✉)
Technische Universität Berlin
Riesenweg 14 i, Freiburg, 79110, Germany
e-mail: deo_pankaj@ymail.com

A. Sayigh (ed.), *Renewable Energy in the Service of Mankind Vol II*,
DOI 10.1007/978-3-319-18215-5_66

66.1 Introduction

Concentrated solar power (CSP) system consists of solar field, power block, thermal storage system and balance of plant. Four technologies, namely parabolic trough, solar tower, linear fresnel and dish stirling are used as solar fields. At the end of year 2012, total installed capacity of CSP was 2.8 GW [1, 2].

Other than power generation, CSP can be used in different industries such as process heat, thermal cooling system and desalination plant. CSP is facing challenges due to its higher project cost as compared to photovoltaic (PV) system. However, technological developments on improving the efficiency and reducing the total project cost are going on for CSP projects to make it more competitive in the market [3].

Conventional power plants cannot be replaced with renewable power plants overnight. Development of renewable energy sources has not reached the stage where it can completely replace the conventional power generation system. Once the renewable systems reach the stage where cost of power generation is less than conventional power plant, then it would be a permanent solution.

During this development stage of CSP system, CSP can join hands with conventional power plants and address the energy issue. If we support coal-based power plants with thermal energy from sun, then we would be able to reduce coal consumption and also carbon emission. It will provide great opportunity to CSP research and development companies for verifying the performance of their systems. It will also encourage financial institution to invest in CSP projects.

CSP project cost can be reduced if CSP plants are installed in coal-based power plants. Research in terms of integration of CSP and conventional power plants is being done for the preheating of feed water in coal power plants [2]. However, this proposed integration design can tap other potential areas of the coal plants for effective integration.

Generally, from the total power generated in a coal-based power plant, 8 % power is consumed to run the auxiliary drives. Drives like boiler feed pump, cooling water pump, condensate water pump, induced draft fan and forced draft fan are the main users of auxiliary power. This power could be generated by CSP solar field which results in reduction of coal consumption and carbon emissions [4, 5].

66.2 Design Concept

CSP-integration method can be applied to existing coal-based power plants located in an area where direct normal irradiation (DNI) is more than or equal to 1900 kWh/m^2/year. Generally, 8 % of total power production would be consumed by auxiliary drives in the coal power plant. Mostly these auxiliaries are pumps and fans. These drives can run on steam energy. In many power plans, boiler feed pump is coupled with small-size steam turbine. The same idea of running the drive with steam energy can be implemented on induced draft fan and force draft fan which are used in boiler. CSP solar field can provide steam and run these auxiliaries which would result in reduction of auxiliary power consumption of coal power plant.

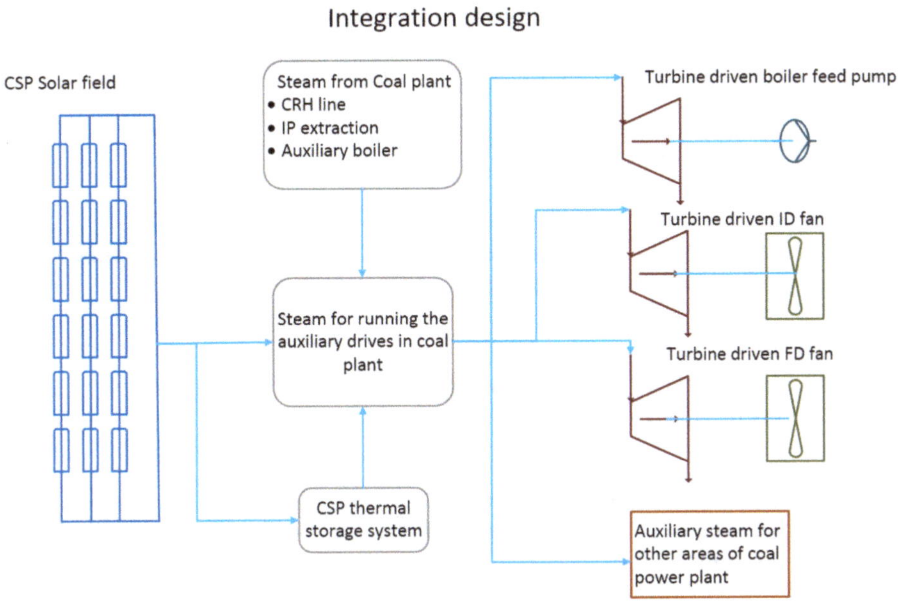

Fig. 66.1 Integration design concept

Figure 66.1 shows that in normal case, steam is provided from cold reheat line or IP extraction line to run auxiliary drives in coal plants. If CSP solar field delivers the required steam to these auxiliary drives, then it can reduce total coal consumption in coal plant or can generate additional electric power from main steam turbine (as additional steam would be admitted in steam turbine). Steam from CSP plant is fed to coal plant due to which there is no need to separate power block and balance of plant to be installed for CSP plant. Nearly 22 % of total CSP project cost is needed for power block and balance of plant installations which could be saved due to this integration design.

This integration design is not restricted to any size of coal power station; however, coal power stations in the range of 1400–2000 MW would be convenient for implementation of this idea as it gives scope for installing 80–120 MW CSP plant, and also sufficient space would be available.

For the case studies, two units of 700 MW in power station are considered. Therefore, 1400 MW power generation unit needs near about 112 MW of auxiliary power consumption. Out of which this integration design can fulfil power consumed by turbine-driven boiler feed pump (TDBFP), induced draft fan and forced draft fan drives (Table 66.1).

66.3 Integration Design

Integration of CSP–coal plant is effectively possible with parabolic trough and linear fresnel as land requirement in these two technologies is comparatively less than solar tower technology. Two case studies are discussed here for the design feasibil-

Table 66.1 Auxiliary drives power consumption in coal power plant

Sr.No	Type of drive	No of units	Capacity in MW	Total Capacity in MW
1	Turbine driven boiler feed pump	4	14	56
2	Force draft fan	4	4.5	18
3	Induced draft fan	4	4.5	18

ity and cost analysis of this design. For that purpose, plant is configured and simulated in System Advisor Model software (SAM).

66.3.1 Integration Design 1 with Linear Fresnel Collector (Without Thermal Storage)

CSP plant capacity: 100 MW Power plant

Coal power plant: 2 × 700 MW Nagpur, India.

Location Details:

Latitude: 21.25 , Longitude: 79.75 , Time zone: GMT 5.5, Wind speed: 1.7 m/s

Annual direct normal radiation: 1959.6 kWh/m^2, Annual global radiation: 2052.7 kWh/m^2

Total land required: 1.35 km^2 (337 acres)

Output: Total annual electricity generation: 193 GWh [6]

66.3.2 Integration Design 2 with Parabolic Trough Collector (Six Thermal Storage)

CSP plant capacity: 100 MW Power plant

Coal power plant: 2 × 700 MW Nagpur, India.

Location Details:

Latitude: 21.25 , Longitude: 79.75 , Time zone: GMT 5.5, Wind speed: 1.7 m/s

Annual direct normal radiation: 1959.6 kWh/m^2, Annual global radiation: 2052.7 kWh/m^2

Total land required: 3.64 km^2 (898 acres)

Output: Total annual electricity generation: 279 GWh [6]

66.4 Results

These two designs were simulated in SAM for cost analysis and energy yield. This integration design eliminates the requirement of power block and balance of plant in CSP project which significantly reduces the overall CSP project cost (20–23 % of total project cost) [4, 7]. The total cost distribution is shown in Fig. 66.2.

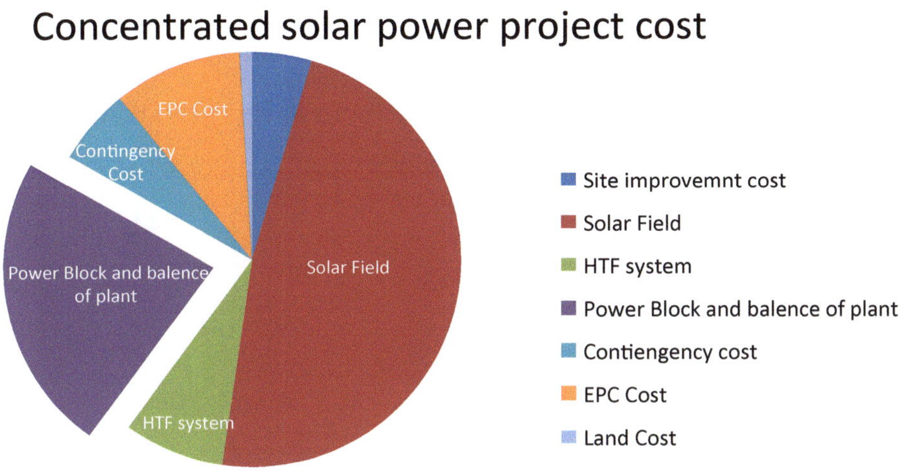

Fig. 66.2 CSP project cost

66.4.1 CSP Project Cost Reduction (As It Does Not Require the Installation of Power Block and Balance of Plant in Integration Design):

Power block and balance of plant mainly comprise steam turbine and its auxiliaries, generator and auxiliaries, condenser, cooling water system and cooling tower. Boiler feed pump (BFP) consumes most of the auxiliary power in coal-based power plant. In most of the recently installed power plants, BFP is coupled with small size steam turbine (TDBFP) and runs on steam energy. CSP can provide the required steam to drive this turbine. So, the steam turbine and its auxiliaries are readily available.

Steam from the exhaust of TDBFP can be fed and condensed in the main condenser of coal plant; therefore there is no need of additional cooling water system. Steam generated from CSP is used to run the steam turbine which is coupled with pumps and fans. So it does not require a separate generator. Basic auxiliaries like compressed air system, de-mineralized water, cooling water, pipe racks and cable trays would be available in conventional power plant, so all these auxiliaries are readily available for CSP.

Normally, coal-based power plants are located in very remote locations. Land acquisition around the running plant would not be a big concern as compared to finding new land for CSP plant.

So to conclude, significant cost reduction is possible due to readily available power block and balance of plant. Moreover, efficiency of this system will be higher than normal CSP plant because it does not convert thermal energy to electrical energy. Thermal energy is utilized to run the pumps and fans only.

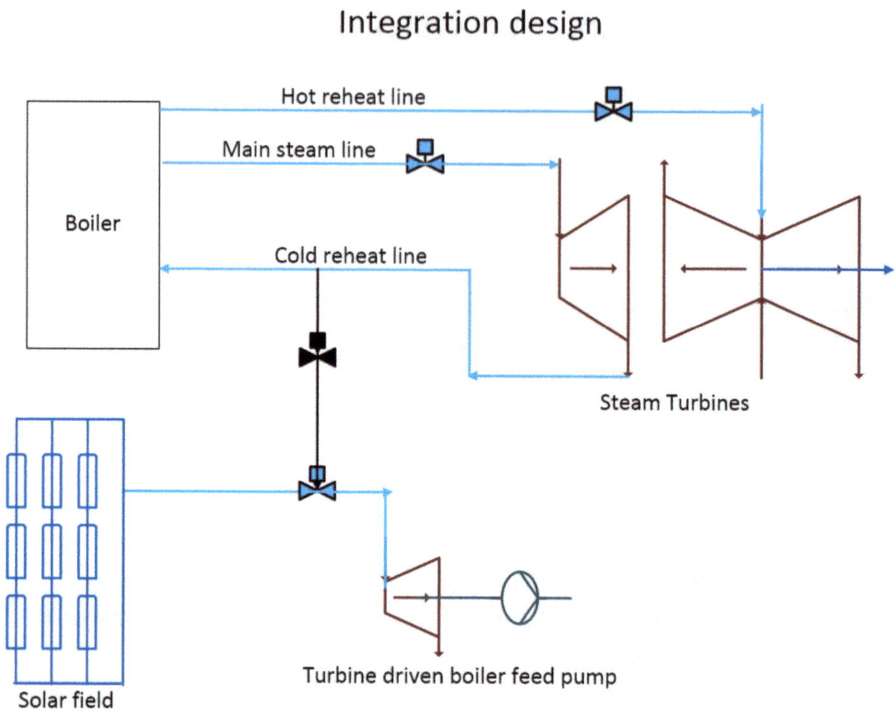

Fig. 66.3 Steam flow in integration design

66.4.2 Additional Cost Benefits

There are many additional cost benefits in different ways. First is in terms of coal saved, as coal plant need not supply steam to run the auxiliary drives any more (as CSP solar field is supplying the same). So coal input to the boiler of coal plant is going to be reduced proportionally.

On the other hand, instead of reducing the coal input to boiler if plant operator decides to run the plant on the same coal input, then additional steam gets admitted to the main steam turbine which results in additional electricity production from coal plant. From Fig. 66.3, it is clear that if we do not extract the steam from the cold reheat line to run the auxiliary drives, then this steam will go back to the boiler of the coal plant to get reheated and admitted into IP turbine. So, selling this additional electricity to grid would be useful for returns on investment calculations for this design.

Cost analysis for Integration design

CSP solar field - Linear Fresnel		
CSP Plant capacity	100	MW (electrical)
Project Cost	282	Millon $
Thermal storage	0	h
Annual Power output	604,2	GWh (thermal)

Coal Plant		
Coal Plant capacity	1400	MW (Electrical)
Import Coal prize for 1 Tonn	65	$
Electricity selling prise in india	5	Rupees
Electricity selling prise in india	0,083	$
Coal Quality	20000	kJ/kg
Generation of 1MW (Electrical)	0,54	Tons coal

Cost of Coal saved		
Annually coal saved due to CSP	104220	Tons coal
Total coal saved for 30 years	3126600	Tons coal
Annually coal cost saved due to CSP	6,77	Millon $
Total coal cost saved for 30 years	203,2	Millon $

Earnings by selling additional Electic energy		
Annual additional power output	193	GWh (electrical
Annual additional earnings from Coal plant	16,0833	Millon $
Total earnings for 30 years	482,5	Millon $

Fig. 66.4 Cost analysis for Linear fresnel [5–7]

66.4.3 Cost Analysis

66.4.3.1 Design 1

Based on the simulation results from social accounting matrices (SAM) for the integration design 2, that is with parabolic trough 193 GWh power would be generated per annum. For this CSP plant installation, the project implementation cost (excluding the costs of power block and balance of plant) would come down to $ 282 million. This cost includes indirect costs like engineering, procurement and construction (EPC) contractor profit and owner's profit.

For generating 100 MW from a coal power plant, 54 tons/hr coal is required (considering 38 % efficiency of the total plant and coal heat value 20000 kJ/kg). So, if CSP plant is used for supplying the steam to auxiliaries, then 0.1 million tons of coal would be saved per year from the coal plant [4, 6].

The calculations for this are shown in Fig. 66.4. Cost of coal in Nagpur (India) for imported coal = $ 65 (4000 INR) per ton (for high heat value 20000 kJ/kg of coal) is considered for the calculations [4, 5].

In 1 year if 0.1 million tons coal would be saved by using this integration method (considering plant efficiency 38 % and coal required per MW = 0.54 tons), then coal of worth $ 203 million would be saved for 30 years of plant life (if we consider cost of coal constant for that time period).

On the other hand, if additional electricity is produced (193 GWh/annum) from the coal plant then $ 482 million would be earned in a life time of a plant (if we consider constant electricity selling price as 8 cents/unit (5 INR/unit) for 30 years) which makes this integration design attractive.

66.4.3.2 Design 2

Based on the simulation results from SAM for the integration design 2, that is with parabolic trough 279 GWh power would be generated per annum. For this CSP

Cost analysis for Integration design

CSP solar field - Parabolic Trough		
CSP Plant capacity	100	MW (electrical)
Project Cost	587	Millon $
Thermal storage	6	h
Annual Power output	858	GWh (thermal)

Coal Plant		
Coal Plant capacity	1400	MW (Electrical)
Import Coal prize for 1 Tonn	65	$
Electricity selling prise in india	5	Rupees
Electricity selling prise in india	0,083	$
Coal Quality	20000	kJ/kg
Generation of 1MW (Electrical)	0,54	Tons coal

Cost of Coal saved		
Annually coal saved due to CSP	150660	Tons coal
Total coal saved for 30 years	4519800	Tons coal
Annually coal cost saved due to CSP	9,79	Millon $
Total coal cost saved for 30 years	293,8	Millon $

Earnings by selling additional Elective energy		
Annual additional power output	279	GWh (electrical
Annual additional earnings from Coal plant	23,25	Millon $
Total earnings for 30 years	697,5	Millon $

Fig. 66.5 Cost analysis for parabolic trough [5–7]

plant installation, the project implementation cost (excluding the costs of power block and balance of plant) would come down to $ 587 million. This cost includes indirect costs like EPC contractor profit and owner's profit.

As per the calculations explained in design 1, 54 tons/hr coal is required (considering 38 % efficiency of the total plant and coal heat value 20000 kJ/kg) for generating 100 MW from coal power plant. So, if CSP plant is used for supplying the steam to auxiliaries, then 0.15 million tons of coal would be saved per year from coal plant [4, 6].

The calculations for this are shown in Fig. 66.5. In 1 year, 0.15 million tons coal per year would be saved if we use this integration method (considering plant efficiency 38 % and coal required per MW = 0.54 tons). Then, coal of $ 293 million would be saved for next 30 years (if we consider cost of coal constant for that time period).

On the other hand, if the additional electricity (279 GWh/annum) is produced from the coal plant during its lifetime, then $ 697.5 million would be earned (if we consider constant electricity selling price as 8 cents/unit (5 INR/unit) for 30 years).

The outcome of this integration can be seen from both ways as coal saving or additional electricity generation depending on the user requirement.

Government subsidies and use of local market for CSP project would further reduce the plant installation cost and make this integration project more attractive. Moreover, in this cost calculations, cost of coal and electricity prices are considered to remain constant for 30 years, but in fact these costs are going to increase day by day which would make this integration design more attractive in terms of returns of investments.

66.5 Conclusion

This proposed integration method reduces the installation cost of CSP plants by 20–22 %. For implementation of this technology, first we have to identify all the potential coal power plants where direct normal irradiation is more than 1900 kWh/m^2.

Based on plot plan of coal power plant, suitable land for solar field could be identified in coal power plant. Generally, coal plants in India are constructed on non-fertile land so if needed plant boundaries can be extended by purchasing the land in affordable prices. Raw water reservoirs in coal plant are of size 0.4–0.6 km^2 (i.e. 100–150 acres) which can store water for 15 days plant capacity [8]. These raw water reservoirs can be covered with concrete slab and the place could be utilized for solar field. This will also reduce the water evaporation losses.

Government policies, such as renewable energy certificates and subsidies for CSP installation, enhance the practicability of this technique. Coal-plant owners need not look for other options to get the renewable energy certificate; they can install renewable energy in-house and earn the certificate. Moreover, this technique can be utilized while designing upcoming coal power plant projects so that space for solar field is considered well in advance for effective implementation of the integration design.

This integration method will encourage CSP power development institutions, power plant owners and financial institutions for installation of new CSP projects due to its reduced cost. On the other hand, it will speed up the CSP research and development work due to the increment in number of CSP installations.

References

1. Schlumberger (2013), LEADING THE ENERGY TRANSITION - Concentrating solar power, SBC Energy Institute, Gravenhage
2. Simbolotti G. (2013), Concentrating Solar Power - Technology Brief, IEA-ETSAP and IRENA.
3. National Renewable Energy Laboratory, US Department of Energy. NREL (Online). http://www.nrel.gov/csp/. Accessed 22 Jan 2014
4. Deo P Own work experience as a project engineer in the field of construction and commissioning of thermal power plants in the period of 2009 to 2013 at Larsen & Toubro Power Ltd. India
5. Coal India-Government Website (Online). https://www.coalindia.in/. Accessed 04 Feb 2014
6. System Advisor Model (SAM) - National Renewable Energy Laboratory
7. Turchi C (2010) Parabolic trough reference plant for cost modeling with the solar advisor model (SAM), National Renewable Energy Laboratory, Colorado
8. Authority CE (2010) Review of land requirement for thermal power stations CENTRAL ELECTRICITY AUTHORITY, New Delhi

Chapter 67
Indoor Experimental Investigations of Two Different Static 3-D Solar Concentrators

Imhamed M. Saleh Ali, K S Reddy and Tapas K Mallick

Abstract Designs and experimental investigations of two different hyperboloid (elliptical hyperboloid concentrator (EHC) and circular hyperboloid concentrator (CHC)) solar concentrators have been reported in this chapter. Indoor tests have been carried out for the two prototypes of solar concentrators. In order to estimate the temperature of the surface area of the receiver tubes for process heat applications, thermal imaging analysis has been carried out and also presented in this chapter. It was observed that the outlet temperature of 93 °C is obtained for the EHC solar concentrator and the outlet temperature of 60 °C is obtained for the CHC solar concentrator, respectively. Indoor experimental temperature distribution of the receiver was carried out for two types of receivers. Temperature measurement was carried out under 'flow' and 'no flow' conditions. Both the EHC and CHC systems were tested and their concentration ratio was found to be 20×. In order to estimate the temperature distribution along the receiver, thermal imaging camera and FLIR Tools software were used to obtain the temperature variation along the receiver at three points. Under 'no flow' conditions, stagnation temperatures were measured at three points (represented as T1, T2 and T3) located on the receiver surface.

Keywords Static concentrator · Concentrating ratio · Elliptical hyperboloid concentrator

I. M. Saleh Ali (✉)
Department of Mechanical Engineering, Sirte University, P.O. Box 674, Sirte, Libya
e-mail: isa6@su.edu.ly

K. S. Reddy
Heat Transfer and Thermal Power Laboratory, Department of Mechanical Engineering, Indian Institute of Technology Madras, 600036 Chennai, India
e-mail: ksreddy@iitm.ac.in

T. K. Mallick
Environment and Sustainability Institute, University of Exeter, Penryn TR10 9FE, UK
e-mail: t.k.mallick@exeter.ac.uk

A. Sayigh (ed.), *Renewable Energy in the Service of Mankind Vol II*,
DOI 10.1007/978-3-319-18215-5_67

67.1 Introduction

Because of the energy crisis, the need to use alternative renewable energy sources such as solar and wind energy has gained lots of momentum in many developing countries. In the past few decades, the use of concentrating solar technology for thermal applications has increased tremendously. Along with the concentrator system, a tracking system is also used to track the sun continuously from east to west, which increases the overall cost of the system. To overcome this, a three-dimensional (3-D) static hyperboloid solar concentrator is designed, developed and applied for the solar thermal applications.

Hassib [1] discussed the geometric analysis of a compound conical concentrator with receivers of various geometries. The shape of the receiver determines the profile of the reflector. It was also shown that the resulting flux distribution determines the shape of the receiver and its position relative to the reflector. El-Refaie [2] described the conical solar energy concentrator with a tubular axial absorber. The effects of apex angle, diameter ratio and truncation ratio on the concentrated power, concentration profile and the reflector surface area have been investigated.

Benitez et al. [3] presented a new method to obtain elliptical ray bundles in a 3-D geometry by reformulating the conditions of a bundle to be considered elliptical and also looked for the shapes and profiles produced by these elliptical bundles using a flow-line design method. The solution of the problem provided may allow for the design of a new set of 3-D concentrators that are 'ideal'. Sellami et al. [4] designed a novel 3-D static concentrator and evaluated the best combination of optical efficiency and acceptance angle. A good agreement between experimental measurements from a manufactured square elliptical hyperboloid and optical simulation results was observed. Ali et al. [5] evaluated the performance of a 3-D elliptical hyperboloid concentrator using a ray-tracing software and carried out the optimisation study of the profile and geometry to improve the overall performance of the concentrator. The optical efficiency was found to be 28 % for a concentration ratio of 20×.

Based on the literature review, it was found that a very limited research has been carried out on hyperboloid concentrators. The idea of creating smooth 3-D elliptical hyperboloid concentrator geometry shows a lot of strengths. In the literature review, it was found that the hyperbolic profile in nonimaging concentrators shows improved performance in acceptance angle and energy collection [6–11]. It was also stated that the 3-D solar concentrator based on the hyperboloid revolution can concentrate all the incoming solar rays [12].

This study presents fabrication of two solar concentrators and indoor experimental analysis of a static 3-D solar elliptical hyperboloid concentrator (EHC) and a static 3-D solar circular hyperboloid concentrator (CHC) for process heat applications. The optical efficiency of 28 % and concentration ratio of 20× were achieved. The indoor testing under solar simulator in Heriot Watt University (HWU) Laboratory was carried out. The stagnation temperature of the receiver corresponds to the measurement of the surface temperature of the receiver at a zero flow rate. The stagnation temperature is measured using the thermocouple fixed at the surface of the receiver. In addition, the experiments were carried out for different receiver

Table 67.1 Specifications of a solar concentrator

Parameters	EHC	CHC
Height of the concentrator	0.4 m	0.4 m
Aperture major axis of the concentrator	0.447 m	0.477 m
Aperture minor axis of the concentrator	0.089 m	0.477 m
Receiver major axis	0.1 m	0.1 m
Receiver minor axis	0.02 m	0.1 m
Acceptance angle	±45°	±15°
Concentration ratio	20×	20×
Reflectivity of the aluminium sheet	Chrome coating, 0.92	Chrome coating, 0.92

volume flow rates to study the performance of the EHC. The experimental study was also carried out to estimate outlet temperature. In order to estimate the temperature distribution along the receivers of EHC and CHC, we used a thermal imaging camera to obtain the temperature variation along the receiver at three points. Under 'no flow' conditions, stagnation temperatures were measured at different points.

67.2 Specification of Two Different Solar Concentrators

Based on the optical analysis of the 3-D EHC [5], the specifications of the two different solar concentrators fabricated are given in Table 67.1. The EHC and CHC with a concentration ratio of 20× and a concentrator height of 0.4 m were fabricated from Star Prototype, China. The prototypes of the EHC and CHC are shown in Fig. 67.1. The schematic of the EHC open cycle is shown in Fig. 67.2.

67.3 Indoor Experimental Testing of a Hyperboloid Concentrator

The temperature distributions for two types of receiver were carried out. Temperature measurement was carried out under 'flow' and 'no flow' conditions. Water was used as heat-transferred fluid and made to pass through the receiver. A receiver was placed under the solar simulator, where average solar radiation of 1200 W/m^2 was obtained. The temperature of the receiver surface was increased and the oil was heated. The heated oil was made to flow through the system by using a peristaltic pump (Watson Marlow 323), which was placed after the receiver, and passed through a water tank. During the experimental measurement, inlet and outlet temperature of the receiver and the temperature of the tube were measured. The temperature measurements were carried out by two T-type thermocouples (for inlet and outlet temperature of the water). In addition to that solar radiation from

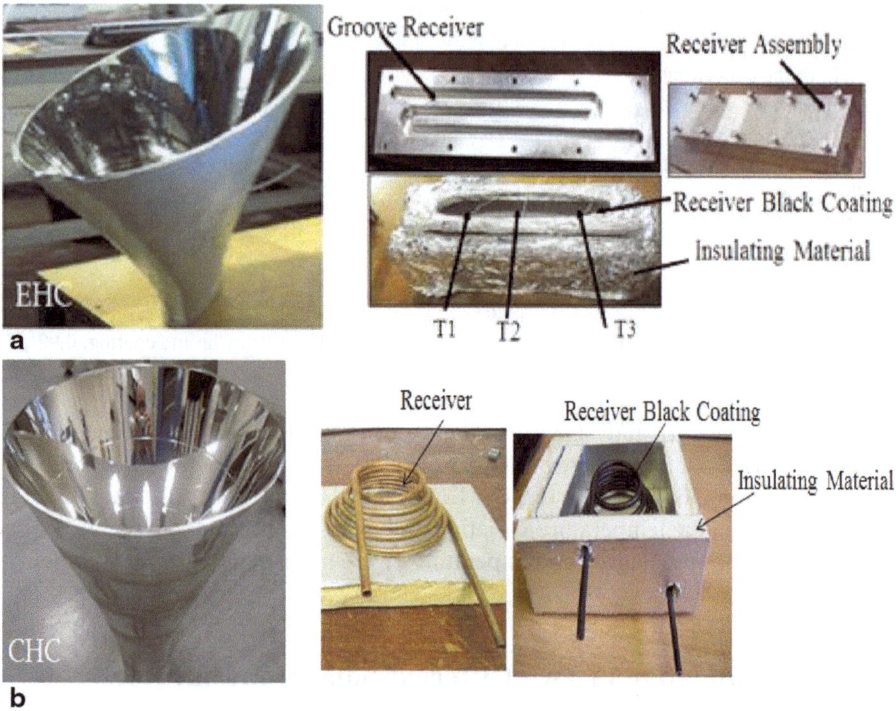

Fig. 67.1 Prototype of elliptical hyperboloid concentrator (*EHC*) and circular hyperboloid concentrator (*CHC*). **a** *EHC* and groove receiver, **b** *CHC* and coil receiver

the simulator was measured. For measuring the solar radiation, a Kipp & Zonen CM21/CM22 pyranometer was used. For collecting all these data (temperature and solar radiation), a Keithley 2700 Multimeter/Data Acquisition System was used, and all data were collected and processed from a computer using an Excel file. The experimental cycle consisted of the elliptical solar concentrator with flat grooved aluminium receiver/copper coil receiver; one end of the receiver is connected to the inlet tank and the other end to the outlet tank. The entire experimental setup for the EHC open cycle is shown in Fig. 67.3. In order to estimate the temperature distribution along the receiver, we used thermal imaging camera and FLIR Tools software to obtain the temperature variation along the receiver at three points. Under 'no flow' conditions, stagnation temperatures were measured at three points (T1, T2 and T3) on the receiver as shown in Fig. 67.1a. The maximum stagnation temperature of 93 °C was observed for solar radiation of 1200 W/m^2, and the average stagnation temperature of 82 °C was observed. The variation of stagnation temperatures with time is shown in Fig. 67.4. The measured stagnation temperatures were also compared with the results of thermal imaging camera. The temperatures from the thermal imaging camera were found to be in close agreement with the estimated stagnation temperature. The photographs of thermal imaging temperature are also

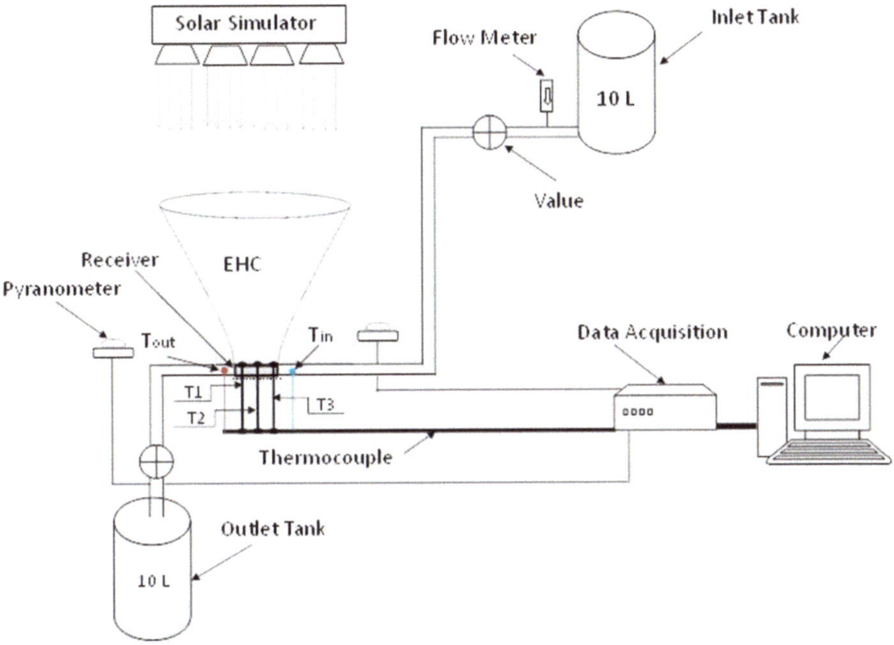

Fig. 67.2 Schematic of the elliptical hyperboloid concentrator (EHC) open cycle

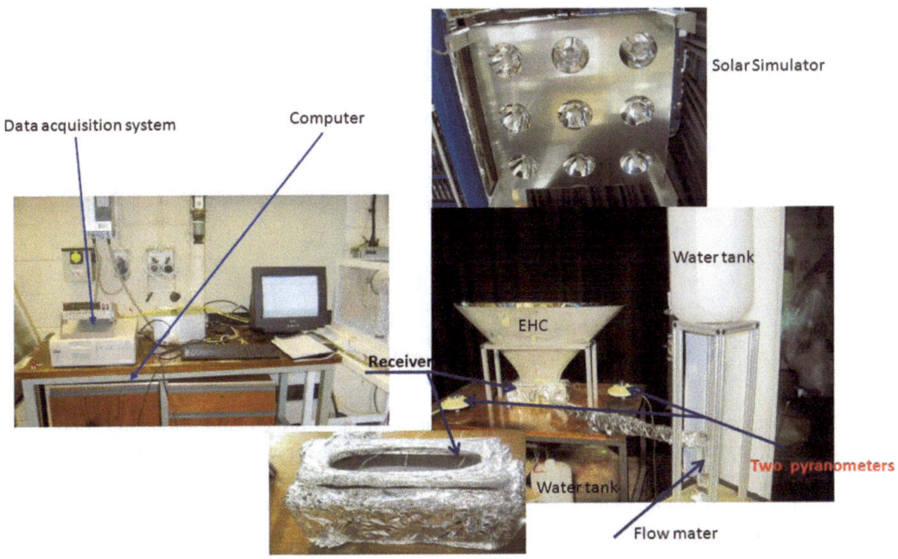

Fig. 67.3 Experimental setup for the EHC open cycle

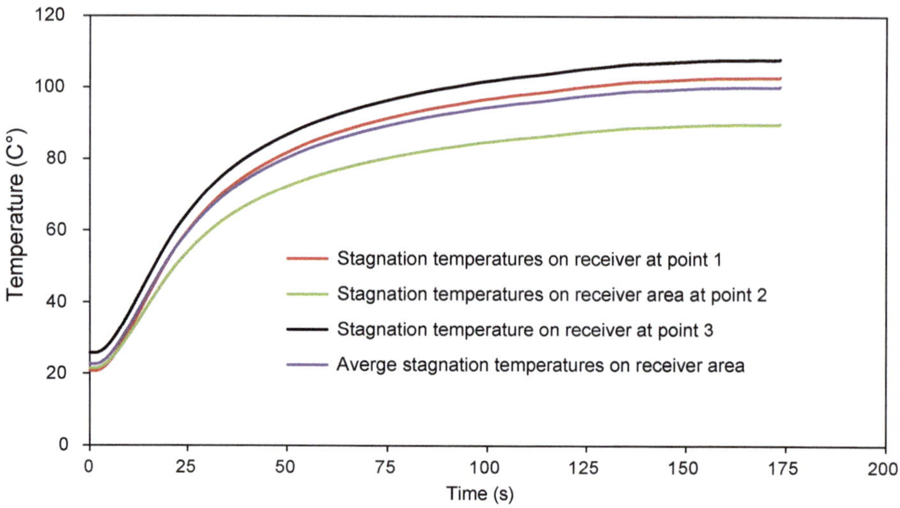

Fig. 67.4 Variation of stagnation temperature with time for EHC

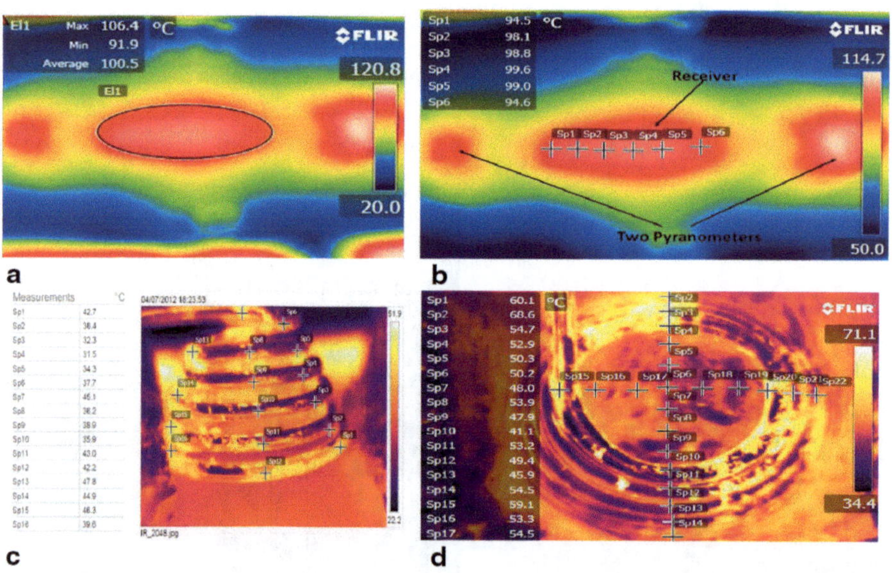

Fig. 67.5 Photographs of thermal imaging temperature of flat grooved aluminium and copper coil receiver where (**a**) the average temperature on the grooved receiver area for EHC, (**b**) temperatures different along the major axis for EHC, (**c**) temperatures different along the vertical axis on coil receiver for CHC, (**d**) temperatures different along the major axis and minor axis on coil receiver for CHC are shown

shown in Fig. 67.5a, 67.5b, 67.5c and 67.5d.Under 'flow conditions', the inlet and outlet temperatures of water were measured when the flow was 0.12 kg/min and the results are shown in Fig. 67.6. As a part of developing the desalination system, the solar concentrator with receiver was constructed and experimental measurements

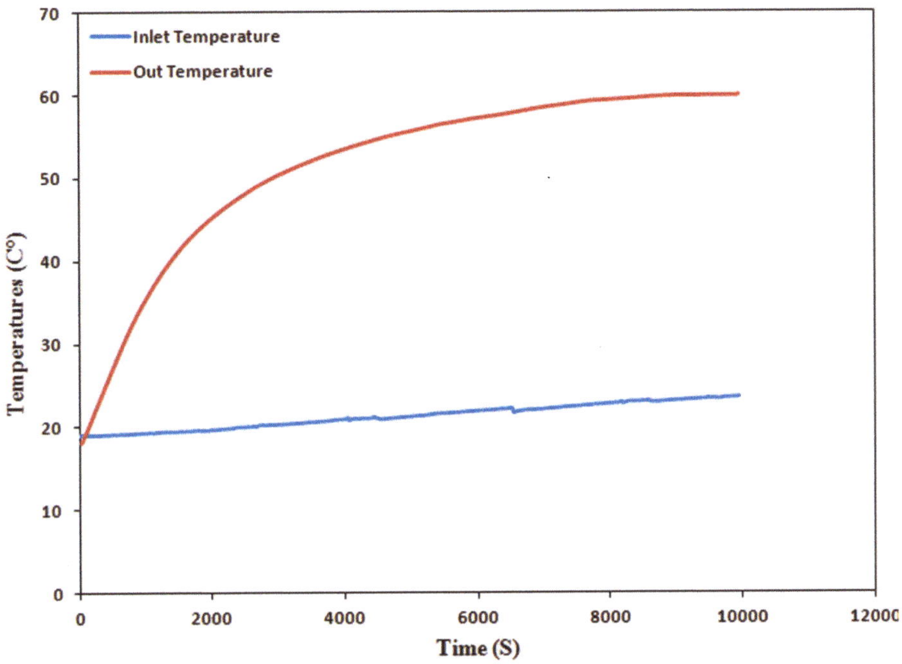

Fig. 67.6 Variation of inlet and outlet temperatures with time for CHC

were taken under 'flow' and 'no flow' conditions. In the flow conditions, oil was allowed to flow through the receiver and heated depending on the solar concentration level. A solar concentrator was placed on the top of the receiver and radiation from the solar simulator reached the receiver through multiple reflections in the concentrator and finally absorbed by the receiver.

67.4 Conclusions

In this chapter, the results of optical analysis of CHC and EHC have been discussed. For indoor characterisation, an in-house built solar simulator was constructed in HWU. The design, fabrication details and indoor characterisation of CHC and EHC with flat grooved aluminium and copper coil receiver have also been discussed. The temperature distributions of the flat grooved aluminium and copper coil receiver were also obtained. An optimal concentration ratio of 20× with an optical efficiency of 28 % was predicted. During this study, an experimental investigation was carried out to verify its operation under indoor test conditions for a 3-D EHC. The maximum stagnation temperature recorded was 93 °C, and the outlet temperature from the system was 60 °C when the flow was 0.12 kg/min.

Acknowledgement We acknowledge the support of Srite University, Libya, and the UKIERI Grant (SA08-061) received for this research.

References

1. Hassib AM (1986) Compound conical concentrators with elliptical receivers. Sol Energy 36:89–92
2. El-Refaie MF (1982) Theoretical analysis of the performance of a conical solar concentrator. Appl Energy 12:37–51
3. Fernandez-Balbuena AA, Garcia-Botella A, Bernabeu E, Vázquez. D (2006) Elliptical concentrators. Appl Opt 45:7622–7627
4. Sellami N, Mallick TK, McNeil DA (2012) Optical characterisation of 3-D static solar concentrator. Energy Convers Manage 64:579–586
5. Saleh Ali IM, Reddy TS, Mallick KS, Tapas K (2013) An optical analysis of a static 3-D solar concentrator. Sol Energy 88:57–70
6. Norton B (2001) Solar process heat: distillation, drying, agricultural and industrial uses, solar energy, the state of the art. James & James, London, pp 477–496 (ISBN 1-902916-23-9)
7. Spate F, Hafner B, Schwarzer K (1999) A system for solar process heat for decentralised applications in developing countries. In: Proceedings of ISES Solar World Congress on CD-ROM, Jerusalem, Israel, 1999
8. Benz N, Gut M RW (1998) Solar process heat in breweries and dairies. In: EuroSun 98. Portoroz, Slovenia
9. Benz N, Gut M (1999) Beikircher, Solar process heat with non-concentrating collectors for food industry. In: Proceedings of ISES Solar World Congress on CD ROM, Jerusalem, Israel, 1999
10. Luminosu I, Zaharie I, Negrea R Ignea D (2008) The study of a static paraboloidal concentrator through the ray-tracing method, incorporated in the southern wall of a building, in: National Forum of Energy-FOREN 2008, Neptun, June 15, 2008
11. Winston R, Welford WT (1979) Geometrical vector flux and some new non-imaging concentrators. J Opt Soc Am 69:532–536
12. Kritchman E (1981) Optimized second stage concentrator. Appl Opt 20:2929–2933

Chapter 68
Effect of Isothermal Dehumidification on the Performance of Solar Cooling System in Tropical Countries

M. M. S. Dezfouli, Kamaruzzman Sopian, Sohif Bin Mat and K. S. M. Sahari

Abstract In recent years, due to the increasing price of energy, and also a high portion of the energy used by conventional air conditioning systems in tropical countries, solar desiccant cooling systems have been identified as energy-efficient cooling systems. The significant impact of the dehumidification process on the performance of a solar cooling system as well as on the thermal comfort condition is because of hot and humid outdoor conditions, and also high latent heat loads of buildings in tropical countries. In this study, to evaluate the isothermal dehumidification effect on the performance of the solar desiccant cooling system, the coefficient of performance (COP) and the zone temperature and humidity ratio of two simulated models (one stage and two stage) of the solar desiccant cooling system have been compared. The Transient Systems Simulation (TRNSYS) modeling validation is carried out by data measurement of a one-stage solar desiccant cooling system which was installed in the technology park of Universiti Kebangsaan Malaysia (UKM) in Malaysia. The capacity of cooling load and also the sensible heat ratio of the zone were 1 ton and 0.25, respectively.

The comparison results between one-stage and two-stage desiccant cooling systems show that under the same operation conditions (regeneration temperature and outdoor conditions) the COP of one-stage and two-stage models were 0.50 and 1.06, respectively, while the supply air of the two-stage model could handle the latent and sensible loads. Therefore, it was achieved that the isothermal dehumidification is a great method to improve the COP of the solar desiccant cooling system, and also indoor comfort conditions in tropical countries.

M. M. S. Dezfouli (✉)
Institute of Product Design & Manufacturing (IPROM), Universiti Kuala Lumpur, 56100 Kuala Lumpur, Malaysia
e-mail: salehi.solar@gmail.com

K. Sopian · S. B. Mat
Solar Energy Research Institute, Faculty of Engineering, Universiti Kebangsaan Malaysia, 43600 Bangi, Selangor, Malaysia
e-mail: k_sopian@yahoo.com

K. S. M. Sahari
Department of Mechanical Engineering, Universiti Tenaga Nasional, Jalan IKRAM-UNITEN, 43000 Kajang, Selangor, Malaysia

A. Sayigh (ed.), *Renewable Energy in the Service of Mankind Vol II*,
DOI 10.1007/978-3-319-18215-5_68

Keywords Isothermal dehumidification · Solar cooling system · Tropical weather

68.1 Introduction

The solar desiccant cooling systems have been well-known as efficient cooling systems in recent years because of the high energy consumption of conventional air conditioning systems. Pennington in 1955 presented the first desiccant cooling system [1]. According to the Pennington cycle, various configurations of solid desiccant cooling cycles have been designed in the world by different researchers. The coefficient of performance (COP) of the desiccant cooling system, regeneration temperature, mass flow rate, fresh air, and relative humidity of the supply air are the important parameters considered by the researchers. Jain et al. [2] evaluated ventilation and recirculation cycles based on weather data in India to determine the effectiveness of evaporative coolers on the COP of a cooling system. Boundoukan et al. [3] evaluated a comparison between ventilation and recirculation procedure modes; critical values were achieved for cycle component performance specifics required to yield a given supply air temperature condition. Kanoglu et al. [4] evaluated the first law and second law for COPs of ventilation and recirculation cycles. Panaras et al. [5] investigated the behavior of the COP of the ventilation and recirculation desiccant cooling systems in terms of the airflow rate. They found that the ventilation cycle presents a higher COP, and the required airflow rate is similar for both cycles. Sphaier et al. [6] studied the simulation of ventilation and recirculation desiccant cooling systems to analyze the impact of cycle component characteristics on the overall system performance. They concluded that, although all components can influence the COP, the COP values of the recirculation cycle are lower than the ventilation cycle. With regard to the increase in temperature of the process air after dehumidification, the performance of solar desiccant cooling systems was limited to the range of 0.4–0.7. Many researchers have attempted to increase the performance of desiccant cooling systems by means of different methods. The isothermal dehumidification was one of the best solutions to improve COP of solar desiccant cooling systems especially in hot and humid areas [7–9]. In fact, isothermal dehumidification is a method of using multiple stages of a desiccant wheel and heat recovery wheel for thermal isolation of process air. Li et al. [10] evaluated the two-stage desiccant cooling system driven by using evacuated glass tube solar air collectors. Their experimental results indicated that the average thermal COP in the cooling cycle was 0.97. Li et al. [11] carried out experimental investigation on a one-rotor two-stage desiccant cooling/heating system driven by solar air collectors. The average thermal COP in the cooling cycle is 0.95 in hot and humid climate conditions.

The present chapter evaluates the effect of isothermal dehumidification on the performance of solar desiccant cooling systems in humid areas by means of experimental and simulation analysis.

68.2 Methodology

The methodology of this study is based on comparison evolution between thermal and isothermal dehumidification systems that are identified as one-stage and two-stage desiccant cooling systems. Generally, the methodology includes two steps. The first step is the experimental setup of one solar desiccant-evaporative cooling to collect data. The second step is the simulation modeling of the installed system and isothermal desiccant cooling system. In fact, to compare the simulation modeling of one- and two-stage solar desiccant cooling, the validation of the Transient Systems Simulation (TRNSYS) has been done by experimental setup. The methodology details of this study are explained in the following sections.

68.2.1 Experimental Setup

As shown in Fig. 68.1, to handle sensible and latent loads of a test room, a solar desiccant-evaporative cooling was installed in a technology park (UKM) in Malaysia. The latent and sensible loads of the test room were 0.8 and 2.6 kW, respectively, while the cooling capacity was 1 ton. According to the American Society of Heating, Refrigerating and Air Conditioning Engineers (ASHRAE) comfort condition, indoor condition designs should consist of a temperature of 25 °C, relative humidity of 50 %, and humidity ratio of 0.0098 kg/kg [12]. According to the Malaysian weather, in outdoor condition design, the temperature should be 30 °C, and the humidity ratio should be 0.0020 kg/kg. The cooling system includes four main parts, namely, Desiccant Wheel (DW) as dehumidifier, a heat recovery wheel, an evaporative cooler as humidifier, and a solar-evacuated tube collector as heat source.

Fig. 68.1 Experimental setup of one-stage desiccant cooling system in technology park of UKM

68.2.2 *Simulation of a Solar Desiccant Cooling System*

68.2.2.1 Thermal Dehumidification Cooling System

According to the configuration of the solar desiccant cooling system which was installed, a simulation model of one-stage desiccant cooling was designed with the TRNSYS software. To find the validation of the simulated model, the experimental data were matched with the simulation results of the one-stage solar desiccant cooling system. Figure 68.2a shows that the solar desiccant cooling is an open cycle that provides supply air to the room from ambient air. Figure 68.2b shows modeling of the solar desiccant cooling system in ventilation mode that was designed by the TRNSYS software. The returning air from the room after the few processes is converted to ambient temperature as exhaust air. Therefore, two kinds of air are present: the process side that produces the supply air, and the regeneration side that releases the returning air from room to ambient temperature. In the process side, first, the ambient air becomes dry because of the DW. Then, the heat recovery wheel cools the air in the process side. In the next step, the air becomes cold because of the evaporative cooler; then, the air moves into the room as supply air. In the regeneration side, the returning air whose latent and sensible loads were left in the room becomes cold because of the evaporative cooler. The heat recovery wheel acts as a heater in the regeneration air side. Then, the heat from the heat exchanger and the heater is transferred to the air. Thus, in the last step in the regeneration side, the hot air absorbs the humidity of the DW and converts it to ambient temperature as exhaust air.

In this simulation, DW is type 683, the evaporative cooler is type 506c, and the heat recovery wheel is type 760b.The heat exchanger is type 91, the evacuated tube solar collector is type 71, the pump is type 3b, the fan is type 112a, the zone load (room) is type 690, and the weather data in Kuala Lumpur (Malaysia) are type 109-TMY2.

68.2.2.2 Isothermal Dehumidification Cooling System

The components of the two-stage desiccant cooling are two DWs, two heat recovery wheels, two evaporative coolers with different capacities, two heat exchangers, two

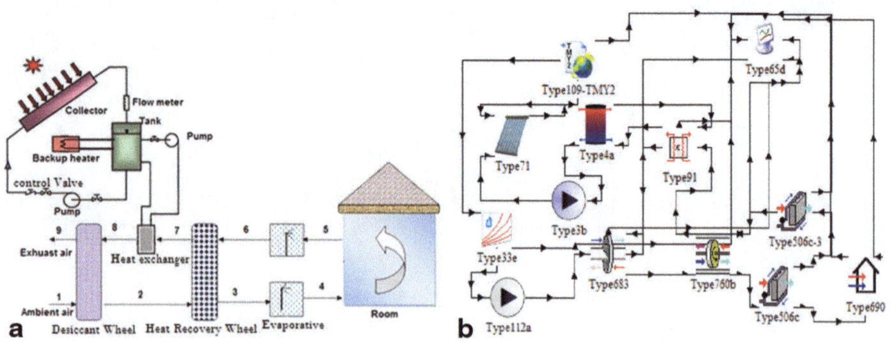

Fig. 68.2 One-stage solar desiccant cooling. (**a**) schematic. (**b**) simulation mode in Transient Systems Simulation (*TRNSYS*)

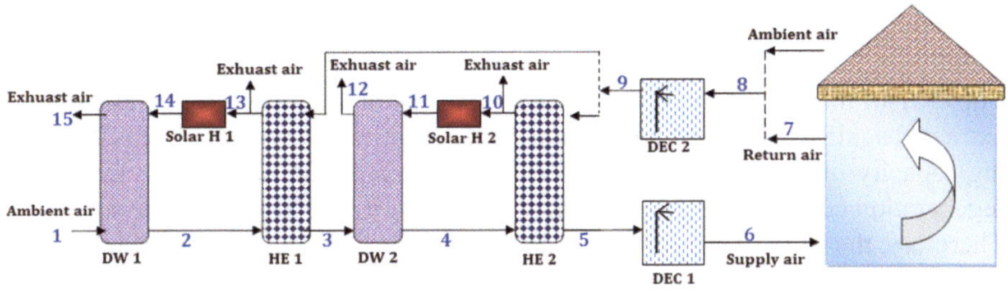

Fig. 68.3 Schematic of two-stage solar desiccant cooling system

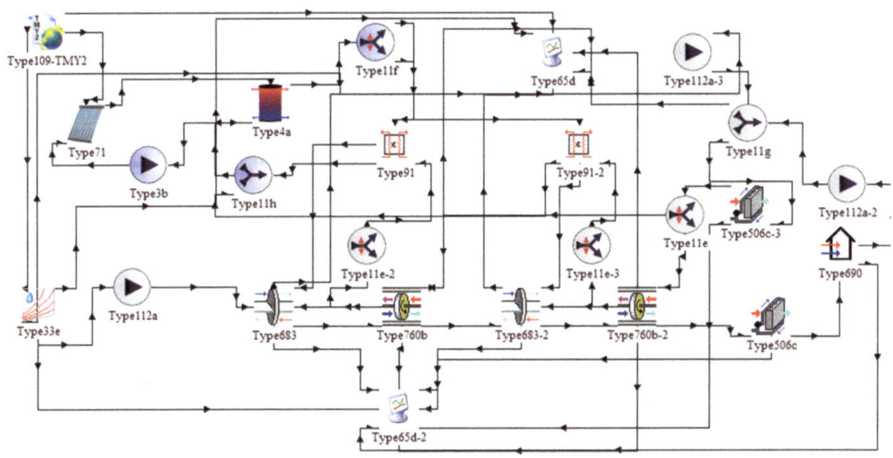

Fig. 68.4 Simulation modeling of two-stage solar desiccant cooling system

auxiliary heaters, and an evacuated tube collector. According to the high dehumidification capacity of the two-stage DW and the humid weather in Malaysia, the set point of dehumidification was adjusted at a 0.005 kg/kg humidity ratio.

Figure 68.3 shows the two-step dehumidification of the process air in the solar desiccant cooling system using evaporative cooling that is an open cycle. In the regeneration side, the returning air is mixed with the ambient air. The air properties of each of the 15 points are characterized by the simulation model. The air properties are explained in the result section. The simulation of the solar desiccant cooling system is shown in Fig. 68.4.

The supply air in the process side is produced from the ambient air by undergoing a few stages, such as the two-step dehumidification in the DWs (2 and 4), two-step cooling in the heat recovery wheels (3 and 5), and one-step cooling in the direct evaporative cooler (6). In the regeneration air side, the returning air (7) is mixed with the ambient air (8) before being cooled in one direct evaporative cooler. Then, the output air from the evaporative cooler (9) is divided into two stages with the same airflow rate. In each stage, air is heated in the heat recovery wheel (10 and 13) and the solar heaters (11 and 14), and humidified in the DWs (12 and 15).

68.2.3 Performance of Solar Desiccant Cooling System

The COP of the solar desiccant cooling system can be calculated by the rate of heat extracted against the rate of heat regenerated. The rate of heat extracted is the cooling capacity of the system that supplies cooling air to the room. The rate of heat regeneration consists of the regeneration heat input by the heater and the solar heat. Therefore, the COP of the system is obtained by the following relationship [13]:

$$COP = \frac{Q_{\mathrm{COOL}}}{Q_{\mathrm{Regeneration}}}. \tag{68.1}$$

68.2.3.1 One-Stage Solar Desiccant Cooling Systems

The COP of the solar desiccant cooling under the ventilation mode can be written as:

$$COP = \frac{\dot{m}_s(h_1 - h_4)}{\dot{m}_r(h_8 - h_7)}, \tag{68.2}$$

where, $\dot{m}_s$ (kg/h) is the mass flow rate of the supply air, $\dot{m}_r$ (kg/h) is the mass flow rate of the regeneration air, and h (kJ/kg) is the enthalpy of air.

68.2.3.2 Two-Stage Solar Desiccant Cooling Systems

The COP of the two-stage solar desiccant cooling can be written as [14]:

$$COP = \frac{\dot{m}_{s1}(h_1 - h_6)}{\dot{m}_{r13}(h_{14} - h_{13}) + \dot{m}_{r10}(h_{11} - h_{10})}, \tag{68.3}$$

where, $\dot{m}_s$ (kg/h) is the mass flow rate of the supply air, $\dot{m}_r$ (kg/h) is the mass flow rate of the regeneration air, and h (KJ/kg) is the enthalpy of air.

68.3 Results and Discussion

The results of this study include the simulation results of the two configurations of the solar desiccant cooling system, namely, the one-stage and the two-stage one, based on the hot and humid weather in Malaysia. According to the main objective of this study, to find the effect of isothermal dehumidification on the performance of solar desiccant cooling systems, the temperature and humidity ratio of the specified points in the four systems were analyzed.

68.3.1 Results of the One-Stage Solar Desiccant Cooling System

Figure 68.5 shows the psychometric chart of the one-stage solar desiccant cooling system in which the air property and process air on all points are specified. The process air and regeneration air sides are indicated by blue and red lines, respectively. Because of the dehumidification process between points 1 and 2, the air temperature increases, whereas the humidity ratio decreases. By the cooling process between points 2 and 3, the air temperature decreases, whereas the humidity ratio of air in the heat recovery wheel remains constant. The temperature decreases, and the humidity ratio increases because of the evaporative cooling process between points 3 and 4. The line between points 4 and 5 shows that the air property changes from supply air to the returning air inside the room. The regeneration side includes the four stages, namely, the evaporative cooling process between points 5 and 6, heating between points 6 and 7, heating process between points 7 and 8, and humidification process between points 8 and 9.

68.3.2 Results of the Two-Stage Solar Desiccant Cooling System

Figure 68.6 shows the psychometric chart of the two-stage solar desiccant cooling system under the ventilation mode. The process air side of this system includes the two-step dehumidification process (1 and 2 and 3 and 4) by the two DWs, the two-step cooling process (2 and 3 and 4 and 5) by the two heat recovery wheels, and the one-stage cooling process (5 and 6) by the evaporative cooler. The process from points 1 to 5 is defined as isothermal dehumidification. The regeneration air side

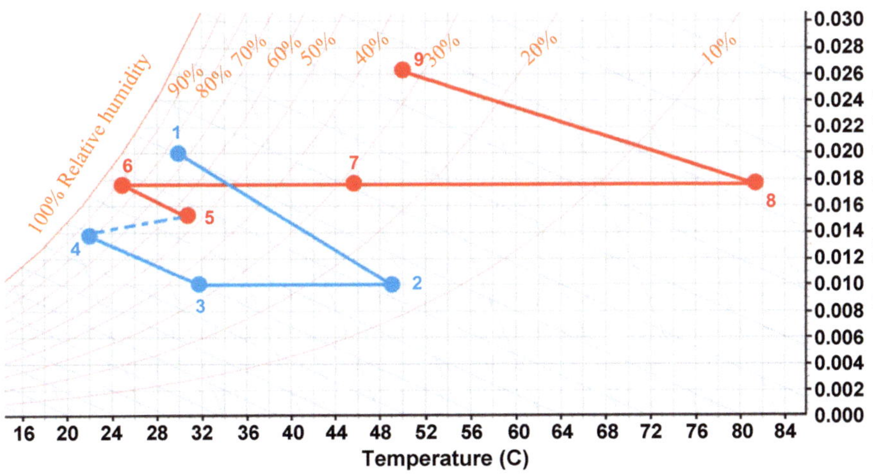

Fig. 68.5 Psychometric chart of the one-stage solar desiccant cooling system under the ventilation mode

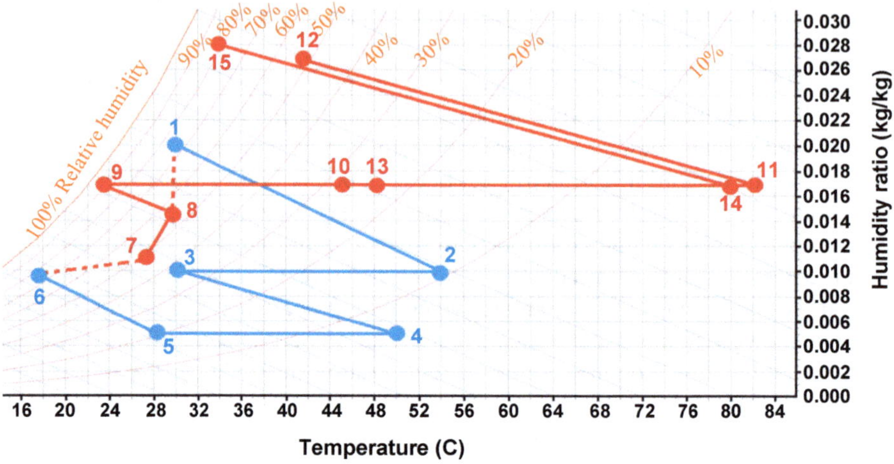

Fig. 68.6 Psychometric chart of two-stage solar desiccant cooling system in ventilation mode

includes a few processes such as the mixing of ambient air and returning air process (7 and 8), evaporative cooling process (8 and 9), two-step heating process (9 and 10 and 9–13) by the heat recovery wheels, two-step heating process (10 and 11 and 13 and14) by the solar heater, and two-step releasing humid air to outdoor (11 and 12 and 14 and 15).

68.3.3 Comparison Results of Thermal and Isothermal Dehumidification Desiccant Cooling Systems

The regeneration temperature, supply air temperature and humidity, and the return air temperature and humidity of two configurations of solar desiccant cooling are shown in Table 68.1. The results show that under the same regeneration temperature for both systems, the temperature and humidity ratio of supply air and return air in two-stage desiccant cooling systems is less than that in one-stage desiccant cooling systems. In two-stage desiccant cooling systems, the humidity ratio can reach 0.005 kg/kg after two steps of dehumidification, while it can reach 0.0010 kg/kg in one-stage desiccant cooling systems. In fact, due to increased moisture removal capacity in two-stage desiccant cooling systems, the effectiveness of evaporative cooling increases; hence, temperature and humidity ratio of the room can touch thermal comfort conditions. The performance amounts of the two mentioned systems are shown in Fig. 68.7. The COP of the two-stage is higher than that of the one-stage desiccant cooling system.

Table 68.1 Air properties of thermal and isothermal solar desiccant cooling systems

Type of configurations	T_{reg} (°C)	Supply air			Return air		
		T (°C)	HR (kg/kg)	RH (%)	T (°C)	HR (kg/kg)	RH (%)
One-stage	81.60	22.00	0.0137	82.5	30.80	0.0152	55.0
Two-stage ventilation	82.10–80.00	17.60	0.0096	76.2	27.30	0.0109	48.0

T temperature, *HR* humidity ratio, *RH* return air temperature

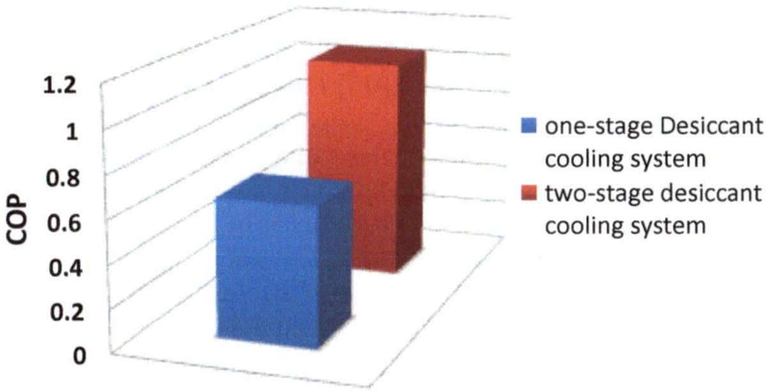

Fig. 68.7 Comparison of performance between one-stage and two-stage solar desiccant cooling configurations

68.4 Conclusions

To find the effect of isothermal dehumidification on the performance of a solid desiccant cooling system, two configurations of one-stage and two-stage desiccant cooling systems have been simulated and compared in this study. The TRNSYS models have been validated by experimental measurement of a solar desiccant cooling system which was installed in the technology park of UKM. Results show that the COP of solar desiccant cooling systems in a two-stage configuration is higher than the one in a one-stage configuration. Also, the temperature and humidity ratio of supply air in two-stage desiccant cooling systems is less than in one-stage systems. Therefore, it was found that the isothermal dehumidification has the positive effect to increase the performance of solar desiccant cooling systems in tropical countries.

Acknowledgments The authors would like to thank the Institute of Product Design & Manufacturing (IPROM) in Universiti Kuala Lumpur and also Solar Energy Research Institute (SERI) in UKM for providing the laboratory facilities and technical support.

References

1. Hürdoğan E et al (2010) Experimental investigation of a novel desiccant cooling system. Energy Build 42(11):2049–2060
2. Jain S, Dhar PL, Kaushik SC (1995) Evaluation of solid-desiccant-based evaporative cooling cycles for typical hot and humid climates. Int J Refrigeration 18(5):287–296

3. Bourdoukan P, Wurtz E, Joubert P (2010) Comparison between the conventional and recirculation modes in desiccant cooling cycles and deriving critical efficiencies of components. Energy 35(2):1057–1067
4. Kanoğlu M, Özdinç Çarpınlıoğlu M, Yıldırım M (2004) Energy and exergy analyses of an experimental open-cycle desiccant cooling system. Appl Therm Eng 24(5):919–932
5. Panaras G, Mathioulakis E, Belessiotis V (2007) Achievable working range for solid all-desiccant air-conditioning systems under specific space comfort requirements. Energy Build 39(9):1055–1060
6. Sphaier LA, Nóbrega CEL (2012) Parametric analysis of components effectiveness on desiccant cooling system performance. Energy 38(1):157–166
7. Dezfouli MMS, Salehi MM et al (2012) Experimental investigation of solar hybrid desiccant cooling system in hot and humid weather of Malaysia. In Proceedings of the 10th International Conference on Environment, Ecosystems and Development (WSEAS'12)
8. Dezfouli MMS, Sohif MAT, Sopian K (2013) Comparison simulation between ventilation and recirculation of solar desiccant cooling system by TRNSYS in hot and humid area
9. Dezfouli MMS et al (2014) Simulation analysis of the four configurations of solar desiccant cooling system using evaporative cooling in tropical weather in Malaysia
10. Sphaier LA, Nóbrega CEL (2012) Parametric analysis of components effectiveness on desiccant cooling system performance. Energy 38(1):157–166
11. Li H et al (2012) Case study of a two-stage rotary desiccant cooling/heating system driven by evacuated glass tube solar air collectors. Energy Build 47:107–112
12. ASHRAE Handbook (1996) HVAC systems and equipment. American Society of Heating, Refrigerating, and Air Conditioning Engineers, Atlanta
13. Daou K, Wang R, Xia Z (2006) Desiccant cooling air conditioning: a review. Renew Sustain Energy Rev 10(2):55–77
14. Ge TS et al (2009) Experimental study on a two-stage rotary desiccant cooling system. Int J Refrigeration 32(3):498–508

Chapter 69
Heat-Driven Heat Pumps—The Future of Domestic Heating in Europe?

R. E. Critoph

Abstract There have been many attempts to commercialize and introduce heat-driven (particularly gas-fired) heat pumps over three decades. There are now three domestic systems on the market [Robur, Vaillant and Viessmann], with others under development.

The different types are reviewed, the markets assessed, and the barriers to wider uptake are discussed. Other options for future heating systems proposed within the UK are a range of electric heat pump developments and fuel cell/micro-CHP units. Moving to an all-electric decarbonised electricity grid is shown to require a vast investment to perhaps triple the capacity of the electricity infrastructure and whilst possible in the long term cannot secure the emission reductions essential in the medium term. The case is proposed for a mixed heating solution with both gas-fired and electric heat pumps, also hybrids being used well in the 2040s. New-build houses will be almost exclusively electric and will need integration with advanced storage to supply domestic hot water. Older properties with higher heat loads will either use hybrid electric heat pump—gas boiler systems or gas-fired heat pumps. The proposed mix, whilst not being the minimal emission route, is much more affordable and a pragmatic solution to domestic heating.

Keywords Heat pump · Adsorption · Domestic

69.1 Introduction

69.1.1 Technologies

In the past, there have been many approaches to the provision of gas-driven heat pumps that have reached varying stages of development. Most, but not all, have not gone beyond RD&D to commercial launch. Systems can be categorised as engine driven or sorption.

R. E. Critoph (✉)
School of Engineering, University of Warwick, Coventry CV4 7AL, UK
e-mail: R.E.Critoph@warwick.ac.uk

A. Sayigh (ed.), *Renewable Energy in the Service of Mankind Vol II*,
DOI 10.1007/978-3-319-18215-5_69

Engine-driven systems can range from conventional gas-fuelled engines driving conventional vapour compression heat pumps to novel engine types, sometimes combined with novel heat pump cycles. In the past, extensive research went into Stirling–Stirling, Rankine–Rankine and other concepts. Whilst some of the more adventurous ideas may still have some potential, the only ones to survive today as commercial systems are conventional engine types (spark ignition or diesel) driving conventional heat pump cycles via open compressors and recovering engine waste heat as useful output. The vast majority of these systems are sized for larger buildings (30 kW+) where a strict maintenance schedule can be enforced and economies of scale work well. There are four major manufacturers, all Japanese, and European sales are estimated at around 10,000 in total [1]. Sales in Japan are an order of magnitude higher. The focus of this work is on domestic applications perhaps extending to light commercial, a size range less suited to the complexity and maintenance of engine-driven systems, although in Japan there are domestic products championed by Osaka Gas. Within the EU, there is consensus that the way forward for domestic units is through sorption systems.

69.1.1.1 Absorption and Adsorption Cycles

What all sorption systems have in common is that they use the ability of a liquid or solid sorbent to ab/adsorb large amounts of refrigerant to be used in a 'thermal compressor', analogous to the mechanical compressor in an electric heat pump. The refrigerant is ab/adsorbed at a low pressure and temperature, then the sorbent is heated (in our case by burning gas), and the refrigerant is then driven out (desorbed) at high pressure. The fundamental thermodynamics are the same whether the refrigerant gas is being absorbed into a liquid or adsorbed into a solid. There are engineering advantages and disadvantages in using solids or liquids, but neither is inherently 'better', and the choice of one or the other in a particular application will depend on the way they are implemented and the resulting efficiency, cost, size, etc.

69.1.1.2 Refrigerants

The refrigerants used by either absorption or adsorption cycles need the same characteristics (high latent heat being the major one), and the three main refrigerants considered for both are the same: water, methanol and ammonia. Water has the highest latent heat by far and so for the same complexity, and cost of machine gives higher COPs [COP is defined as heat output to load/ heat input from gas]. Water also has the benefits of low cost, being non-toxic, environmentally friendly, etc., but has one major disadvantage; it is very low operating pressure. From a practical point of view, it is not possible to boil water in the evaporator of a heat pump at less than about 5 °C. In the case of an air source heat pump, this would restrict the ambient temperature at which any heat pumping could be carried out to more than 10 °C, which is clearly not sensible.

Ammonia has much lower latent heat, but none of the other disadvantages of water. It is the refrigerant of choice in large industrial systems, is not a global warming or ozone-depleting gas and can be used at ambient temperatures below −20 °C. Its higher pressure allows smaller pipe sizes, thereby reducing cost. Its disadvantages are that it is chemically incompatible with copper or brass and that it is toxic; however, neither of these problems is insurmountable, and existing regulations govern the design of refrigerators or heat pumps that use ammonia refrigerant.

Methanol is (in pressure terms) between ammonia and water, and although methanol-based systems are well below atmospheric pressure, it should be possible to use it to extract heat from ambient air at around 0 °C. It is environmentally friendly but has the disadvantage that it decomposes at temperatures beyond about 120 °C, which could be problematic. High-efficiency systems tend to need heat input from the burning fuel at higher than 120°.

69.1.2 Existing and Near-Market Products

There are at present three products on the market and two others at an earlier stage of development. All are being developed or sold by European companies.

69.1.2.1 Vaillant (Information from Ref. [2])

The Vaillant system uses water as the refrigerant, together with a zeolite adsorbent, and consists of the heat pump itself, a solar collector that acts as the low-temperature heat source and a water storage tank. In summer, the solar collectors can provide domestic hot water. It is only intended for use with underfloor heating systems with maximum output temperature of 40 °C, and under these conditions, it has a claimed reduction of annual energy use of 18 % when compared with a condensing boiler. In principle, it could also be used with fan convectors with a flow temperature of 40 °C but perhaps requiring twice or more the wall area occupied by heat emitters compared with conventional radiators supplied by a gas boiler. The initial system sale price was around € 16,000 (Fig. 69.1).

69.1.2.2 Robur (Technical Data from Robur literature, Ref. [3])

The Robur product is an ammonia water absorption heat pump (i.e. ammonia refrigerant, water absorbent), which offers air, water and ground source options. It is a development of technology previously used for air conditioning and thus is comparatively mature. The present air source machine can deliver domestic hot water at 65 °C (gross COP 1.24) and will supply 38 kW to radiators (supply temperature 50 °C) with a COP of 1.52 (gross) and 1.38 (net). This represents a saving of about 40 % in gas consumption compared to a condensing boiler. The unit is a single module intended to be positioned outside the heated building and 854(w) x 1256(d) x 1281(h). An 18-kW unit, more suited to typical UK dwellings, is

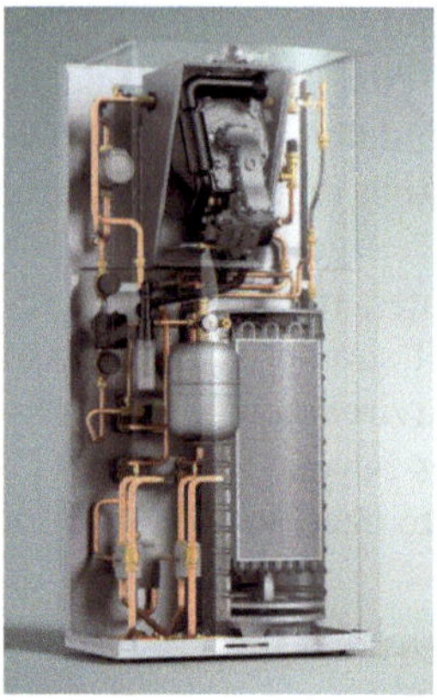

Technical data of zeoTHERM VAS 106/4

Rated heat output range Heating 1,5-10 kW
Rated heat output range d.h.w. 4,2-12,5 kW
Adjustable flow temperature 20-75 °C
Recommended max. flow temperature HC < 40 °C
El. power consumption max. 100 W
Appliance width 772 mm
Appliance height incl. flue outlet 1.700 mm
Appliance depth 718 mm
Transport weight (without casing) 160 kg
Operating weight 175 kg
Integrated controller
zeolite module > no moving parts / no maintenance

Fig. 69.1 Vaillant gas heat pump

Fig. 69.2 Robur gas heat pump

under development. The product is 'badged' by a range of manufacturers including BDR Thermea and Bosch, and the cost of the 40-kW unit is c. £ 12,000 (Fig. 69.2).

69.1.2.3 Viessmann

Viessmann has a water–zeolite system intended for ground heat source and underfloor heating only and is very similar in concept to the Vaillant product. It has

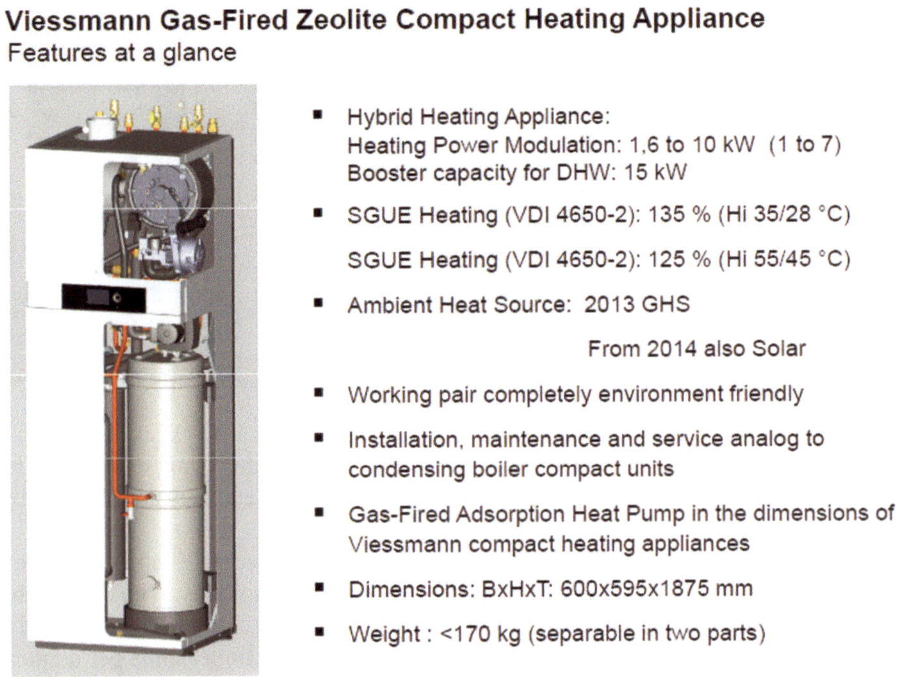

Fig. 69.3 Viessmann gas heat pump

recently had its market launch. The details below are from [4]. Fuel savings are claimed to be 20–25 % compared to a condensing boiler (Fig. 69.3).

69.1.2.4 Sorption Energy

The company (of which the author is a director) is a university spin-out developing an ammonia-active carbon adsorption heat pump. The concept is based on research carried out at the University of Warwick. The product envisaged would consist of an indoor unit of comparable size to a condensing boiler and an external air source evaporator. Emphasis is being placed on the need for low capital cost, low maintenance and installation by non-specialist personnel. Targets include 30 % savings on gas consumption and a better than 3-year payback (Fig. 69.4).

69.2 Market Potential in the Short Term

The information presented below is largely adapted from studies [5, 6] carried out for sorption energy on the market for its heat pump concept, plus the report by Delta-ee [7]. The premise is that in the short term, gas heat pumps must compete with condensing gas boilers rather than electric heat pumps and that any system that has

Fig. 69.4 Sorption energy concept

the potential for major impact must be suitable for retrofit and be air source. Whilst ground source and new-build may be important niche markets, they are dwarfed by the market size of the retrofit market.

The world domestic boiler market is estimated at between 10 and 12 million units per annum. This includes all fuels: gas, oil and electricity. The UK is the biggest market in terms of volume and value, followed by South Korea and Italy. The size of the UK market is around 1.5 million domestic boiler units per annum, and the figure below gives data for 2006 and projections thereafter. UK sales were 1.4 M in 2012 [8].

Historically, regulation has driven technology uptake and witnessed the adoption of Band A condensing boilers influenced by building regulation which saw condensing boiler sales go from zero to about 1 million units over a 6-year period [9].

The route by which boilers pass from manufacturer to private householder is dominated (79 %) by contractor/installers who buy from builders' merchants and this must be the target route if there is to be a widespread uptake of gas heat pump technology. In private replacements, the installer/contractor generally has most influence over the boiler selected as he advises the householder. The factors that influence the installer/contractor are as follows:

- Familiarity with boiler (installed them before, received training from manufacturer, accredited installer).
- Easy and quick to install (weight/fittings). Must be a 'one man fit'.
- Good backup support from manufacturer should there be a problem, e.g. spares available.
- Performance/reliability (if the boiler breaks down, it will be under warranty (2 years generally) from the manufacturer, but the installer will be called back).
- Incentives: Manufacturers sometimes give vouchers to installers for each boiler installed.

Social housing replacements will be determined by the provider's purchasing manager, and decisions are based on 'best lifetime value'. Deals will be in place with OEMs. Large social housing groups generally have one or more boiler brands which they have selected as 'preferred' for installation. The small range of boilers means they are not completely reliant on one supplier, but also do not have a large variety of boilers installed which would be more expensive and complex to maintain. Decisions are made on 'best lifetime value', i.e. minimising capital cost/installation cost plus lifetime maintenance costs.

To summarise, private new-build and retrofit homes are best target markets. Retrofit is by far the largest market and essential if gas heat pumps are to become mass market.

The payback times depend on the size of house and comfort standards since in the domestic range the capital cost of the heat pump is not strongly dependent on the nominal power output. Assuming economies of scale in production, payback times of 3 years seem feasible for larger houses and perhaps 5 years in smaller ones. Renewable heat incentive (RHI) could improve the situation and make gas heat pumps a very attractive replacement for conventional gas boilers, depending on how it might operate. Initially, it would be sensible to target larger owner-occupied homes (4 bed+, 3 m homes or 12 % of the total) where residents tend to have longer investment horizon.

In private new-build (150,000 p.a), gas heat pumps will be most efficient with lower-temperature water output. This requires a well-insulated home with modern radiators or underfloor heating. This means it is well suited to a new-build or major refurbishment.

69.3 Barriers to Development

69.3.1 Specifier/Installer Training

Given the difficulties that have been seen with specification and installation of electric heat pumps, there are obvious pitfalls regarding poor choice of system, controls, etc. Discussions with boiler manufacturers and utilities have lead to the conclusion that it is technically feasible and also necessary that gas heat pumps should be 'black boxes' that can be fitted by a gas safe technician without specialist training beyond a one-day familiarisation course.

69.3.2 Consumer Behaviour

UK consumers are well known to be conservative with respect to new technology and are therefore less likely to switch to new space and water heating systems unless there are clear incentives, such as grant support, to encourage investment in new systems or economic advantages that can be realised in the relative short term. Ref [5] even suggested that in the UK, a sorption heat pump should be sold as a 'super-efficient boiler' rather than as a heat pump in order to avoid scaring customers with new technology. The situation in Germany would of course be reversed.

Only approximately 4–5 % of replacements are motivated by concerns over the efficiency of the old systems.

69.3.3 Capital Cost

Ref [6] suggests that for a mass market consumer product, the payback should be 2–3 years and that for a more niche market, a period of 5 years is acceptable. The Delta-ee study [7] suggests that for a typical heat demand of 18,000 kWh/year, the savings would be more than £ 300/year even with conservative performance estimates. Assuming no difference in installation or maintenance costs and no green deal or other incentives, it can be seen that the excess cost of a gas heat pump over a condensing boiler should ideally be no more than about £ 750, but that up to £ 1500 would still feed a niche market. The present 40-kW Robur unit costs around £ 12,000. A scaled down version produced in larger numbers would presumably come closer to the target, but only Robur could estimate how close. Sorption Energy (who are still several years from market) have studies by Capparo, Angle and Dynamiq, suggesting that the target is feasible, but this has yet to be proven. For the UK market, capital cost is a more critical parameter than operating efficiency.

69.3.4 Life/Reliability

It is difficult to estimate the lifetime or reliability of new products such as these. Whether absorption or adsorption, there are no components inherently subject to wear or corrosion if properly designed. As with all products, there is a trade-off between capital cost and reliability and service life. Life should be at least 10 years, and there is no reason why it should not be twice that.

69.4 Market Potential in the Longer Term

It is a commonly stated criticism of the concept of gas-fired heat pumps that in the longer term, there will be an all-electric renewable energy system; electric heat pumps will be ubiquitous, and there will be no place for burning fuel. Whether this is a likely outcome depends both on the ultimate technical potential of gas heat pumps and on the more difficult questions relating to the energy (electricity, gas and heat) supply grid and its development.

Seasonal COPs of present gas-fired systems range from 1.1 to 1.5, and the theoretical limit could exceed 3, but keeping within what might be possible in systems we can envisage now, a seasonal COP in the UK of 2.0 [gross CV, air to water at 50°C] is certainly not unreasonable. Machines that can achieve these efficiencies will use thermodynamic cycles that are more complex, with cascaded temperature levels, improved heat regeneration, etc., and will probably be larger per unit of heat delivered. At the moment, gas heat pumps must compete with gas boilers rather than electric heat pumps, but in the future, we can imagine that they will have to compete more with advanced electric heat pumps and a COP of 2.0 might be necessary as well as desirable. The TINA Summary Report on Heat [10] suggested that 1.6 was a lower limit to be of interest in 2050.

DECC's document 'The Future of Heating: Meeting the challenge', published March 2013 [11], sees a future for gas heat pumps well beyond the 2030s. Points made include the following:

- Even condensing boilers have a major role in the 2030s.
- Based on the RESOM model, gas will still be used to provide heat in winter, and 'there will be no role for gas boilers in 2050. The ESME results suggest that gas absorption heat pumps (GAHPs) may be taken up as a more efficient solution'.
- 'The modelling is highly sensitive to the assumptions made on the costs and performance of technologies, including technologies which are yet to reach commercial scale deployment such as hybrid systems and domestic GAHPs [gas (ab) sorption heat pumps]. However, the analysis does suggest that there is still no role for standalone gas boilers in 2050, with the least-cost path requiring us to begin to deploy more efficient gas technologies in the 2020s.

The report states that condensing boilers will be in use well in the 2030s, particularly in suburban areas, and suggests that perhaps half will be replaced by 'more efficient systems such as gas absorption heat pumps', with the latter only diminishing in use in the 2040s.

Both the National Grid [12] and Delta-ee [13] have looked at scenarios for 2050 and beyond that encompass a role for both gas and electric heat pumps.

The National Grid GG (Gone Green) scenario energy demand/duration mix has demand for heat being much 'peakier' than demand for electricity. Demand for electricity excluding heating ranges from a maximum of 1110 GWh/day to a minimum of 690 GWh/day (62 % of peak daily heat demand). The heating load ranges from a maximum of 3000 GWh/day to a minimum of 500 GWh/day (17 % of peak daily heat demand). As a result of the large range in the heat demand, it would be very difficult to electrify the entire heat load. Electric heat pumps can be used to supply a significant fraction of the heat demand, but in order to satisfy the entire load, it would be necessary to build between 100 GW and 150 GW of new capacity, most of which will run for half the year or less, with correspondingly unfavourable economics.

Electricity in heat pumps has higher carbon emissions than gas condensing boilers until 2015, and gas heat pumps can decarbonise heat quicker than electrification up to 2030. Dual fuel [electric heat pump + gas or gas heat pump] can deliver similar carbon benefits to full electrification. The all-electric solution needs a 100-GWe capacity, whereas a hybrid or gas heat pump mix needs only 13 GWe. It is anticipated that gas heat pumps would be best used in older, larger, difficult to insulate properties with electric heat pumps in new-build and smaller properties, apartments, etc.

The breakdown of the evolving market is shown in Fig. 56 of the FES [11], below. In domestic and commercial premises, there are marginal improvements in energy efficiency between 2030 and 2050 with most of the more cost-effective refurbishments of the UK's buildings being done by 2030. Between 2030 and 2050, most efficiency gains are made by the changing the way heat is supplied in buildings with a shift to heat pumps for well-insulated new properties with hybrid heat pump (gas or electric) and boiler systems being used in buildings in which the difference in seasonal heat requirements makes them less suited to be supplied by heat pumps alone. Note that gas boilers still make up some 10 % of the mix in 2050 and that gas will also be used in the hybrid heat pump/boiler or gas heat pump systems.

Delta-ee's report, 2050 Pathways for Domestic Heat [13], carried out detailed modelling to 'provide a desk top study on the optimal appliance technology pathways, by property type, based on known and emerging heating technology, required to meet carbon and renewables targets, highlighting the impact on consumers (cost to change and behavioural) and the potential load changes on the gas and electricity distribution networks out to 2050'.

It developed a residential heat model that used a very detailed housing stock model, segmenting the UK housing stock into 35 segments. For each segment, the thermal demand, and how this changes decade by decade to 2050, was defined.

It also used a technology performance model, forecasting future cost and performance of many appliances: gas boiler; gas heat pump; low electrical efficiency micro-CHP; high electrical efficiency micro-CHP; gas boiler + solar thermal; air source heat pump; ground source heat pump; hybrid gas boiler + air source heat pump; biomass boiler; district heating; direct electric (storage) heating.

Finally, it combined these with a customer choice model that incorporated physical fit of different technologies with different parts of the housing stock; customer uptake based on payback and upfront cost; and customer attitudes to different technologies (Fig. 69.5).

This residential heat model was used to determine the future appliance mix under three scenarios:

1. Customer Choice—allowing the customer to choose based on physical fit, customer economics and attitudes. Unsurprisingly, the Customer Choice scenario fails to meet the 2050 carbon reduction targets.
2. Electrification and Heat Networks—Delta-ee defining a pathway where these solutions dominate in 2050. There is a 96 % reduction in carbon emissions (from 2010 to 2020 levels).
3. Balanced Transition—Delta-ee developing a pathway where electric heating, heat networks and gas all play a role in 2050. There is a 90 % reduction in carbon emissions.

The report suggests keeping a variety of options open. The scale of the challenge to decarbonise heat gives lower risks and potentially a lower cost path than pursu-

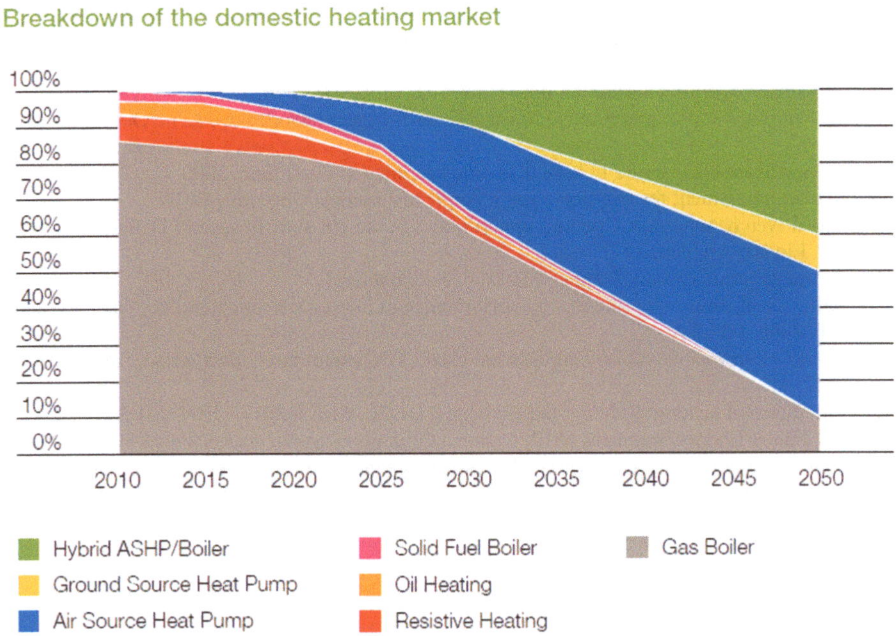

Fig. 69.5 Breakdown of domestic heating market (National Grid, Future Energy Scenarios 2012 [11])

ing a narrower end point. Although BT achieves a 90 % (rather than 96 %) carbon reduction from today to 2050, it has two key benefits:

- It avoids moving an additional 12 million homes completely away from gas—where the highest customer costs are imposed.
- Lower impacts on the energy system—additional peak generation demand grows to 24 GW, rather than 48 GW, as under E&HN. Costs (discounted, opex and capex) to re-enforce the electricity distribution network are € 8 bn lower. Part of the € 4 bn cost to shut down the entire gas network is avoided.

The message from all three sources (DECC, National Grid, and Delta-ee) is similar; there should be a place for gas heat pumps in the longer term to 2050 in a balanced and sustainable yet affordable energy supply and utilisation system.

References

1. Gas heat pumps: product overview, Juliette Promelle, GDF SUEZ, Research division, presented to Marcogaz gas heat pump workshop, 1,2nd December 2011, Paris
2. Vaillant systems with zeoTHERM VAS 106/4, Hendrik Tiemeier, Vaillant, presented to Marcogaz gas heat pump workshop, 1,2nd December 2011, Paris

3. 'Robur—Integrated heating and cooling solutions with absorption heat pumps powered by natural gas and renewable energies', Robur S.p.A. http://www.robur.com/documenti_prodotto/ROBUR_ABSO_EN13_07_c-20130715090407.pdf. Accessed 25 July 2013
4. Viessmann gas driven sorption heat pumps, Dr. Belal Dawoud, Manager R&D „Thermally driven heat pumps", presented to Marcogaz gas heat pump workshop, 1,2nd December 2011, Paris
5. Market assessment in the field of domestic heating, Optimat, June, 2009
6. Market assessment for sorption energy, Angle technology, December 2010
7. Gas-driven heat pumps: Opening opportunities in the UK retrofit sector? Delta-ee Whitepaper, Lindsay Sugden, 2012
8. Private communication, Stephen Marland, National Grid
9. Policy brief: Improving the energy performance of domestic heating and hot water systems, July 2008, DEFRA
10. The low carbon innovation coordination group TINA summary report on heat of September 2012
11. The future of heating: Meeting the challenge, DECC, March 2013, URN: 13D/033
12. UK Future Energy Scenarios 2012, National Grid, http://www2.nationalgrid.com/UK/Industry-information/Future-of-Energy/FES/Documents-archive/. Accessed 17 Aug 2015
13. 2050 Pathways for domestic heat final report, Delta-ee for the gas futures group of the energy networks association, 16th October 2012

Chapter 70
Implementation of a CSP Plant for Localized Energy Supply at Populated Zones in the North of Chile

José M. Lobo, Hugo Osorio and M. Pilar Gárate

Abstract Chile is nowadays facing a complicated energy situation characterized by an energy matrix dominated by fossil sources, unclear energy policies, and rising prices, which causes loss of competitiveness of the national industry and an inefficient electricity grid. These variables, among others, make the national electric system to be characterized by its high economic, environmental, and social costs, for which nonconventional renewable energies (NCRE), such as solar, are seen as a possible solution in the medium term, considering the existing high levels of radiation in certain geographical areas.

This chapter seeks to determine technical and economic feasibility of implementing a central tower concentrated solar power (CSP) plant, technology currently under development for the conversion of solar radiation into electricity, assessing the total supply of a populated zone. To carry out this study, implementation of energy efficiency (EE) measures in certain areas (domestic consumption, industrial, commercial, public services, and transport) will be considered. In terms of the electricity distribution system, the plant will be connected via a substation directly to the populated area under consideration. In other words, CSP design and operation is based on localized supply. This grid connection at distribution level reduces cost and provides secure electricity supply. All these aspects will be part of a business model that, carrying a private and social project evaluation, could make a joint venture between the Chilean state and private companies. Analysis of this model led to the need of the Chilean State to take part on clean and efficient energy investments and the importance of sustainable energy for both actors as an alternative to face a future that is expected to be increasingly dependent on energy consumption.

The systematical and massive implementation of the innovative measures presented in this chapter for populated areas with appropriate technologies for different

J. M. Lobo (✉) · H. Osorio · M. P. Gárate
Departamento de Industrias, Universidad Técnica Federico Santa María,
Av. España 1680, 2340000 Valparaíso, Chile
e-mail: jose.lobo@alumnos.usm.cl

H. Osorio
e-mail: hugo.osorio@usm.cl

M. P. Gárate
e-mail: mariapilar.garate@usm.cl

A. Sayigh (ed.), *Renewable Energy in the Service of Mankind Vol II*,
DOI 10.1007/978-3-319-18215-5_70

meteorological and geographical conditions may contribute to mitigate the growth of the main national electricity grids, reduce losses associated with electric power transmission, diminish unnecessary energy consumption, and therefore decrease environmental impacts associated to them.

Finally, noteworthy is the importance of the new developments achieved through innovation and the opportunities resulting from the reduction in costs through scaling of the industry and constructions that could significantly reduce capital investments required for implementation of these kinds of projects.

Keywords CSP · NCRE · EE · Distributed generation · Localized supply

70.1 Introduction

Industrialization and economic development, as experimented in Chile, requires energy. That is the reason why energy demand is expected to increase at growing rates over the coming years [1]. However, its supply must guarantee sustainable development of the energy industry and therefore of the country. In order to achieve that, technological advances and problems related to scarcity of nonrenewable resources and its environmental and health impacts should certainly be taking into account [2].

The renewable energy sources are native resources that may contribute to face these global issues permanently. The central tower (or central receiver) concentrated solar power (CSP) is currently one of the more attractive NCRE technologies in the long term to meet future energy demand in areas with high levels of solar radiation, due to its advantages in terms of energy storage, social benefits, and its variety of secondary applications [3]. Chile, and more specifically the northern part of the country, has one of the world's largest solar potential, which is currently being exploited by few photovoltaic solar farms. Recently, the first CSP was approved and is expected to start its operation in 2017 [4].

Fossil fuels currently dominate Chilean energy matrix (approximately 70 % of primary energy), which creates a high external dependence because the country does not have reserves of these resources [5]. Besides, rising prices of nonrenewable resources make renewable technologies become increasingly competitive. Consequently, the laws regulating the country's energy sector should aim to promote the implementation of technologies based on the country's natural resources.

It is important to note that the large and growing grids of the electric systems have environmental and economic impacts [6]. Environmental damage in terms of land use, landscape, and thus tourism in Chile, and the consolidation of an inefficient electrical system (since most transmission involves more losses) are the main problems that are causing its expansion. This can be faced by encouraging distributed generation, which means that power generation is located in the vicinity of the consumption points, avoiding use of main transmission systems and thus reducing their future growth.

Although private financial evaluation of NCRE projects currently leads the companies to choose for power plants operating with fossil fuels or hydropower dam (both large scale) mainly due to its lower levelized cost of energy (LCOE) [7], an electrical matrix dominated by these resources does not allow national sustainable development in this field. To achieve this long-term goal, the local authorities should consider an integrated supply and demand management, focusing on the use of clean and renewable resources, efficient planning of the interconnected systems, and efficient energy consumption.

This chapter is structured as follows: In Sect. 70.2, a business model for sustainable energy development in populated areas (called localized supply) is presented, which considers the integrated management of the electricity market aiming for sustainable development. A brief description of a CSP central tower (highlighting the features that make it attractive to be implemented with localized supply) is presented in Sect. 70.3. Section 70.4 describes the methodology used for the economic and financial evaluation under the special scheme imposed by the business model previously discussed, introducing the scenarios and parameters that are considered for the model evaluation. Results and sensitivity analysis of the critical variables are presented in Sect. 70.5. Finally, conclusions about its application in Chile are presented in Sect. 70.6.

70.2 Localized Supply, a Model for Sustainable Energy Development

In order to promote sustainable local energy developments, this chapter introduces an innovative business model for electric generation companies that can address this issue within populated areas aiming to be a "Smart City." This model considers the integrated management of energy resources from three approaches:

- Power supply, renewable, and clean energy: the company takes care of generating electricity with one or more NCRE resources and sell it to a distribution company (associated with a substation that takes care of energy local supply) on a scale such that its nominal power will cover at least the current peak power demand on the location. Clearly, this does not ensure that the populated zone will be fully supplied with renewable energy, but provides a hybrid energy supply in which the excess and deficits are traded in the spot market.
- Distributed generation, efficient electric grid: the NCRE power plant is placed near the consumption point connected to the substation. Generation company should, therefore, avoid the use of main transmission system to dispatch energy to the distribution company.
- Energy efficiency, efficient and conscious consumption: The generation company should be responsible for promoting EE over the town in the main consumption segments: residential, industrial, commercial, public services, and transport. It should encourage the transition through the use of efficient technologies (industrial equipments, construction, home appliances, electric vehicles, etc.) and conscious behavior in the use of resources (audits and programs for cultural change).

It can be seen that these three approaches are finally materialized in the implementation of an NCRE power plant in a convenient location, in addition to the implementation of a plan to stimulate the EE, that consider short-term (investment incentives) and long-term instruments (such as programs of education for sustainable development). Note that solar thermal and geothermal energy plants could provide relatively regular daily supply but may not adjust that easily their capacity to demand growth as photovoltaic and eolic energy plants.

The elements considered in this model of integrated energy management generate benefits to society as a whole. However, private profitability of those initiatives is undermined by the use of technologies that have not achieved the maturity and competitiveness that is expected of them, and by EE plans that involve investments on the one hand and reduces the demand of the populated zone on the other hand. This discussion suggests that the state should subsidize the implementation of this business model given the various social incomes that it causes (which will be presented in Sect. 70.4) and, even more significant, its alignment with not only the national energy strategic plan but also with the ones of developed countries. In this context, the subsidy to energy sustainability is proposed as a payment per unit of NCRE dispatched to the populated area intervened, representing part of the monetary social benefit generated.

70.3 CSP Central Receiver

The CSP central receiver plants are characterized by their circular or oval field of flat mirrors (heliostats) which are individually focused, according to the sun's position, to a receiver located in a tower in the center of the reflective field, where solar radiation is concentrated 600 to 1000 times, making it work at temperatures between 500 and 800 °C. These high temperatures not only allow the use of steam cycles, but also of gas turbines and combined cycle systems, besides the incorporation of energy storage system of up to 15 h, achieving electricity generation even in cloudy conditions or night hours [8] (Fig. 70.1).

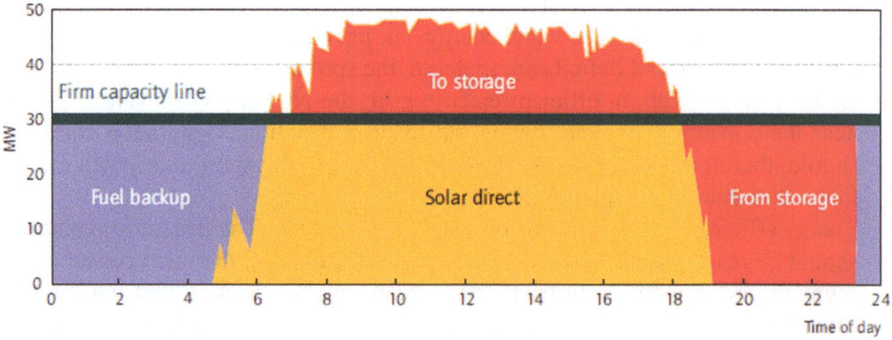

Fig. 70.1 Hybrid operation of a central receiver CSP

This technology seems to be technically favorable to be applied under a scheme of localized supply mainly due to its ability to store energy, as heat to dispatch, in times of absence of direct solar radiation, which increases the amount of energy generated by the plant that will actually be dispatched to the populated zone (see Fig. 70.1). Despite the fact that it is financially more convenient to invest in high-capacity plants (economies of scale), the purpose of the evaluated business model requires some level of parity between electricity generation and consumption in the area under consideration so that distributed generation could be effective. The incentives should therefore only be granted under those conditions.

70.4 Assessment Methodology for Localized Supply from a CSP Plant

To analyze the feasibility of implementing a CSP central tower plant for localized supply of populated centers in northern Chile, it is necessary to know the actual power demand pattern of the zone (which will be modified by the EE measures based on participation of the main areas of consumption), the supply curve of the power plant (differentiating the energy from the sun and back up fossil resource), and its associated costs (both investment and operational). The temporal resolution used in terms of supply and demand is 1 h, in order to represent in a good way the irregularity of renewable energy in their daily patterns of generation.

The demand curves of different electrical substations in Chile and the percentages of total consumption for the main populated areas by sectors (industrial, residential, commercial, and public) are available [9, 10]. Using standardized daily patterns for these sectors, the proportion of demand is estimated for each one of them at each hour of the day, thus facilitating the assessment of EE measures [11]. In this study, the implementation of a central tower CSP for localized supply is assessed for the following populated areas:

- Pozo Almonte: city in the first region, intermediate depression area, peak demand of 4 MW, principally commercial and public consumption (66 %).
- South sector of Antofagasta: city in the second region, coastal area, 14 MW peak demand, mainly residential consumption (47 %).

The CSP plant is simulated with System Advisor Model (SAM), software developed by the NREL which uses a TRNSYS model (transient system simulation tool) to represent the dynamic behavior of a large number of parameters of interest to various NCRE technologies, including the central receiver station [12]. The costs associated with the implementation and operation of the plant have been adapted to Chile using a scaling spreadsheet developed exclusively for SAM. With a TMY3 file (Typical Meteorological Year) of the area where is intended to locally evaluate the technology, and defining certain parameters regarding the capacity and the dispatching pattern allows to obtain the net power generation curve (or supply curve for an average year) and the costs associated with the plant among other results [13].

Investment in EE must be high to achieve the goal of generating sustainable energy areas in the long term. Therefore, we consider as reference, the average investment per capita in developed European countries that are making great efforts in this area (such as France, Italy, Denmark, and Sweden). Since the results of this investment are uncertain, three scenarios referring to different levels of penetration of the EE measures were evaluated: no penetration, realistic, and optimistic, carrying out sensibility analysis for 30 intermediate scenarios in the optimal cases (see Table 70.1). Transport efficiency is considered with a lineal penetration of electric vehicles of 0, 15, and 30 % (for the three scenarios) in the evaluation horizon.

As discussed above, the elements considered in the business model of localized supply, and especially the implementation of EE measures, forces us to analyze the possibility that the state could provide a financial incentive to private investors in such amount that reflects the income generated to the society by its application. Currently in Chile, an application is open to projects regarding the building of a CSP plant which provides a grant of 20 MUSD subject to a minimum capacity of 10 MW nominal [14]. This study considers additionally a subsidy to energy sustainability, so that both actors (state and private investors) may absorb some of the generated profits. In this way, private investors make an economic/financial evaluation of the project, while the state makes an economic/social evaluation of the initiative from their perspective, considering the elements listed in Table 70.2. For private

Table 70.1 Reduction in electricity demand expected in the three EE scenarios

Sector	Energy saves	Energy efficiency measures
Industrial	0 %–6 %–33 %	Energy management systems, production and support equipment, capacity development, self-generation, insulation, etc.
Residential, commercial, and public	0 %–12 %–23 %	Audits, recover the existing buildings, new construction development, capacity development, public lighting

Table 70.2 Financial flows of the generation firm and social flows of the state with localized supply

Actor	Financial/social flows	
Generation company	Investments	CSP plant construction
		Energy-efficiency encouragement
	Operational costs	CSP fixed and variable costs
		Remunerations
	Incomes	Energy and power sales to the populated zone and to the system (spot market)
		Energy sustainability subsidy. "CSP Contest" grants
State	Investments (grants)	"CSP Contest" grants
	Costs (subsidy)	Energy sustainability subsidy
	Social incomes	Reduction in energy consumption by energy efficiency
		Reduce GHG emissions (CERs bonds sale)
		Growing attenuation of main transmission system
		Energy independence (fossil fuel not used in Chile). Local value added (employment creation)

assessment, a discount rate of 8.63 % is used (CAPM), while for social assessment, the rate was 6 % (social discount rate in Chile). In both cases, an evaluation horizon of 20 years was considered. Note that the results of the social assessment were compared with those of a 2013 study obtaining errors less than 20 % [15].

Under those considerations, this chapter economically assesses the implementation of a CSP for localized supply of populated centers in northern Chile with diversity in climatic conditions, population, and energy demand. Economic evaluation is carried out through the calculation of private and social NPV including subsidies, and considering future changes in the investment cost of the plant and in the prices of fossil fuels. Sensibility analysis of the following parameters will determine project feasibility under different scenarios:

- Size of CSP plant: Nominal plant capacity may vary between 4 and 40 MW. As discussed above, small-to-intermediate plant scale is required for local supply.
- Penetration of EE: No EE, realistic, and optimistic scenarios as previously described.
- Financing of CSP plant: The tendency is to finance those projects trough soft loans, high amounts at low interest rate, and extended time horizons. NPV with and without soft financing (50 % of the investment, at a rate of 2 % per year and 20 year term) of the project will be evaluated.

Considering the above-mentioned parameters, a total of 228 scenarios were evaluated. As methodology, we look for those scenarios with global profitability (meaning that the social benefit is enough to generate profits and distribute them between the state and the private through the energy sustainability subsidy) analyzing the maximum private returns that can be achieved in each case (with a subsidy that makes the social NPV equal to zero).

70.5 Results and Sensibility Analysis

As mentioned above, the localized supply business model requires the disposition of the state to provide the necessary resources for its implementation. Figure 70.2 shows how the maximum private profitability varies under a realistic scenario of EE penetration (i.e., the state transfers the entire social benefit to its application). Note that in all the scenarios where a positive private profitability is achieved, it may actually be shared between the two actors involved by decreasing the amount of the energy sustainability subsidy.

As shown in Table 70.3, with the current investment levels of the CSP plant , 60.3 % of the evaluated scenarios generate a global profitability, which increases up to 89.9 % in 2025 considering the learning curves of this technology [3]. Again, the business model is not profitable without fiscal contributions even considering learning curves. The projected price of fossil fuels [16] and its effect on the net economic benefit, keeping constant the other variables in the long term, lead to an increase of 4.6 % of the profitable scenarios in 2030 respect to the actual situation,

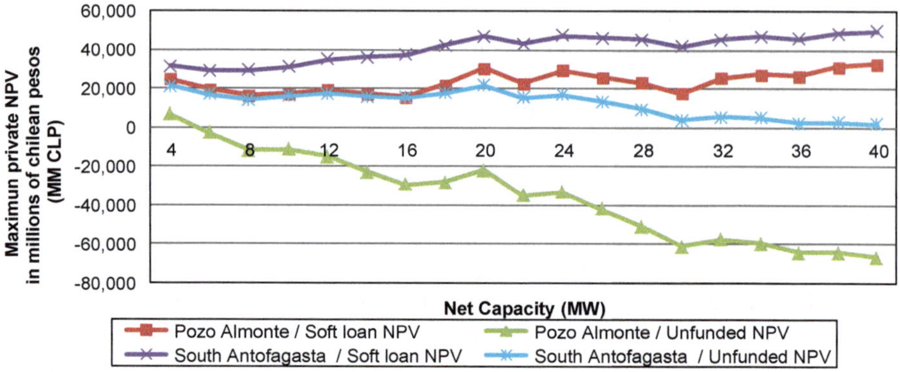

Fig. 70.2 Maximum private profitability for realistic scenarios in both localities

Table 70.3 Percentage of all scenarios evaluated in which overall profitability is generated, considering the projections of the variables analyzed

Variable\year	2013 (%)	2020 (%)	2025 (%)	2030 (%)
CSP investment cost	60.3	81.6	89.9	98.9
Fossil fuel price	60.3	60.3	64.9	64.9

CSP concentrated solar power plant

due to an increase in social income related to energy independence (reduction in fossil resources from the national energy matrix).

In all tested cases, the global profitability of South Antofagasta exceeds that of Pozo Almonte, even if the comparison is made relative to their investment levels. We can affirm that localized supply is more profitable for higher population demand and thus higher net plant capacities under the assessed range. However, it should be notice that a larger demand may difficult the implementation of EE measures, creating more uncertainty on the real social income generated by the project.

Individual analysis of populated zones leads to finding each optimal scenario. Considering the scenarios without soft loan and whit a realistic EE penetration, the optimal cases considers the installation of 4 MW for Pozo Almonte and 20 MW for South Antofagasta, which is close enough to the peak demands of these zones. Some (Key Performance Indicator) KPI of these scenarios are presented in Table 70.4. Furthermore, considering several levels of penetration of energy efficiency (30 interpolations), the proportion of them that generates a global profitability is determined. From the results, we can conclude that the higher the energy demand of the populated zone, less dependent is the global profitability of EE penetration (Table 70.5).

Table 70.4 KPI optimal scenarios with realistic penetration of energy efficiency and pure NPV

KPI\optimal case	Pozo Almonte	Antofagasta Sur
Net electrical output (MW)	4	20
Private maximum NPV (to distribute, MM CLP)	3.570	21.509
Investment (MM CLP)	47.330	125.565
NCRE generation (MWh/year)	35.486	160.918
NCRE distributed generation (%)	98.2%	91.4%
Reduction in GHG emissions (Ton$CO2_{eq}$/year)	34.655	146.520

GHG greenhouse gas

Table 70.5 Percentage of scenarios at different levels of EE in which a global profitability is obtained

Financing\case	Pozo Almonte 4 MW (%)	Antofagasta Sur 20 MW (%)
Unfunded project	58.8	76.5
Soft loan project	82.4	100

EE energy efficiency

70.6 Conclusions

Localized energy supply, as a business model for local energy sustainability, is not financially viable for private investors under any of the scenarios evaluated without external incentives. That means that the participation of the state through grants and/or subsidies would be essential if Chile wants to implement this type of projects. The estimated social benefit of these projects is, under certain conditions, enough to be shared between the state and the private investors, both obtaining profitability from the implementation of this business model.

The projected reductions in CSP installation costs will have an important role in the future development of this market, making viable the business model under almost all scenarios evaluated in this work. Moreover, the projected price of fossil fuels and their effect on net economic benefit increase would not impact significantly the economic feasibility of the implementation of this business model.

Adjusting the capacity of the plant to the demand of the area under study allows higher returns from this model, considering that the subsidy is indexed per unit of NCRE dispatched to the location. Furthermore, the application of localized supply in areas with higher energy demand is more profitable mainly due to the scale factor in construction and equipment costs. By increasing the size of the populated area, the penetration of energy efficiency measures becomes more uncertain and thus affects the social benefit of this component, but also increases the influence of other social factors such as energy independence and local added value neutralizing the effect of such uncertainty.

References

1. O'Ryan R (2008) Diseño de un modelo de proyección de demanda energética global nacional de largo plazo. Universidad de Chile, Santiago, pp 124–137. (Programa de gestión y economía ambiental)
2. King HM (1956) Nuclear energy and the fossil fuels. Shell Development Publications, Texas, pp 1–40
3. Viebahn P, et al (2010) The potential role of concentrated solar power (CSP) in Africa and Europe—a dynamic assessment of technology development, cost development and life cycle. s.l.: Energy Policy
4. Ortega A, Escobar R, Colle S, Luna Abreu S (2009) The state of solar renergy resourse assesment in Chile. Santiago, Chile: ISES Solar World Congress 2009: Renewable Energy
5. Instituto Nacional de Estadísticas (2008) Distribución y consumo energético en Chile
6. Cagna JL (2009) Impacto ambiental de las redes eléctricas. Universidad de Cantabria, Cantabria
7. U.S. Energy Information Administration (2014) Levelized cost and levelized avoided cost of new generation resources in the annual energy outlook 2014. Washington, DC: s.n. pp 6–9. http://www.eia.gov/forecasts/aeo/pdf/electricity_generation.pdf. Accessed Dec 2014
8. Müller-Steinhagen H, Trieb F (2004) Concentrating solar power: a review of the technology. Ingenia, Stuttgard, pp 43–50
9. Centro de Despacho Económico de Carga Retiros de energía a clientes (2012). http://www.cdec-sic-cl/; http://cdec2.cdec-sing.cl/. Accessed Dec 2013
10. Superintendencia de electricidad y combustibles Gobierno transparente (2010). http://www.sec.cl/transparencia/index.html. Accessed Feb 2014
11. Cousiño, José Manuel Infante (2006) Modelo de estimación espacial de consumos eléctricos para la planificación de la expansión de sistemas de transmisión
12. NREL System Advisor Model (2014). http://sam.nrel.gov/. Accessed March 2014
13. Ministerio de Energía Datos solares y meteorológicos de Chile (2012). http://www.minenergia.cl/. Accessed Feb 2014
14. CORFO Concurso Planta de Concentración Solar de Potencia (CSP) (2013). http://www.corfo.cl/programas-y-concursos/programas/concurso-planta-de-concentracion-solar-de-potencia-csp. Accessed Dec 2013
15. NRDC/ACERA/PWC (2013) Beneficios económicos de las energías renovables no convencionales en Chile. Santiago, Chile: s.n., pp 1–4
16. Natural Resources Canada Long term outlook (2010): Crude Oil prices to 2030. https://www.nrcan.gc.ca/energy/publications/markets/6511. Accessed Jan 2014

Chapter 71
Solar Thermal Polygeneration System for Cooling, Fresh Water, and Domestic Hot Water Supply: Experimental Analysis

Gowtham Mohan, Uday Kumar N. T., Manoj Kumar P. and Andrew Martin

Abstract The demands for space air-conditioning and clean drinking water are relatively high in Middle East countries. A sustainable and innovative approach to meet these demands along with the production of domestic hot water is discussed in this chapter. A solar thermal polygeneration (STP) system is designed and developed for the production of chilled water for air-conditioning using absorption chiller, pure water with membrane distiller, and domestic hot water by heat recovery. The STP system has four major components: (i) evacuated tube collector field, (ii) 10TR absorption chiller, (iii) air-gap membrane distillation units, and (iv) heat exchangers integrated together to operate in four different modes for complete solar cooling, cogeneration of pure water and domestic hot water, trigeneration of cooling, pure water and domestic hot water, and cogeneration of cooling and pure water. Experiments on different modes and the analyzed results show the advantages of combined operation through effective utilization of heat lost in the process operation.

Keywords Absorption chiller · Air-gap membrane distillation · Polygeneration · Domestic hot water

U. Kumar N. T. (✉) · G. Mohan · M. Kumar P.
CSEM-UAE Innovation Center LLC, PO Box 31208, Ras Al Khaimah, United Arab Emirates
e-mail: uku@csem-uae.com

G. Mohan · U. Kumar N. T. · A. Martin
Department of Energy technology, KTH Royal Institute of Technology, SE-100 44 Stockholm, Sweden
e-mail: gowtham@kth.se

A. Martin
e-mail: Andrew.martin@energy.kth.se

M. Kumar P.
e-mail: mpo@csem-uae.com

A. Sayigh (ed.), *Renewable Energy in the Service of Mankind Vol II*,
DOI 10.1007/978-3-319-18215-5_71

71.1 Introduction

Electricity demand in United Arab Emirates (UAE) has increased tremendously in past two decades due to rapid industrialization and population growth. Around 30 % of electricity consumption is accounted for providing air-conditioning in the buildings [1] and it is more in summer months. UAE and adjacent countries in Middle East and North Africa (MENA) region do not have adequate natural fresh water resources, and hence most of the fresh water demand is met by fossil fuel-driven sea water desalination technologies. On the other hand, UAE gains abundant solar irradiation with an average global irradiation potential of 600 W/m^2, which could be effectively utilized through solar thermal polygeneration systems for providing air-conditioning in the buildings and fresh water through desalination. Hussain [2] designed and developed a hybrid polygeneration system for utilization in Kuwait, which simultaneously produces power, fresh water, and cooling. The combination of power-reverse osmosis–absorption refrigeration configuration obtained higher fuel savings compared to all other combinations. Picinardi [3] analyzed the performance of cogeneration system for production of cooling and desalination by integrating a single-stage absorption chiller and humidification desalination process. Calise et al. [4] dynamically simulated a solar trigeneration system and analyzed energetically and economically for production of cooling, fresh water, and electricity. The system is modeled with photovoltaic/thermal collectors (PVTs) integrated with absorption chiller and multi-effect desalination system for providing trigeneration.

Membrane distillation (MD) is a thermal-driven desalination process which utilizes low-grade heat energy. The temperature difference between two sides of hydrophobic micro-porous membrane acts as the driving force in the process. Kullab [5] experimentally and numerically analyzed the performance of air-gap membrane desalination (AGMD) system produced by SCARAB development AB for utilization in cogeneration power plants. Burrieza et al. [6] experimentally investigated the performance of AGMD modules for different flow rates and temperatures on the hot and cold side. The potential of solar thermal-driven MD technology as an alternative to the conventional desalinations systems has been researched by several authors. In this chapter, we experimentally analyzed the performance of a flexible solar thermal polygeneration system (STP) to produce simultaneous cooling, pure water, and domestic hot water (DHW) for the weather conditions of UAE. The system is operated for providing cooling for the office cabins in CSEM-UAE during the summer season.

71.2 Description and Integration of Polygeneration System

The schematic sketch of solar thermal polygeneration system investigated in the chapter is shown in Fig. 71.1. The STP system is developed by integrating evacuated tube collectors, single-stage absorption chiller, and membrane distillation unit.

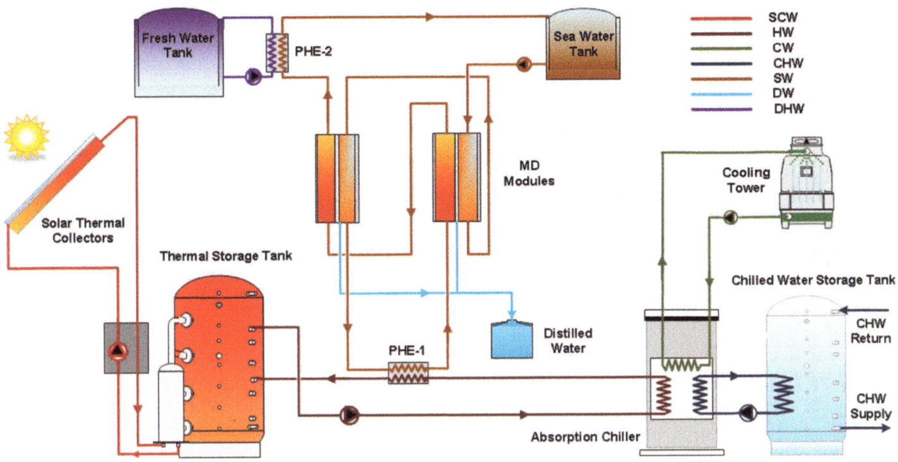

Fig. 71.1 Schematic layout of solar polygeneration system

The system is designed and developed as a flexible polygeneration system in order to operate in different modes for analyzing individual and combined processes. The semicommercial membrane distillation modules developed by SCARAB development AB are utilized in this research.

The evacuated tube collector field collects the thermal energy from solar radiation. The collected thermal energy heats up the fresh water stored in thermal stratified tank, which is then used to drive the polygeneration unit. The hot water from the thermal storage tank is supplied to the absorption chiller for initiating the chilling process. The chilled water produced by the chiller is stored in cold thermal storage tank, which is distributed to the office cabins of CSEM-UAE. The hot water return from absorption chiller is utilized as the heat source for heating the saline water supplied to the membrane distiller using heat exchanger PHE1. The sea water is prefiltered by two micro-filtration cartridges before filling into sea water storage tank. The sea water at ambient temperature is supplied to cold side of MD module, where it gets preheated through internal heat recovery by conduction and latent heat of condensation from hot side of MD and distillate channels, respectively. The preheated sea water exits the MD modules and further gets heated with the heat absorbed from the heat exchanger (PHE1). The hot saline feed water enters the hot side of the MD module where it gets evaporated, passes through a hydrophobic membrane, and condenses on cold side to produce pure distilled water. The hot brine leaving the MD system is passed to a second heat exchanger (PHE2), which extracts the thermal energy for simultaneous production of domestic hot water (DHW). Photographs of the solar thermal polygeneration system installed at CSEM-UAE are shown in Fig. 71.2. The technical data of absorption chiller and membrane distillation systems are shown in Tables 71.1 and 71.2.

Fig. 71.2 Solar thermal polygeneration unit

Table 71.1 Technical data of Yazaki 10TR absorption chiller

Item	Parameter	Unit	Value
Cooling capacity		kW	35.2
Chiller water	Inlet temperature	°C	12.5
	Outlet temperature	°C	7.0
	Rated flow rate	m^3/h	5.47
	Max. operating pressure	kPa	588
Cooling water	Rated inlet temperature	°C	31.0
	Rated outlet temperature	°C	35.0
	Max. operating pressure	kPa	588
	Rated flow rate	m^3/hr	18.4
Hot water	Rated inlet temperature	°C	88
	Rated outlet temperature	°C	83
	Inlet limit	°C	70–95

Table 71.2 Technical data of membrane distillation units

Specification	Value
Membrane area	2.8 m^2
Porosity (ϕ)	0.8
Membrane thickness (b)	0.2 mm
Air-gap length (L)	1 mm
Height of the module	730 mm
Width of the module	630 mm
Thickness of the module	175 mm

71.3 Results and Discussions

The STP system was operated in four different modes to analyze the merits and demerits of integrating different processes together rather than operating individually. The experiments were conducted from 10:00 a.m. to 05:00 p.m. without any auxiliary backup heaters in the sunny weather conditions of United Arab Emirates. The experiments were conducted during peak summer days with peak global radiation varying between 800 and 850 W/m^2 as shown in Fig. 71.3a. The cooling energy produced by

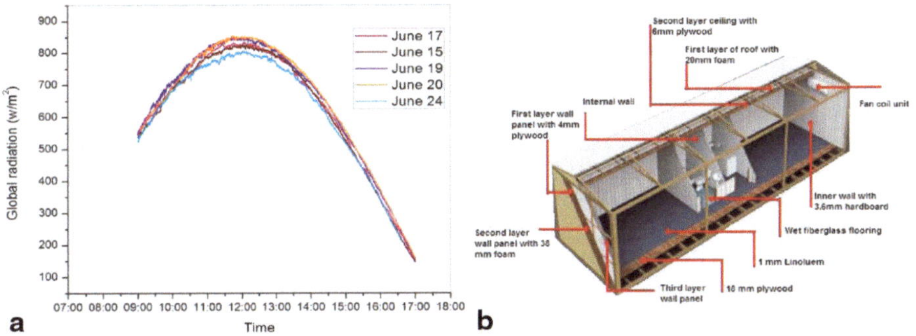

Fig. 71.3 **a** Global irradiation **b** Sectional view of an office cabin

the STP is utilized for providing air-conditioning to the office cabins of CSEM-UAE. A total of eight fan coil units are installed among three office cabins with a total floor area of 91.75 m^2. The sectional view of an office cabin is shown in Fig. 71.3b. The building is constructed with plywood and insulated with thick layer of foam.

(i). Solar cooling

The first mode of operation of the STP system is complete cooling, and it is operated on typical summer days during which both the desalination and domestic hot water supply processes are not operated. The temperature and energy profiles of this mode are shown in Figs. 71.4 and 71.5. As shown in Fig. 71.4, the collector outlet temperature gradually increases from 8:30 A.M. during the charging process of stratified tank. The operation of absorption chiller starts at 10:00 a.m. as the tank top temperature reaches more than 88 °C. Due to steep decrease in hot water supply temperatures, fluctuations in production chilled energy and COP are obtained during the first hour of operation. Throughout the day, the COP of system varies between 0.55 and 0.62 as shown in Fig. 71.5. The refrigeration capacity stabilizes at 25 kW for most parts of the day.

(ii). Cogeneration of distilled water and domestic hot water

In this mode of operation, thermal energy from solar collectors is utilized completely for the MD unit to produce distilled water and the waste heat from MD is recovered for producing hot water for domestic applications. Figure 71.6 shows the temperature profiles and productivity of two-stage MD operation. Saline feed water at conductivities greater than 65,000 μS/cm was distilled using two AGMD modules to produce distilled water at conductivities less than 50 μS/cm and the heat recovered from MD hot side is used to produce DHW at an average of 55 °C. With a feed flow of 1140 l/h, the productivity reached a maximum of 12.5 l/h during noon time and a total of 80 l of pure water is produced for 7 h of daily operation as shown in Fig. 71.7b. Gain-to-output ratio (GOR) is the performance evaluation parameter commonly used in membrane distillation systems. Overall, GOR of two-stage systems is 0.7, which is much higher compared to single-stage system. Compared to single-stage operation, 45 % more productivity is obtained for two-stage process

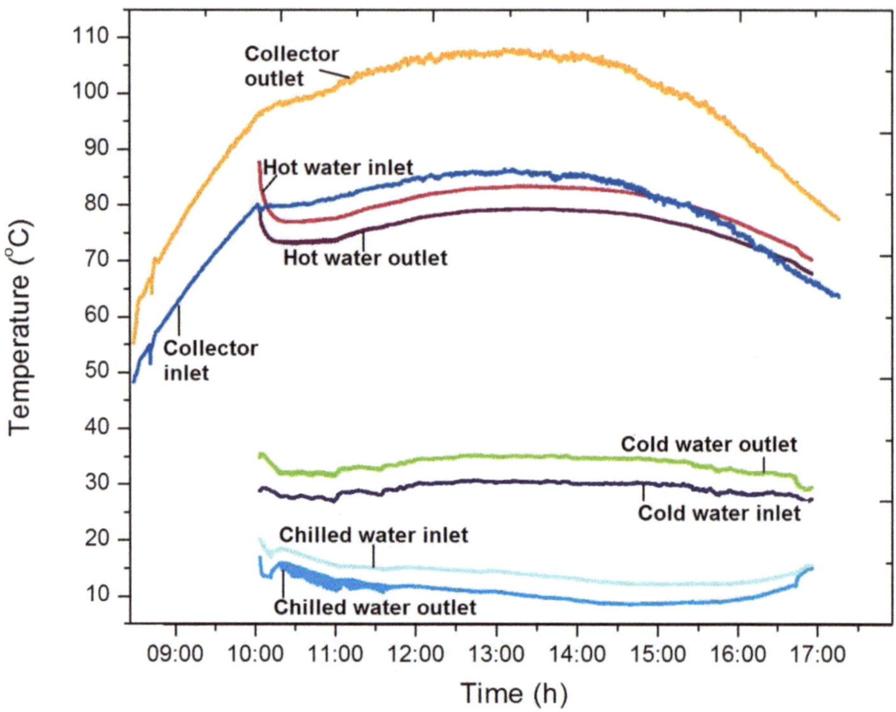

Fig. 71.4 Temperature profile of solar cooling mode

and also heat could be recovered effectively from MD hot side for DHW production. As shown in Fig. 71.7b, from the total useful energy, 50 % energy has been utilized for this mode of operation from which 75 % is recovered for DHW production and the remaining is used for distillate production.

(iii) Trigeneration

This mode is a combination of complete cooling and cogeneration modes. Figures 71.8 and 71.9 summarize the performance of absorption chiller and MD+DHW processes, respectively. As shown in Fig. 71.8, the chilled water is produced at an average of 14 °C when the hot water inlet temperature varies between 70 and 75 °C during the day. Mean chilled energy production of 14 kW is achieved in this mode, which is sufficient to fulfill the cooling demand of two office cabins with COP varying between 0.45 and 0.50. By integration of MD+DHW process with the cooling process, the whole system turns into trigeneration mode. This mode has an advantage of utilizing the total available energy effectively to produce DHW along with cooling and pure water production. In this mode, single-stage membrane distillation unit is integrated with the system, leading to a mean hourly production of 4 l/h with an average delta T of 30 °C between hot and cold sides of MD. As shown in Fig. 71.9, an average energy of 30 kW is consumed by MD process, out of which 25 kW is recovered for domestic hot water production at a mean temperature of 55 °C. In this case, 170 W of thermal energy is used for 1 kg of pure water produc-

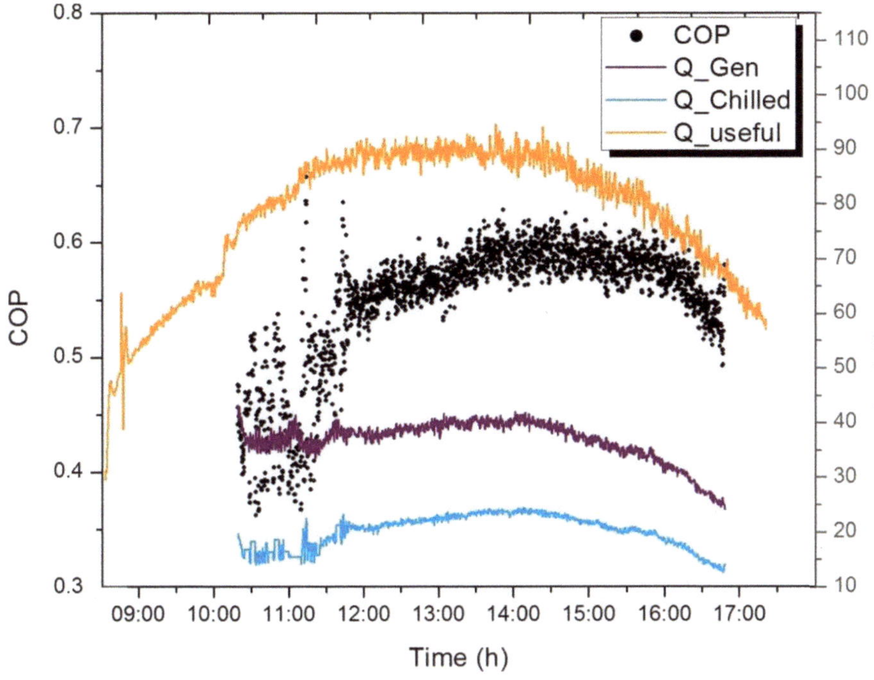

Fig. 71.5 Energy flows of solar cooling mode

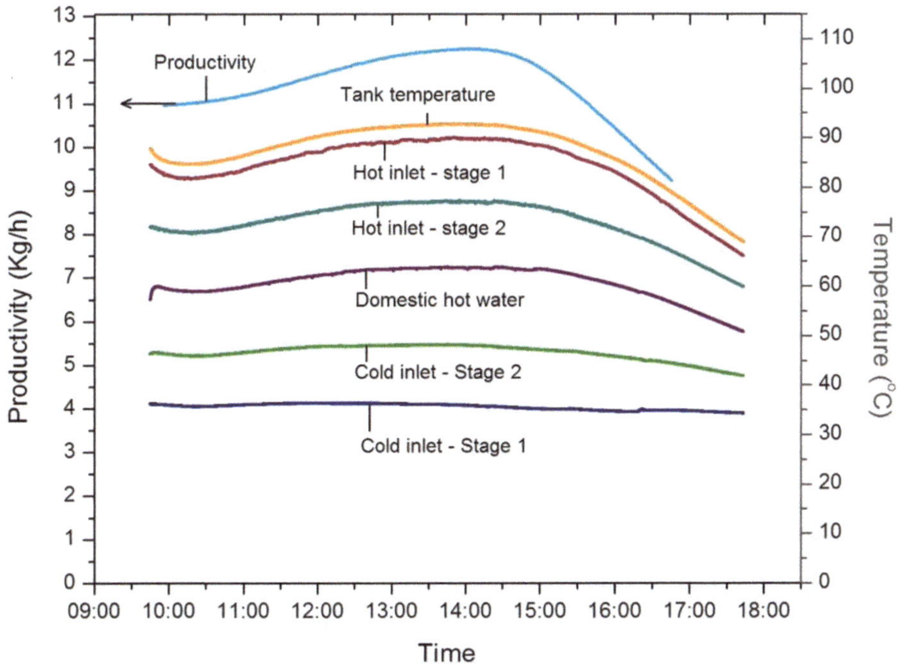

Fig. 71.6 Productivity and temperature profiles distilled water and DHW cogeneration mode

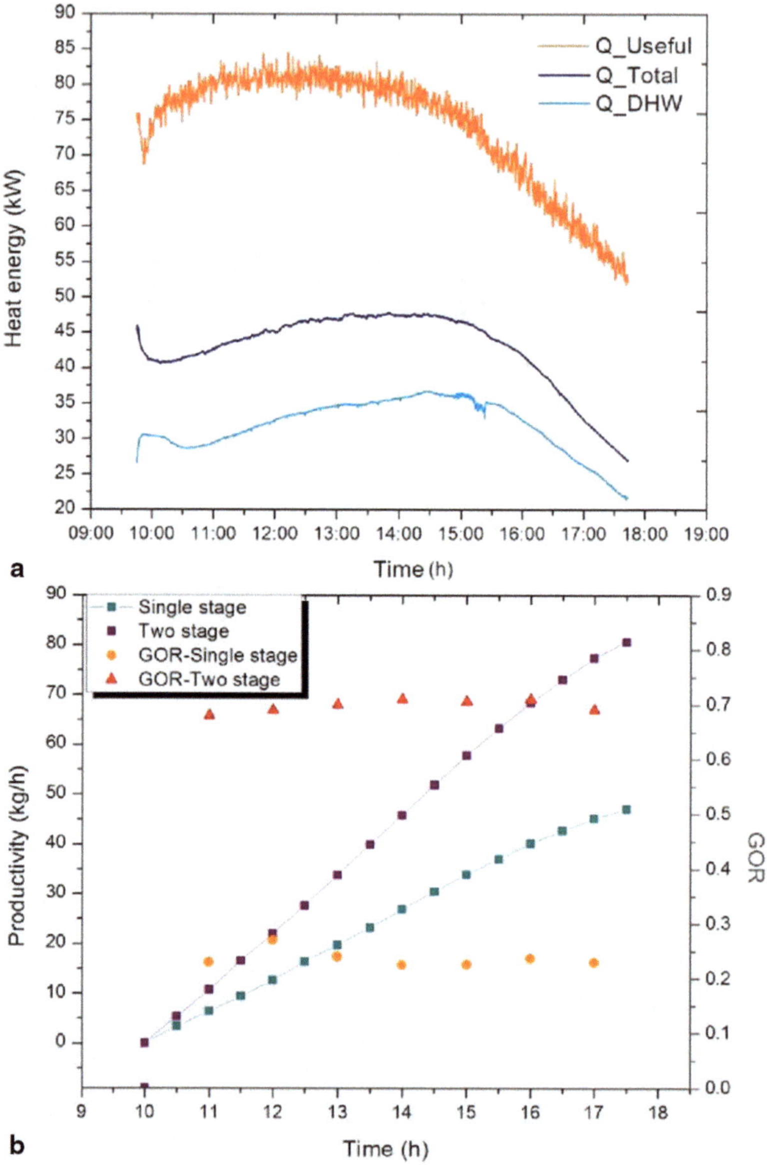

Fig. 71.7 **a** Energy flows in membrane distiller **b** Cumulative productivity and GOR

tion which is five times less than the reported values in the literature [6] without heat recovery option.

(iv) Cogeneration of cooling and desalination

The fourth mode of operation includes a two-stage MD module to increase the pure water production rather than obtaining sufficient temperatures for DHW during heat recovery through single stage. Figure 71.10 shows the performance of two-stage

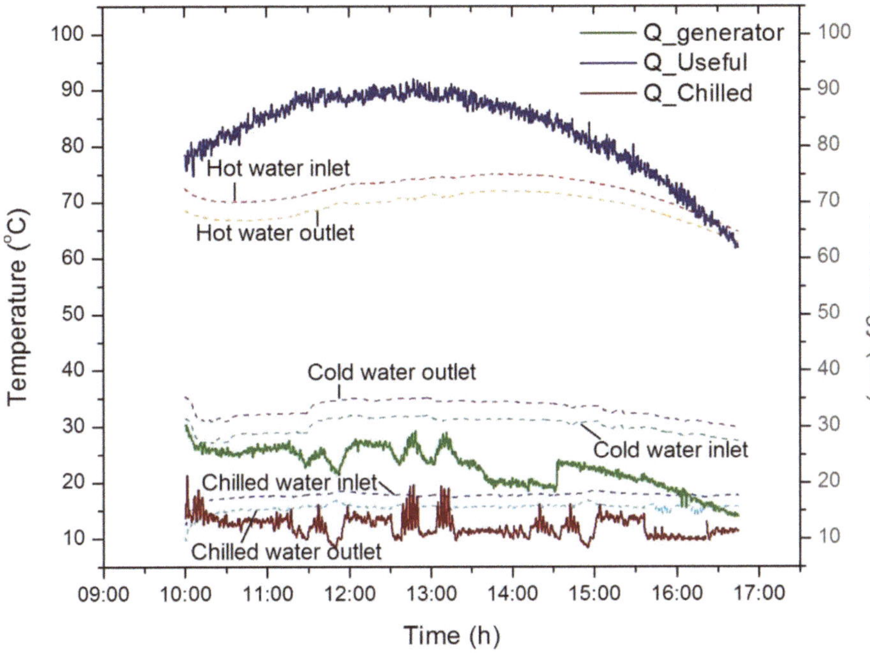

Fig. 71.8 Temperature and energy profile of the absorption chiller in trigeneration mode

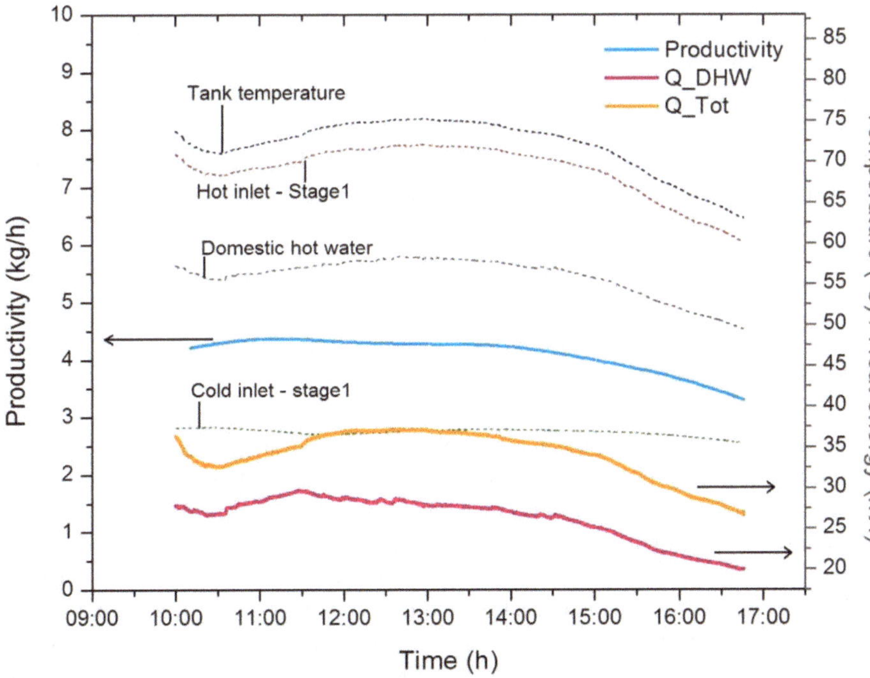

Fig. 71.9 Temperature, energy flows, and productivity of MD system in trigeneration mode

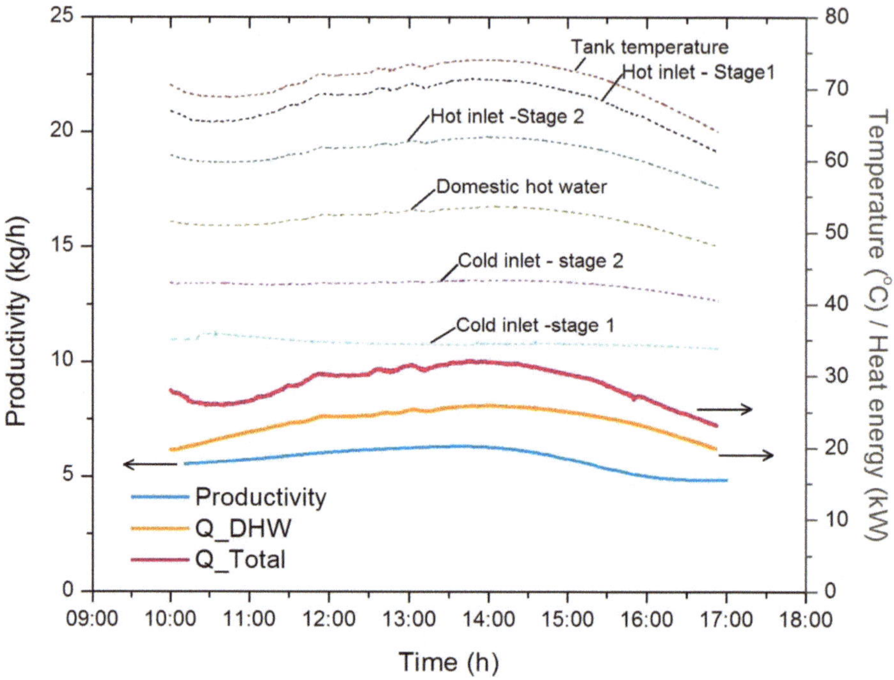

Fig. 71.10 Temperature, energy flows, and productivity of MD system in mode (iv)

MD and DHW system in trigeneration mode. The performance of solar cooling remains similar to the earlier mode of operation. However, compared to the earlier mode of operation, an average of 2 kg/h of distilled water is produced from the two MD modules. Since the DHW could not be obtained at sufficient temperatures (average of 50 °C), this mode could be termed as cogeneration of cooling and desalination instead of trigeneration. This mode is particularly useful in summers during which DHW is not required at high temperatures and also pure water requirement is higher.

71.4 Conclusions

A solar thermal-driven polygeneration system has been developed with a flexibility of operating it in different configurations. The main focus of this chapter is to analyze the advantages of combining different processes together rather than operating individually. Experiments have been carried out on the system consisting of evacuated tube solar thermal collectors, absorption chiller, membrane distillation unit, and heat exchangers for heat absorption and recovery. Operation in different modes shows that cogeneration of pure water and DHW is useful in winter when cooling is not needed, and trigeneration operations for the remaining time expect the summer during which cogeneration of cooling and desalination is more beneficial. In trigeneration mode, heat could be recovered effectively from MD process for DHW production, thus reducing overall energy consumption by MD for kilogram of water production. However, from application viewpoint, capacities of different processes are needed to be optimized to fit into reasonable application.

Acknowledgments The authors would like to extend their gratitude to the Ras Al Khaimah Investment Authority for their financial support. They would also like to thank to KTH Royal Institute of Technology, Stockholm, for the fruitful cooperation in the project.

References

1. Kazim AM (2007) Assessments of primary energy consumption and its environmental consequences in the United Arab Emirates. Renew Sustain Energ Rev 11:426–446
2. Hussain HJ (March 2010) Development of a hybrid powerplant for Kuwait: the simultaneous production of power, fresh water and cooling, Doctoral thesis. Cranfield University, United Kingdom
3. Picinardi A (2011) Cogeneration of cooling energy and fresh water. Doctoral Thesis, University of Bergamo, Italy
4. Calise F, d'Accadia MD, Piacentino A (2014) A novel solar trigeneration system integrating PVT (photovoltaic/thermal collectors) and SW (Seawater) desalination: dynamic simulation and economic assessment. Energy 67:129–148
5. Kullab A (2011) Desalination using membrane distillation: experimental and numerical study. Doctoral Thesis, KTH Royal Institute of technology, Sweden
6. Burrieza EG, Blanco J, Zaragoza G, Alarcón DC, Palenzuela P, Ibarra M, Gerjak W (2011) Experimental analysis of an air gap membrane distillation solar desalination pilot plant system. J Membr Sci 379:386–396

Chapter 72
Solar Thermal Systems for Zero-Energy Buildings: Perspectives and Challenges

Agis M. Papadopoulos

Abstract The implementation of energy-efficiency measures as well as the integration of renewable energy systems, on the move towards the zero (or nearly zero)-energy building is certainly a challenging task. This is even more the case, when buildings are to operate in conditions with increased cooling demand, as it happens in the Eastern Mediterranean, the Middle East and North Africa. The aim of a zero-energy building cannot be achieved without a truly integrated energy design approach, by a combination of two major tools: (a) avoiding, postponing or reducing the generation of heating and cooling loads, by applying the basic principles of building physics and (b) using alternative, renewable sources and systems to produce the heating and refrigeration necessary.

Understanding the requirements of the user and providing adequate solutions remains the single most important prerequisite, if the design had to be successful and just a simple academic exercise. Targeted, interdisciplinary postgraduate courses are needed, with curricula focusing on the utilization of state-of-the-art technology, but also with the holistic approach in the core of their educational attitude, in order to provide architects and engineers with the expertise needed.

Keywords Solar thermal systems · Energy design · Integration · Education

72.1 Introduction

The European Directive on the Performance of Buildings recast (2010/31/EC) is the most recent in a long series of regulatory actions, aiming at the improvement of building's energy behaviour. This necessity to improve the buildings' energy performance became peremptory during the two oil crises in the 1970s, and was

---

A. M. Papadopoulos (✉)
Process Equipment Design Laboratory Department of Mechanical Engineering, Aristotle University Thessaloniki, 54124 Thessaloniki, Greece
e-mail: agis@eng.auth.gr

A. Sayigh (ed.), *Renewable Energy in the Service of Mankind Vol II*,
DOI 10.1007/978-3-319-18215-5_72

expressed in the effort to reduce the demands for heating, ventilation and air-conditioning, without endangering the living standards of the day. However, the phenomenon is neither new nor one-dimensional. In the recent, but deceptively easily forgotten first half of the twentieth century, drastic energy conservation measures were applied affecting the economic and social life in most European countries. At the end of the twentieth century, and especially during the rather careless, at least as far as energy prices were concerned, 1990s, most of the energy conservation actions taken, both on a national and on an international level, had their origins rather in environmental than in purely energy saving motives, an approach that is being reviewed due to the volatility and the overall increase in energy prices after 2003. Finally, the impact of establishing satisfactory indoor air quality conditions had, throughout the twentieth century, a tight relation to the ventilation of buildings and hence to their energy behaviour.

The main conclusion to be drawn from this brief and rather incomplete list of events and developments is that regulating the energy behaviour of buildings has been a goal certainly predating the volatility of the energy markets and has become the driving force behind all actions taken after 1973. The increased environmental concern and sensitivity of the recent years, combined with the technological development in solar energy use and the continuous urbanization, formulate a new framework in which new priorities and considerations arise for the building sector. The rapid industrialization and urbanization occurred during the second half of the twentieth century lead to a dramatic increase of the cities' size and also of their density, consequently leading to an increase in the respective energy consumption. Thus, any aspect of energy efficiency of buildings cannot ignore the particular requirements and constraints of both the urban buildings and the urban microclimate and environment. In this line of thought, Article 9 of the EPBD recast 2010/31/EU foresees that EU Member States shall ensure that (a) by 31 December 2020, all new buildings are nearly zero-energy buildings and (b) after 31 December 2018, new buildings occupied and owned by public authorities are nearly zero-energy buildings.

It is fairly clear that the aim of a zero-energy building cannot be achieved without a truly integrated energy design approach—and also its consequent implementation. Achieving the aim of a zero-energy building can be dealt with by a combination of two major tools: (a) avoiding, postponing or reducing the generation of heating and cooling loads, by applying the basic principles of building physics. This implies the implementation of sound sun-protection schemes, the use of thermal insulation and high-performance glazings, the use of reflective and low-absorbing materials on the building's envelope, reasonable ventilation patterns and the reduction in internal thermal loads production, using high efficiency heating ventilation and air-conditioning systems and utilizing state-of-the-art building automation and control systems and (b) using renewable sources to produce the remaining heating and refrigeration necessary.

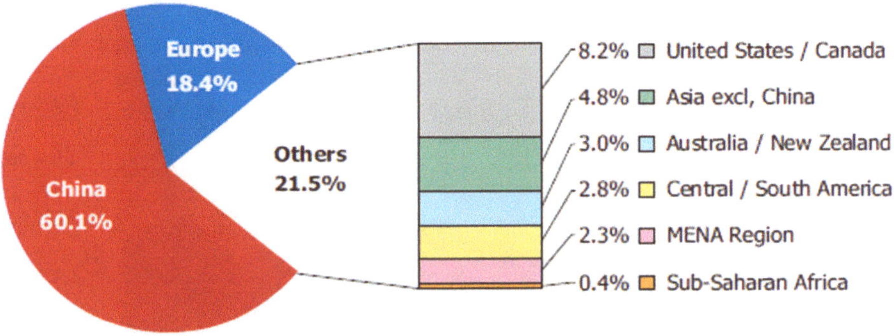

Asia excluding China: India, Japan, Korea South, Taiwan, Thailand
Central / South America: Barbados, Brazil, Chile, Mexico, Uruguay
Europe: Albania, EU 27, Macedonia, Norway, Switzerland, Turkey
MENA Region: Israel, Jordan, Lebanon, Morocco, Tunisia
Sub-Saharan Africa: Namibia, South Africa, Zimbabwe

Fig. 72.1 Installed solar capacities in operation in the end of 2010 [1]

72.2 Solar Thermal Market in Europe

Solar thermal markets, in particular domestic hot water systems, have shown an extremely positive evolution, during the past 20 years. Europe was for at least two decades the leading region globally, with China taking the lead after 2000, as shown in Fig. 72.1.

The European market can be divided into three zones with respect to annually newly installed capacities of glazed collectors: markets above 400,000 m^2, between 400,000 and 200,000 m^2 and below 200,000 m^2. The overall European reliance on Germany is decreasing with Austria, France, Greece, Italy and Spain together accounting for 39 % of the overall European solar market; the other countries now representing 23 % of the market and becoming increasingly relevant, showing a clear trend for fast growth [2] (Figs. 72.2, 72.3, 72.4).

72.3 Solar Thermal Applications in Greece

Greece has significant solar potential and is amongst the most successful countries worldwide in the use of solar thermal energy. The application of active solar systems in Greece started systematically in the mid-1970s. The use of electric heaters in almost every household, in combination with the oil crisis, and the rising price of electricity during this period provided the background for the solar market to develop. Several promotion schemes led towards this direction, at least until 2006; advertising campaigns supported by the Greek government, low-interest loans and

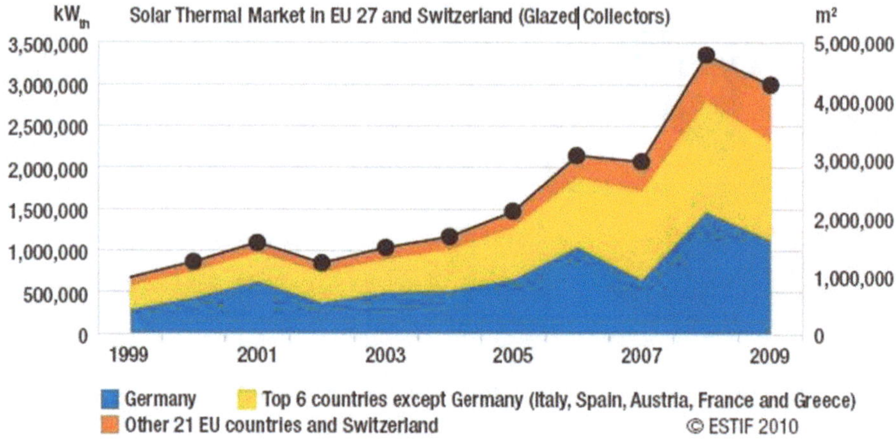

Fig. 72.2 Installed capacity in kW_{th} and area in m^2 for the three zones of the solar thermal market in EU and Switzerland

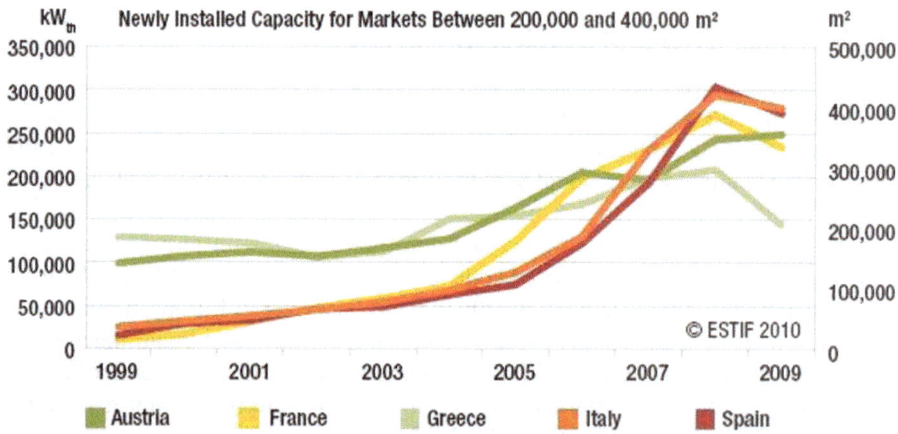

Fig. 72.3 Newly installed capacity in kW_{th} and area in m^2 for the European markets between 200,000 and 400,000 m^2

tax deductions, national standards helped in improving the efficiency and reliability of these products. As a result, Greece ranked for many years in the second place with respect to the per capita-installed thermal systems capacity, with Cyprus being the leader. It is only after 2006 that Austria moved up to the second place, after adopting a generous incentives' scheme [2, 3].

The application of solar systems for satisfying needs in heating and cooling is a strategic option of great importance for the Greek energy system and for the domestic sector and the electricity system in particular. It is also an important choice for the national economy in total, since a very dynamic and competitive industry con-

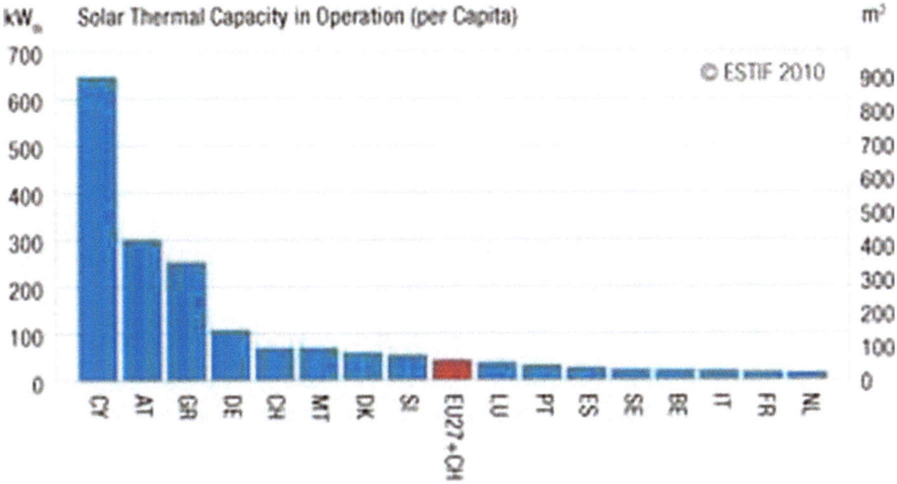

Fig. 72.4 Per capita-solar thermal-installed capacity in Europe

cerning the manufacturing of solar systems has already been developed and needs to be supported.

After 2009, the Greek market has contracted—by almost one third, from approximately 300,000 to 206,000 m^2 and reverted to the level existed in 2004. Following several years of strong growth, the solar market was expected to undergo some form of readjustment; however, this was much more drastic than anticipated, intensified by the financial and economic crisis.

The recession of the Greek economy was mirrored by a contraction of almost 4.2 %, a situation not expected to improve before 2013. With respect to solar systems, the support schemes available for energy-efficiency measures and replacement of older heating equipments proved to be of some help. Together with the steep increase in energy prices due to taxation (oil retail prices almost doubled between 2009 and 2013), a modest annual average growth of 4 % in solar systems sales has been recorded after 2011.

Furthermore, having reached a certain level of expertise and quality, and facing difficulties in the home market, most of the solar systems' manufacturers turned to product development and to exports. The success by the pioneers has motivated other companies as well. By 2011, more than 45 % of the solar collector production of the Greek Solar Industry Association (EBHE) members was exported, compared to less than 5 % in 1991 [4]. Since the beginning of the depression in 2009, exports have covered, at least to some extent, the slump in the domestic market, as shown in Fig. 72.5.

In the history of the Greek solar thermal market, the main solar thermal product was, and still is, the thermosiphonic water heater. The breakdown of solar thermal applications in Greece, according to the European Solar Thermal Industry Federation—ESTIF, showed that almost 98 % of the installed systems are thermosiphonic-type water heaters with closed loop systems which use antifreeze liquid to avoid the

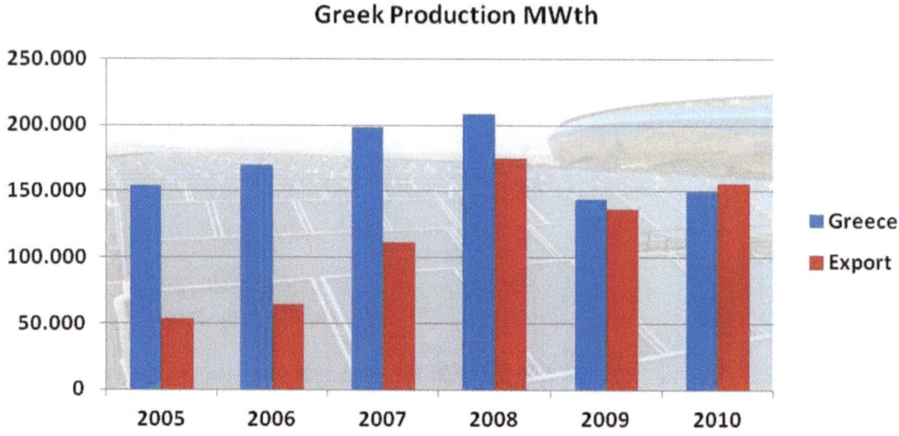

Fig. 72.5 Development of the Greek solar thermal production

freezing of the collector loop. The domination of this system has remained unchallenged over the years. The storage tank can be vertical or horizontal and is mounted higher than the collectors. The average size of thermosiphonic-type systems is 2.4 m^2 collector area and 150 l storage tank. The range of most commonly used systems is between 120 l and 220 l with 1.8–4 m^2 collector surface area. Solar water heating (SWH) systems are designed to deliver the optimum amount of hot water for most of the year. However, in winter, there may sometimes not be adequate solar heat gains to deliver sufficient hot water, particularly in Northern Greece, where the solar radiation and sunshine hours figures are lower [4, 5]. Therefore, all systems are equipped with electric backup heaters and 30 % of the systems are equipped with an additional heat exchanger connected to the fuel or gas-fired boiler of the heating system.

Central solar heating can involve large-scale thermal storage, reaching from diurnal storage to seasonal thermal storage. Central solar systems can also be used for solar cooling in the form of district cooling. In this case, the overall efficiency is high due to the high correlation between the energy demand and the solar radiation. Central systems in Greece cover mostly the need for hot water production in hotels, whilst they are a rather rare sight in residential buildings, where the independence of the thermosiphonic system is a highly rated quality.

Other technologies became popular in the last decade. A solar combi-system provides both solar space heating and cooling, as well as hot water from a common array of solar thermal collectors, usually backed up an auxiliary conventional heat source. Solar-assisted air-conditioning (SAC) technologies require solar heat to drive the cooling process and have proved their efficiency and reliability during the last decade. SAC systems use harmless water-based cooling fluids and much less primary energy than the conventional systems. SAC systems can be used, either as stand-alone systems or with conventional AC, to improve the indoor air quality. Additionally, they cooperate with the already-existing conventional indoor installations [6]. SAC systems installed so far may be classified into closed and open systems. Closed systems are thermally driven chillers, which provide chilled water that is either used in air-handling units to supply conditioned air or is distributed via a chilled water network to the designated rooms to operate decentralized room installations. Technically mature machines for this purpose are mostly absorption chillers and rarely adsorption chillers, which are a few hundred machines worldwide, but of rising interest in SAC. Open systems allow a complete air conditioning by supplying cooled and dehumidified air according to the comfort conditions. The "refrigerant" is spray water, which is in direct contact with the atmosphere. Most common systems are desiccant cooling systems using a rotating dehumidification wheel with solid sorbent [6, 7].

The majority of solar thermal applications in Greece is limited to producing domestic hot water and has not yet been expanded significantly to the area of space heating and cooling. Large collective solar systems are installed mainly in hotels for hot water production and account for almost 1 % of the installed collector area. Finally, space heating, district heating, air conditioning and industrial process heating combined have less than 1 % of the installed collector area [4]. This can be explained by several problems that are identified in the existing support scheme. It is observed that the support scheme existing from 1980 till 2006 favoured only applications such as thermosiphonic solar thermal systems, whilst there is no differentiation of the amount of economic support depending on the type of technology. Additionally, the amount of the tax deduction in the case of central solar systems, combi-systems or solar cooling is negligible and does not give a real incentive to choose such a technology. Investment subsidies in the tertiary and industrial sectors are based on calls and are not available on a constant basis. Another reason for the small penetration of solar thermal applications for space heating and cooling in the market is the fact that such systems require properly trained staff with technical skills, as far as their design, installation and maintenance are concerned. Unfortunately, only few of the market stakeholders have invested in this field, in order to be able to implement such projects [2].

Still, and despite those hurdles, some very interesting demonstration projects have been carried. The most well known is "Solar Village 3", which refers to a settlement of the Greek Worker's Housing Organisation located in the outskirts of Athens, in Lykovrissi; it was built as part of a joint Greek–German demonstration project for solar technologies and was finished in 1989. The settlement has approximately 1700 inhabitants and occupies a total built area of 90.440 m^2; it consists of 25 buildings with 435 dwellings, a solar information centre, a community centre, shops, a cafeteria, a library and the necessary services. The settlement and the buildings have been designed by architect Alexandros Tombazis. The active systems used in the buildings supply energy for space heating and domestic hot water. There are both autonomous and central systems. Thirty-two apartments are without any conventional heating systems (the so-called passive ones), whilst the other ones are served by some of the eight different types of flat plate, vacuum tube and air collectors, an interseasonal heat storage tank of 563 m^3, as shown in Fig. 72.6, a water to water heat pump and oil-fired boilers can work well.

Fig. 72.6 Partial view of Solar Village 3 with the interseasonal storage tank in the front

The project demonstrated that even innovative solar systems can perform well for a period of over 15 years. Unfortunately, it also showed that as soon as the appropriate maintenance was curtailed, the systems deteriorated fast [8–10].

Another, newer interesting project is the solar air—conditioning system at the Center for Renewable Energy Sources (CRES) in Athens. The high solar fraction plant, which supplies space heating and cooling, is installed at an existing office building (426.6 m^2), at the site of CRES and is in operation since December 2011. The solar contribution to the building requirements is designed to reach values over 80 %. The plant design includes the solar thermal collectors (149.5 m^2), the underground thermal energy storage (58 m^3), the absorption cooling machine (35 kW), the cooling tower and the heat pump. In the heating period, the building energy requirements are covered mainly by solar gains coming directly from the solar collectors or the thermal storage. When the hot water's temperature is below 45 °C, the heat pump rises the temperature. The operational principle is depicted in Fig. 72.7.

As the heat pump is driven by solar energy, it operates with increased efficiency. The thermal storage smoothens the time lag between maximum solar radiation and building energy requirements both in a short- and a long-term sense. During low-demand seasons, such as autumn and spring, a large amount of thermal energy is stored, to be recovered at the following heating or cooling period. The performance of the thermal storage is improved through the integration of a heat pump during heating. In cooling period, the building's energy requirements are covered by the absorption chiller, which, driven by hot water at temperatures over 65 °C, produces chilled water at 7 °C. This energy emanates from the collectors when the radiation

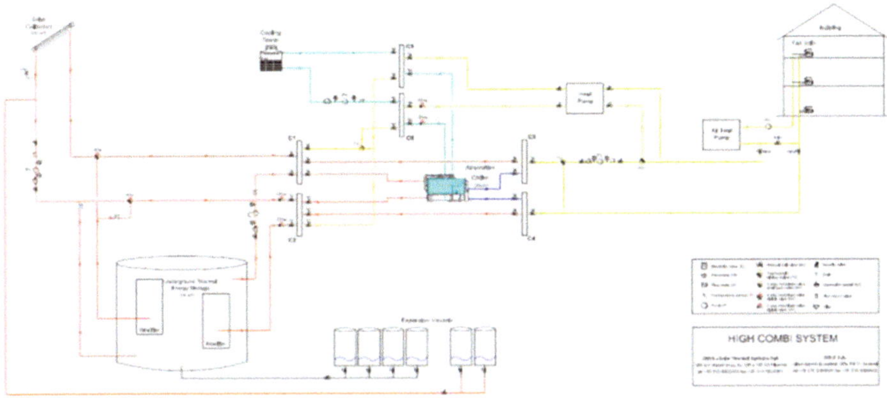

Fig. 72.7 Operational principle of the CRES High-Combi System [11]

is sufficient or from the thermal storage in low-radiation periods, such as morning and evening hours or cloudy hours of the day [11].

There are also several projects for process heating in the dairy, wine, textile dyeing and finishing, rice drying and tannery industry, but the fall beyond this chapter's scope. Still, it has to be noted that solar systems can provide significant savings in those production processes [12].

72.4 Steps into the Future

At the end of the day, it is the integration of the various elements of the building's design that lead to success. The design of the building's envelope is crucial for achieving an indoor space, which effectively responds to environmental changes and is able to reduce their impact. At the same time, an optimized envelope design can ensure significant energy savings in buildings. This is something that can be effectively achieved in the initial stages of the building's design, as it is depicted in Fig. 72.8.

On this basis, the effective thermal protection of the envelope is crucial for allowing a cost optimal use of low-temperature renewable systems and for achieving a high overall energy efficiency of the building. As the dynamic thermal characteristics of the building, namely the time lag and the decrement factor, are affected by the heat storage capability of building elements as well as by the location of thermal insulation. The compatibility between the heat distribution system and the heat generation system is obviously also a dominant factor. The use of low-temperature systems is almost a prerequisite for the successful use of solar or geothermal energy, a case which is clearly demonstrated by the success of underfloor heating systems combined with solar or geothermal-driven hot water systems in Central and Western Europe. One has, however, to keep in mind that such solutions cannot address the

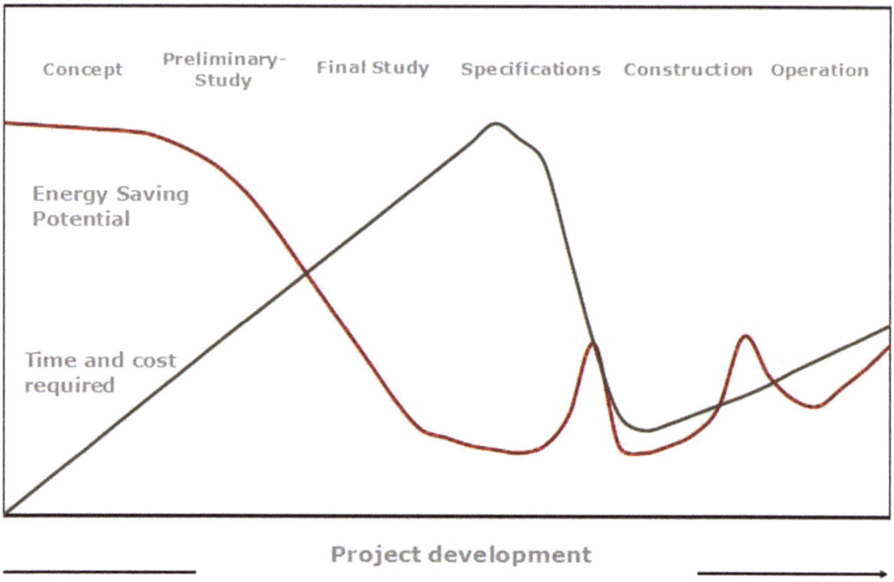

Fig. 72.8 On the effectiveness of incorporating solar (and other RE) systems

cooling and air-conditions problems that dominate the building's energy behaviour in the Mediterranean or the Middle East region.

The problem, however, is much more complicated than this: Ensuring the efficiency and durability of complex, integrated solar systems are not always easy; they call for meticulous design, careful construction and appropriate maintenance. The economics of such systems are also complicated: They depend on their cost—and on the incremental cost they represent—with respect to the purchasing power of the final consumer, but also to the cost of energy conserved, and this is something that varies with respect to conventional energy pricing. A different energy pricing policy is needed, considering the value of energy conserved, and this can only be done if the mid-term marginal cost of power generation is mirrored. They also depend on the capital cost for the investor and on the existence of specific financing tools. This complex interdependence is depicted in Fig. 72.9.

In any case, targeted policies are needed to achieve the goal of zero, or even nearly zero, energy buildings, which include amongst other:

a. The implementation of the existing codes and regulations and the development of more effective regulations that will mirror realities but set ambitious goals. This is clearly a task for the state.
b. Improved and certified expertise is needed for the design and the construction of systems. This is a call for the education of engineers, and hence for the universities, and also for the continuous education and vocational training, which has to be considered in cooperation with professional associations and chambers.

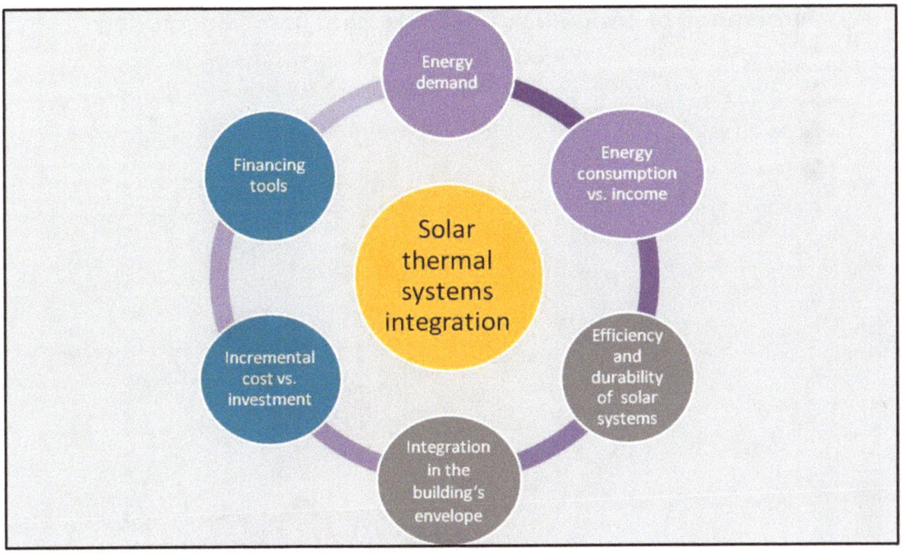

Fig. 72.9 Parameters affecting the integration of solar thermal systems

Considering the aspects of the engineers' education, it is clear that a continuous assessment is needed, followed by a revision and/or adaptation of the courses' syllabi, so that they mirror the truly interdisciplinary nature of the energy design of buildings' scientific field. Such a syllabus may include amongst other elements of architectural design, building construction, thermodynamics, heating, ventilation, air-conditioning, automation and control, investment appraisals, financing, etc. In order to achieve this, the training of future engineers and architects has to become as interdisciplinary as possible, considering traditional elements of architectural design (for the engineers) and thermodynamics (for architects), but also more fields such as materials' science (nanotechnologies), production management (lean production techniques) and logistics to improve the efficiency and cost-effectiveness of systems for everyone.

72.5 Concluding Remarks

During the past 20 years, renewable energy systems for building applications have shown a remarkable evolution. The market has reacted to this, and applications for hot water production and, to some extent, space heating and cooling have become very popular. There have been solar systems working successfully for more than 30 years now, and the solar market is considered to be a mature one, at least in Europe.

If, however, we want to achieve a zero-energy building status, then we need a truly integrated approach. This has to include "classical" solar thermal applications

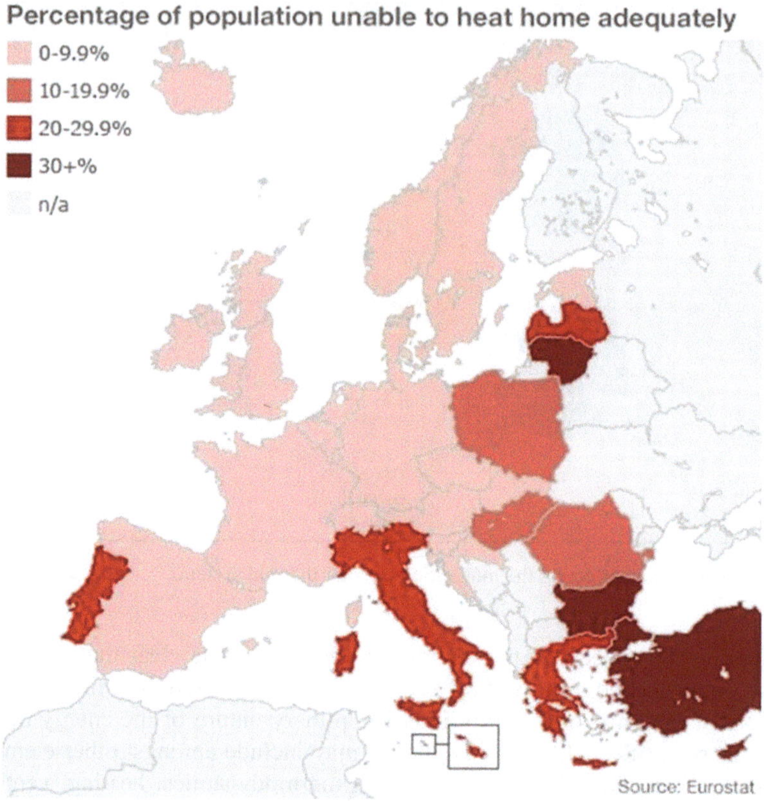

Fig. 72.10 Percentage of population unable to heat their home in 2013 [13]

as hot water production and space heating, but it also has to include space cooling and eventually to be extended to domestic white appliances. In order to achieve this, we need to overcome barriers of not all technical nature. Legislative and regulatory aspects have to be dealt with, in order to produce the necessary framework for the solar systems. Still, in order to capitalize on the technological advance, we need a truly integrated building's energy design, as early as possible; ideally from the conceptual design stage.

Eventually, European cooperation has to be promoted even stronger, to foster the transfer of expertise, as there are interesting differences in the level of progress in the various fields of energy design and of integration of renewable energy systems (Fig. 72.10).

Finally, we cannot neglect the socioeconomic aspects: With energy poverty in Europe reaching record high levels, as shown by latest Eurostat data, providing affordable energy for households is more than an academic exercise; it is a matter of social justice if not of survival.

Acknowledgment The author wishes to acknowledge the European Commission for financing the Energy Efficiency, Renewable Energy and Environmental Impacts master study project ENERESE, JPCR 530194-2012, within the TEMPUS programme.

References

1. Weiss W, Mauthner F (2012) Solar heat worldwide, Edition 2012, AEE INTEC, Austria
2. European Solar Thermal Industry Federation (2011) Solar thermal markets in Europe, trends and market statistics 2010. ESTIF
3. Steinbach J, Ragwitz M (2010)First consultation workshop on A common framework for a coordinated/harmonized policy on renewable heating, report part of the IEE project "Policy development for improving RES-H/C penetration in European Member States (RES-H Policy)"
4. Tsilingiridis G, Martinopoulos G (2010) Thirty years of domestic solar hot water systems use in Greece—energy and environmental benefits—future perspectives. Renew Energy 35:490–497
5. Drosou V, Karagiorgas M, Travasaros C (2011) The Greek solar thermal market and industrial applications, overview of the market situation. Proceedings of world sustainable energy days 2011, conference "Solar Process Heat". Wels, Austria
6. Oxizidis S, Papadopoulos AM (2008) Solar air conditioning: a review of technological and market perspectives. Adv Build Energy Res 2:123–158
7. Eicker U, Schneider D, Schumacher J, Ge T, Dai Y (2010) Operational experiences with solar air collector driven desiccant cooling systems. Appl Energy 87:3735–3747
8. Sotiropoulos VA et al (1992) Final report on the evaluation of active solar systems of Solar Village 3, Aristotle University Thessaloniki, Thessaloniki, Greece
9. Papadopoulos MA et al (1992) Final report on the evaluation of passive solar systems of Solar Village 3, Aristotle University Thessaloniki, Thessaloniki, Greece
10. Tsilingiridis G, Sotiropoulos V (1998) Three-year measurements of ambient and indoor air temperatures in Solar Village, Athens, Greece. Energy Build 28:127–136
11. Tsekouras P, Drosou V, Antonopoulos K, Karytsas C (2012) An innovative high solar fraction heating and cooling plant in Athens—Control strategy and initial measurements, Proceedings of the ASME 2012 Summer Heat Transfer Conference HT2012, July 8–12, 2012, Puerto Rico, USA
12. Tsoutsos T, Karagiorgas M, Zidianakis G, Drosou V, Aidonis A, Gouskos Z, Moeses C (2009) Development of the applications of solar thermal cooling systems in Greece and Cyprus. Fresen Environ Bull 18(7b):1367–1380
13. EUROSTAT, Statistics on population unable to heat their homes adequately. Accessed 1 Apr 2014

Chapter 73
Experimental Investigation of the Effect of Solar Collector's Inclination Angle on the Generation of Thermosiphonic Flow

G. P. Panayiotou, S. A. Kalogirou, G. A. Florides, G. Roditis, N. Katsellis, A. Constantinou, P. Kyriakou, Y. Vasiades, T. Parisis, A. Michaelides and J. E. Nielsen

Abstract Cyprus is currently the leading country in the world with respect to the application of solar water heaters for domestic applications, with more than 93 % of the houses equipped with such a system. The great majority of these solar water heaters are of the thermosiphonic type. Currently, the knowledge about the parameters affecting the 'thermosiphonic phenomenon' is rather poor while on an international level (International Organization for Standardization, ISO, and Comité Européen de Normalisation CEN committees) there is no standard available to test thermosiphon solar collectors. The deeper understanding of the 'thermosiphonic phenomenon' and the identification of the key parameters affecting it is the main aim of a research project currently in process in Cyprus.

In this chapter, the experimental results of the research project are presented. Specifically, a special test rig was set up and equipped with all the sensors necessary to measure all the parameters that are most likely to affect the 'thermosiphonic phenomenon'. All tests were conducted according to ISO 9459-2:1995(E). The system was able to operate in various weather and operating conditions and could

S. A. Kalogirou (✉) · G. P. Panayiotou · G. A. Florides
Cyprus University of Technology, 30 Archbishop Kyprianou Str., 3036 Limassol, Cyprus
e-mail: Soteris.kalogirou@cut.ac.cy

G. Roditis · N. Katsellis · A. Constantinou
Applied Energy Centre, 6 Andreas Araouzos Str., 1421 Nicosia, Cyprus

P. Kyriakou
Cyprus Institute of Energy, Agapinoros 2 & Arch. Makariou III Megaro IRIS, 1st Floor, 1076 Nicosia, Cyprus

Y. Vasiades
Cyprus Organisation for Standardisation, 30 Kosta Anaxagorou & Lemesou Av., 3rd Floor, 2014 Nicosia, Cyprus

T. Parisis · A. Michaelides
RTD Talos, 1 Diogenous Str., Kykkos' Commercial Center, Block A, 4th Floor, 2404 Engomi, Nicosia, Cyprus

J. E. Nielsen
PlanEnergi, A. C. Meyers Vænge 15, 2450 København SV, Denmark

A. Sayigh (ed.), *Renewable Energy in the Service of Mankind Vol II*,
DOI 10.1007/978-3-319-18215-5_73

accommodate the change of inclination of the collector. During the experimental procedure, three different inclination angles of the solar collector were tested in order to evaluate their effect on the generation of thermosiphonic flow. The thermal performance of the collector was calculated both in thermosiphonic operation and also according to EN12975-2:2006 in order to determine the thermal performance at a flow and operation conditions specified by the standard. Finally, a series of correlations were attempted using the experimental results for the thermosiphonic operation of the collector which are the following: (i) the temperature difference of the water at the outlet and the inlet of the collector (ΔT) with solar global radiation, (ii) the water mass flow with the solar global radiation and (iii) the water mass flow with the temperature difference of the water at the outlet and the inlet of the collector. The results of the data analysis showed that the examined parameters were well correlated and also the optimum inclination angle in terms of the highest thermosiphonic flow generation was that of 45°.

Keywords Thermosiphonic phenomenon · Solar water heaters · Cyprus

73.1 Introduction

Cyprus has no natural oil resources at the moment and relies entirely on imported fuel for its energy demands. The only natural energy resource available is solar energy. Cyprus has a very sunny climate with an average annual solar radiation of 5.4 kWh/m^2-day (on a horizontal surface). Solar water heating units are extensively employed in Cyprus. In fact, the total number of units installed is such that make Cyprus a leading country in the world in this area. These units are mostly of the thermosiphonic type. This type of solar water heater consists of two flat-plate solar collectors having an absorber area between 3 and 4 m^2, a storage tank with capacity of 150–180 L and a cold water storage tank, all installed on a suitable frame. An auxiliary electric immersion heater and/or a heat exchanger, for central heating assisted hot water production, are used in winter during periods of low solar insolation.

Because the manufacturing of solar water heaters and mainly that of the thermosiphon type has expanded rapidly in Cyprus, there is a need to study in depth and model this type of systems. It is also required to validate the model using simple physical experiments. In this way, the model can be used to investigate the effect of design changes and therefore improve the solar water heater performance.

There have been extensive analyses of the performance of thermosiphon solar water heaters by numerous researchers [1]. Some of the most important are presented here.

Close [2] made the first published analysis of thermosiphon solar water heater circuit. He presented a mathematical model where mean system temperature and water mass flow rate can be predicted by testing two thermosiphon systems with different characteristics. The results conformed well to those predicted.

Hahne [3] calculated the efficiency and warming up time of flat-plate water collector under steady state and transient state conditions. Comparison of collector performance in summer and winter shows the importance of collector inclination and the effect of pipe spacing.

Morrison and Braun [4] studied system modelling and operation characteristics of thermosiphon solar water heater with vertical and horizontal storage tank. They found that system performance is maximised when the daily collector volume flow is approximately equal to the daily load flow. The very important fact that they found is that the system with vertical tank can perform better than the horizontal one. They found good agreement between the simulation results and experimental data as they compared an efficient numerical simulation model for thermosiphon solar water heaters for two locations. The model they used has also been adopted by the TRNSYS simulation program.

Hobson and Norton [5] made the characteristic curve for an individual directly heated thermosiphon solar energy water heater obtained from the data of a 30-day test period. Using this curve, they calculated the annual solar fraction which agreed well with the corresponding value computed from the numerical simulation. They produced a simple but relatively accurate design method for direct thermosiphon solar water heater.

Kalogirou and Papamarkou [6] studied the modelling of a thermosiphon solar water heating system with a simple model validation in Nicosia, Cyprus. They used two flat-plate collectors with an area of 2.7 m^2 and a storage tank of 150 L modelled using TRNSYS to obtain a detailed analysis and a long-term system performance. They found that the annual solar contribution of the simulated system was 79 %, and during the summer months no auxiliary heating was required as the solar contribution of the simulated system was 100 %. This means that the solar fraction reached 100 %. However, during the summer months the demand for hot water from the storage tank was reduced. Experimentally, they found that there is a decrease in solar radiation during May because of special conditions to the development of clouds encountered in Nicosia. They also made an economic analysis from which they found that the payback time of the system is 8 years, and the present worth of life cycle savings is equal to 161 Cy pounds (€ 275).

Chuawittayawuth and Kumar [7] presented details of experimental observation of temperature and flow distribution in natural circulation solar water heating systems. They found that the temperature values at the riser tubes of the collector inlet are generally much higher than those at the other risers on a clear day, while on cloudy days the temperature is uniform. They concluded that the temperature of water in the risers depends on its flow rate. They also carried out the measurements of the glass temperature.

Jiang et al. [8] found that the thermal performance of a solar thermosiphon system for water heating depends on its design characteristics and manufacturing quality; therefore, they analysed four characteristic parameters which helped them assess not only the solar thermosiphon system performance but also the direction for system performance improvement.

Riahi and Taherian [9] presented a detailed review of other studies and made a detailed analysis by discussing and comparing their results with other studies. They have used the hydrogen bubble method to measure the flow rate as shown by

Bannerot [10]. They collected data for several sunny and cloudy days. Also, they studied the dynamic response of the system to variations in solar insolation. It was found that such systems can provide ample energy to satisfy the demand for hot water.

As can be understood from this literature review, although many scientists have carried out research on various aspects of the thermosiphon system, nobody was concerned with the ability of the system to generate thermosiphon flow and thus the thermosiphon loop which is the main objective of this project. This will show how the performance of such systems can be improved as there is a possibility that not all flat-plate collectors are suitable to be used in thermosiphon solar water heating systems, as it is done today, without too much consideration on the actual performance of the system. In this chapter, the results of the experimental procedure investigating the effect of the solar collector's inclination angle (35, 45 and 55°) to the generation of thermosiphonic flow are analysed in order to define the optimum angle and to correlate the main parameters of such a system such as solar global radiation, temperature difference of the water on the outlet and the inlet of the collector and water mass flow rate.

73.2 Description of the Test Rig

The thermosiphon test rig used for the experimental procedure, shown in Fig. 73.1, was assembled by the staff of Applied Energy Laboratory of the Cyprus Ministry of Commerce, Industry and Tourism and consists of three main components, namely the test rig, the cylinder and the heat sink which are described below.

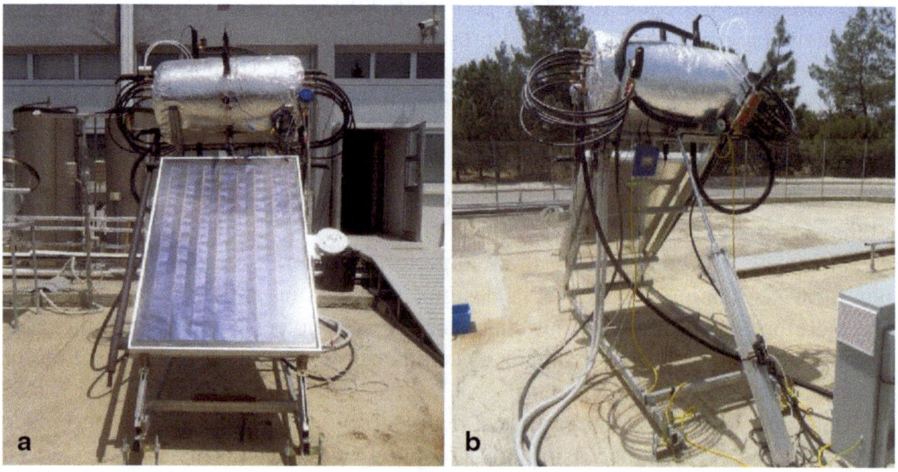

Fig.73.1 Different views of the test rig used for the experiments (**a** front view, **b** rear view)

Test Rig The test rig is made out of galvanised steel, and it was designed to adjust the angle of the collector. For the adjustment of the angle of the collector, a fully automatic pneumatic stroke is installed. The angle of the collector is measured with an inclinometer. In addition, the test rig is designed so that the height between the collector and the cylinder can be adjusted. Additionally, a pyranometer is attached on the test rig in order to measure the global radiation during the test period.

Cylinder The cylinder is made of copper, and it is insulated with natural mineral wool. Sensors are attached to the cylinder for recording the temperature of the water such as: inlet temperature (2 x PT100), outlet temperature (2 x PT100), draw off inlet and outlet (2 x PT100), mixing temperature (2 x PT100), stratification temperature (5 x Thermocouple), inlet and outlet of the heat exchanger (2 x PT100). A pneumatic ball valve is attached at the draw off outlet in order to regulate the flow at the draw off period. In addition, a mixing pump is attached at the back of the cylinder. Furthermore, the cylinder is equipped with a magnetic flow meter for measuring the flow at the exit of the collector, and a second portable ultrasonic flow meter is attached at the inlet pipe of the collector. It also has a differential pressure transmitter for measuring the pressure drop of the collector.

Heat Sink A supply tank is used as heat sink, which provides water to the system. At the exit of the supply tank, there is a three-way valve with an actuator and a pump. This allows regulating the temperature and flow of the water which is provided to the system.

73.3 Methodology

The experimental data presented in this chapter were recorded using Agilent data acquisition equipment between the 29th of May and the 30th of July 2014. The data were recorded from 07.30 to 18.45. The area of the collector used for the experiments was 1.36 m^2, and its orientation was south.

Initially, the solar collector was tested according to EN12975-2:2006 in all inclination angles under investigation (35, 45 and 55°) in order to determine the thermal performance at a flow and operation conditions specified by the standard. According to the standard, the experimental parameters needed to compute the collector's efficiency under constant inlet temperature and mass flow during a test period are the water temperature at the collector inlet, water temperature at the collector outlet, water mass flow, solar global radiation, ambient temperature and wind speed.

Accordingly, the test was repeated with the collector operating thermosiphonically again for all inclination angles examined.

Consequently, the obtained data were then analysed, the average efficiency of each case was calculated and several correlations were attempted which are the following: (i) the temperature difference of the water at the outlet and the inlet of the collector (ΔT) with the solar global radiation, (ii) the water mass flow with the solar global radiation and (iii) the water mass flow with the temperature difference of the water at the outlet and the inlet of the collector. All tests were conducted according to ISO 9459-2:1995(E).

73.4 Results and Discussion

The results concerning the thermal performance of the collector according to EN12975-2 and under thermosiphonic operation are depicted in Fig. 73.2. It can be observed that the thermal performance according to EN12975-2 is slightly higher than that of the thermosiphonic operation, which is rather logical due to the fact that the flow during the thermosiphonic operation is lower than the one according to EN1297-2 since it is naturally created as a result of the temperature difference between the inlet and the outlet water of the collector (thermosiphonic phenomenon).

Consequently, by using the results obtained during the experimental procedure, the thermal performance in terms of the average thermal efficiency of the solar collector for the inclination angles examined was calculated, and the results are presented in Fig. 73.3. The average thermal efficiency is defined as the ratio of the incident solar radiation falling on the collector aperture by the useful energy collected by the circulating fluid (water) for the same time period. As it can be seen by EN12975-2, the thermal performance of the collector is the same for 35 and 45°,

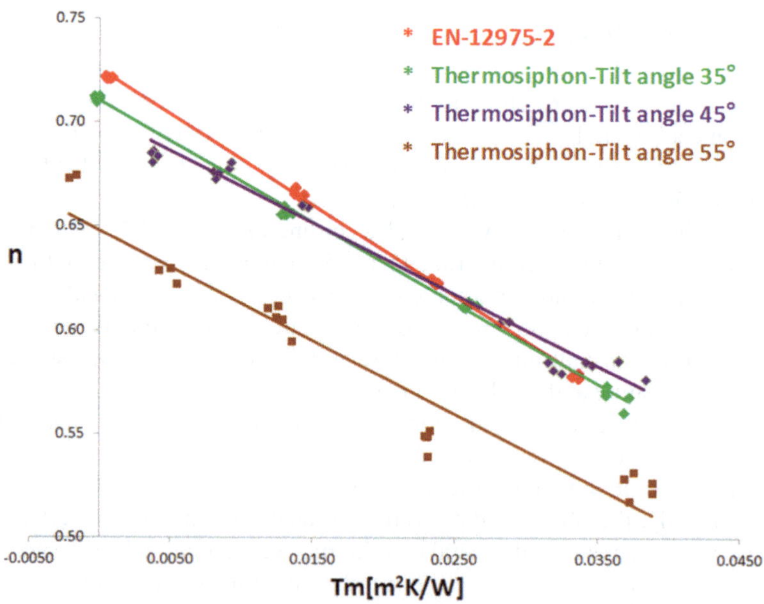

Fig. 73.2 Thermal performance of the solar collector according to EN12975 and under thermosiphonic operation for different inclination angles (35, 45 and 55°)

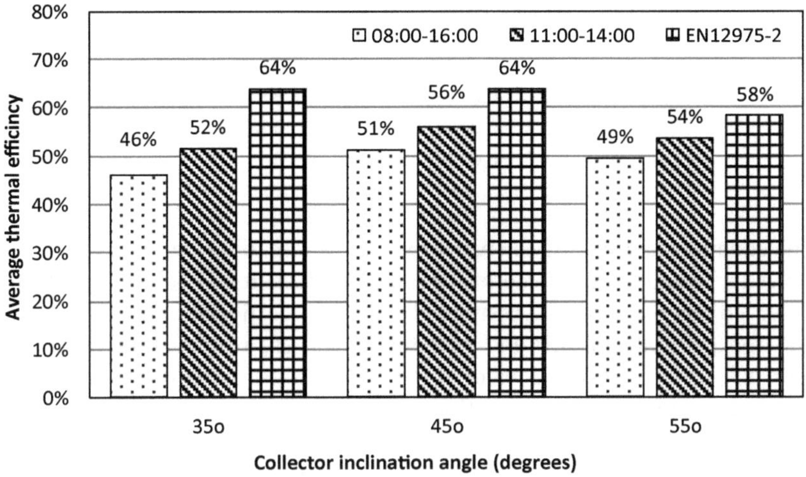

Fig.73.3 Average thermal efficiency of the solar collector for the three inclination angles examined using EN12975-2 and thermosiphonic operation for two time periods (08.00–16.00 and 11.00–14.00)

and it decreases significantly for 55° (from 64% it drops to 58%). On the other hand, when the collector is operating thermosiphonically, the highest energy performance is observed for an inclination angle of 45° in both time periods examined. It should be noted that the thermal performance for the time period 11.00–14.00 is significantly higher than that of 08.00–16.00 (56% instead of 51%) since during that time the solar global radiation is much higher.

Furthermore, several correlations between the main parameters of such a system were attempted, and the resulting equations together with the coefficient of determination (R^2) are presented in Tables 73.1, 73.2, and 73.3. As it can be observed, the examined parameters are very well correlated between them since in most cases the coefficient of determination exceeds 0.9 with only one case where it drops below 0.8.

The equations correlating the water mass flow rate *(m)* to the solar global radiation *(G)* are presented in Table 73.1 where, as can be observed, they are very well correlated since the coefficients of determination (R^2) are more than 0.95. From these results, it can be concluded that water mass flow which is naturally created from the temperature difference of the water at the outlet and the inlet of the collector (thermosiphon phenomenon) is directly correlated to the amount of the incident solar radiation on the collector.

Table 73.1 Equations correlating the water mass flow rate to the solar global radiation

Inclination angle	Equations	Coefficient of determination (R^2)
35°	$m=-9E-08G^2+0.0002G-0.0169$	0.95
45°	$m=-8E-08G^2+0.0002G-0.0084$	0.96
55°	$m=-6E-08G^2+0.0002G+0.002$	0.97

Table 73.2 Equations correlating the water mass flow rate to the temperature difference of the water in and out of the collector

Inclination angle	Equations	Coefficient of determination (R^2)
35°	$m = -0.0006\Delta T^2 + 0.0213\Delta T - 0.0696$	0.93
45°	$m = -0.0006\Delta T^2 + 0.0204\Delta T - 0.0394$	0.81
55°	$m = -0.0003\Delta T^2 + 0.0137T + 0.0196$	0.71

The equations correlating the water mass flow rate to the temperature difference of the water in and out of the collector (ΔT) are presented in Table 73.2. It can also be observed that they are well correlated since the coefficient of determination (R^2) is 0.93 for 35°, 0.81 for 45° and 0.71 for 55°. These results show that when the temperature difference of the water at the inlet and the outlet of the collector (ΔT) increases then the thermosiphonic flow also increases.

The equations correlating the temperature difference between the inlet and the outlet water of the collector with the solar global radiation are presented in Table 73.3. As can be observed, they are also well correlated since the coefficients of determination (R^2) are over 0.81. These results are rather logical since when the incident solar radiation is higher, the temperature difference between the water inlet and outlet also increases.

By using the equations of Tables 73.2 and 73.3 and for a solar global radiation of 1000 W/m^2, the water mass flow rate of the inclination angles examined was calculated in order to define which of these angles results to the generation of higher thermosiphonic flow and thus performs better under thermosiphonic conditions. According to the results, the highest thermosiphonic flow (Fig. 73.4) is generated in the inclination angle of 45° with 35° following and 55° being last (0.129, 0.115 and 0.101 kg/s).

As part of this project, we are going to suggest modifications of the existing standard for the performance of solar collectors (EN12975-2). This will be very similar to the existing standard with the difference that the requirement of the fluid flow rate set point will not be followed, the fluid flow rate stability during the test will have to be extended to $\pm 2\,\%$ (from the existing $\pm 1\,\%$), and the collector incidence angle will not be limited to the requirement of less than 20° but will face south and inclined according to the manufacturer specification, or for the optimum angle for each location, which for Cyprus as proved in this project is 45°.

Table 73.3 Equations correlating the temperature difference of the water in and out of the collector to the solar global radiation

Inclination angle	Equations	Coefficient of determination (R^2)
35°	$\Delta T = -1E-06G^2 + 0.0135G + 2.5611$	0.94
45°	$\Delta T = -6E-06G^2 + 0.0213G - 1.1413$	0.88
55°	$\Delta T = -4E-06G^2 + 0.0166G - 0.6858$	0.81

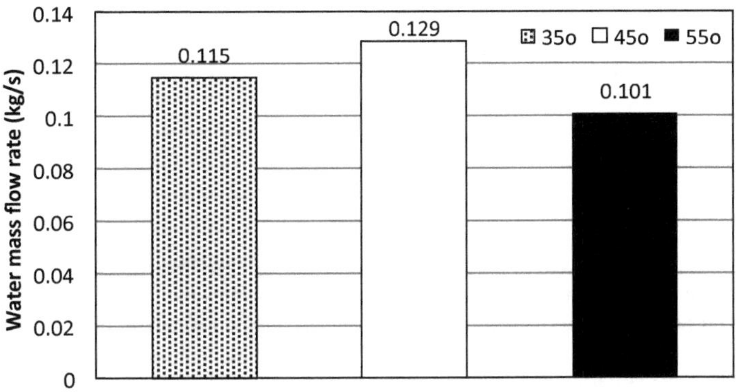

Fig. 73.4 Water mass flow rate for the inclination angles examined calculated using the correlation equations

73.5 Conclusions

This chapter constitutes the experimental data of a research project currently in process in Cyprus which aims to gain deeper understanding of the 'thermosiphonic phenomenon' and the identification of the key parameters affecting it. The main aim of this chapter is to investigate the effect of the solar collector's inclination angle to the generation of thermosiphon flow. Specifically, in this chapter, the angles examined are those of 35, 45 and 55°.

For this purpose, a special test rig was set up and equipped with all the sensors necessary to measure all the parameters that are most likely to affect the 'thermosiphonic phenomenon'. All tests were conducted according to ISO 9459-2:1995(E).

Additionally, a series of correlations were attempted using the data acquired when the collector is operating thermosiphonically for the following parameters: the temperature difference of the water at the inlet and the outlet of the collector, the solar global radiation and the water mass flow rate.

The results of the data analysis showed that these parameters correlate well between them since the coefficient of determination (R^2) is over 0.81 in most cases. Additionally, it was calculated that the thermal performance of a solar collector when operating thermosiphonically reaches its highest value when the inclination angle of the collector is 45°. Also, in this angle the highest water mass flow rate (thermosiphonic flow) is generated.

This work is ongoing, and the remaining objectives are to investigate the thermal performance of the system as a whole and not just the collector, to evaluate the effect of the distance between the top of the collector and the bottom of the storage tank to the performance of the system and also to evaluate another type of solar collector with different characteristics (pipe diameters, etc.) by applying the previously defined optimum conditions.

Acknowledgments This work was carried out as part of a research project co-funded by the Research Promotion Foundation (RPF) of Cyprus under contract ΤΕΧΝΟΛΟΓΙΑ/ΕΝΕΡΓ/0311(ΒΙΕ)09 and the European Regional Development Fund (ERDF) of the EU.

References

1. Kalogirou S (2009) Thermal performance, economic and environmental life cycle analysis of thermosiphon solar water heaters. Sol Energy 83(1):39–48
2. Close DJ (1962) The performance of solar water heaters with natural circulation. Sol Energy 6:33
3. Hahne E (1983) Parameter effects on design and performance of flat plate solar collectors. Sol Energy 34:497–504
4. Morrison GL, Braun JE (1985) System modelling and operation characteristics of thermosiphon solar water heaters. Sol Energy 34:389–405
5. Hobson PA, Norton BA (1989) A design monogram for direct thermosiphon solar energy water heaters. Sol Energy 43:89–95
6. Kalogirou SA, Papamarcou C (2000) Modelling of a thermosiphon solar water heating system and simple model validation. Renew Energy 21:471–493
7. Chuawittayawuth K, Kumar S (2002) Experimental investigation of temperature and flow distribution in a thermosiphon solar water heating system. Renew Energy 26:431–448
8. Jiang DY, Wei H, Hou JX, Jei J (2010) A new performance evaluation method for solar thermosiphon systems. Int J Carbon Technol 5:239–244
9. Riahi A, Taherian H (2011) Experimental investigation on the performance of thermosiphon solar water heater in the south Caspean sea. Therm Sci 15:447–456
10. Bannerot RB, Tu YW, Scott A, Placke G, Poché T (1992) A simple device for monitoring flow rates in thermosyphon solar water heaters. J Solar Energy Eng 114(1): 47–52

Chapter 74
Performance Improvement in a BIST Water Collector: A Parametric Study

Gilles Notton, Christian Cristofari, Fabrice Motte and Jean-Louis Canaletti

Abstract A flat-plate solar collector with high building integration was designed and prototyped. The experimentations showed that the performances of this solar collector can be improved. A numerical thermal model, developed in Matlab® environment using a finite difference model, was validated. Then, a modelling of the complete solar domestic hot water system (solar collector + water storage + piping) was realized. The performance of this system was calculated for various solar collector configurations such as the number and the position of the water pipes, air layer thickness, fluid flow rate, etc. Several solar fractions were used to implement this optimization procedure. An optimized solar collector structure is finally presented.

Keywords Thermal solar collector · Building integration · Modelling · Experimentation · Optimization

74.1 Introduction

A large Web survey on architectural integration of solar technologies (addressed to more than 170 European architects and building professionals) [1] showed that the architectural integration is a major issue in the development and spreading of solar thermal technologies.

First, we present the patented concept and the implemented experiment. Second, we underline the main negative points noted during the experiment. Third, we present the numerical models developed for each part of the thermal system: the first one for the solar collector presented in detail in [2] and a second one for the storage and the thermal loop inspired by the work of Haillot et al. [3]. This numerical model will be used for determining the best configuration of the thermal system and studying the influence of various parameters such as flow rate and air layer thickness. At last, we present a new improved configuration of the solar collector.

G. Notton (✉) · C. Cristofari · F. Motte · J.-L. Canaletti
Research Centre of Vignola, University of Corsica Pasquale Paoli, UMR CNRS 6134, Centre of Vignola, Route des Sanguinaires, F20000 Ajaccio, France
e-mail: gilles.notton@univ-corse.fr

A. Sayigh (ed.), *Renewable Energy in the Service of Mankind Vol II*,
DOI 10.1007/978-3-319-18215-5_74

74.2 Solar Collector and Experiment

The patented concept H2OSS® is invisible from the ground because it is inserted within a drainpipe (Fig. 74.1) which conserves its rainwater evacuation role. The canalizations connecting the house to the heating collector are hidden in the vertical drainpipe. An installation consists of several connected modules. One module is about 1 m length and 0.1 m in width (individual houses), and larger modules can be developed for buildings. The number of modules depends on the drainpipe length. The structure of the collector is composed of glass, an air layer, a highly selective absorber, and an insulation layer (Fig. 74.1). The cold fluid flows from the tank through the inferior insulated tube and then into the upper tube in thermal contact with the absorber.

The experimentation (Fig. 74.3) allowing one to operate closer to the European Standard EN 12975-1 was implemented to test the thermal behaviour, to validate the thermal model, and to improve the performances by some parameter adjustments. Four rows of four thermal modules (1.8 m^2), connected in serial or parallel, are fixed to a solar tracker for better control of the solar intensity and direction (Fig. 74.2).

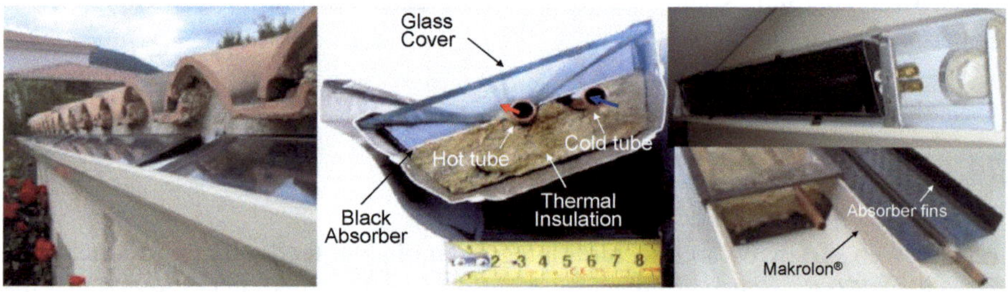

Fig. 74.1 The solar collector H2OSS®

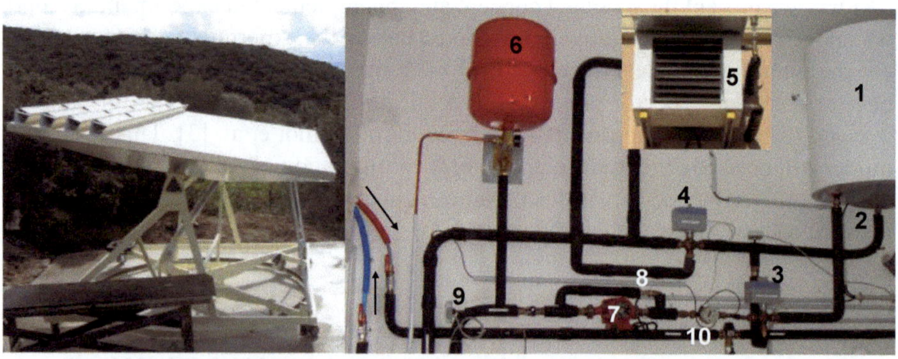

1. Water heater with electrical resistance 2.Temperature sensor (dry cooler gate control). 3. Water heater gate. 4. Dry cooler gate 5. Dry cooler (outdoor). 6. Expansion tank. 7. Pump. 8. Flow rate adjustment. 9. Temperature sensor (water heater gate control. 10. Flow meter

Fig. 74.2 The experiments: the solar tracker with H2OSS modules and the thermal loop

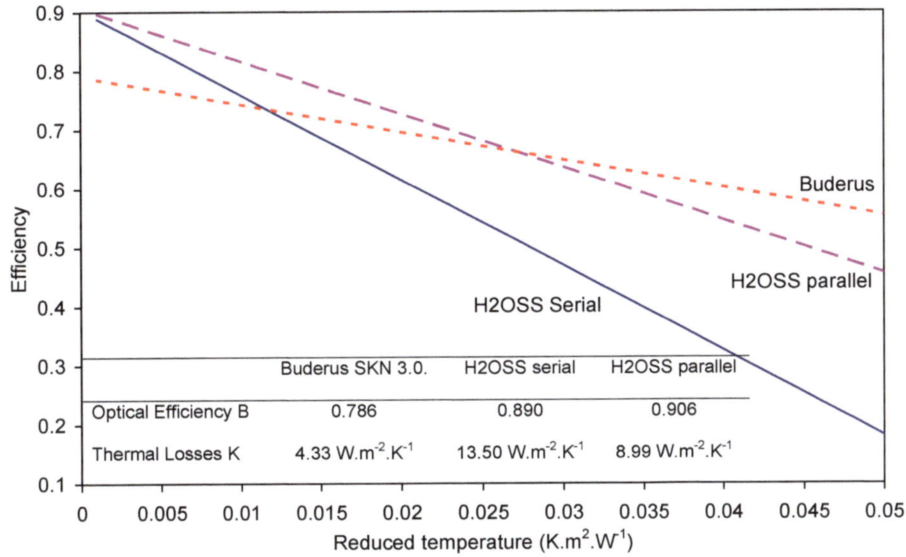

Fig. 74.3 Efficiencies versus the reduced temperature for H2OSS and Buderus SKN. 3.0

We compared the performances of the prototype with a commercial solar collector (Buderus 3.0) (Fig. 74.3); we calculated the optical efficiency and the thermal loss coefficient. Our optical efficiency is high, but the thermal losses are high particularly on the sides which are more important, and the performances decrease rapidly when the reduced temperature increases; the best performance is obtained with a low reduced temperature, that is with a low-input water temperature; it is more interesting to use a water storage tank with a thermal stratification, working at low flow rate with a colder temperature.

74.3 Thermal Models

The solar domestic hot water system is described in Fig. 74.4. The thermal model consists of 2 models:

- One for the H2OSS module, which calculates various temperatures in the solar collector [2]
- One for the hydraulic loop with water storage and distribution circuit developed by Haillot et al. [3]

We developed a bidimensional model with thermal transfers composed of a serial assembling of one-dimensional elementary models. The domain is broken up into elementary isotherm volumes, and for each of the 97 nodes, we write a thermal balance equation using an electrical analogy. All parameters of this model can be changed in such a way that we can estimate the influence of future changes on the

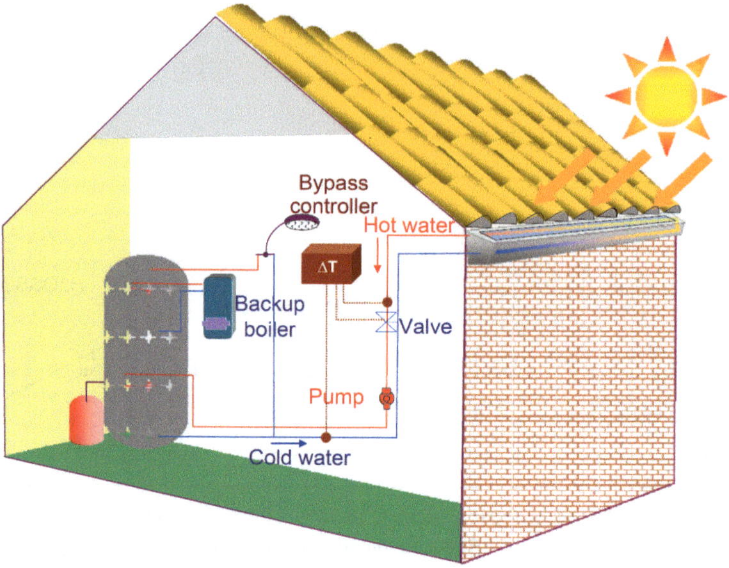

Fig. 74.4 The solar domestic hot water system

performances of solar module. The collectors can be connected in serial or parallel; in the first case, in the flow direction, the output fluid temperature of the first module becomes the input fluid temperature of the next one (Fig. 74.5); in the second case, the output temperature is the same for all the lines of modules and the total water flow rate is the sum of the flow rates of each line.

The experimental validation showed a good accuracy of the model: the relative root mean square errors are around 5 % for the water temperatures and from 4.6 to 10 % for the internal ones [2].

During the experimentation, the hydraulic resistance due to the linear structure was a problem. To reduce it, the solar system must work at low flow rate with these advantages [4]:

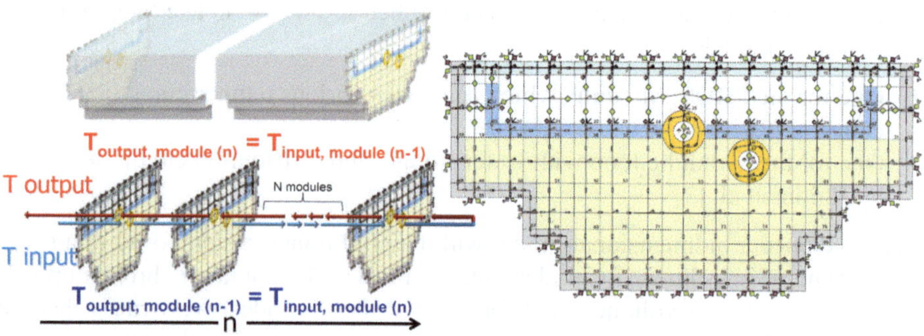

Fig. 74.5 Electrical analogy of the solar thermal collector and serial connexion

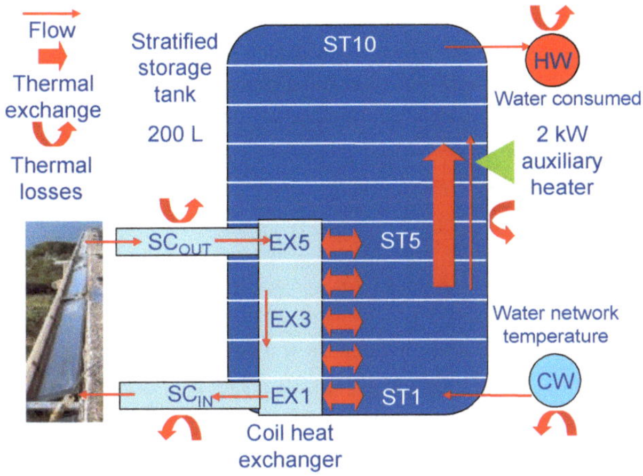

Fig. 74.6 Model node definitions [3]

- Thermal stratification: Using low flow operation results in an increased outlet temperature from the collector and a higher degree of thermal stratification inside the heat storage, the temperature at the top of the storage will be closer to the desired load temperature reducing the auxiliary energy consumption and increasing the solar fraction. The return temperature to the collector will be lowered, and the working periods and the output energy for the collector will be increased.
- Piping: We can use smaller pipes and the heat losses are reduced.
- Pump: The energy consumption of the pump is decreased which is very important in our case.

The thermal loop behaviour is simulated using a numerical code based on a nodal approach [3, 5]. It is divided into 19 nodes: 7 for the fluid circulation, 10 for the storage tank (optimal number of nodes for taking into account optimally the thermal stratification) [4], and 2 for the water inlet and outlet in the storage (Fig. 74.6). The temperature of water at the outlet of the solar collector and the average temperature of the solar absorber are obtained from the modelling of the solar collector.

The energy balance, in 1D, is applied and an iterative method is used to solve the first-order differential equations. A "reversion–elimination mixing algorithm" based on a thermal mix of some storage tank nodes to obtain a correction factor in order to have a positive temperature gradient from the bottom to the top of the tank [6, 7] is used to simulate the thermal stratification. More information on the water utilization profile and thermal regulation are given in [8].

The model of the thermal loop was validated by Haillot et al. [3] from measured data. The comparison between experimental and computed data is conduced to a relative root mean square error of 8.6 % for the yearly average solar fraction; then, the thermal loop model is validated. We plotted in Fig. 74.7 the temperatures ST1–ST10 into the tank, and we see the stratification phenomenon. The two models have good accuracy and can be coupled to simulate the thermal behaviour of the total solar system.

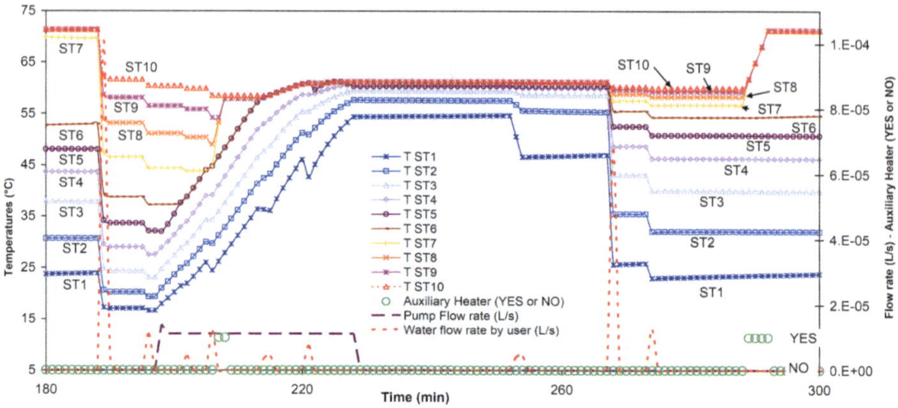

Fig. 74.7 Illustration of the thermal stratification into the storage tank

74.4 Solar Collector Optimization

To choose the optimal configuration, we calculate during each simulation the following data:

- The working time of the pump and its electrical energy consumed
- The working time of the electrical auxiliary heater and its electrical energy consumed
- The thermal energy drawn to the storage tank, that is useful for the user
- The thermal energy produced by the solar system (solar and electrical)
- The part of the thermal energy produced by the solar resource
- The thermal losses by the storage tank and the distribution water network

From these data, we defined three solar fractions:

- SF: the conventional solar fraction, which is the ratio of the total solar energy delivered to the tank $E_{\text{Thermal, Solar}}$ (kWh) and the total energy delivered to the tank E_{Thermal} (kWh). E_{Thermal} is the sum of the solar energy delivered to the tank and the auxiliary energy delivered to the tank.

$$SF = E_{\text{Thermal,solar}} / E_{\text{Thermal}} = E_{\text{Thermal,solar}} / \left(E_{\text{Thermal,solar}} + E_{\text{Electrical,AuxHeat}} \right). \quad (74.1)$$

- SF+: we noted that high hydraulic losses occur in the collector due to serial connexions. These losses impose the use of a high-power electrical pump with a high electrical consumption. It is necessary to take into account this supplementary energy due to the pump working in the solar fraction, and we added the electrical energy of the pump $E_{\text{Electrical,pump}}$ to the electrical energy used for the auxiliary heating:

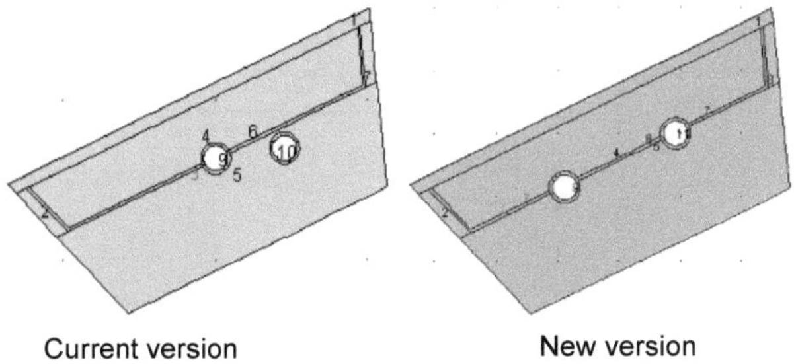

Fig. 74.8 The two versions of the solar collector H2OSS used for optimization

$$SF^{+} = E_{\text{Thermal,solar}} \Big/ \left(E_{\text{Thermal,solar}} + E_{\text{Electrical,AuxHeat}} + E_{\text{Electrical,pump}} \right). \quad (74.2)$$

- SF++: the value of electric power and thermal energy differs due to the form of energy. Electricity is a high-grade form of energy since it is converted generally from thermal energy. We used an overall thermal efficiency [9] $\eta_{\text{Ther-Elec}}$ for converting thermal energy into electricity and taken equal at 0.38. This formulation suggests energy equivalence between electricity and thermal energy with an electrical-to-thermal ratio equal to 2.63 (1/0.38):

$$SF^{++} = E_{\text{Thermal,solar}} \Big/ \left[E_{\text{Thermal,solar}} + \left(E_{\text{Electrical,AuxHeat}} + E_{\text{Electrical,pump}} \right) \Big/ \eta_{\text{Ther-Elec}} \right]. \quad (74.3)$$

We saw during the experiment that our collector has high heat losses due to its particular shape. The objective being to reduce the temperature of the absorber, we tested, after preliminary studies, a new configuration of the H2OSS collector called new version (Fig. 74.8).

Our optimization was realized for a solar system used by four persons in Corsica and composed by 35 serial connected modules (4 m^2) and a tank of 200 L. First, we verified that using 35 serial modules does not produce a temperature saturation, that is the water temperature continues to increase. Figure 74.9 shows, in steady state, the water temperature versus the number of modules (solar irradiance 750 W m^{-2}, ambient temperature 25 °C, wind speed 1 m s^{-1}, flow rate 60 L h^{-1}) for the two versions of solar collectors.

A more rapid saturation occurs for the new version but we can install efficiently, in serial order, 50 modules (50 m of gutter rarely available in a house). We used a pump with electrical power proportional to the flow rate between 30 W for 15 L h^{-1} and 250 W for 200 L h^{-1}. The calculations were realized for winter (January) and summer (July). We varied the fluid flow rate, the air thickness between cover and absorber (reduction in convective losses by the front face), and the insulation thickness (back and lateral thermal losses). The collector dimensions must stay within the commercial standards (the gutter must evacuate rainwater: limits of insulation). As an example, we show in Table 74.1, the monthly energies for the "current" configuration.

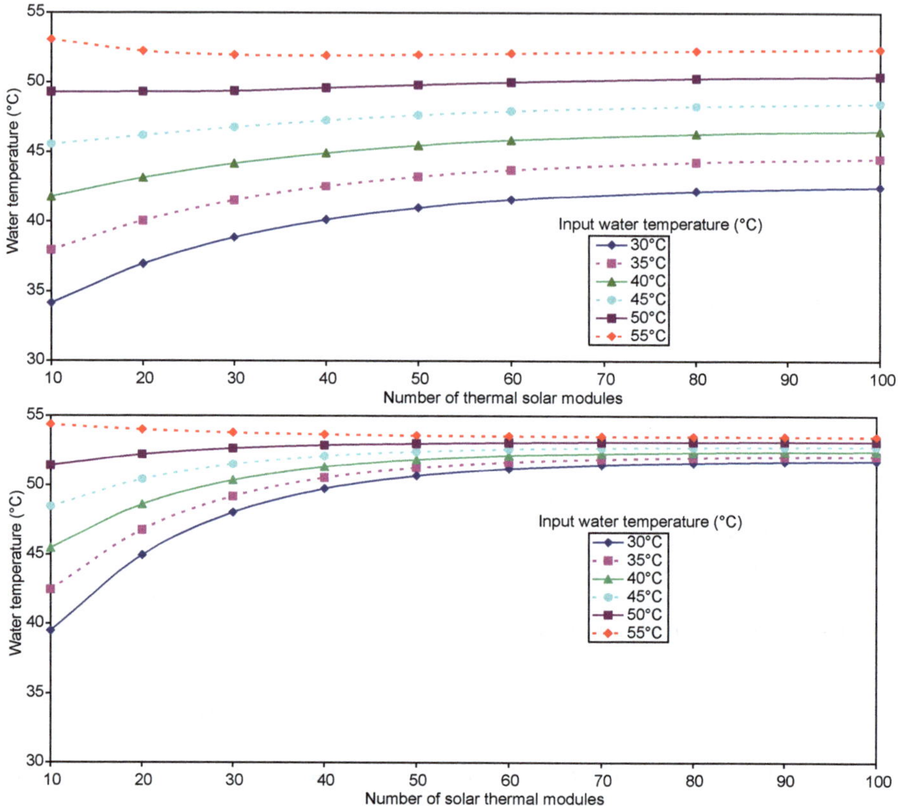

Fig. 74.9 Water temperature versus number of modules for current and new version of solar collector

Influence of the Flow Rate A low flow rate allows a thermal stratification of the storage, a hydraulic losses reduction, and a small pump power and tube diameter [10]. Hollands and Lightstone [11] calculated an annual energy gain of 38 % compared with high flow rate and Cristofari et al. [4] of 5.25 %. The hydraulic losses are high, and the fact to take into account is that the pump consumption in FR+ and FR++ should modify the optimization results compared with the use of FR. We see in Fig. 74.10 that the performances of the new version are better. Then, from now, we will only consider the new version. An optimum flow rate appears with SF+ and SF++ because they are a better indication of the performance level. The optimal flow rates are, respectively, 50 L h^{-1} and 75 L h^{-1} for summer and winter for the new configuration, and around 75 L h^{-1} for the current one. For winter, the performance gaps between a flow rate of 50 L h^{-1} and 75 L h^{-1} is lower than that in summer; thus, the optimum flow rate is taken equal to 50 L h^{-1} and the following optimization calculations will be realized for this flow rate, which is considered as a low flow rate; in fact, we consider as a low flow rate, a value between 7 and 15 L h^{-1} m^{-2} [12], that is for our 4 m^2 between 28 and 60 L h^{-1}.

Table 74.1 Example of calculation of the solar fractions for the current configuration

		January	July
Thermal energy drawn to the storage tank	kWh	182.3	171.0
Thermal losses (storage tank and water distribution circuit)	kWh	27.8	43.9
Thermal energy needs	kWh	210.1	215.0
Thermal energy produced by the solar resource	kWh	51.1	146.0
Running time of the pump	hour	62.5	21.9
Electrical energy for the pump	kWh	6.2	21.9
Running time of the auxiliary electric heater	hour	79	34.5
Electrical consumption of the auxil-iary heater	kWh	159	69
Solar fraction SF	%	24.3	68
Solar fraction SF+	%	23.6	61.7
Solar fraction SF++	%	18.6	38.0

Influence of Air Thickness We calculated the three solar fractions for various air thicknesses (Fig. 74.11). The form of the curve is due to the variation of the convective coefficient in the air layer thickness. It appears in two optimums: 1 cm for winter and 1.125 cm for summer. The optimum is the same whatever the solar fraction used because the pump consumption has small influence. We chose an optimal value of 1.125 cm of air thickness. This result is in accordance with the literature [13] (Fig. 74.10).

Influence of Thermal Insulation We first studied the backside insulation but with the influence being small, the results will not be shown here. The very hot absorbers fins are insulated from the aluminium body by only 2 mm of Makrolon® (polycarbonate product). We added insulation between fins and body without decreasing the collection for thicknesses up to 16 mm for each side, but the maximum usable thickness is 4 mm for a gutter with good rainwater evacuation (Fig. 74.12).

From a thickness of 4–5 mm, the curves tend towards an asymptote. With an insulation of 4 mm on both sides (2 mm more than now), we have an improvement of 1.3 % in winter and 2 % in summer for SF.

74.5 Synthesis and Conclusion

From the previous results, we created an "ideal" H2OSS® solar collector with the characteristics summarized in Table 74.2 for a solar installation for four persons, situated in Corsica and 35 m of solar modules (4 m^2) for water storage of 200 L. To see the consequences of this optimization on the thermal performances, we plotted the efficiency versus the reduced temperature for the current and the new

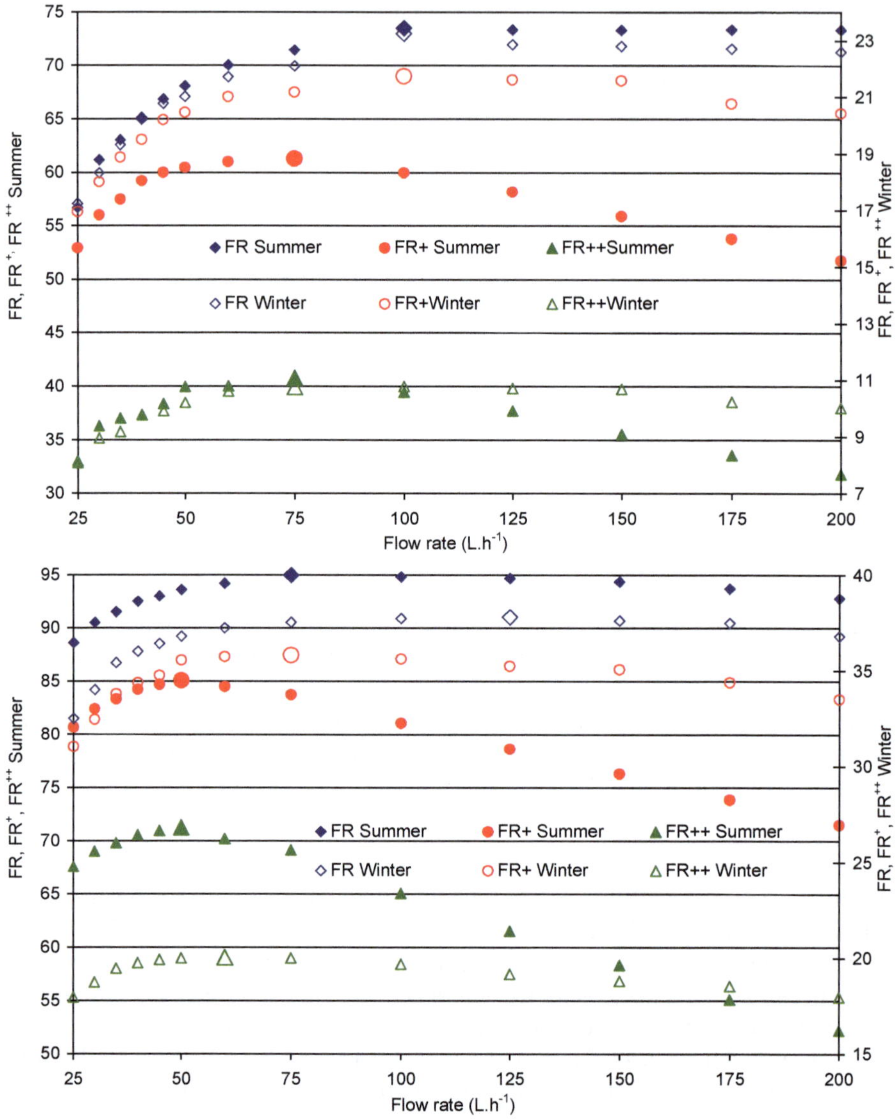

Fig. 74.10 Solar fractions versus water flow rate for the current and new version of the solar collector

optimized version in Fig. 74.13. The thermal loss coefficient decreases from 13.80 to 6.25 W m^{-2} K^{-1} (more than 50 %) but the optical coefficient decreases from 0.903 to 0.780 (−13 %). Despite the improvement of the thermal loss coefficient, it stays higher than that of the Buderus 3.0. (4.3 W m^{-2} K^{-1}) (Fig. 74.13).

The new positioning of the cold tube in the absorber and the optimization showed much better performance than with the actual prototype with an annual fraction passing from 41 % for the current version to 76 % for the new one.

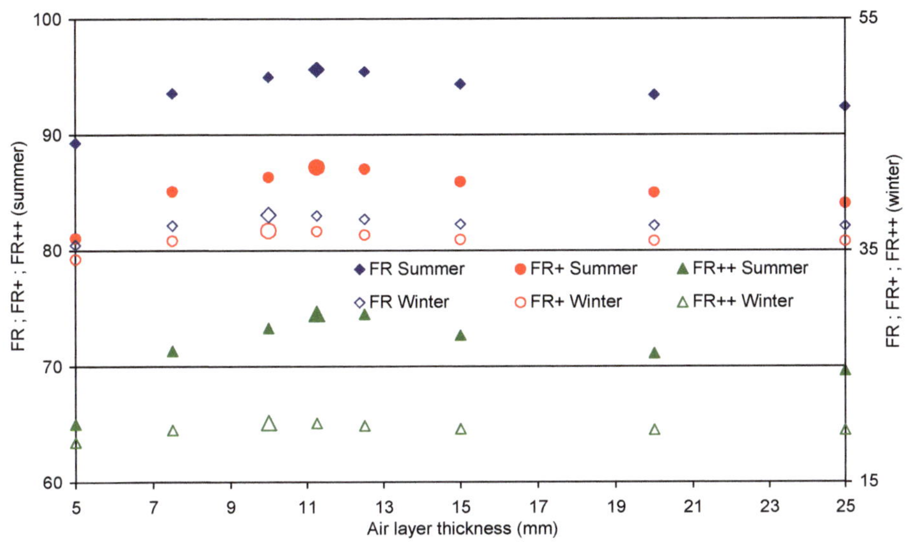

Fig. 74.11 SF, SF+, and SF++ versus the air layer thickness

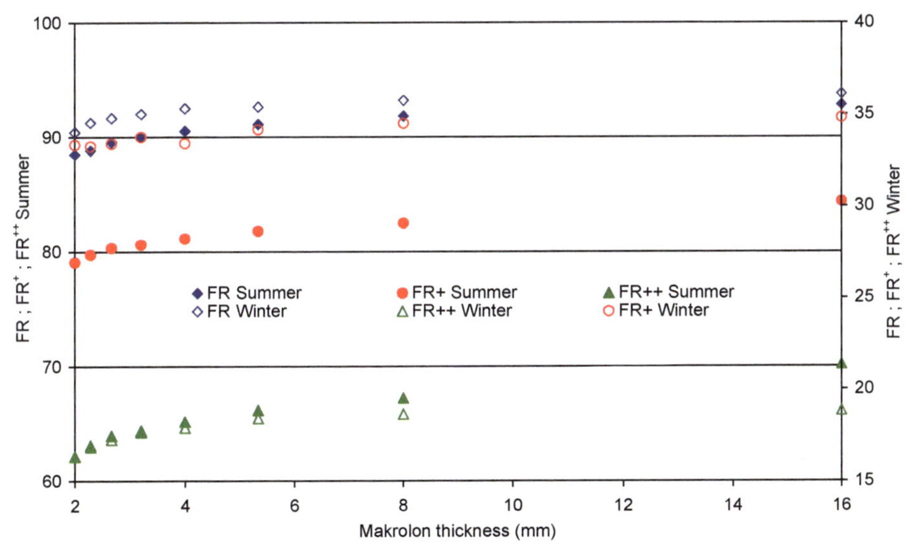

Fig. 74.12 Influence of the Makrolon® thickness between the absorber fins and the aluminium body

Table 74.2 Synthesis of the main optimization results

Parameter	Optimal value
Water tube positions	In the absorber (the cold tube is no longer in the insulation)
Flow rate	50 L h^{-1}
Air layer thickness	1.125 cm
Backside thermal	No more
Makrolon® insulation	+2 mm on both sides

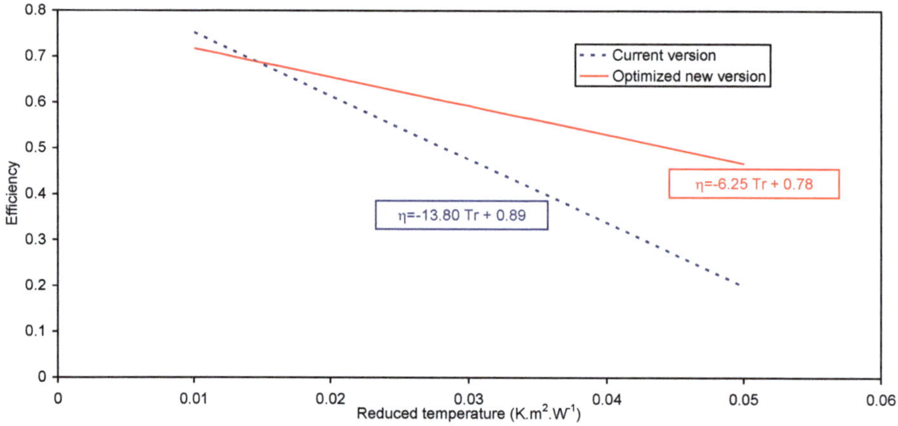

Fig. 74.13 Performance straight lines for current and optimized versions

Acknowledgements The authors thank the Territorial Collectivity of Corsica and OSEO-ANVAR for their financial supports. This work was realized in the frame of the COST Action TU1205 on "Building Integration of Thermal Solar Systems (BISTS)".

References

1. Munari Probst MC, Roecker C (2007) Towards an improved architectural quality of building integrated solar thermal systems (BIST). Sol Energy 81:1104–1116
2. Motte F, Notton G, Cristofari C, Canaletti JL (2013) Design and modelling of a new patented thermal solar collector with high building integration. Appl Energy 102:631–639
3. Haillot D, Nepveu F, Goetz V, Py X, Benabdelkarim M (2012) High performance storage composite for the enhancement of solar domestic hot water systems Part 2: numerical system analysis. Solar Energy 86(1):64–77
4. Cristofari C, Notton G, Poggi P, Louche A (2003) Influence of the flow rate and the tank stratification degree on the performances of a solar flat-plate collector. Int J of Thermal Sci 42(5):455–469
5. Haillot D (2009) Matériaux composites à hautes performances énergétiques pour l'optimisation des chauffe-eau solaires individuels: du matériau au procédé, Phd dissertation, University of Perpignan, France, (in French)

6. Mather D, Hollands KG, Wright J (2002) Single- and multi-tank energy storage for solar heating systems: fundamentals. Sol Energy 73:3–13
7. Nepveu F (2008) Production décentralisée d'électricité et de chaleur par système Parabole/Stirling: Application au système EURODISH, Phd dissertation, University of Perpignan
8. Notton G, Motte F, Cristofari C, Canaletti JL (2014) Performances and numerical optimization of a novel thermal solar collector for residential building. Ren Sust Energy Rev 33:60–73
9. Huang BJ, Lin TH, Hung WC, Sun FS (2001) Performance evaluation of solar photovoltaic/thermal systems. Sol Energy 70:443–448
10. Shah LJ (1999) Investigation and modeling of thermal conditions in low flow SDHW systems, Department of Buildings and energy Technical University of Denmark Report R-034
11. Hollands KGT, Lightstone MF (1989) A review of low flow, stratified-tank solar water heating systems. Sol Energy 43(2):97–105
12. Kenjo L, Inard C, Caccavelli D (2007) Experimental and numerical study of thermal stratification in a mantle tank of a solar domestic hot water system. Appl Therm Eng 11–12:1986–1995
13. Ferahta FZ, Bougoul S, Ababsa D, Abid C (2011) Numerical study of the convection in the air gap of a solar collector. Energy Procedia 6:177–185

Chapter 75
Performance Evaluation of a Novel Solar Industrial Wastewater Treatment Unit for Reuse

Hamdy El-Ghetany and M. Hamdy El-Awady

Abstract In the present study, a novel solar industrial wastewater treatment unit (NIWWTP) was designed, constructed, and tested under actual meteorological conditions of Greater Cairo, Egypt. Nonconventional treatment and decontamination of highly polluted wastewater such as solar degradation is the new approach and currently the focus of intense investigations due to their potential for revolutionary technological applications in the twenty-first century. Industrial wastewater is considered as chemically polluted water that causes most of water-borne diseases. The pollution potential of textile dyes has been primarily prompted due to concerns regarding their possible toxicity and carcinogenicity. Color is the first contaminant to be recognized in wastewater and has to be removed before being discharged into water bodies or on land. This is mainly due to the fact that many dyes are made from known carcinogens and other aromatic compounds. Photocatalytic degradation of dyestuffs from aqueous synthetic sample as well as from real textile wastewater representing industrial color-containing wastewater from a textile company will be used in this study to investigate the performance efficiency of the system under investigation for color removal and decontamination of pollutants. Results showed that the treated water was colorless, with minimum chemical contaminates and pathogenically decontaminated. Experimentally, the system was affected by meteorological conditions such as ambient temperature, wind speed, solar radiation intensity, pollutant concentration in wastewater, quantity of water exposed, and water contact time in the system. The use of solar energy for the water treatment process will provide good potential of using renewable energy on a large scale for a nonconventional wastewater treatment plant using clean energy in the coming decades.

Keywords Solar energy · Photocatalytic degradation · Industrial wastewater treatment

---

H. El-Ghetany (✉)
Solar Energy Department, National Research Centre, Dokki, Cairo, Egypt
e-mail: hamdy.elghetany@gmail.com

M. H. El-Awady
Water Pollution Research Department, National Research Centre, Dokki, Cairo, Egypt
e-mail: hawady@yahoo.com

A. Sayigh (ed.), *Renewable Energy in the Service of Mankind Vol II*,
DOI 10.1007/978-3-319-18215-5_75

Aim of the Work Due to the noticeable shortage of power resources and the high cost of energy supply in developing countries, Egypt has to find out a renewable source with low-cost infrastructure and minimum operation and maintenance facilities. So, this innovative system was manufactured to compensate the gap between the available and the required energies in the industrial zones and to offer a clean technology for the industrial sector.

75.1 Introduction

The pilot plant under investigation represents an independent unique innovative prototype that can be implemented for all categories for industrial wastewater on small and medium scales without need for electric or any other power source, especially in remote and new communities. Wesley [1], declared that due the nature of the industry and the projected uses of the waters of the receiving stream, various waste constituents may have to be removed before discharge, especially dyestuffs, soluble organic pollutants such as phenol, heavy metals, cyanide, and toxic chemicals. These undesirable waste characteristics present esthetic problems even though they may not be particularly deleterious for most water uses. In addition, discharge of industrial effluents from textile industries containing highly colored dyes into aquatic ecosystems and their efficient removal from the textile industry is still a major environmental challenge. Different techniques including almost all the known physical and chemical and biological techniques were described for decolorization, and the final conclusion was that each process alone might not be able to meet the requirements [2]. Moreover, Bizani et al. [3] showed that photocatalytic degradation of two commercial azo dyes in the presence of TiO_2 suspensions were investigated, where decolorization was achieved within 100 min of illumination, while in the presence of TiO_2 and UV, complete color disappearance took place in less than 50 min of light exposure. Also, Sahel et al. [4] showed that photocatalytic degradation of a mixture of two anionic dyes was investigated where the amount of each dye in the mixture was compared individually and mixed with the degradation of both dyes and which had the same discoloration rate for both dyes individually or mixed at low concentration of dyes. Consequently, Ibrahim et al. [5] concluded that anionic acid, reactive, and direct Ciba dyestuffs were effectively removed from their aqueous solutions as well as dye house effluents. Results indicated that the percentage efficiency of color removal, % CR, followed a decreasing order: direct > acid > reactive, respectively. On the other hand, on utilizing cement kiln dust as an adsorbent, for removing color from synthesized disperse and vat aqueous solutions, as well as from textile industrial wastewater, results showed that color in all cases was totally removed from all samples with a percentage reaching 99.9 %, respectively. Moreover, El-Awady et al. [6] concluded that photocatalytic degradation of dyestuffs from aqueous synthetic samples as well as from real textile industrial wastewaters in the presence of an artificial source of ultraviolet irradiation has be carried out. The utility of hydromag–TiO_2 in the photocatalytic degradation of aqueous as well as dyestuff effluents showed a complete destruction of acid, reactive, and vat dyestuffs, at their optimum operating conditions. Accordingly, Sandeep et al. [7] compared the adsorption isotherms and photocatalytic decolorization of two anionic dyes Reactive Red 120 (RR120) and Reactive Black 5 (RB5) in a single-dye solution to that of their mixture. Both dyes showed adherence to Langmuir adsorption isotherms in the single-dye solution.

75.2 Experimental Setup

The solar industrial waste water treatment (IWWT) system has been experimentally installed in the National Research Centre, Dokki, Cairo, Egypt, as shown in Fig. 75.1. The subsequent paragraphs briefly describe the system operation. The industrial wastewater is stored in a chemically polluted elevated water tank that is fixed on a metal frame of 2 m height. The cold contaminated water is then passed from the raw water tank to the solar IWWT unit through a fan coil unit (FCU), which represents air–water heat exchanger to create a cooling effect on the coil surface, leading to humid air coming from the solar IWWT unit to be condensed on the coil surface, and consequently, the condensate water is accumulated in a galvanized steel basin located under the coil. A DC fan is used to convectively force the humid

Fig. 75.1 Photographical view of the novel solar industrial wastewater treatment plant

Fig. 75.2 Photographical view of the metallic basin and glass cover

air across the coil surface. The DC fan is electricity supplied by a solar PV module located beside the solar IWWT unit. This makes the system work independently from the national grid electricity.

The warm contaminated water outlet from the FCU was passed through the solar IWWT unit. The unit is consisted of a black-painted metal basin with an area of 1.0 m2. The wastewater was passed at the desired flow rate through the manufactured ten channels of 1.0 m length each. Therefore, the water sample was exposed to solar radiation in the heated basin with a total stroke of 10 m prior outlet from the solar IWWT unit. Some of the water sample was evaporated due to the heat and mass transfer inside the thermal enclosure. The humid air was then condensed on the inner surface on the truncated pyramidal glass cover. The accumulated condensate was let out from the system through the basin corners.

The Wooden Base The wooden base is manufactured with a dimension of 1 x 1 m and a thickness of 5 cm. It has been designed to have a supporting frame fixed with four wheels to make the system mobile. These wheels are supported by a manual break to fix the system's motion.

Metallic Basin The basin is made from a galvanized steel sheet with an area of 1 m^2 and a thickness of 1 mm. It is designed to be squared shape with a height of 5 cm. Nine galvanized metal strips with a height of 4 cm and a thickness of 0.5 mm are fixed in the basin to form ten identical channels as shown in Fig. 75.2. The basin has two inlet and exit ports, and each port is designed to have two connections, one for the inlet or exit and the other for drain. When the industrial wastewater sample entered the metal basin, it passed through ten channels in a series, each channel of 1 m length, and then, the total water stroke was 10 m with the desired flow rate.

Glass Cover Four strips of transparent glass covers were used to cover the water basin. Each strip was designed to be with a trapezoidal shape with a minor base of 30 cm and a major base of 1 m with a thickness of 4 mm. When the four glass strips were fixed on the basin and sealed from each side by a silicon rubber, they formed a truncated transparent pyramidal shape, its base is 1 × 1 m cross-sectional area and its top is 30 × 30 cm cross-sectional area as shown in Fig. 75.2.

Fan Coil Unit The FCU consisted of three parts: The first one was the electrical fan that was designed to be a DC fan to be powered by the system; the second part was a solar PV module to make the system work standalone in any place regardless of the availability of the national grid electricity. The third part of the FCU is the copper coil to be used as heat exchanger to cool the humid air in the system to be condensed at the coil fins and accumulated in a condensed water basin located under the coil.

75.3 Materials and Methods

75.3.1 Materials

75.3.1.1 Commercial Anionic Dyes

a. *Kiton Scarlet* 4R: It was also known as Victoria Scarlet 3R, strawberry red with molecular formula $C_{20}H_{11}N_2Na_3O_{10}S_3$ and molecular weight 604.473048. The following represents the chemical structure as shown in Fig. 75.3.

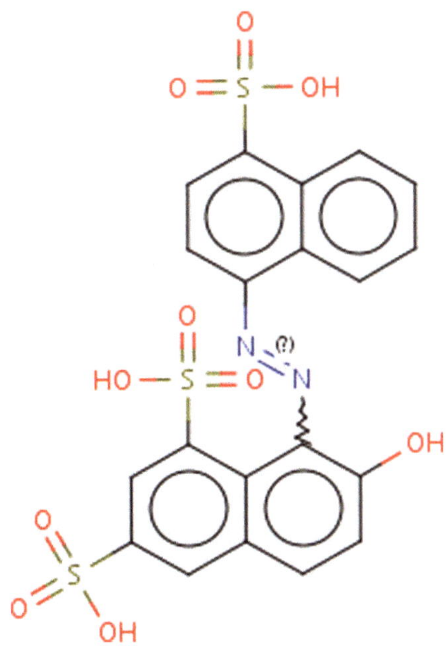

Fig. 75.3 Chemical structure of the Kiton Scarlet 4R dyestuff

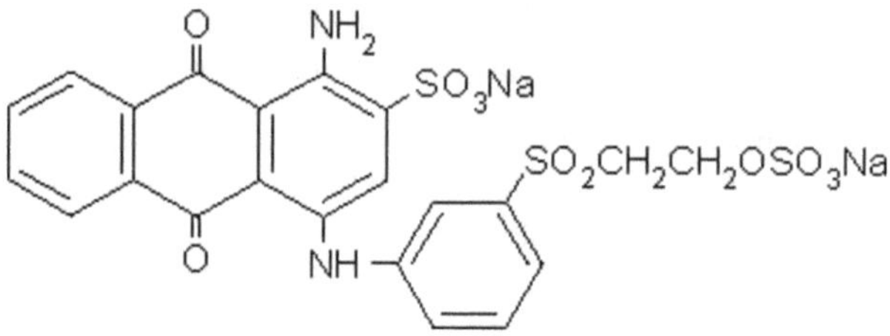

Fig. 75.4 Chemical structure of Reactive Blue No. 19 dyestuff

b. *Reactive Blue* No. 19: [2-anthracene-sulfonic acid, 1-amino-9,10- dihydro-9,10-dioxo-4-((3-((2-(sulfooxy)ethyl) sulfonyl) phenyl)amino)-, disodium salt] (CAS Reg. No. 2580-78-1). The chemical structure of Reactive Blue No. 19 dyestuff is shown in Fig. 75.4.
c. *Spectrophotometer*: UV/VIS Spectrophotometer with an *integrated* printer was used to detect the maximum wavelength (λ max) of both dyestuffs used in the reactor to detect the residual as well as the efficiencies of color removal along with the experimental works.
d. *Raw feeding stock solutions*: About 40 L from each aqueous dye water was freshly prepared with initial concentration of 10 μg/L each by adding chlorine-free tap water to represent the raw aqueous feeding water. The raw water was transferred to the main feeding tank.
e. *Solar unit operation*: Raw feeding water was prepared to adjust the hydraulic retention time, flow rate, sunshine, and sunset, with meteorological data of the system.

75.4 Results and Discussion

75.4.1 Detection of the Maximum Wavelength of Both Dyestuffs

A wavelength scanning was carried out for the two dye solutions; results showed that the maximum wavelength of the Kiton Scarlet was 434 nm, while it was 662 nm for Reactive Blue.

Evaluation of the performance of the novel industrial wastewater treatment plant (NIWWTP) was done for the removal of the two dyes from industrial solutions. Several experiments have been carried out to investigate the system performance. In each experiment, the weather conditions such as ambient dry and wet bulb temperatures and solar radiation intensity were measured. Also, the wavelength scanning of the periodical samples throughout the experiment time schedule was carried out. A sample weather conditions day of experiment is shown in Fig. 75.5.

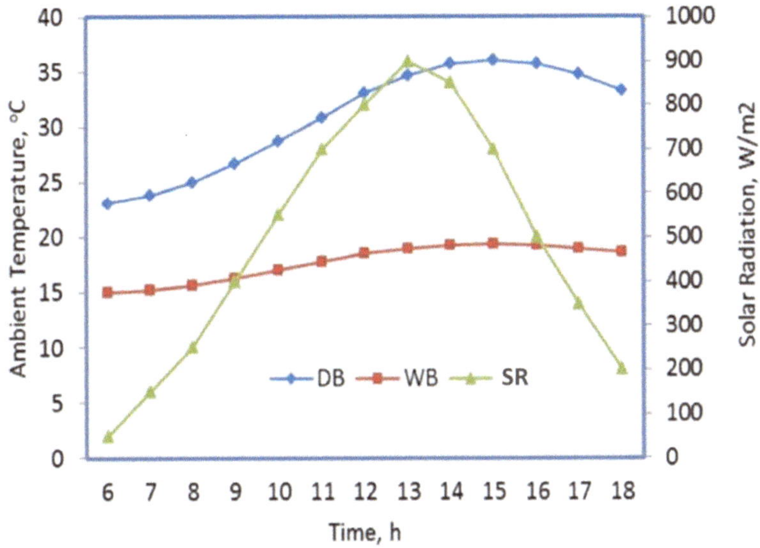

Fig. 75.5 Ambient temperature and solar radiation variation of the test day

75.4.2 Kiton Scarlet and Reactive Blue Dyestuff

- A stock solution was freshly prepared in the feeding tank, while continuous flow was adjusted to fulfill the optimum retention time. Sequential frequent samples were collected along the test period, and each sample was collected after 15 min.
- Raw as well as treated samples were collected and analyzed for the residual dye concentration.
- Results were collected and analyzed.
- Results showed that a gradual reduction was observed during collection time, while complete color removal was observed after 60 min for the first dye (*Kiton Scarlet*) and after 120 min for the second dye (*Reactive Blue*).

All results were recorded for the NIWWTP without any additional parameters. Progress and more efficient removal can be fulfilled with any additional parameters such as a catalyst or other enhancement parameters. The initial dye concentration of the first dye type (Red Kiton Scarlet 4R Acid-Ciba anionic dyestuff) was 4.0994 while in the second dyestuff (Reactive Blue anionic dyestuff) it was 6.21432. This means that two different concentrations have been used to investigate the system capability to treat industrial wastewater.

The percentage of dye removal from the water sample versus the retention time for both heavy concentrated Reactive blue dye stuff and moderate concentration Kiton scarlet 4R red dyestuff is shown in Fig. 75.6. It is clear that the moderate concentration Kiton scarlet 4R red dyestuff was totally removed from the water sample within the retention time of 60 min. while the heavy concentrated Reactive Blue dyestuff was totally removed from the water sample within the retention time

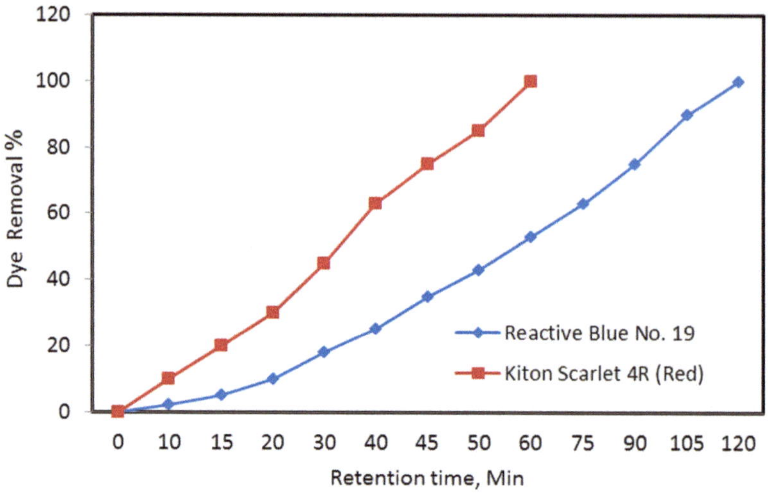

Fig. 75.6 The percentage of dye removal from the water sample versus the retention time

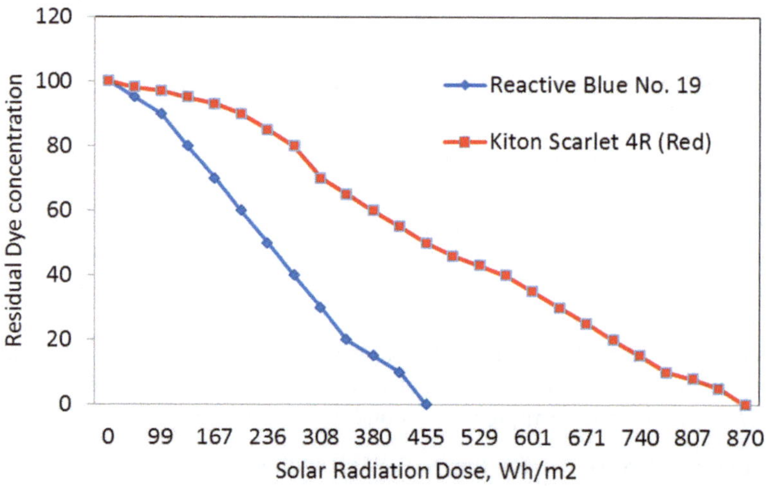

Fig. 75.7 The residual dye concentration from water sample versus solar radiation dose

of 120 min, which means that the higher the sample concentration the higher the retention time needed.

The residual dye concentration of the studied dyestuff is presented versus the exposed solar radiation dose as shown in Fig. 75.7. It is clear that the moderate concentration Kiton scarlet 4R red dyestuff was totally removed from the aqueous sample, when it was exposed to 455 Wh/m^2, while the heavy concentrated Reactive Blue dyestuff was totally removed from the water sample when it was exposed to 870 Wh/m^2, which means that the solar energy can positively contribute to the industrial wastewater treatment, as the daily average solar radiation dose falling

Fig. 75.8 The flow stream of the red anionic dyestuff

Fig. 75.9 The flow stream of the blue anionic dyestuff

on the horizontal surface in Cairo is about 5500 Wh/m^2/day that encourages one to utilize solar energy in the field of industrial water treatment techniques.

The Solar Photo degradation of Commercial anionic dyestuff, namely Diphenyl Red 5B Direct dyestuff, where the fate of color degradation along the reactor basin, was illustrated in Figs. 75.8 and 75.9.

75.5 Conclusion and Recommendations

The given results showed that the system fulfills the following:

75.5.1 Conclusions

1. A remarkable reduction for both dyestuffs.
2. The quality of treated industrial wastewater complies with regulatory guidelines for wastewater disposal to water bodies.

3. The condensate can be safely reused in all categories of industrial sectors.
4. Fully operating with solar energy source.

75.5.2 Recommendations

1. The novel system is a safe way for wastewater treatments as a clean and cheap technology.
2. Reuse of reclaimed water as a nonconventional source of water.

References

1. Wesley WE Jr. (2001) Industrial water pollution control, 3rd edn, McGraw-Hill Science/Engineering/Math
2. Ilyas S, Sultan S, Rehman A (2012) De-colorization and degradation of azo dye, Synozol Red HF6BN, by *Pleurotusostreatus*. Afr J Biotechnol 11(88):15422–15429
3. Bizani E, Fytianos K, Poulios I, Tsiridis V (2006) Photocatalytic de-colorization and degradation of dye solutions and wastewaters in the presence of titanium dioxide. J Hazard Mater 136(1):85–94
4. Sahel K, Perol N, Dappozze F, Bouhent M, Derriche Z, Guillard C (2010) Photocatalytic degradation of a mixture of two anionic dyes: Procion Red MX-5B and Remazol Black 5 (RB5). J Photochem Photobiol A: Chem 212(2–3):107–112
5. Ibrahim NA, El-Awady MH, Abo-Shosha MH (1998) Removal of anionic dyestuffs from aqueous and effluent liquors using cement kiln dust, textile processing, Colour Publications PVT. LTD. Colourage
6. El-Awady MH, Estefan SF, Kamel EM, Abdel Alim IA, Zayed MF (2005) Utility of Hydromag-TiO_2 in the photocatalytic degradation of textile effluent and dyes, Egypt. J Chem 48(3):347–353
7. Sharma SK, Bhunia H, Bajpai PK (2012) Photocatalytic decolorization kinetics and adsorption isotherms of a mixture of two anionic azo dyes: reactive red 120 and reactive black 5. Desalin Water Treat 44:261–268

Chapter 76
Theoretical and Experimental Investigations of Solar Heating Systems Under Specified Output Conditions of Hot Water

W. Tadros, M. Saadeldin and S. A. Hassan

Abstract The theoretical analysis discussed in this work is a suitable mathematical tool by which the performance of the proposed collector can be predicted. The obtained experimental results coincide with the obtained theoretical data from a devised computer program. Controlled output temperature can be obtained from the proposed system. The performance of the tested collector under the proposed intermittent flow conditions is superior to that of the conventional thermosyphone flow collector.

Keywords Solar collector · Thermal efficiency · Thermosyphone · Intermittent flow · Mass flow rate · Solar heating system

76.1 Introduction

Solar water heating is the most promising technique for the utilization of solar energy. Hot water at different temperature levels can be obtained. Solar water-heating systems are mainly composed of a solar flat-plate collector and a hot water storage tank. Hot water flows from the collector to the storage by either natural or forced circulation. In the case of natural circulation, hot water is transferred from the collector to the storage by the thermosyphone phenomena. Hot water entering the storage is substituted by cold water from the same storage to be recirculated through the collector. The main drawback of this type is the night reverse cycle, which results in considerable night losses by sky radiation. In the case of forced circulation, hot water storage can be installed at a level either lower or higher than that of the collector exit. Water is circulated by a pump. A differential thermostat is used to switch off the circulating pump when the temperature of water at the exit of

M. Saadeldin (✉)
Physics Department, Faculty of Science, Cairo University, Giza 12613, Egypt
e-mail: msaadeldin@hotmail.com

W. Tadros · S. A. Hassan
Solar Energy Department, National Research Center, Giza 12622, Egypt

A. Sayigh (ed.), *Renewable Energy in the Service of Mankind Vol II*,
DOI 10.1007/978-3-319-18215-5_76

the collector is lower than that of hot water in the storage tank, which occurs during the cloudy days and during the night-time [1, 2]. In both of the thermosyphone and forced systems, hot water temperature is uncontrolled due to the continuous change of solar intensity from hour to hour and from day to day. If water is required at a temperature higher than that achieved by solar heating, auxiliary electric heaters should be used for achieving the required temperature. Solar water-heating systems are mainly composed of a solar flat-plate collector and a hot water storage. Hot water flows from the collector to the storage by either natural or forced circulation. A thermostat is inserted at the outlet of the collector, which permits the water to flow from the collector to the tank if its temperature is equal to the required value. A solenoid valve is inserted in the connection between the collector and the storage, which is activated by the thermostat.

76.2 Experimental Methods and Results

In this system, cold water is fed to the collector from the main water supply or from a cold water feeding tank that may be used in rural areas where water supply lines are not available. Water content of the collector is subjected to solar energy absorbed by the collector and as a result, its temperature increases. When the water temperature reaches to the required value, at which the thermostat is adjusted, the solenoid valve is activated and water flows from the collector to the storage tank. When the temperature of flowing water decreases than the required value, the thermostat switches off the solenoid valve and the cycle is repeated (Fig. 76.1). It is noticed that this system is suitable to be used in rural hot areas like that in Upper Egypt. The results are classified into two sets. The first set (Fig. 76.2) represents the behavior of the solar collector under the proposed intermittent flow conditions. The second set (Fig. 76.3) represents a comparison between the performance under the proposed intermittent flow conditions and that under conventional flow conditions [3, 4]. The first set of the experimental results is used to evaluate the behavior of the collector under the proposed intermittent flow conditions as will be discussed in the following text. The hourly discharged quantity of hot water per square meter of the absorber area at different output temperatures during summer and winter at dif-

Fig. 76.1 Schematic diagram of a forced-circulation solar water heater

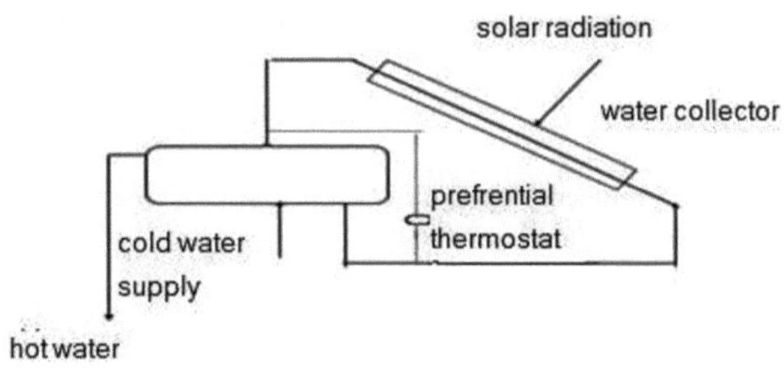

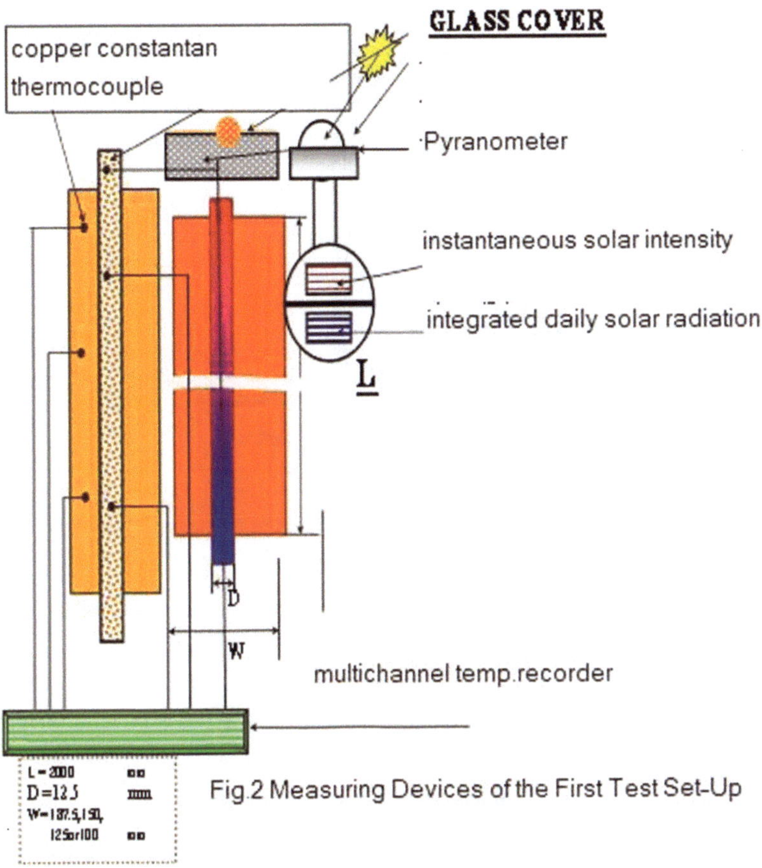

Fig. 76.2 Measuring devices of the first test setup

ferent values of (W/D) is deduced from the measured parameters during the first set of experiments. The maximum temperature of both the plate and water is constant during the time of the measurements. This is because the maximum temperature of water is calculated according to the temperature at which the thermostat is adjusted. Consequently, the maximum plate temperature is limited due to the discharging of hot water and refilling with cold water. It is shown that at the start of the heating cycle, the top loss coefficient is low due to the small difference between the plate and the ambient air temperature. The increase of the plate and water temperature is accompanied with an increase in the top loss coefficient.

For this reason, the behavior of the relation between the top losses and the time is nearly the same as that of the relation between the plate temperature and the time. The daily useful heat increases and consequently the overall efficiency decreases with the increase of the output temperature. This is because the increase in the output temperature results is the increase in the plate temperature. As a result, the thermal losses increase with the increase in the output temperature that, as discussed

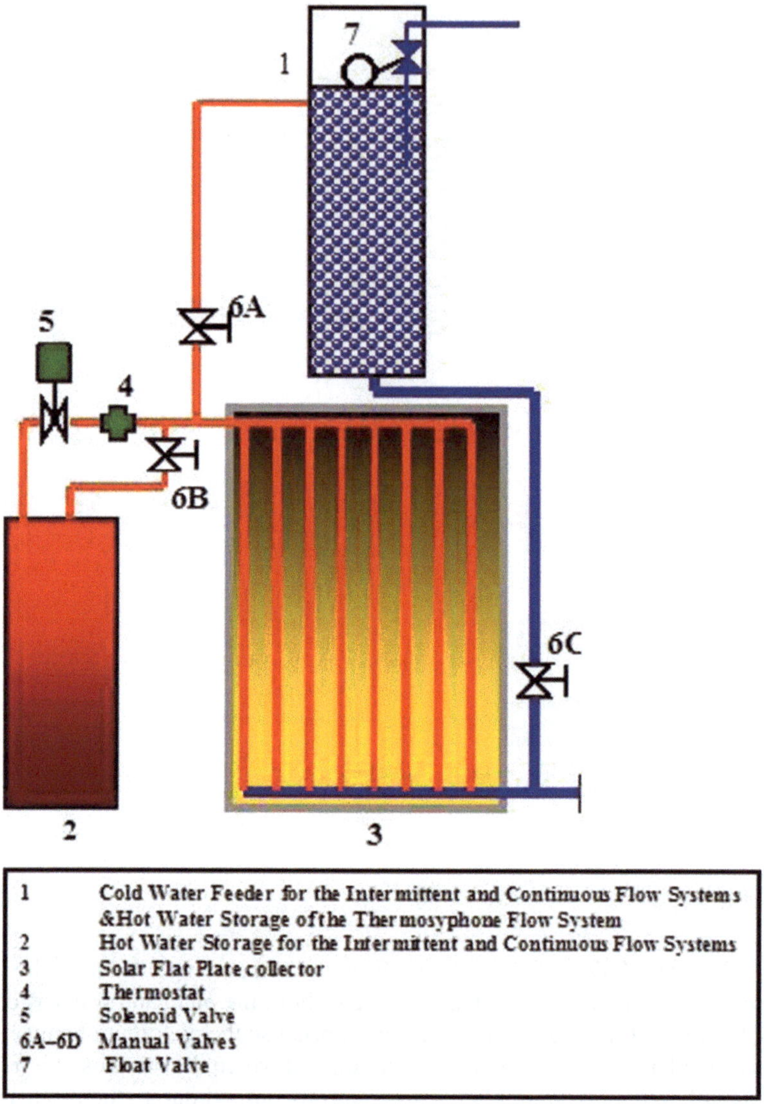

Fig. 76.3 Schematic diagram of the second test

before the increase of (W/D), results in a decrease in the productivity and efficiency of the collector. This is because the increase in the tube pitch results in an increase in the plate temperature and consequently results in the increase in the losses from the collector. During both summer and winter, the performance of the intermittent flow system is better than that of the thermosyphone system. This is because the efficiency of the collector under intermittent flow conditions is higher than before. In addition, the thermal losses in the piping system in the intermittent system are less than those in the thermosyphone system. The proposed collector can be successfully used in solar water-heating systems by which hot water at a prespecified temperature can be obtained. This can be achieved by the use of a thermostat in which hot water is not allowed to be discharged through a solenoid valve, from the collector at lower temperature. The devised computer program, discussed through this chapter, can be used to obtain all of the affecting parameters of the proposed collector at any given operation conditions and any technical data of the collector. The performance of the tested collector under the proposed intermittent flow conditions is better than that of the conventional thermosyphone flow collector. Controlled output temperature can be obtained from the proposed system while in the thermosyphone system, hot water temperature varies with the variation of the following parameters, the climatic conditions under which the system is operated and the rate at which the hot water is discharged from the storage tank that is accompanied with cold water feeding to substitute the discharged water; the proposed system has less construction requirements than the thermosyphone system [5].

Theoretical Analysis

Computer programs are basically devised for complete investigation of the collector performance. A subsidiary program is designed by which solar radiation input to the collector at any location in Egypt can be obtained at any value of the affecting parameters. The obtained results from these programs are presented and interpreted. Two systems that can be theoretically analyzed are the continuous flow and the proposed intermittent flow. The collector efficiency of a flat-plate collector [6] can be written as follows:

$$\eta = F_R\,\alpha\tau\text{-}F_R U_L\left(T_{fi} - T_a\right)/I_T \tag{76.1}$$

Where F_R is collector removal factor which is defined as the ratio of the actual heat gain rate to the gain, T_{fi} is the inlet fluid temperature which is equal to T_{f0} which is the outlet fluid temperature at the end of heating cycle, α & τ are transmissivity and absorptivity product for radiation falling on the collector respectively, I_T is the flux density on the top cover of the collector and U_L is overall heat loss coefficient.

The variation of water temperature rises during each month at different mass flow rates of water (m^*) through the collector is calculated. These results are obtained from the obtained output temperature and the given cold water temperature. The results are presented in Table 76.1. The variation of the collector efficiency with the mass flow rate depends on its heat removal factor as given by Eq. (76.1). It is noted that the heat removal factor is mostly constant during all months. Its value depends mainly on the water flow rate through the collector and the geometry of this collector. For the same geometry the heat removal factor increases with the increase of water mass flow rate. The mean values of the water temperature rise, the output water temperature, and the thermal efficiency of the collector at different mass flow rates during the selected months are summarized in Tables 76.1, 76.2, 76.3, 76.4, and 76.5.

From these tables it can be concluded that the mean output temperatures of the heated water at different months are shown in Table 76.3. From this table, the quantity of heat gain can be calculated. The quantity of heat input to the collector can

Table 76.1 Quantity of discharged water and its output temperature at different flow rates during January

Time	Discharged water in liters at (m^*)				Output temperature (°C) at (m^*)			
	0.005	0.01	0.015	0.02	0.005	0.01	0.015	0.02
7–8	18	36	54	72	18.5	16.3	15.8	15.7
8–9	18	36	54	72	23.7	17.8	17.6	17.7
9–10	18	36	54	72	26.6	19.6	19.8	18.4
10–11	18	36	54	72	28.3	21.5	21.6	19.6
11–12	18	36	54	72	29.2	23.0	22. 9	20.6
12–13	18	36	54	72	29.2	23.3	23.3	20.8
13–14	18	36	54	72	28.2	22.8	22.7	19.9
14–15	18	36	54	72	26.5	21.8	21.7	18.6
15–16	18	36	54	72	24.1	20.3	20.3	17.2
16–17	18	36	54	72	20.6	18.6	18.5	16.5

m^* (kg/m^{2}s) is the mass flow rate

Table 76.2 Quantity of discharged water and its output temperature at different flow rates during June

Time	Discharged water in liters at (m^*)				Output temperature (°C) at (m^*)			
	0.005	0.01	0.015	0.02	0.005	0.01	0.015	0.02
7–8	18	36	54	72	30.9	28.4	27.6	26.9
8–9	18	36	54	72	38.5	32.5	29.8	28.6
9–10	18	36	54	72	44.5	35.9	31.9	30.3
10–11	18	36	54	72	49.5	38.4	33.6	31.8
11–12	18	36	54	72	52.7	39.7	34.8	32.9
12–13	18	36	54	72	53.5	39.4	35.0	33.0
13–14	18	36	54	72	51.6	38.2	34.2	32.1
14–15	18	36	54	72	47.6	36.4	32.7	30.9
15–16	18	36	54	72	42.3	34.2	30.6	29.6
16–17	18	36	54	72	35.3	31.8	28.5	28.4

m^* (kg/m^{2}s) is the mass flow rate

Table 76.3 Daily quantity of discharged water and its output temperature at different mass flow rates and months

Month	$m^*=0.005$		$m^*=0.01$		$m^*=0.015$		$m^*=0.02$	
	Q	T	Q	T	Q	T	Q	T
Jan.	180	254.9	360	205	540	204.2	720	184.4
Apr.	180	385.3	360	289.5	540	283.3	720	249.3
Jun.	180	466.4	360	354.9	540	310.7	720	304.5
Oct.	180	398.5	360	305.2	540	275.4	720	260.5

m^* (kg/m^{2}s) is the mass flow rate

Q total daily quantity of heated water in liters, T output temperature

Table 76.4 Thermal efficiency of the collector at different mass flow rates and months

Month	Thermal efficiency			
	$m^*=0.005$	$m^*=0.01$	$m^*=0.015$	$m^*=0.02$
January	0.46	0.48	0.71	0.59
April	0.49	0.45	0.61	0.44
June	0.41	0.53	0.43	0.50
October	0.55	0.49	0.44	0.40

m^* (kg/m²s) is the mass flow rate

Table 76.5 Productivity of the intermittent collector at different values of $(T_{fo}-T_{fi})$ at $D_o=0.0125$ m and $W=0.1$ m during October

Temperature	Water at $(T_{fo}-T_{fi})$ of	Discharge difference	Local timer
30	20	10	
4.4	5.4	18.5	8–9
11.0	15.9	39.2	9–10
12.0	22.2	51.3	10–11
13.7	24.5	56.8	11–12
14.5	25.5	59.4	12–13
13.4	24.5	56.8	13–14
10.1	19.3	45.3	14–15
6.1	13.2	33.4	15–16
85.2	150.5	309.7	Total discharged water in liters
10.7	12.6	13.00	Total heat gained in MJ
0.27	0.29	0.30	Overall efficiency %

be obtained from Fig. 76.3. From these quantities, the thermal efficiency of the collector under different conditions is calculated and presented in Table 76.4. The thermal efficiency of the collector increases with the increase of the mass flow rate, and the collector is continuously fed with cold water at a temperature of T_{FI}. As a result, the efficiency will be decreased with time due to the increase of T_{fi}, as shown by Eq. (76.1). The results are in good agreement with previous results obtained by many researchers. This means that the computer program can be used for studying the performance of the continuous flow solar collectors. From Table 76.5 and other results taken at different months it can be shown that the performance of the collector under the proposed intermittent flow conditions is close to that under the continuous flow conditions at low specific flow rates. The increase of the flow rate results in an increase of the efficiency and a decrease in the output temperature of water when the cold water temperature is constant. It should be taken into consideration that the relation between the productivity of the continuous flow collector is not linearly proportional to the temperature rise of water. This is due to the effect of the change of the mass flow rate on the plate temperature and the heat removal factor. The increase of the plate temperature increases the thermal losses whereas the increase of the heat removal results in an increase of the thermal efficiency of the collector. As a result, the relation between the change of the collector efficiency and the change of its productivity is nonlinear.

A comparison between the performance of the collector under the proposed intermittent flow conditions and that under the conventional flow conditions has been made, and the following results are obtained from the measurement of the second set of experiments. Figure 76.5 shows the net heat gained from the system during summer and winter at following conditions, when the collector is operated under intermittent flow conditions at output temperatures of 40, 50, and 60 °C. When the collector is operated under thermosyphone system with a storage tank of a volume of 100 L. The heat gained by the collector in each case is calculated as follows, when it is operated under intermittent flow conditions, the heat gain is equal the total mass of the discharged water multiplied by $[(T_o - T_i) * C_p]$. When it is operated under thermosyphone conditions the hot water storage is discharged after the sunset and the temperature top of each pitch of 5 L is measured. The total heat gain is equal to $[5(T_{op} - T_{ii}) * C_p]$ where T_{op} is the temperature of each discharged petch and T_{ii} is the temperature of the cold water in the storage tank at the start of the experiment at sunrise.

From Fig. 76.4 it can be seen that, during summer and winter the heat gained and the overall efficiency of the collector under intermittent flow conditions is higher than those of the thermosyphone collector at all values of the output temperature of the intermittent collector. This is because the mean plate temperature of the thermosyphone collector is higher than that of the intermittent flow one. This is due to the continuous discharge of hot water from the intermittent collector and refilling it with cold water. The mean temperature of the obtained hot water from the thermosyphone of a storage tank of 100 L per one square meter of the collector is equal to 45 °C in summer and is equal to 32 °C in winter. When the thermostat of the intermittent flow collector is adjusted at 45 °C during summer, the daily quantity of discharged water is equal to about 122 L. During winter, when the thermostat is adjusted at 32 °C, the daily quantity of discharged water is equal to about 110 L. This means that during both summer and winter, the performance of the intermittent flow system is better

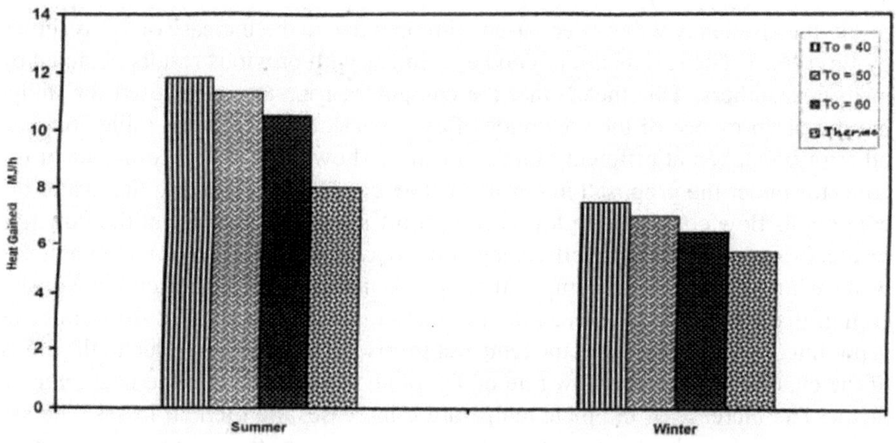

Fig. 76.4 Heat gained by one square meter of the collector at different output temperatures during summer and winter, T_i in summer=250 C and T_i in winter=150 C

than that of the thermosyphone system. This is due to the higher efficiency of the collector under intermittent flow conditions than that of the thermosyphone collector, as mentioned before. In addition, the thermal loss in the piping system in the intermittent system is less than that in the thermosyphone system. [7, 8]

76.3 Conclusions and Future Work

From the theoretical analysis and the experimental investigation of the proposed collector under intermittent flow conditions, it is concluded that

(1) The proposed collector can be successfully used in solar water-heating systems by which hot water at a prespecified temperature can be obtained. This can be achieved by using a thermostat by which hot water is not allowed to be discharged through a solenoid valve from the collector at lower temperature.
(2) The theoretical analysis, discussed in this chapter, is a suitable mathematical tool to predict the performance of the proposed collector.
(3) The devised computer program, through this chapter, can be used to obtain all of the affecting parameters of the proposed collector at any given operation conditions and any technical data of the collector.
(4) The obtained experimental results coincide with the obtained theoretical data derived from the devised computer program.
(5) The performance of the tested collector under the proposed intermittent flow conditions is superior to that of the conventional thermosyphone flow collector. This is because:

   - Continuous cyclic discharging of hot water from the collector and refilling it with cold water prevent the continuous increase of the absorber plate temperature as in the case of the thermosyphone system. As a result of this phenomenon, thermal losses from the collector are limited and most of the absorbed solar energy is transferred to water and the overall efficiency of the system is increased.
   - The piping system of hot water in the proposed system is shorter than that in the conventional thermosyphone system.
   - No fluctuation in the temperature of hot water, in the storage tank, is observed while in the thermosyphone system, the discharging of hot water is accompanied with mixing of cold water with hot water in the storage tank.

References

1. Kaptan IN, Killic A (1996) Theoretical and experimental investigation of a novel built-in-storage solar energy water heater. Sol Energy 57:393–400
2. Mohamed AA (1997) Integrated collector storage tank system with thermal diode. Sol Energy 61:211–218

3. Khattab NM (1976) Factors affecting the performance of solar water heaters, MSc Thesis, Faculty of Science, Cairo University
4. Hassan SA (2003) Theoretical and experimental investigation of solar heating systems at specified output conditions of hot water, PhD Thesis, Faculty of Science. Cairo University, Egypt
5. Tadros WH, Soliman SH (1989) Effect of solar radiation on performance of flat plate collector. J Inst Eng India 69:31–34
6. Iqbal M (1983) An introduction to solar radiation. Academic Press, New York
7. Jerome E, Johnson Ed D (2009) The development of a solar thermal water purification, heating, and power generation system Madison, WI March
8. Werner W (2003) Solar heating system for houses: a design handbook for solar combisystem

Chapter 77
Thirty-Five Years of Experience with Research, Promotion and Use of Solar Cookers to Save Conventional Fuels and Reduce Carbon Emission

Shyam S. Nandwani

Abstract Since 1979, the author has designed, studied and used various models of solar cookers—from thermal food warmer in 1979, to PV multi-voltage cooker in 2010 and to single-axis tracker for solar cookers in 2012. Besides publishing the results in technical journals and national and international conferences, these have been promoted through lectures in communities, educational and cultural centres, TV, radio, newspapers, workshops and seminars in Costa Rica and many other countries. For one of the models of solar cooker, the author got a patent in 1984.

In the present work, different models and some promotional activities, especially at educational institutes, to warm the lunch for students are mentioned.

Keywords Solar cooker · Solar oven · Solar hybrid oven · Solar cooker-cum-dryer · PV solar cooker · Energy saving

77.1 Introduction

Food provides nutrients to grow and run our body. Thus we are what we eat. Therefore, if we improve the quality of cooking, we improve the quality of life. What could be better than cooking with the blessing of the sun.

The concept of hot box for cooking is more than 230 years old [1]. In 1776, Benedict de Saussure, a Swiss alpinist, using a simple box with various glasses on the top and solar energy, could attain a temperature of around 90–100 °C and could bake some fruits. Thus, solar cooking is not new; however, its acceptance and massive use is a tough process.

Unlike many authors who start their studies on solar cooker because of need of firewood for 2500 million poor people, cutting firewood, deforestation, walking long distance with risk to carry it, lung problem due to smoke and global warming

S. S. Nandwani (✉)
P.O. Box 728, 3000 Heredia, Costa Rica
e-mail: snandwan@yahoo.com

A. Sayigh (ed.), *Renewable Energy in the Service of Mankind Vol II*,
DOI 10.1007/978-3-319-18215-5_77

Fig. 77.1 A simple low-cost solar food warmer. (March 1979)

and so on, I started it for the first time in February 1979 at my house in Costa Rica because of family necessity.

The main source of electricity in this country was, and still is, hydropower (about 80 %). Like most people, I was using an electric range for cooking at my house. At the beginning of 1979, due to high solar radiation (as usual), there was a shortage of water in rivers; this resulted in a shortage of electricity. Owing to this crisis, national electricity utility (ICE), in February 1979, decided to ration electricity for 2 days a week, from 7 am to 5 pm.

However, the literature mentions that water and food pasteurize at 65–70 °C and food cooks at about 75–82 °C. Thus, I made a simple solar hot box at house just to warm food, cooked the previous night [2].

A. Food Warmer

As shown in Fig. 77.1 it is an insulated wooden box and a galvanized iron metal sheet painted black, glass wool as heat insulation and one window glass on top to reduce the heat losses. The low-wavelength solar radiation passes through transparent glass and is absorbed by the black paint. With this the metallic plate is heated. Now the metallic hot plate also emits the radiation but in the higher wavelength region (heat). The same window glass on the top of the box impedes the transmission of heat radiation, as a result the air in the box is heated and can be used for warming food.

The maximum air temperature in the solar oven reached up to 80–90 °C and 3–4 meals were heated during mid day in 40–60 min. Materials cost was approximately US$ 15–20. Our purpose was solved. However, I got interested in this practical device and thought of continuing making different models to cook meals and also started disseminating/promoting solar cookers in and out of Costa Rica.

For the past 35 years I have designed, constructed, studied and published various types of cookers—hot box types, hybrid solar electric (110 V AC and 12 V DC), solar electric microwave oven, cooker with dryer, cooker with water heater/pasteurizer, and so on. Although the details of cookers can be obtained from various articles [3, 4], we can mention in brief some of our cookers with corresponding references.

Fig. 77.2 A conventional hot box oven with wood (**a**) and with stainless steel (**b**)

B. Hot Box Oven for Cooking
After seeing the success with food warmer, it was decided to improve the box and convert it into a solar food cooker. Two changes were made in the solar food warmer mentioned already. First, we added another window glass on the top of the box to reduce heat losses, and second, we used a reflector to increase the solar radiation [5]. The reflector can be adjusted (recommended) every hour or two. Figure 77.2a and b show the improved solar oven constructed in 1982 and it is still in use at the author's house.

With these two changes, the maximum plate temperature was increased to 150 °C with no food inside and to 110–120 °C during cooking. Normally, 3–4 meals can be cooked in 3–4 h.

This has been used for cooking, baking, and roasting food (rice, beans, vegetables, lentils, meat, deserts, etc.), and also for purifying tap/river water from the microbiological point of view (pasteurization).

Once convinced with its working, another cooker was made with stainless steel (Fig. 77.2b) as an outer box to increase the durability of the oven due to high humidity and rain in Costa Rica. In 1984, we got the national patent (no. 2367) for this solar oven. This type is the most common type of solar cooker used worldwide.

Some Problems/Inconveniences Depending on the quantity of food and change in climate, the food may or may not be always cooked properly. Although partially cooked food could be completely cooked on a conventional stove and cooking could be finished, however practically no housewife will love this idea and this could be the sufficient reason to get frustrated. Fortunately, this happened with my wife and it compelled me to look for a solution. Although heat storage cookers are studied [6], they are very expensive. Thus in 1986, I preferred to make a hybrid cooker. Both solar-electric ([7], Fig. 77.3a) and solar-LPG gas ([8], Fig. 77.3b) hybrid cookers were made but I will mention only the solar electric oven that I have found to be more convenient and we use it at home whenever required.

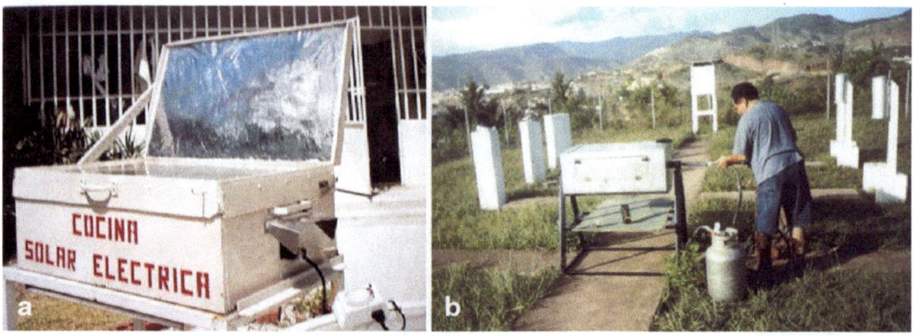

Fig. 77.3 Hybrid solar electric oven (**a**) and hybrid solar gas oven (**b**)

C. Hybrid (Sun–AC Electricity) Oven

As shown in Fig. 77.3a, it is similar to a hot box oven, but the black metallic plate has an electrical resistance beneath and a thermostat to regulate the temperature [8]. The hybrid cooker kept in the sun is also connected with electricity, and the thermostat is fixed at 90–100 °C. If solar radiation alone enables the system to reach this temperature, electricity will not be consumed/passing. If the solar intensity falls below this, and plate temperature does not attain the required temperature, electricity is connected automatically to the electric plate and is disconnected if the solar intensity is increased.

With this device your meal is cooked at the right time with reduced consumption of conventional fuel, as electricity in this case.

D. Electric Storage Model: Multi-voltage Oven

In the hybrid model mentioned already, the electricity from the conventional source (110 V AC) was used, thus it could be used only at places connected to the electric grid. There are many families/places (like farm house, ecological cabins) that are not connected to the electric grid. To satisfy their needs, we designed another model using electricity generated from photovoltaic panels and solar energy.

Here photovoltaic panels are used to convert solar energy in DC electricity. This is then stored in deep-cycle 12 V batteries. The stored electricity is directly used for cooking or converted in 110 (or 220) V AC through an inverter, which can be used efficiently for running conventional a microwave oven [9] or a 200–350 W low-wattage cooker [10]. Figure 77.4 shows the final model. Depending on the connections of the resistance (series/parallel), the oven could be used with 12 V DC or 110 V AC.

E. Multipurpose Oven

Cooking needs higher temperature, and thus requires good solar radiation. Depending on the climate/solar radiation at a given place, solar cookers can be used for 6–9 months in a year. On the other hand, water heating and drying need less temperature and thus the process can be achieved even with low solar radiation. It means if the same cooker (with some modification) is used for both purposes—cooking and water heating (or drying)—it can be used for longer period (9–10 months). With this objective, two multipurpose devices—cooker-cum-water heater [11] and cooker-cum-dryer [12]—were designed, made and studied. Here we mention only cooker-cum-dryer.

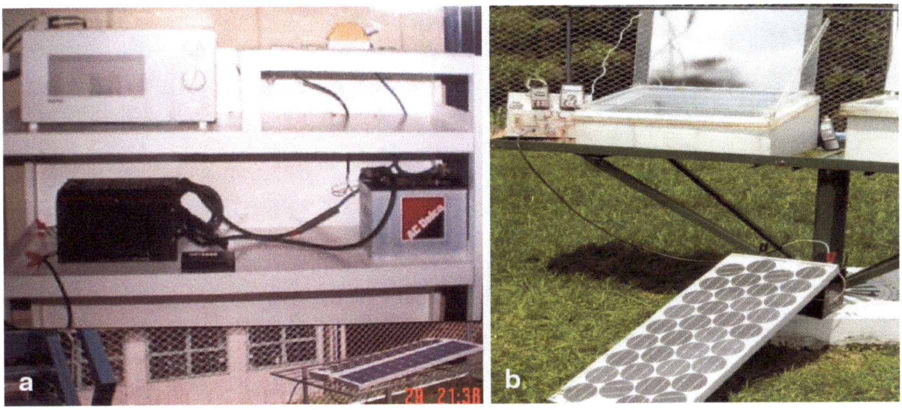

Fig. 77.4 A PV solar oven, microwave (**a**) and a low-wattage multi-voltage oven (**b**)

Fig. 77.5 A hybrid solar electric dryer (**a**) and in the mode of drying (**b**)

As can be seen in Fig. 77.5a, the model is conceptually similar to a conventional solar oven, except that the box has two holes in front of the oven for the entrance of ambient air as well as another three holes at the back of the oven for the exit of hot, humid air. The metallic tray(s) with the products to be dried are kept at the top of the absorbing plate. The holes are closed for cooking and opened for drying. Here the metallic plate also has electric resistance as well as a thermostat to regulate the plate temperature. For water heating, the metallic plate is replaced by a rectangular tank for storing water (capacity about 20–30 L).

The real solar drying has been done with various domestic products, such as tomatoes, coriander, onions and pineapple (Fig. 77.5b).

More solar ovens are studied and used but are not described here due to space constrains. These are concentrating cooker [3], short-term heat storage cooker based on sensible heat [13] and cooker based on phase change [14], two-compartment oven, mainly for academic purpose [14], and portable PV solar oven [15].

Fig. 77.6 Use of solar ovens for warming lunch at one educational centre, Costa Rica (2014)

Dissemination Activities In addition to publish articles for my academic career and get promotion, somehow I also wanted that more people should use the solar cooker whenever possible. Thus I thought of organizing workshops to teach how to make it, use it properly and cook food on it not only in Costa Rica but in other 10 countries too whenever requested. Some of our participants have made solar cookers for their personal use; some have even made the foundation to promote solar cooking and even some sell solar cookers based on the requirement of their clients. Maybe about 500 solar ovens have been made individually, without any government incentive.

In recent years, directors of some schools and colleges have asked for solar cookers for warming lunch of the students during mid day with the intention to reduce the use of microwave ovens and consumption of electricity. Even some institutions and foundations, including mine, have earned national and international awards for research on using solar cookers. Figure 77.6 shows just one of the solar ovens (1.2 m^2) designed by us and installed in May 2014 at one college in San Jose to heat 50–60 lunch boxes.

77.2 Advantages

Solar cookers have many advantages such as saving energy and money and improving health of persons and the planet [4, 16].

Combining the solar cooker with the conventional fuel stove, users will need less fuel (especially firewood) and not require to walk daily, could be once every third or fourth day to look for firewood. Reduced use of firewood for cooking will lead to less deforestation.

However, using one family solar cooker (0.20–0.25 m^2 of absorbing area) for 7–8 months in a year can save about 700–900 kWh electricity or 750 kg firewood per year. This can reduce the emission of about 1–1.25 t of CO_2 per year [16].

77.3 Conclusions

There is no doubt that solar cooking has many advantages; however, we cannot forget that it has some limitations too. On the basis of my 35 years of experience, there are various steps that could be undertaken to increase the awareness and use of solar cookers. Some of these could be:

- Adopting different medium of communications to inform, show and convince people about the use of solar cookers,
- Designing the solar cooker appropriately for meeting the needs of the people and communities for which they are intended,
- Making both commercial and DIY (do-it-yourself) models available,
- Teaching the buyers the proper use of solar cookers, not only selling,
- Informing the users about both the benefits and limitations of solar cookers, including very clearly that the device is intended to save and not to substitute conventional fuels.

Thus, with the combined efforts of designers, manufacturers, promoters and government incentives and so on, the use of solar cooker could be increased.

Acknowledgements The author is thankful to Solar Cookers International, Sacramento, CA, USA; the Abdus Salam International Center for Theoretical Physics, Trieste, Italy; Red Iberoamericana de Coccion Solar de Alimentos, Salta, Argentina and World Renewable Energy Network, England, for invitations at their meetings, where good information was exchanged among participants of different countries for research and dissemination. The author is also thankful to Technicians Carlos Delgado for constructing some of these cookers and Guillermo Torres for making some measurements.

References

1. Cookers International. Sacramento, CA, USA Non Government Organization. Solar. www.solarcooking.org, www.solarcookers.org
2. Nandwani SS (1979) Sunny visións—getting a charge from Old SUN. Report on this solar food warmer. *The TICO TIMES* (Central American News Paper, www.ticotimes.net), Costa Rica, May 4, p 21
3. Nandwani SS (2011) Solar cooker and dryers to conserve human and planet health, Encyclopedia of sustainability science and technology, in the section solar thermal energy for New Springer Encyclopedia, March 15, 2011
4. Nandwani SS (1993\2004) Book, La Cocina/Horno Solar, Hagala Usted mismo, 2nd revised edn. Editorial Fundación Universidad Nacional, Costa Rica, p 100, 120
5. Nandwani SS (1988) Experimental and theoretical analysis of simple solar oven in the climate of Costa Rica I. Sol Wind Tech 5(2):159–170 (Pergamon Press)
6. Rommel M, Stahl W, Wittwer V (1986) A solar cooker based on a flat plate collector with oil storage. Sun Work Eur. April(1):3–5.
7. Nandwani SS (1989) Design, construction and experimental study of electric cum solar oven. Sol Wind Tech 6(2):149–158

8. Flores MA, Flores AE, Nandwani SS (2002) Diseño, Construcción y Funcionamiento de un Horno Solar Híbrido CPC. Presented at XI Congreso Iberico e VI Congresso Ibero—Americano de Energía Solar, held at Vilamoura (Algarve), Portugal, Sept 29–Oct 2, Published in the CD Rom, pp 128–137
9. Currin C, Nandwani SS, Marvin A (1994) Preliminary study of solar microwave oven. In: Nandwani SS, Pejack E, Blum B (eds) Proceedings of the second world conference on solar cookers' use and technology, July 12–15. Universidad Nacional, Costa Rica, pp 149–158
10. Nandwani SS (2008) Design, construction and study of a hybrid and dual voltage solar cooker in the climate of Costa Rica. Renew Energy (Annual Edition):103–105 (Sovereign Publications)
11. Nandwani SS, Otoniel FG (1993) Experimental study of solar oven cum water heater and solar oven cum Drier- II. In: Nandwani SS, Pejack E, Blum B (eds) Proceedings of the Second World Conference on Solar Cookers—Use and Technology, July 12–15. Universidad Nacional, Costa Rica, pp 273–284
12. Nandwani SS (2007) Design, construction and study of a hybrid solar food processor in the climate of Costa Rica. Renew Energy 32(3):427–441
13. Nandwani SS (1990) Cheap solar oven with heat storage, preliminary study for Costa Rica's climate, the Heliograph, Royal Institute of Technology, Stockholm, No. 2, pp 4–13
14. Nandwani SS, Steinfart J, Henning HM, Rommel M, Wittwer W (1997) Experimental study of multipurpose solar hot box at Freiburg, Germany. Renew Energy 12:1–20
15. Nandwani SS (2006) Estudio de una Estufa Solar Portatil y Hibrida—I. HIDRORED (Spanish Journal, Peru) Oct 20:1–6
16. Nandwani SS (1996) Solar cookers—cheap technology with high ecological benefits. J Ecol Econ 17:73–81

Chapter 78
Development of an Efficient Low- and Medium-Temperature Vacuum Flat-Plate Solar Thermal Collector

G. S. F. Shire, R. W. Moss, P. Henshall, F. Arya, P. C. Eames and T. Hyde

Abstract Production of heat accounts for over half of our overall primary energy consumption in domestic and industrial applications. Despite the great scope for deployment of solar thermal collectors to provide low- and medium-temperature heat, there is relatively little uptake of this technology. The requirements for heat provision are studied, and the desired characteristics of potential solutions considered. Application areas are discussed in addition to the potential for system integration. An assessment is made of the shortcomings of solar thermal collectors and the requirements for new technologies suggested. This leads to a design approach for a collector that is effective across a range of applications and provides further supplementary benefit for system or building integration.

A vacuum flat-plate (VFP) solar thermal collector is proposed as a solution to the requirements for domestic and industrial heat at the low- to medium-temperature range. VFP solar thermal collectors have several potential advantages over other collector types: they may deliver heat efficiently at 150–250 °C, and they may be made much thinner than the existing collectors, offering new application opportunities in industrial process heat and for building integration. VFP collectors achieve this by combining the optical properties of flat-plate (FP) collectors with the heat loss characteristics of evacuated tubes. This chapter considers the development of a micro/mini-channel absorber plate and vacuum enclosure for a VFP collector.

Keywords Renewable energy · Solar · Thermal · Heat transfer · Vacuum

G. Shire (✉) · R. Moss
School of Engineering, University of Warwick, Coventry CV4 7AL, Coventry, UK
e-mail: G.S.F.Shire@warwick.ac.uk

P. Henshall · P. Eames
Centre for Renewable Energy Systems Technology, Loughborough University, Loughborough, UK

F. Arya · T. Hyde
School of the Built Environment, University of Ulster, Coleraine, UK

A. Sayigh (ed.), *Renewable Energy in the Service of Mankind Vol II*,
DOI 10.1007/978-3-319-18215-5_78

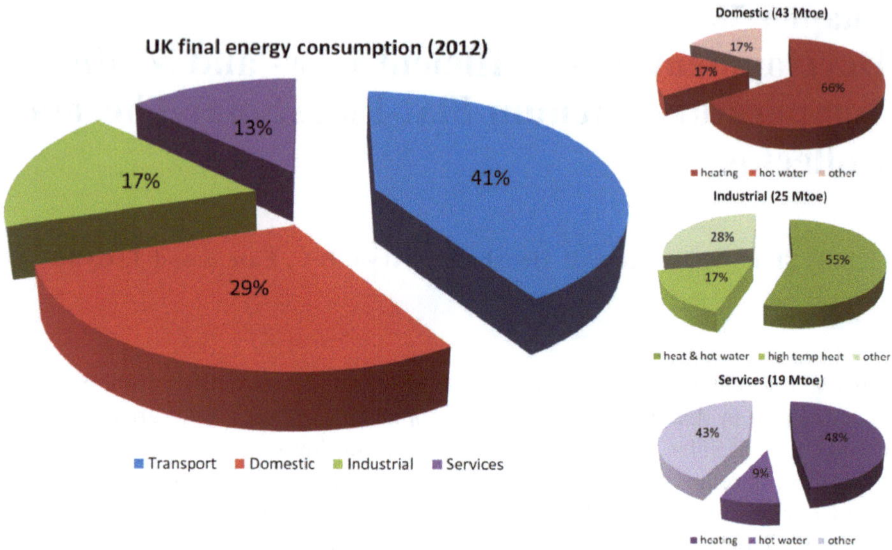

Fig. 78.1 UK final energy consumption by sector [3]

78.1 Introduction

The extraction of energy from fossil fuels and the consequential impact on global climate change is perhaps the most significant issue of our age and has been the subject of major intergovernmental talks since the Rio Earth Summit of 1992. Most nations have agreed targets for reducing emissions of carbon dioxide, and there is significant pressure on developed nations to lead the way in this regard since they are principally responsible for the great majority of historic emissions. The UK has pledged to reduce overall emissions of greenhouse gases by 80 % from 1990 levels by 2050, and this was established in the Climate Change Act [13]. If we are to develop an effective strategy for achieving our targets, we must first consider our current energy needs, where this energy is being used and in what form.

Provision of heat is a large component of our energy needs and should not be overlooked. In fact, 40 % of all fossil fuels are burnt for low-temperature heating purposes [12]. Efficient, low-cost generation of renewable low- and medium-temperature heat is crucial to meeting our obligations to reduce emissions. The most recent data from the Department of Energy and Climate Change [3] show that provision of heat still accounts for 44 % of UK's total primary energy use. Figure 78.1 illustrates how this consumption of energy is split between various sectors and how it is utilised within each sector (other than transport).

From Fig. 78.1, it can be seen that the domestic, industrial and services sectors use almost 70 % of the total energy consumed to provide heating and hot water. Although space heating is a seasonal demand, the summer requirements for heat are comparable to the total electrical demand in the UK.

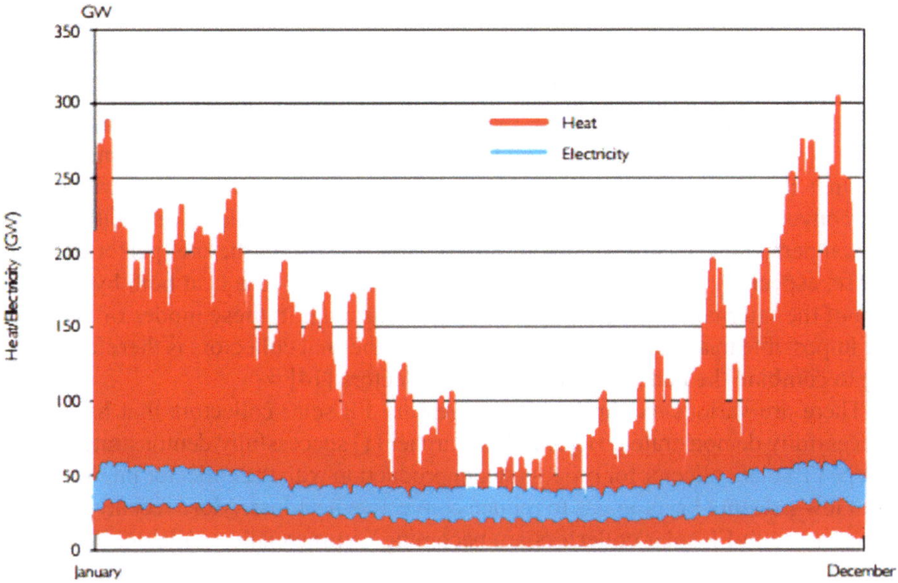

Fig. 78.2 Comparison of heat and electricity demand variability across a year (domestic and commercial), 2010 [2]

Figure 78.2 shows the seasonal variation of the heat demand compared to the electrical demand in the UK for 1 year. There has been some talk of this heat demand being provided by heat pumps utilising renewable electricity. However, it is clear from this figure that electricity cannot be used in order to provide all our heating needs. Grant Wilson et al. [7] showed that even a partial electrification of domestic heating demand will have serious implications for UK's ageing electrical transmission and distribution networks. Shifting just 30 % of only the non-daily metered heat demand onto electrically driven heat pumps with a coefficient of performance of 3 would result in a 25 % increase in daily peak electricity demand. This daily peak is also a significant underestimate of the variability in instantaneous power demand on the electrical network. Even a simple analysis demonstrates that significant electrification of heating will not be practicably achievable with the existing electricity grid.

There is clearly a need for renewable heat that is not delivered through the electricity grid. Although seasonal, a significant proportion of our energy needs could be delivered by solar thermal systems; MacKay (2009) estimates that solar thermal systems could provide 10 % of our total energy needs. Despite the seasonality of space heating, more than half of industrial energy consumption in the UK is required for heating, and generating hot water. This heat is in large part required for process hot water use in the paper, chemicals and food processing industries, for which demand is spread throughout the year and solar thermal collectors would be appropriate [9].

78.2 Background

Conventional solar thermal collectors are available in two forms: glazed flat-plate (FP) design and evacuated tube (ET) design. The former is simpler to construct but suffers from higher heat loss by conduction and convection through the internal air volume; the latter incorporates an insulating vacuum that eliminates conduction and convection within the collector glazing. FP collectors typically exhibit superior optical performance compared with ET collectors, but their thermal performance is worse, especially at elevated temperatures. FP collectors lose heat both by convection of the internal air (or gas) and conduction through it; these modes of heat loss are impossible in a vacuum. Research into solar thermal collectors is therefore seeking to combine the benefits of ET and FP collectors [14].

There are various examples of low-pressure FP solar collectors that have been successfully demonstrated. Benz and Beikircher [1] successfully demonstrated a prototype FP solar collector for process steam production with the collector interior filled with low-pressure krypton gas to reduce convective heat loss. More recently, a number of vacuum FP solar collectors are becoming commercially available. Both SRB Energy and TVP Solar are developing solar thermal collectors that utilise vacuum technologies developed at the European Organization for Nuclear Research (CERN). It is anticipated that a vacuum flat-plate (VFP) solar collector will exhibit greater efficiencies at higher temperatures in comparison to both ET and FP collectors.

The concept of employing an evacuated or low-pressure enclosure to enhance the thermal performance of FP solar collectors is a concept which dates back to the 1970s [6]. At this time, FP solar collectors were limited in their achievable performance, with efficiencies usually less than 40 % for absorber plate temperatures greater than 100 °C. Eaton and Blum [6] suggest that the use of only a moderate vacuum environment (150–3500 Pa) between the absorber plate and enclosure glass cover is sufficient to allow the collector to efficiently operate at temperatures exceeding 150 °C. Achieving higher temperatures would allow FP collectors to be considered for process heat applications. The moderate vacuum pressure range, while being sufficient to effectively suppress convective heat transfer between the absorber plate and the collector glass, still allows for gas conduction heat transfer to take place. Gas conduction can account for several Watts of total power loss from a solar collector [1]. Subsequently, it is desirable to attain vacuum pressure between the plate and glass cover of less than 0.1 Pa in order to fully suppress both convection and gas conduction processes and maximise collector thermal performance.

78.3 Collector Design

Attaining and maintaining enclosure pressures below 0.1 Pa for an adequate product lifetime represents a significant engineering challenge for an FP collector geometry. This is especially the case when the vacuum layer volume is very small, as in the case of vacuum glazing, which typically employs vacuum layers less than 0.5 mm

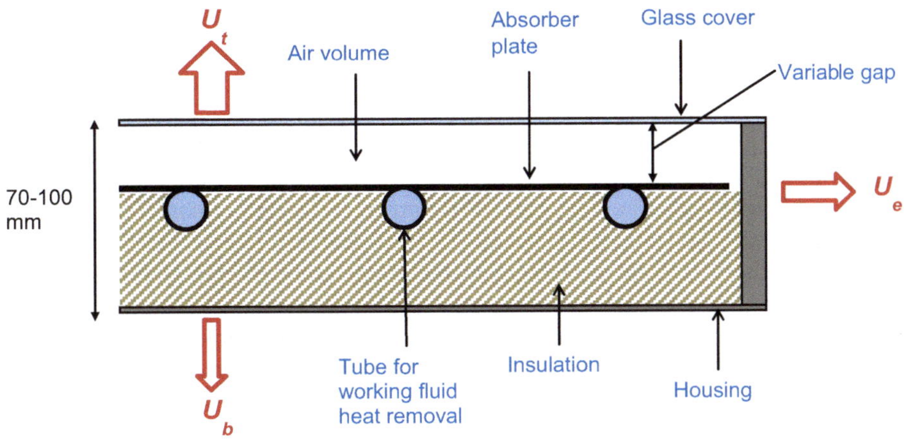

Fig. 78.3 Conventional configuration of an FP collector [3]

thick [5]. To this end, the design of the evacuated enclosures should protect the glass cover and sealing material from the large stresses imposed by atmosphere pressure forces and stresses due to differential thermal expansion among the various enclosure components [8].

Conventional FP solar collectors are typically configured as depicted in Fig. 78.3.

As can be observed in Fig. 78.3, the absorber plate and heat removal tubes are insulated on the rear side of the collector with an air gap between the absorber and glass cover on the front side. For such collectors, convective heat loss between the absorber and glass cover can be significant. Typically, heat loss from the collector is characterised by the collector overall loss coefficient (U_L), where U_L is calculated as:

$$U_L = U_t + U_b + U_e, \quad (78.1)$$

where U_t is the top loss coefficient, U_b is the back loss coefficient and U_e is the edge loss coefficient (see Fig. 78.4).

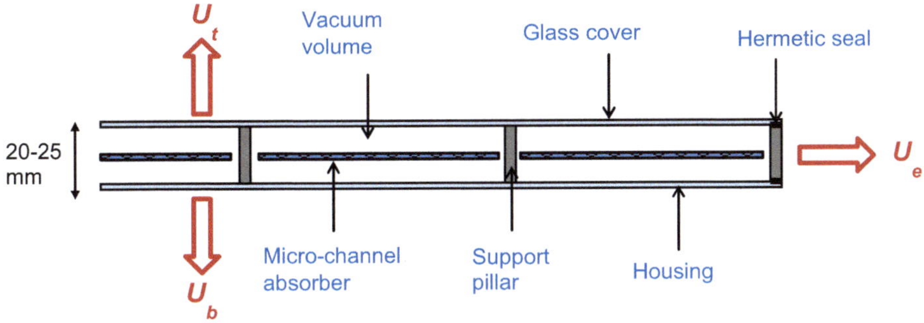

Fig. 78.4 VFP configuration

The VFP configuration under consideration in this chapter is depicted in Fig. 78.4. In this VFP configuration, the absorber plate is suspended within the housing such that it is completely surrounded by vacuum, suppressing convective heat loss. An array of support pillars is required to allow the glass cover and housing to resist atmospheric pressure loading resulting from the evacuated volume within the collector. In this VFP configuration, it is expected that U_t will be similar in magnitude to U_b, with only radiative heat transfer occurring between the absorber plate and glass cover/back housing. Subsequently, a large decrease in heat loss is expected, resulting in the VFP collector being capable of operating efficiently at higher temperatures.

The absorber plate of the collector is that of a micro-channel or flooded plate design, which offers several benefits to the design and operation of such a collector. The absorber plate can be made very thin, allowing the vacuum enclosure to be of a reduced height: this offers new potential applications in building integration. Because there are no long fins in a flooded plate design, the absorber plate has a very uniform temperature, and consequently, the heat removal factor for the absorber plate is very high. This improves thermal efficiency but has to be balanced against the pumping power required to extract that heat [11].

The vacuum insulation layer within the enclosure and surrounding the solar absorber can be very thin while remaining effective, thus allowing the collector itself to be only slightly deeper than the depth of the solar absorber plate, as no bulky backing insulation is required. These characteristics greatly enhance the potential for collectors to be integrated into the facades of commercial buildings.

78.4 Collector Performance

Using a 1-D FP solar collector modelling methodology, as described by Duffie and Beckman [4], values of heat loss coefficients were estimated for both configurations of collector as shown in Table 78.1. The general characteristics of the two configurations were based on the Kingspan Solarmax FN 2.0 FP collector [10], for which an air gap of 28 mm and a glass cover emissivity of 0.915 are assumed. Subsequently, both configurations were modelled, and efficiency curves have been plotted as shown in Fig. 78.5.

Figure 78.5 shows the modelled performance of the FPW21 collector as well as a Kingspan Solarmax DF 100−20 ET solar collector for comparison [10]. The

Table 78.1 Estimated heat loss coefficients for FP and VFP collectors at 120 °C

Heat loss coefficient	FP	VFP
U_t (W/m²K)	3.9	0.8
U_b (W/m²K)	0.4	0.8
U_e (W/m²K)	0.4	–
U_L (W/m²K)	4.7	1.6

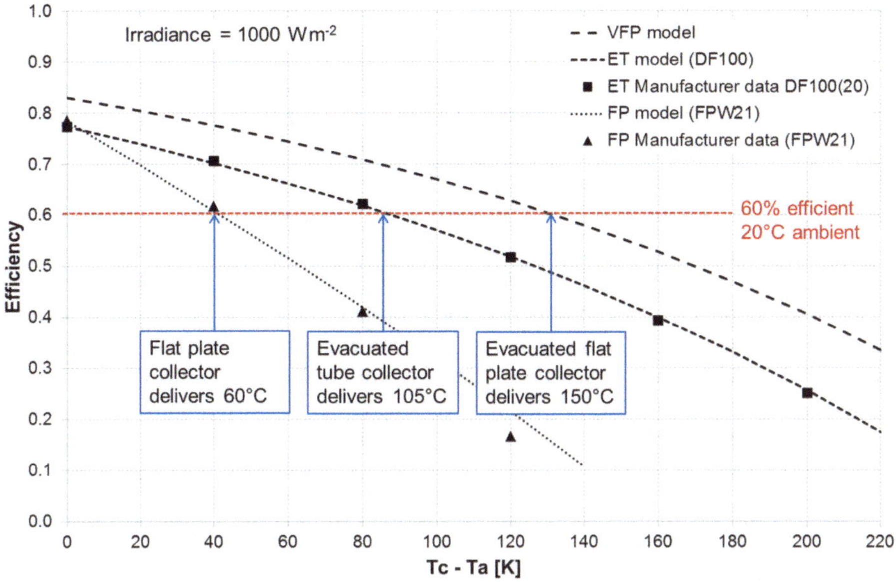

Fig. 78.5 Efficiency curves for FP, ET and VFP collectors (EN12975-2:2006)

modelled performance of the ET collector includes a shape factor to take into account the effects of a cylindrical enclosure on radiative heat transfer and cover area. As can be seen in Fig. 78.5, the model is consistent with manufacturer performance data and so gives some confidence in the predicted performance of the VFP collector.

When using the same modelling method to predict performance, the VFP system gives much greater efficiencies at higher average absorber plate temperatures (T_c). From this analysis it is estimated that at an ambient temperature of 20 °C, the FP collector could heat domestic hot water at 60 °C with an efficiency of 60 %. Under the same environmental conditions, an ET collector could heat water at 105 °C with similar efficiency while a VFP collector could operate with efficiencies greater than 60 % at mean plate temperatures up to 150 °C.

This suggests that a VFP collector is suitable for a wide range of applications such as domestic hot water/space heating and process heat production. For domestic hot water/space heating, a VFP collector could efficiently provide heat up to ~ 100 °C to a suitable heat storage system.

For process heat applications, VFP collectors could provide heat up to 200 °C with efficiencies greater than 50 %. This would satisfy several industrial heating requirements such as chemical heat treatments and industrial steam washing [9]. Furthermore, VFP collectors could provide this level of performance, at peak daily insolation, without the need for a solar concentrator and associated tracking systems.

78.5 Conclusions

The design of a novel vacuum FP solar thermal collector has been outlined. The features of this design allow the collector to be constructed with a low profile that would have significant appeal for householders for domestic application, and this has greatly enhanced the potential for building integration. Building integration offers benefits of effective cost reduction as well as secondary benefits of improved building energy performance and reduced heating demand. This collector has been shown to have the potential to provide hot water and heating with high efficiency.

Such collector arrays could provide medium temperature (150 °C) heat that can be utilised for absorption cooling or desiccant air-conditioning systems and perhaps even for electrical power generation. The collectors would also improve the thermal performance of the building, reducing excess solar gain during summer days and the heat loss from buildings during cooler months. As well as saving energy and reducing payback, this would improve the thermal comfort for occupants.

Acknowledgements The authors are grateful to the Engineering and Physical Sciences Research Council (EPSRC) for its funding of this work.

References

1. Benz N, Beikircher T (1999) High efficiency evacuated flat-plate solar collector for process steam production. Sol Energy 65:111–118
2. Department for Energy and Climate Change (DECC) (2012) The future of heating: a strategic framework for low carbon heat in the UK. London
3. Department for Energy and Climate Change (DECC) (2013) Energy consumption in the United Kingdom. London
4. Duffie J, Beckman W (2006) Solar engineering of thermal processes, 3rd edn. Wiley, New York
5. Eames P (2008) Vacuum glazing, current performance and future prospects. Vacuum 82(2008):717–722
6. Eaton C, Blum H (1975) The use of moderate vacuum environments as a means of increasing the collection efficiencies and operating temperatures of flat-plate solar collectors. Sol Energy 17:151–158
7. Grant Wilson IA, Rennie AJR, Ding Y, Eames PC, Hall PJ, Kelly NJ (2013) Historical daily gas and electrical energy flows through Great Britain's transmission networks and the decarbonisation of domestic heat. Energy Policy 61(2013):301–305
8. Henshall P, Moss R, Arya F, Eames P, Shire S, Hyde T (2014) An evacuated enclosure design for solar thermal energy applications. Grand Renewable Energy Conference, Tokyo, Japan
9. Kalogirou S (2003) The potential of solar industrial process heat applications. Appl Energy 76:337–361
10. Kingspan (no date) Thermomax technical design guide, Kingspan
11. Moss RW, Shire S (2014). Design and performance of evacuated solar collector microchannel plates. EuroSun Conference, Aix-les-Bains, France
12. Nera Economic Consulting (2009) Renewable heat technologies for carbon abatement: characteristics and potential, final report to the Committee on Climate Change, July 2009
13. Parliament of the United Kingdom (2008) Climate Change Act 2008, Chap. 27
14. Zambolin E, Del Col D (2010) Experimental analysis of thermal performance of flat plate and evacuated tube solar collectors in stationary standard and daily conditions. Sol Energy 84:1382–1396

Chapter 79
Analytical Modelling of Integrated Solar Drying System

M. A. Karim and Zakaria Mohd. Amin

Abstract The drying of fruit and vegetables is a subject of great importance. Dried fruit and vegetables have gained commercial importance, and their growth on a commercial scale has become an important sector of the agricultural industry. However, food drying is one of the most energy intensive processes of the major industrial process and accounts for up to 15 % of all industrial energy usage. Due to increasingly high electricity prices and environmental concern, a dryer using traditional energy sources is not a feasible option anymore. Therefore, an alternative/renewable energy source is needed. In this regard, an integrated solar drying system that includes highly efficient double-pass counter flow v-groove solar collector, conical-shaped rock-bed thermal storage, auxiliary heater, the centrifugal fan and the drying chamber has been designed and constructed. Mathematical model for all the individual components as well as an integrated model combining all components of the drying system has been developed. Mathematical equations were solved using MATLAB program. This paper presents the analytical model and key finding of the simulation.

Keywords Solar drying · V-groove collator · Rock-bed thermal storage · Conical tank

Nomenclature

D_h, D'_h	Hydraulic diameter of first and second pass (m)
$h_{1,2,3,4}$	Convection heat transfer coefficients (W/m^{2}K)
h_{rs}	Glass cover to sky radiative heat transfer coefficient (W/m^{2}K)
$h_{r21,r23}$	Radiative heat transfer coefficient (W/m^{2}K)
h_w	Wind convection heat transfer coefficient (W/m^{2}K)
H_c	Gap between v-groove absorber and glass cover (m)
H_g	Height of v-groove (m)
I	Solar radiation (W/m^2)
k	Thermal conductivity of air (W/mk)

M. A. Karim (✉) · Z. M. Amin
Mechanical Engineering Discipline, Science and Engineering Faculty, Queensland University of Technology, Brisbane, Australia
e-mail: azharul.karim@qut.edu.au

A. Sayigh (ed.), *Renewable Energy in the Service of Mankind Vol II*,
DOI 10.1007/978-3-319-18215-5_79

k_i	Insulation thermal conductivity (W/mk)
L	Length of the collector (m)
m	Air mass flow rate (kg/m^{2}s)
n_c	Efficiency of the collector
N	Number of glass cover
$Q_{1,2}$	Heat transferred to the air in first and second pass (W/m^2)
$S_{1,2}$	Solar radiation absorbed by glass cover and absorber plate (W/m^2)
$T_{1,2,3,4}$	Mean temperatures of surfaces (K)
T_a	Ambient temperature (K)
$T_{f1,f2}$	Mean fluid temperature (K)
T_s	Sky temperature (K)
T_{fi}	Initial air temperature (K)
T_o	Output air temperature (K)
U_b	Bottom heat loss coefficient (W/m^{2}K)
U_t	Top heat loss coefficient (W/m^{2}K)
U_f	Air velocity in the collector (m/s)
V	Wind velocity (m/s)
W	Width of the collector (m)
x	Insulation thickness (m)
α_1	Absorptivity of glass cover
ε_1	Emissivity of glass cover
τ_1	Transmittance of glass cover
σ	Boltzmann constant (5.67×10^{-8} W/m^{2}K^4)
ρ	Air density (kg/m^3)
$\varnothing$	Tilt angle of the collector (degrees)
μ	Dynamic viscosity (kg/ms)
A	Area (m)
h_v	Volumetric heat loss coefficient (W/m^3 °C)
L	Length (m)
$\dot{m}$	Mass flow rate (kg/ms)
N	Number of bed elements
T_{amb}	Ambient air temperature (°C)
$T_{a,m}$	Inlet air temperature to bed element (°C)
$T_{b,m}$	Mean temperature of bed element (°C)
$T_{a,m+1}$	Outlet air temperature of bed element (°C)

79.1 Introduction

Traditional sun drying is a labour-extensive, time-consuming slow process with requirement of large area [1]. Also, quality of the product cannot be maintained as there is no control on the drying process. On the other hand, the use of mechanical drying using fossil fuel or electricity will solve quality problem that occurs during traditional sun drying but is highly energy intensive and expensive. Furthermore, most small farmers cannot afford this technology and cost of fossil fuel [2]. These drawbacks can be solved by the use of solar dryer. However, initial cost of solar dryers can be considered high and therefore payback period should be low to make this technology attractive. To make the payback period shorter, the solar dryer needs to be very efficient [3].

To successfully develop an efficient solar drying system, it is important to design an air collector with high efficiency since it is one of the main components and would lead to a better performance of the system [4]. Flat plate air collectors are widely used; however, out of the three collector plates (namely flat plate, v-corrugated and finned air collectors) studied in [5], v-corrugated collector has higher efficiency, thus considered to be better for the solar drying system. The efficiency is further increased in double-pass operation, and optimal flow rate is determined to be 0.035 kg/m^2 s [5]. A mathematical model was also developed and compared the results with the experiments [6]. Systems using solar energy only operate during hours of adequate sunlight; therefore, to increase the reliability of the system, thermal storages need to be used. Thermal energy stored may be used later when solar energy is not available or sufficient. The packed bed storage system consisting of loosely packed rock material is considered most suitable for solar drying [7]. Literature shows that conical-shaped thermal storage tank is suitable for small-scale operation.

In this research, an integrated solar drying system is designed and constructed. Mathematical models for the major component and the entire system were developed. The models were solved using MATLAB program.

79.2 Solar Drying System

The principal components of solar drying systems are solar air collector which is used for heating ambient air, a drying unit where air extracts moisture from the product and the air handling unit which circulates the air. A thermal storage tank can be added to store energy for later use, as in Fig. 79.1.

As mentioned above, the solar air collector being investigated is a double-pass counter flow v-groove air collector in which the inlet air initially flows at the top part of the collector and changes direction once it reaches the end of the collector and flows below the collector to the outlet. As shown in Fig. 79.1, the outlet air from the solar collector then goes to the branched section in which the air flow is divided. The amount of air flow in each branch depends on the current setting of the valves and condition of the air. For example, if the output air temperature of the solar collector exceeds the current requirement at the drying chamber, higher percentage of the air will go through the thermal storage tank. Where there is no drying, all the air will go to the thermal storage. Control of air flow leads to a more flexible system, which can operate at optimum conditions. At times of low sunlight or when it is night-time, the stored energy in the thermal storage will be discharged, thus allow-

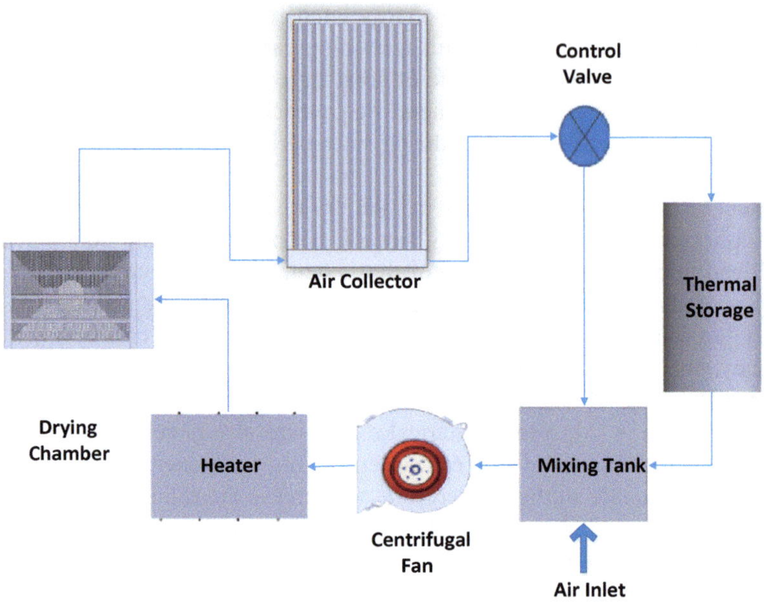

Fig. 79.1 Solar drying system components

ing the hot air to be used for drying. The auxiliary heater at the inlet of the drying chamber is used for back up heating or for maintaining a constant air temperature at the inlet of the drying chamber.

79.3 Mathematical Modelling

79.3.1 Solar Air Collector

A readily available analytic model for the double-pass counter flow v-groove solar air collector is not found in the literature. Thus, to investigate the solar collector performance and calculate its operating parameters such as temperature, development of the model from basic heat balance equations is necessary. The solution involves iterative matrix method to calculate output parameters. Figure 79.2 shows the collector cross section and thermal network of the double-pass V-groove collector.

Energy balance equations based on the thermal network shown in Fig. 79.2 for the glass cover, first-pass fluid, absorber plate, second-pass fluid and back plate are given in the equations below.

The energy balance in the top plate is given by

$$S_1 + h_{r21}(T_2 - T_1) + h_1(T_{f1} - T_1) = U_T(T_1 - T_a)$$

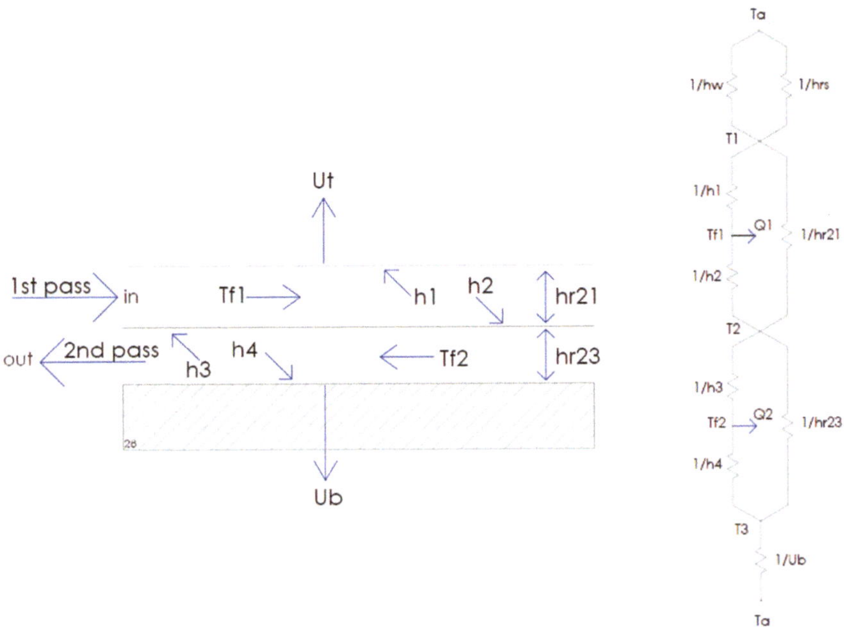

Fig. 79.2 Cross section and thermal network of a double-pass v-groove solar air collector

For the fluid's first pass,

$$h_2(T_2 - T_{f1}) = h_1(T_{f1} - T_1) + Q_1$$

Energy balance in the absorber plate is

$$S_2 = h_3(T_2 - T_{f2}) + h_2(T_2 - T_{f1}) + h_{r23}(T_2 - T_3) + h_{r21}(T_2 - T_1)$$

For the fluid's second pass,

$$h_3(T_2 - T_{f2}) = h_4(T_{f2} - T_3) + Q_2$$

For the bottom plate,

$$h_4(T_{f2} - T_3) + h_{r23}(T_2 - T_3) = U_b(T_3 - T_a)$$

These equations are then arranged in terms of temperature and placed in a 5×5 matrix in the form of

$$[A][T] = [B]$$

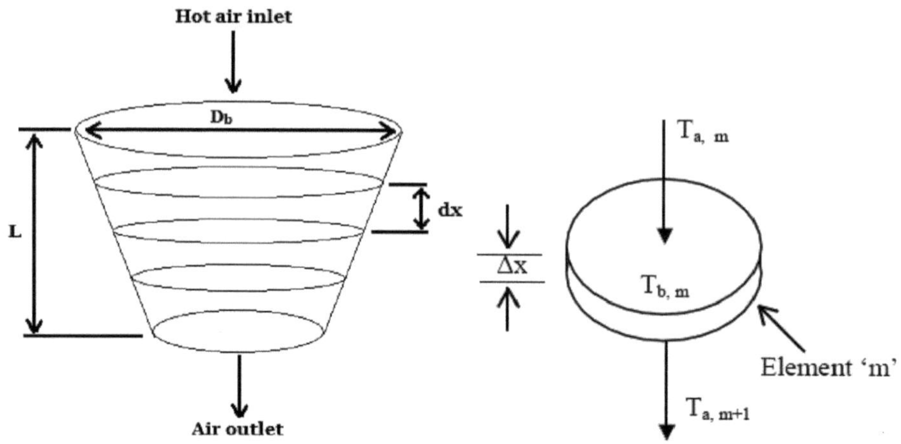

Fig. 79.3 Conical thermal storage tank (*left*) and element 'm' of the tank (*right*)

The mean temperature can be determined by using array division

$$[T]=[A]^{-1}[B]$$

79.3.2 Thermal Storage Tank

To model the storage tank mathematically, the length (height) L of the storage tank is divided into several nodal elements dx as depicted in Fig. 79.3. The value of dx is small and only exaggerated in the figure.

The following governing temperatures are used to evaluate the temperature distribution for air and solid in the thermal storage tank

$$T_{a,m+1}=T_{b,m}+(T_{a,m}-T_{b,m})\exp(-\phi_1)$$

$$\phi_1=\frac{h_v AL}{N(\dot{m}_a Cp_a)}=\frac{NTU}{N};\ N=\frac{L}{\Delta x}$$

$$T_{b,m(t+\Delta t)}=T_{b,m(t)}+\left[\phi_2\left(T_{a,m}-T_{a,m+1}\right)-\phi_3\left(T_{b,m}-T_{amb}\right)\right]\Delta t$$

$$\phi_2=\frac{\dot{m}_a Cp_a N}{(\rho_r Cp_r)AL(1-\epsilon)};\quad \phi_3=\phi_2\left(\frac{U\Delta A_m}{\dot{m}_a Cp_a}\right)$$

Neglecting loss to the surroundings, $T_{b,m(t+\Delta t)}$ becomes

$$T_{b,m(t+\Delta t)}=T_{b,m(t)}+\left[\phi_2\left(T_{a,m}-T_{a,m+1}\right)\right]\Delta t$$

79.3.3 Auxiliary Heater

Air flows through the heater with a known mass flow rate and inlet temperature. Assuming a constant heat transfer and negligible heat loss to the environment, the performance of the auxiliary heater is determined by the simple equation

$$\dot{Q} = \dot{m}C_p[T_2 - T_1]$$

79.3.4 Drying Chamber

79.3.4.1 Material Model

The moisture balance equation is $\dfrac{\partial M}{\partial t} + u\dfrac{\partial M}{\partial x} = D_{\text{eff}}\dfrac{\partial^2 M}{\partial x^2}$

Temperature balance equation is $\dfrac{\partial T}{\partial t} + u\dfrac{\partial T}{\partial x} = D_{\text{eff}}\dfrac{\partial^2 T}{\partial x^2}$

With initial condition at t=0, $M = M_0$ and $T = T_0$

Boundary condition at x=0, $\dfrac{\partial M}{\partial t} = 0$ and $\dfrac{\partial T}{\partial t} = 0$

Effective diffusivity is $\dfrac{D_{ref}}{D_{eff}} = \left(\dfrac{b_0}{b}\right)^2$

Thickness ratio is obtained by the following equation

$$b = b_0\left[\frac{\rho_w + M_w\rho_s}{\rho_w + M_0\rho_s}\right]$$

79.3.4.2 Equipment Model

The energy balance at the drying chamber is

$$\frac{\partial T_a}{\partial z} = \frac{\rho_S\left(C_s + C_m M\right)(1-\in)}{(G_0 C_{pa})}\frac{\partial T}{\partial t} + \frac{\partial M}{\partial t}h_{fg}\frac{\rho_s(1-\in)}{G_0 C_{pa}}$$

The moisture balance is

$$\frac{\partial Y}{\partial z} = \frac{\rho_s(1-t)}{G_0}\frac{\partial M}{\partial t}$$

Boundary and initial conditions are as follows:
At x=0 and t=0,

$$T_a = T_0 \text{ and } Y = Y_0$$

$$\frac{dT_a}{dt} = 0 \text{ and} \frac{dY}{dt} = 0.$$

At x>0 and t=0,

$$T_a = T_0 \text{ and } Y = Y_0$$

The differential equations are discredited and written in finite difference form before performing simulation.

79.4 Results and Discussion

The entire solar drying system is simulated in MATLAB. The initial key parameters are presented in Table 79.1.

The figures below show how the temperature changes as it progresses through each component. In Fig. 79.4, the output air temperature in the double-pass v-groove air collector is plotted against time. It also shows the temperature of the input air. It can be seen from the graph the change in temperature of the air once it passed through the air collector. The maximum output temperature occurs at 1:00 pm, which is approximately 315 K.

In the exit of the solar air collector, there is a branch. For the current simulation, 100% air passed through the storage tank. From Fig. 79.5, it can be seen that in the first few hours, the output air temperature of the thermal storage is very low as energy is used to heat the rocks and energy is stored in the rocks. Then, as time progress, the output air temperature in the thermal storage increases. It is interesting to see that, around 5 pm, when the solar radiation is quite low, output temperature from the storage is still quite high and the thermal storage tank is charged during the day. In Fig. 79.6, it can be seen that the input temperature from storage overlaps the output temperature. This is because in the current operation, all the air flows to the thermal storage and no air flow from the collector goes to the mixing tank. If the proportion is changed, then the output temperature will be somewhere in between the two air temperatures from the collector and storage tank. In Fig. 79.7, the auxiliary heater increased the air temperature before it goes to the drying chamber.

Then, in Fig. 79.8, the temperature change through the system is shown. It can be seen that the air temperature increases by approximately 40 °C, which is a desirable result. The temperature range of the air going to the drying chamber is ideal for drying apple as determined by [8]; thus, for the simulation study, the air temperature

Table 79.1 Variables considered in drying tests.

Mass flow rate (kg/s)	0.1
Drying duration (h)	7
Sample (apple) weight (kg)	1.72
Number of samples (apple)	45
Sample thickness (mm)	10
Heater size (kW)	3
Air velocity (m/s)	0.61

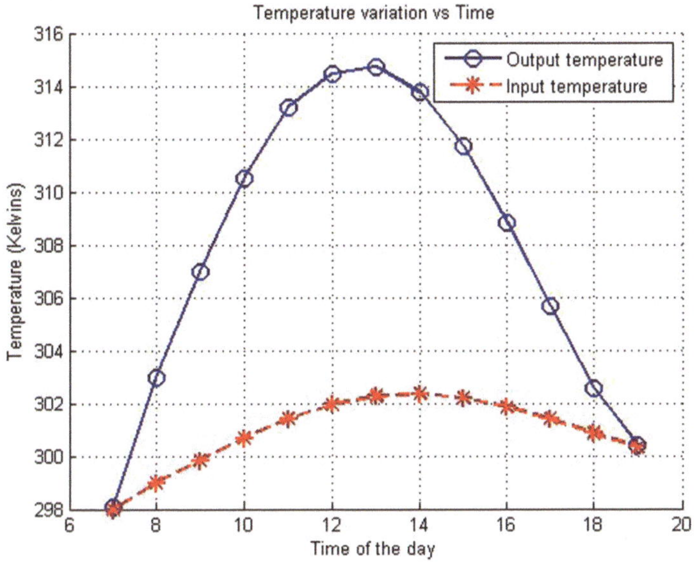

Fig. 79.4 Solar air collector inlet and outlet air temperature

entering the drying chamber may be 60 − 80 °C for faster drying time and better quality product.

The hot air coming out from the heater goes to drying chamber, as represented in the set-up schematic in Fig. 79.1. The drying curves of the material (apple) are presented in Figs. 79.9 and 79.10. For both the figures, initial rate of moisture removal and rate of temperature rise are high. It can also be seen that both moisture removal rate and temperature rise stabilize after 2 pm. Thus, these curves follow the normal drying trend of apple [8].

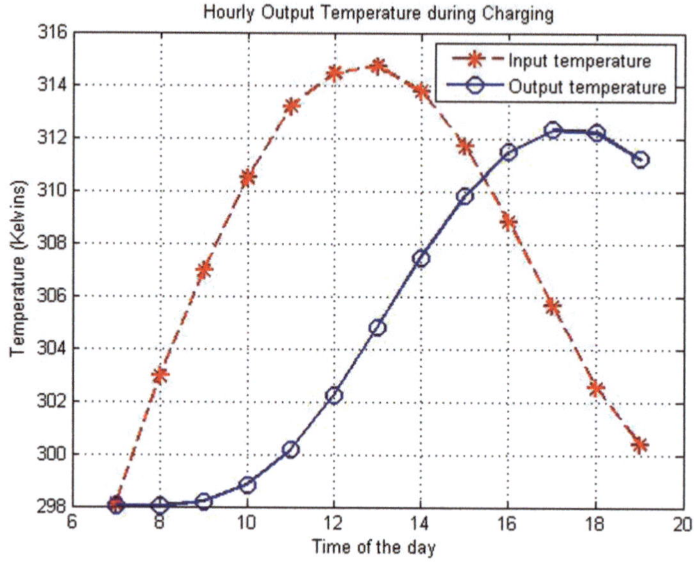

Fig. 79.5 Rock-bed storage tank inlet and outlet temperature

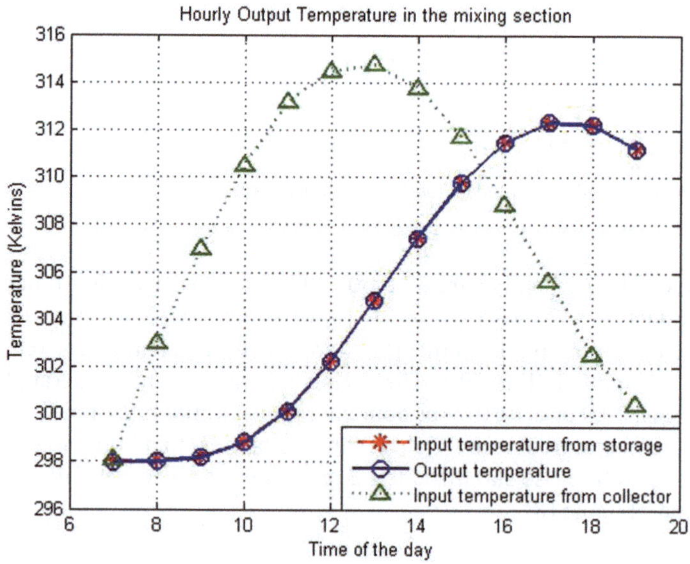

Fig. 79.6 Outlet temperature from mixing chamber

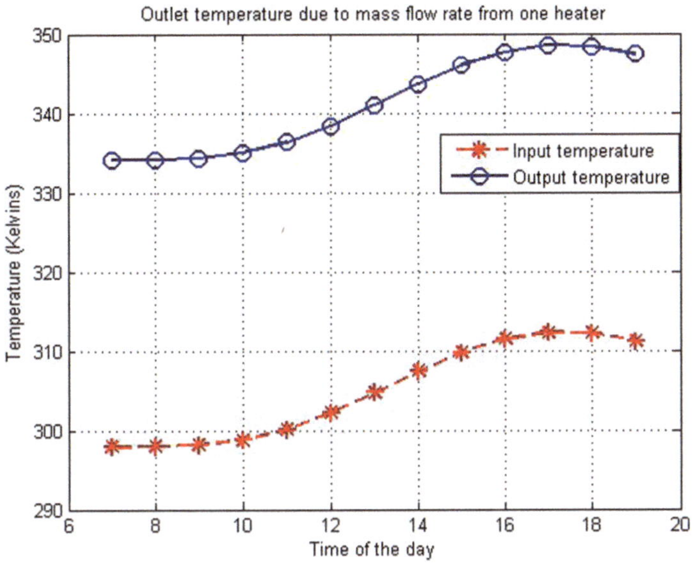

Fig. 79.7 Output temperature from the auxiliary heater

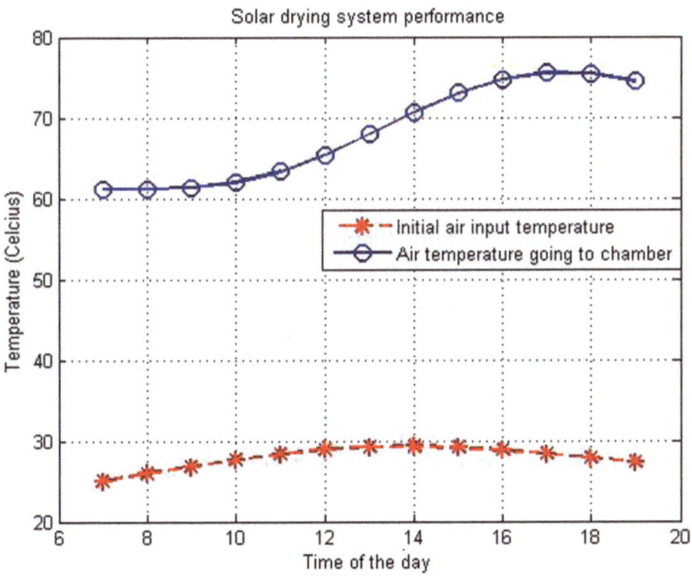

Fig. 79.8 Total increase in temperature of the solar drying system

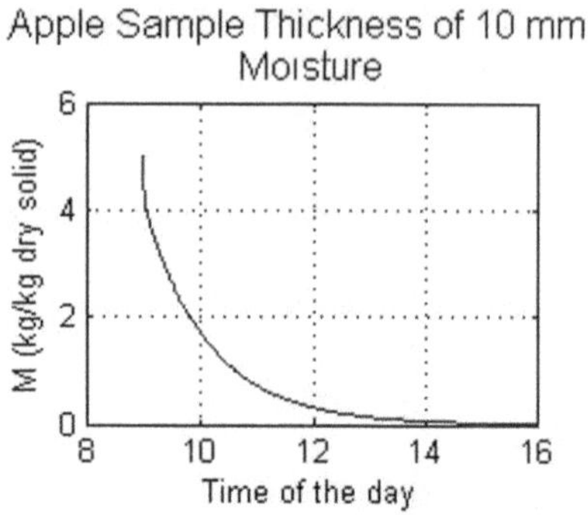

Fig. 79.9 Drying moisture curve

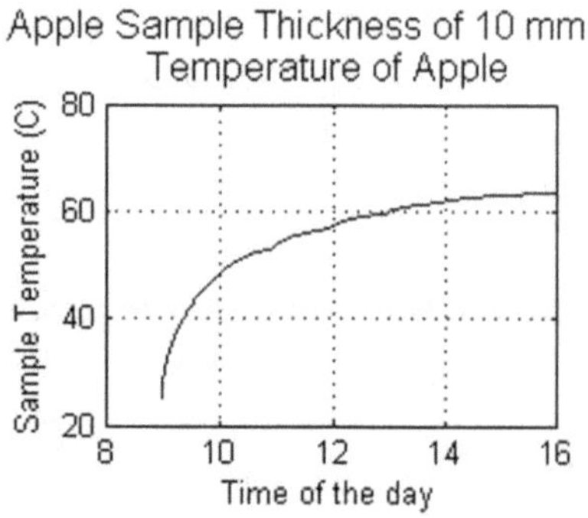

Fig. 79.10 Drying material temperature curve

79.5 Conclusion

The mathematical model was able to predict the mean temperature, the instantaneous air temperature, the output air temperature and efficiency of any component of the drying system. The thermal performance of the thermal storage tank, drying chamber and heater was determined using the mathematical model developed, which were then interconnected by MATLAB codes to simulate the whole system. The parameters determined by the model are considered very important to design any drying system for specific drying purposes. Present study provided the much needed integrated model which is absent in the literature. The temperature range in which the system will run is between 60 and 80 °C. This is the ideal drying temperature for apple. Along with an air velocity entering the drying chamber at 1–2 m/s and air humidity as low as possible, the drying time can be greatly reduced and the product quality is improved.

References

1. Shalaby SM, El-Sebaii AA (2012) Solar drying of agricultural products: a review. Renew Sustain Energy Rev 16:37–43
2. Chua KJ, Mujumdar AS, Hawlader MNA, Chou SK, Ho JC (2001) Convective drying of agricultural products. Effect of continuous and stepwise change in drying air temperature. Dry Technol 19(8):1949–1960
3. VijayaVenkataRaman S, Iniyan S (2012) A review of solar drying technologies. Renew Sustain Energy Rev 16:2652–2670
4. Karim MA, Hawlader MNA (2006) Performance investigation of flat plate, v-corrugated and finned air collectors. Energy 31:452–470
5. Karim MA, Hawlader MNA (2006) Performance evaluation of v-groove solar air collector for drying applications. Appl Therm Eng 26:121–130
6. Karim MA, Perez E, Amin ZM (2014) Mathematical modelling of counter flow v-grove solar air collector. Renew Energy J 67:192–201
7. Karim MA, Amin ZM (2014) Mathematical modelling of rock-bed solar thermal storage tank. Solar 2014 Conference, Melbourne, Australia, May-2014
8. Menges HO, Ertekin C (2006) Mathematical modeling of thin layer drying of Golden apples. J Food Eng 77:119–125

Chapter 80
Ejector Chillers for Solar Cooling

Adriano Milazzo and Federico Mazzelli

Abstract Ejector chillers are being studied at Department of Industrial Engineering of Florence (DIEF) since 2000, both theoretically and experimentally. This chapter discusses the application of solar-powered chillers in air conditioning and details the fundamental parameters. This technology is not far from being a technically feasible alternative to commercially available single-stage absorption chillers, but obviously a huge effort is still needed to improve its performance and gain access to the market.

Keywords Solar cooling · Ejectors · Natural refrigerants

80.1 Introduction

Solar cooling technologies may be classified according to the solar energy harvesting mode (concentrating or non-concentrating) or according to the energy conversion performed (electric or thermal). According to a recent review [1], at present the cheapest option is solar electric, that is photovoltaic (PV) panels coupled with commercial vapour compression chiller. However, this result is largely due to the recent dramatic decrease in PV cost and the large production volumes that make vapour compression chillers very inexpensive, notwithstanding their inherent complexity. Under a long-term perspective, it could be worth to concentrate some research effort on systems that, though more expensive at current prices, are based on a small number of simple components.

Solar thermal technologies have also seen a significant decrease in their cost, mainly due to the large amount of collectors manufactured and installed in China. With respect to concentrating solar thermal collectors, non-concentrating collectors have a simpler structure, do not require a sun-tracking system and can be effective in locations with high share of diffused radiation. Evacuated tube collectors, thanks to their reduced heat loss towards the environment, perform better at relatively high temperatures and have reached a high market share.

A. Milazzo (✉) · F. Mazzelli
Department of Industrial Engineering (DIEF), University of Florence, Florence, Italy
e-mail: adriano.milazzo@unifi.it

A. Sayigh (ed.), *Renewable Energy in the Service of Mankind Vol II*,
DOI 10.1007/978-3-319-18215-5_80

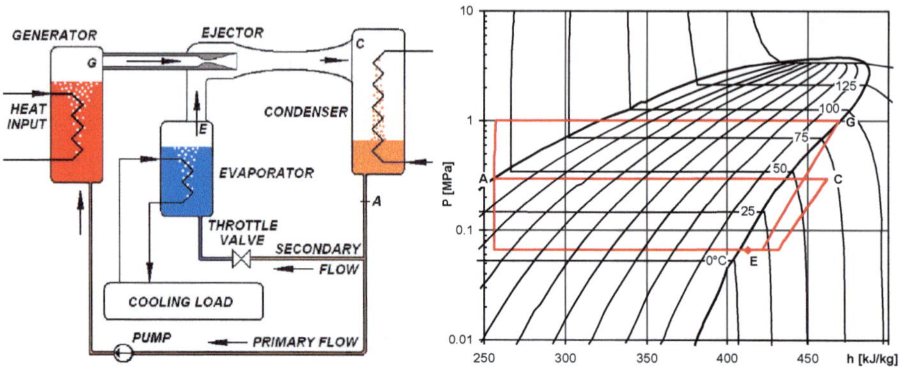

Fig. 80.1 Basic scheme of ejector chiller and thermodynamic cycle (working fluid R245fa)

Once an amount of thermal energy is given, heat powered cooling may be obtained through various operating principles. Absorption cooling is by far the most consolidated option and its convenience is not questioned here. However, ejector chillers may represent an interesting alternative, due to their simple and inexpensive construction, simplicity of operation and maintenance and potential for improvement.

In its basic configuration (Fig. 80.1), an ejector chiller comprises three heat exchangers, an ejector, a pump and an expansion valve. The working fluid exiting condenser at point A divides into a primary and a secondary flow. The first runs through a thermal engine cycle AGC, converting the heat input received at generator into kinetic energy through the primary nozzle within the ejector. The ejector replaces the compressor in the refrigeration cycle AEC, where the secondary flow runs through a conventional throttle valve and an evaporator. The two cycles share the condenser that must therefore discharge the sum of their thermal powers. System performance is measured by the ratio between cooling and motive power, that is:

$$COP = \frac{Q_f}{Q_{mot} + W_{pump}} = \omega \frac{h_E - h_A}{h_G - h_A} \tag{80.1}$$

$\omega = \dot{m}_s / \dot{m}_p$ being the "entrainment ratio" between secondary and primary flow rates.

In Eq. 80.1, mechanical power and thermal power are simply added, notwithstanding their different thermodynamic and economic value. Further discussion on this point will follow.

For given fluid and operating temperatures, the enthalpy differences experienced by the primary and secondary flow are fixed, so the system coefficient of performance (COP) is proportional to ω, which depends on the ejector geometry, operating conditions and working fluid.

Ejectors have a long history and relatively widespread applications but, quite surprisingly, their operation still offers matter of discussion and their design is largely a trial-and-error process.

Use of computational fluid dynamics (CFD) is widely described in the relevant literature [2, 3], but any CFD analysis must be initiated by a preliminary design, which is often done empirically. A set of prescriptions for steam ejector design, for example, is given by ESDU [4]. A significantly improved approach relies on thermodynamic models, which should account for the specific fluid features and boundary conditions.

Most authors recommend the so-called constant pressure mixing scheme, which dates back to the work of Keenan et al. [5]. This scheme is substantiated in a very simple geometry, with a conical inlet section, a cylindrical mixing chamber and a conical diffuser. A normal shock is thought to exist, at least in a one-dimensional analysis, somewhere in the cylindrical mixing chamber. Many refinements have been proposed from the original Keenan's model, from the early effort of Munday and Bagster [6] to the successful model proposed by Huang et al. [7], used by many authors. All these thermodynamic models can give, at most, the area of a few flow sections within the ejector, without any attempt to specify their distance. In other words, they actually are "zero-dimensional". Furthermore, they rely on empirical efficiencies for the fundamental components and hence need experimental validation. Last but not least, ideal gas behaviour is assumed for the working fluid, claiming that pressure within the mixing chamber is very low.

As far as we know, the only attempt to set a really one-dimensional design criterion is the "constant rate of momentum change" (CRMC) formulated by Eames [8]. This criterion suggests designing the ejector by imposing a constant rate for the momentum reduction along the diffuser. This reduction rate should be low enough to give a reasonable diverging angle at the diffuser exit. Furthermore, a gradual transition should occur in the diffuser throat between the supersonic and the subsonic flow. Obviously, this is a very difficult target that may be reached only for a single value of the inlet conditions. No attempt is made by the CRMC precept to describe the mixing process, which is inherently two-dimensional. Again, the working fluid is treated as an ideal gas.

According to our experience, the CRMC concept needs further refinement. The mixing zone must be described accounting for the gradual momentum exchange between the motive and the entrained flow. The throat needs to be shaped and possibly contoured through a smooth transition with the mixing zone and the diffuser. This latter should be carefully designed in order to recover as much pressure as possible. Even the primary nozzle, which is often taken for granted, must be scrutinized in terms of exit diameter (i.e. exit pressure) and angle.

All these issue, once properly addressed, can bring significant improvement in the ejector entrainment ratio, which is the fundamental performance parameter for the whole cooling system. However, the other components, that is heat exchangers, expansion valve and generator feed pump, should not be neglected. It is worth noting that a significant part of the system cost is represented by the heat exchangers, so that their optimization in terms of temperature differences and pressure losses can greatly improve the economic viability of the system.

The feed pump is the sole electric load within the ejector cooler, and its power, even if by far smaller than the thermal power, is actually the only operating cost when the system is powered by renewable energy. Therefore, the pump must be as efficient as possible. Furthermore, as it drives the fluid from a condenser, it must be highly resistant to cavitation.

Last but not least, the working fluid selection is an important and non-trivial task. The fluid must run through a combined engine and refrigeration cycle, covering a wide range of temperatures. Fluid charge is higher in comparison with vapour compression refrigerators, which makes the use of costly or dangerous fluids impractical. Some fluids experience huge pressure variations when brought from the condenser to the generator temperature, worsening the pump operation.

80.2 Fluid Selection

Favourable features of a working fluid for an ejection refrigerator are as follows:

- Zero ozone depletion potential (ODP) and low global warming potential (GWP)
- Low flammability and toxicity
- High latent heat and density at generator, condenser and evaporator temperatures

Ejection chillers may be perfectly fluid tight (the feed pump may be driven through a magnetic coupling); hence, fluid leakage can be moderate or null. However, the fluid amount per unit power is large, yielding high damage in case of accidental release. In Europe, F-gas regulations will ban the use of refrigerants with GWP>2500 by 2017. This value may hence be set as a threshold.

A set of candidate fluids is listed in Table 80.1. Once ammonia is excluded for its toxicity and CO_2 for its critical point below the condenser temperature, the first and rather obvious choice is water. Costless and absolutely safe, it has a very high latent heat throughout the above-specified range of temperatures. Icing problems may be encountered due to the rather high triple point, but the major drawback is the very low pressure and density of steam at cold temperatures.

Table 80.1 Relevant data for a set of non-toxic and zero-ozone depletion potential (ODP) fluids

Fluid	M_{mol} [kg/kmol]	T_{crit} [K]	P_{crit} [MPa]	N.B.P.[a] [K]	Expansion	GWP	Safety
Water	18.015	647.1	22.064	373.12	W	0	A1
R600a	58.122	407.81	3.629	261.4	D	20	A3
R134a	102.03	374.21	4.059	247.08	W	1300	A1
R152a	66.051	386.41	4.517	249.13	W	120	A2
R236ea	152.04	412.44	3.502	279.34	D	1200	A1
R245ca	134.05	447.57	3.925	298.28	D	640	A1
R245fa	134.05	427.16	3.651	288.29	D	950	A1
R365mfc	148.07	460.0	3.266	313.3	D	890	A1
R1234yf	114.04	367.85	3.382	243.66	D	4	A2L
R1234ze	114.04	382.52	3.636	254.19	D	6	A2L
R1233zd	130.5	438.75	3.772	195.15	D	<5	A2L

M_{mol}, T_{crit}, P_{crit} and N.B.P. from NIST Refprop; Global warming potential (GWP) from [13]
W wet, *D* dry expansion
[a] Normal boiling point (P=101.3 kPa)

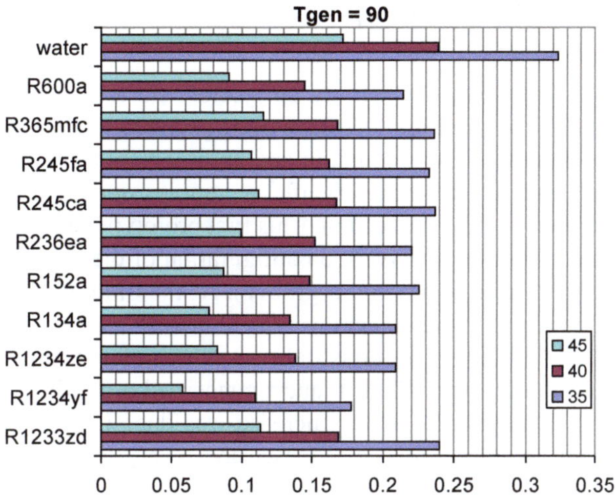

Fig. 80.2 Coefficient of performance (COP) of ejector cycle working with different fluids—model results for T_{gen} = 90 °C and T_{eva} = 5 °C

A second possibility is given by hydrocarbons, which play a central role in the domestic refrigeration market. Isobutane (R600a) is taken as an example of this class of fluids that share low GWP and high COP in vapour compression cycles. Isobutane has a "dry expansion", that is its entropy decreases along the upper limit curve, which is useful to avoid condensation within the ejector. The obvious burden of hydrocarbons is flammability, which may represent a serious problem as the fluid charge increases.

A third group collects the fluorocarbons, which may have zero or low flammability and favourable thermodynamic properties, but generally high GWP. Some of them have dry expansion. Among fluorocarbons, R134a has a reasonable cost and is well known in the refrigeration industry, but has high saturation pressure at generator temperature and imposes a high compression ratio on the feed pump. R134a has a relatively high GWP, but its value is below the 2500 threshold.

R152a is interesting in terms of very low GWP, but is flammable. R236ea, R245ca and R245fa are quite similar: all have GWP<2500 and dry expansion. R365mfc has the highest critical temperature, acceptable GWP and dry expansion. The three fluoro-olefins at the bottom are promising alternatives to fluorocarbons with low GWP and generally low flammability, though likely to have high cost in the near term.

Our research group set up a comprehensive thermodynamic model of an ejector refrigerator [9] that may work with different fluids accounting for their real gas behaviour through NIST Refprop functions. Ongoing experimental activity [10] allows a model validation/calibration on R245fa. COP values calculated by the model for all mentioned fluids at T_{gen} = 90 °C, T_{eva} = 5 °C and T_{cond} = 35, 40 and 45 °C are shown in Fig. 80.2.

A part from the absolute values, which depend on the specific working conditions, the thermodynamic advantage of water is evident. This will orient our future efforts, as will be discussed later on.

80.3 Experimental Activity

As mentioned, our activity on ejectors started in the early 2000s. Even if the results shown in Fig. 80.2 have been gathered much later, water was selected as the working fluid for economic and environmental reasons. A prototype 5-kW chiller was built at our laboratory after a complex optimization process [11] and a few experiments were carried out. This first prototype featured an innovative double-stage structure, with an inverted configuration of the second stage (i.e. the primary flow entered from a peripheral groove). At the time, the CRMC concept was unknown and the mixing chambers were cylindrical.

In year 2010, an industrial research partner (Frigel S.p.A.—Florence) asked for a new prototype. Their request concerned a possible industrial use, so that an order of magnitude power increase was due. This forced to abandon water, which would require a huge condenser and a very low pressure evaporator, in favour of a synthetic fluid. After an extensive search, R245fa was selected for its favourable thermodynamic features. The prototype was designed using a first version of the above-mentioned thermodynamic simulation code and a first testing campaign was performed [12]. The rather unsatisfactory results, together with the first CFD results, suggested two further designs, and at last, the system reached the present working configuration (Fig. 80.3). The Frigel prototype is significantly more powerful with respect to common laboratory scale (90 kW_{th} at generator, 40 kW cooling power). It is designed to work on hot water (90 °C) as thermal input, producing

Fig. 80.3 40-kW prototype ejector chiller—the generator, not visible, is below the evaporator

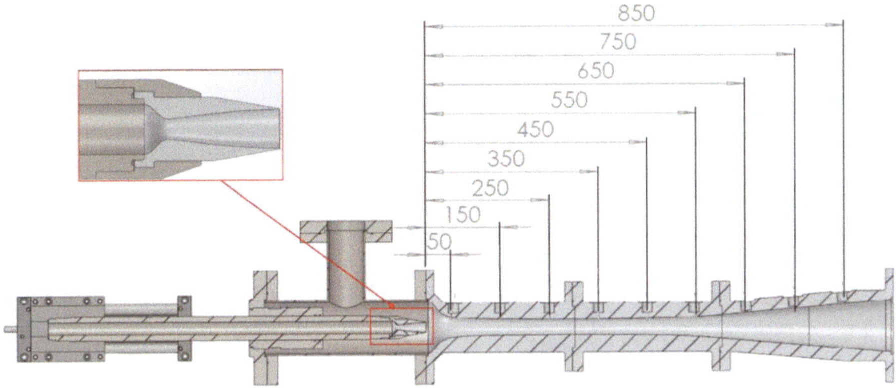

Fig. 80.4 Ejector with detail of the primary nozzle

chilled water at 7 °C. The condenser is cooled by ambient temperature water coming from an external, humidified air cooler.

An interesting feature of this prototype is the vertical layout, allowing a 2-m head between condenser (on top) and pump (at the bottom—see Fig. 80.3). This latter is a multistage centrifugal unit, selected for its resistance to cavitation. The ejector is fitted with nine static pressure ports and a movable primary nozzle (Fig. 80.4), in order to optimize the relative position between nozzle and mixing chamber.

Piezoelectric pressure transducers and 4-wire Pt100 temperature sensors are placed in all significant points. Water flow rates are measured on all water streams, and R245fa flow rates are calculated accordingly. Pump flow rate may be varied through an inverter. The prototype has been subjected to a number of tests for 5–6 h each, that is a duration representative of a day of operation for an air-conditioning system. Operation has always been very stable, even in the presence of swift variations of the condenser water temperature.

A sample of the prototype performance is shown in Fig. 80.5. Experimental points are shown with their uncertainty intervals together with CFD results to be discussed later. The measured COP is well above 0.4 when evaporating at 5 °C and around 0.55 when evaporating at 10 °C. These levels are not too far from those obtained by single-effect absorption chillers and match or exceed the performance obtained by other authors [13]. Note the decrease in performance experienced above a well-recognizable condenser temperature (critical temperature), which is typical of supersonic ejectors. Note also the heavy impact of the evaporating temperature.

80.4 Numerical Activity

Since the beginning of the activity with Frigel, it was evident how thermodynamic modelling was useful as a first design tool, but needed substantial refinement by more detailed analysis in order to reach good results. A commercial CFD tool

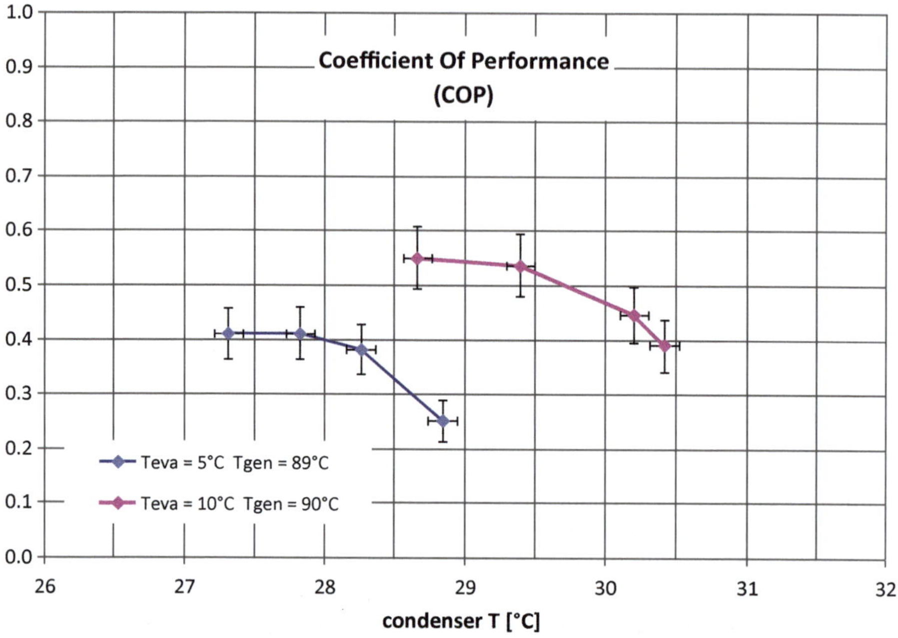

Fig. 80.5 Ejector chiller coefficient of performance (COP) as a function of operating temperatures

(Fluent) was selected, for its easy use and ability to model various fluids. At first, the R245fa was modelled through embedded NIST fluid properties, in order to make fluent results fully comparable with thermodynamic simulations. However, numerical stability problems and computational time soon suggested using a Peng Robinson equation of state, which was checked for very close resemblance with NIST data. An axial-symmetric domain was used. Due to the high directionality of the flow (axial velocity component always much greater than the transversal component), a structured grid is selected. This allows for longitudinal stretching of the elements, thus reducing the grid size. Complete grid independence was reached at 80,000 elements.

The lines shown in Fig. 80.5 demonstrate a very close agreement between CFD and experimental results. Furthermore, as shown in Fig. 80.6, CFD analysis gives a deep insight into the ejector behaviour. Only the supersonic flow field is shown for clarity. It is clear how the supersonic flow reaches the mixing chamber wall somewhat after the throat (note the continuous shape obtained through CRMC criterion). In this way, the entrainment ratio becomes independent from the downstream conditions, as shown also in Fig. 80.5 by the practically horizontal shape of the COP curves before the condenser critical temperature/pressure. As the condenser pressure rises, the shock train marking the transition to subsonic flow moves to the left and this "sealing" effect is lost, making the entrainment ratio dependent on the exit pressure.

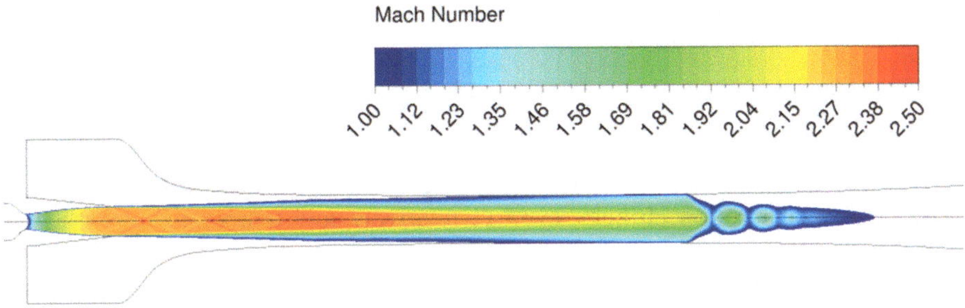

Fig. 80.6 Contour map of Mach > 1 zone—$T_{eva} = 5\,°C$

At very high condenser temperature, the flow may even reverse from the condenser towards the evaporator, but even when this condition has been reached for testing purpose, the normal prototype operation was easily restored, as soon as normal working conditions were re-established.

In the present configuration, entrainment of the secondary flow and mixing are satisfactorily performed, though an improved design of the inlet part will be pursued. What still needs a major redesign is the final part of the diffuser (not shown in Fig. 80.6), where a recirculation zone limits the pressure recovery and hence lowers the critical condenser temperature. This would impair the system utilization in hot/humid climates, where demanding external conditions claim for higher condenser temperatures. Therefore, this aspect will form our priority in future CFD optimization.

The reliability of CFD simulation must be partly ascribed to the favourable characteristics of R245fa. Dry expansion avoids any phase change throughout the numerical domain, making the simulation much easier. Unfortunately, R245fa has a relatively high GWP and this can make its usability incompatible with coming regulations. Alternative fluids are proposed, like above-mentioned R1233zd, but their cost and commercial widespread availability are questionable. Furthermore, thermodynamic simulations show that water, besides being costless and absolutely safe, could give significant COP improvements (35 %, according to data shown in Fig. 80.2). Therefore, next CFD simulations will be made on steam. This will greatly increase the complexity of the simulation, due to condensation along the expansion. Previous analysis [14] and literature results show that condensation does occur, even if its inception is hard to locate within fast expanding ejector flow, where metastable conditions may last much more than expected. Even if commercial CFD codes (including Fluent) have specific models for condensing steam, their scope is mainly limited to steam turbine analysis and their use near 0 °C may be limited. Homogeneous condensation may be very fast and cause abrupt changes in the flow field. A close scrutiny and careful experimental validation will be needed in order to have a reliable tool for optimization.

80.5 Concluding Remarks

From the discussion and the results presented above, we may conclude what follows:

- Ejectors may well surpass COP=0.5 and hence be competitive with single-stage H_2O-LiBr absorption chillers, even at relatively low temperatures of the hot source as those obtained by evacuated tube solar collectors.
- Apparently the best working fluid is water, as it is safe and costless and offers significant higher COP in the relevant working conditions.
- A huge effort is needed to improve the design tools: a detailed and well-calibrated thermodynamic model must give a physically consistent shaping to the various parts of the ejector and subsequent CFD analysis must refine and optimize this shape, accounting for all two- or three-dimensional effects. This may be particularly difficult for water, due to phase change effects.
- Significant performance improvements can be gained if evaporation temperature is set to somewhat higher values. This is feasible through a proper dimensioning of the air-conditioning system in terms of components and flow rates. An evaporation temperature of 10 °C instead of 5 °C would increase COP by almost 40 %.

A further point can be added: using water as the working fluid, high liquid density and low pressure allow for gravity feeding of generator, once it is placed at suitably lower elevation than the condenser [15]. Considering as an example a saturation temperature of 90 °C at generator and 35 °C at condenser, a height of 6.6 m would be sufficient to overcome the pressure difference, that is the layout would be compatible with a normal house, the generator being placed in the basement and the condenser on the roof. This would further reduce the system cost (investment and operation) and complexity (completely sealed circuit with no moving parts).

References

1. Kim DS, Infante Ferriera CA (2014) Solar refrigeration options—a state-of-the-art review. Int J Refrig 31:3–15
2. Ruangtrakoon N, Thongtip T, Aphornratana S, Sriveerakul T (2013) CFD simulation on the effect of primary nozzle geometries for a steam ejector in refrigeration cycle. Int J Therm Sci 63:133–145
3. Hemidi A, Henry F, Leclaire S, Seynhaeve J, Bartosiewicz Y (2009) CFD analysis of a supersonic air ejector. Part II: relation between global operation and local flow features. Appl Therm Eng 29:1523–1531
4. ESDU Ejectors and jet pumps computer program for design and performance for compressible gas flow (2015) ESDU 92042. www.esdu.com
5. Keenan H, Neumann EP, Lustwerk F (1950) An investigation of ejector design by analysis and experiment. J Appl Mech Trans ASME 72:299–309
6. Munday JT, Bagster DF (1977) A new ejector theory applied to steam jet refrigeration. Ind Eng Chem Process Des Dev 16:442–449
7. Huang BJ, Chang JM, Wang CP, Petrenko VA (1999) A 1-D analysis of ejector performance. Int J Refrig 22:354–364

8. Eames IW (2002) A new prescription for the design of supersonic jet-pumps: the constant rate of momentum change method. Appl Therm Eng 22:121–131
9. Grazzini G, Milazzo A, Paganini D (2012) Design of an ejector cycle refrigeration system. Energy Convers Manage 54:38–46
10. Mazzelli F, Milazzo A, (2015) Performance analysis of a supersonic ejector cycle working with R245fa. Int J Refrig 49:79–92
11. Grazzini G, Rocchetti A (2002) Numerical optimization of a two-stage ejector refrigeration plant. Int J Refrig 25:621–633
12. Eames IW, Milazzo A, Paganini D, Livi M (2013) The design, manufature and testing of a jet-pump chiller for air conditioning and industrial application. App Therm Eng 58:234–240
13. Chen X, Omer S, Worall M, Riffat S (2013) Recent developments in ejector refrigeration technologies. Renew Sustain Energy Rev 19:629–651
14. Grazzini G, Milazzo A, Piazzini S (2008) Prediction of condensation in steam ejector for a refrigeration system. Int J Refrig 34:1641–1648
15. Grazzini G, D'Albero M (1998) A jet-pump inverse cycle with water pumping column. IIF-IIR—Gustav Lorentzen Conference—Oslo, Norway—1998

Chapter 81
Modelling Conical Rock-Bed Solar Thermal Storage Tank

M. A. Karim and Zakaria Mohd. Amin

Abstract An important application of solar thermal storage is for power generation or process heating. Low-temperature thermal storage in a packed rock bed is considered the best option for thermal storage for solar drying applications. In this chapter, mathematical formulations for conical have been developed. The model equations are solved numerically for charging/discharging cycles utilizing MATLAB. Results were compared with rock-bed storage with standard straight tank. From the simulated results, the temperature distribution was found to be more uniform in the truncated conical rock-bed storage. Also, the pressure drop over a long period of time in the conical thermal storage was as low as 25 Pa. Hence, the amount of power required from a centrifugal fan would be significantly lower. The flow of air inside the tank is simulated in SolidWorks software. From flow simulation, 3D modelling of flow is obtained to capture the actual scenario inside the tank.

Keywords Solar energy · Rock bed · Conical storage tank

Nomenclature

A_s	Surface area, m^2
Cp_a	Specific heat of air, J/kg°C
D_s	Diameter of bed storage, m
D_r	Diameter of rock element, m
F_r	Heat removal factor
f	Friction factor
G	Air mass flow rate, kg/m^2s
H	Latent heat, kJ/kg
h	Heat loss coefficient, W/m^2K
h_v	Volumetric heat loss coefficient, $W/m^{3}°C$
I	Solar insolution, W/m^2
k_a	Thermal conductivity of air, W/mK
k_r	Thermal conductivity of rock element, W/mK

M. A. Karim (✉) · Z. M. Amin
Mechanical Engineering Discipline, Science and Engineering Faculty,
Queensland University of Technology, Brisbane, Australia
e-mail: azharul.karim@qut.edu.au

A. Sayigh (ed.), *Renewable Energy in the Service of Mankind Vol II*,
DOI 10.1007/978-3-319-18215-5_81

L Length, m
L_c Hydraulic diameter, m
m Mass, kg
$\dot{m}$ Mass flow rate, kg/s
N Number of bed elements
Nu Nusselt number
ΔP Pressure drop, Pa

81.1 Introduction

Efforts of rational and effective energy management as well as environmental considerations increased the interest in utilization of renewable energy sources, especially solar energy. Solar energy systems only operate during hours of adequate solar radiation, and therefore, to increase the reliability of the system, thermal storages are used. There are many situations where available energy is wasted because it is in the wrong place and/or at the wrong time. There is therefore a need for energy storage to overcome this problem [1]. A wide range of techniques can be used for this purpose including electrical storage, compressed air storage, hydrogen storage, and (thermal) storage of heat transfer fluid. All such methods involve high-capital investment, and therefore, a cheaper alternative has been widely exploited in which thermal energy is stored in a suitable medium.

The packed-bed systems generally consist of loosely packed rock material that allows thermal energy to be stored. Heat from the solar air collector is passed through the rocks in the storage where air is usually passed from the top of the bed to the bottom. To utilize the heat from the system, the air flow is reversed to take the energy from hot rocks. The performance of the rock-bed storage depends, among other factors, on the geometry of the bed. According to Zanganeh A. et al. [2], the thermal performance of the cylindrical rock-bed storage can be improved by changing the geometry of the storage. They performed experimental test between the cylindrical and truncated conical thermal storage. This led them to discover that the pressure drop in the conical thermal storage was lower in long periods of charging.

Karim et al. [3] studied the temperature gain of a single rock-bed element over a period of 6 h. Singh et al. [4, 5] proposed a simulation model for the thermal behaviour of the packed bed. They ran simulations for bed elements of different sphericity and void factors. However, their study is only limited to cylindrical storage tanks.

Although conical tanks are reported as more efficient in experimental studies, no mathematical equations are properly developed for describing the behaviour of this tank. In this study, mathematical model has been developed for cylindrical thermal storage tanks.

81.2 Mathematical Model

As shown in Fig. 81.1, the length of the storage tank chamber is divided into a number of nodal elements. Considering the temperature variation along the length, the following assumptions were made: (1) properties of both solid and fluid are constant, (2) no heat loss and mass transfer to the surroundings, and (3) conduction from the fluid to the rocks was negligible. Based on the assumptions, the proposed energy balance equation for the system yields

$$(\dot{m}_a Cp_a)\, dT_a = (\dot{m}_a Cp_a)(T_{a,m} - T_{a,m+1}) \tag{81.1}$$

$$dT_a = (\dot{m}_a Cp_a)(T_{a,m} - T_{a,m+1}) / (\dot{m}_a Cp_a). \tag{81.2}$$

Solving the Eq. (81.2), the air temperature ($T_{a,m+1}$) at the exit of each element can be determined [6].

$$T_{a,m+1} = T_{b,m} + (T_{a,m} - T_{b,m})\exp(-\phi_1) \tag{81.3}$$

where

$$\phi_1 = \frac{h_v AL}{N(\dot{m}_a Cp_a)} = \frac{NTU}{N}; \quad N = \frac{L}{\Delta x}. \tag{81.4}$$

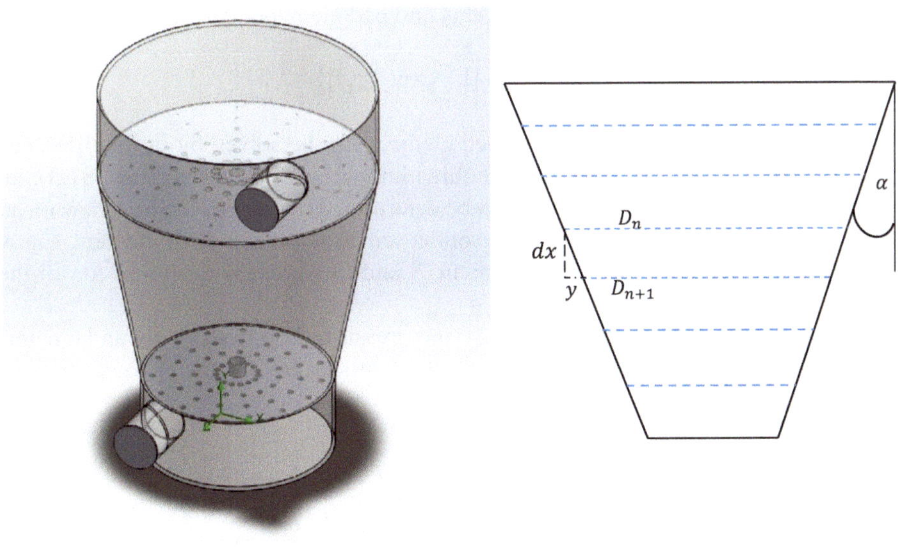

a Isometric View **b** 2D view

Fig. 81.1 Schematic of conical thermal storage tank. **a** Isometric view, **b** 2D view

For each element in the storage, the energy balance equation for a time increment yields

$$(\rho_r Cp_r)(1-\varepsilon) A\Delta x\left(\frac{dT_{b,m}}{dt}\right) = (\dot{m}_a Cp_a)(T_{a,m} - T_{a,m+1}) - (U\Delta A_m)(T_{b,m} - T_{amb}). \tag{81.5}$$

The finite difference form is given as

$$T_{b,m(t+\Delta t)} = T_{b,m(t)} + [\phi_2(T_{a,m} - T_{a,m+1}) - \phi_3(T_{b,m} - T_{amb})]\Delta t \tag{81.6}$$

where

$$\phi_2 = \frac{\dot{m}_a Cp_a N}{(\rho_r Cp_r) AL(1-\varepsilon)} \tag{81.7}$$

$$\phi_3 = \phi_2\left(\frac{U\Delta A_m}{\dot{m}_a Cp_a}\right). \tag{81.8}$$

However, neglecting heat loss to the surroundings, the finite difference expression yields

$$T_{b,m(t+\Delta t)} = T_{b,m(t)} + [\phi_2(T_{a,m} - T_{a,m+1})]\Delta t. \tag{81.9}$$

According to Singh et al. [5], this method provides a stable prediction of the bed element temperatures for time increments and bed elements with a relationship:

$$\Delta t \le \{\phi_2[1-\exp(-\phi_1)]\}. \tag{81.10}$$

The above expression for the mean bed element can be solved by finite difference method, with all bed element temperatures known at $t = 0$. Using Eq. (81.3), an outlet air temperature from the bed can be determined. From Eq. (81.6), a new mean bed temperature can be obtained. The outlet temperature from bed element 1 now becomes the inlet temperature to element 2, and this process continues for all the elements in the rock-bed storage.

According to Zanganeh G. et al. [2], the pressure drop in the bed can be determined from the following equation with the temperature change and geometry of the rocks and bed.

$$P = \frac{LG^2}{\rho d}\left(A \times \frac{(1-\varepsilon)^2}{\varepsilon^3 \psi^2}\left(\frac{\mu}{Gd}\right) + B\left(\frac{1-\varepsilon}{\varepsilon^3 \psi}\right)\right) + \rho g L\left(\frac{\Delta T}{T}\right) \tag{81.11}$$

where ψ is the sphericity of the rocks, and for randomly shaped gravel with similar size, A and B are 217 and 1.83, respectively [2]. The temperature difference ΔT is the difference in the inlet and outlet temperature of the node. Hence, the pressure is determined for every nodal element for the time step selected.

For the energy gained within the thermal storage, the following expression can be used for every nodal element in the time frame. The temperature difference used is the increase in temperature of the bed element.

$$\dot{Q}_2 = \dot{m}_a Cp_a (T_{b,m(t+\Delta t)} - T_{b,m}). \tag{81.12}$$

Hence, the summation of the energy in the thermal storage would yield the total amount of energy over the period of charging. For the development of an optimal conical thermal storage shown in Fig. 81.1, the equations shown previously for a cylindrical storage can be utilized. However, as the diameter of each node changes throughout the system, the following process was adopted to consider the changes. To determine change in the diameter of each node, the change in the horizontal direction is expressed as (Fig. 81.1)

$$y = 2 \times dx \times \tan \tan(\alpha). \tag{81.13}$$

Hence, the average area can be determined by finding the average diameter of the nodal element. Therefore, the diameter due to change of angle of the conical thermal storage yields

$$D_{n+1} = D_n - y. \tag{81.14}$$

Therefore, the average diameter and the area between the two nodal elements yield

$$D_{avg} = \frac{D_n + D_{n+1}}{2}, \tag{81.15}$$

$$A = \frac{\pi D_{avg}^2}{4}. \tag{81.16}$$

With the change in area, the surface area of each node is given as

$$SA_{node} = 2 \times \frac{\pi \left(D_n^2 - D_{n+1}^2 \right)}{4} + \pi (D_n - D_{n+1}) dx. \tag{81.17}$$

81.3 Results and Discussion

The performance of packed rock-bed storage was determined under changing ambient temperature conditions as with the solar air collector. This affected the mass flow rate of the system, as a constant inlet temperature to the rock-bed storage was

Table 81.1 Parameters

Length (m)	1
Diameter (m)	1.2
Volume (m^3)	0.78
Inclination angle, α (°)	12
Void fraction	0.45
Charging air temperature, C	80
Discharging air temperature, C	30

considered. With a time step of 5 s, the thermal performance of the rock bed over a period of 8 h is determined. The general parameters used in the simulation programme are shown in Table 81.1.

Temperature distribution across the storage for 8 h of charging is presented in Fig. 81.2. As observed in Fig. 81.2a after a charging period of 1 h, the first element has a mean temperature of 73 °C. As time progresses, the remaining elements are charged. After a period of 8 h, the storage tank can be considered completely charged with approximately half of the system with a mean bed temperature of 80 °C and a minimum bed temperature of about 76 °C.

Similarly, with the discharging of the thermal storage for 8 h as shown in Fig. 81.2b, the entire system is nearly discharged over this period. For discharging of the thermal storage, fresh air enters from the bottom of the thermal storage and exits from the top. As shown in Fig. 81.2, as the ambient air enters the thermal storage, heat is transferred from the rocks to the air. Thus, after 1 h of discharging, the first bed element is still above an ambient air temperature of 25 °C. As the air flows across the thermal storage, the temperature of rocks is gradually reduced as shown over the 8-h period. After the last hour of discharging, the final bed temperature in

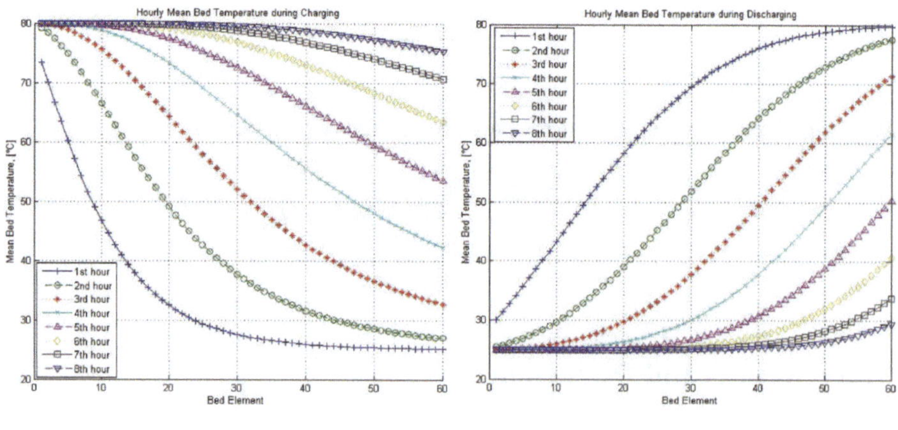

a **Charging** **b** **Discharging**

Fig. 81.2 Temperature distribution during 8 h of charging and discharging for conical thermal storage. **a** Charging, **b** discharging

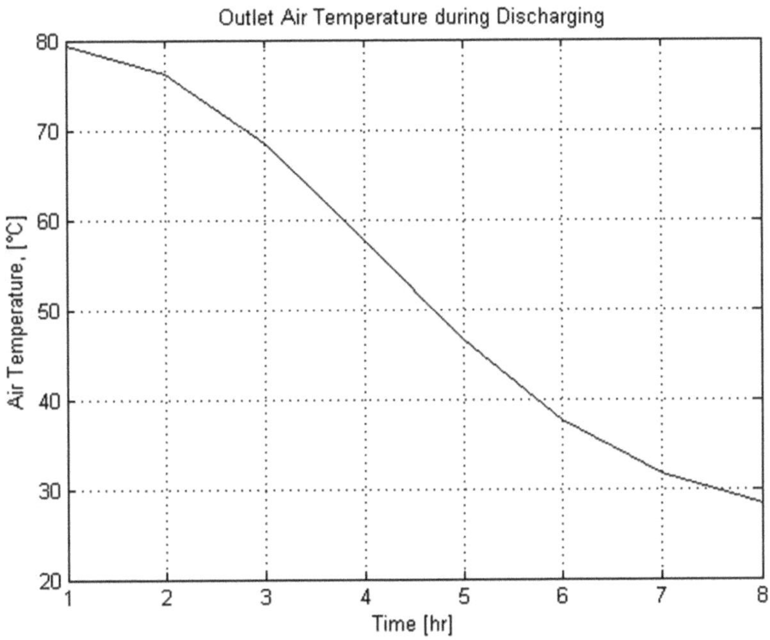

Fig. 81.3 Outlet air temperature during discharging for conical thermal storage

the thermal storage is approximately 40 °C. Therefore, a longer discharging period is required to remove the remaining energy in the thermal storage.

Because of the design of the conical tank, higher temperature is exposed to bigger area and lower temperature is exposed to smaller area. Hence, the system is more efficient for the same aspect ratio of length and diameter. During discharge, heated air is removed as ambient air is forced into the system. In Fig. 81.3, the average outlet temperature of the tank for 8 h of discharging is presented. As can be seen, decrease in the outlet temperature is minimum in the 1st hour, and after that, fall in the outlet temperature is steep. In the 8th hour, the outlet air temperature gradually decreases to below 30 °C.

In order to analyse the performance of thermal storage, the amount of energy stored is required. The total amount of energy stored over 8 h of charging is shown in Fig. 81.4. With an initial temperature of 25 °C and an inlet air temperature of 80 °C, the total energy accumulated by the thermal storage tank is approximately 1040 kW-h (= 130 kW × 8 h). Therefore, approximately 1000 kW-h of energy is available to be used during periods of low solar radiation or night-time operation.

As the pressure inside the tank is affected by the temperature of the air, Fig. 81.5 shows the pressure drop during charging and discharging, respectively. As observed, the initial pressure drop in the system is approximately 2.5 Pa, for an outlet temperature of 25 °C. However, as the outlet air temperature within the thermal storage increases, the pressure drop also increases. Hence, the pressure at the end of

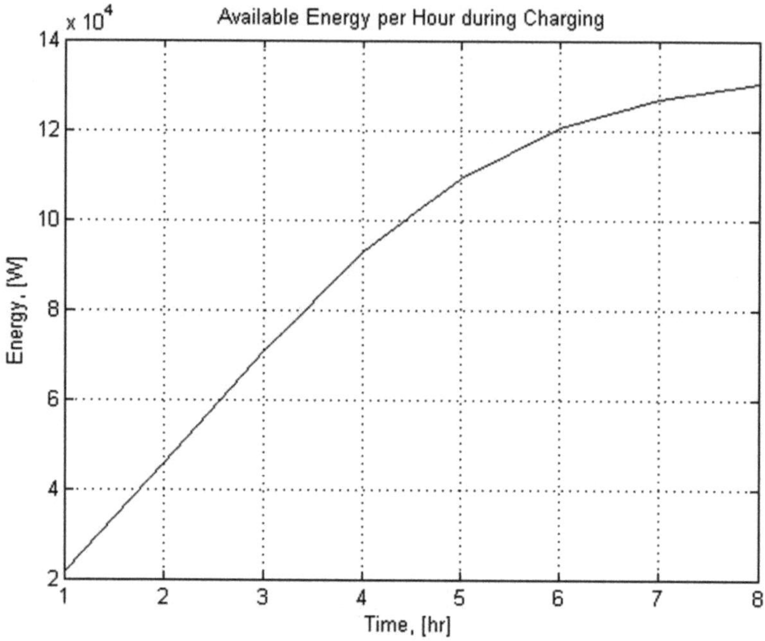

Fig. 81.4 Available energy during charging over 8 h

the 8-hh charging is about 25 Pa. This pressure drop may then be used to determine the required power of the fan/blower for the system.

To investigate the behaviour of airflow inside the storage tank, SolidWorks flow simulation is used, as shown in Fig. 81.6. A 3D modelling of flow was obtained to capture the actual flow behaviour of air inside the tank. As can be seen, the velocity reaches its maximum value in the outlet reaching 7.3 m/s. The velocity around the inner wall is around 1.6 m/s and gets weaker as it moves towards the middle.

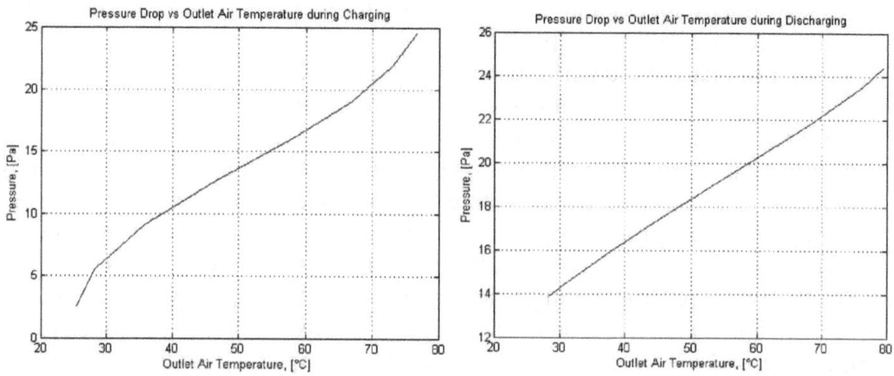

Fig. 81.5 Pressure drop during charging *(right)* and discharging *(left)* for outlet air temperature

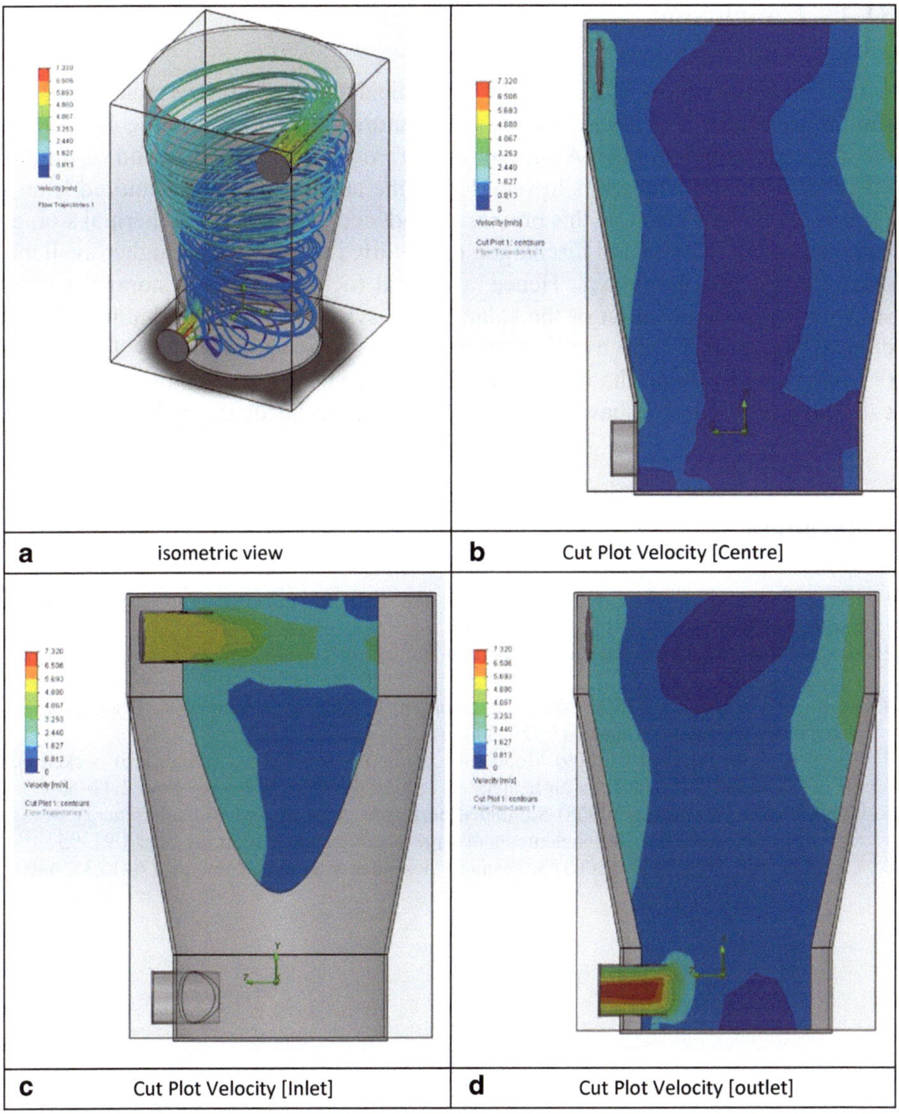

Fig. 81.6 Air flow inside the conical storage tank. **a** Isometric view, **b** cut plot velocity *(centre)*, **c** cut plot velocity *(inlet)*, **d** cut plot velocity *(outlet)*

As observed during discharging, as the area at the bottom of the conical thermal storage is smaller than the top, the initial pressure drop for discharging is larger than during charging. With an initial pressure drop of 14 Pa, the maximum pressure drop in the thermal storage during discharging would be approximately 24.2 Pa.

81.4 Conclusion

A mathematical model is developed for conical thermal storage tank. From the simulated results, it is found that the temperature distribution is more efficient in the conical thermal storage. Also, the pressure drop over a long period of time in the conical thermal storage is lower. Hence, the amount of power required from a centrifugal fan to overcome this pressure is reduced. As the conical thermal storage tank has better charged and discharged capability, this results in a higher available energy output for the system. Hence, a conical rock-bed thermal storage is more suitable for the application of the solar drying system. With a maximum pressure drop of about 24 Pa, the available energy stored was approximately 1040 kW-h (= 130 kW × 8 h). From the flow simulation, it was observed that transactional circular flow is enabling to have uniform air distribution inside the tank.

References

1. Karim MA (2011) Experimental investigation of a stratified chilled water thermal storage system. Appl Therm Eng 31(11–12):1853–1860
2. Zanganeh A et al (2012) Packed-bed thermal storage for concentrated solar power. Solar Energy I(86):3084–3098
3. Karim MA, Hawlader MNA (2004) Development of solar air collectors for drying applications. Energy Convers Manage 45:329–344
4. Singh R, Saini R, Saini J (2009) Models for predicting thermal performance of packed bed energy storage system for solar air heaters—a review. Open Fuels Energy Sci J 2:47–53
5. Singh R, Saini R, Saini J (2008) Simulated performance of packed bed solar energy storage system having storage material elements of large size. Open Fuels Energy Sci J I:91–95
6. Crandall DM, Thacher EF (2003) Segmented thermal storage. Solar Energy 77(4):435–440

Chapter 82
Reducing Biomass and the Use of Kerosene for Cooking in Indonesia to Support the Global Efforts to Reduce CO_2 Emission

Herliyani Suharta and A.M. Sayigh

Abstract We discuss Indonesia's efforts in poverty alleviation within the context of national energy. The statistical data regarding provincial poverty lines, the number of people who live below this line, the energy type used by low-income families and the gender situation are given. Several approaches to energy provision for firewood consumers are described in the global context. It started with a field research initiative until the formation of a national program for socio-economic development including World Bank involvement. Various demands such as electricity development, overcoming climate change disasters and infrastructure reconstruction led to funding scarcity. We describe ideas/initiatives to obtain funding by showing the benefits of reducing CO_2 emission. The estimated annual global costs of adapting to climate change and the Cancun Agreement are briefly described within the funding scarcity context.

Keywords Biomass · CDM · Climate change · Kerosene · LPG · Poverty line · Renewable firewood · Sun cooking

82.1 Background

The Handbook of Energy and Economy Statistic of Indonesia 2012 [1] published by the Ministry of Energy and Mineral Resources (KESDM in bahasa) shows the National Primary Energy Provision in 2011 was 1,490,771 thousand barrels of oil equivalents (BOE). This consisted of natural gas (458,952 × 10^3 BOE), coal (334,143 × 10^3 BOE), crude oil (327,422 × 10^3 BOE), biomass (280,171 × 10^3 BOE), other fuel or "minyak bakar" (200,795 × 10^3 BOE), liquefied natural gas (LNG) (197,244 × 10^3 BOE), biofuel (46,676 × 10^3 BOE), hydro power (31,269 × 10^3 BOE), other petroleum

H. Suharta (✉)
Energy Technology Center (B2TE)-BPPT, PUSPIPTEK, Serpong, Tangerang, Banten 15314, Indonesia
e-mail: herli@iptek.net.id

A. Sayigh
World Renewable Energy Network (WREN), BN2 1YH, PO Box 362, Brighton, United Kingdom

A. Sayigh (ed.), *Renewable Energy in the Service of Mankind Vol II*,
DOI 10.1007/978-3-319-18215-5_82

products ($27{,}029 \times 10^3$ BOE), liquefied petroleum gas (LPG) ($17{,}564 \times 10^3$ BOE), geothermal ($16{,}494 \times 10^3$ BOE), and electricity (1558×10^3 BOE). KESDM Data and Information Center stated that firewood and charcoal used in households were classified as biomass energy. This makes biomass the fourth largest primary energy need after natural gas, coal and crude oil.

If energy transformation, own use, losses and statistical discrepancies are omitted from the National Primary Energy Provision, the final energy consumption becomes $1{,}114{,}767 \times 10^3$ BOE. This consist of consumption by industry ($359{,}687 \times 10^3$ BOE), households ($320{,}369 \times 10^3$ BOE), transportation ($277{,}405 \times 10^3$ BOE), non-energy use ($98{,}413 \times 10^3$ BOE), commercial ($34{,}077 \times 10^3$ BOE), and other sectors ($24{,}816 \times 10^3$ BOE). Household energy is the second largest consumer.

"Statistics Indonesia" published in 1999 [2] showed 52.54 % of households in Indonesia use firewood. This fell to 47.49 % in 2009 [3], and to 36.35 % in 2012 [4]. People use firewood as their economic conditions are limited. The black bar in Fig. 82.1 is the percentage of households that use kerosene, which is almost zero in Java and Bali. The use of LPG for cooking increased sharply after 2009 because of government interventions (see Sect. 3.4).

Firewood is used by 81.26 % of households in East Nusa Tenggara (83.51 % in 2009), 51.51 % in Central Kalimantan (60 % in 2009), 51.29 % in West Sumatera (56 % in 2009), and 43.16 % in South Kalimantan (50 % in 2009).

The main reason why households do not cook is to save money or that they do not have enough money. In East Nusa Tenggara, for example, the population in September 2010 was 4,683,827 (4,619,700 in 2009); 98,050 people in urban areas live below the income of 321,163 IDR/capita/month, while 911,100 people in rural areas live below the income of 234,141 IDR/capita/month.

Cooking with firewood is often carried out indoors with lack of efficient ventilation. The efficiency of this process varies widely, from 10 % for a well-made fire to 40 % for a custom-designed charcoal stove [5]. In barren areas and in the era of climate change, the demand for firewood and the slow growth of trees are not in balance, leading to a scarcity of firewood. Efficient cooking technology reduces firewood combustion which means reducing green house gas (GHG) emission, smoke, and toxic particles inhaled by the people cooking and their children. Furthermore, it saves human energy in collecting firewood in the harsh climate of barren areas where firewood is scarce.

Smoking-related diseases are a major worldwide health issue. Across the developing world, 1.6 million children die annually because of exposure to smoke from indoor cooking. Exposure to smoke throughout their life is reported to have a detrimental impact on the growth and development of 100 million children. This is a key barrier to successful socio-economic development and to making progress towards the Millennium Development Goals [6]. However, the program to reduce biomass for cooking has to compete with many other development programs such as the national electricity program. Helping the victims of natural disasters including those caused by climate change are urgent, while infrastructure reconstruction also needs a huge amount of national funding.

National Electricity Development to Compare the Urgency. PT. PLN (the state electricity company) stated that national peak load is 29,500 MW, and the installed

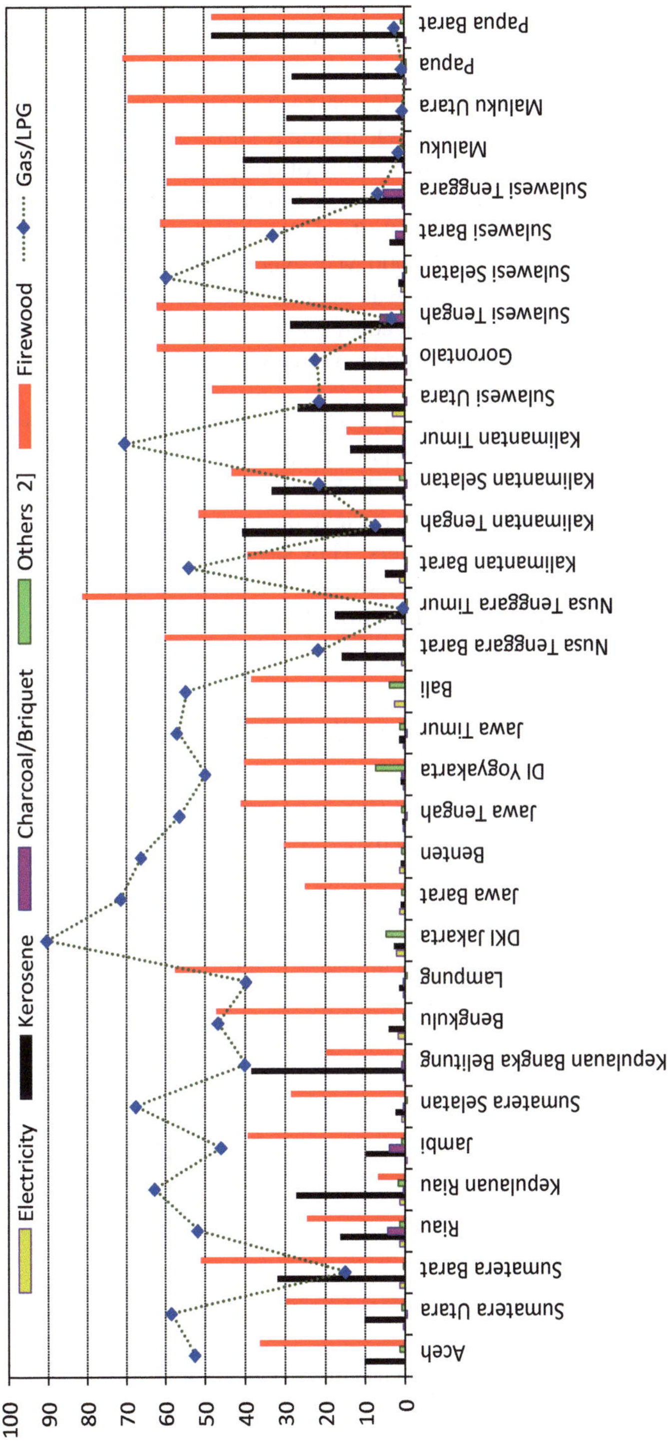

Fig. 82.1 *Red bar* is the percentage of households that used firewood by province in 2012 [4]

capacity is 34,000 MW [7]. The reserve is only 15.25 %, while the demand for national electricity is growing by 9 % a year. The growth rate for outside Java is approximately 12–18 %.

Oil replacement in power plants decreases the capacity and life time of the power plants. Ngurah Adyana, Java-Bali Operational Director of PT. PLN stated "For a good security, the reserve should be 30 % of the peak load of each operation area (wilayah)." An electricity crisis has emerged in North Sumatera. It is predicted that this will also occur in Java by 2018 [8, 9]. Jusuf Kala, Vice President 2004–2009 stated "Indonesia should build 10,000 MW every 3 year. The first 10,000 MW program was announced in 2007 and was planned to finish in 2010/2011. The second 10,000 MW program was announced in 2008 and was planned to finish in 2014. However, the first program was produced only 8000 MW, while the second program is not realized yet" [7].

There is a systematic effort in electricity development and there is a subsidy for electricity. Those who use massive amounts of electricity (industries and commercial sectors) receive more subsidy benefits. As they are producers, they can decide the price of their products to retrieve the money back. On the other hand, the majority of households are consumers. They save when the electricity price is high. Households that have no electricity do not benefit from electricity subsidies and those who have no car do not benefit from fuel subsidies.

82.2 International Fund for Climate Change Adaptation

The Copenhagen Accord states that CO_2 concentration in the troposphere must not exceed 450 parts per million (ppm) by 2100, so that the increase in the global temperature should be below 2°C [10]. As only 50 ppm are left for 87 years of actions (2100−2013), this means humans are allowed to emit only 1 ppm per 1.7 years or less. Figure 82.2 shows that CO_2 increased from 386 ppm in 2009 to 400 ppm on 2 May 2013 (or 3.5 ppm a year). This indicates that CO_2 sequestration is not balanced with anthropogenic GHG emission.

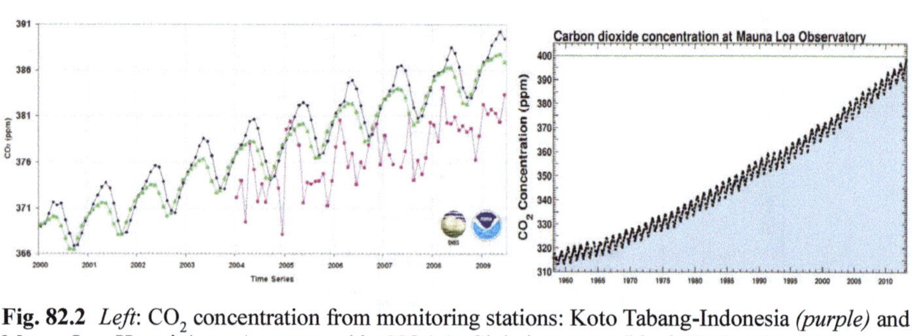

Fig. 82.2 *Left*: CO_2 concentration from monitoring stations: Koto Tabang-Indonesia *(purple)* and Mauna Loa-Hawai *(green)*, processed by NOAA. Global average *(black)* [11] *right*: CO_2 concentration profile from Mauna Loa monitoring station shows 400 ppm on 2 May 2013

The International Energy Agency (IEA) reported that emission in 2010 from fossil burn had reached 30.6 Giga tCO_2equivalent. The total including land use, land-use change and forestry (LULUCF) reached 50 Giga tCO_2Eq. At this rate, the IEA predicted that CO_2 concentration will reach 420 ppm in 2017 [12]

The Cancun Agreement [13] suggests further discussion as to whether the global average temperature rise should be decreased to 1.5°C at the next Conference of Parties meeting in 2015 in Paris. It reports a series of UN decisions on the scheme to help poor countries in cutting their carbon emission and to adapt to climate change, to share and to develop green technology, to consider "a climate risk insurance facility" to help poor countries to cope with extreme weather impacts, and to pay poor countries not to chop down trees.

Figure 82.2 left shows the warming of the climate system is "unequivocal." Low data from KotoTabang indicates the impact of sequestration of Indonesia's forests. Therefore, the Indonesian government has taken tight measures to protect its rain forests and run national afforestation and reforestation programs across the country.

Parry et al. reported [14] on the annual global costs of adapting to climate change estimated by the UN Framework Convention on Climate Change (UNFCCC) as shown below:

- Water: The UNFCCC estimate only USD 11 billion, excluding costs of adapting to floods, and no costs for transferring water within nations from surplus to deficit areas.
- Health: The UNFCCC estimate only USD 5 billion for assessing malaria, diarrhea and malnutrition. This covers only 30–50 % of the global total disease burden and excludes developed nations,
- Infrastructure: The UNFCCC estimate only USD 8−130 billion. This investment must increase in order to reduce poverty and thus avoid continuing high levels of vulnerability to climate change. This upgraded infrastructure could be eight times more costly than the estimate.
- Coastal zones: The UNFCCC estimate only USD 11 billion, excluding increased storm intensity. They used low predictions of sea level rise published in the 2007 IPCC report. Including these points, the costs could be three times greater than the estimate.
- Ecosystems: The UNFCCC excluded the costs of protecting ecosystems and the services they can provide for society. This could cost over USD 350 billion.

82.3 Various Cooking Technologies That Use Solar Energy, Renewable Biomass and LPG

82.3.1 Solar Box Cookers Dissemination in Indonesia

Various designs of solar box cookers have been assessed since 1992. The introduction of sun cooking is seen as an alternative way to reduce smoke inhaled by

Fig. 82.3 Solar box cooker: model HS 5521

firewood consumers, who are a low income society with low education. Community education, on a gender equity basis, aims to improve their knowledge for finding possible solutions to their energy limitation and teach them how to use and make solar cookers. The development and field testing of user-made designs has been in use since 1995 combined with the involvement of international volunteers to widen the global impact across. The design shown in Fig. 82.3 was the product of training.

The temperature inside the cooking chamber is 202–204°C without load. The cooking time depends on the design, the cooking pot used, the quantity of food to be cooked, the air temperature, wind, latitude and the amount of solar energy.

Food cooks faster between 10:00 and 14:00 and slower between 8:00 and 10:00 and between 14:00 and 16:00. Sun cooking usually avoids stirring or turning the food since opening the cooking chamber will release the heat and slow the cooking process. The food is chopped in small pieces for faster cooking. Different foods can be cooked simultaneously depending on the size of the cooking chamber. A family may use one or more solar cookers. Compared to cooking over a fire, the food does not have a smoky flavor or pollution.

The Indonesian Sun Cooking project received a 5-year field research grant from 1994 to 1998 from The Center for Field Research in Earthwatch Institute, Massachusetts, USA. The project continued without this support until 2004 [15–21]. Slow acceptance, however, is not a failure. As children watch their parent using a solar cooker, they will grow up and may adapt to new ideas/methods. In the longer term, greater understanding might arise from multiple sites and could help to shape the fundamental policy changes that are required. The field research findings into the broader impacts within the policy elite might help shift the policy debates toward providing facilities and funding.

82.3.2 Concentrated Cooker for the Clean Development Mechanism (CDM) Solar Cooker Project in Aceh

The CDM provides a financial mechanism where the government and private sectors can invest in greenhouse gas (GHG) mitigation projects in Annex I countries. Proved emission reductions will be credited to the country that invests to provide necessary technology and the Certified Emission Reduction (CER) credit will be counted towards their national CO_2 emission reduction target. The participating developing country gains technology for the target user that has no purchasing ability. In this global understanding and agreement, the dissemination of solar cookers was arranged under the CDM initiative.

At a meeting with Dieter Seifert at his exhibition booth at Umbutu Village, Johannesburg in July 2002, Herliyani suggested expanding solar cooking activity in Indonesia. Funding uncertainty drove her to write a proposal called "Global Solar Cooker Programme," which was submitted to Alluminium Canada (ALCAN)—in London in August 2002; the proposal provided the name of the patent holder for solar cooker K14. The idea was presented at the World Renewable Energy Network meeting in Brighton, UK, in October 2002.

Project design document (PDD): 0218-CDM Solar Cooker Project Aceh 1 dated 10 October 2005, category SSC 1c for switching non-renewable biomass/non-sustainable harvested firewood. Methodology used: CDM Project AMS-1C ver.6—Thermal energy for user. Date of request for registration was 7 January 2006–5 February 2006. UNFCCC registration on 6 February 2006. Host party: Indonesia. The project sites are in Sabang Islands and Badar in South East Aceh. A thousand sets containing a solar concentrator cooker K14 plus a wonder box were transferred to the target users with one set for each household. The life span of the K14 is >10 years. The wonder box is for heat retention cooking (simmering) and to keep the food hot during meal times.

It was the first registered CDM project for Indonesia and for Germany. ALCAN is an investor in the CDM project. The funding was managed solely by the German company as they wanted to acheive 100% CER. It is the "first" project that represents the global importance of incorporating CDM finance into a community educational program to ensure the best use of the cooking devices. The target users who used solar cookers have made an effort to contribute to reducing their CO_2 emission ([19, 22–24]; Fig. 82.4).

It is true that the Indonesian government is willing to improve the energy infrastructure, to improve the living conditions of poor people and to keep the environment green. On the other hand, there are companies in industrialized countries that are still producing CO_2 during their industrial processes for profit, while simultaneously looking for CER to fulfill their CO_2 reduction obligation through their involvement in the CDM project. Although these two contrasting standards of living meet in the CO_2 trading arena, we are all hoping to clean the environment.

However, we need to know if the targets of CO_2 emission reduction have been achieved in this balance [6].

Fig. 82.4 Solar concentrator cooker K14

In the era of climate change, the increase in global temperature caused more sea water evaporation. Thick clouds hanging in the sky above the Indonesian archipelago caused intermittent solar insolation, while on a clear day, the weather is very hot and the sun irritates the skin. This might affect sun cooking.

82.3.3 CDM Cook Stove Project Kupang

The CDM project was designed to distribute 30,600 sets of an efficient stove "Save80" plus a wonder box to households in Kupang, East Nusa Tenggara. The project provides thermal energy to replace the use of kerosene or other fossil fuels for cooking. The stove Save80 (see Fig. 82.5), which uses small pieces of firewood

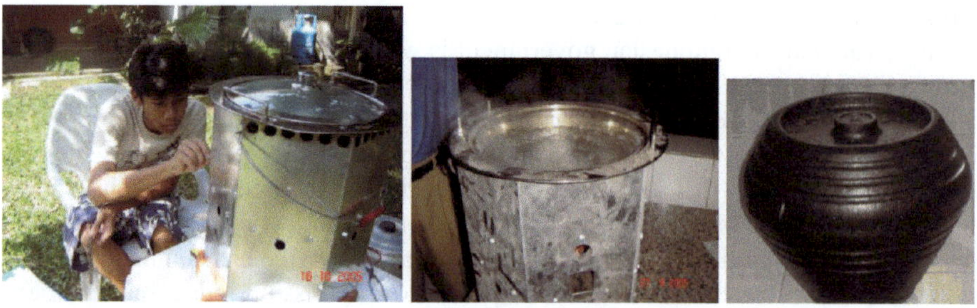

Fig. 82.5 Save80 stove, boiling water inside the stove and the wonder box

that are categorized as renewable, is suitable to prepare all usual dishes. The lifetime if used appropriately is > 12 years; the metallic parts in contact with the flames is made of stainless steel. A wonder box is for heat retention cooking (simmering) in order to save fuel. In a field test on 26 September 2005, the water temperature in the pot filled with 6 L of boiling water and stored in the wonder box remained hot at 70 °C after 7 h.

The overall efficiency of the Save80 is approximately 45 %. As the nominal effective thermal power is 1.45 kW, the total thermal power from 30,600 Save80 stoves distributed by the project will be 44.4 $MW_{thermal}$. As it is below the limit of small-scale CDM projects, that is, 45 $MW_{thermal}$, the simplified methodology for small-scale CDM projects applies.

The PDD was completed in January 2006. The preliminary meeting with TUEV Sued-Munich took place in February 2006. The validation process and site visit was in May 2006. The promotion of Save80 in Kupang City Hall was performed on 31 May 2006. The preliminary field test was carried out after October 2006. National approval was authorized on 27 December 2006. The registration to UNFCCC in Bonn followed. Production of 30,600 Save80 stoves was planned to be completed by June 2007.

The mean emission reduction from one “Save80” stove in combination with a wonder box is estimated to be 1.487 tCO_2Eq. For 30,600 stoves, it is estimated to be 42,379 tCO_2Eq./year. The emission reduction is assumed constant during the crediting period. Thus, the total emission reduction over the 10-year crediting period is 423,790 tCO_2Eq.

The crediting period is 1 July 2007 until 30 June 2017. The German company wanted 100 % CER.

The project barriers and risks written in the PDD are lack of acceptance by the users, and financial barriers. To overcome the first barrier, an approach was incorporated into community education that was designed to ensure the best use of the cooking devices. Random visits during monitoring, performed by skilled personnel, provide opportunities to correct mistakes and deficiencies. The second barrier is a reason why the project was proposed under CDM. Without CDM credits the entire project will not be self-supporting; therefore, pre-financing was needed until the PDD was published on the UNFCCC website. However, as the CDM Cook Stove Project Kupang could not get investment, all efforts were in vain; there was no field research and no CERs to payback the capital invested [27–29].

82.3.4 LPG to Replace Kerosene for Cooking

The program “LPG to Replace Kerosene for Cooking” in Indonesia started in 2007 in response to the Presidential Decree No.104/2007. This program distributed a start-up package containing “a set of cylinders filled with 3 kg LPG, a burner and accessories” to replace the kerosene burners. Millions of sets were distributed in 21 provinces—Aceh, North Sumatera, Riau, Riau Island, Jambi, South Sumatera,

Fig. 82.6 LPG 12-kg cylinder *(blue)*, and 3-kg cylinder *(green-melon)* [35]

Lampung, DKI Jakarta, West Java, Banten, Central Java, DI Yogyakarta, East Java, Bali, West Kalimantan, South Kalimantan, East Kalimantan, North Sulawesi, Gorontalo, West Sulawesi, South Sulawesi. It began by distributing 4,030,683 sets in 2007, 15,407,777 in 2008; 24,156,307 in 2009; 4,239,078 in 2010; and 5,283,834 in 2011 [30].

According to the data shown in Fig. 82.1, the next program should be in West Sumatera, South East Sulawesi, Central Sulawesi and East Nusa Tenggara, Central Kalimantan, Maluku, Maluku Utara, Papua and West Papua.

In October 2013, the price of a 3-kg cylinder of LPG at PT. Pertamina gas station was 12,750 IDR; 13,250 IDR at an agency, and 14,500 IDR in the town of Pangkalan. In the market, it often sold at 15,000 IDR [31]. "The need of subsidized LPG in January 2014 was 6,193 tonnes" according to Rudy Biantoro, PT.Pertamina Marketing Operation Regional VI Kalimantan [32].

The price of non-subsidized LPG in 12-kg cylinder in October 2009 was 5850 IDR/kg. However, it often sold at 75,000 IDR.

On 1 January 2014, the price was raised to 9809 IDR/kg, or 117,708 IDR/12-kg cylinder [33]. However, it sold at 120,000–130,000 IDR.

The price was raised after BPK (Indonesian Supreme Audit Board) found that PT. Pertamina lost 7.73 trillion IDR between 2001 and October 2012 [34]. Since the price of non-subsidized LPG in 12-kg cylinders increased, some of the consumers changed to buying subsidized LPG in 3-kg cylinders allocated for disadvantaged people. This caused subsidized LPG in 3-kg cylinders to become scarce and the price increased to 17,000 IDR Fig. 82.6.

82.3.5 Green Stoves for East Nusa Tenggara Province

On my first visit to Kupang in 1993, I saw a cluster of Kosambi trees and Lontar trees on the way to Oesapa; however, in 2005, those clusters were no longer there.

Fig. 82.7 **a** 3-stone fireplace. **b** Renewable harvested firewood for cooking with the green stove. **c**: The green stove is used to boil 5 L of water [5]. Portability and self-regulating air supply are the advantages

Kosambi wood has high calorific value. The population of Kupang city depends on kerosene; however, firewood is a buffer to get their energy need. The 3-stone fireplace (see Fig. 82.7a) is widely used, although logging of trees is forbidden. Firewood is transported from green areas and is traded. People buy it from retailers. Inefficient use of firewood often destroys green areas. However, some experts think "it may be a minor cause of deforestation;" therefore, it is negligible compared to the deliberate destruction of clearing land for agricultural use [36].

"Getting kerosene is becoming more and more difficult. We could not buy direct as take and pay. We would queue to get 10 L kerosene per week. Often we left a small plastic drum (jirigen in Bahasa) at the retailer shop and would take the filled drum once per week. If we could not get kerosene we used our firewood stock collected from our garden. From May until November we use firewood to boil water. We only need to put three stones for a fireplace in our garden. We can save 65,000 IDR per month. The price of subsidized kerosene is 3000 IDR/L. This means we save 48 % of our energy expenditure. Lately, we only get 5 L kerosene for our weekly needs. The firewood becomes scare. We use whatever dry wood we can find. The famous Kosambi wood is difficult to find today" Ignas, sms communication on 20 July 2012.

Firewood consumers in East Nusa Tenggara suffer from limited kerosene supply and limited biomass resources.

Overcoming the firewood crisis is one of the greatest challenges of humanity.

The program of "LPG to Replace Kerosene for Cooking" for East Nusa Tenggara is still at a waiting list status.

In Kupang, 12-kg LPG plus cylinder is sold at 850,000–1,000,000 IDR. Changing an empty 12-kg cylinder for a full cylinder costs 160,000-190,000 IDR. Only several shops sell these goods and their stock is only 10–20 tubes [37]. Therefore, only 0.37 % of the population use gas for cooking (see Fig. 82.1).

After the "CDM Cook Stove Project Kupang" could not get investment, various stoves designs, their performances and prices were assessed to find low-cost projects for wider reach and impact. A simple stove was bought in Nepal from Sanu Kaji Shrestha. After testing using small pieces of dry branches with a diameter of < 2 cm (see Fig. 82.7b), we called it Tungku Hemat Energi in Bahasa or green stove. It can also use larger pieces of chopped dead wood. The stove can boil 5 L of water from 27 °C within 30 min and needs 350 g of small pieces of dry wood. The user can get complete combustion easily which leads to less smoke [38–40].

A survey in 2004 showed that the average need of kerosene in middle to upper level society was 1.5 L a day, with a maximum of 2 L. The average need of kerosene in lower level society was 0.5 L a day, with a maximum of 1 L. The average need is 1–1.5 L of kerosene a day, and 1 L a day is sufficient to obtain a reduction in CO_2 emission [5].

A survey in 1998 which investigated the demand for firewood [38, 41] reported that

- Families buy an average of 300 kg/month of firewood.
- Families that collect firewood do so 9 times a month; each collection is 15–20 kg or 135–180 kg a month.

The luxury of bathing in warm water and the ability to buy raw cooking food dictate this difference in demand.

The average demand for firewood is 217.5 kg/month or 7 kg/day.

The green stove needs 350 g of wood to boil 5 L of water in 30 min. This means the stove should be used for 5.279 h a day to compete with a kerosene brazier that needs 1 L of kerosene a day. If this operating time is used to boil the water, the need for small pieces of firewood is (5.279×60 min)/30 min × 350 g = 3.689 kg. The green stove is saving (3311/7000 × 100 % =) 47.3 % of firewood per day.

This means a reduction of almost half of the environmental burden, reducing smoke inhalation and reducing CO_2 emission.

This green stove fulfills all principles of frugal engineering—robustness, portability, simplicity, matching the local culture and offers the possibility of mega-scale production. A large manufacturer was approached to support the production for wider distribution. In July–August 2012, a combustion chamber was made of stainless steel that could resist temperature up to 1500 °C and could resist heat corrosion for 5 years, but still remained at a cheap price.

The stoves will be transferred to people via training to demonstrate its capability for efficient cooking using firewood. Recruiting female participants in the "training-for-trainers" program and for monitoring will be carried out by a local women's organization (PKK) that is promoting how to cook local foods in an efficient way.

Ideas for Obtaining Funding Years passed by, but no funding for the action described in Sect. 3.3 was available. As Save80 could not easily be made locally, a proposal was made to use easier technology that could be made locally. The proposal is not registered as a CDM project; however, the idea of counting towards the CER would increase the value of the project.

The estimated CO_2 emission reduction per green stove is 0.9297 $tCO_2Eq./year$ [5].

Note: The first year of counting is 2012 as the preliminary proposal was made in 2012 (see Table 82.1).

If we distributed 10,000 green stoves in the first year, the total CO_2 emission reduction is estimated to be approximately 9297 tCO_2Eq./year. If the price of CER is 10 USD/t, this project will create a CER of 92,970 USD in the second year (2013). The stoves that have been implemented in the first year will still give a CER in the following years which is assumed constant. Therefore, the distribution of 50,000 stoves will give a CO_2 emission reduction of approximately 46,500 tCO_2Eq./year. The CER that can be harvested in 2017 will be 465,000 USD at 100 % abatement success. After 2017, the accumulated CER generated by the distribution of 50,000 green stoves is 1,393,820 USD. In the worst scenario, where the abatement success is 50 %, while the CER price is rated at only 2 USD, this action will still give a CER of 139,448 USD. If the CER value can be reduced in advance, there will be funding to start this project.

It is necessary to bring the field findings to the attention of international policy makers in order to break the funding scarcity in helping the poor to live in barren areas of the world. Therefore, this proposal was submitted to the Indonesia Clean Energy Development (ICED) Program (USAID), a strong funding holder, on 20 April 2013, but it was not successful.

82.3.6 Efforts initiated by The World Bank

The World Bank in cooperation with The Ministry of Energy and Mineral Resources of Indonesia (Ditjen EBTKE-KESDM) arranged the "Multi-Phase Clean Stove Initiatives" in Indonesia. The first phase consisted of 3 stages—consultation workshops held in May, July and October 2012 [42, 43]:

The first meeting discussed the location of the existing activities and the development of various stoves. It was found that there was minimal market penetration for improved stoves. The second meeting discussed the intervention strategies.

The third meeting discussed further steps to prepare Phase II, and discussions on stove standards, test procedures and the institutes who carry out the testing, funding mechanisms such as targets, stove production, and the monitoring and verification of possible supply lines. The participants included provincial and regional government personnel, experts, private stake holders and various nongovernment organizations. The World Bank offered a loan. International financial business meetings took place regarding poverty alleviation and environment development by the Indonesian Government. The question being asked was whether this could be part of the solution discussed in the Cancun Agreement

Table 82.1 The estimated amount of CO_2 emission reductions and the possible CER harvested during the project period [5]

Year	Stoves imple-mented	Abatement success				Various CER price/tonne: 10 USD/t; 5 USD/t and 2 USD/t								
		100 %	75 %	50 %	25 %	10 USD/t	5 USD/t	2 USD/t	10 USD/t	5 USD/t	2 USD/t	10 USD/t	5 USD/t	2 USD/t
		Tonnes of CO_2 e.g.				USD	USD	USD	USD	USD	USD	USD	USD	USD
						100 % abated			75 % abated			50 % abated		
2012	2012: 10,000	0	0	0	0	0	0	0	0	0	0	0	0	0
2013	2013: 10,000	9297 for 10,000 stoves	6973	4638	2,319	92,970	46,380	18,594	69,730	34,865	13,946	46,380	23,190	9,276
2014	2014: 10,000	18,594 for 20,000 stoves	13,945	9,297	4,649	185,940	92,970	37,100	139,450	69,725	27,890	92,970	46,485	18,594
2015	2015: 10,000	27,891 for 30,000 stoves	20,918	13,945	6,473	278,910	139,450	46,500	209,180	104,590	41,836	139,450	69,725	27,890
2016	2016: 10,000	37,100 for 40,000 stoves	27,825	18,594	9,287	371,000	185,940	74,200	278,250	139,125	55,650	185,940	92,970	37,188
2017	2017: 10,000	46,500 for 50,000 stoves	34,875	23,250	11,625	465,000	232,500	93,000	348,750	174,375	69,750	232,500	116,250	46,500
	Total 50,000					1,393,820	697,240	269,394	1,045,360	522,680	209,072	697,240	348,620	139,448

Minimum capital need 5 × 100,000 USD/year or 500,000 USD for the 5-year project period
If the abatement success reaches 100 %, at CER price of 5 USD/t, the project has shown pay back
If the abatement success reaches 75 %, at CER price of 5 USD/t, the project will still be able to pay back
If the abatement success reaches 50 %, the CER price should be 10 USD/t, the project will be able to pay back

82.4 Gender Consideration

In 2010, the Indonesian population was 237,641,326. There were 101.4 males for every 100 females. The number/proportion of females in various ranks of the civil service in 2000 is shown in Fig. 82.8 left [2].

Although the number of females at the lowest rank is low, the number of females at rank IIB to IIIB is significant. However, the decline in numbers at the higher ranks means less access to decision making. The patriarchal system and the societal mindset dictates that women perform domestic chores, nurturing tasks, and support their husband; societal perceptions and biases are a stigma to women. Cooking is a woman's task including firewood gathering and all tasks related to cooking. Married women in low-income families endure many stigmas as well as having to manage with a limited amount of money. It is difficult for them to get involved in activities outside the home, as these tasks take up all their time.

Male and female civil servants on the same rank receive the same salary; however, because of the many stigmas facing women, they are unlikely to further their career. Free competition in the job market meant women were left far behind. Although women have broad technological knowledge and experience, and are particularly beneficial in certain situations, their knowledge is often not promoted, leading to fewer advantages for women in society.

In 2013, the gender situation improved (see Fig. 82.8 right).

A regulation was introduced stating that the highest rank in the civil service can be achieved via structural career and via functional career. There is a method to count each functional rank to analogy to the rank of the civil servant. Today, many women have working hard on the detail in every functional career. This regulation

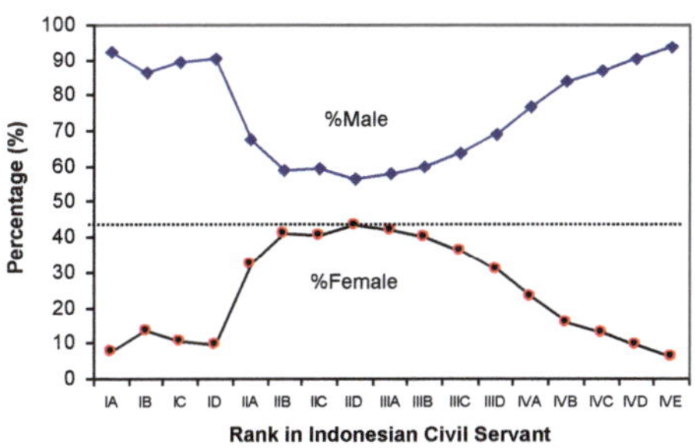

Fig.100. X-profile of Male-Female existance in various ranks of Indonesian civil servant.
In the rank IIB until IIIB the presence of female is significant. However, it declines at the higher ranks.
Source: Statistik Indonesia 1999, BPS (2000: p75).

Occupation	Male	Female	Total	F/T (%)	M/T(%)
Specific Functional	935838	1316252	2252090	58.45	41.55
General Functional	1225945	751485	1977430	38	62
Structural	170766	67696	238462	28.39	71.61
5th echelon	2744	1240	3984	31.12	68.88
4th echelon	112734	53991	166725	32.38	67.62
3rd echelon	44633	10868	55501	19.58	80.42
2nd echelon	10110	1490	11600	12.84	87.16
1st echelon	545	107	652	16.41	83.59
Total	2332549	2135433	4467982	47.79	52.21

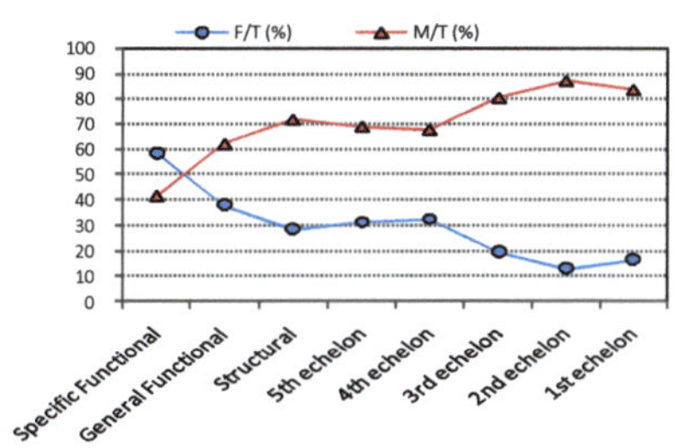

Fig. 82.8 *Left*: The number/proportion of females in various ranks of the civil service in Indonesia [2]; the highest rank is IVE *right*: The number of civil servants by occupation and sex [4]. The first rank is the highest structural career

has led to an increasing number of women occupying the highest ranks in the civil service, and the gap is shortening as shown in Fig. 82.8 left.

Although the majority of women in this situation do not cook with firewood, they are aware of the hardship and difficulties that women have to face every day, especially those with low incomes.

82.5 Concluding Remarks

The use of various cooking technologies has been rated appropriately. However, barriers such as lack of awareness and inappropriate government policy stand in the way. In many seminars that discuss technology dissemination, or poverty alleviation, it is often stated that "giving is not educating people" or "we are not social agents but businesses." On the other hand, the poverty line of rural areas in some provinces is very low. East Nusa Tenggara, for example, where 81.26% of households use firewood, the poverty line is 234,141 IDR/capita/month, or 7553 IDR/day. If one person works, but he or she has two children, the 7553 IDR is only for food. As the government support is only for basic schooling, most of the children leave school as no funding is available for higher education; this is probably a similar situation in many places in the world.

One of the resources of forests is firewood. Forests are important for maintaining global balance. Forests act as "carbon sinks." They conserve soil and water, control avalanches and desertification, stabilize sand dunes and protect coastal areas from flood. It is good idea if the dissemination and education approach for various cooking technologies can support reducing tree felling in order to support the national afforestation and reforestation widespread programs throughout the country.

Annual global costs of adapting to climate change should include the costs of protecting ecosystems. Therefore, the previously described proposals can be supported. It provides an opportunity for change with wider community participation, as well as promoting renewable energy deep into the root of the nation—the families. The efforts and the willingness to initiate change need to be appreciated. If we fully involve the public in the process of science we shall not only give them understanding, we shall give the world a future. Courage is needed to raise optimism.

References

1. KESDM (2012) Handbook of energy and economy statistic of Indonesia 2012
2. BPS (2000) Statistic Indonesia 1999.BPS Jakarta, Indonesia, p 75
3. BPS (2009) Statistic Indonesia 2009. BPS Jakarta, Indonesia, p 82, 182–184
4. BPS (2013) Statistic Indonesia 2013. BPS Jakarta, Indonesia, Table 4.3.5, Table 2.3.2
5. Suharta H, Sayigh A (2012) Green stove dissemination via community education to face firewood scarcity and financing possibility through C-trading. ISESCO J Sci Technol 8(14):70–85

6. Suharta H (2009) Brief assessment on CDM solar cooker project Aceh 1, Indonesia, submitted to *Asia Pacific Tech Monitor* (a web-based, on-line bi-monthly journal featuring articles on different aspects of technology development and transfer). The United Nations Asian and Pacific Centre for Transfer of Technology (UN-APCTT), New Delhi, India
7. EVY/MAR/AOI (2014) Proyek Pembangkit Terlambat Dibangun, Kompas daily news, head line, 27 February 2014
8. Evy (2014) Kondisi Pasokan Listrik Merisaukan, Kompas daily news, 22 March 2014, p 19
9. Kala MJ (2014) Solusi untuk Kelistrikan Kita, Wakil Presidem RI (2004–2009), Kompas daily news, 3 Maret 2014, p 6
10. United Nations Climate Change Conference—COP15/MOP5 (2009) Copenhagen Accord, Copenhagen, Denmark, 7–18 Dec 2009
11. Aldrian E (2009) Short communication, director of center for climate change and air quality, meteorological climatological and geophysical agency (BMKG), Jakarta, Indonesia
12. International Energy Agency (2011) World energy outlook, November 2011
13. United Nations Climate Change Conference—COP16/MOP6 (2010) Cancun Agreement, Cancun, Mexico, 29 Nov—10 Dec 2010
14. Parry M et al (2009) A new report on costs of adapting to climate change significantly underestimated, published by International Institute for Environment and Development (IIED) and the Grantham Institute for Climate Change at Imperial College London, UK, 26 August 2014. List of authors are: Martin Parry (Imperial College London), Nigel Arnell, Richard Tiffin and Tim Wheeler (Reading Univ.), Pam Berry (Oxford Univ.), David Dodman and David Satterthwaite (IIED), Chris Hope (Cambridge Univ.), Sam Fankhauser (London School of Economics), Sari Kovats (London School of Hygiene and Tropical Medicine), Robert Nicholls (Southampton Univ.)
15. Suharta H, Fahrudin D (1995, 1996, 1997, 1998) Solar oven, the alternative solution of fuelwood shortage for cooking in remote & barren area in Indonesia, in *Indonesia sun cooking expedition briefing*. Earthwatch Institute, Massachusetts
16. Suharta H, Abdullah K, Sayigh AM (1998) The solar oven: development and field testing of user-made design in Indonesia. J Sol Energy 64(4–6):121–132
17. Suharta H, Sena PD, Sayigh AM, Komarudin (1998) The social acceptability of solar cooking in Indonesia. J Renew Energy 16(1999):1151–1154 (AM Sayigh (ed) Pergamon Press, UK)
18. Suharta H, Sayigh AM, Komarudin, Abdullah K (1999) An educational approach in a cross cultural context to promote solar cookers in Indonesia. In: proceeding of the World Renewable Energy Congress, Juni 1999, Kuala Lumpur, Malaysia, pp 265–270
19. Suharta H, Sayigh A (2005) Cooking with the sun, RENEWABLE ENERGY 2005. An official publication of the World Renewable Energy Network—affiliated to UNESCO. Sovereign Publications Ltd, London, pp 64–68
20. Suharta H, Sayigh A, Nasser SH (2005) Sun cooking: the best practice in Indonesia, ISESCO science and technology vision. J ISESCO Cent Promot Sci Res 1(1):69–75
21. Suharta H, Sayigh AM, Nasser SH (2006) Comparing a way to calculate the heat loss coefficient of solar flat plate collector. Journal Ilmiah Teknologi Energi 1(2) (Pebruari B2TE–BPPT, Jakarta, Indonesia)
22. Suharta H et al (2006) Clean development mechanism solar cooker project Aceh 1, Indonesia, proceeding of solar cookers and food processing international conference, 12–16 July 2006. Granada, Spain
23. Seifert D, Suharta H, Gadhia D, Anderson J, Vílchez M (2006) Clean development mechanism (CDM)—a powerful instrument to fulfil the UN millenium goals—experiences, visions and suggestions. In: proceeding of solar cookers and food processing international conference, 12–16 July 2006. Granada, Spain
24. Suharta H (2009) FUTURE PROSPECT OF CDM Solar Cooker Project Aceh 1, Indonesia, Jurnal Ilmiah Teknologi Energi, Vol 1, No 8, Februari 2009, 1–15. ISSN 1858–3466, Jakarta, Indonesia
25. Seifert D (2002) Test of K14 at 48.2 NL and 12.7 EL on 1 October 2002, place of test D-84524 Neuotting, Germany. Email: bdiv.seifert@t-online.de

26. Solar Gis (2013) World map of global horizontal irradiation, Geo Model Solar. (Retrieved on 15 March 2014)
27. Suharta H, Seifert D, Sayigh AAM, Mathew K (2006) Cook stove SAVE80 for clean development mechanism Project Kupang 1, East Nusa Tenggara, Indonesia. In: proceeding the World Renewable Energy Congress, 5–8 Feb 2007, Perth, Australia
28. Suharta H, Sayigh AAM (2007) Can clean development mechanism is used to finance a project that exerting efforts to achieve the UN millennium development goals?, Jurnal Ilmiah Teknologi Energi, B2TE, vol. 1, no. 4, Februari 2007, 67–77. ISSN 1858–3466, Jakarta, Indonesia. Presented in WRERCE 2007, Grand Hyatt, 4–7 November 2007, Jakarta, Indonesia
29. Project Developers (2005) Clean development mechanism cook stove Project Kupang 1, Indonesia, simplified project design document for small-scale project activities version 02 (01/07/2006), UNFCC web
30. Anonimous, slide presentation that quoted data from Directorate General of Oil and Gas (2012) Overview of LPG Conversion Program (4). Consultation Workshops 2, World Bank-DitJen EBTKE, 12 July 2012, Jakarta
31. Wijayanto N (2013) Harga Elpiji 3 kg naik…., Sindonews.com, 1 October 2013 at 14:00 WIB
32. Necolsen G (2014) Cara Pertamina Hadapi Panic Buying Elpiji 3 kg, Tribunenews.com, 6 January 2014 at 6:00 WIB
33. PT.Pertamina Persero (2014) Pertamina Resmi Menaikkan Harga Elpiji 12 kg, Tribunenews.com, Tanjung Redep, 1 January 2014 at 00:00 WIB
34. Sinaga A (2014) Harga Elpiji 3 kg dopasaran ikutan naik, Sindonews.com, 3 Januari 2014 at 10:41 WIB
35. Dony (2014) Penurunan kembali harga LPG 12 kg rugikan agen, Pertamina cuek, Lensaindonesia.com, 11 January 2014 at 18:57 WIB, Editor Suyono, retrieved 13 March 2014
36. Wikipedia—the free encyclopedia (2006) Biofuel. (retrieved 20 April 2006)
37. Dientje A (2014) LPG tube situation in Kupang, short communication by mobile phone on 12 March 2014
38. FAO: Wood Fuel Surveys. Annex III—(a) Measuring cooking fuel economy, 1. Introduction *http://www.fao.org/documents/show_cdr.asp?url_file=/docrep/Q1085E/q1085e0b.htm*
39. Suharta H (2008) "Berbagai briket sampah alam untuk bahan bakar Tungku Hemat Energi (THE) S1 dan S2" Jurnal Ilmiah Teknologi Energi, Vol. 1, No. 6, Februari 2008. The Center of Energy Technology (B2TE), BPPT, Jakarta, p 42–57
40. The Center of Energy Technology (2009) Pengujian Tungku Hemat Energi (THE) S1. The test result was signed on 12 January 2009. B2TE, BPPT, Jakarta, Indonesia
41. Suharta H, Abdullah K, Sayigh AM (1998) The solar oven: development and field testing of user-made design in Indonesia. J. Sol Energy 64(4–6):121–132
42. Kusdiana D (2012) Plan of GOI towards clean household cooking solution, presented in 1st consultation workshop on clean stove initiative in Indonesia, Ditjen EBTKE, 7 May 2012
43. Zhang Y (2012) Indonesia clean biomass cookstoves program *draft* program document—key elements for discussion, Energy Economist, the World Bank, October16th, 2012

Chapter 83
The Solar Greenhouse: A Strategy for Energy Savings in Social Housing in Milan

Valentina Dessi'

Abstract The recently updated Lombardy Region law no. 39/2004 and the new Milan city building code stimulate more and more the use of passive climate control strategies to reduce thermal energy consumption. Solar greenhouse is one of the strategies promoted by regulations both for new and retrofit buildings.

This chapter compares solar greenhouse regulation requirements (orientation, natural ventilation, openings, shading devices, etc.) with the Milan city public housing policies and programmes. In particular, it focuses on the large public building stock produced between the post-second war period and the 1980s, which has a very low energy performance and therefore high retrofit potential.

Both the regional law and the building code consider greenhouse retrofitting by closing balconies or integrating building elements, such as the "loggia". The solar greenhouses are feasible only if they reach at least a 10% energy savings, calculated in accordance with the UNI EN ISO 13790/2008 (energy performance of buildings—calculation of energy use for space heating and cooling).

The closing of balconies and loggias, especially in social housing, is an action that has often been made to increase the space available in the residence. However, the closing is often individual unauthorized action followed by subsequent amnesty and payment, which legalizes the increased volume and the facade change, independently by any energy performance evaluation.

The consequence of individual uncontrolled actions is the change of facades, which often radically modify the visual impact of the residential complex, out of any facade image of the entire building.

This chapter analyses a case study of a 1940–1950s social housing building in Milan.

The selected case is a linear building with a mix 1/2/3-bedroom flats which have the requirements of the legislation to implement the bioclimatic greenhouses. Over the years, many flats became private property and many of the loggias were closed actually becoming real bioclimatic greenhouses. This chapter specifies the actual energy savings and takes into account the potential visual impacts that would occur with an action controlled and bound by guidelines provided by the local authorities and by the Lombardy public housing company (Aler).

V. Dessi' (✉)
Dept DAStU, Politecnico of Milano, via Bonardi 3, Milano, Italy
e-mail: valentina.dessi@polimi.it

A. Sayigh (ed.), *Renewable Energy in the Service of Mankind Vol II*,
DOI 10.1007/978-3-319-18215-5_83

Keywords Solar greenhouse · Social housing · Energy retrofit · Evaluation tool · Building code

83.1 Introduction

The European 2020 climate and energy package (reduction in CO_2 emissions, energy conservation, and use of resources from renewable energy sources) has triggered a major rethink on how to design, more sustainable, less energy consumers, more comfortable, and environmental friendly settlements. Our environment responsibility as well as the users' needs is changing, and consequently, the way we design our homes is changing. After the European directives for energy savings have been implemented in Italy, everyone is now convinced of the inevitable and without delay necessity to design environmentally sustainable buildings through bioclimatic approach. Since at least a decade, the topic of energy efficiency and sustainability is clearly present in the architectural debate and in competitions especially when proposed by the public administration. This is a very important aspect because it is the public authority that—in addition to verify the design and the process of construction—it also has the role and the ability to drive a process and therefore serves as a guide and example for a correct way to build in terms of sustainability and energy efficiency.

If this is now clear thanks to new realizations, it is a bit more difficult to observe it in existing buildings renewal. Moreover, if we look at the scope of public housing, we see that in new constructions, sustainable buildings, in terms of environment social and economic, are still seen as the result of experiences combining research and design and therefore not yet well assimilated in the normal praxis of the design.

In recent years, several Italian regions gave funds to managing bodies (through tax breaks for the financing of interventions or fully funded) to improve public residential buildings (ERP—*Edilizia Residenziale Pubblica*) environmental performance. The interventions are generally limited to reduce losses from external envelope (walls and windows), to improve thermal boilers efficient, to integrate of solar technologies—thermal and PV systems—and to use of geothermal energy.

On the issue of energy improvement of ERP, the contribution focuses on an element that crosses three main architectural aspects: energy needs, life quality, and visual perception.

The element taken into consideration is the bioclimatic greenhouse, a passive solar heating system made by a glazed space (i.e. a loggia) which in winter converts solar energy into heat, reduce leakage, and consequently reduce the energy requirements for space heating. The study is based on an extensive reading of the ERP buildings in Milan to understand which of these buildings have the morphological requirements to design the integration of greenhouses, on an observation of the users' behaviours. As a result, the study is a proposal to use the greenhouse as a strategy that can reduce the heating demand. The aim is to propose the use of this element that meets the needs of the inhabitants and of the public body in the belief that a lead intervention can prevent uncontrolled negative modifications of the facades and can instead give a new interpretation of the possible operation to get a more sustainable building. The proposal is conceived in terms of unitary intervention by the public entity, as well as single operation undertaken by the resident in his home (e.g. in cases where public buildings are partly sold to tenants and so there is the coexistence of owners and tenants).

83.2 The Bioclimatic Greenhouse

The easiest way to create a bioclimatic greenhouse in a flat is to close the loggia (or balcony). On the other hand, the closing of the loggia is a common practice especially in residential public buildings, where the inhabitants have consolidated over time, creating volumes in various ways, with the aim of obtaining a wider space without additional charge. It is indeed an unauthorized increase of volume, which modifies the facade of the building, making worse, generally, the overall image.

However, when properly designed, greenhouse becomes a part of the formal language of architecture and the building interprets an energetic phenomenon that characterizes the environmental behaviour of that particular flat or building. There are now many interesting examples of multistorey buildings, which can be studied or simply observed. Since the 1970s, Italians have before observed examples of the pioneers, especially in northern and central Europe (Austria and the Netherlands in particular) and then, many years later, found inspiration. Among the most significant works, it is sufficient to recall the architecture of the Gneiss Moss solar settlement at Salzburg in Austria in the 1990s by architect George W. Reinmeberg. In the settlement of Gneis Moss, the outside air at 2 °C is channelled into homes through the solar greenhouses—where even with little sun—warms up to about 12 °C; it reduces the contribution of heating to 10 °C instead of 18 °C, and it makes possible to save about 20 % thermal energy. In addition, the greenhouse becomes a liveable and enjoyable space in all seasons and it represents a particular linguistic element where the greenhouses are extruded in a vertical section. This element was designed to optimize the shading and cooling in summer [5] (Fig. 83.1).

The other pioneer who is always mentioned when talking about solar architecture is the architect Thomas Herzog who we can cite, for his houses, as well as the participation in the design of solar neighbourhoods, such as the Solar City. Thomas Herzog is the leader of the design team that in 2005 won a competition sponsored by the city of Rome for the construction of two interventions of public housing in the "*PDZ Lunghezzina 2*" (Fig. 83.2).

The proposal aims to achieve three goals, namely to reduce the energy requirements and to ensure thermal comfort conditions thanks to the use of a lot of bioclimatic devices, including the bioclimatic greenhouse; all included at a cost of 1000 €/m^2 to achieve low-cost architectures. Particular importance in the project was the presence of the loggias greenhouse inside the flats, equipped with shading systems for the period that does not require heating.

Fig. 83.1 Southern view of the building of Gneiss Moss III, Salzburg, and a view of one of the greenhouses of the building III. (Source: Malighetti Laura, "Una serra in facciata", in Il Sole 24 Ore—Arketipo, no. 5, 2006)

Fig. 83.2 An image of the winning project for new public residential buildings in Rome

83.3 The Reference Norm in Milan

In 2004, the Lombardy Region issued the law n. 39 "Regulations for energy efficiency in buildings and for the reduction of pollutant and greenhouse emissions[1]".

[1] In January 2014, the Regional Law 39/04 changed. The attached document concerns the "criteria for the recognition of the role of bioclimatic greenhouses and loggias, in order to make them equivalent to technical volumes".

Since 2007, the building code of Milan takes into account for the first time the bioclimatic greenhouse specifying some additional requirements compared to the regional legislation. In July 2014, the new building code was published. On this issue, it changes, in part, the previous regulation. In order to be considered technical volumes and therefore not counted in the volumetric calculations, the bioclimatic greenhouses (Article 133) must meet a number of conditions. In particular, in the case of housing redevelopment, residents have the option of closing an intermediate space between the outside and the house if it is south-oriented (up to a deviation of 45°), and if in winter this space is free from shading. This evaluation requires a site analysis, which mainly affects the distance from the facing buildings and the presence of any obstacles on the horizon that could prevent the solar radiation of winter.

The greenhouse should be integrated into the building structure, it should not be wider than 1.5 m, and it cannot have direct access to the property (it must be separated by at least one window or French window). For the glass-to-floor-area ratio, the glazed surface of the greenhouse must be greater than 1/10 of the surface of the floor of the heated space. The surface of the greenhouse cannot be larger than 10 % of the flat. Regarding the relationship with the flat, the greenhouse must have a window, facing the one of the flat (easily) openable for at least one third of its surface, with a U value of the glass of at least 1.5 W/m^2 K (i.e. at least a double glass), and must be equipped with a shading device to screen at least 70 % of the summer solar radiation.

Last compulsory requirement for the greenhouse and perhaps the most difficult to achieve, especially in case of ERP in Milan among 1940 and 1970, with very high needs, is the energy savings of at least 10 % of the heat demand. For this evaluation, it is necessary to calculate or use specific programmes that assess the energy performance of the greenhouse.

83.4 The Case of Varesina District in Milan

The case study is a building in a settlement in the north-west semi-suburban area of Milan. This is the first public housing neighbourhood after the World War II designed by architects Diotallevi and Marescotti and built between 1945 and 1950. Their aim, through the study of economic public housing, was to combine the reconstruction of housing with the social and political reconstruction. Over time, a large number of houses have been sold and became private property (Fig. 83.3).

The settlement was built on two parallel streets and is composed by couples of in-line four-storey buildings. One side faces to the common garden, and the other side faces to the gardens pertaining to the inhabitants of the ground floor.

The buildings are 16 with 640 of one-, two-, and three-bedroom flats, one living room with two expositions (south-east and south-west), a small kitchen, one bathroom, and one south-east-oriented loggia.

Fig. 83.3 Aerial view of the settlement *(left)*, view of the facade *(middle)*, and original aerial perspective of one flat of the building *(right)*. Original drawing of the plant of a two-bedroom flat *(below on the right)*

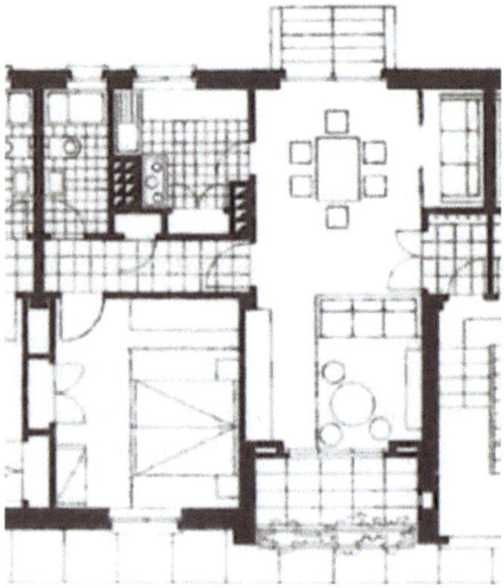

Many of the loggias have been windowed by inhabitants, without the necessary permits, and without even a criterion oriented to the preservation of the unitary image of the facade; the sole aim was expanding their own confined space to be used in various ways and maybe in different ways depending on the season.

This is a building that has all the requirements to be energetically improved with a series of interventions, ranging from classics such as the replacement of the existing single glazed windows with low E double-glazing windows, the addition of an insulating coat (the U value of the external walls is currently around 1.1 W/m^{2}K), and the enhancing of the efficiency of the heating system nowadays consisting in a three boilers thermal station which serves the entire group of 640 housing, that is obsolete and very expensive to keep active.

This chapter aims to evaluate the contribution of the solar greenhouse in flat of about 60 m^2, composed by a living room, one bedroom, one kitchen, and one bathroom. The loggia, which overlooks the south-east side, is about 4 m long and 1.5 m deep, perfectly meets the requirements specified in the Article 133 of the new building code. At this point, it is opportune to check the condition of the surroundings, in order to verify whether the building in front and the possible presence of

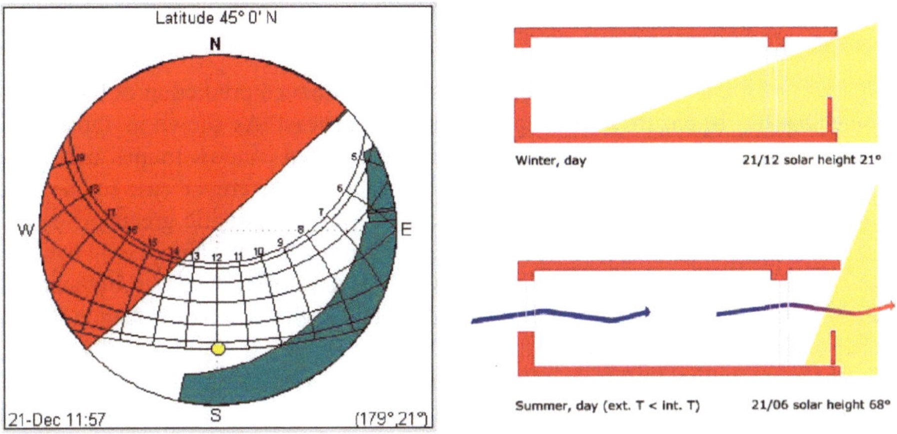

Fig. 83.4 Study of the obstacles on the horizon inside the loggia *(left)* and energy flows in winter and summer during sunny days *(right)*

the evergreen plants could screen the greenhouse in the winter season. To assess the obstacles on the horizon, it is enough to draw a shadow mask. Taking the test on a point inside the loggia emerges that in the winter months, the loggia is properly reached by the solar radiation. In addition, the vegetation, quite lush in summer, is not an obstacle in winter (Fig. 83.4).

The glazing of the greenhouse will be a double glass openable for at least 1/3 of the surface to provide ventilation indoor and in the greenhouse during the summer.

In summer, the shading will have to screen at least 70 % of the solar radiation. In this case, a simple check can be done graphically, with a series of operating diagrams. As we can see, the ceiling of the loggia prevents the radiation inside the property and it is therefore not necessary to use tents or brise soleil sun shading system. In addition, the double facade allows the cross-ventilation in summer, because the window is able to capture the wind from the west (Fig. 83.5).

Fig. 83.5 View from inside the solar greenhouse in winter season (closed loggia) with deciduous trees that allow the solar radiation access *(left)*. View from inside with the closed French window between the heated space and the greenhouse *(central photograph)*. View from the greenhouse to the house *(right)*

83.5 Results

This part of the chapter concerns the energy performance prediction of the greenhouse that has to ensure energy savings of 10% at least. As shown in Table 83.1, the flat without the greenhouse with the original wooden window frames and single glazing and not insulated external walls was taken as a reference case to compare with the different variations and the presence of the bioclimatic greenhouse. The heating energy demand of the reference case is about 246 KWh/m^{2}a, i.e. in band G, the last category of energy certification (G$\geq$175 Kwh/m^{2}a). The requirement has been calculated with the software CENED, binding in the Lombardy Region for the release of energy certificate.

As shown in Table 83.1, the closing of the loggia with a single glazing, which is in fact the typical do-it-yourself operation of the inhabitants, it is possible to save about 16%, and it turned out to be sufficient in terms of energy as the new building code requires a minimum of 10% (the previous one required a 5% contribution).

However, if we use the double glazing in the greenhouse, we would certainly have a lower solar contribution, compensated by a greater buffer effect, which would still meet the requirement, and in fact, the reduction would be about 19%. It is quite evident the energy demand reduction when no other measure is taken into account in the building envelope.

When in the flat the low-E double-glazing windows replace the single glazing, the energy demand reduction due to the greenhouse effect is about 2%. In other words, in a small flat, the effects due to a very efficient greenhouse or the improvement of the glazing are very similar (see cases 3 and 6 in Table 83.1).

The heating contribution of the greenhouse is not so evident when some measures in the opaque envelope have been already taken into account. A minimum operation was simulated, considering 4 cm of insulation (polystyrene) in the external

Table 83.1 Energy demand and percentage (%) of energy saving compared to the reference case. Different configurations of the flat (Kwh/m^{2}a) have been considered: flat with single or double glazing, with or without greenhouse; greenhouse single or double glazing; external walls without insulation (U value = 1.1 w/m^{2}K) and with two thickness of insulation (4 and 10 cm, i.e. = 0.59 and 0.25 w/m^{2}K). On the right, the home energy performance rating chart in Lombardy Region

	House glazing	Greenhouse glazing	Uvalue	E D	Saving
			W/m^{2}K	Kwh/m^{2}a	%
1	SG	No GH	1.1	246	
2	SG	SG	1.1	206	16
3	SG	DG	1.1	200	19
4	SG	No GH	0.59	217	12
5	SG	No GH	0.25	194	21
6	DG	No GH	1.1	199	19
7	DG	SG	1.1	201	18
8	DG	DG	1.1	196	20
9	DG	No GH	0.59	170	31
10	DG	SG	0.59	170	31
11	DG	DG	0.59	167	32
12	DG	No GH	0.25	148	40
13	DG	SG	0.25	147	40
14	DG	DG	0.25	144	41

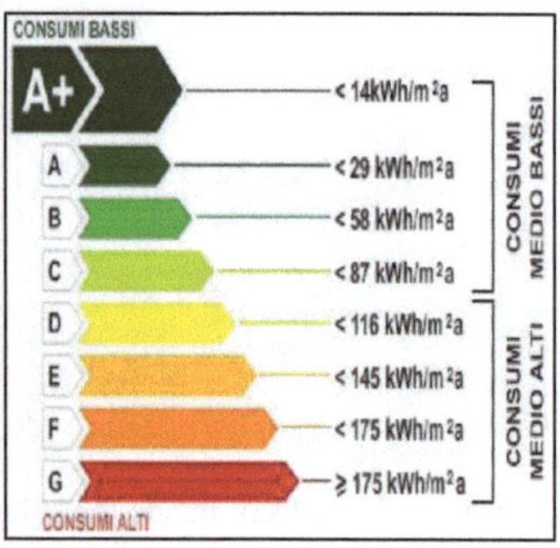

GH greenhouse, *SG* single glazing, *DG* double glazing, *ED* energy demand

wall (i.e. U value from 1.1 w/m^2K become 0.59) and a second one that considers 10 cm of insulation (U value of 0.25 w/m^2K). In the first case, the outcome was the improvement compared with the reference case of 12 %, while in the second case, the improvement was about 21 % (cases 4 and 5 in Table 83.1). It is obvious that important improvements occur when both transparent and opaque surfaces are modified. In this case, the energy demand would be 40 % lower than in reference case (case 12 in Table 83.1).

At this point, with an energy demand of 148 kwh/m^2a, a minimum contribution would come from the solar greenhouse, 1 % with single glazing and 2 % with a double glazing in the greenhouse. Comparing it with the reference case, the reduction would be about 41 % (cases 1 and 14 in Table 83.1).

Due to the ease and the bearable cost of the realization,[2] the greenhouse could be considered a first step of environmental refurbishment also because it would be well accepted by the inhabitants (both tenants and owners). This can also represents a first step to stimulate an environmental consciousness where it does not exist yet (at least till when many conventional central heating plants are yet in use and the European directive 2012/27/EU on energy efficient, which considers the installation of individual consumption meters to measure the consumption of heat or cooling or hot water for each unit, is far from been implemented). This set of strategies make the flat move from band G to band E of the home energy performance rate.

The last aspect concerns the visual impact of the interventions, which is currently "out of control" because no indication is given about the uniformity of interventions when the single individual makes these. One of the few indications comes from recent legal judgements. According to these, the solar greenhouse in a condominium is considered regular if it respects the architectural decoration of the building, if not harmful to the neighbourhood, and finally, if it guarantees equal use of the common good.

The choice of the topic of public housing is also based on need to establish clear guidelines, which are not only based on performance but also on geometry or visual perception.

Here, it makes no sense to proceed with a specific project for the selected case study, but it is more useful to look at the examples that have common features with the case study but also with the public residential buildings that have the functional requirements for the realization of the greenhouse.

In the first case, in Fig. 83.6, the loggias were windowed; each loggia has its own system of glazing and shading, but they are the same all over the building. This means that the closure of the loggias may be not concurrent, but can follow different timings. The example also includes the glazing of the external surface of the stairwell. Here, we can have different possibilities. We can replace the masonry wall with an openable glass to prevent overheating in summer, or, instead of removing the original wall, we can realize a Trombe wall, which heats the air in the stairwell or in the greenhouses.

[2] If we consider a price around 200 €/m^2, the construction costs of the greenhouse could be around € 1500.

Fig. 83.6 Sketches with different configuration of the greenhouse: glazing and shading device for each loggia of the building

Figure 83.7 shows another aspect instead, namely the wooden shielding solar, which has the possibility to be moved along the facade according to the seasonal or daily needs. They give a horizontal rhythm quite stiff, while the vertical scanning is rather variable, not fixed, but in any case, they characterize very clearly the facade, more than the greenhouses themselves.

The Fig. 83.7 on the right shows how the fixed screens can be implemented or minimized depending on the season, and in particular, the arrangements they assume are of three types: summer, when the shading device has to screen much direct radiation as possible; winter, when, on the contrary, it allows passage of as much solar radiation as possible; and a greater variability in the intermediate season, still uncertain, with days that may require shielding or others that require solar radiation.

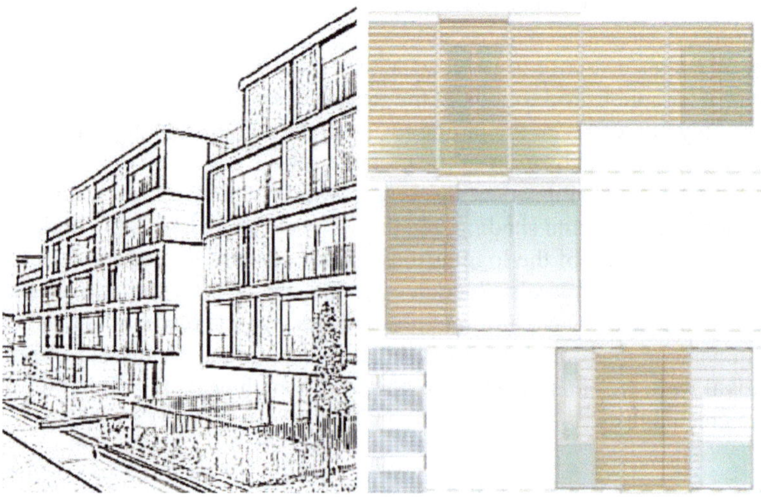

Fig. 83.7 Moveable shading device along the whole facade and different seasonal way to use them

Fig. 83.8 Greenhouses full height along the facade, in alternance with masonry parts

The glazing should not necessarily affect the entire facade. The example on the right shows how the alternating opaque/transparent even with high percentages of transparent surfaces compared to the opaque can respect the decorum of the building. Unlike the other examples, this is an operation that cannot be realized in different timings, but it is a single intervention because it is a unique greenhouse. Inside, there may be a separation (in the case of loggias that serve single flats) or not (in case there are common open areas or the stairwell), but in each case, the image consists in a single element in the facade which alternates with the opaque element, the masonry (Fig. 83.8).

83.6 Conclusions

With this proposal, at least three goals would be achieved: from one hand energy savings beyond the minimum required by the regulations with a consequent decreasing in the amount of CO_2 emissions in the atmosphere, a real increase in living space (at least for certain periods of the year), aspect which affects more than the others the current inhabitants, and the relatively low cost of the operation.

The last point concerns the visual impact of an action that is not currently subject to compliance with guidelines to protect the image of the building.

Following the closure of the loggias and other retrofit interventions, the image of the building could be very different from the original, but it is necessary to maintain a unified image. Intervention should therefore be recognized as energy retrofits, and it should "tell" the environment operation of the architecture, not only made of matter but also of energy flows that can help achieve low energy conditions of comfort which more and more people are looking for in their own homes. A successful intervention, which meets the favour of the people, is undoubtedly the best incentive to continue along this path.

References

1. Battisti A, Herzog T, Tucci F (2012) Sperimentazioni di housing sociale tra efficienza energetico-ambientale e basso costo. Techne 4:343–354. University Press, Firenze
2. McCullagh JG (1978) The solar greenhouse book. Rodaus Press, Emmaus
3. Pugliese R (2005) La casa popolare in Lombardia*: 1903–2003*. Unicopoli, Milano
4. Reinberg GW (1998) Architecture by Georg W. Reinberg. Alinea international, Firenze
5. Sartogo F (2004) L'Architettura Solare di Reinberg. Intervista a Georg W. Reinberg. AR 52:4–11. Prospettive ed. Roma
6. Zappone C (2012) La serra solare. Esselibri, Napoli

Chapter 84
Solar-Assisted Ultra-supercritical Steam Power Plants with Carbon Capture and Storage

Giorgio Cau, Daniele Cocco and Vittorio Tola

Abstract This chapter focuses on the evaluation of the potential benefits arising from the integration of concentrating solar systems into coal-based ultra-supercritical (USC) power plants with post-combustion CO_2 capture (PCC). In order to offset the efficiency penalty introduced by CO_2 removal, the USC-PCC plant was integrated with a concentrating solar field with direct steam generation based on parabolic trough and linear Fresnel collectors. The performance of the solar-assisted USC-PCC power plant was evaluated by means of specifically developed simulation models by using data sets for a typical meteorological year for the sites of Cagliari (Sardinia, Italy) and the North Western Australia Coast. A preliminary cost analysis was also carried out.

Keywords USC · CO_2 removal · Solar integration · Linear Fresnel collectors · Parabolic trough collectors

84.1 Introduction

As known, concentrating solar power (CSP) plants are one of the most interesting solar conversion technologies. Today, the current CSP world generating capacity is around 3800 MW and is rapidly increasing. More than 1500 MW of additional capacity is currently under construction and more than 4000 MW is under development [1]. For CSP plants, different options are available for solar field, power block, heat transfer fluid (HTF), and thermal energy storage (TES) [2, 3]. Currently, parabolic trough collectors (PTCs) are the most commercially proven technology, especially in integration with steam Rankine cycles, the use of thermal oil as HTF and molten salts for the TES section. Linear Fresnel collectors (LFCs) may be a viable alternative to PTCs owing to their lower land requirements and lower capital costs, although their optical efficiency is lower than that of PTCs [4, 5]. Moreover,

V. Tola (✉) · G. Cau · D. Cocco
DIMCM, Department of Mechanical, Chemical and Materials Engineering, University of Cagliari, Via Marengo 2, Sardinia, 09123 CA, Italy
e-mail: vittorio.tola@dimcm.unica.it

A. Sayigh (ed.), *Renewable Energy in the Service of Mankind Vol II*,
DOI 10.1007/978-3-319-18215-5_84

direct steam generation (DSG) is a very interesting option to raise the performance of CSP plants based on steam Rankine cycles [6]. Finally, one of the most interesting options to reduce the electricity production costs is the integration of the solar field with a conventional power plant fed by fossil fuels through hybrid CSP plants. Some examples of hybrid CSP plants can be found in North Africa and Middle East countries, where solar system has been integrated with combined cycles [7–9]. Solar contribution in hybrid CSP plants allows to reduce fuel consumption and therefore CO_2 emissions.

Obviously, even with hybrid CSP plants, a substantial reduction of CO_2 emissions can only be achieved with the integration of carbon capture and storage (CCS) systems. Among the different CCS technologies, PCC processes, especially chemical absorption with amine-based solvents, appear today the most suitable option for coal-based power plants [10]. In power generation plants, CO_2 capture, transport, and storage introduce large energy penalties, mainly due to the large heat consumption required for solvent regeneration, which is usually carried out by supplying low-pressure steam (in particular, to remove 90 % of the CO_2 produced, the target of most CCS projects, a penalization of 10–11 % points on net efficiency is expected) [11, 12]. Therefore, solar-assisted post-combustion carbon capture (SPCC) systems can be a suitable option to mitigate the energy penalties introduced by CO_2 removal.

SPCC systems have been studied in several papers [13–15] and all studies demonstrated that solar integration improves the economic feasibility of CCS, especially for decreasing collector costs and increasing CO_2 emission prices. The integration of concentrating solar collectors in SPCC plants can be carried out by means of two main approaches: (i) production of low-pressure steam for the solvent regeneration process and (ii) production of high- or intermediate-pressure (IP) steam for the high-pressure (HP) and IP turbines. From a thermodynamic point of view, the latter is undoubtedly the preferred approach [16], although the best option also depends on climatic conditions and solar collector technology.

In this framework, this study evaluates the performance of solar-assisted USC steam power plants integrated with a post-combustion CO_2 removal process (PCC). USC systems represent the state of the art in the field of coal-fired power plants, with conversion efficiencies above 45–46 %, more than 5 % points greater than those of conventional steam power plants [17]. The USC-PCC power plant was integrated with a solar field based on LFC or PTC with direct steam production. TES is a very challenging task in DSG solar collectors and, therefore, this option was not considered in this study, as in most hybrid solar/fossil CSP projects.

The comparative analysis aims to evaluate the increase in USC annual energy production and global efficiency due to the solar energy contribution and was carried out by considering two different locations: Sardinia in the Mediterranean Sea and North Western Australia Coast. Finally, a preliminary assessment of the energy production cost was also included.

84.2 Configuration and Performance of the Solar Field

As mentioned, two different options, namely PTC and LFC, were compared here. For both options, the solar field is based on DSG and is composed of several lines of linear collectors connected in parallel to achieve the required thermal power output. Each PTC and LFC collector line includes several modules connected in series, and Table 84.1 shows the main geometrical and performance parameters assumed in this study [5, 18]. The performance of the solar field was evaluated on a yearly basis by means of a specifically developed simulation model starting from hourly data of direct normal irradiation (DNI), solar position, and air temperature. In particular, the thermal power output Q_{FLD} transferred to the steam was evaluated hour by hour according to the following equation:

$$Q_{FLD} = m_{STM} \cdot (h_{out} - h_{in}) = Q_{SOL} \cdot \eta_{OPT,R} \cdot \mathrm{IAM}(\theta) \cdot \eta_{END} \cdot \eta_{CLN} - Q_{THR} \quad (84.1)$$

where Q_{SOL} is the solar power input (i.e., the product of DNI and collecting area A_C), $\eta_{OPT,R}$ is the reference optical efficiency, IAM(θ) is the incidence angle modifier, η_{END} is the end-loss optical efficiency, η_{CLN} is the surface cleanliness efficiency, and Q_{THR} is the receiver thermal losses. Figure 84.1 shows the two IAM components as a function of the longitudinal and transversal components θ_L and θ_T of the solar incidence angle θ. The end-loss optical efficiency was evaluated as a function of collector length, focal height, and longitudinal component θ_L. Thermal losses were evaluated according to the specific correlations as a function of the difference

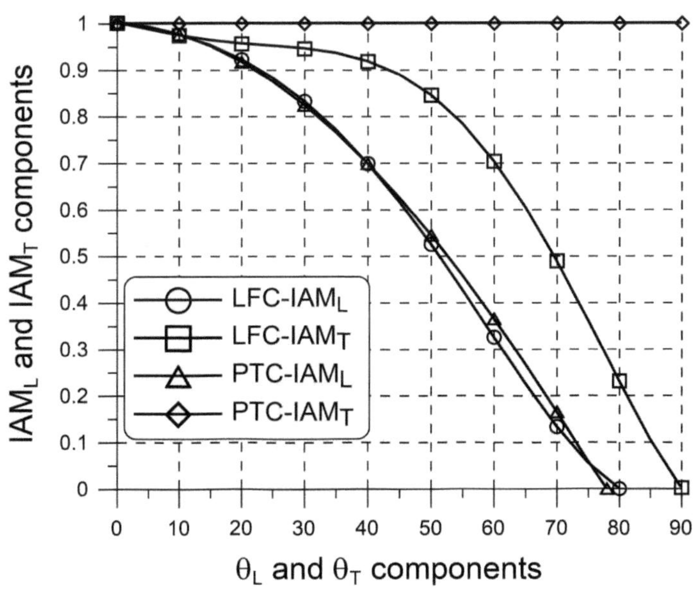

Fig. 84.1 Longitudinal and transversal IAM components

Table 84.1 Main characteristics of solar collectors

	LFC	PTC
Module length/width	44.8/16.56 m	150/5.77 m
Focal length (F)	7.4 m	1.71 m
Module collecting area (AC)	513.6 m^2	865.5 m^2
Modules per line	16	4
Reference optical efficiency, EVA/SH section	0.67/0.65	0.75/0.75
Cleanliness efficiency	0.98	0.98
Receiver losses, EVA	$0.056\Delta T + 0.000213\Delta T^2$	$0.00047\Delta T^2$
Receiver losses, SH	$0.013\Delta T + 6.2510^{-10}\Delta T^4$	$0.00047\Delta T^2$

ΔT between average steam temperature in the receiver tube and ambient temperature. Similarly, piping thermal losses were evaluated as a function of the difference between steam temperature and ambient temperature. A more detailed description of the simulation model can be found in [19].

As shown in Fig. 84.2, the most interesting areas in the world for CSP plants are located in North Africa, South Africa, Middle East, India, Australia, North America, and South America. Today, CSP projects are commercially developed for at least 2000 kWh/m^2/year of DNI availability, although with lower costs of the solar field future plants may be built in areas with at least 1800 kWh/m^2/year [20]. For this reason, the present comparative study was carried out by considering two different sites, with high and low DNI availability. In particular, the site with high DNI availability refers to meteorological data of the North Western Australia Coast, while the site with low DNI refers to meteorological data of Sardinia (Italy). The data set for a typical meteorological year was obtained by the Meteonorm software [21] and includes DNI, solar azimuth and elevation, air temperature, relative humidity, and wind velocity. Table 84.2 summarizes the most important meteorological data of the two sites and the corresponding design conditions assumed for the solar field.

For the two collector technologies and the two sites, Table 84.3 shows the main performance of a single line. In particular, Table 84.3 shows the reference values of thermal power output and line efficiency (i.e., the values referred to the design conditions of Table 84.2), as well as the annual average values of thermal energy

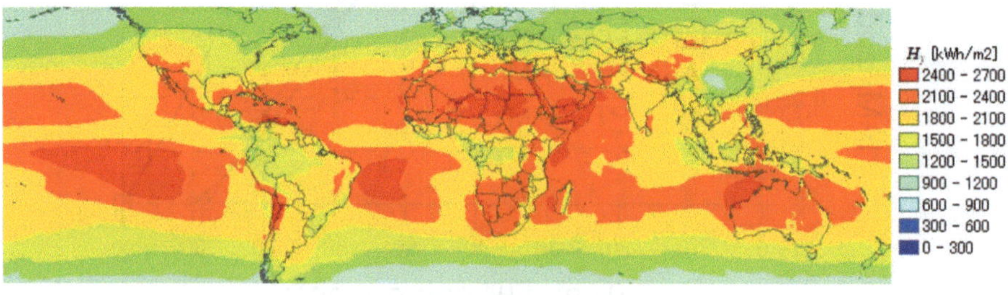

Fig. 84.2 Distribution of the annual DNI around the world

Table 84.2 Meteorological data for the sites of Sardinia and Australia and solar field design assumptions

		Sardinia	Australia
Available DNI	kWh/m^{2}year	1720	2774
Average ambient temperature	°C	17.2	24.5
Average wind velocity	m/s	3.96	3.80
Design DNI	W/m^2	800	900
Design elevation/azimuth angles	°	74.2/0.0	89.6/0.0
Design ambient temperature	°C	30.0	35.0

production and solar line efficiency (the ratio of annual thermal energy production and annual available DNI). Obviously, the different values of the reference power output depend on the different collecting areas of PTC and LFC and the different DNI. The reference conversion efficiency increases by about 6–7 % switching from Sardinia to Australia mainly due to the higher value of the design DNI and by about 10 % substituting LFC with PTC, due to the higher optical efficiency. Table 84.3 demonstrates that PTC in high DNI sites allows to achieve the better conversion efficiencies.

As mentioned, the solar field includes several collector lines and Fig. 84.3 relates reference power output and solar field collecting area for both the collector technologies located in Sardinia and Australia. The corresponding number of collector lines can be easily calculated by the ratio of solar field area and line collecting area. The reference power output obviously increases linearly with the solar field area, and for a given thermal power output the steam mass flow produced depends on the input and output enthalpy values. In particular, with the assumptions of this study, every megawatt of thermal power input allows to produce 0.371 kg/s of steam at 500 °C and 80 bar. Overall, a reference thermal power output of 200 MW requires a solar field with a collecting area of about 300,000 m^2 for PTC installed in Australia (which corresponds to about 900,000 m^2 of land area) but over 400,000 m^2 for LFC installed in Sardinia (which corresponds to about 750,000 m^2 of land area).

Table 84.3 Performance of a single LFC and PTC line

		LFC		PTC	
		Sardinia	Australia	Sardinia	*Australia*
Collecting area	m^2	8217.6	8217.6	3462.0	3462.0
Reference thermal power	MW	3.905	4.697	1.814	2.194
Reference efficiency	%	59.40	63.51	65.49	70.43
Available DNI	GWh/year	14.12	22.79	5.95	9.60
Thermal energy production	GWh/year	5.10	10.1	2.93	5.60
Average efficiency	%	36.11	44.30	49.26	58.33

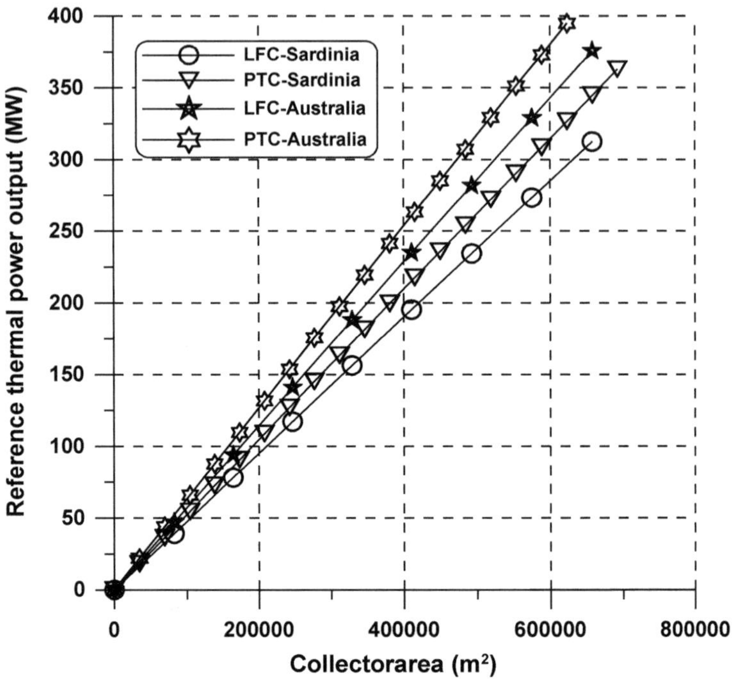

Fig. 84.3 Reference thermal power output

84.3 Configuration and Performance of the USC-PCC Power Plant

The reference configuration of the USC section is based on a typical 600–800 MWe unit with a coal power input of 1.5 GW. In particular, a reheated Rankine cycle with three steam turbines (HP, IP, and low-pressure (LP) turbines) and eight regenerative steam extractions was considered here. Figure 84.4 shows a simplified scheme of the USC power plant. The flue gas treatment section is based on a conventional "high-dust" configuration and includes an SCR DeNOx reactor, an electrostatic precipitator, and a low-temperature flue gas desulfurization system. The CO_2 removal section is based on a chemical absorption process with amine-based solvents, which is currently considered the most suitable option for CO_2 capture from flue gases [22]. The CO_2 capture section includes an absorption column, where most of the CO_2 contained in the flue gas is absorbed by the aqueous methyl-ethanolamine (MEA) solution, and a regeneration column, where the absorbed CO_2 is released. Furthermore, the CO_2 capture section is integrated with a proper conditioning and compression section to match the pressure (11 MPa) and CO_2 purity (above 99.5 %) required for CO_2 transport and storage. Table 84.4 shows the main operating parameters of both USC plant and CO_2 capture section.

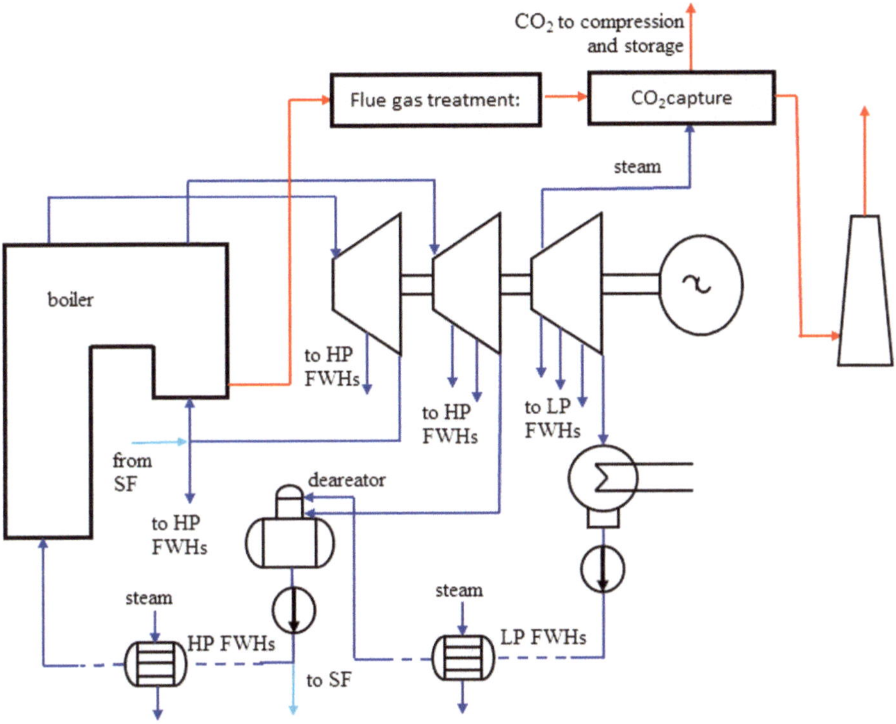

Fig. 84.4 Simplified scheme of the USC-PCC power plant

Table 84.4 Operating parameters of the USC-PCC plant

Fuel chemical power input	1500	MW
Coal lower heating value	25.0	MJ/kg
SH/RH steam temperature	600/610	°C
SH/RH steam pressure	26.8/6.4	MPa
Condenser pressure	4.4	kPa
Boiler feedwater temperature	309	°C
Flue gas temperature	120	°C
Solvent/gas mass ratio		4.4
MEA mass fraction	%	30
CO_2/MEA molar ratio		0.28
Reboiler specific thermal energy	MJ/kg_{CO2}	3.45

The performance analysis of the USC power plant was carried out by using simulation models specifically developed through Aspen-Plus and Gate-Cycle software platforms. In particular, Aspen-Plus models are used to simulate the flue gas treatment and the CO_2 removal processes, while Gate-Cycle models are used to simulate the power generation section. A more detailed description of the simulation models,

Table 84.5 Main performance of the USC power plant with and without CCS

		USC	USC-PCC
Fuel chemical power input	MW	1500.0	1500.0
Gross power output	MW	672.0	577.1
Power required for flue gas treatment	MW	14.0	14.0
Power required for CO_2 removal	MW	–	4.2
Power required for CO_2 compression	MW	–	56.5
USC net power output	MW	658.0	502.4
USC net efficiency	%	43.87	33.49
Heat required for solvent regeneration	MW	–	451.5
CO_2 specific emissions	g/kWh	790.0	103.5

as well as a detailed analysis of the performance of USC power plants integrated with CO_2 capture systems, can be found in [23].

Table 84.5 shows the main performance of the USC power plant without CO_2 removal and the USC-PCC power plant considered here. As demonstrated in Table 84.5, CO_2 removal largely reduces the USC performance mainly due to both the heat required by the solvent regeneration process and the power required by the CO_2 compression system. In particular, to remove 90 % of the CO_2 the solvent regeneration process requires about 450 MWt, supplied by a hefty steam extraction from the low-pressure (LP) turbine (about 200 kg/s at 3.3 bar). The latter steam extraction reduces the USC power output by about 95 MW and the mass flow of the LP steam turbine from 315.9 to 152.6 kg/s. Moreover, the CO_2 removal section requires more than 60 MW for powering CO_2 compressors, solvent pumps, and process auxiliaries. Overall, integration with the CO_2 removal section reduces the USC power output by about 155 MW, with a corresponding efficiency penalty of about 10.5 % points (from 43.87 to 33.49 %). Obviously, the CO_2 specific emissions greatly benefit from the introduction of the PCC system (which decreases from 790 to about 100 g/kWh).

84.4 Performance of the Integrated USC-PCC Power Plant

The performance of the USC-PCC plant reported in Table 84.5 refers to design conditions and without the contribution of the solar field. During periods of solar energy availability, IP steam produced by the solar field increases the mass flow rate of both IP and LP steam turbines and, therefore, leads to the off-design operation mode of the USC plant, with a corresponding efficiency penalty. Moreover, the increase in the LP turbine mass flow leads to higher condenser thermal loads and, therefore, higher cooling water requirements or higher cooling water outlet temperatures. Figure 84.5 shows the increase of the condenser pressure and mass flow (in percentage with respect to design conditions) as a function of the solar field

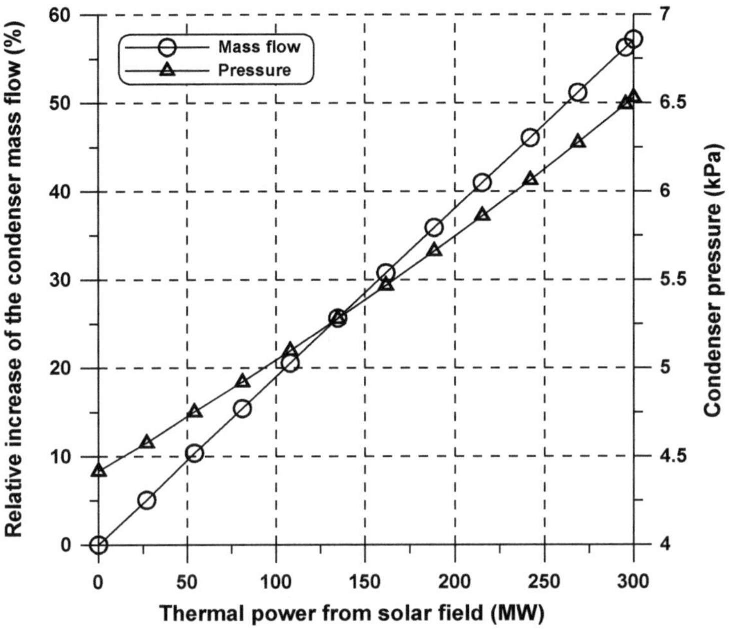

Fig. 84.5 Mass flow and pressure of the condenser as a function of the solar field power output

power output. The condenser mass flow greatly increases with solar power production contribution. To avoid an excessive increase in the condenser thermal load, a maximum solar contribution of 300 MW was considered in this study. The increase in steam pressure and mass flow leads to corresponding changes in the turbine power output owing to the sliding pressure operating mode of the steam turbine. Figure 84.6 shows the USC-PCC net power output as a function of the collecting area and the thermal power produced by the solar field for the two different collectors and the two different locations considered here. As shown in Fig. 84.6, the net power output (and therefore the power gain due to the solar field contribution) linearly increases with the solar contribution for all cases and the highest values refer to the PTC installed in Australia. Obviously, the annual average performance of the power plant is greatly influenced by the annual DNI availability. During most of the solar field operating hours, the DNI is below its design value (800 and 900 W/m^2 for Sardinia and Australia, respectively), due to cloudiness and nights. Moreover, concentrating solar collectors require a minimum available DNI (100–150 W/m^2) for their operation. Therefore, the USC power plant operates without solar energy integration, and therefore at reference conditions, for a large share of the annual operating hours.

Table 84.6 summarizes the annual performance of the USC-PCC options studied here. For comparative purposes, a collecting area of about 330,000 m^2 was considered (the corresponding reference power output of the solar field ranges from 156 to 208 MW, depending on collector technology and location). In particular,

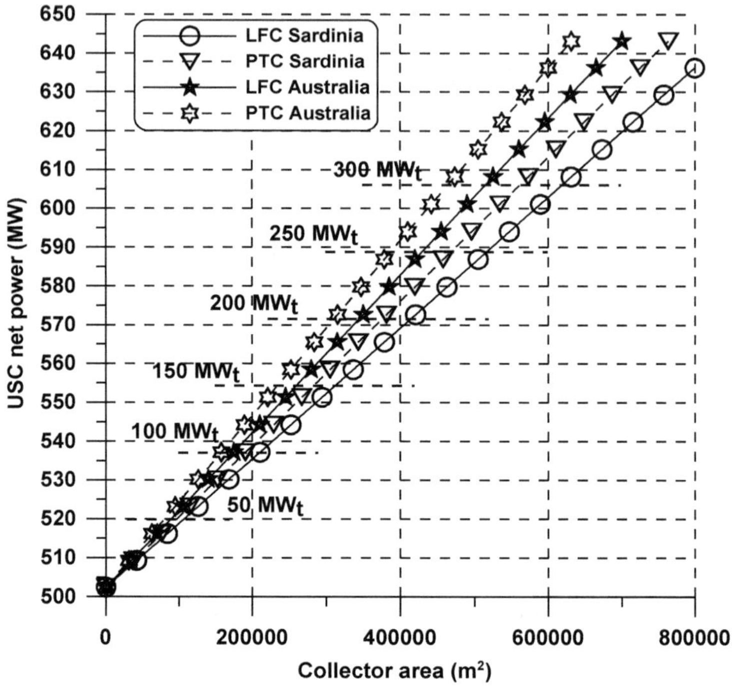

Fig. 84.6 USC-PCC power output as a function of the collecting area and the power output of the solar field

Table 84.6 Performance of the USC-PCC integrated with solar field

		LFC		PTC	
		Sardinia	Australia	Sardinia	Australia
Number of lines		40	40	95	95
Collecting area	m^2	328,704	328,704	328,890	328,890
Reference thermal power	MW	156.20	187.88	172.33	208.43
Increase in steam mass flow	kg/s	57.92	69.67	63.90	77.29
Increase in energy production	%	1.41	2.80	2.20	4.23
USC-PCC-CSP efficiency	%	34.03	34.57	34.23	34.91
Solar field efficiency	%	36.11	44.30	49.26	58.33
Average solar to electrical conversion efficiency	%	12.55	15.46	17.16	20.41

Table 84.6 reports the increase in the annual electricity production (percentage of the corresponding value of the same power plant without solar integration), the average conversion efficiency (with respect to coal LHV), the solar field efficiency, and the average conversion efficiency of solar energy into electrical energy.

By considering, for simplicity, a plant availability of 100 % (8760 operating hours), the net electricity production of the USC power plant (658 MW) would be 5764 GWh/year. For the USC-PCC power plant, the energy requirements related to the CO_2 removal section reduce the annual energy production to 4399 GWh (−23.7 %). With the integration of a 330,000 m^2 solar field, Table 84.6 shows that the solar energy contribution to annual electricity production is quite low in Sardinia (1.41 % for LFC and 2.20 % for PTC) but it is about twice in Australia (2.80 % for LFC and 4.23 % for PTC). Obviously, a larger collecting area would increase annual solar contribution. As shown in Table 84.6, the average conversion efficiency of the USC-PCC power plant, calculated as the ratio of the annual energy production and fuel energy input (13,140 GWh/year for all cases), ranges from 34.0 to 34.9 %. Since the efficiency of the USC-PCC power plant is 33.49 %, Table 84.6 demonstrates that solar field integration makes it possible to compensate for the efficiency penalty only to a small extent owing to the limited solar energy contribution. Once again, the best performance can be achieved by PTC located in Australia (the efficiency increases by more than 1.4 % point) and the worst performance can be achieved by LFC in Sardinia (0.5 % of efficiency increase). The average solar to electrical conversion efficiency was calculated as the ratio of electricity production gain due to solar field integration and solar energy input, and therefore it includes both solar field and steam cycle efficiencies. Solar conversion efficiency is quite low for LFC, especially in Sardinia (12.55 %), while better values can be obtained with PTC in Australia (20.41 %).

84.5 Preliminary Economic Analysis

A preliminary economic analysis was carried out to compare the energy production cost of the reference USC power plant (with and without the PCC section) and the solar-assisted configurations studied here. Solar integration improves electricity production and conversion efficiency of USC-PCC power plants but obviously requires higher capital costs. In particular, the economic analysis evaluates the levelized cost of energy (LCOE) of the USC and USC-PCC power plants and the marginal levelized cost of energy ($LCOE_M$) of the solar-assisted USC-PCC configurations. The latter was calculated as the ratio of the increase in the annual cost of the solar-assisted plants with respect to the reference USC-PCC plant and the corresponding increase in annual electricity production due to solar energy contribution. Both $LCOE$ and $LCOE_M$ are calculated with reference to the International Energy Agency (IEA) simplified methodology [24], according to:

$$\begin{aligned} LCOE &= \frac{C_T}{E_E} = \frac{TCI \cdot FCR + C_{O\&M} + C_F}{E_E} \\ LCOE_M &= \frac{\Delta C_T}{\Delta E_E} = \frac{TCI_{SF} \cdot FCR + C_{O\&M,SF}}{\Delta E_E} \end{aligned} \quad (84.2)$$

Table 84.7 Main assumptions for the economic analysis

USC specific capital investment	1300 €/kW
USC-PCC specific capital investment	2470 €/kW
O&M costs of USC (% of TCI)	3%
O&M costs of USC-PCC (% of TCI)	2.5%
Coal price	75 €/t
Solar field specific capital investment	100–260 €/m²
Specific land cost	2 €/m²
O&M costs of solar field (% of TCI)	1.5%
Real debt interest	7%
Annual insurance rate	1%
Operating lifetime	20 years

where *TCI* is the total capital investment, *FCR* is the fixed charge rate, $C_{\mathrm{O\&M}}$ are the operating and maintenance costs, C_{F} is the cost of the fuel, and E_{E} is the annual energy production. The subscript SF indicates the solar field. The main economic assumptions are reported in Table 84.7. In particular, on the basis of the available data, for the USC power plant, a specific capital investment of 1300 €/kW was assumed while that of the USC-PCC plant is almost twice [25]. The FCR depends on the annual real debt interest (7% in this case) and the operating lifetime (20 years). TCI_{SF} was estimated starting from a specific cost in the range of 100–300 €/m², while the annual operating and maintenance costs are assumed to be 1.5% of the TCI_{SF}.

Figure 84.7 shows the $LCOE_M$ for the four different cases analyzed and the *LCOE* for both the reference USC (4.56 c€/kWh) and USC-PCC (8.27 c€/kWh) power plants as a function of solar field cost. It is obvious from Fig. 84.7 that $LCOE_M$ always increases with the solar field cost and $LCOE_M$ values for the LFC case located in Sardinia always exceed the *LCOE* of the two reference power plants (USC with or without CO_2 removal) even for the lowest solar field costs considered here. For the same location, the more efficient PTC operates with lower solar energy production costs, especially in the case of reduced solar field costs. In particular, the $LCOE_M$ is lower than the *LCOE* of the reference USC-PCC power plant for solar field costs of about 120 €/m². For power plants located in Australia, both collectors operate with $LCOE_M$ lower than the *LCOE* of the reference USC power plant without CO_2 removal in case of very low solar field costs (lower than 110–120 €/m²). For the Australia site, $LCOE_M$ values lower than the *LCOE* of the USC-PCC can be reached for a solar field cost of about 200 €/m² (PTC) and 240 €/m² (LFC). For the same solar field cost, PTC solutions assure an $LCOE_M$ lower than that of LFC, especially in Sardinia and for major solar field cost, although the cost of LFCs is typically lower than the cost of PTCs.

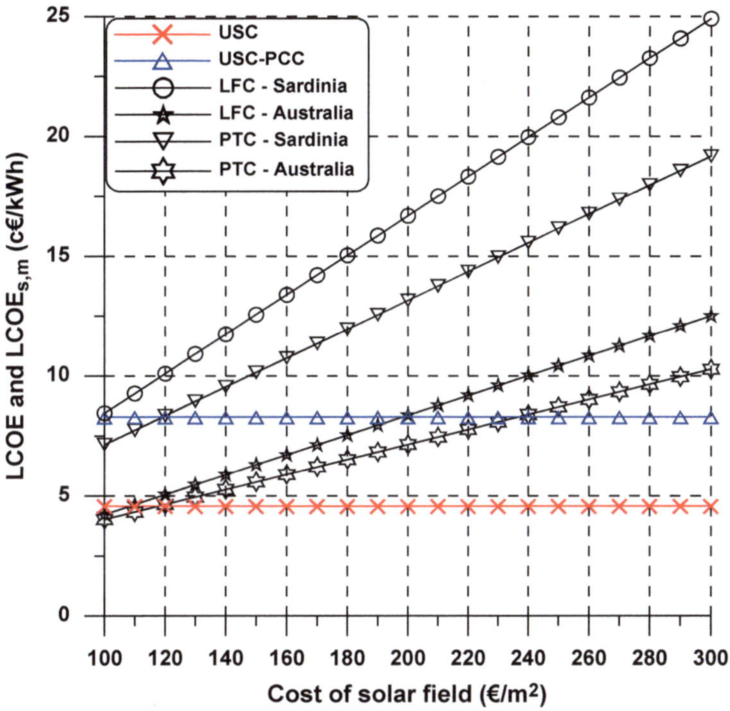

Fig. 84.7 LCOE and $LCOE_M$ as a function of solar field cost

84.6 Conclusions

This study demonstrates that concentrating solar collectors with DSG can be an interesting solution for hybrid power generation plants based on fossil fuels. In particular, steam generation from solar energy enhances electricity production from coal and partially offsets the energy consumption due to CO_2 removal in USC power plants. Comparison of solar fields based on different collector technologies shows that PTCs lead to better performance than LFCs. Moreover, electricity production from solar energy greatly depends on DNI availability and, therefore, on the installation site (the Australia site here considered leads to a solar contribution almost twice than that of the Sardinia site). Overall, in hybrid USC-PCC power plants, solar contribution to electricity production remains remarkably lower (2–4 % on average) than coal contribution, even with large solar fields. What is more is that the introduction of the steam produced by the solar field alters the mass flow and energy balance of the steam power plant (in particular, turbines, heat exchangers, and condenser) and therefore the solar contribution cannot be too high (a maximum value of 300 MW was assumed here).

Finally, the results of a preliminary economic analysis demonstrate that solar-assisted configurations based on the most efficient collector technologies (PTC) installed in sites with high DNI availability may operate with competitive electricity production costs. Less efficient collector technologies (LFC) installed in sites with low DNI availability require reduced solar field costs to be competitive.

References

1. http://www.solarpaces.org. Accessed 31 July 2014
2. Reddy VS, Kaushik SC, Ranjan KR, Tyagi SK (2012) State-of-the-art of solar thermal power plants—a review. Renew Sustain Energy Rev 27:258–273
3. Tian Y, Zhao CY (2013) A review of solar collectors and thermal energy storage in solar thermal applications. Appl Energy 104:538–553
4. Giostri A, Binotti M, Silva P, Macchi E, Manzolini G (2013) Comparison of two linear collectors in solar thermal plants: parabolic trough versus Fresnel. J Sol Energy Eng 135(1):011001
5. Morin G, Dersch J, Platzer W, Eck M, Häberle A (2012) Comparison of linear Fresnel and parabolic trough collector power plants. Sol Energy 86:1–12
6. Feldhoff JF, Schmitz K, Eck M, Schnatbaum-Laumann L, Laing D, Ortiz-Vives F, Schulte-Fischedick J (2012) Comparative system analysis of direct steam generation and synthetic oil parabolic trough power plants with integrated thermal storage. Sol Energy 86:520–530
7. Behar O, Kellaf A, Mohamedi K, Belhamel M (2011) Instantaneous performance of the first integrated solar combined cycle system in Algeria. Energy Procedia 6:185–193
8. Montes MJ, Rovira A, Munoz M, Martinez-Val JM (2011) Performance analysis of an integrated solar combined cycle using direct steam generation in parabolic trough collectors. Appl Energy 88(9):3228–3238
9. Nezammahalleh H, Farhadi F, Tanhaemami M (2010) Conceptual design and techno-economic assessment of integrated solar combined cycle system with DSG technology. Energy Procedia 6:186–194
10. Lim Y, Kim J, Jung J, Lee CS, Han C (2013) Modeling and simulation of CO_2 capture process for coal-based power plants using amine solvent in South Korea. Energy Procedia 37:1855–1862
11. Goto K, Yogo K, Higashii T (2013) A review of efficiency penalty in a coal-fired power plant with post-combustion CO_2 capture. Appl Energy 111:710–720
12. Tola V, Pettinau A (2014) Power generation plants with carbon capture and storage: a techno-economic comparison between coal combustion and gasification technologies. Appl Energy 113:1461–1474
13. Mokhtar M, Ali MT, Khalipur R, Abbas A, Shah N, Al Hajaj A, Armstrong P, Chiesa M, Sgouridis S (2012) Solar-assisted post-combustion carbon capture feasibility study. Appl Energy 92:668–676
14. Zhao Y, Hong H, Zhang X, Jin H (2012) Integrating mid-temperature solar heat and post-combustion CO_2-capture in a coal-fired power plant. Solar Energy 86:3196–3204
15. Qadir A, Mokhtar M, Khalilpour R, Milani D, Vassallo A, Chiesa M, Abbas A (2013) Potential for solar-assisted post-combustion carbon capture in Australia. Appl Energy 111:175–185
16. Jamel MS, Abd Rahman A, Shamsuddin AH (2011) Advances in the integration of solar thermal energy with conventional and non-conventional power plants. Renew Sustain Energy Rev 20:71–81
17. Yang Y, Wang L, Dong C, Xu G, Morosuk T, Tsatsaronis G (2013) Comprehensive exergy-based evaluation and parametric study of a coal-fired ultra-supercritical power plant. Appl Energy 112:1087–1099

18. Novatec Solar GmbH. (2014) Solar Boiler performance based on Fresnel Collector Technology. www.novatecsolar.com. Accessed 15 July 2014
19. Cau G, Cocco D (2014) Comparison of medium-size concentrating solar power plants based on parabolic trough and linear Fresnel collectors. Energy Procedia 45:101–110
20. Breyer C, Knies G (2009) Global energy supply potential of concentrating solar power. Proceedings of the SOLARPACES conference 2009, Berlin, September 15–18, 2009
21. Meteonorm (2010) Version 6.0, Meteotest, Bern (CH)
22. Khalipour R, Abbas A (2011) HEN optimization for efficient retrofitting of coal-fired power plants with post-combustion carbon capture. Int J Greenhouse Gas Control 5:189–199
23. Cau G, Tola V, Deiana P (2013) Comparative performance assessment of IGCC and USC plants integrated with CO_2 capture systems. Fuel 116:820–833
24. International Energy Agency (IEA) and OECD Nuclear Energy Agency (NEA) (2010) Projected costs of generating Electricity—2010 Edition, Report IEA-NEA, ISBN 978-92-64-08430. www.iea.org. Accessed 15 July 2014
25. U.S. Department of Energy (2010) Cost and performance baseline for fossil energy plants—volume 1: bituminous coal and natural gas to electricity. Technical report DOE/NETL-2010/1397. http://www.energy.gov. Accessed 15 July 2014

Chapter 85
A Modular Linear Fresnel Reflecting Solar Concentrator for Low-Enthalpy Processes

O. A. Jaramillo, J. O. Aguilar, R. Castrejón-García and N. Velázquez

Abstract One of the aims of solar-thermal engineering is to develop 'medium temperature collectors' in the range of 70–250 °C that are suitable for applications such as cleaning, drying, evaporation, distillation, pasteurization, sterilization and cooking, as well as applications with low-temperature heat demand and high consumption rates (domestic hot water, space heating and swimming pool heating), and heat-driven refrigeration and cooling. In this work, we present an applied research on the development of a modular linear Fresnel reflecting solar concentrator (MLFRC) for heating water and steam generation at low enthalpy in the temperature range of 70–110 °C. We present a simple mathematical method to determine the distance among the reflecting elements and the angle for each one in order to avoid shading between the elements during the period of operation. We designed a non-imaging CPC cusp-type collector with a cylindrical receiver as a secondary concentrator. We also evaluated the thermal performance of MLFRC according to the Standard ANSI/ASHRAE 93:2003.

Keywords Modular linear Fresnel concentrator · Water and steam at low enthalpy · Concentrated solar energy

O. A. Jaramillo (✉)
Instituto de Energías Renovables, Universidad Nacional Autónoma de México,
Privada Xochicalco s/n, Temixco, MOR 62580, Mexico
e-mail: ojs@ier.unam.mxe; oajaramillo@gmail.com

J. O. Aguilar
División de Ciencias e Ingeniería, Universidad de Quintana Roo,
Boulevard Bahía s/n, Chetumal, Q Roo 77019, Mexico

R. Castrejón-García
Dirección de Desarrollo Tecnológico, Instituto Nacional de Astrofísica, Óptica y Electrónica,
Luis Enrique Erro # 1, Tonantzintla, PUE 72840, Mexico

N. Velázquez
Instituto de Ingeniería, Universidad Autónoma de Baja California, Blvd. Benito Juárez y
Calle de la Normal s/n, Mexicali, BC 21280, Mexico

A. Sayigh (ed.), *Renewable Energy in the Service of Mankind Vol II*,
DOI 10.1007/978-3-319-18215-5_85

85.1 Introduction

The development of linear Fresnel technology has been carried out during several decades. The first meaningful linear Fresnel collector was patented and prototyped by Francia in Italy in 1964 [1]. An innovative commercial prototype based on the linear reflector technology was made by Solar Heat and Power Company [2]. Because of the demand for renewable energy, which plays a significant role in the utility-scale power supply market, the current linear Fresnel technology is being targeted at high-temperature power cycles. Solar plants using linear Fresnel technology have been commercially tested and/or deployed on a utility scale around the world [3]. Currently, Novatec Solar [4] and Areva Solar [5] are leaders in the development of a high-temperature linear Fresnel collector. But some studies have also addressed on low- or medium-temperature heat generation, given that this technology has wide applications in building cooling and heating, industrial process, water treatment, etc. [6,7]. A complete discussion about history, state of the art and future of this technology can be found in reference [8].

This work presents the design and evaluation of a modular linear Fresnel reflecting solar concentrator (MLFRC), which was designed to obtain hot water and low-enthalpy steam at 1.5 atm in the range of 70–110 °C. The MLFRC uses an array of low-profile flat reflectors and a mobile receiver assembly that includes a linear receiver tube and a secondary concentrator. The prototype was evaluated according to the ANSI/ASHRAE 93:2003 [9], since this standard is widely known and provides test methods for determining the thermal performance of solar energy concentrators that use single-phase fluids and have no significant internal energy storage. During the evaluation of MLFRC, both the thermal efficiency and the time constant for heating and cooling were obtained.

85.2 Design of the MLFRC

MLFRC prototype has a rim angle of 45°, a length of 2.44 m and an aperture area close to 5 m^2. The frame is made by using structural steel hollow sections, conventional flat mirrors are assembled to the frame, and a mobile compound parabolic concentrator made of aluminium is used as a secondary concentrator (see Fig. 85.1). In the construction of the MLFRC, there was no need for sophisticated machinery or skilled labour; and only hand tools were required for its assembly. Since the MLFRC is for low-enthalpy steam generation and hot water supply, it was designed with an unshielded receiver to reduce both production and transportation costs. A simple solar tracking system is used when the concentrator is operated in the north-to-south direction.

To perform the optical design of MLFRC, the following assumptions were made:

- All reflecting surfaces are specular and free from deformations.
- Each reflector is pivoted at the centre of its longitudinal axis.

Fig. 85.1 Modular linear Fresnel reflecting solar concentrator

- The movement axes of each reflector are in the same plane.
- Reflecting elements do not shade each other during the period of operation from 9:00 to 15:00 h (solar time).
- The solar acceptance angle is $\xi = 32' = 4.65\text{mrad}$ and it is independent of the atmospheric conditions.
- The sun passes through the zenith.
- All mirror elements have equal width W and constant length l.
- The system uses a non-imaging CPC cusp-type collector as a secondary concentrator.

Figure 85.2 shows a schematic diagram of the geometry of MLFRC where the reflecting elements have the same width and their movement is pivoted at the centre of its longitudinal axis.

The angle Ψ_i of each reflecting element is calculated for redirecting the direct radiation (that arrives with an angle θ) towards the focal plane that is located at a distance f from the reflecting element placed in the centre of MLFRC. Considering the initial conditions of operation at 9:00 am ($\theta_s = 45°$):

$$L_0 = 0 \tag{85.1}$$

$$\Psi_0 = 22.5° \tag{85.2}$$

As shown in Fig. 85.2, right side reflectors are closer to the horizontal position, while the left side elements are closer to the vertical position. Note that the elements of the left side of Fig. 85.2 exhibit greater shadow projection than the elements on the right side, and this is due to the inclination of the reflectors themselves. To calculate the distance L_i among elements and their angles Ψ_i, we used a set of equations that were taken and modified from reference [10], namely,

$$\Psi_i = \frac{1}{2}\arctan\left[\frac{L_{i-1}}{f} + \frac{W}{2f}\left(\frac{\sin\Psi_{i-1} + \sin\Psi_i}{\tan(90° - \theta_s - \xi)} + \cos\Psi_{i-1} + \cos\Psi_i\right)\right] + \frac{\theta_s}{2} \tag{85.3}$$

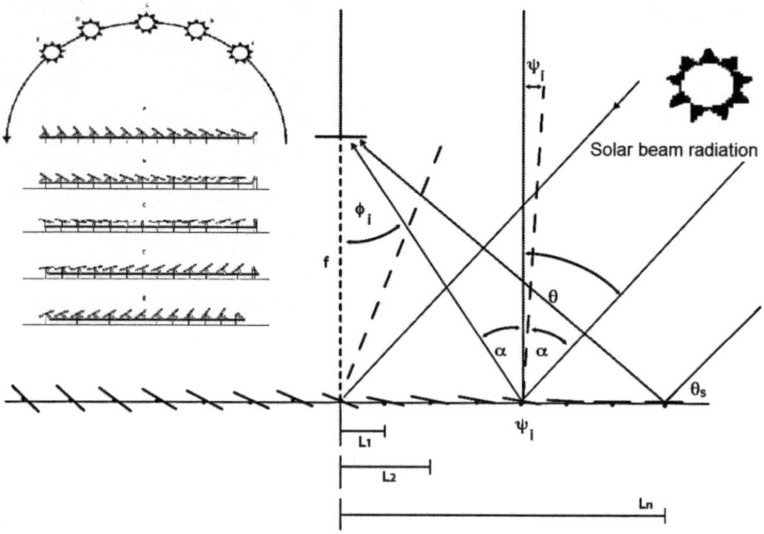

Fig. 85.2 Schematic representation of the geometry of MLFRC

$$L_i = L_{i-1} + \frac{W}{2}\left(\frac{\sin\Psi_{i-1} + \sin\Psi_i}{\tan(90° - \theta_s - \xi)} + \cos\Psi_{i-1} + \cos\Psi_i\right) \quad (85.4)$$

where f is the focal distance, W is the width of the reflecting element and n is the number of reflecting elements of the left side of MLFRC, as shown in Fig. 85.2. Equations (85.3–85.4) are solved iteratively for $(1 \le i \le n)$ taking into account the initial conditions, Eqs. (85.1) and (85.2).

By symmetry, the spacing among the reflecting elements on the left side is the same as the spacing of the right side of MLFRC, that is to say:

$$L_l = L_r \quad (85.5)$$

while the angles of the reflecting elements on the right side are determined by

$$\Psi_r = \frac{\theta_s - \arctan\left[\frac{L_r}{f}\right]}{2} \quad (85.6)$$

where the subscript is established as $(1 \le r \le n)$.

In order to determine the length Rpl, that is the width of the receptor at the receiving plane, it is necessary to consider the projection of the last element at the left side of the MLFRC, as shown in Fig. 85.3.The width R_{pl} is calculated as

$$R_{pl} = r_l + r_c + r_r \quad (85.7)$$

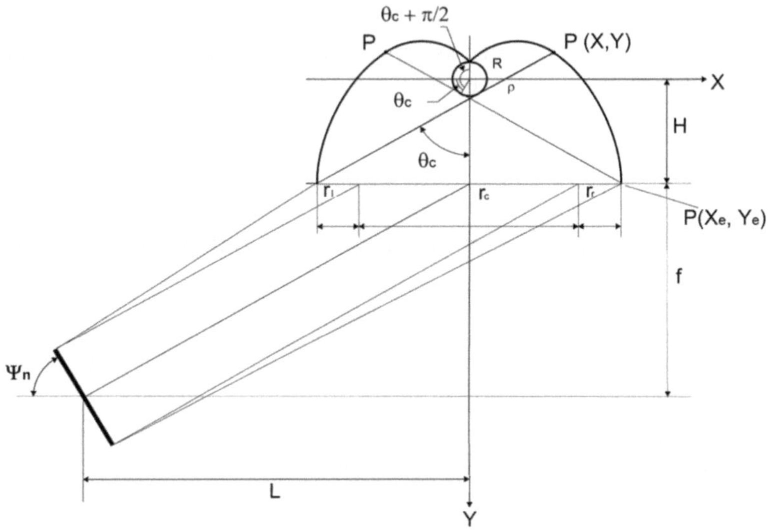

Fig. 85.3 Parameters involved in the design of the Cusp-CPC

where

$$r_l = \left(f + \frac{W}{2}\sin\Psi_n\right)\left(\tan\left[\arctan\left[\frac{L_n}{f} + \xi\right]\right] - \tan\left[\arctan\left[\frac{L_n}{f}\right]\right]\right) \quad (85.8)$$

$$r_c = W\cos\Psi_n + W\sin\Psi_n\tan\left[\arctan\left[\frac{L_n}{f}\right]\right] \quad (85.9)$$

$$r_r = \left(f - \frac{W}{2}\sin\Psi_n\right)\left(\tan\left[\arctan\left[\frac{L_n}{f}\right]\right] - \tan\left[\arctan\left[\frac{L_n}{f} - \xi\right]\right]\right) \quad (85.10)$$

The width R_{pl} must match with the width of the secondary concentrator. For example, in this work we considered the using of a non-imaging CPC cusp-type collector as a secondary concentrator that is shown in Fig. 85.3.

In a non-imaging CPC cusp-type collector with a cylindrical receiver, the radius, R, and the acceptance half angle, θ_c, the distance, ρ, along a tangent from the receiver to the curve, are related to the angle θ(outlined between the radius to the bottom of the receiver and the radius to the point of tangency, T), by the following expressions for the two sections of the curve [11].

$$\rho = R\theta, |\theta| \le \theta_c + \frac{\pi}{2} \quad (85.11)$$

$$\rho = R\left(\frac{\theta + \theta_c + \frac{\pi}{2} - \cos(\theta - \theta_c)}{1 + \sin(\theta - \theta_c)}\right), \theta_c + \frac{\pi}{2} \le \theta \le \frac{3\pi}{2} - \theta_c \quad (85.12)$$

The two expressions for $\rho(\theta)$ are equivalent at point P (see Fig. 85.3), where $\theta = \theta_c + \frac{\pi}{2}$. The curve is generated by incrementing θ, calculating ρ and then calculating the coordinates, X and Y, by the following expressions [11]:

$$\begin{aligned} X &= R\sin\theta - \rho\cos\theta \\ Y &= -R\cos\theta - \rho\sin\theta \end{aligned} \tag{85.13}$$

The coordinate X_e at the end of the cusp-CPC is the half of r_c plus the distance of r_i, that is:

$$X_e = \frac{r_c}{2} + r_l \tag{85.14}$$

On the other hand, it is possible to use the following set of coupled equations to calculate the radius R of the cylindrical receiver,

$$\frac{r_c}{2} + r_l = R\sin\left(\frac{3\pi}{2} - \theta_c\right) - \rho_e\cos\left(\frac{3\pi}{2} - \theta_c\right) \tag{85.15}$$

$$Y_e = -R\cos\left(\frac{3\pi}{2} - \theta_c\right) - \rho_e\sin\left(\frac{3\pi}{2} - \theta_c\right) \tag{85.16}$$

$$\rho_e = R\left(\frac{2\pi - \cos\left(\frac{3\pi}{2} - 2\theta_c\right)}{1 + \sin\left(\frac{3\pi}{2} - 2\theta_c\right)}\right) \tag{85.17}$$

Equations (85.15–85.17) is a system with three equations and three unknowns (R, Y_e and ρ_e), where Y_e and ρ_e are the extreme values in the cusp-CPC geometry, as shown in Fig. 85.3.

Figure 85.3 shows a full, untruncated curve, which is the mathematical solution for a reflector shape with the maximum possible concentration ratio. The total concentration ratio C for the MLFRC, including the secondary reflector, is established by the ratio between the aperture area A_a and the receptor area A_r,

$$C = \frac{A_a}{A_r} = \frac{\left(\sum_{i=1}^{n} W\cos\left[\Psi_i - \frac{\theta_s}{2}\right]\right) \times l}{\pi R \times l} = \frac{\sum_{i=1}^{n} W\cos\left[\Psi_i - \frac{\theta_s}{2}\right]}{\pi R} \tag{85.18}$$

One important aspect of the linear Fresnel collector is the optical characterization, which was performed by using the simulation software TracePro [12]. Considering the path of each ray interacting with the reflecting surfaces of the collector, the flux energy arriving on the absorber can be thus obtained. To carry out the ray-tracing

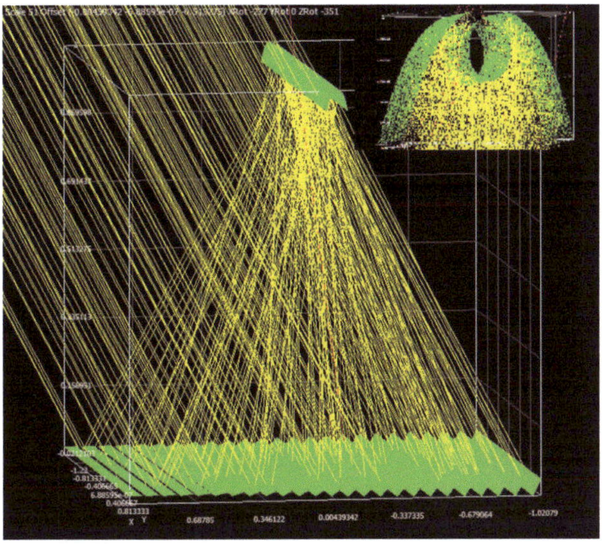

Fig. 85.4 Simulation of the MLFRC by using Sol TracePro

simulation process, we first established a 3D model of the concentrator and cavity receiver system in CAD and then transferred to TracePro, where the surface properties such as reflectivity ρ and absorptivity α were set into the simulation process. Figure 85.4 shows the results of the ray-tracing method implemented in TracePro. As can be noted, the reflecting elements do not shade each other and the secondary concentrator collects the solar beam in the receiver tube. The main results of design MLFRC are reported in Table 85.1.

85.3 Thermal Performance of MLFRC

The thermal performance of MLFRC was evaluated experimentally according to the ANSI/ASHRAE 93:2003 standard [9]. The evaluation was conducted to determine the time constant and the thermal instantaneous efficiency using water as the thermal fluid at a flow rate of 5 l/m. During evaluation a direct solar radiation of 876 W/m^2 was measured.

85.3.1 Time Constant

The collector time constant is the time required for the fluid leaving the MLFRC to achieve 63.2 % of its steady-state change following change in irradiance. This parameter determines the time response of the solar collector in order to evaluate the transient behaviour of the collector and select the correct time intervals for the quasi-steady-state or steady-state efficiency tests. Figure 85.5 illustrates the

Table 85.1 Key features of MLFRC

i	Ψ_i	L_i (cm)	i	Ψ_i	L_i(cm)	Parameter value
−12	45.0°	−99.96	1	20.2°	7.92	$W = 6.00$ cm
−11	43.7°	−91.44	2	17.9°	15.96	$f = 99.96$ cm
−10	42.3°	−82.92	3	15.7°	24.11	$C = 20.87$
−9	40.8°	−74.41	4	13.5°	32.35	$r_c = 8.49$ cm
−8	39.2°	−65.92	5	11.4°	40.66	$r_l = 0.47$ cm
−7	37.4°	−57.46	6	9.4°	49.04	$r_r = 0.45$ cm
−6	35.6°	−49.04	7	7.5°	57.46	$\theta_c = 45°$
−5	33.6°	−40.66	8	5.8°	65.92	$R = 1.06$ cm
−4	31.5°	−32.35	9	4.2°	74.41	$\rho_e = 7.73$ cm
−3	29.3°	−24.11	10	2.7°	82.92	$P(X,Y) = (2.52, -1.02)$
−2	27.0°	−15.96	11	1.3°	91.44	$P(X_e, Y_e) = (4.72, 6.23)$
−1	24.8°	−7.92	12	0.0°	99.96	$\rho = 0.9$
0	22.5°	0.0				$\alpha = 0.95$

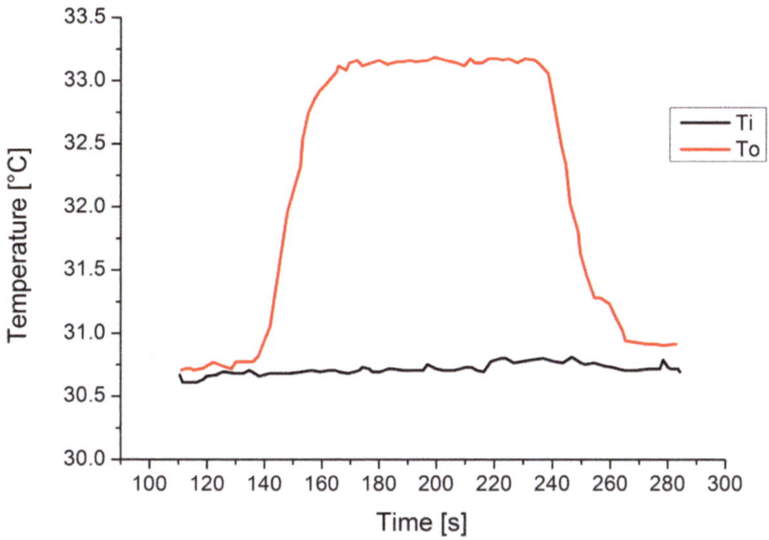

Fig. 85.5 Evaluation of the time constant of MLFRC

Table 85.2 The time constant for heating and cooling of MLFRC

Time constant	
Cooling	Heating
18.0 s	14.5 s

exponential nature of heating and cooling time and the time constants are reported in Table 85.2.

85.3.2 Thermal Efficiency

The ANSI/ASHRAE 93:2003 standard [9] requires efficiency tests to be run for a period equal to one time constant or 5 min, whichever is larger. The thermal instantaneous efficiency η_g was evaluated experimentally by considering a time period of 5 min:

$$\eta_g = \frac{\dot{m} C_P (T_o - T_i)}{A_a G_b} \tag{85.19}$$

where T_i and T_o are the inlet and outlet temperatures, respectively, $\dot{m}$ is the mass flow rate, C_P is the specific heat, A_a is the aperture area of the collector and G_b is the direct solar irradiance component in the aperture plane of the collector. The

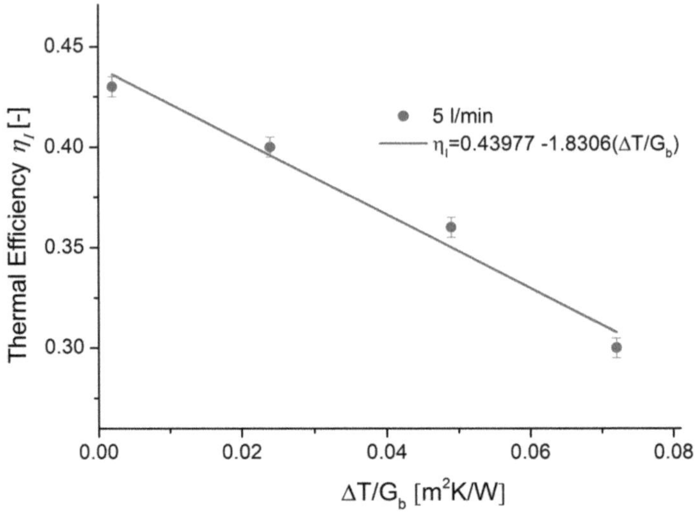

Fig. 85.6 Thermal efficiency data and best-fit curve for the MLFRC

inlet temperature T_i is restricted to an interval occurring from 20 to 90 °C in order to avoid the phase change in water that is used as thermal fluid.

On the other hand, the thermal efficiency of the concentrators from the First Law is given by,

$$\eta_I = F_R\left(\eta_o - \frac{U_L}{C}\left(\frac{\Delta T}{G_b}\right)\right) \tag{85.20}$$

where $\Delta T = T_i - T_a$ is the temperature rise across the receiver and T_a is the ambient temperature. Note that Eq. (85.20) has the form $y = b + mx$, which can help to experimentally obtain the heat removal factor F_R and the overall heat loss coefficient U_L. Note that $F_R U_L / C$ is the slope of the line and $F_R \eta_o$ is the *y*-intercept. It is important to indicate that for a collector operating under steady irradiation and fluid flow rate, $F_R \eta_o$ and $F_R U_L / C$ are nearly constant. Therefore, Eq. (85.20) plots as a straight line on a graph of efficiency versus the heat loss parameter $\Delta T / G_b$.

The efficiency curve for MLFRC is shown in Fig. 85.6. The points show the thermal instantaneous efficiency, and a straight line of best fit can be drawn between those points to obtain the thermal efficiency for the solar concentrators. Note that the linear fit instead of a second-degree fit is acceptable in the present case involving solar concentrators with low temperature. The values of the thermal efficiency η_I , the heat removal factor F_R and the overall heat loss coefficient U_L are reported in Table 85.3.

Table 85.3 Thermal performance of MLFRC

$\eta_o[-]$	0.54
$\eta_i[-]$	$\eta_i = (0.43977 \pm 0.00895) - (1.8306 \pm 0.19805)$ $(\Delta t/G_b)$
m [l/min]	5.0
$C[-]$	20.87
$F_R[-]$	0.81
U_L [W/m²K]	47.16
$F_R\eta_o[-]$	0.43977
F_RU_L/C [W/m²K]	1.8306

85.4 Conclusions

We designed, developed and evaluated an MLFRC with a rim angle of 45°, an aperture area close to 5.0 m² and a concentration ratio close to 20. The materials used in its construction and the method of manufacture were cheap and did not require complex manufacturing systems. The frame was made using structural steel hollow sections; conventional flat mirror reflector strips were assembled to the frame and a compound parabolic concentrator made from aluminium was used as a secondary concentrator. In the construction of the MLFRC, there was no need for sophisticated machinery or skilled labour; and during the assembly, only hand tools were required. Since the MLFRC is for low-enthalpy steam generation and hot water, it was designed with an unshielded receiver to reduce both production and transportation costs. A simple solar tracking system is used when it is oriented in the north-to-south direction.

To determine the dimension of the MLFRC, we developed a simple mathematical method on the basis of simple geometrical optics, and an on-imaging CPC cusp-type collector with a cylindrical receiver was designed and developed as a secondary concentrator to increase the concentration ratio of the MLFRC. To evaluate the optical performance of the MLFRC, we implemented a ray-tracing simulation process in TracePro. Results of the simulation showed that the reflecting elements do not shade each other and the secondary concentrator collects the solar beam on the receiver tube with an optical efficiency estimated close to 54 %.

The thermal performance of the MLFRC was tested on the basis of ANSI/ASHRAE 93:2003 standard. Peak efficiency close to 45 % was assessed. The efficiency curve was established as a straight line on the basis of the best fit for the thermal instantaneous efficiency.

Our results show that the MLFRC provides an adequate thermal efficiency and enough structural resistance. Because of its modular design, several MLFRC can be connected in series, providing an increment in the thermodynamic availability of the fluid. Given the geographical position of Mexico, solar heat for industrial processes is very promising.

Acknowledgements The authors thank José de Jesús Quiñones Aguilar, Jorge Alejandro Monreal Cruz and Álvaro Junior Cepeda Cifuentes for their technical assistance in the development of the device and evaluation. They also thank Dr. Maximiliano Valdez González for his technical support in the network management and software implementation. This work was supported by PAPIIT-UNAM under the project IT100514.

References

1. Silvi C (2009) The pioneering work on linear Fresnel reflector concentrators in Italy. In: Presented at the 15th SolarPACES International Symposium, Berlin, Germany
2. Mills D, Morrison GL, Pye JD, Le Lievre P (2006) Multi-tower line focus Fresnel array project. J Sol Energy Eng 128:118–120
3. Mills D, Morrison G (2000) Compact linear Fresnel reflector solar thermal power plants. Sol Energy 68(3):262–283
4. Solar N (2013) http://www.novatecsolar.com/. Accessed 7 Nov 2013
5. Solar A (2013) http://www.areva.com/EN/solar-220/arevasolar.html/. Accessed 15 Nov 2013
6. Wagner M, Zhu G (2012) A direct-steam linear Fresnel performance model for NREL's system advisor model. In: Presented at the ASME 2012 6th Int. Conf. Energy Sustainability, San Diego, CA
7. Velázquez N, García-Valladares O, Sauceda D, Beltrán R (2010) Numerical simulation of a linear Fresnel reflector concentrator used as direct generator in a Solar-GAX cycle. Energy Convers Manage 51(3):434–445
8. Zhu G, Wendelin T, Wagner M, Kutscher C (2013) History, current-state and future of linear Fresnel concentrating solar collectors. Sol Energy 103:639–652
9. ANSI/ASHRAE Standard 93 (2003) Methods of testing to determine the thermal performance of solar collectors
10. Mathur SS, KandpaL TC, Negi BS (1991) Optical design and concentration characteristics of linear Fresnel reflector solar concentrators II. Mirror elements of equal width. Energy Convers Manage 31(3):221–232
11. Kalogirou S (1997) Design, construction, performance evaluation and economic analysis of an integrated collector storage system. Renew Energy 12(2):179–192
12. Pro T (2014). http://www.lambdares.com/. Accessed 7 Feb 2014

Chapter 86
Exergetic, Environmental and Economical Analysis of a Cogeneration Plant Connected to a District Heating Network.

K. Sartor and P. Dewallef

Abstract Based on actual operational and economic data of an existing Rankine cycle cogeneration plant connected to a district heating network, a detailed economic and exergetic evaluation is carried out in order to study the influence of the network temperature level of the cogeneration plant efficiency, namely the sum of the electrical and thermal efficiency.

Keywords Energetic analysis · CHP · DHN · Economic analysis · Environmental analysis

86.1 Introduction

During the design of a cogeneration plant intended to feed a district heating network, the selection of the water temperature level of a district heating network (DHN) is an important step as it is directly linked to the power consumed by the circulation pumps and the heat losses to the environment. If the DHN is fed by natural gas boilers, the influence of the temperature level is not that important as the natural gas boiler efficiency varies slightly between 97 and 105 % (based on LHV) [1]. Lately, the use of low-temperature DHN (the return water temperature is lower than 50 °C) enabled the exploitation of latent due to flue gas condensation. However, this type of DHN, so-called 4th-generation DHN, is not widespread nowadays and most of the DHNs in operation have supply temperature around 80 °C [2]. However, if a cogeneration plant based on a Rankine cycle is used to feed the DHN, lowering the network temperature enables an increase in the electricity production. Indeed, the use of back-pressure steam turbine allows the production of high temperature steam supplied in a heat exchanger heating the DHN water by condensing the steam. Setting the temperature of the DHN together with the heat exchanger efficiency fixes the steam saturation temperature and consequently the level of back pressure. The

K. Sartor (✉) · P. Dewallef
Department of Aerospace and Mechanics, Thermodynamics Laboratory, University of Liège, Chemin des Chevreuils 7, B-4000 Liège, Belgium
e-mail: kevin.sartor@ulg.ac.be

A. Sayigh (ed.), *Renewable Energy in the Service of Mankind Vol II*,
DOI 10.1007/978-3-319-18215-5_86

lower the temperature requirement of the DHN is, the lower the pressure level of the back pressure and the higher the electricity production.

In this matter, the cogeneration efficiency defined as the sum of the electrical efficiency and thermal efficiency is often misleading. Indeed, increasing the level of temperature required by the DHN increases the back pressure, which, in turn, decreases the electrical efficiency. However, the thermal energy not converted into electricity is available as useful thermal energy. As a result, changing the level of water temperature into the DHN slightly affects the cogeneration efficiency. As it will be shown herein, the cogeneration efficiency is a poor indication of the plant performance. To underline this fact, an economic study is carried out to check the influence of the DHN temperature on the cost of thermal energy generated by the plant.

To support the discussion, an existing cogeneration plant connected to a DHN installed on the University Campus in Liège (Belgium) is used as an application test case. Based on actual operational and financial data, a detailed economic, environmental and energetic evaluation is carried out that is supported by a calibrated simulation model of the whole installation (this model is detailed in other contributions by the authors [3, 4]) in order to study the influence of the network temperature level on the CHP plant performance. In a second step, the simulation model is used to perform an exergetic analysis intended to give physical insights to the economical study. This results in a broader discussion on the several methods available to assess the quality of cogeneration plants and the opportunity to develop such a technology together with district heating networks.

86.2 Problem Statement

When using a CHP to feed a DH network, a conventional goal is to optimize the conversion efficiency (η_{CHP}) of the primary energy into useful energy, namely electricity and heat. If P_{prim} is the rate of primary energy consumption and *Pelec* (*Pheat*) is the electricity (thermal) production, the electrical (thermal) efficiency is defined by $P_{elec}=\eta_{elec/}P_{prim}$ $(P_{heat}=\eta_{heat/}P_{prim})$. Consequently, the cogeneration efficiency is $\eta_{CHP}=\eta_{heat}+\eta_{elec}$. Similar indicator may be defined from an environmental point of view by noting that the combustion of 1 kilowatt-hour (kWh) of natural gas releases 251 g of CO_2, while the same primary energy of biomass releases between 0 and about 30 g of CO_2 depending on the supply chain [5]. Biomass fuels may be of very different types, and it is beyond the scope of the present contribution to exhaustively list them, and it will be considered herein that the biomass fuel generates 30 g of CO_2 per kWh of primary energy which corresponds to wood pellets [6]. Most of the European countries have policies to reward the use of biomass fuels under the form of subsidies or premium on the energy selling price [7, 8]. Other methodologies are available (e.g. life cycle analysis) that take into account the whole production process from the construction of the plant to its decommissioning to assess the actual saving in terms of CO_2 emissions. In the framework of the present study, these more advanced methodologies do not bring any significant change in the conclusions and it is decided to stick to the simple approach.

In addition to the energetic analysis, an exergetic analysis of the whole system is carried out. the exergy e is defined as $e=(h-h_0)-T_0(s-s_0)$ where h stands for the enthalpy, s for the entropy, and the subscript 0 for the reference conditions, namely a pressure of 101 kPa and an absolute temperature of 288.15 K [9].

Besides the energetic and environmental aspects, the most critical and decisive criteria for the development of such technologies is the cost of the energy. The problem of optimizing a CHP plant connected to a DH network is considered herein as designing the CHP plant minimizing the cost of heat (COH) supplied to the user. To do so, a cost model for the heat production is set up that takes into account the cost of capital (i.e. the initial investment), the fuel costs, the operation and maintenance cost as well as the selling price of the electricity produced. The previous consideration translates the fact that the plant is supposed to be heat-driven, the electricity being considered as a by-product sold to the grid operator or consumed locally.

The cost model per unit of thermal energy used herein is derived similarly to the one defined for electricity in [10]. According to this model, the cost of heat is expressed as:

$$COH = \underbrace{\frac{1}{\xi_{DHN}}}_{\text{DHN losses}} \left[\underbrace{\frac{C \cdot \psi + U_{\text{fix}}}{T_e P_{\text{i,th,CHP}}}}_{\text{Fixed cost}} + \underbrace{\frac{y_{\text{f}}}{\bar{\eta}_{\text{th,CHP}}} + u_{\text{var}}}_{\text{Variable cost}} - \underbrace{(y_{\text{e}} + T_{\text{cv}} \cdot y_{\text{cv}}) \cdot \frac{\bar{\eta}_{\text{el,CHP}}}{\bar{\eta}_{\text{th,CHP}}}}_{\text{Electricity selling price}} \right] \tag{86.1}$$

In the above formula, C represents the initial investment cost and ψ is the annuity factor taking into account the present value of money ($C\psi$ is the annual repayment for the initial investment expressed in $year^{-1}$) assessed through:

$$\psi = \frac{d}{1-(1+d)^{-N}} \tag{86.2}$$

where d is the discounting rate per year and N the number of years for which the installation is used (e.g. the life time of the plant). $P_{i,\ th,chp}$ is the installed thermal power of the CHP plant in MW, and τ_e is the equivalent utilization time at rated power output. τ_e embeds the availability factor of the plant. y_f is the cost of fuel in €/MWh, U_{fix} is the fixed cost of operation, maintenance and administration in €/year, and u_{var} is the variable cost of operation, maintenance and repair in €/MWh. $\bar{\eta}_{th,\ chp}$ and $\bar{\eta}_{e,\ chp}$ are, respectively, the annual average thermal and electricity efficiencies yet taking into account the start/stop procedures (if any) and the part load efficiency. y_e is the price of electricity in €/MWh, while τ_{cv} and y_{cv} are, respectively, the number of green certificates per MWh of electricity produced[1] and the unit price of a green certificate. The term $\tau_{cv} y_{cv}$ is replaced by the premium on the electricity selling when feed-in tariffs are used instead.

[1] For the Walloon region of Belgium one green certificate is granted for every 456 kg of CO_2 saving. A maximum of 2 green certificates are allowed per MWh of electricity produced.

The determination of C, U_{fix}, u_{var} , d and N is not within the scope of the present contribution and reliable estimates can be found, e.g., in [1, 10]. Neither the influence of fuel cost y_f nor the one of the supporting policies $\tau_{cv}y_{cv}$ will be long discussed herein and representative value of the market in Belgium will be used, as it is relatively straightforward for the reader to include relevant data into the above model.

The determination of T_e, $\overline{\eta}_{e,\, chp}$ and $\overline{\eta}_{th,\, chp}$ is not straightforward and very often overlooked as these values strongly depend upon the size of the CHP plant and the evolution in time of the heat demand. Studying how the CHP plant matches the heat demand and assessing the resulting performance are not the main subject of the present contribution, and it is treated in [3]. Herein, the focus is set on the influence of water temperature level of the DHN on the final cost of energy. As this temperature level has a direct influence on the heat losses, a transport efficiency ξ_{DHN} is introduced in Eq. 86.1 to assess the influence of these heat losses and to express the cost of heat per unit of heat delivered to the customer. It is defined as the ratio of the heat delivered to the consumer to the heat produced by the plant. As previously said, this temperature level has a significant influence on the global CHP plant annual efficiency if it is based on a Rankine cycle: assuming the plant is made of one back-pressure steam turbine, the condenser located downstream of the steam turbine exchanges heat to the DHN. The electricity production of this steam turbine is being expressed as $P_{elec} = \varepsilon\, \eta_{alt} \cdot \dot{M}_{steam} \cdot (h_{su} - h_{ex,\, s})$ where h_{su} and $h_{ex,s}$ stand for the supply and isentropic exhaust enthalpy of the steam, ηalt the alternator efficiency, ε the isentropic steam turbine efficiency, and $\dot{M}_{steam}$ the steam mass flow rate. Depending on the supply DHN water temperature, the requirements on the exhaust steam vary: the higher the supply water temperature is, the higher the required steam temperature and pressure are. The electricity production decreases when the DHN temperature increases. This aspect is discussed in the following by taking into account the modification of the thermodynamic cycle during its design phase to minimize the cost of heat.

In the case in which the heat delivered by the CHP plant for the base load and by a backup generation for the peak load (e.g. a natural gas boiler), the average cost of heat must be weighted proportionally to the production of the heat sources. If the costs of heat generated through relation (1) are denoted, respectively, COH_{chp} and COH_{bck} for the CHP plant and the backup boiler and Θ is the ratio of the thermal energy generated by the CHP plant to the total thermal energy for the considered time interval, the average cost of heat is assessed through the following equation:

$$COH = \theta \cdot COH_{\mathrm{CHP}} + (1-\theta) \cdot COH_{\mathrm{bck}} \qquad (86.3)$$

Note that, in the case of natural gas boilers, the last term in Eq. 86.1 is cancelled.

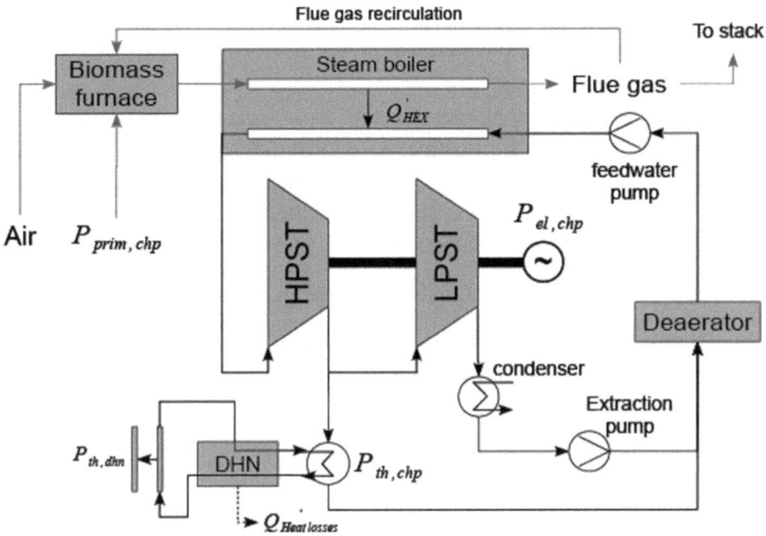

Fig. 86.1 Schematic of CHP plant

86.3 Simulation Model

The simulation model is used to assess the exact amount of heat supplied by a CHP plant to a DH network whose heat profile is known. Through the calibrated simulation model, the net electricity production of the plant is known for both rated and part load operation together with the heat losses related to the DHN. The considered biomass CHP plant consists of a biomass furnace connected to a boiler made of an economizer, an evaporator and superheater section. The steam is expanded into two successive steam turbines to produce electricity. After the steam expansion at the high pressure steam turbine, a portion of the steam is extracted to supply heat to the DHN through a heat exchanger. A schematic of the cycle is shown in Fig. 86.1.

For the sake of simplicity, a brief description of the model is provided in this section but more information can be found in [3, 4] to which the interested reader is referred to. The complete simulation model of the plant is an aggregation of basic components modelled by a zero-dimensional (i.e. input–output) approach (biomass furnace, steam boiler, steam turbines, district heating network) verifying the conservation of mass, energy and momentum. The biomass combustion model is handled through a general biomass composition $C_m H_n O_x N_y S_z$ (where the subscripts are the ratio between wet basis mass fraction of each component to its molar mass) able to assess the composition of 15 species of the flue gas products, namely H_2, O_2, H_2O, CO, CO_2, OH, H, O, N_2, N, NO, NO_2, CH_4, SO_2, SO_3.

Steam turbines are modelled by the Stodola line [11] where the steam turbine is considered as a nozzle whose mass flow rate depends upon the inlet pressure and temperature and the outlet pressure:

$p\frac{\dot{m}_{st}\sqrt{T_{su,st}}}{\sqrt{p^2_{su,st}-p^2_{ex,st}}}=K$ where the parameter K is constant for a wide range of operation and therefore can be determined from nominal operation. $\dot{m}_{st}$ stands for the steam mass flow in the turbine, p for the pressure, and T for the absolute temperature. The subscripts *st* and *ex* stand, respectively, for supply and exhaust. Concerning the DHN heat losses, their assessment rely on the resolution of a steady-state two-dimensional heat conduction–convection problem. This enables the calculation of a global heat transfer coefficient, $\Lambda(\frac{W}{K})$ used to assess the heat losses as a function of ambient temperature and DHN water temperature level as $\dot{Q}_{losses,\ DHN}=\Lambda\ (T_{fd}-T_{ambient})$ where T_{fd} is an average temperature of the water circulating in the DHN.

86.4 Application Test Case

The aforementioned simulation model is applied to a typical district heating application available on the University campus in Liège (Belgium). The installed network has a total length of 10 km and distributes pressurized hot water at about 125 °C, on average, to approximately 70 buildings located in the University campus representing a total heat area of about 470,000 m^2. Buildings are very different in nature, namely classrooms, administrative offices, research centers, laboratories and a hospital. The hospital represents about 25 % of the total heated area and requires steam for the kitchen and air humidity control system that justified this water temperature level. The effective peak power of the network is around 56 MW for a total of 61,000 MWh per year.

While all the buildings are heated between 4:00 and 20:00, the hospital needs heating and steam 24 h a day, 365 days per year. The DH network is operating since the 1960s and was originally operated from natural gas boilers. In order to cope with the restrictions in terms of CO_2 emissions, the University of Liège decided to invest in a biomass CHP plant whose purpose is to feed the base load heat demand of the campus. The CHP plant has started full operation in 2012. It is made of a moving grid biomass boiler with nominal primary power of 12 MW supplying steam to a back-pressure turbine and a condensing turbine with nominal power of 2.4 MW. The extracted steam is condensed in a heat exchanger feeding the DH network with a nominal power of 7 MW. The remaining thermal power needed by the DH network is provided by two natural gas boilers with a total installed power of 54 MW. The flue gas at the exhaust of the furnace passes successively through an evaporator (platen), screen tubes, two superheaters, one evaporator and four economizers. Exhaust gases are filtered before being directed to the stack. The steam cycle is representative of a traditional cycle with extraction turbines, and the steam generated has a temperature of about 420 °C. The district heating network is divided into

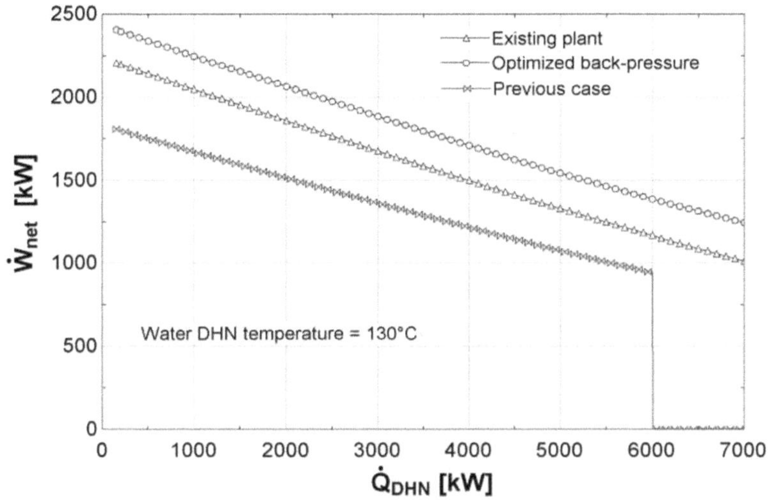

Fig. 86.2 Electricity production for the three cases studied [kW]

twenty-three sections having the same geometric characteristic but pipe diameters ranging from 50 to 350 mm. The insulation used is mineral wool with an identified thermal conductivity of 0.047 W m^{-1} K^{-1}.

With respect to [3, 4], the CHP plant studied herein is slightly different as an exhaust gas recirculation at the level of the primary air was added together with a modification of the steam cycle. Indeed, instead of injecting saturated steam from the steam drum to heat the deaerator, a portion of the extracted steam is now used which improves the conversion efficiency.

86.5 Results

The existing plant being already built, it is not possible to change the level of back pressure. In order to underline the potential of optimizing the level of DHN temperature, the simulation model is used to study the influence of the turbine back pressure on the whole-plant efficiency and generation costs. In the following figures, the results corresponding to the existing plant are referred to as existing plant, while those related to the simulated one are referred to as optimized back-pressure. The net electricity production is sketched for these two situations as a function of the heat delivered to the DHN (Fig. 86.2).

When the temperature of the DHN is modified, the cogeneration efficiency as well as the CO_2 emissions can be assessed and compared to the separated generation of heat and power. They are represented for full load conditions in Figs. 86.3a and b. Representing these results at full load makes the cogeneration efficiency directly proportional to the electrical efficiency.

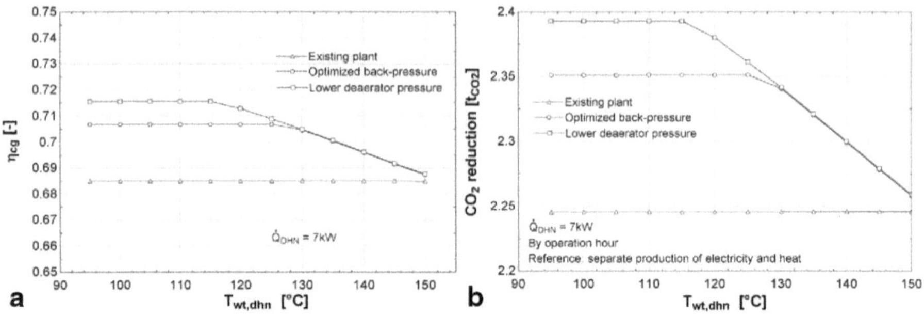

Fig. 86.3 cogeneration efficiency η_{CHP} and CO_2 emissions as a function of the water temperature supplied to the DHN for rated CHP heat output

For the existing plant, the cogeneration efficiency is strictly constant as the back pressure is fixed, and this sets the steam mass flow rate at a constant value (as the heat demand is constant) and keeps the cogeneration efficiency constant. When the back pressure is optimized to the DHN temperature level, the cogeneration efficiency increases. Additionally, it can be seen that the cogeneration efficiency increases when the level of temperature decreases up to a point where the efficiency remains constant. This saturation of the cogeneration efficiency is due to the fact that a minimum of steam extraction is required to maintain the level of temperature in the deaerator. In order to go a step further, a third test case is developed where the deaerator temperature can be decreased down to 115 °C.

This test case is referred to as lower deaerator pressure in Figs. 86.3a and b. It has to be noted that variations of the efficiency appear important due to the scale used in the figures yet they are limited to 3 % which is relatively low for the large range of DHN temperature studied.

When looking at the exergetic efficiency of the plant taking into account the exergy content of the hot water and the net electricity production, it can be seen from Fig. 86.4a that the exergetic efficiency of the plant increases with

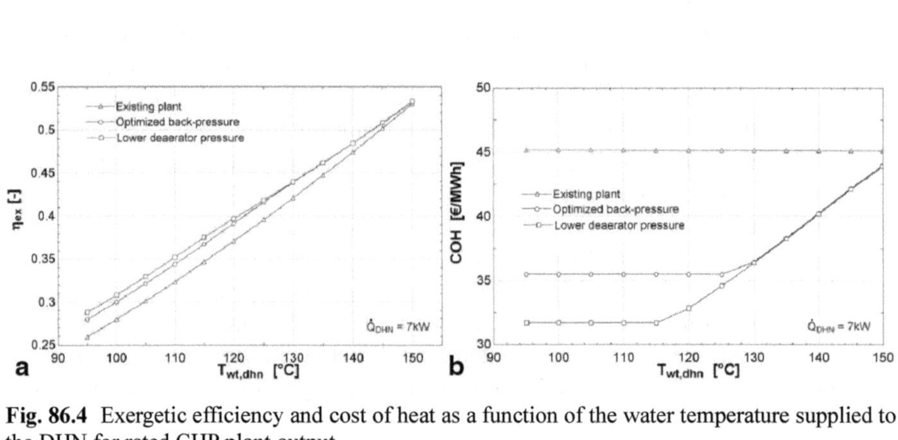

Fig. 86.4 Exergetic efficiency and cost of heat as a function of the water temperature supplied to the DHN for rated CHP plant output

Table 86.1 Energy-based COH comparison for two cases investigated for a DHN temperature level of 130 °C for the current load profile of the ULg

Configuration	Current	Optimized pressure
τ_e [-]	0.6613	0.6613
η_e Annual mean electricity efficiency [-]	0.1070	0.1251
$\eta_{\bar{th}}$ Annual mean thermal efficiency [-]	0.4412	0.4412
Fixed costs [€/MWh]	25.27	25.27
Variable costs [€/MWh]	87.18	87.19
Electricity production [€/MWh]	−69.61	−81.41
DHN costs [€/MWh]	2.25	2.25
Total CHP plant COH [€/MWh]	45.10	33.30
Global application COH [€/MWh]	**52.7**	*44.84*

the DHN temperature. This is a straightforward consequence from the exergetic efficiency definition. As the energy of the hot DHN water increases with the level of temperature, the energy produced increases. However, when taking a look at Fig. 86.4b, the cost of the generated heat increases with an increasing DHN temperature level which tends to underline, as others studies [8, 12–14], an incorrect definition in the current subsiding policies for the CHP plant. In Eq. 86.1, the yearly average of the thermal and electrical efficiencies is calculated from the instantaneous efficiencies at full load obtained from the model and multiplied by 0.98 and 0.92 to take into account the reliability and the availability of the plant. All the values used to compute the COH defined in Eq. 86.1 were extracted from a previous study detailed in [3].

For the existing plant, the cost of heat is constant whatever the level of temperature is (it is only influenced by the level of heat transport losses), but when the back pressure is optimized, lower generation costs can be obtained. Furthermore, when the deaerator temperature is adapted, another cost reduction can be achieved. For a deeper understanding of the cost structure, the COH can be also detailed for the existing plant and optimized back pressure for the actual annual load profile to focus on the difference between raw CHP plant performances and the influence of the annual load of a DHN (Table 86.1). The indicated costs correspond to the COH computed through Eq. 86.3 for a backup natural gas boiler with COH of 65.37 € per MWh (see [3]).

Concerning the optimized pressure configuration, it leads to higher annual average thermal and electricity efficiencies and a higher electricity production (in absolute value) and therefore a reduced COH.

To compensate the subsiding policies inconsistency previously identified and to assess correctly energy systems performance, several indicators could be used [8, 12, 13, 15, 16]. The alternative way proposed in this study is to assess the COH considering that the efficiencies defined in Eq. 86.1 are replaced by the thermal and electrical *exergetic* efficiencies. The exergy-based COH is represented in Fig. 86.5 with plain symbols, while the energy-based COH is represented with empty symbols as a reminder.

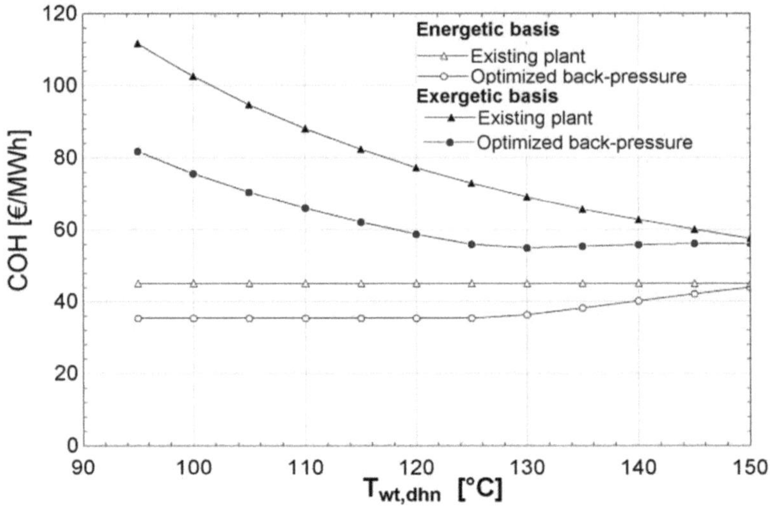

Fig. 86.5 Comparison of the COH energy and exergy-based COH as a function of the water temperature supplied to the DHN for rated CHP plant output

For the existing plant, the exergy-based COH decreases when the water temperature increases, while the exergetic efficiency increases due to the higher water exergy content. When the back pressure is optimized and is not restricted by any external stress, in the range of water temperature of 125–150 °C, the exergy-based COH is quite constant (about ± 1 € per MWh of exergy compared to ± 10 per MWh of energy). In the other range of temperature, when no improvement is feasible, it follows the same trend that of the existing case.

This constancy reflects that the available exergy is used in a better way for each DHN temperature level. Indeed, the exergy degradation that occurs in the steam extraction valve decreases through the back-pressure adjustment. Therefore, the exergy analysis seems to be a better indicator to rank energy systems such as CHP plants. It should be promoted in the subsiding policies, while the cogeneration efficiency, which is generally used, is not a good indicator to correctly assess the CHP plant performance, while the production costs increase with the exergetic efficiency.

86.6 Conclusions

This contribution extends previous studies on the energetic, environmental and economical analysis of a CHP plant connected to a district heating network considering the exergy aspect. The exergy analysis is a lean indicator to show the irreversibilities of a system and which is proposed as an indicator to rank the CHP plant. Several cases are studied: an existing plant and two alternatives of improvement which could be investigated during the CHP plant design process.

For the existing CHP plant, the energy-based COH, the CO_2 emissions reduction and the cogeneration efficiency of the CHP plant remain constant whatever is the DHN temperature level, while the exergetic efficiency increases with this temperature leading to a lower exergy-based COH.

When the back pressure can be adjusted, i.e. an improvement is performed, a reduction in CO_2 emissions (0.15 t of CO_2 avoided per hour of production) and higher cogeneration efficiency (about 3 %) can be achieved but they are quite limited for the large range of the DHN temperature level considered. When this back-pressure adjustment is not constrained by any external stress and safety margin, the energy-based COH increases significantly with the DHN temperature level, while the exergetic efficiency increases, which seems to be a paradox (about 35 % or 13 € per MWh of energy). On the other hand, the related exergy-based COH is quite constant which means the *available exergy* is used in a better way for each DHN temperature level. In the other temperature range, the COH (energy and exergy-based) follows the same trends that in the existing plant, while there is no improvement anymore.

These examples tend to underline that the current subsiding policies, energy-based, do not encourage the thermodynamic improvements and are not a good indicator to rank the CHP plant performance. Therefore, modified subsiding policies and indicators, such as the exergy system analysis studied herein, have to emerge to rank correctly the CHP plant and assess others' systems which could be connected like DHN.

As discussed briefly, the authors wish also to remind the major aspect of the load profile on the energy-based COH, while a large difference between the raw and the annual COH is shown: an increase of 17 % or 8 € per MWh of energy for the annual COH.

References

1. Energi Styrelsen (2012) Technology data for energy plants: generation of electricity and district heating, energy storage and energy carrier generation and conversion. Technical report, Energi Styrelsen
2. Skagestad B, Mildenstein P (1999) District heating and cooling connection handbook. Technical report, IEA
3. Sartor K, Quoilin S, Dewallef P (2014) Simulation and optimization of a {CHP} biomass plant and district heating network. Appl Energy 130(0):474–483
4. Sartor K, Restivo Y, Ngendakumana P, Dewallef P (2014) Prediction of {SOx} and {NOx} emissions from a medium size biomass boiler. Biomass Bioenerg 65(0):91–100. (21st European Biomass Conference)
5. Edenhofer O, Pichs-Madruga R, Sokona Y, Seyboth K, Arvizu D, Bruckner T, Christensen J, Devernay J-M, Faaij A, Fischedick M, Goldstein B, Hansen G, Huckerby J, Ja¨ ger-Waldau A, Kadner S, Kammen D, Krey V, Kumar A, Lewis A, Lucon O, Matschoss P, Maurice L, Mitchell C, Moomaw W, Moreira J, Nadai A, Nilsson LJ, Nyboer J, Rahman A, Sathaye J, Sawin J, Schaeffer R, Schei T, Schlo¨ mer S, Sims R, Verbruggen A, von Stechow C, Urama K, Wiser R, Yamba F, Zwickel T. (2011) Ipcc special report on renewable energy sources and climate change mitigation—complete report. 06/2011

6. SGS (2012) Biomass verification procedure—energy and carbon balance form. Technical report, COFELY Services
7. Moiseyev A, Solberg B, Maarit A, Kallio I (2014) The impact of subsidies and carbon pricing on the wood biomass use for energy in the {EU}. Energy 76(0):161–167
8. Ertesvg IS (2007) Exergetic comparison of efficiency indicators for combined heat and power (chp). Energy 32(11):2038–2050
9. Kotas TJ (1985) Chapter 2–basic exergy concepts. In: Kotas TJ (ed) The exergy method of thermal plant analysis. Butterworth-Heinemann, London, pp. 29–56
10. Bachmann R, Nielsen H, Warner J (1999) Combined—cycle gas & steam turbine power plants. Pennwell Books, Tulsa
11. Lowenstein ALC (1927) Stodola. Steam and gaz turbines, vol 1, 6th edition. McGraw-Hill, New-York
12. Liao C, Ertesvg IS, Zhao J (2013) Energetic and exergetic efficiencies of coal-fired {CHP} (combined heat and power) plants used in district heating systems of china. Energy 57(0):671–681
13. Lythcke-Jørgensen C, Haglind F, Clausen LR (2014) Exergy analysis of a combined heat and power plant with integrated lignocellulosic ethanol production. Energy Convers Manag 85:817–827
14. Verbruggen A, Dewallef P, Quoilin S, Wiggin M (2013) Unveiling the mystery of combined heat and power (cogeneration). Energy 61(0):575–582
15. Dincer I, Rosen MA (2007) Exergy, environment and sustainable development. In: Dincer I and Rosen MA (ed), {EXERGY}. Elsevier, Amsterdam, 36–59
16. Holmberg H, Tuomaala M, Haikonen T, Ahtila P (2012) Allocation of fuel costs and co2-emissions to heat and power in an industrial {CHP} plant: case integrated pulp and paper mill. Appl Energy 93(0):614–623. ((1) Green Energy; (2) Special Section from papers presented at the 2nd International Enery 2030 Conf.)

Chapter 87
Investigation on the Thermal Characteristics of a Bi-fluid-Type Hybrid Photovoltaic/Thermal (PV/T) Solar Collector

Hasila Jarimi, Mohd Nazari Abu Bakar, Norain A. Manaf, Mahmod Othman and Mahadzir Hj Din

Abstract Known as a photovoltaic/thermal (PV/T) solar collector, this type of hybrid solar collector involves the integration between conventional PV cells and a solar thermal component. Up to now, the research in this field has tended to focus on either air or water as the working fluid. When both water and air are utilised under the same PV/T solar collector, it is seen as more appealing due to its space-saving design and range of potential applications. Known as a bi-fluid-type PV/T solar collector, this type of hybrid system is the subject of this research. For collector optimisation purposes, thermal modelling is considered important and hence, 2-dimensional (2-D) steady-state energy balance equations are derived to model the bi-fluid PV/T system during the simultaneous mode of fluid operation. However, when the fluids are to be operated simultaneously, the equations are not solvable analytically, and hence the numerical method is employed. Following this, the common useful thermal characteristic parameters of the collector, that is, the collector efficiency factor F', heat removal F_R and also the overall heat transfer loss coefficient U_l cannot be determined analytically. Nevertheless, these quantities can be evaluated experimentally and graphically through computer simulations using the derived 2-D steady-state analysis. Theoretically, using the results obtained from the modelling, suitable curves are plotted of which the values of F_R for the hybrid solar collector during independent mode of fluid operation for water and air and for bi-fluid configuration are computed as 0.59, 0.71 and 0.82 respectively. Meanwhile, at

H. Jarimi (✉) · N. A. Manaf
Faculty of Applied Sciences, Universiti Teknologi MARA Shah Alam, 40450 Shah Alam, Selangor, Malaysia
e-mail: H.Jarimi@gmail.com

M. N. Abu Bakar · M. Hj Din · N. A. Manaf · H. Jarimi
Faculty of Applied Sciences, Universiti Teknologi MARA Perlis, 02600 Arau, Perlis, Malaysia
e-mail: mohdnazari@perlis.uitm.edu.my

M. Othman
Faculty of Computer Sciences and Mathematics, Universiti Teknologi MARA Perlis, 02600 Arau, Perlis, Malaysia

A. Sayigh (ed.), *Renewable Energy in the Service of Mankind Vol II*,
DOI 10.1007/978-3-319-18215-5_87

zero reduced temperature, the predicted thermal efficiencies are at 42.8, 51.3, and 59.2 % respectively. Experiments are then performed for steady-state analysis under the solar simulator at the Solar Energy Research Lab UiTM Perlis, Malaysia. The results obtained are found to be in good agreement with the values predicted by the mathematical model. From this study it can be concluded that the utilisation of both water and air as the working fluid under the same hybrid solar collector is seen as promising in order to optimise solar energy utilisation. The validated 2-D steady-state analysis is very useful in order to predict the collector performance under the influence of important environmental and design parameters. This study contributes as the starting platform in a research on a bi-fluid-type hybrid PV/T solar collector covering both theoretical and experimental studies.

Keywords Hybrid · Bi-fluid (PV/T) · Thermal parameters · 2-D steady state

87.1 Introduction

Manifesting itself in the form of low-grade energy by a conventional thermal solar collector and high-grade energy via a photovoltaic (PV) panel, solar energy is by far seen as the largest source of renewable energy. Typically made from crystalline and amorphous types of semiconductors, the electrical temperature dependency of PV cells has been a point of attention. Following this, methods to improve the performance of PV cells have been actively investigated. Since the 1970s, research on the integration between a conventional PV panel and a thermal solar collector has been actively conducted [1]. This type of hybrid solar collector is known as a 'photovoltaic/thermal (PV/T) solar collector' [1]. Designed as the cooling element to the PV panel, in addition to the improved electrical energy performance, useful thermal energy is produced by the thermal component of the solar collector via its working fluid. When compared to the conventional thermal solar collector and PV panel, the electrical and thermal performance is not as good [2]. However, when analysed as a PV/T solar collector as a whole, both energy yields (electrical and thermal) are of the same surface area. This implies that the thermal efficiencies of the combined collector are higher in comparison to the side-by-side arrangement [2].

The research so far has tended to focus on either air or water as the working fluid. When both fluids are utilised as the working fluids, the collector is known as bi-fluid PV/T solar collector, which was first introduced by Tripanagnostopoulos [3]. However, the research focused mainly on experimental studies of the collector during independent mode of fluid operation. Followed by [4], a bi-fluid PV/T solar collector was developed and analysed theoretically using 2-dimensional (2-D) steady-state analysis. However, the design of the collector is such that the position of air and water alternates, hence the 2-D analyses for the water and air components were carried out separately.

Abu Bakar et al. [5] have proposed a novel design of a hybrid PV/T solar collector which integrates the use of both water and air as the working fluids. In the

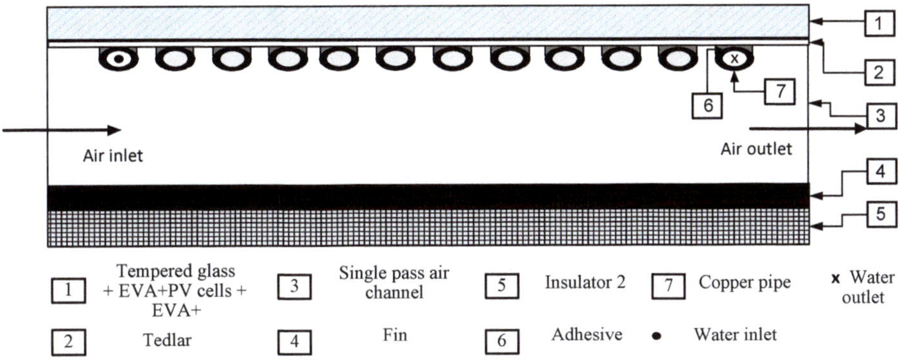

Fig. 87.1 Cross-sectional side view of the designed collector

study, where the simultaneous utilisation of fluids is the highlight, a serpentine-shaped copper tube as the water heat exchange system has been integrated with a conventional PV module and a single pass air channel as the air heat exchange system. 2-D steady-state analysis is then developed and validated to simulate the collector's performance [5]. As discussed by Jarimi et al. [6], the performance of the collector is then further improved by the researchers by introducing the use of a heat transfer enhancement technique via the use of a set of fins parallel to the direction of the air flow with a slight modification to the mathematical model.

Up to now, for a hybrid (PV/T)-type solar collector, no specific standards have been published to evaluate the thermal performance of this type of hybrid system [7, 8]. To address its performance, as reported by Amrizal et al. [7], the similar behaviour of the thermal performance between the PV/T collectors and the conventional thermal module is observed, and hence the available standards for solar thermal and PV devices are officially referred to. The common thermal characteristic parameters of the collector, that is, the heat removal F_R, collector efficiency factor F' and also the overall heat transfer loss coefficient U_l are useful to analyse the thermal performance of the collector. These quantities can be evaluated by solving thermal equations analytically or experimentally [9]. In this study, following the developed 2-D mathematical model as discussed in [5], a slight modification to the model suitable for the finned air channel configuration as shown in Fig. 87.1 is discussed. The thermal characteristic parameters are then evaluated graphically through computer simulations using the derived 2-D steady-state analysis. Experiments have been conducted at the Solar Energy Research Lab, UiTM Perlis, Malaysia to validate the simulation results obtained in the study.

87.2 Experimental Study

In order to evaluate the performance of the bi-fluid-type hybrid solar collector, a prototype and indoor test-rig facilities have been fabricated and set up at the Solar Energy Research Lab, UiTM Perlis, Malaysia. The prototype is non-optimised

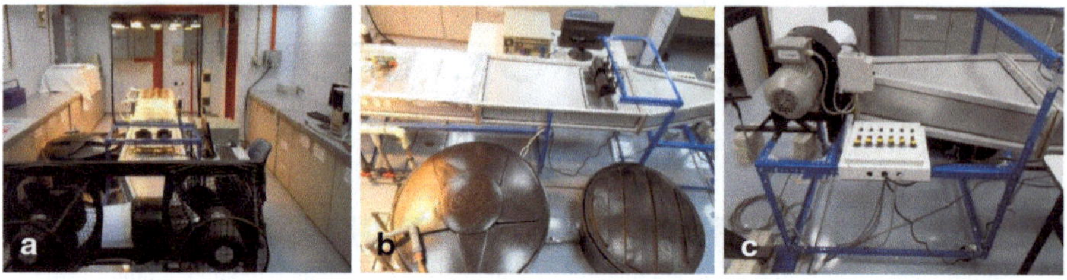

Fig. 87.2 **a** Indoor experimental set-up, **b** water and air heat exchange supply system and **c** power supply and control system

and fabricated to validate the simulation results from the developed 2-D mathematical model for the collector under study. The indoor test-rig facilities, shown in Fig. 87.2, consist of three important sections, namely solar simulator, air and water heat exchange supply system and the power supply, and control system.

Many low-cost solar simulators suitable for solar collector testing were fabricated as early as 1974 [10, 11]. According to [10, 12–14], a low-cost simulator, which consists of a 3-phase array of 30 quartz-halogen tungsten lamps installed on a lamp casing with a reflector arranged on a steel support structure 1.4 × 1.4 m in size, is fabricated for the collector indoor testing. Capable of producing solar intensities up to 800 W/m^2, the intensity of the lamps is made to vary using the United Automation CSR2-E series power regulator installed at the power supply control box for the solar simulator. A calibrated Eppley Black & White 8–48 pyranometer is used to measure the incident solar radiation.

Water is pumped into the collector by an SP737 Tideway electric submersible water pump from the primary water tank whose speed is controlled using a United Automation CSR2-E series power regulator. In order to fix the temperature of the water at a constant temperature, a secondary water tank is used to discard the heated water from the panel. Prior to entering the collector, the water flow rate is first measured using a Blue-White F-440 inline flow meter. Meanwhile, air is pumped into the collector using an air blower, which consists of a unit of fan driven by a Siemens 3-phase induction AC motor connected to an AC driver to control the air flow rate into the collector. The air flow rate is measured using an SDL 350 hot-wire thermo-anemometer from Extech Instruments.

The temperatures of the PV module both at the top and back surface, back plate and fluids are all measured using a K-type thermocouple, which was calibrated prior to the collector testing. All the thermocouples and the pyranometer are connected to the channels of Advantech DAQ modules (ADAM) for data logging via a PC. The electrical characteristics of the PV module for indoor testing are measured using a VS-6810 IV tracer. The readings are recorded at a time interval of 1 s and are averaged over the period during which the system is in steady state. The combination of both heat exchange systems into the collector leads to a new test-rig design which must be tailored to the dual heat exchange demands. Therefore, the fabrication of the test-rig facilities discussed in this study is a good starting platform to further develop experimental research on a bi-fluid-type PV/T solar collector.

87.3 Mathematical Modelling

For a practical heat conduction problem which involves complicated geometries with complex boundary conditions or variable properties, the thermal balance equations associated with the heat transfer cannot be solved analytically [15]. In this study, when both fluids are to be operated simultaneously, the system is considered complex and complicated [5, 6]. By referring to Craft [16], the system is analysed numerically, in which a 2-D steady-state analysis with a finite difference method is employed with temperature nodes defined using the procedure as discussed by Abu Bakar et al. [5]. Interested readers can refer to [5, 6] for detailed discussion on the mathematical model.

Since the collector designed in this study is the flat-plate-type collector (FPC), regardless of the temperature gradient across the PV cells which acts as the collector absorber, the average temperature is still used to determine the performance of the collector [8]. Therefore, by modifying the famous Hottel-Whillier-Bliss (H-W-B) equations, the total useful thermal energy produced by the collector can also be expressed in terms of the heat removal factor F_R, collector efficiency factor F' and also the overall heat loss coefficient U_l.These parameters are useful since the total useful energy can be predicted by knowing the inlet fluid temperature [9]. However, since the energy balance equations developed are not solvable analytically using the developed model, the values of the thermal characteristic parameters of the collector can be computed graphically using the computer simulations. As discussed in Duffie and Beckman [9], by definition, the useful thermal energy produced by a collector with fluid f, can be expressed in terms of the fluid heat removal factor F_R, collector aperture A_c, the overall heat loss coefficient U_l and the fluid input temperature T_{fi} as follows:

$$Q_{TH} = \dot{m}_f C_{pf}(T_{fo} - T_{fi}) = A_c F_R[S_c - U_l(T_{fi} - T_a)]. \tag{87.1}$$

The actual fraction of the thermal energy absorbed, S_c, is given by the following expression in terms of the available global solar radiation G, transmittance–absorptance factor $\alpha_c\tau_g$, packing factor PF_c, and electrical efficiency η_{el}:

$$S_c = \alpha_c\tau_g PF_c(1-\eta_{el})G. \tag{87.2}$$

The electrical efficiency of the collector is modelled as a function of temperature based on Florschuetz [17] equations as follows:

$$\eta_{el} = \eta_{ref}(1 - \beta_{ref}(T_{pv} - T_{ref})), \tag{87.3}$$

with η_{ref}, β_{ref}, and T_{ref} as the electrical efficiency, temperature coefficient and the PV cell temperature, respectively. The value for η_{ref} is obtained experimentally as 0.04 at $T_{ref} = 40^{\circ}$C.

In general, (87.1) is useful to describe the collector performance when the fluids are operated independently. In order to describe the collector's thermal performance during simultaneous mode of fluid operation, a slight modification to (87.1) is made by assuming that at steady state the total energy produced by the collector can be expressed in terms of the average total heat removal factor of the collector F_{Rtotal} and the average overall heat loss coefficient $\overline{U}_1$, in which the calculation includes the total energy loss when air T_{f1} and water T_{f2} are utilised simultaneously as shown in (87.4):

$$\begin{aligned} Q_{\mathrm{TH,tot}} &= \dot{m}_{\mathrm{f1}} C_{\mathrm{pf1}} (T_{\mathrm{f1o}} - T_{\mathrm{f1i}}) + \dot{m}_{\mathrm{f2}} C_{\mathrm{pf2}} (T_{\mathrm{f2o}} - T_{\mathrm{f2i}}) \\ &= A_{\mathrm{c}} S_{\mathrm{c}} (F_{\mathrm{Rtotal}}) - A_{\mathrm{c}} \overline{U_1} F_{\mathrm{Rtotal}} ((T_{\mathrm{f1i}} - T_{\mathrm{a}}) + (T_{\mathrm{f2i}} - T_{\mathrm{a}})). \end{aligned} \tag{87.4}$$

Following the work in [5], the expression for the overall thermal efficiency is then given by

$$\begin{aligned} \eta_{\mathrm{TH,tot}} &= \frac{A_{\mathrm{c}} S_{\mathrm{c}} (F_{\mathrm{Rtotal}}) - A_{\mathrm{c}} \overline{U_1} F_{\mathrm{Rtotal}} ((T_{\mathrm{f1i}} - T_{\mathrm{a}}) + (T_{\mathrm{f2i}} - T_{\mathrm{a}}))}{G A_{\mathrm{c}}} \\ &= \frac{S_{\mathrm{c}} (F_{\mathrm{R,tot}})}{G} - (\overline{U_1} F_{\mathrm{R\,total}}) \left(\frac{\Delta \mathrm{T}}{G} \right). \end{aligned} \tag{87.5}$$

From here, by following the standard thermal characteristic analysis of a thermal solar collector, sets of data from experiments and computer simulations are obtained. Later, graphs of thermal efficiency against $\frac{\Delta T}{G}$ with $\Delta T = (T_{\mathrm{fi}} - T_{\mathrm{a}})$ and $\Delta T = ((T_{\mathrm{f1i}} - T_{\mathrm{a}}) + (T_{\mathrm{f2i}} - T_{\mathrm{a}}))$ for the independent and simultaneous mode of fluid operation, respectively, are plotted to evaluate the collector performance. The gradient and the interception of the graph reflect the performance of the collector in which the indication of useful energy absorbed and lost can be obtained from the graph interception, that is, $\alpha_{\mathrm{c}} \tau_{\mathrm{g}} P F_{\mathrm{c}} (1 - \eta_{\mathrm{el}}) F_{\mathrm{R,tot}}$, and gradient, that is, $\overline{U_1} F_{\mathrm{R_{total}}}$, respectively.

In addition, the energy produced by the collector can also be represented using the collector efficiency factor F' as follows [18]:

$$Q_{\mathrm{TH}} = A_{\mathrm{c}} F' [S_{\mathrm{c}} - U_1 (\overline{T}_{\mathrm{f}} - T_{\mathrm{a}})]. \tag{87.6}$$

Equation (87.6) is then modified to obtain an expression for the bi-fluid configuration during simultaneous mode of fluid operation:

$$Q_{\mathrm{THtotal}} = A_{\mathrm{c}} S_{\mathrm{c}} (F'_{\mathrm{total}}) - A_{\mathrm{c}} \overline{U_1} F'_{\mathrm{total}} ((\overline{T}_{\mathrm{f1}} - T_{\mathrm{a}}) + (\overline{T}_{\mathrm{f2}} - T_{\mathrm{a}})). \tag{87.7}$$

As considered in [18], the mean fluid temperature in the equation can be replaced by the arithmetic average of inlet and outlet fluid temperature. In (87.7) above, the mean average of air and water temperatures is then computed as follows:

$$\overline{T}_{\mathrm{f1}} = \frac{T_{\mathrm{f1i}} + T_{\mathrm{f1o}}}{2} \text{ and } \overline{T}_{\mathrm{f2}} = \frac{T_{\mathrm{f2i}} + T_{\mathrm{f2o}}}{2}.$$

Therefore, the expression for the overall thermal efficiency can be expressed as

$$\eta_{\mathrm{TH,tot}} = \frac{S_{\mathrm{c}}(F'_{\mathrm{total}})}{G} - \left(\overline{U}_{\mathrm{l}}\, F'_{\mathrm{total}}\right)\left(\frac{\left(\frac{T_{\mathrm{f1i}} + T_{\mathrm{f1o}}}{2} - T_{\mathrm{a}}\right) + \left(\frac{T_{\mathrm{f2i}} + T_{\mathrm{f2o}}}{2} - T_{\mathrm{a}}\right)}{G}\right)$$
$$= \frac{S_{\mathrm{c}}(F'_{\mathrm{total}})}{G} - \left(\overline{U}_{\mathrm{l}}\, F'_{\mathrm{total}}\right)\left(\frac{\Delta\overline{T}_{\mathrm{f}}}{G}\right). \quad (87.8)$$

The value of F' can be computed from the interception of thermal efficiency against the $\frac{\Delta\overline{T}_{\mathrm{f}}}{G}$ curve with $\Delta\overline{T}_{\mathrm{f}} = (\overline{T}_{\mathrm{f}} - T_{\mathrm{a}})$ and $\Delta\overline{T}_{\mathrm{f}} = (\overline{T}_{\mathrm{f1}} - T_{\mathrm{a}}) + (\overline{T}_{\mathrm{f2}} - T_{\mathrm{a}})$ for the independent and simultaneous mode of fluid operation, respectively.

87.4 Results and Discussion

The experiments are performed under the fabricated solar simulator at the Solar Energy Research Lab UiTM Perlis, Malaysia. With the solar simulator fixed at an intensity of 800 W/m^2, wind speed at 1.0 ms^{-1} and mass flow rate at 0.02 kg/s for each of the fluids, the hybrid collector is tested and the performance is analysed for both thermal and electrical performance at steady-state condition. In addition, the average PV cell temperatures are estimated using the measured temperatures of the back surface of Tedlar in accordance with the guidelines given in [8]. Using the 2-D steady-state analysis, computer simulation is performed using MATLAB to simulate the collector performance with the important parameters; average solar intensity *G*, wind speed and ambient temperature T_{a} are based on the recorded experimental data.

The computer simulation and experimental results as illustrated in Fig. 87.3a, b show the graph of thermal efficiency plotted against the ratio $\frac{\Delta T}{G}$ and $\Delta\overline{T}_{\mathrm{f}} / G$ for the independent mode and simultaneous mode of fluid operation, respectively. Due to the experimental limitation and also to avoid unwanted effects caused by the overheated collector under the solar simulator, the inlet-ambient temperature difference for water and air will not exceed 20 °C. Therefore, for the theoretical analysis, the range of temperature difference is set at a higher value in order to get more accurate results for the thermal characteristics analysis. As can be seen from the plotted curves, the 2-D mathematical modelling is concluded to be in a good agreement with the experimental results. The equations that describe the curves presented in Fig. 87.3a, b are summarised in Table 87.1. From the equations, the heat removal F_{R}, collector efficiency F' and also the overall heat transfer loss coefficient U_{l}

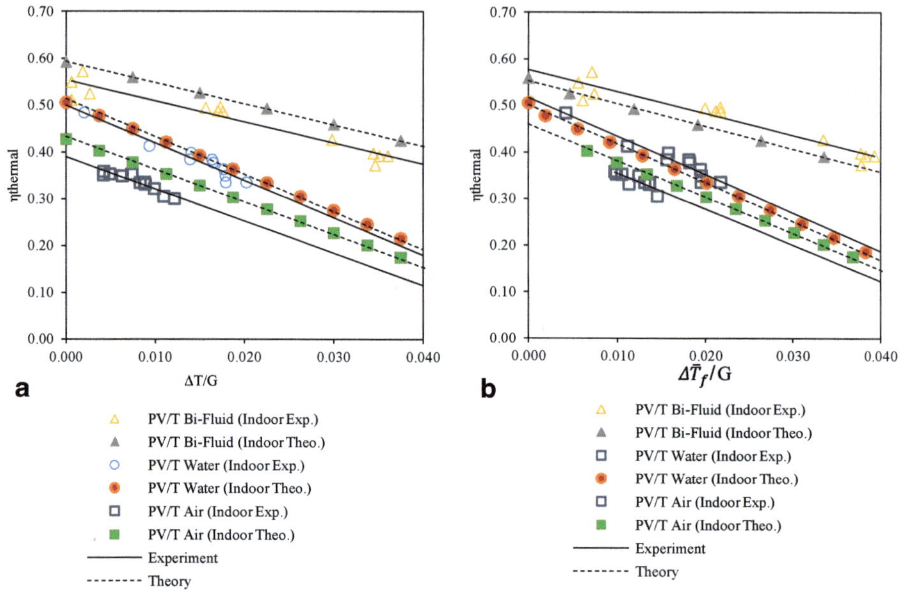

Fig. 87.3 **a** Variation of thermal efficiency with $\dfrac{\Delta T}{G}$ and **b** variation of thermal efficiency with $\Delta \overline{T}_{\mathrm{f}} \,/\, G$

Table 87.1 Summary of the steady-state performance equations

PV/T type	Experimental	Predicted
PV/T air	$\eta_{th(exp)}=-6.8639\ \Delta T/G+0.3907$	$\eta_{th(theory)}=-6.6808\ \Delta T/G+0.4280$
PV/T water	$\eta_{th(exp)}=-7.9907\ \Delta T/G+0.5000$	$\eta_{th(theory)}=-8.0380\ \Delta T/G+0.5131$
PV/T bi-fluid (total)	$\eta_{th(exp)}=-4.4481\ \Delta T/G+0.5529$	$\eta_{th(theory)}=-4.4934\ \Delta T/G+0.5924$
PV/T air	$\eta_{th(exp)}=-7.7373\ \Delta\overline{T}_f/G\ +0.4326$	$\eta_{th(theory)}=-7.8148\ \Delta\overline{T}_f/G\ +0.4600$
PV/T water	$\eta_{th(exp)}=-8.2497\ \Delta\overline{T}_f/G\ +0.5168$	$\eta_{th(theory)}=-8.3303\ \Delta\overline{T}_f/G\ +0.5010$
PV/T bi-fluid (total)	$\eta_{th(exp)}=-4.7112\ \Delta\overline{T}_f/G\ +0.5773$	$\eta_{th(theory)}=-4.8629\ \Delta\overline{T}_f/G\ +0.5524$

when the collectors are operated independently and simultaneously are computed and summarised in Table 87.2. From the computed results, it can be concluded that when operated independently, the collector's performance is considered satisfactory with its efficiency at zero reduced temperature obtained theoretically at 51.3 and 42.8 % for PV/T water and air, respectively. Meanwhile, experimentally, the values obtained are 50.0 and 39.1 %, respectively.

Table 87.2 Computed thermal characteristics

PV/T Type		F_R	U_l (W K m²)	F'
PV/T Bi-fluid (Total)	Experiment	0.79	5.63	0.84
	Theory	0.82	5.45	0.89
PV/T (Water)	Experiment	0.71	11.25	0.73
	Theory	0.71	11.30	0.74
PV/T (Air)	Experiment	0.56	12.26	0.63
	Theory	0.59	11.27	0.69

In addition, the simultaneous utilisation of air and water improves the collector's thermal performance, and this is reflected by an increase in the value of F_R and F' which can reach up to 0.79 and 0.84, respectively, with the efficiency at zero reduced temperature reaching up to 55.3 %. From the computed overall heat loss coefficient using experimental results it can also be concluded that the simultaneous use of both fluids complements one another such that in comparison to the values during independent mode, $\overline{U_l}$ is the lowest at 5.63 W/K m². This lowest value is also reflected by the low average temperature of the PV panel. It is important to note here that thermal performance is analysed at the non-optimised design parameters. Therefore, a better performance is expected when the collector is to be operated at its optimum condition.

87.5 Conclusions

Theoretical analysis based on 2-D steady-state analysis is developed and used to predict the thermal characteristics of a bi-fluid-type hybrid solar collector during independent and simultaneous mode of fluid operation. Mathematical expressions that define the thermal characteristics are derived. Suitable curves are then plotted accordingly using the data obtained from the computer simulations and experimental results. The mathematical model is then validated against the experimental results obtained from the collector indoor testing. It can be concluded that the utilisation of both fluids in the same hybrid collector improves the collector performance, and this is reflected by the increase in the average heat removal factor value and also the decrease in the overall heat loss coefficient. Therefore, it can be concluded that the bi-fluid-type PV/T solar collector designed in this study has many potential applications in various fields with space-saving features for installation.

Acknowledgement This work is funded by the Malaysian Fundamental Research Grant Scheme (FRGS) 600-RMI/ST/FRGS 5/3/Fst (160/2010) and the Solar Energy Research Lab, Universiti Teknologi Mara (UiTM) Perlis, Malaysia.

References

1. Chow TT (2010) A review on photovoltaic/thermal hybrid solar technology. Appl Energy 87:365–379
2. PVT ROADMAP (2006) PVT ROADMAP A European guide for the development and market introduction of PV-Thermal technology
3. Tripanagnostopoulos Y (2007) Aspects and improvements of hybrid photovoltaic/thermal solar energy systems. Sol Energy 81:1117–1131
4. Assoa YB, Menezo C, Fraisse G, Yezou R, Brau J (2007) Study of a new concept of photovoltaic–thermal hybrid collector. Sol Energy 81:1132–1143
5. Abu Bakar MN, Othman M, Din M Hj, Manaf NA, Jarimi H (2014) Design concept and mathematical model of a bi-fluid photovoltaic/thermal (PV/T) solar collector. Renew Energy 67:153–164
6. Jarimi H, Abu Bakar MN, Manaf NA, Othman M, Din M (2013) Mathematical modelling of a finned bi-fluid type photovoltaic/thermal (PV/T) solar collector. In: Clean Energy and Technology (CEAT), 2013 IEEE Conference on, 2013, pp. 163–168
7. Amrizal N, Chemisana D, Rosell JI (2013) Hybrid photovoltaic–thermal solar collectors dynamic modeling. Appl Energy 101:797–807
8. PV catapult (2005) PVT performance measurement guidelines—guidelines for performance measurements of liquid-cooled non-concentrating PVT collectors using c-Si cells. Report D8–6, PV-Catapult.
9. Duffie JA, Beckman WA (2006) Solar engineering of thermal processes. Wiley, New Jersey
10. Garg HP, Shukla AR, Madhuri I, Agnihotri RC, Chakravertty S (1985) Development of a simple low-cost solar simulator for indoor collector testing. Appl Energy 21:43–54
11. Abu Bakar MN, Othman MY Hj (2013) Teknologi pengumpul suria fotovolta terma. Penerbit Universiti Kebangsaan Malaysia, Malaysia
12. Othman MY, Yatim B, Sopian K, Abu Bakar MN (2007) Performance studies on a finned double-pass photovoltaic-thermal (PV/T) solar collector. Desalination 209:43–49
13. Hussain F, Othman MYH, Yatim B, Ruslan H, Sopian K, Anuar Z, et al (2011) Fabrication and irradiance mapping of a low cost solar simulator for indoor testing of solar collector. J Sol Energy Eng 133:044502–044502
14. Agrawal S, Tiwari GN, Pandey HD (2012) Indoor experimental analysis of glazed hybrid photovoltaic thermal tiles air collector connected in series. Energy Build 53:145–151, 10
15. Çengel YA, Ghajar AJ, Kanoğlu M (2011) Heat and mass transfer: fundamentals and applications. McGraw Hill, USA
16. Craft TJ (2010) Finite difference schemes [PDF document]. Lecture Notes Online Web site: http://cfd.mace.manchester.ac.uk/twiki/pub/Main/TimCraftNotes_All_Access/cfd1-findiffs.pdf Accessed 18 May 2014
17. Florschuetz LW (1979) Extension of the Hottel-Whillier model to the analysis of combined photovoltaic/thermal flat plate collectors. Sol Energy 22:361–366
18. Okeefe MJ, Francey JLA (1985) A modified solar collector flow factor. Aust J Phys 38:233–237

Chapter 88
A Modified Solar/Gas Thermodynamic Hybridization Scheme in ISCC Plants for Reducing the Air-Cooled Condenser Power Consumption

Fouad Khaldi and Mounir Aksas

Abstract The cooling systems for most existing and under construction integrated solar combined cycle (ISCC) power plants are based on air-cooled condensers (ACCs). This condensing technology limits the cycle efficiency and requires additional power consumption to run fans. The present study aims at investigating the thermodynamic feasibility of replacing the conventional bottoming steam cycle by a binary cycle, which is based on the coupling between an air bottoming cycle (ABC) and organic Rankine cycle (ORC). The advantage of the novel solar/gas thermodynamic hybridization scheme is that the ABC gets rid of the condensing process, and the ORC is more efficient for recovering heat at low/medium temperatures. The study is based on making some thermodynamic modifications on the first Algerian ISSC power plant, which is used as a reference case study. The thermodynamic simulations performed by the flow sheet software Cycle-Tempo demonstrate that the new proposed thermodynamic layout can reduce the condenser (CD) heat duty and power consumption by about 20 %. The study analyzes the sensitivity of the performance of the new thermodynamic concentrating solar power (CSP) conception applied to air temperature and heat transfer fluid (HTF) mass flow rate.

Keywords Integrated solar combined cycle plant · Organic Rankine cycle · Air bottoming cycle · Air-cooled condenser · Condenser heat duty and power consumption · Thermodynamic simulations

Nomenclature

ABC	Air bottoming cycle
AC	Air compressor
ACC	Air-cooled condenser

F. Khaldi (✉) · M. Aksas
Department of Physics, Faculty of Sciences, University of Batna, Batna, Algeria
e-mail: fouadkhaldi@gmail.com

M. Aksas
e-mail: m_aksas@hotmail.com

A. Sayigh (ed.), *Renewable Energy in the Service of Mankind Vol II*,
DOI 10.1007/978-3-319-18215-5_88

AREC	Air recuperator
ATC	Air topping cycle
CC	Combustion chamber
CD	Condenser
CH	Chiller
CSP	Concentrating solar power
DB	Duct burner
DE	Deaerator
DNI	Direct normal irradiance, W/m^2
DR	Drum
GTPP	Gas turbine power plant
ECO	Economizer
EVA	Evaporator
GT	Gas turbine
HRSG	Heat recovery steam generator
HRVG	Heat recovery vapour generator
HTF	Heat transfer fluid
ISCC	Integrated solar combined cycle
LHV	Lower heating value, J/kg
ORC	Organic Rankine cycle
PMP	Pump
REC	Recuperator
RG	Regenerator
SBC	Steam bottoming cycle
SF	Solar field
SHE	Superheater
SSG	Solar steam generator
ST	Steam turbine
SVG	Solar vapour generator

88.1 Introduction

With intensive research activities, concentrating solar power (CSP) can have a strong opportunity to become a key source of renewable energy for the future. Therefore, for moving progressively from gas-fuel electricity to solar electricity, it has been demonstrated that the hybridization of solar/gas is the most suitable option [1]. Furthermore, from both economic and technical views, the best thermodynamic hybridization scheme involves implementing parabolic trough solar energy in integrated solar-combined cycle power plants [2]. Most integrated solar combined cycle (ISCC) power plants are located in arid areas. Thus, because of limited water availability the cooling system is based on air-cooled condensers (ACCs). The ACC technology has some disadvantages on the ISCC plant performance [3–5]. The first reason is relative to the inherent under-performance of the bottoming steam cycle when the condensing process temperature is high. The second reason is the additional auxiliary power requirements necessary for running fans. Some innovative thermodynamic schemes for bottoming cycles could be a promising solution for this issue [6]. Several studies [7–9] demonstrated the feasibility of air bottoming cycle (ABC), which completely eliminates the condensing process. The ABC has technical and economic advantages compared with steam bottoming cycle (SBC) when combined with small- and medium-scale gas turbines (GTs). Organic Rankine cycles (ORCs) based on toluene are a competitive option when combined with high-efficiency GTs with low exhaust temperatures [10]. A number of studies [11–14] proposed new thermodynamic schemes and examined the influences of several designing and operating parameters on the performance of ORC-based CSP plants.

The purpose of this chapter is to propose a new thermodynamic scheme for the hybridization of solar/gas to reduce the power consumption of the ACCs. The case study is based on making some thermodynamic modifications in the existing ISCC power plant in Algeria [15].

The idea of the chapter is based on the coupling of three thermodynamic cycles: gas turbine air bottoming cycle/organic Rankine cycle (GT-ABC/ORC). Indeed, the SBC is replaced by a binary cycle: ABC/ORC. The performance sensitivity of the new thermodynamic cycle to air temperature and heat transfer fluid (HTF) mass flow is analyzed.

The thermodynamic simulations were performed by the flow sheet program "Cycle-Tempo."

88.2 Reference Case Study: The Gas Turbine–Steam Bottoming Cycle–Concentrating Solar Power (GT-SBC-CSP) Thermodynamic Layout

The first Algerian ISCC power plant is located at the Algerian's largest natural gas field, Hassi R'Mel. The design of the power plant considered baseline direct normal irradiance (DNI) estimated at 751 W/m^2, air ambient at 0.928 bars and 35 °C, with relative humidity at 24 %. At design running conditions, the full capacity of the plant is 160 MW with a solar energy share of near 14 % (22 MW). The thermal efficiency of the cycle is 56 %, and the solar-to-electricity efficiency is about 17 % [15].

The flow diagram of the plant, under the acronym GT-SBC-CSP, is shown in Fig. 88.1. The plant is composed of a power block and solar field (SF). The power block is a conventional combined cycle power plant with two 40-MW SGT-800 gas turbine power plants (GTPPs) and an 80-MW SST-900 steam turbine (ST). The isentropic efficiencies of the air compressor (AC), GT and ST are 0.86, 0.88, and 0.90, respectively. The power block also contains two identical single-pressure heat recovery steam generators (HRSGs) with supplementary firing and no reheats. The flow diagram shows the power block with only 1 GTPP and 1 HRSG. The HRSG

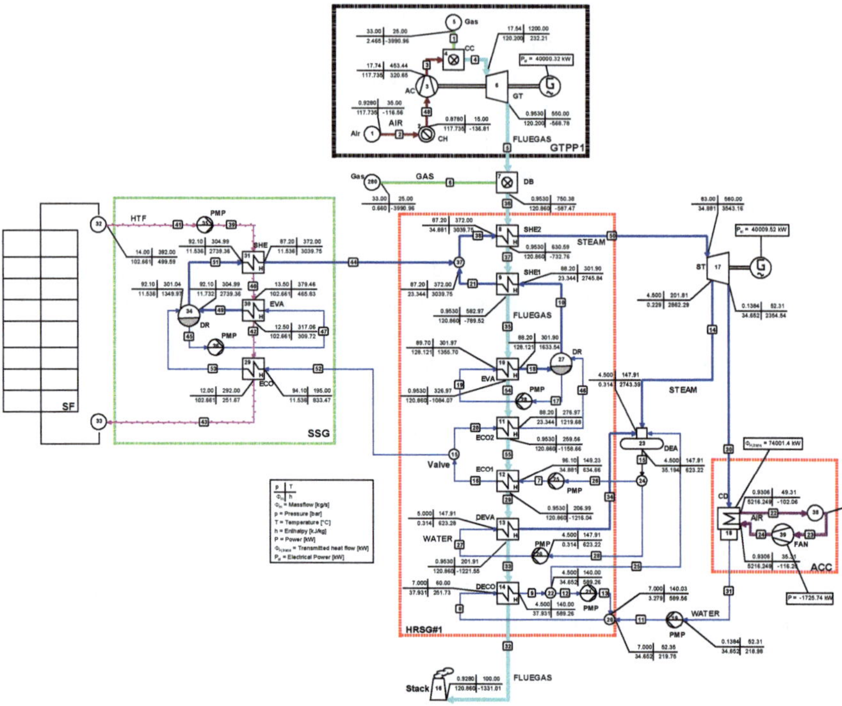

Fig 88.1 Flow diagram of the GT-SBC-CSP layout

is equipped with low-pressure economizer and evaporator (DECO and DEVA), two economizers (ECO1 and ECO1), an evaporator (EVA), two super heaters (SHE1 and SHE2) and a duct burner (DB)-integrated downstream of the GT exit.

The HRSG is with an approach temperature of 25 °C and a pinch temperature of 25 °C. The solar steam generator (SSG) supplies about 50 MW of heat at a temperature of 393 °C. It is the assembling of an economizer (ECO), an EVA with a drum (DR), and a superheater (SHE). The HTF used in the SF is synthetic oil Therminol PV-1. The SSG is with an approach temperature of 4 °C and a pinch temperature of 12 °C. The GTPP is fuelled with natural gas of which lower heating value (LHV) =45773 kJ/kg. The two HTSGs and the SSG together produce about 70 kg/s of superheated steam (superheated to 258 °C), at 83 bars and 560 °C, of which 67 % (34.9 kg/s) is produced by the HRSGs and 33 % (11.5 kg/s) is produced by the SSG.

The GTPP is supported by 4 Sadinter chiller (CH) unites, which contribute to boost the GTPP by reducing intake air temperature to 15 °C. The SPX dry cooling system, equipped with 15 fans, provides saturated water at 52 °C and at about 0.14 bars, with a condenser (CD) heat transfer duty of about 147.6 MW. The fans consume 3.45 MW of electricity.

The design parameters of the power plant are detailed in [15].

88.3 The GT-ABC/ORC-CSP Thermodynamic Layout

As shown in Fig. 88.1, in the new thermodynamic layout, under the acronym GT-ABC/ORC-CSP, only the SBC is modified in comparison with the GT-SBC-CSP layout. Indeed, for both layouts, the two GTPPs reject about 242 Kg/s of exhausted gas at 750 °C, and the SSG/SVG supplies about 50 MW of heat at a temperature of 393 °C. However, for the GT-ABC/ORC-CSP scheme the SBC is replaced by a binary cycle—ABC/ORC. The ABC comprises an AC, a GT, a generator (G) and an air recuperator (AREC). This cycle is chosen to be without intake air cooler and intercooler. The ORC, using toluene, comprises two identical HRVGs, a turbine, a toluene–toluene regenerator and an air–toluene recuperator. Each HRVG has an ECO, an EVA with a DR and a SHE. The regenerator (RG) is used to augment the cycle efficiency by using the sensible heat, containing toluene, after expansion. This is because toluene is a dry fluid with positive slope of the saturated vapour curve. The ABC recovers thermal energy through the AREC, from 242 kg/s of flue gas discharged from the GTPPs at 750 °C. The ORC recovers thermal energy from three streams. Indeed, the ORC recovers some heat through the HRVGs from 240 kg/s of air at 415 °C rejected from the ABC. It also recovers some thermal energy through the recuperator (REC) from 242 kg/s of flue gas leaving the AREC at 321 °C. This recovered heat is used to preheat toluene before being sent into the SVG. Through the SVG, toluene recovers energy from 205 kg/s of HTF at 393 °C. The GT-ABC/ORC-CSP cycle rejects 240 kg/s of pure air at 128 °C and 142 kg/s of flue gas at 129 °C to the environment (Fig. 88.2).

The two HRVGs and the SVG together produce 292 kg/s of slightly superheated toluene vapour (superheated by 2 °C), at 35.8 bars and about 309 °C [16], of which 41.5 % (121 kg/s) is produced by the HRVGs and 58.5 % (171 kg/s) is produced by the SVG.

With regard to the ABC, the ambient temperature and pressure are 35 °C and 0.928 bars, respectively. The relevant air mass flow rate is estimated at 240 kg/s, and the optimum pressure ratio in the compressor is 4.8. The compressor and turbine have isentropic efficiencies of 86 and 88 %, respectively. The thermal effectiveness of the AREC is 90 %.

In the ORC, the isentropic efficiency of the turbine is 90 %. The thermal effectiveness of the RG is 73 %, and the thermal effectiveness of the REC is 65 %. The HRVG is with an approach temperature of 23 °C and a pinch temperature of 12 °C.

Considering ambient temperature at 35 °C, the ACC provides saturated toluene at 0.135 bars and 52.3 °C. Cycle-Tempo computes the CD heat transfer duty based on the assumptions that the initial temperature difference T_{ITD} (vapour-to-ambient) is 17 °C, and the temperature difference at the hot side of the CD ΔT_{out} is 3 °C. It also computes the total power consumption of all the fans supposing that for every fan, the isentropic efficiency is 0.80, the air pressure ratio is 1.0028, and the mechanical efficiency is 0.94 [17].

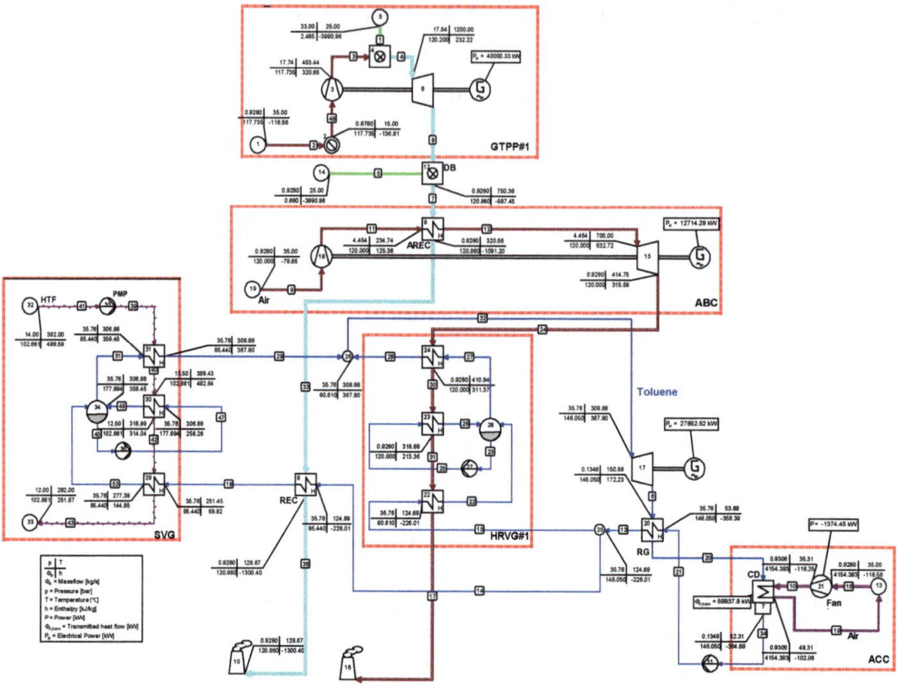

Fig 88.2 Flow diagram of the GT-ABC/ORC-CSP layout

88.4 Results and Discussion

88.4.1 Power Consumption of the ACC

As noted in Fig. 88.1, the nominal capacity of the GT-ABC/ORC-CSP cycle is 161.1 MW, with a thermal efficiency of about 56%. The binary bottoming cycle ABC/ORC generates a total of 81.1 MW. The ABC generates 25.4 MW (31.32%) and the ORC generates 55.7 MW (68.68%). The CD heat transfer duty is about 117.9 MW and the power consumption of the fans is 2.75 MW. In comparison to the existing reference case, the GT-SBC-CSP layout, the new thermodynamic conception (GT-ABC/ORC-CSP scheme) has almost the same nominal capacity with the same thermal efficiency, but the CD heat transfer duty and power requirement of the fans decrease by about 20%. This is because a significant share in nominal capacity of the GT-ABC/ORC-CSP scheme is produced by the ABC, which does not require the condensing process. The reduction in CD heat transfer duty has an impact not only on power consumption but also on the size of CD heat exchange area, on the size and number of fans, and in consequence on the initial investment cost of the ACC.

Besides, there are some advantages associated to the new hybridization scheme; the GT-ABC/ORC-CSP is technically less complex compared with the GT-SBC-CSP such as it needs less number of heat exchangers, there is no need to deaerator (DE), and no need to preheating loop.

88.4.2 Effect of Air Temperature

In this section, we compare the sensitivities of the performances of both cycles, GT-SBC-CSP and GT-ABC/ORC-CSP, with air temperature. The performance evaluation is made in terms of nominal capacity, cycle efficiency, CD heat duty and fans power consumption. The evaluation is considered at the design conditions, except that the air volumetric flow rate in the GTPP is fixed at the design value, whereas air mass flow rate varies against air temperature. Indeed, AC runs at constant volumetric flow rate, and then, with the increase in air temperature, air density decreases, and as a result, the admission of air mass flow rate into the compressor decreases. This effect impacts the performance of the power plant.

For the two cycles, over the air temperature range 10–50 °C, the nominal capacity and cycle efficiency are almost the same and decrease with similar trends. Averagely, for both cycles, for every 1 °C temperature increase, the nominal capacity decreases by approximately 1.17 MW and the cycle efficiency decreases by about 0.12 point.

For both cycles, there is proportionality between the CD heat duty and fans' power consumption; this is a way they both decrease in a trend similar with air temperature. This is because the production of steam and toluene vapour decreases with air temperature. The GT-ABC/ORC-CSP cycle is more sensible, that is, with increasing air temperature, it requires less and less heat duty for condensing, and consequently, it requires less and less power for running fans compared with the GT-SBC-CSP cycle. Averagely, for every 1 °C temperature increase, the ACC power consumption decreases by 9 kW for the GT-ABC/ORC-CSP scheme and decreases by 5 kW for the GT-SBC-CSP scheme; however, the CD heat duty decreases by about 0.47 MW and 0.27 MW, respectively. The new hybridization reduces the ACC power consumption by 18–25 %, over the examined air temperature range 10–50 °C (Fig. 88.3).

88.4.3 Effect of HTF Mass Flow Rate

In the SSG/SVG, the steam/vapour mass flow rate is more or less proportional to the HTF mass flow rate. The HTF mass flow rate varies following the DNI intensity, in other words, according to time during day and climate conditions. The effect of DNI on the performances of two cycles, GT-SBC-CSP and GT-ABC/ORC-CSP, is taken into account indirectly *via* the HTF mass flow rate. For both cycles, the full nominal capacity and cycle efficiency increase with HTF mass flow rate. However, for GT-SBC-CSP, the variations are slightly more rapid. At low HTF mass flow rate, the GT-ABC/ORC-CSP is more efficient, but at some critical value of HTF

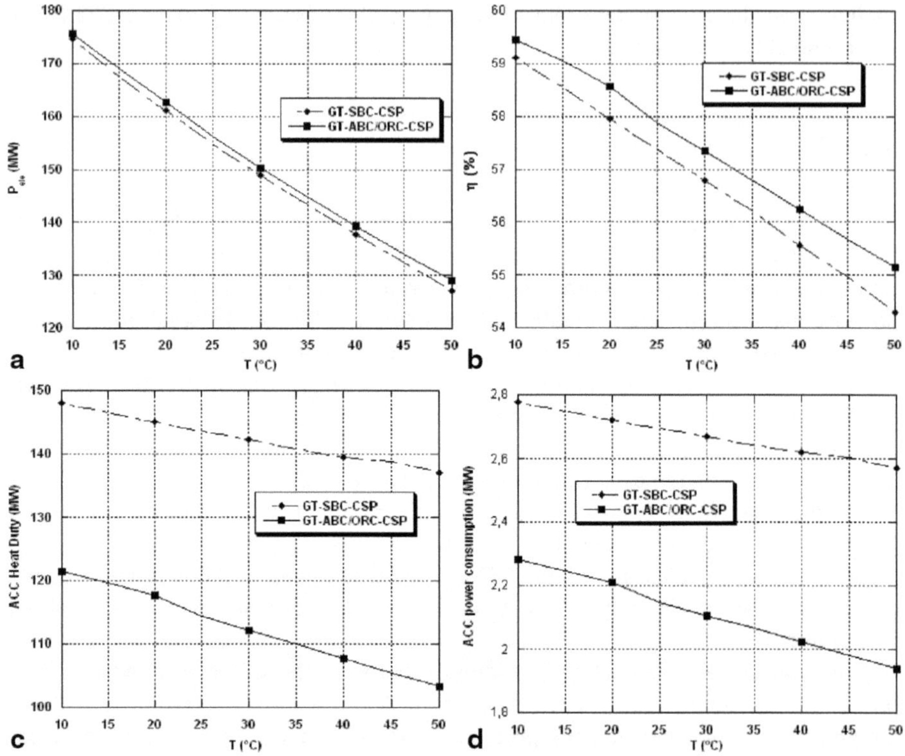

Fig. 88.3 Effect of air temperature on the performance of the GT-SBC-CSP and the GT-ABC/ORC-CSP: **a** Nominal capacity, **b** Cycle efficiency, **c** ACC heat duty and **d** ACC power consumption

mass flow rate, around 205 kg/s, the GT-SBC-CSP becomes more efficient. From 105 to 305 kg/s, the nominal capacity of the GT-SBC-CSP increases from 150 to 170 MW, whereas that of the GT-ABC/ORC-CSP increases from 153 to 169 MW. The cycle efficiency varies from 49 to 63 % for the first cycle and from 51 to 52 % for the second cycle. Averagely, for every 1 kg/s increase in HTF flow mass rate, the nominal capacity increase by 0.20 MW for the GT-SBC-CSP and by 0.16 MW for the GT-ABC/ORC-CSP. In the same order, the cycle efficiencies increase by 0.11 and 0.14 points, respectively.

The increase of HTF mass flow rate increases the CD heat duty and consequently increases the ACC power consumption for both cycles with almost the same pace. The GT-ABC/ORC-CSP keeps its superiority in terms of lower requirements in CD heat duty and fans power consumption. Averagely, for every 1-kg/s increase in HTF flow mass rate, the fans power consumption increases by 8 kW and the CD heat duty by 0.35 MW for both cycles (Fig. 88.4).

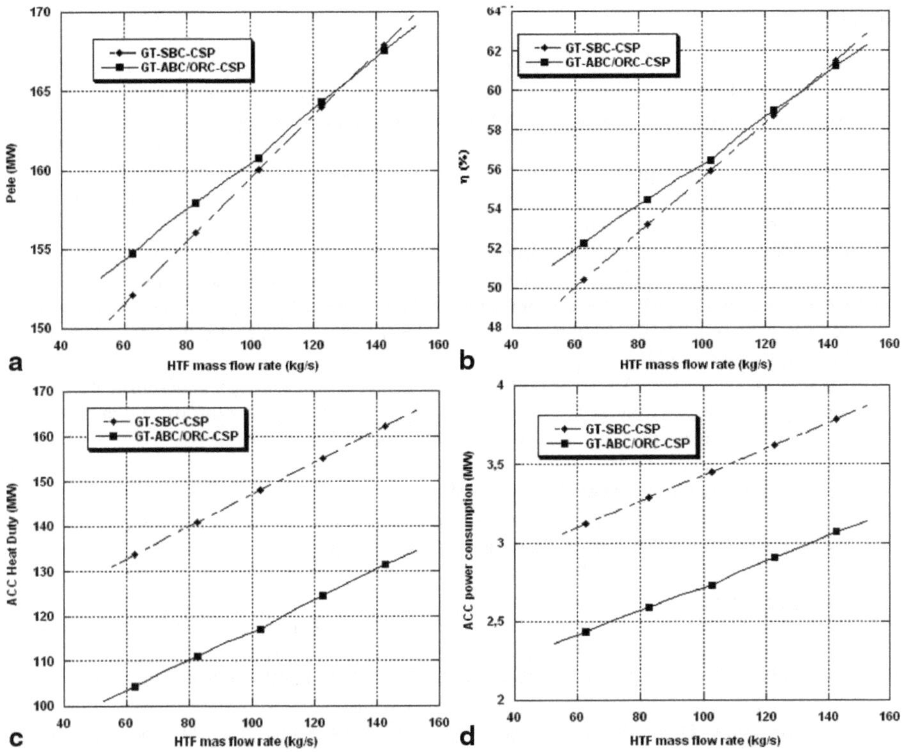

Fig. 88.4 Effect of HTF mass flow rate on the performance of the GT-SBC-CSP and the GT-ABC/ORC-CSP: **a** Nominal capacity, **b** Cycle efficiency, **c** ACC heat duty and **d** ACC power consumption

88.5 Conclusions

The purpose of the present study is to make some thermodynamic modifications on an existing ISCC power plant in operation in Algeria. A binary cycle, ABC/ORC, is proposed instead of the well-recognized SBC. The thermodynamic simulations demonstrate that the novel solar–gas thermodynamic hybridization scheme (GT-ABC/ORC-CSP) reduces, at design running conditions, the heat duty and power consumption of the ACC by 20 %. The performance sensitivities of the novel thermodynamic layout and existing reference case to air temperature and HTF mass flow rate are quasi-similar. Averagely, for every 1 °C temperature increase, the full nominal capacity of the cycle decreases approximately by 1.17 MW, and the cycle efficiency decreases by about 0.12 point. The GT-ABC/ORC-CSP layout reduces the ACC power consumption by 18–25 %, over the examined air temperature range 10–50 °C. On the other hand, averagely, for every 1 kg/s increase in HTF flow mass rate, the nominal capacity increases by 0.16–0.20 MW and the cycle efficiency increases by 0.11–0.14 points, respectively.

References

1. Peterseim JH, White S, Tadros A, Hellwig U (2013) Concentrated solar power hybrid plants, which technologies are best suited for hybridisation? Renew Energy 57:520–532
2. Pihl E, Spelling J, Johnsson E (2013) Thermo-economic optimization of hybridization options of solar retrofitting of combined-cycle power plants. J Solar Energy Eng 136:021001
3. Deng H, Boehm RF (2011) An estimation of the performance limits and improvement of dry cooling on trough solar thermal plants. Appl Energy 88:261–223
4. Poullikkas A, Hadjipaschalis I, Kourtis G (2013) A comparative overview of wet and dry cooling systems for Rankine cycle based CSP plants. Trends Heat Mass Transf 13:27–50
5. Blanco-Marigorta AM, Sanchez-Henríquez MV, Peña-Quintana JA (2011) Exergetic comparison of two different cooling technologies for the power cycle of a thermal power plant. Energy 36:1966–1972
6. Bianchi M, De Pascale A (2011) Bottoming cycles for electric energy generation: parametric investigation of available and innovative solutions for the exploitation of low and medium temperature heat sources. Appl Energy 88:1500–1509
7. Najjar YSH, Zaamout MS (1996) Performance analysis of gas turbine air-bottoming combined system. Energy Convers Manag 37:399–403
8. Bolland O, Førde M, Hånde B (1996) Air bottoming cycle: use of gas turbine waste heat for power generation. J Eng Gas Turbines Power 118:359–367
9. Ghazikhani M, Khazaee I, Abdekhodai E (2014) Exergy analysis of gas turbine with air bottoming cycle. Energy 72:599–607
10. Chacartegui R, Sanchez D, Munoz JM, Sanchez T (2009) Alternative ORC bottoming cycles FOR combined cycle power plants. Appl Energy 86:2162–2170
11. Canada S, Cohen G, Cable R, Brosseau D, Price H (2005) Parabolic trough organic Rankine cycle solar power plant. NREL/CP-550-37077, January 2005
12. Hen Y-L, Mei D-H, Tao W-Q, Yang W-W, Liu H-L (2012) Simulation of the parabolic trough solar energy generation system. Appl Energy 97:630–641
13. Al-Sulaiman FA (2014) Exergy analysis of parabolic trough solar collectors integrated with combined steam and organic Rankine cycles. Energy Convers Manag 77:441–449
14. Al-Sulaiman FA (2013) Energy and sizing analyses of parabolic trough solar collector integrated with steam and binary vapor cycles. Energy 58:561–570
15. Khaldi F (2012) Energy and exergy analysis of the first hybrid solar-gas power plant in Algeria. In: Proceedings of ECOS2012, the 25th International Conference Efficiency, Cost, Optimization, Simulation and Environmental Impact of Energy Systems, June 26–29, Perugia, Italy
16. Rayegan R, Tao YX (2011) A procedure to select working fluids for solar organic Rankine cycles (ORCs). Renew Energy 36:659–670
17. Wagner MJ, Gilman P (2011) Technical manual for the SAM physical trough model. Technical Report, NREL: TP-5500–51825, June 2011

Index

A. Sayigh (ed.), *Renewable Energy in the Service of Mankind Vol II*,
DOI 10.1007/978-3-319-18215-5

MIX
Papier aus verantwortungsvollen Quellen
Paper from responsible sources
FSC® C105338

If you have any concerns about our products,
you can contact us on
ProductSafety@springernature.com

In case Publisher is established outside the EU,
the EU authorized representative is:
Springer Nature Customer Service Center GmbH
Europaplatz 3, 69115 Heidelberg, Germany

Printed by Libri Plureos GmbH
in Hamburg, Germany